Geophysical Monograph Series

Geophysical Monograph Series

Geophysical Monograph 289

Distributed Acoustic Sensing in Borehole Geophysics

Yingping Li
Robert Mellors
Ge Zhan
Editors

This Work is a co-publication of the American Geophysical Union and John Wiley and Sons, Inc.

ADVANCING EARTH AND SPACE SCIENCES

WILEY

This edition first published 2025

Published under the aegis of the AGU Publications Committee

Matthew Giampoala, Vice President, Publications
Steven A. Hauck, II, Chair, Publications Committee
For details about the American Geophysical Union visit us at www.agu.org.

Registered Office
John Wiley & Sons, Inc., 111 River Street, Hoboken, NJ 07030, USA

Editorial Office
111 River Street, Hoboken, NJ 07030, USA

For details of our global editorial offices, customer services, and more information about Wiley products visit us at www.wiley.com.

Wiley also publishes its books in a variety of electronic formats and by print-on-demand. Some content that appears in standard print versions of this book may not be available in other formats.

Library of Congress Cataloging-in-Publication Data Applied for:
Hardback: 978-1-394-17924-4

Cover Image: Borehole monitoring system with fiber-optic cables (red lines) and active and passive sources (blue circles) overlaying waveforms recorded by distributed fiber optic sensing technology

Cover Design: Wiley

Cover Illustration: © photographer3431/Getty Images

Background: Waveform image courtesy of Teck Kean Lim, Toshiaki Kobayashi, Akira Fujimoto, Sergei Tcherkashnev, and Michael Mondanos

Set in 10/12pt Times New Roman by Straive, Pondicherry, India

CONTENTS

Part IV Borehole Distributed Acoustic Sensing Monitoring

LIST OF CONTRIBUTORS

Rufat Aghayev
Silixa Ltd.
Elstree, UK

Jonathan Ajo-Franklin
Department of Earth, Environmental and Planetary Sciences
Rice University
Houston, Texas, USA

Shujaat Ali
SLB
Houston, Texas, USA

Andrey Bakulin
Formerly
EXPEC Advanced Research Center
Saudi Aramco
Dhahran, Saudi Arabia

Presently
Bureau of Economic Geology
University of Texas at Austin
Austin, Texas, USA

Audrius Berzanskis
MagiQ Technologies Inc.
Somerville, Massachusetts, USA

Falko Bethmann
Geo-Energie Suisse
Zurich, Switzerland

Pierre Bettinelli
SLB
Kuala Lumpur, Malaysia

Riley Blais
Department of Environmental Engineering and Earth Sciences
Clemson University
Clemson, South Carolina, USA

Guido Blöcher
Helmholtz Centre Potsdam
GFZ German Research Centre for Geosciences
Potsdam, Germany

Fateh Bouchaala
Department of Earth Sciences
Khalifa University of Science and Technology
Abu Dhabi, UAE

Alex M. Brisbourne
British Antarctic Survey
Natural Environment Research Council
Cambridge, UK

Xiaohui Cai
Department of Earth, Energy, and Environment
University of Calgary
Calgary, Alberta, Canada

Zhonglin Cao
BGP Inc
China National Petroleum Corporation
Zhuozhou, China

Hui Chen
BGP Inc
China National Petroleum Corporation
Zhuozhou, China

Yuanzhong Chen
School of Information and Communication Engineering
University of Electronic Science and Technology of China
Chengdu, China

and

Optical Science and Technology (Chengdu) Ltd.
Chengdu, China

Ze Chen
Silixa Ltd.
Elstree, UK

Arthur Cheng
Earth Environmental Science Programme
The Chinese University of Hong Kong
Hong Kong, China

Yun-Ze Cheng
Institute of Earth Sciences
Academia Sinica
Taipei, Taiwan

Caleb Christensen
MagiQ Technologies Inc.
Somerville, Massachusetts, USA

Olivia Costantino
Department of Environmental Engineering and Earth Sciences
Clemson University
Clemson, South Carolina, USA

Scott DeWolf
Department of Environmental Engineering and Earth Sciences
Clemson University
Clemson, South Carolina, USA

Ben Dyer
Geo-Energie Suisse
Zurich, Switzerland

Mattew V. Eaid
Chevron
Houston, Texas, USA

David Eaton
Department of Earth, Energy and Environment
University of Calgary
Calgary, Canada

Spadavecchia Emmanuel
ENI Natural Resources
Milan, Italy

Mahmoud Farhadiroushan
Silixa Ltd.
Elstree, UK

Akira Fujimoto
Technology & Research Center
Japan Organization for Metals and Energy Security
Chiba, Japan

Yoshihiro Fujise
WELMA Co., Ltd.
Fukuoka, Japan

Brian N. Fuller
Sterling Seismic & Reservoir Services
Littleton, Colorado, USA

Shigetsugu Furuya
Idemitsu Kosan Co., Ltd.
Tokyo, Japan

Leonid N. Germanovich
Department of Environmental Engineering and Earth Sciences
Clemson University
Clemson, South Carolina, USA

Domenico Giardini
Institute of Geophysics
ETH Zurich
Zurich, Switzerland

Arran Gillies
Silixa Ltd.
Elstree, UK

Pavel Golikov
EXPEC Advanced Research Center
Saudi Aramco
Dhahran, Saudi Arabia

Yves Guglielmi
Energy Geosciences Division
Lawrence Berkeley National Laboratory
Berkeley, California, USA

Jing Guo
Center for Optical Materials Science and Engineering Technologies
Clemson University
Clemson, South Carolina, USA

Jakob B. U. Haldorsen
MagiQ Technologies Inc.
Somerville, Massachusetts, USA

Yoko Hasada
Engineering Advancement Association of Japan
Tokyo, Japan

and

Daiwa Exploration and Consulting Co., Ltd.
Tokyo, Japan

Guangming He
BGP Inc
China National Petroleum Corporation
Zhuozhou, China

Marian Hertrich
Swiss Seismological Service
ETH Zurich
Zurich, Switzerland

Chet Hopp
Energy Geosciences Division
Lawrence Berkeley National Laboratory
Berkeley, California, USA

Guangmin Hu
School of Resources and Environment
University of Electronic Science and Technology of China
Chengdu, China

Qi Hu
Department of Energy Science and Engineering
Stanford University
Stanford, California, USA

Liwei Hua
Center for Optical Materials Science and Engineering Technologies
Clemson University
Clemson, South Carolina, USA

and

Department of Environmental Engineering and Earth Sciences
Clemson University
Clemson, South Carolina, USA

Hsin-Hua Huang
Institute of Earth Sciences
Academia Sinica
Taipei, Taiwan

and

Department of Geosciences
National Taiwan University
Taipei, Taiwan

Lianjie Huang
Los Alamos National Laboratory
Los Alamos, New Mexico, USA

Thomas S. Hudson
Department of Earth Sciences
University of Oxford
Oxford, UK

and

Department of Earth Sciences
ETH Zurich
Zurich, Switzerland

Kristopher A. Innanen
Department of Earth, Energy, and Environment
University of Calgary
Calgary, Alberta, Canada

Ge Jin
Department of Geophysics
Colorado School of Mines
Golden, Colorado, USA

Junzo Kasahara
Engineering Advancement Association of Japan
Tokyo, Japan

and

Center for Integrated Research and Education of Natural Hazards
Shizuoka University
Shizuoka, Japan

Scott D. Keating
Department of Earth and Planetary Sciences
ETH Zurich
Zurich, Switzerland

J. Michael Kendall
Department of Earth Sciences
University of Oxford
Oxford, UK

Manish Lal Khaitan
SLB
London, UK

Timothy Kneafsey
Energy Geosciences Division
Lawrence Berkeley National Laboratory
Berkeley, California, USA

Toshiaki Kobayashi
Technology & Research Center
Japan Organization for Metals and Energy Security
Chiba, Japan

Charlotte M. Krawczyk
Helmholtz Centre Potsdam
GFZ German Research Centre for Geosciences
Potsdam, Germany

and

Department of Applied Geophysics
Technische Universität Berlin
Berlin, Germany

Chin-Shang Ku
Institute of Earth Sciences
Academia Sinica
Taipei, Taiwan

Federica Lanza
Swiss Seismological Service
ETH Zurich
Zurich, Switzerland

Donald C. Lawton
Department of Earth, Energy, and Environment
University of Calgary
Calgary, Canada

and

Carbon Management Canada
Calgary, Alberta, Canada

Joel Le Calvez
SLB
Sugar Land, Texas, USA

Ariel Lellouch
School of the Environment and Earth Sciences
Tel Aviv University
Tel Aviv, Israel

David Li
Los Alamos National Laboratory
Los Alamos, New Mexico, USA

Xiang Li
BGP Inc
China National Petroleum Corporation
Zhuozhou, China

Yingping Li
BlueSkyDAS LLC
Sugar Land, Texas, USA

and

Department of Earth and Atmospheric Sciences
University of Houston
Houston, Texas, USA

Yunyue Elita Li
Department of Earth, Atmospheric, and Planetary Sciences
Purdue University
West Lafayette, Indiana, USA

Teck Kean Lim
Consultant, Geophysicist/Geoscientist
Tokyo, Japan

Chin-Jen Lin
Institute of Earth Sciences
Academia Sinica
Taipei, Taiwan

Kou-Fong Ma
Institute of Earth Sciences
Academia Sinica
Taipei, Taiwan

and

Department of Earth Sciences
National Central University
Taoyuan, Taiwan

and

Earthquake-Disaster & Risk Evaluation and Management Center
Taoyuan, Taiwan

Yuanyuan Ma
Department of Earth, Energy and Environment
University of Calgary
Calgary, Canada

and

Department of Earth, Environmental, and Planetary Sciences
Rice University
Houston, Texas, USA

Kelly MacDougall
Department of Earth, Energy and Environment
University of Calgary
Calgary, Canada

Vikram Maji
Department of Earth, Energy and Environment
University of Calgary
Calgary, Canada

and

Department of Earth Science
University of Bergen
Bergen, Norway

Alejandro Martinez
SLB
Houston, Texas, USA

Evgeniia Martuganova
Department of Geoscience & Engineering
Delft University of Technology
Delft, The Netherlands

and

Helmholtz Centre Potsdam
GFZ German Research Centre for Geosciences
Potsdam, Germany

Jack Maxwell
Silixa Ltd.
Elstree, UK

Peter Meier
Geo-Energie Suisse
Zurich, Switzerland

Robert Mellors
Scripps Institution of Oceanography
University of California San Diego
La Jolla, California, USA

Donald Richard Metzbower
MagiQ Technologies Inc.
Somerville, Massachusetts, USA

Hitoshi Mikada
Department of Civil and Earth Resources Engineering
Kyoto University
Kyoto, Japan

Craig Milne
Silixa Ltd.
Elstree, UK

Francesco Miranda
ENI Natural Resources
Milan, Italy

Takashi Mizuno
SLB
Sugar Land, Texas, USA

Rob Moak
Department of Environmental Engineering and Earth Sciences
Clemson University
Clemson, South Carolina, USA

Michael Mondanos
Silixa Ltd.
Elstree, UK

Lawrence C. Murdoch
Department of Environmental Engineering and Earth Sciences
Clemson University
Clemson, South Carolina, USA

Yanrui Ning
Department of Geophysics
Colorado School of Mines
Golden, Colorado, USA

Ben Norden
Helmholtz Centre Potsdam
GFZ German Research Centre for Geosciences
Potsdam, Germany

Hiroshi Ohnuma
Engineering Advancement Association of Japan
Tokyo, Japan

Tom R. Parker
Silixa Ltd.
Elstree, UK

Daniel Pérez
Silixa Ltd.
Elstree, UK

Grant Plunkett
Department of Environmental Engineering and Earth Sciences
Clemson University
Clemson, South Carolina, USA

Olga Podgonova
SLB
Cambridge, Massachusetts, USA

Antonio P. Rinaldi
Swiss Seismological Service
ETH Zurich
Zurich, Switzerland

Michelle Robertson
Energy Geosciences Division
Lawrence Berkeley National Laboratory
Berkeley, California, USA

Verónica Rodríguez Tribaldos
GFZ German Research Centre for Geosciences
Potsdam, Germany

and

Energy Geosciences Division
Lawrence Berkeley National Laboratory
Berkeley, California, USA

Seda Rouxel
Department of Earth Sciences
Khalifa University of Science and Technology
Abu Dhabi, UAE

Rebecca O. Salvage
Department of Earth, Energy and Environment
University of Calgary
Calgary, Canada

and

Natural Resources Canada
Geological Survey of Canada
Calgary, Canada

Ali Sayed
SLB
Houston, Texas, USA

Francisco Serbeto
Geo-Energie Suisse
Zurich, Switzerland

Sergey Shatalin
Silixa Ltd.
Elstree, UK

Masahiro Shimazaki
Idemitsu Kosan Co., Ltd.
Tokyo, Japan

Ilya Silvestrov
EXPEC Advanced Research Center
Saudi Aramco
Dhahran, Saudi Arabia

Robert Smith
EXPEC Advanced Research Center
Saudi Aramco
Dhahran, Saudi Arabia

Robert R. Stewart
Department of Earth and Atmospheric Sciences
University of Houston
Houston, Texas, USA

Po-Li Su
Institute of Earth Sciences
Academia Sinica
Taipei, Taiwan

Sergei Tcherkashnev
ASTO Geophysical Consulting Pty Ltd.
Perth, Australia

Katinka B. Tuinstra
Swiss Seismological Service
ETH Zurich
Zurich, Switzerland

Manuel Useche
SLB
Bogota, Colombia

Chaoyi Wang
Department of Earth, Energy and Environment
University of Calgary
Calgary, Canada

Fei Wang
BGP Inc.
China National Petroleum Corporation
Chengdu, China

Stefan Wiemer
Swiss Seismological Service
ETH Zurich
Zurich, Switzerland

Mark E. Willis
Halliburton
Houston, Texas, USA

En-Shih Wu
Institute of Oceanography
National Taiwan University
Taipei, Taiwan

Junjun Wu
Optical Science and Technology (Chengdu) Ltd.
Chengdu, China

Xiang Wu
Halliburton
Houston, Texas, USA

Yongji Wu
Center for Optical Materials Science and Engineering Technologies
Clemson University
Clemson, South Carolina, USA

Shujun Xia
BGP Inc.
China National Petroleum Corporation
Chengdu, China

Hai Xiao
Center for Optical Materials Science and Engineering Technologies
Clemson University
Clemson, South Carolina, USA

Jinling Xiong
Dagang Oilfield
China National Petroleum Corporation
Tianjin, China

Jizhong Yang
State Key Laboratory of Marine Geology
Tongji University
Shanghai, China

and

Shanghai Sheshan National Geophysical Observatory
Shanghai, China

Gang Yu
BGP Inc
China National Petroleum Corporation
Zhuozhou, China

and

Optical Science and Technology (Chengdu) Ltd.
Chengdu, China

Alba Zappone
Swiss Seismological Service
ETH Zurich
Zurich, Switzerland

and

Department of Mechanical Engineering
ETH Zurich
Zurich, Switzerland

Ge Zhan
TGS-NOPEC Geophysical Company
Houston, Texas, USA

Hao Zhang
Institute of Geophysics
Shanghai Petroleum Corporation Ltd.
Shanghai, China

Yu Zhang
BGP Inc
China National Petroleum Corporation
Zhuozhou, China

Yishui Zhao
Optical Science and Technology (Chengdu) Ltd.
Chengdu, China

Yingcai Zheng
Department of Earth and Atmospheric Sciences
University of Houston
Houston, Texas, USA

Bing Zhou
Department of Earth Sciences
Khalifa University of Science and Technology
Abu Dhabi, UAE

Xuran Zhu
Center for Optical Materials Science and Engineering Technologies
Clemson University
Clemson, South Carolina, USA

Jingjing Zong
School of Resources and Environment
University of Electronic Science and Technology of China
Chengdu, China

PREFACE

Distributed acoustic sensing (DAS) is a crucial technology within the distributed fiber optic sensing (DFOS) family. DAS utilizes lasers to detect acoustic or vibration signals in real-time along an optical fiber cable that can extend for tens of kilometers. A DAS system consists of an optical fiber cable and an optoelectronic device known as an interrogator unit (IU). The IU sends laser pulses into the fiber and detects coherent backscattered Rayleigh waves caused by imperfections within the optical fiber. Changes in the backscattered waves over time are used to measure the dynamic strain rate or strain at all points along the fiber cable. DAS systems record acoustic and vibration signals that affect the fiber with high resolution in both temporal and spatial domains and over a broad frequency spectrum. The first publication utilizing Rayleigh backscatter-DAS (RBS-DAS) was in 2011, and since then, DAS technology has been applied to diverse fields and topics, including geophysics, seismology, hydrology, glaciology, the energy industry, geotechnical engineering, environmental protection, and geohazard mitigation and prevention. This has led to numerous significant research publications and practical applications, both in academia and industry.

In 2021, the American Geophysical Union (AGU) published one of the first comprehensive books on DAS titled *Distributed Acoustic Sensing in Geophysics: Methods and Applications* (Geophysical Monograph Series Volume 268). This publication documented advancements in DAS technology and its geophysical applications from both academia and industry, and it sought to raise awareness and attract potential users from various fields. The monograph was divided into four main topics: (1) DAS concepts, principles, and instrumentation; (2) DAS applications in the energy and resource industries; (3) DAS monitoring applications; and (4) DAS environmental applications. Additionally, it featured a review chapter summarizing the historical development of and advances in DAS applications up to 2020. However, the rapid advancements in DAS technologies and geophysical applications since then require updated publications to stay current with the latest developments. Notably, there has been an explosion of innovative work focused on DAS applications in borehole geophysics, which inspired this volume.

Borehole geophysics involves measuring the physical properties of subsurface rock formations, imaging subsurface structures, and monitoring subsurface activities with instruments deployed in wellbores drilled to various depths. For seismology, borehole deployments are especially useful, as sensors placed in the wellbore are closer to these subsurface targets and are subject to less noise compared to surface seismic sensors. This leads to high-quality measurements and images and enables borehole sensors to effectively monitor with a high signal-to-noise ratio. Unfortunately, deep boreholes often face high-temperature and high-pressure conditions. Conventional seismic sensors—such as geophones, seismometers, and accelerometers—which contain electronic components, cannot reliably function over extended periods under these conditions, making continuous long-term monitoring difficult or even impossible. However, borehole DAS systems, using optic fiber cables installed in the wellbore and an IU on the surface, can reliably operate in these harsh environments due to the durability of optic fibers. DAS fibers can also be deployed outside the well casing, which permits seismic monitoring during ongoing well operations. Consequently, the DAS borehole system is a viable alternative or complement to conventional borehole seismic sensors.

In planning this book, we reached out to many DAS and borehole specialists. The consensus within the community was that a book devoted to borehole DAS was critically needed, and we received enthusiasm and interest in contributing chapters to the proposed book. With encouragement and support from AGU and Wiley for a follow-up to the 2021 volume, we submitted a proposal for a monograph titled *Distributed Acoustic Sensing in Borehole Geophysics*. The proposal received highly favorable feedback, and we are grateful to the three anonymous reviewers for their positive comments and constructive suggestions.

Distributed Acoustic Sensing in Borehole Geophysics includes 31 chapters and focuses on four main topics: (1) borehole DAS instrumentation and modeling; (2) borehole DAS acquisition and processing; (3) borehole DAS imaging and inversion; and (4) borehole DAS monitoring. In the introductory chapter, we provide an overview of the chapters in each of the four parts, along with a summary of the development history and primary geophysical applications of DAS technologies, including fiber Bragg grating-quasi distributed acoustic sensing (FBG-qDAS) and RBS-DAS.

Part I, Borehole Distributed Acoustic Sensing Instrumentation and Modeling, includes five chapters. These chapters discuss an integrated DFOS system for wellbore and reservoir monitoring using different modes of fibers in the same fiber cable. This is a novel approach for measuring distributed strain using microwave photonics interferometry, a qDAS system for obtaining three-component (3C) measurements, 3D fiber installation of DAS radar for 3C DAS measurements, and modeling of DAS and geophone waveforms to enhance the DAS-vertical seismic profiling (VSP) processing method in complex geological settings.

Part II, Borehole Distributed Acoustic Sensing Acquisition and Processing, contains 10 chapters detailing various projects related to DAS acquisition and processing. These include joint borehole-surface DAS acquisition and processing of DAS-VSP data both onshore and offshore; examples of hybrid wireline-fiber cables for acquiring DAS-VSP data; borehole DAS projects in geothermal energy exploration in Japan; borehole DAS applications in glacial environments; 3D DAS-VSP acquisition and processing for gas hydrate exploration in Alaska; DAS measurements for attenuation in carbonate rocks; a DAS processing method focused on DAS-to-velocity transformation and Q estimation to improve DAS images of subsurface structures; an automatic first-arrival time-picking approach for DAS-VSP data based on the differential-phase-spectral slope method; and an evaluation of the potential of downhole DAS technology to detect and locate distant microseismicity.

Part III, Borehole Distributed Acoustic Sensing Imaging and Inversion, encompasses seven chapters. These chapters cover a rapid ray-based DAS-VSP imaging method using both PP and PS waves; a workflow for imaging subsurface structures by maximizing the utility of borehole DAS data; a time-lapse DAS-VSP method for detecting diffracted wavefields from induced fractures by stimulations; and the processing and imaging results of four-well tubing-strapped 3D DAS-VSP data acquired under flowing conditions. Additionally, this part discusses the integration of existing geological and geophysical data with images derived from 3D DAS-VSP data acquired using the hybrid wireline DAS system at a geothermal site. It also showcases notable images of subsurface structural details in the high-temperature Takigami geothermal field and describes a full-waveform inversion method applied to both borehole DAS and accelerometer data to produce high-quality subsurface images.

Part IV, Borehole Distributed Acoustic Sensing Monitoring, comprises seven chapters. It begins with a review discussing the potential, capabilities, and limitations of downhole DAS within a broad seismological context. This part also includes a case study demonstrating that borehole DAS systems designed for geothermal energy exploration can effectively monitor earthquakes in earthquake-prone areas. It provides an excellent example of monitoring earthquakes and ground motions and analyzing ambient noise with borehole DAS data acquired in two wells situated across an active fault zone. Furthermore, it introduces a source-free DAS logging method that utilizes borehole DAS ambient noise root mean square amplitudes to monitor, detect, and locate fractures and faults. This part also introduces low-frequency DAS (LF-DAS) and its advantages for subsurface imaging and qualitative interpretations. It demonstrates the feasibility of using temporary and retrievable optical fibers to monitor hydraulic fracturing operations and analyze microseismic and LF-DAS signals. Lastly, it presents DAS/distributed strain sensing monitoring results from four fluid injection experiments conducted in two different underground research laboratories.

The final chapter of this volume summarizes the key findings and outlines potential directions for future research. Supplementary reading materials are included throughout this monograph to enrich and complement the content.

This monograph would not have been possible without the invaluable contributions of the 130 authors from around the world (see List of Contributors), and we are deeply grateful for their efforts. We also acknowledge the 60 volunteer reviewers who generously dedicated their time and expertise to ensuring each chapter met the highest standards of quality. We would also like to thank AGU and Wiley for their support in bringing this monograph to fruition. Special thanks to Dr. Taka'aki Taira, the AGU subject editor for this volume, who meticulously reviewed all chapters, provided constructive comments, and ultimately approved them. Our appreciation extends to the Editor in Chief for AGU Books, Kate Lajtha, as well as Jenny Lunn and Dara Liling from AGU Publications. The team at Wiley, especially Summers Scholl, Kimberly Monroe-Hill, Lesley Fenske, Naveen Kumaran Shanmugam, Tiffany Taylor, Rajalaxmi Rajendrasingh, Noel McGlinchey, Keerthana Govindarajan, Nathanael Jude McGavin, Vignesh Lakshmikanthan, and Gunasekaran Arumugam, are recognized for their exceptional organization, management, cover design, and permission clearance. We also thank Robert Mellors, Ge Zhan, Teck Kean Lim, Dana Jurick, Mahmoud Farhadiroushan, and others for providing their borehole DAS photos and images as candidates for the book cover design.

This monograph will serve as a comprehensive handbook for a diverse audience interested in borehole DAS technologies and their geophysical applications. We anticipate its appeal to a broad readership, including geophysicists, seismologists, geologists, geoscientists, environmental scientists, and graduate and undergraduate students in geophysics and geoscience, all with an interest in DAS borehole geophysical applications. We believe this book will offer a valuable platform for both scientific and industrial communities to exchange cutting-edge advances in borehole DAS technology.

Yingping Li
BlueSkyDas LLC, USA
and
University of Houston, USA

Robert Mellors
University of California San Diego, USA

Ge Zhan
TGS-NOPEC Geophysical Company, USA

1

Recent Advances in Distributed Acoustic Sensing for Borehole Geophysics

Yingping Li[1,2], Robert Mellors[3], and Ge Zhan[4]

ABSTRACT

This comprehensive monograph on distributed acoustic sensing (DAS) in borehole geophysics spans six parts, each delving into distinct aspects of DAS technology and applications. This chapter serves as an introduction, outlining the principles of DAS within distributed fiber-optic sensing (DFOS) and its recent advancements, emphasizing its capabilities in measuring acoustic/vibration signals with high resolution along long fiber-optic cables. It discusses the challenges and advantages of DAS in borehole geophysics and its applications for recording seismic measurements, monitoring downhole activities, detecting microearthquakes, and assessing seismic risks. This chapter also traces the evolution of DAS technology from fiber Bragg grating DAS to Rayleigh backscattering-based DAS (RBS-DAS), showcasing its expanding global use in various geophysical projects. The subsequent parts of this book delve into instrumentation and modeling (Part I), acquisition and processing (Part II), imaging and inversion techniques (Part III), and applications across earthquake seismology, fracture zone detection, active fault monitoring, and low-frequency strain monitoring (Part IV). The last chapter in this volume summarizes the conclusions for Chapters 2–30 and looks ahead to future endeavors in DAS technology development and borehole DAS geophysical applications. All chapters are authored by experts in the field, providing a comprehensive overview of the current state, challenges, and potential of DAS technology in borehole geophysics and related disciplines.

1.1. INTRODUCTION

This monograph, titled *Distributed Acoustic Sensing in Borehole Geophysics*, serves as an introduction to the technologies of distributed acoustic sensing, with an emphasis on recent applications in borehole geophysics. DAS is a subset of distributed fiber-optic sensing (DFOS), which utilizes an entire fiber-optic cable, extending up to several tens of kilometers, to continuously measure physical signals at high spatial and temporal resolution. DFOS technologies harness Rayleigh, Raman, and Brillouin backscattering of laser pulses to enable distributed measurements of acoustic/vibration, static strain, and temperature, known respectively as distributed acoustic sensing (DAS), distributed strain sensing (DSS), and distributed temperature sensing (DTS). Among these, DAS stands out as a key member of the DFOS technology family and is becoming widely used in borehole geophysics.

DAS technology utilizes a fiber-optic cable and an optoelectronic device, known as the interrogator unit, to measure acoustic or vibration signals by sensing the dynamic strain rate along the fiber's axial direction.

[1] *BlueSkyDAS LLC, Sugar Land, Texas, USA*

[2] *Department of Earth and Atmospheric Sciences, University of Houston, Houston, Texas, USA*

[3] *Scripps Institution of Oceanography, University of California San Diego, La Jolla, California, USA*

[4] *TGS-NOPEC Geophysical Company, Houston, Texas, USA*

Distributed Acoustic Sensing in Borehole Geophysics, Geophysical Monograph 289, First Edition.
Edited by Yingping Li, Robert Mellors, and Ge Zhan.

DOI: 10.1002/9781394179275.ch01

The DAS observation system, the interrogator, consists of a laser light source sending pulses through the fiber cable and a photoelectric signal detector that receives and analyzes coherent Rayleigh backscatter from the pulses generated by imperfections inside the optical fiber. DAS instruments can capture sound and vibration signals along tens of kilometers (or more) of sensing fiber cable with a spatial resolution ranging from 0.25 to tens of meters and across a wide frequency band from millihertz to tens of kilohertz. The DAS technologies have various applications in seismic measurements, monitoring downhole activities, detecting earthquakes, and assessing seismic risks. This book primarily delves into the application of DAS technologies in borehole geophysics, although we will also touch on DSS and DTS applications (e.g., Hartog, 2017) in certain chapters to complement the discussion on DAS technology.

Borehole geophysics, a specialized branch of geosciences, entails measuring physical properties within Earth's subsurface rock formations using instruments deployed in wells drilled to different depths. Borehole seismic sensors are employed to deduce the properties of the media immediately surrounding the borehole and to obtain high signal-to-noise ratio (SNR) measurements of signals from more distant sources. This is because the background ambient noise at depth tends to be significantly quieter compared to noise near the surface, where environmental and human activities generate more substantial disruptions.

Challenges in borehole geophysics arise for two reasons: (a) making measurements during borehole operations, such as production or stimulation; and (b) sustaining functionality in harsh environmental conditions with a limited geometry. The conventional method for seismic measurements in a borehole involves using strings of geophones lowered down the well, which can interfere with other operations, or using a dedicated monitoring well. In contrast, the fiber-optic technology for DAS measurements can be installed either outside the casing, allowing measurements without obstructing the borehole, or along tubing inside the well. Additionally, the fiber cable is relatively inexpensive and highly resistant to extreme temperatures and pressures, enabling long-term and continuous monitoring of signals.

The enduring characteristics of optic fibers under high pressure and temperature in harsh downhole environments make DAS an appealing geophysical tool for borehole applications. It enables the measurement of acoustic, vibrational, and related signals in a relatively quiet environment, facilitating continuous monitoring of time-lapse variations in diverse geophysical parameters of rock formations. Moreover, it aids in assessing seismic risks arising from natural and induced earthquakes.

Hill et al. (1978) were pioneers in fabricating fiber Bragg gratings (FBGs) on small segments of fiber cores using a visible laser along the fiber. Subsequently, Meltz et al. (1989) developed the ultraviolet (UV) side-writing process, enabling compact spatial fabrication of FBGs. This breakthrough paved the way for developing various FBG-based distributed sensing (FBG-DxS) sensors capable of measuring strain, temperature, and pressure (Kersey et al., 1992, 1997; Udd, 1991), where "x" represents one of the physical parameters being measured.

The Bragg reflection wavelength, λ_{BR} (or f_{BR}, frequency), is determined by the grating period (Λ), also known as the index modulation. FBG sensors measure strain by detecting the frequency or wavelength shift of the Bragg reflection spectrum due to small variations in Λ caused by applied strain. Udd et al. (1998, 2000) introduced a multiaxis fiber grating technique to measure multidimensional strain fields. Kersey (2000) summarized the development of various FBG sensors for measuring parameters such as strain, pressure, temperature, acceleration, multiphase flow, vibration, and acoustic fields for borehole monitoring in oil and gas reservoirs. Although FBGs can be continuously distributed along a fiber, in practice, FBG sensors with groups of gratings are typically made in a "discrete" manner. When multiple FBG sensors are created along a very long optical fiber, discrete fiber Bragg grating sensors (FBGSs) enable DFOS measurements. Therefore, we refer to "discrete acoustic sensing" using the FBG technique as quasi-distributed acoustic sensing (qDAS) to differentiate it from backscattering-based DAS.

Bostick (2000) documented the inaugural implementation of an FBG-qDAS system in borehole seismic applications within the oil and gas industry. This pioneering initiative took place during a seismic field experiment in December 1999, conducted in a 132-meter-deep test well at the Devine Test Site in Texas, USA. The experiment aimed to demonstrate the effectiveness of the FBG-qDAS system equipped with three-component (3C) acceleration sensors, showcasing its performance against conventional geophones across various parameters such as directional response, sensitivity, signal-to-noise ratio, hodogram, amplitude, and phase of seismic events. The integration of fiber-optic accelerometer data into particle velocity seismograms demonstrated alignment with geophone data as well as differentiation of geophone data into particle acceleration. Moreover, alongside monitoring ambient noise and drilling-bit noise, the experiment encompassed various analyses concerning dynamic range, bandwidth, directionality, and fundamental seismic imaging capabilities. This involved the utilization of dynamite shot holes and a weight drop source in offset and walkaway VSP (vertical seismic profiling) survey geometries. The results revealed a considerable dynamic range, ensuring reliable detection of both compressional and shear waves, along with a broader bandwidth compared to geophones. The validation of the FBG-qDAS system's performance through these experimental results suggests its potential for widespread adoption within the oil and gas industry, attributed to its durability, longevity, and reliability, even in harsh environmental conditions.

The success of the initial experiment sparked a series of tests worldwide on FBG-qDAS systems in both onshore and offshore wells. These tests aimed to explore effective installation methods such as passive mandrel, active clamp, and tube convey, as documented by Hornby et al. (2003, 2005) and Knudsen et al. (2006). They also focused on acquiring offset and walkaway VSP surveys (Bostick et al., 2003; Hornby et al., 2005), obtaining time-lapse seismic images (Blanco et al., 2006; Keul et al., 2005; Knudsen et al., 2003), monitoring downhole noise from flow (Knudsen et al., 2006), and induced micro-seismicity from fracturing (Bostick & Travis, 2014; Gaston & Bostick, 2011). Furthermore, these methods found applications in geothermal fields (Paulsson et al., 2014; Rassenfoss, 2014) and carbon capture and storage (CCS) projects (Gaston & Bostick, 2012; Lescanne et al., 2011; Prinet et al., 2013).

In October 2006, an FBG-qDAS system was installed in a production well in the deep-water Mars field in the Gulf of Mexico (GOM), comprising 11 vertical and 4 horizontal qDAS acceleration sensors. This installation aimed to utilize seismic sources for a 3D VSP survey with a 30-level geophone array, assessing the system's feasibility for 3D and 4D FBG-qDAS surveys. Subsequently, two simultaneous ocean bottom node (OBN)-qDAS-VSP surveys were conducted near the Mars field in 2007 and 2010. These surveys yielded high-quality, high-frequency 3D seismic images (Hornby & Burch, 2008) and 4D seismic images (Hornby et al., 2012; Wu et al., 2011) of target reservoirs. Although only waveforms from the 11 vertical sensors were utilized, the qDAS-VSP images with high-frequency content were comparable to those obtained by OBN data. However, the limited number of sensors and shallow depth coverage constrained the outcomes of the 4D FBG-qDAS surveys. Readers interested in FBG-qDAS technology can refer to review papers by Eriksrud and Kringlebotn (2013) for oil and gas applications and Allwood et al. (2017) for more general applications for further reading.

The optical time domain reflectometer (OTDR) (Barnoski & Jensen, 1976; Barnoski et al., 1977) plays a crucial role as a quality control (QC) and quality assurance (QA) tool used to detect imperfections in fiber resulting in Rayleigh backscattering (RBS). The OTDR can be seen as the predecessor of interrogators for Rayleigh backscattering-based DAS systems (Juskaitis et al., 1994; Shatalin et al., 1998). DAS systems, which rely on RBS (Hartog & Gold, 1984; Healey, 1985), utilize distributed points along the fiber cable, spaced as close as 0.25 to 1 m apart, covering distances of up to 40 km or more. These systems detect vibration signals by measuring the axial strain rate or strain along the fiber, which is influenced by the surrounding medium. The RBS-DAS technology effectively repurposes the otherwise detrimental Rayleigh scatters, which typically degrade transmitted laser signals in telecommunications, turning them into backscattering reflectors for sensing vibration or acoustic signals. This technological evolution represents a significant leap forward in the field.

The initial applications of RBS-DAS technologies in borehole geophysics on land were reported in DAS-VSP surveys, with the goal of enhancing velocity models and improving subsurface imaging (Mateeva et al., 2012; Mestayer et al., 2011; Miller et al., 2012). These technologies have been utilized for various purposes, including monitoring downhole activities (Bateman et al., 2013; Denney, 2012; Johannessen et al., 2012; Koelman, 2011, 2012a, 2012b; Parker et al., 2012; Van Der Horst et al., 2013; Xiao et al., 2013), detecting microearthquakes induced by fracture stimulations (Grandi et al., 2013; MacPhail et al., 2012; Molenaar et al., 2011, 2012; Ugueto et al., 2014; Webster et al., 2013), and carbon capture, utilization, and storage (CCUS) projects (Barberan et al., 2012; Cocker et al., 2014; Correa et al., 2017; Daley et al., 2013; Götz et al., 2015; Harris et al., 2016; Mestayer et al., 2011), as well as geothermal projects (Feigl et al., 2017; Jousset et al., 2016; Lord et al., 2016; Mellors et al., 2019; Reinsch et al., 2015; Wang et al., 2014).

After successful trials of RBS-DAS applications on land, a sparse walk-above DAS-VSP survey was conducted in the North Sea in late 2011 (Madsen et al., 2012, 2013). Additionally, the first dual-well 3D DAS-VSP survey in deep water was conducted simultaneously with an ocean bottom seismometer (OBS) survey in the GOM in 2012 (Mateeva et al., 2013a, 2013b). Analysis of the dual-well 3D DAS-VSP data (Li et al., 2015; Wong et al., 2015; Wu et al., 2014, 2015) showed that DAS-VSP data can effectively enhance velocity models through travel-time tomography, resulting in significantly improved migration images of subsurface structures for both OBS and DAS-VSP surveys. Since then, numerous marine 3D DAS-VSP surveys have been conducted and processed globally (e.g., Abdul Rahim et al., 2017; Ball et al., 2017; Gerritsen et al., 2016; Hance et al., 2016; Jiang et al., 2016; Li et al., 2018; Nizkous et al., 2015; Zhan et al., 2015, 2016). Li et al. (2019) and Duan et al. (2020) utilized a 3D DAS-VSP acquired inside a salt pile in the GOM to delineate salt boundaries using a salt proximity migration imaging method (Wu et al., 2012).

Because repeatable 3D DAS-VSP surveys can be acquired cost-effectively, they serve as valuable tools for monitoring variations in subsurface reservoirs and optimizing oil and gas production. Multiple repeated 3D DAS-VSP surveys conducted in deep-water environments and on land have demonstrated that time-lapse (4D) DAS-VSP surveys can produce "4D" seismic images, revealing both temporal and spatial fluctuations in subsurface reservoirs (Hill, 2014; Kiyashchenko et al., 2013; Mateeva et al., 2013a, 2015). Researchers have also utilized time-lapsed DAS-VSP methods to monitor CO_2 injection processes (e.g., Daley et al., 2016; Harris et al., 2016; Pevzner et al., 2017) and detect temporal-spatial variations in physical properties in subsurface formations caused by hydraulic fracturing treatments (e.g., Byerley et al., 2018; Meek et al., 2017; Teff et al., 2016).

The rapid advancement of RBS-DAS technologies and their expanding applications in geophysics have led to a notable increase in the volume of articles published in international peer-reviewed journals and conferences. According to Li et al. (2021), the number of such articles rose from fewer than 10 in 2011 to over 60 in 2014. However, there was a noticeable slowdown in 2015, coinciding with a major downturn in the oil and gas industry. Nevertheless, the numbers rebounded, with around 100 articles in 2016 and over 200 papers in 2018, attributed to substantial contributions from both DAS instrument manufacturers and service providers, as well as an increased number of users from academia and industry.

At the 2017 AGU Fall Meeting held in New Orleans, scientists and engineers from academia and industry gathered to promote the applications of RBS-DAS in geophysics, seismology, geosciences, and geo-engineering. The event featured 19 papers across two main DAS sessions and six DAS presentations in other sessions. The majority of attendees involved in the development and application of DAS technology expressed a desire for a DAS AGU monograph. Such a publication would serve to advance DAS techniques and geophysical applications by promoting broader public awareness and education. This led to the publication of the first AGU DAS monograph, titled *Distributed Acoustic Sensing in Geophysics: Methods and Applications*, featuring 17 chapters. The realization of this publication was facilitated by the contributions of 74 authors, 53 peer reviewers, and AGU and Wiley staff. Nevertheless, the publication process encountered substantial delays as a result of the COVID-19 pandemic.

The first AGU DAS book encompassed nine chapters on DAS borehole geophysical applications; however, it did not comprehensively cover all significant aspects of the field or reflect recent advancements. Driven by increasing interest in the topic, approval was granted for a second AGU DAS monograph focusing solely on borehole DAS, which is this volume. This book includes 31 chapters organized into four parts: borehole DAS instrumentation and modeling, borehole DAS acquisition and processing, borehole DAS imaging and inversion, and borehole DAS monitoring, as well as introductory and concluding chapters authored by the editors.

The current chapter serves as a brief guide for readers, outlining the challenges faced by the authors and their methodologies for overcoming them. The subsequent parts cover instrumentation and modeling (Part I), acquisition and processing (Part II), imaging and inversion techniques (Part III), and applications in earthquake seismology, fracture zone detection, active fault monitoring, and low-frequency strain monitoring (Part IV). Each part features chapters authored by experts in the field, providing a thorough overview of the current state, challenges, and potential of DAS technology in borehole geophysics and related disciplines. In the concluding chapter, the authors discuss the key research findings and potential future directions of study in the field. Although it is not feasible to cover all significant developments in borehole DAS techniques and applications, the book strives to include important or milestone literature as recommended further reading materials to enhance its comprehensiveness.

1.2. BOREHOLE DAS INSTRUMENTATION AND MODELING

Part I of this book comprises five chapters, three of which focus on DAS instrumentation and the remaining two of which delve into DAS modeling.

Simultaneously measuring multiple physical parameters of subsurface formations—such as dynamic strain rate (DAS), static strain (DSS), and temperature (DTS)—using different modes of optic fibers within the same fiber cable offers significant scientific merits and economic advantages. In Chapter 2, Farhadiroushan et al. (2024) review recent developments in integrated DSS technology for monitoring wellbores and reservoirs using the full backscatter spectrum, including Rayleigh, Brillouin, and Raman scatters. They employed a precision-engineered fiber cable with a lower noise floor and an interferometry technique to measure the Brillouin frequency shift, achieving integrated strain measurements with a fine resolution of 1 μm. Temperature effects can be measured and compensated for using the Raman DTS within the same fiber cable.

Real-time and continuous monitoring and characterization of hydrological and chemical properties of percolating water along deep sections of the vadose zone and underlying saturated zone are crucial for minimizing potential subsurface pollution of underground water. Hua et al. (2024), in Chapter 3, present a novel approach for measuring distributed strain known as coherence-length-gated microwave photonics interferometry (CMPI), which has provided new insights into hydrologic processes within the vadose and saturated zones. CMPI measures strain by detecting optical interference phase changes in the microwave domain, achieving a resolution of 10nε when employing centimeter-scale spatial resolution. The technique was employed to measure strain induced by minor periodic variations in air pressure (amplitude of 25 Pa and frequency of 4 Hz) in a sand-filled laboratory column simulating barometric loading in the vadose zone. The CMPI distributed strain measurement system significantly enhances our capability to continuously monitor and characterize the vadose and saturated zones.

Haldorsen et al. (2024), in Chapter 4, point out the significance of downhole DAS systems positioned in close proximity to targets, enabling near-continuous monitoring of mechanical disturbances arising from hydrocarbon production, fluid flow, and induced or natural earthquakes. This monitoring serves the purpose of optimizing oil and gas production, evaluating associated risks, and providing early warnings of potential critical situations. However, existing RBS-DAS systems only capture disturbances along a single component of axial strain rate along the fiber, leading to ambiguities in measuring azimuthal and incidence angles of incident seismic waves. In this chapter, Haldorsen et al. present a hybrid fully optic qDAS receiver array system with a wide band and wide aperture, comprising a microphone, a hydrophone, and a three-component accelerometer. This configuration aims to offer vector information of the incoming wavefield without relying on downhole electronic units, ensuring reliable and continuous operation even in wells exceeding 200 °C. The optical accelerometer, a key component of the system, is constructed by winding optical fiber around a mass spring capable of flexing in only one direction. As the spring flexes, the fiber extends, and the qDAS interrogator measures the extension ΔL of the fiber, which is proportional to the acceleration of the mass.

Li et al. (2024c), in Chapter 5, emphasize the necessity of employing 3C seismic instruments due to the vector nature of seismic wavefields, particularly in borehole seismic applications. These applications include VSP 3C-3D (three-dimensional) vector migration imaging, measuring both P and S wave velocity profiles, pinpointing the locations of microseismic events, measuring shear-wave splitting for medium anisotropic parameters, and 3D salt proximity, among others. However, current RBS-DAS systems only measure a single-component axial strain rate or strain, making them inadequate for many of these seismic tasks. Despite the significant advantages of DAS systems over conventional borehole seismic instruments, they suffer from three major disadvantages: single-component measurement, fiber directivity, and low SNR. In this chapter, the authors propose a solution in the form of 3D DAS "radar," which involves the installation of a fiber cable in three dimensions to acquire 3C DAS data without ambiguity regarding the azimuthal and incidence angles of incident seismic waves.

High-quality borehole seismic imaging of steeply dipping salt flanks is essential for accurately locating exploration and production wells near the salt boundary in deep-water oil and gas fields, such as the GOM. Understanding the differences in migration images obtained from emerging DAS systems and conventional geophones is critical. In Chapter 6, Zong et al. (2024) use numerical simulations of a subset of the SEAM (SEG advanced modeling) model, which features a steeply dipping salt flank, to study the theoretical wavefields and their corresponding images for both the strain rate field (analogous to DAS signals) and the particle velocity field (analogous to geophone signals). They compare the simulated DAS data to both single-component and multicomponent geophone data. The walkaway VSP geometry is utilized due to its advantages in observing wave propagation characteristics. As current seismic signal processing technologies are primarily developed for standard geophone data, these studies provide important references for correlating DAS strain rate measurements with geophone particle velocity data.

1.3. BOREHOLE DAS ACQUISITION AND PROCESSING

Part II comprises 10 chapters on borehole DAS acquisition and processing.

Acquiring high-quality seismic reflection data in regions with complex near-surface geological structures poses significant technical challenges using conventional seismic survey methods. In Chapter 7, Bakulin et al. (2024) investigate the Smart DAS uphole acquisition system, a pioneering advancement tailored for imaging and monitoring such complex near-surface environments. By integrating surface seismic advantages with borehole geophysics, this system establishes a 3D imaging network using strategically positioned DAS receivers in shallow upholes. Through synthetic and field studies conducted in desert environments, the researchers confirm the system's superior SNR and exceptional repeatability. Furthermore, they outline crucial best practices for uphole drilling and fiber-optic cable installation, which are imperative for technical success and economic feasibility. Importantly, the flexible uphole placement of the Smart DAS system enables imaging over any desired extent, thus overcoming the limitations typically associated with VSP surveys.

Simultaneously acquiring surface and borehole seismic data using seismic nodes and DAS fibers, along with shared seismic sources, offers significant technical and economic advantages. Chapter 8 by Yu et al. (2024) details the acquisition of 3D DAS-VSP data through downhole armored optical cables, conducted alongside an onshore 3D seismic data project in the Dagang Oilfield and 3D OBN data projects in the East China Sea and Abu Dhabi. When collecting surface 2D/3D seismic data, simultaneous acquisition of 2D/3D DAS-VSP data can provide comprehensive, high-resolution structural images around the borehole. Integrating arrival times from both surface and borehole seismic data can refine the velocity model for improved subsurface imaging. Utilizing joint seismic data processing and imaging methods driven by borehole data can significantly enhance the quality of 2D or 3D surface seismic imaging. The chapter also discusses the workflow for processing 3D DAS-VSP data and its integrated interpretation.

The hybrid fiber-wireline cable stands out as an exceptionally effective method for deploying fiber optics in DAS-VSP surveys. These cables ingeniously incorporate optical fiber strands alongside the copper lines of traditional electric wireline cables. This integration of optical fibers into a hybrid optical-electrical logging cable enables the acquisition of seismic data during any logging intervention within the well. Notably, this setup eliminates rig time when VSPs are acquired using the hybrid cable, allowing other tools to remain stationary for their respective operations. This innovative technology holds immense promise for revolutionizing borehole seismic operations, particularly in the exploration and appraisal of oil and gas fields. Moreover, various types of wireline cables and slick lines used for cased-hole interventions now integrate optical fibers as well. Although primarily intended for interventions in producing wells, these cables still facilitate the acquisition of seismic measurements afterward. Martinez et al. (2024), in Chapter 9, provide compelling examples of acquiring DAS-VSP data using the hybrid logging cable technique, establishing it as a standard measurement in exploration and appraisal campaigns. The authors underscore the successful utilization of DAS technology in hybrid cables across diverse environments, including open-cased holes, sub-vertical to deviated well trajectories, shallow and deep wells, various tool combinations, and tension profiles. Notably, hybrid cable DAS-VSP surveys streamline measurements and imaging of the subsurface, not only during production but also throughout the exploration phase, all without the need to install permanent fiber cables in the wells.

Geothermal energy, a form of renewable energy, is gaining attention, with supercritical water emerging as a promising future energy source due to its significantly higher geothermal production capacity resulting from its very high enthalpy. Targeting depths deeper than ordinary geothermal reservoirs, particularly supercritical water zones located 3–4 km near or below the brittle-to-ductile transition zone, presents unique challenges. Traditional borehole geophysical tools often fail to perform adequately under such high-temperature conditions. In Chapter 10, Kasahara et al. (2024b) utilize fiber-optic DAS and DTS systems deployed in high-temperature geothermal boreholes to conduct geophysical explorations across six geothermal fields in Japan. Employing full-waveform inversion (FWI) on deep PP reflections, the authors produce high-quality images of deep structures alongside temperature-depth profiles. Synthetic DAS waveforms aid in interpreting observations, and the authors also investigate potential temperature effects on seismic waves. By integrating DAS and DTS measurements with existing geological, geophysical, and drilling data, the authors refine and update geothermal subsurface structures.

Seismology has long been applied in glacial environments and is now a common tool for studying processes and changes in the cryosphere due to rapid climate warming. In Chapter 11, Brisbourne et al. (2024) provide an overview of DAS applications in glacial settings. Despite its recent introduction in these challenging environments, early studies have demonstrated the feasibility and potential advantages of DAS technology. Particularly in borehole settings, where DAS offers significant advantages, this method has facilitated observations that would have been impractical with traditional seismic instruments. Borehole DAS has been instrumental in detecting seismic anisotropy in deeper ice layers, identifying basal seismic activity and low-amplitude volcanic tremors that might otherwise go unnoticed, and quantifying variations in seismic reflectivity due to climatic influences on grain size. The authors emphasize the critical role of proper coupling during data acquisition and noise reduction during data processing. Additionally, they note that the relatively straightforward velocity structure of the cryosphere, compared to more complex geological environments, presents an ideal setting for developing and validating DAS data processing techniques applicable beyond glaciological research.

The 3D DAS-VSP technique has been instrumental in gas hydrate exploration. Kean Lim et al. (2024) in Chapter 12 delve into the processing and fault interpretation carried out for the largest known onshore 3D DAS-VSP survey at a gas hydrate research well Hydrate-01 on the North Slope of Alaska. The DAS data was recorded via a single-mode fiber-optic cable installed in the well with 1,701 shots of the active seismic source Vibroseis. Because DAS is a unique omnidirectional single-component sensor, a tailored processing approach was developed to provide accurate and high-resolution reflected P-wave images. The processed high-quality 3D DAS-VSP volume clearly revealed minor faults that were absent from the existing 3D surface seismic volume. These results were pivotal in establishing trajectories for production and monitoring wells. This study underscores the critical role of integrating advanced technologies like DAS in geophysical investigations. Such integration not only enhances subsurface imaging but also improves decision-making processes in hydrocarbon exploration and production.

Due to its significant sensitivity to the physical characteristics of geological formations, seismic wave attenuation is a key parameter for further study. Nevertheless, it is challenging to accurately estimate and comprehend the process of seismic attenuation in complex media, like carbonate rocks. Bouchaala et al. (2024), in Chapter 13, aim to clarify how seismic attenuation is affected by the petrophysical characteristics of carbonate rocks. In addition, this chapter looks into appropriate methods for all kinds of seismic data—from seismic to ultrasonic scales, particularly DAS data. These multiscale analyses of seismic attenuation enable us to better understand the attenuation mechanisms in carbonate rocks. The authors demonstrate that attenuation is highly sensitive to the petrophysical characteristics of carbonate rocks. The study observed several distinctions from sandstones, such as the fluid saturation sensitivity of the compressional to the shear attenuation ratio. It has been noted that there are several attenuation processes indicating the complexity of carbonate rocks, which is related to their intricate pore structure. Nonetheless, the precision and vertical spatial resolution of attenuation must be balanced. Although high-density DAS data indeed provides a high spatial resolution of attenuation, comprehensive studies are required to improve the accuracy of attenuation measurements.

Chapter 14 by Sayed et al. (2024) explores the application of DAS in borehole geophysics, focusing on the estimation of seismic quality factor (Q), the DAS-to-velocity transform, and the imaging process. The authors discuss the relationship between axial strain and particle velocity for body waves and evaluate the effect of typical variation in local velocities on the estimation of Q using synthetic data. The study compares different methods of imaging using DAS data, highlighting the differences in frequency and amplitude response between DAS and geophone sensors and the effects of scaling and correction factors on the data. It emphasizes the importance of gauge length correction and the effectiveness of the DAS-to-velocity transform in producing consistent and accurate images. This chapter highlights the importance of DAS-VSP applications and their capacity to significantly improve the accuracy of seismic imaging for subsurface structures when compared to images generated using traditional 3C geophone data.

Accurately, precisely, and efficiently picking first-arrival times is an important task for many active and passive borehole seismic surveys. This is particularly crucial for borehole DAS applications with a very high density of fiber sensors, such as 1 m of channel space over a fiber cable of more than a few kilometers. Li et al. (2024a), in Chapter 15, develop an automatic first-arrival time-picking method based on the differential-phase-spectral slope method (DPSSM). For two seismic events, or the same event recorded by two nearby sensors with highly similar waveforms, the differential phase spectrum, $\Delta\Phi$ (f), is proportional to the differential time, Δt, between the two events. Thus, the differential time can be determined by matching the slope in the plot of the differential phase spectra versus the frequency. The authors automated this method and applied it to analyze two zero-offset DAS-VSP common-shot gathers obtained from two near-vertical geothermal wells at the Utah FORGE (Frontier Observatory for Research in Geothermal Energy) site. The two wells contain 1,010 and 1,206 DAS channels, respectively, with a DAS channel space of 1 m. Using the automatically picked first-arrival times, the authors generate interval and average P-wave velocity profiles and find that these profiles are comparable to those obtained from sonic logs and velocity surveys with a perforate shot and some deeper-induced earthquakes. The authors demonstrate that DPSSM can be an effective tool to accurately and quickly pick hundreds to thousands of first-arrival picks on DAS-VSP data and expect to further develop this method for locating microearthquakes and updating velocity models with travel-time tomography.

In Chapter 16, Lanza et al. (2024) evaluate the potential of downhole DAS technology to detect and locate distant microseismicity. Identifying and locating microseismicity at remote distances is crucial for reservoir monitoring both during and after operations, as it allows for the assessment of geomechanical conditions and the detection of pre-existing, favorably oriented faults that could affect the stimulation region. The authors leverage different scales with examples from the full-scale Utah FORGE site, USA, and the mesoscale Bedretto Underground Laboratory, Switzerland. They compare performances in terms of data quality, SNR, detectability, and the applicability of a newly developed location methodology for out-of-network seismicity (HADES-R).

1.4. BOREHOLE DAS IMAGING AND INVERSION

Part III presents seven chapters on borehole DAS imaging and inversion.

Chapter 17 by Chen et al. (2024) delves into a rapid ray-based borehole seismic imaging method and showcases its applications in P-wave and converted wave imaging of subsurface structures utilizing DAS-VSP survey data. The authors employ a multitemplate fast-advancement algorithm (MSFM) to compute the travel timetable. By allowing for flexible migration apertures, their method amalgamates the benefits of both the VSP-CDP transform and depth migration techniques. This approach concentrates on imaging near the reflection path, diminishes migration artifacts, and tackles the issue of uneven coverage imaging amplitude by considering the fold of coverage during the imaging process. The authors illustrate satisfactory DAS-VSP P-wave and converted wave images achieved through the grid ray tracing imaging method with both numerical synthetics and field data examples.

Effectively leveraging borehole DAS-VSP data for subsurface imaging projects poses a significant technical challenge for borehole geophysicists. Mizuno et al. (2024), in Chapter 18, aim to tackle this challenge by developing a workflow to maximize the utility of borehole DAS data. The authors delve into two key aspects: DAS-VSP survey designs and DAS data processing procedures essential for obtaining accurate subsurface images. The authors emphasize the critical role of survey design, highlighting its ability to facilitate large-scale DAS surveys in a quantitative and automated manner. They propose that correcting DAS data responses within the imaging algorithm is pivotal in achieving high-quality DAS images comparable to those produced by 3C borehole geophones. Furthermore, the authors explore the generalized radon transform (GRT) migration problem, an extension of Kirchhoff migration, as part of their study.

According to theoretical and seismic modeling studies, rock fractures with crack widths as small as a millimeter are expected to generate diffracted wavefields when stimulated by seismic waves with frequencies below 100 Hz. If these modeling predictions hold true, diffracted wavefields from such thin fractures could potentially form the basis of a time-lapse seismic method for analyzing reservoir fracture patterns and behavior over the course of the reservoir's life cycle. In Chapter 19, Fuller (2024) utilizes Kirchhoff depth migration images derived from time-lapse borehole DAS-VSPs obtained in a well that underwent hydraulic stimulation. The objective is to determine whether diffracted wavefields from fractures are observable in the time-lapse borehole DAS data by summing along anticipated time-offset curves of diffractions. The time-lapsed DAS-VSP surveys were conducted using Vibroseis sources at the surface, with DAS fiber cable cemented behind casing in the fracture stimulation well. A baseline survey was completed prior to hydraulic stimulation, and subsequent surveys were conducted after stimulation, flow back, and several weeks of production. The positioning of DAS fiber in the horizontal leg of the stimulation well aimed to capture any potential diffraction, given its proximity to both natural and hydraulically stimulated fractures in the reservoir. Fuller examines the 2D Kirchhoff depth migration images generated from the time-lapse DAS-VSP dataset to determine if diffracted wavefields from thin fractures within the reservoir are detectable, thus providing insights into characterizing reservoir fracture patterns.

Although capturing tubing-strapped fiber 3D DAS-VSP data in flowing wells significantly reduces the cost of fiber deployment and data acquisition, processing the recorded data can be challenging due to the poor SNR. Ali et al. (2024) discuss in Chapter 20 the processing and imaging of four-well tubing-strapped 3D DAS-VSP data acquired under flowing conditions. The coupling of fiber to the formation varies widely in such acquisitions, leading to significant spectrum overlaps between the VSP signal and flow noise. Obtaining a relative amplitude image in deviated wells requires crucial DAS directivity correction. Multiwell DAS geometries present additional challenges, including inconsistent lighting, misaligned reflectors across wells due to inaccurate migration models, and varying illumination angles.

The Groß Schönebeck Geothermal Research Laboratory in Germany serves as a European testing center for low-permeability deep geothermal reservoir development. Various geophysical research programs, including extensive 3D surface seismic and 3D VSP utilizing a wireline DAS technique, were conducted at the center in 2017. The hybrid wireline DAS-VSP data were successfully collected in two geothermal boreholes at depths of approximately 4.2 km, marking one of the earliest publicly accessible instances of wireline DAS operation. Martuganova et al. (2024), in Chapter 21, focus on analyzing and integrating both existing geological and new geophysical data, particularly 3D DAS-VSP images derived with the hybrid wireline DAS system. This chapter illustrates that reservoir properties may be derived with high resolution even under challenging conditions such as elevated temperature, pressure, and salinity.

Kasahara et al. (2024a), in Chapter 22, provide a notable example of the application of DAS and DTS technology to obtain subsurface structural details in high-temperature geothermal fields for commercial purposes. The DAS and DTS surveys were conducted in 2022 at the Takigami geothermal field in northeastern Kyushu, Japan. Optical fiber cable was deployed down to a depth of 2,000 m in the TP-2 geothermal well with 12 seismic sources near the Takigami field. The authors measure the borehole temperature profile using DTS and construct a subsurface P-wave velocity depth profile using DAS. They obtain migration images of subsurface structures using the reflected seismic waves recorded by the DAS system.

The FWI technique is crucial for uncovering subsurface structures under complex geological conditions. To make seismic acquisition more cost-effective, DAS provides dense data sampling, making it a practical alternative to traditional sensors. Cai et al. (2024), in Chapter 23, combine DAS-VSP and accelerometer data in elastic FWI acquired at the Newell County Facility in Alberta, Canada. The authors aim to assess field velocities by implementing multiazimuth FWI and comparing the inversion results across two azimuths within the full dataset. They use a single-parameter compressional velocity (Vp) parameterization based on well-log analysis and an effective source scheme to improve the reliability of elastic FWI with DAS and accelerometer data. The developed depth registration scheme, based on cross-correlation scans of the two datasets, could yield reliable depth estimates for each channel of DAS data. DAS, combined with acceleration measurements from accelerometer data, has resulted in improved accuracy in subsurface imaging.

1.5. BOREHOLE DAS MONITORING

Part IV encompasses seven chapters focusing on borehole DAS monitoring. Four chapters delve into DAS borehole monitoring of earthquakes and ambient noise for fracture zones. The remaining three chapters explore borehole monitoring of induced microseismicity by fracture stimulation and low-frequency DAS (LF-DAS) events.

DAS systems installed in wellbores have been utilized to monitor downhole activities, including various flows, fractures, and microearthquakes induced by fracture stimulations for oil and gas production, CCUS, and geothermal energy development. However, downhole fibers have thus far been underutilized in earthquake seismology studies. In a brief review in Chapter 24, Lellouch (2024) discusses the potential, capabilities, and limitations of downhole DAS in a broad seismological context. The spatial continuity of DAS measurements is especially crucial for local earthquakes, whereas for distant events, the lower noise levels at depths may play a significant role.

Japan is one of the most earthquake-prone and volcanic countries, boasting abundant geothermal energy resources. In Chapter 25, Hasada et al. (2024) evaluate optical fiber DAS systems through complementary observations with surface seismometers. The authors obtain high-quality DAS records of seismic signals using DAS interrogators and optic fibers deployed in existing high-temperature geothermal wells. Through a comparison of DAS and surface seismometer records, the authors identify P-to-S and S-to-P converted waves. Observing these converted waves via DAS proves valuable for estimating seismic impedance contrasts in the subsurface of geothermal fields, which could be indicative of the fluid contents of geothermal reservoirs. During relatively short periods of DAS geothermal exploration experiments in the hot wellbores across different geothermal fields in Japan, it is not surprising that numerous natural earthquakes are recorded, underscoring the potential for DAS systems, primarily employed in geothermal resource exploration, to function as DAS networks for monitoring natural earthquakes and volcanic tremors and thus providing early warnings for geohazard risks posed by large, damaging earthquakes and violent volcanic eruptions.

Active faults are challenging to monitor in situ with high resolution due to their unknown depths or heavily weathered outcrops. In Chapter 26, Huang et al. (2024) address this challenge by using an optic fiber array installed in two boreholes located on both sides of an active fault, with DAS interrogators on the surface, to monitor ground motions at various depths across the fault zone. The Milun Fault, ruptured in the 2018 ML6.4 Hualien earthquake, was drilled through by the Milun Fault Drilling and All-inclusive Sensing project (MiDAS), creating a unique cross-fault fiber array for downhole earthquake monitoring and active fault detection. The high-fidelity DAS system records abnormally large DAS amplitudes, enabling the authors to identify a 20-meter-thick major fault zone structure consistent with well logging core samples. Variations in subsurface velocity structures over time are mapped using a strain-amplitude-based method. The authors anticipate broader applications of this method in geoenergy, groundwater, and natural earthquake monitoring to mitigate geohazards in earthquake-prone areas.

Accurately locating and monitoring fracture zones around boreholes is crucial for developing energy resources and subsurface waste storage. In Chapter 27, Li et al. (2024b) developed a source-free logging method based on borehole DAS ambient noise analyses to identify and locate fractures around injection wells. The authors analyze 48 hours of ambient noise data and construct DAS ambient noise root mean square (RMS) depth profiles for both sedimentary and granite formations and compare them with the associated wireline logging data. They apply this method at a geothermal site in the Utah FORGE. Their analysis results for RMS amplitudes of DAS ambient noise indicate strong correlations between noise peaks and physical properties, such as low velocity, low density, high porosity, and low Poisson ratio, corresponding to fracture zones in both granite and sedimentary formations. Therefore, the DAS noise RMS peaks can be used to identify the location of fractures, interfaces, and faults. Such a source-free logging method using borehole DAS ambient noise is expected to have broader applications in oil, gas, and geothermal energy exploration and development, as well as CCUS.

Low-frequency DAS (LF-DAS) is an emerging, valuable technology utilizing fiber-optic cables to monitor subsurface strain changes with remarkable precision. Its high spatial and temporal resolution enables comprehensive subsurface strain monitoring, facilitating qualitative and quantitative assessments across hydraulic fracturing, fracture monitoring, geomechanical modeling, and downhole pressure monitoring. In Chapter 28, Ning and Jin (2024) introduce LF-DAS and its advantages for subsurface imaging, followed by discussions on sensing cable deployment methods, qualitative interpretations of LF-DAS data for various oil-field applications, geomechanical modeling, and quantitative investigation. Furthermore, they delve into the emerging utilization of Rayleigh frequency shift (RFS)-based DSS technology for strain monitoring during production. Throughout the chapter, ongoing research efforts and prospects for LF-DAS are emphasized, making it a valuable resource for geoscientists and industry practitioners.

In Chapter 29, Ma et al. (2024) detail two pioneering DAS monitoring projects in Canada known as the Canadian Dip-in DAS (CanDiD) projects. These initiatives involved deploying temporary and retrievable optical fibers to monitor hydraulic fracturing operations. The primary objective of CanDiD was to assess the efficacy of retrievable optical fibers for fracturing monitoring through the analysis of microseismic and LF-DAS signals. The authors discovered that the SNR of LF-DAS signals resembled that of cemented-in DAS fibers. However, microseismic signals recorded by the retrieval fiber exhibited a higher fraction of coherent noise compared to cemented fibers. Additionally, wireline-deployed DAS captured numerous fiber waves induced by wireline slippage, with their spatiotemporal distribution aligning with fracture hits observed in LF-DAS data. These findings suggest that dip-in DAS systems hold significant potential for mapping fracture geometry without the necessity of installing permanent optical fibers behind casing, offering a cost-effective and flexible alternative.

DAS and DSS are emerging as excellent tools for characterizing and monitoring the geomechanical response of rock masses during the injection and/or production of fluids in subsurface operations. However, due to the relative novelty of this application, approaches are still under development. In Chapter 30, Rodríguez Tribaldos et al. (2024) evaluate the potential of these technologies and the impact of various experimental components (e.g., cable type, installation conditions, choice of fiber-optic technology) on the results by analyzing DAS and DSS monitoring data from four fluid injection experiments conducted in two different underground research laboratories (URLs). Their observations are compiled into a list of lessons learned, many of which are transferable to field-scale investigations.

1.6. SUMMARY

The first chapter of the book serves as an introduction to DAS in borehole geophysics, focusing on recent applications and advancements in DAS technology. It explains the principles behind DAS and its role within DFOS, highlighting its ability to measure acoustic/vibration signals with high spatial and temporal resolution along fiber-optic cables. This chapter discusses the challenges and advantages of using DAS in borehole geophysics, detailing its applications in seismic measurements, monitoring downhole activities, detecting microearthquakes, and assessing seismic risks. It also mentions the evolution of DAS technology, from FBG-qDAS to RBS-DAS, and its expanding use in various geophysical projects globally, including oil and gas reservoir monitoring, CCS projects, and geothermal applications. This chapter concludes by highlighting the growth of research and publications in the field, leading to the development of dedicated monographs and conferences focused on advancing DAS techniques and applications in geophysics and related disciplines.

Part I encompasses five chapters focusing on DAS instrumentation and modeling for borehole applications. Chapter 2, by Farhadiroushan et al. (2024), explores the latest advancements in integrated DSS technology for well and reservoir monitoring, employing different modes of optic fibers within the same cable for simultaneous DAS, DSS, and DTS measurements. Hua et al. (2024), in Chapter 3, introduce CMPI for hydrological process monitoring. In Chapter 4, Haldorsen et al. (2024) present a hybrid 3C qDAS receiver array system for continuous downhole monitoring without any downhole electronic units. Li et al. (2024c), in Chapter 5, emphasize the need for 3C DAS instruments and propose 3D DAS "radar" for improving and enhancing 3C DAS observations. Chapter 6, by Zong et al. (2024), conducts numerical simulations to compare DAS strain rate measurements with geophone particle velocity data, focusing on steeply dipping salt flank imaging in deep-water oil and gas fields like the GOM.

Part II delves into various aspects of borehole DAS acquisition and processing across 10 chapters. It begins with Chapter 7, by Bakulin et al. (2024), which explores the Smart DAS uphole acquisition system designed for imaging and monitoring complex near-surface environments. Chapter 8, by Yu et al. (2024), discusses the advantages of simultaneous acquisition of surface and borehole seismic data using DAS fibers, showcasing its potential for high-resolution imaging. Martinez et al. (2024), in Chapter 9, highlight the effectiveness of hybrid fiber-wireline cables for DAS-VSP surveys, particularly in exploration and appraisal campaigns. Chapter 10, by Kasahara et al. (2024b), focuses on utilizing DAS and DTS systems in high-temperature geothermal boreholes, offering insights into geophysical explorations and deep structure imaging. Brisbourne et al. (2024), in Chapter 11, explore DAS applications in glacial settings, emphasizing its role in studying cryosphere processes. Chapter 12, by Kean Lim et al. (2024), delves into gas hydrate exploration using 3D DAS-VSP techniques, showcasing the integration of DAS in geophysical investigations. Bouchaala et al. (2024), in Chapter 13, investigate seismic wave attenuation in carbonate rocks using DAS data, providing insights into attenuation mechanisms. Sayed et al. (2024), in Chapter 14, discuss the application of DAS in borehole geophysics, focusing on seismic quality factor estimation and imaging processes. Lastly, Chapter 15, by Li et al. (2024a), introduces an automatic first-arrival time-picking method for DAS-VSP data, demonstrating its effectiveness in accurately and efficiently picking first-arrival times for seismic surveys. Lanza et al. (2024), in Chapter 16, assess the potential of downhole DAS technology for detecting and locating distant microseismic events, highlighting its importance for reservoir monitoring and comparing its effectiveness across different scales and sites.

Part III encompasses several chapters on borehole DAS imaging and inversion techniques. It begins with Chapter 17, by Chen et al. (2024), which introduces a rapid ray-based borehole seismic imaging method for P-wave and converted wave imaging using DAS-VSP survey data. Chapter 18, by Mizuno et al. (2024), focuses on maximizing the utility of borehole DAS data through survey design and data processing, including addressing the GRT migration problem. Fuller (2024), in Chapter 19, explores diffracted wavefields from fractures in time-lapse borehole DAS-VSPs, examining their potential in reservoir fracture analysis. Ali et al. (2024), in Chapter 20, discuss processing challenges and imaging techniques for tubing-strapped 3D DAS-VSP data, emphasizing the importance of meticulous processing for quality imaging. Martuganova et al. (2024), in Chapter 21, analyze wireline DAS-VSP data from a geothermal research center, demonstrating the feasibility of high-resolution data acquisition in challenging conditions. Lastly, Chapter 22, by Kasahara et al. (2024a), showcases DAS and DTS technology applications in high-temperature geothermal fields, and Chapter 23, by Cai et al. (2024), presents elastic FWI using DAS-VSP and accelerometer data, emphasizing the technique's cost-effectiveness and reliability in subsurface imaging under complex geological conditions.

Part IV delves into the applications of borehole DAS technology across seven chapters, covering earthquake seismology, fracture zone detection, active fault monitoring, and low-frequency strain monitoring. Chapter 24, by Lellouch (2024), offers a thorough review of borehole DAS applications in earthquake seismology, emphasizing its potential benefits alongside strategies to mitigate higher noise levels. In Chapter 25, Hasada et al. (2024) evaluate optical fiber DAS systems in geothermal wells, showcasing their efficacy in earthquake and volcanic tremor monitoring, providing valuable insights for seismic hazard assessment in seismically active regions. Chapter 26, by Huang et al. (2024), focuses on active fault monitoring using DAS technology, demonstrating its capability to detect ground motions along depths across active fault zones, aiding geohazard mitigation in earthquake-prone areas. Li et al. (2024b), in Chapter 27, introduce a source-free logging method based on borehole DAS ambient noise analyses for identifying fractures around injection wells, offering a cost-efficient and accurate means of fracture detection for energy exploration. Chapter 28, by Ning and Jin (2024), discusses LF-DAS technology's qualitative and quantitative assessments in subsurface strain monitoring, showcasing its potential in various oil-field applications. Ma et al. (2024), in Chapter 29, detail two pioneering DAS monitoring projects known as the CanDiD projects, showcasing the potential of retrievable optical fibers for fracture mapping and subsurface monitoring. Lastly, Chapter 30, by Rodríguez Tribaldos et al. (2024), evaluates DAS and DSS technologies' potential in subsurface operations, emphasizing ongoing research needs to optimize approaches and address challenges, consolidating observations from fluid injection experiments. Overall, this chapter offers a detailed exploration of DAS technology's evolving role in subsurface monitoring and geomechanical response characterization, showcasing its potential across diverse applications in energy exploration, seismic hazard assessment, and geohazard mitigation.

ACKNOWLEDGMENTS

We extend our gratitude to all of the contributing authors whose efforts have brought this monograph to fruition. The authors are particularly thankful to Dr. Taka'aki Taira, an AGU Books Editorial Board member, for his meticulous review of this manuscript and for providing valuable feedback and suggestions to enhance the chapter. We also express our appreciation to the rest of the AGU Books Editorial Board and the editorial team at Wiley for their ongoing support of this monograph.

REFERENCES

Abdul Rahim, M. F., Ghazali, A. R., Hardy, R. J. J., Bourdon, L. M., El Kady, M. N., Muhammed, M. D. D., & Konuk, T. (2017). *Estimation of anisotropy in the presence of gas cloud using 3D DAS VSP.* Paper presented at the Asia

Petroleum Geoscience Conference and Exhibition. Expanded Abstracts, 43580.

Ali, S., Miranda, F., Emmanuel, S., & Podgornova, O. (2024), Multiwell distributed acoustic sensing 3D vertical seismic profile under flowing conditions: toward solving the processing and imaging challenges. This volume: Chapter 20.

Allwood, G., Wild, G., & Hinckley, S. (2017). Fiber Bragg grating sensors for mainstream industrial processes. *Electronics*, *6*(4), 92. doi:10.3390/electronics6040092

Bakulin, A., Silvestrov, I., Smith, R., & Golikov, P. (2024), Smart distributed acoustic sensing uphole acquisition system: bridging the gap between surface seismic and borehole geophysics for imaging and monitoring in complex near-surface environments. This volume: Chapter 7.

Ball, M., Mathieson, A., & Soulas, S. (2017). *The evolution of life of field seismic data on the Clair Field.* Paper presented at the 1st EAGE Workshop on Practical Reservoir Monitoring. Expanded Abstracts, We PRM 09. doi:10.3997/2214-4609.201700018

Barberan, C., Allanic, C., Avila, D., Hy-Billiot, J., Hartog, A., Frignet, B., & Lees, G. (2012). *Multi-offset seismic acquisition using optical fiber behind tubing.* Paper presented at the 74th EAGE Conference and Exhibition. Expanded Abstracts, Y003. doi:10.3997/2214-4609.20148798

Barnoski, M. K., & Jensen, S. M. (1976). Fiber waveguides: a novel technique for investigating attenuation characteristics. *Applied Optics*, *15*(9), 2112–2115. doi:10.1364/AO.15.002112

Barnoski, M. K., Rourke, M. D., Jensen, S. M., & Melville, R. T. (1977). Optical time domain reflectometer. *Applied Optics*, *16*, 2375–2379. doi:10.1364/AO.16.002375

Bateman, K., Molenaar, M. M., & Brown, M. D. (2013). *Lessons learned from shell's history of casing conveyed fiber optic deployment.* Paper presented at the SPE Unconventional Resources Conference, SPE-167211-MS. doi:10.2118/167211-MS

Blanco, J., Knudsen G. S., & Bostick, F. X. (2006). *Time-lapse VSP field test for gas reservoir monitoring using permanent fiber optic seismic system.* Paper presented at the 76th SEG Annual International Meeting. Expanded Abstracts, 3447–3451. doi:10.1190/1.2370250

Bostick, F. X. (2000). *Field experimental results of three-component fiber-optic seismic sensors.* Paper presented at the 70th SEG Annual International Meeting. Expanded Abstracts, 21–24. doi:10.1190/1.1815889

Bostick, F. X., Knudsen, S., Nakstad, H., Blanco, J., & Mastin, E. (2003). *Permanently installed fiber optic multi-station 3-C in-well seismic trial at Izaute Field.* Paper presented at the 65th EAGE Conference and Exhibition. doi:10.3997/2214-4609-pdb.6.F45

Bostick, T., & Travis. P. (2014). *Optical multi-component borehole seismic systems continue to deliver unique permanent downhole data.* Paper presented at the 76th EAGE Conference and Exhibition. Expanded Abstracts, 413–414. doi:10.3997/2214-4609.20140609

Bouchaala, F., Rouxel, S., & Zhou, B. (2024), Potential of seismic attenuation for exploring complex media: a focus on carbonate rocks. This volume: Chapter 13.

Brisbourne, A., Hudson, T., & Kendall, M. (2024). Borehole and surface applications of distributed acoustic sensing for characterization of the cryosphere and glacial environments. This volume: Chapter 11.

Byerley, G., Monk, D., Aaron, P., & Yates, M. (2018). Time-lapse seismic monitoring of individual hydraulic frac stages using a downhole DAS array. *The Leading Edge*, *37*, 802–810. doi:10.1190/tle37110802.1

Cai, X., Kristopher, I., Mattew, E., Scott, K., Hu, Q., & Lawton, D. (2024), Multiazimuth elastic full-waveform inversion of fiber-optic and accelerometer vertical seismic profiling data: a field example from the Carbon Management Canada Newell County Facility, Alberta, Canada. This volume: Chapter 23.

Chen, Y., Zong, J., Xia, S., Yu, G., Fei, W., Hao, Z., & Hu, G. (2024), Distributed acoustic sensing-vertical seismic profiling P-P and P-S wave imaging method: a fast ray-based approach through a multitemplate fast-advancement algorithm. This volume: Chapter 17.

Cocker, J. D., Herkenhoff, E. F., Craven, M. E., Nemeth, T., Daley, T. M., White, D., & Strudley, A. (2014). *Simultaneous acquisition of DAS and conventional down-hole geophone array at Aqusitore, Canada.* Paper presented at the 76th EAGE Conference and Exhibition. Extended Abstracts, WS15-C03. doi:10.3997/2214-4609.20140616

Correa, J., Egorov, A., Tertyshnikov, K., Bona, A., Pevzner, R., Dean, T., et al. (2017). Analysis of signal to noise and directivity characteristics of DAS VSP at near and far offsets — A CO2CRC Otway Project data example. *The Leading Edge*, *36*(12), 994a1–994a7. doi:10.1190/tle36120994a1.1

Daley, T. M., Freifeld, B. M., Ajo-Franklin, J., Dou, S., Pevzner, R., Shulakova, V., et al. (2013). Field testing of fiber-optic distributed acoustic sensing (DAS) for subsurface seismic monitoring. *The Leading Edge*, *32*, 699–706. doi:10.1190/tle32060699.1

Daley, T. M., Miller, D. E., Dodds, K., Cook, P., & Freifeld, B. M. (2016). Field testing of modular borehole monitoring with simultaneous distributed acoustic sensing and geophone vertical seismic profiles at Citronelle, Alabama. *Geophysical Prospecting*, *64*(5), 1318–1334. doi:10.1111/1365-2478.12324

Denney, D. (2012). Distributed acoustic sensing for hydraulic-fracturing monitoring and diagnostics. *Journal of Petroleum Technology*, *64*(3), 68–74. SPE-0312-0068-JPT. doi:10.2118/0312-0068-JPT

Duan, Y., Li, Y., Kryvohuz, M., Mateeva, A., & Chen, T. (2020). *3D salt-boundary imaging with transmitted waves in DAS VSP data acquired in salt.* Paper presented at the SEG Technical Program. Expanded Abstracts, 370–374. doi:10.1190/segam2020-3428055.1

Eriksrud, M., & Kringlebotn, J. T. (2013). Fiber optic sensor technology for oil and gas applications. In A. Cutolo, et al. (Eds.), *Photonics for safety and security* (pp. 309–333). World Scientific. doi:10.1142/9789814412971_0014

Farhadiroushan, M., Parker, T. R., Sergey Shatalin, S., Chen Z., Gillies, A., Aghayev, R., Pérez, D., Maxwell, J., & Milne, C. (2024), Integrated distributed strain sensing for wellbore and reservoir monitoring. This volume: Chapter 2.

Feigl, K. L., Patterson1, J., Parker, L., Reinisch, E. C., Zeng, X., Cardiff, M. A.,et al. (2017). *Characterization of material properties at Brady Hot Springs, Nevada by inverse modeling of data from seismology, geodesy, and hydrology*. Paper presented at the AGU Fall Meeting, Abstract H41O-01.

Fuller, B. (2024), Seismic imaging of reservoir fractures using distributed acoustic sensing data. This volume: Chapter 19.

Gaston, G., & Bostick, T. (2011). Practical downhole production and reservoir monitoring with optical sensors. *First Break, 29* (7), 86–88. doi:10.3997/1365-2397.29.7.51917

Gaston, G., & Bostick. T. (2012). *Permanent downhole optical sensing continues to deliver value - well into the 4th dimension*. Paper presented at the 74th EAGE Conference and Exhibition. Expanded Abstracts, Y007. doi:10.3997/2214-4609.20148802

Gerritsen, S., Ernst, F., Field, C., Abdullah, Y., Daud, D. N. P. H., & Nizkous, I. (2016). Velocity model building challenges and solutions in a SE Asian basin: beyond reflection tomography. *First Break, 34*(10), 91–97. doi:10.3997/1365-2397.34.10.86978

Götz, J., Lüth, S., Henninges, J., & Reinsch T. (2015). *Using a fibre optic cable as distributed acoustic sensor for vertical seismic profiling at the Ketzin CO_2 storage site*. Paper presented at the 77th EAGE Conference and Exhibition. Expanded Abstracts, Th P2 13. doi:10.3997/2214-4609.201413363

Grandi, S. K., Webster, P., Hornman, K., Lumens, P.G.E., Franzen, A., Kindy, F., et al. (2013). *Microseismic applications using DAS*. Paper presented at the 4th EAGE Passive Seismic Workshop. Expanded Abstracts, PS02. doi:10.3997/2214-4609.20142338

Haldorsen, J., Christensen, C., Metzbower, R., & Berzanskis, A. (2024), A fiber-optic, multicomponent sensor for borehole-seismic formation imaging and for monitoring of natural or induced seismicity. This volume: Chapter 4.

Hance, T., Jiang, T., Zhan, G., Kjos, E., Geetan, R., Soulas, S., & Thomas, I. (2016). *Learnings from distributed acoustic sensing data processing for seismic applications - a case study from the North Sea*. Paper presented at the 78th EAGE Conference and Exhibition. Extended Abstracts, Th STZ2 04. doi:10.3997/2214-4609.201601601

Harris, K., White, D., Melanson, D., Samson, C., Daley, T. M., & T.M. (2016). Feasibility of time-lapse VSP monitoring at the Aquistore CO_2 storage site using a distributed acoustic sensing system. *International Journal of Greenhouse Gas Control, 50*, 248–260. doi:10.1016/j.ijggc.2016.04.016

Hartog, A. H. (2017). *An introduction to distributed optical fibre sensors*. Boca Raton, FL: CRC Press.

Hartog, A. H., & Gold, M. P. (1984). On the theory of backscattering in single-mode optical fibers. *Journal of Lightwave Technology, 2*(2), 76–82. doi:10.1109/JLT.1984.1073598

Hasada, Y., Kasahara, J., Mikada, H., & Ohnuma, H. (2024), Investigation of distributed acoustic sensing characteristics by complementary observations of distributed acoustic sensing in geothermal wells and by surface seismometers. This volume: Chapter 25.

Healey, P. (1985). Statistics of Rayleigh backscatter from a single-mode optical fibre. *Electronics Letters, 21*(6), 226–228. doi:10.1049/el:19850161

Hill, D. J. (2014). *Case study of a multi-well 4D DAS-VSP*. Paper presented at the 76th EAGE Conference and Exhibition. Extended Abstracts, WS15-B02. doi:10.3997/2214-4609.20140612

Hill, K. O., Fujii, Y., Johnson, D. C., & Kawasaki, B. S. (1978). Photosensitivity in optical fiber waveguides: application to reflection filter fabrication. *Applied Physics Letters, 32*(10), 647–649. doi:10.1063/1.89881

Hornby, B. & Burch, T. (2008). *Passive "drive by" imaging in a deepwater production well using permanent borehole seismic sensors*. Paper presented at the 78th SEG Annual International Meeting. Expanded Abstracts, 349–352. doi:10.1190/1.3054820

Hornby, B., Barkved, O., Sugianto, H., & Nahm, J. (2012). *4D monitoring with permanent borehole seismic sensors: progress, challenges, and the road ahead*. Paper presented at the 82nd SEG Annual International Meeting. doi:10.1190/segam2012-1519.1

Hornby, B. E., Bostick, F. X., Williams, B., Lewis, K., & Garossino, P. (2005). Field test of a permanent in-well fiber-optic seismic system. *Geophysics, 70*(4), E11–E19. doi:10.1190/1.1993709

Hornby, B. E., Williams, B. A., Lewis, K. A., Garossino, P. G., & Bostick, F. X. (2003). *Field test of a permanent in-well fiber optic seismic system*. Paper presented at the 73rd SEG Annual International Meeting. Expanded Abstracts, 2219–2222. doi:10.1190/1.1817788

Hua, L., Murdoch, L., Zhu, X., Germanovich, L., Wu, Y., Guo, J., Blais, R., Plunkett, G., Costantino, O., DeWolf, S., Moak, R., & Xiao, H. (2024), A microwave photonics optical fiber method for measuring distributed strain for hydrologic applications in the vadose and saturated zones. This volume: Chapter 3.

Huang, H. H., Ma, K. F., Wu, E. S., Cheng, Y. Z., Lin, C. J., Ku, C. S., & Su, P. L. (2024), Spatiotemporal monitoring of a frequent-slip fault zone using downhole distributed acoustic sensing at the MiDAS Project. This volume: Chapter 26.

Jiang, T., Zhan, G., Hance, T., Sugianto, S., Soulas, S., & Kjos, E. (2016). *Valhall dual-well 3D DAS VSP field trial and imaging for active wells*. Paper presented at the 86th SEG Annual International Meeting. Expanded Abstracts, 5582–5586. doi:10.1190/segam2016-13871754.1

Johannessen, K., Drakeley, B., Farhadiroushan, M. (2012). *Distributed acoustic sensing - a new way of listening to your well/reservoir*. Paper presented at the SPE Intelligent Energy International. Extended Abstracts, SPE 149602. doi:10.2118/149602-MS

Jousset, P., Reinsch, T., Henninges, J., Blanck, H., & Ryberg, T. (2016). Strain and ground-motion monitoring at magmatic areas: ultra-long and ultra-dense networks using fibre optic sensing systems. General Assembly European Geosciences Union. *Geophysical Research Abstracts, 18*, EGU2016-15707

Juskaitis, R., Mamedov, A. M., Potapov, V. T., & Shatalin, S. V. (1994). Interferometry with Rayleigh backscattering in a single-mode optical fiber. *Optics Letters, 19*(3), 225–227. doi:10.1364/OL.19.000225

Kasahara, J., Hasada, Y., & Mikada, H. (2024a), Geothermal reservoir characterization using distributed acoustic sensing from vertical seismic profiling in six geothermal fields in Japan. This volume: Chapter 10.

Kasahara, J., Hasada, Y., Furuya, S., Shimazaki, M., Mikada, H., Ohnuma, H., & Fujise, Y. (2024b), distributed acoustic sensing and distributed temperature sensing geophysical study in the Takigami geothermal field in Northeastern Kyushu, Japan. This volume: Chapter 22.

Kean Lim, T., Fujimoto, A., Kobayashi, T., Tcherkashnev, S., & Mondanos, M. (2024). STW stratigraphy test well (Hydrate-01) distributed acoustic sensing 3D vertical seismic profiling processing. This volume: Chapter 12.

Kersey, A. D. (2000). Optical fiber sensors for permanent downwell monitoring applications in the oil and gas industry. *IEICE Transactions on Electronics*, *E83-C*(3), 400–404.

Kersey, A. D., Berkoff, T. A., & Morey, W. W. (1992). High resolution fiber Bragg grating based strain sensor with interferometric wavelength shift detection. *Electronics Letters*, *28*(3), 236–238. doi:10.1049/el:19920146

Kersey, A. D., Davis, M. A., Patrick, H. J., LeBlanc, M., Koo, K. P., Askins, C. G., et al. (1997). Fiber grating sensors. *Journal of Lightwave Technology*, *15*(8), 1442–1463. doi:10.1109/50.618377

Keul, P. R., Mastin, E., Blanco, J., Maguérez, M., Bostick, F., & Knudsen, S. (2005). Using a fiber-optic seismic array for well monitoring. *The Leading Edge*, *24*(1), 68–70. doi:10.1190/1.1859704

Kiyashchenko, D., Lopez, J., Berlang, W., Birch, B., Zwaan, M., Adawi, R., et al. (2013). Steam injection monitoring in a field of South Oman with borehole seismic – from single pattern to field scale surveillance with DAS. *The Leading Edge*, *32*, 1246–1256. doi:10.1190/tle32101246.1

Knudsen, S., Havsgård, G. B., Berg, A., & Bostick, F., (2006). *Flow-induced noise in fiber-optic 3-C seismic sensors for permanent tubing-conveyed installations.* Paper presented at the 68th EAGE Conference and Exhibition. doi:10.3997/2214-4609.201402113

Koelman, J. V. (2011). Fiber-optic sensing technology providing well, reservoir information – anyplace, anytime. *Journal of Petroleum Technology*, *63*(7), 22–23. doi:10.2118/0711-0022-JPT

Koelman, J. V., Lopez, J. L., & Potters, H. (2012a). *Fiber optic technology for reservoir surveillance.* Paper presented at the International Petroleum Technology Conference, IPTC-14629-MS. doi:10.2523/IPTC-14629-MS

Koelman, J. V., Lopez, J. L., & Potters, H. (2012b). *Optical fibers: the neurons for future intelligent wells.* Paper presented at the SPE Intelligent Energy International, SPE 150203. doi:10.2118/150203-MS

Knudsen, S., Havsgard, G. B., Berg, A., Nakstad, H., & Bostick, T. (2003). *Permanently installed high-resolution fiber optic 3C/4D seismic sensor systems for in-well imaging and monitoring applications.* Paper presented at the SPIE Sixth Pacific Northwest Fiber Optic Sensor Workshop. doi:10.1117/12.544383

Lanza, F., Tuinstra, K., Rinaldi, A. P., Rodriguez Tribaldos, V., Hopp, C., Bethmann, F., Dyer, B., Meier, P., Serbeto, F., Hertrich, M., Domenico, G., & Wiemer, S. (2024), A prospective evaluation of borehole distributed acoustic sensing technologies for detecting and locating out-of-network microseismicity. This volume: Chapter 16.

Lellouch, A. (2024), Seismological applications of downhole distributed acoustic sensing. This volume: Chapter 24.

Lescanne, M., Hy-Billiot, J., Aimard, N., & Prinet, C. (2011). The site monitoring of the Lacq industrial CCS reference project. *Energy Procedia*, *4*, 3518–3525. doi:10.1016/j.egypro.2011.02.279

Li, D., Huang, L., Zheng, Y., & Li, Y. (2024a), Automatic picking of distributed acoustic sensing vertical seismic profiling first-arrival traveltimes using a differential-phase-spectral slope method. This volume: Chapter 15.

Li, D., Huang, L., Zheng, Y., & Li, Y. (2024b), Monitoring granitic and sedimentary rocks at the Utah FORGE site using borehole distributed acoustic sensing ambient noise. This volume: Chapter 27.

Li, Y., Duan, Y., Chen, T., & Johnson, D. (2019). *Delineating salt boundaries using 3D DAS-VSP acquired inside salt in the Gulf of Mexico.* Paper presented at the 89th SEG Annual International Meeting. Expanded Abstracts, 5355–5359. doi:10.1190/segam2019-w1-02.1

Li, Y., Karrenbach, M., & Ajo-Franklin, J. B. (2021). A literature review: distributed acoustic sensing (DAS) geophysical applications over the past 20 years. Chapter 17. In Y. Li, M. Karrenbach, & J. B. Ajo-Franklin (Eds.), *Distributed acoustic sensing in geophysics: Methods and applications*, Geophysical monograph 268 (pp. 229–291). Hoboken, NJ: John Wiley & Sons, Inc. doi:10.1002/9781119521808.ch17

Li, Y., King, K., Wong, W., Chen, T., & Wu, H. (2018). *Diagnosing azimuthal variations of velocity model uncertainties using 3D DAS-VSP in GOM.* Paper presented at the 88th SEG Annual International Meeting. Expanded Abstracts, 5511–5515. doi:10.1190/segam2018-w20-05.1

Li, Y., Li, D., Huang, L., & Zheng, Y. (2024c), Three-component distributed acoustic sensing arrays with three-dimensional fiber cable deployment. This volume: Chapter 5.

Li, Y., Wu, H., Wong, W., Hewett, B., Liu, Z., Mateeva, A., & Lopez, J. (2015). *Velocity analysis and update with 3D DAS-VSP to improve borehole/surface seismic images.* Paper presented at the 85th SEG Annual International Meeting. Expanded Abstracts, 5285–5289. doi:10.1190/segam2015-5864923.1

Lord, N. E., Zeng, X., Fratta, D., Feigl, K. L., H. F. Wang, H. F., & PoroTomo Team. (2016). *Multispectral analysis of surface wave (MASW) analysis of near-surface structure at Brady Hot Springs from active source and ambient noise using a 8700-meter distributed acoustic sensing (DAS) array.* Paper presented at the AGU Fall Meeting. Abstract S13B-2559. https://agu.confex.com/agu/fm16/meetingapp.cgi/Paper/190705

Ma, Y., Wang, C., Maji, V., Salvage, R., MacDougall, K., & Eaton, D. (2024), Recent advances in the use of temporary optical fiber deployment for downhole hydraulic fracture monitoring. This volume: Chapter 29.

MacPhail, W.F.P., Lisoway, B., & Banks, K. (2012). *Fiber optic distributed acoustic sensing of multiple fractures in a horizontal well.* Paper presented at the SPE Hydraulic Fracturing Technology Conference, SPE-152422-MS. doi:10.2118/152422-MS

Madsen, K. N., Parker, T., & Gaston, G. (2012). *A VSP field trial using distributed acoustic sensing in a producing well in the North Sea.* Paper presented at the 74th EAGE Conference and Exhibition. Expanded Abstracts, Y006. doi:10.3997/2214-4609.20148801

Madsen, K. N., Thompson, M., Parker, T., & Finfer, D. (2013). A VSP field trial using distributed acoustic sensing in a producing well in the North Sea. *First Break*, *31*(11), 51–56. doi:10.3997/1365-2397.2013027

Martinez, A., Useche, M., & Sayed, A. (2024), Distributed acoustic sensing–acquired wellbore seismic data using hybrid wireline cable: field data examples. This volume: Chapter 9.

Martuganova, E., Norden, B., Blöcher, G., & Krawczyk, C. M. (2024), Enhanced geothermal reservoir characterization using wireline distributed acoustic sensing. This volume: Chapter 21.

Mateeva, A., Hornman, J. C., Hatchell, P., Potters, H., & Lopez, J. (2015). *Frequent seismic monitoring for pro-active reservoir management*. Paper presented at the 85th SEG Annual International Meeting. Expanded Abstracts, 4817–4821. doi:10.1190/segam2015-5899850.1

Mateeva, A., Lopez, J., Mestayer, J., Wills, P., Cox, B., Kiyashchenko, D., et al. (2013a). Distributed acoustic sensing for reservoir monitoring with VSP. *The Leading Edge*, *32*(10), 1177–1300. doi:10.1190/tle32101278.1

Mateeva, A., Mestayer J., Yang Z., Lopez J., Wills P., Roy J., & Bown, T. (2013b). *Dual-well 3D VSP in deepwater made possible by DAS*. Paper presented at the 83rd SEG Annual International Meeting. Expanded Abstracts. 5062–5066. doi:10.1190/segam2013-0667.1

Mateeva, A., Mestayer, J., Cox, B., Kiyashchenko, D., Wills, P., Lopez, J., et al. (2012). *Advances in distributed acoustic sensing (DAS) for VSP*. Paper presented at the 82nd SEG Annual International Meeting. Expanded Abstracts, 1–5. doi:10.1190/segam2012-0739.1

Meek, R., Woller, K., George, M., Hull, R., Bello, H., & Wagner, J. (2017). *Time-lapse imaging of a hydraulic stimulation using 4D vertical seismic profiles and fiber optics in the Midland Basin (Part B)*. Paper presented at the SPE/AAPG/SEG Unconventional Resources Technology Conference, URTEC-2695394-MS. doi:10.15530/URTEC-2017-2695394

Mellors, R. J., Sherman, C., Fu, P., McLennan, J., Morris, J., Ryerson, F., & Morency, C. (2019). *Potential use of distributed acoustic sensors to monitor fractures and microseismicity at the FORGE EGS site*. Paper presented at the 44th Workshop on Geothermal Reservoir Engineering, SGP-TR-214.

Meltz, G., Morey, W. W., & Glenn, W. H. (1989). Formation of Bragg gratings in optical fibers by a transverse holographic method. *Optics Letters*, *14*(15), 823–825. doi:10.1364/ol.14.000823

Mestayer, J., Cox, B., Wills, P., Kiyashchenko, D., Lopez, J., Costello, M., et al. (2011). *Field trials of distributed acoustic sensing for geophysical monitoring*. Paper presented at the 81st SEG Annual International Meeting. Expanded Abstracts. 4253–4257. doi:10.1190/1.3628095

Miller, D., Parker, T., Kashikar, S., Todorov, M., & Bostick, T. (2012). *Vertical seismic profiling using a fiber-optic cable as a distributed acoustic sensor*. Paper presented at the 74th EAGE Conference and Exhibition. Expanded Abstracts, Y004. doi:10.3997/2214-4609.20148799

Mizuno, T., Ali, S., Bettinelli, P., & Le Calvez, J. (2024), Distributed acoustic sensing vertical seismic profiling imaging workflow: transitioning from the geophone era. This volume: Chapter 18.

Molenaar, M. M., Hill, D. J., Webster, P., Fidan, E., & Birch, B. (2011). *First downhole application of distributed acoustic sensing (DAS) for hydraulic fracturing monitoring and diagnostics*. Paper presented at the SPE Hydraulic Fracturing Technology Conference, SPE 140561. doi:10.2118/140561-MS

Molenaar, M. M., Hill, D. J., Webster, P., Fidan, E., & Birch, B. (2012). *First downhole application of distributed acoustic sensing (DAS) for hydraulic fracturing monitoring and diagnostics*. Paper presented at the SPE Drilling & Completion, SPE-140561-PA. doi:10.2118/140561-PA

Ning, Y., & Jin, G. (2024), Enhancing unconventional reservoir development through low-frequency distributed acoustic sensing technology. This volume: Chapter 28.

Nizkous, I., Gerritsen, S., Kiyashchenko, D., & Joinson, D. (2015). *Distributed acoustic sensing (DAS) VSP for imaging and velocity model building*. Paper presented at the International Petroleum Technology Conference, IPTC-18483-MS. doi:10.2523/IPTC-18483-MS

Parker, T., Shatalin, S. V., Farhadiroushan, M., Kamil, Y. I., Gillies, A., Finfer, D., & Estathopoulos, G. (2012). *Distributed acoustic sensing - a new tool for seismic applications*. Paper presented at the 74th EAGE Conference and Exhibition. Expanded Abstracts, Y002. doi:10.3997/2214-4609.20148797

Paulsson, B.N.P., Toko, J. L., Thornburg J. A., & He R. (2014). *Fiber optic vector sensors*. Paper presented at the 76th EAGE Conference and Exhibition. Expanded Abstracts, 415–416. doi:10.3997/2214-4609.20140610

Pevzner, R., Urosevic, M., Popik, D., Shulakova, V., Tertyshnikov, K., Caspari, E., et al. (2017). 4D surface seismic tracks small supercritical CO_2 injection into the subsurface: CO2CRC Otway Project. *International Journal of Greenhouse Gas Control*, *63*, 150–157. doi:10.1016/j.ijggc.2017.05.008

Prinet, C., Thibeau, S., Lescanne, M., & Jacques Monne, J. (2013). Lacq-Rousse CO_2 capture and storage demonstration pilot: lessons learnt from two and a half years monitoring. *Energy Procedia*, *37*, 3610–3620. doi:10.1016/j.egypro.2013.06.254

Rassenfoss, S. (2014). In search of better reservoir imaging using fiber optic receivers downhole. *Journal of Petroleum Technology*, *66*(06), 58–65. doi:10.2118/0614-0058-JPT

Reinsch, T., Henninges, J., Götz, J., Jousset, P., Bruhn, D., & Lüth, S. (2015). *Distributed acoustic sensing technology for seismic exploration in magmatic geothermal areas*. Paper presented at the World Geothermal Congress 2015. https://gfzpublic.gfz-potsdam.de/pubman/item/item_1182900

Rodríguez Tribaldos, V., Hopp, C., Rinaldi, A. P., Tuinstra, K., Lanza, F., Ajo-Franklin, J., Kneafsey, T., Robertson, M., Guglielmi, Y., & Zappone, A. (2024), Use of distributed acoustic sensing and distributed strain sensing technologies for geomechanical characterization of rock mass response in mesoscale experiments in underground laboratories: lessons learned. This volume: Chapter 30.

Sayed, A., Khaitan, M., Martinez, A., & Stewart, R. R. (2024), Distributed acoustic sensing to velocity transform for vertical seismic profiling applications. This volume: Chapter 14.

Shatalin, S. V., Treshikov, V. N., & Rogers, A. J. (1998). Interferometric optical time-domain reflectometry for distributed optical-fiber sensing. *Applied Optics*, *37*(24), 5600–5604. doi:10.1364/AO.37.005600

Teff, J., Silver, K., Langton, D., Kahn, D., Borell, J., & Chavarria, J. A. (2016). *A technical comparison of downhole methods through fiber optic VSP, in the Eagle Ford Formation.* Paper presented at the SPE/AAPG/SEG Unconventional Resources Technology Conference, URTEC-2462817-MS. doi:10.15530/URTEC-2016-2462817

Udd, E. (1991). *Fiber optical sensors: An introduction for engineers and scientists.* Hoboken, NJ: John Wiley & Sons.

Udd, E., Schulz, W. L., Seim, J. M., Haugse, E. D., Trego, A., Johnson, P. E., et al. (2000). Multidimensional strain field measurements using fiber optic grating sensors. *SPIE Proceedings*, *3986*, 254–262. doi:10.1117/12.388113

Udd, E., Schulz, W. L., Seim, J. M., Nelson, D. V., & Makino, A. (1998). Transverse fiber grating strain sensors based on dual overlaid fiber gratings on polarization-preserving fibers. *SPIE Proceedings*, *3330*, 253–263. doi:10.1117/12.316981

Ugueto, G. A., Ehiwario, M., Grae, A., Molenaar, M., Mccoy, K., Huckabee, P., & Barree, B. (2014). *Application of integrated advanced diagnostics and modeling to improve hydraulic fracture stimulation analysis and optimization.* Paper presented at the SPE Hydraulic Fracturing Technology Conference, SPE-168603-MS. doi:10.2118/168603-MS

Van Der Horst, J., Den Boer, H., In 't Panhuis, P., Kusters, R., Roy, D., Ridge, A., & Godfrey, A. (2013). *Fibre optic sensing for improved wellbore surveillance.* Paper presented at the International Petroleum Technology Conference, IPTC-16873-MS. doi:10.2523/IPTC-16873-MS

Wang, H. F., Lord, N. E., Chalari, A., Lancelle, C., Baldwin, J. A., Castongia, E., et al. (2014). *Field trial of distributed acoustic sensing using active sources at Garner Valley, California.* Paper presented at the 2014 AGU Fall Meeting, NS41C-07.

Webster, P., Wall, J., Perkins, C., & Molenaar, M. (2013). *Micro-seismic detection using distributed acoustic sensing.* Paper presented at the 83rd SEG Annual International Meeting. Expanded Abstracts, 2459–2463. doi:10.1190/segam2013-0182.1

Wong, W., Wu, H., Li, Y., Hewett, B., Liu, Z., Yang, Z., & Wills, P. (2015). *Experimental processing of dual-well 3D DASVSP simultaneously acquired with OBS in deep water environment in GOM.* Paper presented at the 3rd EAGE Workshop on Borehole Geophysics, Expanded Abstracts, BG18. doi:10.3997/2214-4609.201412197

Wu, H., Kiyashchenko, D., & Lopez, J. (2011). Time-lapse 3D VSP using permanent receivers in a flowing well in the deepwater Gulf of Mexico. *The Leading Edge*, *30*(9), 1052–1058. doi:10.1190/1.3640529

Wu, H., Li, Y., & Hewett, B. (2012), *Offset VSP P wave and converted S wave Salt Proximity Migration Study.* Paper presented at the 82nd SEG Annual Meeting, Technical Program. Expanded Abstracts, 1–5. doi:10.1190/segam2012-1392.1

Wu, H., Wills, P., Li, Y., Wong, W., Hewett, B., & Liu, Z. (2014). *3D DAS-VSP processing and learning – Velocity diagnosis and update for seismic imaging improvement.* Paper presented at the 76th EAGE Conference and Exhibition, Workshops. Extended Abstracts, WS15-D04. doi:10.3997/2214-4609.20140620

Wu, H., Wong, W., Yang, Z., Wills, P., Lopez, J., Li, Y., et al. (2015). Dual-well 3D vertical seismic profile enabled by distributed acoustic sensing in deep water Gulf of Mexico. *Interpretation*, *3*, SW11–SW25. doi:10.1190/INT-2014-0248.1

Xiao, J., Farhadiroushan, M., Clarke, A., Khalifa, Q., Mulhem, A., Reyes, H. F., et al. (2013). *Inflow monitoring in intelligent wells using distributed acoustic sensor.* Paper presented at the SPE Middle East Intelligent Energy Conference and Exhibition, SPE-167447-MS. doi:10.2118/167447-MS

Yu, G., Xiong, J., Liu, C., Wu, J., Chen, Y., Zhao, Y., Cao, Z., He, G., Zhang, Y., Li, X., & Chen, H. (2024), Joint surface and borehole DAS-VSP data acquisition and processing. This volume: Chapter 8.

Zhan, G., Kommedal, J., & Nahm, J. (2015). *VSP field trials of distributed acoustic sensing in Trinidad and Gulf of Mexico.* Paper presented at the 85th SEG Annual International Meeting. Expanded Abstracts, 5539–5543. doi:10.1190/segam2015-5876420.1

Zhan, G., Li, Q., Nahm, J., Kommedal, J., & Konkler, J. (2016). Field trials of distributed acoustic sensing for reservoir delineation with VSP. *CSEG Recorder*, *41*(6), 34–37.

Zong, J., Yang, J., Li, Y. E., Cheng, A., Wu, X., & Willis, M. (2024), Comparing distributed acoustic sensing and geophone vertical seismic profiling imaging: acquisition efficiency of distributed acoustic sensing versus the multicomponent advantage of geophone data. This volume: Chapter 6.

Part I

Borehole Distributed Acoustic Sensing Instrumentation and Modeling

2

Integrated Distributed Strain Sensing for Wellbore and Reservoir Monitoring

Mahmoud Farhadiroushan, Tom R. Parker, Sergey Shatalin*, Ze Chen, Arran Gillies, Rufat Aghayev, Daniel Pérez, Jack Maxwell, and Craig Milne

ABSTRACT

Distributed fiber-optic sensing, where the fiber is permanently installed along the wellbore, offers many advantages for subsurface reservoir monitoring. Using multiple fibers housed in a cable, the full backscatter spectrum can be used to combine distributed acoustic sensing (DAS), distributed strain sensing (DSS), and distributed temperature sensing (DTS) along the entire wellbore. DAS utilizes Rayleigh backscatter along standard single-mode fiber. However, more recently, a precision-engineered fiber with brighter low-loss scatter centers and an advanced optoelectronic interrogator offers a step-change in DAS performance, with a 20 dB (100 times) lower noise floor compared to that achieved with the standard fiber. The precision-engineered fiber can also be used as a "fiber ruler" where the distances between the scatter centers along the fiber are measured to evaluate the changes in the integrated strain. The strain measurements can be calibrated with a new DSS where an interferometric technique is used to measure the Brillouin frequency shift with a fine resolution of 1 $\mu\varepsilon$. Temperature effects can be measured and compensated for using a Raman DTS.

2.1. INTRODUCTION

Distributed fiber-optic sensing, where the fiber is permanently installed along the wellbore, offers many advantages for subsurface monitoring. Distributed acoustic sensing (DAS) utilizes Rayleigh backscatter light and allows seismic data to be acquired along a continuous length of fiber, providing a high-definition seismic measurement along the entire wellbore (Parker et al., 2014). In addition, advances in precision-engineered fiber with brighter scatter centers and a new optoelectronic interrogator offer a step-change in DAS performance, with a 20 dB (100 times) lower noise floor compared to that achieved with standard fiber. This enables accurate measurement of seismic velocity changes in the formation as well as monitoring minute ground motions (on the picometer scale) and the gradual buildup of low-frequency strain along the wellbore.

There is a growing interest in monitoring subsurface reservoir changes, such as subsidence and caprock integrity, during fluid injection and production operations. DAS has been used for seismic measurements and microseismic event detection. The low-frequency response of DAS provides a highly sensitive technique for monitoring slow strain changes due to poroelastic pressure buildup (Richter et al., 2019) or pressure drawdown in the reservoir. Time-shift mapping of seismic data has also been used to study reservoir changes, including subsidence

Silixa Ltd., Elstree, UK

* *This chapter is dedicated to Dr. Sergey Shatalin, who was the Chief Scientist of Silixa from 2008 until he passed away on 30 December 2022.*

Distributed Acoustic Sensing in Borehole Geophysics, Geophysical Monograph 289, First Edition.
Edited by Yingping Li, Robert Mellors, and Ge Zhan.

DOI: 10.1002/9781394179275.ch02

monitoring. Seismic waves propagating through the formation experience different travel times as the stress and strain fields change due to pressure depletion in the reservoir. The strain time shift is accumulated due to changes in the rock dimensions and variations in the rock density, which is typically a factor of 5 to 10 times greater than the dimensional effect, with the ratio between the two referred to as the "R" factor (Hatchell & Bourne, 2005; Røste et al., 2005).

For long-term monitoring, two new distributed sensing capabilities have been developed: (a) interferometric distributed strain sensing (IDSS) based on measuring the frequency of Brillouin backscatter light with high accuracy and (b) a novel optical "fiber ruler" that precisely measures the separation of scatter centers across different sections of engineered fiber.

These distributed strain measurements can be used to monitor changes in strain along the wellbore, such as those resulting from reservoir pressure changes due to injection and production operations. Temperature effects can be measured and compensated for using a Raman DTS.

We briefly outline the principle of the natural elastic and inelastic scattering in optical fiber, namely the Rayleigh, Brillouin, and Raman mechanisms (Hartog, 2017). The elastic scattering mechanism may be enhanced using different fiber core dopants, writing continuous chirped fiber Bragg gratings along the fiber, and introducing precision-engineered scatter centers. We also highlight the advantages of precision-engineered fiber by controlling the phase and amplitude of the scatter centers to achieve shot-noise performance with two orders of magnitude acoustic sensitivity improvement compared to that of standard fiber.

The rock "R" factor—or, as we propose here, the rock index factor "RI"—can be mapped by combining the distributed DAS seismic time shift with strain measurements using LF-DAS, IDSS, and the fiber ruler. A DTS measurement can be used for temperature measurements and, in addition, for compensating strain data for temperature variations.

2.2. DISTRIBUTED SENSORS BASED ON OPTICAL TIME DOMAIN TECHNIQUE

Figure 2.1 shows the principle of distributed optical fiber sensor, where an optoelectronic interrogator is connected to a length of sensing fiber that can be up to tens of kilometers. A laser pulse is launched into the sensing fiber, and as the laser pulse propagates through the fiber, it is perturbed by different scattering mechanisms (Hartog, 2017). A portion of the scattered light is captured by the same sensing fiber and then returned to the optoelectronic interrogator. The sensing location can be determined by measuring the time of flight of the backscatter light. For a silica glass fiber with a refractive index of about n = 1.5, the time of flight for the laser pulse to the sensing location and the return of the backscatter light back to the interrogator is about 10 ns/m (two-way propagation).

2.3. OPTICAL FIBER SCATTERING MECHANISM

Figure 2.2 shows the scattering mechanisms that a laser pulse undergoes as it propagates along a length of fiber. The small variations in refractive index along the fiber act as tiny reflectors, which result in Rayleigh scattering

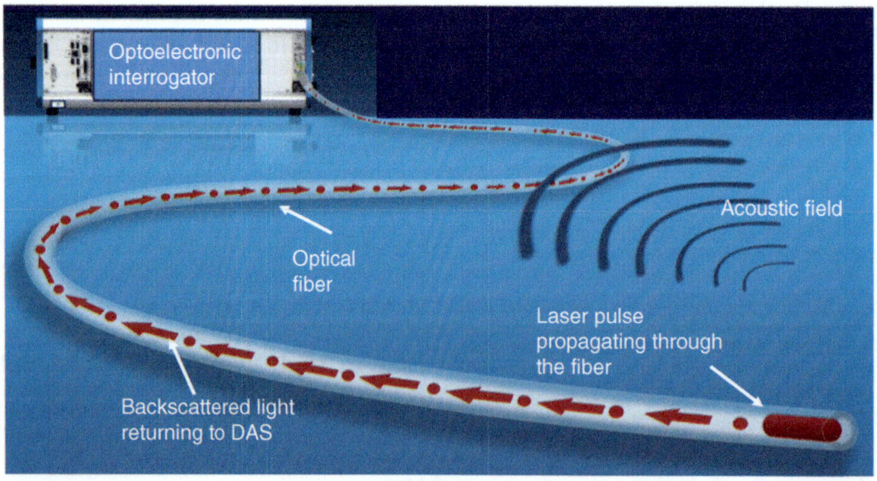

Figure 2.1 Distributed fiber-optic sensor where a laser pulse is launched into the optical fiber and the backscatter light along the fiber is returned to the optoelectronic interrogator. The backscatter light along the fiber is used to measure the acoustic vibration, strain, and temperature along the sensing optical fiber.

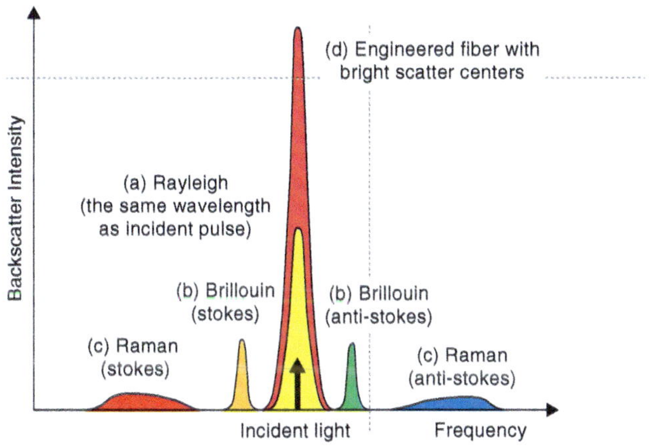

Figure 2.2 Optical fiber scatter mechanisms: (a) elastic Rayleigh scattering, (b) Brillouin and (c) Raman in-elastic Stokes and anti-Stokes scattering, (d) engineered fiber with bright scatter centers.

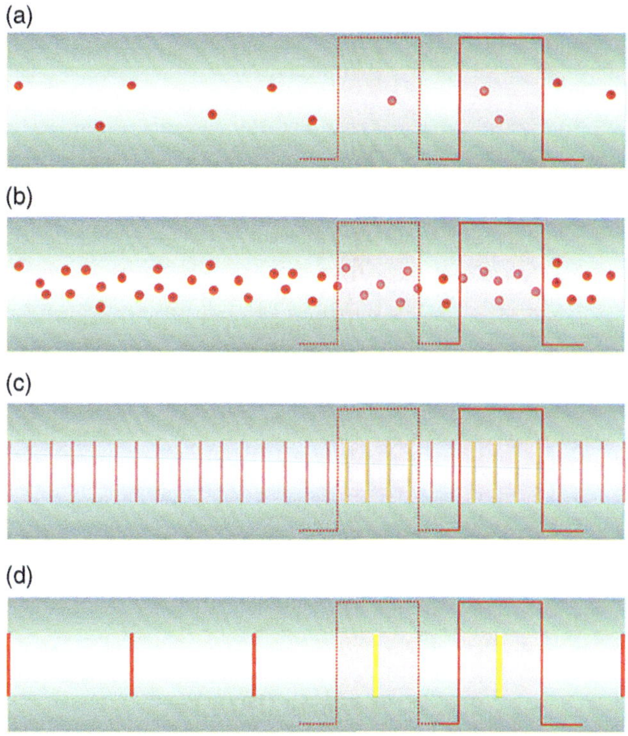

Figure 2.3 Different manufacturing techniques for enhancing the backscattered light along a fiber. The positions of the laser pulse illuminating the sensing fiber along its length are shown by the two red rectangular pulses. (a) Standard fiber with small random fluctuations of the refractive index resulting in Rayleigh backscatter light with varying amplitude and phase. (b) Highly doped fiber with a higher Rayleigh scatter cross-section. (c) Continuous chirped fiber Bragg gratings to enhance backscattering compared to background Rayleigh. (d) Precision-engineered fiber with bright backscatter centers to control the amplitude and phase of reflected laser pulses.

at the same frequency as the laser pulse. The backscattered Rayleigh signal is utilized in DAS (Shatalin et al., 2021). This is elastic scattering where the optical frequency of the laser light and backscatter light are the same.

Brillouin and Raman scattering are inelastic scattering and are generated as the result of the acoustic and optical phonon energy in the fiber, respectively. The frequency of the Stokes scattered light is downshifted, and the anti-Stokes scattered light is upshifted compared to the laser pulse. The Brillouin backscatter light has been used for distributed temperature and strain sensing (DTSS) (Parker et al., 1998), and the Raman backscatter light is used for DTS (Dakin et al., 1985).

2.4. DAS WITH DIFFERENT TYPES OF FIBERS

The elastic scattering mechanism may be enhanced using different fiber manufacturing processes, as indicated in Figure 2.3. In standard fiber (see Figure 2.3 (a)), a small fraction of the scattered light is captured back along the fiber. The scattering mechanism may be enhanced by increasing the doping of the core material, as indicated in Figure 2.3(b). However, this technique also increases the fiber attenuation and limits the sensing range. More recently, an advanced technique has been used to create a continuous series of fiber Bragg gratings along the fiber to significantly enhanced the reflected light, as indicated in Figure 2.3 (c). This type of enhanced backscatter fiber can offer some benefits for long-range sensing applications (Westbrook et al., 2020). However, the additional phase variation between multiple fiber Bragg gratings can limit the full benefits of the enhanced backscatter light to further improve the sensitivity in near-range sensing applications.

The precision-engineered fiber, indicated in Figure 2.3 (d), provides bright scatter centers where the optical amplitude and phase of the return signal along the sensing fiber are precisely controlled. In this case, we can reach shot-noise limited performance with the new DAS interrogator, resulting in a 20 dB (100 times) improvement in performance down to sub-picostrain /$\sqrt{\text{Hz}}$ resolution. Figure 2.4 is a plot of the frequency response of the DAS system (with 10 m gauge length) using the standard single-mode fiber (green) and the precision-engineered fiber (red). In this case, the noise level of –80 **dB rel rad** $\sqrt{\text{Hz}}$ for standard single-mode fiber corresponds to 2.5 pε/$\sqrt{\text{Hz}}$, and –100 **dB rel rad/** $\sqrt{\text{Hz}}$ corresponds to 0.25 pε/$\sqrt{\text{Hz}}$.

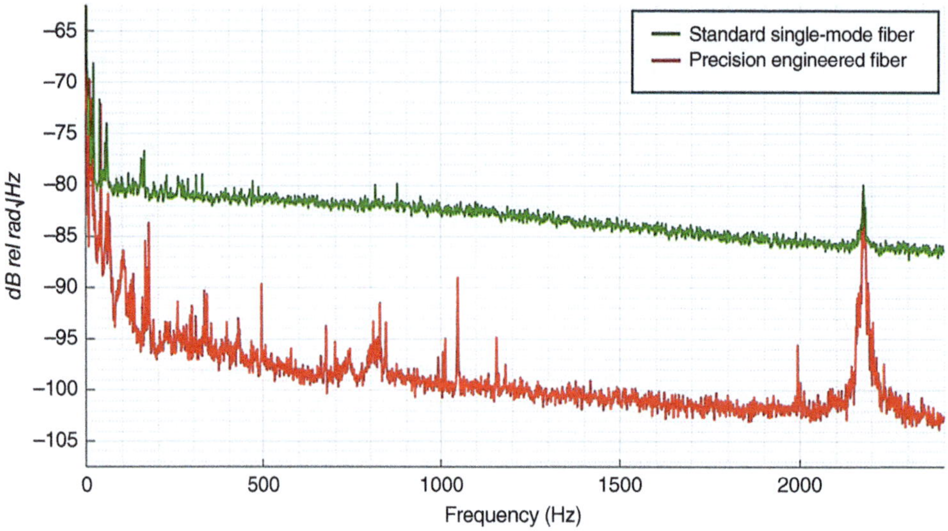

Figure 2.4 The frequency spectra response of the DAS system (10 m gauge length) using the standard single-mode fiber (green) and precision-engineered fiber (red).

2.5. LOW-FREQUENCY DAS

The DAS system measures the strain rate as a function of time t between two scatter centers at position z, which can be expressed as

$$\text{strain} - \text{rate} = \dot{\varepsilon}(t, z) = (\partial/\partial t)(\partial/\partial z)u(t, z) \quad (2.1)$$

where $\dot{\varepsilon}$ is strain rate and u is distance along the fiber. By integrating the strain rate, we can measure very small strain changes over a period of time to measure the low-frequency strain profile with fine resolution, as indicated in equation (2.2):

$$\varepsilon(z) = \int_{t_1}^{t_2} \dot{\varepsilon}(t, z)dt \quad (2.2)$$

Figure 2.5 shows the displacement sensitivity of the DAS system with standard fiber and the precision-engineered fiber compared to a seismometer and geophones. The finer sensitivity of the DAS system with precision-engineered fiber is especially significant at frequencies below 1 Hz compared to the seismometer.

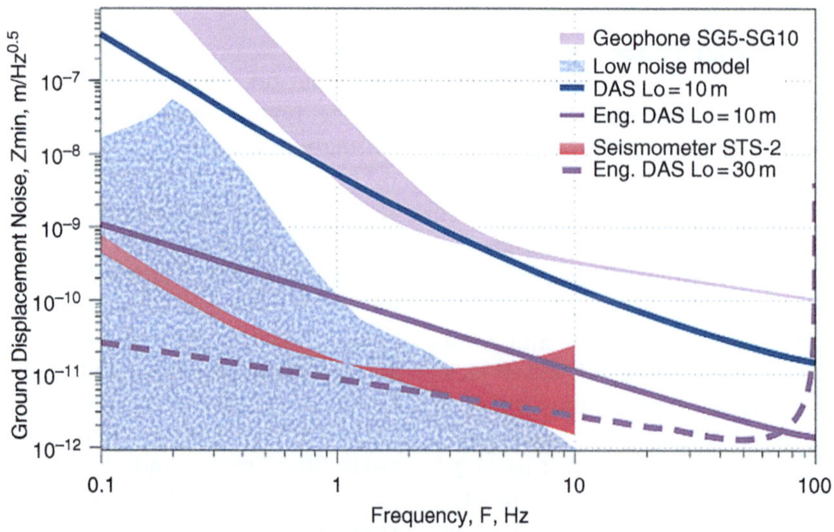

Figure 2.5 Displacement noise comparison of DAS with and without engineered fiber with seismometer and geophone. L_0 denotes gauge length.

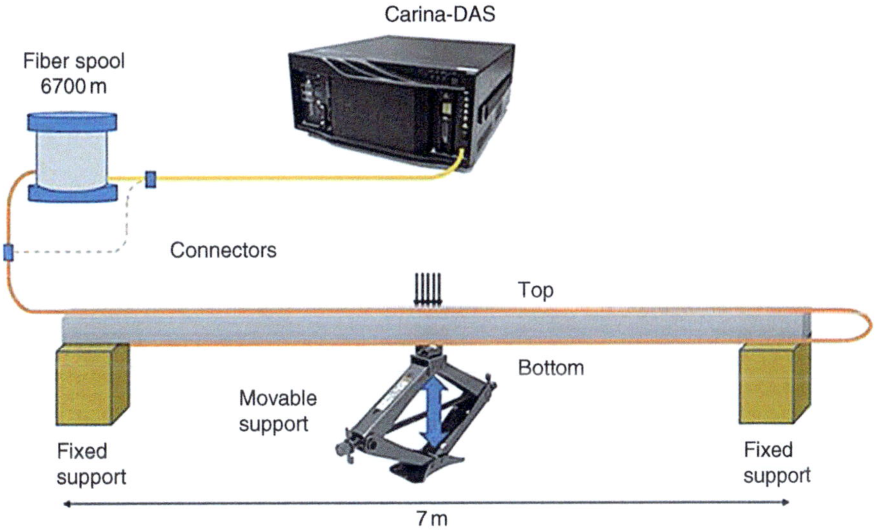

Figure 2.6 The experimental setup for low-frequency strain testing using an engineered fiber with 25 cm gauge length glued directly on the top and the bottom of the pipe with a square cross-sectional area.

Figure 2.6 shows a three-point bending flexural strain experimental setup for low-frequency DAS measurements using a precision-engineered fiber with 25 cm gauge length glued directly on the top and bottom of a pipe with a square cross-sectional area. The center position of the pipe is moved up and down in steps of about 10 με every 10 s over a period of 120 s. Figure 2.7(a) shows the 2D space–time waterfall plot of the strain rate along the pipe, and Figure 2.7(b) shows the strain-rate amplitude in the middle of the pipe at the top and bottom of the pipe. Figures 2.7(c) and 2.7(d) show the strain corresponding to the integrated strain-rate data shown in Figures 2.7(a) and 2.7(b), respectively. The top fiber undergoes compression and the bottom fiber undergoes tension when the center of the pipe is bent downward. Figure 2.7(e) shows the strain profile along the top and the bottom of the pipe. As can be seen, the low-frequency DAS measurements can provide a fine strain resolution of less than a nano-strain (nε) down to 8 mHz frequency, which can be used to monitor the onset of minute structural deformations.

2.6. IDSS

The frequency of the Brillouin backscatter light changes with strain, as indicated in Figure 2.8. We use an interferometric technique similar to the DAS system to measure the frequency of the Brillouin backscatter light. In this case, the instantaneous frequency of the backscatter light, which changes linearly with strain, is converted to a highly sensitive phase shift referenced to an unstrained fiber section.

Figure 2.9(a) shows the experimental setup to evaluate the sensitivity of the IDSS. A 5 m section of the fiber is passed over a low-friction pulley system, and different weights are attached to one side of the fiber while the other end is fixed. Figure 2.9(b) shows the strain measurement along a 5 km fiber. Comparing the variation of the signal over the 50 m unstrained sections of the fiber, a strain resolution of ±1 με is achieved with 60 s integration time. The strain variation of 55 με is observed around the 5 km reel of the fiber due to the variation of the spooling tension.

The temperature dependence of the Brillouin frequency shift can be compensated using a DTS measurement. The compensated IDSS data can be used to provide a periodic strain baseline calibration for the low-frequency DAS strain measurements. This is very useful for monitoring the buildup of strain over a long period of time, such as for subsurface subsidence monitoring.

2.7. FIBER RULER

Using a new optical fiber ruler instrument, we can measure the integrated strain over distances spaced between any scatter center pairs. The experimental setup for the fiber ruler is shown in Figure 2.10. Using a programmable optical delay line, we can vary the optical delay between the scatter centers within 8 km of the lead fiber compared to the scatter centers within 1 km of fiber after the optical delay line. As shown in Figure 2.11, we

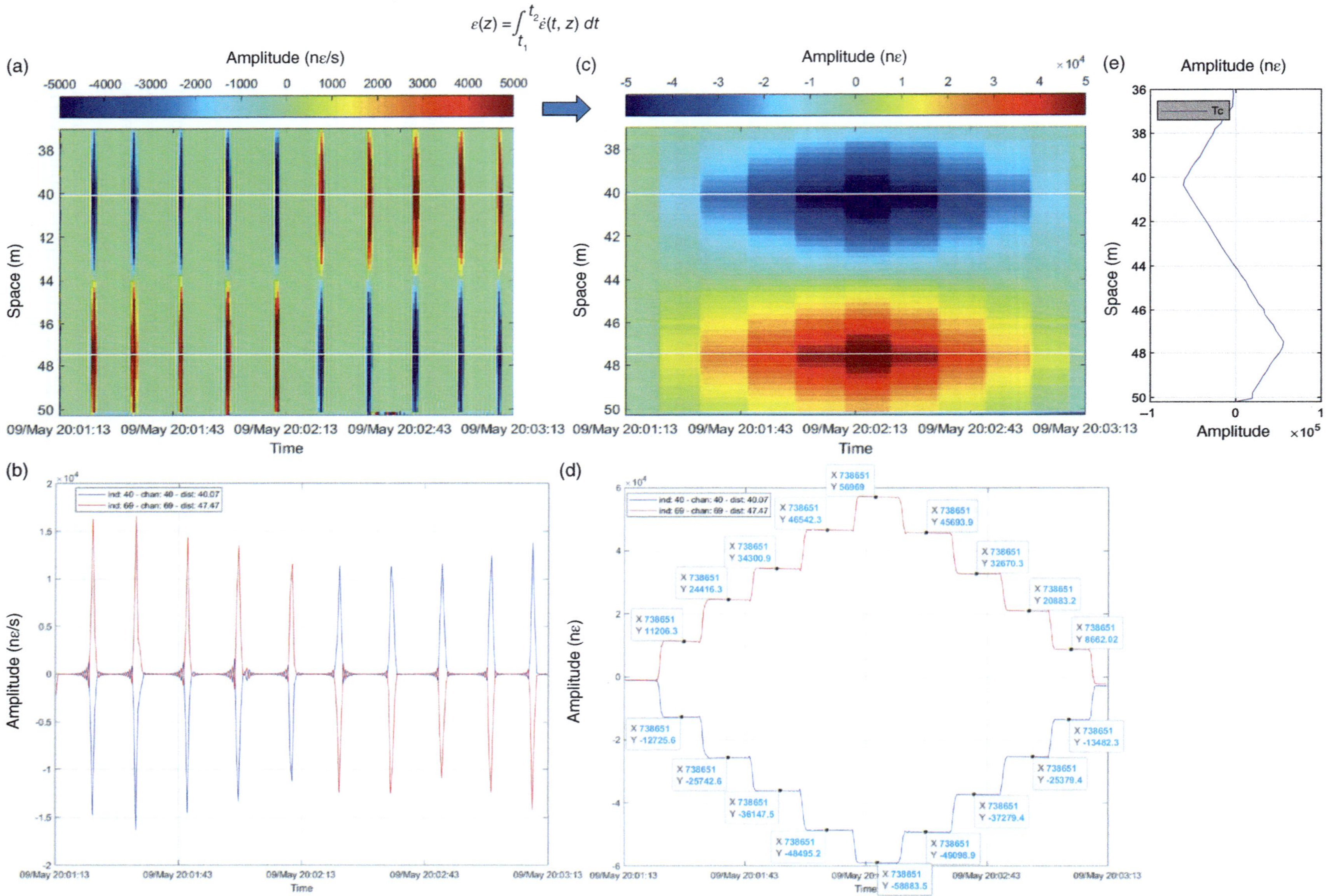

Figure 2.7 Low-frequency DAS strain measurement (over 120 s) using an engineered fiber with 25 cm gauge length. (a) 2D space–time of strain rate. (b) Amplitude of the strain rate in the middle of the pipe for the top and bottom fiber. (c) 2D space–time of converted strain rate to strain. (d) Amplitude of strain in the middle of the pipe for the top and bottom fiber. (e) Strain profile along the top and bottom of the pipe.

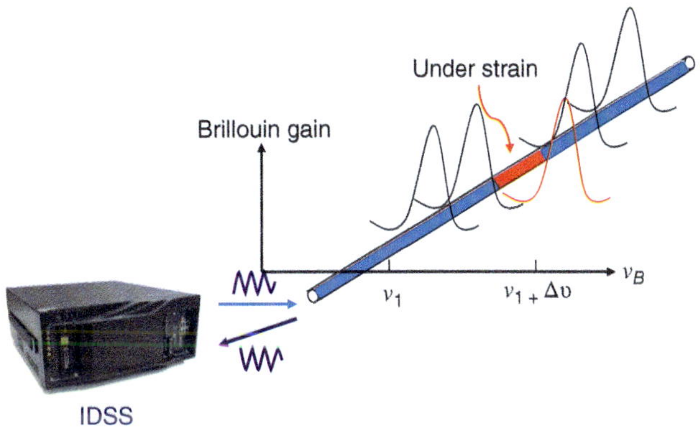

Figure 2.8 IDSS. The frequency of the Brillouin scatter light shifts with strain, and its instantaneous frequency is measured using an interferometric technique similar to the DAS system.

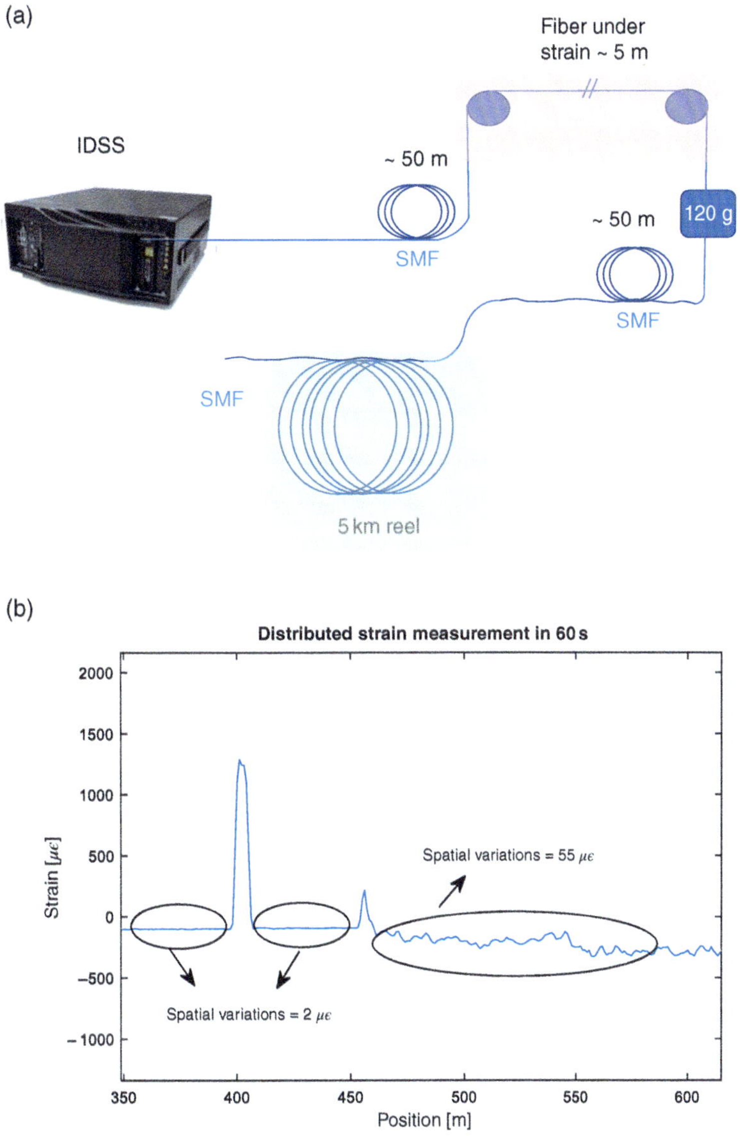

Figure 2.9 (a) IDSS test setup with a calibrated strain applied to a 5 m section of the standard single-mode fiber. (b) The strain profile with distance along the test fiber.

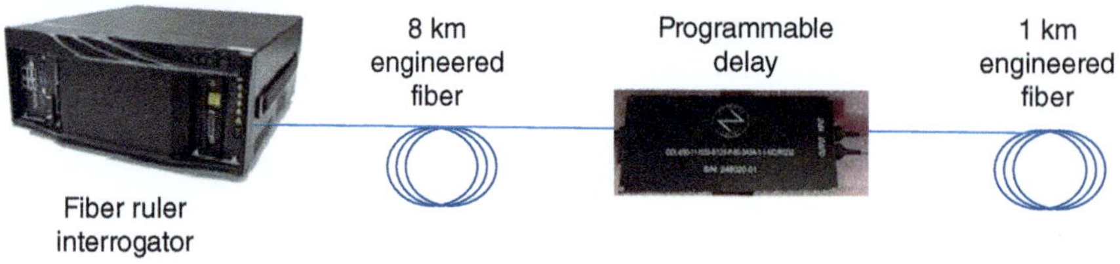

Figure 2.10 Fiber ruler using engineered fiber to measure the optical distance between two scatter centers. A programmable optical delay line is used to change the optical delay between two scatter centers.

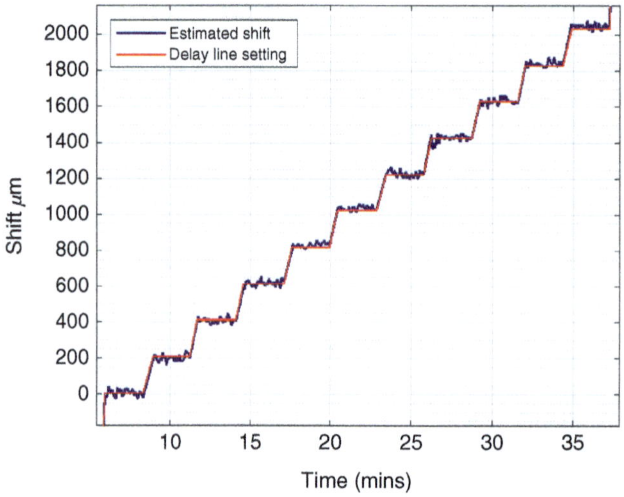

Figure 2.11 Fiber ruler experimental test results showing the estimated optical shift with the programmable delay line setting.

have demonstrated a distance resolution of 5 μm for a 1-minute measurement time, which translates to

- 1 με resolution for scatter centers 5 m apart
- 0.1 με resolution for scatter centers 50 m apart
- 0.01 με resolution for scatter centers 500 m apart

The thermal variations are also corrected using Raman distributed temperature measurements (DTS).

2.8. DISTRIBUTED ROCK INDEX (OR "R") FACTOR

As indicated in Figure 2.11, the bright scatter centers along the engineered fiber can be projected and referenced along the entire wellhole structure. The position of the scatter centers provides precise depth calibration that can be used to significantly improve the time-shift resolution for time-lapse seismic measurements.

Using a rock strain model, Hatchell and Bourne (2005) and Røste et al. (2005) proposed that the fractional change in the travel-time, $\frac{\Delta t}{t}$, of seismic waves may be expressed as a function of the vertical strain ε_z along the wellbore

$$\frac{\Delta t}{t} = (1 + R)\varepsilon_z \tag{2.3}$$

where the dimensionless value R denotes the ratio between travel-time changes due to velocity changes and path length changes.

According to 4D seismic data and a rock strain sensitivity model, in a typical field, the estimated accumulated strain can be 12 με/year with an R factor of 5 to 15; see Davies et al. (2020).

Table 2.1 shows an example of the strain-rate resolution required over a period of 1 year down to a 1-second measurement time.

Table 2.1 Strain rate.

Strain rate		Period
30	με	/year
2.5	με	/month
83	nε	/day
3.5	nε	/hour
57	pε	/minute
~1.0	pε	/sec

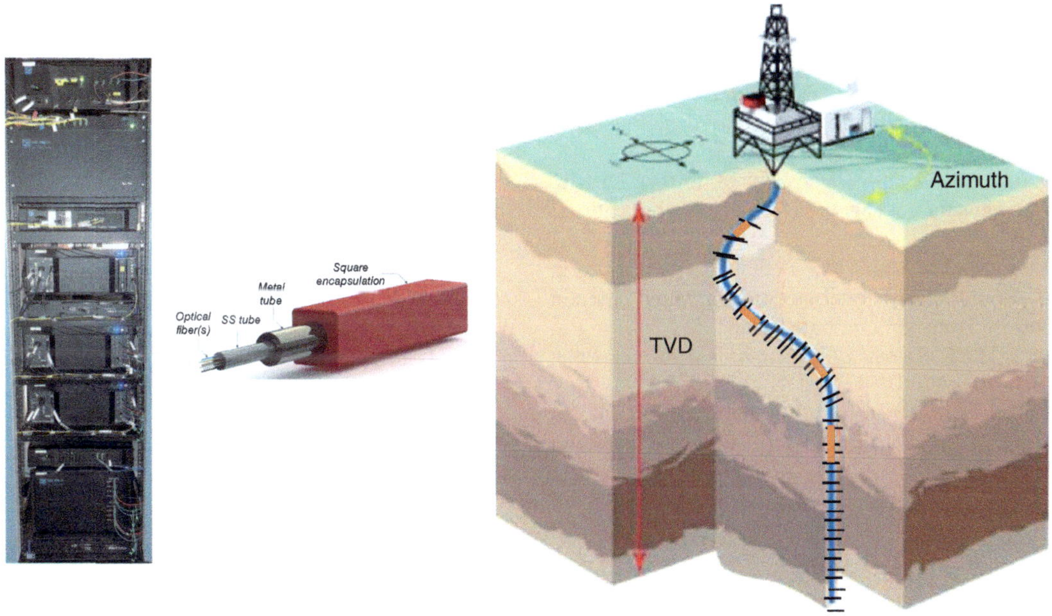

Figure 2.12 Integrated distributed optical fiber sensor for reservoir and subsidence monitoring.

Utilizing engineered fiber, the DAS system provides the high sensitivity required to track integrated low-frequency strain over a period of several months. At the same time, the buildup of strain may induce microseismic events that can be recorded over the wide aperture of engineered fiber using DAS.

Seismic data and strain measurements can be combined to map the distributed "R" factor or, as referred to here, the distributed "rock index," as expressed in the modified equation (2.2):

$$\Delta\nu_z(t)/\nu_z = -R_z(t).\varepsilon_z(t) \tag{2.4}$$

2.9. SUMMARY AND CONCLUSION

The distributed optical fiber sensor system, as shown in Figure 2.12, utilizes the full backscatter spectrum of single and multimode fibers plus scatter centers along an engineered fiber that can be all encapsulated within a single cable to provide integrated high-performance distributed measurements for

- Repeatable DAS vertical seismic profiling (VSP) and accurate time-shift seismic strain measurements
- Strain-rate measurement down to sub-picostrain per second (<1pε/s), where the strain-rate measurement can be integrated to evaluate low-frequency strain buildup with fine resolution
- Measuring direct strain with a fine resolution of 1 με using IDSS and to provide a periodic strain base-line strain calibration for low-frequency DAS measurements
- Measuring integrated strain in time and distance using a new fiber ruler technique that can also be combined with IDSS
- Measuring the temperature along the wellbore using high-resolution DTS and also compensating strain data for the temperature cross-sensitivity
- Measuring slow strain buildup over a long period of time, such as for subsurface subsidence monitoring
- Measuring the distributed rock index by combining seismic time-shift data and the strain profile along the wellbore

ACKNOWLEDGMENTS

We would like to thank all our collaborators, field engineers, and production teams in testing and preparing the equipment for field trials. In particular, we would like to thank Mr. Boris Chorbadzhiev and Plamen Atanassov for their valuable help in building the low-frequency DAS experimental setup.

REFERENCES

Dakin, J., Pratt, D., Bibby, G., & Rose, J. (1985). Distributed optical fiber Raman temperature sensor using a semiconductor light source and detector. *Electronics Letters*, *21*(13), 569–570.

Davies K., et al. (2020). *Fiber optics for sub-surface strain forecasting*. Paper presented at the First EAGE Workshop on Fiber Optic Sensing.

Hartog, H. A. (2017, April). *An introduction to distributed optical fiber sensors* (1st ed.). CRC Press. ISBN: ISBN-10: 1482259575, 7.

Hatchell, P., & Bourne, S. (2005). Rocks under strain: strain-induced time-lapse time shifts are observed for depleting reservoirs. *The Leading Edge*, *24*(12), 1222–1225.

Parker, T., Shatalin, S., & Farhadiroushan, M. (2014). Distributed Acoustic Sensing – a new tool for seismic applications. *First Break*, *32*, 61–69.

Parker, T. R., Farhadiroushan, M., Feced, R., & Handerek, V. A. (1998). Simultaneous distributed measurement of strain and temperature from noise-initiated Brillouin scattering in optical fibers. *IEEE Journal of Quantum Electronics* 1998, *34*(4), 645–659.

Richter, P., Parker, T. R., Woerpel, C., Wu, W., Rufino, R., & Farhadiroushan, M. (April, 2019). Hydraulic fracture monitoring and optimization in unconventional completions using a high-resolution engineered fiber-optic Distributed Acoustic Sensor. *First Break*, *37*(4), 63–68.

Røste, T., Stovas A., & Landrø, M. (2005). *Estimation of layer thickness and velocity changes using 4D prestack seismic data.* Paper presented at the 67th Conference and Exhibition, EAGE. Extended Abstracts, C010.

Shatalin S., Parker T. R., Farhadiroushan M., 2021, High definition seismic and microseismic data acquisition using distributed and engineered fiber optic acoustic sensors, Chapter 1, doi:10.1002/9781119521808.ch1

Westbrook, P. S., Feder, K. S., Kremp, T., Monberg, E. M., Wu, H., Zhu, B., et al. (2020). Enhanced optical fiber for distributed acoustic sensing beyond the limits of Rayleigh backscattering. *iScience*, *23*, 101137.

3

A Microwave Photonics Optical Fiber Method for Measuring Distributed Strain for Hydrologic Applications in the Vadose and Saturated Zones

Liwei Hua[1,2], Lawrence C. Murdoch[2], Xuran Zhu[1], Leonid N. Germanovich[2], Yongji Wu[1], Jing Guo[1], Riley Blais[2], Grant Plunkett[2], Olivia Costantino[2], Scott DeWolf[2], Rob Moak[2], and Hai Xiao[1]

ABSTRACT

Distributed strain measurements appear to hold significant potential for monitoring hydrologic processes. Coherence-length-gated microwave photonics interferometry (CMPI) is a distributed sensing technique that measures strain in an optical fiber by reading optical interference phase changes in the microwave domain. The technique provides 10 nε resolution when centimeter spatial resolution is applied. CMPI was used to measure the strain caused by small periodic variations in air pressure (25 Pa amplitude and 4 Hz) in a sand-filled laboratory column used to represent barometric loading in the vadose zone. The strain varied as a periodic function, with an amplitude that decreased and a phase that increased with depth. The distribution of amplitude and phase of the strain depended on the water saturation, permeability, and presence of a barrier at the top of the sand. A theoretical analysis suggests that the air pressure diffusivity estimated from pressure data is similar to the diffusivity estimated from the distributed strain. These results indicated that the CMPI-distributed strain measurement system could be used to improve monitoring and characterization of the vadose and underlying saturated zone.

3.1. INTRODUCTION

Understanding changes in water content and pore pressure in geologic media is a central theme to much of hydrology and related fields, and this has motivated the development of a wide range of monitoring methods (Dobriyal et al., 2012; Lekshmi et al., 2014). The typical approach is to use a sensor that measures a local property, such as a transducer to measure the fluid pressure or a capacitance gauge to measure dielectric properties. Changes in pore pressure or water content deform porous media, and measurements of this deformation have potential as an alternative monitoring signal (Murdoch et al., 2020, 2023). One example of this approach is the use of interferometric synthetic aperture radar (InSAR) or global positioning system (GPS) to measure deformation at the ground surface caused by fluid pumping or injection in underlying aquifers or reservoirs (Galloway & Burbey, 2011; Khakim et al., 2012; Monfared & Rothenburg, 2011; Vasco et al., 2022).

It is also possible to measure strain in the subsurface using discrete or distributed sensors (Murdoch et al., 2020). The deformation that can be resolved by subsurface sensors is many orders of magnitude higher than InSAR or GPS, providing a mechanism for detecting

[1]*Center for Optical Materials Science and Engineering Technologies, Clemson University, Clemson, South Carolina, USA*

[2]*Department of Environmental Engineering and Earth Sciences, Clemson University, Clemson, South Carolina, USA*

Distributed Acoustic Sensing in Borehole Geophysics, Geophysical Monograph 289, First Edition.
Edited by Yingping Li, Robert Mellors, and Ge Zhan.

DOI: 10.1002/9781394179275.ch03

effects of small changes in pressure or water content. Distributed strain sensors typically involve uniaxial measurements along an optical fiber (Becker et al., 2017, 2023; Sun et al., 2020; Xue & Hashimoto, 2017), although combining multiple fibers together in a helical wrap can provide some strain components normal to the axis of the fibers (Murdoch et al., 2020). Optical fiber can also be packaged in discrete instruments to measure multiple components of the strain tensor with high precision (Murdoch et al., 2023).

Many applications have been focused on evaluating strains caused by pressure changes in the saturated zone. Well tests used to characterize aquifers or reservoirs cause tilt or strain that can be measured and inverted to help the characterization process (Becker et al., 2017; Cappa et al., 2007; Fabian & Kümpel, 2003; Guglielmi et al., 2014; Hisz et al., 2013; Jin et al., 2021; Karasaki et al., 2000; Murdoch et al., 2020; Schweisinger et al., 2009, 2011; Sun et al., 2020, 2021; Svenson et al., 2008; Vasco et al., 2001; Xue & Hashimoto, 2017). Leaks of CO_2 may also cause a strain signal that could be used to help locate the leak and characterize its severity (Rutqvist & Tsang, 2002; Xue & Hashimoto, 2017).

Deformation also occurs in the vadose zone in response to hydrologic processes or surface loads (Keller et al., 2012; Murdoch et al., 2015). Changes in pore pressure in an underlying aquifer will cause characteristic patterns of deformation in the vadose zone (Murdoch et al., 2015). Evaluating small strains in the subsurface can be complicated by changes in barometric pressure. An increase in barometric pressure acting on the ground surface causes negative vertical strain (compression), but an increase in the pressure in pore space causes a positive strain (tension). The net observed strain is a combination of these two offsetting effects. Changes in water content will increase the weight of shallow soils, compressing the underlying formation (Marin et al., 2010; Van Der Kamp & Schmidt, 1997), and will also affect the strain by altering the air phase's relative permeability.

We measured vertical strain in the vadose zone with the intent of estimating changes in water content in overlying soils due to rainfall or evaporation (Murdoch et al., 2015). We found that this process was affected by strains caused by barometric pressure loading in the vadose zone. In a related process, diffusion of air pressure through the vadose zone is widely recognized to affect water levels in wells in unconfined aquifers (Quilty & Roeloffs, 1991; Rasmussen & Crawford, 1997; Rojstaczer, 1988; Rojstaczer & Riley, 1990; Weeks, 1979; Yusa, 1969).

This motivated us to seek a method that could measure the strain caused by barometric pressure fluctuations as a function of depth. Our initial motivation was to improve the interpretation of the strain data in Murdoch et al. (2015), but this raised an even more intriguing possibility: if the strain caused by barometric pressure fluctuations was affected by moisture content, then could it perhaps be used to monitor moisture content changes in the vadose zone? This would have important applications in hydrology, but it would require measuring small, distributed strains with high spatial and temporal resolution (nε strain resolution with centimeter spatial resolution and 100 Hz sampling).

Fiber Bragg gratings (FBGs) are commonly used optical fiber sensors for measuring strain. These sensors use gratings etched over a short length of fiber to measure strain, and multiple gratings along the same fiber can measure strain at different locations. However, FBGs have a limited resolution of approximately 0.1 microstrain, which may not be sufficient for detecting subtle changes. Additionally, the localized nature of FBG measurements may not capture strain variations occurring at larger scales (López-Higuera et al., 2011). Time-domain reflectometry distributed sensors, which rely on Brillouin peak shifting in the light signal, provide continuous sensing along the entire length of the fiber using standard telecoms optical fiber. However, the strain resolution of Brillouin peak shifting is limited to 5 microstrain due to low strain sensitivity. These methods also require wavelength scanning and have relatively longer measurement times (Bao & Chen, 2012). Phase optical time-domain reflectometry (ΦOTDR) and its modified version, distributed acoustic sensing (DAS), offer event detection and strain measurements based on weak reflector arrays. DAS, particularly used for measuring seismic signals in the oil industry, provides strain-rate measurements in a wide frequency range. However, DAS requires expensive equipment and relies on integration of strain-rate data over a few-meter baseline, which may not be ideal for applications requiring sharper spatial resolution. Considerable computational processing is necessary for strain calculation and resolution improvement, making calibration and validation challenging (Paitz et al., 2021).

Overall, the measurement requirements we have mentioned exceeded the capabilities of available technology at the time, so we evaluated the feasibility of developing a new distributed strain sensing technique using microwave photonics (Hervás et al., 2018; Hua et al., 2017). CMPI measures strain directly, so the integration step required by DAS is avoided. Moreover, the strain measurements between CMPI reflectors can be calibrated and both the location and spatial resolution are known. This makes it possible for CMPI to measure sharper spatial resolution with a clearly defined uncertainty, which is a significant advantage over DAS when making critical measurements. The interrogator required for CMPI can be fabricated from common, off-the-shelf components.

It would be possible for investigators to make their own interrogator for CMPI, but some expertise in optics would be required, and we expect that most users would prefer to buy a functional interrogator.

The objective of this chapter is to describe a microwave photonics technique for measuring distributed strain using optical fibers with applications to hydrologic sensing. The first part of the chapter describes the optical sensing method, named coherence-length-gated microwave photonics interferometry (CMPI) (Hua et al., 2017, 2021a, 2021b; Zhu et al., 2021). The method provides high-sensitivity measurements by recording optical interference signals from time-multiplexed interferometers in the microwave domain. This is followed by a section describing experiments where the CMPI method was used to measure small strains caused by changes in air and water pressure in a laboratory column configured like the vadose zone. We also describe a field application of the CMPI system used to characterize strain during a well test.

This project was conducted as a collaboration between investigators with backgrounds in optics, electrical engineering, hydrology, and geomechanics. This interdisciplinary collaboration has led to a new measurement technique, CMPI, that has generated new insights into hydrologic processes. Our intent with this chapter is to explain relevant details required to implement CMPI and use it for hydrologic applications in order to motivate future collaborations.

3.2. DISTRIBUTED SENSING USING OPTICAL FIBERS WITH MICROWAVE PHOTONICS

Microwave photonics modulates light in the microwave band and then synchronizes the optical detection system with the microwave modulation frequency (Hervás et al., 2018). As a result, the system has a high SNR and thus an improved detection limit compared to the conventional time-domain main measurement-based detection systems. Microwave photonics-based sensing technologies have been demonstrated to measure temperature (Huang et al., 2015; Tian et al., 2022), strain (Hua et al., 2015), and distance (Dong et al., 2018), among others. Due to their extensive measurement range and relatively low implementation cost, microwave photonics sensing technologies have emerged as promising contenders for detecting hydrologic and geophysical processes (Bogris et al., 2022; Dauson et al., 2022).

The phase of the microwave-modulated light can be measured and Fourier transformed to find the time-of-arrival information for distributed sensing. The highly stable microwave phase determines locations along the signal path and thus enables a measurement range of more than 10 km (Clement et al., 2019). Frequency modulation in the gigahertz bandwidth enables spatial resolution in the centimeter range (Huang et al., 2014). Distributed systems based on this approach have been developed for interrogating Rayleigh scattering (Liehr et al., 2010), FBG arrays (Cheng et al., 2015; Clement Bellido et al., 2018; Ricchiuti et al., 2014), and interferometers (Hua et al., 2017; Huang et al., 2014; Wu et al., 2021; Zhu & Huang 2023).

The CMPI method provides high-sensitivity sensing by recording optical interference signals from time-multiplexed interferometers in the microwave domain. The coherence length of the light source is carefully chosen or controlled to eliminate crosstalk or gate the signal between adjacent interferometers such that distributed sensing can be achieved. A frequency-time joint method is used to separate interferometers in the time domain. As a result, a maximum measurable frequency of tens of kilohertz for distributed strain sensing can be obtained (Hua et al., 2021a). The CMPI interrogation mechanism can be combined with low-coherence interferometry (LCI) to achieve fully distributed sensing (Zhu et al., 2021). The CMPI sensors are typically constructed using single-mode fiber (SMF) with weak reflectors or broadband FBGs. However, in cases where optimal performance is not essential, it is still feasible to utilize regular SMF, albeit with compromised system performance.

The following section outlines the components and architecture of the CMPI system along with the theoretical basis of the method. It describes how the CMPI method acquires signals from distributed interferometers. The next section outlines the approach for determining strain from the interferometer data and, in particular, describes how CMPI can measure strain from static to acoustic frequencies. The section concludes with a description of the CMPI performance characteristics and trade-offs affecting spatial resolution, number of sensing units, strain resolution, frequency constraints, and other characteristics.

3.2.1. Architecture and Theoretical Basis of CMPI

A typical CMPI system includes a microwave source, a vector microwave detector, a light source, an electro-optical modulator, a photodetector (PD), and distributed sensors formed by interferometers (Figure 3.1(a)). A continuous-wave (CW) laser with a center frequency of ω is used as the light source. The light intensity is modulated by a microwave signal. The light launches into an optical fiber sensing network and is reflected by reflectors with respective time delays τ. Any two reflections that have a lag $(\tau_{k2} - \tau_{k1})$ much smaller than the coherence time of the laser produce optical interference, so the responsive light paths form an interferometer IF_k. Light reflections

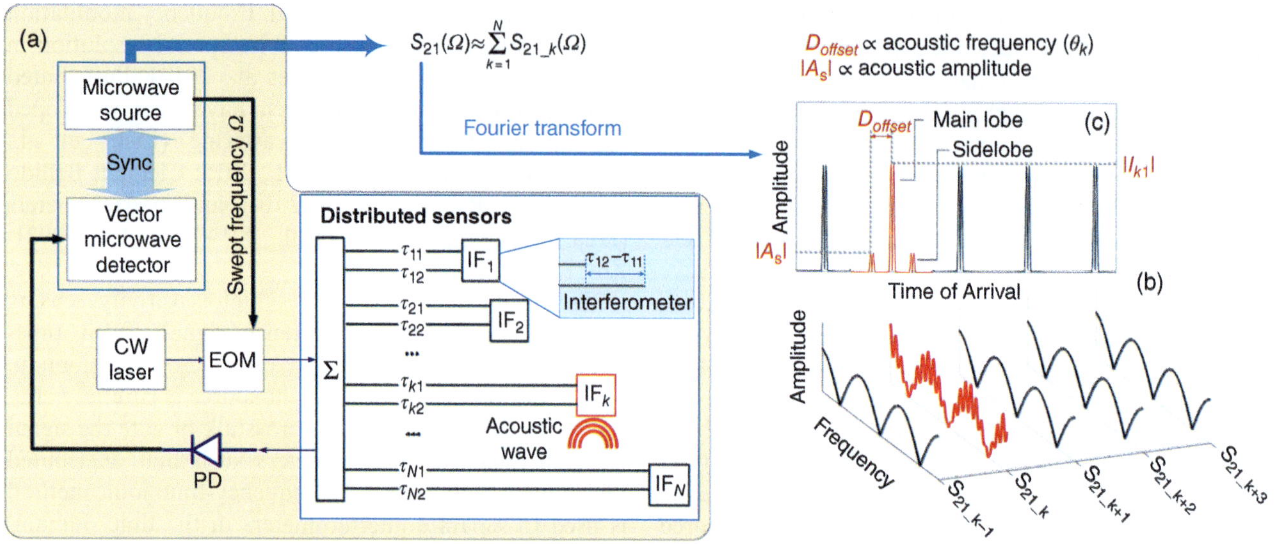

Figure 3.1 (a) Schematic of the CMPI system for distributed strain sensing and distributed acoustic sensing. PD, photodetector; EOM, electro-optic modulator; CW laser, continuous-wave laser. (b) The amplitude of the decomposed frequency domain signal from each interferometer. (c) The amplitude of the time domain signal. L. Hua, X. Zhu (2021) / MDPI / CC BY 4.0.

are injected into a PD, which converts the light power into electric voltage and sends it back to the vector microwave detector. The frequency of the microwave signal is swept over a designed range. At each modulation frequency Ω, synchronized detection is conducted, and the amplitude variation and phase shift of the modulation envelope are measured.

The optical path difference (OPD) for each interferometer, IF_k ($k = 1, 2., N$) is designed to be much smaller than the coherence length of the light source (Lc); the separation distance between adjacent interferometers is designed to be much larger than Lc. Therefore, the optical interference only occurs within each interferometer, and it is negligible among different interferometers. The complex frequency response of the system S_{21} can be approximately expressed as

$$S_{21}(\Omega) \approx \sum_{k=1}^{N} S_{21\_k}(\Omega), \tag{3.1}$$

where $S_{21\_k}(\Omega)$ is the frequency response from the kth interferometer, which is expressed as

$$\begin{aligned} S_{21\_k}(\Omega) = m\big[&\left(A_{k1}{}^2 + A_{k1}A_{k2}\cos\Delta\phi_k\right)e^{-\tau_{k1}\Omega} \\ &+ \left(A_{k2}{}^2 + A_{k1}A_{k2}\cos\Delta\phi_k\right)e^{-\tau_{k2}\Omega}\big], \end{aligned} \tag{3.2}$$

where m is a constant decided by the modulation depth of the electro-optic modulator (EOM), A_{k1} and A_{k2} are the amplitudes of the two reflected waves, and $\Delta\phi_k$ is the optical phase difference between the two waves. When the interference phase of the kth interferometer varies sinusoidally as a function of time with frequency Θ_k, amplitude δ_k, and the initial phase Φ_0, we can express it as

$$\Delta\phi_k = \Delta\phi_{k_{\text{static}}} + \delta_k \cos(\Theta_k t + \Phi_0), \tag{3.3}$$

where $\Delta\phi_{k_{\text{static}}}$ is a constant within each S_{21} measurement window, $\delta_k \cos(\Theta_k t + \Phi_0)$ is the dynamic phase, and t is the time variable. Assume the microwave modulation frequency is linearly scanned from Ω_{start} with a step of $\Delta\Omega$ and a constant sampling time Δt, we can express t as

$$t = \frac{\Delta t(\Omega - \Omega_{\text{start}})}{\Delta\Omega}. \tag{3.4}$$

When $\delta_k \ll 1$, $\cos\Delta\phi_k$ can be simplified through Fourier–Bessel expansion and linear approximation and expressed as

$$\begin{aligned} \cos\Delta\phi_k \approx \cos\Delta\phi_{k_{\text{static}}} - \delta_k \sin\Delta\phi_{k_{\text{static}}} \\ \cos\left(\frac{\Theta_k\Delta t}{\Delta\Omega}\Omega + \Phi_{\text{cons}}\right), \end{aligned} \tag{3.5}$$

where

$$\Phi_{\text{cons}} = \Phi_0 - \frac{\Theta_k \Delta t}{\Delta\Omega}\Omega_{\text{start}}. \tag{3.6}$$

Substituting equation (3.5) into equation (3.2), the result indicates the amplitude of the microwave interferogram S_{21_k} is modulated by a sinusoidal signal that has a frequency of $\frac{\Theta_k \Delta t}{\Delta\Omega}$ as illustrated in Figure 3.1(b). The inverse Fourier transform of $S_{21}(\Omega)$ is expressed as

$$F(t_z) = m\sum_{k=1}^{N} F_k(t_z), \tag{3.7}$$

which includes "Static" and "Dynamic" components as

$$F_k(t_z) = \{I_{k1}\delta(t_z - \tau_{k1}) + I_{k2}\delta(t_z - \tau_{k2})\}_{\text{Static}}$$
$$- \{A_s\delta(t_z - \tau_{k1} + D_{\text{offset}}) + A_s^*\delta(t_z - \tau_{k1} - D_{\text{offset}})$$
$$+ A_s\delta(t_z - \tau_{k2} + D_{\text{offset}}) + A_s^*\delta(t_z - \tau_{k2} - D_{\text{offset}})\}_{\text{Dynamic}} \tag{3.8}$$

where

$$I_{k1(2)} = A_{k1(2)}^2 + A_{k1}A_{k2}\cos\Delta\phi_{k\_\text{static}} \tag{3.9a}$$

$$A_s = \delta_k A_{k1}A_{k2}\sin\Delta\phi_{k\_\text{static}}\, e^{j\Phi_{\text{cons}}}; \text{and } D_{\text{offset}} = \Theta_k\frac{\Delta t}{\Delta\Omega}. \tag{3.9b}$$

The "Static" component in equation (3.8) includes two time pulses (main lobes) that are generated by the two reflected waves from interferometer IF$_k$, and they are centered at τ_{k1} and τ_{k2}, respectively. The amplitude of the main lobe $|I_{k1(2)}|$ is determined by the average interference phase $\frac{L_k\omega}{c}$ within the S_{21} measurement window.

The "Dynamic" component is formed by time pulses generated by the dynamic interference changes. Those pulses occur at both sides of the respective main lobes (side lobe, Figure 3.1(c)). The offset between a side lobe and the corresponding main lobe (D_{offset}) is proportional to the acoustic frequency Θ_k with a scaling coefficient $\Delta t/\Delta\Omega$. The amplitude of the dynamic interference phase change δ_k is proportional to the amplitude of the sidelobes $|A_S|$.

The average interference phase can be calculated using the chirp effect of the EOM to perform quadrature-phase demodulation (Hua et al., 2021b). This is done by tuning the EOM bias to add opposite phase shifts of $\pi/4$ to the optical interference phase, so the peak values of the two main lobes are in quadrature as $A_{k1}^2 + A_{k1}A_{k2}$ cos $(\Delta\phi_{k\_\text{static}} + \pi/4)$ and $A_{k2}^2 + A_{k1}A_{k2}\cos(\Delta\phi_{k\_static} - \pi/4)$. The standard quadrature-phase demodulation method can be used to calibrate the conic coefficients (A_{k1}^2, A_{k2}^2, $A_{k1}A_{k2}$, and the quadrature error) and calculate the interference phase (Požar et al., 2011).

$S_{21\_k}(\Omega)$ can be extracted from $S_{21}(\Omega)$ through Fourier transform of the gated time-domain signal as

$$S_{21\_k\_g}(\Omega) = F[F(t_z)g_k(t_z)] \tag{3.11}$$

where $g_k(t_z)$ is a time-domain gate function that selects the time pulses generated by the kth interferometer. The temporal signal can be reconstructed by applying a high-pass filter to $|S_{21\_k\_g}(\Omega)|$. The time is linearly converted from Ω using equation (3.4), and the amplitude is also corrected by using the calculated interference phase.

The individual reflected wave from one interferometer cannot be resolved in the time domain when the microwave bandwidth becomes small. Consequently, equation (3.8) can be approximated as

$$F_k(t_z) \approx \{(I_{k1} + I_{k2})\delta(t_z - \overline{\tau_k})\}_{\text{Static}}$$
$$-\{2A_s\delta(t_z - \overline{\tau_k} + D_{offset}) + 2A_s^*\delta(t_z - \overline{\tau_k} - D_{offset})\}_{\text{Dynamic}} \tag{3.12}$$

where $\overline{\tau_k} = (\tau_1 + \tau_2)/2$.

3.2.2. CMPI for Strain Measurement

CMPI-distributed sensors are created with cascaded weak reflectors in SMF. The reflectors are normally fabricated using femtosecond laser micromachining by focusing light into the fiber core, which creates reflectivity from −35 to −45 dB (Song et al., 2016). Any two reflectors with a separation distance smaller than the coherence length of the light source (Figure 3.2(a)) form an intrinsic Fabry–Perot interferometer (IFPI). The strain applied to the IFPI changes the distance between the reflector pair and changes the interference phase, which is expressed as (Butter & Hocker, 1978)

$$d\Delta\phi_k = \frac{2n\omega(1 - P_{\text{eff}})}{c}\Delta l_k\varepsilon \tag{3.13}$$

where c is the speed of light, n is the refractive index of the fiber, ω is the frequency of the optical light, P_{eff} is the effective strain-optic coefficient, and Δl_k is the cavity length of IFPI. Static strain changes can be measured by monitoring the static interference phase change. Dynamic strain results in an interference phase change at the strain change frequency, which can be calculated from the $D_{\_\text{offset}}$. The amplitude and phase of the dynamic strain can be calculated using A_s. The temporal strain can be recovered through time-domain gating and Fourier transform, as shown in equation (3.11). The sensitivity of the strain is $\frac{2n\omega(1 - P_{eff})}{c}\Delta l_k$, which is linearly proportional to Δl_k.

CMPI provides high strain sensitivity and a large dynamic frequency measurement range. A set of lab tests has been performed to demonstrate strain sensing using an array of weak reflectors in optical fiber. In one setup, a weak reflector array was prestrained and fixed on two linear translation stages. Strains were applied to the sensor array by moving the stage axially. Results show that the interference phase increases linearly as a function of the applied strain. The characterized static strain resolution was 0.6 με for a 1.5 cm cavity IFPI sensor, and the resolution increased to 10^{-8} ε when the cavity length was 1 m (Hua et al., 2017, 2021b). In another setup, the fiber reflector array was wrapped around a piezoceramic cylinder (PCC), which was driven by a waveform generator to create acoustic signals. Acoustic signals of frequency up to 15.6 kHz were detected. The experimental results show that the sensing system can locate an acoustic

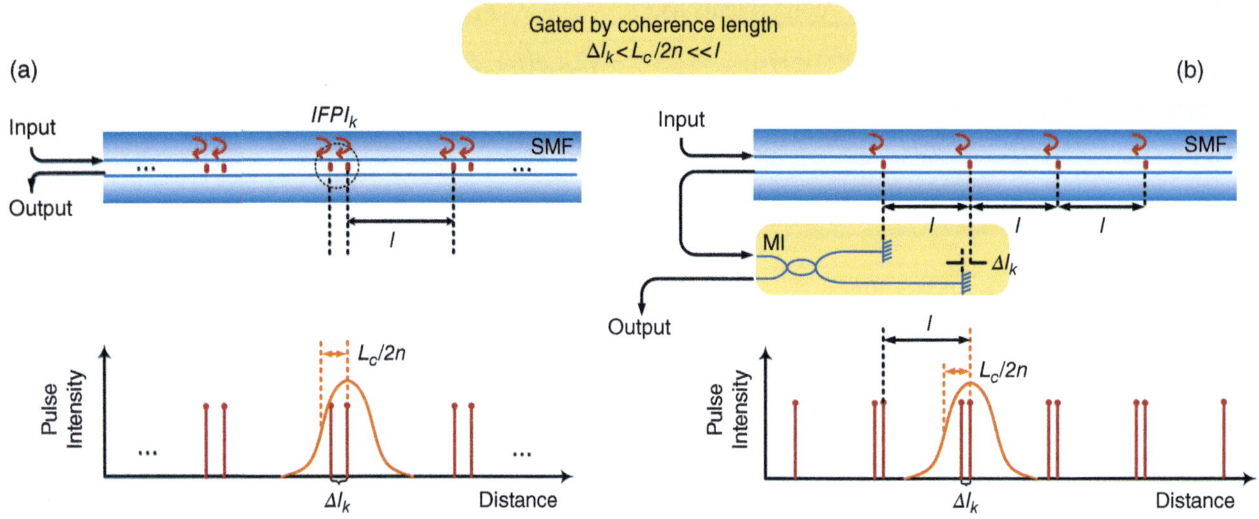

Figure 3.2 Schematic of the interferometers formed by cascaded weak reflectors (red dots) in optical fiber (blue band). (a) Cascaded IFPIs with a separation distance of l. (b) LCI from a weak in-fiber reflector array and a local Michelson interferometer (MI). L_c is the coherent length of the light source, and n is the refractive index of the optical fiber.

source. The recovered acoustic signals match the excitation signals in frequency, amplitude, and phase (Hua et al., 2021a).

One major concern is the crosstalk among CMPI sensors. To avoid crosstalk, the separation between the IFPIs needs to be larger than the coherence length, as illustrated in Figure 3.2. Therefore, the regions between adjacent IFPIs are insensitive to the strain, and those regions are so-called dark zones. Dark zones reduce the number of sensor units, increase the cost of the sensor array, and make the sensor package complicated.

LCI has been adapted into the original CMPI system to tackle the dark-zone problem. LCI was originally used for absolute remote measurement of displacement (Rao & Jackson, 1996), in which a local receiving interferometer (LRI) is connected inline with the remote sensing interferometer. The OPD of LRI is adjusted to match that of the sensing interferometer within the coherence length of the source, such that the interference occurs at the output of the LRI.

A local Michelson interferometer (MI) is implemented as the LRI in the CMPI architecture (Figure 3.2). Cascaded IFPIs are formed by two weak reflectors with a similar cavity length, l. The cavity length is designed to match the OPD of the MI, and Δl represents the maximum matching deviation. The coherent length of the light source is larger than the maximum matching deviation and much smaller than the cavity length ($\Delta l < L_c/2n \ll l$). The interference occurs between the pair of lightwaves: one reflected by the first reflector and the long arm of the MI; the other reflected by the second reflector and the short arm of the MI. As such, the OPD and the separation distance of these combined interferometers are still Δl and l. Dark zones along the sensing fiber are eliminated using this approach. If there are N weak reflectors along the fiber, this configuration creates $N - 1$ low-coherent interferometers. The relationship between interference phase change and the applied strains becomes

$$d\Delta\phi_k = \frac{2n\omega\left(1 - P_{eff}\right)}{c} l\varepsilon. \tag{3.14}$$

Here, the local MI, as an LRI, is used to enable the IFPIs, which have a similar cavity length as the arm length difference of the MI. As such, we can also increase the arm length difference of the MI to il, where i is an integer, so the IFPIs with a cavity length of il are enabled. The relationship between interference phase change and the applied strains becomes

$$d\Delta\phi_{k\_i} = \frac{2n\omega\left(1 - P_{eff}\right)}{c} il\varepsilon \tag{3.15}$$

Equations (3.14) and (3.15) indicate that the strain sensitivity is proportional to the MI cavity length. Consequently, the number of sensors decreases to $N - i$. MI works as a low pass filter to enable the average strain reading over the MI cavity length. The spatial resolution is always decided by the separation between the adjacent reflectors. One can use this feature to trade off the system specs in terms of the number of sensing units, cutoff spatial frequency, and strain sensitivity.

3.2.3. CMPI System Hardware

Hardware for the CMPI system includes three primary components: an interrogator, a reference MI, and a sensing fiber (Figure 3.3). A low-coherence light source (LCS) is used as the light carrier. The light was intensity-modulated by a microwave signal via an EOM (Lucent, X2623Y). The modulation signal is generated by a vector network analyzer (VNA). The bias voltage of the EOM is provided by an external DC power supply. An inline polarization controller (Thorlabs, US) is used to optimize the modulation depth of the EOM. The microwave-modulated light output from the EOM is first amplified by an Erbium-doped fiber amplifier 1 (EDFA1) and then launched into the distributed sensors. Reflected signals from the distributed fiber sensors are then coupled into the MI, which is made of a coupler and two Faraday rotation mirrors (FRMs). The MI was well confined and stabilized in a black box using epoxy to minimize the effects of environmental vibrations. The output light is amplified by the EDFA2 and filtered by the BPF. A high-speed photodetector detects the filtered signal and passes over the converted electrical signal to VNA through port 2.

The CMPI interrogator can be assembled using readily available, off-the-shelf components, which typically cost between $15,000 and $20,000. The MI can be fabricated in-house using fiber couplers and Faraday mirrors, which are also common off-the-shelf components (Zhu et al., 2021). For the sensing fiber, we have utilized reflector arrays on a regular SMF. These reflectors were fabricated through femtosecond laser micromachining, where light was focused into the fiber core, resulting in reflectivity ranging from −35 to −45 dB (Song et al., 2016). Alternatively, broadband weak FBGs can also serve as the sensing fiber, and these are available through a few vendors. The cost of FBG arrays is proportional to the number of FBGs included. For instance, a 100 weak FBG array typically costs between $200 and $2,000.

To demonstrate the concept, a CMPI system was built in-house and validated by applying known strain to the sensor. A l = 10 cm MI was used to select reflectors every 10 cm apart. In the static strain testing, an SMF that has six weak reflectors 10 cm apart was attached along a cantilever beam. Strains along the cantilever beam were measured and compared with the theoretical value. The results show that the system provides a static strain resolution of 95 nε at a spatial resolution of 10 cm (Zhu et al., 2021). To test the dynamic strain sensing, a piece of SMF with an array of 12 weak reflectors 5 cm apart was wrapped on two PCCs driven at frequencies of 1,700 and 3,430 Hz as acoustic sources. The experimental results show that the sensing system can locate multiple acoustic sources. The system resolves 20 nε when the spatial resolution is 5 cm. The recovered acoustic signals match the excitation signals in frequency, amplitude, and phase (Hua et al., 2021a).

3.2.4. System Specification Trade-offs and Limitations

Dynamic strain (acoustic signal) is sampled during microwave frequency scanning, so the sampling rate is determined by the frequency scanning sampling time

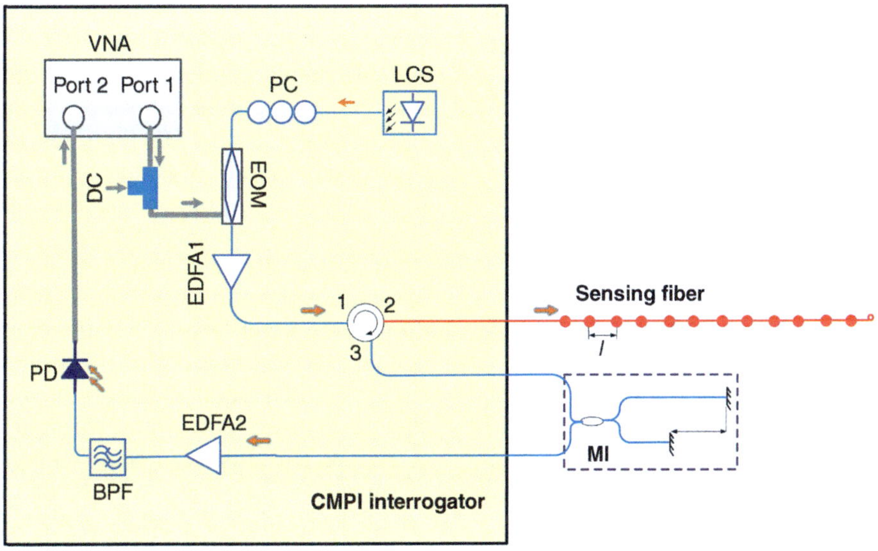

Figure 3.3 Hardware used in the CMPI system includes an interrogator, Michelson interferometer (MI), and sensing fiber. The interrogator includes a low-coherence light source (LCS), a vector network analyzer (VNA), a laser, an electro-optical modulator (EOM), a photodetector (PD), and two Erbium-doped fiber amplifiers (EDFAs).

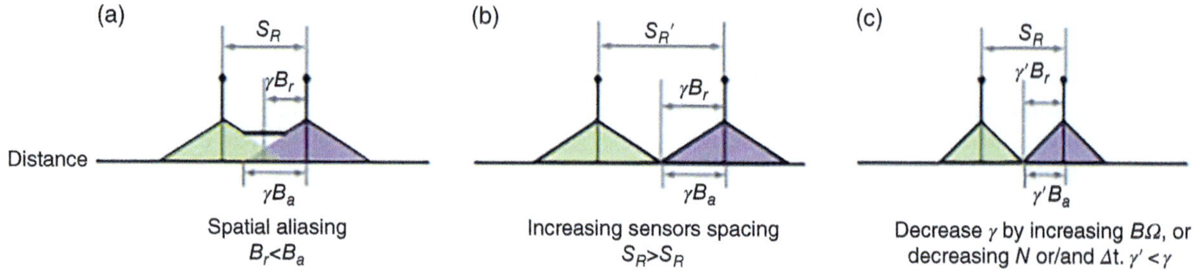

Figure 3.4 Relationship between the spatial resolution (S_R) and reception bandwidth (B_r). (a) Spatial aliasing when $S_R < 2\gamma\ B_a$. (b) Increase S_R or (c) decrease γ to avoid spatial aliasing. Adapted from Hua et al. (2021a).

(Δt). Therefore, the maximum measurable frequency is $\Theta_{\max} = 1/2\Delta t$. An acoustic signal centered at any frequency between DC and $\Theta_{\max}$ can be detected. However, the sidelobes generated by different sensors may overlap when the acoustic signal bandwidth (B_a) approaches the upper limit, as shown in Figure 3.4(a). These overlaps (spatial aliasing) cause crosstalk among sensors, and more sophisticated signal processing methods are required to identify the frequency components and recover the temporal signal. The reception bandwidth that avoids spatial aliasing (B_r) is expressed as

$$B_r = \frac{S_R}{2\gamma}, \quad \text{where } \gamma = \frac{cN\Delta t}{2B_\Omega n}. \tag{3.16}$$

The reception bandwidth can be increased by fabricating reflectors with larger intervals (Figure 3.4(b)), decreasing γ (Figure 3.4(c)) by either increasing the frequency scanning bandwidth (B_Ω), decreasing the scanning points (N), or reducing the sampling time (Δt). These four parameters also affect other aspects of system performance, such as the range sampling resolution (Sr), maximum measurement range ($L_r = \frac{cN}{2B_\Omega n}$), and signal-to-noise ratio (inversely proportional to Δt). The trade-offs among the different aspects of system performance can be tuned to satisfy the requirements of different applications.

The maximum number of sensing units N_C in the system is the ratio of the maximum measurement range and the spatial resolution ($N_c = L_r/S_R$). Using equation (3.16) gives

$$N_c = (2B_r\Delta t)^{-1} \tag{3.17}$$

Equation (3.17) indicates that the number of sensing units is inversely proportional to the reception bandwidth and Δt. The loss from the sensing cable can also limit the N_c, but this factor is minor when the reception bandwidth is greater than several hertz.

There is a time delay between the waveform reading and the occurrence due to the frequency sweeping and signal postprocessing. The delay is usually from milliseconds to seconds, which can be minimized by choosing a small number of modulation frequencies.

It is worth noting that temperature variations would also cause the interference phase change. The strain-temperature cross-talk of a typical fused silica SMF is about 9.57 με/°C (Song et al., 2016). The strain-temperature cross-sensitivity needs to be considered when applying the sensor and system in real applications.

3.3. STRAINS CAUSED BY SMALL FLUCTUATIONS IN AIR PRESSURE, AN APPLICATION OF CMPI

Barometric pressures fluctuate due to atmospheric tides, heating and cooling of air from daily solarization, short- and long-term fluctuations from storm systems and weather patterns, and annual fluctuations from seasonal changes in the atmosphere, among other factors (Kuang et al., 2013; Massmann & Farrier, 1992). These effects cause diurnal to semidiurnal fluctuations (12 or 24 hr. period) of roughly +/− 300 Pa near the equator, and the magnitude decreases with latitude and can be nearly zero at latitudes above 60° (Massmann & Farrier, 1992). Variations of 1,000 Pa to several thousand Pa can occur over several hours during severe storms or during several days in response to weather fronts.

Changes in barometric pressure have long been known to affect the water level in wells (Jacob, 1940); increases in barometric pressure acting on the ground surface cause a drop in the water level in a well completed in a confined aquifer, and vice versa. This effect is more complicated in an unconfined aquifer where the barometric pressure change diffuses downward through the vadose zone (Davis & Rasmussen, 1993; Quilty & Roeloffs, 1991; Rasmussen & Crawford, 1997; Rojstaczer, 1988; Rojstaczer & Riley, 1990; Weeks, 1979; Yusa, 1969), which damps the effect of barometric pressure change on the water level. Interestingly, the connection between the diffusion of air pressure through the vadose zone and water level changes in wells was recognized (Tolman, 1937) long before methods for analyzing this process were available.

Pressure diffusion through the vadose zone also affects strain in the vadose zone. We showed that changes in water content correlated with strains in the underlying soil, but the effect was complicated by changes in barometric pressure (Murdoch et al., 2015). Our initial approach was to account for changes in barometric pressure by assuming it only acted at the ground surface. This explained some effects, but the correlation between barometric pressure and strain changed with time, and this complicated the interpretation. The change in correlation between barometric pressure and strain in the subsurface appeared to be a result of changes in the air pressure diffusivity caused by changes in water content. Barometric pressure could diffuse relatively deep into the subsurface when the water content was low, but pressure diffusion decreased as the water content increased because the gas phase pressure diffusivity (and relative permeability) decreased to essentially zero. This seemed to be a reasonable explanation, but little work was available describing the effects of barometric pressure on strain in the vadose zone.

The objective of the experiments described here is to evaluate the feasibility of measuring and evaluating strains caused by barometric pressure fluctuations. This would build on our earlier work, where we needed to improve the barometric calibration of in situ strain data. However, it became apparent that the effects of barometric pressure diffusing into the vadose zone would be controlled by variations in pneumatic diffusivity, which depends on water content and relative permeability. As a result, a secondary motive is to evaluate the sensitivity of distributed strain to basic vadose zone processes, such as changes in water content.

3.3.1. Experimental Methods

The approach was to evaluate the strains and pressure in a column in the laboratory before proceeding to the field. A cylindrical PVC pipe 0.2 m in diameter and 1.03 m long was fitted with flanges at the end and tapped holes along the side (Figure 3.5). The tapped holes were used to secure air pressure transducers, and the distance from the top of the cylinder was used as a reference (Figure 3.5) to locate the transducer (Costantino, 2021). Water could be injected or drained through a fitting in the bottom flange. A pipe fitting in the upper flange was attached to a plate bolted to the face of an 8-in. acoustic speaker wired to a signal generator (Figure 3.5).

The strain was measured in the sand using an optical fiber interrogated with a CMPI system, as described above. The optical fiber was an SMF with reflectors spaced 5 cm apart. The reflectors were fabricated using femtosecond laser micromachining, which created reflectivity from −35 to −45 dB (Song et al., 2016). A 10-cm-long cavity MI was used to select reflectors every 10 cm apart to form an IFPI.

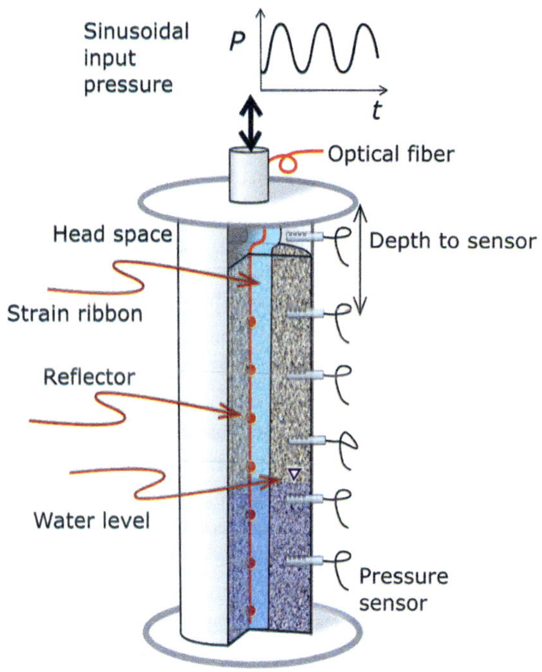

Figure 3.5 Schematic of the sand-packed column with a sinusoidal air pressure applied to the headspace in the laboratory experiments. A strain ribbon containing an optical fiber with a uniformly spaced reflector extends along the axis of the column to measure strain.

The fiber was packaged to improve the frictional strain transfer with the porous media and to provide protection from breakage during deployment. The packaging consisted of laminating the fiber between two layers of 100-μm-thick polyester film with a contact adhesive. This was accomplished using a purpose-built device that dispensed the optical fiber between two rolls of 3M 396 tape (Plunkett, 2021). This created a uniform ribbon 2.5 cm wide with the optical fiber along its axis laminated between two pieces of film for a total thickness of 200 μm. The coefficient of friction between the strain ribbon and sand is 0.35 +/− 0.1, and the axial compliance of the ribbon is 80 με/N, according to measurements described in Plunkett (2021).

Air pressure was measured with Bourns BPS110 Series analog low-pressure sensors rated to measure between 1 and 6.89 kPa with 0.25% accuracy and a total error band of 1.5%. The transducers were attached to a short piece of vinyl tubing attached to a brass tube inserted through the wall of the column (Figure 3.5). The transducer was mounted slightly above the brass tube so that air was trapped next to the transducer when the water level in the column was above the level of the transducer. This protected the transducer, which is designed to measure air pressure and is damaged when exposed to water. It also caused the air pressure transducers to continue to measure pressure changes when the sand was completely saturated, even though the air pressure in the column would have been zero. These data are omitted to avoid confusion.

Experiments were conducted by configuring the column and attaching the speaker to a signal generator and attaching the pressure transducers to a data acquisition system (LabJack T7). The signal generator was set up to drive the speakers with a 4 Hz sinusoid. The frequency of 4 Hz was selected to ensure that the wavelength of the pressure wave is slightly longer than the column height, thereby enabling differentiation of the phase at adjacent sensors. Additionally, this helps minimize the difference between phases to below 2π, thus avoiding any potential ambiguity. The strain was measured using the CMPI interrogator, as outlined earlier, and the data were sampled and stored in MATLAB. The VNA measured the amplitude and phase of the electrical signal at the modulation frequency, which was set to 40,001 sampling points in the microwave band from 0.3 kHz to 9 GHz. The intermediate frequency bandwidth (IFBW) was set to 10 kHz, and the sampling time for each modulation frequency was set as 1 ms. The bandwidth for each sensing section is solely dependent on the sampling time, which was 500 Hz for each sensing unit. The LabJack was triggered by the VNA to start the pressure signal sampling. The pressure signal was applied for several minutes and then sampled over approximately 10 s. The resulting S_{21} and pressure data were saved to a local drive and processed offline.

The experimental program includes six configurations of different water content and materials in the column. The baseline configuration was created by attaching the strain ribbon to the center of the bottom flange and then tensioning the ribbon with a 10 N weight. A 2-cm-thick layer of gravel was placed at the bottom of the column, and the rest was filled with dry, medium-grained, well-sorted sand to approximately 7 cm from the top. This created a headspace between the upper flange and the top of the sand. The brass tubes used for air pressure measurements were in place as the column was filled. The strain ribbon extended along the axis of the column (Figure 3.5), and the measurement fiber passed through a pressure seal in the upper flange.

A suite of the following six configurations was tested.

1. **Dry**: This is the baseline outlined previously—the column is filled with dry, medium-grained sand. All configurations used this sand except the Silt-to-48 configuration.
2. **Fill to 42**: Water was injected and filled the column to a depth of 42 cm from the top of the column. The point of zero pressure (PZP) in the column was measured at 42 cm using an external standpipe. A capillary fringe is expected to extend approximately 10 cm above the PZP (Plunkett, 2021). The PZP is similar to the water table in an unconfined aquifer.
3. **Drained**: Water in the Fill-to-42 experiment was drained, and the test was repeated.
4. **Fill to 29**: Water was injected into the column to fill it to a depth of 29 cm. This is the level of the uppermost strain and pressure sensors.
5. **Barrier**: A thin plastic film was placed across the column and sealed to the sides, creating a barrier to airflow. A few centimeters of sand was placed on top of the film to hold it firmly against the underlying sand. The film is flexible, so it can transmit effective stress to the underlying sand, but it reduces the transmission of air pressure.
6. **Silt to 48**: The upper 48 cm of sand was removed from the column and replaced with a mixture of sand and silt. This created a two-layer system with the permeability of the upper layer less than that of the underlying layer.
7. **Fill and Drain**: The speaker was removed, and the top of the column was vented. Water was injected at a constant rate using a peristaltic pump until water was ponded on the sand. The pump was then turned off, the system was allowed to equilibrate, and then the pump was reversed and the water removed.

3.3.2. Results

Pressure and strain were monitored as functions of depth during the experiments, and the results were compiled as time series, which were then processed and summarized as variations in amplitude and phase delay. The summaries were used to facilitate a comparison between the experimental configurations.

Dry Experiment

The pressure at the sensor closest to the speaker and located in the headspace (depth = 6 cm from the top of the column) fluctuates with the greatest amplitude, approximately 55 Pa from peak to trough (Figure 3.6(a)), and the amplitude progressively decreases approximately as a negative exponential function of depth with a semilog slope of $s_p = -1.2$ 1/m (Figure 3.7(a)). The maximum pressure is 10 Pa at a depth of 67 cm (Figure 3.6(a)).

The pressure in the sand is also periodic, and the signal is delayed behind the pressure in the headspace (Figure 3.6(a)). The delay relative to the pressure in the headspace, P_r, increases with depth at a gradient of 130 ms/m (Figure 3.8(a)), which indicates an air pressure phase velocity of approximately v_{pp} = 8 m/s. The period is 0.25 s, which implies a wavelength of λ_p = 2 m.

The vertical strain also varies as a periodic function of time (Figure 3.6(b)) with the same period as the pressure. The largest strain amplitude is 200 nε, and it occurs at the

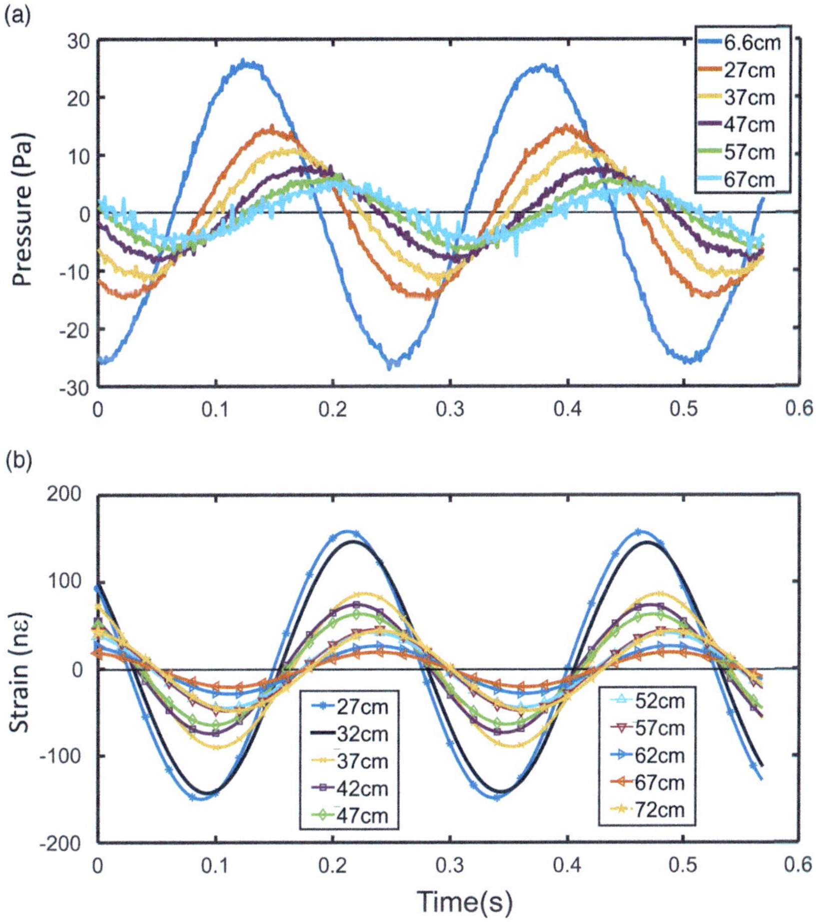

Figure 3.6 Air pressure (a) and strain (b) at different depths as a function of time when applying a sinusoidal pressure at the top of the column of dry sand. The air pressure at 7 cm is used as a reference, P_r.

sensor closest to the top of the column (27 cm). The strain amplitude decreases with depth and is 25 nε at the lowest strain sensor (72 cm in Figure 3.6). The strain amplitude attenuates as approximately a negative exponential with a semilog slope of $s_\varepsilon = -2.0$ 1/m.

The strain minima at the shallowest sensor occur slightly before the pressure maxima in the headspace, and the strain minima occur progressively later as the depth of the strain sensor increases (Figure 3.6). When the air pressure loads only the surface of the sand and does not pressurize the sand (e.g., if the sand was impermeable), the air pressure and strain should be exactly opposite in sign; an increase in air pressure in the headspace is associated with a decrease in strain (compression), and tension occurs when the air pressure decreases. We will use the air pressure in the headspace, P_r, as a reference, and a positive strain delay will be taken as the time when a strain minimum occurs after the pressure maxima in the headspace.

The strain delay is negative at the shallowest sensor, and it increases with depth, according to the time series (Figure 3.6). The strain delay is approximately linear with depth, and the gradient is 85 ms/m (Figure 3.8(b)), which is flatter than that of the pressure. It follows that the phase velocity, $v_{p\varepsilon}$, of the strain is $v_{p\varepsilon}$ = 12 m/s, and the strain wavelength is λ_ε = 3 m.

Loading on the surface of the sand will create an elastic *p* wave that propagates downward. According to Barrì et al. (2012), the velocity of *p* waves in the sand is approximately 150 m/s, which is an order of magnitude faster than the phase velocity of the strain. As a result, it appears that the strain waves are a result of the pressure propagating into the sand, even though the phase velocity of the strain is faster than that of the pressure.

Fill-to-42 Experiment

The time series for this and the other experiments are similar to Figure 3.6 with variations in amplitude and delay that appear to result from the configuration of the experiment. The amplitude of the pressure in the upper part of the Fill-to-42 experiment is greater, and then amplitude decreases more slowly with depth than in the Dry experiment (Figure 3.7(a), (b)). The slope of the pressure delay in this experiment is slightly less than in the Dry experiment (Figure 3.7(a), (b)). The pressure amplitude decreases sharply below the PZP. The pressure data where the column is saturated are uncertain because the measurements were made with transducers designed to measure air pressure.

The volume of air-filled pores in this experiment is significantly smaller than in the Dry experiment. The smaller volume apparently lessened the decrease in pressure with depth compared to the Dry experiment.

The amplitude of the strain signal above the PZP is greater than in the Dry experiment, but the amplitude of the strain decays sharply below the PZP (Figure 3.7 (b)). This is consistent with the pressure amplitude, which is relatively large in the zone above the PZP.

The delay of the strain above the PZP decreases with depth (Figure 3.8). Everywhere else in this experiment and in the other experiments, the strain delay increases

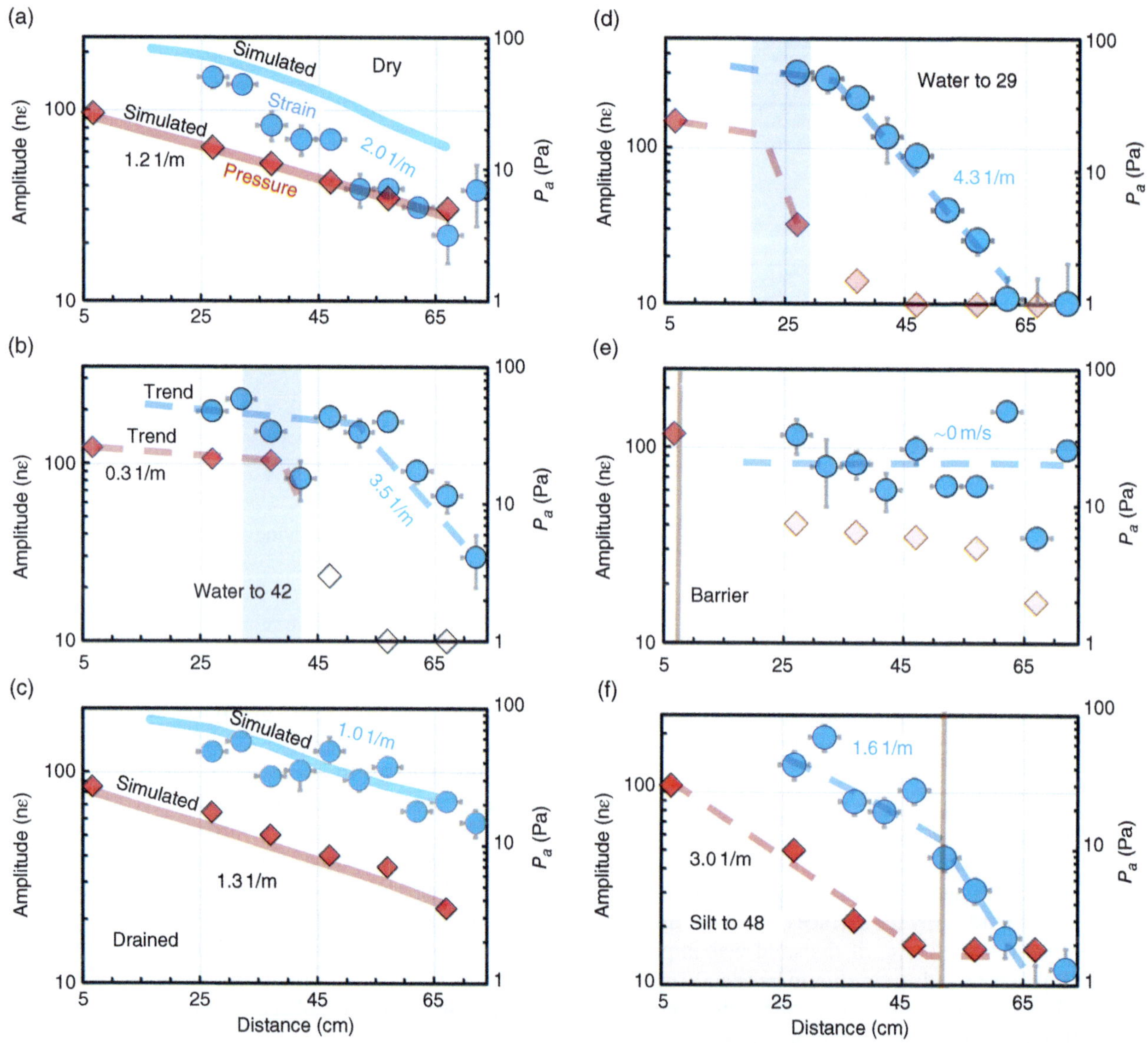

Figure 3.7 Amplitude of periodic pressure (red) and strain (blue) signals as functions of distance in different configurations of a sand-filled column. Surface of the sand at x = 6 cm. (a) Dry sand. (b) Water filled to 42 cm. (c) Water drained after filling. (d) Water filled to 29 cm. (e) Barrier at surface of sand. (f) Silty sand to 48 cm underlain by sand. Numbers are semilog slopes. Data shown as symbols, numerical simulation as thick solid lines in Dry and Drained plots, trend of data estimated manually as thick dashed line. Vertical bands are water level or contact between sand and silty sand. Lightly shaded red symbols are uncertain pressure measurements.

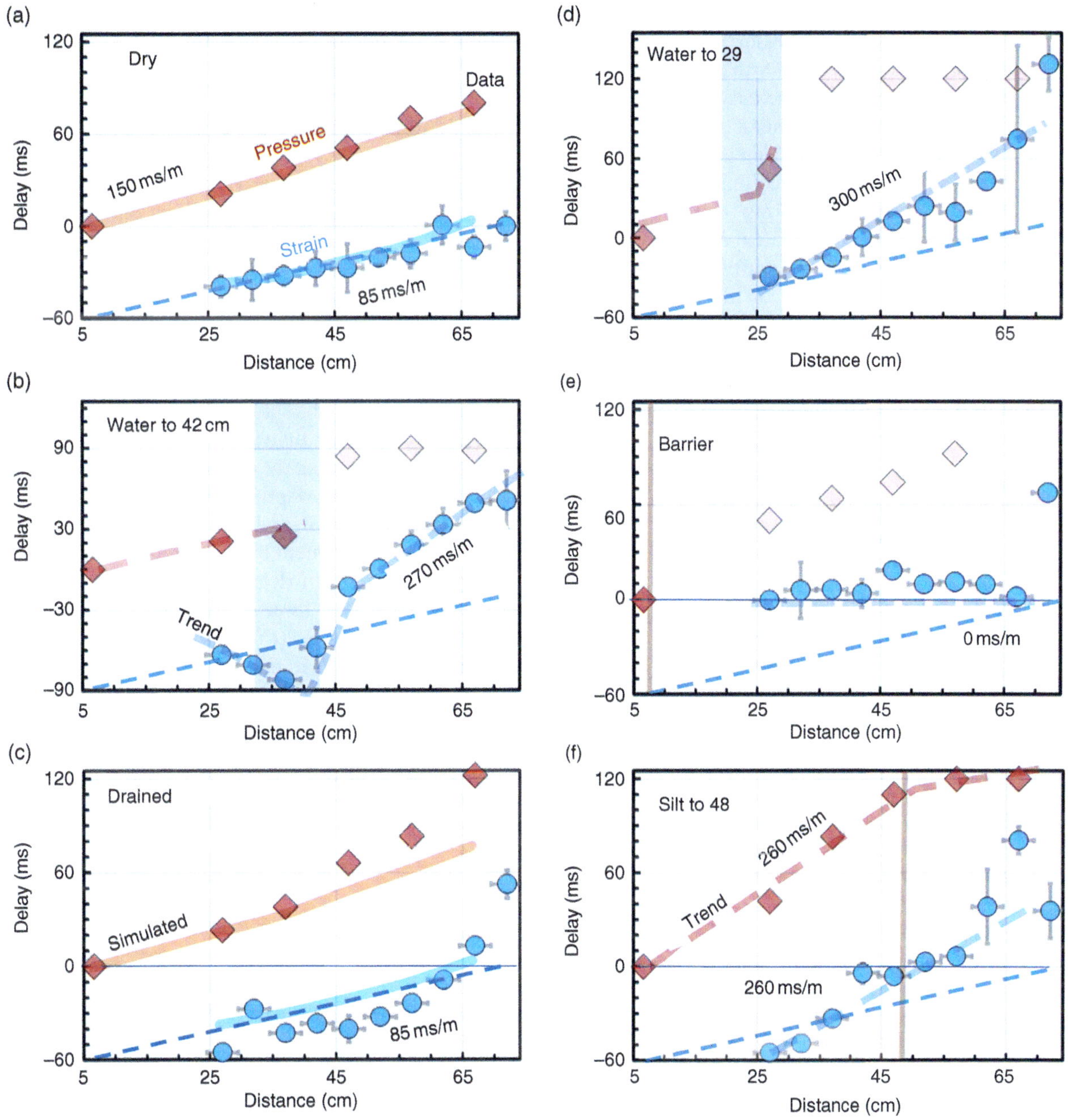

Figure 3.8 Delay of the periodic pressure (red) and strain (blue) signals as functions of distance in different configurations of a sand-filled column. The surface of the sand at x = 6 cm. (a) Dry sand. (b) Water filled to 42 cm. (c) Water drained after filling. (d) Water filled to 29 cm. (e) Barrier at the surface of sand. (f) Silty sand to 48 cm underlain by sand. Best-fit slopes as shown. Data are shown as symbols, an analytical solution for Dry baseline as a thin dashed line, numerical simulation as thick solid lines in Dry and Drained plots, and trend of data estimated manually as thick dashed line. Vertical bands are water level or contact between sand and silty sand. Lightly shaded red symbols are uncertain pressure measurements.

with depth, so this effect is unusual. Below the PZP, the strain delay increases sharply with depth at a slope of 270 ms/m ($v_{p\varepsilon}$ = 4 m/s, λ_ε = 1 m). This indicates that the velocity of the strain in saturated conditions slows to approximately one-third that of dry conditions.

Drained Experiment

The water that was injected to partially fill the column in the Fill-to-42 experiment was drained, and another experiment was conducted. The amplitude and delay of the pressure in the upper half of the column during the Drained experiment are essentially the same as during the Dry experiment (Figures 3.7 and 3.8). The pressure delay in the lower half of the column is longer than in the Dry experiment, probably because the saturation there is higher.

The strain delay rate is also similar to that of the Dry experiment in the upper half and greater than the Dry experiment in the lower half of the column. The saturation in the lower half of the column during the Drained experiment is greater than in the Dry experiment, and this would have reduced the pneumatic diffusivity. This could explain the increase in the delay of the pressure in the lower half of the experiment.

The amplitude of the strain decreases with depth in the Drained experiment, but it does so at a semilog slope that is less than in the Dry experiment (Figure 3.7(c)).

Fill-to-29 Experiment

This experiment involved filling the column so the PZP was at 29 cm, so it is similar to the Fill-to-42 Experiment. The pressure in the uppermost pressure sensor (27 cm) was less than, and the delay was significantly greater than, in the previous experiments. The PZP was below the 27-cm pressure sensor, but the saturation would have been elevated in the zone up to 10 cm above the PZP because of capillarity (Plunkett, 2021), and this likely affected the pressure amplitude and delay.

The strain amplitude attenuates with a semilog slope of approximately $s_\varepsilon = -4.3$ 1/m, which is the steepest of any of the experiments. The strain delay is also the steepest of any experiments, approximately 300 ms/m ($v_{p\varepsilon}$ = 3 s/m and λ_ε = ¾ m). These results are similar to the amplitude decay and delay in the saturated region of the Fill-to-42 experiment. They indicate that going from dry to saturated conditions slows the velocity, reduces the wavelength, and increases the amplitude attenuation of the strain.

Barrier Experiment

The Barrier experiment was conducted by sealing a thin, compliant plastic film to the walls of the column near the top of the sand. The film was covered with a few centimeters of sand to hold it in place. The film was intended to transmit stresses but prevent air pressure from propagating through the sand.

The pressure is attenuated with depth much more in this experiment than in the others (Figure 3.7(d)). Interestingly, a periodic pressure response occurs even with the barrier in place. This may have been because of small unrecognized leaks in the film, or because the film was displaced enough to transmit a slight change in pressure.

The strain response in this experiment is significantly different from the responses in the other experiment. The strain amplitude (60–90 nε) and phase (+/− 10 ms) are somewhat variable, but they are essentially independent of depth (Figures 3.7(d) and 3.8(d)). To account for a possible delay resulting from synchronization between the strain and pressure measurements, an assumed offset was applied to the strain delay in Figure 3.8(d). This was done because the delay could have affected the accuracy of the dataset. Nevertheless, it is clear that the increase in delay with depth that occurred in the other experiments was absent in this one.

Silt to 48 Experiment

The sand was removed from the column and replaced with dry silty sand with a lower permeability than the original sand. The strain ribbon was tensioned in place, and the silty sand was packed in lifts as the column was filled.

The air pressure in the silty sand attenuates more rapidly and the delay is greater than in the sand. This effect appears to change when crossing the contact between the silty sand and the sand at 48 cm depth (Figures 3.7(f) and 3.8(f)).

The strain in the silty sand attenuates more rapidly than in the Drained experiment but less rapidly than in the Dry experiment. The strain delay in the silty sand increases with depth two to three times faster than in sand (Figures 3.7(f) and 3.8(f)). This indicates that $v_{p\varepsilon}$ is approximately 4.5 m/s in the dry silty sand, but it is much faster, 12 m/s, in the dry and drained sand.

Fill and Drain Experiment

This experiment involved injecting water into the bottom of the column at a constant rate with a peristaltic pump until a small volume of water ponded on the surface of the sand and then reversing the pump to remove the water. The strain was measured between 10 reflectors spaced 5 cm apart in a strain ribbon along the axis of the column.

There is a systematic variation in strain with time as the column fills, ponding occurs, and then is drained (Figure 3.9). Early in the experiment, a few με of tensile strain occur in the lower part of the column and compressive strain occurs in the top. Strain above approximately 25 cm increases by 5–10 με and reaches a maximum at times that increase progressively with height above the column. The maximum strain at each reflector pair occurs approximately when the water level or PZP in the column reaches the reflector pair. For example, at the reflector pair between 40 and 45 cm, the strain increases and reaches a maximum at approximately 1,300 s, and the scaled water volume at this time is at 50 cm, slightly above the top reflector (Figure 3.9).

The scaled water volume is roughly equal to the height of the PZP in the column. The PZP is analogous to the water table in an aquifer. A pressure transition zone extends 10–15 cm above the PZP (Plunkett, 2021) and is similar to the capillary fringe above an aquifer. The results during the filling of the column indicate that the sand is

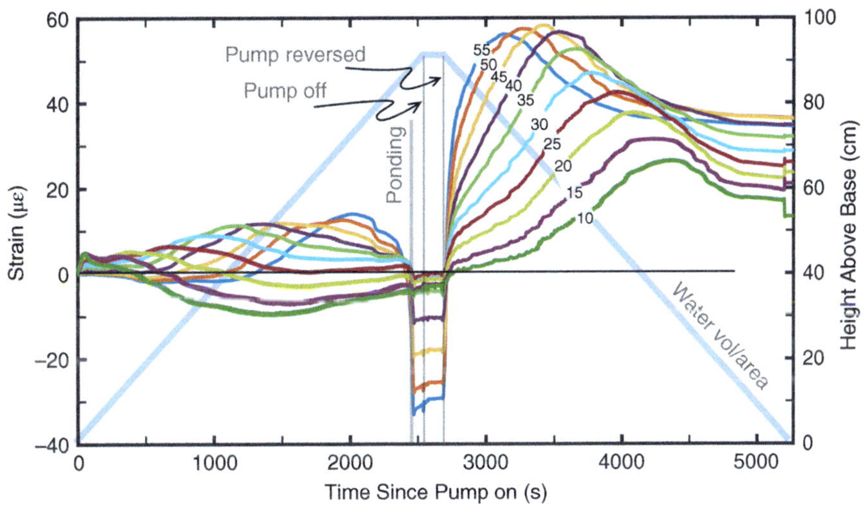

Figure 3.9 Axial strain measured with a strain ribbon and CMPI at different elevations (numbers on colored lines are cm above the datum) in a sand-filled column as functions of time when the column is filled and drained at a constant rate. Ponding occurs approximately 90 s before the pump is turned off. Injected water volume is scaled to the cross-sectional area and porosity to give an effective height above the base of the column (blue line).

compressed slightly above a rising water table, but then it starts to expand as the pore pressure increases in the capillary fringe. The sand expands as the pore pressure increases, and then it decreases slightly as the PZP rises above the upper reflector pair (Figure 3.9).

The process just outlined occurs progressively upward as the pore pressure increases, but a striking change occurs as the column becomes filled with water (Figure 3.9). A slight increase in compression occurs as the capillary fringe starts to interact with the ground surface, but then a large and rapid increase in compression occurs when ponding starts to occur at the ground surface. The greatest compressive strain is 40–50 με and occurs at the upper reflector pair, but compression occurs throughout the upper half of the column when ponding occurs. The compression abruptly stops when water covers the surface of the column, and the strain rate essentially goes to zero before the pump is turned off (Figure 3.9).

An abrupt tensile strain occurs when the pump is reversed and the surface pond is fully drained (Figure 3.9). The strain at the upper reflector pair increases by more than 80 με at the onset of draining. It reaches a peak and then decreases, and this pattern is repeatedly progressively downward as the column is drained (Figure 3.9).

The sharp compression as the column filling nears completion was surprising, but it appears to be related to capillary exclusion (Philip et al., 1989; Plunkett, 2021) of the rising pressure transition zone. Capillary exclusion causes the ground surface to behave as a no-flow boundary, and this increases the pore pressure as the pressure transition zone interacts with the free surface. The pressure increases by approximately 1,000 Pa from this effect, according to the measurements in a similar experiment by Plunkett (2021) The observed compression is consistent with a load of 1,000 Pa acting on the upper boundary, suggesting that the pressure caused by capillary exclusion causes the compression observed in the experiments.

The experimental data described here and shown in Figures 3.6–3.9 indicate that changes in pore pressure representing fluctuations of barometric pressure or changes in the height of the water table cause strains that can be characterized with distributed strain measurements made with CMPI. These results suggest that distributed strains could offer insights into hydrologic processes in the vadose zone.

Analysis

The strain and pressure in the lab experiments were analyzed using a simple analytical solution that describes the basic process, and a numerical solution that has the potential to include more details. Both analyses are one dimensional with a pressure that varies periodically, with amplitude A_p and period λ at one end. The analytical solution assumes the pressure and strain are uncoupled and the properties of the porous media are homogeneous and isotropic. We assume the pressure variations can be described using the diffusion equation with uniform and constant pressure diffusivity, D. This requires that the change in pressure is small enough to ignore fluid compressibility. The pressure is uncoupled from the strain rate in the solid because of the high compressibility of the air in the pores. This boundary value problem is solved by Carslaw and Jaeger (1986) for an analogous problem in heat conduction. The pressure in the half-space, P, is given in dimensionless form as

$$P^* = \frac{P}{A_p} = e^{-z^*} sin(t^* - z^*) \tag{3.18}$$

where the characteristic time and length

$$t_c = \frac{\lambda}{2\pi} \tag{3.19}$$

$$L_c = \sqrt{\frac{\lambda D}{\pi}} \tag{3.20}$$

are used to define

$$t^* = \frac{t}{t_c} = \frac{t}{\frac{\lambda}{2\pi}} = \frac{2\pi t}{\lambda} \tag{3.21}$$

$$z^* = \frac{z}{L_c} = \frac{z}{\sqrt{\frac{\lambda D}{\pi}}} = z\sqrt{\frac{\pi}{\lambda D}} \tag{3.22}$$

The total strain is the sum of the strain caused by the pressure loading at the ground surface and the strain caused by the change in pore pressure. The strain caused by the pressure loading at the ground surface is

$$\varepsilon_s = -B \sin(t^*) \tag{3.23}$$

where

$$B = \frac{A_p(1+\nu)(1-2\nu)}{E(1-\nu)} \tag{3.24}$$

and E is Young's modulus and ν is Poisson's ratio of the porous medium. The strain caused by the pore pressure change is

$$\varepsilon_p = B\alpha_b e^{-z^*} \sin(t^* - z^*) \tag{3.25}$$

where α_b is the Biot–Willis coefficient. We will assume $\alpha_b = 1$, which is reasonable for materials like soils with a bulk modulus much less than the grain modulus, and saturation effects on the apparent α_b will be ignored. The total strain is the sum of the two components, so in dimensionless form,

$$\varepsilon^* = \frac{\varepsilon_s + \varepsilon_p}{B} = e^{-z^*} \sin(t^* - z^*) - \sin(t^*). \tag{3.26}$$

The trigonometric identity for an arbitrary phase shift, φ_ε, gives

$$\varepsilon^* = e^{-z^*} \sin(t^* - z^*) - \sin(t^*) = c_\varepsilon \sin(t^* + \varphi_\varepsilon) \tag{3.27}$$

where c_e is the observed strain amplitude. It follows that

$$c_\varepsilon = \sqrt{e^{-2z^*} - 2e^{-z^*}\cos z^* + 1} \tag{3.28}$$

and the amplitude of the pressure is $c_p = e^{-z^*}$.

The delay of the strain also follows from equation (3.27). Changing the sign to write the phase as a delay in keeping with the experimental data,

$$\varphi_\varepsilon = -\tan^{-1}\left(\frac{e^{-z^*}\sin z^*}{1 - e^{-z^*}\cos z^*}\right). \tag{3.29}$$

It is apparent that the delay of the strain is approximately linear for $z^* < 1$ (dashed line in Figure 3.10). The slope is ½ at $x^* = 0$, and the slope of a best-fit line

$$\varphi_\varepsilon \approx \frac{\pi}{4}\left(\frac{1}{2}z^* - 1\right) \tag{3.30}$$

is $\pi/8$. For comparison, the delay for the pressure is (Figure 3.10)

$$\varphi_p = z^*. \tag{3.31}$$

The experimental results for the delay appear to be generally consistent with equations (3.30) and (3.31). The strain and pressure delays for uniform sand are approximately linear with depth. The slope of the pressure delay is steeper than that of the strain, and the strain delay is less than that of the pressure (Figures 3.8 and 3.10).

Substituting and then solving for D in terms of air pressure phase velocity,

$$D_{vp} = \frac{\lambda v_p^2}{4\pi} \tag{3.32}$$

and in terms of the strain phase velocity

$$D_{ve} = S_1 \frac{v_\varepsilon^2 \lambda}{4\pi} \tag{3.33}$$

where the coefficient is $0.15 < S_1 < 0.25$ using the different approximations for the slope of the phase velocity. D can also be estimated from the semilog slope of pressure decay

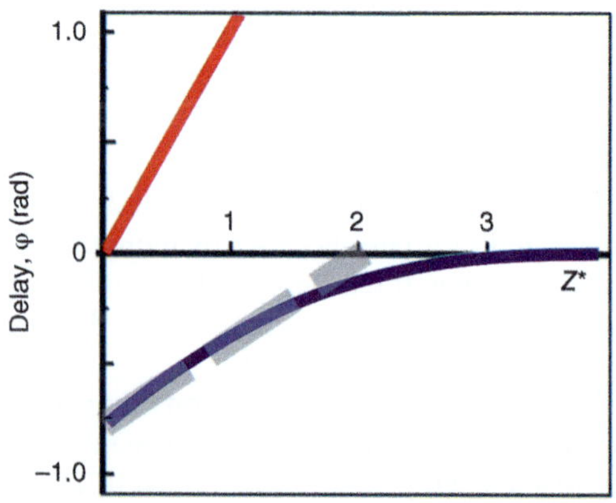

Figure 3.10 Phase delay in radians of the strain (blue) and pressure (red) relative to the pressure at the ground surface. Positive delay means the signal arrives after and negative means it arrives before the reference signal. Phase lag of the strain can be approximated by the gray dashed line for $z^* < 1$.

$$D_{sp} = \frac{0.59}{s_p^2 \lambda}. \quad (3.34)$$

Using values for the air pressure from the Dry and Drained experiments, $v_p = 7.6$ m/s gives $D_{vp} = 1.1$ m^2/s, and $s_p = 1.3$ 1/m gives $D_{sp} = 1.4$ m^2/s. The phase velocity of the strain is $v_e = 12$ m/s, which gives 0.4 m^2/s < D_{ve} < 0.7 m^2/s.

The values of D determined using the air pressure (1.1–1.4 m^2/s) are slightly greater than values estimated from the strain delay, but the values are remarkably similar. The values differ by approximately a factor of 2, which could be because friction on the wall of the column slowed v_e compared to the semi-infinite case in the analytical solution.

Numerical Simulation. Results from the analytical solution are encouraging, but they omit a variety of factors that limit the application of the results to uniform materials. In many cases, field conditions would be characterized by soil materials with different textures and water pore pressure. This will cause the soil properties to vary in space. Changes in air pressure cause changes in water pressure, which may also cause the properties to change—most of the relevant properties in the analysis are functions of pore pressure. It follows (Costantino, 2021) that these conditions are described using two nonhomogeneous diffusion equations for the liquid and gas phases,

$$-\nabla \cdot D_l \nabla (P_l + \rho_l g z) + \frac{\partial P_l}{\partial t} = -\Gamma_l \quad (3.35a)$$

$$-\nabla \cdot D_g \nabla \left(P_g + \rho_g g z\right) + \frac{\partial P_g}{\partial t} = -\Gamma_g \quad (3.35b)$$

where the diffusivities of the liquid and gas pressure are, respectively,

$$D_l = \frac{k k_{rl}}{\mu (S_l S_{\sigma l} - C_m)} = \frac{k k_{rl}}{\mu (S_l S_{\sigma l} + |C_m|)} \quad (3.36a)$$

$$D_g = \frac{k k_{rg}}{\mu \left(S_g S_{\sigma g} - C_m\right)} = \frac{k k_{rg}}{\mu \left(S_g S_{\sigma l} + |C_m|\right)} \quad (3.36b)$$

with the terms on the right side serving to couple changes the liquid and gas phase pressures

$$\Gamma_l = \frac{C_m}{(S_w S_\sigma - C_m)} \frac{\partial P_g}{\partial t} \quad (3.37a)$$

$$\Gamma_g = \frac{C_m}{\left(S_g S_\sigma - C_m\right)} \frac{\partial P_l}{\partial t}. \quad (3.37b)$$

P_n is the pressure of the nth phase, where $n = l$ for the liquid and $n = g$ for the gas phases, respectively. ρ is the fluid density, μ is fluid viscosity, g is gravity, k and k_r are the intrinsic and relative permeabilities, S_n is the saturation of the nth phase, the elastic storage term is $S_{\sigma n} = \phi \beta_n + \left(\frac{1}{K} - \frac{1}{K_s'}\right)$, and C_m is the specific moisture capacity (derivative of the soil water retention curve with respect to pressure). Constitutive functions describing water content, k_r, and C_m as functions of capillary pressure given by van Genuchten (1980) were used here.

Stresses are determined using momentum conservation for quasistatic conditions

$$\frac{\partial \sigma_{ij}}{\partial x_i} = -F_i, \quad (3.38)$$

where σ_{ij} is the total stress tensor and F_i is the body force vector. The ij subscript assumes Einstein notation, where repeated subscripts indicate summation. The total stress depends on the effective stress on the solid, σ_{eij}, and the effective pore pressure, P (Borja, 2006, equation 3.42),

$$\sigma_{ij} = \sigma_{eij} - \alpha P \quad (3.39)$$

where α is the Biot–Willis parameter. The effective pressure in the unsaturated zone is weighted between the water and the air pressures (Borja, 2006, equation 3.24; Lu et al., 2010, equation 3.18)

$$P = S_e P_l + (1 - S_e) P_g, \quad (3.40)$$

where S_e is the effective liquid saturation. The material parameters are characterized by constitutive functions of the capillary pressure. Young's modulus was characterized as a power function given by Lu and Kaya (2014). Poisson's ratio was described using the function given (Kumar Thota et al., 2021).

The governing equations were discretized over a 2D region 0.2 m wide × 1 m high bounded on the sides and bottom by no-flow conditions for air and water and no mechanical displacement. These conditions assumed the PVC pipe used to contain the soil was rigid. We conducted tests where the PVC pipe was included explicitly, and there was negligible difference with the case that used the zero displacement boundary, although including the PVC pipe explicitly increased the run time. The top boundary was loaded by a periodic air pressure at 4 Hz, and the bottom and sides were assumed impermeable to air. The top and sides were assumed to be impermeable to water, and the water pressure at the bottom boundary was assumed to be fixed at 0 because there was a thin layer of water that could not be completely drained. Gravity drainage to equilibrium was used to establish initial conditions for the water pressure. The initial air pressure was assumed to be pneumatostatic. Initial stresses were in gravity equilibrium. The region was discretized with a rectangular, mapped finite element mesh with 800 cells vertically and 12 cells horizontally (each mesh element is 1.2 mm × 17 mm). The governing equations for air and water pressure and their boundary conditions were fully coupled and solved using the Galerkin finite element method. The resulting pressures were then used to solve for the displacements and strains, also using the Galerkin finite element method. The solution was implemented and the results rendered using Comsol Multiphysics software.

Results. The results were rendered along a line at the center axis of the region resembling the experiments. Parameters were adjusted manually so the results were similar to results from the Dry and Drained experiments. The simulations were intended to provide insights into the general behavior of the experiments, so a formal parameter estimation process was unnecessary.

The resulting strain in the experiments has components caused by the surface load and the pore pressure (Figure 3.11). The component from the surface load at 0.2 m depth fluctuates between +/− 400 nε, and the pressure fluctuates between +/− 25 Pa. The phase of the surface load component is always delayed by ½ period relative to P_r (pressure in the headspace). However, the pressure load component of strain is delayed only slightly relative to P_r, and its magnitude is slightly less than the surface load component. The result of adding the two components is that the resulting strain minima occur before the maxima in P_r (Figure 3.11(a)).

The phase of the two components moves closer together at 0.4 m, and they are nearly in phase from one another at 0.6 m (Figure 3.11(b), (c)) The magnitude of the strain from surface load also decreases with depth. This is because the lower boundary condition prevents displacement. The resulting total strain at 0.4 m is approximately half of that from the 0.2 m depth, whereas the resulting total strain at 0.6 m is barely fluctuating at all.

Interactions between the surface and pressure loads cause the time series of the strain to lead to the pressure, and the results of the simulations resemble those of the experiments (Figures 3.6 and 3.12). The simulations of the pressure and strain magnitudes and delays are also similar to the experimental data (Figures 3.7(a), (c) and 3.8(a), (c)).

A key aspect controlling the strain response is the pore pressure that is created by the fluctuating air pressure at the ground surface. The simulations show that the air pressure perturbation is transferred to the water pressure and continues to propagate downward, causing a strain response. The pressure propagation depends on the air and water pressure diffusivity D_w, which varies with depth and saturation. The air pressure diffusivity, D_a, is approximately 1.4 m²/s in dry sand, and the D_w is approximately 0.12 m²/s in saturated conditions (Figure 3.13). Interestingly, both D_a and D_w decrease and are small in the transition zone at the water table, suggesting the water table would impede downward propagation of barometric pressure.

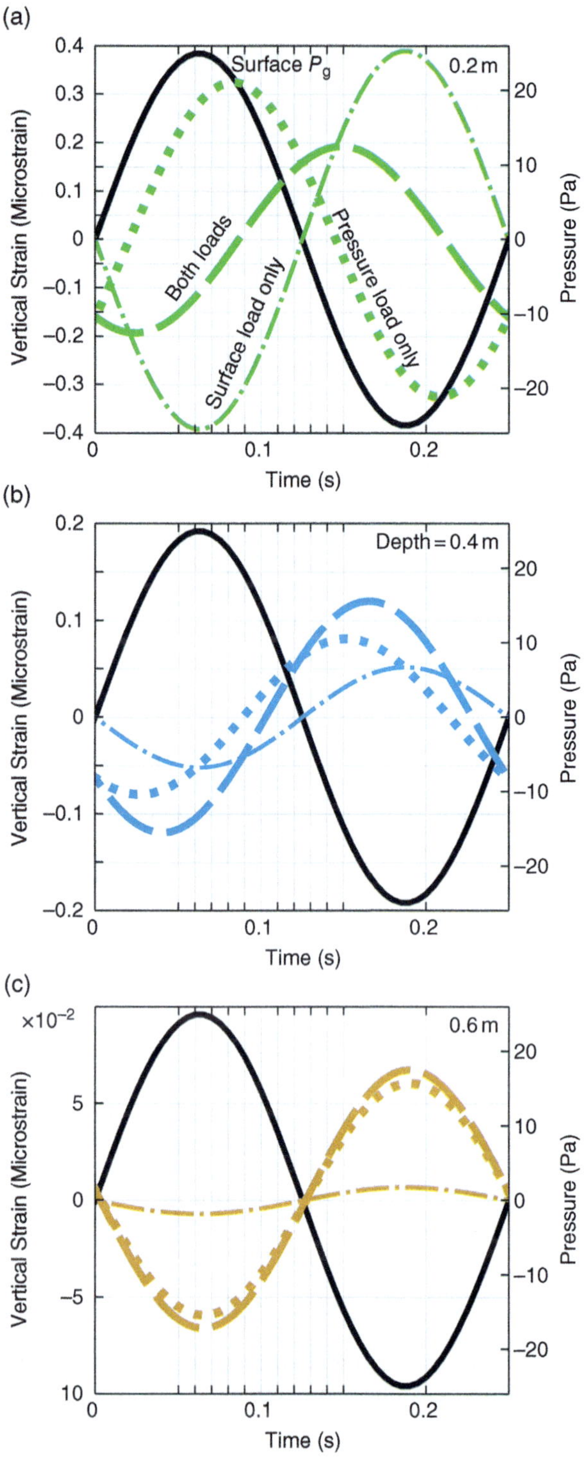

Figure 3.11 Components of the vertical strain from an air pressure fluctuation on the surface. The black solid line is P_r, the pressure at the ground surface used for reference; dashed-dot line is the strain component from surface loading only; dotted line is the strain component from pressure in the porous media only; and thick dashed line is the sum of the two components. From numerical simulations based on equations (3.19)–(3.24).

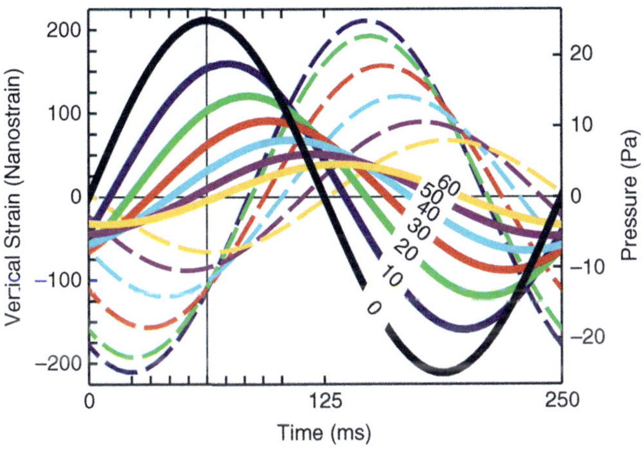

Figure 3.12 Pressure (thick solid lines) and strain (dashed lines) at different depths in cm (numbers) in dry sand using numerical simulation based on equations (3.35)–(3.40).

The numerical analysis is capable of representing the partially saturated conditions and resulting nonlinear material properties in the other experiments, and analyses of those experiments are ongoing.

Discussion

The experiments and simulations outlined here indicate that strains in soil are closely related to small periodic changes in air pressure and saturation at the centimeter-to-meter scale. Strain amplitudes in the sand caused by small fluctuations in air pressure (+/− 25 Pa) were as small as 10 nε during a 4 Hz signal, and spatial variations in strain were detected with a 5 cm spacing. The effects of fluctuating barometric pressure on water levels are widely known, but to our knowledge, the effects of barometric pressure on strain in partially saturated media have not been documented. One reason for this is that detecting these signals challenges the available measurement capabilities (strain resolutions in the nε range, sampling rates of at least 100 Hz, and spatial resolutions at the cm scale in a package that could be deployed in the field). The development of the CMPI methods with optical fiber packaged in a strain ribbon has made it feasible to measure these signals.

Characterizing the Vadose Zone. The experimental results show that a periodic fluctuation of air pressure at the ground surface will cause a strain signal that penetrates into and responds to the properties of an underlying porous medium. The phase velocity (inverse of s_p) of the strain is roughly twice that of the pressure under dry conditions (Figure 3.8(a), (c)), and it decreases by a factor of roughly one-third when the sand becomes saturated (Figure 3.8(a), (c) and (b), (d)). D_w is expected to be significantly less than D_a, according to Figure 3.13 and equation 3.21, and this may explain why the strain phase velocity is slower in saturated than dry conditions.

The effects of air pressure diffusing into the subsurface are illustrated by comparing results from the Barrier experiment (Figures 3.7 and 3.8(e)) and the other experiments (Figures 3.7 and 3.8). Strain delay and magnitude are variable but essentially independent of depth in the Barrier experiment where a membrane was used to seal the downward diffusion of pressure but allow stress to be transmitted. In contrast, the delay and magnitude of the strain varied with depth in the other experiments where the pressure barrier was omitted. According to the theoretical analyses (e.g., equation 3.6), strain is independent of depth when only the surface load is considered. This is consistent with the experiments and confirms that systematic variations of strain with depth are due to pressure variations.

The amplitude of the strain varies with saturation, with smaller strain magnitudes occurring in dry conditions and larger strains in saturated conditions (Figure 3.7(a), (c) and (b), (d)). The strain magnitude will be affected by both the pressure magnitude and the elastic modulus, and both of these parameters are functions of pore pressure (equation 3.20; Borja, 2006; Lu & Kaya, 2014).

The experiments conducted here used pressure fluctuations that are smaller and faster than fluctuations of barometric pressure, where diurnal variations of several hundred Pa are common, and larger variations in the kPa range occur over longer periods. The characteristic length of a diurnal variation in the dry sand used in the experiment would be more than 100 m, and it would be more than 10 m when the pressure diffusivity was two orders of magnitude less than in the experiments. This suggests that strain variations caused by variations in barometric pressure could extend to significant depths in the vadose zone. We have shown that the strain signal is affected by the pore pressure and water content, and this raises an important point for both the potential and the challenges of interpreting strains caused by barometric pressure variations. It appears that variations in strain caused by barometric pressure fluctuations could be used to continuously interrogate changes in pore pressure and properties in the vadose zone, which would have wide applications for environmental characterization and monitoring. However, the quantitative interpretation would likely require inverting numerical analyses similar to those described previously. The key material parameters in that analysis are all nonlinear functions of pore pressure, which will raise challenges during the inversion process. An inversion workflow described by Roudini et al. (2023) that uses a proxy model to reduce the number of simulations required for inversion is a possible path forward for addressing these challenges.

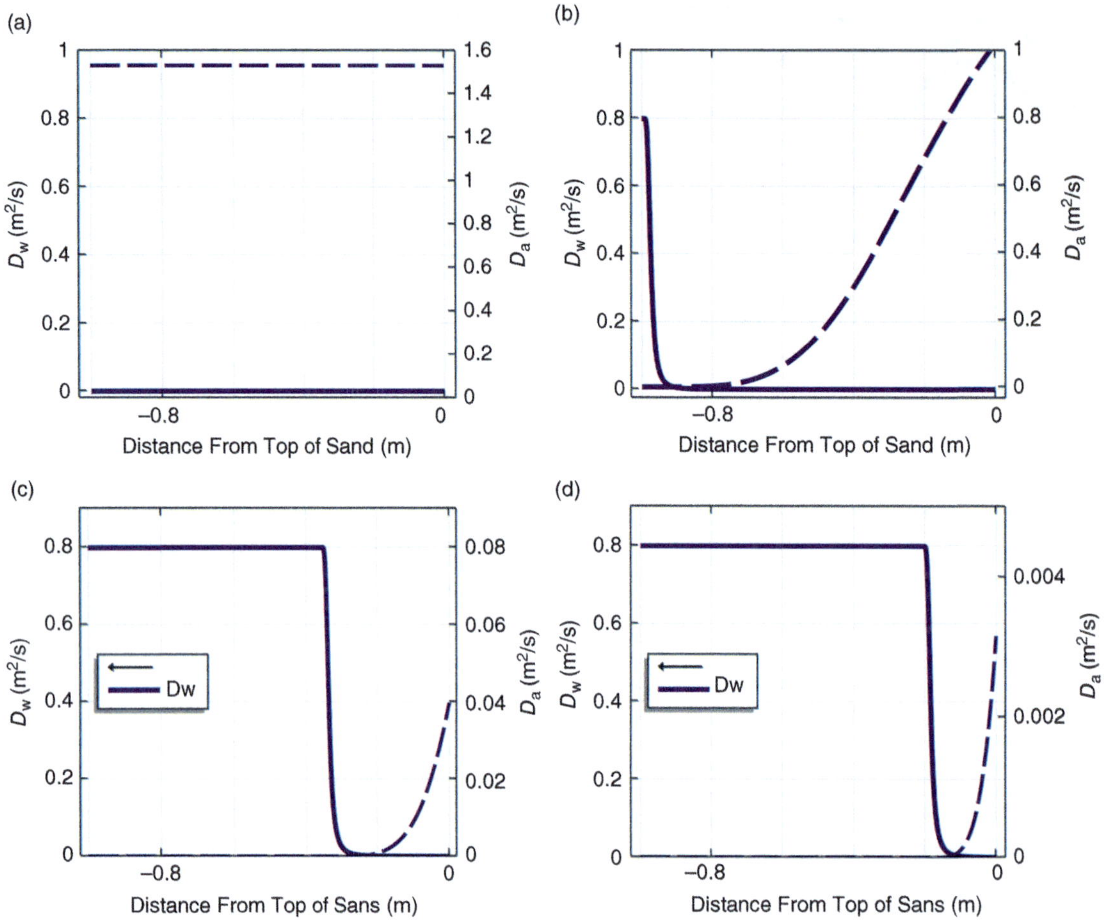

Figure 3.13 Initial pressure diffusivity for water, D_w, and air, D_a, as a function of depth for different initial conditions. (a) Dry. (b) Gravity drainage. (c) PZP at d = 0.35 m. (d) PZP at d = 0.2 m. From numerical simulations using equations 3.35–3.40.

Characterizing the Saturated Zone. We are optimistic that measurements of strain made with optical fibers could have important applications for monitoring processes in the vadose zone, but there are also important potential applications in the saturated zone. The Fill and Drain experiments show a strong and distinctive strain signal associated with a rising water table that reaches the ground surface. In addition, we have used the CMPI technique with a strain ribbon in the field to monitor transient well tests. One experiment was conducted by deploying a strain ribbon along the wall of a borehole and then filling a polyethylene sleeve in the borehole with sand to couple the ribbon to the wall of the boring (Figure 3.14). Becker et al. (2017) describe a similar technique for pressing a fiber cable against the wall of a borehole with a water-filled sleeve. The strain ribbon was deployed in the vadose zone from the ground surface to a depth of 7 m in saprolite, and weak reflector pairs were placed every 0.85 m along the fiber. Each weak reflector pair forms a 15 cm cavity strain meter. This allowed the vertical strain to be measured every 0.85 m of depth (Figure 3.14(a), (b)). The strain ribbon was at a radial distance of $r = 4.4$ m.

The water table of the unconfined aquifer in the saprolite was at a depth of approximately 9 m, and the saturated zone was bounded below by low permeability crystalline rock at a depth of approximately 30 m. A pumping well was drilled through the saturated zone to the top of the crystalline rock. We conducted a constant-rate pumping test for 6 hr, and we monitored the strain in the vadose zone and the drawdown in monitoring well WL-5 at $r = 5$ m m from the pumping well.

The drawdown in the monitoring well increased sharply soon after the pump was started, and the rate of increase diminished with time in a pattern (Figure 3.14(c)) typical of pumping tests in an unconfined aquifer (Neuman, 1975). The vertical strain in the vadose zone increased as soon as pumping started, and then the rate diminished

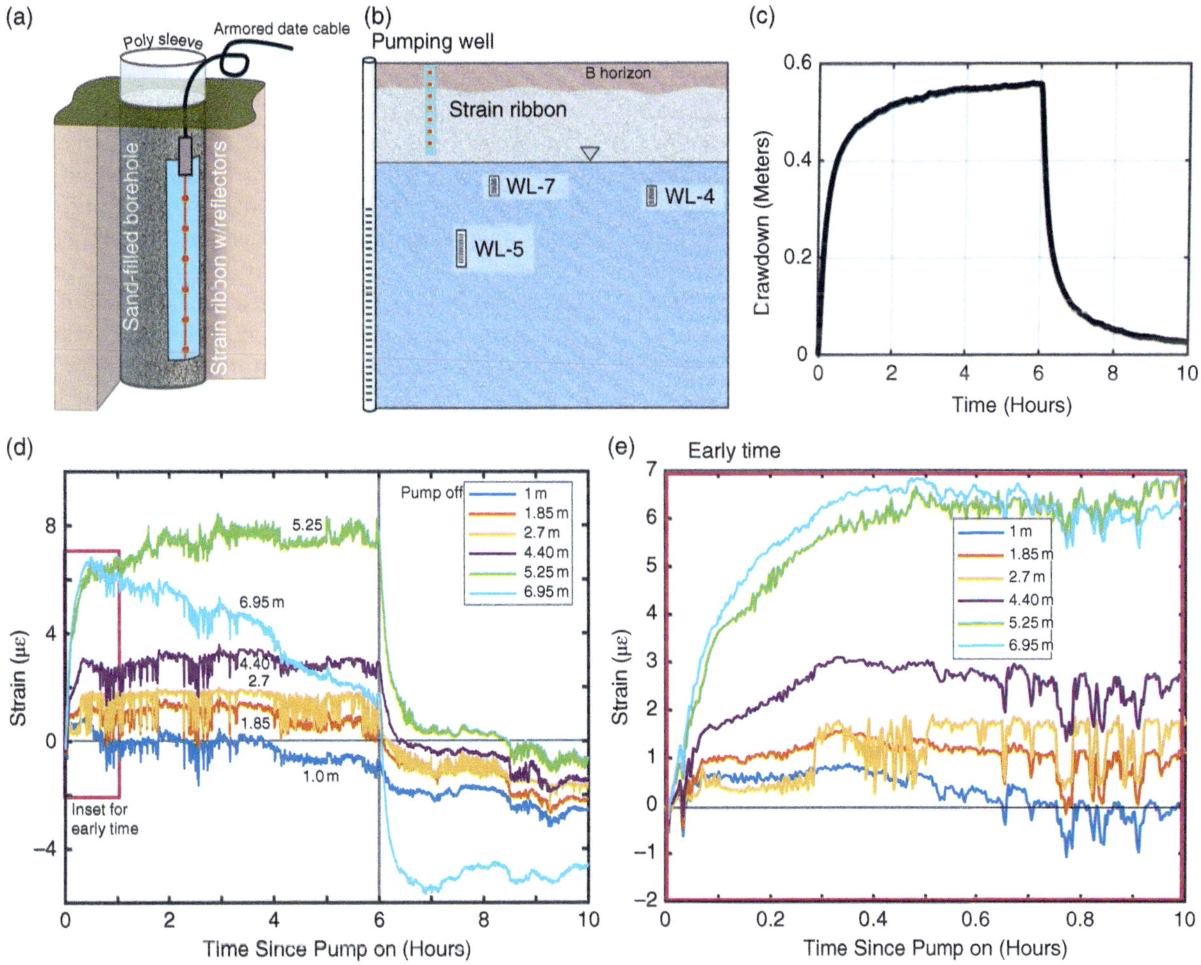

Figure 3.14 Field test of a CMPI strain ribbon used to characterize a pumping test. (a) Schematic of the strain ribbon pressed against the wall of a borehole by a sand-filled polyethylene sleeve. (b) Cross-section of the field site. (c) Drawdown at monitoring well WL-5. (d) Vertical strain at different depths below the ground surface during pumping and recovery after the pump is turned off. The red box frames the data in panel (e). (e) Strains as a function of time during the first hour of the pumping test. Strain data at 3.55 and 6.95 m are omitted from plots due to large demodulation error.

with time in a pattern that resembles the drawdown (Figure 3.14(d), (e)). The strain is greatest at the bottom of the ribbon near the water table, where it reaches a maximum of approximately 7 με. The strain decreases upward, but the CMPI ribbon resolves a clear strain signal that reaches a maximum of approximately 0.6 με at a depth of 1 m (Figure 3.14(e)).

Poroelastic simulations of pressure and strain in the vicinity of a well being pumped at a constant rate in an unconfined aquifer were conducted using methods outlined earlier and in Murdoch et al. (2020) and configured to the field site. Those analyses show that the strain signal resembles the drawdown and that the strain magnitude increases with depth. An exception occurs near the water table where pumping drops the pore pressure, increasing compression and reducing strain magnitude resembling the blue curve from 6.95 m depth in Figure 3.14.

Similarities between the observed strains and simulations during pumping tests suggest that strain data in the vadose zone could be used with inversion methods to estimate properties of underlying unconfined aquifers. This has important implications for characterizing aquifers because it suggests that measurements of strain in the vadose zone could be used instead of, or to supplement, the pressures that are typically used to interpret well tests. This builds on findings by Murdoch et al. (2021, 2023) that show the strain in the overburden above a confined aquifer can be analyzed to estimate the saturated hydraulic diffusivity in the aquifer. The data in Figure 3.14 suggest that vertical strain near the water table is affected by upward propagating pressure changes caused by pumping in the underlying aquifer (blue curve from 6.95 m depth in Figure 3.14), whereas overlying strains that are above the extent of the changing pressure are unaffected.

Pressure propagation in the vadose zone will be affected by constitutive functions that vary with pore pressure. This implies that distributed strain measurements in the vadose zone during pumping tests could be used to help characterize the vadose zone constitutive functions. In situ characterization of vadose zone constitutive functions is challenging, so this approach could be a step forward in the characterization of hydrologic processes associated with shallow aquifers and the vadose zones that overlie them.

3.4. CONCLUSION

Distributed strain and acoustic measurements outlined throughout this AGU volume have provided new insights into geophysics. In this chapter, we have shown how an approach for measuring distributed strain, called CMPI, has enabled new insights into hydrologic processes in the vadose and saturated zones.

CMPI measures strain by reading optical interference phase change in the microwave domain. A local MI was implemented to eliminate the dark zones for high spatial resolution sensing. A frequency-to-time-domain mapping signal processing method was developed to increase the sensor bandwidth for DAS. Sensing performance of centimeter spatial resolution, up to 10nε sensitivity, and kilohertz strain measurement rate was achieved.

CMPI was used to measure the strain caused by small periodic variations in air pressure (25 Pa amplitude and 4 Hz) in a sand-filled laboratory column used to represent pressure changes in the vadose zone caused by barometric fluctuations. The amplitude of the pressure decreased as an exponential function of depth, and the phase delay of the pressure increased as a linear function of depth in the experiments. The amplitude and phase delay of the strain followed similar functions of depth in dry sand. Strain amplitudes decreased from roughly 200 nε at shallow depths to roughly 10 nε near the bottom of the column. The phase velocity of the pressure in dry sand was approximately 7.6 m/s, and the velocity of the strain was faster, approximately 12 m/s. Saturating the sand reduced the phase velocity of the strain to 3–4 m/s, and the velocity increased to 12 m/s when the water was drained. Changing the material from a permeable, clean sand to a lower permeability silty sand reduced the strain phase velocity from 12 to 4.5 m/s. The strain varied periodically but was essentially independent of depth when a pressure barrier was placed at the top of the sand.

The results of the experiments show how the strain distribution with depth caused by air pressure fluctuation is affected by the saturation and permeability distribution in the underlying porous media. An analytical solution used to interpret the data gives similar results for the air pressure diffusivity estimated from the air pressure and the strain, suggesting that distributed strain could be used to characterize vadose zone properties. Inverting a numerical analysis that includes variable material properties appears to offer a path forward for more detailed interpretation. These results will help future investigations account for the effects of barometric pressure on data from strainmeters, and they suggest that the constant interrogation by a fluctuating barometric pressure may lead to a new strategy for monitoring the vadose zone.

ACKNOWLEDGMENTS

This work was funded by the U.S. Department of Energy National Energy Technology Laboratory under grant DE-FE028292. We appreciate the support from the U.S. DOE, but the views and conclusions expressed in this document are solely those of the authors and do not necessarily reflect those of the U.S. DOE. We appreciate input from Yang Song, Jincheng Lei, and Daniel Hestenes for the sensor fabrication and packaging.

AVAILABILITY STATEMENT

Data used in the creation of section "Characterizing the Vadose Zone" were stored at NETL'S Energy Data eXchange (Hua et al., 2023). Those data were collected from March 18 through March 22, 2020, at the hydromechanical test facility at the Clemson University Simpson Station research facility in Central, South Carolina, USA. This includes ambient data in addition to a pumping test performed between 14:31 and 20:31 EST on 22 March 2020 from Pumping Well 2 (PW2). Figures were made with MATLAB, available under the MATLAB license at www.mathworks.com.

REFERENCES

Bao, X., & Chen, L. (2012). Recent progress in distributed fiber optic sensors. *Sensors (Basel, Switzerland)*, *12*(7), 8601–8639. doi:10.3390/s120708601

Barrì, J., Bordes, C., Brito, D., Sénéchal, P., & Perroud, H. (2012). Laboratory monitoring of P waves in partially saturated sand. *Geophysical Journal International*, *191*, 1152–1170. doi:10.1111/j.1365-246X.2012.05691.x

Becker, M. W., Ciervo, C., Cole, M., Coleman, T., & Mondanos, M. (2017). Fracture hydromechanical response measured by fiber optic distributed acoustic sensing at milliHertz frequencies. *Geophysical Research Letters*, *44*(14), 7295–7302. doi:10.1002/2017GL073931

Becker, M. W., Harris, B., & Pevzner, R. (2023). Characterization of aquifer poroelastic response to impulse and oscillatory well pressure using distributed acoustic sensing. *Geophysical Research Letters*, *50*(1), e2022GL100904. doi:10.1029/2022GL100904

Bogris, A., Nikas, T., Simos, C., Simos, I., Lentas, K., Melis, N. S., et al. (2022). Sensitive seismic sensors based on microwave frequency fiber interferometry in commercially deployed cables. *Scientific Reports*, *12*(1), 14000. doi:10.1038/s41598-022-18130-x

Borja, R. I. (2006). On the mechanical energy and effective stress in saturated and unsaturated porous continua. *International Journal of Solids and Structures*, *43*(6), 1764–1786. doi:10.1016/j.ijsolstr.2005.04.045

Butter, C. D., & Hocker, G. B. (1978). Fiber optics strain gauge. *Applied Optics*, *17*(18), 2867–2869. doi:10.1364/AO.17.002867

Cappa, F., Guglielmi, Y., & Virieux, J. (2007). Stress and fluid transfer in a fault zone due to overpressures in the seismogenic crust. *Geophysical Research Letters*, *34*(5). doi:10.1029/2006GL028980

Carslaw, H. S., & Jaeger, J. C. (1986). *Conduction of heat in solids* (2nd ed.). Oxford University Press.

Cheng, R., Xia, L., Yan, J., Zhou, J., Wen, Y., & Rohollahnejad, J. (2015). Radio frequency FBG-based interferometer for remote adaptive strain monitoring. *IEEE Photonics Technology Letters*, *27*(15), 1577–1580. doi:10.1109/LPT.2015.2406112

Clement Bellido, J., Peralta, J. H., Madrigal, J. M., Vicente, H. M., Penalva, G. T., Fernandez-Pousa, C. R., & Maicas, S. S. (2018). Fast incoherent OFDR interrogation of FBG arrays using sparse radio frequency responses. *Journal of Lightwave Technology*, *36*(19), 4393–4400. doi:10.1109/JLT.2018.2821199

Clement, J., Maestre, H., Torregrosa, G., & Fernández-Pousa, C. R. (2019). Incoherent optical frequency-domain reflectometry based on homodyne electro-optic downconversion for fiber-optic sensor interrogation. *Sensors*, *19*(9), 2075.

Costantino, O. (2021). *Evaluating the feasibility of measuring and analyzing strain caused by changes in barometric pressure in the vadose zone (Master of Science)*. Clemson: Clemson University.

Dauson, E., Donahue, C., DeWolf, S., Hua, L., Xiao, H., Murdoch, L., & Johnson, P. (2022). Damage detection in a laboratory-scale wellbore applying time reversal and nonlinear elastic wave spectroscopy (TR NEWS). *NDT and E International*, *126*, 102573. doi:10.1016/j.ndteint.2021.102573

Davis, D. R., & Rasmussen, T. C. (1993). A comparison of linear regression with Clark's method for estimating barometric efficiency of confined aquifers. *Water Resources Research*, *29*(6), 1849–1854. doi:10.1029/93WR00560

Dobriyal, P., Qureshi, A., Badola, R., & Hussain, S. A. (2012). A review of the methods available for estimating soil moisture and its implications for water resource management. *Journal of Hydrology*, *458–459*, 110–117. doi:10.1016/J.JHYDROL.2012.06.021

Dong, H., Liu, S., Yang, L., Peng, J., & Cheng, K. (2018). Optical fiber displacement sensor based on microwave photonics interferometry. *Sensors (Switzerland)*, *18*(11), 3702. doi:10.3390/s18113702

Fabian, M., & Kümpel, H.-J. (2003). Poroelasticity: observations of anomalous near surface tilt induced by ground water pumping. *Journal of Hydrology*, *281*(3), 187–205. doi:10.1016/S0022-1694(03)00234-8

Galloway, D. L., & Burbey, T. J. (2011). Review: regional land subsidence accompanying groundwater extraction. *Hydrogeology Journal*, *19*(8), 1459–1486. doi:10.1007/s10040-011-0775-5

Genuchten, M. T. V. (1980). A closed-form equation for predicting the hydraulic conductivity of unsaturated soils. *Soil Science Society of America*, *44*(5), 892–898. doi:10.2136/sssaj1980.03615995004400050002x

Guglielmi, Y., Cappa, F., Lancon, H., Janowczyk, J. B., Rutqvist, J., Tsang, C. F., & Wang, J. S. Y. (2014). ISRM suggested method for step-rate injection method for fracture in-situ properties (SIMFIP): using a 3-components borehole deformation sensor. *Rock Mechanics and Rock Engineering*, *47*(1), 303–311. doi:10.1007/s00603-013-0517-1

Hervás, J., Barrera, D., Madrigal, J., & Sales, S. (2018). Microwave photonics filtering interrogation technique under coherent regime for hot spot detection on a weak FBGs array. *Journal of Lightwave Technology*, *36*(4), 1039–1045. doi:10.1109/JLT.2018.2793161

Hisz, D. B., Murdoch, L. C., & Germanovich, L. N. (2013). A portable borehole extensometer and tiltmeter for characterizing aquifers. *Water Resources Research*, *49*(12), 7900–7910. doi:10.1002/wrcr.20500

Hua, L., Song, Y., Huang, J., Lan, X., Li, Y., & Xiao, H. (2015). Microwave interrogated large core fused silica fiber Michelson interferometer for strain sensing. *Applied Optics*, *54*(24), 7181–7187.

Hua, L., Song, Y., Cheng, B., Zhu, W., Zhang, Q., & Xiao, H. (2017). Coherence-length-gated distributed optical fiber sensing based on microwave-photonic interferometry. *Optics Express*, *25*(25), 31362–31376.

Hua, L., Zhu, X., Cheng, B., Song, Y., Zhang, Q., Wu, Y., et al. (2021a). Distributed acoustic sensing based on coherent microwave photonics interferometry. *Sensors*, *21*(20), 6784. doi:10.3390/s21206784

Hua, L., Zhu, X., Dewolf, S., Lei, J., Zhang, Q., Murdoch, L., & Xiao, H. (2021b). Phase demodulation by frequency chirping in coherence microwave photonic interferometry. *IEEE Journal of Selected Topics in Quantum Electronics*, *27*(2), 1–9. doi:10.1109/JSTQE.2020.2975575

Hua, L, Riley, B., Murdoch, L. C., & DeWolf, S. (2023). PW2 pumping: March 18 thru March 22, 2020. https://edx.netl.doe.gov/dataset/pw2-pumping-march-18-thru-march-22-2020

Huang, J., Lan, X., Luo, M., & Xiao, H. (2014). Spatially continuous distributed fiber optic sensing using optical carrier based microwave interferometry. *Optics Express*, *22*(15), 18757. doi:10.1364/OE.22.018757

Huang, J., Lan, X., Song, Y., Li, Y., Hua, L., & Xiao, H. (2015). Microwave interrogated sapphire fiber Michelson interferometer for high temperature sensing. *IEEE Photonics Technology Letters*, *27*(13), 1398–1401. doi:10.1109/LPT.2015.2422136

Jacob, C. E. (1940). On the flow of water in an elastic artesian aquifer. *Eos, Transactions, American Geophysical Union, 21* (2), 574–586. doi:10.1029/TR021i002p00574

Jin, G., Ugueto, G., Wojtaszek, M., Guzik, A., Jurick, D., & Kishida, K. (2021). Novel near-wellbore fracture diagnosis for unconventional wells using high-resolution distributed strain sensing during production. *SPE Journal, 26*(05), 3255–3264. doi:10.2118/205394-PA

Karasaki, K., Freifeld, B., Cohen, A., Grossenbacher, K., Cook, P., & Vasco, D. (2000). A multidisciplinary fractured rock characterization study at Raymond field site, Raymond, CA. *Journal of Hydrology, 236*(1–2), 17–34. doi: 10.1016/S0022-1694(00)00272-9

Keller, T., Arvidsson, J., Schjønning, P., Lamandé, M., Stettler, M., & Weisskopf, P. (2012). In situ subsoil stress-strain behavior in relation to soil precompression stress. *Soil Science, 177* (8), 490–497. doi:10.1097/SS.0b013e318262554e

Khakim, M. Y. N., Tsuji, T., & Matsuoka, T. (2012). Geomechanical modeling for InSAR-derived surface deformation at steam-injection oil sand fields. *Journal of Petroleum Science and Engineering, 96–97*, 152–161. doi:10.1016/j.petrol.2012.08.003

Kuang, X., Jiao, J. J., & Li, H. (2013). Review on airflow in unsaturated zones induced by natural forcings. *Water Resources Research, 49*(10), 6137–6165. doi:10.1002/WRCR.20416

Kumar Thota, S., Duc Cao, T., & Vahedifard, F. (2021). Poisson's ratio characteristic curve of unsaturated soils. *Journal of Geotechnical and Geoenvironmental Engineering, 147*(1), 04020149. doi:10.1061/(asce)gt.1943-5606.0002424

Lekshmi, S. S. U., Singh, D. N., & Baghini, M. (2014). A critical review of soil moisture measurement. *Measurement, 54*, 92–105. doi:10.1016/J.MEASUREMENT.2014.04.007

Liehr, S., Nöther, N., & Krebber, K. (2010). Incoherent optical frequency domain reflectometry and distributed strain detection in polymer optical fibers. *Measurement Science and Technology, 21*, 017001. doi:10.1088/0957-0233/21/1/017001

López-Higuera, J. M., Cobo, L. R., Incera, A. Q., & Cobo, A. (2011). Fiber optic sensors in structural health monitoring. *Journal of Lightwave Technology, 29*(4), 587–608. doi:10.1109/JLT.2011.2106479

Lu, N., & Kaya, M. (2014). Power law for elastic moduli of unsaturated soil. *Journal of Geotechnical and Geoenvironmental Engineering, 140*(1), 46–56. doi:10.1061/(asce)gt.1943-5606.0000990

Lu, N., Godt, J. W., & Wu, D. T. (2010). A closed-form equation for effective stress in unsaturated soil. *Water Resources Research, 46*(5). doi:10.1029/2009wr008646

Marin, S., van der Kamp, G., Pietroniro, A., Davison, B., & Toth, B. (2010). Use of geological weighing lysimeters to calibrate a distributed hydrological model for the simulation of land–atmosphere moisture exchange. *Journal of Hydrology, 383*(3–4), 179–185. doi:10.1016/J.JHYDROL.2009.12.034

Massmann, J., & Farrier, D. F. (1992). Effects of atmospheric pressures on gas transport in the vadose zone. *Water Resources Research, 28*(3), 777–791. doi:10.1029/91WR02766

Monfared, K. A., & Rothenburg, L. (2011). Ground surface displacements and tilt monitoring for reconstruction of reservoir deformations. *International Journal of Rock Mechanics and Mining Sciences, 48*(7), 1113–1122. doi:10.1016/j.ijrmms.2011.07.003

Murdoch, L. C., Freeman, C. E., Germanovich, L. N., Thrash, C., & Dewolf, S. (2015). Using in situ vertical displacements to characterize changes in moisture load. *Water Resources Research, 51*(8), 5998–6016. doi:10.1002/2015WR017335

Murdoch, L. C., Germanovich, L. N., DeWolf, S. J., Moysey, S. M. J., Hanna, A. C., Kim, S., & Duncan, R. G. (2020). Feasibility of using in situ deformation to monitor CO_2 storage. *International Journal of Greenhouse Gas Control, 93*, 102853. doi:10.1016/j.ijggc.2019.102853

Murdoch, L. C., Germanovich, L. N., Roudini, S., DeWolf, S. J., Hua, L., & Moak, R. W. (2021). A type-curve approach for evaluating aquifer properties by interpreting shallow strain measured during well tests. *Water Resources Research, 57*(9), e2021WR029613. doi:10.1029/2021WR029613

Murdoch, L. C., DeWolf, S., Germanovich, L. N., Moysey, S., Hanna, A., Roudini, S., & Moak, R. (2023). Using the shallow strain tensor to characterize deep geologic reservoirs. *Water Resources Research, 59*, e2022WR032920. doi:10.1029/2022wr032920

Neuman, S. P. (1975). Analysis of pumping test data from anisotropic unconfined aquifers considering delayed gravity response. *Water Resources Research, 11*(2), 329–342. doi:10.1029/WR011I002P00329

Paitz, P., Edme, P., Gräff, D., Walter, F., Doetsch, J., Chalari, A., et al. (2021). Empirical investigations of the instrument response for distributed acoustic sensing (DAS) across 17 octaves. *Bulletin of the Seismological Society of America, 111*(1), 1–10. doi:10.1785/0120200185

Philip, J. R., Knight, J. H., & Waechter, R. T. (1989). Unsaturated seepage and subterranean holes: conspectus, and exclusion problem for circular cylindrical cavities. *Water Resources Research, 25*(1), 16–28. doi:10.1029/WR025i001p00016

Plunkett, G. (2021). *A novel technique for distributed measuring of strain in the vadose zone (Master of Science)*. Clemson: Clemson University.

Požar, T., Gregorčič, P., & Možina, J. (2011). A precise and wide-dynamic-range displacement-measuring homodyne quadrature laser interferometer. *Applied Physics B: Lasers and Optics, 105*(3), 575–582. doi:10.1007/s00340-011-4512-5

Quilty, E. G., & Roeloffs, E. A. (1991). Removal of barometric pressure response from water level data. *Journal of Geophysical Research, 96*(B6), 10209–10218. doi:10.1029/91JB00429

Rao, Y. J., & Jackson, D. A. (1996). Recent progress in fibre optic low-coherence interferometry. *Measurement Science and Technology, 7*(7), 981–999. doi:10.1088/0957-0233/7/7/001

Rasmussen, T. C., & Crawford, L. A. (1997). Identifying and removing barometric pressure effects in confined and unconfined aquifers. *Ground Water, 35*(3), 502–511. doi:10.1111/J.1745-6584.1997.TB00111.X

Ricchiuti, A. L., Hervas, J., Barrera, D., Sales, S., & Capmany, J. (2014). Microwave photonics filtering technique for interrogating a very-weak fiber Bragg grating cascade sensor. *IEEE Photonics Journal, 6*(6), 1–10. doi:10.1109/JPHOT.2014.2363443

Rojstaczer, S. (1988). Determination of fluid flow properties from the response of water levels in wells to atmospheric loading. *Water Resources Research*, *24*(11), 1927–1938. doi:10.1029/WR024i011p01927

Rojstaczer, S., & Riley, F. S. (1990). Response of the water level in a well to Earth tides and atmospheric loading under unconfined conditions. *Water Resources Research*, *26*(8), 1803–1817. doi:10.1029/WR026I008P01803

Roudini, S., Murdoch, L. C., Shojaei, M., & DeWolf, S. (2023). Proxy-based Bayesian inversion of strain tensor data measured during well tests. *Geomechanics for Energy and the Environment*, , *36*, 100506. doi:10.1016/j.gete.2023.100506

Rutqvist, J., & Tsang, C. F. (2002). A study of caprock hydromechanical changes associated with CO_2-injection into a brine formation. *Environmental Geology*, *42*(2–3), 296–305. doi:10.1007/s00254-001-0499-2

Schweisinger, T., Svenson, E. J., & Murdoch, L. C. (2009). Introduction to hydromechanical well tests in fractured rock aquifers. *Ground Water*, *47*(1), 69–79. doi:10.1111/j.1745-6584.2008.00501.x

Schweisinger, T., Svenson, E. J., & Murdoch, L. C. (2011). Hydromechanical behavior during constant-rate pumping tests in fractured gneiss. *Hydrogeology Journal*, *19*(5), 963–980. doi:10.1007/s10040-011-0728-z

Song, Y., Hua, L., Lei, J., Zhang, Q., Liu, J., Ye, L., & Xiao, H. (2016). An IFPI temperature sensor fabricated in an unstriped optical fiber with self-strain-compensation function. *Journal of Sensors*, *2016*. doi:10.1155/2016/6419623

Sun, Y., Xue, Z., Hashimoto, T., Lei, X., & Zhang, Y. (2020). Distributed fiber optic sensing system for well-based monitoring water injection tests—A geomechanical responses perspective. *Water Resources Research*, *56*(1), e2019WR024794. doi:10.1029/2019WR024794

Sun, Y., Xue, Z., Hashimoto, T., & Zhang, Y. (2021). Optically quantifying spatiotemporal responses of water injection-induced strain via downhole distributed fiber optics sensing. *Fuel*, *283*, 118948. doi:10.1016/J.FUEL.2020.118948

Svenson, E., Schweisinger, T., & Murdoch, L. C. (2008). Field evaluation of the hydromechanical behavior of flat-lying fractures during slug tests. *Journal of Hydrology*, *359*(1–2), 30–45. doi:10.1016/j.jhydrol.2008.06.004

Tian, X., Wang, Y., Zhu, D., & Shi, J. (2022). Radio-frequency intensity interrogation for temperature measurement. *IEEE Photonics Technology Letters*, *34*(18), 973–976. doi:10.1109/LPT.2022.3195960

Tolman, C. F. (1937). *Ground water*. McGraw-Hill.

Van Der Kamp, G., & Schmidt, R. (1997). Monitoring of total soil moisture on a scale of hectares using groundwater piezometers. *Geophysical Research Letters*, *24*(6), 719–722. doi:10.1029/97GL00521

Vasco, D. W., Karasaki, K., & Kishida, K. (2001). A coupled inversion of pressure and surface displacement. *Water Resources Research*, *37*(12), 3071–3089.

Vasco, D. W., Kim, K. H., Farr, T. G., Reager, J. T., Bekaert, D., Sangha, S. S., et al. (2022). Using Sentinel-1 and GRACE satellite data to monitor the hydrological variations within the Tulare Basin, California. *Scientific Reports*, *12*(1), 3867. doi:10.1038/s41598-022-07650-1

Weeks, E. P. (1979). Barometric fluctuations in wells tapping deep unconfined aquifers. *Water Resources Research*, *15*(5), 1167–1176. doi:10.1029/WR015I005P01167

Wu, N., Xia, M., Wu, Y., Li, S., Qi, R., Huang, Y., & Xia, L. (2021). Microwave photonics interrogation for multiplexing fiber Fabry-Perot sensors. *Optics Express*, *29*(11), 16652. doi:10.1364/oe.424059

Xue, Z., & Hashimoto, T. (2017). Geomechanical monitoring of caprock and wellbore integrity using fiber optic cable: strain measurement from the fluid injection and extraction field tests. *Energy Procedia*, *114*, 3305–3311. doi:10.1016/j.egypro.2017.03.1462

Yusa, Y. (1969). The fluctuation of the level of the water table due to barometric change. *Special Contributions, Geophysical Institute, Kyoto University*, *9*, 15–28.

Zhu, C., & Huang, J. (2023). Multiplexed optical fiber tip refractive index sensor interrogated by microwave photonics. *Optics and Lasers in Engineering*, *163*, 107477. doi:10.1016/j.optlaseng.2023.107477

Zhu, X., Hua, L., Lei, J., Tang, J., Murdoch, L., & Xiao, H. (2021). Microwave-photonic low-coherence interferometry for dark zone free distributed optical fiber sensing. *Optics Letters*, *46*(5), 1173–1176. doi:10.1364/ol.414699

4

A Fiber-Optic, Multicomponent Sensor for Borehole Seismic Formation Imaging and for Monitoring of Natural or Induced Seismicity

Jakob B. U. Haldorsen, Caleb Christensen, Donald Richard Metzbower, and Audrius Berzanskis

ABSTRACT

The material presented is taken from work by the authors either alone, together, or with other coauthors. The objective of the present work is not to present new research, but to bring together pieces already published. We argue for the use of three-component sensors for borehole seismic work and describe a range of novel optical point sensors, including a hydrophone and a three-component accelerometer, that can be used in a borehole seismic sensor array. A relatively sparse array of wide-bandwidth, optical three-component accelerometers can be used to supplement fiber-optical distributed acoustic sensing (DAS), where the point sensors compensate for the inability of the DAS sensors to measure the direction of particle motion in three-dimensional space. A hybrid fully optical receiver array provides vector information about the wave field and will be a wide-band, wide-aperture sensor array without any electronics at or near the sensors. With carefully selected components, this hybrid array will not be adversely affected by temperatures exceeding 250 °C in geothermal and hot oil and gas wells.

4.1. INTRODUCTION

The material presented in this chapter is based on work by the authors on fiber-optic point sensors and on the use of three-component (3C) sensors for borehole seismic applications. The objective of the present work is not to present new research but to bring together in a coherent way work already published. The referenced material has been published by the present authors over several decades, either together, alone, or with other coauthors. The published research will be referenced at the appropriate places in the text.

When considering which sensor technology to use, one has to weigh the costs of the measurement technology against the objectives for acquiring the data. The analyses of the available technologies will normally end up in a compromise, and the compromise will result in reduced capabilities. It is vital to determine that the reduced capabilities do not unduly hurt the objectives for acquiring the data and that the measurements still result in sufficiently actionable information.

Geophones, hydrophones, and accelerometers respond in different ways to particle motion and vibrations and are used by geophysicists to extract information about materials and structures invisible to the naked eye—either to locate and understand sources of acoustoelastic waves or for structural imaging using reflections of these waves.

Mechanical disturbances in the subsurface may be related to hydrocarbon production, fluid flow, or injection and flow of CO_2 for storage. These disturbances can generate microseismic ruptures. To see the weakest of these, one will have to utilize sensors in wells close to the rupture. Borehole sensors installed close to the reservoir make it possible to continuously monitor for induced or natural microseismicity and will allow early warnings of developing potentially critical situations.

MagiQ Technologies Inc., Somerville, Massachusetts, USA

Distributed Acoustic Sensing in Borehole Geophysics, Geophysical Monograph 289, First Edition.
Edited by Yingping Li, Robert Mellors, and Ge Zhan.

DOI: 10.1002/9781394179275.ch04

Hydrophones measure dynamic pressure and are not sensitive to the direction of particle motion, as pressure changes depend on changes in the local particle density, which depends only on the speed at which particles move away from each other and not on the direction they are moving in. Modern marine seismic towed arrays use a mixture of hydrophones and multicomponent accelerometers (Frømyr, 2004). A pressure differential across the array will induce transverse motion, which is detected by the accelerometers. Combining pressure measurements with the measurements of the direction of transverse motion captures the 3C–3D direction of propagation for the wave field. In shallower waters, the propagating wave field couples to Scholte interface waves at the seafloor, which can be measured with multicomponent seafloor cables or nodes.

Commercially available distributed acoustic sensing (DAS) technology (Farhadiroushan et al., 2009) transforms a fiber-optical cable into an acoustic sensor array capable of near-continuous measurements. The technology allows measuring changes in the laser light reflected back from points along an optical fiber. These changes are caused by changes in the induced strain along the fiber at a location determined from the time of the reflection. The changes in strain can be interpreted as indications of passing seismic waves. However, the strain rate of the fiber is coupled significantly only to the component of particle motion along the fiber and to a much lesser degree to particle motion transverse to the fiber. As there is nothing but glass in the well, a DAS array has a very long life expectancy even when exposed to persistent high temperatures. This is a significant advantage for monitoring geothermal and hot oil and gas wells.

We believe that DAS arrays should be complemented by 3C sensors, and we will describe optical point sensors developed by MagiQ Technologies that can replace electrically powered geophones for this purpose. With optical 3C point sensors, one can build a fully optical hybrid sensor cable, including DAS and sparsely distributed optical 3C sensors. This hybrid array will have a similarly long life expectancy as a pure DAS cable, but the information added by the transverse components of the optical 3C sensors allows a significantly better description of seismic wavefields.

To stress the importance of 3C sensors for achieving the objectives for seismic imaging, we will start with a synoptic review of seismic imaging, followed by a similarly synoptic review of the consequences these objectives have for the acquisition and processing of borehole seismic data, in particular borehole seismic data acquired for monitoring microseismic activities in and around an active subsurface field. Finally, we will describe a set of fiber-optic sensors that MagiQ Technologies has developed for borehole seismic applications, which are well suited for combining with single-component DAS arrays.

4.2. SEISMIC IMAGING

The seismic imaging technique sends acoustic waves into a formation and records the signal returned after the waves are scattered by inhomogeneities in the formation. The goal of many seismic investigations is to map the inhomogeneities and use the map to identify resources, hazards, or other structures of interest. In other cases, the properties of the source are of interest, for example, in identifying causes of induced seismicity.

The scattered wavefield at a particular location is the convolution of the earth's response with the incoming field. To find the earth response, one will need to reconstruct both the source and the scattered wavefield at every location and deconvolve the reconstructed scattered wavefield by the reconstructed source wavefield (Claerbout, 1985).

The accuracy of a seismic image is closely related to the ability the data provide for accurately reconstructing the location of a point source (Miller et al., 1987). With a single-component sensor array, the source reconstruction relies entirely on the differences in arrival times at adjacent sensors. These traveltime differences also allow the estimation of the component of the local elastic propagation velocities along the axis of the array. With a 3C sensor array, one can combine the traveltime differences with the measured direction of the particle motion vector. This allows the determination of the following:

1. Whether the different wavefield components are compressional (P) or shear (S) waves (Leaney & Esmersoy, 1989)
2. The direction the different wavefield components are propagating
3. The velocities at which they are propagating

Conventionally, three- or four-component (3C, 4C) electrically powered sensors are used to measure the local particle motion vector. Fiber-optic DAS technology eliminates the need for electrical power and electronics to digitize and transmit the data back for recording, but the insensitivity of the DAS cable to the transverse components means that the DAS cable on its own is only good for making traveltime measurements.

4.2.1. Borehole Seismic Data

Vertical seismic profiling (VSP) provides images of a geological formation, using controlled seismic sources at the surface and receivers in a well close to the formation target. The receivers will record the wavefield incident from the source and—some time later—the wave field reflected back from the formation. Estimates of these two wavefields are used to generate detailed and high-resolution three-dimensional (3D) images of the formation. Although inherent in most standard processing of VSP data, the relationship between incoming and reflected wavefields at the level of the borehole receivers is formalized in the interferometric approach to seismic data analysis (Bakulin & Calvert, 2004).

With 3C sensors and a carefully formulated process, formation images can be generated using only local formation properties (Haldorsen & Jahren, 2020). These images are less dependent on detailed knowledge of the formation between the physical source and the borehole seismic sensors. Knowing the direction of propagation and the P- and S-wave velocities, one can determine the distance back along the propagation direction to the location where the compressional and shear waves coincide in time and space (Haldorsen, 2002; Haldorsen & Jahren, 2020; Haldorsen et al., 2013). The 3C sensors provide more information, allowing the generation of more accurate seismic images.

Placing the sensors closer to the target means greater details in the formation image. This is important when the objective is to monitor changes related to production or injection of fluid or CO_2. However, the lateral extent of the image will be limited because the sensors are close to the target. To counteract this, it is desirable to combine sensors close to the target with sensors further away, but instrumenting the well to the surface would require a large number of sensors. The conventional requirement for avoiding spatial aliasing at least two measurement points per wavelength for the shortest usable wavelength. A wavefield at 75 Hz, propagating at 2,250 m/s, will have a wavelength of about 30 m; and to avoid spatial aliasing, the distance between the individual sensors in the array would have to be 15 m or less. If the target formation is at a depth of 3,500 m, at least 230 sensors are needed. An S-wave—moving much slower and have significantly shorter wavelength—would require many more sensors. However, Haldorsen (2021) shows that the requirements for dense spatial sampling can be relaxed for an array of 3C sensors—substantially reducing the cost of instrumenting a deep well for acoustic monitoring.

4.2.2. Locating Microseismic Sources

Haldorsen and Jahren (2020) introduced a concept of "fat-ray" imaging for borehole seismic data from 3C sensors, and Haldorsen (2022) applied this concept, in combination with the process described by Haldorsen et al. (2013), to the problem of locating microseismic sources. This section gives a summary of these applications.

Haldorsen et al. (2013) estimate the signatures of P- and S-waves emitted from a prospective source location by:

1. Projecting the wavefield recorded at each 3C sensor onto and perpendicular to the rays connecting each sensor to the prospective source location
2. Aligning the projected data by shifting them by the modeled P- and S-wave traveltimes
3. Averaging the projected and aligned data

For the prospective source location to be good, the reconstructed P- and S-wave signatures must be coherent and simultaneous. Their mutual coherency is used as a measure of the likelihood that the location is a source hypocenter.

The estimates from the individual 3C sensors would give an accurate distance from the sensor to the hypocenter location, each effectively with a cosine tapered smearing over ±45° sectors in azimuth and polar angles. The final location of the source is determined from a combination of the spread in polar angles and the curvature of the wavefront. With a reasonable aperture of the array, smearing along the array will be significantly reduced; however, smearing in the azimuthal direction, perpendicular to the axis of the array, will remain.

Haldorsen et al. (2013) appear to avoid the azimuthal smearing by first finding the maximum azimuthal energy and then proceed to estimate the depth and the distance to the event hypocenter at this azimuth; however, for a general 3D estimation of hypocenter locations, the cosine tapered angular smearing remained a problem. To overcome essentially the same problem with VSP reflection imaging, Haldorsen and Jahren (2020) introduced the concept of "fat rays" that only allowed contributions to the image from data vectors that deviate less than a few degrees (2° to 5°) from the ray that connects the prospective location of a reflection to the 3C sensor that recorded the data vector. This approach enabled formal 3D imaging to locate points of reflection or conversion of components of the borehole seismic data.

Haldorsen (2022) shows that "fat-ray" imaging helps resolve the hypocenter locations for several overlapping microseismic sources. This was demonstrated using synthetic data generated by microseismic sources emitting both P- and S-waves. For simplification, the formation was set to be homogenous and isotropic, with a compressional velocity of 2,750 m/s and a shear velocity of 1,588 m/s.

The synthetic data were recorded by a 450 m linear sensor array in a vertical well. To avoid spatial aliasing of S-waves below 120 Hz, the distance between the sensors should not exceed 6.5 m. At a sensor interval of 45 m, any S-wave at frequencies above 18 Hz and P-wave above 30 Hz will be spatially aliased.

The data were generated by three sources, all at the same depth but at different azimuthal directions and different distances from the well. One source was at +45° azimuth, 200 m away; a second source was at +15° azimuth, 150 m away; and a third source was at −15° azimuth, 100 m away. The source signatures for both the P- and S-waves were 120 Hz Ricker wavelets, with significant energy at frequencies beyond 250 Hz.

Figure 4.1 shows the synthetic data recorded by 101 single-axis sensors at regular 5 m intervals. This dense spatial sampling is adequate for S-waves at frequencies

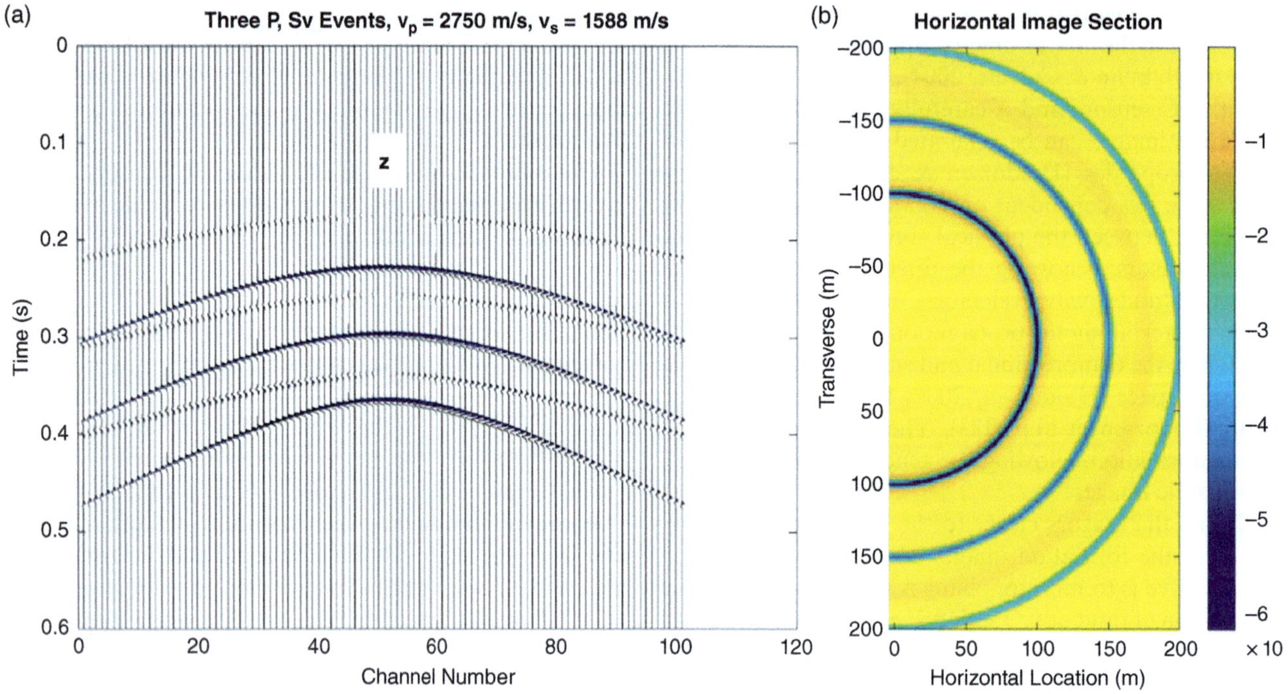

Figure 4.1 (a) Synthetic data for three overlapping microseismic events recorded by 101 vertical-component-only receivers at 4.5 m intervals. All three P and S sources are located at the depth of the middle receiver, a first source 200 m away from the well in the azimuthal direction of 45° North from East, a second 150 m away at 15° North, and a third 100 m away at 15° South. The source signatures are 120 Hz Ricker wavelets. (b) A horizontal section through the 3D image cube at the depth of the sources. The image is obtained without any constraints from horizontal-axis receivers. Haldorsen (2022) / with permission of the Society of Exploration Geophysicists..

below 160 Hz. The figure also shows a horizontal section through the 3D image generated from the data—at the depth of the three events.

The 450 m array is long enough and the spatial sampling dense enough to give accurate estimates of the depths of the events, and the time delays between the P- and S-wave arrivals give accurate estimates of the distances away from the well. However, because the data contain no information about the azimuth, the image is rotationally symmetric around the linear array. At any azimuth angle, there are three events at 100, 150, and 200 m from the well. There also are potential issues with deciding which of the wave modes should be labeled "P-wave" and which should be labeled "S-wave." However, for this modestly complicated geometry, it appears that there are no visible artifacts from labeling ambiguities.

The data in Figure 4.2 were recorded by 11 3C sensors at regular 45 m intervals. With classical spatial sampling limits, any S-wave at frequencies above 18 Hz and P-wave above 30 Hz will be spatially aliased. However, spatial aliasing is about the non-uniqueness of wavefield propagation direction for a given frequency, and it appears that the direction of particle motion measured by the 3C sensors resolves the ambiguity, and the "fat-ray" tapered weights give sharply focused images of the sources without the artifacts normally associated with spatial aliasing.

4.3. HYBRID SENSOR ARRAY

The data acquisition required for repeat (4D) VSP is both costly and time-consuming, in particular offshore where the cost of rig time has to be factored in. Because of this, the number of acquired 4D VSP surveys is very limited. New sensor technologies for borehole applications aim at reducing the cost, and/or extending the lifetime, and/or improving the performance and flexibility of the instrumentation used for imaging and monitoring.

Surveillance data should be acquired frequently. Changes in the data would allow assessing the effects on the reservoir of either production or injection. Dynamic, high-resolution 3D images would show fluid migration pathways in and near the reservoir. Installing a sensor array permanently in a well will increase the initial costs but significantly reduce the cost of repeat surveys. With an array of acoustic sensors permanently installed in the borehole, 3D VSP data can be acquired on-demand and 4D images of the reservoir around the well can be generated in a timely and cost-effective

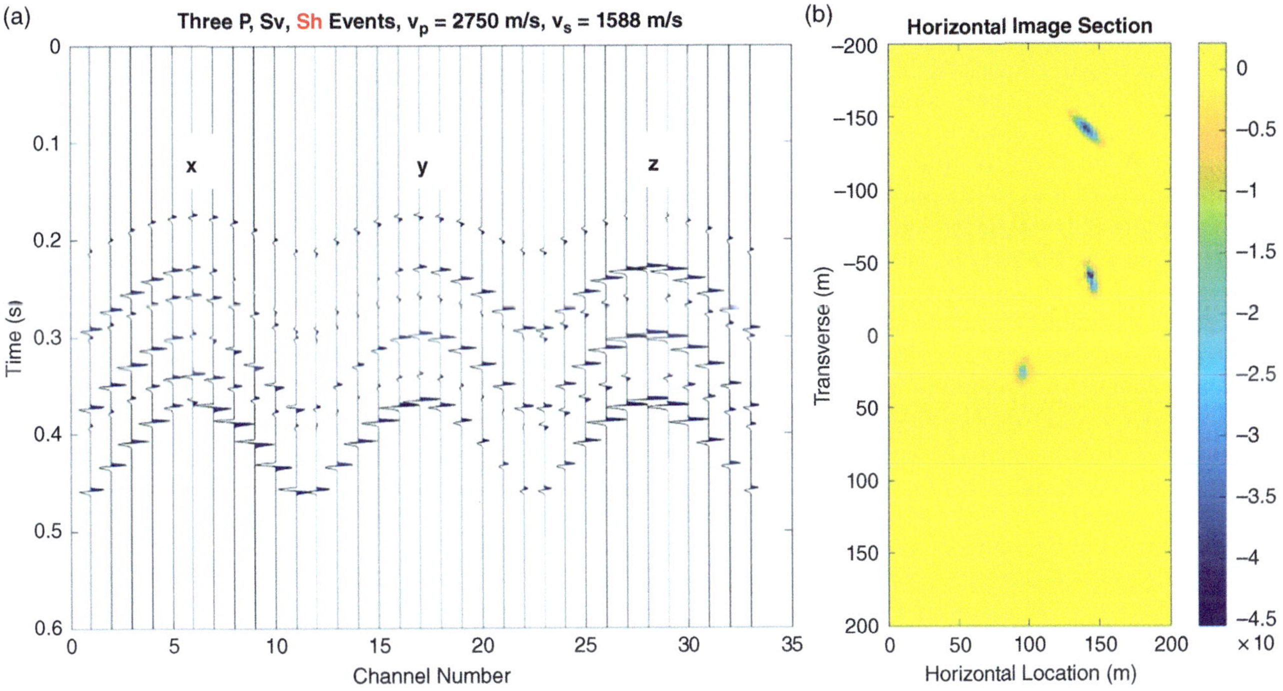

Figure 4.2 (a) Synthetic data for three overlapping microseismic events recorded by 11 3C receivers at 45 m intervals. The locations of the three P and Sv sources are the same as in Figure 4.1. (b) A horizontal section at the known depth of the sources, through the 3D image cube obtained with ±2° "fat-ray" imaging from this 11-element sparse array of 3C sensors. Haldorsen (2022) / with permission of the Society of Exploration Geophysicists.

manner. These images will provide information on sweep efficiencies, thief zones, potential zones of early breakthrough, injection conformance, and potential identification of well safety issues, among other applications that contribute to improving well efficiency and safety.

In addition to the purchasing costs for the sensor system, the choice of sensors for permanent installation in a well may be severely limited by the required temperature, completion complexity, and well architecture. Commercially available DAS interrogators transform a fiber-optical cable into a sensor array capable of near-continuous measurements of the component of an acoustic field along the axis of the optical fiber. A DAS sensor array is attractive but has fundamental limitations for single-well monitoring as it gives no indication of the direction of propagation for the incident wavefield.

Combining the DAS array with 3C sensors for a hybrid sensor array, the azimuthal indetermination of the DAS data would be totally resolved by the 3C sensors. Traditionally, borehole seismic surveys have used arrays of 40–100 3C sensors spaced at regular 15 m intervals. Placing the array about midway between the formation target and the surface source gives a reasonable lateral image aperture. The reduced propagation paths give improved resolution compared to surface seismic data. Combining 3C sensors near the target with a DAS sensor array extended to the surface may be a better choice, from both a technical and a commercial point of view (Haldorsen et al., 2018).

Permanently installed hybrid sensor arrays would significantly improve the versatility and lower the cost for both single and repeat VSP acquisitions. The array would give data that can be used for wide-aperture imaging and data that are adequate for acoustic monitoring. If desirable for particular applications, the 3C sensors can be supplemented with hydrophones, providing point measurements of dynamic pressure. This is commonly done with seafloor arrays (Frømyr, 2004), and doing this may improve the separation of P-waves, S-waves, and interface waves.

Geophone nodes are mechanically robust, but because of the electronics, they are not resistant to long-term exposure to high temperatures. A slim, optical 3C accelerometer has been designed by MagiQ Technologies for installation in narrow borehole annuli and can be used for a fully optical hybrid sensor array. This optical hybrid sensor array would have no downhole electronics, and there would be no need for downhole power—thereby eliminating the risk of failure of electronics left in the well for a long period of time. This would obviously significantly increase the expected lifetime of the sensor array in high-temperature wells. An optical hybrid sensor array is expected to be operational at temperatures exceeding 200 °C by a sizeable margin.

4.4. FIBER-OPTIC POINT SENSORS

A DAS interrogator transforms a fiber-optic cable into a sensor array capable of near-continuous measurements of the component of an acoustic field along the axis of the optical fiber. A DAS array is attractive but has fundamental limitations for single-well monitoring as the sensors are only sensitive to the component along the fiber axis of the motion induced by the passing wavefield, and therefore give no indication of the direction of propagation for the passing wavefield. Geophone sensors are mechanically robust, but because of the electronics, they are not resistant to long-term exposure to high temperatures.

A fiber-optic sensor for measuring a particular material property will depend on technology that translates a change in that material property into a change in the strain of an optical fiber. If the fiber is wound around a cavity that will contract and expand in response to the pressure, the readings would be proportional to pressure. By winding an optical fiber around a mass-spring that is free to flex in only one direction, one can make an optical sensor that measures motion in the direction the mass-spring can flex. As the spring flexes, the fiber extends or contracts, and using an interrogator similar to a DAS interrogator will give readings that are proportional to one component of the acceleration vector. Three of these devices can together measure the full particle motion vector.

Based on these principles, MagiQ Technologies has developed a range of fiber-optic point sensors for different applications (Haldorsen et al., 2021a, 2021b). These sensors can replace electric geophones—and are capable of withstanding temperatures well in excess of 200 °C. The fiber-optic sensors can be used in combination with DAS for a hybrid, full-featured, fully optical sensor array. This fiber-optic hybrid array will not require electronics near the sensors, will have a long life expectancy, and will deliver comparable or better performance than an array of electronic sensors.

In the following, we describe the sensors and their associated interrogator, and describe laboratory and field data recorded by these sensors.

4.4.1. The Interrogation System

The sensors are read by a system somewhat similar to DAS interrogators, measuring changes in fiber strain between Fiber Bragg gratings (FBGs) at each sensor. Figure 4.3, from Haldorsen et al. (2021a), illustrates one manifestation of the highly multiplexed architecture for the optical sensing system. The system includes splitters directing a measured fraction of the laser pulses to

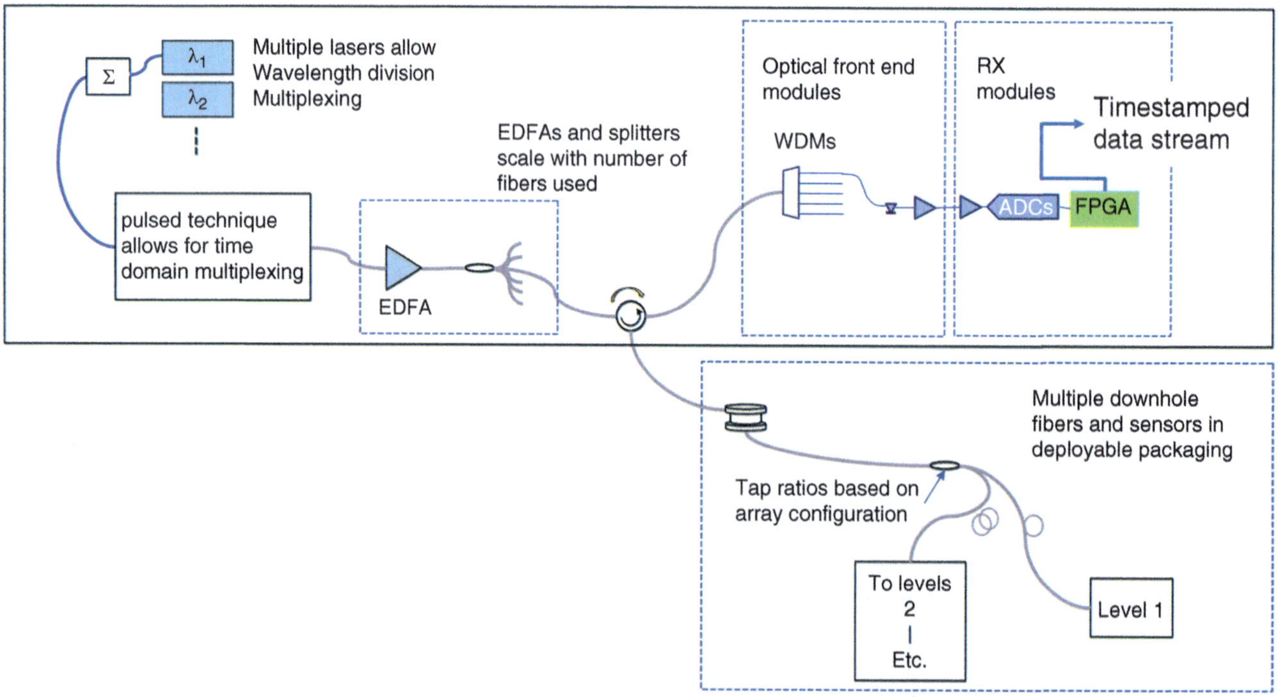

Figure 4.3 The highly multiplexed system architecture for the interrogation. The proprietary interferometry technique developed by MagiQ Technologies is particularly robust against laser noise and transmission-line pickup. The interrogator works with a single fiber per 3C sensor, each fiber extending to 10–20 additional 3C sensors. Different levels of redundancy can be obtained by different patterns of interleaving sensor strings. The architecture allows hundreds or thousands of sensors in a single network. Haldorsen et al. (2021) / with permission of the Society of Exploration Geophysicists.

each sensor. Using modulated laser beams with relatively short pulses at different frequencies, the system allows for time and wavelength division multiplexing. Laser interferometry (Hornby et al., 2005) allows very accurate estimates of changes in fiber strain between adjacent FBGs.

Including segments of nonresponsive reference fiber in each sensor allows for canceling laser noise and transmission line pickup. Up to 20 3C sensor nodes can be supported on a single fiber. The architecture is flexible and scalable and can accommodate a rather large number of sensors in the array.

4.4.2. The Fiber-Optic 3C Accelerometer

Several manufacturers offer multicomponent optical seismic tools at different stages of development. A large array of ocean-bottom multicomponent sensors has been deployed over Jubarte, a deep-waters Brazilian offshore field (Maas et al., 2015). The sensor array, OptoSeis, developed by Petroleum GeoServices and now owned by Geospace, is reported to give substantially good data. The OptoSeis sensor measures the changes in strain of a fiber attached to the two sides of a flexing plate. This system is reportedly being adapted for borehole deployment. Another system is the FOSVS sensor array developed by Paulsson Inc. (Paulsson et al., 2019). This system appears to be similar to the early prototype of the "Alpha" sensor (Figure 4.4, Haldorsen et al., 2021a) with respect to both size and the use of FBGs.

The fiber-optic accelerometer described by Haldorsen et al. (2021a) has been optimized for sometimes conflicting requirements regarding size, aspect ratio, deployment, and response. Two of the designs intended for borehole applications are shown in Figure 4.4. The "Alpha" sensor, with a form factor of an approximate cube with sides of 2.5 in. (0.063 m), was designed for standard borehole seismic applications. The "Beta" sensor was designed for seismic monitoring in a permanent or semipermanent installation. For this, the sensor needed to fit into a narrow borehole annulus as part of the well completion. The "Beta" sensor is cylindrical in shape with a length of about 3.5 in. (0.088 m) and a diameter of 0.85 in. (0.021 m).

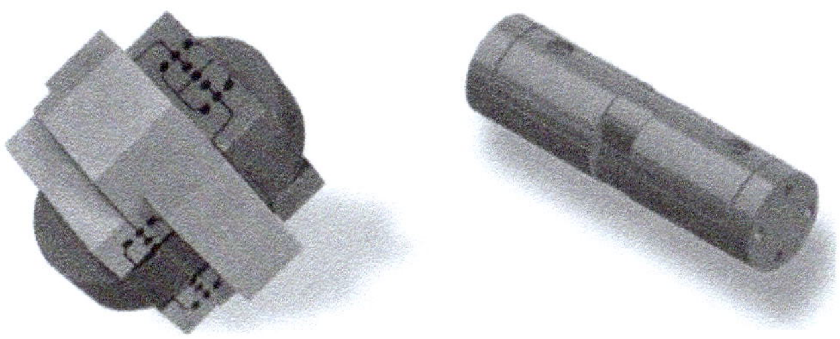

Figure 4.4 The MagiQ optical accelerometer is built on a mass-spring that is free to extend in one direction only. The "Alpha" sensor (left), approximately shaped like a cube with a side of 2.5 in. (0.063 m), was designed for acquisition of borehole seismic data in a standard geometry. The "Beta" sensor (right), shaped like a cylinder of length about 3.5 in. (0.088 m) and diameter 0.85 in. (0.021 m), is designed for permanent or semipermanent installation in a narrow borehole annulus as part of the completion of a well. Haldorsen et al. (2021) / with permission of the Society of Exploration Geophysicists.

Field tests indicated that the "Alpha" sensor had very good sensitivity and bandwidth. However, the system was found to be undesirably sensitive to laser vibrations, something that is also commonly seen with DAS arrays (Haldorsen et al., 2018). After fundamentally changing the interrogation architecture and adding new real-time signal processing steps, data acquired with the new sensor no longer show any signs of noise that can be attributed to vibrations of either the interrogator or the laser.

Figure 4.5 shows the 3C GeoLite sensor. This is assembled from three "Beta"-style sensors, each sensitive to motion in three mutually perpendicular directions. Adding a "fiber-management" module, the assembly fits into a tube with an outer diameter of about 1.1 in. (0.028 m) and a length of about 2 ft (0.6 m). As in this figure, the "Beta" sensor can be enclosed in a rigid housing; however, a flexible housing will allow a large array to be spooled on a reel, which would greatly simplify the deployment. On the other hand, a rigid housing helps the vector fidelity of the sensor and also simplifies the assembly.

Laboratory Data

In this section, we will discuss laboratory data obtained to demonstrate the functionality of the directional GeoLite sensor.

The resonance frequency for the flexing springs in the optical accelerometers is adjusted based on the

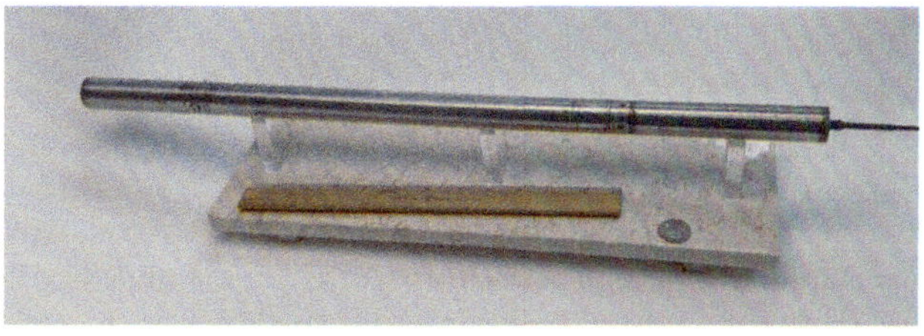

Figure 4.5 The GeoLite sensor, an assembly of three "Beta" sensors where the mass-springs are free to move in mutually perpendicular directions. Adding a "fiber-management" module, the assembly fits into a tube with an outer diameter of 1 in. (0.025m), and a length of about 2 ft (0.6 m). A U.S. quarter coin and a 1 ft (0.3 m) long ruler are included for scale. Haldorsen et al. (2021b) / with permission of IEEE.

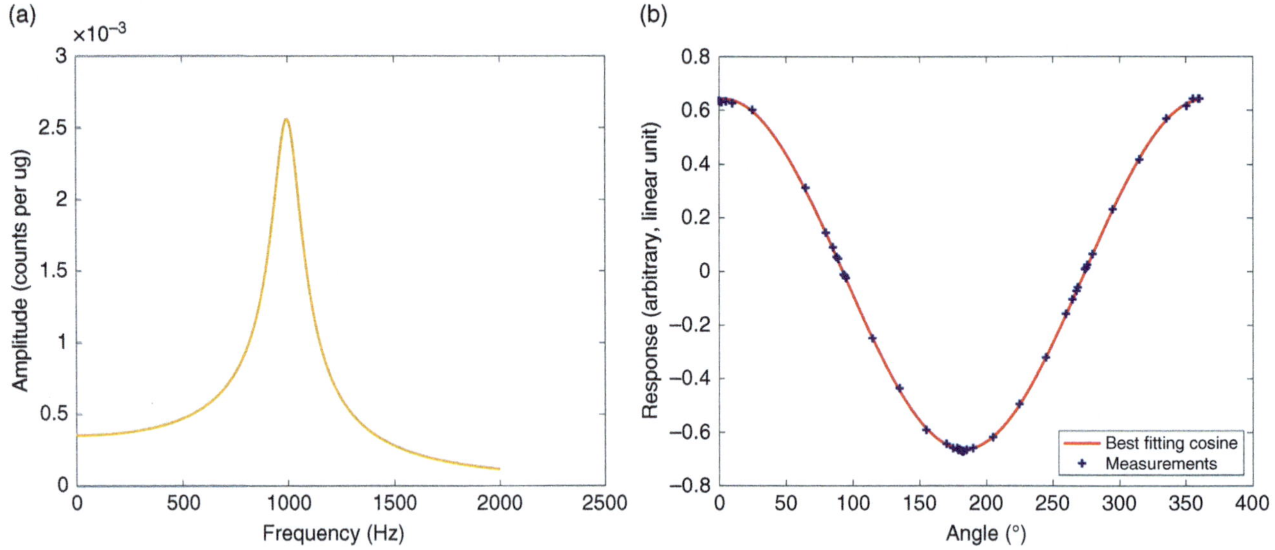

Figure 4.6 (a) The frequency-dependent sensitivity for the calibrated transverse and axial individual optical accelerometers. Dividing the raw frequency-dependent amplitudes by these converts the raw measurements to μg. The sensitivity curves, plotted in a linear scale, are calibrated to reference accelerometers. (b) The angular response of the optical accelerometer measured in the lab with a source at 100 Hz, compared to a cosine function, consistent with a projection of the signal onto the axis of the sensor. Haldorsen (2022) / with permission of the Society of Exploration Geophysicists.

requirements from the specific application. Figure 4.6 shows the response curves for the GeoLite sensors at frequencies from DC and 2 kHz, with a resonance frequency at about 1,100 Hz. The sensitivity curves for all three components are calibrated using reference accelerometers.

Figure 4.6 also shows the angular response of the transverse sensor, measured in the laboratory at 100 Hz using a linearly polarized source and a controlled geometry. The response is seen to agree well with the theoretical cosine function from the projection of the transmitted compressional signal onto the axis of the accelerometer. If the measurements by one sensor vary as $A_x(t) = A(t)\cos\phi$, where ϕ is the direction to the source, adding another sensor rotated by $\frac{\pi}{2}$ radians relative to the first sensor, this new sensor will measure $A_y(t) = A(t)\ \cos\left(\phi + \frac{\pi}{2}\right) = A(t)\ \sin\phi$. Thus, with $A_y(t) = A_x(t)\tan\phi$, a hodogram plotting $A_y(t)$ versus $A_x(t)$ will show a straight line pointing in the direction to the source.

The *cross-axis rejection*, the sensitivity of a sensor to a motion perpendicular to its presumed sensitive axis, was measured to 34 dB for the transverse sensor and 33 dB for the axial sensor. This compares to typical values of around 20 dB for geophones (Frømyr, 2004). It should be pointed out that the deployment method and clamping performance may also affect directionality.

In Figure 4.7, we show the laboratory setup and the results from a comparison between the "Beta" optical accelerometer and a conventional reference accelerometer. The two sets of sensors had the sensitive axes aligned. The figure shows the waveforms produced by each sensor in response to a signal from a controllable signal generator. The visually very similar waveforms have both been filtered to the frequency band from 10 to 300 Hz.

Field Data

In this section, we will discuss field deployments of the optical directional GeoLite accelerometer. The tests, reported by Haldorsen et al. (2021b) and Haldorsen et al. (2021a), were done in order to demonstrate that the sensors give data comparable to data recorded by conventional sensors employed by potential end users.

In January 2020, two GeoLite 3C accelerometers were used to acquire VSP data in a research well at the Aramco Research Center in Houston, TX (Haldorsen et al., 2021a). The vertical well (Figure 4.8(b)) was air-filled, 455 m deep, with a diameter of 7 in. (0.176 m).

One of the GeoLite sensors was enclosed in a rigid housing and the other in a flexible housing. Each was strapped to a conventional 3C borehole geophone tool by a number of metal and plastic straps. The arrangement (Figure 4.8 (a)) had the optical tools coupled to the well casing through the body of a geophone tool. The quality of this coupling relies on the rigidity of the strapping both along and transverse to the tool.

Nominally, the vibroseis source (Figure 4.8(c)) generated a signal at frequencies from 8 to 120 Hz. The

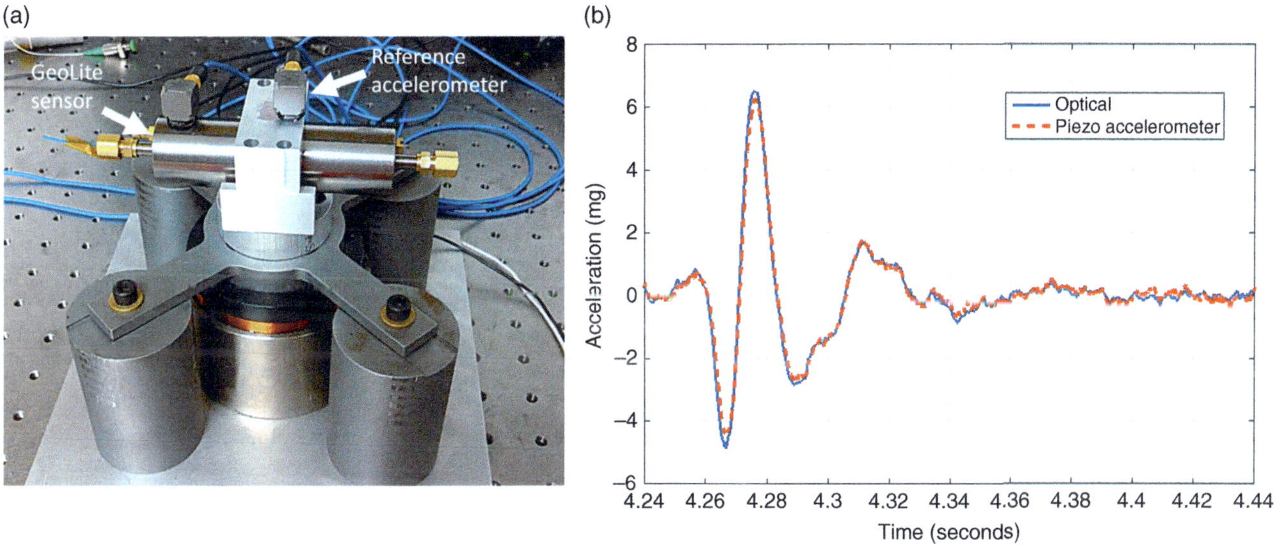

Figure 4.7 (a) Laboratory setup for comparing prototype optical and reference accelerometers. (b) Waveforms produced by each sensor. Haldorsen (2022) / with permission of the Society of Exploration Geophysicists.

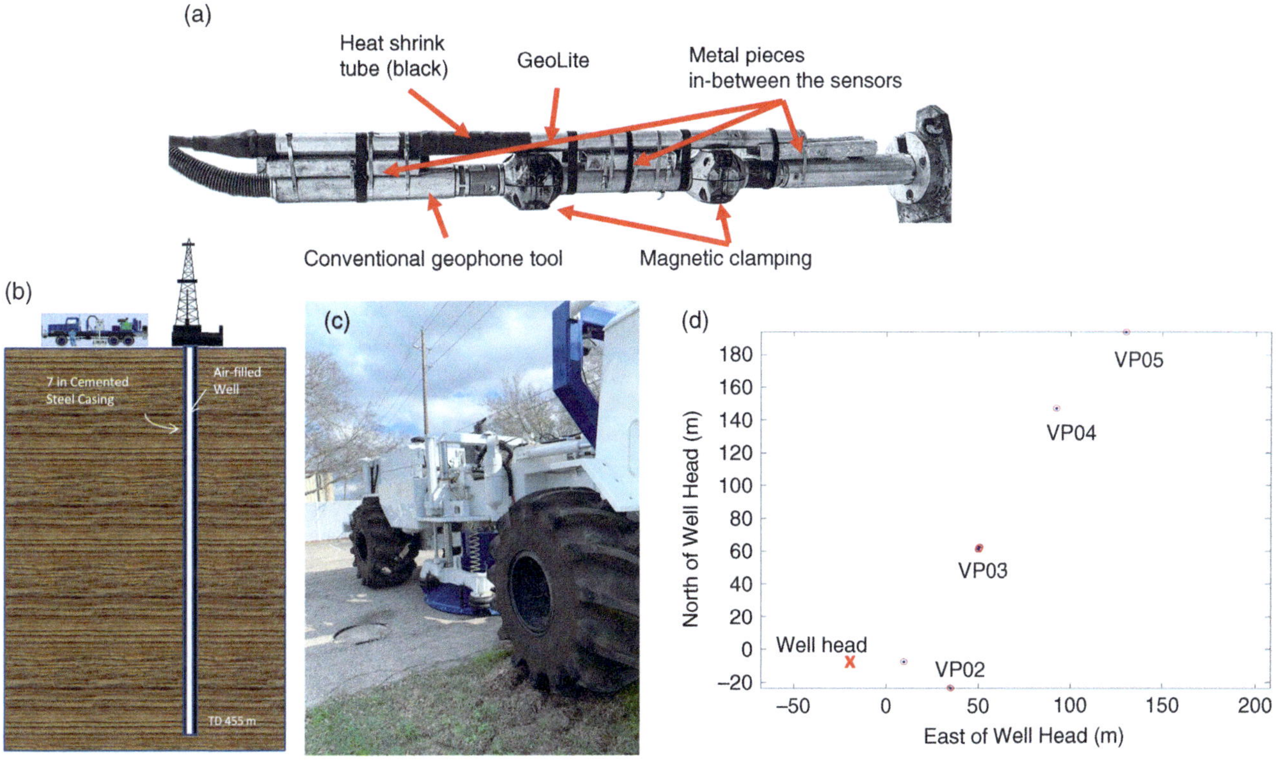

Figure 4.8 (a) Each GeoLite sensor was strapped to a conventional 3C borehole geophone tool by a number of metal and plastic straps. The arrangement had the optical GeoLite tools coupled to the well casing through the body of a geophone tool. The quality of this coupling relies on the rigidity of the strapping both along and transverse to the tool. (b) A sketch of the Saudi Aramco Research well in Houston, TX. The 455 m deep well is vertical, and the 7 in. (0.176 m) casing is cemented. The VSP data were recorded at 5 m intervals at depths from 60 to 150 m. (c) A picture of the vibroseis truck. The dark blue structure near the lower center of the picture is the baseplate, from which the source signal is transmitted. (d) The full data set included 1167 individual VSP-type data records from the four different source locations VP02, VP03, VP04, and VP05, each indicated in the figure by a small red circle with a black point in the center. The well head for the vertical well where the borehole seismic receivers were deployed as indicated in the figure by a red x. Haldorsen et al. (2021) / with permission of the Society of Exploration Geophysicists.

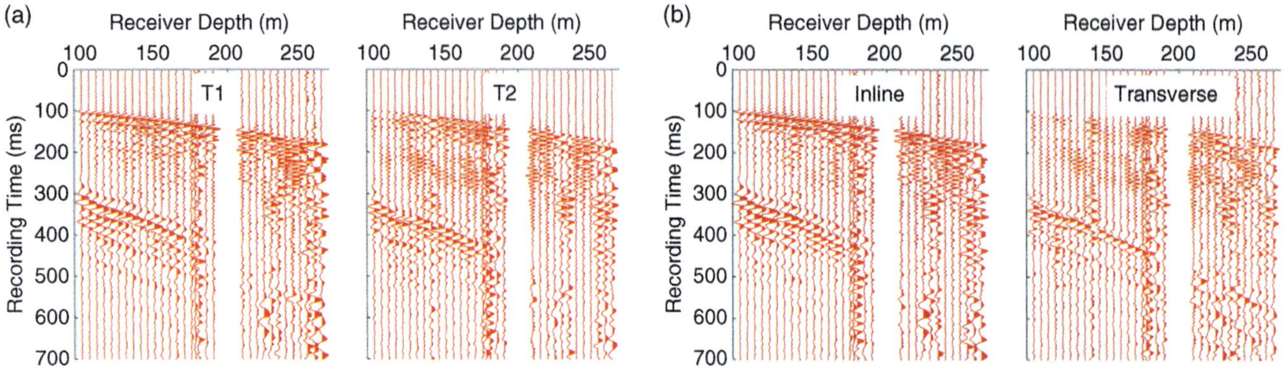

Figure 4.9 Stacked GeoLite VSP data from source location VP03 (Figure 4.8). VP03 is about 72 m from wellhead. (a) Raw horizontal components, T1 and T2, arbitrary orientation. (b) The two horizontal components have been rotated to have the 2nd component (labeled "Inline") for each trace point along the horizontal component of the ray connecting the source and receiver, by minimizing the energy in the direct compressional arrival in trace 3 (labeled "Transverse"). Haldorsen et al. (2021) / with permission of the Society of Exploration Geophysicists.

duration of the signal was 12 s, and the length of the recorded data was 16 s. The vibroseis moved between four different source locations, VP02 VP03, VP04, and VP05, indicated in Figure 4.8(d). Each frequency sweep was repeated 8 times as the sensors were moved at 5 m intervals over the depth range of 60 to 160 m. The full data set included 1,167 individual data records and took four days to acquire.

A VSP tool rotates unpredictably when moved between settings in the well. As a result, the orientation of the transverse components is not known. To align the data, the transverse components are normally rotated mathematically around the vertical component. This rotation process is set to minimize the energy on one of the transverse components, having the other component point along the vertical plane through the locations of both source and sensor. Figure 4.9 shows stacked data recorded by the horizontal components, both raw and aligned. The vibroseis was at location VP03, about 72 m from the wellhead. Figure 4.10 compares stacked and aligned 3C data recorded by the optical tool to those recorded by the geophone tool, also for source location VP03. They both look like VSP data, and the differences between them can most likely be attributed to the different coupling of the two tools, with the optical sensor riding "piggyback" on the geophone, as shown in Figure 4.8(a). In this configuration, the fiber-optic sensor will be coupled to the well casing through the body of the

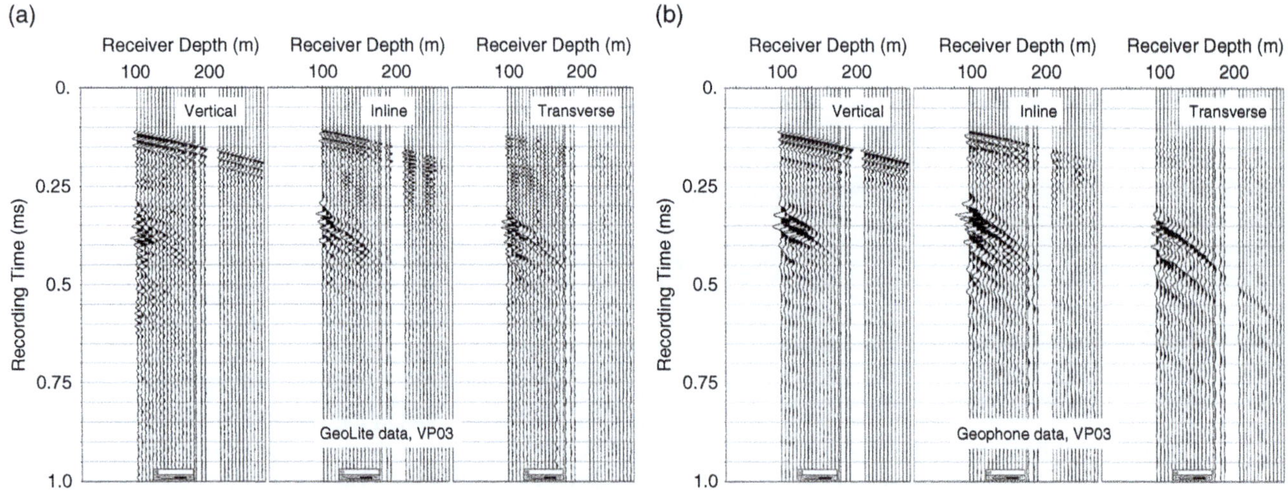

Figure 4.10 A comparison between 3C data acquired by (a) the prototype optical tool and (b) the conventional electrical borehole seismic geophone tool for vibrator location VP03. The data are oriented like in Figure 4.9. Haldorsen et al. (2021) / with permission of the Society of Exploration Geophysicists.

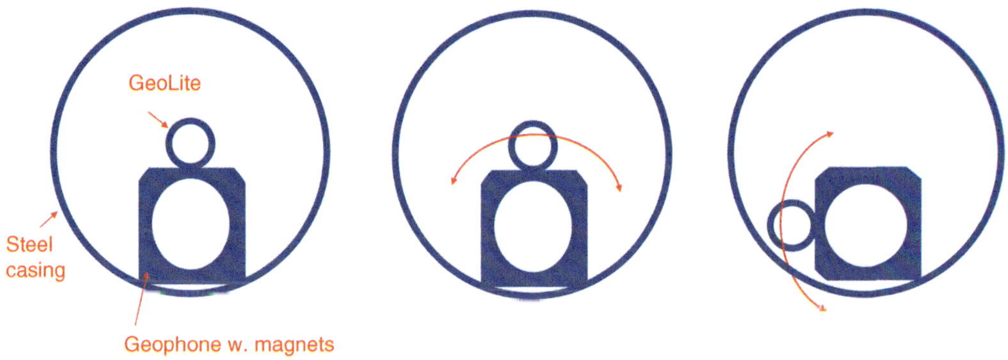

Figure 4.11 Illustration showing the cross-section of the well with the GeoLite sensor riding "piggyback" on the geophone tool as shown in Figure 4.8(a). In this configuration, the fiber-optic sensor will be coupled to the well casing through the body of the geophone. Depending on the tightness and elasticity of the straps used to fasten one sensor to the other, the "multistory" arrangement may wobble. This will affect the transverse components the most.

geophone. Depending on the tightness and elasticity of the straps used to fasten one sensor to the other, the "multistory" arrangement may wobble—which will affect the transverse components the most. If the tool wobbles, as illustrated in Figure 4.11, the traditional tool orientation using the direct P-wave arrivals will leave some energy on the transverse components. Although the vector fidelity of the fiber-optic sensor is demonstrated by laboratory measurements shown in Figure 4.6, the fact that significant energy remains on the transverse channels after the tool orientation—as apparent in Figure 4.9—means the direct P-arrivals are not recorded by the sensors as linearly polarized.

Field Data at High Frequencies

The vibrator was set to generate energy at frequencies from 8 to 120 Hz. The frequency response of the hydraulic system driving the vibrator baseplate is not strictly linear—neither is the transfer of energy from the baseplate to the soil. As a result, the signal transmitted into the soil contains energy at multiples of the intended instantaneous driving frequency. Figure 4.12 shows time–frequency analyses of the raw, uncorrelated z-axis data from two colocated geophone and GeoLite tools. These analyses both show a "baseband" of intended source energy from 8 to 120 Hz over 12 s, as well as higher harmonics at frequencies exceeding 200 Hz. Using the technique suggested by

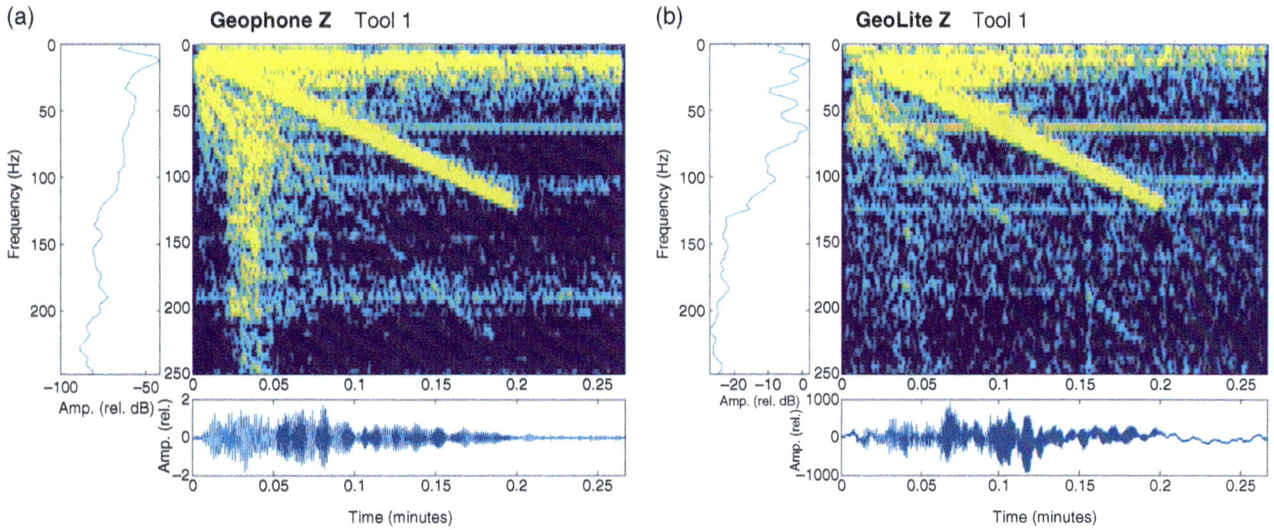

Figure 4.12 Time-frequency analyses of the raw, 20 s, uncorrelated z-axis data from two colocated sensor tools, (a) from the geophone tool and (b) from the GeoLite tool. These analyses both show a baseband of intended source energy from 8 to 120 Hz over 12 s, as well as higher harmonics at frequencies exceeding 200 Hz. The analyses also reveal somewhat different exposure of the two tools to environmental noise.

Haldorsen et al. (2004) for processing VSP data, the high-frequency energy in the harmonics could potentially be used for imaging. The analyses also reveal that the two tools are subjected to somewhat different environmental noise.

Field Data at Really Low Frequencies

Figure 4.13 shows the 16 s raw data and 1 s correlated data from the two GeoLite 3C sensors for two records acquired 6 min apart. Neither the source nor the receivers had been moved between the two recordings. For the earlier record, we see very strong acoustic energy at frequencies at and below 1.5 Hz. After correlation with the 8–120 Hz intended signature sweep, the data with or without this low-frequency energy were indistinguishable. At some measure, the similarity of the correlated data demonstrates the linear response of the optical sensor.

The recording system for the optical sensors records a continuous stream of data. This allowed extracting a 72 s continuous segment that included the 16 s record with the low-frequency energy in Figure 4.13. The 72 s data segment is shown in Figure 4.14, together with a low-frequency spectral analysis of the data. The traces appear to contain energy extending to DC, well below the lower end of 8 Hz for the nominal source signature. Haldorsen et al. (2021a) suggest that the strong, amplitude-modulated energy below 1.5 Hz could be from standing waves in the air-filled well. The surface waves will push at the surface end of the well pipe, pumping energy into the air-filled well as in an "organ pipe." The standing waves in an open, air-filled pipe will be resonating at

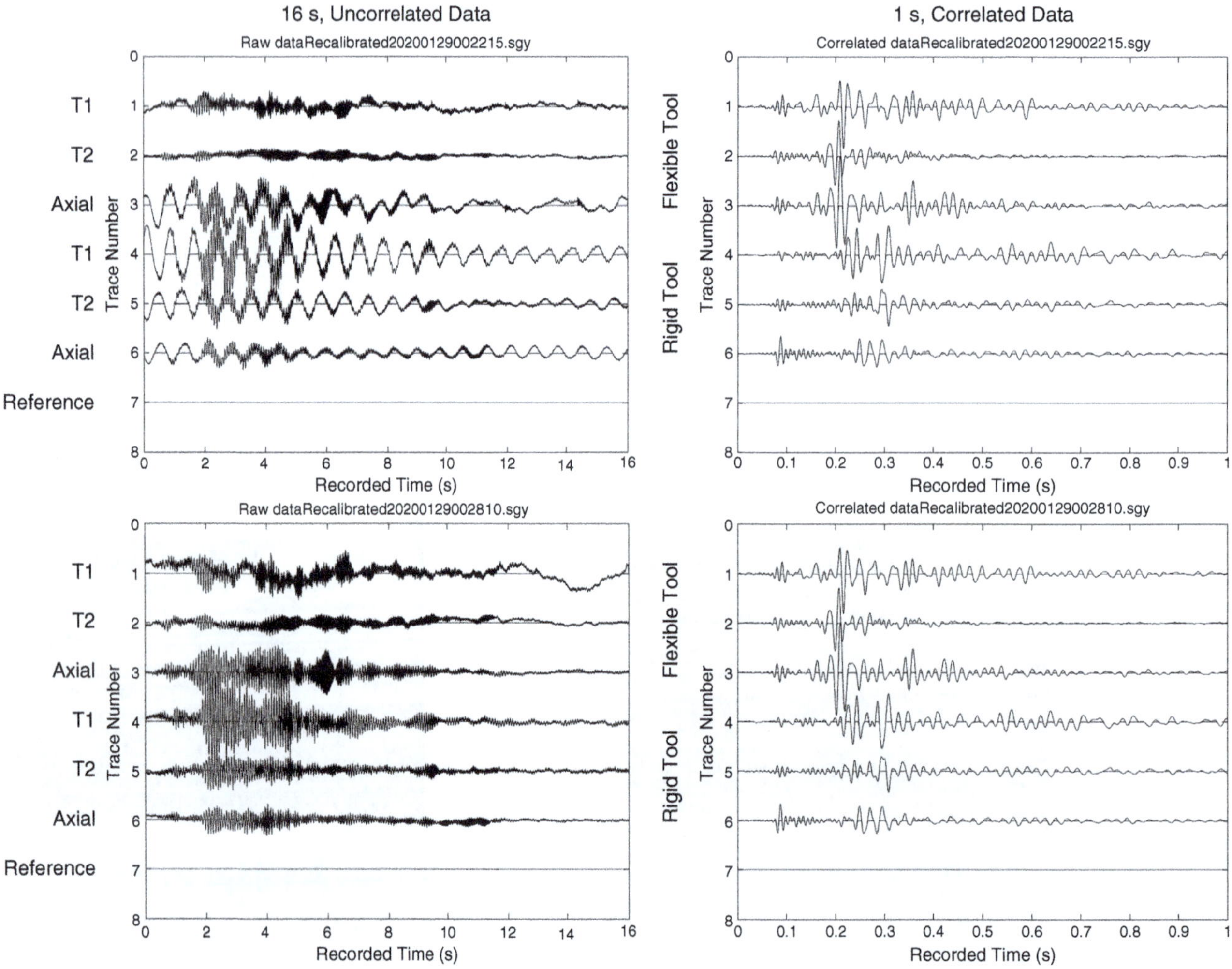

Figure 4.13 Raw data at 16 s and correlated data at 1 s for two records acquired 6 minutes apart. The source or receivers were not moved between the two records. The sequence of the six traces is, from the top: Tool 1 (Transverse 1, Transverse 2, Axial), Tool 2 (Transverse 1, Transverse 2, Axial). For the first record, we see very strong acoustic energy at frequencies at and below 1.5 Hz. Haldorsen et al. (2021) / with permission of the Society of Exploration Geophysicists.

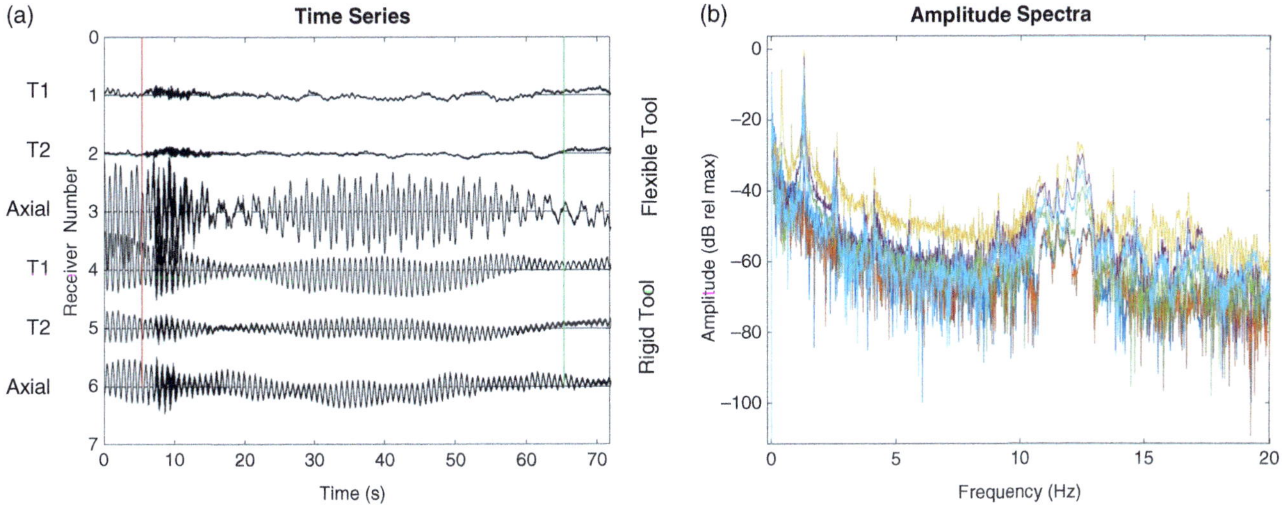

Figure 4.14 (a) A section of 72 s of raw GeoLite data; (b) Low-frequency spectral analysis of the data. Trace sequence in (a), starting from the top (six traces): Tool 1 (Transverse 1, Transverse 2, Axial), Tool 2 (Transverse 1, Transverse 2, Axial). The traces obviously contain significant energy below the lower limit of 8 Hz for the nominal source signature and extending to DC. We believe that energy at and below 1.5 Hz could be from a standing wave in the air-filled well, with the well acting as an organ pipe. Haldorsen et al. (2021) / with permission of the Society of Exploration Geophysicists.

frequencies $f_n = \frac{(2n-1)v}{4L}$, where L is the length of the air-filled well (455 m) and v is the speed of sound in air at 343 m/s. The fundamental resonance frequency for the open pipe is 0.19 Hz, with harmonics at 0.57 Hz, 0.95 Hz, and 1.33 Hz. The amplitude spectra in Figure 4.14 indicate high amplitudes at around these frequencies.

4.4.3. A Fiber-Optic Pressure Sensor

To construct a pressure sensor, the optical fiber is wound around a flexible cavity that expands and contracts in proportion to the external pressure. Figure 4.15 from Haldorsen et al. (2021b) shows the optical hydrophone cast into a polyurethane form that is shaped for low-resistance towing. Although this hydrophone is not designed for borehole applications, with some well-understood changes it could be, and it could be added to the 3C fiber-optic sensor tool. The hydrophone would provide point measurements of dynamic pressure, which may allow better separation of P-waves, S-waves, and interface waves.

The optical hydrophone is designed for very broadband response. Figure 4.16 compares data recorded by these hydrophones to data recorded by reference U.S. Navy H48 hydrophone at frequencies from 0.2 to 100 Hz. These measurements were made in collaboration with Naval Undersea Warfare Center (NUWC), Division Newport's Underwater Sound Reference Division (USRD).

Figure 4.15 In the hydrophone, the optical fiber is wound around a flexible cavity that is allowed to expand and contract in proportion to the pressure. In this display, the hydrophone is cast into a polyurethane form that is shaped for low-resistance towing. Haldorsen et al. (2021) / with permission of IEEE.

Although the low-frequency roll-off of the reference sensor starts at about 10 Hz, the fiber-optic hydrophone provides data to the very low end of the spectrum.

Figure 4.17 shows a comparative analysis of data recorded by the reference sensor and the fiber-optic hydrophone. For these data, the source swept through frequencies from 200 to 1,600 Hz. The analyses show that the reference sensor and the fiber-optic sensor provide essentially the same measured values at all frequencies, and in this figure, the two plots overlap to the extent that they look like a single curve.

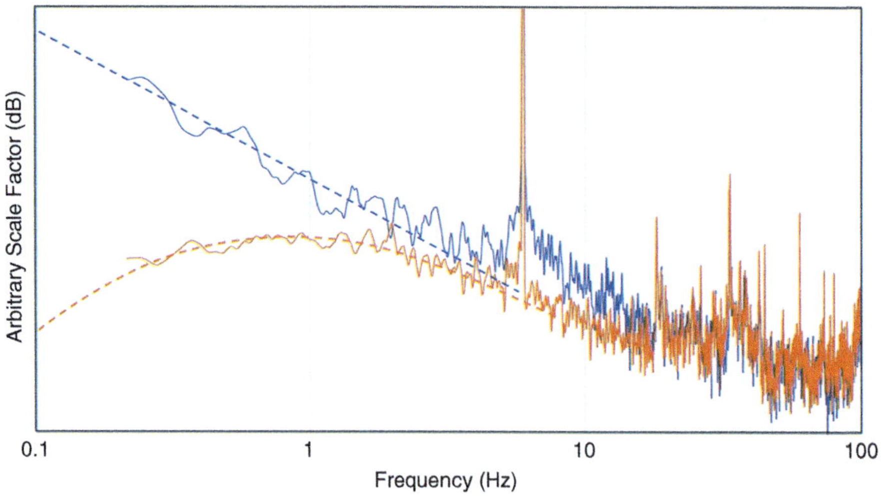

Figure 4.16 The fiber-optic hydrophone is designed to measure acoustic disturbances at very low frequencies. Here we compare the fiber-optic hydrophone measurements (in blue) to those of a reference hydrophone (in red) at frequencies from 0.2 to 100 Hz. The fiber-optic sensor provides data to the very low end of the spectrum, whereas the data from the reference sensor appear to start rolling off at about 10 Hz. Haldorsen et al. (2021b).

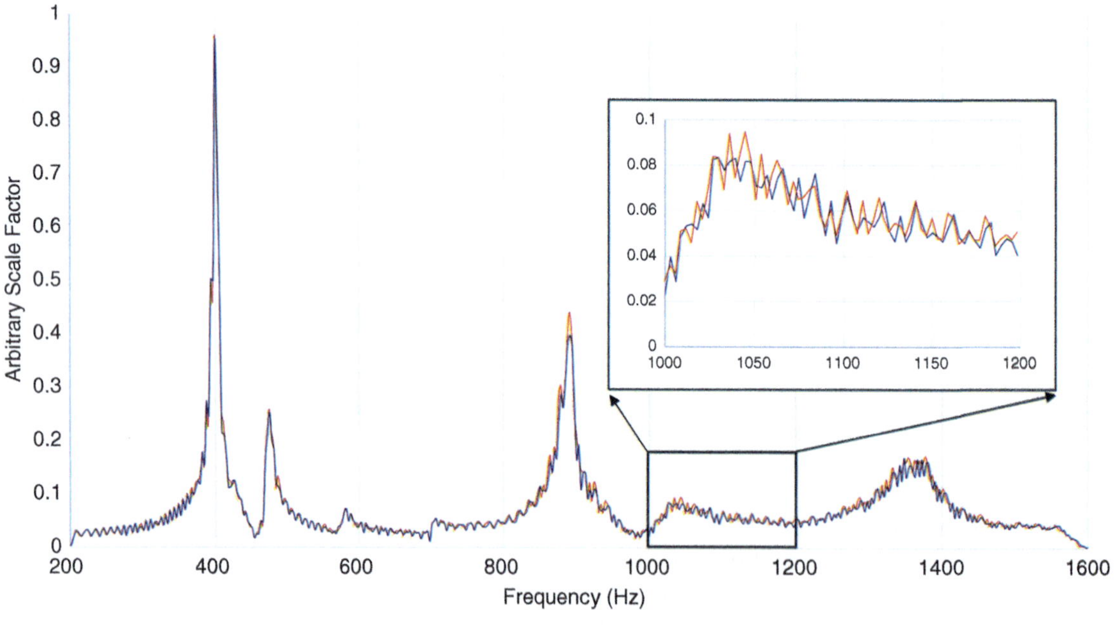

Figure 4.17 Comparing the fiber-optic hydrophone to a reference sensor for a source sweeping from 200 to 1,600 Hz. The fiber-optic sensor (in blue) and the reference sensor (in orange) provide essentially the same measured values over all frequencies, and the two plots overlap to the extent that it looks like one curve. The insert is a zoom-in from 1,000 to 1,200 Hz to show that there are two curves on the plot. Haldorsen et al. (2021b).

4.5. CONCLUDING REMARKS

Haldorsen (2021, 2022) shows that with 3C sensors, the distance between sensors can be increased beyond the limits of half a wavelength of the maximum usable frequency without generating artifacts normally associated with spatial aliasing. A near-continuous single-axis DAS array has no spatial-aliasing problems but lacks information about transverse components of particle motion. Adding a single 3C sensor would solve the azimuthal ambiguity of a DAS array—at least when only a single microseismic source is active within the sensitive area around the array.

A less radical approach than this is to supplement a fiber-optical DAS array with a relatively sparse array of wide-bandwidth, optical 3C accelerometers, where the fiber-optic point sensors compensate for the inability of the DAS sensors to measure the direction of particle motion in 3D space.

As it will have no electrical or electronic components, a fully optical hybrid DAS/3C array will have significantly improved reliability and lifetime in a high-temperature environment, which would otherwise challenge the temperature limits of current state-of-the-art electrical sensors. Such applications would include geothermal wells. For this, or for life-of-field monitoring of oil and gas reservoirs, removal of electrical components would eliminate electrical or electronic failures and at the same time lower the cost of the downhole installed hardware.

This would give data that are adequate for acoustic monitoring, and significantly improve the versatility and lower the cost of both single and repeat VSP acquisitions. Without downhole electronics, there is no need for downhole power—and the risk of failure of electronics left in the well for a long period of time is eliminated, significantly increasing lifetime in high-temperature wells. The hybrid optical sensor array is expected to be operational at temperatures exceeding 200 °C by a sizable margin.

The acquisition system for the hybrid optical array will be simpler and enable "always-on" monitoring of induced seismicity, giving information about changes in the stress and pressure in the reservoir. In relation to CO_2 storage, on-demand active seismic surveys allow both short-term and long-term monitoring of the CO_2 plume.

ACKNOWLEDGMENTS

This material has been produced with support in part by the U.S. Department of Energy, Office of Science, National Energy Technologies Laboratory, under Award Number DE-SC00015781, and by the U.S. Department of Defense (NORTHCOM) under Award Number HQ003419C0045.

REFERENCES

Bakulin, A. & Calvert, R. (2004). *Virtual source: new method for imaging and 4D below complex overburden.* Paper presented at the 74th SEG Annual Meeting and International Exposition.

Claerbout, J. (1985). *Imaging the Earth's interior.* Blackwell Science Inc.

Farhadiroushan, M., Parker, T. R., & Shatalin, S. (2009). Method and apparatus for optical sensing. *U.S. Patent No. 9,541,425.* Washington, DC: U.S. Patent and Trademark Office

Frømyr, E. *Recent advances in multicomponenet seismic.* Paper presented at the 5th Conference & Exposition on Petroleum Geophysics, Hyderabad-2004, 2004.

Haldorsen, J. B. U. (2002). *Converted-shear and compressional images using projection imaging.* Paper presented at the 64th Meeting of the European Association of Geoscientists and Engineers.

Haldorsen, J. B. U. (2021). Spatial aliasing and three-component seismic sensors. *Geophysics, 86*(4):V255–V267.

Haldorsen, J. B. U. (2022). *Microseismic imaging with three-component borehole sensors.* Paper presented at the 92nd SEG Annual Meeting.

Haldorsen, J. B. U., Borland, W., & Heijna, H. B. (2004). *Extended high-frequency processing of vibrator vsp data.* Paper presented at the 66th Meeting of the European Association of Geoscientists and Engineers.

Haldorsen, J. B. U., Brooks, N. J., & Milenkovic, M. (2013). Locating microseismic sources using migration-based deconvolution. *Geophysics, 78*(5):KS73–KS84.

Haldorsen, J. B. U., Christensen, C., Metzbower, D. R., Berzanskis, A., Machnizh, J., Bergery, G., et al. (2021a). A three-component optical sensor for borehole-seismic applications. *The Leading Edge, 40*(6):447–453.

Haldorsen, J. B. U., Christensen, C., Metzbower, D. R., & Berzanskis, A. (2021b). *Fiberoptic wide-band acoustic multi-component sensors.* Paper presented at the Oceans Conference and Exhibition, San Diego.

Haldorsen, J. B. U., Christensen, C., Metzbower, R., & Berzanskis, A. (2022). *Why you should wish you had three-component optical borehole sensors in your sensor array.* Paper presented at the SEG Summer Research Workshop, Stanford University, CA.

Haldorsen, J. B. U., Jahren, L., Milenkovic, M., & Hilton, T. (2018). *Vertical seismic profiling using a hybrid tool, combining three-component geophones with das sensors.* Paper presented at the 80th Meeting of the European Association of Geoscientists and Engineers.

Haldorsen, J. B. U. & Jahren, L. (2020). Imaging below a complex overburden with borehole-seismic data. *Geophysics, 85* (3), S135–S150.

Hornby, B. E., Bostick, F. X., Williams, B. A., Lewis, K. A., & Garossino, P. G. (2005). Field test of a permanent in-well fiber optic seismic system. *Geophysics, 70* (4):E11–E19.

Leaney, W. S. & Esmersoy, C. (1989). *Parametric decomposition of offset VSP wave fields.* Paper presented at the 59th SEG Annual Meeting and International Exposition.

Maas, S. J., Bunn, J. B., Metzbower, D. R., Smith, A. B., & Stickler, D. (2015). *Jubarte PRM system manufacturing and installation lessons learned.* Paper presented at the 3rd EAGE Workshop on Permanent Reservoir Monitoring.

Miller, D., Oristaglio, M., & Beylkin, G. (1987). A new slant on seismic imaging: classical migration and integral geometry. *Geophysics, 52*(7):943–964.

Paulsson, B., Wylie, M., Thornburg, J., He, R., Hardiman, H., & Patel, D. (2019). *A fiber optic single well seismic system for geothermal reservoir imaging and monitoring.* Paper presented at the 44th Workshop on Geothermal Reservoir Engineering at Stanford University.

5

Three-Component Distributed Acoustic Sensing Arrays with Three-Dimensional Fiber Cable Deployment

Yingping Li[1,2], David Li[3], Lianjie Huang[3], and Yingcai Zheng[2]

ABSTRACT

We design two fiber distributed acoustic sensing (DAS) arrays to acquire three-component (3C) DAS array data. These two arrays are a cross array and a cross-concentric circle (CCC) array at the surface, with part of the fiber cable installed in a vertical wellbore. We calculate traveltimes from seismic sources to all fiber sensors of the DAS arrays and construct differential traveltime (DTT) profiles. We calculate the source azimuth and incidence angle from the spatial distribution of the differential times and then use the resulting angles to rectify the raw DAS amplitudes distorted by the fiber directivity. We create horizontal and vertical hodograms that are comparable to those of a 3C geophone at the DAS array center using the recovered amplitudes. Our DTT and waveform modeling results show that a 3D optimally distributed fiber cable can effectively acquire 3C DAS array data. Employing beamforming and coherent stacking techniques, we can significantly increase the signal-to-noise ratio (SNR) of the DAS system. This 3C DAS array overcomes three disadvantages of conventional DAS measurements: single component, fiber directivity, and lower SNR. Our 3C DAS array can also be used to monitor and study earthquakes and explosions at local, regional, and teleseismic distances.

5.1. INTRODUCTION

Distributed acoustic sensing (DAS) is a technology that uses a fiber-optic cable to measure seismic (strain) signals. A DAS system includes a fiber-optic cable connected to an optoelectronic device containing a laser light source and an optical detector (interrogator unit, IU) to measure coherent Rayleigh backscattering generated by imperfections in the optical fiber (Mestayer et al., 2011; Hartog, 2017). Changes in the return time of the scattered light allow us to make continuous and real-time measurements of small strain changes in the material around the fiber with a high spatial resolution in a wide frequency range. These properties make DAS an appealing tool in both earthquake seismological and exploration geophysical research (Li et al., 2021, 2022b). In contrast to conventional three-component (3C) geophones, DAS does have some significant drawbacks, including single-component measurement, fiber directionality, and low intrinsic signal-to-noise ratios (SNR). We note that the DAS measurement of the strain or strain rate is the average over a length, called the gauge length, which is related to the DAS spatial resolution.

In Figure 5.1, we compare single-component DAS recordings with 3C geophone seismograms of an M = −0.72 microearthquake induced by fracture stimulations at a geothermal site in Utah FORGE during a fracture stimulation project in 2019 (Moore et al., 2020). A DAS system with a 1-km-long fiber-optic cable

[1]*BlueSkyDas LLC, Sugar Land, Texas, USA*
[2]*Department of Earth and Atmospheric Sciences, University of Houston, Houston, Texas, USA*
[3]*Los Alamos National Laboratory, Los Alamos, New Mexico, USA*

Distributed Acoustic Sensing in Borehole Geophysics, Geophysical Monograph 289, First Edition.
Edited by Yingping Li, Robert Mellors, and Ge Zhan.

DOI: 10.1002/9781394179275.ch05

cemented outside the well casing and a geophone array with 12-level 3C geophones is deployed in the observation well to simultaneously monitor the induced microearthquakes by the fracture stimulations. The channel spacing of DAS fiber is 1 m with a gauge length of 10 m, and the geophone depth interval is 30 m. The time domain sampling rate for both systems is 0.5 ms (Li et al., 2022a). Figure 5.1(a) shows 30-channel DAS data with channel depths ranging from 829.2 to 858.4 m. We are able to pick the first arrival times of both P and S waves manually or automatically using an LTA/STA method or other appropriate approaches, but the high noise amplitudes of some traces prevent us from measuring the times reliably and accurately.

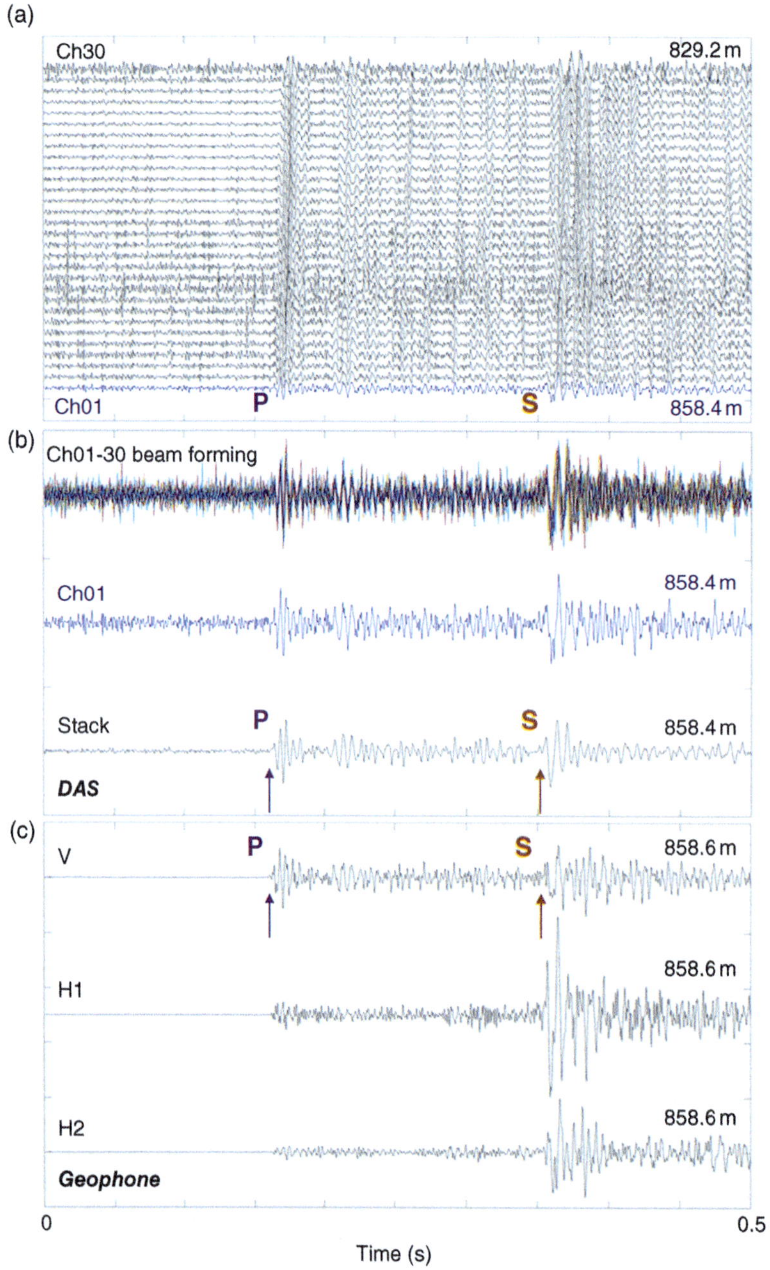

Figure 5.1 (a) DAS waveforms of a M = −0.72 earthquake recorded by 30 DAS channels with a channel space of 1 m show P and S waves and background noise. (b) Coherent waveform beamforming of 30 traces with both P and S wave windows compared a single trace of Ch01 and a stacking trace. (c) 3C geophone recording of the same seismic event.

We select time windows around the P and S waves of channel 01 to cross-correlate with the other 29 channels, to beamform the coherent waveforms by time-shift, and to construct a stack trace (Figure 5.1(b)). Compared to the raw DAS trace of channel 01, the coherent waveform stacking trace increases SNR by a square root of 30 theoretically and is comparable to the vertical component waveform recorded by a 3C geophone at the same depth (Figure 5.1(c)). The DAS first-arrival waveforms of P and S waves are very similar to those of vertical component recording by a 3C geophone, and the first arrival times of both P and S waves can be reliably and accurately picked from the stack DAS trace. This example not only illustrates the power of beamforming a stack of coherent waveforms to significantly increase the SNR of DAS data to improve measurement of first arrival times for both P and S waves but also indicates that DAS data is only one component compared to the 3C geophone data (Figures 5.1(b) and 5.1(c)).

Because seismic wavefields are vectors, 3C geophones, seismometers, or accelerometers are necessary in many seismological studies, particularly in borehole seismic applications like vertical seismic profiling (VSP). These borehole applications include 3C–3D (three-dimensional) vector migration imaging, measuring P and S wave velocity profiles, pinpointing the locations of microseismic events using both P and S waves, measuring shear-wave splitting for medium anisotropic parameters, and 3D salt proximity, among others. Unfortunately, fiber DAS data are measurements of a single-component axial strain or strain rate. Strain (or strain rate) is a tensor rather than a vector.

Helically wound fiber cables (Hornman, 2017; Innanen, 2017; Kuvshinov, 2016; Ning & Sava, 2018) are broadside sensitive and have been developed and field-tested for creating 3C DAS measurements. Helically wound fiber cables are typically deployed in trenches or shallow horizontal boreholes, mainly for surface seismic applications (Mateeva et al., 2021). A subsurface chemical explosion at the Nevada National Security Site (NNSS) was recorded by both helical fibers on two nearby boreholes and a surface helical fiber (Abbott et al., 2019; Mellors et al., 2021). Several field tests have indicated that the special cables do record broadside seismic energy. However, fully recovering 3C DAS measurements is still a technical challenge.

DAS systems based on fiber Bragg grating (FBG) techniques are typically considered "quasi" distributed acoustic sensing (qDAS) systems (Li et al., 2021). The FBG-qDAS system with the multiaxis FBG enables 3C measurement of seismic signals. 3C borehole seismic systems with various qDAS technologies have been developed and are commercially available (Haldorsen et al., 2021; Paulsson et al., 2014, 2015; Zhang et al., 2022), but sensor numbers are limited. Reducing the relatively expensive cost to build, install, and maintain qDAS systems is a tremendous technical and financial challenge.

Conventional seismic arrays (Rost & Thomas, 2002; Schweitzer et al., 2012) deployed at the surface in a regular geometry pattern, such as a cross (LASA = Large Aperture Seismic Array and YKA = Yellowknife Seismological Array) or concentric circles (NORES and ARCES, two small-aperture arrays with a diameter of only 3 km in southern and northern Norway; Schweitzer et al., 2012), have been widely used to determine the directions of coming seismic waves and to beamform their coherent waveforms for increasing SNR, and have played important roles in monitoring, detecting, locating, identifying, and distinguishing earthquakes, explosions, and other seismic events for the CTBT (Comprehensive Nuclear Test-Ban Treaty). In several DAS applications in earthquake seismology, surface DAS fiber cables are deployed in different geometric shapes, such as in the east–west and north–south trenches (Ajo-Franklin et al., 2015; Xu et al., 2021), triangles (Castongia et al., 2017; Walter et al., 2020), and five fiber cables in a perpendicular cross with another four fiber cables (Ajo-Franklin et al., 2017). Wang et al. (2018) deployed a DAS fiber network with a zigzag shape at the surface and installed a fiber cable in a shallow vertical well at one end of the DAS network at Brady Hot Springs, Nevada. The DAS seismic networks with zigzag and other irregular geometry shapes have been used to conduct beamforming to increase SNR for the DAS observation systems (Ichinose et al., 2022; Näsholm et al., 2022; van den Ende & Ampuero, 2021).

In this study, we design two 3C fiber sensor DAS arrays: (a) a cross array and (b) a cross-concentric circular array, both at the surface with a fiber cable in a vertical well, to acquire 3C DAS array data of microearthquakes induced by fracture stimulations in geothermal reservoirs and unconventional oil and gas reservoirs. We examine if the fiber cable installed in the special geometry shapes distributed in three dimensions can optimally record 3C DAS array data. We compute traveltimes from a seismic source to all fiber sensors of the DAS arrays and generate profiles of differential traveltimes in both map view and vertical cross-sectional view. We estimate the source azimuth and incidence angle using the profiles of DTTs and use the resulting two angles to correct DAS raw amplitudes distorted by the fiber directivity, recovering the true amplitudes of seismic waves that are similar to those recorded by a 3C geophone. We then use the recovered DAS amplitudes to construct horizontal and vertical hodograms and show that those hodograms are equal to or comparable to the particle motion plots of a 3C geophone at the DAS array center ($x = 0$, $y = 0$).

We generate synthetic waveforms of acoustic response and DAS amplitudes for a given seismic source using a finite-difference method (Gao & Huang, 2017) and two P-wave velocity models: a constant velocity model and a layered velocity model. We find that both DTT modeling and synthetic waveform amplitudes of fiber sensors in the DAS array are feasible to estimate the azimuth and incidence angle of the seismic source. Our modeling results indicate that such a 3D distributed fiber cable with a special geometry shape can be used to acquire 3C DAS array data. We discuss some features and advantages of the 3C DAS fiber array. We should be able to significantly increase the SNR of the DAS system using beamforming (e.g., Li et al., 2020) and coherent signal stacking techniques, thus decreasing the magnitude detection threshold of microseismic events using as many fiber sensors as possible in the DAS array. This 3C DAS array enables us to overcome three major disadvantages of current DAS measurements, including single-component, fiber directivity, and low SNR. The 3C DAS fiber array discussed here is hopefully expected to be expanded to a large-scale 3C DAS array for monitoring and studying local, regional, and even teleseismic events, such as explosions and earthquakes (Kennett, 2022; Mellors et al., 2021).

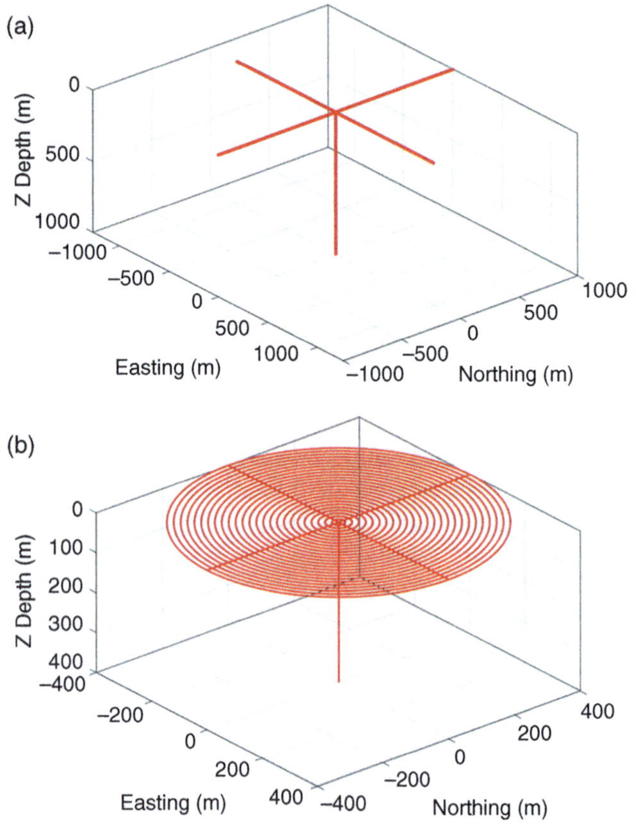

Figure 5.2 (a) Layout of a 3C DAS array consisting of a fiber-optic cable installed with a portion in a cross (orthogonal linear) shape at the surface and a portion in a vertical wellbore. (b) Layout of a 3C DAS array with a fiber-optic cable deployed with a portion in the crossed lines (a Cartesian coordinate system) and 25 concentric rings (a polar coordinate system) at the surface and a portion in a vertical wellbore.

5.2. FIBER CABLE DEPLOYMENT GEOMETRIES OF DAS ARRAYS AND MODELING METHODS

This study examines two experimental DAS fiber-optic cable layout designs: (a) a cross array, a fiber-optic cable deployed in a cross shape at the surface and in an underground vertical well (Figure 5.2(a)); and (b) a cross-concentric circle (CCC) DAS fiber array, a fiber-optic cable installed a portion in the crossed lines (a Cartesian coordinate system) and 25 concentric rings (a polar coordinate system) at the surface and a portion in a vertical wellbore (Figure 5.2(b)).

5.2.1. Fiber Cable Deployment Geometries of DAS Arrays

Figure 5.2(a) shows the first DAS fiber array layout design with a portion of fiber-optic cable installed in a cross (orthogonal linear) shape at the surface and a portion of the fiber cable cemented outside the well casing. This ensures the best coupling with subsurface geologic formations. For near-surface deployment, the fiber cable is typically buried or cemented in a trench. The ground fiber cable is 2 km in length in both the north–south and east–west directions. The fiber cable in the vertical well is 1 km in length. The total length of the cable is 5 km. For field deployment, we recommend using a 10-km fiber cable to increase the redundancy and reliability of the DAS array system. The sensor (or channel) spacing and gauge length of a DAS observation system can typically be selected. In our modeling, we use a gauge length and sensor spacing of 10 m and a total fiber cable length of 5 km. This DAS cross-array system contains 401 fiber sensors at the surface and 100 fiber sensors in the vertical well.

In the second configuration, we employ a layout of a CCC DAS array as shown in Figure 5.2(b). Compared to the 3C DAS cross array in Figure 5.2(a), we add a fiber cable along 25 concentric rings at the surface to construct this DAS CCC array. The radii of these 25 concentric circles range from 16 to 400 m, with an increment of 16 m. This DAS CCC array allows DAS fiber sensors to be distributed along three axes of a Cartesian coordinate system as well as along radial directions in a polar coordinate system. For a given sensor spacing (also gauge length) of 8 m, the circle portion of this fiber DAS array with a cable length of about 32.7 km contains 4,084 fiber sensors, and the cross portion of the fiber DAS array with a cable length of 2 km comprises 253 positions of fiber sensors. If we would like to build a duplicate fiber cable for redundancy along the three axes of the cross DAS subarray, we can add an additional 2 km of cable with 253 fiber sensors to the DAS array system. Nevertheless, we do not need this redundancy and only use 4,335 fiber sensors for our modeling.

5.2.2. DTT Modeling Method

For a seismic array to record high-quality array data, the geological condition of the subsurface beneath any good seismic array should ideally be as homogeneous as possible (Rost & Thomas, 2002). A high-quality fiber DAS array should also be installed in a geological homogeneous site, following the best practice of seismic array deployment. Therefore, in our modeling of traveltimes and DTTs for DAS fiber cable deployment geometries in Figures 5.2(a) and 5.2(b), we use a simplified velocity model for the medium beneath the DAS arrays. This model has a homogeneous P-wave velocity of 2,000 m/s in a half-space. Although this velocity model is rather simple, we can still use it to illustrate the concept of how DTTs from the fiber sensors can be used to determine the source azimuth and the angle of the incident P wave.

We first carry out travel-time modeling using the DAS cross array with 401 fiber sensors at the surface and 100 fiber sensors in the vertical borehole (Figure 5.3(a)). For a given seismic source (Figure 5.3(a)), we calculate the traveltimes t_i of the P wave to each of the fiber sensors in the cross array using a constant velocity model of V_p = 2,000 m/s,

$$t_i = D_i/V_p, \tag{5.1}$$

where D_i is a hypocentral distance from the source to each fiber sensor. We subtract the traveltime t (x = 0, y = 0) from the source to a fiber sensor at the center of DAS array (x = 0, y = 0) from the traveltime t_i at each fiber sensor to obtain differential traveltimes (DTT), dt_i,

$$dt_i = t_i\,(x, y) - t\,(x = 0, y = 0). \tag{5.2}$$

Arrival times and traveltimes may not be the same values if the source origin time is unknown. However, the differential time for arrival time and traveltime are the same because the source origin time is subtracted and canceled. Therefore, we use the term *differential time* without distinguishing arrival or traveltimes hereafter.

We plot the differential times, dt_i, for each fiber sensor of the DAS cross array in map view in Figure 5.3(b) and cross-sections in Figures 5.3(c) and 5.3(d). Fiber sensors with negative differential times are plotted in red, whereas fiber sensors with positive differential times are in blue. Negative differential times indicate that the seismic wave arrives at the fiber sensors marked in red earlier than that arriving at the reference fiber sensor at the array center (x = 0, y = 0). It shows the quarter-half-space in which the source lies; the actual differential times, of course, can determine the azimuth and incidence angles of the seismic source (Figures 5.3(b) and 5.3(d)). We construct isolines for fiber sensors with the same differential times and plot them in a map view (Figure 5.3(b)) and in a cross-section (Figure 5.3(d)) in the incidence plane to estimate the source azimuth and the angle of incidence. However, for a given dt, we can project the real fiber sensors to positions along the azimuth to create "virtual" fiber sensors there and construct DTT isolines in Figure 5.3(d) in the incidence plane. This result demonstrates the advantage of the DAS cross-array in determining the directions of incident seismic waves.

Figure 5.4(a) depicts a DAS CCC array at the surface with 4084 fiber sensors along 25 concentric rings and 201 fiber sensors at the DAS cross array using a 35 km fiber. For a seismic source at an azimuth of 45° with a hypocentral distance of 9 km and a constant P-wave velocity of 2000 m/s, we calculate traveltimes, t_i, and DTTs, dt_i, for each fiber sensor of the DAS CCC array and display their contour maps in Figures 5.4(b) and 5.4(c), respectively. Both contour maps clearly indicate that fiber sensors with faster arrival times (negative dt_i) with respect to the array center origin are in the northeast quadrant, whereas fiber sensors with slower arrivals (positive dt_i) with respect to the array center origin are in the southwest quadrant, implying the seismic source is in the northeast direction.

We plot the positions of fiber sensors with negative and positive (i.e., sign) differential times in negative (red) and positive (blue), respectively, in Figure 5.4(d), and find a boundary between the negative and positive (red and blue) sensors that divides the DAS array into two semicircles. The normal of the negative–positive (red–blue) boundary points toward the seismic source. Therefore, we can quickly and accurately estimate the source azimuth. Alternatively, we can search for positions of fiber sensors having the minimum and maximum values in differential time on each ring. We can accurately calculate the source azimuth by fitting a straight line passing through the two extreme locations. When the source azimuth is accurately determined, we can estimate the incidence angle by generating isolines of DTTs from sensors in the vertical well and in the incident plane.

5.3. DTT MODELING RESULTS

5.3.1. Modeling Results for the DAS Cross Array

We first apply our traveltime modeling method to a DAS cross-array with four seismic sources in four quadrants and two sources on the north and east axes (Figure 5.5(a)). We calculate the DTT of the six sources to each fiber sensor of the DAS cross-array and plot DTTs to every fiber sensor for each of the six seismic sources in

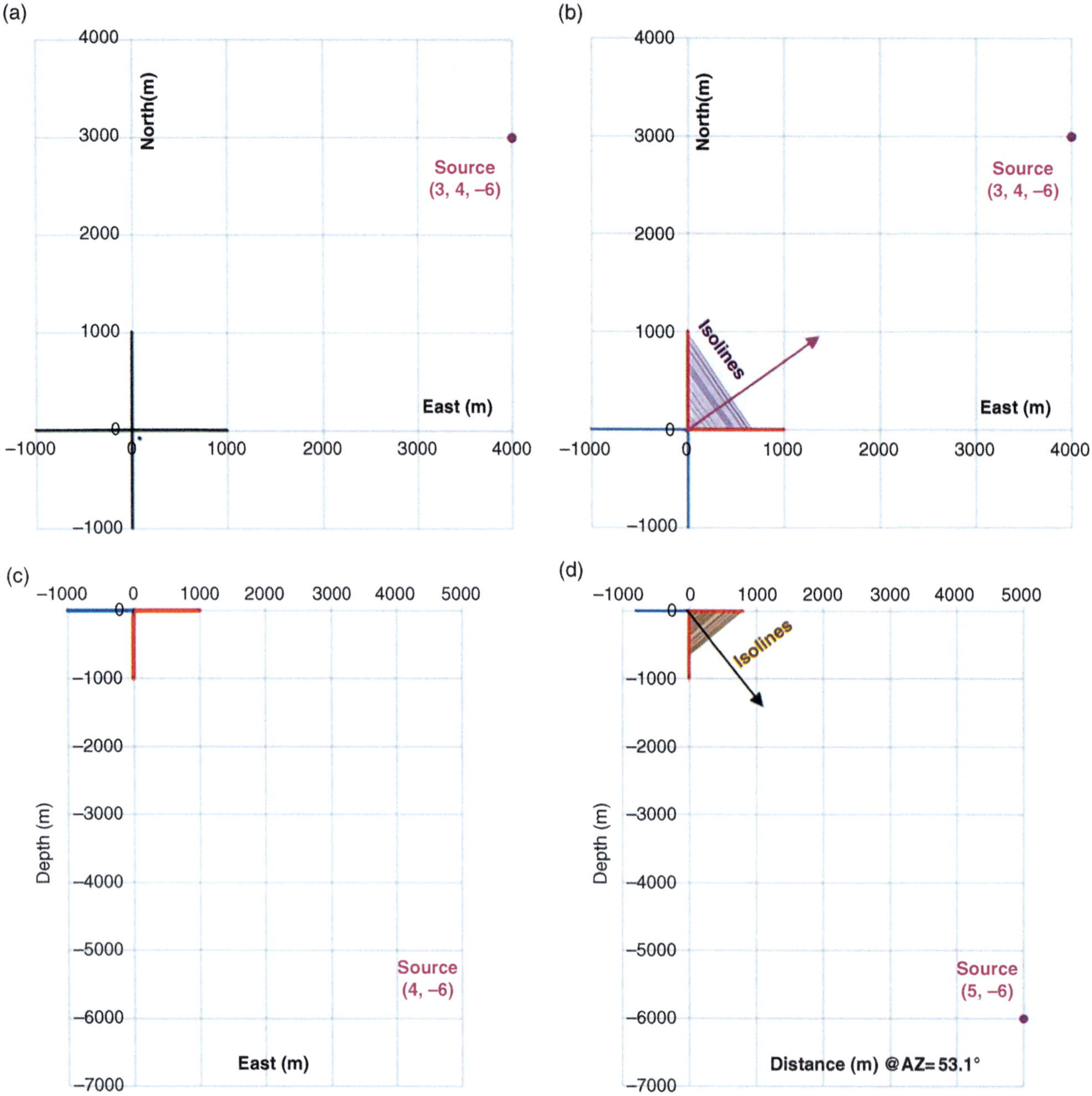

Figure 5.3 (a) A map view showing a DAS cross array with 401 fiber sensors along a 2 km fiber-optic cable at 10 m spacing in both the north and east directions and a seismic source (magenta dot). The numbers in parentheses are the coordinates of the seismic source in km. The x and dot at (0, 0) indicate the fiber going into and out of a vertical well. (b) The negative values (red) of DTTs (dt_i) for "sensors" along both the north and east axes indicate that seismic waves are coming from the northeast corner. These isolines of DTT (purple lines) can be used to estimate the source azimuth. (c) An east-depth cross-section displays fiber sensors with negative (red) and positive (blue) DTT and a seismic source on this plane. (d) A cross-section along the source azimuth of 53.1° shows fiber "sensors" with negative dt_i (red), the seismic source position, and isolines of DTT (brown lines) that can be used to estimate incidence angle.

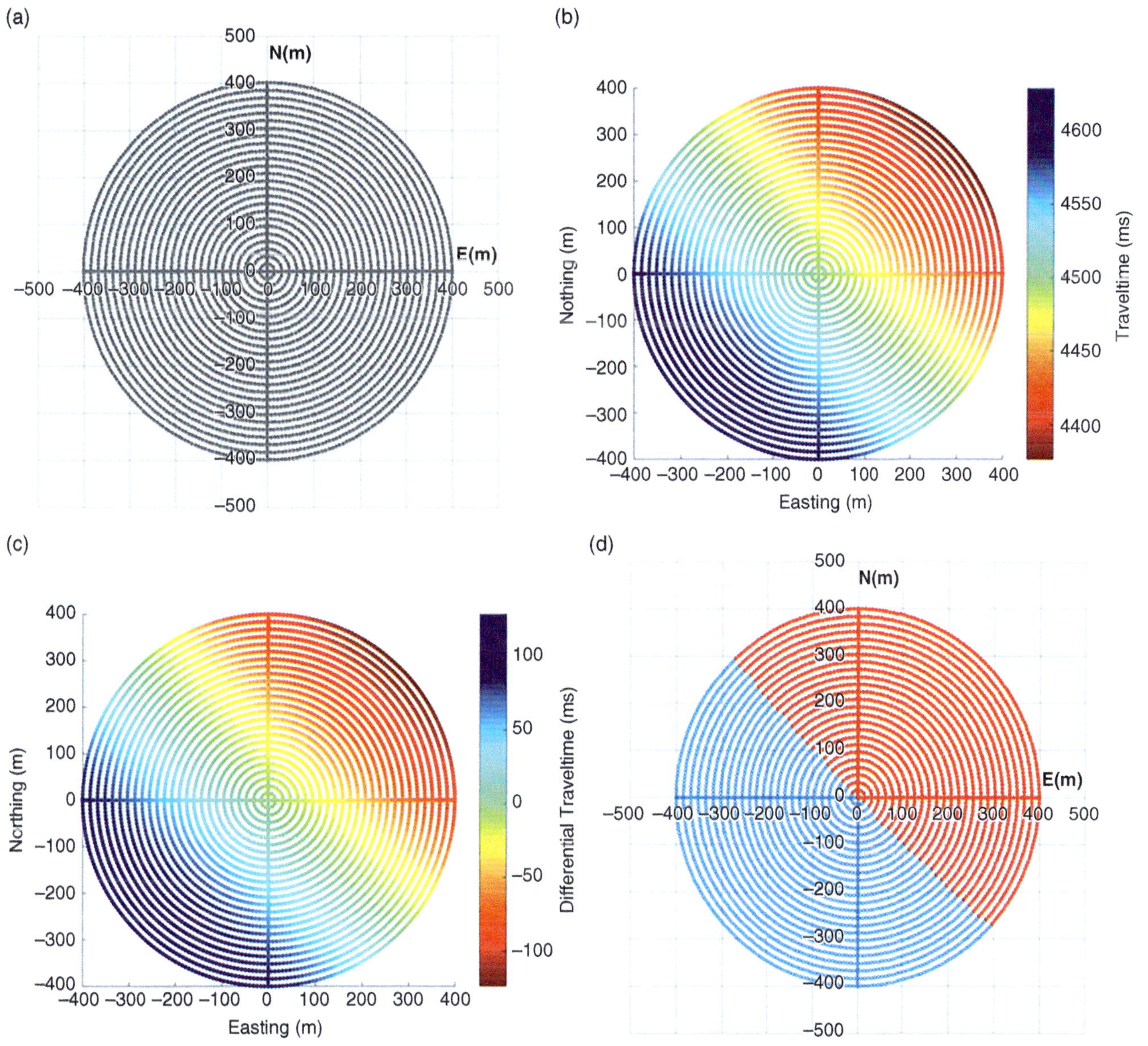

Figure 5.4 (a) A map view of a DAS fiber array along a cross and 25 concentric rings with 201 and 4,084 fiber sensors with a sensor spacing of 8 m. The radii of the circle rings range from 16 to 400 m. (b) Travel-time contours from a source located at a distance of 5,600 m to the DAS array center (x = 0, y = 0) at an azimuth of 45°. (c) Contours of DTTs ($dt_i = ti - t(x = 0, y = 0)$) indicate that dt values increase from northeast to southwest. (d) Fiber sensors with negative (red dots) and positive (blue dots) DTTs (dt_i). The red north and east axes indicate the source is in the first quadrant (northeast), and the normal of the red half circles (or the bisector) is the source azimuth of about 45°.

Figure 5.5(a). Fiber sensors with negative and positive differential times are plotted as red and blue dots, respectively. If we observe that red sensors appear only along the north (or east) axis, then the source is in the north (or east) direction. When red sensors appear along any two axes, we can easily determine which quadrant the seismic source originates from. Modeling results for this group of seismic sources show that our method can distinguish azimuthal directions of seismic sources.

Figure 5.5(b) is a map view showing the locations of a DAS cross array at (x = 0, y = 0) and seven seismic sources (magenta dots) in the first (northeast) quadrant, which result in red fiber sensors appearing along the north and east axes. We use this group of sources to demonstrate how to estimate source azimuths using isolines in map view. Isolines of differential times for seven seismic sources with different azimuths are plotted in Figure 5.5(b). We find that the normal vectors of these isolines point to

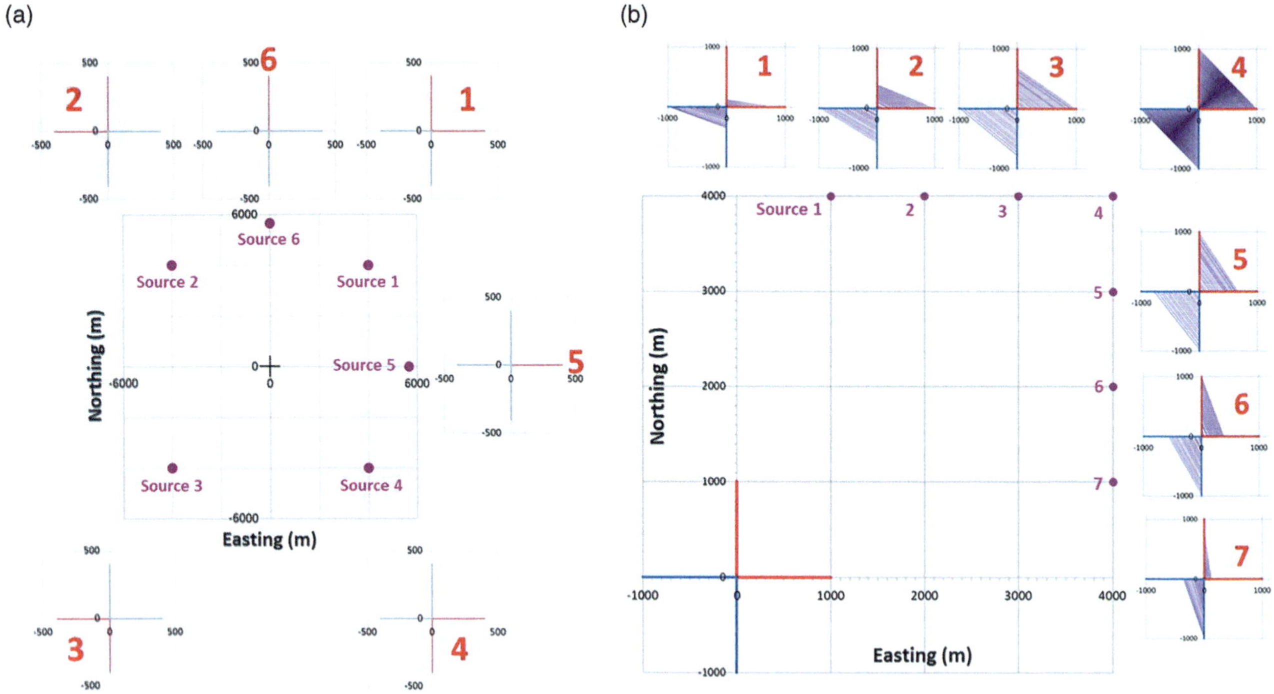

Figure 5.5 (a) A map view showing the locations of a cross DAS array, four seismic sources (magenta dots) in four quadrants, and two sources (magenta dots) on the north and east axes. Corresponding distributions of DTTs (dt_i) at each fiber sensor of the DAS cross array are shown with source numbers. For sources 5 and 6, the fiber sensors with negative dt_i values (red points) are only along one axis. Therefore, we indicate that the sources are located in the east and in the north. Fiber sensors with negative dt_i values (red points) along any two axes demonstrate that the seismic waves come from the sources in four different quadrants. (b) A map view of locations for the DAS cross-fiber array and seven sources (magenta dots) in the first (northeast) quadrant with all red sensors along the north and east axes. Corresponding plots of isolines (purple lines) of DTTs (dt_i) for seven numbered sources in the first quadrant. The normals of the isolines can be used to estimate the azimuths of the seismic sources.

the corresponding seismic sources. The greatest number of isolines corresponds to the source with a 45° azimuth. The number of isolines then gradually decreases as the source azimuths decrease or increase and reaches a minimum when the azimuth approaches either the east or north axes. Therefore, errors for estimating a source azimuth using only isoline normal are smallest at 45° and then increase to a maximum when the source azimuth approaches the north and east axes. However, when the source azimuths are close to the north or east axes, red color fiber sensor distributions will accurately point to the sources with smaller errors (sources 5 and 6 in Figure 5.5(a)).

After we obtain a reliable estimate of a seismic source azimuth, we can define a vertical incidence plane for the P wave and construct isolines of DTTs, dt_i, to estimate an incidence angle (with respect to the vertical direction) of the seismic wave. For seismic sources along the west–east or south–north axes, it is straightforward. We can use red fiber sensors along the axes and red fiber sensors in the vertical well to generate isolines of DTTs in the incidence plane to estimate the incident angles. For seismic sources along the other azimuths, there are no real fiber sensors physically along those azimuths in the DAS cross-array. We need to use real red fiber sensors in the vertical well and "virtual" red fiber sensors to construct isolines of DTTs in the incidence plane to estimate the incident angles. For a given dt_i, we can project real fiber sensors along the (east or north) axes to the radial direction at the azimuth to create "virtual" fiber sensors, as long as the assumption of a homogeneous medium is still valid.

Figures 5.6(a) to 5.6(d) show four seismic sources with different incidence angles and corresponding dt_i isolines on the incident planes. The normal vectors of these isolines point toward the seismic sources. When the incidence angle is about 45° (Figure 5.6(a)), the error of the estimate is the smallest. For the isoline normal method, the estimated errors increase when the incident angle gradually reduces toward 0° or 90° (Figure 5.6(d)). For the cross array, if we observe all red fiber sensors in the vertical well and all blue fiber sensors at the surface, then the seismic source is right beneath the vertical well.

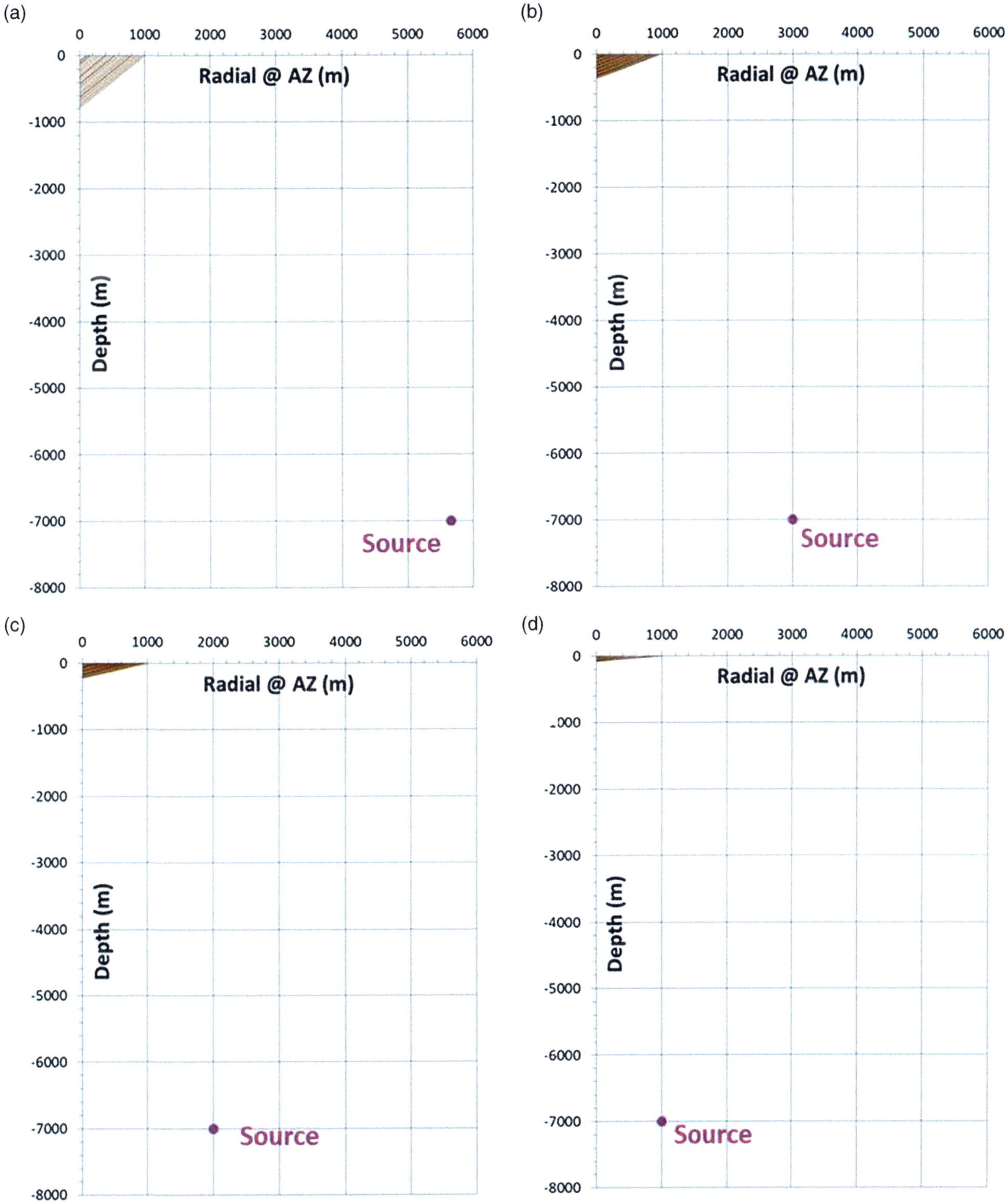

Figure 5.6 Cross-sections in incidence planes showing four seismic sources (magenta dots) and isolines (brown lines) of DTTs (dt_i) for the four sources with distinct incidence angles. Normals of the isolines can be used to derive the incident angles of seismic waves.

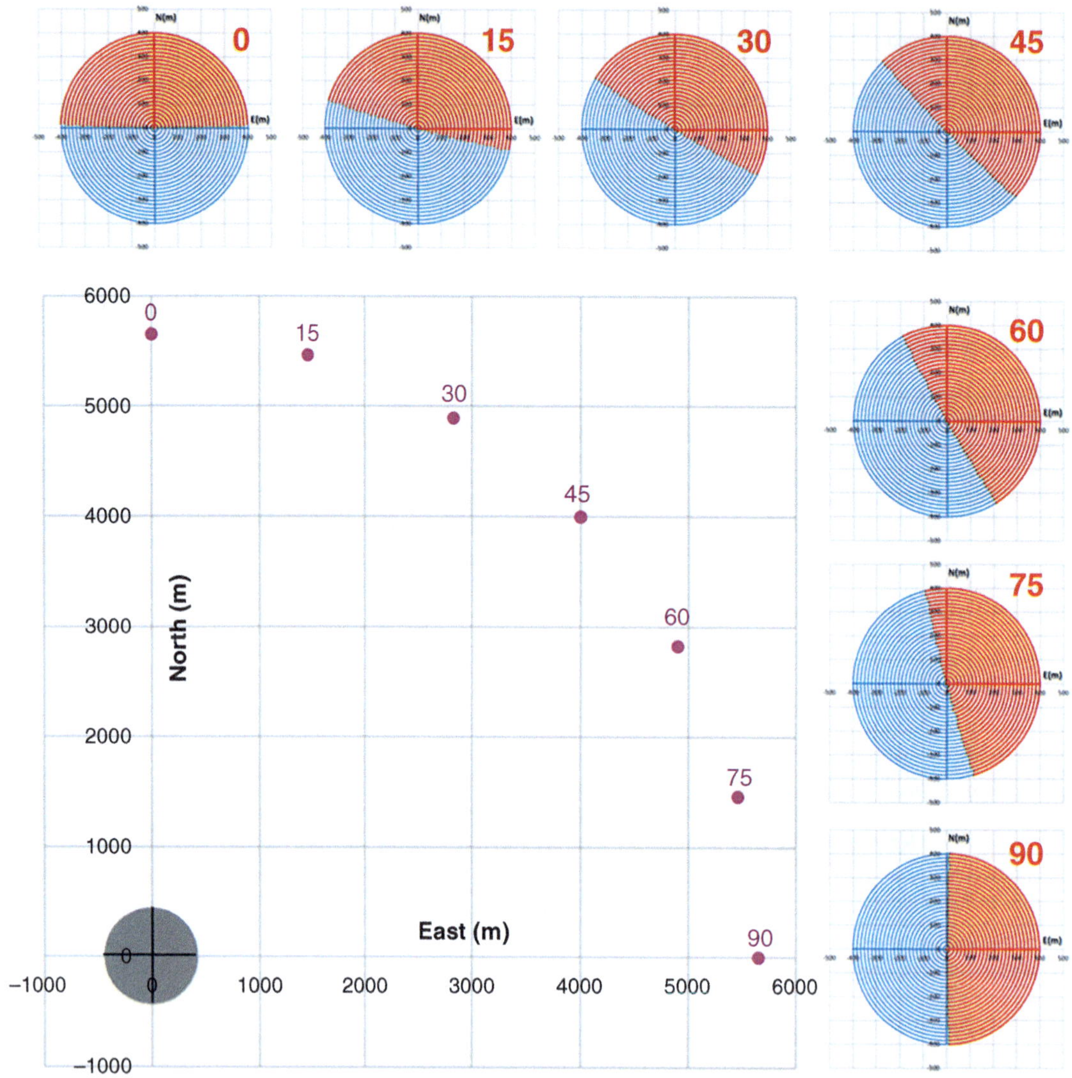

Figure 5.7 A map view showing the locations of a DAS CCC array and seven seismic sources (magenta dots) in the first (northeast) quadrant at different azimuths from 0° to 90° with an increment of 15°. Plots of negative (red) and positive (blue) DTTs (dt_i) at each fiber sensor of the DAS array for corresponding sources demonstrate that normals of two-color DTT circles point to those seismic sources, providing a robust approach to quickly estimating the source azimuths.

5.3.2. Modeling Results for a CCC DAS Array

We present the modeling results of the CCC configuration (Figure 5.2(b)) to demonstrate how to take advantage of the DAS CCC array to determine the source azimuth and the angle of the incident seismic wave. Figure 5.7 shows the locations of a DAS CCC array at (0, 0) and seven seismic sources (magenta dots) with different azimuths in the first (northeast) quadrant. We calculate DTTs, dt_i, using seven seismic source locations for each fiber sensor of the DAS CCC array and plot the negative (red) and positive (blue) dt_i distributions for each of the seven sources in Figure 5.7. The red–blue plots show a clear boundary between negative and positive dt_i values. The normals of the red–blue boundaries point to the seismic sources. For each concentric circle, we can search for the positions of two fiber sensors with the minimum and maximum dt_i values. A line connecting these two points approximately lines up with the source azimuth. Given the density of fiber sensors at the surface, the uncertainty of estimating the source azimuth using this method can be smaller than 1°. The modeling results demonstrate that the normals of a bisector of two-color circles point to the seismic source azimuth. Our results show that this fiber geometry is a robust approach to quickly and accurately determining the azimuths of seismic sources. Note that the DAS cross-array results in large uncertainties for estimating smaller azimuths close to the cross axes and needs virtual fiber sensors along the radial directions. We solve these issues using the DAS concentric circle subarray.

We use the next two numerical modeling examples to illustrate in detail our method for estimating the source azimuth and the angle of the incident seismic wave using a DAS CCC array. Figure 5.8(a) is a map view showing locations of a DAS CCC array and a seismic source at coordinates (4 km, 4 km, −7 km) with a hypocentral distance of 9 km, an azimuth of 45°, and an incident angle of 38.9°. We plot negative (red) and positive (blue) DTTs in Figure 5.8(b). We observe that the normal of the red–blue boundary points to the seismic source.

We search for a pair of two fiber sensors (red and blue with negative and positive values) having the minimum and maximum differential times in each ring of the concentric circle DAS subarray (Figure 5.8(b)). We use straight lines connecting all pairs of two-color sensors with negative (red) and positive (blue) dt_i values to derive an estimate of the source azimuth to be 44.5° ± 1.8°. Using the data of the outer rings only (ring number > 11), we obtain a better estimated azimuth of 44.9° ± 0.3° because there are fewer fiber sensors in the inner rings. Alternatively, we can use all pairs of those red–blue sensors on all rings to fit a straight line to obtain an estimate of the source azimuth of 44.8° ± 0.2°. When we fit a straight line with only the red–blue sensor pairs on the outer rings, from 12 to 25 rings, the estimated source azimuth is 45.0° ± 0.1°. We can also calculate the average value and one standard deviation of the normals of isolines in the cross-subarray to estimate the source azimuth to be 45.0° ± 0.0° (Figure 5.8(c)) because the source is symmetrical to both axes.

When the source azimuth of about 45° is determined, we construct an incidence plane with red fiber sensors along both radial and vertical directions on this incidence plane with a seismic source (Figure 5.8(d)) and generate isolines of DTTs, dt_i, (Figure 5.8(e)) to estimate the incidence angle. From an average normal of the isolines shown in Figure 5.8(e), we obtain an estimate of the incident angle of 38.8° ± 0.3°.

Figure 5.8(f) shows a seismic source at coordinates (1 km, 1.75 km, −6 km) with a magenta dot and a DAS CCC array at (0, 0) with a plot of two-color circles for the DTT. The hypocentral distance from the source to the center of the DAS CCC array is 6.33 km, with a source azimuth of 29.7° and an incident angle of 18.6°. We identify each pair of fiber sensors having the minimum (red) and maximum (blue) dt_i values for each ring of the DAS CCC array and plot the red–blue pairs with purple circles (Figure 5.8(g)). At each ring, seismic wave energy arrives at the red fiber sensor first with a minimum dt_i value but arrives at the blue fiber sensor last with a maximum dt_i value. We use each pair of the two-color fiber sensors to obtain an estimate of the source azimuth, resulting in the estimated azimuth being 29.8° ± 1.4° for all rings. We obtain a better estimate for the source azimuth of 29.7° ± 0.3° when we use only the data of the outer rings (ring number > 11) of the DAS array. We also use all the purple points in Figure 5.8(g) to fit a straight line to obtain the source azimuth of 30.0° ± 0.6°. If we use only purple points in outer rings (ring 12 to ring 25) to fit straight lines, we derive the source azimuth of 29.78° ± 0.04°, reducing the uncertainty significantly. Alternatively, the normal vector of DTT isolines from the cross-subarray can also be calculated to estimate the source azimuth of 29.1° ± 1.9° (Figure 5.8(h)), but with relatively larger uncertainty.

When the source azimuth of about 29.8° is determined, we construct an incidence plane with red fiber sensors along both radial and vertical directions (Figure 5.8(i)). We generate isolines of DTTs, dt_i, (Figure 5.8(j)) to estimate the incidence angle. From an average normal of the isolines shown in Figure 5.8(j), we obtain an estimate of the incident angle of 18.0° ± 0.6°. As the incident angle decreases to a small angle very close to the vertical axis, uncertainty for estimating the incident angle increases. There is room to reduce uncertainties for estimating the incident angle by increasing the sampling rate in the time domain and decreasing the fiber sensor spacing of the DAS recording system.

We conduct uncertainty analysis for estimating the seismic source azimuth using the DAS CCC array. We estimate the azimuth of a seismic source shown in Figure 5.8(a) using an individual pair of two fiber sensors having the minimum and maximum DTTs (dt_i) on each ring of the concentric circle DAS subarray and plot the estimated azimuth values in Figure 5.9(a). An estimated azimuth averaged over 25 red–blue fiber sensor pairs of all rings is 29.8° ± 1.4°, with a relative error of 4.7%. Estimated azimuths with the red–blue sensor pairs of inner rings (ring numbers < 12) of the DAS array are much more scattered than those estimated with the pairs on the outer rings. We exclude those fiber sensor pairs on the inner rings and calculate an average azimuth of 29.7° ± 0.3° with a relative error of only 1%. The mean values and ± 1 standard deviation for all pairs and outer pairs only are shown in Figure 5.9(a). Two mean values are comparable to each other and very close to the true azimuth of 29.74°, but the standard deviation for all pairs is much larger than that of the outer rings because of large estimated errors from the inner rings.

Figure 5.9(b) shows and compares the uncertainties of estimated azimuths with straight-line fitting methods

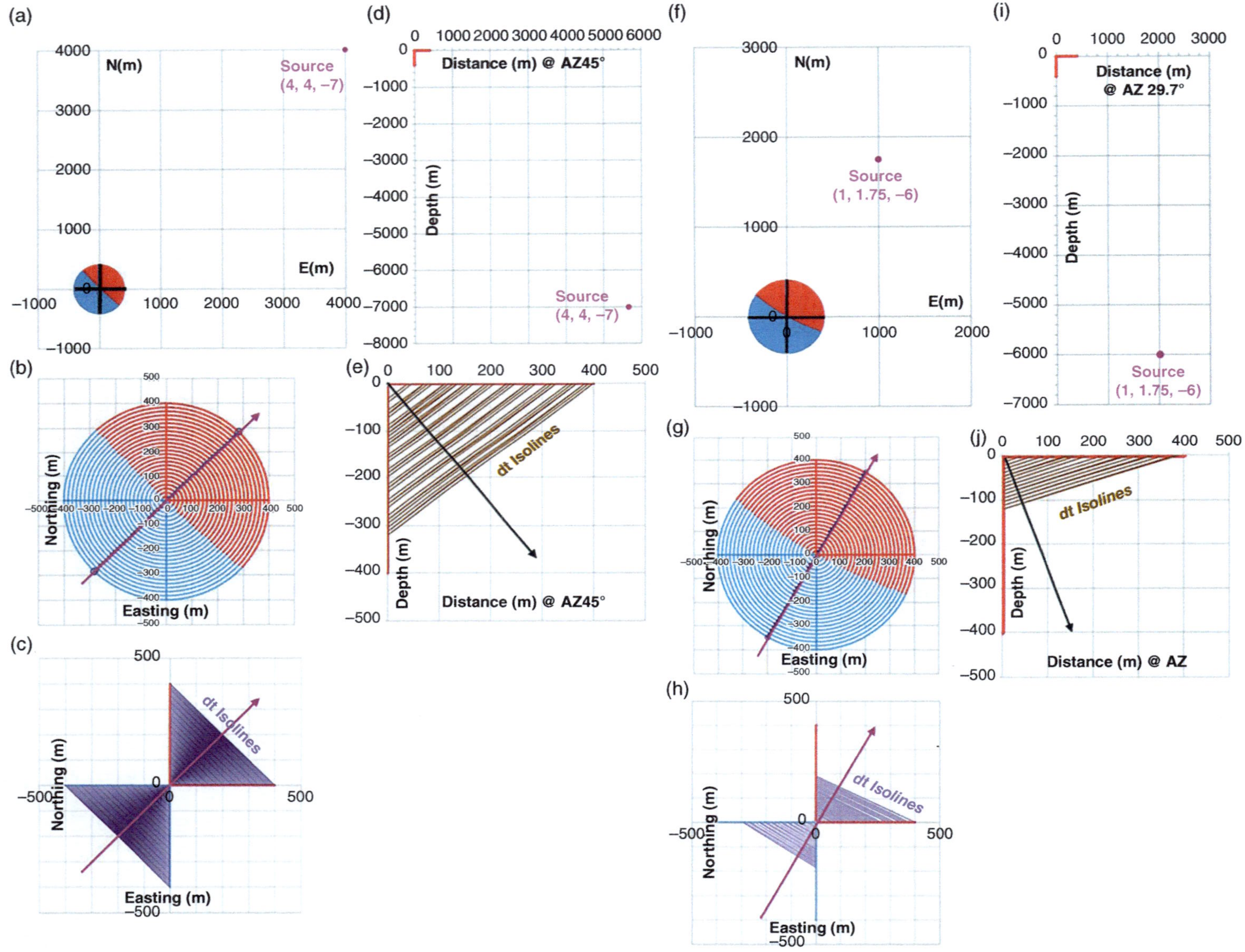

Figure 5.8 (a) A map view showing the DAS CCC array, a seismic source at an azimuth of 45° (magenta dot). The numbers in parentheses are the coordinates of the seismic source in kilometers. (b) A plot of the negative (red) and positive (blue) DDTs (dt_i) of all sensors in the array. The purple dots show the minimum and maximum dt_i values for each ring. We can calculate the source azimuth by fitting a straight line to those dots. (c) Isolines of DTTs from the cross-subarray can be used to estimate the source azimuth too. (d) A cross-section of the incidence plane at an azimuth of 45° showing the locations of a seismic source and DAS sensors along the vertical and radial axes with negative dt_i values (red points). (e) Normal of isolines of negative dt values on the cross-section used to estimate the incident angle of the seismic wave. (f) A map view showing the DAS CCC array, a seismic source at an azimuth of 29.7° (magenta dot). The numbers in parentheses are the coordinates of the seismic source in kilometers. (g) A plot of the negative (red) and positive (blue) dt_i (DTT) sensors. The purple dots show the minimum and maximum dt values for each ring. We can calculate the source azimuth by fitting a straight line to those dots. (h) Isolines of DTTs from the cross-subarray can be used to estimate the source azimuth too. (i) A cross-section of the incidence plan at an azimuth of 29.7° showing the locations of a seismic source and DAS sensors along the vertical and radial axes with negative dt_i values (red points). (j) Normal of isolines of negative dt_i values on the cross-section used to estimate the incident angle of the seismic wave. Li et al. (2023).

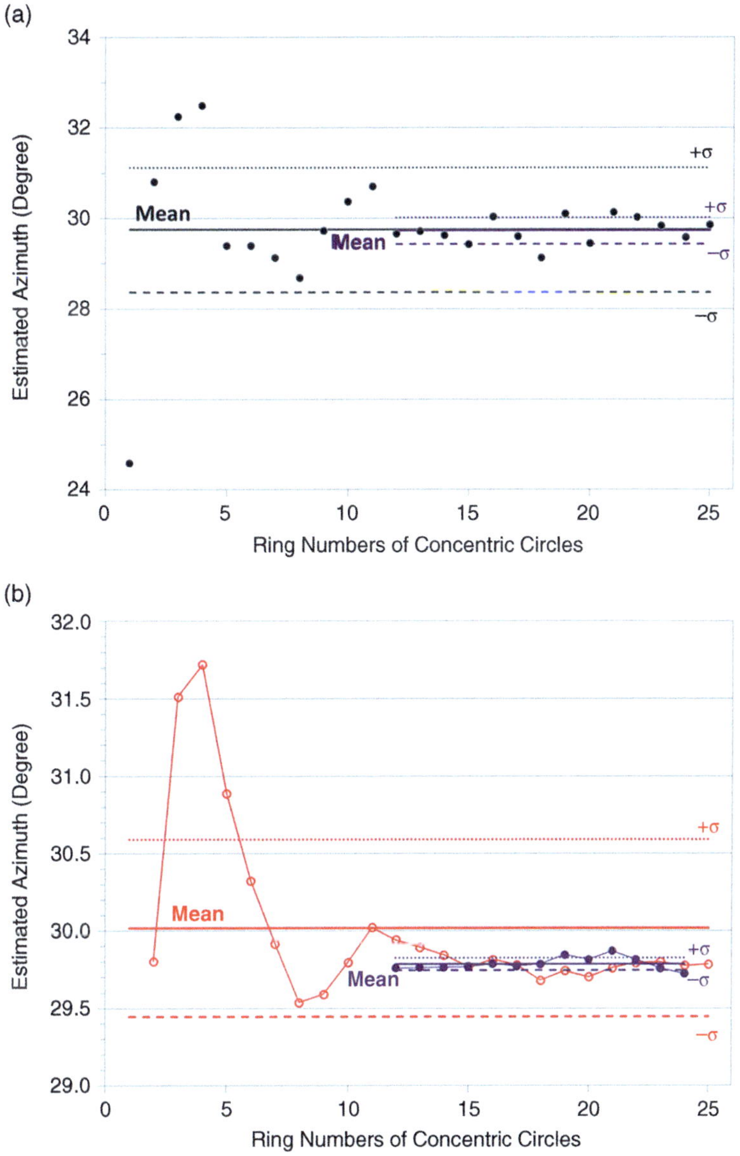

Figure 5.9 Uncertainty assessment of estimating a source azimuth using DTTs (dt_i) of a DAS CCC array. Solid lines are mean values; dotted and dashed lines are ±1 standard deviation (±σ). (a) Black dots are estimated source azimuths using each pair of red–blue fiber sensors with the minimum and maximum dt_i values on each ring. Purple lines are results using pairs of outer rings only. (b) Red curves and lines are the results of fitting a straight line using all pairs of the red–blue sensors on the rings. Purple lines are results obtained using the red–blue fiber sensors on the outer rings only.

using different combinations of all red–blue fiber sensor pairs having the minimum and maximum dt_i values on each ring of the DAS CCC array. The red curve with open circles shows estimated azimuths from our first fitting approach, in which we fit a straight line starting with the inner pairs. Ring number 5 on the horizontal axis of Figure 5.9(b) represents a fitting result using fiber sensor pairs from ring 1 to ring 5, and ring number 25 means that all fiber sensor pairs from ring 1 to ring 25 are used to fit a straight line. Red flat solid, dotted, and dashed lines are the mean value of estimated azimuths and ± 1 standard deviation. The average azimuth estimated with this method is 30.0° ± 0.6°. However, the average azimuth estimated using this fitting method, excluding data from the inner rings (ring number < 12), is 29.79° ± 0.07°, which is very close to the true source azimuth of 29.74°.

We also fit straight lines for estimating source azimuths using the fiber sensor pairs, with the minimum and maximum dt_i values only on the outer rings of the concentric circle DAS subarray. We fit a straight line, starting with the red–blue fiber pairs on the outermost ring of the DAS subarray. For this straight-line fitting practice, ring numbers 12 (or 24) on the horizontal axis of Figure 5.9(b) represent fitting a straight line using the fiber sensor pairs on rings 12 (or 24) to 25. Estimated azimuths with this fitting method (purple dots in Figure 5.9(b)) using only data from the outer rings are very stable, with an estimated azimuth of 29.79° ± 0.04°. Purple flat solid, dotted, and dashed lines are the mean values of estimated azimuths and ± 1 standard deviation. This example confirms that the red–blue fiber sensor pairs of all outer rings are most useful for accurately and reliably estimating the azimuth of a seismic source, as expected.

We find that the cross-subarray has a simple geometry, is easy to deploy, and can get estimated angles in a cost-effective way. However, uncertainties of estimated azimuthal and incidence angles vary with the angles because the number of isolines changes significantly for different seismic sources. Limited numbers of sensors in the cross-subarray alone will not enable effective beamforming and coherent signal stacking to improve the SNR. In contrast, the CCC array overcomes the disadvantages of the cross-array alone. The uncertainties of the estimated angles by the CCC array are smaller and more stable. More than 4,000 fiber sensors enable optimal beamforming and coherent signal stacking to significantly improve the SNR and optimize the DAS array performance.

Our robust method accurately determines the source azimuth and incidence angle with small and acceptable uncertainties, as demonstrated by two numerical modeling examples in Figure 5.8 and an uncertainty assessment for estimating the source azimuth in Figure 5.9. We expect such results because we take advantage of the thousands of fiber sensors in the DAS CCC array, which has a favorable geometrical shape compared with conventional seismic arrays with a limited number of seismometers. We show that we can reliably estimate the source azimuth and the angle of the incident seismic wave by analyzing the DTTs of a DAS CCC array. We will focus on our second task, which is to address the issue of how to correct DAS amplitudes distorted by the fiber directivity effects using the reliably estimated source azimuth and incidence angles.

5.4. CORRECTING DAS AMPLITUDES DISTORTED BY FIBER DIRECTIVITY

It is well known that a fiber DAS system records seismic signals by detecting strain (or strain rate) along the axial length of the fiber. The theoretical DAS sensitivity of an incident P wave varies with the incidence angle, θ, as a function of $\cos^2\theta$. When a P wave is incident along the fiber length ($\theta = 0°$), the DAS sensitivity is 1. However, at an incidence angle of $\theta = 90°$, the DAS sensitivity is 0. This phenomenon is known as fiber response directivity, and it distorts recorded DAS amplitudes. Because a source azimuth and incidence angle can be determined by the DAS CCC array, it provides an opportunity for us to correct the effects of fiber directivity and recover "true" amplitudes of the incidence vector wave field, such as those recorded by a 3C geophone, for the entire fiber DAS array.

We generate synthetic signal amplitudes with a minimum phase wavelet from a seismic source shown in Figure 5.8(a), recorded by a 3C geophone at the center ($x = 0$, $y = 0$) of the DAS CCC array. Figures 5.10(a) and 5.10(b) show the horizontal and vertical hodograms (or particle motion plots) of the signal recorded by the 3C geophone. The 3C geophone hodograms indicate that the seismic source has an azimuth of 29.7° and an incidence angle of 18.6°. For the given seismic source position, we calculate azimuths and incidence angles for all fiber sensors of the DAS fiber cross-subarray and use $\cos^2\theta$ of the two angles to generate distorted DAS amplitudes caused by the fiber directivity effect at fiber sensors on the cross DAS subarray. We use the synthetic geophone amplitude as a reference for the true amplitude of wave fields.

We can use DAS amplitudes of any symmetric fiber sensor pair on the two orthogonal axes, such as [($x = 0$ m, $y = 40$ m), ($x = 40$ m, $y = 0$ m)] in a map view or [($x = 40$ m, $z = 0$ m), ($x = 0$ m, $z = -40$ m)] in the incidence plane to generate horizontal and vertical hodograms. Figures 5.10(c) and 5.10(d) show the DAS amplitude hodograms in a map view and in the incidence plane constructed using the amplitudes of 10 symmetric pairs of fiber sensors on the axes, compared with those of a 3C geophone at the center of the DAS CCC array. We observe that the DAS amplitude hodograms are significantly different from those of the 3C geophone, demonstrating the severe DAS amplitude distortion effects caused by fiber directivity.

We use a source azimuth of 29.8° and an incidence angle of 18° to make $\cos^2\theta$ corrections to the DAS amplitudes. Figures 5.10(e) and 5.10(f) depict horizontal and vertical hodograms created with corrected DAS amplitudes and the two estimated angles to each sensor on the axes. Figure 5.10(e) shows horizontal hodograms generated with the exact angle $\cos^2\theta$-corrected DAS amplitudes using 10 fiber sensor pairs at different locations on the axes that are perfectly matched with that of the 3C geophone at the center ($x = 0$, $y = 0$) of the DAS CCC array. Figure 5.10(f) shows vertical hodograms plotted with the exact angle-corrected DAS amplitudes on the incidence plane that match each other perfectly. However, they have a systematic error of 1.2° compared with the

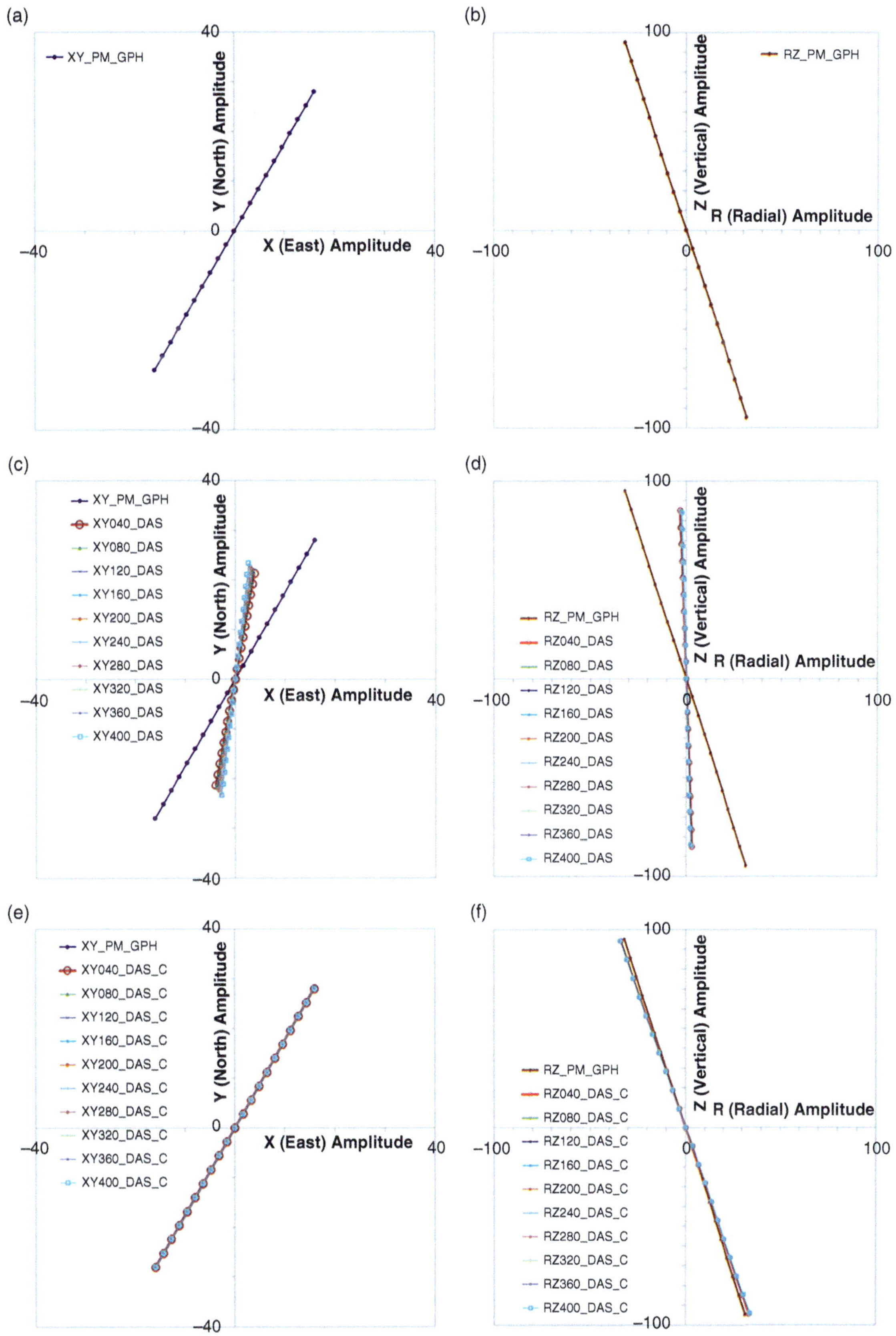

Figure 5.10 (a) Horizontal (purple line) and (b) vertical (brown line) hodograms (particle motion plots) using synthetic amplitudes of a 3-C geophone at the center (x = 0, y = 0) of the fiber DAS array for a seismic source shown in Figure 5.10 with an azimuth of 29.7° and an incidence angle of 18.6°. (c) Horizontal and (d) vertical hodograms, constructed using synthetic DAS amplitudes for the same seismic source, significantly deviate from those of the 3C geophone because DAS amplitudes are severely distorted by the fiber directivity. (e) Horizontal and (f) vertical hodograms, generated with the DAS amplitudes corrected for the fiber directivity effect, perfectly match those of the 3C geophone at the DAS array center (x = 0, y = 0). Li et al. (2023).

incident angle of the 3C geophone (Figure 5.10(b)). We attribute this systematic error to the uncertainty of our estimated incidence angle from DAS array analysis. We use an incident angle of 18° rather than 18.6° in our DAS amplitude $\cos^2\theta$ corrections.

We demonstrate that analyzing the DTTs of a DAS CCC array can accurately estimate the source azimuth and incidence angle with a homogenous velocity model. We can use the resulting source azimuth and incident angle to recover "true" amplitudes by applying $\cos^2\theta$ corrections and obtain hodograms using the symmetric fiber sensor pairs on the axes of the DAS cross-subarray with the corrected DAS amplitudes. We determine the source azimuth and the angle of the incident seismic wave using the horizontal and vertical hodograms and compare them with those estimated from analyzing DAS array DTTs. We can validate our robust array methods by deploying 3C geophones colocated with the fiber sensors on a real DAS CCC array. The example demonstrates that DAS hodograms with corrected amplitudes can potentially be utilized as 3C DAS data for further seismological studies, including seismic event location and shear-wave splitting analysis. We also note that amplitude corrections of fiber directivity responses ($\sin 2\theta$ rather than $\cos^2\theta$) to S waves should be treated differently.

5.5. SYNTHETIC WAVEFORM MODELING WITH TWO VELOCITY MODELS

We have demonstrated some encouraging results from our DTT modeling of the DAS CCC array. However, the constant velocity model we used for DTT modeling is too simplified to represent a more realistic earth velocity model. The first critical question is: are the modeling results still valid for a more realistic P-wave velocity model? The second challenge issue is: how do distorted DAS amplitudes of seismic waves recorded by the DAS CCC array affect the accuracy of first arrival picks and errors of estimated source azimuth and incidence angles? Thus, synthetic waveform modeling of the DAS CCC array using a finite-difference method will help to address those issues.

Figure 5.11 shows two P-wave velocity models: a constant P-wave velocity of 2,000 m/s and a layered velocity model. We generate synthetic waveforms of acoustic responses to the seismic source in Figure 5.8(a) for all fiber sensors of the DAS CCC array using these velocity models (Figure 5.11) and an improved rotated staggered-grid finite-difference method with fourth-order temporal accuracy (Gao & Huang, 2017). The recordings of acoustic waveforms should be similar to those obtained by hydrophones at the same positions of the fiber sensors in the DAS array.

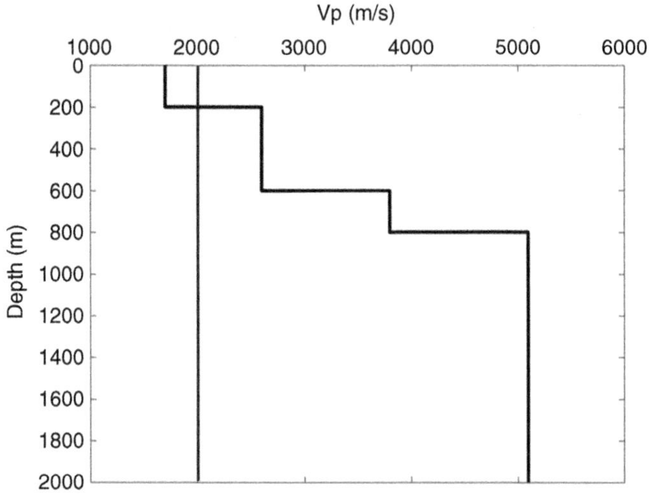

Figure 5.11 Two P-wave velocity models: a constant velocity model of Vp = 2,000 m/s and a layered velocity model, used for waveform modeling with a finite-difference method.

We show the resulting synthetic acoustic waveforms and traveltimes (red line) of the fiber sensors on each of 25 concentric rings and along the west–east, south–north, and vertical axes, derived for the constant velocity model and the layered velocity model in Figures 5.12(a) and 5.12(b), respectively. We observe relatively simple acoustic waveforms for the constant velocity model (Figure 5.12(a)) but more complex acoustic waveforms for the layered model (Figure 5.12(a)), as expected. However, no matter which velocity model is used, two synthetic acoustic waveforms display a common feature. For each ring of the DAS CCC fiber array, especially the outer rings, fiber sensors with the fastest and slowest traveltimes (marked by magenta arrows) indicate the azimuth of the seismic source. We calculate DTTs for both velocity models and plot two-color DTT distributions in Figures 5.12(c) and 5.12(d) for comparison. The source azimuth estimated from both two-color DTT plots is almost identical. These examples of the modeling of synthetic waveforms and traveltimes illustrate that our previous conclusions based on the modeling of DTTs only with a constant velocity model are still valid for a general layered velocity model.

We multiply $\cos^2\theta$ by the synthetic acoustic waveforms to add the fiber directivity effect and create synthetic waveforms with DAS amplitudes that should be recorded at all fiber sensors (channels) on the 25 concentric circles and along the west–east, south–north, and vertical axes of the DAS CCC array. We plot synthetic waveforms with DAS amplitudes and modeled traveltimes (red lines) as a function of channel number for a constant velocity model in Figure 5.12(e) and for a layered velocity model in Figure 5.12(f), respectively. In addition to those

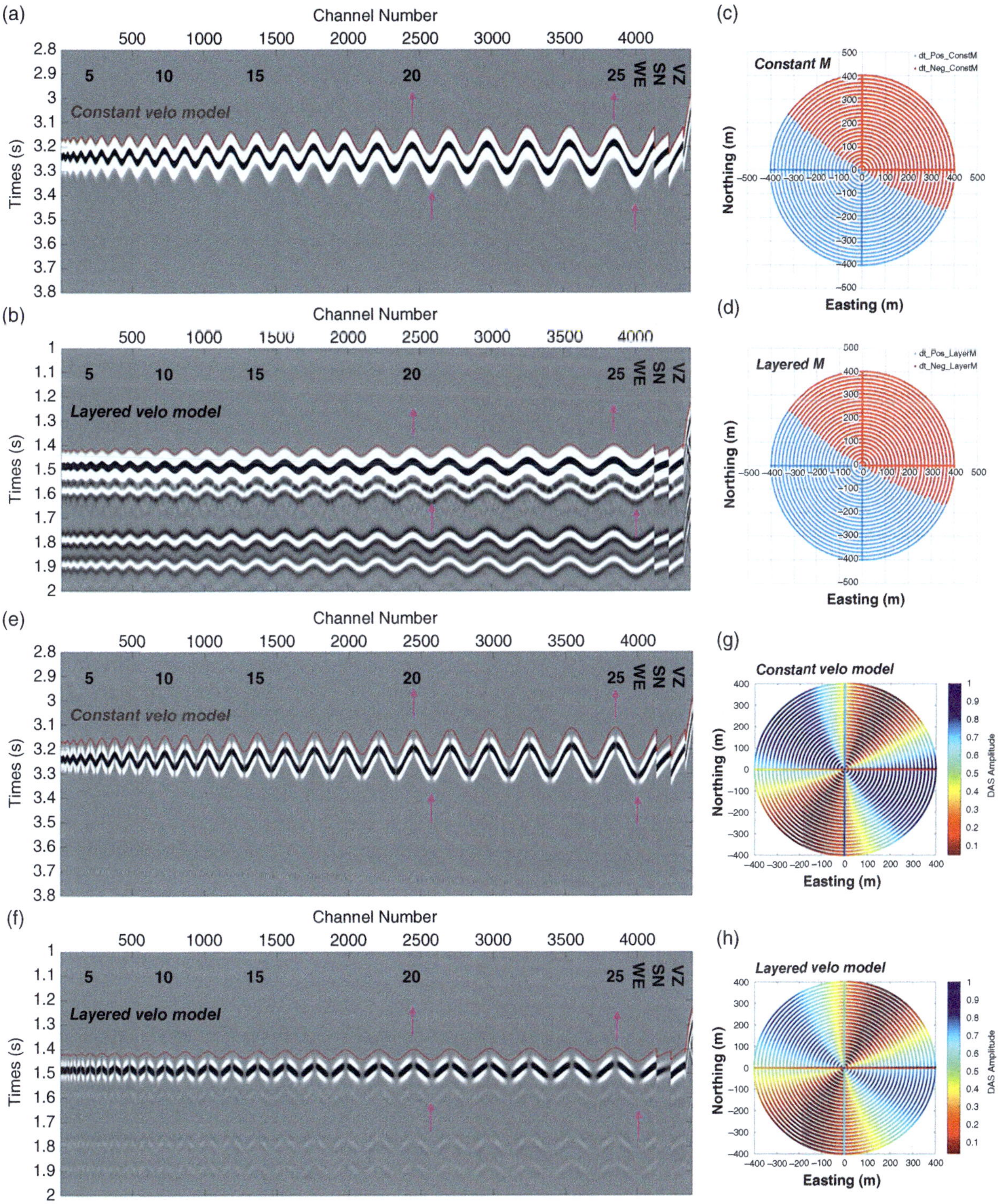

Figure 5.12 Synthetic waveforms of acoustic responses to a seismic source shown in Figure 5.8(a) for all fiber channels (sensors) at 25 concentric rings and along the west–east, south–north, and vertical axes, calculated using (a) the constant velocity model and (b) the layered velocity model. Synthetic first arrival times are indicated by red lines, and the concentric rings are marked with black numbers. Magenta arrows indicate fiber sensors of the fastest and slowest traveltimes of each ring that can be used to estimate the source azimuth. (c) and (d) are two-color plots of DTT distributions with negative (red) and positive (blue) DTT values generated using both the constant and the layered velocity models. Normals of both red-colored half-circles point to the seismic source. Synthetic waveforms of the DAS responses with fiber directivity to a seismic source in Figure 5.8(a), for all fiber channels (sensors) at 25 concentric rings and along the west–east, south–north, and vertical axes, calculated using (e) the constant velocity model and (f) the layered velocity model. Synthetic first arrival times are indicated by red lines, and the concentric rings are marked with black numbers. Magenta arrows indicate fiber sensors of the fastest and slowest traveltimes of each ring that can be used to estimate the source azimuth. Distributions of the synthetic amplitudes of P-wave first arrivals are shown in (g) for a constant velocity model and in (h) for a layered velocity model. Fiber sensors with minimum DAS amplitudes of first P-wave arrivals are along the source azimuth, whereas those with maximum amplitudes of first P-wave arrivals are in a direction perpendicular to the source azimuth. Channel 1 starts from the inner circle for our modeling.

common features shown in Figures 5.12(a) and 5.12(b) used for estimating a source azimuth, we note that synthetic DAS waveforms display periodical variations of DAS amplitudes with the minimum DAS amplitudes along the radial direction toward (or away from) the seismic source.

The theoretical DAS sensitivity of an incident plane seismic P wave with an incidence angle θ of 90° will be zero. However, axial strain (or strain rate) for a perpendicular incident spherical or cylindrical wave is small but not zero, which is proved from theoretical modeling by Martin (2018) and is supported by DAS directivity measurements (Correa et al., 2017; Willis et al., 2021; Wu et al., 2017). The first arrival times of P waves with nonzero DAS amplitudes along the radial direction may be measurable, but the measurement will suffer larger pick errors due to the poor SNR. We can estimate the first arrival times for those radial sensors by fitting the trend of the first arrivals (Figures 5.12(e) and 5.12(f)).

We utilize the DAS amplitudes of P-wave first arrivals from all fiber sensors to generate DAS amplitude distribution plots for a constant velocity model (Figure 5.12(g)) and a layered velocity model (Figure 5.12(h)). The DAS amplitude distribution plots of P-wave first arrivals for both velocity models indicate that fiber sensors with the lowest first arrival amplitudes point to the seismic source, and fiber sensors with the highest first arrival amplitudes are perpendicular to it. In addition to the distributions of DTT (or moveout trends of traveltimes) from the DAS CCC array, DAS amplitude distribution plots of the first arrivals can also be used to estimate the seismic source azimuth.

The theoretical fiber angular response to incidence S waves is proportional to sin2θ rather than $\cos^2\theta$ (Bakku, 2015; Wu et al., 2017). Therefore, DAS amplitude distributions of incident S waves from an earthquake source at the concentric circle subarray should have the maximum (minimum) amplitudes along the directions of 45° (90°) symmetric to the source azimuth direction. Waveform modeling of DAS amplitudes and traveltimes of S waves from seismic sources involves calculating elastic responses with more computation power, and it is beyond the scope of our current study. However, it is an interesting and important topic for the 3C DAS array investigation to follow in the near future.

Calculated with the constant and layered velocity models shown in Figure 5.11, enlarged P-wave synthetic waveforms of DAS responses to a seismic source (Figure 5.8(a)) and traveltimes (red lines) for all fiber sensors (channels) at ring 25 and along the west–east, south–north, and vertical axes are shown in Figures 5.13(a) and 5.13(b), respectively. We observe that the DAS amplitudes of P waves for the 25th ring of the CCC array decrease around fiber sensors with the fastest and slowest traveltimes, implying the seismic source azimuth, too. However, we find that the dimmed DAS amplitudes will result in large uncertainties for the first arrival picks. We may fit trends of the first arrival times using those of nearby fiber sensors at the concentric rings to get an estimate of the traveltimes.

Alternatively, traveltimes of real fiber sensors on the west–east, and south–north axes of the cross-subarray can be used as traveltimes of virtual sources that are projected in the radial direction. The traveltimes (and/or DTT) of fiber sensors (real or virtual) along the radial direction and vertical axis are important to obtain a reliable estimate of the incidence angle on the vertical incident plane. Slope variations of the first arrival times for fiber sensors on the vertical axis for the layered velocity model indicate changes in incidence angles at different depths, whereas an unchanged slope of the first arrival times for fiber sensors on the vertical axis calculated using the constant velocity model shows a consistent incidence angle. The DAS amplitudes of P-wave first arrivals on the three orthogonal axes without periodic dimmed amplitudes are suitable to generate horizontal and vertical hodograms after correcting for fiber directivity effects. The concentric circle and cross subarrays of the DAS CCC array are complemented by each other for 3C DAS array data analyses.

Using the constant and layered velocity models, we integrate traveltimes, DTTs, and DAS amplitudes of P-wave first arrivals to create comprehensive graphs versus fiber channels (sensors) of the DAS CCC array. The results are shown in Figures 5.14(a) and 5.14(b), respectively. Utilizing all measurable parameters from the 3C DAS array, the distinctive distribution patterns of the comprehensive plots serve as a visualization analysis tool for the 3C DAS array data to quickly estimate the source azimuth and incidence angle. This favorable 3D distribution of fiber sensors enables it to flexibly process and rearrange 3C DAS array data. It may serve as the DAS radar so that radar antenna theory and phase-controlled processing methods can be easily implemented for seismic studies using the DAS radar. Various DAS array datasets shown in Figures 5.14(a) and 5.14(b) may provide suitable seeds for machine learning processing of 3C DAS array data.

5.6. DISCUSSION

We design the 3C DAS arrays consisting of a 2D DAS fiber sensor array at the surface, either a cross array or a CCC array, and a fiber cable in an underground vertical well as the third component to record a vector field of seismic waves.

For modeling DTTs, we only need to assign coordinates for each of the fiber sensors in the fiber DAS arrays.

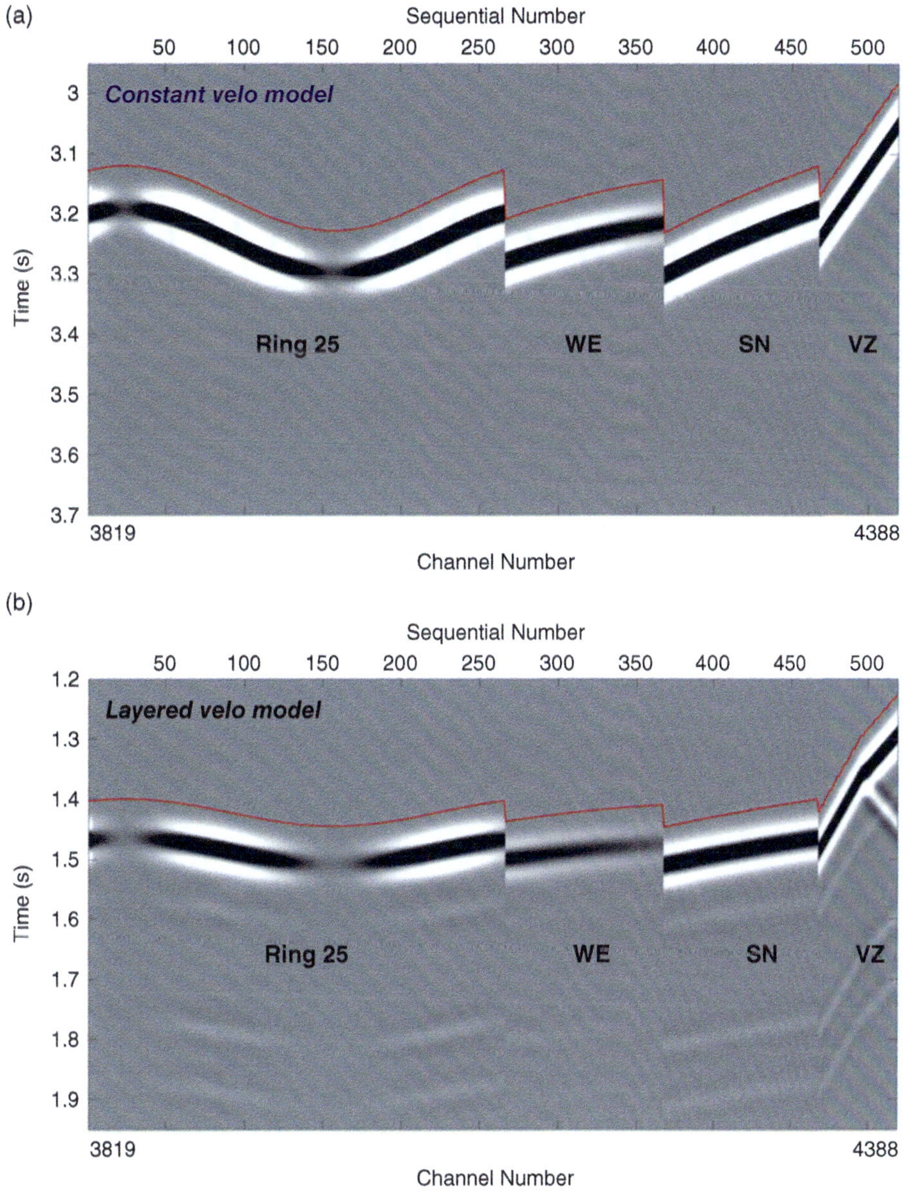

Figure 5.13 Enlarged synthetic waveforms of DAS responses with fiber directivity to a seismic source in Figure 5.10(a) for all fiber channels (sensors) at ring 25 and along the west–east, south–north, and vertical axes, calculated using (a) the constant velocity model and (b) the layered velocity model and a finite-difference method. Note variations in DAS amplitudes along the 25th ring.

However, we require layout schematics to build a real fiber sensor DAS array using an entire fiber cable of about 37 km long to ensure including some redundant fiber sensors. Figure 5.15 depicts a conceptual diagram showing how to deploy a DAS CCC array using a continuous fiber cable. One end of the fiber cable connects with an IU at the north, and the fiber DAS array starts at arrow 1, and then the fiber cable goes into the vertical wellbore at X 2 and returns to the surface at solid circle 3. The cable goes to the west (arrow 4) and takes a U-turn back to the east (arrows 5 and 6) and a U-turn again to the west (arrow 7). When the cable reaches the DAS array center (x = 0, y = 0), it turns to the south (arrow 8) and backs to the north (arrow 9). The cable starts the first circle (arrow 11) at arrow 10 until reaching the end of the last ring (17) at square 18.

The installed fiber cable should have single-mode (SM) common and engineering fibers, at least two for each type, to ensure redundancy of fibers inside the cable and to compare and evaluate the performance of both common

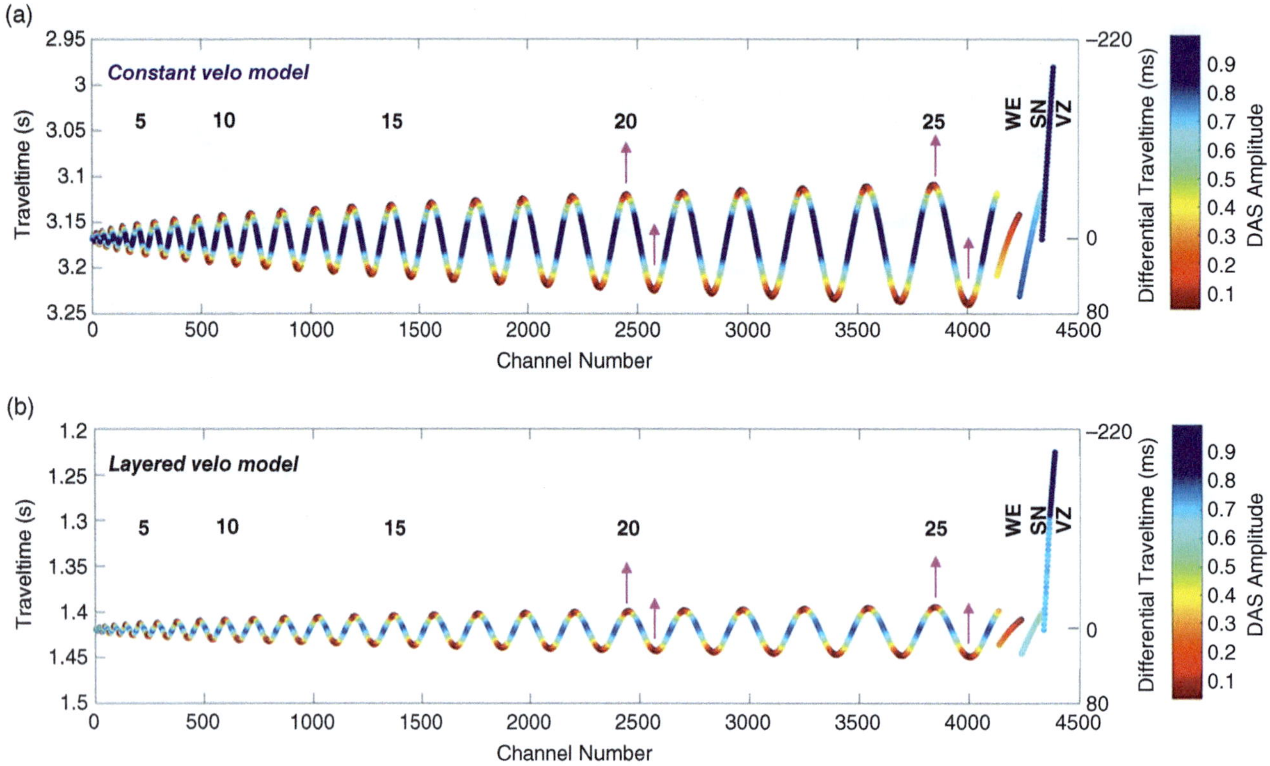

Figure 5.14 Modeling of traveltimes, different traveltimes, and synthetic DAS amplitudes of first arrivals are plotted against all fiber channels of the DAS CCC array for (a) the constant velocity model and (b) the layered velocity model. Comprehensive analysis of traveltimes, DTTs, and amplitudes of DAS data acquired from the CCC fiber array enables us to estimate the source azimuth and incidence angle.

and engineering fibers. In boreholes, DAS monitoring of microseismic events, a sensor spacing of 1 m, a gauge length of 10 m, and a time sampling rate of 0.5 ms are typically used. In our modeling of the DTTs of the DAS arrays, we adopt a sensor spacing of 8–10 m, a gauge length of 8–10 m, and a 0.5 ms time sampling rate. For such a DAS array, it is expected to produce SEG-Y files of 32 TB per day for continuous recording. Thus, preprocessing and trigger algorithms are required to store the data for seismic events only. Nevertheless, in a real DAS fiber array, most IUs can adjust all these parameters based on the specific purposes and objectives of the DAS arrays. The radius of each concentric circle, the number of rings, the total fiber cable length of the fiber DAS array, sensor spacing, and gauge length are all adjustable and subject to trade-offs to satisfy different seismic monitoring purposes.

This kind of installation of a fiber cable in Figure 5.15 ensures the fiber sensors repeat five times at the DAS array center (x = 0, y = 0) with double fiber sensors along the axes of the north–south, west–east, and up–down directions. Redundant colocated fiber sensors along a fiber cable running in opposite directions not only provide redundant first arrival picks of incoming seismic waves at the same positions but also use more fiber sensors to time-shift and stack for increasing SNR.

Beamforming methods for DAS arrays with irregular geometry shapes have been evaluated by van den Ende and Ampuero (2021) and Ichinose et al. (2022) using DAS waveforms of earthquakes recorded by a DAS network at Brady Hot Springs, Nevada (Wang et al., 2018). An example in Figures 5.1(a) and 5.1(b) also shows the power of a beamforming method to increase SNR significantly. Therefore, beamforming techniques should work here for the designed 3C DAS CCC array. Waveform cross-correlations for coherent waveform beamforming may reveal if there are potential differences in the stacking of fiber sensors in the inline or cross-line directions with different fiber directivity effects.

Given more than 4,000 fiber sensors in the DAS CCC array, it enables us to employ beamforming techniques on the 3C DAS array to significantly increase SNR by a factor of the square root of N, where N is the number of coherent signals stacked, and thus partially resolve a major issue of low detectability for small-magnitude seismic events because of the low SNR for non-engineering

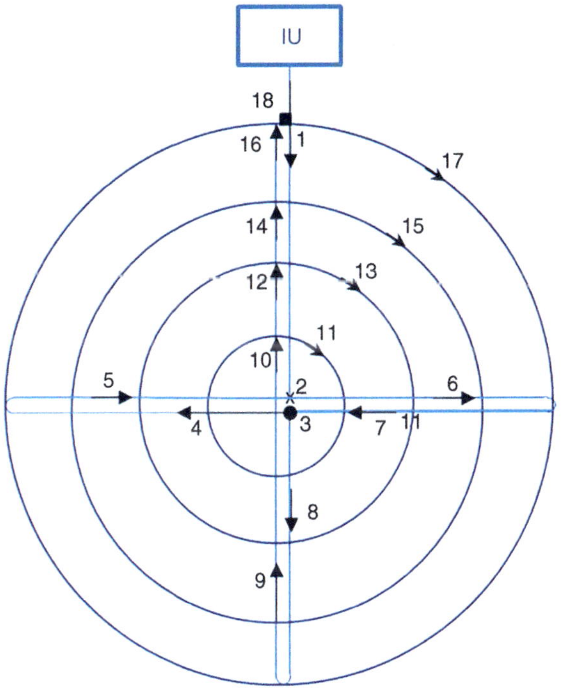

Figure 5.15 A conceptual diagram showing how to deploy a DAS CCC array using a continuous fiber cable. The fiber cable passes through the array center (x = 0, y = 0) five times and twice along the axes of the cross to increase redundancy at key locations of fiber sensors. IU is a DAS interrogator unit. The black x indicates the fiber cable getting into the vertical well, and the black solid circle represents the fiber cable being out of the vertical well. Numbered arrows illustrate the installation path of the fiber cable.

DAS fibers. We can use the frequency-wavenumber (f-k) analysis to estimate the slowness vector (Näsholm et al., 2022) for the DAS CCC array with regular geometry. An array transfer function can enable us to evaluate the sensitivity and resolution of a 3C fiber sensor DAS array for seismic signals with different frequency contents and slowness, providing some clues for improving our design of 3C fiber DAS arrays at different scales.

Although our waveform and travel-time modeling of the DAS CCC array provide promising theoretical results, we face some technical challenges in the implementation of a real DAS array. As with any high-quality seismic arrays, we need to build the DAS array in geological formations as homogeneous as possible. However, inevitable localized media abnormalities may cause local variations in traveltimes and DAS amplitudes through the amplifying and attenuation effects. The fiber sensors of the DAS CCC array with those local variations should be detectable using calibrated active seismic sources around and on the array and excluded from further analysis and processing. The calibrating active sources also play important roles in accurately determining the positions of fiber sensors (channels) in the DAS CCC array. The calibrating active sources with S_H and S_V components can be used to examine if the medium in the vicinity of the 3C DAS CCC array is as isotropic as possible; thus, the 3C DAS array can be potentially used to analyze shear-wave splitting for anisotropy of the medium between the seismic sources and the DAS array.

Fiber coupling conditions with the media will be another great technical challenge for the 3C DAS arrays. Cunningham et al. (2023) reported a 5.16-km-long SM fiber cable deployed in two L shapes in a map view and a vertical plane at the SURF (Sanford Underground Research Facility) in Lead, South Dakota. This quasi-3D DAS array recorded 18 teleseismic events with a magnitude >6 in 2022 and 2023, including Turkey's M = 7.8, 7.5, and 6.7 earthquakes in February 2023. It is necessary to adjust DAS amplitudes for different segments of two L-shape fiber cables before utilizing the 3D DAS recordings as 3C DAS array data because the coupling conditions for each segment of the fiber cable differ.

Similarly, for our designed 3C DAS CCC array, surface portions of the fiber cable will be buried in the trenches, whereas the vertical portion of the cable will be cemented outside the well casing. Surface and vertical portions of the fiber cable apparently have different coupling conditions, resulting in recorded 3C DAS amplitudes not in 1 : 1 : 1 scale. Ambient noises at the surface are typically noisier than those in the underground wellbore. Thus, the SNR for fiber sensors at the surface will be worse than that in the wellbore. Many more surface fiber sensors in our designed DAS CCC array allow us to significantly improve SNR for the surface fiber sensors using beamforming techniques. We should calibrate and correct DAS amplitudes recorded by different portions of the fiber cable using various active sources around the DAS array and passive sources from local, regional, and teleseismic earthquakes and explosions and compare them with recordings from 3C geophones or seismometers.

The 3C DAS CCC array is designed for 3C–3D migration imaging of subsurface structures, updating velocity models using active seismic sources, monitoring microearthquakes induced by fracture stimulations in the vicinity of oil, gas, and enhanced geothermal system (EGS) reservoirs, and detecting time-lapse variations during geologic carbon storage. We can use such a 3C DAS array with the dense distribution of fiber sensors to derive anisotropic parameters of fracture zones, obtain high-precision relative event locations, and retrieve source time functions of microearthquakes. This 1-km-scale fiber sensor DAS array can be extended to a large-scale DAS array to study local, regional, and teleseismic earthquakes and explosions. Our seismological studies can take advantage of numerous seismic phases detectable by the DAS array at regional and teleseismic distances. A smaller 3C DAS array with a diameter of a few hundred meters in marine may serve as a new DAS seismic tool, similar to "streamers," P-cables, or an array of OBSs (ocean bottom seismometers), for deep-water exploration and for monitoring and locating marine mammals and various other marine and sub-marine activities.

5.7. CONCLUSIONS

We have designed two 3C fiber sensor DAS arrays to record a vector field of seismic waves. The two 3D DAS fiber arrays include a cross array and the CCC array at the surface, with part of the fiber cable installed in a vertical well. We model DTT using these 3D fiber sensor DAS arrays and seismic sources at different locations with different azimuths and incident angles. Our DTT modeling results indicate that we can estimate the source azimuth and the angle of the incident seismic wave with small uncertainties (<0.1° for the azimuth and about 0.6° for the incident angle). We recover the true amplitude of the seismic wavefield by correcting raw DAS amplitudes distorted by fiber directivity using two estimated angles. Using the recovered DAS amplitudes, we generate horizontal and vertical hodograms (particle motion plots) that are consistent with the particle motions of a 3C geophone at the DAS CCC array center ($x = 0$, $y = 0$). The source azimuth and the incident angle measured from the DAS particle motions are close to or equal to the two angles estimated from the DTT modeling.

The synthetic DAS amplitudes and traveltimes from the finite-difference waveform modeling demonstrate that we can estimate P-wave source azimuths with distribution patterns of both DAS amplitude and DTT using two different velocity models. Dimmed P-wave DAS amplitudes at some fiber sensors along the radial direction, which are distorted by fiber directivity, actually indicate the source azimuth direction already. However, this results in poor picks of P-wave arrival times for fiber sensors with dimmed amplitudes. We use the trends of the P-wave first arrival time along the fiber rings, or "virtual sensors," projected by the sensors along the axes to estimate arrival times. The symmetric features of DTTs about the sensors with dimmed amplitudes could provide better estimates of the arrival times. Then we can use the DTT estimates on the radial and vertical components to determine the incidence angle.

We can process the 3C DAS array data using techniques for processing conventional seismic array data, such as beamforming, f-k analysis for slowness vectors, coherent signal stacking, and so on. Given the large number of fiber sensors in the 3C DAS array, a stacking technique can significantly increase the SNR of DAS observation systems and reduce the threshold for detecting smaller seismic events. The 3C DAS array technology partially overcomes three major disadvantages of conventional DAS measurements, including single-component, fiber directivity, and low SNR. The 3C DAS array data can enhance applications of the DAS technology in oil and gas, geothermal energy, and geologic carbon storage. The 3C DAS array data aids in imaging subsurface structures with 3C–3D migration, timely monitoring and accurately locating microearthquakes, studying fracture zone and reservoir anisotropy, and monitoring CO_2 injection time-lapse variations. Such 3C DAS arrays can also be developed for large-scale applications, such as those used for monitoring earthquakes and explosions at local, regional, and teleseismic distances.

ACKNOWLEDGMENTS

The authors would like to thank managing editor R. Mellors and two anonymous reviewers for their careful review and constructive comments that led to significant improvement of the manuscript. We also appreciate Professors C. Thurber, F. Wang, and A. Lellouch, who read the original manuscript and provided valuable feedback and helpful discussions. This work was supported by the U.S. Department of Energy (DOE) through the Los Alamos National Laboratory (LANL), which is operated by Triad National Security, LLC for the National Nuclear Security Administration (NNSA) of U.S. DOE under Contract No. 89233218CNA000001, and through the University of Houston under Contract No. DE-EE0008764.

AVAILABILITY STATEMENT

The DAS seismic data used in this chapter are publicly available and can be downloaded from the Utah FORGE Data Distribution at https://constantine.seis.utah.edu/datasets.html.

REFERENCES

Abbott, R., Mellors, R., & Pitarka, A. (2019). *Distributed acoustic sensing observations and modeling of the DAG series of chemical explosions*, Paper presented at the CTBT Science and Technology Conference. https://ctnw.ctbto.org/ctnw/abstract/32643

Ajo-Franklin, J., Lindsey, N., Dou, S., Daley, T. M., Freifeld, B., Martin, E. R., et al. (2015). *A field test of distributed acoustic sensing for ambient noise recording*. Paper presented at the 85th SEG Annual International Meeting. Expanded Abstracts, 2620–2624. doi:10.1190/segam2015-5926936.1

Ajo-Franklin, J. B., Dou, S., Daley, T., Freifeld, B., Robertson, M., Ulrich, C., et al. (2017). *Time-lapse surface wave monitoring of permafrost thaw using distributed acoustic sensing and a permanent automated seismic source*. Paper presented at the 87th SEG Annual International Meeting. Expanded Abstracts, 5223–5227. doi:10.1190/segam2017-17774027.1

Bakku, S. K. (2015). *Fracture characterization from seismic measurements in a borehole* (PhD thesis). Cambridge, MA: MIT.

Castongia, E., Wang, H. F., Lord, N., Fratta, D., Mondanos, M., & Chalari, A. (2017). An experimental investigation of distributed acoustic sensing (DAS) on lake ice. *Journal of Environmental and Engineering Geophysics*, *22*, 167–176. doi:10.2113/JEEG22.2.167

Correa, J., Egorov, A., Tertyshnikov, K., Bona, A., Pevzner, R., Dean, T., et al. (2017). Analysis of signal to noise and directivity characteristics of DAS VSP at near and far offsets — a CO2CRC Otway Project data example. *The Leading Edge*, *36* (12), 994a1–994a7. doi:10.1190/tle36120994a1.1

Cunningham, E., Fratta, D., Lord, N., Chavarria, A., Thurber, C., & Wang, H. (2023). *Observations of regional and teleseismic events at SURF*. Paper presented at the Distributed Acoustic Sensing RCN Workshop, Madison, Wisconsin, USA.

Gao, K., & Huang, L. (2017). An improved rotated staggered-grid finite-difference method with fourth-order temporal accuracy for elastic-wave modeling in anisotropic media. *Journal of Computational Physics*, *350*(1), 361–386. doi:10.1016/j.jcp.2017.08.053

Haldorsen, J. B. U., Christensen, C., Metzbower, D. R., Berzanskis, A., Machnizh, J., Bergery, G., et al. (2021). A three-component optical sensor for borehole-seismic applications. *The Leading Edge*, *40*(6), 447–453. doi:10.1190/tle40060447.1

Hartog, A. H. (2017). *An introduction to distributed optical fibre sensors*. Boca Raton, FL: CRC Press.

Hornman, J. C. (2017). Field trial of seismic recording using distributed acoustic sensing with broadside sensitive fibre-optic cables. *Geophysical Prospecting*, *65*(1), 35–46. doi:10.1111/1365-2478.12358

Ichinose, G. A., Mellors, R. J., Barno, J. G., & Gok, R. (2022). Comparisons between array derived dynamic strain rate (ADDS) and fiber-optic distributed acoustic sensing (DAS) strain rate. *Journal of Geophysical Research: Solid Earth*, *127*, e2022JB025101. doi:10.1029/2022JB025101

Innanen, K. A. (2017). *Parameterization of a helical DAS fibre wound about an arbitrarily curved cable axis*. Paper presented at the 79th EAGE Conference and Exhibition. Expanded Abstracts, We A5 15. doi:10.3997/2214-4609.201701202

Kennett, B. L. N. (2022). The seismic wavefield as seen by distributed acoustic sensing arrays: local, regional and teleseismic sources. *Proceedings of the Royal Society A*, *478*, 47820210812202210812. doi:10.1098/rspa.2021.0812

Kuvshinov, B. N. (2016). Interaction of helically wound fibre-optic cables with plane seismic waves. *Geophysical Prospecting*, *64*(3), 671–688. doi:10.1111/1365-2478.12303

Li, D., Huang, L., Zheng, Y., Li, Y., Wannamaker, P., & Moore, J. (2022a). Feasibility of source-free DAS logging for next-generation borehole imaging. *Scientific Reports*, *12*, 11910. doi:10.1038/s41598-022-16027-3

Li, D., Tian, X., Hu, H., Tang, X.-M., Fang, X., & Zheng, Y. (2020). Gaussian beam imaging of fractures near the wellbore using sonic logging tools after removing dispersive borehole waves. *Geophysics*, 85, D133–D143. doi:10.1190/geo2019-0104.1

Li, Y., Karrenbach, M., & Ajo-Franklin, J. B. (2021). A literature review: distributed acoustic sensing (DAS) geophysical applications over the past 20 years. Chapter 17. In Y. Li, M. Karrenbach, & J. B. Ajo-Franklin (Eds.), *Distributed acoustic sensing in geophysics: Methods and applications*, AGU geophysical monograph 268 (pp. 229–291). The American Geophysical Union and John Wiley & Sons. doi:10.1002/9781119521808.ch17

Li, Y., Karrenbach, M., & Ajo-Franklin, J. (2022b). Using sound and vibration signals to understand the subsurface. *Eos*, *103*. doi:10.1029/2022EO215001

Li, Y., Li, D., Huang, L., & Zheng, Y. (2023). *3D installation of fiber cables for acquiring 3C DAS array data*. Paper presented at the SEG Technical Program. Expanded Abstracts, 1787–1791. doi:10.1190/image2023-w04-01.1

Martin, E. (2018). *Passive imaging and characterization of the subsurface with distributed acoustic sensing* (dissertation). Stanford University.

Mateeva, A., Duan, Y., Kiyashchenko, D., & Lopez, J. (2021). On the surmountable limitations of distributed acoustic sensing (DAS) vertical seismic profiling (VSP) – depth calibration, directionality, and noise. Chapter 6. In Y. Li, M. Karrenbach, & J. B. Ajo-Franklin (Eds.), *Distributed acoustic sensing in geophysics: Methods and applications*, AGU geophysical monograph 268 (pp. 81–91). The American Geophysical Union and John Wiley & Son. doi:10.1002/9781119521808.ch6

Mellors, R. J., Abbott, R., Steedman, D., Podrasky, D., & Pitarka, A. (2021). Modeling subsurface explosions recorded on a distributed fiber optic sensor. *Journal of Geophysical Research: Solid Earth*, *126*, e2021JB022690. doi:10.1029/2021JB022690

Mestayer, J., Cox, B., Wills, P., Kiyashchenko, D., Lopez, J., Costello, M., et al. (2011). *Field trials of distributed acoustic sensing for geophysical monitoring*. Paper presented at the 81st SEG Annual International Meeting. Expanded Abstracts, 4253–4257. doi:10.1190/1.3628095

Moore, J., McLennan, J., Pankow, K., Simmons, S., Podgorney, R., Wannamaker, P., et al. (2020). *The Utah Frontier Observatory for Research in Geothermal Energy (FORGE): a laboratory for characterizing, creating and sustaining enhanced geothermal systems*. Paper presented at the 45th Workshop on Geothermal Reservoir Engineering, Stanford University, CA, USA.

Näsholm, S. P., Iranpour, K., Wuestefeld, A., Dando, B. D. E., Baird, A. F., & Oye, V. (2022). Array signal processing on distributed acoustic sensing data: directivity effects in slowness space. *Journal of Geophysical Research: Solid Earth*, *127*, e2021JB023587. doi:10.1029/2021JB023587

Ning, L. C. I., & Sava, P. (2018). Multicomponent distributed acoustic sensing: concept and theory. *Geophysics*, *83*(2), P1–P8. doi:10.1190/GEO2017-0327.1

Paulsson, B. N. P., Thornburg, J. A., & He, R. (2015). *A fiber-optic borehole seismic vector and acoustic sensor system for geothermal site characterization and monitoring*. Paper presented at the World Geothermal Congress, Melbourne (Australia).

Paulsson, B. N. P., Toko, J. L., Thornburg, J. A., & He, R. (2014). *Fiber optic vector sensors*. Paper presented at the 76th EAGE Conference and Exhibition. Expanded Abstracts, 415–416. doi:10.3997/2214-4609.20140610

Rost, S., & Thomas, C. (2002). Array seismology: methods and applications. *Reviews of Geophysics*, *40*(3), 2.1–2.27. doi:10.1029/2000rg000100

Schweitzer, J., Fyen, J., Mykkeltveit, S., Gibbons, S. J., Pirli, M., Kühn, D., & Kværna, T. (2012). Seismic arrays. In P. Bormann (Ed.), *New manual of seismological observatory practice 2 (NMSOP-2)* (pp. 1–80). Deutsches GeoForschungsZentrum GFZ. doi:10.2312/GFZ.NMSOP-2_ch9

van den Ende, M. P. A., & Ampuero, J.-P. (2021). Evaluating seismic beamforming capabilities of distributed acoustic sensing arrays. *Solid Earth*, *12*, 915–934. doi:10.5194/se-12-915-2021

Walter, F., Gräff, D., Lindner, F., et al. (2020). Distributed acoustic sensing of microseismic sources and wave propagation in glaciated terrain. *Nature Communications*, *11*, 2436. doi:10.1038/s41467-020-15824-6

Wang, H. F., Zeng, X., Miller, D. E., Fratta, D., Feigl, K. L., Thurber, C. H., & Mellors, R. J. (2018). Ground motion response to an ML 4.3 earthquake using co-located distributed acoustic sensing and seismometer arrays. *Geophysical Journal International*, *213*, 2020–2036. doi:10.1093/gji/ggy102

Willis, M. E., Ellmauthaler, A., Wu, X., & LeBlanc, M. J. (2021). Important aspects of acquiring distributed acoustic sensing (DAS) data for geoscientists. Chapter 2. In Y. Li, M. Karrenbach, & J. B. Ajo-Franklin (Eds.), *Distributed acoustic sensing in geophysics: Methods and applications*, AGU geophysical monograph 268 (pp. 33–44). The American Geophysical Union and John Wiley & Sons. doi:10.1002/9781119521808.ch2

Wu, X., Willis, M. E., Palacios, W., Ellmauthaler, A., Barrios, O., Shaw, S., & Quinn, D. (2017). Compressional- and shear-wave studies of distributed acoustic sensing acquired vertical seismic profile data. *The Leading Edge*, *36*(12), 987–993. doi:10.1190/tle36120987.1

Xu, T., Feng, S., Li, F., Ma, L., & Yang, K. (2021). Distributed acoustic sensing system based on phase-generated carrier demodulation algorithm, Chapter 4. In Y. Li, M. Karrenbach, & J. B. Ajo-Franklin (Eds.), *Distributed acoustic sensing in geophysics: Methods and applications*, AGU geophysical monograph 268 (pp. 57–64). The American Geophysical Union and John Wiley & Sons. doi:10.1002/9781119521808.ch4

Zhang L., Jing Q., Bing Han, B., Rao Y., Li, W., Xin, P., et al. (2022), *3C fiber-optic quasi-distributed acoustic sensing (3C-QDAS)*. Paper presented at the Eighth Symposium on Novel Photoelectronic Detection Technology and Applications, 12169AV. doi:10.1117/12.2626266

6

Comparing Distributed Acoustic Sensing and Geophone Vertical Seismic Profiling Imaging: Acquisition Efficiency of Distributed Acoustic Sensing Versus the Multicomponent Advantage of Geophone Data

Jingjing Zong[1], Jizhong Yang[2,3], Yunyue Elita Li[4], Arthur Cheng[5], Xiang Wu[6], and Mark E. Willis[6]

ABSTRACT

Fiber-optic distributed acoustic sensing (DAS) has revolutionized conventional seismic acquisition by supplying high-resolution data, efficient acquisition, and the potential for permanent monitoring. Further investigation into comparing DAS recordings with standard geophone responses when applying the conventional seismic signal processing methods is of interest. Whereas geophones record the particle velocity of the ground motion, DAS interrogator units measure the phase changes in backscattered laser lights at impurities along the fiber, which are related to its average axial strain rate over a gauge length. We use numerical simulations of the SEAM model under a vertical seismic profiling (VSP) acquisition geometry to study their theoretical wavefields and corresponding images. Compared to the single-component particle velocity data, the strain rate field demonstrates directional sensitivity: waves polarized along/orthogonal to the extension of the fiber have the maximum/minimum sensitivity. In a walkaway VSP experiment with sufficient shot coverage, however, the image of the strain rate data is comparable to that of the single-component particle velocity data. Multicomponent geophone records provide a vector field, allowing for a more complete wavefield analysis and reconstruction for an enhanced final image compared to the DAS data. Understanding these differences is key for effective imaging evaluation and survey design.

6.1. INTRODUCTION

The ultimate goal of seismic measurements is to assess ground motion responding to seismic wave propagation (Lindsey et al., 2020). Recent advances in fiber-optic distributed acoustic sensing (DAS) technology have enabled more efficient and intense geophysical surveys including environmental sensing (e.g., Fang et al., 2020; Zhu et al., 2021), geohazard monitoring (e.g., Hadley et al., 2008), as well as land and borehole seismic prospecting (e.g., Daley et al., 2013; Zhan, 2020). Essentially, the DAS interrogator unit (UI) repurposes a fiber cable into a single-component sensor by measuring the phase shifts between consecutive laser light pulses backscattered at numerous reflectors (i.e., impurities or small changes in the refractive index of the glass) distributed along the fiber, as the response of its dynamic strain rate related to the ground deformation (Krohn et al., 2014; Li et al., 2021). On the other hand, the standard geophone

[1]School of Resources and Environment, University of Electronic Science and Technology of China, Chengdu, China
[2]State Key Laboratory of Marine Geology, Tongji University, Shanghai, China
[3]Shanghai Sheshan National Geophysical Observatory Shanghai, China
[4]Department of Earth, Atmospheric, and Planetary Sciences, Purdue University, West Lafayette, Indiana, USA
[5]Earth Environmental Science Programme, The Chinese University of Hong Kong, Hong Kong, China
[6]Halliburton, Houston, Texas, USA

Distributed Acoustic Sensing in Borehole Geophysics, Geophysical Monograph 289, First Edition.
Edited by Yingping Li, Robert Mellors, and Ge Zhan.

DOI: 10.1002/9781394179275.ch06

measures the particle velocity of the ground motion induced by seismic waves with a known instrumental response function. The advantages of DAS include denser spatial sampling at meter scale (Lindsey et al., 2017), a broader frequency band (the longest period field signal observed at 200 s by Yu et al., 2019), extensive receiver array coverage for tens of kilometers (Hartog, 2020), and the potential for permanent installation/monitoring without interfering with other activities in the well. These advantages allow rapid uptake of DAS technology in compensating for or replacing some conventional geophone surveys. For incidence, DAS has been widely deployed in vertical seismic profiling (VSP) surveys for around a decade and has achieved some promising results (e.g., Mateeva et al., 2012, 2017; Zhan et al., 2019).

Continuous efforts are being made to explore the correspondence between geophone particle velocity measurements and DAS strain rate responses so that optimal survey design and seismic signal processing methods can be accommodated. Mateeva et al. (2014) found that the polarity of DAS signals is independent of wave incident direction but has an opposite polarity of the reflected wave to the geophone data. Willis et al. (2016) and Wu et al. (2017) quantitatively studied the kinematics (time) and dynamics (amplitude) of the DAS signal in a VSP configuration and provided an empirical relationship for the DAS first-arrival amplitude as a function of the incident angle of seismic energy. Lindsey et al. (2020) extended the quantitative analysis of DAS amplitude to a broader frequency band and provided an empirical calibration methodology to recover seismic energy. Dean et al. (2017) investigated the directionality of DAS responses on wavenumber and the angle of incidence. Martin et al. (2021) reviewed theoretical geophone and DAS responses for body waves and surface waves and extraction of coherent signals via ambient noise interferometry. As current seismic signal processing technologies are mostly developed for standard geophone data, these studies provide important references for correlating DAS strain rate measurements to geophone particle velocity data.

In this chapter, we aim to understand the first-order ramifications of switching from geophone particle velocity field responses to DAS strain rate field recordings in migrated images. To achieve this, we use numerically simulated data to compare wavefield characteristics under a VSP survey configuration. We extend the analysis by comparing the final images of DAS, the single-component geophone, and the separated P-wavefield from multicomponent geophone data. We employ a subset of the SEAM model featuring a steeply dipping salt flank. Both vertical and deviated wellbores proximal to the salt dome are designed to capture different ranges of incidence angles for salt flank reflection. To focus our comparison on image differences primarily caused by the nature of different sensing quantities, we make two simplifications: (a) the effect of the gauge length, which has been actively studied by many researchers (e.g., Dean et al., 2017; Mizuno et al., 2019), is not considered in this test—only point-wise axial strain rate along the fiber is simulated for comparison; and (b) because it is not the objective of this study to consider the role of noise in their respective images, we simulate idealized noise-free signals for imaging comparison. It should be noted that the signal-to-noise ratio of field DAS recording can be lower than geophone data, depending on the interrogator design (Sidenko et al., 2021) and coupling conditions (Daley et al., 2013).

Part of this work was presented at the 2020 SEG Complex Overburden and Imaging Workshop (Wu et al., 2020).

6.2. NUMERICAL SIMULATED WAVEFIELDS

We simulate the strain rate and particle velocity wavefields for a subset of the SEAM model, which contains 1,600 × 2,000 model grids with a 2.5 m grid interval. The P- and S-wave velocity models are displayed in Figures 6.1a and 6.1b, respectively. Both vertical and deviated wellbores are designed with the wellhead located at $x = 2{,}700\ m$ to accommodate a wider range of incidence wave angles. The receiver lines for both vertical and deviated wells cover the depth from 0 to 1,880 m, as indicated by dotted lines. For the deviated well, the bottom receiver is laterally offset by 1,000 m from the wellhead location. The geophone and DAS configurations have receiver intervals of 20 and 2.5 m, respectively. We place sources (denoted by purple stars) at the top of the model from 0 to 4,000 m in the x direction, with an interval of 20 m. The source function is a Ricker wavelet at a center frequency of 20 Hz. We simulate a total recording time window of 3 s at a sampling rate of 0.25 ms.

Assuming that a vertical fiber along the wellbore is deployed, the axial strain rate $\dot{\varepsilon_{zz}}$ along the vertical direction (Z direction) can be written as

$$\dot{\varepsilon}_{zz} = \frac{\varepsilon_{zz}}{\partial t} = \frac{\partial \dot{u}_z}{\partial z}$$

where ε_{zz} and u_z represent strain and displacement in the Z direction, respectively. The strain rate at a point in the Z direction ($\dot{\varepsilon}_{zz}$) can be regarded as a spatial derivative of the particle velocity field along the Z direction ($\dot{u}_z$).

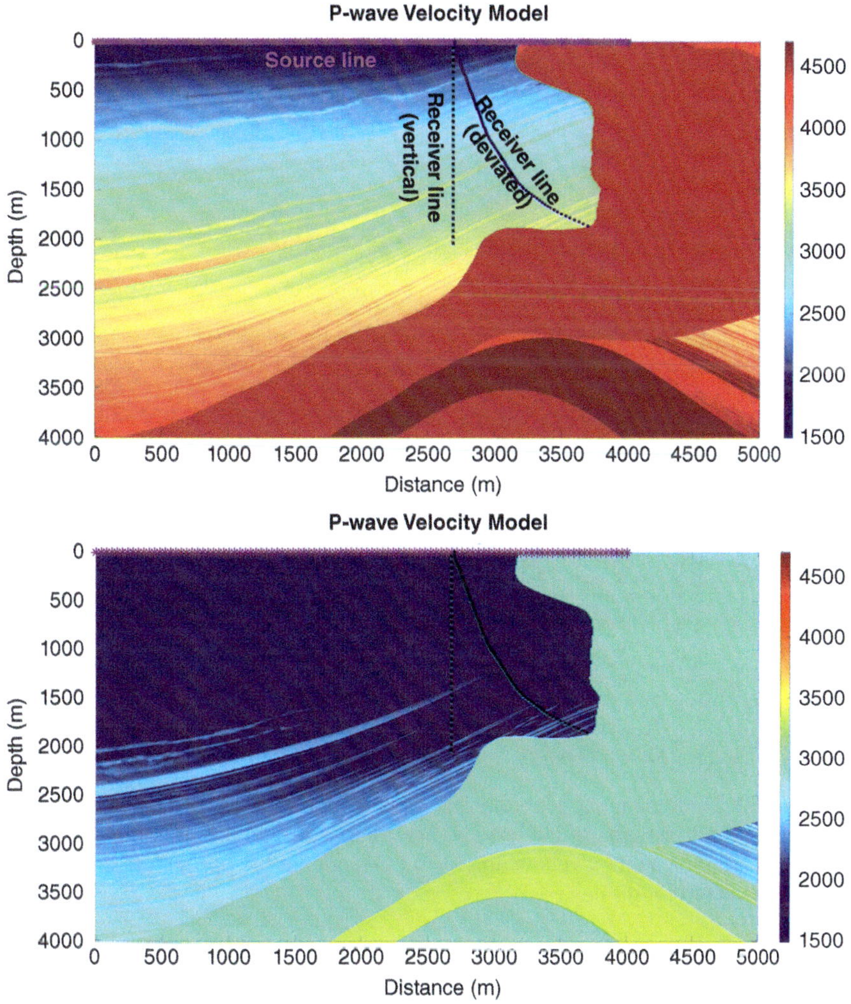

Figure 6.1 The SEAM model is partially displayed and utilized for numerical simulations. Two different receiver array configurations (vertical and deviated, indicated by dotted lines) are used to compare the received wavefields. The source line, represented by purple stars, remains unchanged for both configurations.

6.2.1. Simulated Wavefields in a Vertical Well

The *Z*-component geophone wavefields are simulated and displayed in Figure 6.2. The amplitude of each individual shot gather is normalized in a range between –1 and 1 for display. Four example shots from either side of the wellbore with varying offsets are selected to demonstrate near-, middle-, and far-offset wavefield characteristics. The offset for shots located on the left side of the wellbore is defined to be negative, and for the ones located on the right, it is positive. In the top-right corner of Figure 6.2, the model is displayed with four yellow arrows marking example shot locations at –2,700, –1,200, –300, and 900 m, respectively. The strain fields are displayed in Figure 6.3 for comparison. P-transmissions and major P-reflections are marked by solid and dashed blue lines, respectively. Major converted S-transmissions and S-reflections are marked by solid magenta lines and dashed red lines, respectively. The move-outs (apparent velocity) of different events on both fields are largely identical. However, it is noticed that some portions of the events are weak or even missing in the strain field, as marked by blue boxes. Strain sensing has directional P-wave sensitivity where the wave-particle motion parallel to the fiber has maximized strain sensitivity and particle motion orthogonal to the fiber cannot introduce significant axial strain. Herein, the missing portions are mostly coincident

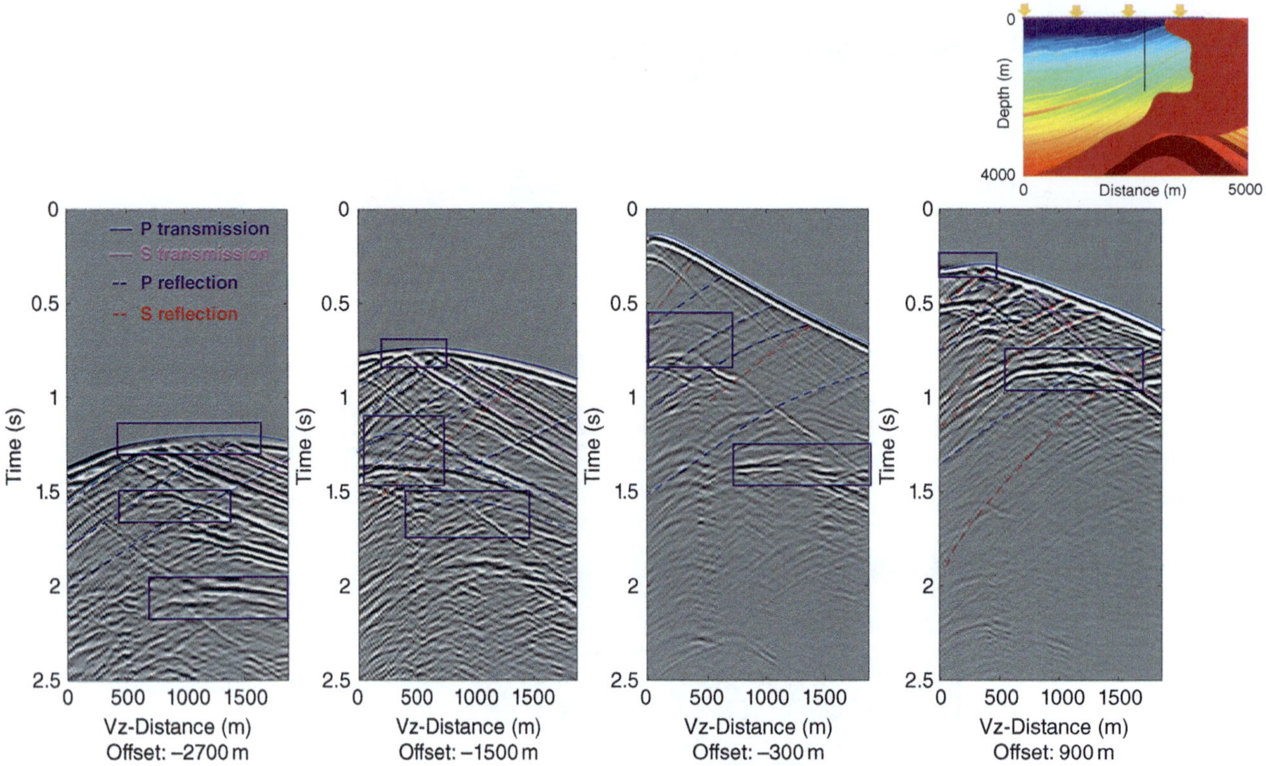

Figure 6.2 The simulated Z-component geophone shot gathers for offsets at –2,700, –1,200, –300, and 900 m, respectively. Blue boxes mark the area of interest and are compared with the strain field the DAS shot gathers. Example shot locations are indicated by yellow arrows in the model in the upper-right corner. The amplitude of each shot gather is normalized for display.

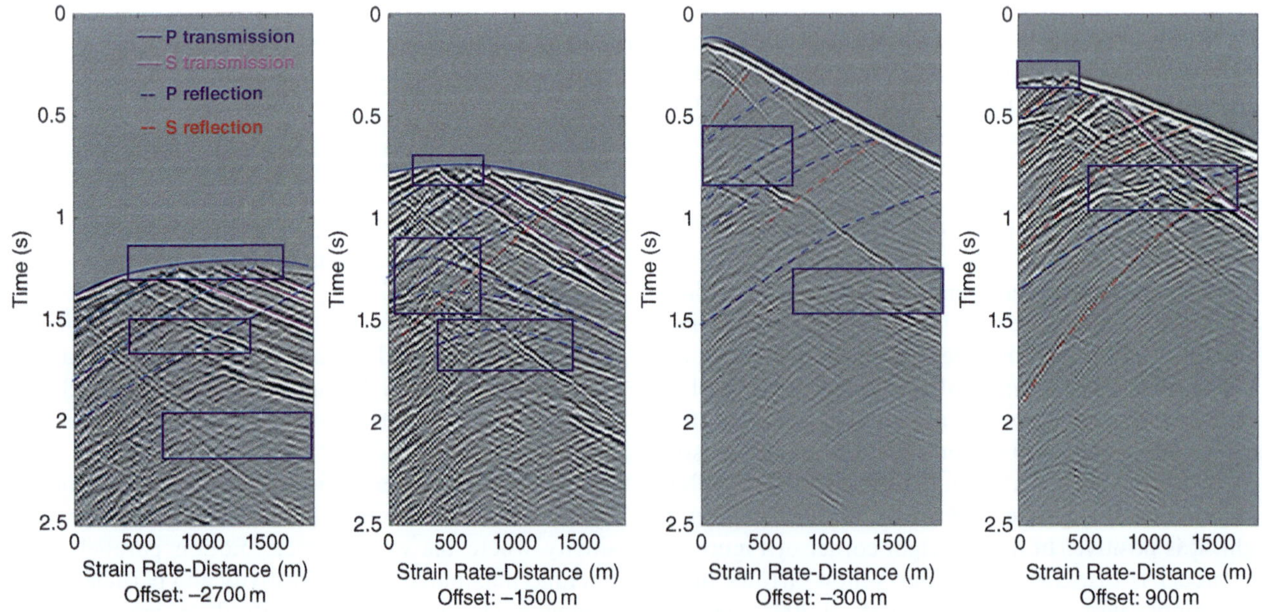

Figure 6.3 The simulated strain field DAS shot gathers for offsets at –2,700, –1,200, –300, and 900 m, respectively. Blue boxes mark the area of interest and are compared with Z-component geophone data. The amplitude of each shot gather is normalized for display.

with the flattened portion of the events that are associated with large P-indecent angles (large apparent velocity). The energy of S-waves, on the other hand, appears to be relatively better captured than the P-waves at far-offset shot locations (offset of –2,700 and –1,200 m) in the strain rate field, as denoted by dashed purple lines. It might result that at large S-incident angles, the particle motion of the S-wave, orthogonal to its propagation direction, is more aligned with the S-wave sensitivity of the fiber.

In this study, we focus on acoustic images. Therefore, we separate P-wavefields for the multicomponent particle velocity data (from both vertical and horizontal components) using Helmholtz decomposition through the divergence operator (Morse & Feshbach, 1954; Huang & Greenhalgh, 2021), as displayed in Figure 6.4. Combining data from multiple components, more complete P-wavefields are revealed compared to those in the Z-component. Red boxes mark some evident P-wave reflections that are weak or even absent in the strain rate field. Considering that the receiver array is somewhat parallel to the steeply dipping salt flank, the marked events are likely to be P-wave reflections at the salt flank with large incident angles for shots to the left of the borehole. The capability of capturing them might directly affect the final images.

During practical field applications, P-wave reflection wavefields of the geophone data are often obtained from a wave-by-wave approach that includes (a) hodogram rotations among three components to analyze and redistribute the primary wavefield energy, which is not viable for DAS data; (b) a ray-tracing-based time-variant orientation to separate the reflection P- and S-wavefields; and (c) deconvolution and a combination of various filtering to further enhance the primary P-wavefield. In terms of optimal wavefield analysis and decomposition, the multicomponent vector geophone field has irreplicable advantages in allowing for a more comprehensive reconstruction of the desired wavefield compared to single-component DAS data.

6.2.2. Simulated Wavefields in a Deviated Well

One of the most-discussed challenges of the chosen SEAM model is to image the steep salt flank. As discussed earlier, the vertical receiver array would tend to pick up P-reflections at the salt flank at large incident angles under the current geometry. To alleviate the effects of directional sensitivity of axial strain rate sensing in illuminating the steep salt flank, we design a deviated wellbore to allow different ranges of incident angles. For the deviated observation system where the coordinate system is counterclockwise rotated by angle θ, the particle velocity field can be represented by multiplying a rotating matrix. In the

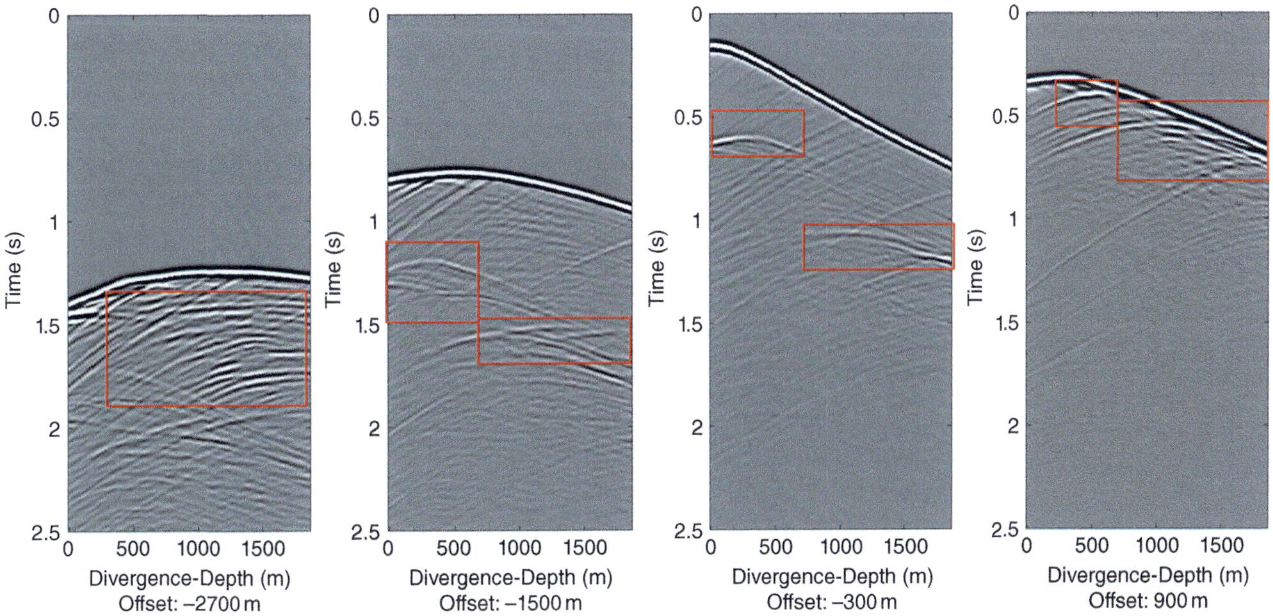

Figure 6.4 Separated P-wave (divergence field) shot gathers obtained from horizontal X- and vertical Z-component particle velocity data for the following offsets: –2,700, –1,200, –300, and 900 m. Red boxes indicate areas where P-wave reflections are weak or even absent in both the Z-component particle velocity data and the strain rate field. The amplitude of each shot gather has been normalized for display.

depth-distance space (Z–X plane), the rotated coordinate system can be written as follows:

$$Rot\,\dot{u} = \begin{bmatrix} \cos\theta & -\sin\theta \\ \sin\theta & \cos\theta \end{bmatrix} \begin{bmatrix} \dot{u}_x \\ \dot{u}_z \end{bmatrix}$$

The strain rate in the rotated system, on the other hand, is a tensor quantity and can be calculated by multiplying a rotation matrix and its transpose on either side of the strain rate tensor in its original coordinate system (Martin et al., 2018):

$$Rot\,\dot{\Sigma} = \begin{bmatrix} \cos\theta & \sin\theta \\ -\sin\theta & \cos\theta \end{bmatrix} \begin{bmatrix} \dot{\varepsilon}_{xx} & \dot{\varepsilon}_{xz} \\ \dot{\varepsilon}_{xz} & \dot{\varepsilon}_{zz} \end{bmatrix} \begin{bmatrix} \cos\theta & -\sin\theta \\ \sin\theta & \cos\theta \end{bmatrix}$$

The simulated Z-component particle velocity field and the strain rate field are displayed in Figures 6.5 and 6.6, respectively. The strain rate field appears to be better comparable with single-component particle velocity in this deviated case than vertical receiver geometry, especially for the near-offset shot. The deficiency of some P-transmission (solid blue lines) and reflection events (dashed blue lines) with large incident angles, although less obvious, can be observed (marked by blue boxes). More coherent and stronger S-wave events, especially at far-offset shots, are observed in the point-wise strain rate data. We further obtain the divergence field and display it in Figure 6.7. Similar to the case in the vertical receiver array acquisition configuration, the divergence field contains a more complete P-wave field combining information from multiple components. Red boxes mark areas where P-wave reflections are weak or even absent in both the single-component particle velocity field and the strain rate field.

6.3. COMPARISON OF GEOPHONE AND DAS IMAGES

The main objective of this study is to compare acoustic images obtained from geophone and DAS data. We image the single-component geophone records, DAS strain rate data, and separated P-wavefields from multicomponent geophone data using the reverse-time-migration (RTM) algorithm. A smoothed version of the true SEAM model velocity is employed and displayed in Figure 6.8. The smoothing windows for vertical and horizontal directions are 500 and 200 m, respectively. Acoustic RTM images for the vertical receiver array using data from P-wavefields separated from multicomponent particle velocity records, single-component particle velocity records, and point-wise strain rate data are shown in Figure 6.9(a)–(c), respectively. For the single-component particle velocity and strain

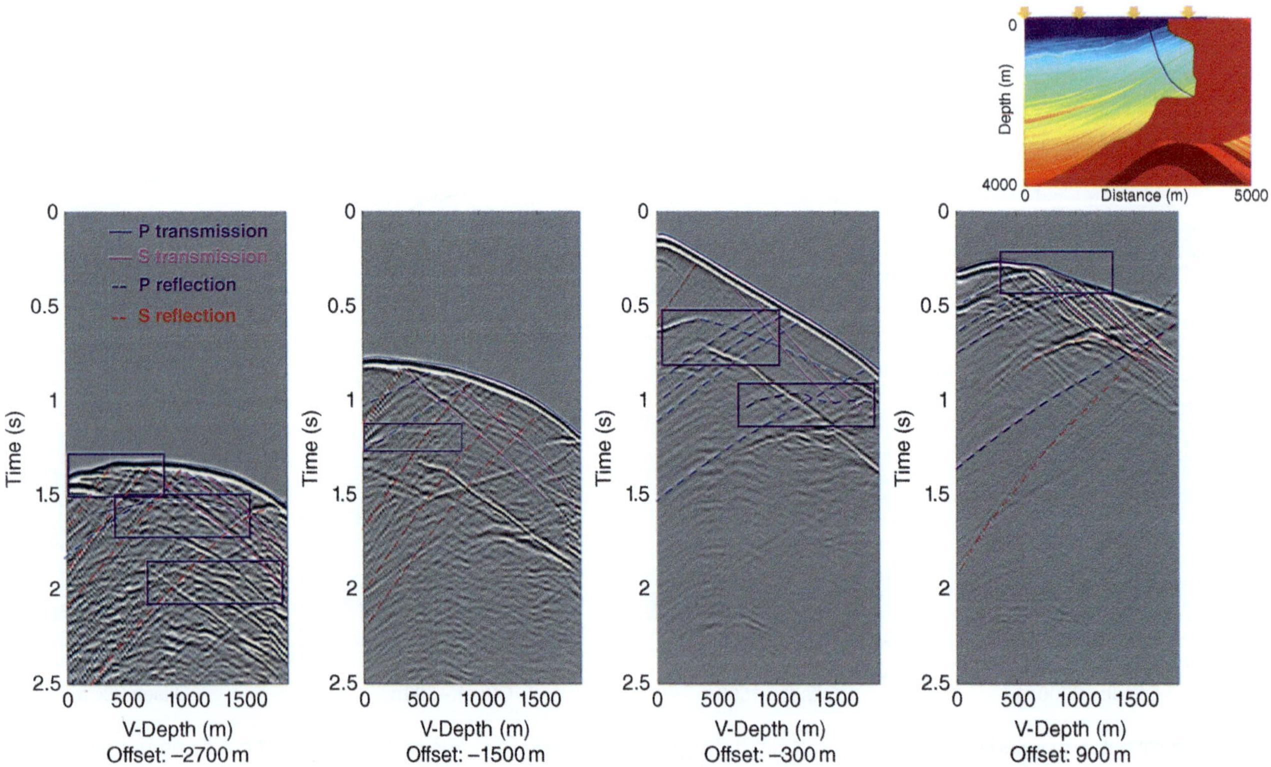

Figure 6.5 Simulated strain field DAS shot gathers for offsets at −2,700, −1,200, −300, and 900 m, respectively. Blue boxes mark the area of interest and are compared with the Z-component geophone data. The amplitude of each shot gather is normalized for display.

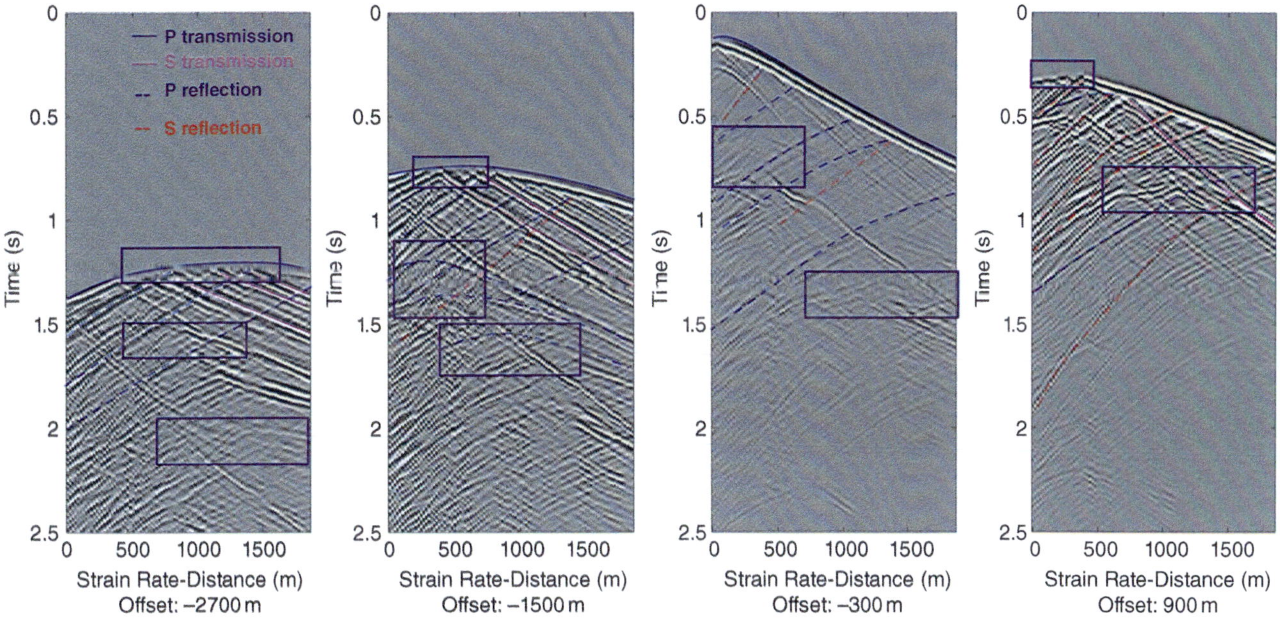

Figure 6.6 Simulated strain field DAS shot gathers for offsets at −2,700, −1,200, −300, and 900 m, respectively. Blue boxes mark the area of interest and are compared with the *Z*-component geophone data. The amplitude of each shot gather is normalized for display.

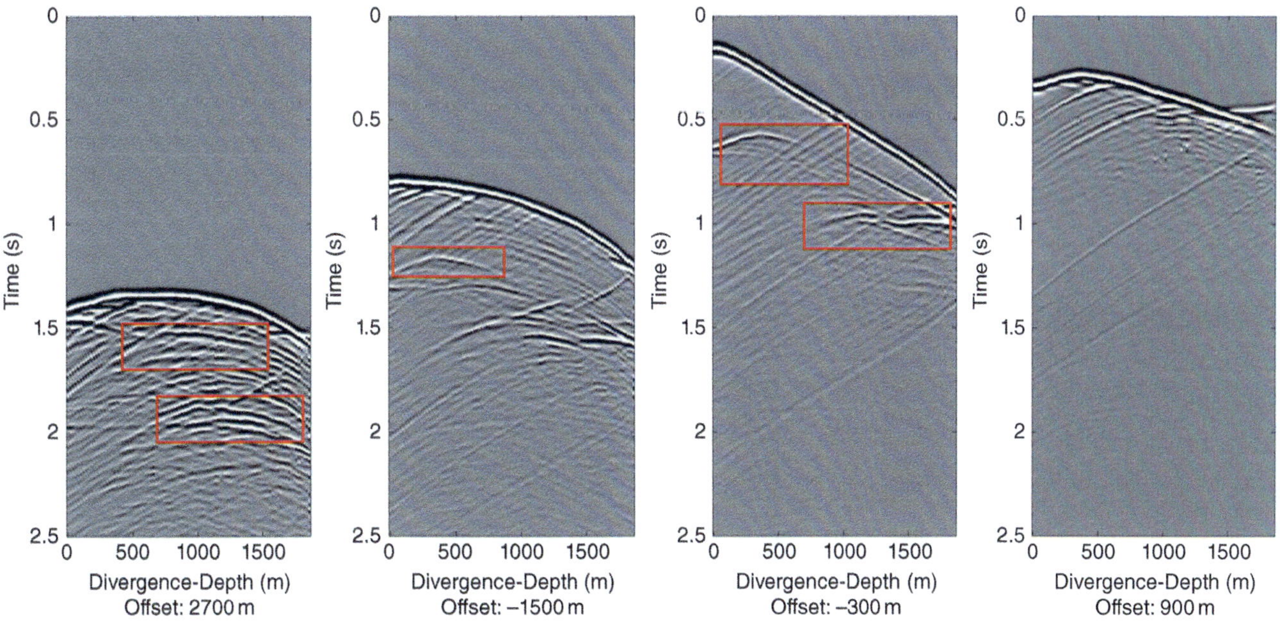

Figure 6.7 Separated P-wave (divergence field) shot gathers obtained from horizontal *X*- and vertical *Z*-component particle velocity data for the following offsets: −2,700, −1,200, −300, and 900 m, respectively. Red boxes mark the areas where P-wave reflections are weak or even absent in both the single-component particle velocity field and the strain rate field. The amplitude of each shot gather is normalized for display.

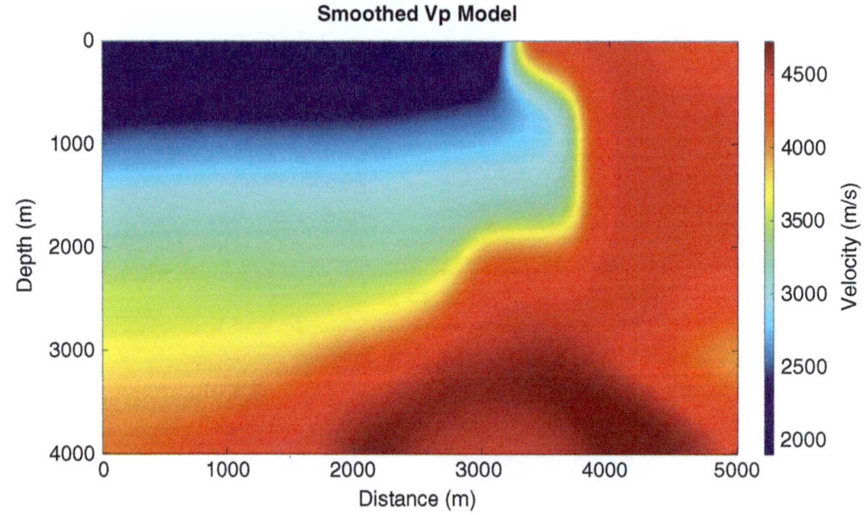

Figure 6.8 The true SEAM model is smoothed for imaging. The smooth windows for vertical and horizontal directions are 500 and 200 m, respectively.

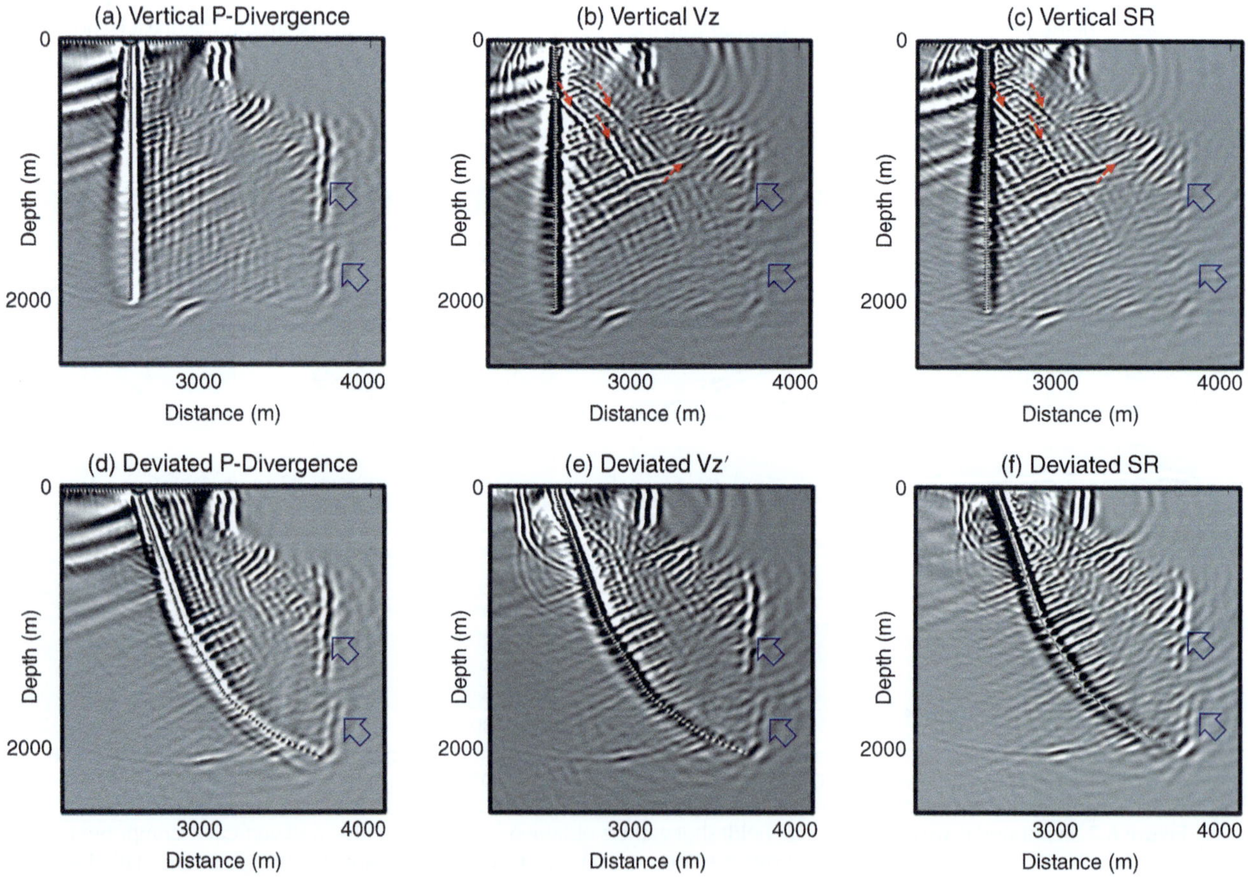

Figure 6.9 RTM images for the vertical receiver array using data from (a) P-wavefield separated from multicomponent particle velocity records, (b) single-component particle velocity records, and (c) point-wise strain rate data, and for the deviated receiver array using data from (d) P-wavefield separated from multicomponent particle velocity records, (e) single-component particle velocity records, and (f) point-wise strain rate data.

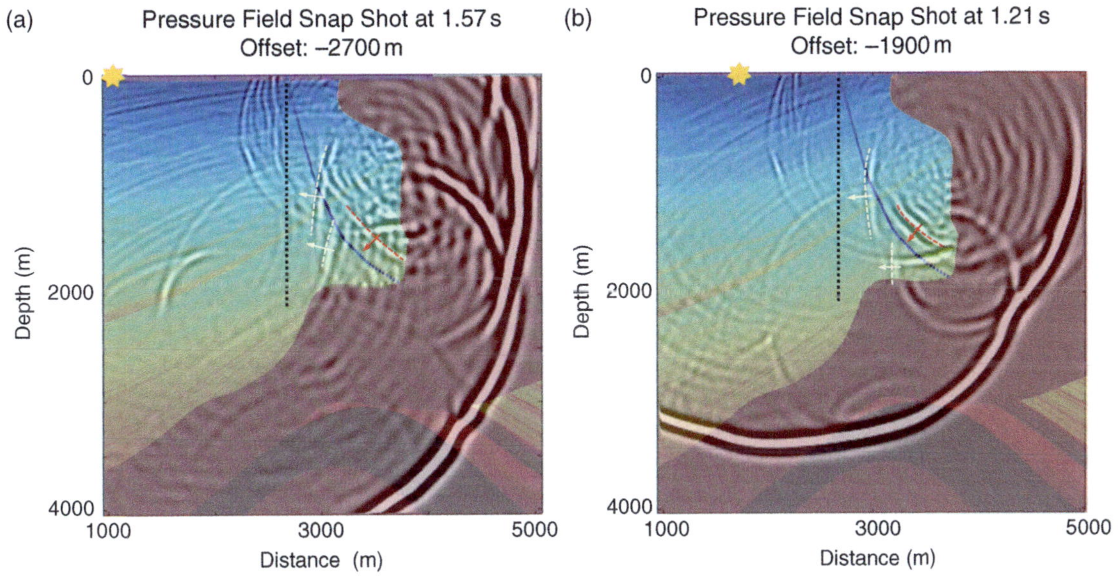

Figure 6.10 Pressure field snapshots plotted over the true model for (a) offset at –2,700 m at time 1.57 s and (b) offset at –1,900. The shot locations are marked by yellow stars. The wavefronts of the reflections at the salt flank and the salt overhang are marked by dashed white lines and dashed red lines, respectively. Schematic wave propagation directions are denoted by white and red arrows.

rate data, we image raw records without separating shear modes, allowing for a comparison of the resultant artifacts. A Laplacian filter was applied to suppress low-wavenumber artifacts caused by direct arrivals and diving waves. The image of the divergence field appears to have the most complete delineation of the steep salt flank (indicated by blue arrows). The images of the single-component particle velocity field and the strain rate field are largely similar. Both images suffer greatly from imaging artifacts. For incidence, the strong events indicated by red arrows are possibly induced by converted S-wave modes. Neither of them has a well-resolved salt flank. The strain rate image has even less focused energy of the salt flank compared with particle velocity. Based on the earlier wavefield analysis, it can be explained by the observation that the strain rate field has less sensitivity to the high-incidence-angle P-wave reflected at the steep salt boundary. Such a situation is relieved significantly under the deviated receiver array configuration. The acoustic RTM images for the vertical receiver array using data from P-wavefield separated from multicomponent particle velocity records, single-component particle velocity records, and point-wise strain rate data are shown in Figure 6.9(d)–(f), respectively. As pointed out by blue arrows, the salt flank is much better resolved in both the single-component particle velocity image (Figure 6.9e) and the strain rate image (Figure 6.9c). In addition, the two images behave similarly. We plot two pressure field snapshots for far-offset shot locations at –2,700 and –1,900 m in Figure 6.10(a), (b), respectively. For illustration purposes, we mark the reflection wavefronts of the steeply dipping flank with dashed white lines. The corresponding wave propagation direction is denoted by white arrows. The inclination of the receiver array enables detection of P-wave reflections at the salt flank with smaller incidence angles than those by the vertical receiver array. This results in better sensitivity to P-reflections and, thus, an improved image of the salt flank. However, none of the tested geometries could resolve the salt overhang. This is because, on the one hand, the P-wave reflection, whose wavefront is marked by a dashed red line, arrives at the deviated receiver at a high incidence angle. On the other hand, it further propagates and reflects at the bottom salt extension before being picked up by the vertical receiver array. As a result, the primary reflections are significantly scattered, and the salt overhang images are unfocused.

6.4. DISCUSSION AND CONCLUSION

We compare the theoretical strain rate responses to single-component and multicomponent particle velocity measurements to evaluate the behavior of analog fiber-optic DAS data in comparison to geophone signals using conventional seismic processing techniques. The numerical simulation in this study serves the primary goal of analyzing the directional sensitivity of different measuring quantities in DAS and geophone responses in their acoustic images. Various instrumental factors (e.g., gauge length, instrumental responses at varying frequencies, and fiber coupling condition), although not considered in this study, play critical roles in the actual field acquisition and are actively studied.

The current analysis, which is conducted under a VSP geometry, suggests that noise-free point-wise strain rate data is insensitive to waves polarized orthogonally to the fiber compared to the single-component particle velocity data. For a 2D walkaway VSP survey with hundreds of shots distributed from near to far-offset locations, however, the final images of the strain rate data and the single-component particle velocity data behave similarly, given the greatly improved angular coverage from densely distributed shots. Considering that the limited length of the geophone receiver array in a conventional VSP acquisition process would require repetitive shots at each shot location to obtain sufficient depth aperture, DAS VSP acquisition offers advantages in both cost efficiency and source wavelet consistency.

Yet compared to multicomponent geophone records, which provide a vector field of wave propagation, the single-component DAS VSP strain rate field captures only a subset of the wavefield information, resulting in a less complete image. A better survey design that honors the directional sensitivity of DAS measurements can help alleviate this issue. In our test, we observe an enhanced salt flank image of the strain rate data in the deviated receiver array arrangement compared to the vertical array. On the other hand, multicomponent recordings allow for wave polarization analysis based on hodogram distribution, which is absent in the single-component DAS VSP data. Hence, in areas with complex subsurface geometries, wavefield analysis can be more challenging for DAS VSP data. Future research on extracting multimode wavefields from single-component DAS data is of interest.

ACKNOWLEDGMENTS

The authors thank Huaigu Tang and Xinding Fang for contributing the original modeling codes and helpful discussions. Jingjing Zong is funded by the National Natural Science Foundation of China (grant nos. 42474171 and 42104130), the China Postdoctoral Science Foundation (grant no. 2021M690536), and the Fundamental Research Funds for the Central Universities. Jizhong Yang is supported by the National Natural Science Foundation of China (grant no. 42004096), Shanghai Sheshan National Geophysical Observatory (No. SSOP202205), and the Fundamental Research Funds for the Central Universities.

REFERENCES

Daley, T. M., Freifeld, B. M., Ajo-Franklin, J., Dou, S., Pevzner, R., Shulakova, V., et al. (2013). Field testing of fiber-optic distributed acoustic sensing (DAS) for subsurface seismic monitoring. *The Leading Edge*, *32*(6), 699–706.

Dean, T., Cuny, T., & Hartog, A. H. (2017). The effect of gauge length on axially incident P-waves measured using fibre optic distributed vibration sensing. *Geophysical Prospecting*, *65*(1), 184–193.

Fang, G., Li, Y. E., Zhao, Y., & Martin, E. R. (2020). Urban near-surface seismic monitoring using distributed acoustic sensing. *Geophysical Research Letters*, *47*(6), e2019GL086115.

Hadley, C., Peters, D., Vaughan, A., & Bean, D. (2008). *Gumusut-Kakap project: geohazard characterization and impact on field development plans.* Paper presented at the International Petroleum Technology Conference, European Association of Geoscientists & Engineers.

Hartog, A. H. (2020). Distributed sensors in the oil and gas industry. In I. del Villar & I. R. Matias (Eds.), *Optical fibre sensors* (pp. 151–191). Hoboken, NJ: John Wiley & Sons.

Huang, X., & Greenhalgh, S. (2021). A finite-difference iterative solver of the Helmholtz equation for frequency-domain seismic wave modeling and full-waveform inversion. *Geophysics*, *86*(2), T107–T116.

Krohn, D. A., MacDougall, T., & Mendez, A. (2014). *Fiber optic sensors: Fundamentals and applications* (pp. 233–310). Bellingham, WA: Spie Press.

Li, Y., Karrenbach, M., & Ajo-Franklin, J. B. (2021). A literature review: distributed acoustic sensing (DAS) geophysical applications over the past 20 years. In Y. Li, et al. (Eds.), *Distributed acoustic sensing in geophysics: Methods and applications* (pp. 229–291). American Geophysical Union.

Lindsey, N. J., Martin, E. R., Dreger, D. S., Freifeld, B., Cole, S., James, S. R., et al. (2017). Fiber-optic network observations of earthquake wavefields. *Geophysical Research Letters*, *44*, 11792–11799.

Lindsey, N. J., Rademacher, H., & Ajo-Franklin, J. B. (2020). On the broadband instrument response of fiber-optic DAS arrays. *Journal of Geophysical Research: Solid Earth*, *125*, e2019JB018145. doi:10.1029/2019JB018145

Martin, E. R., Lindsey, N. J., Ajo-Franklin, J. B., & Biondi, B. L. (2021). Introduction to interferometry of fiber-optic strain measurements. In Y. Li, et al. (Eds.), *Distributed acoustic sensing in geophysics: Methods and applications* (pp. 111–129). American Geophysical Union.

Martín-Martín, A., Orduna-Malea, E., Thelwall, M., & López-Cózar, E. D. (2018). Google Scholar, Web of Science, and Scopus: a systematic comparison of citations in 252 subject categories. *Journal of Informetrics*, *12*(4), 1160–1177.

Mateeva, A., Lopez, J., Chalenski, D., Tatanova, M., Zwartjes, P., Yang, Z., et al. (2017). 4D DAS VSP as a tool for frequent seismic monitoring in deep water. *The Leading Edge*, *36*(12), 995–1000.

Mateeva, A., Lopez, J., Potters, H., Mestayer, J., Cox, B., Kiyashchenko, D., et al. (2014). Distributed acoustic sensing for reservoir monitoring with vertical seismic profiling. *Geophysical Prospecting*, *62*, 679–692.

Mateeva, A., Mestayer, J., Cox, B., Kiyashchenko, D., Wills, P., Lopez, J., et al. (2012). *Advances in distributed acoustic sensing (DAS) for VSP*. Paper presented at the SEG Technical Program. Expanded Abstracts, 1–5.

Mizuno, T., Leaney, S., Calvez, J. L., Naseer, F., & Khaitan, M. L. (2019). *The significance of gauge length in particle velocity estimation from DAS data: VSP and microseismic cases*. Paper presented at the SEG Technical Program. Expanded Abstracts.

Morse, P. M., & Feshbach, H. (1954). Methods of theoretical physics. *American Journal of Physics*, *22*(6), 410–413.

Sidenko, E., Pevzner, R., Bona, A., & Tertyshnikov, K. (2021). *Experimental comparison of directivity patterns of straight and helically wound DAS cables*. Paper presented at the EAGE Annual Conference & Exhibition, Amsterdam.

Willis, M. E., Barfoot, D., Ellmauthaler, A., Wu, X., Barrios, O., Erdemir, C., et al. (2016). Quantitative quality of distributed acoustic sensing vertical seismic profile data. *The Leading Edge*, *35*(7), 605–609.

Wu, X., Willis, M. E., Palacios, W., Ellmauthaler, A., Barrios, O., Shaw, S., & Quinn, D. (2017). Compressional-and shear-wave studies of distributed acoustic sensing acquired vertical seismic profile data. *The Leading Edge*, *36*(12), 987–993.

Wu, X., Willis, M., Zong, J., & Li, Y. E. (2020). *Comparing DAS and geophone VSP imaging – Wider depth aperture and acquisition efficiency of DAS versus the multi-component advantage of geophone data*. Paper presented at the 2020 SEG Complex Overburden and Imaging Workshop, Society of Exploration Geophysicists.

Yu, C., Zhan, Z., Lindsey, N. J., Ajo-Franklin, J. B., & Robertson, M. (2019). The potential of DAS in teleseismic studies: insights from the Goldstone experiment. *Geophysical Research Letters*, *46*(3), 1320–1328.

Zhan, G., Li, Y., Tura, A., Willis, M., & Martin, E. (2019). Introduction to special section: distributed acoustic sensing and its oilfield potential. *Interpretation*, *7*(1).

Zhan, Z. (2020). Distributed acoustic sensing turns fiber-optic cables into sensitive seismic antennas. *Seismological Research Letters*, *91*(1).

Zhu, T., Shen, J., & Martin, E. R. (2021). Sensing Earth and environment dynamics by telecommunication fiber-optic sensors: an urban experiment in Pennsylvania, USA. *Solid Earth*, *12*(1), 219–235.

Part II

Borehole Distributed Acoustic Sensing Acquisition and Processing

7

Smart Distributed Acoustic Sensing Uphole Acquisition System: Bridging the Gap Between Surface Seismic and Borehole Geophysics for Imaging and Monitoring in Complex Near-Surface Environments

Andrey Bakulin[1], Ilya Silvestrov[2], Robert Smith[2], and Pavel Golikov[2]

ABSTRACT

Imaging and monitoring onshore reservoirs or gas storages with reflection seismology is a challenging task, primarily due to near-surface noise, scattering, and seasonal changes. Traditional surface seismic methods often suffer from the dominance of near-surface energy, masking deeper reflections. Meanwhile, multiwell 3D vertical seismic profiling (VSP), although effective, is limited by its dependence on a few deep wells, making it susceptible to variable illumination and flow noise. To address these challenges, we propose a Smart DAS uphole acquisition system. This system integrates multiple shallow vertical arrays (50–500 m) spaced 100–1000 m apart, connected via long distributed acoustic sensing (DAS) cables. This innovative hybrid approach enhances signal-to-noise ratio and data repeatability, free from the issues associated with flow noise and deep well dependencies. Through elastic synthetic and field tests, we demonstrate that vertical-array data can produce high-quality images comparable to those obtained through surface seismic methods. The system's ability to provide superior near-surface characterization and robust reflection imaging makes it highly effective for imaging beneath complex overburdens. Moreover, the Smart DAS uphole system's scalability and improved repeatability offer a cost-effective alternative to traditional buried point-sensor systems, making it suitable for extensive 3D monitoring of onshore reservoirs and gas storages.

7.1. INTRODUCTION: WHY VERTICAL ARRAYS?

Surface seismic (Figure 7.1(a)) allows the densest and most flexible placement of sensors at the surface of the Earth. However, although such acquisition offers the best way to deploy sensors, it suffers two significant limitations:

1. Most recorded energy is associated with near-surface arrivals such as ground roll, guided waves, refracted waves, and so on, and as a result, reflected energy from targets of interest is often hidden behind near-surface noise, requiring dense trace acquisition and sophisticated and massive multichannel processing to uncover them (Meunier, 2011). Consequently, land seismic crews reach hundreds of thousands of channels to address this issue.
2. The repeatability of the data recorded with surface sensors is typically poor (Bakulin et al., 2012; Schissele et al., 2009), making them inefficient for seismic monitoring. Furthermore, even when permanently placed, surface sensors are subject to daily/seasonal changes that are difficult or impossible to remove in processing.

[1] *Bureau of Economic Geology, University of Texas at Austin, Austin, Texas, USA*

[2] *EXPEC Advanced Research Center, Saudi Aramco, Dhahran, Saudi Arabia*

Distributed Acoustic Sensing in Borehole Geophysics, Geophysical Monograph 289, First Edition.
Edited by Yingping Li, Robert Mellors, and Ge Zhan.

DOI: 10.1002/9781394179275.ch07

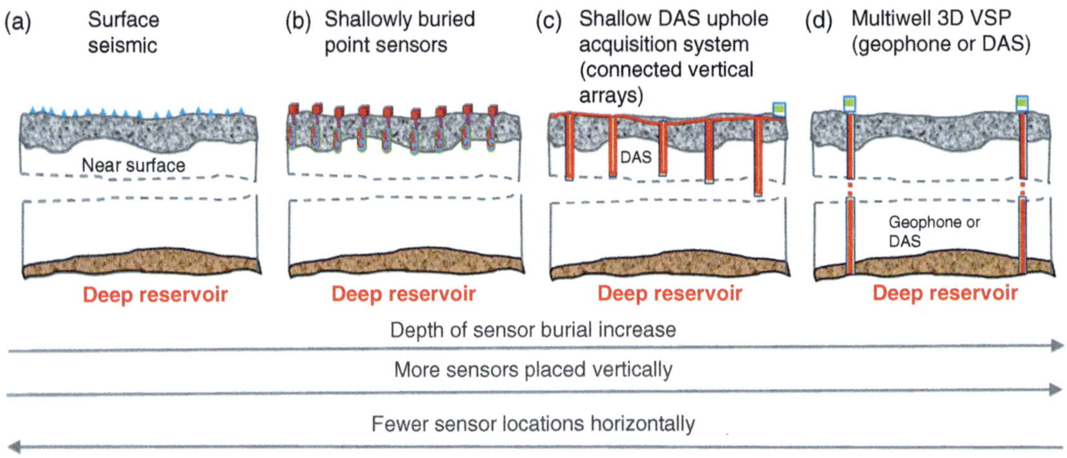

Figure 7.1 Spectrum of seismic acquisition configurations from surface to downhole: (a) surface seismic, (b) shallowly buried point sensors, (c) shallow DAS uphole acquisition system with connected vertical DAS arrays, (d) multiwell 3D VSP using deep wells.

Buried receivers may significantly reduce the amount of recorded near-surface energy (Bakulin et al., 2015, 2017; Ikelle & Wilson, 1999; Meunier et al., 2001). As a result, reflections of interest can often be observed on prestack records (Berron et al., 2015; Meunier et al., 2001). In addition, a better signal-to-noise ratio implies that less sensor density is needed to obtain seismic images (Meunier, 2011).

In addition, the burial of seismic sensors, even at a shallow depth, significantly improves repeatability. Bakulin et al. (2012) compared the repeatability of surface and shallowly buried sensors at various depths. In a 2D configuration, normalized root-mean-square (NRMS) repeatability was 53% for permanent surface sensors, whereas it was reduced to 24% for sensors at 10 m and 17% for geophones at 30 m depth of burial.

Data quality and repeatability improvements due to burial gave rise to the acquisition of shallowly buried point sensors (Figure 7.1(b)), which became popular for monitoring (Bakulin et al., 2012; Jervis et al., 2018; Meunier et al., 2001; Roach et al., 2015; Shulakova et al., 2015). Burial is relatively shallow, ranging from a few meters (Shulakova et al., 2015) to a few tens of meters (Bakulin et al., 2012; Jervis et al., 2018; Roach et al., 2015), and is typically implemented through shallow vertical drilling and connecting point sensors to the surface. A shallowly buried array with a point sensor is also used for microseismic monitoring of gas storage (Barnavol & Payre, 2017) and other passive surveys, such as hydraulic fracture monitoring (Snelling & Taylor, 2013). Improved data quality allows for sparser horizontal sampling, as seen in Figure 7.1(b) when compared to Figure 7.1(a), which represents surface seismic. In acquisition geometry, a survey with shallowly buried point sensors is still closest to the surface seismic because the density of vertical locations remains high, whereas the burial depth stays low.

On the other side of the acquisition spectrum is borehole vertical seismic profiling (VSP), which usually utilizes one or several injector/producer boreholes penetrating the deep target of interest (Figure 7.1(d)). Although geophone acquisition is feasible for 3D VSP, it becomes much less convenient for 4D VSP because permanently placing geophones behind tubing is costly and complex, especially for larger channel counts. Distributed acoustic sensing (DAS) rapidly becomes the preferred acquisition technology for multiwell 3D and 4D VSP because it enables permanent, low-cost, and intervention-less instrumentation of active deep wells (Kiyashchenko et al., 2020; Mateeva et al., 2013).

Between shallowly buried point sensors (Figure 7.1(b)) and multiwell 3D VSP (Figure 7.1(d)), there is a continuum of options where multiple intermediate-depth observation boreholes are instrumented with vertical sensor arrays. Placement in observation boreholes provides all the benefits of buried sensors and eliminates the potentially strong fluid flow noise present in VSP. In addition, shallow holes allow complete flexibility to place them at required locations for imaging and monitoring. This could be important in the case of long horizontal injectors/producers when the actual space to be monitored may be out of reach for DAS VSP using fiber installed in vertical sections. Deeper observation holes permit larger vertical antennas, which allow for greater horizontal separation (compared to shallowly buried point sensors from Figure 7.1(b)). However, it is typically impractical to drill deep observation wells. As an alternative, ultra-shallow boreholes (5–50 m) used to deploy shallowly buried arrays with point sensors (Figure 7.1(b)) can be cost-effectively extended to around 400–500 m. This provides an opportunity to instrument a more considerable vertical extent and increase the hole spacing, thus reducing the surface hole density.

Similar to deep VSP, conventional point sensor arrays are not cost-effective, even for shallow configurations. The Microhole project demonstrated (nonconnected) instrumentation of five microholes with 20-level geophone arrays (Majer, 2007). The achieved channel count and hole density were low and insufficient for at-scale imaging or monitoring. Likewise, Meunier et al. (2001) instrumented several 175 m holes with independent 23-level 4C arrays and used a sparser 400–500 m between antennas targeting reservoir monitoring use cases. We believe DAS instrumentation is the most efficient and fit-for-purpose for implementing vertical arrays (Figure 7.1(c)), offering several advantages. First, it allows an order of magnitude increase in the channel count, with DAS cables offering meter-like sampling. Second, it enables efficient connection of many holes on a single fiber (with connecting cables trenched at the surface) and uses a single recorder for all of them. Bakulin et al. (2017) first introduced this configuration of connected vertical DAS arrays, which they dubbed the Smart DAS uphole acquisition system (Bakulin et al., 2018a). This system is an interconnected network of multiple Smart DAS upholes that can be acquired with a single DAS interrogator.

This chapter analyzes the Smart DAS uphole acquisition system in detail. We begin by briefly describing the practical acquisition aspects critical for implementation. We then dive into a synthetic case study that was used as preparation for the subsequent field tests. We focus on understanding vertical DAS array data and utilizing it for subsurface imaging and monitoring of deep targets. Next, we speculate on the 3D implementation of the Smart DAS uphole acquisition. Finally, we summarize various near-surface characterization techniques enabled by the new acquisition system.

7.2. SEISMIC ACQUISITION ASPECTS

Seismic acquisition with the Smart DAS uphole acquisition system requires efficient drilling and robust installation of fiber-optic cables. Drilling accounts for most of the acquisition cost, making it a critical factor in optimizing the survey design. Although portable rigs for drilling ultra-shallow (0–50 m) and shallow (0–500 m) boreholes are often comparable in cost, targeting 500 m or more depth requires significantly more expensive drilling equipment, such as used for water wells. As a result, shallow drilling (0–500 m) may provide an optimal window for buried seismic acquisition for the following reasons:

- Multiple low-cost portable rigs can drill a grid of wells, enabling sufficient density of shallow holes.
- Shallow holes are decoupled from producers/injectors, avoiding any interference with subsurface activities while allowing flexibility to move the shallow hole locations away from any current or future surface obstructions.
- Shallow survey design (Figure 7.1(c)) remains closer to carpet-like surface seismic (Figure 7.1(a)) than multi-well 3D VSP (Figure 7.1(d)), which has variable illumination and associated artifacts of borehole imaging.
- Increasing the depth of shallow holes allows larger horizontal well spacing with a simple rule of thumb that hole spacing comparable to hole depth could deliver similar subsurface coverage/illumination to surface seismic (Ikelle & Wilson, 1999). This rule applies when channel spacing is similar between seismic and vertical arrays. In addition, denser channel spacing with DAS increases cross-hole spacing proportionately (Ikelle & Wilson, 1999). This could significantly minimize surface imprint and simplify logistics compared to ultra-shallow but dense buried point-sensor surveys (Figure 7.1(b)).

Significant acquisition experience has accumulated in the seismic industry for "uphole surveys" (Cox, 1999), which deliver essential information to build and calibrate near-surface models. The conventional approach involves drilling shallow holes and measuring the traveltime from a surface energy source to a single downhole geophone. The sensor is moved from the bottom to the top (Figure 7.2(a)), with traveltimes recorded at 2–4 m spacing below 50 m and 1–2 m in the ultra-shallow part (0–50 m). Uphole surveys are particularly beneficial in environments with a complex near surface (Cox, 1999), such as in desert environments (Ley et al., 2003). However, conventional uphole surveys often suffer from hole collapses due to unstable or unconsolidated near-surface sediments, resulting in loss of equipment and costly redrill. For this reason, surveys are usually restricted to zero-offset acquisition because of the limited time open boreholes can remain stable for surveying. To overcome some of the limitations of the conventional uphole system, Bakulin et al. (2017, 2018a) introduced the Smart DAS uphole concept (Figure 7.2(c)). As soon as the hole is drilled, the fiber is lowered into place, and the hole is backfilled or cemented. DAS turns the entire vertical length of the uphole into a sensor with meter-like sampling, often with a separate downgoing and upgoing fiber for redundancy. As a result, a zero-offset survey can be acquired with a single shot. Figure 7.2(d) shows a consistent source waveform with depth on Smart DAS uphole data instead of a variable and jumpy waveform obtained with conventional uphole using a single moving geophone and sequential source excitations (Figure 7.2(b)).

Daisy-chaining Smart DAS upholes on a single fiber, we arrive at a Smart DAS uphole acquisition system shown in Figure 7.3. Every shot at the surface is now recorded by all DAS channels located along the fiber cable. They can be divided into downhole "vertical" channels (along the upgoing and downgoing legs of the fiber)

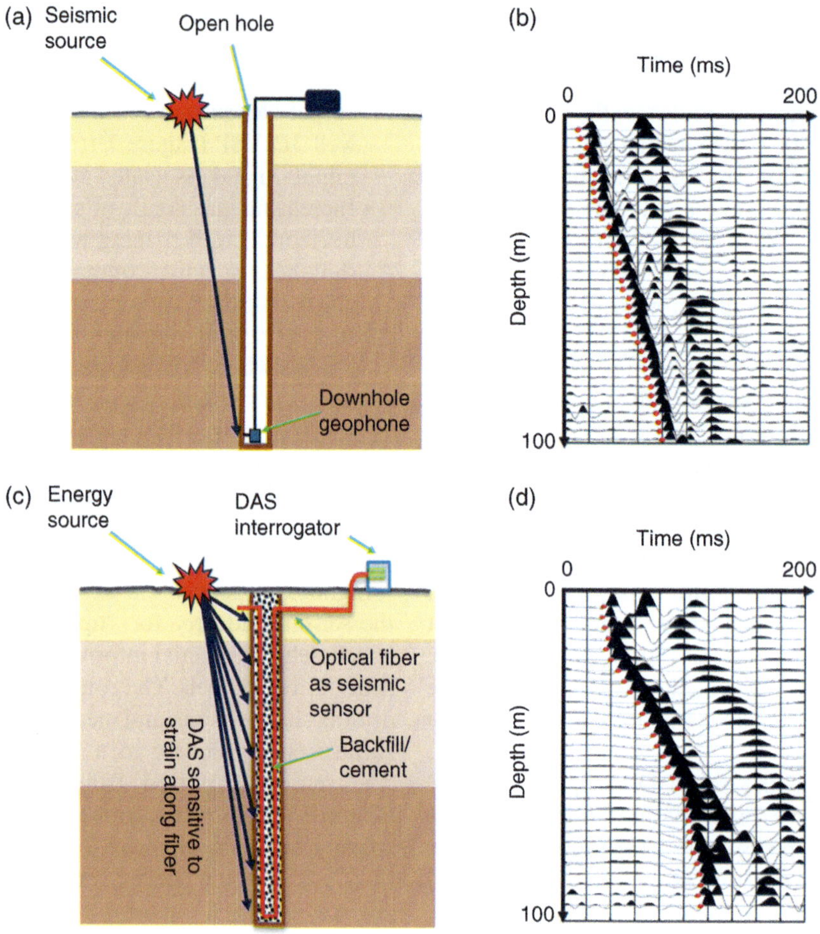

Figure 7.2 Comparing conventional (a, b) and Smart DAS uphole (c, d) acquisition systems and examples of data. Smith et al. (2019) / with permission of OnePetro.

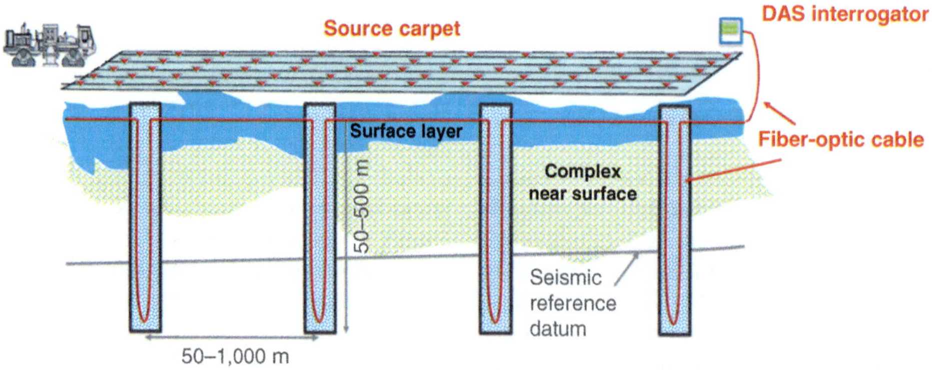

Figure 7.3 More detailed schematic of Smart DAS uphole acquisition system shown in 2D. Daisy-chained Smart DAS upholes from Figure 7.2(c) represent connected vertical arrays. Adapted from Bakulin et al. (2017).

and "horizontal" surface channels along the connecting segments. Such a system not only fulfills the original objectives of the upholes to calibrate the vertical near-surface velocity but also becomes an efficient vertical-array imaging system from Figure 7.1(c).

Although conceptually simple and efficient, practical implementation for such acquisition requires significant drilling and fiber installation optimization. Bakulin et al. (2022a) presented a version of high-productivity seismic acquisition for Smart DAS uphole acquisition. Most

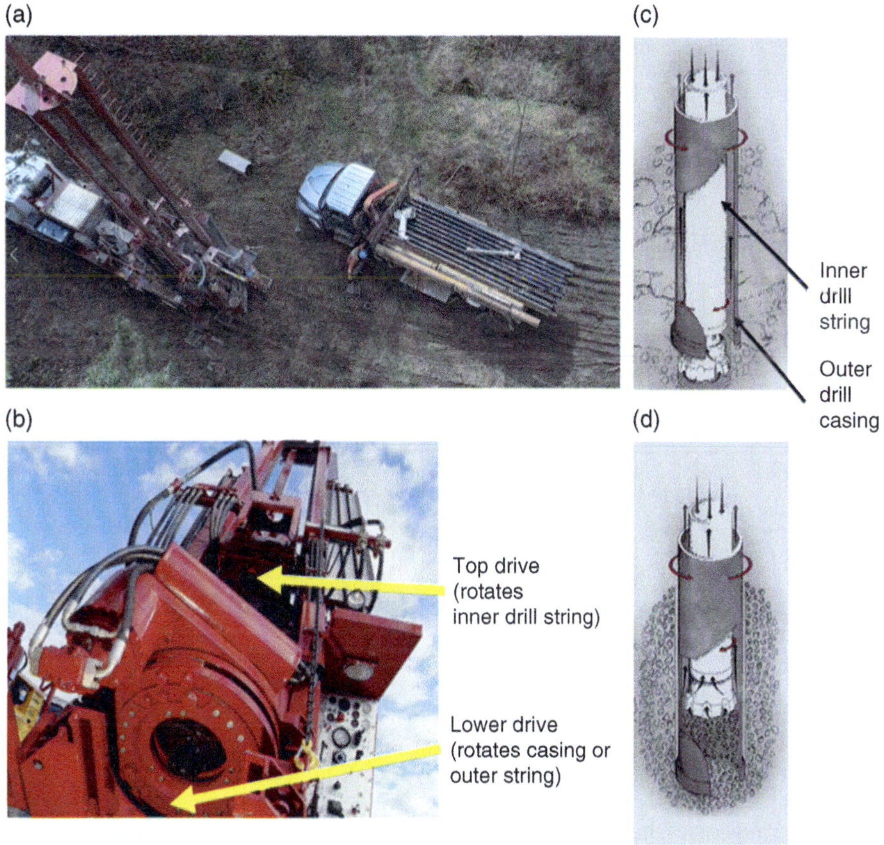

Figure 7.4 Optimal drilling equipment for Smart DAS uphole acquisition: (a) drone view of the portable dual rotary rig with support trucks carrying drill pipes; (b) two drives independently rotate inner and outer drill strings; (c) inner drill string is advanced in competent formation; (d) outer casing is ahead when drilling through unconsolidated sediments, karsts, or loss-circulation zones. Bakulin et al. (2022a) / with permission of the Society of Exploration Geophysicists.

nonproductive time in shallow drilling is associated with unconsolidated shallow sediments, deeper karsts, and low-circulation zones. Dual rotary drilling (Figure 7.4) was demonstrated as a fit-for-purpose technique to address all these challenges efficiently. The portable rig has two top drives (Figure 7.4(b)) that independently advance the inner drill string and outer casing (Figures 7.4(c),(d)). In competent formations, the inner drill string is advanced first (Figure 7.4(c)). In contrast, when going through unconsolidated sediments or karsts, the outer casing is pushed ahead, preventing collapse and maintaining circulation through the inner annulus. In summary, dual rotary versatility enables high-productivity drilling through any near surface.

In addition, dual rotary drilling enables robust fiber installation, as shown in Figure 7.5. With conventional single rotary drilling, as soon as the drill string is removed, there is a high chance of collapse that can ruin the subsequent installation. With dual rotary drilling, the outer drill casing can be left in place as deep as needed to contain the unconsolidated geology and de-risk the fiber installation (Figure 7.5(b)). The DAS fiber cable is deployed outside the grouting PVC pipe, serving as a carrier (Figure 7.5(d)). The carrier pipe with fiber is lowered inside the protective outer casing, eliminating the risk of collapse (Figure 7.5(b)). As soon as the fiber reaches the bottom of the hole, cement is pumped through the grouting pipe and fills the hole from the bottom up (Figure 7.5(b)). During cementing, the outer casing is removed (Figure 7.5(c)) to allow the best possible coupling in the shallowest section. At this point, even if the sediment collapse starts to happen, the fiber is safely in place and will not be damaged. Cementing is essential for two reasons: first, to provide good coupling of the DAS sensor to the formation, and second, to avoid parasitic tube waves that could be a significant source of downhole noise (Meunier et al., 2001).

Another option is to leave the protective casing in place. However, experimental data suggest that uncemented casing generally leads to poor coupling (Hardage, 2000). Bakulin et al. (2022a) have confirmed poor coupling for short-term measurements (~ hours and days after installation) with casing left by dual rotary. They concluded that pulling out the protective casing (Figure 7.5(c)) is required

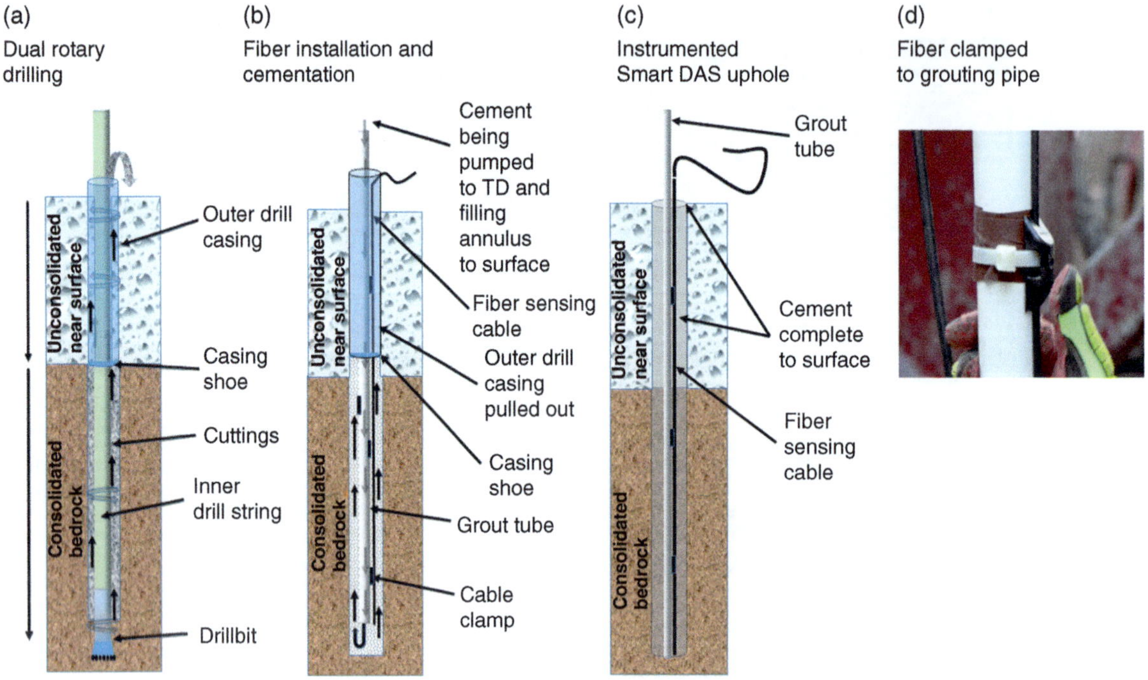

Figure 7.5 Stages of the protected fiber installation using dual rotary setup: (a) dual rotary drilling showing how outer drill casing prevents collapses of unconsolidated sediments; (b) lowering and cementing grouting pipe with fiber inside protective outer casing (TD, total depth); (c) cementing to the surface and removing protective outer casing; (d) photo showing fiber clamping to the carrier grouting pipe made of PVC. Bakulin et al. (2022a) / with permission of the Society of Exploration Geophysicists.

for Smart DAS upholes to achieve high-quality data in the shallowest part of the near surface. Other longer-term studies noted that better seismic bonding might evolve with time because drilling mud and rock cutting could better fill the annulus between the casing and the formation and lead to some solidification (Goetz, 2013). In a monitoring use case, a change of coupling over time between the uncemented casing and formation observed by Goetz (2013) may introduce an additional source of nonrepeatability that is hard to correct in processing. Finally, it is essential to remember that coupling concerns only apply to the portion of the DAS uphole coinciding with the limited extent of the uncemented outer casing typically used only for the shallowest, most unconsolidated part. Some monitoring scenarios may not require sensors in the shallowest portion of the hole. For example, Meunier et al. (2001) only instrumented a depth range of 60–175 m with geophones and hydrophones, suggesting that the shallowest sensors could be excluded because of lower repeatability. If we replicate this practice with DAS, the outer casing could remain in place, making the installation process even simpler.

Segments of fiber between the upholes can be efficiently trenched using cable-plowing machines available for laying telecommunications fiber. Figure 7.6 shows one example of a machine used in a desert environment and anticipated for surface seismic with DAS (Bakulin et al., 2020). Shallow burial (10–20 cm) could be acceptable for short-time characterization applications (Bakulin et al., 2022b; Nap et al., 2020). However, deeper burial may be required for monitoring use cases to avoid exposure of the cables to the surface over time. For instance, a 100–120 cm burial depth was used in a desert environment with shifting sand and still exposed some cables to the surface over a year or more (Jervis et al., 2018).

7.3. SYNTHETIC CASE STUDY

Before diving into the field experiment, let us perform a synthetic imaging study using a 2D line consisting of only 13 vertical arrays (Figure 7.7(a)). We assume that only vertical arrays are used for reflection imaging to match the field example where straight fiber was used in the acquisition. The connecting segments are ignored as they are sensitive to mostly horizontal motions due to the directivity of the DAS fiber. We honor the fiber directivity in the synthetic exercise but ignore gauge-length averaging and associated array filtering (Bakulin et al., 2020). So, DAS channels in the holes are approximated by point vertical displacements in finite-difference modeling. Such simplification is not essential for this conceptual

Figure 7.6 Cable plowing machines for automated trenchless deployments of horizontal segments for Smart DAS uphole acquisition system: (a) cable spool is carried in the front of the machine; (b) blade runs through the soil at a preselected depth to create a trench in which the cable is placed by feeding it down a chute located on the back of the blade; placement is done before the soil has time to fill the trench resulting in "trenchless" installation. Bakulin et al. (2020) / with permission of the Society of Exploration Geophysicists.

imaging exercise and could be easily upgraded using the scheme described by Egorov et al. (2021).

7.3.1. Illumination and Angular Coverage Achieved by Vertical Arrays

If we focus on the illumination of a deep reflector, there are two competing effects. Increasing the spacing between vertical arrays reduces illumination. In contrast, increasing the instrumented extent of the vertical arrays raises subsurface illumination. Balancing these competing effects, it is possible to obtain vertical-array configurations with illumination similar to surface seismic (Ikelle & Wilson, 1999).

For instance, they demonstrated that vertical arrays spaced at 500 m and a depth of 500 m could lead to an angular coverage similar to the surface seismic provided both have an equal channel spacing of ~25 m. They suggested that the two factors controlling the imaging resolution (without showing any imaging examples) are the total angular coverage and the density of sampling within this angular coverage. Furthermore, they highlighted that finer channel sampling along the vertical array could further boost angular sampling density. Figure 7.7 shows the illumination exercise contrasting 13 vertical arrays at 200 m depth and 400 m cross-hole spacing with the surface seismic using source/receiver sampling of 10 m. Figure 7.7(b) (red line) confirms that Smart DAS uphole acquisition and 2D surface seismic could achieve a similar fold when DAS channel spacing is 5 m (one-half of the 2D seismic sampling). If this sampling is halved to 2.5 m, the fold can be doubled (blue line in Figure 7.7(b)). Conversely, if the spacing is doubled to 10 m, the fold is halved (green line in Figure 7.7(b)). Considering that fine DAS sampling (0.5–2 m) is easily achieved, this opens an opportunity to either increase the fold or reduce the cross-hole spacing even further. Figure 7.7(c) validates Ikelle and Wilson's (1999) conjecture that similar average angular coverage could be achieved with fixed vertical arrays and 2D surface seismic spreads. What was not exposed in the previous studies is the inhomogeneity of the angular coverage for a fixed subsurface point (Figure 7.7(d)). Indeed, the central subsurface point from Figure 7.7(a) exhibits illumination in three ranges of angles (analyzed with a one-degree bin): around 0° (below the central uphole), 18–26°, and 32–52°. In contrast, 1–18 and 26–32° represent nonilluminated zones. If we use a 10° bin, the illumination distribution becomes much smoother and approaches 2D seismic (Figure 7.7(d), dashed red line). Please note that the number of occurrences for the 10° curve was multiplied by 1/10 to display it in the same plot as the 1° curve. When we explore the illumination of other subsurface points (not shown), they have teeth and notches moving along the angle scale. When the entire spread is analyzed, the average distribution is nearly identical to surface seismic except for minor jagging (Figure 7.7(c)).

The signal-to-noise ratio of vertical-array data is typically higher than surface seismic. Redundant illumination coverage mainly reduces noise during migration or stacking (Meunier, 2011). Therefore, with better data quality, comparable images could be obtained with less fold or angular sampling, implying that even sparser arrays in Smart DAS uphole acquisition could still be feasible for

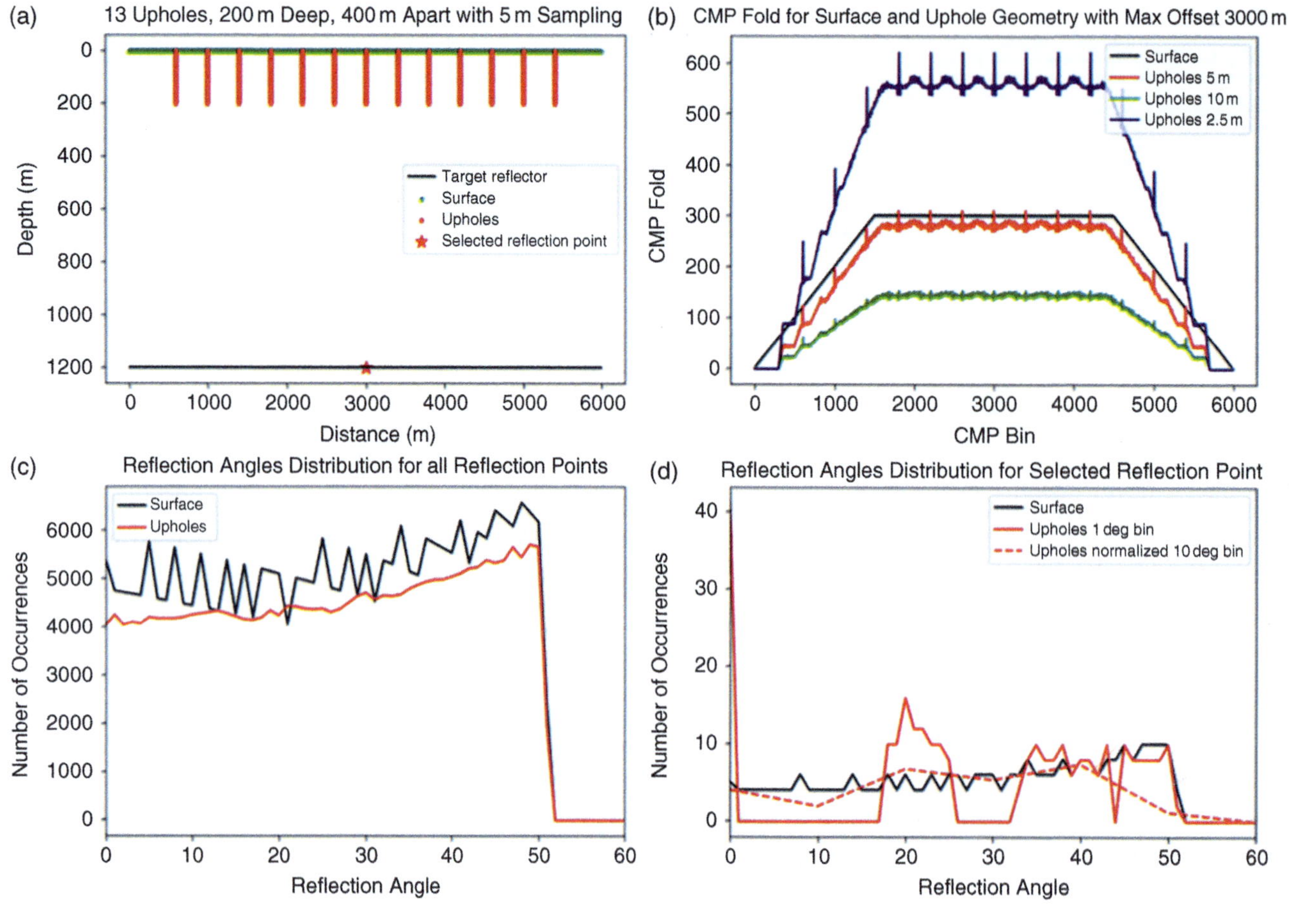

Figure 7.7 Synthetic case study showing deep imaging with Smart DAS uphole acquisition systems containing 13 DAS upholes at 200 m depth, spacing of 400 m, and vertical DAS channel spacing of 5 m: (a) acquisition geometry along with the target reflector at 1,200 m (surface sources are in green, and DAS receivers are in red); (b) fold achieved with a base channel sampling of 5 m (red) can be doubled with 2.5 m sampling (blue) or halved with 10 m (green); (c) distribution of the illumination angles (binned with 1°) for the target reflector achieved with the entire spread of the Smart DAS uphole system from (a); (d) distribution of illuminating angles for a fixed subsurface point from (a) when averaged with one-degree (solid red) and 10° (dashed red) bins (scale for the dashed red line is 10 times the shown scale). Black lines in (b)–(d) denote corresponding quantities for a fixed 2D surface seismic spread with coincident source and sensor sampling of 10 m that cover the same range as the green line in (a). Although the illumination of fixed subsurface points consists of teeth and notches (d), the illumination for 10° bins or average for the entire spread c is always comparable to surface seismic.

robust imaging and monitoring. Successes of monitoring with limited geophone arrays (Meunier et al., 2001) and SeisMovie monitoring with sparse, shallowly buried point sensors and sources (Berron et al., 2015) suggest the potential viability of such a proposition. Let us perform initial validation of imaging with vertical arrays with synthetic and field data.

7.3.2. Depth Imaging with Vertical Arrays

Let us now perform imaging of the realistic synthetic data. Figure 7.8 displays the subsurface model inspired by the field experiment shown in the following sections. The target of interest is a low-relief structure with a closure of 25 m at a depth of around 1,200 m. The overburden consists of highly contrasting high- and low-velocity layers typical for the desert environment. The vertical velocity profile for the overburden is compiled from a combination of checkshot, log, and uphole information in the area. Although the deeper layers conform to the structure, the near surface above the seismic reference datum at 100 m contains superimposed heterogeneities (Figure 7.8) borrowed from the SEAM Arid model (Oristaglio, 2012). Only P-wave velocity is shown; however, the model is elastic to replicate ground roll and other near-surface arrivals contaminating the reflections.

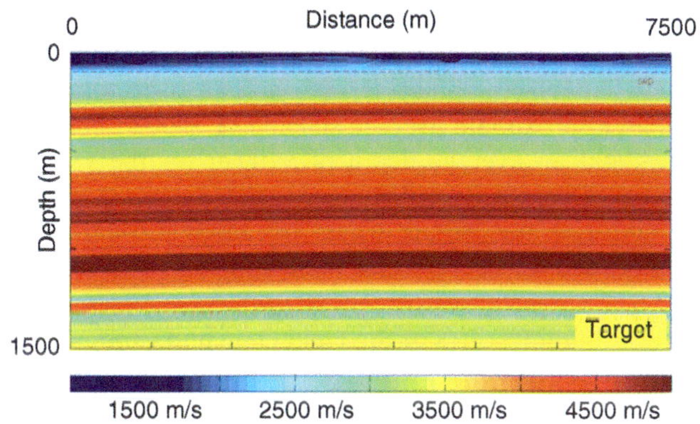

Figure 7.8 Compressional velocity for a 2D synthetic model with a 25 m low-relief target at 1,200 m depth. The model is inspired by a velocity profile from an actual onshore field and contains alternating high- and low-velocity layers with high contrast. Near-surface layer has heterogeneities borrowed from the SEAM Arid model. Adapted from Oristaglio (2012).

Figure 7.9(a) represents typical common-receiver gathers (CRG) in this model. The first observation is that reflected events are seen in the central parts of the gathers, so DAS cables in vertical wells should have excellent sensitivity to reflections. The second observation is that CRGs from receivers closest to the surface show the most significant amount of ground roll and near-surface arrivals, whereas deeper data display more unobscured reflections. The third observation is that even CRG at 160 m (Figure 7.9(a)) still contains significant trains of horizontally propagating energy with low apparent velocities, as typically observed in a highly contrasting layering environment. Thus, buried-receiver data requires noise-removal processing typical for surface seismic data.

After linear noise removal, we performed acoustic reverse-time migration (RTM) of the vertical-array day and arrived at the image shown in Figure 7.9(b). Please note that spacing between upholes was doubled to 400 m, whereas their depth is kept at 200 m to simulate a scenario closer to the field experiment. Although some imaging artifacts due to sparse uphole spacing are present in the shallowest part of the image, the target of interest at 1,200 m is well-imaged with the correct velocity model. These results confirmed that acoustic RTM of the elastic vertical-array data could lead to robust images of the deeper subsurface similar to surface seismic.

7.4. FIELD EXPERIMENT FROM A DESERT ENVIRONMENT

7.4.1. Summary of the Field Acquisition

Let us now review the imaging results from a field case study in a desert environment. Acquisition details are provided by Bakulin et al. (2017). Figure 7.10 summarizes the achieved acquisition geometry with six connected Smart DAS upholes shown in red. This early experiment preceded the development of the advanced high-productivity acquisition described in the previous section. Only 6 out of 10 upholes are connected, and uphole depth is variable. These are manifestations of Smart DAS uphole acquisition challenges when the conventional single rotary drilling is utilized. Fiber installation in open holes without

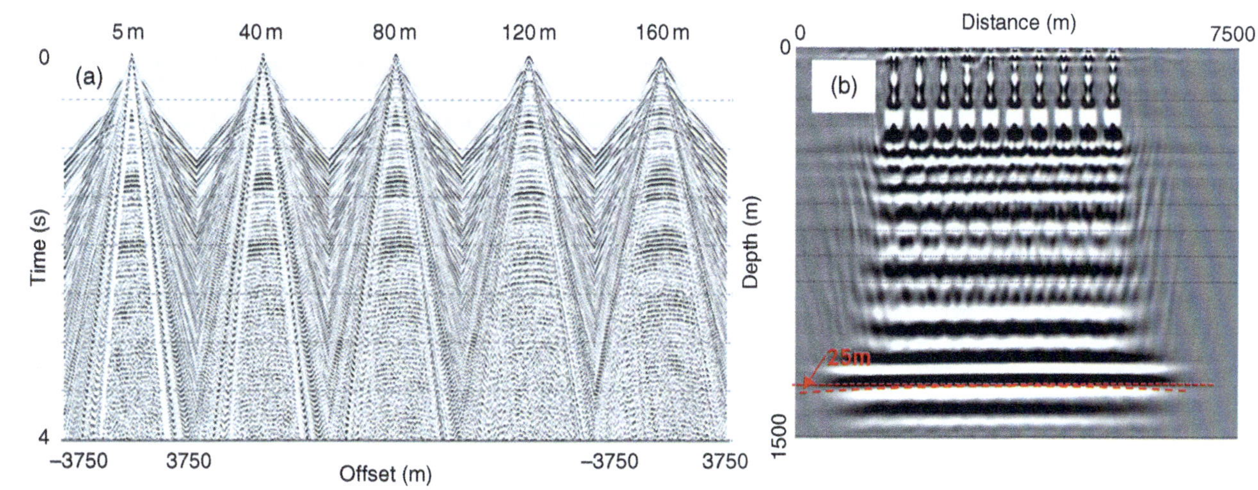

Figure 7.9 Synthetic prestack and imaged data for 10 upholes at 200 m, spaced at 400 m, and vertical channel spacing of 10 m: (a) common-receiver gather simulated with receivers at different depths (5–160 m) using elastic modeling; (b) acoustic RTM of Smart DAS uphole data after noise removal. An accurate image of the low-relief structure (closure of 25 m) is obtained at a depth of around 1,200 m. Observe cleaner reflection data with increased receiver depth in (a) due to the reduced energy of near-surface arrivals such as ground roll. The maximum offset is 3,750 m on all gathers. Adapted from Bakulin et al. (2018b, 2018c).

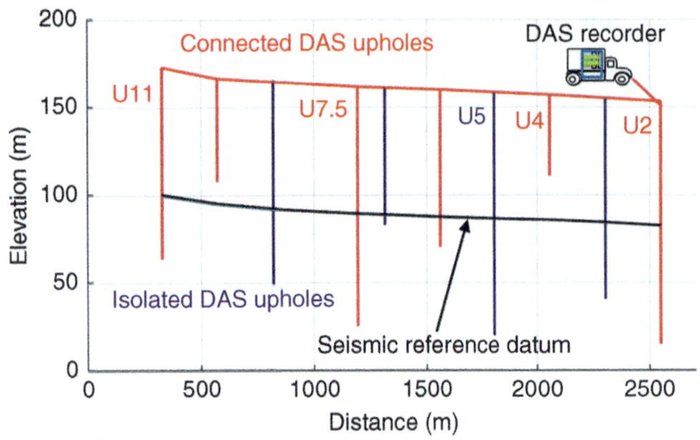

Figure 7.10 Configuration of Smart DAS uphole acquisition system used in this study. Six connected upholes are shown in red and follow the daisy-chained structure from Figure 7.3. Isolated DAS upholes are shown in blue. Adapted from Bakulin et al. (2017).

a protective outer casing from Figure 7.5(b) remained risky and prone to collapses and fiber damage, preventing uniform depth of the upholes. A much more robust acquisition is now feasible with dual rotary drilling and protected fiber installation, as described earlier.

DAS fiber delivered 1,180 channels spaced at 4 m with 299 channels inside vertical arrays and the remaining horizontal channels at the surface. Several 2D lines with vibroseis sources acquired with a 10 m inline and 10 m crossline sampling resulted in 2,843 vibrator points. We use only downhole channels for reflection imaging due to their strong sensitivity to near-vertical P-wave reflections. The achieved Smart DAS uphole acquisition system has an approximate average spacing of ~400 m between holes and an average depth of ~100 m for the vertical arrays. Densely sampled horizontal sections record ground roll for surface-wave inversion and refracted energy used for near-surface modeling.

7.4.2. Characteristics of the Field Data and Comparison with Modeling

Figure 7.11(a) shows an example of common-shot gather acquired using six connected upholes. This gather displays the entire fiber line (surface and downhole) with 4 m sampling. High data quality is observed with different segments of the fiber detecting various elastic waves. Horizontal surface sections capture refracted events and well-sampled surface waves. Reflection events are not visible on surface cables due to their predominantly horizontal directivity. Such an interpretation is supported by comparison with elastic modeling shown in Figure 7.12. We use a synthetic model from Figure 7.8, resembling the velocity model of this field. First, we obtain excellent general agreement between field and synthetic data, although we have not attempted any inversion and simply tried to replicate a realistic vertical 1D velocity profile. Second, we observe energetic near-surface arrivals dominating field and synthetic gather, suggesting that elastic modeling is a must to replicate those features on DAS records.

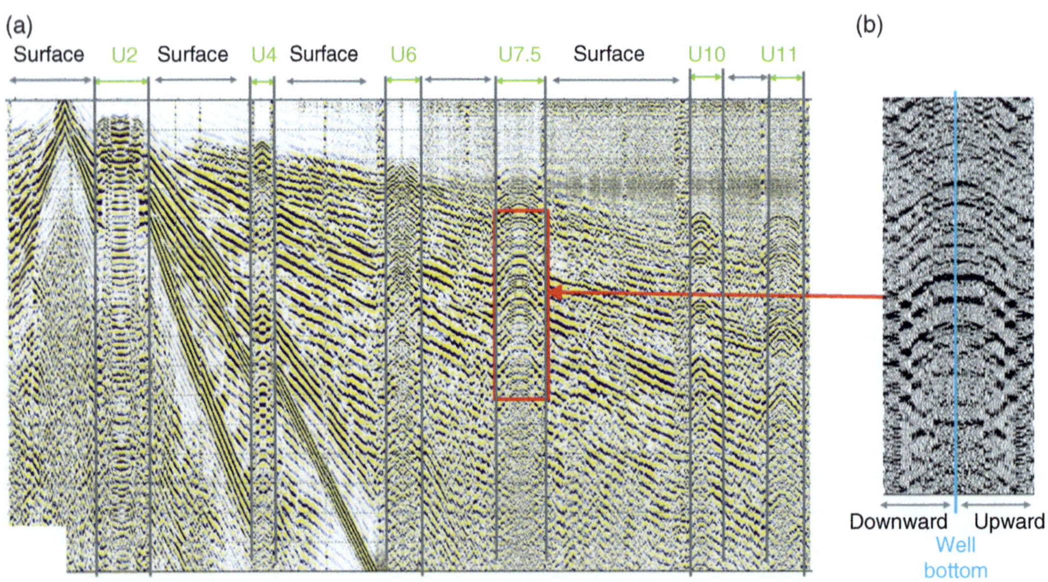

Figure 7.11 Raw common-shot gather acquired using the Smart DAS uphole acquisition system: (a) gather consisting of interleaved horizontal and vertical arrays recorded simultaneously; (b) vertical array (Figure 7.3) contains both downward and upward legs of the fiber, resulting in a symmetric subgather. Adapted from Bakulin et al. (2018b).

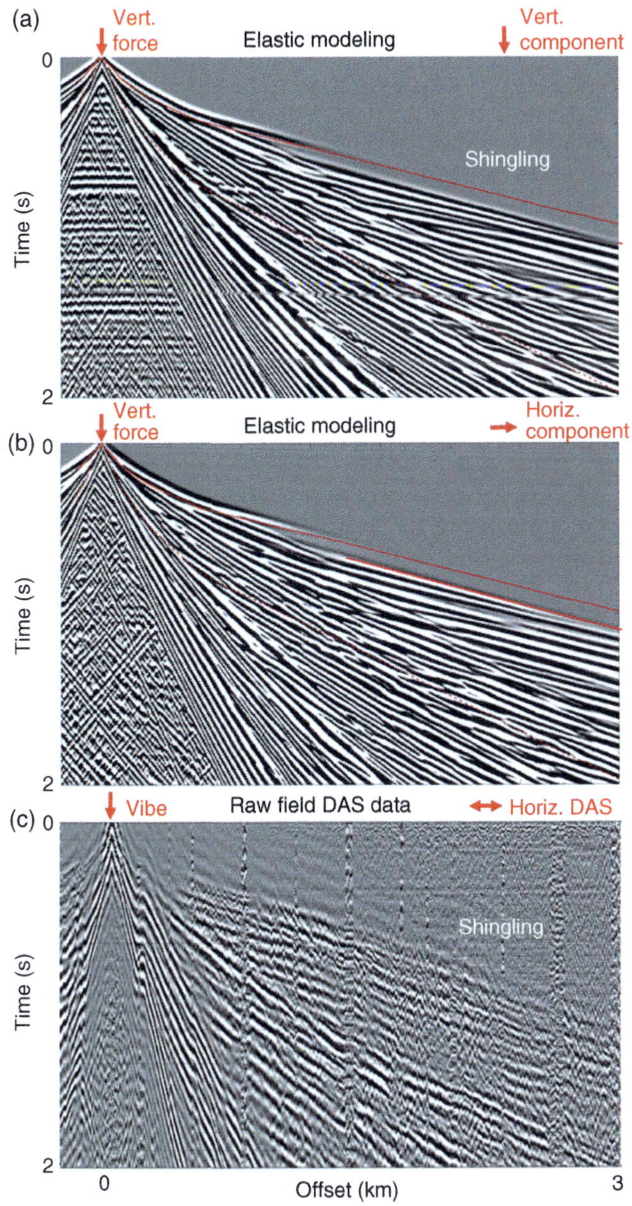

Figure 7.12 Comparison of synthetic data and field data. Synthetic data is generated using elastic modeling and using the realistic model from Figure 7.8. (a) Common-shot gather simulating vertical component. (b) Common-shot gather simulating horizontal component. (c) Field common-shot gather obtained with horizontal DAS fiber trenched at 1 m depth. Adapted from Bakulin et al. (2017).

Third, we conclude that the DAS field gathers (Figure 7.12(c)) are closer to horizontal component recording (Figure 7.12(b)) due to their primarily axial directivity. Indeed, vertical component recordings (Figure 7.12(a)) are expected to reveal some reflections at least near-vertical angles. However, such reflections are not recorded by horizontal DAS cables (Figure 7.12(c)) and are similarly absent in the simulated recording of inline components (Figure 7.12(b)).

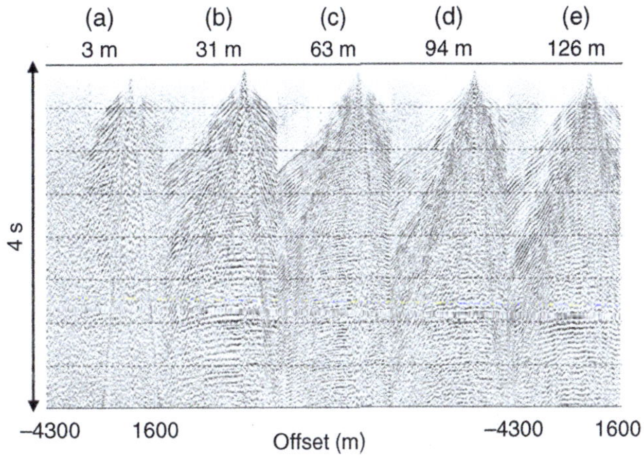

Figure 7.13 Field CRG with DAS channels at different depths: (a) 3 m, (b) 31 m, (c) 63 m, (d) 94 m, (e) 126 m. Please note the lowest signal-to-noise ratio on the shallowest channel (a) with no visible reflections. Generally, cleaner reflection data is observed at more considerable depths. Adapted from Bakulin et al. (2018c).

In contrast, vertical arrays exhibit reflected events needed for deep imaging. Please note that each uphole contains downgoing and upgoing legs of the fiber, resulting in a symmetric gather (Figure 7.11(b)). In the imaging exercise, we use channels from both legs to increase the channel count and improve angular coverage (Ikelle & Wilson, 1999), and increase the signal-to-noise ratio.

Modeling results suggest increasing receiver depth should improve the signal-to-noise ratio and make more reflections visible (Figure 7.9(a)). Field data from Figure 7.13 shows a similar trend, although perhaps with a more complex noise pattern than on synthetic. Suppose we perform a more focused comparison of modeled and field data similar to the previous surface data, but now done on common-receiver gather. In that case, we also observe surprisingly good agreement between signal and noise observed on the field and synthetic gathers (Figure 7.14). More complex noise events seen at near offsets on the field data (Figure 7.14(b)) may require a more detailed near-surface model to replicate them. Nevertheless, the target reflector from 1,200 m depth (red arrow in Figure 7.14) is robustly detected on the raw, unprocessed vertical DAS channel. Therefore, reflections are robustly captured on vertical DAS arrays (Figure 7.14(b)) but not on horizontal fiber segments (Figure 7.12(c)).

7.4.3. Processing and Imaging of Vertical-Array Data

We performed a crossline stack of three shot lines to further increase the signal-to-noise ratio for imaging. Such a step suppresses the crossline noise while enhancing the

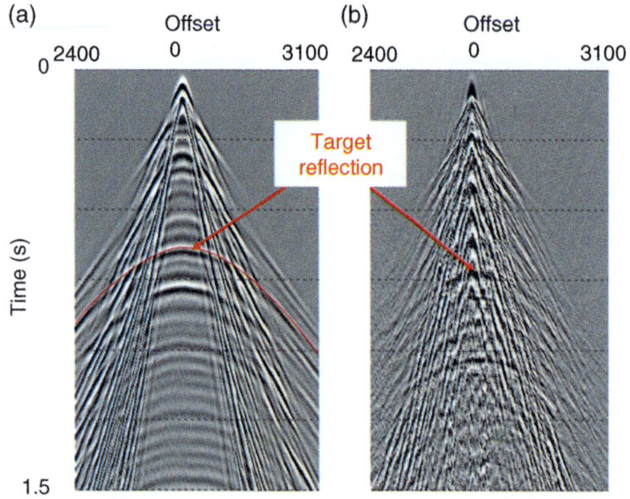

Figure 7.14 Comparison of synthetic (a) and raw field (b) CRG with DAS channels 130 m (uphole U7.75 in Figure 7.10). We observe clear subsurface reflections without processing despite more complex noise patterns on field data near offsets. This agreement confirms the excellent sensitivity of vertical DAS channels to subsurface reflections. Adapted from Bakulin et al. (2017).

inline propagating events. We emphasized that vertical-array data is still more similar to surface seismic. Also, shallow data requires significant time processing before imaging. To use conventional processing tools, the data was projected from downhole locations to the surface (Bakulin et al., 2018b). Such projection can be achieved via different methods: using direct arrival traveltimes (Fuller & Sterling, 2010), wave-equation redatuming (Al-Ali & Verschuur, 2006), or interferometric approaches (Schuster, 2009). After careful numerical testing, we determined that conventional static time shifts are acceptable for this dataset due to very slow velocities in the near surface. Time shifts were directly measured from traditional zero-offset DAS uphole surveys at each of the six locations (Bakulin et al., 2017). After applying time-shift to each DAS trace, we obtain redatumed data with the surface acquisition geometry. Redatumed surface data were subject to a typical time processing sequence for land seismic data. Figure 7.15 demonstrates the impact of random and linear noise removal on vertical-array data. Dense spatial sampling (10 m) on CRG enables the robust suppression of ground roll and other linear noise (see Figures 7.15(a) and 7.15(b)). After noise removal, we observe robust prestack reflections suitable for deep imaging.

Common-midpoint gathers obtained after redatuming DAS data allowed us to update stacking velocities initially taken from legacy seismic. Revised velocities improved the coherency of the common-depth-point(CDP) stack. In addition, surface-consistent deconvolution improved the recovery of higher frequencies (Bakulin et al., 2018d). Finally, surface-consistent residual static correction compensated for high-frequency static anomalies on the CDP gathers. The CDP stack of vertical DAS array data is shown in Figure 7.16(a). Despite shorter DAS arrays (on average ~ 100 m versus 200 m in the synthetic study in Figure 7.9(b)), we still obtain a reasonable image of the target at 1,200 m and deeper reflectors. Shallow imaging artifacts are apparent in the field and synthetic images but primarily healed well above the shallow target.

7.4.4. Comparison with Surface Seismic

Let us compare data and images from Smart DAS uphole acquisition with 2D surface seismic. To make a valid comparison, we have selected a similar 2D subset from the legacy 3D seismic data. It consists of one receiver line and a single parallel shot line with 60 m spacing for both. First, we compare prestack data recorded with

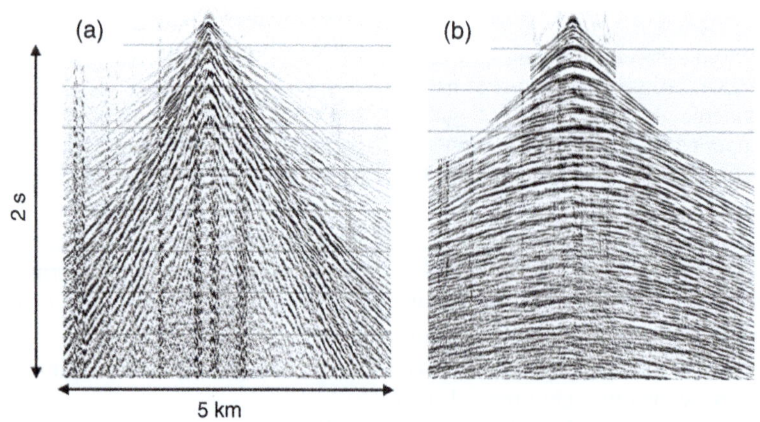

Figure 7.15 Common-receiver gather recorded by vertical DAS channel at 125 m: (a) before processing, (b) after linear and random noise removal. Adapted from Bakulin et al. (2018b).

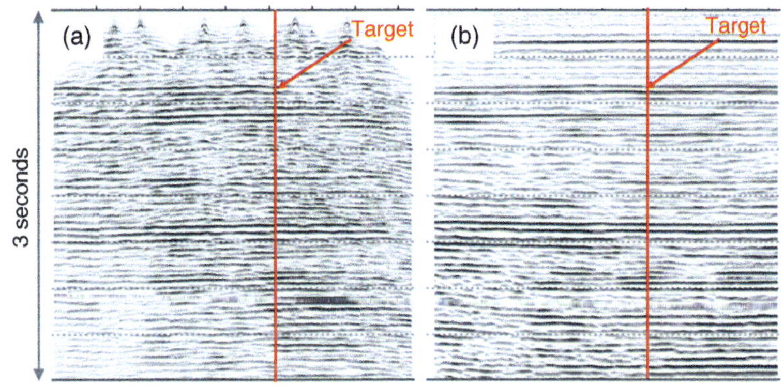

Figure 7.16 Stack sections obtained using various data: (a) vertical DAS array data, (b) conventional 2D seismic. Adapted from Bakulin et al. (2018b).

surface geophones and shallow vertical DAS channels. Figure 7.17(a), (b) depicts raw CRG without any processing. Both gathers reveal similar reflectors; however, the DAS data contain more linear noise and appears to have a higher spatial resolution. If we recall that surface data were acquired with 72-geophone arrays and 5-vibrator source groups, it explains why a significant amount of linear noise is filtered by field arrays (Figure 7.17(a)). The geophone array is areal and extensive, 55 m × 55 m (Figure 7.17(a)). In contrast, a single DAS channel averages strain over the gauge length of only 7 m or about 1/8 of the linear size of the geophone array. In this regard, the DAS channel behaves more like a single sensor and naturally preserves all the linear noise, including ground roll with the lowest apparent velocities (Figure 7.17(b)). More "resolved" nature stems from the finer shot spacing of 10 m compared to 60 m in surface seismic. To make a more fair comparison, we perform linear noise removal (similar to field array, but 2D) and decimate DAS gather to 60 m shot spacing. The resulting gather (Figure 7.17(c)) is more comparable to the geophone record (Figure 7.17 (a)), with both having little linear noise and a similar level

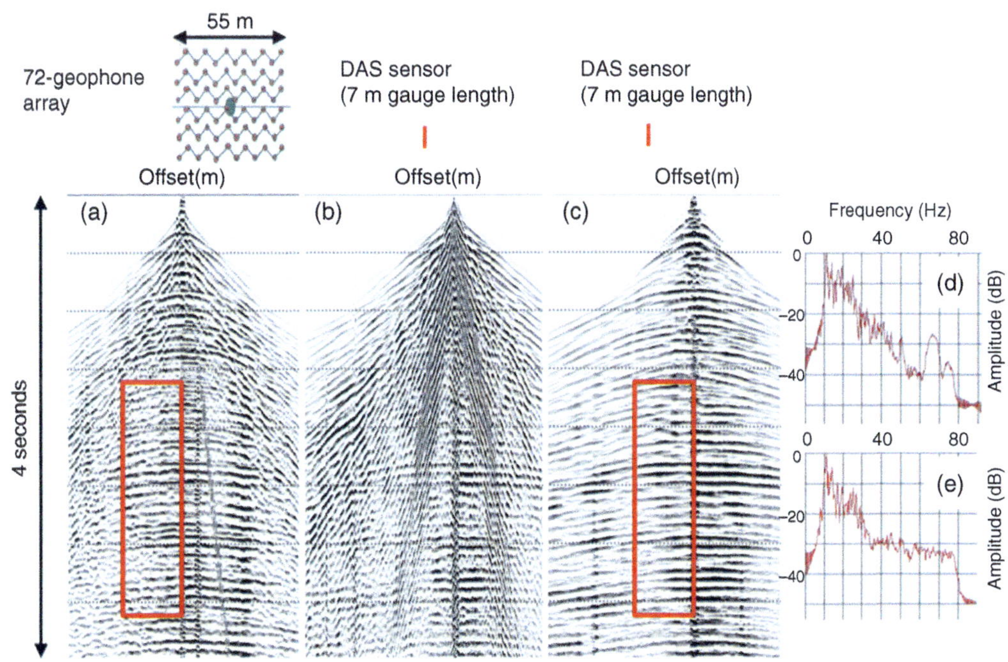

Figure 7.17 Comparison of prestack data obtained with 2D surface seismic and Smart DAS uphole acquisition system: (a) common-receiver gather from 2D surface seismic with 72-geophone array; (b) raw DAS gather from a single receiver at 15 m depth and shot spacing of 10 m; (c) same DAS gather after removing linear noise and decimating to 60 m source spacing. After noise removal, amplitude spectra of surface geophone data (d) and shallow DAS data (e) exhibit similar behavior. Adapted from Bakulin et al. (2017).

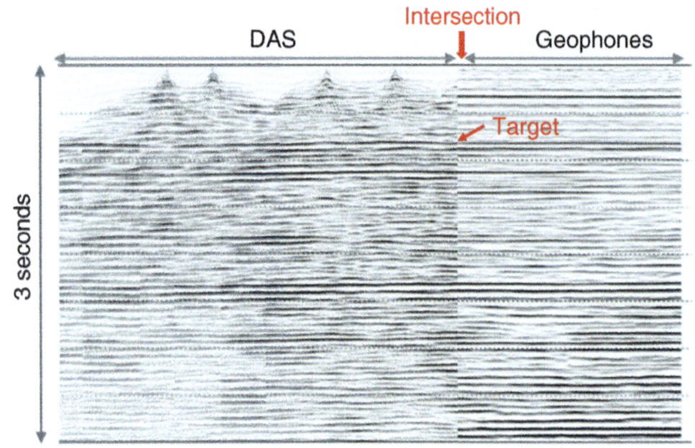

Figure 7.18 Comparison of 2D stack image obtained with vertical DAS arrays (left) and surface seismic (right). The red arrow marks the intersection point between two lines running at 45° to each other. Observe good ties between a shallow target as well as deep reflectors. Adapted from Bakulin et al. (2017).

of smoothness and exhibiting almost one-to-one correspondence between key reflectors. Because both acquisitions employed a linear sweep of 8–80 Hz, we can also compare their amplitude spectra (computed inside the red box in Figure 7.17(a), (c)). Spectra appear quite similar (Figure 7.18(e), (d)), validating the vertical directivity and broadband nature of DAS channels in the upholes. We have not done any conversion between DAS and geophone data in this study. Yet the close similarity between the two types of data is perhaps because the geophone array and DAS cable perform similar spatial averaging of the recorded wavefield quantities.

Let us now compare images obtained with geophone and DAS data. To perform the unbiased comparison, we have applied the same time processing to both datasets and used the same legacy stacking velocities. Figure 7.16 (b) shows seismic stacks obtained with legacy geophone data. Despite some minor differences in data character caused by the uneven distribution of DAS receivers, both images reveal identical targets and similar structures (Figure 7.16). Because 2D lines from DAS and geophone acquisition run 45° to each other, Figure 7.18 shows a view merged at the intersection point with a perfect tie between the two images. Shallow and deep reflectors are well-matched at the target and across an entire interval of 3 s.

7.4.5. Depth Imaging of DAS Vertical-Array Data and Image Sensitivity to the Near-Surface Velocity

The earlier synthetic exercise (Figure 7.9) suggested that depth imaging of sparse vertical-array data could be feasible if a velocity model in depth is obtained. Let us evaluate the effectiveness of depth imaging, determine sensitivity to the near-surface velocity, and compare the results with time imaging. Because good-quality data with vertical DAS arrays were recorded down to 3–4 s, the original synthetic model created before the acquisition (Figure 7.8) was no longer satisfactory in total depth. Moreover, it requires updating in the shallow part with new DAS uphole data. Therefore, we built a new 1D deep velocity that absorbed new data from DAS upholes and legacy data from check shots, sonic logs, and root mean square (RMS) stacking velocities. Despite being 1D and smooth, the new model shown in Figure 7.19(a) captures most details of vertical layering, especially multiple velocity inversion characteristics for the desert environment. To evaluate the sensitivity of the images to the near-surface velocity, we constructed two models that are primarily identical except for the shallowest near surface (Figure 7.19(b)). One is based on legacy uphole information from the areas available before the experiment (blue line in Figure 7.19(b)). In contrast, another is built using the average near-surface velocity as measured from Smart DAS upholes of this field experiment. With the velocity model in hand, migration can use the actual receiver positions in depth and avoid time shifts to the surface used for time imaging. In essence, we now could replicate RTM migration applied to synthetic data in Figure 7.9(b).

Figure 7.20 displays RTM depth images obtained with these two different models. Deep images (Figure 7.20(a), (b)) are broadly similar, but updated near-surface velocity improved the coherency of shallow and deep reflectors. It also produced somewhat cleaner quiet areas in the middle section with fewer artifacts. Zooming into the top part of the section containing the target (Figure 7.20(c), (d)) presents us with more tangible evidence of improved coherency, highlighted by the blue arrows. Although differences between the two velocity versions are very minute and localized (Figure 7.19(b)), they suggest significant image sensitivity to the near-surface velocity when using vertical-array data. Observe that the strong reflector below the target experiences change of shape between Figure 7.19(c) and 7.19(d). The significant impact of near-surface velocity on low-relief structures with a vertical closure of fewer than 60 m (or 30 ms) is well-known in time imaging (Bakulin et al., 2017). Nosjean et al. (2017) presented an example where different near-surface models could lead to reservoir volumes varying by a factor of three. Likewise, depth imaging suffers from a similar sensitivity, as supported by examples in this and other studies. In addition to imaging potential approaching surface seismic, the Smart DAS uphole acquisition system has another unmatched capability to accurately and directly characterize near-surface velocity (Bakulin et al., 2017). When dealing with complex shallow overburden, both imaging and near-surface description represent a perfectly

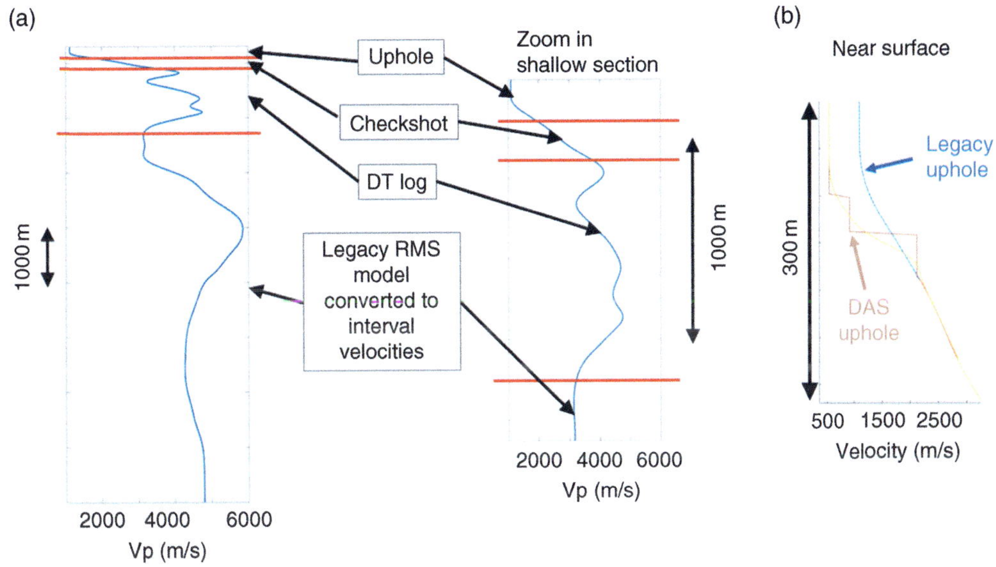

Figure 7.19 Composite 1D depth velocity model (a) built from uphole, checkshot, logs, and legacy stacking velocities. The updated near-surface section (b) exhibits reduced velocity in the shallowest portion as measured by the Smart DAS uphole of this survey.

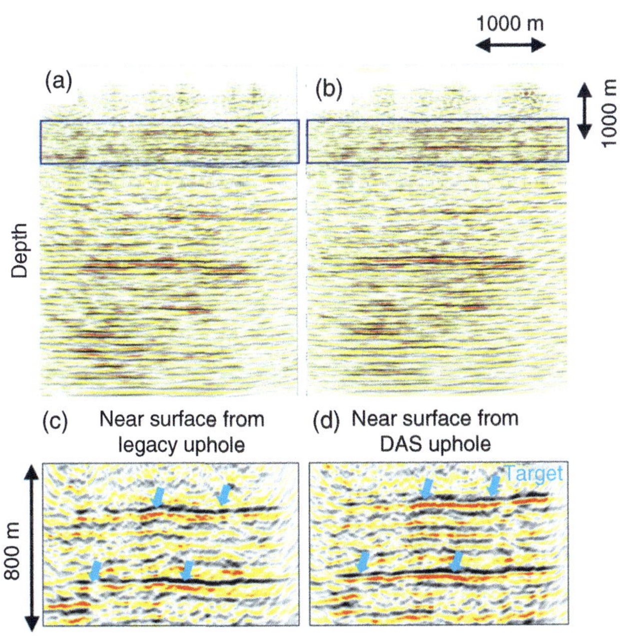

Figure 7.20 RTM depth images (top row) and zooms (bottom row) using different versions of the 1D velocity model from Figure 7.19: (a) image using higher velocity in the near-surface as estimated from legacy uphole (blue curve in Figure 7.19(b)); (b) image using lower velocity in the near-surface as measured from DAS uphole of this survey (orange curve from Figure 7.19(b)). Image zoom using updated lower velocity from DAS uphole (d) exhibits improved coherency of the shallow target and the next reflector compared to the image using velocity from legacy uphole (c).

complementary pair. The following section discusses various approaches to constructing accurate near-surface models from Smart DAS acquisition data. We also explain the benefits of Smart DAS acquisition for the monitoring below near surface experiencing seasonal changes.

The last thing is to compare the time and depth imaging results side by side. We convert the RTM image from depth to time and compare it with the CDP time stack (Figure 7.21). Both images depict identical structural

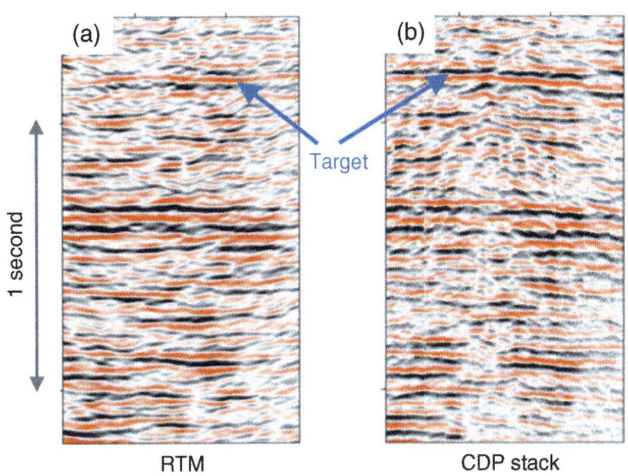

Figure 7.21 Comparison of depth (a) and time (b) images after the RTM image was converted to time for ease of benchmarking. Observe good agreement between two utterly different imaging approaches.

information, as expected for the case of relatively flat geology. Somewhat different amplitude signatures are expected due to significantly different imaging approaches. The good overall agreement suggests that both imaging approaches are robust for vertical-array data. Depth migration, such as RTM, allows handling the exact acquisition geometry with buried receivers. It could be the preferred approach when the velocity model could be constructed from the surface to the target of interest. Although we review the construction of the near-surface models from Smart DAS acquisition, building a deeper velocity model remains a topic that is less explored in the literature and requires further research.

7.5. NEAR-SURFACE CHARACTERIZATION USING THE SMART DAS UPHOLE ACQUISITION SYSTEM

Smart DAS uphole acquisition can benefit areas with complex near surface (Bakulin et al., 2017; Smith et al., 2019a). In this case, both vertical and horizontal arrays could enable very efficient near-surface modeling methods with much-reduced uncertainty compared to surface acquisition. However, although vertical DAS arrays have excellent sensitivity to reflections, surface arrays are akin to horizontal geophones (Figure 7.12). Because horizontal geophones are not optimal for recording reflections and refractions with near-vertical polarization, the obvious extension is to use a spiral or other omnidirectional cables for surface segments (Hornman, 2017). Figure 7.22 summarizes four main avenues of near-surface modeling in a graphic form, whereas Table 7.1 captures each technique's advantages and disadvantages. Let us review them in more detail.

7.5.1. Smart DAS Upholes

The Smart DAS uphole survey (Figure 7.22(a)) is akin to a mini-checkshot or mini zero-offset VSP (Figure 7.2(c)). This is the only direct method to measure near-surface velocity (Cox, 1999). It is considered the most accurate and often used to calibrate other indirect measurements. Although for competent sediments, one can record higher frequencies, in the desert environment, the frequency range of uphole surveys is similar to surface seismic (0–90 Hz) due to the highly attenuative nature of the shallow near surface. Smart DAS upholes have a significant advantage over conventional uphole surveys with geophones discussed earlier and elaborated in other publications utilizing the same dataset (Bakulin et al., 2017, 2018a; Smith et al., 2019a). Upholes are local measurements around the well. Therefore, they may not capture the global trend if drilled through a shallow anomaly. Nevertheless, a dense kilometer-size grid of upholes can directly and robustly capture the long-wavelength component of the static corrections for time imaging (Bakulin et al., 2017) or enable an accurate depth model for depth imaging from uphole interpolation alone. However, Bakulin et al. (2018a) have

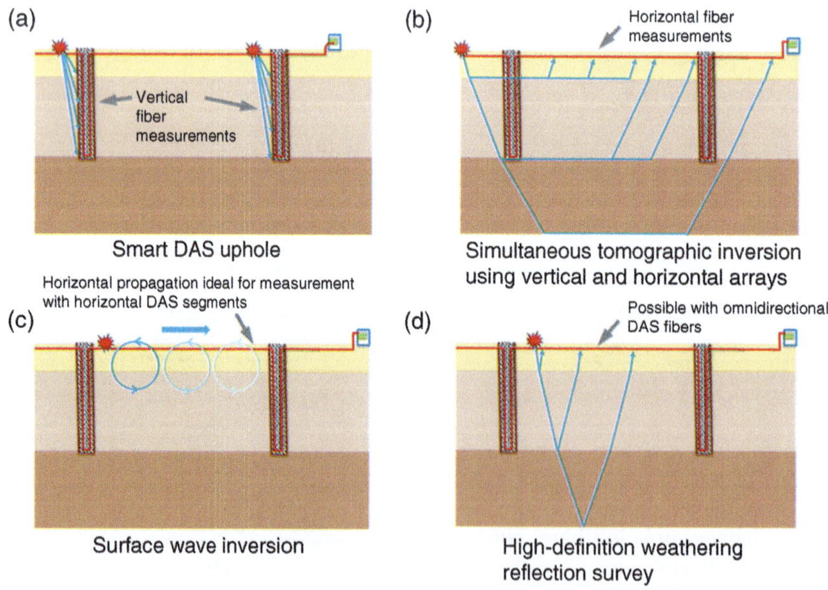

Figure 7.22 Near-surface characterization applications enabled by Smart DAS uphole acquisition: (a) uphole measurement; (b) refraction tomography using densely sampled horizontal and vertical arrays; (c) surface-wave inversion; (d) high-definition (HD) weathering surveys (requires omnidirectional cables at the surface). Smith et al. (2019) / with permission of OnePetro.

Table 7.1 Summary of near-surface characterization techniques enabled by Smart DAS uphole acquisition system.

Method	Advantages	Disadvantages
Smart DAS upholes	• Direct velocity measurement • Velocity inversions can be captured • Can be extended to walk away and multiazimuth surveys to characterize anisotropy	• Local measurement next to the borehole • Restricted to a maximum depth of upholes • Possible issues for shallow measurements
Simultaneous tomographic inversion using vertical and horizontal arrays	• Sensitive to a large volume of the subsurface • Can resolve velocity inversions • Can obtain deeper information than possible with upholes	• Dependent on the quality of data and first-break picks • Non-uniqueness of inversion • Measures near-horizontal velocity • Long wavelength less certain
High-definition surface-wave inversion	• High sensitivity to the very shallow part • Shear velocity estimation	• Not sensitive to a deeper part • Needs picking of higher modes for P-wave velocity inversion • Relies on the 1D approximation
High-definition weathering reflection survey	• Measures the same range of angles as deep reflection data • Consistent with reflection imaging	• Requires very dense acquisition • Requires omnidirectional cables for surface segments

Smith et al. (2019) / with permission of OnePetro.

shown that a 7 m gauge length could be too large for low-velocity near surface, leading to attenuation of higher frequencies (~50–70 Hz) and reducing the survey resolution. Luckily, DAS acquisition allows changing gauge length in the interrogator box itself. Bakulin et al. (2020) advocated using multiscale DAS measurements with several different gauge lengths. Alfataierge et al. (2021) have numerically tested these ideas and discussed optimal acquisition parameters for DAS uphole surveys.

7.5.2. Simultaneous Tomographic Inversion Using Vertical and Horizontal Arrays

Refracted arrivals can be recorded even with straight fibers used at the surface, although some preprocessing such as supergrouping or interferometric enhancement is usually required (Smith et al., 2019a). Using omnidirectional cables, refraction data quality improves significantly and no longer requires enhancement (Bakulin et al., 2022b). As a result, simultaneous tomographic inversion of first breaks or early waveforms could be performed using both vertical and horizontal arrays (Figure 7.22(b)). Alshuhail et al. (2019) used synthetic data with velocity inversions to demonstrate a much-improved estimation of static corrections with refraction tomography when vertical-array segments were included. Likewise, Egorov et al. (2022) have applied 2D Bayesian tomography and proved a considerable reduction in uncertainty in the presence of vertical arrays. Figure 7.23 demonstrates typical improvements and reductions in uncertainty at a single vertical location of the SEAM Arid model (Oristaglio, 2012). Direct arrivals and upgoing refractions from multiple shots to the vertical arrays serve as strong subsurface constraints supplementing the information obtained with the surface channels. Their impact is particularly critical when reconstructing models with multiple velocity inversions typical for the desert environment, such as the SEAM Arid model. Figure 7.23 shows that velocity inversion at ~200 m cannot be reliably mapped from surface data (Figure 7.23(a)). In contrast, 300 m vertical arrays allow robust identification of the velocity inversion with small uncertainty (Figure 7.23(c)). Similar improvement is expected for a full-waveform inversion and would be the subject of future studies.

7.5.3. High-Definition Surface-Wave Inversion

3D surface seismic has a typical inline sampling of 12.5–25 m, which is not optimal for surface-wave inversion. Also, data are acquired with small geophone arrays, reducing the spatial resolution of the inversion. Horizontal segments from Smart DAS acquisition enable "zoom into the near surface" through a meter-like sampling and small gauge length. Figure 7.24 demonstrates surface-wave inversion on one of the gathers from the field dataset record with horizontal DAS channels (Bakulin et al., 2020). Similar to the gauge-length effect on uphole data with a very slow near surface, this figure highlights that a 7 m gauge length is suboptimal for surface-wave inversion. Blue arrows on the modeled gather (Figure 7.24(e)) identify dimming energy on the

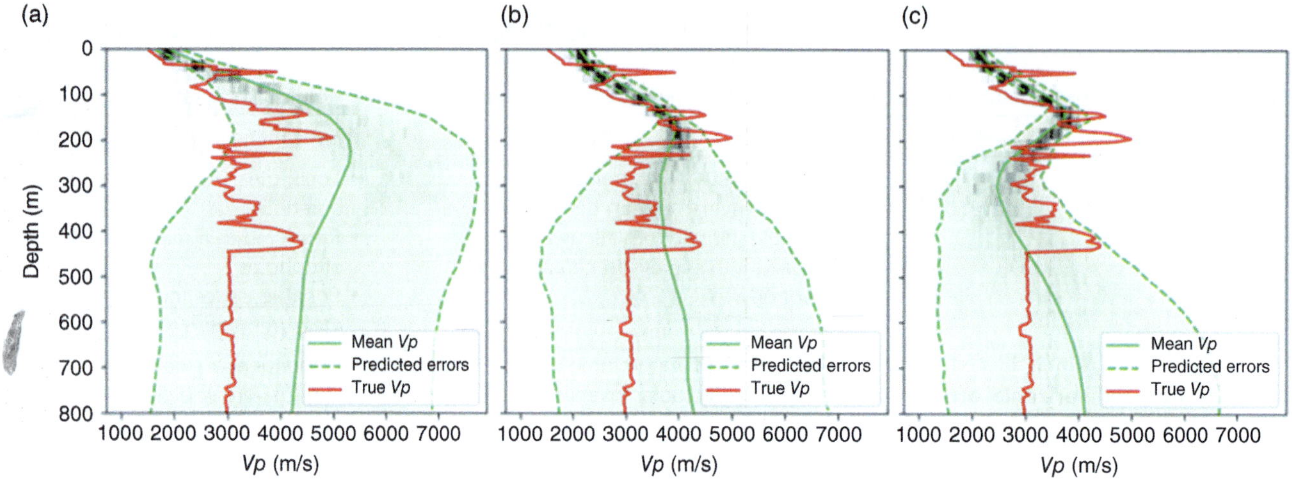

Figure 7.23 Vertical profiles from Bayesian tomography approach applied to synthetic SEAM Arid dataset: (a) surface data only; (b) surface data combined with vertical arrays down to 150 m; (c) surface data combined with vertical arrays down to 300 m depth. The red curve is the true velocity from the model; green solid and dotted lines denote the mean and three standard deviations, respectively, obtained from Bayesian tomography. Observe a significant uncertainty reduction when adding vertical-array data and increasing their depth. Egorov et al. (2022) / with permission of EAGE.

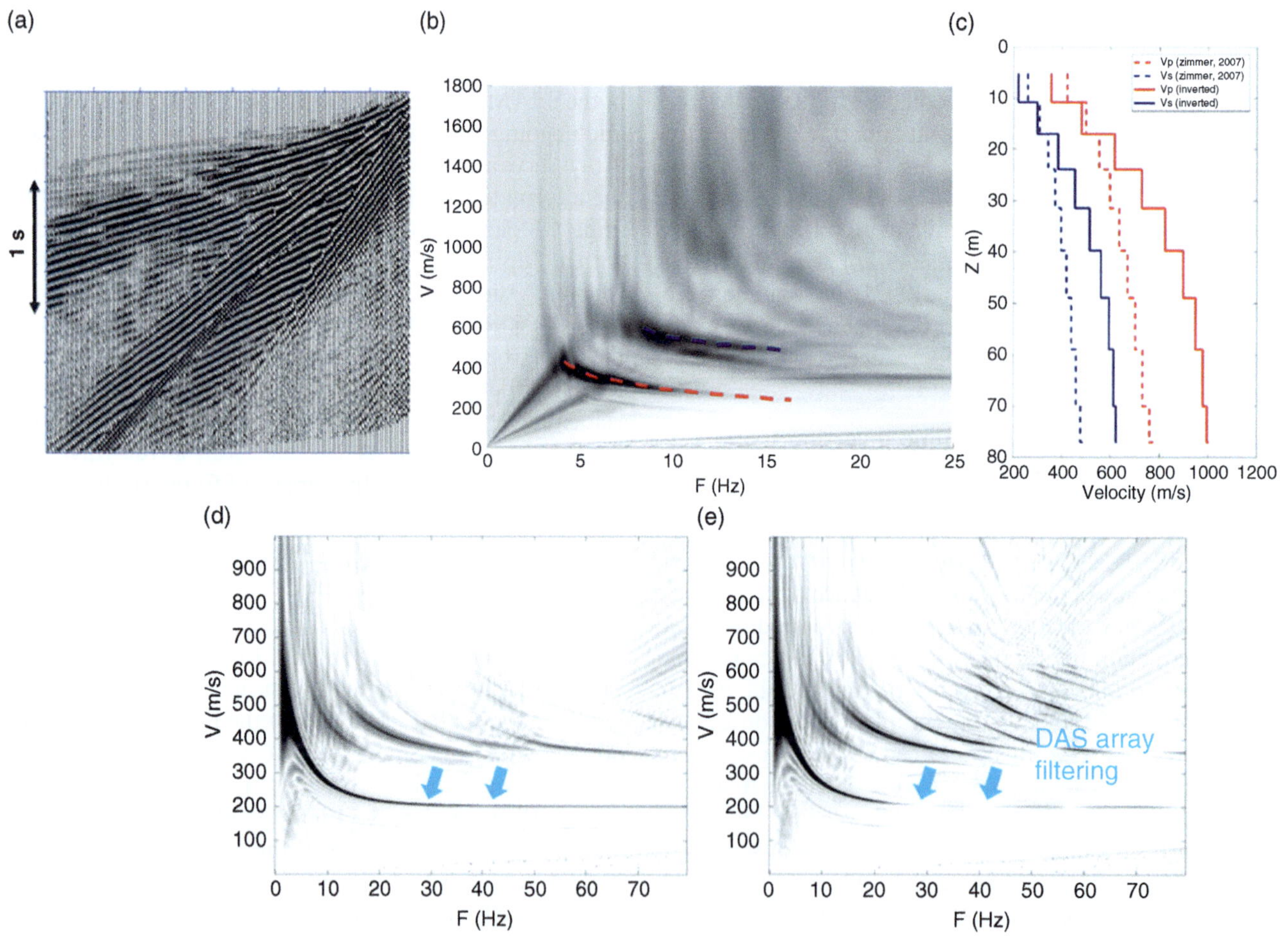

Figure 7.24 Surface-wave inversion DAS data from surface cables: (a) common-receiver gather with 4 m sampling of horizontal DAS channels; (b) dispersion curves picked on (a); (c) velocity profiles after inversion using different relations between P- and S-wave velocities. Reduced energy above 17 Hz in (a) can be explained by DAS array filtering as replicated with synthetic velocity-frequency panels: (d) dense single geophone data at 1 m sampling and (e) simulated DAS data with a field geometry (gauge length 7 m, channel spacing 4 m). Note the low shear-wave velocity of 200 m/s. Bakulin et al. (2020) / with permission of the Society of Exploration Geophysicists.

dispersion curves caused by DAS array filtering. Similar dimming is seen in the field data (Figure 7.24(b)). The lack of energy dimming on the point-sensor synthetic (Figure 7.24(e)) suggests that the use of a smaller gauge length (1–2 m) could be optimal to characterize very slow shear velocities (100–200 m/s) during surface-wave inversion. In summary, multiscale DAS measurements with multiple gauge lengths could enable "near-surface zoom" using surface-wave inversion. Characterizing the shallowest near-surface could benefit velocity model building for imaging and assist with removing near-surface seasonal variations occurring during monitoring (Smith et al., 2020).

7.5.4. High-Definition Weathering Reflection Surveys

Weathering reflection surveys (WRSs) with small receiver sampling of 1–2 m can characterize near surface using reflection-based velocity analysis (Martin et al., 2009). Although production data sampling (12.5–25 m) is insufficient, creating "reflection blind spots" in the near surface, densely sampled DAS data can overcome this limitation. Bakulin et al. (2020) simulated an example of WRS data with DAS using a realistic synthetic model from a desert environment. Figure 7.25 shows data demonstration from the field experiment described above using the shallowest DAS channel from a vertical array at 7 m depth. Raw gather exhibits intense linear events, completely masking shallow reflectors (Figure 7.25(a)). However, densely sampled data enables efficient linear noise removal to eliminate the most energetic noise and reveal robust shallow reflectors (Figure 7.25(b)). The shallowest recognizable reflector is at 150 ms, corresponding to approximately 60 m depth, an extremely shallow depth not achievable with seismic production surveys. Although straight DAS cables deployed at the surface are not suitable for WRS surveys, helical cables with a 30° or 60° wrapping angle could deliver robust shallow reflectors required for WRS with meter-like sampling (Bakulin et al., 2022b; Hornman et al., 2013).

7.6. TOWARD 3D IMPLEMENTATION

7.6.1. Characterization and Imaging Below the Complex Near Surface

Although most discussions and examples stayed in the context of the 2D acquisition, extension to 3D is relatively straightforward. Figure 7.26 shows two sample layouts using eight independent fibers connecting various upholes

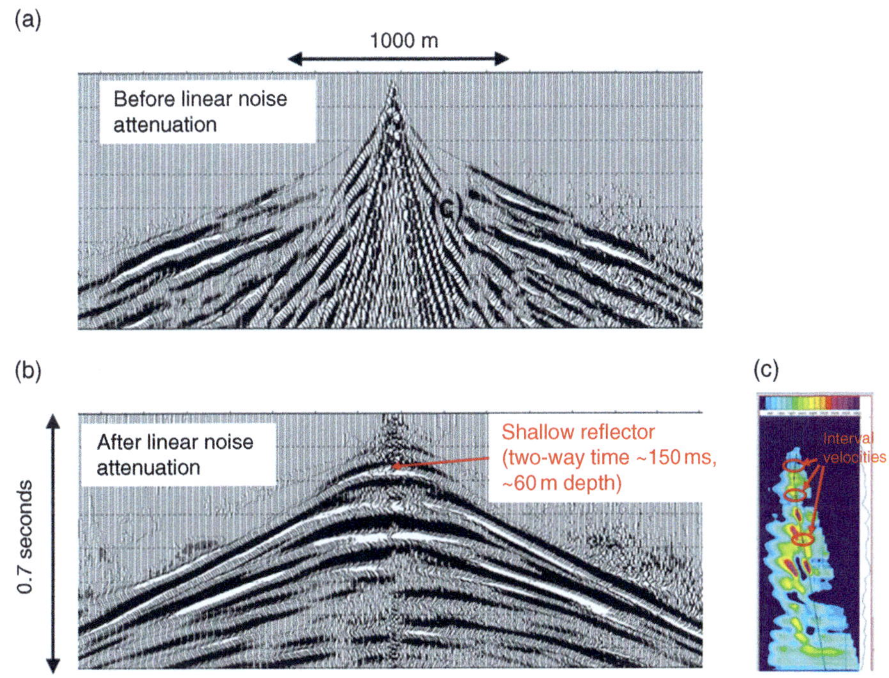

Figure 7.25 Shallow reflection can be uncovered from well-sampled DAS data: (a) common-receiver gather recorded by the shallow channel at 7 m depth in Smart DAS uphole U4 (Figure 7.10); (b) same as (a), but after linear noise removal; (c) semblance panel obtained from cleaned data (b) allowing to pick normal moveout velocities. The first recovered reflector is at 150 ms, allowing velocity characterization of the shallow near-surface layer at a depth of ~60 m. Adapted from Smith et al. (2019a).

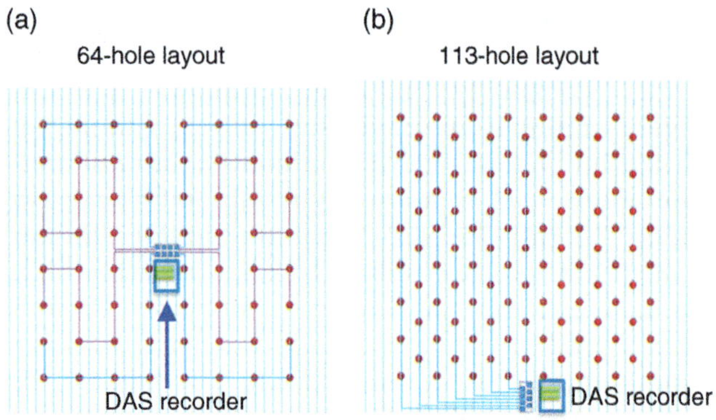

Figure 7.26 Possible 3D layouts of 3D Smart DAS uphole acquisition: (a) 64-hole design with central position of the DAS recording unit; (b) 113-hole design with side position of the DAS recording unit. Each acquisition assumes that eight independent fibers are used to connect different upholes. Recording at the central site requires eight co-located interrogators with single fibers or a single unit taking all eight fibers.

and running to a central recording location. Unlike geophones, DAS recording leads to the gradual deterioration of the signal-to-noise ratio along the cable associated with the optical attenuation (Bakulin et al., 2020). As a result, in practice, fiber length of limited length is selected depending on the acceptable acquisition goals. The typical fiber length currently used for seismic acquisition is around 10 km. Assuming 500 m holes and 500 cross-hole distance, 10 km fiber can be sufficient to connect 10 Smart DAS upholes. Super-interrogators taking 16–48 fibers are feasible, or existing DAS boxes can be stacked (Bakulin et al., 2020). Similar to the 2D case (Ikelle & Wilson, 1999), symmetric placement of vertical DAS arrays in both horizontal directions should lead to angular illumination comparable to 3D seismic with the orthogonal acquisition. Likewise, trading surface receiver locations for downhole ones should result in a significant increase in signal-to-noise ratio due to the reduced amount of near-surface energy on the buried channels (Bakulin et al., 2015, 2017; Ikelle & Wilson, 1999; Meunier et al., 2001). Therefore, even with a lower trace density, a Smart DAS uphole acquisition system may deliver similar image quality compared to surface seismic. Smart DAS acquisition could be a method of choice in areas with a complex near surface, such as a desert environment (Ley et al., 2003), where the success of the imaging hinges on the accuracy of the near-surface reconstruction (Bakulin et al., 2017). Other potential benefiting areas are mineral and geothermal exploration in crystalline rocks (Eaton et al., 2003; Schreiter et al., 2015) that exhibit additional scattering challenges caused by seismic speckle noise (Bakulin et al., 2022c). In this case, vertical arrays could improve the characterization of small-scale heterogeneities and mitigate them during imaging.

7.6.2. Monitoring Below a Complex and Changing Near Surface

Buried sensors and vertical arrays possess much higher repeatability (Bakulin et al., 2012; Meunier et al., 2001; Schreiter et al., 2015; Smith et al., 2020), making such a configuration also a preferred choice for monitoring use cases. Smith et al. (2019b) have presented a decimation exercise from an actual carbon-dioxide-enhanced oil recovery (CO2-EOR) project done with the 3D installation of 1,000 shallow-buried geophones. They demonstrated that the sparsifying surface locations for the buried grid with single sensors (Figure 7.1(b)) have quickly made repeatability unacceptable for sensitive monitoring. This result suggests that acquisition density (or angular coverage) is as essential for repeatability as for image signal-to-noise ratio. In contrast, significantly sparser Smart DAS uphole acquisition (Figure 7.1(c)) enables increased acquisition or trace density using deeper and well-sampled DAS arrays. The result is a much lighter 3D installation with fewer holes but deeper antennas than reported by buried point-sensor surveys (Jervis et al., 2018; Roach et al., 2015). Still, it could achieve robust imaging and monitoring in the most challenging conditions, such as the desert environment.

With significantly higher data quality of vertical-array data, it becomes feasible to reduce hole density even further. For example, Meunier et al. (2001) used 400–500 m cross-hole spacing with 175 m vertical arrays. Such data may no longer be suitable for 3D imaging; however, it could deliver sufficient monitoring information to assist reservoir production or monitor gas storage. In this case, a permanent buried source could also be used to improve repeatability at the expense of shrinking the monitoring locations. Berron et al. (2015) presented a system with 1,490 buried point hydrophones at 20 m and 40 sources at 25 m depth. Such a system with shallowly buried point sensors achieved sensitive monitoring of steam-assisted oil recovery over a 1 × 1 km pad. However, a similar system failed to deliver results in more challenging desert environments with deeper reservoirs (Berron et al., 2012). Desert environments require significantly higher acquisition and trace density for imaging, even with buried receivers (Bakulin et al., 2012). This is precisely where vertical arrays with deeper depth and higher channel count could provide an edge. However, no monitoring tests have been reported thus far in the desert environment with shallow DAS vertical arrays, which remains a topic for future studies.

7.7. DISCUSSION

The Smart DAS uphole acquisition system was initially designed for desert environments. However, it may find many applications in other onshore settings. Shallow drilling is done with portable rigs that require a relatively small footprint. After installation is completed, the Smart DAS uphole consumes little surface imprint. A careful cementation procedure was designed to satisfy the environmental requirements similar to plugged and abandoned wells, preventing cross-communication between shallow aquifers. For monitoring applications, the role of surface connecting cables is mainly to reduce the number of interrogators used for acquisition. If permanent trenching is not feasible for whatever reason, then surface extension cables could be used, similar to one reported by AlDawood et al. (2023). For 4D seismic surveys that need to be repeated every few years, such an option is a feasible, less-intrusive alternative. Splicing is required, but if carefully executed, it results in a manageable optical loss that could be tolerated. As Bakulin et al. (2017) demonstrated, splicing could be avoided for the trenched version. There are existing commercial solutions to execute fiber U-turns at the bottom of the upholes with a diameter smaller than 6 cm and loss of less than 0.3 dB. The use of special bend-improved single-mode fibers can further reduce the diameter and loss to ~0.1 dB. The longevity of the fiber-optic cables has been well demonstrated for downhole conditions. Shallow holes may use more cost-effective cables with less protection, but their longevity for long-term monitoring requires further investigation for various environments.

7.8. CONCLUSIONS

The Smart DAS uphole acquisition system utilizes a network of interconnected shallow upholes instrumented with DAS fiber to enable seismic imaging and monitoring below the complex and changing near-surface. Our 2D field experiments demonstrate robust time and depth imaging, which compares well with conventional surface seismic despite uneven uphole spacing and significant cross-hole distance. This can be attributed to the higher signal-to-noise ratio and fine channel sampling of the DAS array.

The horizontal and vertical components of the DAS arrays provide multiple opportunities for near-surface characterization techniques. The additional effort of drilling and placing DAS fibers inside the shallow subsurface enables a very sharp geophysical "zoom" or "X-ray" of the near surface. In cases when the shallow overburden represents a significant challenge, Smart DAS acquisition could provide better imaging results driven by higher-resolution near-surface models and improved data quality produced by the buried receivers.

The Smart DAS uphole acquisition system also has the potential to scale permanent buried-receiver surveys for seismic monitoring, overcoming the limitations of cost and scalability associated with buried geophone arrays. We present possible 3D implementations of the Smart DAS uphole acquisition that can be used for time-lapse seismic, supported by recent developments in DAS sensing. The buried nature of the DAS upholes limits 4D noise caused by changing near-surface conditions, and the dense channel spacing enables larger uphole spacing to make the system cost-efficient.

The Smart DAS uphole acquisition system bridges the gap between surface seismic and borehole geophysics. It blends elements of both in terms of acquisition as well as processing. With rising demand to efficiently monitor subsurface CO_2 and H_2 storage at scale, in addition to oil and gas reservoirs, the Smart DAS uphole acquisition system represents a viable candidate for onshore reservoirs.

ACKNOWLEDGMENTS

We want to thank the numerous current and former colleagues whose efforts were invaluable for this project. In particular, we appreciate the contributions of Kevin Erickson, Emad Hemyari, Mustafa Al-Ali, Abdullah AlRamadhan, Youngseo Kim, Anton Egorov, and Ezzedeen Alfataierge.

AVAILABILITY STATEMENT

The data used in this article contains confidential information and is not available for public release.

REFERENCES

Al-Ali, M. N., & Verschuur, D. J. (2006). An integrated method for resolving the seismic complex near-surface problem. *Geophysical Prospecting*, *54*, 739–750.

AlDawood, A., Shaiban, A., Alfataierge, E., & Bakulin, A. (2023). *Walkaway DAS VSP simultaneously acquired between two deep wells*. Paper presented at the 84th EAGE Conference and Exhibition.

Alfataierge, E., Alsharif, M., Egorov, A., & A. Bakulin (2021). *Optimal acquisition parameters for smart DAS uphole survey in a desert environment*. Paper presented at the 91st Annual International Meeting, SEG. Expanded Abstracts, 493–497.

Alshuhail, A., Silvestrov, I., Aldawood, A., Bakulin, A., & Alawaji, M. (2019). *Near-surface characterization using vertical array seismic data from smart DAS upholes*. Paper presented at the 89th Annual International Meeting, SEG. Expanded Abstracts, 2893–2897.

Bakulin, A., Burnstad, R., Jervis, M., & Kelamis, P. (2012). *Evaluating permanent seismic monitoring with shallow buried sensors in a desert environment*. Paper presented at the SEG Annual Meeting.

Bakulin, A., Jervis, M., Colombo, D., Tsingas, C., & Y. Luo (2015). *Bring geophysics closer to the reservoir — A new paradigm in reservoir characterization and monitoring*. Paper presented at the 85th Annual International Meeting, SEG.

Bakulin, A., Golikov, P., Smith, R., Erickson, E., Silvestrov, I., & Al-Ali, M. (2017). Smart DAS upholes for simultaneous land near-surface characterization and subsurface imaging. *The Leading Edge*, *36*(12), 1001–1008.

Bakulin, A., Golikov, P., Smith, R., Erickson, K., Silvestrov, I., & Al-Ali, M. (2018a). *Smart DAS uphole acquisition system for near-surface characterization and imaging*. Paper presented at the SEG Technical Program. Expanded Abstracts, 201–205.

Bakulin, A., Golikov, P., Erickson, K., Silvestrov, I., Kim, Y.S., Smith, R., & Al-Ali, M. (2018b). *Seismic imaging of vertical array data acquired using smart DAS uphole acquisition system*. Paper presented at the SEG Technical Program. Expanded Abstracts, 4050–4054.

Bakulin, A., Golikov, P., Smith, R., Erickson, E., Silvestrov, I., & Al-Ali, M. (2018c). *Reflection seismic imaging using Smart DAS upholes*. Paper presented at the 80th EAGE Conference and Exhibition. Expanded Abstracts, Th P7 12.

Bakulin, A., Golikov, P., Smith, R., Erickson, E., Silvestrov, I., & Al-Ali, M. (2018d). *Advances in near-surface characterization and deep imaging with Smart DAS upholes*. Paper presented at the SPE Kingdom of Saudi Arabia Technical Symposium and Exhibition, Dammam, Saudi Arabia. SPE 192310-MS.

Bakulin, A., Silvestrov, I., & Pevzner, R. (2020). Surface seismics with DAS: an emerging alternative to modern point-sensor acquisition. *The Leading Edge*, *39*, 808–818.

Bakulin, A., Alfataierge, E., Berquist, R., & Hermal, F. (2022a). *Acquisition trial of deep Smart DAS uphole: evaluating high-productivity drilling with dual rotary and fiber cable deployment*. Paper presented at the SEG Technical Program. Expanded Abstracts, 602–606.

Bakulin, A., Alfataierge, E., Pevzner, R., & Tertyshnikov, K. (2022b). *Evaluating HD weathering surveys and surface seismic with DAS in a sand dune environment*. Paper presented at the SEG Technical Program. Expanded Abstracts, 621–625.

Bakulin, A., Neklyudov, D., & Silvestrov, I. (2022c). Multiplicative seismic noise caused by small-scale near-surface scattering and its transformation during stacking. *Geophysics*, *87*(5), V419–V435.

Barnavol J. & Payre, X. (2017). *Case study: Lacq pilot CO2 storage in old gas field - microseismic monitoring*. Paper presented at the 79th EAGE Conference and Exhibition, Workshop Programme, WS02 A03.

Berron, C., Forgues, E., Bakulin, A., Jervis, M., & Burnstad, R. (2012). *Effects of complex near surface on 4D acquisition with buried source and receiver*. Paper presented at the 82nd Annual International Meeting, SEG.

Berron, C., Michou, L., De Cacqueray, B., Duret, F., Cotton, J., & Forgues, E. (2015). *Permanent, continuous & unmanned 4D seismic monitoring: Peace River case study*. Paper presented at the SEG Technical Program. Expanded Abstracts, 5419–5423.

Cox, M. (1999). *Static corrections for seismic reflection surveys*. SEG.

Eaton, D. W., Milkereit, B., & Salisbury, M. H. (2003). *Hard-rock seismic exploration*. Society of Exploration Geophysicists. ISBN 9781560801146

Egorov, A, Chavara, M., Alfataierge, E., & Bakulin, A. (2021). *Realistic modeling of surface seismic and VSP using DAS with straight and shaped fibers of variable gauge length*. Paper presented at the 91st Annual International Meeting, SEG. Expanded Abstracts, 437–441.

Egorov, A., Silvestrov, I., & Bakulin, A. (2022). *Effect of vertical arrays on near-surface velocity and statics uncertainty*. Paper presented at the 83rd EAGE Conference and Exhibition.

Fuller B. N. & Sterling, J. M. (2010). *Method for processing borehole seismic data. US Patent No. 7746726 B2*. Washington, DC: US Patent and Trademark Office.

Goetz, J. (2013). *Borehole seismic monitoring of CO2 storage within a saline aquifer at Ketzin, Germany* (PhD thesis). Berlin, Germany: Technical University of Berlin.

Hardage, B. (2000). *Vertical seismic profiling: Principles*. Handbook of geophysical exploration: Seismic exploration (3rd ed., Vol. *14*, p. 552pp). Amsterdam; New York: Pergamon.

Hornman, J. C. (2017). Field trial of seismic recording using distributed acoustic sensing with broadside sensitive fibre-optic cables. *Geophysical Prospecting*, *65*(1), 35–46.

Hornman, K., Kuvshinov, B., Zwartjes, P., & Franzen, A. (2013). *Field trial of a broadside-sensitive distributed acoustic sensing cable for surface seismic*. Paper presented at the 75th EAGE Conference & Exhibition, Tu 04 08.

Ikelle, L. T., & Wilson, R. J. (1999). Potential impacts of vertical cable (VC). *The Leading Edge*, 1154–1157.

Jervis, M., Bakulin, A., & Smith, R. (2018). Making time-lapse seismic work in a complex desert environment for CO2EOR monitoring — design and acquisition. *The Leading Edge*, *38*, 598–606.

Kiyashchenko, D., Mateeva, A., Duan, Y., Johnson, D., Pugh, J., Geisslinger, A., & Lopez, J. (2020). Frequent 4D monitoring with DAS 3D VSP in deep water to reveal injected water-sweep dynamics. *The Leading Edge*, *39*(7), 471–479.

Ley, R., Bridle, R., Amarasinghe, D., Al-Homaili, M., Al-Ali, M., Zinger, M., & Rowe, W. (2003). *Development of near surface models in Saudi Arabia for low relief structures and complex near surface geology*. Paper presented at the SEG Technical Program. Expanded Abstracts, 1992–1995.

Majer, E. (2007). *The application and use of microholes for vertical seismic profiling* (NETL DOE project FEW ESD04–006, 2004–2007). National Energy Technology Laboratory. Retrieved from https://netl.doe.gov/node/3729

Martin, F., Blake, B., Colombo, C., & Elagrari, F. (2009). *Weathering reflection survey (WRS) method for statics computation in desert operations*. Paper presented at the 71st EAGE Conference and Exhibition. Extended Abstracts, P214. doi:10.3997/2214-4609.201400148.P214

Mateeva, A., Mestayer, J., Yang, Z.Lopez, J., Wills, P., Roy, J., & Bown, T. (2013). *Dual-well 3D VSP in deepwater made possible by DAS*. Paper presented at the SEG Technical Program. Expanded Abstracts, 5062–5066.

Meunier, J. (2011). *Seismic acquisition from yesterday to tomorrow: 2011 distinguished instructor course*. Distinguished instructor series, No. 14. SEG and EAGE.

Meunier, J., Huguet, F., & Meynier, P. (2001). Reservoir monitoring using permanent sources and vertical receiver antennae: the Cere-la-Ronde case study. *The Leading Edge*, *20*, 622–629.

Nap, A., Edme, P., Schmelzbach, C., Paitz, P., & Robertsson, J. O. A. (2020). *Surface seismic with distributed acoustic sensing: is trenching worthwhile?* Paper presented at the NSG2020 26th European Meeting of Environmental and Engineering Geophysics, European Association of Geoscientists and Engineers.

Nosjean, N., Hanot, F., Gruffeille, J. P., & Miquelis, F. (2017). Statics: from imaging to interpretation pitfalls and an efficient way to overcome them. *First Break*, *35*, 71–78.

Oristaglio, M. (2012). SEAM phase II—land seismic challenges. *The Leading Edge*, *31*, 264–266.

Roach, L. A., White, D. J., & Roberts, B. (2015). Assessment of 4D seismic repeatability and CO_2 detection limits using a sparse permanent land array at the aquistore CO_2 storage site. *Geophysics*, *80*(2), WA1–WA13.

Schissele, E., Forgues, E., Echappé, J., Meunier, J., De Pellegars, O., & Hubans, C. (2009). *Seismic repeatability—is there a limit?* Paper presented at the 71st EAGE Conference and Exhibition incorporating SPE EUROPEC 2009.

Schreiter, L., Hloušek, F., Hellwig, O., & Buske, S. (2015). Characterization of seismic reflections from faults in a crystalline environment, Schneeberg, Germany. *Geophysical Prospecting*, *63*(4), 1015–1032.

Schuster, G. T. (2009). *Seismic interferometry*. Cambridge University Press.

Shulakova, V., Pevzner, R., Dupuis, C., Urosevic, M., Tertyshnikov, K., Lumley, D., & Gurevich, B. (2015). Burying receivers for improved time-lapse seismic repeatability: CO2CRC Otway field experiment. *Geophysical Prospecting*, *63*(1), 55–69.

Smith, R., Bakulin, A., & Silvestrov, I. (2019a). *Smart DAS uphole acquisition system for near-surface model building: results from the first successful field tests*. Paper presented at the SPE Middle East Oil and Gas Show and Conference, Manama, Bahrain, SPE-195154-MS.

Smith, R., Bakulin, A., & Jervis, M. (2019b). *Performance of a hybrid seismic monitoring system with buried receivers for an onshore carbonate reservoir – current status and way forward*. Paper presented at the 89th Annual International Meeting, SEG. Expanded Abstracts, 5234–5239.

Smith, R., Bakulin, A., Jervis, M., & AlRamadhan, A. (2020). Comprehensive seismic monitoring of an onshore carbonate reservoir: a case study from a desert environment. In *Active Geophysical Monitoring* (2) Chapter 6.4 ed., pp. 581–503). Elsevier.

Snelling, P., & Taylor, N. (2013). Optimization of a shallow microseismic array design for hydraulic fracture monitoring – a horn river basin case study. *CSEG Recorder*, *38*(3).

8

Joint Surface and Borehole Distributed Acoustic Sensing Vertical Seismic Profiling Data Acquisition and Processing

Gang Yu[1,2], Jinling Xiong[3], Yuanzhong Chen[2,4], Junjun Wu[2], Yishui Zhao[2], Zhonglin Cao[1], Guangming He[1], Yu Zhang[1], Xiang Li[1], and Hui Chen[1]

ABSTRACT

Joint surface and borehole seismic exploration is a 2D or 3D surface and borehole seismic exploration method for simultaneous acquisition of onshore or offshore seismic and vertical seismic profiling (VSP) data using the same sources. When acquiring 2D or 3D seismic data in the field, simultaneously acquired 2D or 3D DAS VSP data can provide full-well high-resolution structural images around the borehole and enhance 2D or 3D seismic data processing significantly. In addition, the subsurface accurate time-depth transform relationship, formation velocity, deconvolution operator, spherical diffusion compensation factor, Q value, anisotropy parameters, and other reservoir formation information in a certain range around the borehole or boreholes can be obtained directly from the 2D or 3D DAS VSP data. These can be used to significantly enhance the surface 3D seismic data processing, such as velocity model calibration and modification, static correction, deconvolution, demultiple processing, high-frequency restoration, anisotropic migration, and Q compensation or Q migration. This chapter describes the 3D DAS VSP data acquired by a downhole armored optical cable simultaneously with an onshore 3D seismic data acquisition project in Dagang Oilfield and 3D OBN data acquisition projects in the East China Sea and Abu Dhabi. The 3D DAS VSP data processing workflow and integrated interpretation are also discussed in this chapter.

8.1. INTRODUCTION

Distributed acoustic sensing (DAS) uses an optical fiber cable as a sensor for acoustic or seismic signals and is increasingly recognized as a viable alternative to downhole three-component (3C) geophone arrays for DAS vertical seismic profiling (VSP) data acquisition (Chen et al., 2012). The method to deploy optical fibers into a well can be a cable-based deployment or as part of a well completion string. Boreholes can be retrofitted with a fiber-optical cable by cementing it behind the casing, clamping it on injection or production tubing, pumping it inside the tubing (Mateeva et al., 2014), or using a hybrid fiber-optical wireline cable inside the casing (Frignet & Hartog, 2014; Hartog et al., 2014). Some tests have used optical fiber in an armored wireline clamped with magnets to the inside of the casing (e.g., La Follett et al., 2014). This can dramatically reduce the operating time required to complete full well coverage DAS VSP data acquisition as well as offering the opportunity to

[1]*BGP Inc, China National Petroleum Corporation, Zhuozhou, China*
[2]*Optical Science and Technology (Chengdu) Ltd., Chengdu, China*
[3]*Dagang Oilfield, China National Petroleum Corporation, Tianjin, China*
[4]*School of Information and Communication Engineering, University of Electronic Science and Technology of China, Chengdu, China*

Distributed Acoustic Sensing in Borehole Geophysics, Geophysical Monograph 289, First Edition.
Edited by Yingping Li, Robert Mellors, and Ge Zhan.

DOI: 10.1002/9781394179275.ch08

achieve much higher spatial coverage than is typical using conventional methods such as downhole 3C geophone arrays. The DAS interrogator is a sensing system that uses optical fiber both as a sensing element and a transmission signal medium. DAS VSP surveys have the advantages of full-borehole coverage, high-density sampling, high efficiency, high temperature and high pressure resistance, and low operation cost. It has become an important development direction for the borehole seismic technique (Li et al., 2010; Sun et al., 2009).

Many DAS VSP field tests and case studies have been published (Daley et al., 2013; Hartog et al., 2014; Madsen et al., 2013; Mateeva et al., 2014; Parker et al., 2014; Lesnikov and Allanic, 2014; Daley et al., 2016). These case studies represent a wide range of different source types, source-receiver geometries, and cable deployment methods. The use of fiber-optic cable for DAS VSP data acquisition has become a well-established application for DAS. Recording borehole seismic data using a DAS cable instead of a downhole 3C geophone array has advantages, especially for VSP surveys. In boreholes with preinstalled optical cables for temperature or pressure measurements, DAS VSP on demand or time-lapse DAS VSP can be acquired without well intervention (Mateeva et al., 2012, 2014). In comparison to downhole 3C geophone array recordings, the DAS cable covers the entire borehole, and full-borehole DAS VSP data can be recorded with a single shot on the surface. A DAS VSP survey using fiber-optical cable deployed to the total depth (TD) does not need to move the tool string and repeat surface shots many times like a conventional VSP survey by using a downhole 3C geophone array (Mateeva et al., 2014). This technique makes it economically feasible to interrogate multiple boreholes with the same source on the surface. Downhole armored optical fiber cable can be used to simultaneously acquire 3D surface seismic data and 3D DAS VSP data using the same surface seismic sources. In recent years, borehole seismic surveys have demonstrated the use of the DAS method for 3D DAS VSP data acquisition and comparison with conventional 3D VSP data acquired using downhole 3C geophone arrays (Mateeva et al., 2017; Correa et al., 2018). Borehole 3D DAS VSP data can also provide full-borehole coverage 3D VSP imaging and enhance surface seismic data processing significantly. Seismic imaging and time-lapse 4D reservoir monitoring can give us critical reservoir information to guide the placement of production and water flood injection boreholes in our high-value reservoirs.

Because the downhole 3C geophone array is located near the reservoir, the reservoir formation average and interval velocity, absorption coefficient, attenuation factor, deconvolution operator, anisotropy parameters, and other information can be obtained directly from 3D VSP data around the borehole in a certain range. At the same time, it is possible to obtain high-resolution reflection imaging around and near the borehole with a high fold number. 3D VSP can also be used for amplitude variation with offset (AVO) analysis, reservoir fracture or fault prediction, anisotropy analysis, and reservoir characterization (Chen et al., 2012; Jiang et al., 2016).

Compared with surface seismic exploration, VSP has the advantages of an accurate time-depth relationship, accurate formation information around the borehole, and rich wavefield information. The reflected wave observed by VSP has a short propagation path and is less affected by the low-velocity zone near the surface and surface environmental noise, so VSP data have the advantages of high resolution and high signal-to-noise ratio (SNR).

Borehole and surface joint seismic exploration is a 3D seismic exploration method formed by the combination of surface seismic and VSP surveys simultaneously (Jiang et al., 2016). Using 3D DAS VSP data and boreholes, accurate time-depth relationships, formation velocities, deconvolution operators, true amplitude recovery (TAR) values, Q values, and anisotropy parameters around the wellbore can be obtained. These parameters can be used to significantly enhance surface 3D seismic data processing. Borehole-driven surface seismic data processing includes velocity model calibration and modification, static correction, deconvolution, demultiple processing, high-frequency restoration, anisotropic migration, Q compensation or Q migration, and so on. Through borehole-surface joint seismic exploration, the accuracy and quality of seismic data volume can be improved, the SNR and resolution of reflected waves of target layers can be enhanced, and the ability of the data to characterize the geological target can be increased (Yu et al., 2020).

In this chapter, we first present a case study of onshore 3D seismic data acquisition and 3D borehole seismic data acquisition in two boreholes, simultaneously conducted in Dagang Oilfield near Bohai Bay in China. 3D surface seismic data acquisition covered an area of 200 km^2, and two boreholes (B-01 and B-02) were used to acquire 3D borehole seismic data simultaneously when the crew was acquiring 3D surface seismic data. Apart from the joint 3D ocean bottom node (OBN) and 3D DAS VSP surveys in the East China Sea, the world's largest jointly acquired high-density 3D OBN and 3D DAS VSP survey using six and seven wells in two locations in the Middle East is also described in this chapter.

8.2. JOINT 3D SEISMIC DATA AND 3D VSP DATA ACQUISITION

8.2.1. Joint Onshore 3D Seismic and 3D DAS VSP Data Acquisition

3D seismic data were acquired with a bin size of 10 m × 20 m, source and receiver line spacing of 200 m, receiver

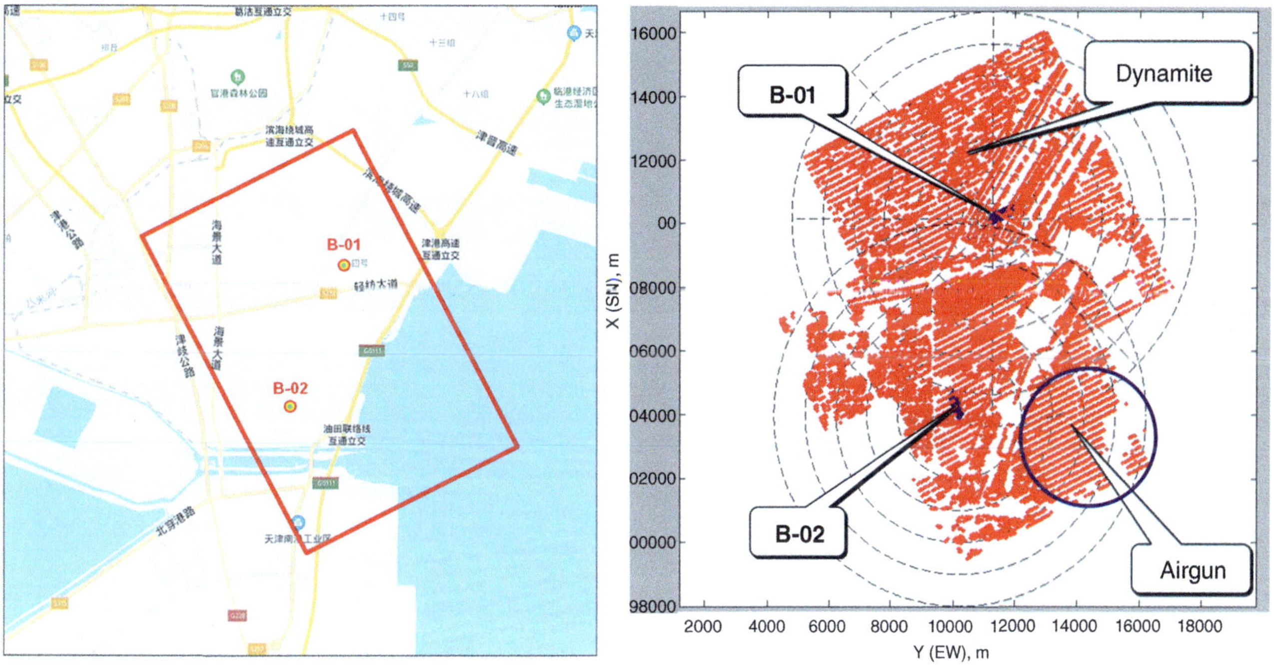

Figure 8.1 Simultaneous 3D surface seismic and borehole seismic survey area map and borehole locations (left), and actual source distribution map (right). Dynamite sources are located onshore, and the lower-right corner within the blue circle is the shallow water (offshore) airgun source location.

spacing of 20 m, shot spacing of 40 m, minimum offset of 22 m, and maximum offset of 4,646 m. The first borehole, B-01, and the second borehole, B-02, have a depth of 3,500 m and 2,770 m, respectively. An armored single-mode optical cable was deployed inside the casing of borehole B-01 from wellhead to TD, and an 80-level 3C downhole geophone array was deployed inside the casing of borehole B-02 from 500 m to 2,080 m. A total of 8,167 surface shots were recorded by the armored optical cable inside borehole B-01 and covered the source area of 74 km^2, and 7,913 surface shots were recorded by the 80-level 3C downhole geophone array and covered an area of 54 km^2. The downhole armored optical cable inside borehole B-01 has a receiver spacing of 2 m (total 1,750 levels), and the 80-level 3C downhole seismic array has a level spacing of 20 m. The total downhole seismic data covered the source area of 127.8 km^2, with a total of 16,080 effective surface shots.

Figure 8.1 shows the simultaneous 3D surface seismic and borehole seismic survey area map and borehole locations (left) and the actual source distribution map (right). About 75% of the survey area is onshore, and 25% of the survey area is in the shallow water area (Bohai Bay). Therefore, both dynamite and airgun sources were used to acquire 3D surface seismic and borehole seismic data simultaneously. The dynamite sources are located onshore, and the airgun sources are located at the lower-right corner of the source map. The total number of shots for the 3D surface seismic survey was 41,990, including 12,580 offshore airgun shots (right in Figure 8.1). Table 8.1 is a summary of the 3D surface seismic survey acquisition parameters. For the VSP survey shown in Figure 8.1, borehole B-01 covered a source area of 74 km^2, and only 8,167 shots were recorded; borehole B-02 covered a source area of 54 km^2, and only 7,913 shots were recorded. About 25% of the sources in the center were recorded by both borehole receiving sensors (optical cable and 3C downhole geophone array).

Table 8.1 3D surface seismic survey acquisition parameters.

Content	Onshore survey parameters	Offshore survey parameters
Observation geometry	24L5S400T	12L10S400T
Inline observation geometry	3990-3910-20-10-3990	3990-3910-20-10-3990
Tracks number	9,600	4,800
Shot line spacing	200 m	200 m
Shot lines	24	12
Shot interval	40 m	40 m
Receiver interval	20 m	20 m
Bin size	10 m × 20 m	10 m × 20 m
Fold number	240	240
Minimum offset	22 m	22 m
Maximum offset	4,646	4,646
Inline roll	200 m	200 m
Xline roll	200 m	200 m
Aspect ratio	0.6	0.6
Trace density	1,200,000/km^2	1,200,000/km^2
Shot density	125/km^2	250/km^2

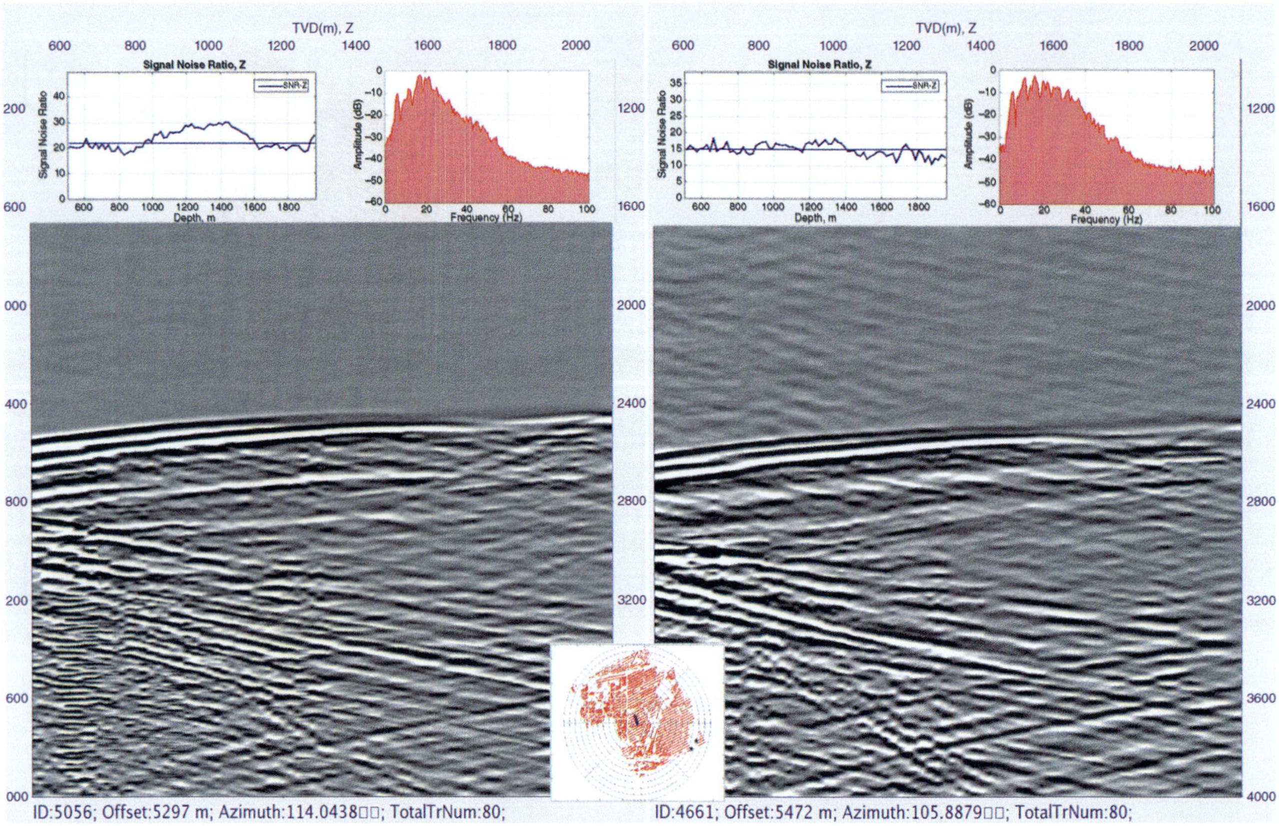

Figure 8.2 Raw CSG data comparison using a dynamite source (left) and an airgun source (right) with the downhole 3C geophone array. The amplitude spectra and calculated SNR are shown at the top of the data plot.

Figure 8.2 shows the raw common shot gather (CSG) data comparison using a dynamite source (left) and an airgun source (right) with the same downhole geophone array. The amplitude spectra and the calculated SNR are shown at the top of the data plot (Figure 8.2). The airgun source data have a broader frequency range than the dynamite source, and the dynamite source data have higher SNR than the airgun sources.

The DAS VSP data acquisition project used an interrogator from the Institute of Semiconductors, Chinese Academy of Sciences. A gauge length of 8 m was used for DAS VSP data acquisition. The DAS interrogator has a strain sensitivity of 0.02 nε/sqrtHz@10 Hz, a minimum spatial sampling interval of 0.25 m, a bandwidth of 1 ~ 2,500 Hz, and a sensing distance of 10,000 m.

Figure 8.3 shows the raw CSG data comparison of the downhole 3C geophone array data and DAS VSP data. Above the shot gather data are the amplitude spectra and calculated SNR of the 3C geophone data and DAS VSP data. The DAS VSP data have richer, lower frequency content than the downhole 3C geophone array.

After the review and analysis of the recorded borehole seismic data, we can draw the following conclusions:

1. The borehole seismic data recorded by the 80-level downhole 3C geophone array in borehole B-02 have high SNR, so they can be used for VSP imaging processing (Figure 8.3, top).
2. The borehole seismic data recorded by the armored optical fiber cable in borehole B-01 have lower SNR and broad frequency range but clear first arrival on the downgoing wave, so they can be used for anisotropic velocity calculation (Figure 8.3, bottom).
3. The DAS VSP data from borehole B-01 have stronger background noise, cable slapping, ringing noise, abnormal channels, and so on, so they need to be removed during subsequent data processing stages.
4. The 3C borehole seismic data from borehole B-02 have rich wavefield information and are suitable for wavefield separation with effective amplitude-preserved processing steps.

8.2.2. Joint 3D OBN and 3D DAS VSP Data Acquisition in the East China Sea

The joint OBN and 3D DAS VSP survey-related publications (Zwartjes et al., 2017; Mateeva et al., 2013a, 2013b) provide us with very valuable information about the capabilities of offshore 3D DAS VSP data. High-quality 3D VSP images obtained around the DAS boreholes by Wu et al. (2015) gave us the confidence and

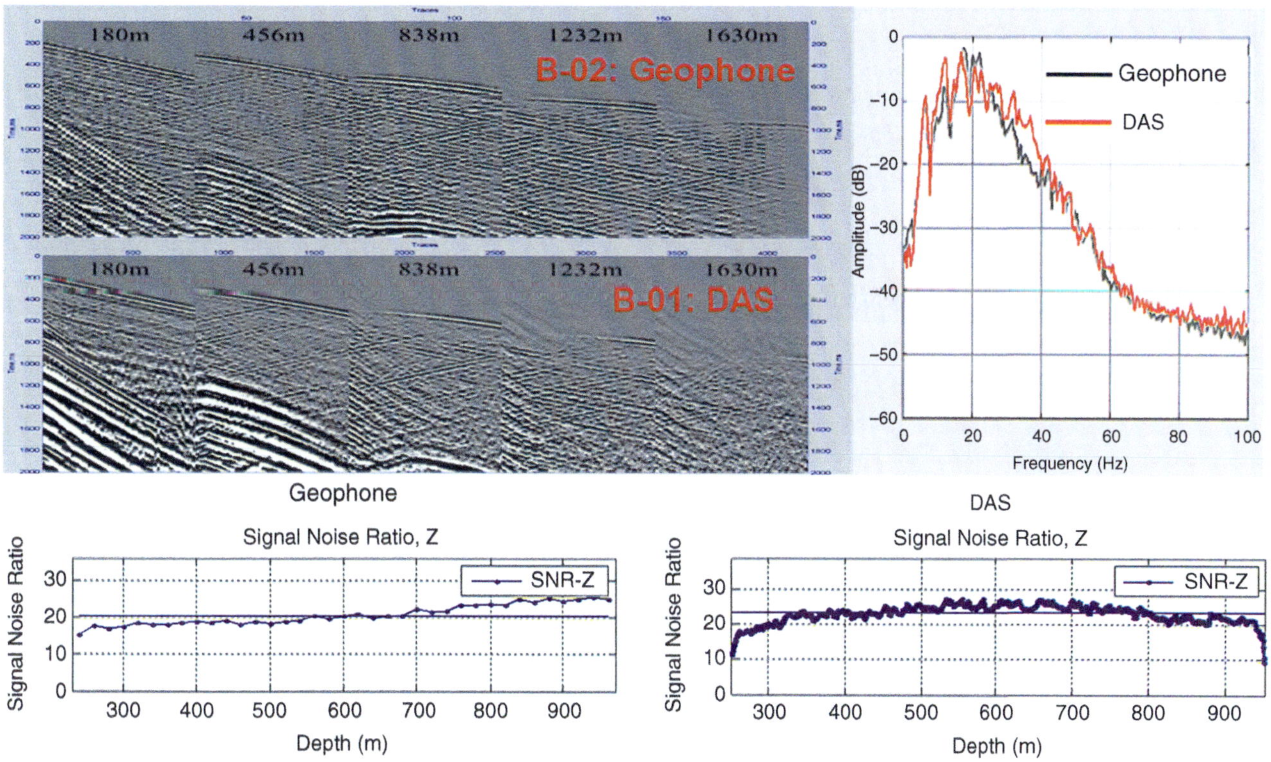

Figure 8.3 Raw CSG data comparison of downhole 3C geophone array data (B-02) and DAS VSP data (B-01). The right side of the shot gathers data and shows the amplitude spectra of geophone data and DAS data. The calculated SNR of geophone data and DAS data are shown at the bottom of Figure 8.2.

decision support to conduct an offshore 3D DAS VSP survey simultaneously with a newly planned high-density OBN survey in the East China Sea.

The Pinghu Oil and Gas Field is in the middle of the Pinghu fault structural belt on the west side of the Xihu Sag in the Shelf Basin of the East China Sea, bounded by the Pinghu main fault on the west. The main fault of Pinghu is NNE trending and has an extension length of more than 100 km, which plays an important role in controlling the formation of the Pinghu Formation in the Pinghu Oil and Gas Field. The field is composed of eight large and small structures, including Fangheting, Bajiaoting, Wanghuting, and Shuangzhaoting, among which Fangheting, Bajiaoting, and Wanghuting are the main structures, all of which are controlled by drilling and tested for industrial oil and gas flow.

Figure 8.4 shows the regional geology and structure map of the Xihu Sag. The Xihu Sag is about 550 km long from north to south and 110 km wide from east to west and covers an area of about 5.9×10^4 km^2. The main sedimentary layers are tertiary strata, and the thickness of the deposits is predicted to be greater than 12,000 m. The structural area of the Xihu Sag is generally divided into three major zones: the western slope area, the central depression area, and the eastern fault terrace area. The Fangheting structure is the largest structure in the Pinghu Oil and Gas Field. The Pinghu Formation trap is an anticline, and the trap area is also large. Its east and west flanks are complicated by the second and first faults, respectively. The Bajiaoting structure is in the descending wall of the Pinghu main fault to the north of the Fangheting structure. The Huagang Formation trap is also a complete anticline, trending northeast, and there are two high points. Borehole Pinghu No. 2 is in the northwest direction of the high point. The Wanghuting structure is located to the west of the Fangheting structure. The Wangyi fault descends the wall. It is a local structure developed on the nose-shaped uplift extending from the Haijiao bulge to the sag. It is divided into three fault blocks by two small faults. At present, the main oil and gas fields and boreholes are distributed in the western slope area, the central inversion structural belt in the central depression area, and the western sub-sag area. There are currently about 125 exploration and evaluation boreholes completed. The degree of exploration is relatively low. Figure 8.5 is a cross-section of the Fangheting structure and Huxinting structure along seismic survey line GP14-534.

1. In 2007, the client acquired a 3D OBC dataset over the Pinghu Oil and Gas Field. The low data quality of the existing vintage ocean bottom cable (OBC) seismic survey restricts further in-depth reservoir

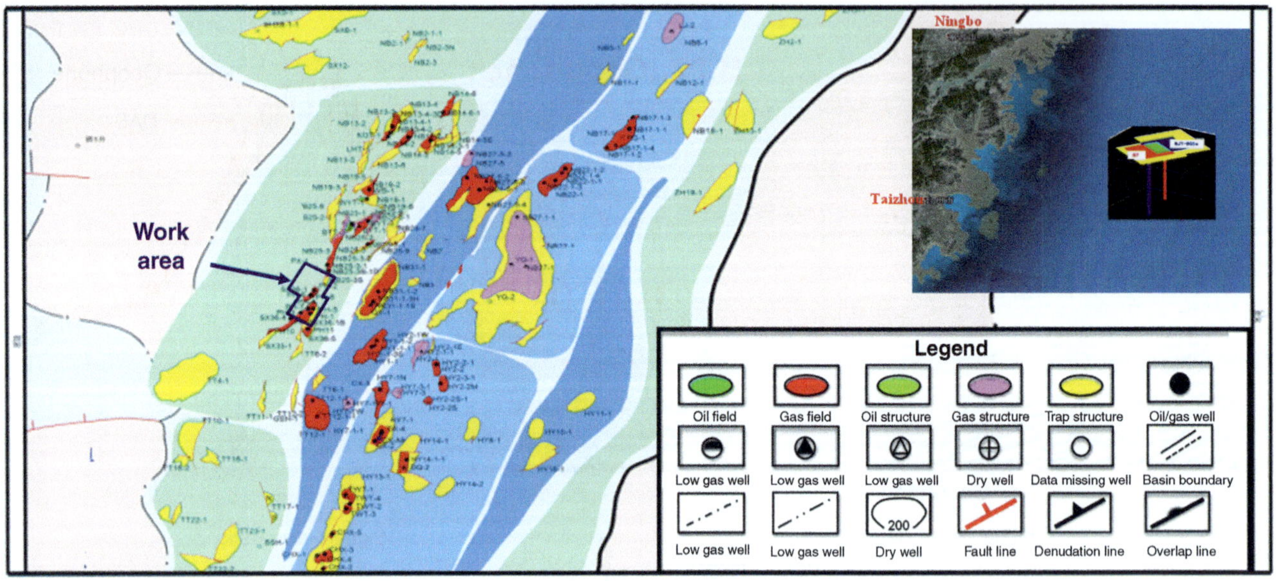

Figure 8.4 Regional geology and structure map of the Xihu Sag. The blue box is the area studied.

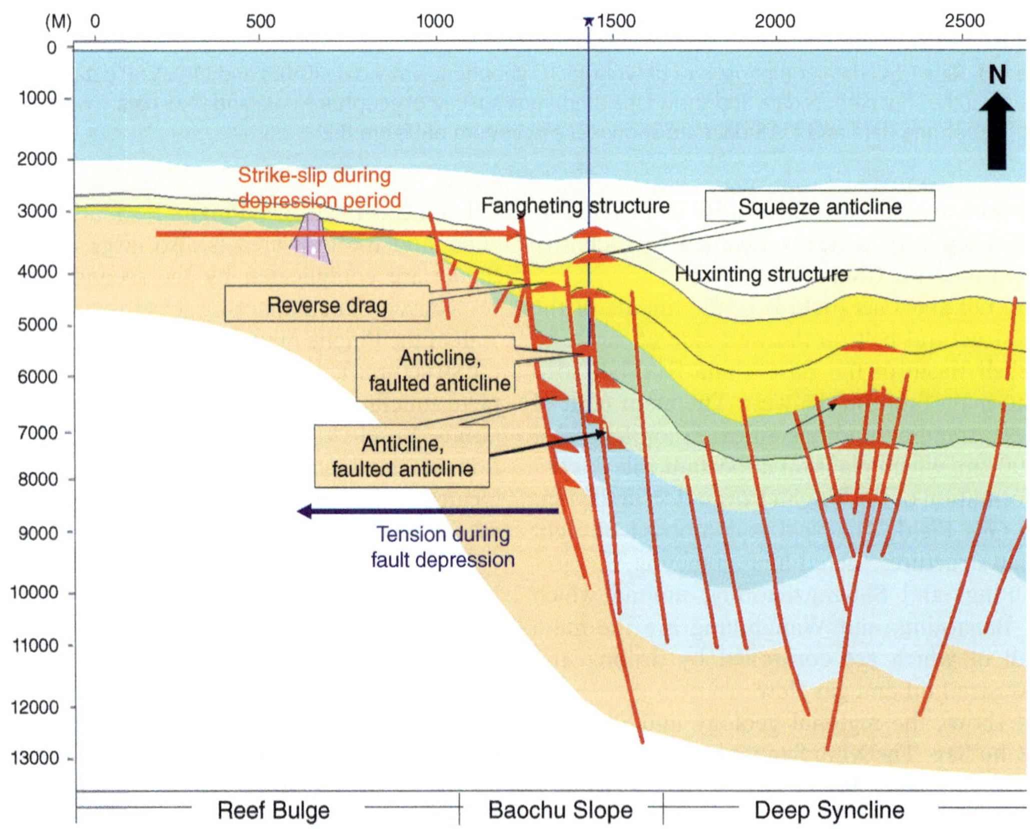

Figure 8.5 Cross-section of the Fangheting structure and Huxinting structure along the seismic survey line of GP14–534.

evaluation and field development. The main problems faced during evaluation and development are as follows:Target layer is deeply buried (more than 3,300 m), and the imaging quality in the middle and deep layers is poor, which makes it difficult to meet the geological requirements for fine structural interpretation and identification of dominant reservoirs.

2. The lateral variation of the reservoir is large, and reservoir prediction is difficult.
3. The vintage OBC data have poor imaging accuracy and low SNR.
4. Noise interference of the original seismic data is relatively serious, and the main structure has a blank area of seismic data.

Therefore, the oil company decided to conduct a new round of 3D seismic exploration using OBN in this area to enable more accurate structural mapping and reservoir characterization. The exploration area is located about 280 km southeast of Zhoushan City, Zhejiang Province, in the Pinghu Oil and Gas Field of the East China Sea. The sea floor of the work area is relatively flat, and the water depth is between 80 and 100 m. There are two operating platform areas in the data acquisition area, namely the Bajiao Pavilion and the Fanghe Pavilion main platform. There are also oil and gas pipelines on the sea floor in the survey area. The client required high-quality OBN seismic data for detailed characterization of deep oil and gas reservoirs, prediction of the properties of lithologic oil and gas reservoirs, fracture distribution in the fractured oil and gas reservoirs, and accurate structural evaluation of remaining oil and gas reservoirs in low-relief structures. The data acquisition requirements are to (a) improve the energy and SNR of deep formation data and (b) optimize the OBN survey design around the platform area to reduce the blank area of seismic data.

Shown in Figures 8.6 and 8.7, the Pinghu 3D OBN seismic data acquisition project is summarized in Table 8.2. The OBN survey designed total shots were 261,129, but the actual number of shots completed was 261,514. There were 322 shots more shots added horizontally across the platform. The total number of deployed nodes is 18,504, including 2,009 nodes deployed again (twice) in the second stage of the survey.

Figure 8.8 shows the borehole casing structure and armored optical cable (blue line) for boreholes G7 (left) and G5 (right). Both boreholes are deviated wells and have multiple casing strings. For well G7, the water depth is 88 m, the completed well depth is 4,270 m, and the production packer was set at 3,595 m. For well G7, the water depth is 81.2 m, the completed well depth was 4,270 m, and the target layer maximum well deviation was 11.4° ~ 36.4°.

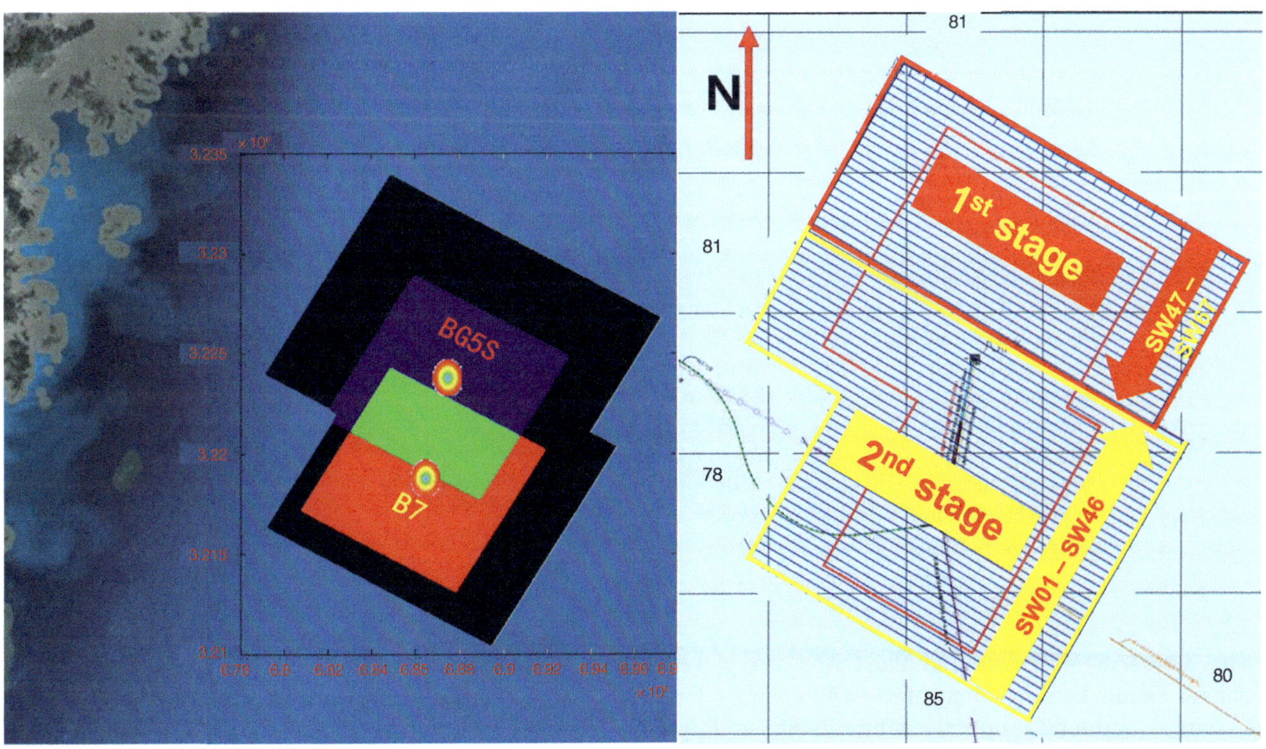

Figure 8.6 (Left) Joint 3D OBN and 3D DAS VSP shots map (black, OBN shots cover area; red, well B7 shots cover area; blue, well B5 shots cover area; green, both wells joint shots cover area). (Right) OBN acquisition shot swath stage plan.

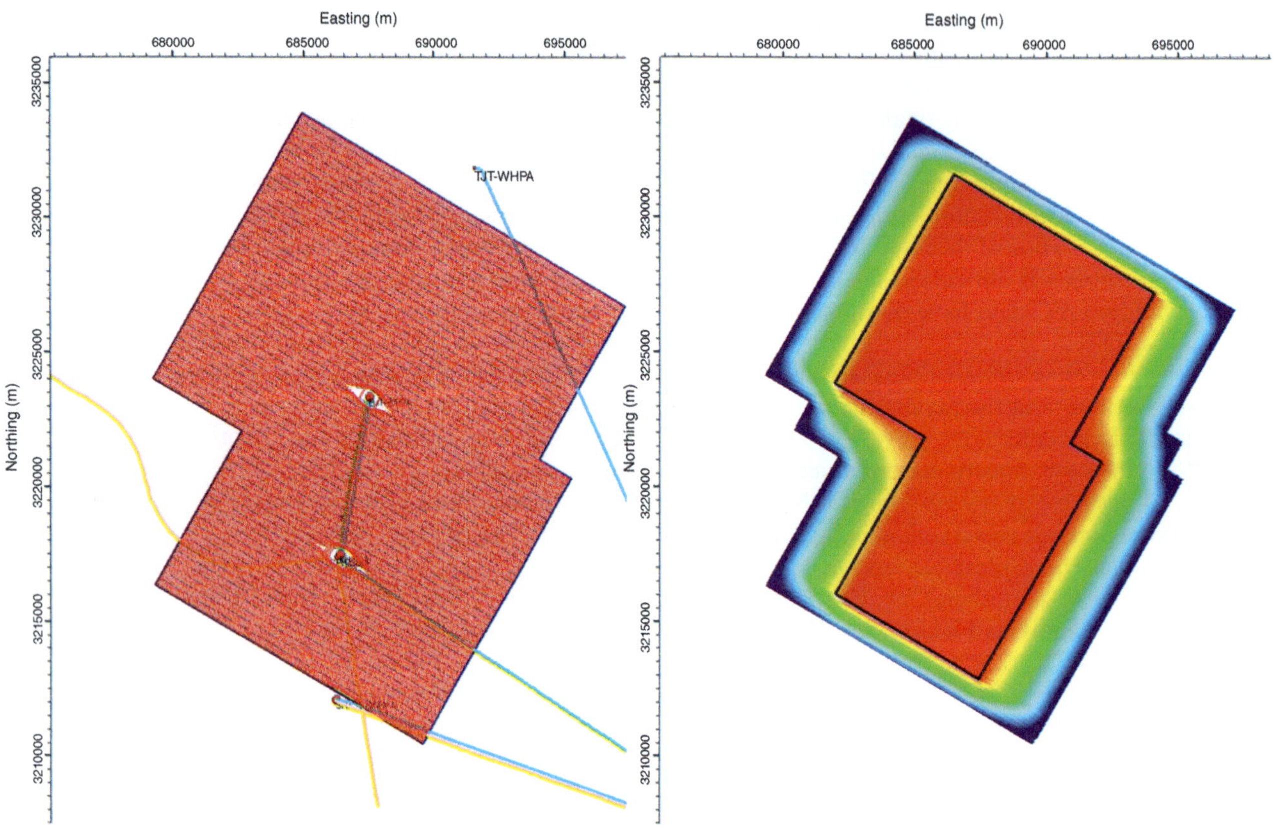

Figure 8.7 Actual OBN survey shot map (left) and full-coverage fold map (right).

Table 8.2 Pinghu 3-D OBN Seismic Data Acquisition Parameters.

Items	Survey parameters	Remarks
Shot swath	67	
Total shots	261,129	261,514 (completed)
Full coverage area	125.21 km^2	
Deployed nodes	18,504	2,009 (repeatedly deployed)
Recording channels	1,920	240 × 8
Receiver spacing	50 m	
Minimum offset	25.67 m	
Maximum offset	6,482.12 m	
Bin size	2.5 m × 25 m	
Fold number	640	8 (h) × 80 (v)
Inline shot spacing	37.5 m	
Crossline shot spacing	50 m	

The 3D DAS VSP data and the OBN survey were acquired simultaneously using the same shots. Two self-coupled armored optical cables with multiple high-temperature-rating single-mode fibers were deployed inside the casing of both wells B5 and B7. The length of the downhole armored optical cable was 3,085 and 3,537 m in wells B5 and B7, respectively. Each armored optical cable was connected to two uDAS interrogators on each platform to record two sets of 3D DAS VSP data, and the two datasets were stacked together for enhanced SNR processing. Figure 8.9 shows pictures of the 3D DAS VSP data recording interrogator uDAS system (left), data storage disk

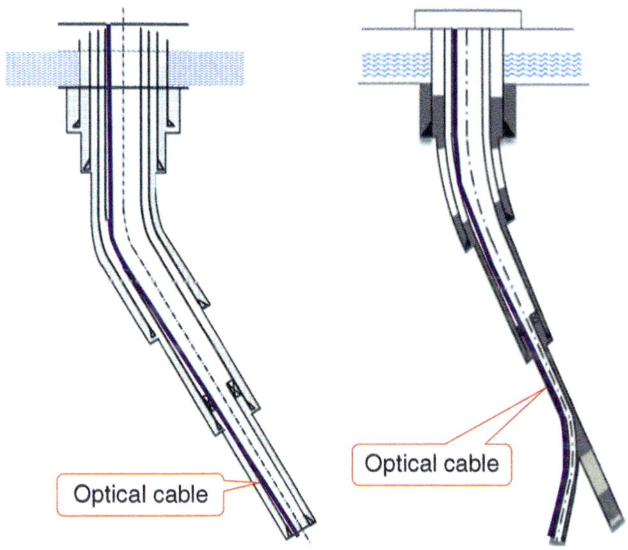

Figure 8.8 Borehole casing structure and armored optical cable (blue) for boreholes G7 (left) and G5 (right).

array (middle), and armored optical cable drum next to the wellhead (right) for 3D DAS VSP data acquisition. The downhole receiver spacing is 1 m from the sea floor to the TD. The armored optical cable recorded 62,473 shots inside well B5 and 48,411 shots inside well B7, respectively. 3D DAS VSP data were recorded continuously 24 hours/day without affecting OBN data acquisition, and the data were segmented according to the GPS time of each airgun shot after 3D DAS VSP data acquisition was completed.

Joint 3D OBN and 3D DAS VSP Data Acquisition in the Middle East

The world's largest 3D DAS VSP data acquisition as of this writing was completed in October 2022 through the joint OBN and 3D DAS VSP data acquisition project. The density of the airgun source carpet of the ongoing high-density OBN survey is 25 m × 25 m, and the OBN spacing is 150 m × 25 m. By using the airgun sources of an ongoing high-density OBN data acquisition project offshore of Abu Dhabi, high-density OBN data and 3D DAS VSP data were acquired simultaneously in a total of 13 wells in two different locations. At each location, all the production and injection wells have armored optical cable preinstalled outside the production or injection tubing, and the optical cable was strapped on the tubing with clampers at 10 m spacing.

At location 1, we have simultaneously completed 3D DAS VSP data acquisition from six wells, and the cumulative shots in each well total 406,309 and cover an airgun source carpet area of 154 km^2. At location 2, 3D DAS VSP data has been simultaneously acquired for seven wells, and the cumulative shots in each well total 431,706 and cover an airgun source carpet area of 157 km^2.

The OBN acquisition employed a pull/push outside shooting methodology with two source fleets operating at a separation of 6 km so that each source point was fired and recorded twice by each DAS cable in the borehole. Each source fleet comprised multiple vessels, with each vessel firing independently but observing a minimum separation of 4 km from any other source vessel. The total number of OBN shots acquired by two passes of air gun source shooting exceeded 2.4 million shots for 3D DAS VSP data acquisition at location 1 and about 3 million shots at location 2. Figure 8.10 shows the survey geometry of the joint high-density OBN and 3D DAS VSP data acquisition project. It is a plane view of the well trajectory projection of the six wells in location 1 and seven wells in location 2. Zipper 2 and Zipper 3 are the two OBN source sweeps around locations 1 and 2. Figure 8.11 is a survey geometry display of the 3D DAS VSP data acquisition in location 1. A plane view of the well trajectory of the six wells in location 1 is shown

Figure 8.9 Pictures of the 3D DAS VSP data recording interrogator uDAS system (left), data storage disk array (middle), and armored optical cable drum next to the wellhead (right).

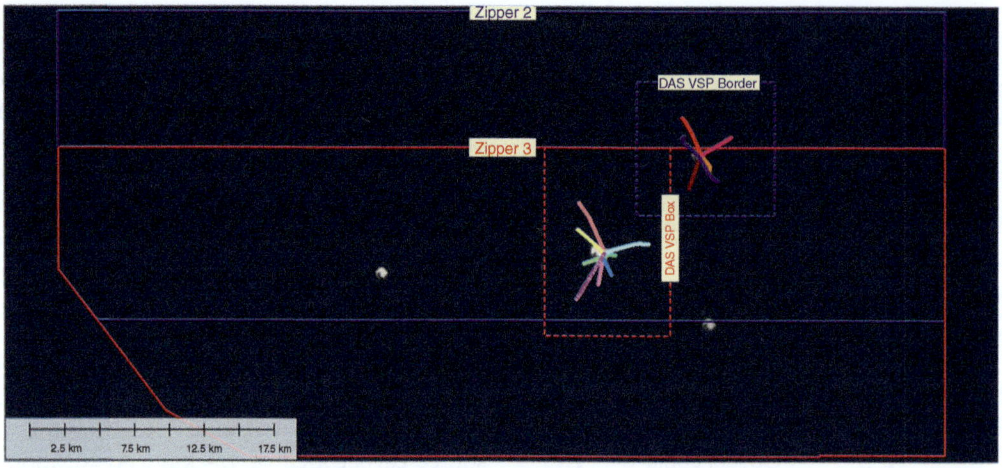

Figure 8.10 The survey geometry of the joint high-density and 3D DAS VSP data acquisition project. Plane view of the well trajectory of the six wells in location 1 and seven wells in location 2. Zipper 2 and Zipper 3 are the two OBN source sweeps around locations 1 and 2.

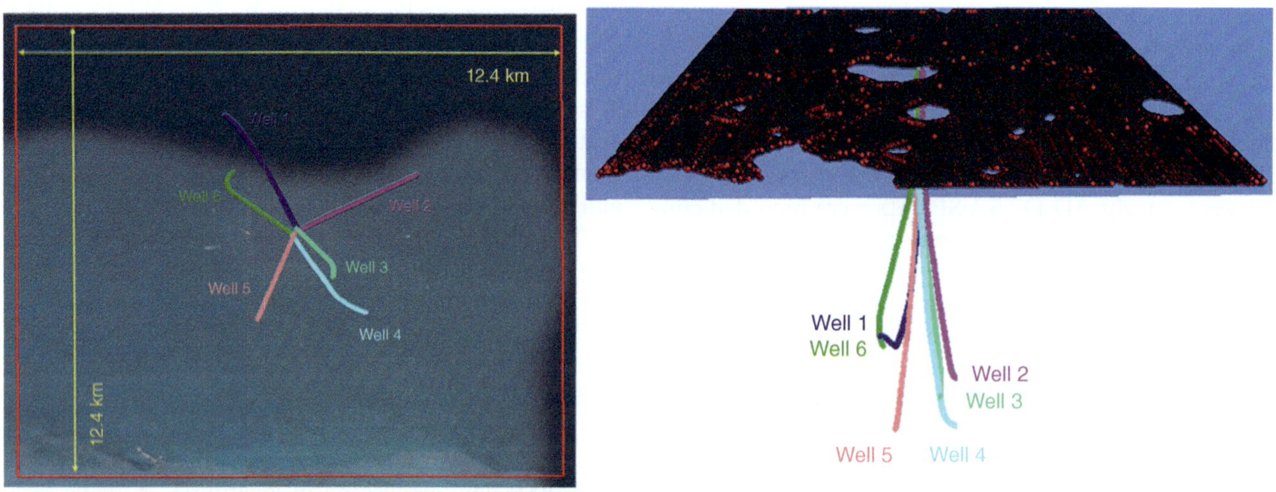

Figure 8.11 The survey geometry of the 3D DAS VSP data acquisition project in location 1. (Left) Plane view of the well trajectory of the six wells in location 1; (right) 3D view of the source distribution and the well trajectory with an armored optical cable in six wells at location 1.

on the left, and a 3D view of the source distribution and the well trajectory of the six wells in location is shown on the right. As shown in Figure 8.10, all 13 wells in both locations are highly deviated, horizontal wells, but the armored optical cables strapped on the outside of the production or injection tubing using metal clampers with 10 m spacing are only installed from the wellhead to the turning point of the horizontal section of the well. The downhole receiver spacing of the 3D DAS VSP data recorded is 1 m, the sampling rate is 1 ms, and the record length of the 3D DAS VSP data is 10 s. The original 3D DAS VSP data were recorded continuously 24/7 in the field. Each OBN shot time and the shot GPS coordinates were used to segment the continuously recorded 3D DAS VSP data into the CSG data with a 10 s record.

8.3. 3D SURFACE SEISMIC DATA AND DAS VSP DATA PROCESSING

8.3.1. Enhanced 3D Seismic Data Processing Using Jointly Acquired DAS VSP Data

Borehole seismic data can provide a higher-resolution structural image around the borehole than that of the surface seismic as its receivers are located inside a borehole and near the reservoir. Apart from its structural imaging capabilities, the formation average and interval velocity, absorption, attenuation, deconvolution operator, anisotropy parameters, and other reservoir information in a certain range around the wellbore can be obtained directly from the VSP data.

Because DAS VSP data are different from the VSP data recorded by the 3C downhole geophone array, the data processing workflow is different between the two datasets. DAS VSP data are the fiber strain, not the particle velocity at each depth location, so the following special data processing procedures were applied: (a) conversion of displacement data to velocity data, (b) optical cable slapping and ringing noise removal, and (c) group forming to increase SNR. After these special processing steps on the DAS VSP data, the typical VSP data wavefield separation and imaging processing steps can be conducted.

The formation average and interval velocity, absorption, attenuation, deconvolution operator, anisotropy parameters, and other information obtained from the VSP data around the well can be used to enhance the surface seismic data processing, and we call this step "borehole-driven 3D surface seismic data processing." It includes velocity model calibration and modification, statics correction, deconvolution processing, demultiple processing, high-frequency restoration, anisotropic migration, Q compensation or Q migration, and so on. The workflow of borehole-driven 3D surface seismic data processing using parameters extracted from the borehole seismic data is shown in Figure 8.12. The most noticeable enhancement of surface seismic data processing is the use of the correct anisotropic parameters and Q volume, as well as a more accurate velocity model for 3D anisotropic and Q migration.

Figure 8.13 compares the surface seismic prestack depth migration (PSDM) image, downhole geophone array walkaway VSP image from well B-02, and walkaway DAS VSP image from well B-01. The frequency range of the three different imaging results is increased from 60 Hz for the surface seismic PSDM data to 70 Hz for well B-02 downhole geophone array walkaway VSP imaging, and further to 85 Hz for well B-01 walkaway DAS VSP imaging. One can easily identify subseismic faults and pinch-out structures from the DAS VSP imaging using higher-frequency and higher-resolution imaging data, but these fine structures are not visible from the surface seismic PSDM and downhole geophone array VSP images.

Seismic exploration practice shows that elastic media rarely exist in the subsurface, and subsurface media exhibit viscoelastic behavior. The viscosity of the subsurface

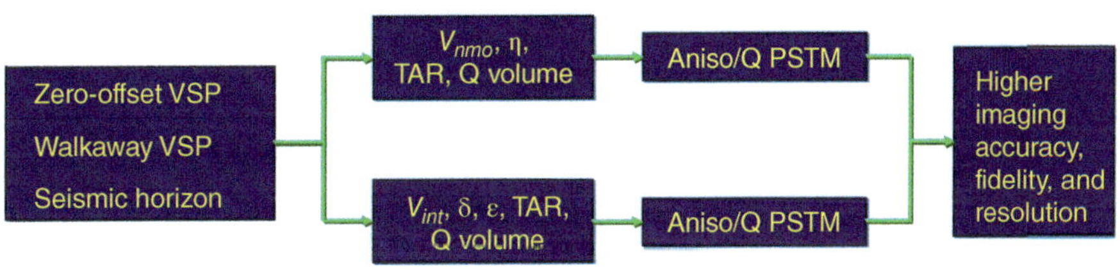

Figure 8.12 Workflow for borehole-driven surface seismic data processing enhancement using parameters extracted from the borehole seismic data.

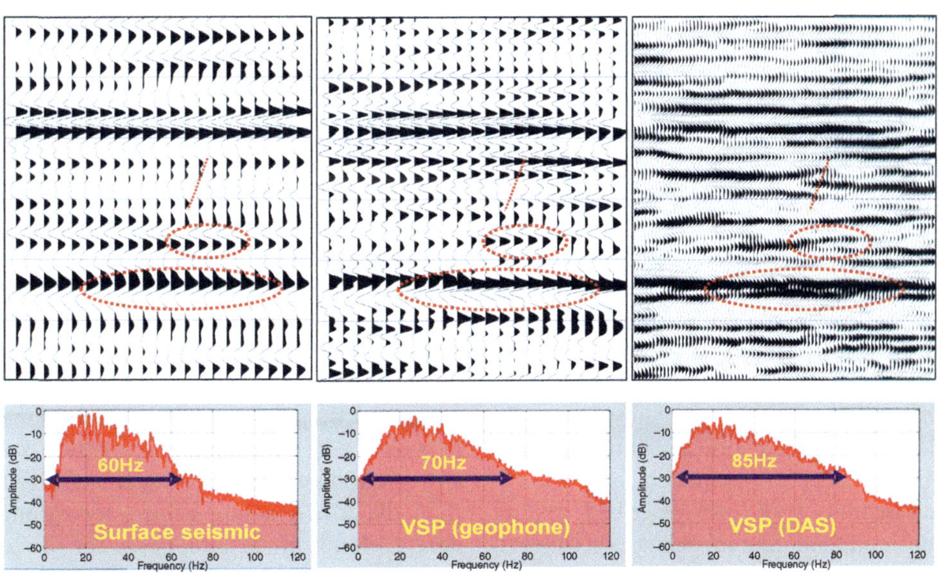

Figure 8.13 Surface seismic PSDM image (left), downhole geophone array walkaway VSP image (center), and walkaway DAS VSP image (right) and their amplitude spectrum display.

medium consumes the energy of the seismic wave, attenuates the amplitude of the seismic signal, distorts the phase of the seismic wavelet, and reduces the resolution and imaging accuracy of the seismic data. Absorption decreases the amplitude of waves propagating in the Earth. In addition, it narrows the bandwidth and modifies the phase. Obtaining true-amplitude migration and acceptable resolution in dissipative media requires a migration algorithm that corrects these effects (Causse and Ursin, 2000). The earliest viscoelastic compensation appeared in a one-dimensional form. This is the reverse Q filtering technique that is still widely used in the industry. However, the attenuation of viscoelastic seismic waves is related not only to the propagation time but also to the frequency of the wavelet, the distribution of the velocity field, the ray path, and the subsurface geological structure. Obviously, one-dimensional compensation cannot fully restore the original subsurface characteristics of the seismic wave field.

Dai and West (1994) proposed an inverse Q migration method to image subsurface structures from seismic reflection data in an anelastic media. The downward extrapolation of surface seismic data is based on a highly accurate numerical solution for a wave system in the space-frequency domain in an inhomogeneous viscoacoustic medium. They used a pseudospectral method to complete the post-stack depth migration of viscous acoustic waves and developed viscoelastic compensation from one dimension to two dimensions. Mittet and Sollie (1995) used explicit convolution filtering in the frequency space to complete the prestack depth migration of viscous acoustic waves and numerically analyzed the sensitivity of the macro-Q model error based on the migration results. They presented a method for accommodating absorption and dispersion effects in depth migration schemes. Extrapolation operators that compensate for absorption and dispersion are designed using an optimization algorithm. The design criterion is that the wavenumber response of the operator should equal the true extrapolator. To better adapt to lateral changes in velocity and better suppress the evanescent wave energy, Causse and Ursin (2000) used a reverse time migration to complete the prestack depth migration of the viscous acoustic wave. They used a stable amplifying finite-difference wavefield extrapolation that compensated for the attenuating propagation of wavefields in the real Earth. Numerical tests on synthetic VSP data illustrated how dissipation destroyed the amplitude, phase, and resolution in the migration images and how their algorithm corrected these effects. The tests also showed that the method was stable with respect to noise and did not require a very detailed macro model for dissipation.

However, the research work just described mainly focused on migration methods and modeling experiments, and few examples of viscoelastic migration techniques have been applied in industry. There are two main factors hindering the industrial application of viscoelastic migration technology. One is that it is difficult to estimate a relatively accurate attenuation parameter model using actual seismic data. There are currently at least 10 methods known to the authors to estimate attenuation parameters from seismic data, but the noise immunity and stability of these methods cannot meet the requirements of commercial seismic data processing. The second is the effect of high-frequency noise on viscoelastic migration. Viscoelastic migration amplifies high-frequency energy. When high-frequency signals are amplified, high-frequency noise is amplified at the same time. The effect of this situation is especially serious on deep-layer imaging. When an elastic wavefield propagating in a viscoelastic medium is attenuated, high-frequency noise in the deep layer is often amplified tens of thousands of times, which seriously reduces the SNR of seismic records after viscoelastic migration.

The spectral ratio method is the most-used approach to estimate the Q factor from VSP data. Dean and Correa (2017) described a methodology that can be used to correct the frequency spectra and thus successfully estimate unbiased Q values from DAS VSP data. For both wells B-01 and B-02, the spectral ratio method was used to derive the Q value from the surface to TD. Zero-offset DAS VSP data and geophone recorded VSP data were used to calculate the Q value for wells B-01 and B-02 separately. Because well B-01 is far from the coastline, the armored optical cable in this well did not record any 3D DAS VSP data from the offshore airgun source. As the gauge length in the DAS method has an attenuative effect on the high frequencies of the DAS VSP data, the Q value from the longer-gauge-length DAS VSP data is the average Q value within the formation of the longer gauge length. In practice, one would prefer to use short-gauge-length DAS VSP data to calculate the Q value if possible. Because we only recorded the DAS VSP data using an 8 m gauge length, the choice of using short-gauge-length DAS VSP data to calculate the Q value was not available.

VSP data from five existing wells within the 3D surface seismic survey area were used to derive Q values from wellhead to TD along each borehole (Figure 8.14). These Q values were used to build the mid-deep layer Q-field data volume for the entire 3D seismic data coverage area, and uphole data were used to establish the near-surface 3D Q-field data volume (Figure 8.15 and Figure 8.16). Then we used our tomographic inversion code Q_{fit} to optimize the initial Q-field. The basic principle of Q tomographic inversion is to invert the Q value according to the frequency anomaly of seismic data. From the perspective of inversion, this method seeks the relative Q variation of anomalies against a background rather than the true absolute values, for example where there are gas clouds in the data. From the perspective of migration, because there are many causes of energy loss, Q migration is one of the means for amplitude compensation, but not the only method. Our tomographic inversion

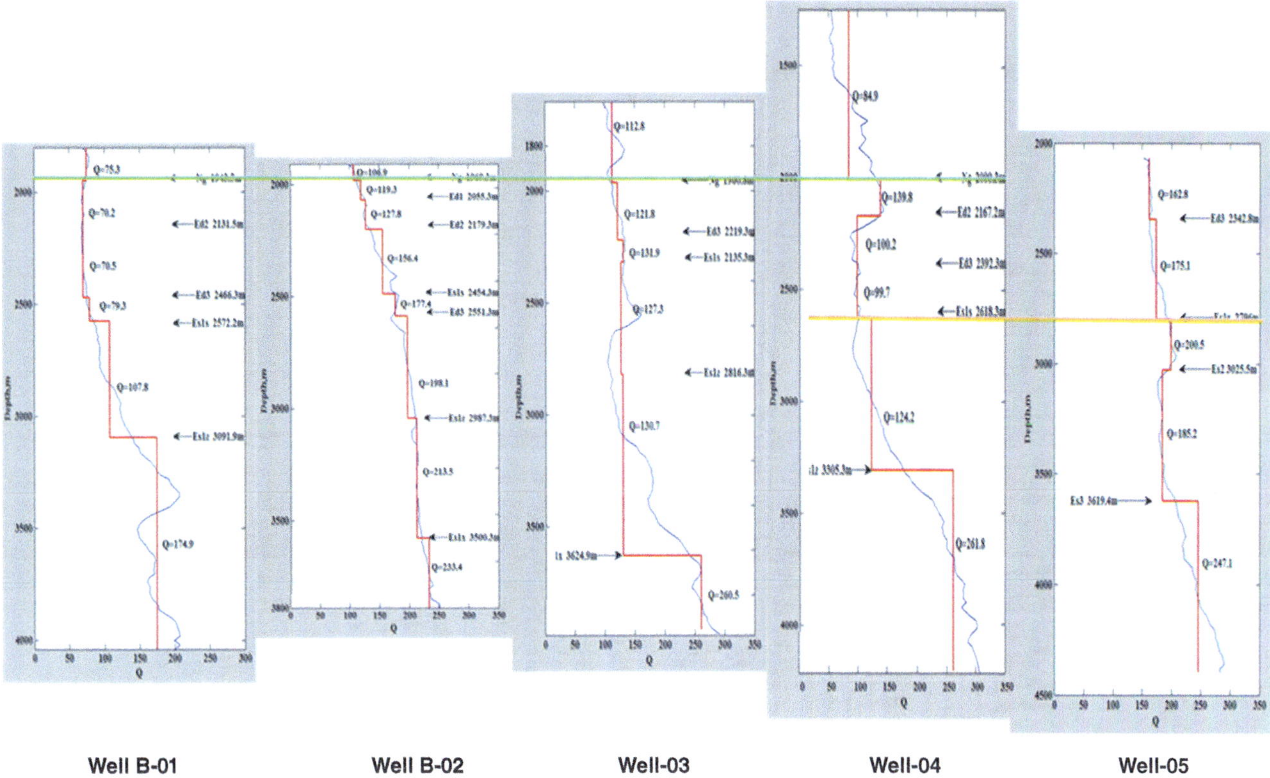

Figure 8.14 VSP-data-derived Q values (thin blue curve) and horizon tops (thick green and orange lines) in five wells within the 3D surface seismic survey area.

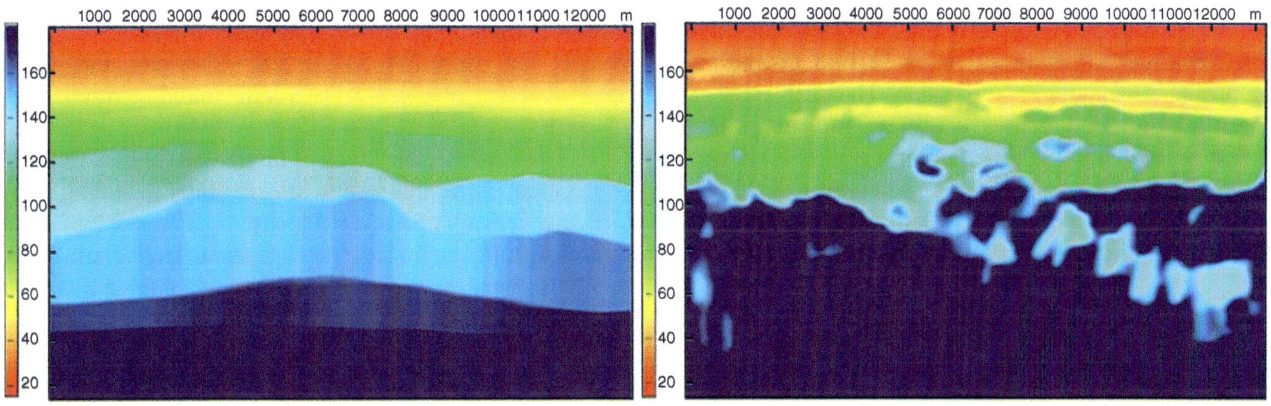

Figure 8.15 The changes in the Q-field before Q tomographic inversion (left) and after Q tomographic inversion (right).

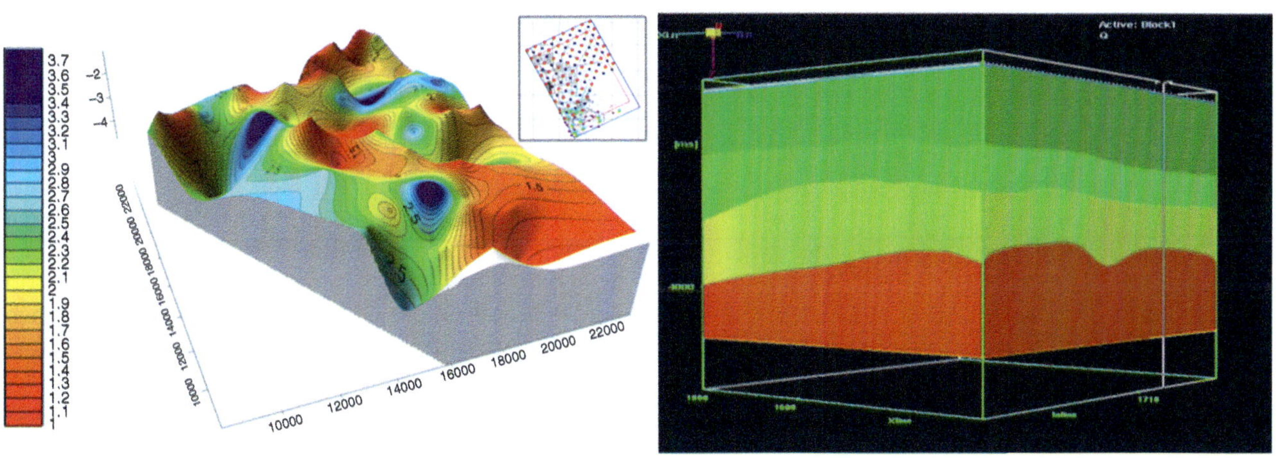

Figure 8.16 (Left) Colored Q value of near-surface formation; the *Z* axis is the thickness of the near-surface low-velocity zone. (Right) Mid-deep formation Q-field data volume from the VSP data in five wells.

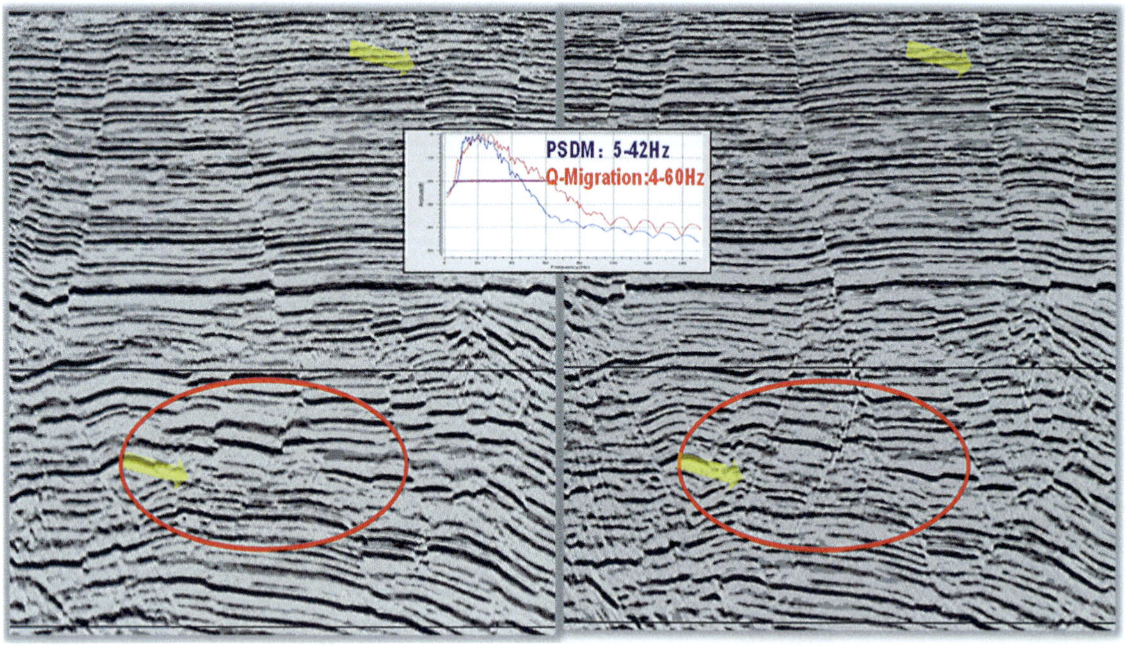

Figure 8.17 Image comparison between conventional migration (left) and Q migration (right).

code Q_{fit} has the inversion function of quality factor (attenuation factor) for 2D and 3D isotropic media. The "essence" of Q inversion is to find Q anomalies according to frequency anomalies. Therefore, the algorithm can only find areas of Q outliers and cannot obtain the real absolute Q value completely. As shown in Figure 8.15, one can see the changes in the Q-field before Q tomographic inversion (left panel) and after Q tomographic inversion (right panel).

Figures 8.17 and 8.18 show the comparison between conventional migration and Q migration in the section and coherency time slice, respectively. Q migration can eliminate the influence of formation attenuation on vertical and horizontal formation resolution and improve imaging accuracy. Both Figures 8.17 and 8.18 show much sharper and more focused faults and fracture zones within the reservoir. We built a new 3D reservoir structural model with much finer details based on the Q migration and coherence data (Figure 8.19). After careful interpretation of the newly built reservoir structural model, three new borehole locations were recommended to the client. The final drilling results on these three new boreholes are very successful, with a high volume of commercial oil flow.

8.3.2. 3D DAS VSP Data Processing for the East China Sea

3D DAS VSP data were acquired by a downhole armored optical cable simultaneously with an ongoing OBN data acquisition project in the East China Sea. Upgoing wave imaging, common receiver gather optimization imaging, and downgoing wave imaging were conducted on this 3D DAS VSP dataset. Integrated interpretation of the 3D DAS VSP imaging results was also performed.

The geological objectives of this project are as follows:

1. Obtain accurate P-wave velocity data, including average velocity and layer velocity.
2. Calibrate accurately each seismic reflection layer with local geological formation.
3. Provide formation attenuation parameter Q, TAR, and deconvolution parameters.
4. Undertake multiple wave analysis of newly acquired OBN data.
5. Extract accurately the anisotropy parameters and borehole-driven processing parameters for migration; it is expected that the resolution of the new migration data will be improved by 5–10 Hz.
6. Perform high-resolution 3D DAS VSP imaging data around the borehole.

The 3D DAS VSP data quality analysis is shown in Figure 8.20, including dominant frequency, SNR, energy, and background noise level. Figure 8.20 shows that the overall data quality is good: (a) single-shot raw data from near to far offset from both wells B7 and B5 have strong energy, high SNR, high dominant frequency, and broad bandwidth, and the 3D DAS VSP data have rich low-frequency energy with clear and continuous first arrivals and can be picked up easily; (b) the 3D DAS VSP data have different background noise and resonance

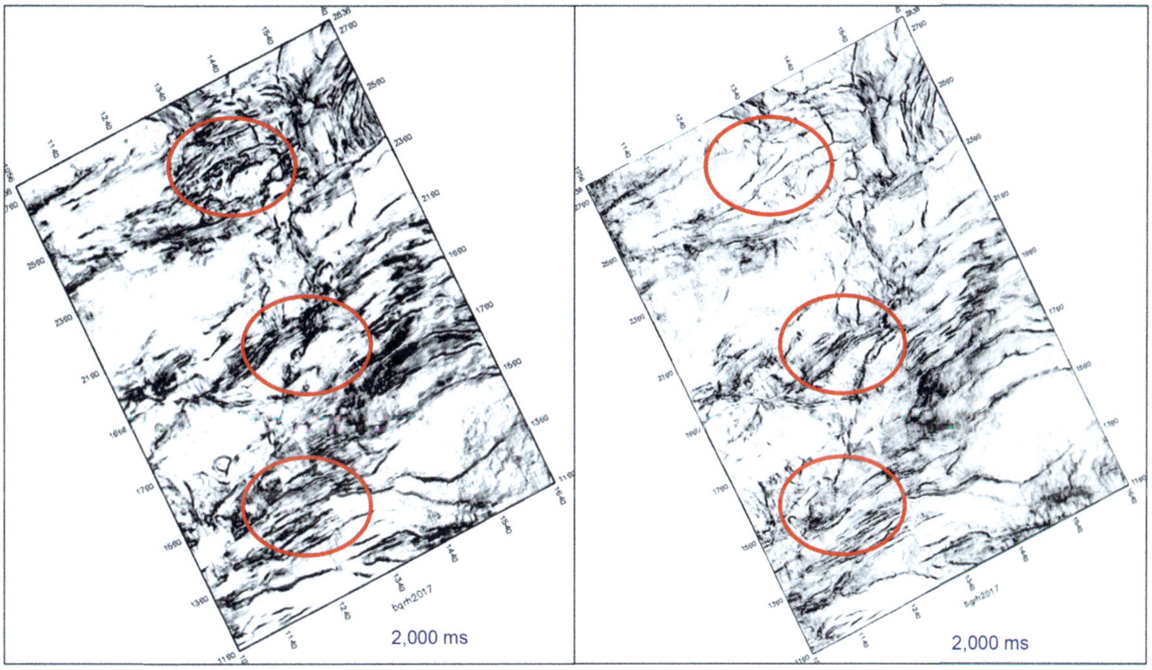

Figure 8.18 Comparison between the 2,000 ms time slice of coherence data from conventional migration (left) and Q migration (right).

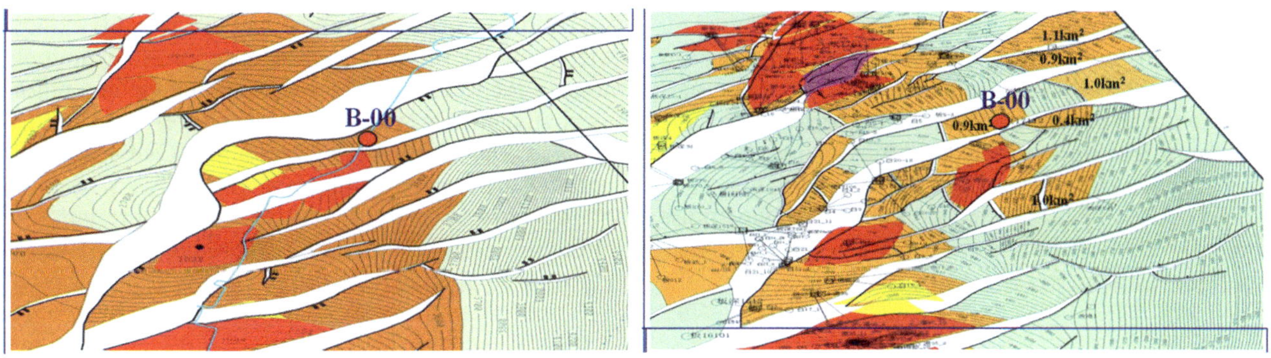

Figure 8.19 Structural maps for the old reservoir top (left) and the new reservoir top (right).

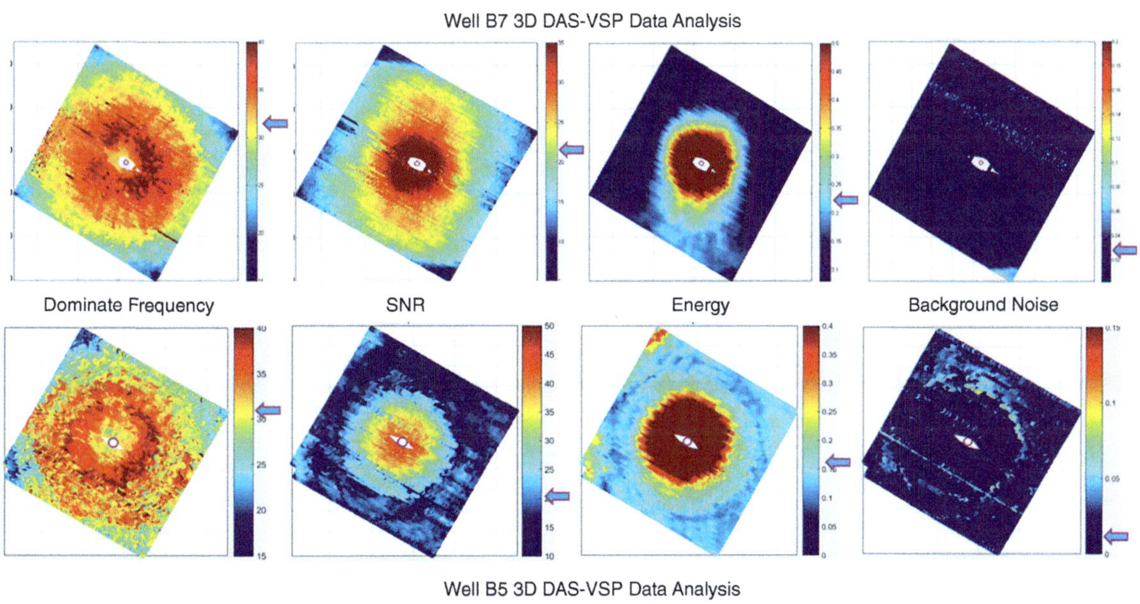

Figure 8.20 Raw 3D DAS VSP data quality analysis (dominant frequency, SNR, energy, and background noise level) for well B7 (top) and well B5 (bottom). Yu et al. (2022).

interference, and some well sections have inevitable interference caused by the surging of seawater; (c) the 3D DAS VSP data have abundant wave fields, and the multiple waves are relatively well developed; (d) there is no difference between records obtained during multiple armored optical cable entries, indicating the stability of the acquisition equipment and that the parameter settings are not abnormal; and (e) all of the well high-density data meet the requirements of borehole seismic parameter calculation and accurate 3D DAS VSP data imaging around the borehole.

The 3D DAS VSP data have the advantages of high density and full well bore coverage, but because of the poor coupling between the armored fiber-optic cable and the casing wall on some parts of the borehole section, the ringing noise from the cable is strong. In Figure 8.21(a), the apparent velocity of the ringing noise is similar to the effective signal in the time domain, and it appears as a periodic pulse signal in the spectrum and periodic in the autocorrelation spectrum. It is difficult to effectively remove such noise with conventional denoising methods.

We used a ringing noise attenuation technique based on a Tau-P transform. The Tau-P transform is first used to transform DAS VSP data into the Tau-P domain, and then the linear prediction method is used to predict the periodic ringing noise in the data. Finally, the predicted ringing noise is subtracted from the original data.

Figure 8.21 shows the effect of removing ringing noise in the 3D DAS VSP data from the Pinghu area. After using the ringing noise attenuation technique of the Tau-P transform, the ringing noise has been effectively attenuated and suppressed (Figure 8.21(b)), and no effective reflection information has been removed from the removed noise data (Figure 8.21(c)).

The conventional 3D VSP data migration processing steps include observation system definition, preprocessing, first-arrival picking, static correction, amplitude compensation, deconvolution, wavefield separation, velocity model building, and upgoing wave migration processing (Li et al., 2010). After denoising processing, wavelet deconvolution, vector wavefield separation, anisotropic velocity modeling, and correction of seismic migration velocity field, upgoing wave imaging, common receiver gather optimization imaging, and downgoing wave imaging of the 3D DAS VSP data are performed. The main processing steps are (a) suppressing ringing noise and abnormal amplitude interference, laying the foundation for the separation of upgoing and downgoing wavefields; (b) deconvolution processing, using downgoing waves to obtain wavelets for bubble suppression and wavelet zero-phase processing; (c) wavefield separation, separating the upgoing and downgoing wavefields to expand the effective bandwidth of the data; (d) shallow water multiple wave suppression, suppressing multiple waves of the water layer and highlighting the real reflection characteristics of the formation; and (e) 3D DAS VSP data imaging, using the angle domain Gaussian beam migration method to process the 3D DAS VSP data to obtain high-precision 3D DAS VSP imaging results.

In comparison with surface seismic data, the VSP wavefield is more abundant, with not only upgoing reflected

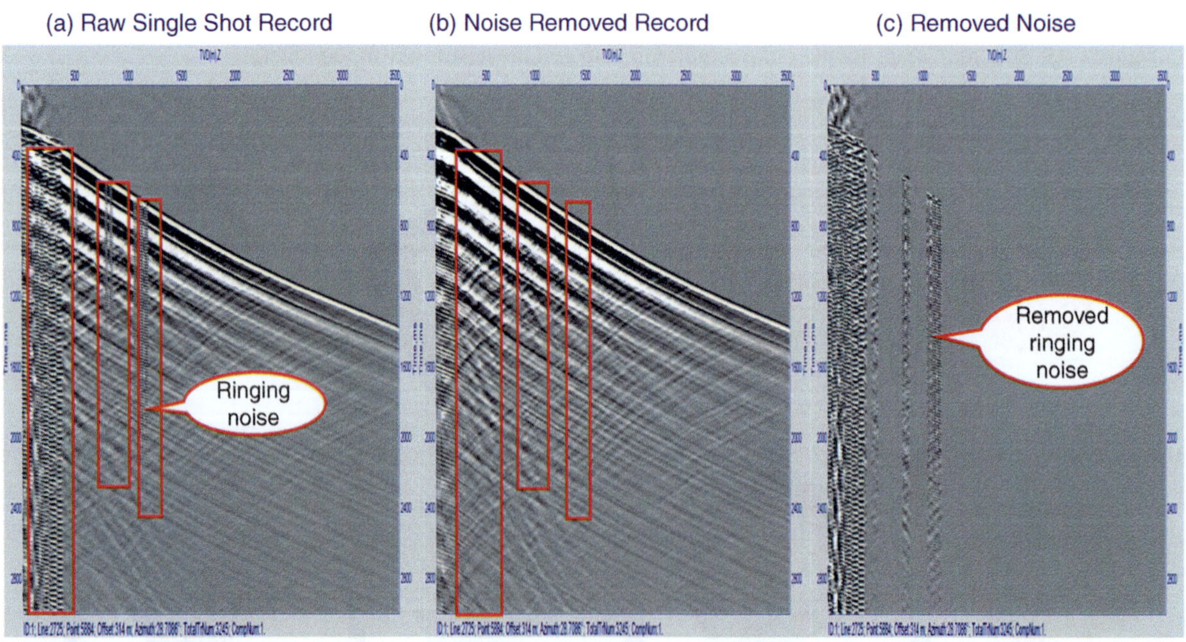

Figure 8.21 Cable ringing noise in 3D DAS VSP data removal: (a) raw single-shot record; (b) ringing noise removed record; (c) removed ringing noise. Yu et al. (2022).

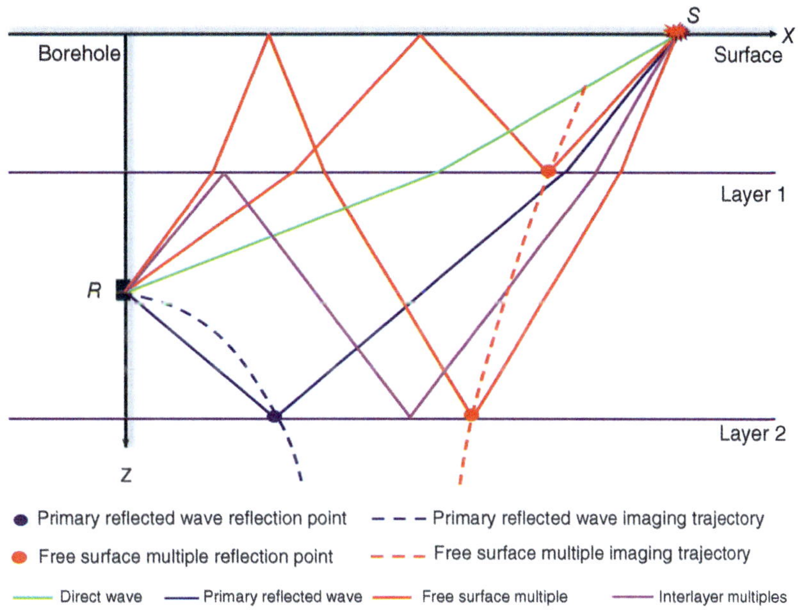

Figure 8.22 The propagation paths of different wave modes recorded by a VSP survey. Yu et al. (2022).

waves but also downgoing reflected waves with wider illumination (Figure 8.22). Normally, the free-surface multiples, interlayer multiples, etc., are included in the downgoing reflected waves. In Figure 8.22, the model has two reflective layers and one observation well, S is the source, and R is the receiver. The propagation path of the direct wave, primary reflected wave, free-surface multiples, and interlayer multiples are along green, blue, red, and pink lines, respectively. The primary reflected wave is propagated from shot point S to the reflection point of the reflective layer to receiving point R. The free-surface multiples are propagated from shot point S to the reflection point of the reflection layer, from the reflection point of the reflection layer to the reflection point of the free surface (surface), and then to receiving point R. Receiver R can receive not only free-surface multiples from the underlying formation but also free-surface multiples from the overlying formation. In comparison with the positions of the reflection points, it shows that the lateral distribution of the multiple reflection points of the free surface is wider, and the vertical distribution is shallower. After the imaging processing of the downgoing reflected wave, the reflection imaging above receiving point R can be achieved, and the lateral imaging range has been expanded.

The free surface multiple migration method for the free-surface multiple (Liu et al., 2011, Liu et al., 2012, Wang, et al., 2014, Liu et al., 2015) is used in the 3D DAS VSP data migration processing. The Fourier integral equation of one-way wave field continuation from the scalar wave equation was derived by Ferguson and Margrave (2005), and they gave an explicit expression of the angular frequency domain. Based on the single-pass prestack depth migration of the primary reflected wave method and the propagation characteristics of free-surface multiples, the migration can be realized only by modifying the wavefield continuation method. The one-way wavefield continuation imaging method of free-surface multiple waves can be described as follows:

Step 1: Rearrange the Z components of the VSP data into the CRG domain dataset.

Step 2: Perform forward continuation of the source wavelet to the surface. Let the source wavelet originate at the receiver point R in the well. The source wavelet wavefield extends upward continuation layer by layer with step Δz, and the continuation formula from z to $z - \Delta z$ is as follows:

$$\psi_s(x, z + \Delta z, \omega) = \psi_s(x, z, \omega)e^{-ik_z\Delta z} \tag{8.1}$$

where $\psi_s(x, z, \omega)$ is the source wavefield in the angular frequency domain at depth z, and k_z is the wave number in the z direction. The wavefield continuation operator can be the Fourier operator of a split-step, Fourier finite-difference operator, generalized screen operator, and so on.

Step 3: Continue one step in the reverse direction. The CRG wavefield extends downward from the surface layer by layer with step Δz according to equation (8.8.2):

$$\psi_r(x, z + \Delta z, \omega) = \psi_r(x, z, \omega)e^{ik_z\Delta z} \tag{8.2}$$

where $\psi_r(x, z, \omega)$ is the free surface multiple wavefield in the angular frequency domain at depth z, and k_z is the wave number in the z direction.

Step 4: Continue forward one step. The source wavelet wavefield of the second step is extended downward from the surface layer by layer with step Δz according to equation (8.8.3):

$$\psi_r(x, z - \Delta z, \omega) = \psi_r(x, z, \omega)e^{ik_z\Delta z} \tag{8.3}$$

where $\psi_r(x, z, \omega)$ is the free-surface multiple wavefield in the angular frequency domain at depth z, and k_z is the wave number in the z direction.

Step 5: Extract the imaging volumes. Cross-correlate the wavefields extended in steps 3 and 4, and take the zero-time imaging volume. Repeat steps 3, 4, and 5 until the maximum depth is reached.

Because both boreholes B5 and B7 are deviated and the airgun source has to be restricted at 500 m from the platform, 3D DAS VSP upgoing reflection wave imaging has a very limited aperture and coverage area, and downgoing multiple waves were used to conduct the 3D DAS VSP data migration to increase the imaging coverage area. According to the characteristics of downgoing multiples in the offshore 3D DAS VSP data, the offshore 3D DAS VSP downgoing multiple migration method has been innovatively developed, which significantly expands the 3D DAS VSP imaging aperture and improves overall 3D DAS VSP imaging quality. Figure 8.23 shows the downgoing multiple reflection wave imaging processing workflow of the 3D DAS VSP data.

Because the resolution of the downgoing multiple wave image is low, Q migration of the downgoing multiple waves was introduced to minimize the influence of absorption attenuation and improve imaging resolution. Figures 8.24 and 8.25 show the migration results of two sets of 3D DAS VSP data for wells B5 and B7: one is the depth migration based on upgoing wave data only (left), and the other is the depth migration based on downgoing multiple wave data (right); an updated velocity field using anisotropic velocity modeling was the final migration velocity model (Sun et al., 2009).

The 3D view of the downgoing multiple reflection wave migration result of the Pinghu 3D DAS VSP from well B7 is shown in Figure 8.26. This 3D imaging data volume will be used to guide the 3D structure interpretation and generate the 3D attribute data volume for reservoir characterization around the wellbore. Because the downgoing multiple reflection wave migration expanded the coverage area of the 3D DAS VSP data, this migration result can be used to map the detailed reservoir boundary and characterize fluid distribution near the bottom of the wellbore.

The comparison of newly acquired 3D OBN data imaging, 3D DAS VSP upgoing wave imaging, and 3D DAS VSP downgoing multiple reflection wave imaging in well

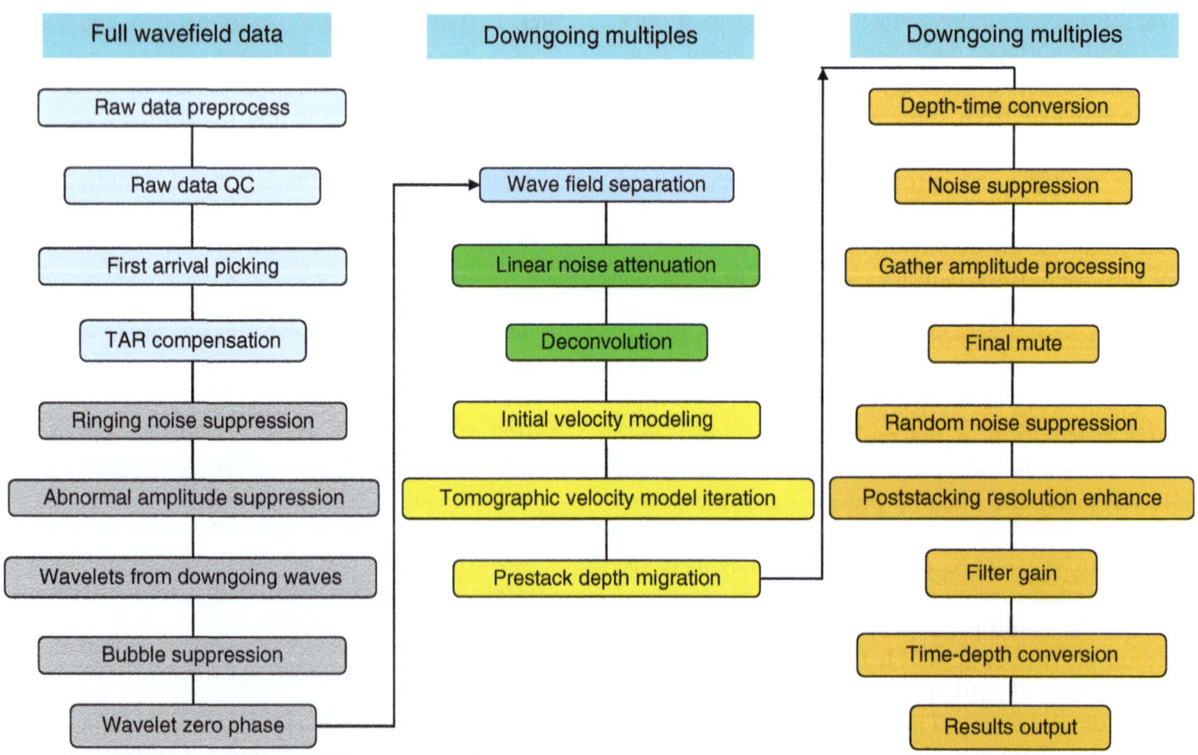

Figure 8.23 The migration processing workflow of 3D DAS VSP data downgoing multiple reflection waves. Yu et al. (2022).

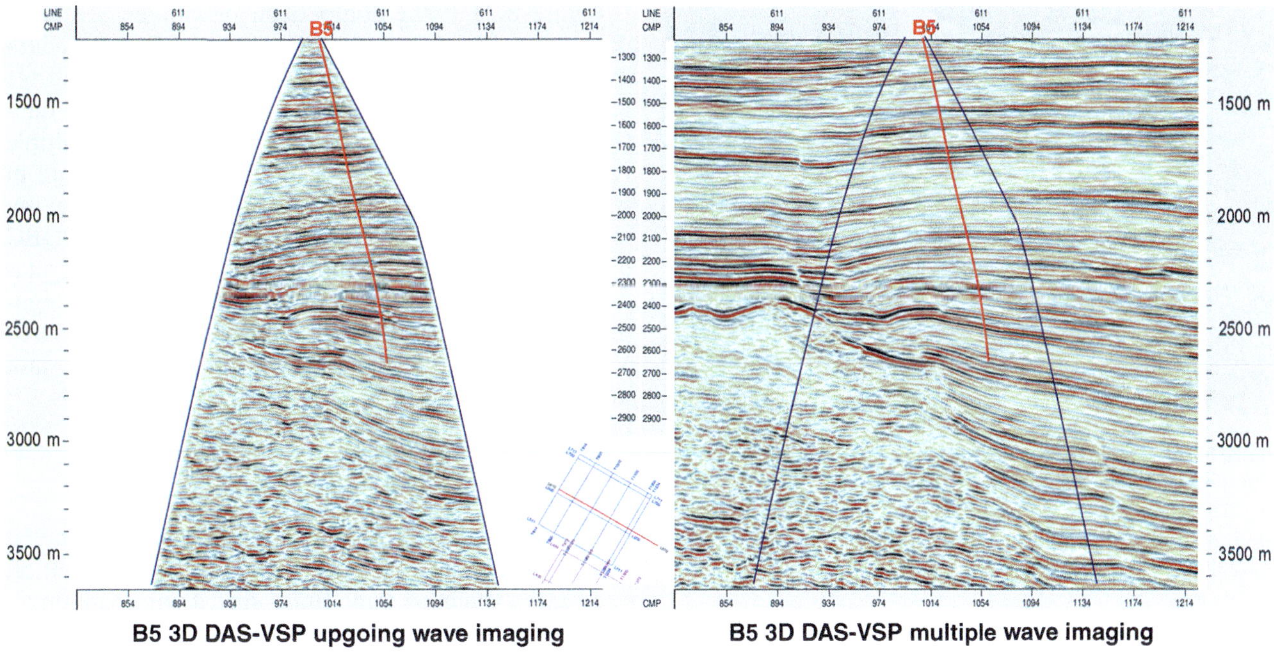

Figure 8.24 Imaging comparison for 3D DAS VSP data from borehole B5: upgoing reflection wave migration (left) versus downgoing multiple reflection wave migration (right). Yu et al. (2022).

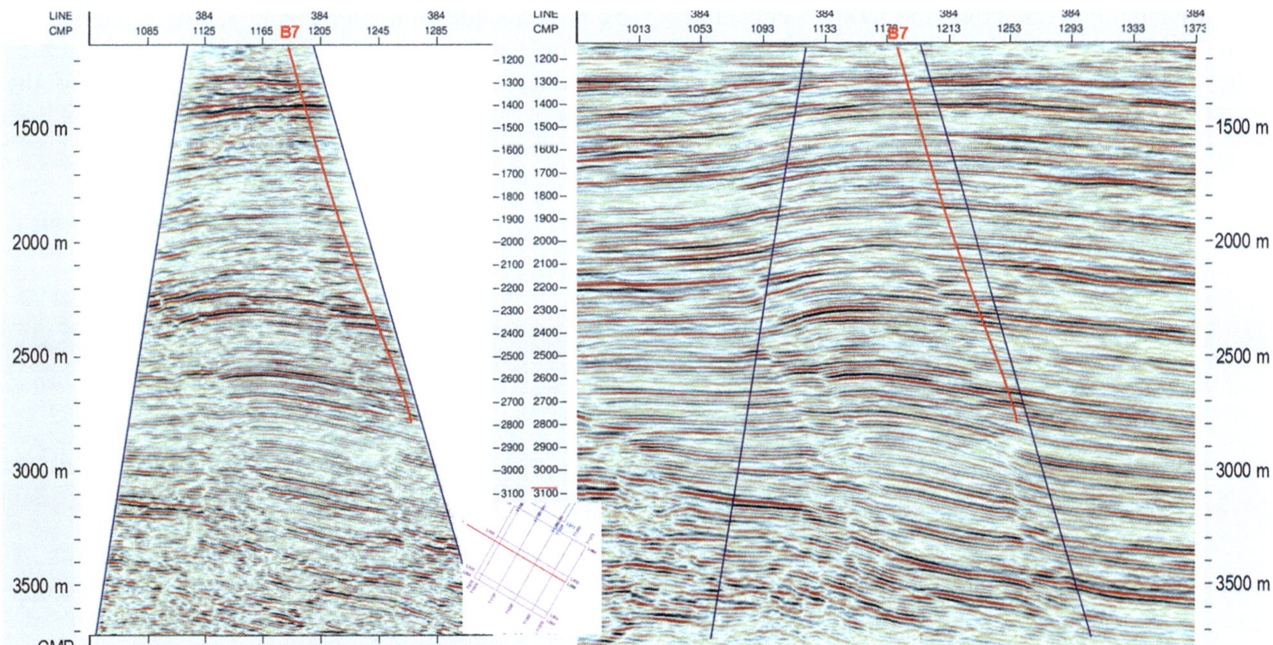

Figure 8.25 Imaging comparison for 3D DAS VSP data from borehole B7: upgoing reflection wave migration (left) versus downgoing multiple reflection wave migration (right). Yu et al. (2022).

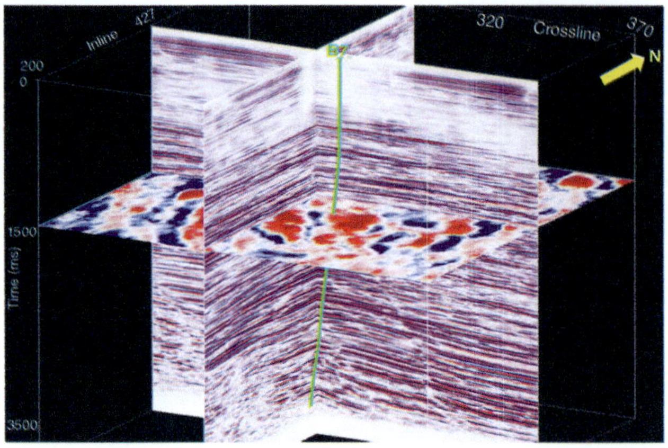

Figure 8.26 3D view of the downgoing multiple reflection wave migration result of the Pinghu 3D DAS VSP data from borehole B7. This imaging can be used to guide 3D structure interpretation and identify small geological features around the wellbore. Yu et al. (2022).

G7 is shown in Figure 8.27. The blue lines (Figure 8.27(b)) define the coverage area of the 3D DAS VSP upgoing wave migration. The yellow lines (Figure 8.27(c)) are the coverage area of the 3D DAS VSP downgoing multiple reflection wave migration. When inserting the 3D DAS VSP migration results into the 3D OBN data migration results, most of the reflection events match very well, as shown in Figure 8.27(c).

Figure 8.28 is the comparison of vintage 3D OBC data migration (Figure 8.28(a)), new 3D OBN data migration (Figure 8.28(b)) and the downgoing multiple reflection wave migration of the 3D DAS VSP data (Figure 8.28(c)) in borehole B5. The downgoing multiple reflection wave migration of the 3D DAS VSP data in Figure 8.28(c) shows significant migration quality improvement in comparison with the vintage 3D OBC data migration. The tracing of the reservoir and formation boundaries becomes much easier and less ambiguous in both the new OBN data migration and 3D DAS VSP data migration than in the vintage 3D OBC data migration.

8.3.3. Integrated Interpretation

Apart from the 3D DAS VSP data processing, detailed reservoir structural interpretation, reservoir formation boundary tracing, and fluid distribution mapping around both wellbores B5 and B7 are also conducted. The joint wellbore and seismic formation top identification and calibration based on accurate VSP time-depth relationship constraint is shown in Figure 8.29. Figure 8.29(a) is the synthetic record and corridor stack in comparison with the VSP section and OBN section at the borehole location, and Figure 8.29(b) is the formation top identification based on the corridor stack of VSP data and synthetic records. The results of the

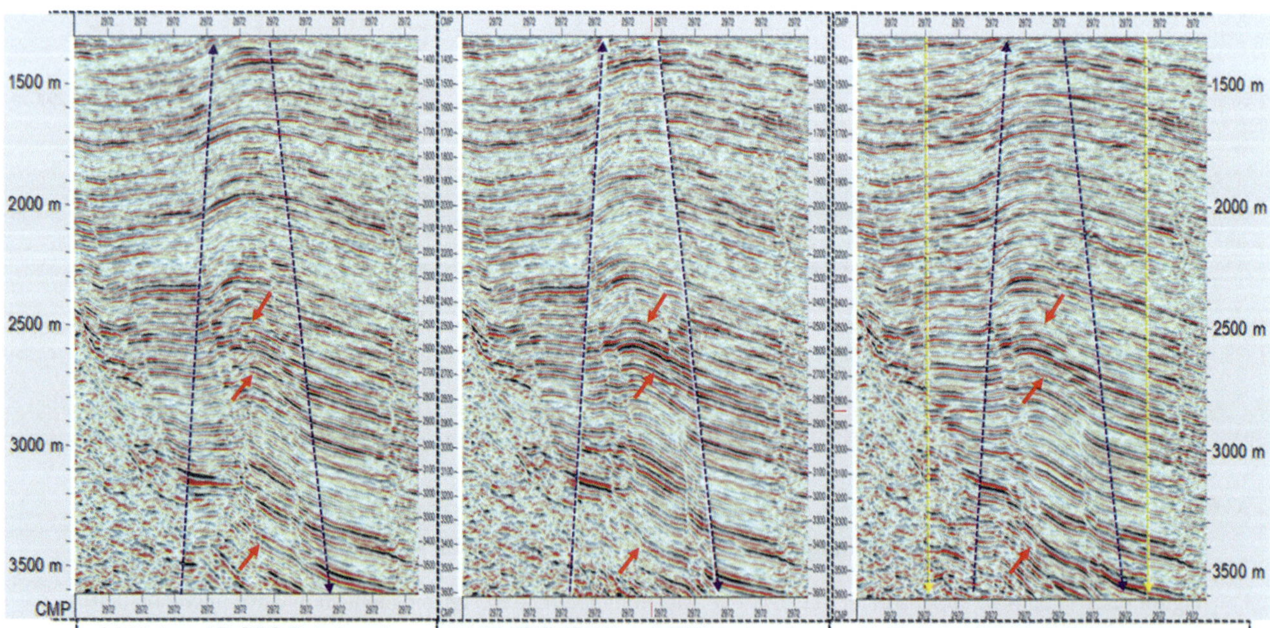

Figure 8.27 Comparison of vintage 3D OBC data migration, 3D DAS VSP upgoing wave migration, and 3D DAS VSP downgoing multiple reflection wave migration in borehole B7. Yu et al. (2022).

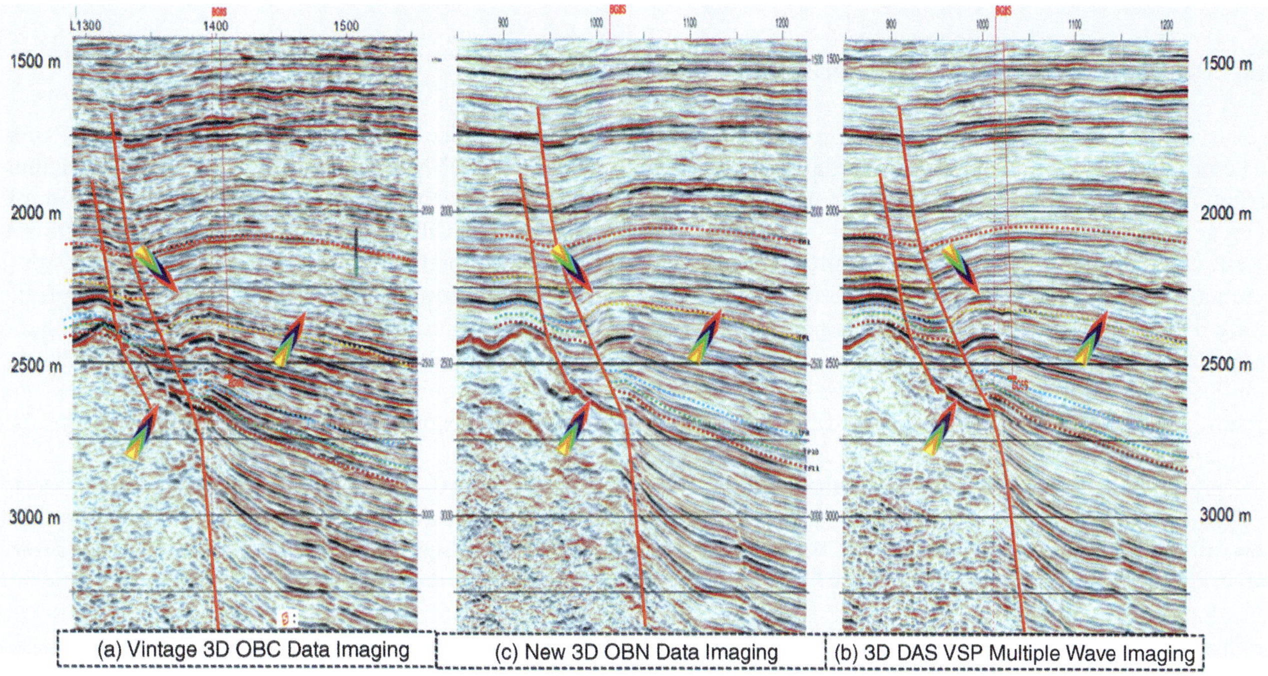

Figure 8.28 Comparison of vintage 3D OBC data migration, new 3D OBN data migration, and 3D DAS VSP downgoing multiple reflection wave migration in borehole B5. Yu et al. (2022).

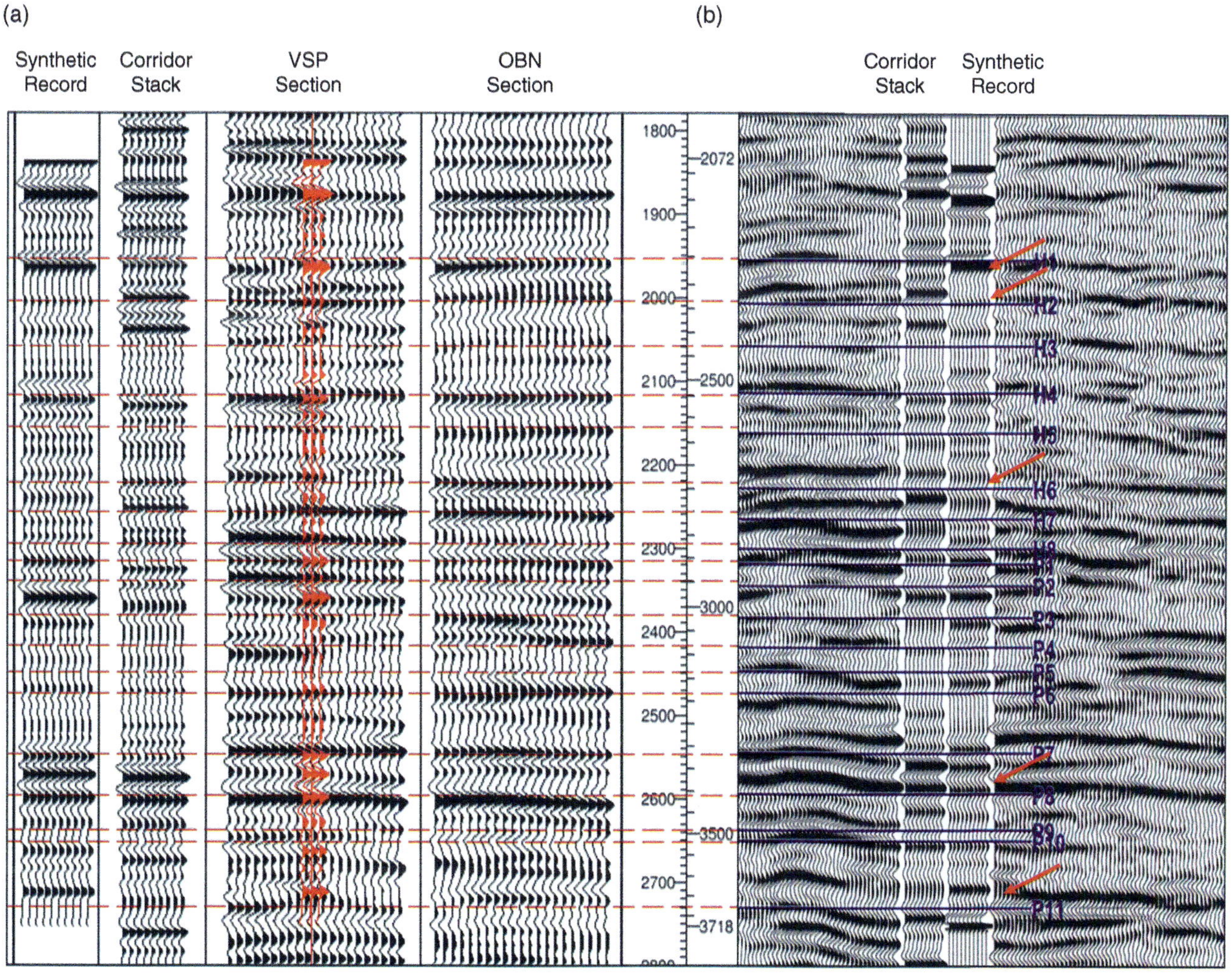

Figure 8.29 Joint borehole and seismic formation top calibration based on the accurate 3D DAS VSP data time-depth relationship constraint. (a) synthetic record and corridor stack in comparison with the 3D DAS VSP section and OBN section at the borehole location (well–seismic tie); (b) formation top identification and calibration based on the corridor stack of 3D DAS VSP data and the synthetic record. Yu et al. (2022).

well–seismic tie at the well location and the results give us the confidence to use the 3D DAS VSP data for detailed structure interpretation and reservoir formation mapping around the borehole area.

The calibration results of the vintage OBC data with the synthetic data are not very good in the study area. By using the accurate time-depth relationship constraints of multiwell 3D DAS VSP data, the accuracy of the calibration has been significantly improved. The consistency of seismic horizon recognition is confirmed by the well-to-well lateral comparison, which is the foundation for the interpretation of the seismic horizon in the study area.

Figure 8.30 shows the comparison of the reservoir structure interpretation map based on the vintage 3D OBC data (left) and the new 3D DAS VSP data (right). The major structure features of the reservoir did not change much between the interpretations from two different datasets, but there are differences in the small structural features between the two interpretations, such as in the areas pointed out by blue arrows.

Figure 8.31 shows the time slice at 2,800 ms for both 3D DAS VSP imaging data and OBN imaging data, as well as the curvature slice at 2,800 ms for both 3D DAS VSP imaging data and OBN imaging data. The 3D DAS VSP imaging data shows more detailed geological and reservoir features, which will enhance the reservoir characterization around both borehole areas.

The joint multiwell 3D DAS VSP data inversion profile and the distribution of the sand bodies of the Pinghu Group along the profile connecting the four wells are shown in Figure 8.32. The amplitude properties and waveform indicators are used in the inversion of the 3D DAS VSP data to generate the inversion profile in Figure 8.32 (top), and the inversion results were used to analyze and predict the lateral thickness distribution of sand bodies, as seen in Figure 8.32 (bottom). The formation with red in the inversion profile indicates the highest hydrocarbon saturation rate, and the formation with yellow has a lower hydrocarbon saturation rate. It is very valuable for the client to know the 3D lateral thickness distribution of sand bodies along with hydrocarbon saturation rate prediction within the sand bodies, and the client can plan an enhanced oil and gas recovery strategy to boost the production of this mature oil and gas field.

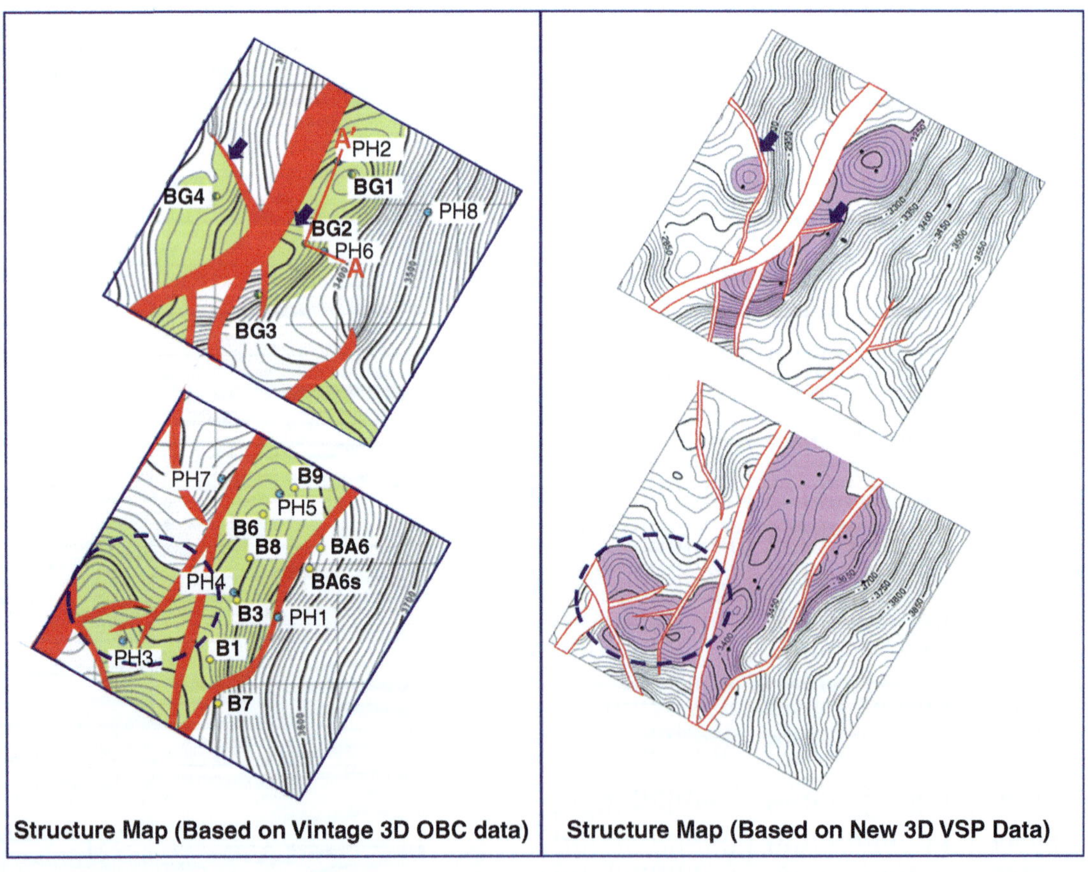

Figure 8.30 Comparison of reservoir structure interpretation ma. Structure map based on vintage 3D OBC data (left) and structure map based on new 3D DAS VSP data (right). Yu et al. (2022).

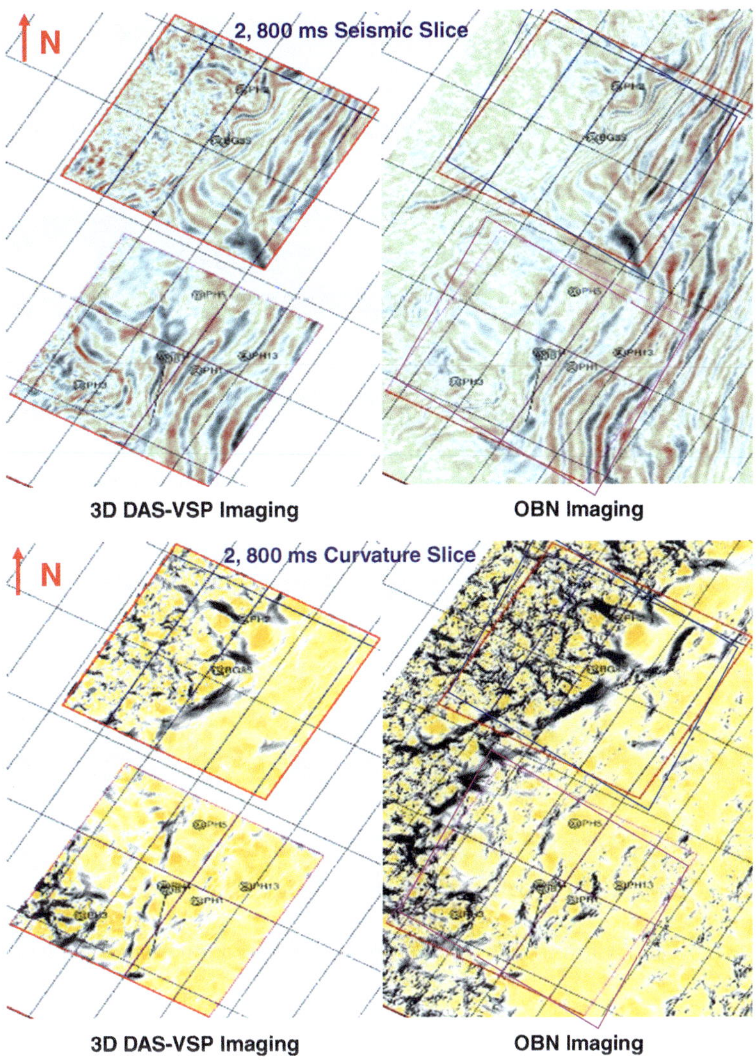

Figure 8.31 The time slice at 2,800 ms for both 3D DAS VSP imaging data and OBN imaging data (top), and the curvature slice at 2,800 ms for both 3D DAS VSP imaging data and OBN imaging data (bottom).

8.3.4. 3D DAS VSP Data Processing in the Middle East

Table 8.3 lists the 3D DAS VSP data processing workflow. There are three major data processing steps: (a) data QA/QC and signal processing, (b) velocity model building, and full waveform inversion (FWI) and imaging.

Key Data Processing Techniques

3D DAS VSP Data Deblending Processing. Because the high-density OBN data were acquired using multiple source boats at the same time, the simultaneously acquired 3D DAS VSP CSG data records have multiple shots in a 10 s record, and the deblending process needs to be conducted on the 3D DAS VSP raw data. The sparse inversion method using all the shots within the source carpet was used for deblending the blended acquired 3D DAS VSP data:

$$d = \boldsymbol{\Gamma m} \tag{8.4}$$

Here, d is the blended acquisition data, m is the effective signal without blending, and Γ is a blended operator.

$$J(m) = \|d - \boldsymbol{\Gamma m}\|_2^2 + \lambda\|\boldsymbol{Fm}\|_0 \tag{8.5}$$

Here, $J(m)$ is the objective function, F is the three-dimensional fast Fourier transform operator, and λ is the regularization parameter, which controls the error term and the weight constraint term.

The objective function of the minimization problem can be solved by the iterative shrinkage threshold method: apply the L_0 norm sparse constraint on the signal model in the 3-D frequency-wavenumber-wavenumber (FKK) domain to minimize the objective function.

The minimization problem of the objective function can be solved by the iterative shrinkage threshold method:

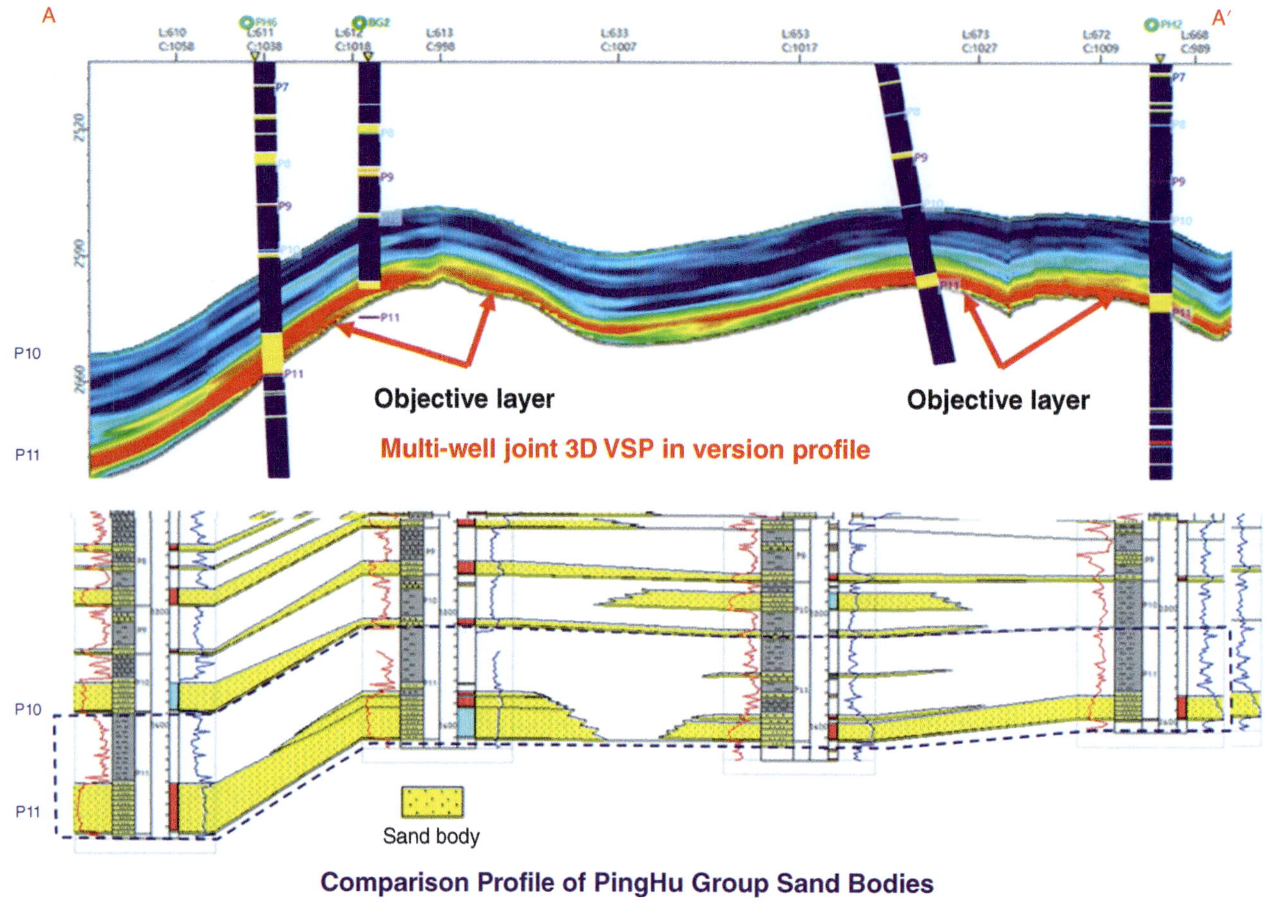

Figure 8.32 Multiwell joint 3D DAS VSP data inversion profile (top); red indicates the highest hydrocarbon saturation rate. Comparison profile of Pinghu Group sand bodies (bottom). Yu et al. (2022).

$$\boldsymbol{m}_{i+1} = \boldsymbol{F}^{-1}\boldsymbol{T}_\tau \boldsymbol{F}\left[\boldsymbol{\Gamma}^{\mathrm{H}}\boldsymbol{d} - \left(\boldsymbol{\Gamma}^{\mathrm{H}}\boldsymbol{\Gamma} - \boldsymbol{I}\right)\boldsymbol{m}_i\right] \tag{8.6}$$

T_τ is the threshold operator of the transformation domain. Here, we used a hard threshold to process the data whose amplitude is greater than the given threshold value τ.

Figure 8.33 shows the 3D DAS VSP data deblending processing results before deblending (left) and after deblending (center), and the removed blending noise (right).

Preprocessing and Denoising. **Spherical Diffusion Compensation (TAR Compensation).** The TAR factor extracted from VSP data by the true amplitude recovery factor method is used for the spherical diffusion amplitude compensation of 3D DAS VSP data. The comparison of before-TAR compensation and after-TAR compensation is shown in Figure 8.34.

Abnormal Amplitude Suppression. Abnormal amplitude suppression processing is used for surge noise attenuation. Surge noise is a type of noise caused by wind and waves and has the characteristics of low frequency (generally below 20 Hz) and large amplitude value. The processing steps we adopted are "simultaneously identifying surge noise on multiple traces and individually denoising the surge noise from each trace." This technique automatically identifies strong energy interference in seismic records at different frequency bands, determines the spatial position of noise, and suppresses it at the time-varying domain (manner) and the space-varying domain (manner) according to the user-defined threshold and attenuation coefficient. The recognition parameter of the surge noise used in the calculation is the transverse weighted median of the data envelope. The frequency division processing method can improve the fidelity of the denoising process. This method is usually used to suppress the interference of strong energy seismic waves and surface waves near the source location. The processing procedures are used in different frequency ranges in the time–space domain. First the surge noise is automatically identified according to the suppression threshold value on a multiple-trace basis, and then the noise energy is suppressed according to the attenuation factor trace by trace. Figure 8.35 displays the results of abnormal amplitude suppression.

Table 8.3 3D DAS VSP data processing workflow.

Processing steps	Processing functions
1. Data QA/QC and signal processing	(a) Load raw data and define geometry (b) 3D DAS VSP data deblending processing (c) Compute derivative of DAS VSP data (d) Generate common receiver gathers (e) Analyze and QC data (f) Pick first break and QC (g) Correct tidal statics and water column using water velocity (h) Process corridor stack section with near offset data (i) Calibrate depth using synthetic records (j) Estimate Q value using zero-offset VSP data and synthetic records (k) Correct depth for 3D DAS VSP data (l) Edit bad traces, if any (m) Correct instrument signature, if provided (n) Estimate source signature (o) Apply 3D deghosting technique to remove source side ghost (p) Apply debubble (q) Apply source designature/zero phasing (r) Apply model-based multiple prediction and removal (s) Apply 3D Tau-P predictive deconvolution for further demultiple (t) Suppress noise if needed (u) Separate up/down wavefield with 3D Tau-P transformation (v) Suppress converted shear wave (w) Attenuate linear noise (such as tube wave, etc.)
2. Velocity model building	(a) Build initial velocity model using legacy velocity/VSP moveout analysis (b) Build initial anisotropy model using well velocity and VSP first-break information (c) Apply FWI to update the velocity and anisotropy model, using data from step 1g (d) Use reverse time migration or Kirchhoff migration to QC velocity quality (e) Invert Q model through FWI
3. FWI and imaging	(a) Run high-frequency FWI with primaries and multiples to output a high-resolution velocity and impedance model and convert to generate a high-resolution reflectivity image • Wave propagation engine in FWI includes anisotropy and Q • Optionally use FWI to invert Q (b) Use least-squares multiple migrations to image VSP data: test imaging multiples using DIC. In our experience, least-squares migration provides a higher broader-band image with much less crosstalk. (c) Conventional imaging: • Option 1: Apply Gaussian beam migration with Q compensation to image primary data, both up and down waves (data from step 1x), and generate angle gathers • Option 2: Apply reverse time migration with Q compensation to image primary data, both up and down waves (data from step 1w) (d) Apply proper post-migration processing to enhance imaging quality and deliver the final products

Signal Deconvolution. **Bubble Suppression.** The airgun (array) is the most frequently used source for offshore seismic surveys. When the airgun is excited, high-pressure air is rapidly released into the water and quickly forms spherical bubbles. The bubbles continue to expand and compress in the water to form multiple bubble oscillations. Each bubble expansion will produce a seismic wave. Due to multiple bubble expansions and contractions, the airgun wave will oscillate many times as it propagates through the water, which reduces the resolution of seismic data. The bubble suppression method is to input the wavelet with a bubble effect, suppress the wavelet with a bubble effect, calculate the shaping filter operator, and use the simulated far-field wavelet to identify the bubble effect and suppress it. Before bubble suppression and the after-bubble suppression result are shown in Figure 8.36.

Ghost Wave Suppression in the FK Domain. The free-surface reflection (ghost wave) suppression operator is calculated in the FK domain to eliminate free-surface reflection, which suppresses the free-surface reflection of both receivers and sources. We can suppress free-surface

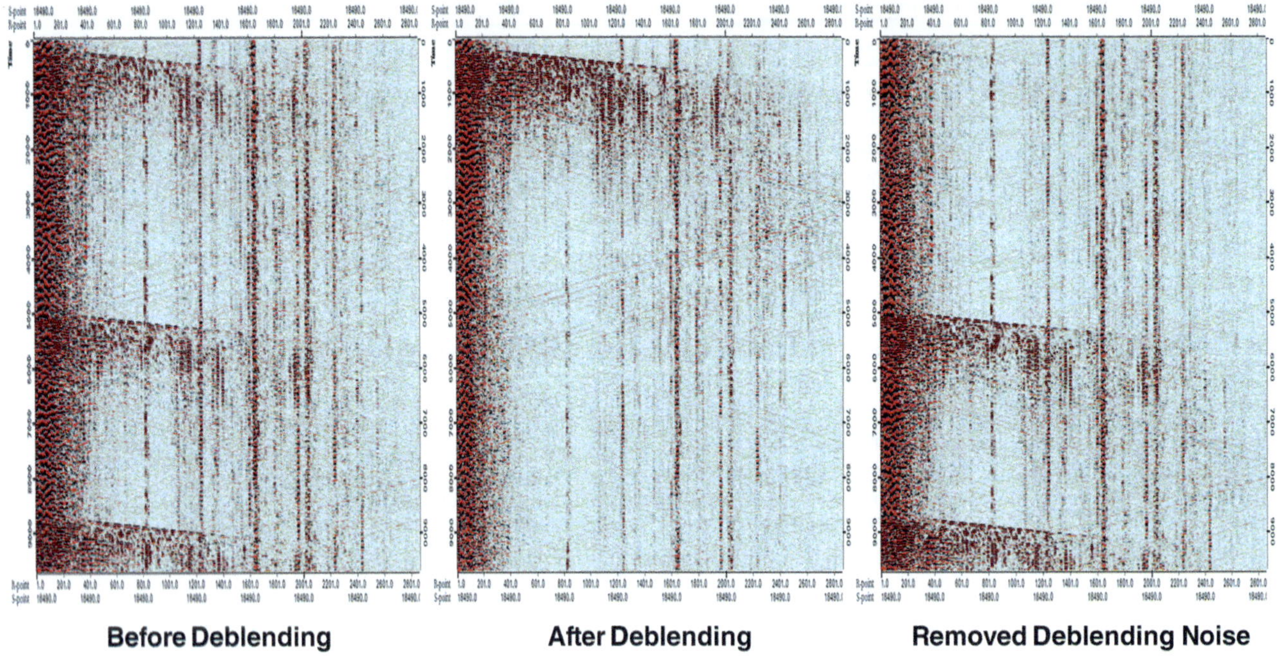

Figure 8.33 Sparse inversion-based CSG data deblending processing results.

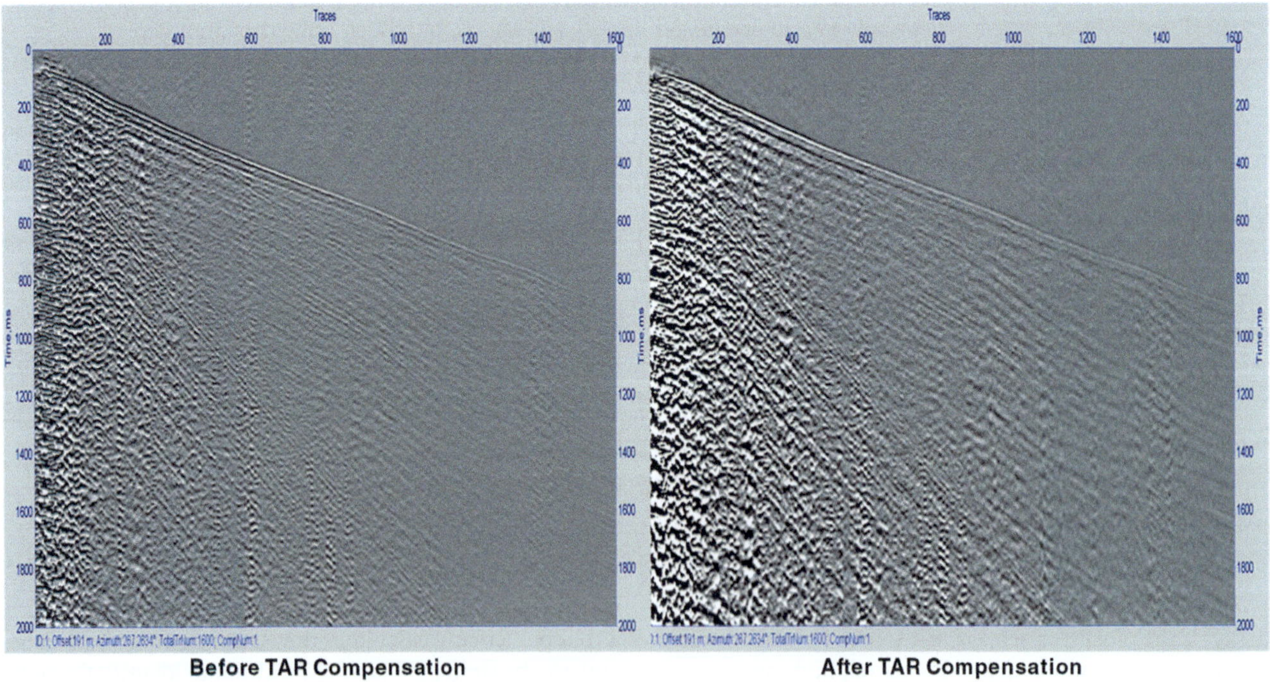

Figure 8.34 Comparison before (left) and after (right) spherical diffusion compensation.

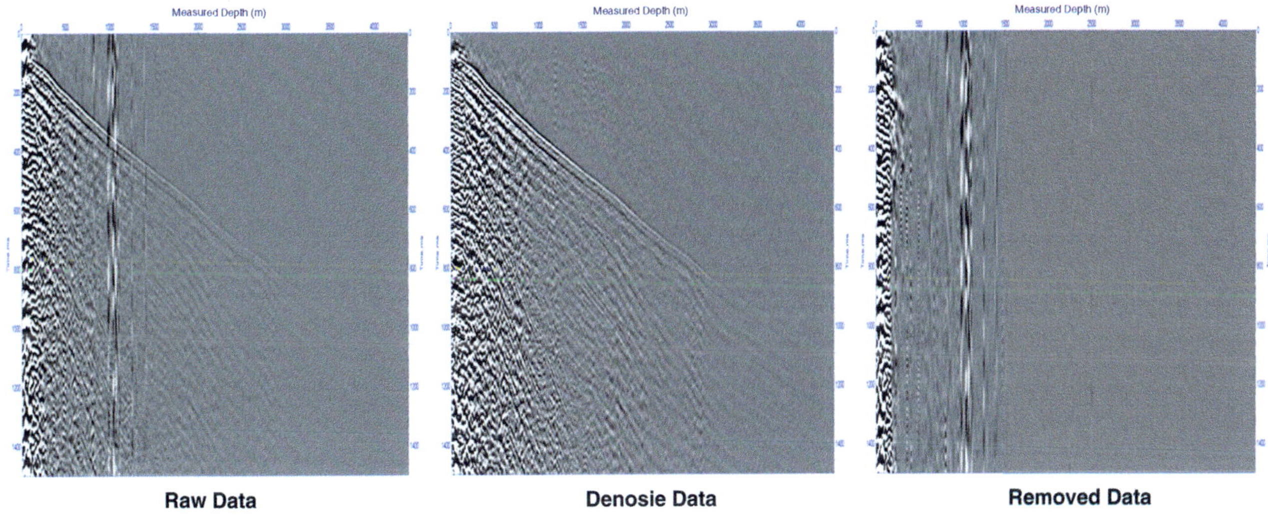

Figure 8.35 Removing abnormal amplitude interference from raw 3D DAS VSP data.

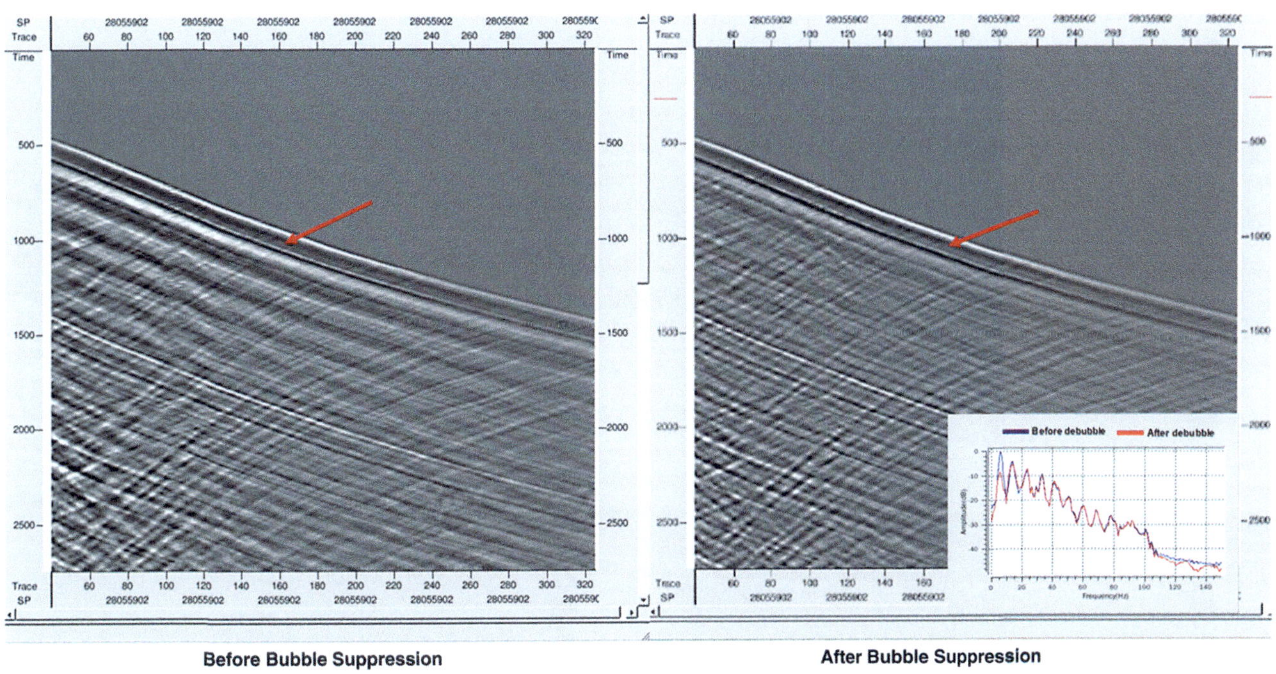

Figure 8.36 Comparison of before bubble wave suppression (left) and after bubble suppression (right).

reflection by giving the depth of the geophone/source or letting the software module automatically scan the best geophone/source depth to suppress free-surface reflection. Figure 8.37 shows the ghost wave suppression processing workflow. Figure 8.38 shows the comparison of before ghost wave suppression (left) and after ghost wave suppression (right).

Wavefield Separation. The Tau-P transform is used to separate upgoing and downgoing wavefields. The VSP data have obvious linear characteristics, and the apparent velocity of upgoing and downgoing wavefields are in opposite directions. The linear radon transform (Tau-P) can be considered for wavefield separation. The common FK filter changes the shape and frequency of the

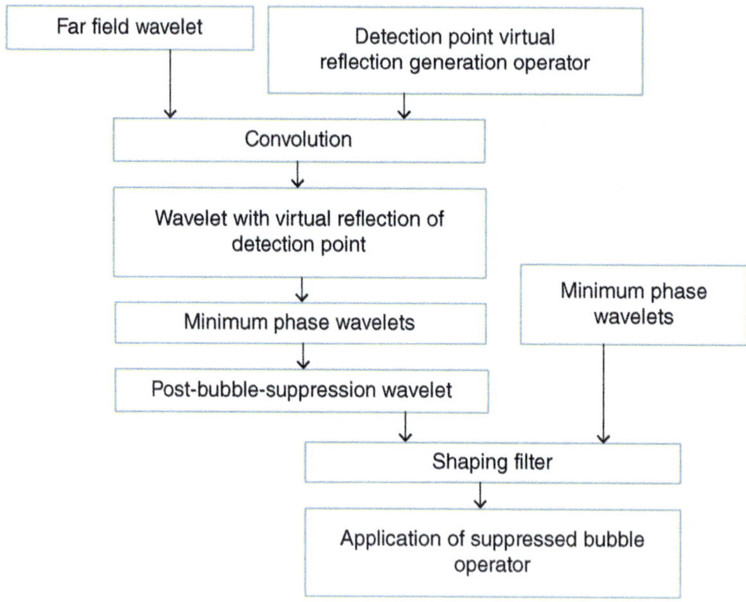

Figure 8.37 Ghost wave suppression processing workflow.

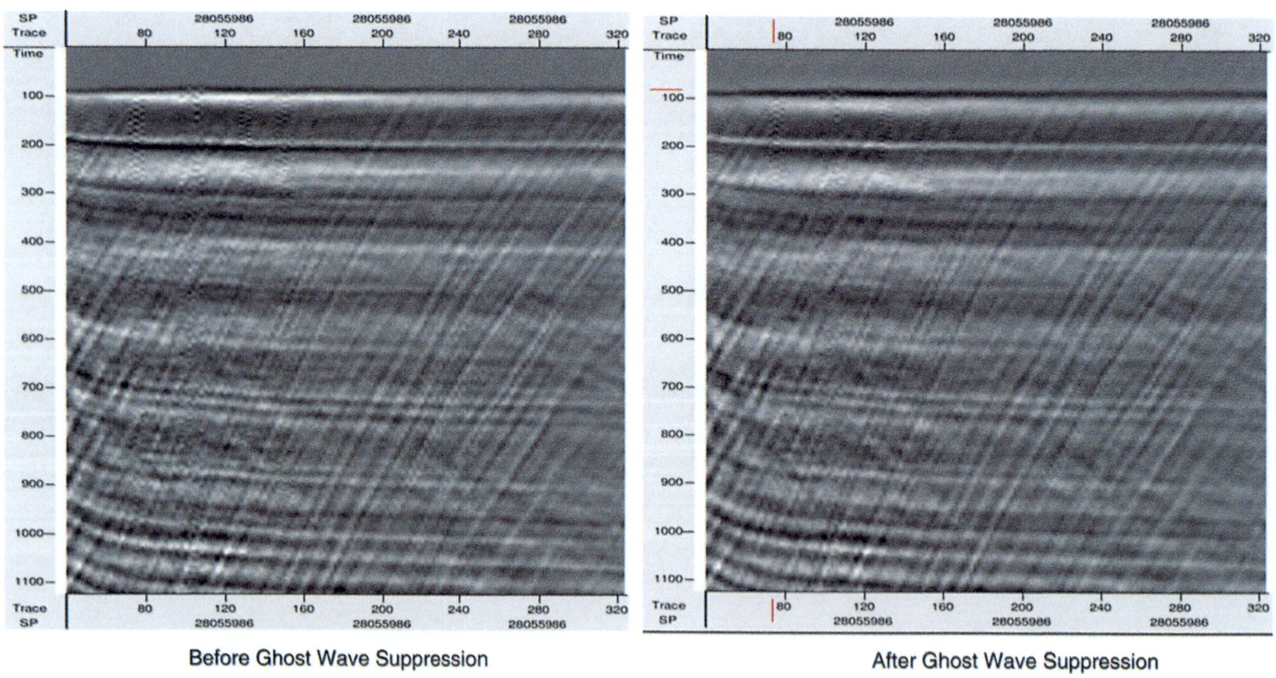

Figure 8.38 Comparison before (left) and after (right) ghost wave suppression.

wavefield greatly, which is not conducive to the fidelity, high frequency, and high precision of DAS VSP data. The processing flow of DAS VSP data using the Tau-P transform and FK filtering has good results, and the upgoing and downgoing wavefield shape and frequency are well preserved (Figure 8.39).

Remove Multiples and Increase Frequency. **Time-Varying Wavelet Deconvolution.** Wavelets are extracted from the downgoing wave measured directly by VSP data, and time-varying wavelet deconvolution is carried out to suppress multiples and improve the resolution of the VSP data.

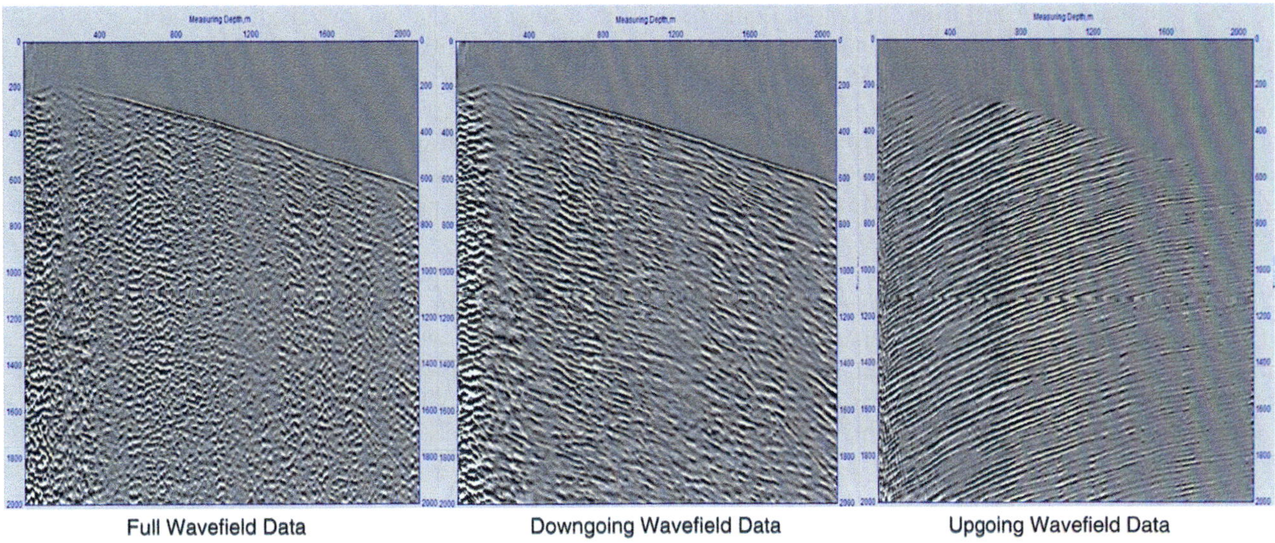

Figure 8.39 Effect of downgoing wavefield (center) and upgoing wavefield (right) separation.

Tau-P Deconvolution. This method can conduct predictive deconvolution on the seismic data after Tau-P transformation to suppress multiples. In the case of nonzero offset, multiples of different orders in seismic records are aperiodic. Currently, it is difficult to suppress multiples only by predictive deconvolution. However, after seismic records containing multiples are transformed into the Tau-P domain, the time difference of multiples is the same (periodic) on the same P-value channel. Therefore, predictive deconvolution can be used to suppress multiples (Figure 8.40).

Velocity Model Building and Image Processing

For the 3D DAS VSP data processing workflow, there are three major data processing steps: (a) data QA/QC and signal processing; (b) velocity model building; and

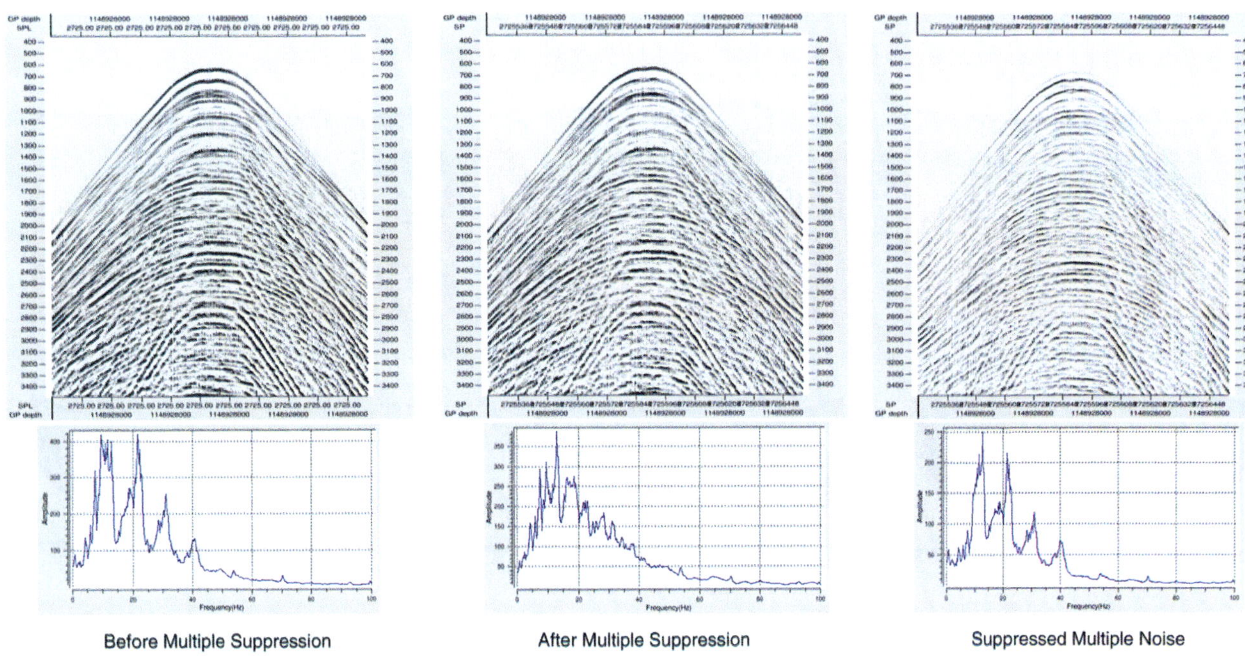

Figure 8.40 Result of multiple suppression: (left) before multiple suppression; (center) after multiple suppression; (right) suppressed multiple noise.

(c) FWI and imaging. Apart from the conventional processing steps of 3D DAS VSP data, deblending processing of multiwell blended acquired 3D DAS VSP data using multiple airgun sources, a special ringing noise removal procedure, joint domain full waveform inversion (JDFWI) for velocity model updating, and one-way wave equation multiple migration (OWEMM) methods were used to generate final imaging.

VSP First Break Tomography Modeling. Conventional 3D VSP data first-break tomography modeling has some problems, such as insufficient near surface ray coverage, low shallow inversion accuracy, poor shot points, static correction effect, and so on. How to increase the number of near-surface coverage and improve the inversion accuracy and shot point static correction quality is the key problem that must be solved for 3D VSP data high-quality imaging. Given the particularity of 3D DAS VSP data, the DAS VSP Fresnel first-break tomography technique was developed to improve the uniformity of VSP first-break path coverage and increase the stability of the inversion solution. 3D DAS VSP borehole seismic combined Fresnel volume tomographic inversion is proposed, utilizing the characteristics of the development and dense arrangement of direct waves in the vertical direction at the first break of the 3D DAS VSP data and the characteristics of uniform distribution in the lateral direction and high precision of the gyratory wave at the first break of the surface. This technique fully combines the advantages of the two types of data and establishes a joint objective function for tomographic inversion, thereby improving the accuracy of the shallow, medium, and deep velocity models. A velocity model was built for generating synthetic VSP data using 2,001 shots on the surface with 12.5 m shot spacing and 241 downhole geophones with spacing of 12.5 m and depth from 200 to 3,200 m. The specific first-break tomographic modeling results and ray coverages are shown in Figure 8.41.

VSP FWI. We have extended our proprietary FWI to VSP data, which takes advantage of both refractions and reflections to invert a high-resolution velocity model. The wave propagation engine of FWI is acoustic, adapting to isotropic, VTI (vertical transversely isotropic), TTI (tilted transversely isotropic), and orthorhombic anisotropy media, and with Q compensation. Instead of inverting velocity, FWI could also be used to generate high-frequency velocity and impedance perturbation data and provide broadband imaging products. Figure 8.42 shows the true velocity model (left) and VSP FWI inverted velocity using forward modeling VSP data (right).

JDFWI. The client provided a vintage 3D velocity model from an early OBN survey covering the 3D DAS VSP data area for us to build an initial 3D velocity model for 3D DAS VSP data migration. The acoustic logging data from all the wells and the JDFWI were used to calibrate and update the initial 3D velocity model for 3D DAS VSP data migration.

Extracting the travel time information of field data lies at the heart of the velocity model building and the application of FWI. To achieve this, researchers have made

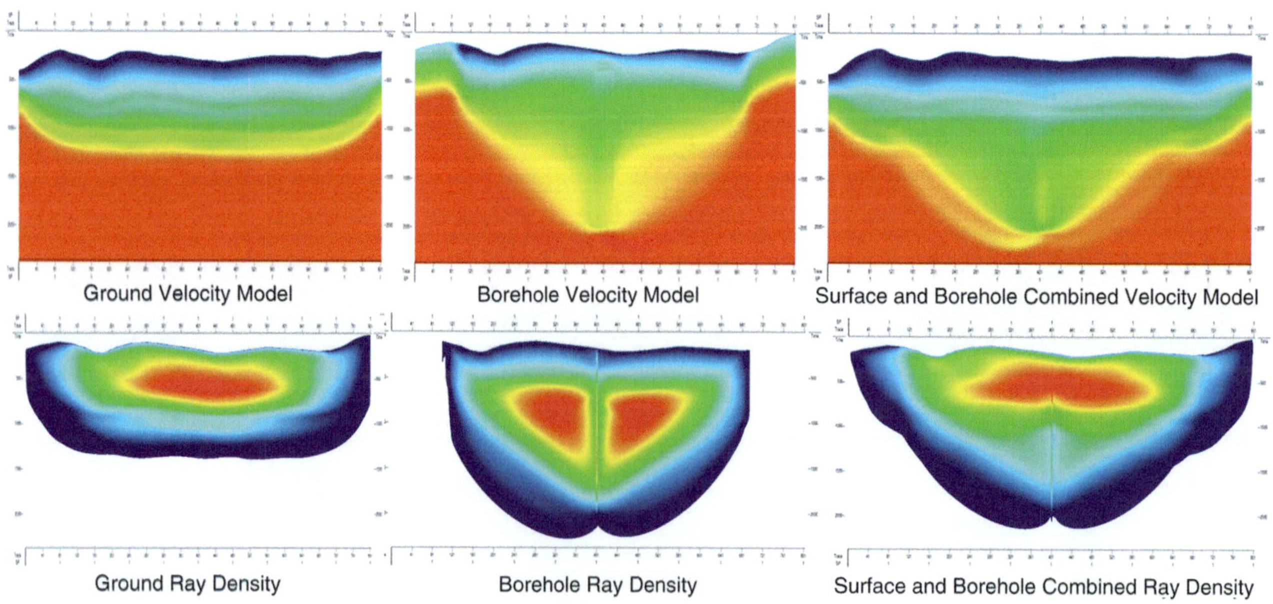

Figure 8.41 Comparison of surface seismic and borehole-surface seismic data combined with Fresnel tomography inversion results (top) and ray coverages (bottom).

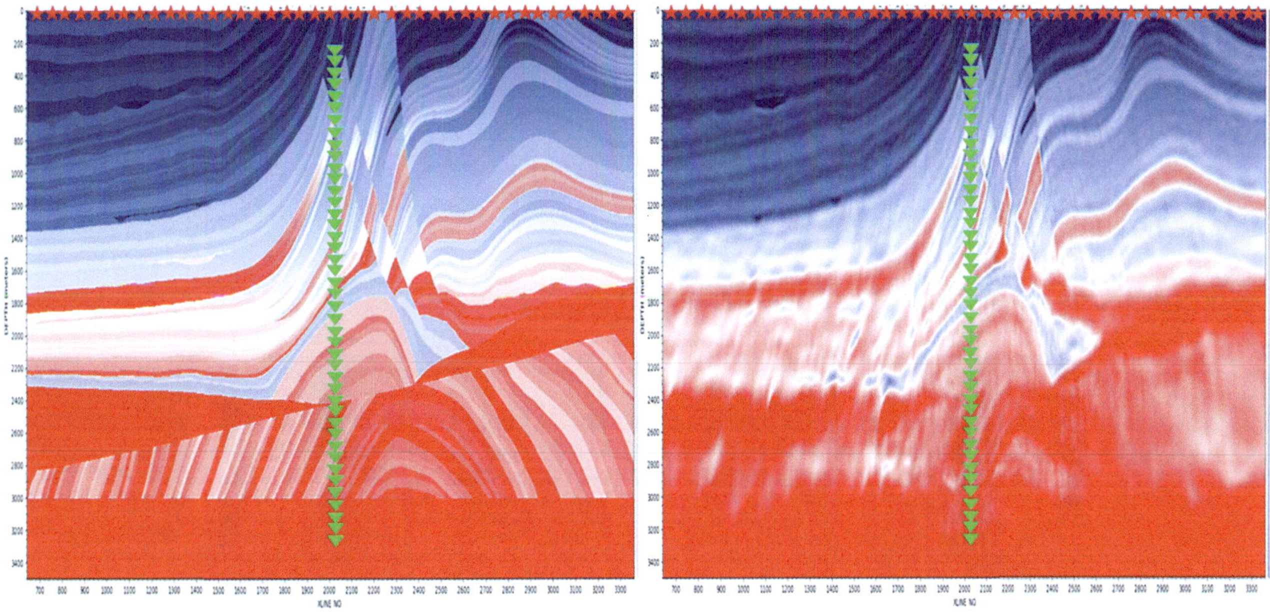

Figure 8.42 A synthetic example of using VSP FWI to invert velocity.

many influential advancements such as adaptive waveform inversion (Warner & Guasch, 2016), dynamic warping FWI (Ma & Hale, 2013), time-lag FWI (Luo & Schuster, 1991), and optimal transport FWI (Yang et al., 2018). All of these methods are formulated in the conventional data domain only, and their success relies on the extraction of the travel time misfit between the field and synthetic data, which may not be trivial when the geology is complicated. An alternative method is based on the delay time common image gather (DTCIG) to directly measure the travel time inaccuracy in the model domain, which is intuitive in applications and more sensitive to velocity inaccuracy. Combining our model domain velocity inaccuracy measure with the conventional data misfit, we derive the formulation of JDFWI, which minimizes the travel time misfit in both data and model domains. Compared to conventional FWI, applications to both synthetic and field data demonstrate that the proposed JDFWI can overcome the cycle-skipping issue with the new model domain travel time misfit term while still retrieving high-resolution complex velocity models. It is thus a more stable and practical solution to our velocity model building application with the eased requirements of a reliable initial model or low-frequency data.

Conventional FWI is a data-driven method that seeks optimal velocity models minimizing the following least-squares misfit objective function between the simulated data $\boldsymbol{d}(\boldsymbol{x}_r;\, \boldsymbol{t};\, \boldsymbol{x}_s)$ and the field data $\boldsymbol{D}(\boldsymbol{x}_r;\, \boldsymbol{t};\, \boldsymbol{x}_s)$ (recorded at receiver locations $\boldsymbol{x}_r$ due to a specified source location $\boldsymbol{x}_s$):

$$E(v) = \iiint (D - d(v))^2 dt dx_s dx_r \qquad (8.7)$$

Here, in a simplified case of the acoustic, isotropic, and constant density wave equation, the simulated data d is predicted using the velocity models $v(x)$ by forward propagating wavefield d as follows:

$$\begin{cases} \left(\dfrac{1}{v(x)^2}\dfrac{\partial^2}{\partial t^2} - \nabla^2\right) p(x;t;x_s) = \delta(x - x_s)S(t), & (8.8a) \\ d(x_r;t;x_s) = p(x_r;t;x_s), & (8.8b) \end{cases}$$

where $S(t)$ is the source wavelet and ∇^2 denotes the Laplacian operator. The nonlinear state-space response, the curse of dimensionality, and imperfect low-frequency content in both the model and data often increase the risk of our conventional FWI being trapped in the local minimum of the objective function in equation (8.7). Compared to the initial vintage 3D velocity model from an early OBN survey in Figure 8.43 (top), the JDFWI inverted velocity (Figure 8.43 (bottom)) has diving wave energy focused at the zero-lag delay time at all the penetrated depths. This confirms the quality of the proposed JDFWI inversion statistically (Duan et al., 2023, Zhang et al., 2023).

With this observation, to overcome the longstanding cycle-skipping issue, travel time misfit in the model domain could be an important addition to our conventional data-driven FWI objective function from equation (8.1). Following the computational cost and velocity sensitivity reported in Sava and Fomel, 2006,

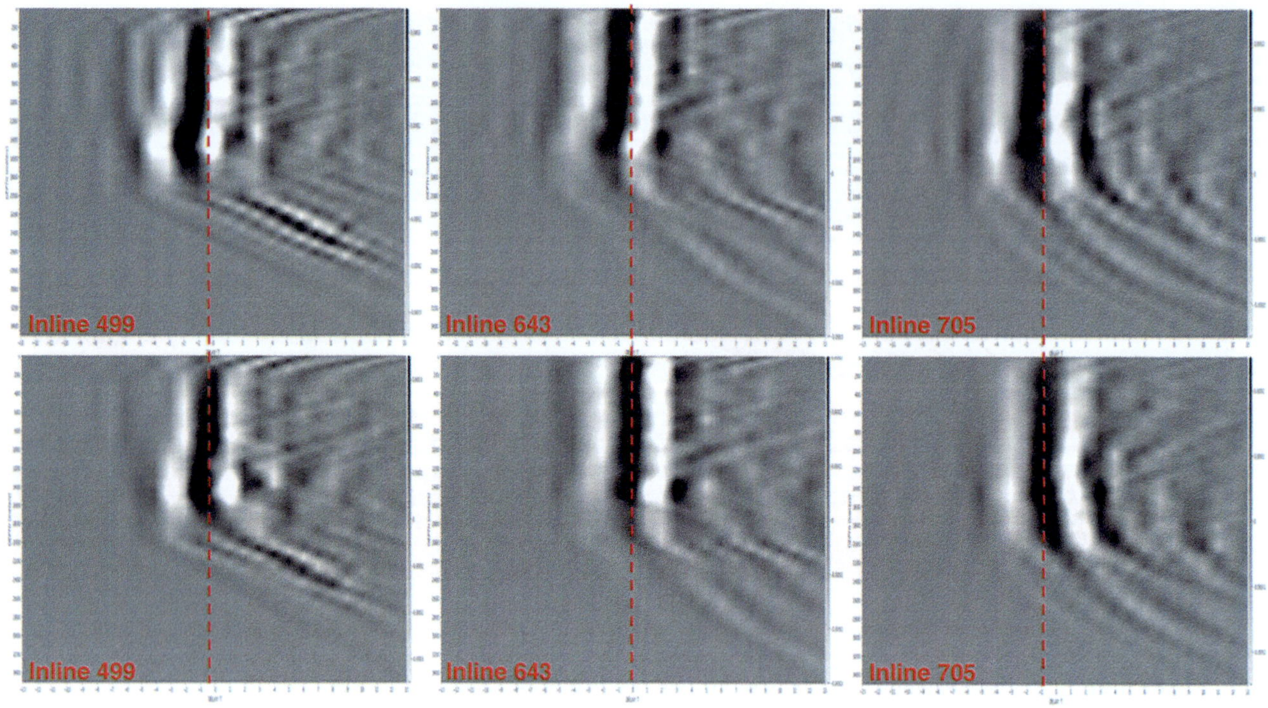

Figure 8.43 Delay-time gather display: (top) initial vintage 3D velocity model, IL 415; (bottom) JDFWI model, IL 415. Red lines indicate the zero-lag ($\tau = 0$) delay time.

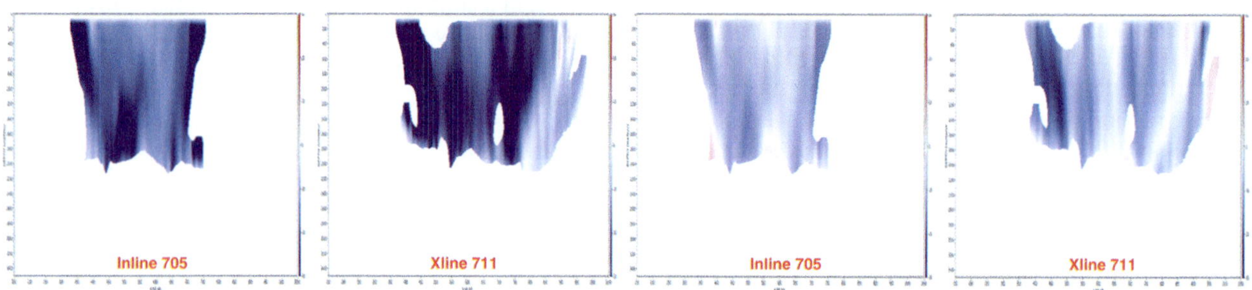

Figure 8.44 3D DAS VSP FWI model building: (left) delay time from the legacy model on inline 706 and xline 711; (right) delay time from the FWI model on the same lines.

we employ DTCIGs to form our JDFWI. Figure 8.44 shows the delay time from the legacy velocity model on inline 706 and xline 711 (left) and the delay time from the FWI model on the same lines (right), and Figure 8.45 shows the delay time from the legacy model on $Z = 250$ m and $Z = 1{,}000$ m (left) and delay time from the FWI model on the same depths (right). The velocity model has been significantly improved in Figures 8.44 and 8.45 by using the JDFWI processing method.

One-Way Wave Equation Multiple Migration (OWEMM). Surface-related multiple-imaging tech enhances signal quality, offering improved convergence and angle info versus standard primary imaging (Whitmore et al., 2010). But it is complex due to crosstalk noise from receiver data-driven forward wavefields. Lu et al. (2022) found a deconvolution imaging condition (DIC) to limit crosstalk. Still, it cannot fix illumination issues from acquisition geometry, also causing instability in null space (Whitmore et al. 2010).

In contrast, least-squares migration works better, iteratively using migration/demigration operators to lessen crosstalk (Li et al., 2022; Herrmann and Li, 2012; Witte et al., 2017). It is more stable than DIC, as no stabilizer is needed for operator evaluation. Most solvers need adjoint links between forward/adjoint operators. DIC in migration needs to change modeling operator and requires data preprocessing (Lu et al., 2022). To avoid

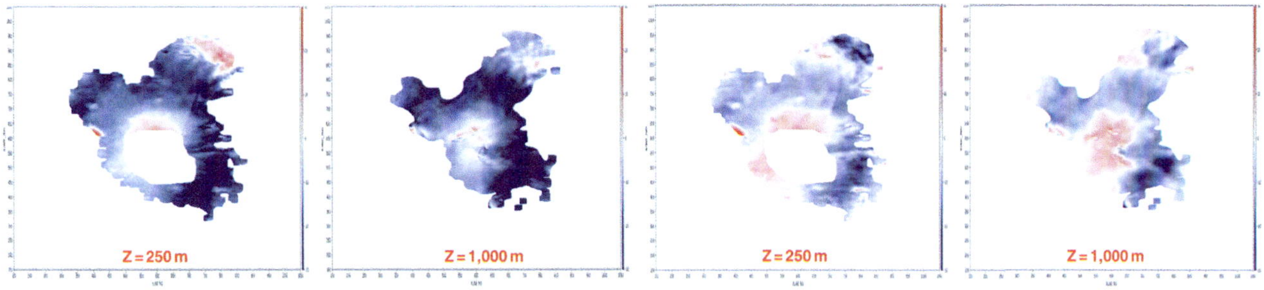

Figure 8.45 3D DAS VSP FWI model building: (left) delay time from the legacy model on $Z = 250$ m and $Z = 1000$ m; (right) delay time from the FWI model on the same depths.

these complexities, we use first-order steepest descent with DIC for gradient preconditioning. Synthetic and real DAS VSP examples prove our method's feasibility even without complex optimizations.

We employ the one-way wave propagator for our least-squares migration due to two main advantages over the two-way kernel. First, the one-way kernel is notably more efficient, enabling us to reach ultra-high frequencies. Second, it avoids imaging crosstalk from internal multiples and high-angle refraction waves. These factors are particularly beneficial when aiming for precise imaging in a more accurate velocity model. This leads us to adopt the monochromatic one-way-based surface-related multiple modeling operator.

Having the linear modeling operator and its adjoint in hand, we can address the least-squares migration challenge through diverse linear optimization solvers. For example, quadratic solvers like LSQR enhance convergence rates. However, most of these solvers require an adjoint test to align the migration operator and its adjoint. This poses a challenge when integrating geophysical prior knowledge, such as gradient preconditioning, as adjustments to the operators can disrupt their adjoint relationships, rendering linear solvers ineffective. To overcome this obstacle, we adopt the gradient descent method for solving the least-squares migration problem. This technique incorporates the DIC within the migration operator, enhancing convergence behavior.

Adhering to the adjoint relationship for linear operator inversion, the migration operator in the equation must employ the cross-correlation imaging condition (CCIC) for gradient computation. However, this introduces the wavelet spectrum from both input and source data into the gradient spectrum. Prior works (Li et al., 2022; Herrmann & Li, 2012) suggest that least-squares migrations ideally eliminate this wavelet spectrum. But this effectiveness might diminish with a limited iteration count, particularly for faint target events or when handling noisy datasets.

To enhance convergence, we employ the DIC, which normalizes the gradient using the l2 norm square of the background wavefield $U(w)$ at each frequency. The DIC acts as a preconditioner for the migration operator, reducing the influence of the source signature in the input data. This strategy resembles an approximate Hessian inversion applied to the gradient. Consequently, we adopt the steepest descent method with modified gradients, achieving improved convergence behavior without necessitating an exact adjoint relationship of the operators.

We have developed multiple-imaging and least-squares multiple-imaging techniques using a one-way wave equation as the wave propagation engine. Numerical experiments show that even with the DIC, crosstalk noise still appears in the multiple images. However, least-squares multiple migrations successfully remove the crosstalk artifacts and provide high-quality broadband multiple images, as shown in Figure 8.46 (location 1) and Figure 8.47 (location 2).

The multiwell multiple migration expanded the imaging aperture and minimized the typical smile effect in the VSP data migration. The imaging results of the multiwell 3D DAS VSP data will be used to generate high-accuracy and high-resolution 3D structure images with a higher number of folds for planning infilled and horizontal borehole drilling.

Multiple reflections contain subsurface reflectivity information that often complements that found in primary reflections. A general linear least-squares joint imaging of multiples and primaries (LSJIMP) inversion method can be used to simultaneously suppress crosstalk noise and combine pegleg multiples and primaries in a prestack sense. In general, LSJIMP is compatible with a wide variety of prestack imaging methods and can be extended to jointly image primaries and other embedded wave modes such as shear-wave conversions (Brown & Guitton, 2005). Primaries and multiples are imaged jointly using least-squares reverse time migration (LSRTM). Joint LSRTM uses multiple energy as a signal while removing crosstalk from the image. A class of multiple scattered events can be imaged using the two-way propagator in modeling and migration. Such events can scatter off sharp interfaces in the migration velocity as well as the sea surface.

Marine seismic surveys in shallow water regions typically suffer from acquisition striping and poor shallow

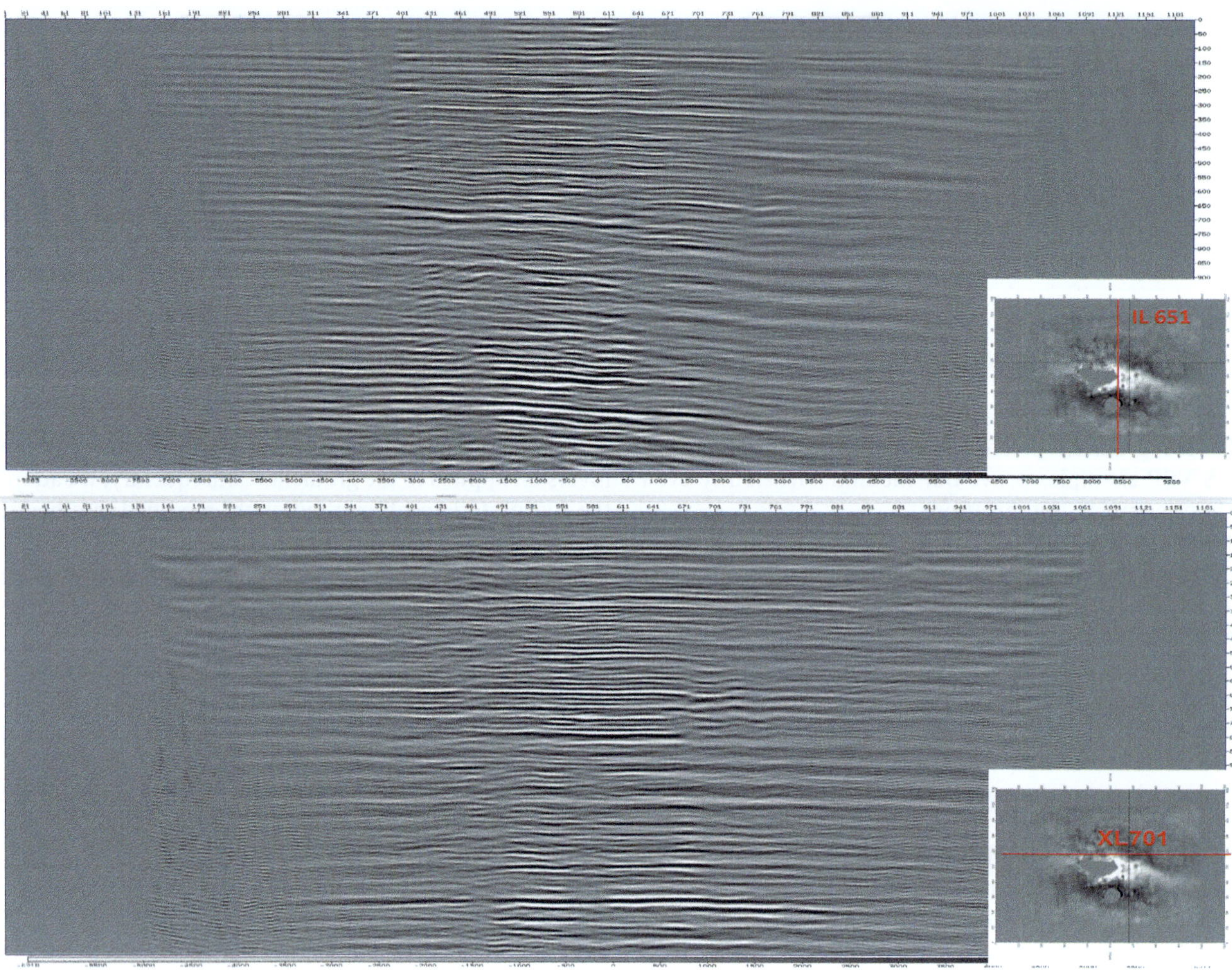

Figure 8.46 Multiwell 3D DAS VSP data multiple migration results (top, IL 651; bottom, XL701) at location 1.

resolution. Multiple imaging has been discussed in the literature for several years as a processing-based approach to this problem. Least-squares wave equation multiple migration (LS-WEMM) results reduce acquisition striping and improve spatial resolution compared with primary imaging. The LS-WEMM method was described by Poole (2019):

$$(t, x, y) = (t, x, y)^{-1}\Phi(f, x, y, z)D_{\Phi}(f, x, y, z)r(x, y, z) \tag{8.9}$$

In equation (8.9), the reflectivity (x, y, z), is derived such that when it is convolved with the forward extrapolated downgoing wavefield, $D_{\Phi}(f, x, y, z)$, and further extrapolated to the surface using operator $\Phi(f, x, y, z)$, the upgoing multiple wavefield, $m(t, x, y)$, is produced to compare with recorded multiples in a least-squares sense. Confidence weights (t, x, y) limit the cost function to recorded data locations within the mute. Spatial coordinates are given by (x, y, z), t and f refer to time and temporal frequency, respectively, and the inverse Fourier transform, F^{-1}, converts data from the frequency domain to the time domain. In practice, recorded data are used as a substitute for the recorded multiples. The reflectivity, derived using the steepest descent iterative solver, represents the subsurface to predict the following order of arrivals from the downgoing wavefield. For example, primaries will predict first-order multiples, first-order multiples will predict second-order multiples, and so on (Poole, 2021).

We have developed multiple-imaging and least-squares multiple-imaging techniques using a one-way wave equation as the wave propagation engine. Numerical experiments show that even with the DIC, cross-talk noise still appears in the multiple images. However, least-squares multiple migration successfully removes the cross-talk artifacts (Figure 8.48) and provides high-quality broadband multiple images (Figure 8.49).

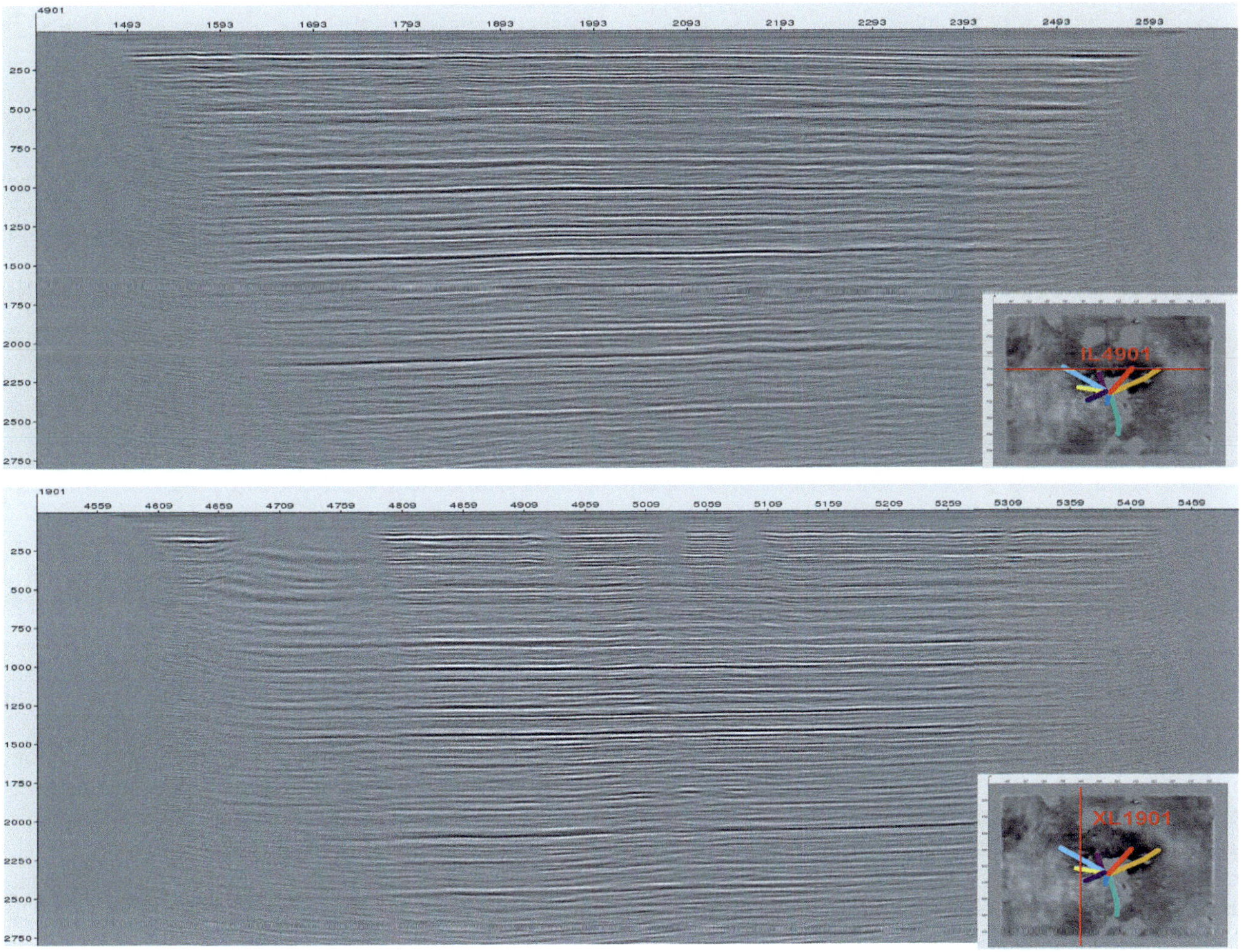

Figure 8.47 Seven-well 3D DAS VSP multiple data migration results along IL 4901 (top) and XL 1901 (bottom) from location 2.

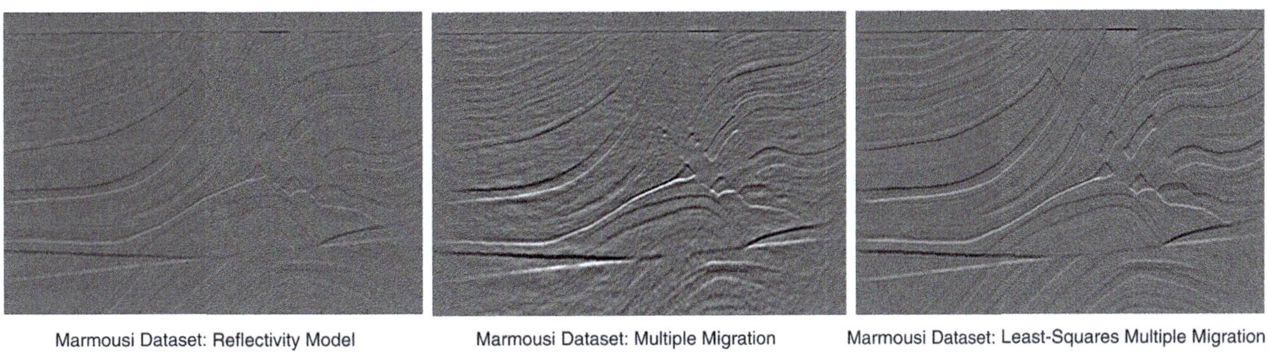

Figure 8.48 (Left) Multiple images using DIC from a single shot record. Note that significant crosstalk artifacts still appear. (Middle) Image from least-squares multiple imaging. (Right) True reflectivity model.

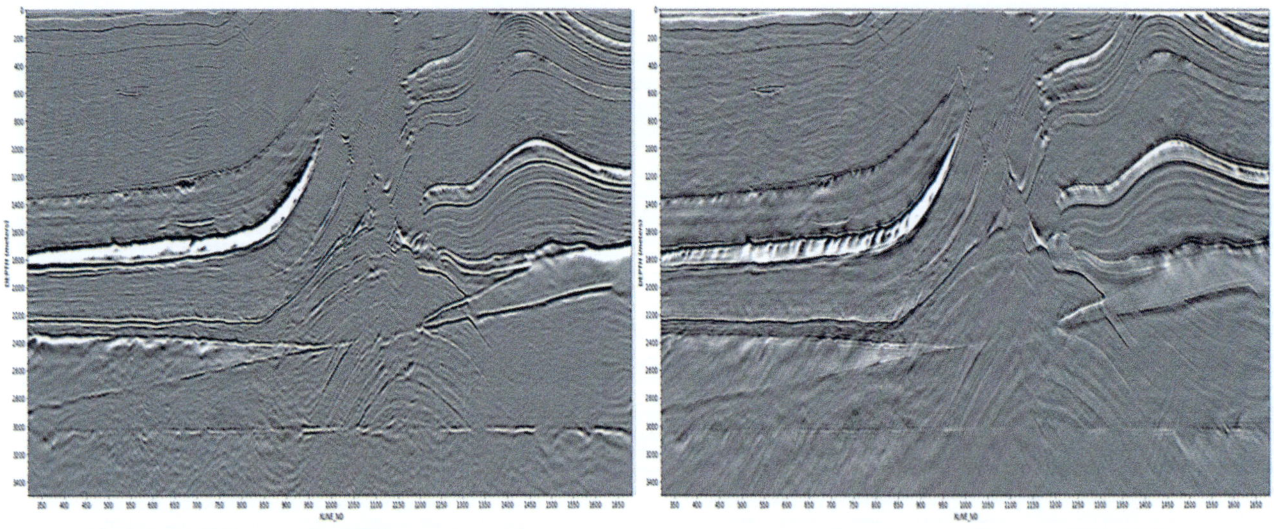

Figure 8.49 (Left) Multiple images using VSP FWI inverted velocity. (Right) Least-squares multiple imaging using VSP FWI inverted velocity.

Gather Processing After Migration. **CADZOW.** Filter noise suppression CADZOW filtering was independently discovered several times and so is also known as singular spectrum analysis (SSA) or the caterpillar method (Cadzow, 1988; Trickett, 2002). Prestack random noise suppression is important for improving AVO and azimuthal (AVAZ) analysis, but it can also improve the stacked section by exploiting the larger volume of data available at the prestack stage, allowing for improved velocity analysis and statics correction, and by cleaning up multiples prior to their removal. Based on the prestack stochastic noise suppression technique of high-dimensional matrix decomposition, effective signal and random noise are separated by the high-dimensional matrix rank reduction decomposition method (Trickett & Burroughs, 2009). Figure 8.50 shows the results before and after prestack random noise attenuation. The SNR of the gathers is significantly improved, and random noise is greatly suppressed (Figure 8.50 (right)).

Spectrum Whitening. Spectrum whitening is used to control the amplitude spectrum of input seismic traces within the frequency band of signals prescribed by users to the same level by time-variable gain and suppress the amplitude out of the frequency band. This process is called "time-variable spectrum whitening" or "time-variable amplitude compensation." Delicate processing of time-variable spectrum whitening can be fulfilled by defining a comparatively small frequency increment. Also, defining a bigger frequency increment can yield a good result for general processing. Seismic data after time-variable spectrum whitening are helpful for velocity analysis and residual static corrections. A 3D DAS VSP profile before and after spectral whitening is shown in Figure 8.51.

Random Noise Suppression. The post-stack random noise attenuation method is a prediction denoising technique in the F-XYO domain based on 3D frequency space. It assumes that the effective wave in the seismic record is predictable in the F-XYO domain, whereas the random noise does not have such characteristics. The multitraces complex least-square principle is used to obtain the 3D prediction operator, and the prediction operator is used to predict and filter the 4D seismic data volume of the frequency component to attenuate random noise. Figure 8.52 shows a 3D DAS VSP profile before and after post-stack random noise attenuation.

8.4. CONCLUSIONS

DAS is increasingly recognized as a viable alternative to geophone arrays for the acquisition of borehole seismic or DAS VSP data. When acquiring surface seismic data in the field, simultaneously acquired DAS VSP data in available wells can provide full-well VSP imaging and enhance surface seismic data processing significantly. As the 3D VSP detectors are located near the reservoir, the formation average and interval velocity, absorption, attenuation, deconvolution operator, anisotropy, and other information around the wellbore in a certain range can be obtained directly from the VSP data. At the same time, it is possible to obtain high-resolution reflection imaging near the well with a higher number of folds. The 3D VSP can also be used for AVO analysis, reservoir fracture prediction,

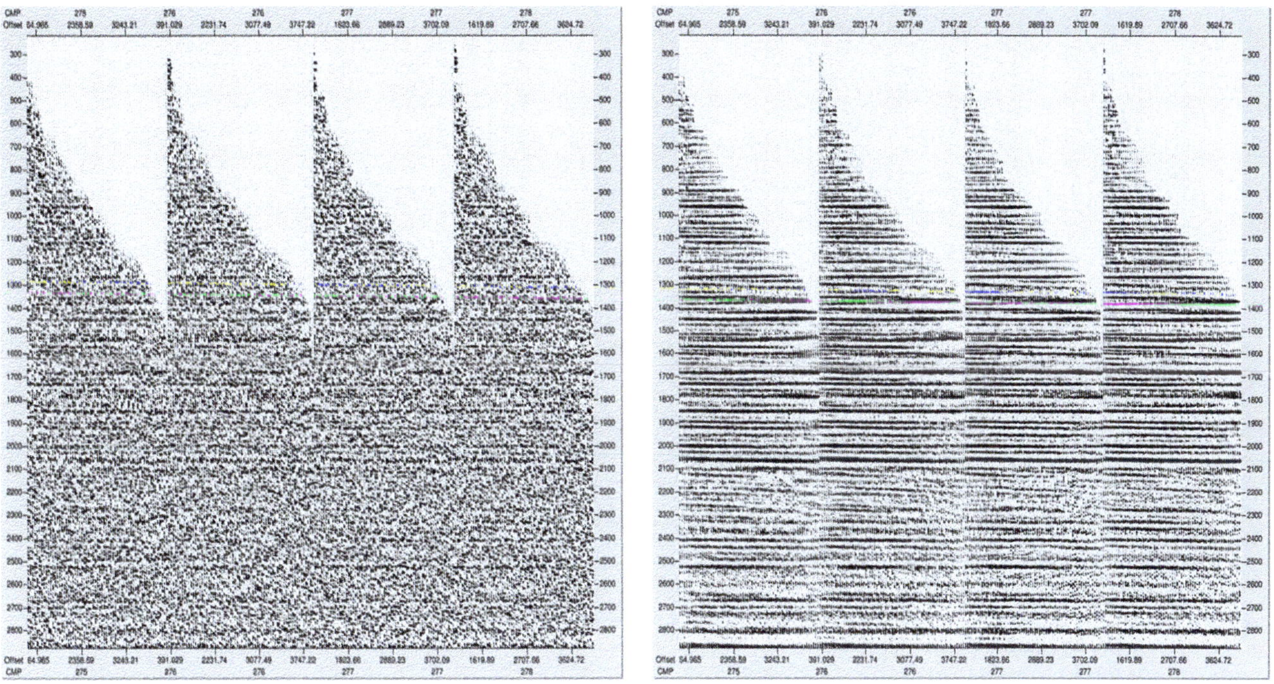

Figure 8.50 Before prestack random noise attenuation (left) and after prestack random noise attenuation (right).

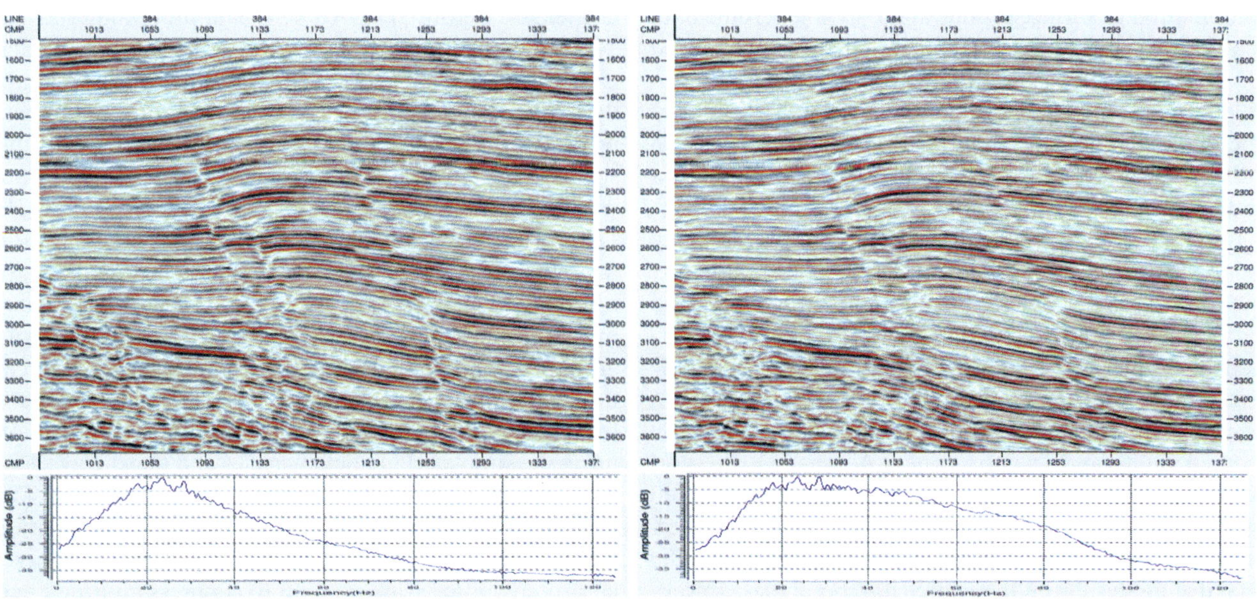

Figure 8.51 A 3D DAS VSP profile before and after spectral whitening.

anisotropy analysis, and reservoir characterization. The information obtained around the well from the VSP data can be used to enhance surface seismic data processing. It includes velocity model calibration and modification, extracting deconvolution operator, demultiple processing, anisotropic migration, Q compensation or Q migration, and so on. The attributes of the amplitude-preserved imaging data can be used to identify fluid in the reservoir formation and trace the reservoir layer distribution around the boreholes for optimization of field development.

The borehole and surface joint seismic exploration technique is a 3D seismic exploration method formed by the

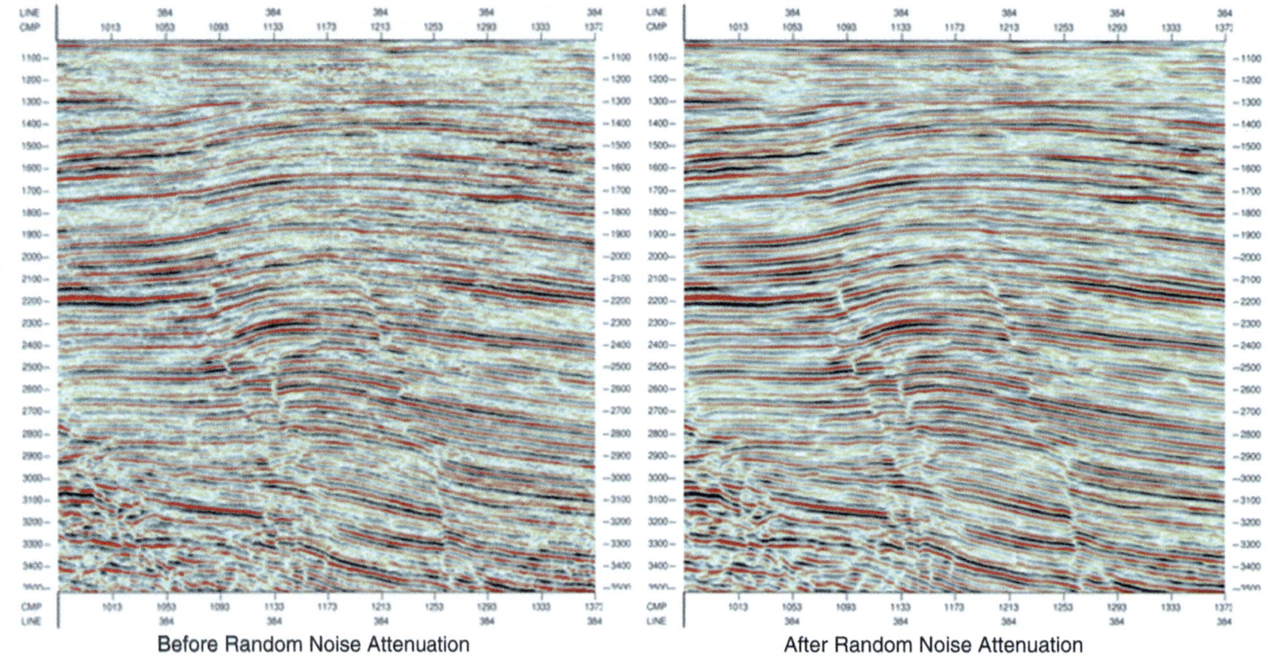

Figure 8.52 A 3D DAS VSP profile before and after post-stack random noise attenuation.

combination of surface seismic and borehole seismic data acquisition simultaneously. In this chapter, we introduced 3D DAS VSP data simultaneously acquired by a downhole armored optical cable along with a 3D surface seismic data acquisition project in Dagang Oilfield, an OBN data acquisition project in the East China Sea and in the Middle East.

Borehole-driven seismic data processing, especially Q migration in the Dagang project, has significantly enhanced imaging quality. Q migration was conducted with accurate Q-field data obtained from the joint inversion of both near surface 3D Q-field data from uphole data and mid-deep layer Q-field data from VSP data. The Q migration results show much sharper and more focused faults and fracture zones. A new 3D reservoir structural model was built with much finer details based on the Q migration and coherence data, and three new borehole locations were recommended to the client. The final drilling results on these three new boreholes are very successful, with a high volume of commercial oil flow.

In the East China Sea project, 3D DAS VSP downgoing multiple reflection wave imaging shows significant imaging quality improvement in comparison with vintage 3D OBC data imaging. Tracing of the reservoir and formation boundary becomes easier and more confident in both new OBN data imaging and 3D DAS VSP data imaging than in the vintage 3D OBC data imaging.

The processing of the world's largest multiwell 3D DAS VSP dataset provided us with unique opportunities and challenges. Before 3D DAS VSP data processing started, 24/7 continuous recorded raw data was split using each OBN shot time and shot GPS coordinates into CSG data with a 10 s record. Because multiple airgun source boats were used to acquire the OBN and 3D DAS VSP data, sparse inversion with all the shots within the source carpet was used for the deblending of the blended acquired 3D DAS VSP data. The spherical diffusion compensation method using the TAR factor extracted from VSP data is used to compensate for the 3D DAS VSP data. Abnormal amplitude suppression is adopted to remove wild amplitude interference using the "multitrace identification and single-trace denoising" idea. The signal deconvolution technique was used to perform bubble suppression using the simulated far-field wavelet and ghost wave suppression in the FK domain on the 3D DAS VSP data. The positive and negative Tau-P transform is used to separate upgoing and downgoing wavefields. Time-varying wavelet deconvolution is carried out to suppress multiples and improve the resolution of the 3D DAS VSP data. The Tau-P deconvolution method is also used to conduct predictive deconvolution on the VSP data after Tau-P transformation to suppress multiples. VSP first-break tomography and FWI are used to build the velocity model, and the least-squares multiple-imaging technique is used to migrate the multiwell 3D DAS VSP data. After image processing, the gather processing steps of the CADZOW filter noise suppression, spectrum whitening, and random noise suppression were applied to the imaging data before the final stack. The processing results of the multiwell 3D DAS VSP data will be used to generate high-accuracy and high-resolution 3D structure images with a higher number of folds for planning infilled and horizontal borehole drilling.

ACKNOWLEDGMENTS

The authors wish to thank the Dagang Oilfield of CNPC, Shanghai Petroleum, and Natural Gas Co., Ltd., ADNOC, BGP, and Optical Science and Technology (Chengdu) Ltd. for permission to publish the data; and they acknowledge the support of BGP's research project "Borehole DAS VSP Key Techniques Research Project" (06-01-2021) and CNPC's research project "Downhole Optical Fiber Intelligent Monitoring Project" (2020F-44).

AVAILABILITY STATEMENT

The data and software used in this chapter are proprietary and are not available to the public.

REFERENCES

Brown, M. P., & Guitton, A. (2005). Least-squares joint imaging of multiples and primaries. *Geophysics*, *70*(5), S79–S89.

Cadzow, J. (1988). Signal enhancement – a composite property mapping algorithm. *IEEE Transactions on Acoustics Speech, and Signal Processing*, *36*, 49–62.

Causse, E., & Ursin, B. (2000). Viscoacoustic reserve-time migration. *Journal of Seismic Exploration*, *9*(2), 165–183.

Chen, Y. Z., Li, Y. P., Zhao, H., Jin, Q. H., Wang, X. M. and Liang, X. (2012). Improving the precision of surface seismic data processing by walkaway-VSP. Paper presented at the 74th EAGE Conference & Exhibition incorporating SPE EUROPEC, A032.

Correa, J., Egorov, A., Tertyshnikov, K., Bona, A., Pevzner, R., Dean, T., et al. (2017). Analysis of signal to noise and directivity characteristics of DAS VSP at near and far offsets - A CO2CRC Otway Project data example. *The Leading Edge*, *36*, 994a1–994a7.

Correa, J., Pevzner, R., Popik, S., Tertyshnikov, K., Bona, A., & Freifeld, B. (2018). *Application of 3D VSP acquired with DAS and 3C geophones for site characterization and monitoring program design: preliminary results from stage 3 of the CO2CRC Otway project*. Paper presented at the SEG International Exposition and 88th Annual Meeting.

Dai, N. & West, G. F. (1994). *Reserve Q migration*. Paper presented at the SEG International Exposition and 66th Annual Meeting.

Daley, T., Freifeld, B., Ajo-Franklin, J., Dou, S., Pevzner, R., Shulakova, V., et al. (2013). Field testing of fiber-optic distributed acoustic sensing (DAS) for subsurface seismic monitoring. *The Leading Edge*, *32*(6), 699–706.

Daley, T. M., Miller, D. E., Dodds, K., Cook, P., & Freifeld, B. M. (2016). Field testing of modular borehole monitoring with simultaneous distributed acoustic sensing and geophone vertical seismic profiles at Citronelle, Alabama. *Geophysical Prospecting*, *64*(5), 1318–1334.

Duan, L., Zhang, Y., Chen, F., Han, W., Zhou, Z., & Song, Q. (2023). *Joint domain full waveform inversion*. Paper presented at the 84th EAGE Annual Meeting, Vienna, Austria.

Ferguson, R. J., & Margrave, G. F. (2005). Planned seismic imaging using explicit one-way operators. *Geophysics*, *70*(5), 101–109.

Frignet, B. & Hartog, A. (2014). *Optical vertical seismic profile on wireline cable*. Paper presented at the SPWLA 55th Annual Logging Symposium.

Hartog, A., Frignet, B., Mackie, D., & Clark, M. (2014). Vertical seismic optical profiling on wireline logging cable. *Geophysical Prospecting*, *62*(4), 693–701.

Herrmann, F. J., & Li, X. (2012). Efficient least-squares imaging with sparsity promotion and compressive sensing. *Geophysical Prospecting*, *60*(4), 696–712.

Jiang, T., Zhan, G., Hance, T., Sugitanto, S., Soulas, S. & Kjos, E. (2016). *Valhall dual-well 3-D DAS VSP field trial and imaging for active wells*. Paper presented at the SEG International Exposition and 86th Annual Meeting.

La Follett, J., Wyker, B., Hemink, G., Hornman, K., Lumens, P., & Franzen, A. (2014). *Evaluation of fiber-optic cables for use in distributed acoustic sensing: commercially available cables and novel cable designs*. Paper presented at the SEG International Exposition and 84th Annual Meeting.

Lesnikov, V. & Allanic, C. (2014). *DAS VSP acquisition - perspectives and challenges*. Paper presented at the 76th EAGE Conference and Exhibition.

Li, X., Song, Q., Song, J., Tran, A. and Zhou, Z. [2022] Preconditioned Linearized Inversion with the Inverse Scattering Imaging Condition. 2022(1), 1–5.

Li, Y. P., Chen, Y. Z., Peng, J., & Jin, Q. H. (2010). *Walkaway VSP multi-wave imaging over a gas cloud area*. Paper presented at the 72nd EAGE Conference & Exhibition incorporating SPE EUROPEC.

Liu, Y. K., Chang, X., Jin, D. G., He, R. Q., Sun, H. C., & Zheng, Y. C. (2011). Reverse time migration of multiples for subsalt imaging. *Geophysics*, *76*(5), 209–206. doi:10.1190/geo2010-0312.1

Liu, S. W., Wang, H. Z., Chen, S. C., & Liao, J. P. (2012). Joint imaging method of VSP upgoing and downgoing reflection wave. *Chinese Journal of Geophysics*, *55*(9), 3126–3133. doi:10.6038/j.issn.0001-5733.2012.09.030

Liu, Y. K., Hu, H., Xie, X. B., Zheng, Y. C., & Li, P. (2015). Reverse time migration of internal multiples for subsalt imaging. *Geophysics*, *80*(5), S175–S185. doi:10.1190/geo2014-0429.1

Lu, S., Wu, H., Dong, X., Zhong, T., Qiu, L., Li, X., et al. (2022). Building adjoint operators for least-squares migration using acoustic wave equation. *Geophysics*, *2022*(ja), 1–64.

Luo, Y., & Schuster, G. T. (1991). Wave-equation traveltime inversion. *Geophysics*, *56*, 645–653.

Ma, Y., & Hale, D. (2013). Wave-equation reflection traveltime inversion with dynamic warping and full-waveform inversion. *Geophysics*, *78*, 223–233.

Madsen, K., Thompson, M., Parker, T., & Finfer, D. (2013). A VSP field trial using distributed acoustic sensing in a producing well in the North Sea. *First Break*, *31*, 51–56.

Mateeva, A., Lopez, J., Chalenski, D., Tatanova, M., Zwartjes, P., Yang, Z., et al. (2017). 4D DAS VSP as a tool for frequent seismic monitoring in deep water. *The Leading Edge*, *36*, 995–1000.

Mateeva, A., Lopez, J., Mestayer, J., Wills, P., Cox, B., Kiyashchenko, D., et al. (2013a). Distributed acoustic sensing for reservoir monitoring with VSP. *The Leading Edge*, *32*, 1278–1283. doi:10.1190/tle32101278.1

Mateeva, A., Lopez, J., Potters, H., Mestayer, J., Cox, B., Kiyashchenko, D., et al. (2014). Distributed acoustic sensing for reservoir monitoring with vertical seismic profiling. *Geophysical Prospecting*, *62*, 679–692.

Mateeva, A., Mestayer, J., Cox, B., Kiyashchenko, D., Wills, P., Lopez, J., et al. (2012). *Advances in distributed acoustic sensing (DAS) for VSP*. Paper presented at the SEG International Exposition and 82nd Annual Meeting.

Mateeva, A., Mestayer, J., Yang, Z., Lopez, J., Wills, P., Roy, J., & Bown, T. (2013b). *Dual-well 3D VSP in deepwater made possible by DAS*. Paper presented at the 83rd Annual International Meeting, SEG. Expanded Abstracts, 5062–5066. doi:10.1190/segam2013-0667.1

Mittet, R., Sollie, R., & Hokstad, K. (1995). Prestack depth migration with compensation for absorption and dispersion. *Geophysics*, *60*(5), 1485–1494.

Parker, T., Shatalin, S., & Farhadiroushan, M. (2014). Distributed acoustic sensing – a new tool for seismic applications. *First Break*, *32*, 61–69.

Poole, G. (2019). *Shallow water surface related multiple attenuation using multi-sailline 3D deconvolution imaging*. Paper presented at the 81st EAGE Conference and Exhibition. Expanded Abstracts, Tu R1 5.

Poole, G. (2021). *Least-squares multiple imaging constrained jointly by OBN and towed-streamer data*. Paper presented at the 82nd EAGE Conference and Exhibition. Expanded Abstracts, 1–5. doi:10.3997/2214-4609.202113070

Sava, P., & Fomel, S. (2006). Time-shift imaging condition in seismic migration. *Geophysics*, *71*, 209–217.

Sun, X. G., Lin, Y., Gao, J., Sun, Z. S., & Lin, J. X. (2009). *Anisotropic parameter calculation by walkaway VSP and full azimuth seismic data*. Paper presented at the SEG International Exposition and 79th Annual Meeting.

Trickett, S. (2002). *F-x eigenimage noise suppression*. Paper presented at the 72nd Annual International Meeting, SEG. Expanded Abstracts, 2166–2169.

Trickett, S. & Burroughs, L. (2009). *Prestack rank-reducing noise suppression: theory*. Paper presented at the SEG Technical Program. Expanded Abstracts, 3332–3336. doi:10.1190/1.3255552

Wang, Y. B., Chang, X., & Hu, H. (2014). Simultaneous reverse time migration of primaries and free-surface related multiples without multiple prediction. *Geophysics*, *79*(1), 1–9.

Warner, M., & Guasch, L. (2016). Adaptive waveform inversion: theory. *Geophysics*, *81*, 429–445.

Whitmore, N. D., Valenciano, A. A., Sollner, W. & Lu, S. (2010). *Imaging of primaries and multiples using a dual-sensor towed streamer*. Paper presented at the SEG Technical Program. Expanded Extracts, 3187–3192.

Witte, P. A., Yang, M., & Herrmann, F. J. (2017). *Sparsity-promoting least-squares migration with the linearized inverse scattering imaging condition*. Paper presented at the EAGE Annual Conference.

Wu, H., Wong, W., Yang, Z., Wills, P., Lopez, J., Li, Y., et al. (2015). Dual-well 3D vertical seismic profile enabled by distributed acoustic sensing in Deepwater Gulf of Mexico. *Interpretation*, *3*, SW11–SW25. doi:10.1190/INT-2014-0248.1

Yang, Y., Engquist, B., Sun, J., & Hamfeldt, F. (2018). Application of optimal transport and the quadratic Wasserstein metric to full-waveform inversion. *Geophysics*, *83*, 43–62.

Yu, G., Liu, H. B., Cao, Z. L., Li, J. J., Li, J. G., Chen, Y. Z., et al. (2022). Joint OBN and 3D DAS-VSP data acquisition and processing in the East China Sea. *First Break*, *40*(9), 37–48. doi:10.3997/1365-2397.fb2022073

Yu, G., Xiong, J. L., Wu, J. J., Chen, Y. Z., & Zhao, Y. S. (2020). Enhanced surface seismic data processing using simultaneous acquired DAS-VSP data. *First Break*, *38*(6), 29–36. doi:10.3997/1365-2397.fb2020039

Zhang, Y., Duan, L., Han, W., Chen, F., Chen G., & Zhou, Z. (2023). *Overcoming cycle-skipping issue using joint-domain full waveform inversion*. Paper presented at the 18th International Congress of the Brazilian Geophysical Society, Rio de Janeiro, Brazil.

Zwartjes, P., Mateeva, A., Tatanova, M., Chalenski, D., Yang, Z., Lopez, J., et al. (2017). *4D DAS VSP in deepwater – proof of concept and next steps*. Paper presented at the 87th SEG Annual Meeting. Expanded Abstracts, 5802–5807.

9

Distributed Acoustic Sensing Acquired Wellbore Seismic Data Using Hybrid Wireline Cable: Field Data Examples

Alejandro Martinez[1], Manuel Useche[2], and Ali Sayed[1]

ABSTRACT

Hybrid wireline cables represent a significant advancement in the deployment of fiber optics within the context of well logging operations. These cables integrate fiber optic strands with the copper lines found in conventional electric wireline cables. When optical fibers are part of a hybrid optical-electrical logging cable, seismic data can be acquired during any logging intervention, improving efficiency and data collection. This integration reduces the acquisition time for Vertical Seismic Profiling (VSP) to mere minutes, compared to the hours or days required by traditional methods. Additionally, rig time is eliminated if VSPs are conducted using hybrid cables while other tools remain stationary during their operations. This chapter centers on VSP jobs recorded with hybrid optical-electrical hepta-cables, highlighting their potential to revolutionize borehole seismic operations, particularly in well exploration and appraisal. Furthermore, various wireline cables and slicklines used in cased-hole interventions now incorporate optical fibers, allowing seismic measurements to be gathered concurrently with other interventions. This technology offers a transformative approach to conducting borehole seismic operations, enhancing efficiency and data quality. Drastic reduction of operating time benefits environmental aspects such as lowering both sound exposure levels detrimental to marine fauna and CO_2 emissions.

9.1. INTRODUCTION

In geophysical projects in the energy industry, distributed acoustic sensing (DAS) technology uses optical fiber as a vibration sensor for seismic acquisitions, which is deployed along portions or the entire well, either permanently or temporarily. The principle of DAS, discussed at length in previous chapters, is that a short pulse of light travels down an optical fiber, and as the pulse of light propagates, a small amount of energy scatters at each elemental section of the fiber. A small fraction of the scattered light is guided back to the launching end and captured by an optical interrogator at the surface. The returning scattered light is acquired as a function of the time elapsed from launching the probe pulse. The timing of a sample of the acquired trace is related to the location in the fiber where the light was scattered because the group velocity of light in the fiber is known (Frignet & Hartog, 2014; Hartog, 2017; Hartog et al., 2014).

The most common deployments of optical fibers in the industry are in well completion, either cemented behind the casing (Olofsson & Martinez, 2017) or strapped around the production tubing (Ali et al., 2021; Duque et al., 2023b; Ferla et al., 2018). Typically, these fibers have been used for surveillance and production monitoring, mainly using distributed temperature sensing (DTS) applications. In these environments, DAS is increasingly becoming adopted for seismic measurements around production or injection monitoring: for example, in time-lapse vertical seismic profiles (Mateeva et al., 2017),

[1]*SLB, Houston, Texas, USA*
[2]*SLB, Bogota, Colombia*

Distributed Acoustic Sensing in Borehole Geophysics, Geophysical Monograph 289, First Edition.
Edited by Yingping Li, Robert Mellors, and Ge Zhan.

DOI: 10.1002/9781394179275.ch09

crosswell tomography (Nakatsukasa et al., 2018), strain monitoring (Starr & Jacobi, 2017), microseismic monitoring (Le Calvez et al., 2022), as well as in the well integrity space, such as in leak detection (Cherubini et al., 2023).

Hybrid wireline cable is a highly effective method of deploying fiber optics. The term *hybrid* in this context refers to the fact of combining fiber-optic technology within a conventional logging cable. These cables are built so that the fiber-optic strands are bundled with the copper (electric) lines of an otherwise conventional electric wireline cable. When optical fibers are deployed as part of a hybrid optical–electrical logging cable, any logging intervention in the well can be leveraged to acquire seismic data. This opens a new avenue of well seismic measurements in the exploration and appraisal of oil fields. In this novel approach, a hybrid logging cable becomes the seismic sensing element so that the velocity measurement along the entire well is conducted efficiently. The acquisition time for checkshots and zero-offset vertical seismic profiling (ZOVSP), which are routinely performed in the industry, is reduced to minutes instead of hours or days required by conventional methods. Rig time is eliminated if VSP is acquired with the hybrid cable while other tools remain stationary as part of their own operation (Kimura et al., 2018; Martinez et al., 2018). Sometimes, geophone arrays are required for specific objectives requiring three-component (3C) data, and simultaneous conveyance with DAS hybrid cable allows obtaining additional data above the geophone tools without incurring additional operating time. The approach of acquiring VSP with hybrid cable is covered in this chapter.

Other conveyances, such as coiled tubing or carbon rods featuring optical fiber, offer alternatives for collecting well seismic measurements (Danardatu et al., 2014; Kimura et al., 2019; Kobayashi et al., 2020). Several types of wireline cables and slicklines used for cased-hole interventions now feature optical fibers, and although they are mostly geared for interventions in producing wells, seismic measurements can still be acquired at the back of these operations. Also, drop-down dissolvable optical fibers that are deployed by gravity in wells present opportunities for footprint reduction of surface equipment and potential cost optimization of DAS interventions (Bostick et al., 2019a, 2019b). More recently, surface DAS deployments are gaining traction in the industry, especially for time-lapse monitoring and microseismic evaluation (Bakulin et al., 2019; Hornman et al., 2013).

Moreover, the use of DAS is also expanding to the new energy field, such as carbon capture storage (CCS), hydrogen storage, and drilling for helium, lithium, and geothermal (enhanced geothermal systems [ESGs]).

The focus of the discussion herein is on VSP jobs recorded using hybrid optical–electrical heptacables or temporary interventions in well logging operations. This is a technology that can dramatically change the way borehole seismic operations are conducted, particularly exploration and appraisal of oil and gas fields.

9.2. DESCRIPTION OF THE TECHNOLOGY

A hybrid logging cable has been designed for VSP acquisition using DAS technology (Varkey et al., 2008). As shown in Figure 9.1, the hybrid optical heptacable is an innovative way to acquire borehole seismic data during any other descent scheduled in a wellbore, either in an open hole or cased hole (e.g., reservoir pressure/sampling, petrophysical measurements, casing evaluation, etc.). This technology provides access to any well at any time to collect borehole seismic measurements when performing well logging operations. The fiber optic is bundled with electrical lines, which turn the entire logging cable into a seismic receiver. This simplifies the planning around VSP surveys, and it significantly reduces the overall cost of collecting such information because a dedicated logging run for seismic acquisition is eliminated.

The advantage of this technology, which allows the acquisition of borehole seismic data with DAS, is clear. The operator can obtain well seismic information using minimum rig time (Useche et al., 2020) to zero additional rig time when the tools conveyed with the hybrid cable are required to remain stationary as part of the operations. Some examples of DAS acquisition with hybrid cables are when collecting reservoir samples (Guerra et al., 2020), sidewall coring (Kimura et al., 2017), calibrating nuclear magnetic tools (Martinez et al., 2018), acquiring well integrity logs in cased-hole environments (Martinez et al., 2021), and even during hybrid DAS geophone operations in the case of complex seismic jobs such as 3D VSP (Kimura et al., 2018).

Much experience has been gained in the last decade on DAS VSP, and the technology is rapidly advancing as well. Virtually all kinds of VSP under diverse well conditions have been acquired using DAS (Figure 9.2). Table 9.1 presents a small sample of actual jobs, which is not intended to be comprehensive. Details of each job can be found in the reference listed.

The results obtained from these surveys range from the simple time-depth pairs to calculate interval velocities and use in synthetic seismograms to a corridor stack for an advanced well tie, vertical transverse isotropy (VTI) model calibration, and 3D imaging in complex geologies.

There are many optical interrogators in the market that employ different implementations of the physics of measurement, as well as the quality and specifications of the recorded data. Also, the technology evolves rapidly, so any attempt to leave printed references of current

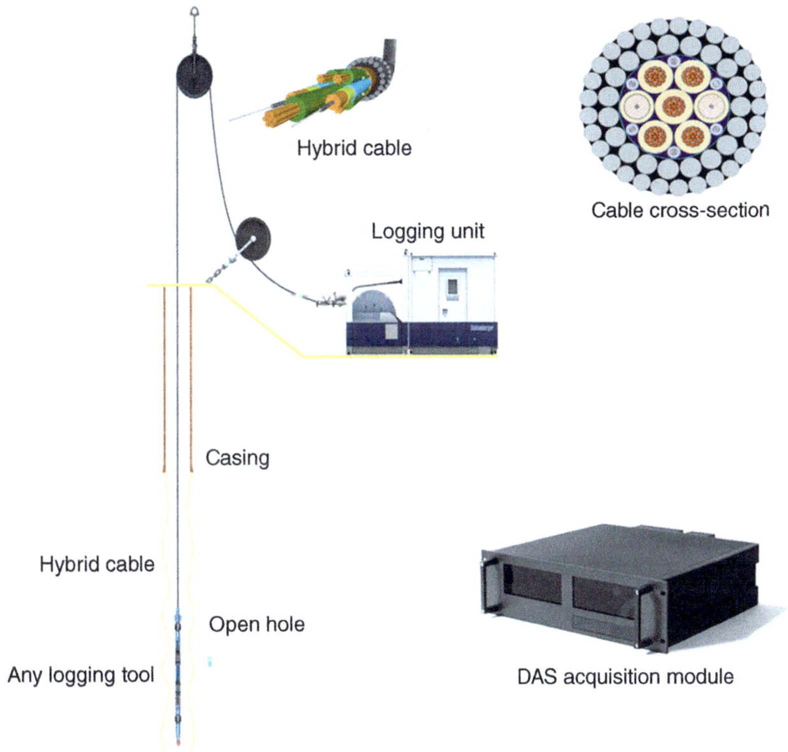

Figure 9.1 Hybrid optical logging cable operation for DAS VSP acquisitions. Martinez et al. (2021) / Society of Exploration Geophysicists.

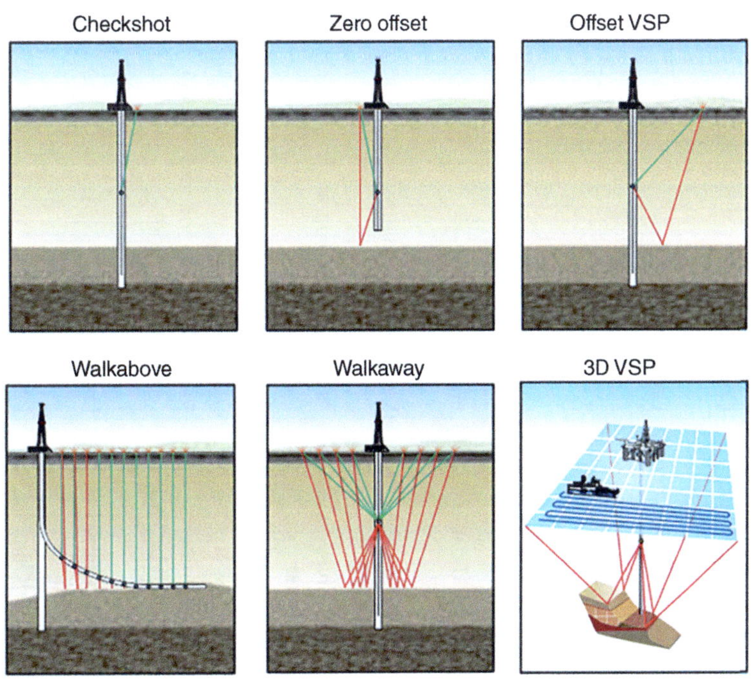

Figure 9.2 Schematic representation of VSP geometries of jobs. Adapted from Martinez and Jones (2010).

Table 9.1 DAS borehole seismic examples in the literature featuring different survey geometries and well environments.

Application/ Geometry	Conveyance	Objective	EOF[a] (m)	Max deviation/ type of well[b]	Environment	Tool combination	Seismic source	Reference
Checkshot	DAS hybrid cable	Time-depth curve and velocity	1,328	18° OH	Land	Magnetic resonance and formation tester	Airgun	Kimura et al. (2017)
Zero offset	DAS hybrid cable	Corridor stack	3,115	4° OH	Land	Magnetic resonance	Airgun	Martinez et al. (2018)
Zero offset	DAS hybrid cable and geophone	Corridor stack	1,435	18° CH	Land	Cement evaluation	Vibrator	Martinez et al. (2021)[c]
Zero-offset lookahead	DAS hybrid cable and geophone	Corridor stack	1,775	Vertical OH	Offshore	Formation tester	Airgun	Guerra et al. (2020)[c]
Zero offset	DAS hybrid cable and geophone	Corridor stack	>5000	5° OH	Offshore	Sonic	Airgun	Rossi et al. (2023)
Zero offset	DAS hybrid cable	Imaging	>4200	40° OH	Offshore	Coring	Airgun	Rossi et al. (2023)
Walkabove	DAS hybrid cable	Velocity and imaging	3,700	Deviated OH	Land	Resistivity imager	Vibroseis	Useche et al. (2020)
Zero offset and walkaway	Hybrid survey DAS and geophone	Anisotropy calibration and imaging	>3000	16° CH	Offshore	Seismic-geophone	Airgun	Horne et al. (2020)[c]
3D VSP	Hybrid survey DAS and geophone	Imaging	>3000m [9000ft]	55° OH/CH	Offshore	Seismic-geophone	Airgun	Kimura et al. (2018)
Walkaway and fixed-offset VSP	Permanent fiber cemented behind casing	Velocity information and depth calibration	2,655	Vertical CH	Land	NA	Dynamite and vibroseis	Olofsson and Martinez (2017)
Walkaway Walkaround	Permanent fiber in completions	Imaging	3,300	Vertical CH	Land	N/A	Vibroseis	Ferla et al. (2018)[c]
3D VSP	Permanent fiber in completions	Imaging	2,700	Deviated CH	Offshore	N/A	Airgun	Ali et al. (2021)

[a] *EOF, end of fiber.*
[b] *OH, open hole; CH, cased hole.*
[c] *Includes partial or complete comparison with geophone data.*
Adapted from Martinez (2022).

specifications will soon be outdated. However, if the reader is interested in such information, the specification of the equipment used in the examples to be discussed next can be found at www.slb.com (both optical interrogator and hybrid logging cable).

For the discussion that follows, field examples acquired in land wells are showcased, mostly using vibrators as the seismic source, except for two jobs in which airguns deployed in a portable acquisition tank were used. Examples of VSP surveys acquired in offshore wells are also discussed. The analysis and jobs presented herein correspond to VSP surveys acquired with hybrid logging cables.

9.3. ANATOMY OF A DAS RECORD ACQUIRED WITH HYBRID LOGGING CABLE

When the cable remains stationary, as part of regular logging operations (i.e., downhole tool calibrations) or by request, seismic data can be acquired with no additional rig time (Kimura et al., 2017) or with minimum operational time on the order of minutes. Several shots of the seismic source, typically about 30 to 100, are required to produce a stack with good signal-to-noise ratio (SNR). This represents significantly less time than conventional seismic acquisitions with conventional geophone-based technology.

Figure 9.3 shows an example of raw stack DAS of 50 shots acquired in a 3,000-m-deep well with 18° deviation and a combination of open- and cased-hole sections. Parameters of the acquisition were gauge length (GL) 10.2 m, trace output every 2.04 m, 1 ms sampling rate (SR), and pulse frequency (PRF) of 5,000 Hz. DAS recording was performed while the cable was under full down tension while conveying a nuclear magnetic resonance (NMR) tool. Some elements can be seen immediately. (a) Clear first breaks can be seen along the wellbore, except in areas marked as poor coupling or near casing shoes where conventional borehole seismic acquisition may also suffer. Near surface is also a place where the quality of the seismic measurement is affected by the presence of multiple casings. These are all areas where the cable may not be properly touching the borehole walls. (b) Even with minimum deviation, DAS data of good quality can be obtained. This is because in many operations, the tension in the logging cable is reduced, forcing the cable to coil up and be in contact with the borehole. Depth correction is made assuming linear distribution of the stretch,

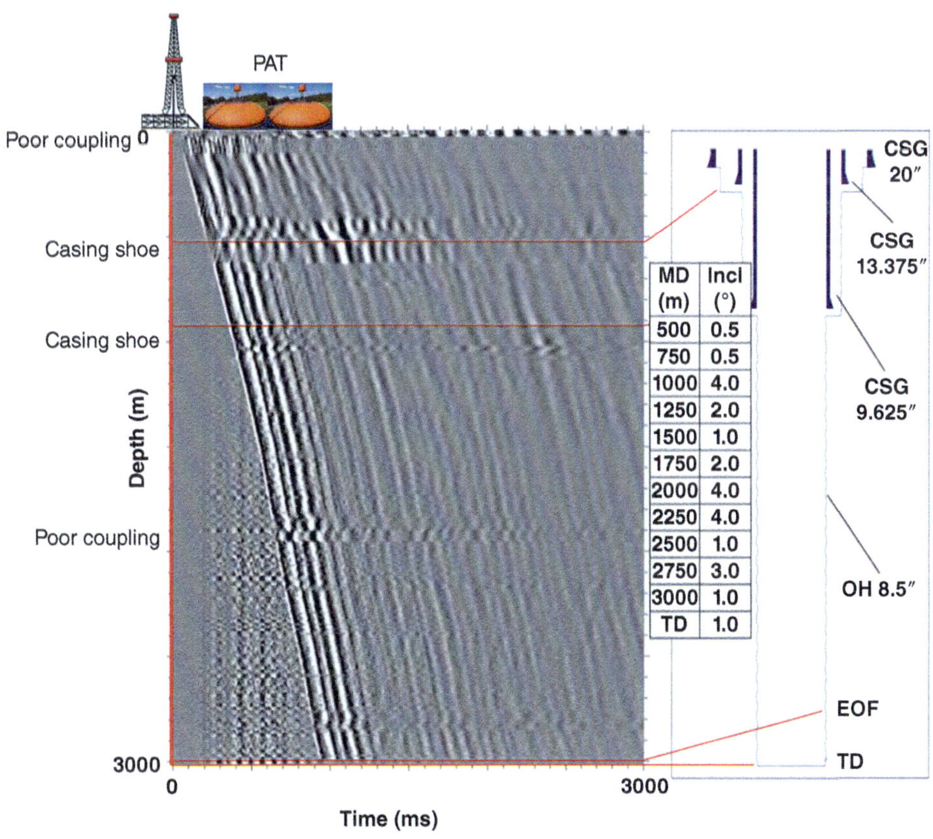

Figure 9.3 Raw stack DAS data. Dual portable airgun tanks were used to deploy each parallel-gun cluster, 350 in^3 total volume. Martinez et al. (2018) / Society of Exploration Geophysicists.

which is accurate in most cases (Guerra et al., 2020). (c) Noise artifacts like those observed with vertical moveout and arriving before the first breaks can be attributed to equipment noise. In this case, the cause is excessive vibration of the optical interrogator unit at the surface. (d) Source multiples that travel along the first breaks are a good quality control check for SNR and overall wavefield behavior. In this job, a large airgun cluster deployed in two portable tanks was used as the seismic source (Martinez et al., 2018). (e) Even on a raw stack wavefield like the one displayed directly from acquisition, there is already evidence of reflections. (f) The end of fiber (EOF) can be clearly observed at the deepest part of the record. This is where the optical fiber terminates, and it coincides with the top of the tool string immediately below the cable. This position can be accurately matched in depth with standard wireline techniques backed up by precise determination of fiber length measured optically. (g) Data in open- and cased-hole sections are of equal quality, provided acoustic coupling is achieved. (h) Seismic source parameters and equipment used in DAS acquisitions are identical to those used for geophone acquisition.

9.4. P-WAVE VELOCITY INFORMATION

Most of the value of any VSP survey resides in the time-depth pairs and calculated interval velocities. The interval velocities are an essential tool for correcting sonic logs for dispersion effects (Martinez & Jones, 2010; Stewart et al., 1984), which are then used to derive synthetic seismograms for well-tie purposes.

In the example shown in Figure 9.4, a checkshot survey was recorded in Brazil using DAS. The operator wanted time-depth information at discrete markers in the well. Initially, the survey was to be conventional, but it was acquired with DAS instead during a wireline descent consisting of NMR and reservoir sampling tools. During the calibration of the NMR, which is an intrinsic part of the logging operation, the tools must remain stationary. In this case, the tuning process took approximately 20 min, and this very same moment was used to take 50 shots for a stack. The source used in this case was a dual-airgun cluster deployed in a portable tank. This is one of the first documented well seismic records obtained with zero additional rig time. Acquisition parameters were GL 10.2 m and SR 1 ms, and cable under tension.

The distributed acoustic measurement with a spatial output of 2.04 m was decimated with coarser traces, as shown, partly due to acoustic coupling noise and because of the need only for a few specific checkshot points. The data were sufficient to effectively provide velocity characterization and to aid in the well tie using the synthetic seismogram shown in Figure 9.5.

In another example, an operator interested only in time-depth information acquired a DAS survey during a petrophysical logging run. As in the previous example, the seismic source used the acquisition tank technology; in this case, a cluster of tanks provided increased energy output because the well was deeper. The cable remained

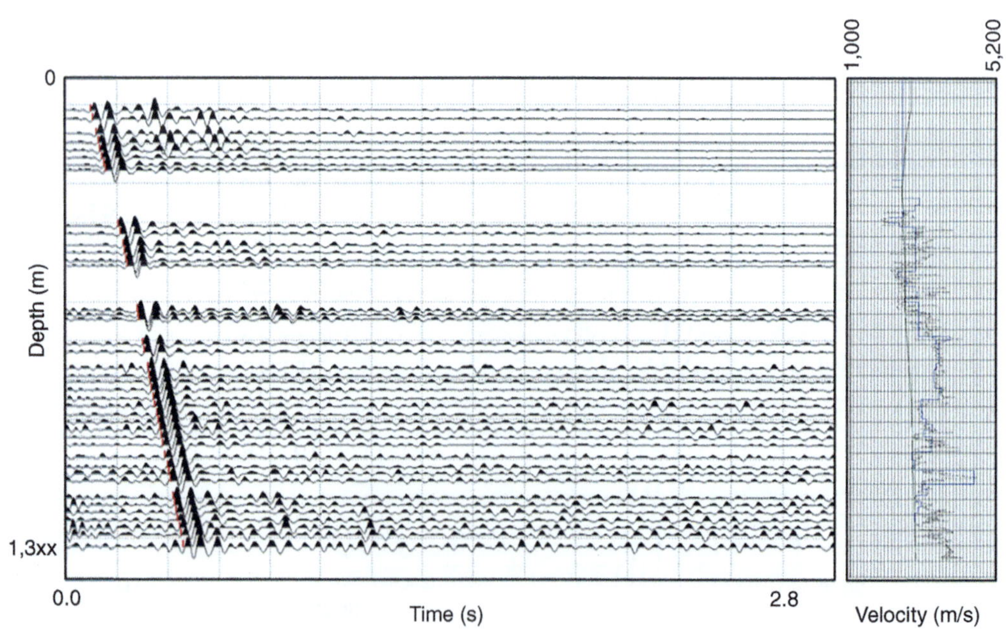

Figure 9.4 Time-depth curve and velocity analysis derived from checkshot and sonic data (right). Blue curve is interval velocity from VSP; black curve from sonic. First breaks (red) are plotted with inverse polarity, as expected from airgun data (left). Adapted from Martinez et al. (2018).

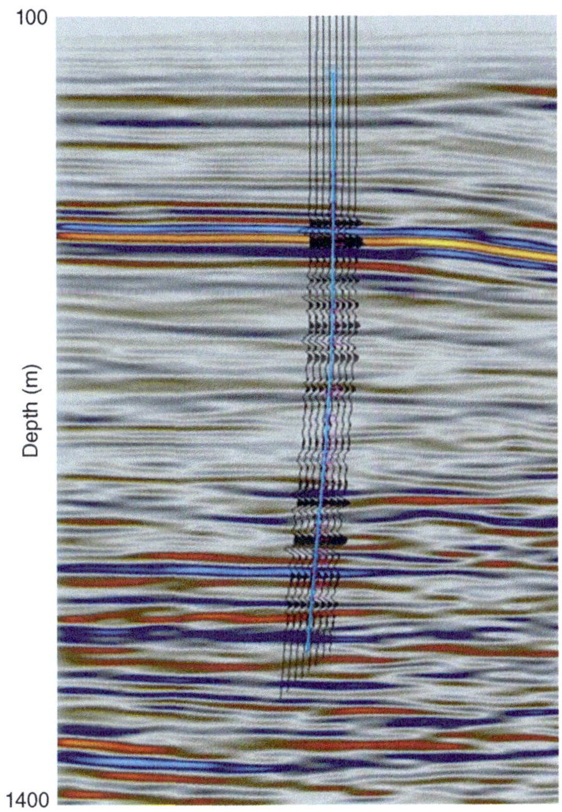

Figure 9.5 Well-tie analysis using synthetic seismogram overlying 2D surface seismic line (normal polarity: an increment in acoustic impedance is a red peak). Alvopetro Energy Ltd.

stationary, and the seismic acquisition added 45 min to the overall logging operation. Figure 9.6 shows that the main objective of obtaining time-depth data was achieved successfully; interval velocities derived from sonic and seismic data correlate very well. It is important to highlight that although the objective was a simple checkshot, the same dataset was used to process a full VSP corridor stack. With this technology, obtaining such a corridor is only a processing effort and not an acquisition one using the same input dataset, in contrast to conventional VSP acquisition. In this case, the extra effort to acquire data for VSP could have been more than 5 hr. In addition, besides the correlation between sonic and seismic, DAS provided velocity information almost all the way to the surface, where sonic was not recorded.

9.5. S-WAVE VELOCITY INFORMATION

An additional value of the DAS datasets is converted S-wave velocities calculation when present in the wavefield because typical denser trace output can minimize spatial aliasing effects. This sometimes creates confusion for some VSP users who question the existence of shear waves in one-component (1C) data such as DAS. S waves can also be present in the vertical axis or *Z* component of geophone tools. In the absence of a fit-for-purpose S-wave seismic source (i.e., shear vibrators), the P-wave to S-wave conversion (PS) is often found in VSP surveys due to

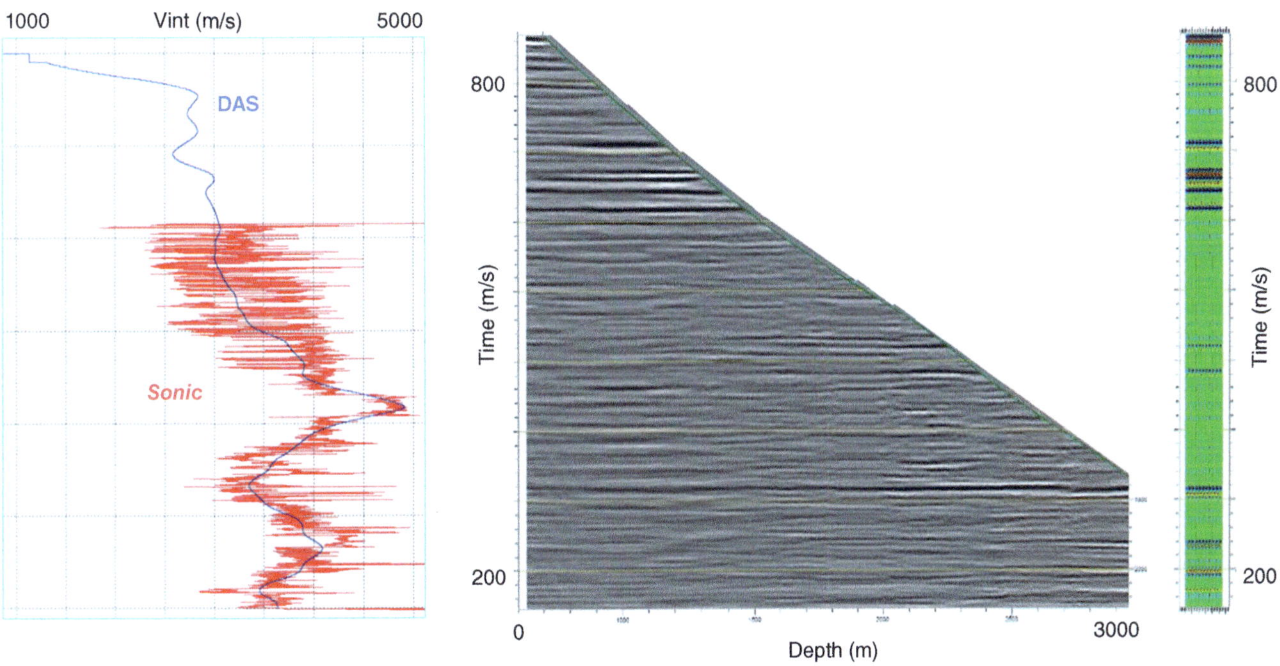

Figure 9.6 (Left) Velocity analysis from sonic and seismic. (Center) Deconvolved upgoing wavefield. (Right) Corridor stack generated. Adapted from Martinez et al. (2018).

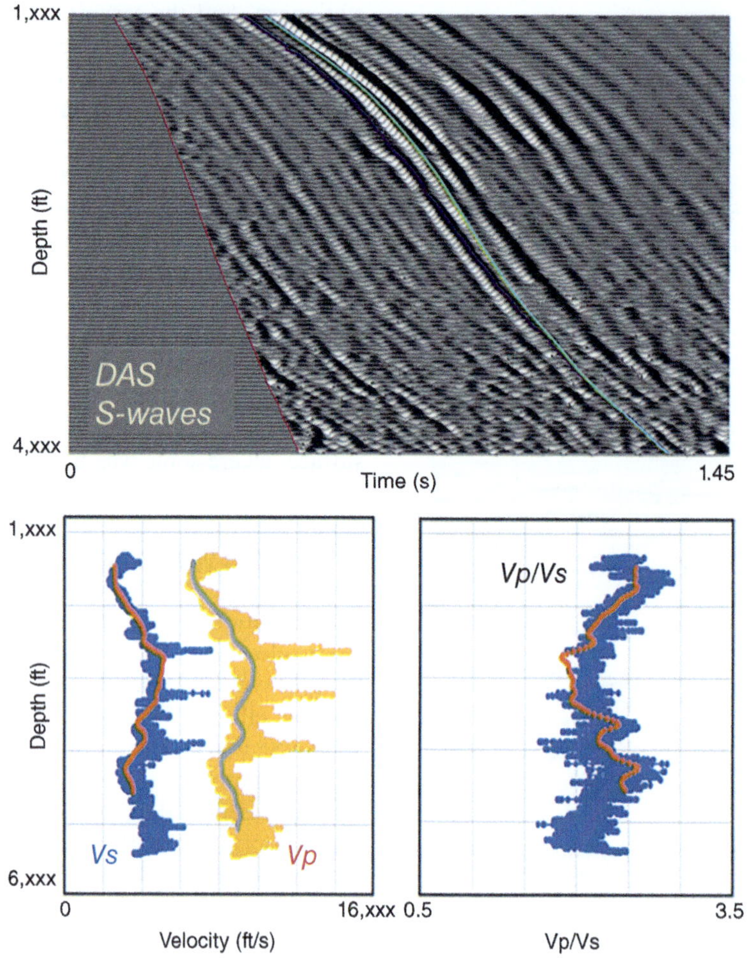

Figure 9.7 (Top) Clearly shows the S-wave downgoing trend corresponding to the residual after first wavefield separation (blue/green curve). (Bottom) Velocity analysis from sonic and seismic data for both P and S waves. Martinez et al. (2020) / Society of Exploration Geophysicists.

high-impedance contrast in the formation caused by features such as weathering layers, salt boundaries, and others. The fine sample spacing of DAS traces helps prevent spatial aliasing, especially in low-velocity shallow deposits. Figure 9.7 shows the time picking of S waves in a DAS downgoing residual wavefield after direct P-wave subtraction. P and S velocities, as well as the independent Vp/Vs ratio derived from DAS data, correlate very well with monopole and dipole sonic data, respectively.

This type and amount of data would require considerable time and effort to be acquired with conventional tools, which would challenge its acquisition as a nonprimary objective. The result is that the S-wave information often is neglected. With DAS, S waves can be utilized more routinely if they are present in the acquired wavefield. These data could be of importance for both localized geomechanical and improved resolution of well tie-analysis if the surface seismic analysis is making use of shear waves.

An example of S waves extracted from a DAS VSP comes from a deviated well in Colombia. The operator wanted to acquire a rig-source VSP in the final section of the well; however, the well construction at the time of intervention would have casing extending all the way from the bottom of the 9½ in. intermediate section, causing poor seismic response. The plan adopted was to perform an intermediate logging intervention. In this case, DAS data were obtained while conveying a sonic-caliper combo tool. The data shown Figure 9.8 are from a 100-shot stack collected in approximately 50 min. Excellent data quality was recorded along most of the wellbore, which had a maximum deviation of 30°. However, it can be noted that the S-shape transitioned from vertical to maximum deviation and then back to a vertical trajectory. The source employed in this acquisition was vibroseis sweeping linearly frequencies 8–96 Hz for 12 s and SR 1 ms. The downhole recording parameters were GL 10.2 m, trace output every 2.04 m, 1 ms SR, and PRF

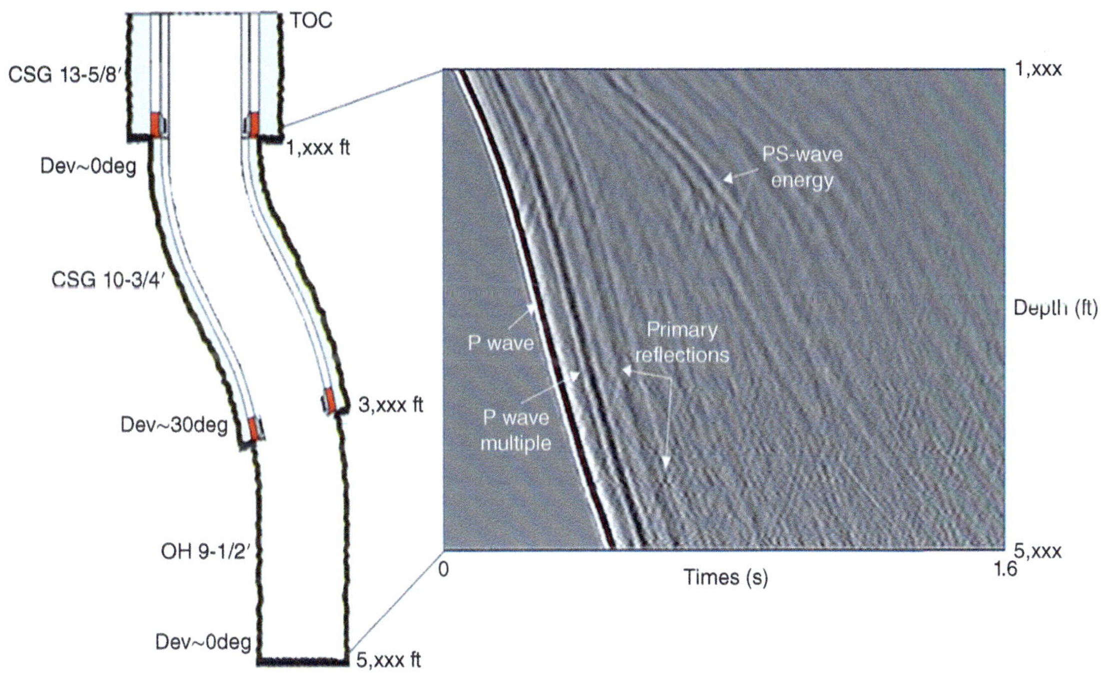

Figure 9.8 DAS in deviated well showing clear PS-wave energy. Martinez et al. (2020) / Offshore Technology Conference.

5,000 Hz. DAS recording was performed while the cable was under full down logging tension.

9.6. COMPARISON OF DAS AND GEOPHONE

The next two studies highlight the results of DAS ZOVSP surveys in cased-hole environments. VSP surveys were acquired while conveying ultrasonic-imager and cement-bond tools with a hybrid optical heptacable. The ZOVSP objectives were general well-tie analysis and, if possible, generating a seismic image near the well. A one-to-one comparison was made with conventional geophone tools in a separate descent.

The geophone examples shown here were in fact acquired with geophone accelerometers (GACs). Characteristics of the GAC can be found in Martinez and Jones (2010). To compare DAS and GAC, both measurements need to be converted to velocity.

The first example features a zero-offset VSP in an 18° deviated cased-hole well. It was recorded to provide high-quality velocity tie-in analysis in a thin sandstone layered reservoir with intercalations of claystones and shales and strong lateral variability as observed in nearby wells.

The seismic source was located with an offset of 511 ft from the wellhead at 90° azimuth using a linear 8–96 Hz, 12 s-long sweep. In the case of geophone data, a four-level tool and five good shots were recorded per stack, which took nearly 14 hr to record. DAS data were acquired while conveying an ultrasonic imager for casing evaluation, with a 75-sweep stack within 1 hr of total operation.

The raw stack wavefields of both DAS and GAC measurements are displayed in Figure 9.9. It can be appreciated that DAS is richer in low frequencies and has a 90° phase rotation compared to GAC, which falls into the normal comparison scenario using raw data. Geophone data is sampled at 50 ft, whereas DAS data is kept at 15 ft (with 30 ft GL).

The conversion of GAC acceleration data to velocity is simply done by applying a sensor transform. DAS was converted from native strain to velocity following (Sayed et al., 2020). This conversion is done in two steps, where first gauge-length corrections corresponding to the normalized sinc function in the spatial frequency domain to yield axial strain (Figure 9.10), followed by converting axial strain to velocity (Figure 9.11).

Frequency spectra of the two datasets after conversion to velocity can be seen in Figure 9.12. DAS data seem to have a broader bandwidth in the range of 8–60 Hz. The ability to reconstruct frequencies higher than 60 Hz for DAS may be compromised due to coupling noise.

So far, DAS data show characteristics similar to geophone data, which gives assurance that the VSP processing should render similar results in both cases. The data processing workflows of both types of data are nearly identical, as shown in Figure 9.13. Some differences in the DAS processing are as follows. (a) The process of

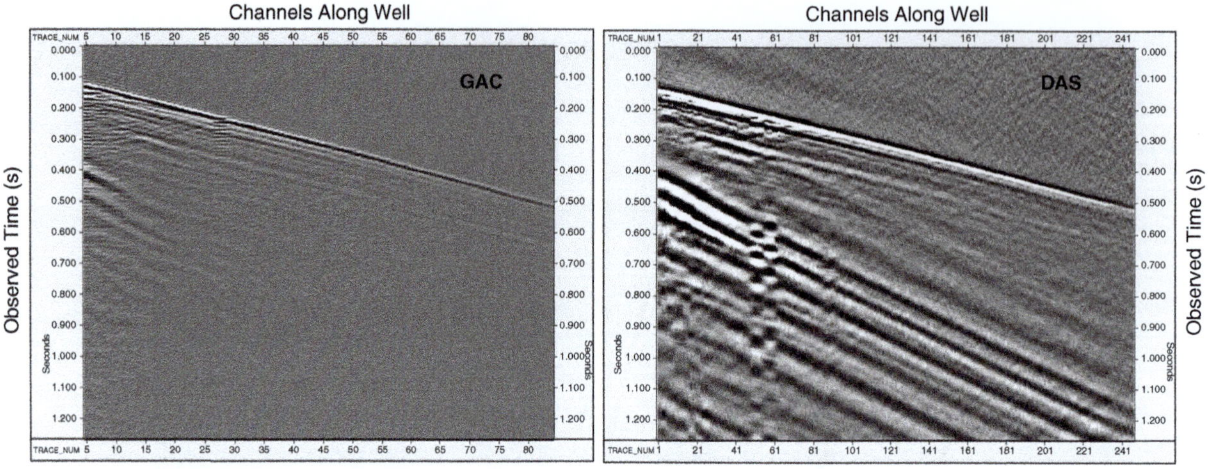

Figure 9.9 Raw stack of GAC (left) and DAS (right) acquisitions. Clear direct arrivals, P, and converted waves are seen in both wavefields. Frequency and phase are different between GAC and DAS measurements. Khaitan et al. (2021) / European Association of Geoscientists & Engineers.

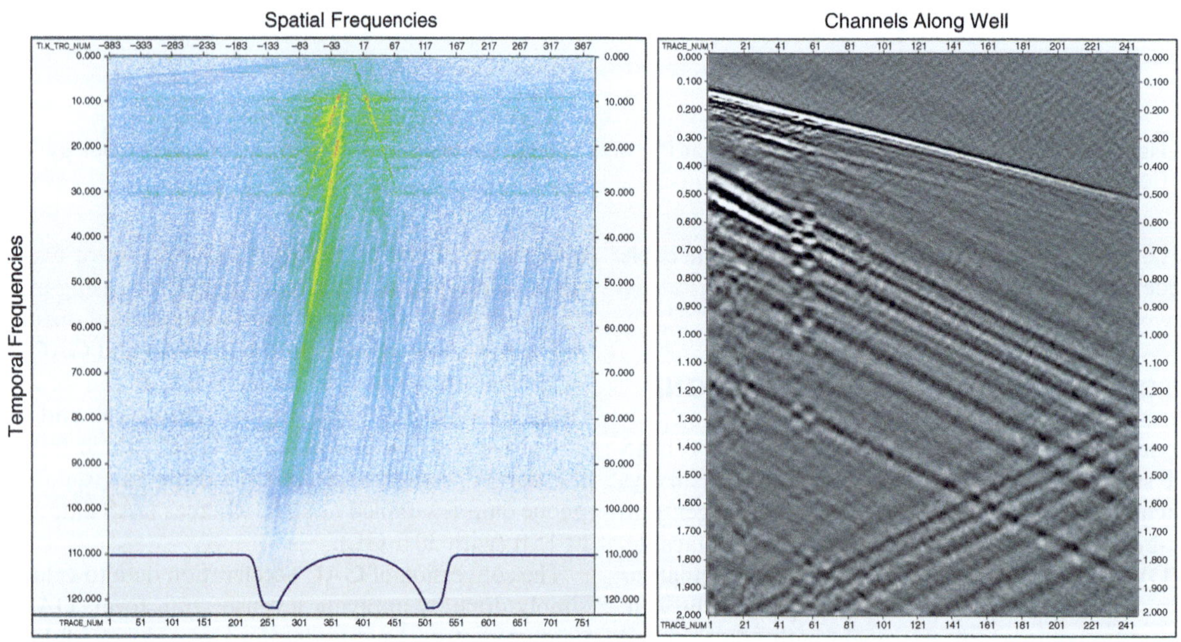

Figure 9.10 (Left) FK spectrum of raw DAS stack. (Right) Axial strain after applying gauge length corrections factors: blue line overlying FK spectrum. Khaitan et al. (2021) / European Association of Geoscientists & Engineers.

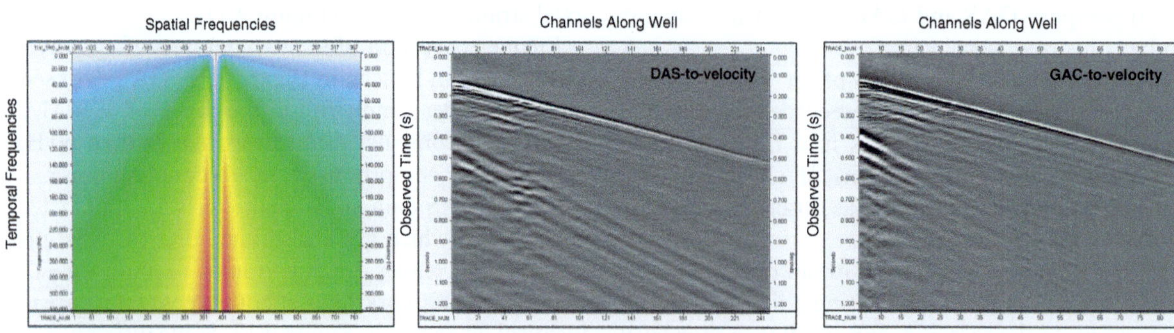

Figure 9.11 Scaling factors in the FK spectrum (left) applied to axial strain to obtain velocity (center). Velocity transformed from GAC measurement shown on the right. Khaitan et al. (2021) / European Association of Geoscientists & Engineers.

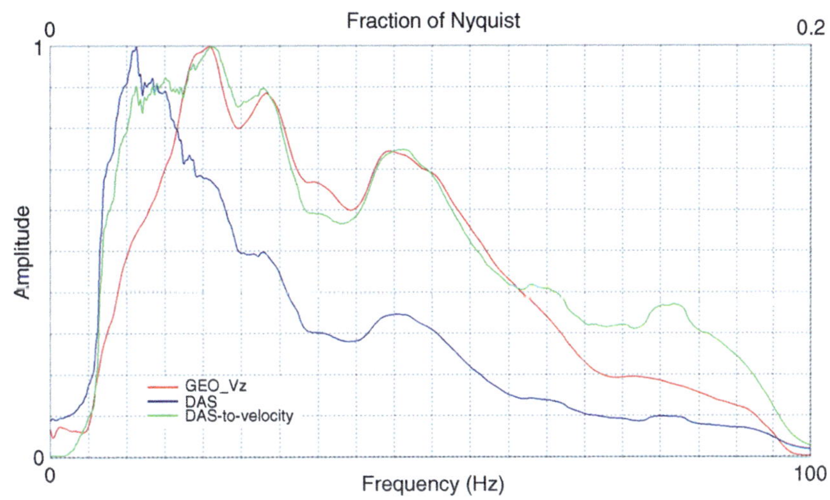

Figure 9.12 Spectrum of DAS strain data (blue), DAS to velocity (green), and GAC acceleration to velocity (red).

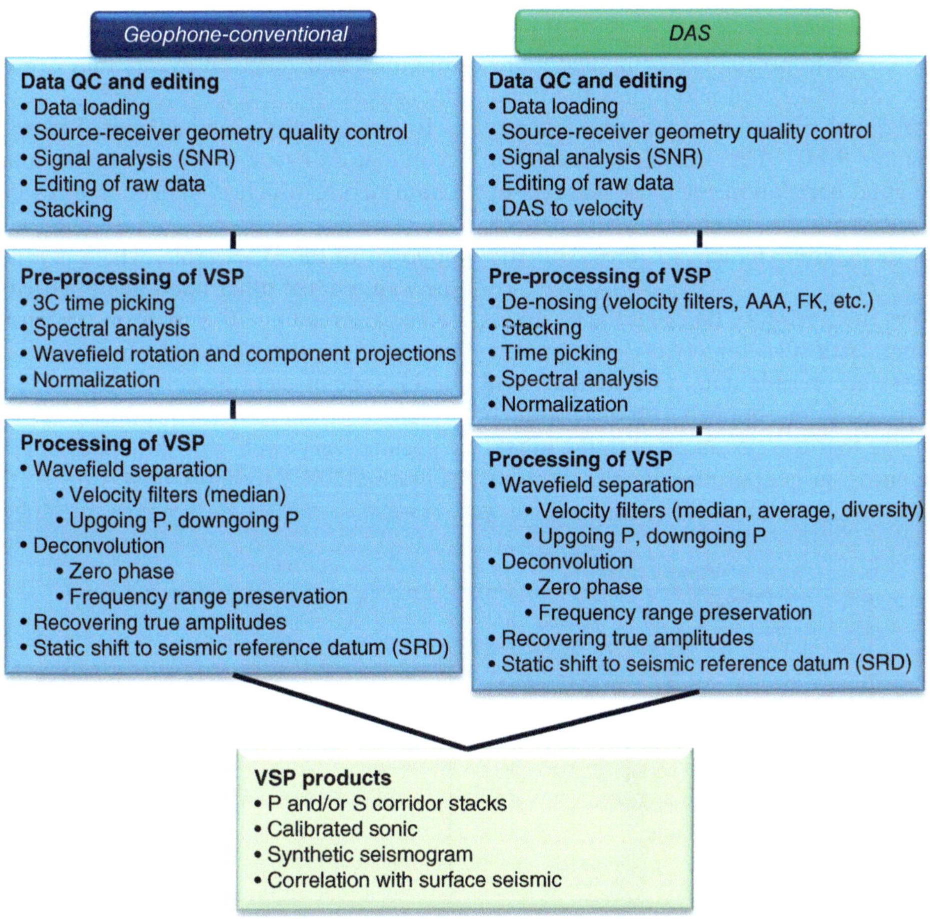

Figure 9.13 Typical ZOVSP data processing chains used for geophone (left) and DAS (right) data.

conversion of DAS to velocity differs (Sayed et al., 2020, 2021, 2022), as explained in the previous section. (b) Denoising is performed because the data usually present bands of poor coupling that can be effectively improved with anomalous amplitude attenuation (AAA) filters commonly used in surface seismic signals (Lawrence et al., 2021; Omega-SLB, 2022); FK filters are sometimes needed in case of tube-wave contamination, especially in land VSP operations. (c) In the case of the geophone data, the difference in preprocessing relates to dealing with time picking the horizontal components if they are to be used. (d) For time picking in DAS, a travel-time smoothing inversion is used to handle densely spaced traces and produce stable interval velocity profiles (Lizarralde & Swift, 1999). (e) During wavefield separation, the velocity filters may be the same; however, due to the higher spatial density of DAS, more traces are typically used than in geophones: for example, 20 versus 7 or 9. Likewise, at the time of generating the corridor stack, a higher number of traces (at the bottom of the well) is used in DAS.

Both VSP datasets are of good quality, and the traveltimes obtained are nearly identical. The computed interval velocities track each other well and follow the trend of a much higher frequency variation of the sonic interval velocity log (Figure 9.14). The corridor stacks in Figure 9.15 show good correlation for most markers of interest, perhaps except for the basement, where neither method shows a conclusive match to the synthetic seismogram.

Figure 9.16 shows the three-way tie using the two corridor stacks, along with the migrated VSP images obtained and the surface seismic data. For comparison, the displays show the corridor obtained with one technology next to the image that was produced with the other. Except for differences in amplitudes in the bottom marker, as mentioned before, the tie-in is of exceptionally good quality.

In the second example, acquisition with conventional geophones took 16 hr, rendering 73 VSP traces every 50 ft, using a multilevel tool to expedite the operation. DAS acquisition was performed in 1.2 hr, recording 411 traces over the same well interval, representing nearly 95% operation efficiency. Figure 9.17 shows a direct comparison of the volume of VSP data acquired with both technologies. Both datasets are of good quality; however, the DAS data show better spatial sampling in the top 2,000 ft. of the borehole.

The consistency of first breaks and similar character of the wavefields is exceptional, with differences of ±1 ms within the accuracy of time picking. Figure 9.18 shows the comparison of the traveltimes; the curves are so close that it is virtually impossible to see the small differences. A plot creating an arbitrary merge of the wavefields obtained with both technologies is presented to allow visual inspection of the continuity of the time breaks and the remarkable uniformity of features such as downgoing multiples and reflected events that can be traced across datasets.

In terms of frequency content, Figure 9.19 shows that geophone data tend to preserve higher amplitudes at maximum depth, which is understandable because the tools are strongly anchored against the borehole wall, as opposed to the DAS cable, which is just in contact with the wall. On the other hand, the low-frequency end tends to be preserved by DAS much more constantly across the measured depths. This is an observation commonly made in this type of recording.

The corridor stacks produced with both datasets correlate very well for the main markers of interest (Figure 9.20). Kirchhoff migration of the datasets was

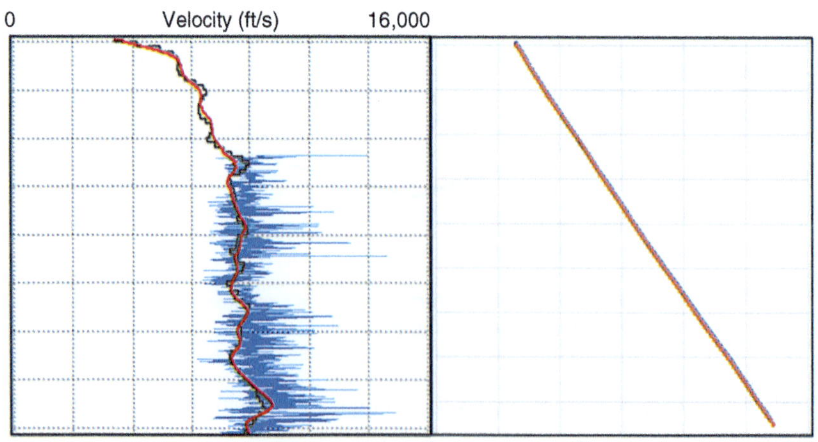

Figure 9.14 (Left) Interval velocities from geophone (black), DAS (red), and sonic data (blue). (Right) Observed transit times from geophone tools and DAS with less than 1 ms difference. Martinez et al. (2021) / Society of Exploration Geophysicists.

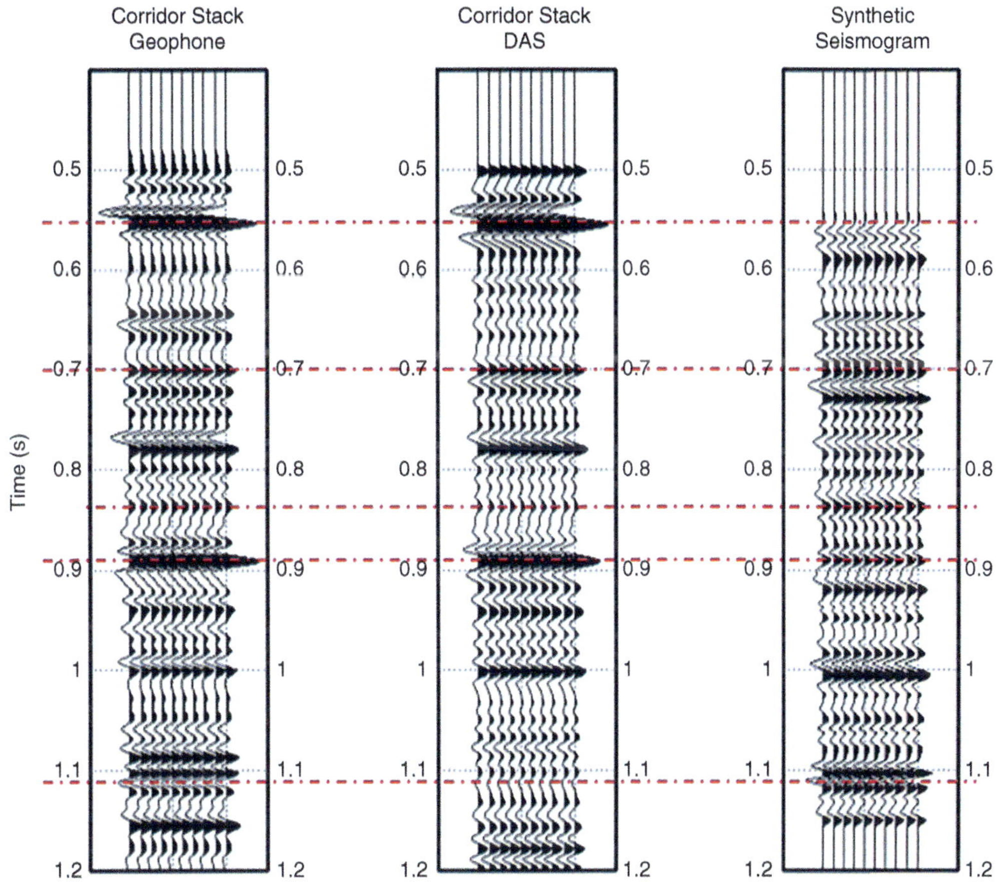

Figure 9.15 Corridor stacks derived from geophone and DAS data, compared with synthetic seismograms. Martinez et al. (2021) / Society of Exploration Geophysicists.

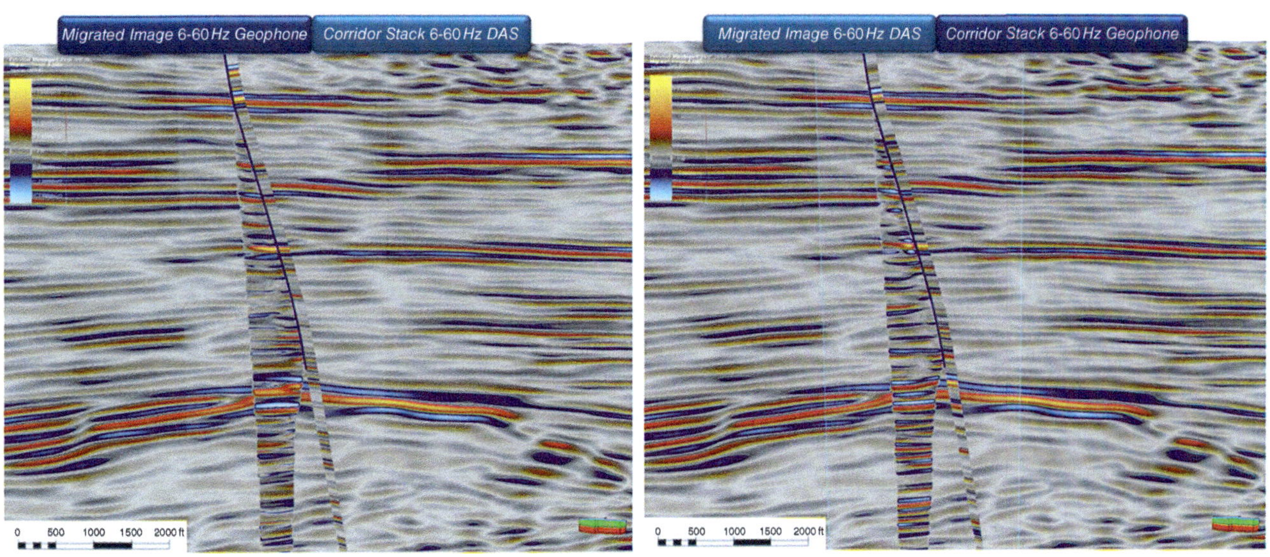

Figure 9.16 (Left) Integrated results from migrated image of geophones, DAS corridor stack overlying surface seismic. (Right) Migrated image of DAS, geophone corridor stack compared to surface seismic. Martinez et al. (2021) / Society of Exploration Geophysicists.

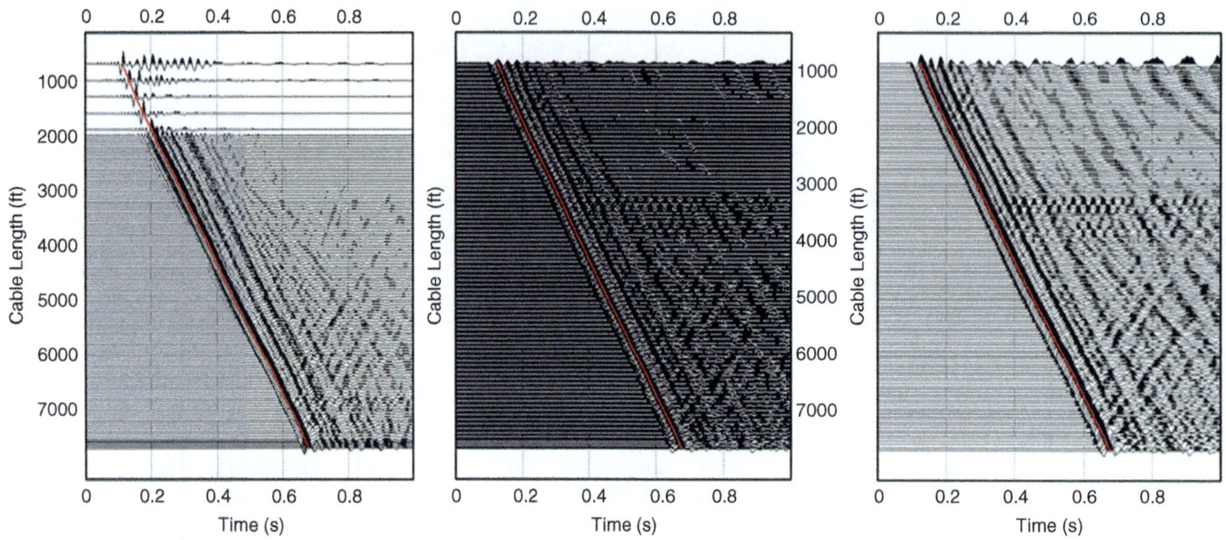

Figure 9.17 Geophone and DAS data. (Left) Geophone vertical component spaced 50 ft., 73 traces. (Center) DAS discretized in 411 traces spaced every 16 ft. (Right) DAS data decimated to 134 traces spaced 50 ft. First breaks indicated with red line. Duque et al. (2022) / American Association of Petroleum Geologists.

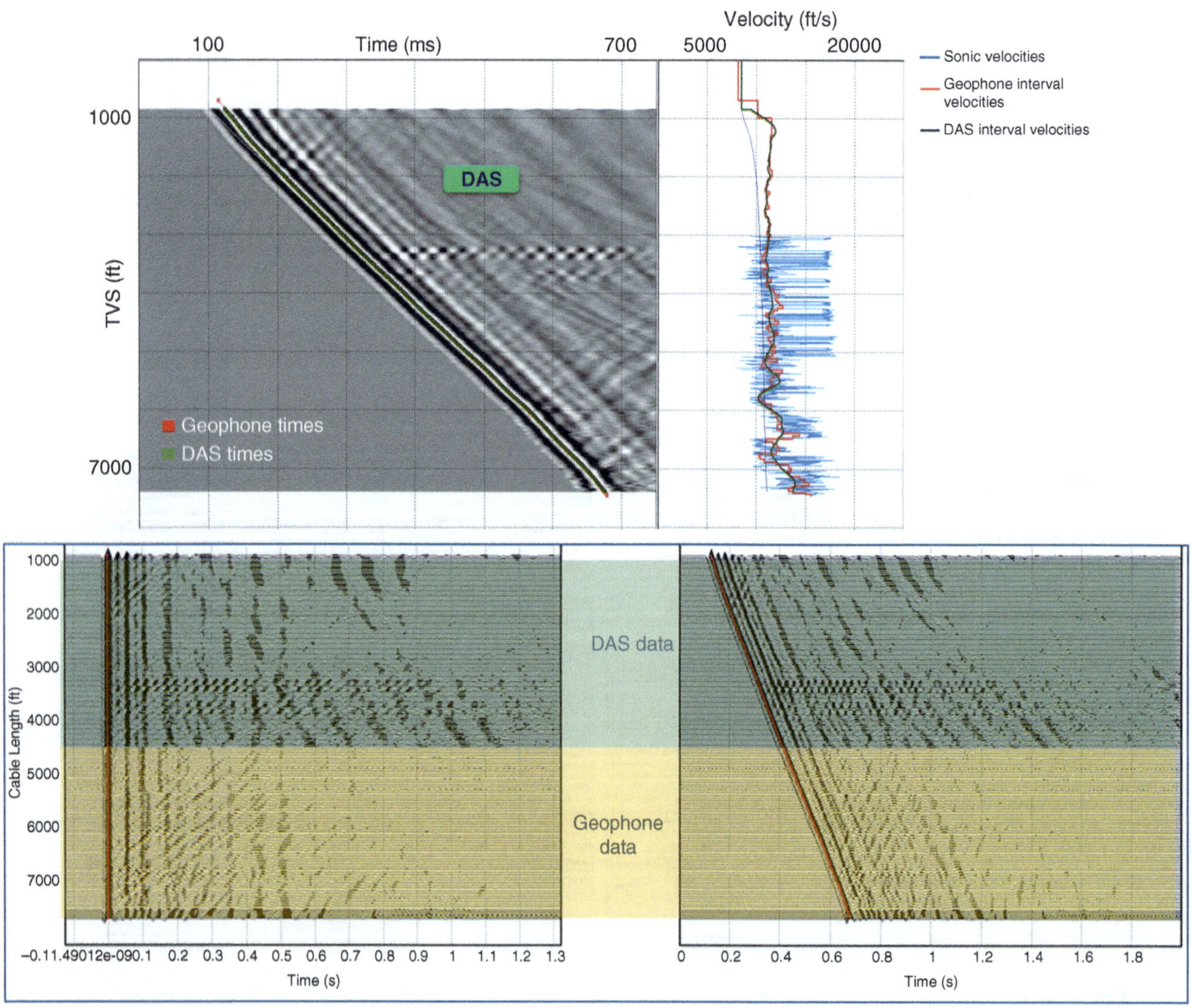

Figure 9.18 Geophone and DAS transit times comparison. (Top-left) DAS downgoing wavefield with first breaks of DAS (green) and geophone (red). (Top-right) Interval velocities calculated independently with DAS (black), geophone (red), and sonic (blue) data. (Bottom) Merge of DAS and geophone datasets at an arbitrary depth to show continuity of first breaks and coherency of wavefields. Helman et al. (2023) / Society of Petroleum Engineers.

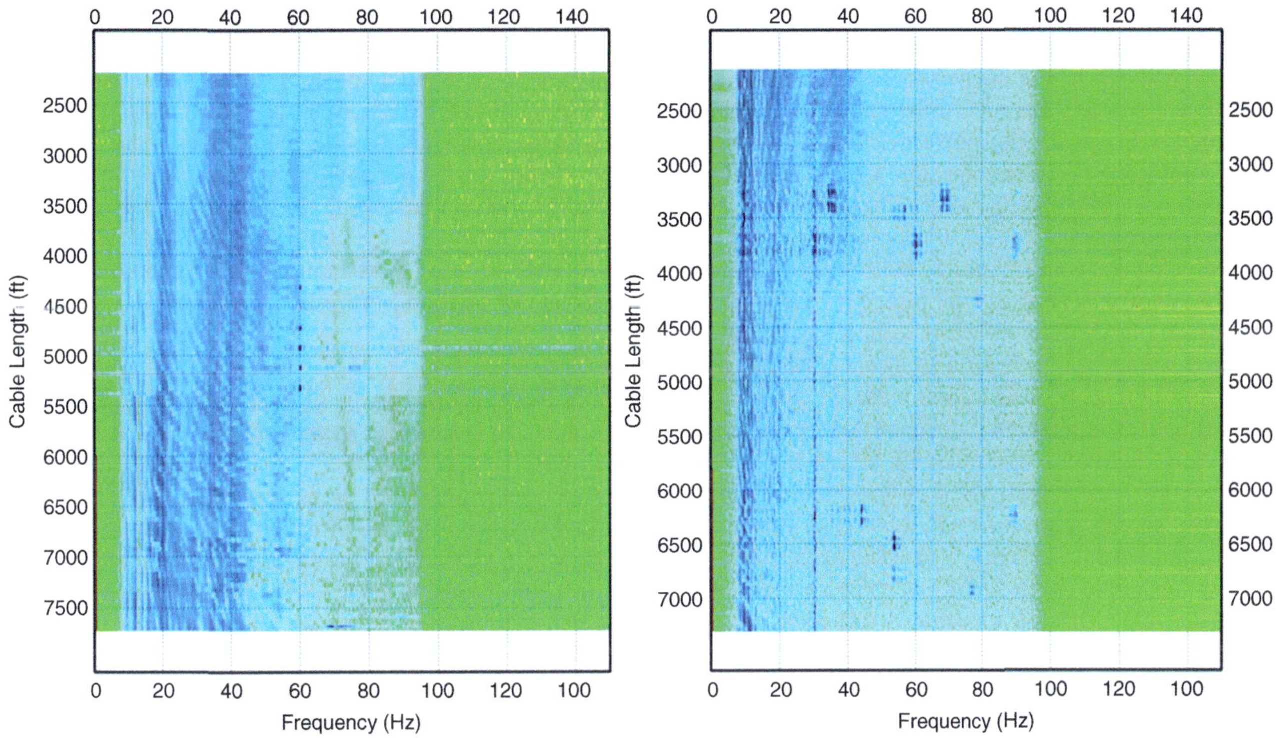

Figure 9.19 Geophone (left) and DAS (right) frequency spectra. Helman et al. (2023) / Society of Petroleum Engineers.

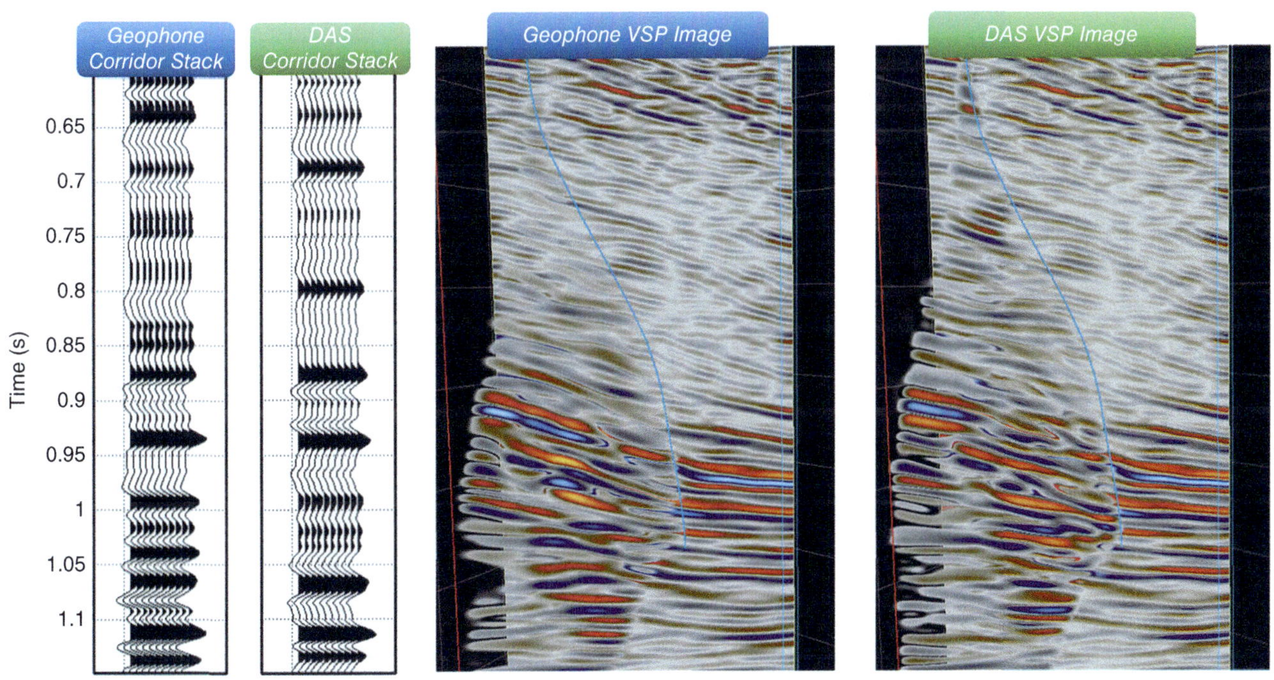

Figure 9.20 (Left) Corridor stacks of geophone and DAS data. (Right) Migrated images of geophone and DAS datasets. Distance from wellhead to target offset is approximately 1,750 ft. Duque et al. (2022) / American Association of Petroleum Geologists.

performed with the same parameters except for migration aperture, which, in the case of geophone data, was 8°, whereas in DAS, it was opened to 10°. The images produced have subtle differences inherent in the frequency content of both wavefields. However, the DAS image provided confidence in capturing shallow events and the lateral continuity of targeted events. The operator used the DAS image to further the interpretation of local faulting in the area, which could be hinted at in the surface seismic data and confirmed by the VSP image (Figure 9.21).

Note that DAS data with this technology can also be collected while conveying conventional seismic tools. Acquiring both DAS and geophone data can serve the purpose of extending the aperture of recording for travel-time estimation and imaging in 2D/3D VSP surveys. Hybrid surveys (DAS plus geophones) are also utilized in specific cases when 1C DAS and 3C measurements complement each other. Hybrid surveys could be used, for example, in geologically complex areas where wavefields are expected to contain out-of-plane reflections that cannot be adequately separated in the main energy wavefield by using only one component or when DAS can be challenged by directivity effects. Also, specific objectives such as horizontal transverse isotropy (HTI) anisotropy characterization may require 3C data. In fact, for the two examples discussed in this section, the logging runs with the geophone tool were also conveyed with the fiber-optic cable, and DAS data were also recorded prior to the acquisition of the geophone survey.

There are comparisons in the literature of VSP data acquired concurrently with geophones and DAS in open-hole and offshore environments using hybrid optical logging cables (Guerra et al., 2020; Horne et al., 2020). In addition, an interesting comparison of geophone data with DAS recorded in permanent fibers deployed in completions can be found in Ferla et al. (2019).

9.7. DAS VSP IN SLANT OR HIGHLY DEVIATED WELLS

High well deviation introduces potential challenges related to fiber response directivity. Vertical or deviated wells in complex geological settings could encounter similar complications. In such cases, full 3D ray-trace modeling will help in calculating the angles of both the direct and reflected arrivals at the receivers. By doing this, a survey configuration can be recommended where the resulting VSP may involve more than one source location.

The next example showcases a highly deviated appraisal well drilled onshore for which surface seismic amplitudes helped define the prospect. A high-density

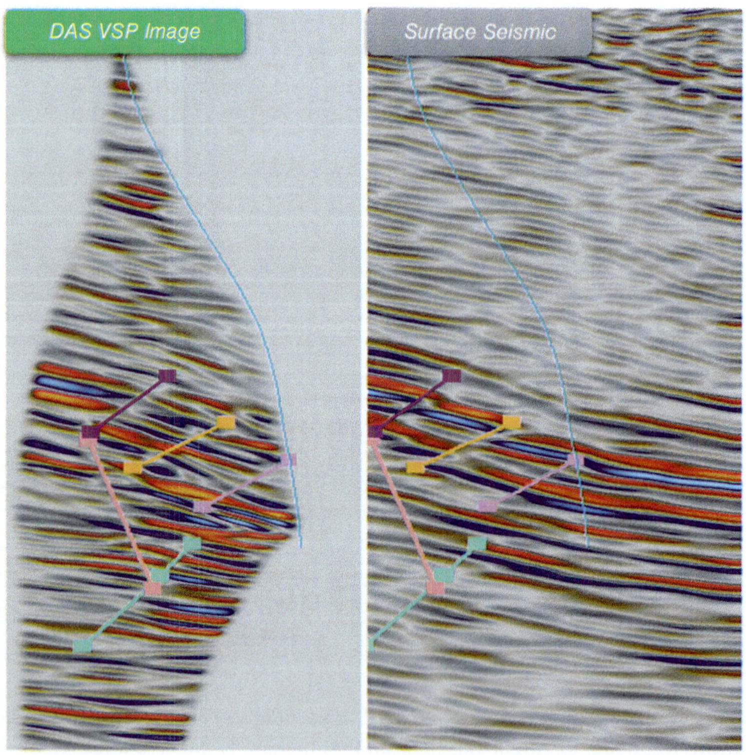

Figure 9.21 Faulting as interpreted from DAS VSP and surface seismic data. Distance from wellhead to target offset is approximately 1,750 ft. Duque et al. (2022) / American Association of Petroleum Geologists.

checkshot survey was recommended to deliver reliable time-depth velocity. Due to the well deviation, ray-tracing modeling was carried out, and two source positions were shown as the best solution to ensure vertical incidence of the P-wave energy for a reliable time-depth relation. Also, the incident angles of the reflected rays were checked so that the response of the fiber was within the acceptable polarization angles to reconstruct the signal for imaging purposes (i.e., $\cos^2\varphi$ of amplitude decay).

Figure 9.22 shows the summary of the ray-trace modeling performed and the raw data obtained in relation to the well structure. Data were of excellent quality for almost the entire well. Note that the seismic data did not reach the total depth of the well due to wireline tools getting held up during the descent.

The velocity analysis and subsequent VSP processing render very good results. The match between seismic and sonic velocities, as well as the synthetic seismogram and migrated images obtained, were of excellent quality (Figure 9.23). An important part of the objective for this well, as it was drilled on the edges of the surface seismic cube and velocity uncertainty was high, was to confirm the prognosis of target depth. The resulting migrated image shows a very good tie-in of the target horizon (thick event in red).

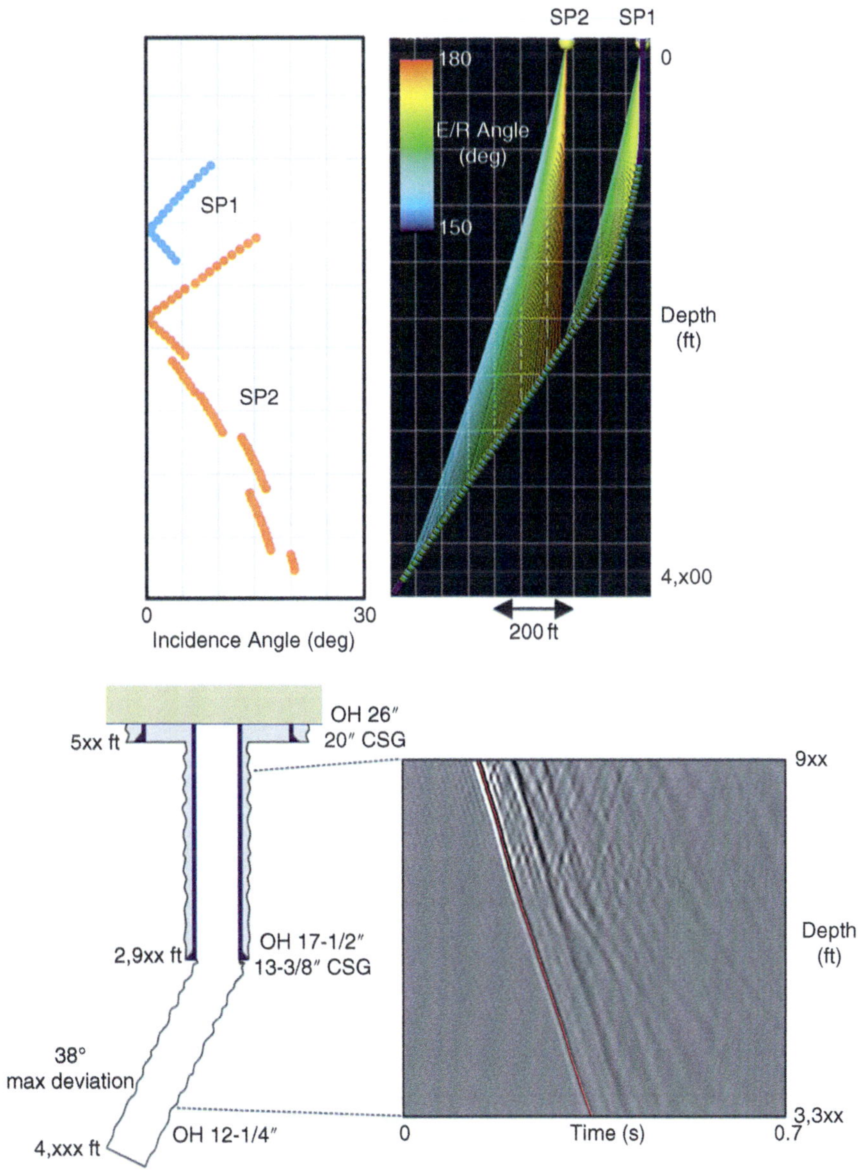

Figure 9.22 (Top) Ray-tracing modeling indicating (left) angles of direct arrivals (typically kept within 20° to honor vertical incidence). (Right) Emerging/reflected ray angles arriving at receivers. In this case, the angles were all kept within 30–40°. (Bottom) DAS raw data quality obtained along the wellbore. Useche et al. (2020) / Society of Exploration Geophysicists.

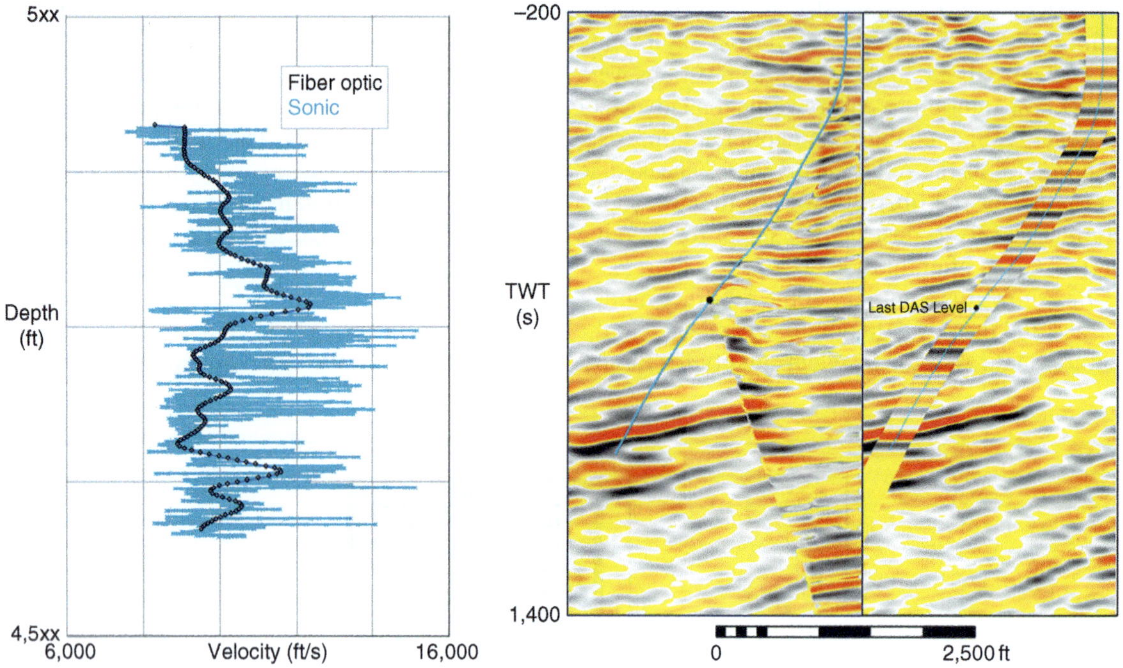

Figure 9.23 (Left) Sonic and seismic velocity comparison. (Right) Synthetic seismogram and migrated image superimposed on surface seismic. Useche et al. (2020) / Society of Exploration Geophysicists.

Another example of DAS acquisition in slant wells comes from an offshore open-hole well with a 40° deviation in deep-water Brazil. VSP was recorded while conveying a sidewall coring tool. DAS data were acquired with 1 ms SR, 2.55 m spatial sampling, GL 10.2 m, and pulse-width resolution of 10 KHz. A large-volume gun cluster of 1,200 in^3 was used as a seismic source, hanging from the rig's crane. The results of the job in terms of well-tie and imaging were quite remarkable, as shown in Figures 9.24 and 9.25.

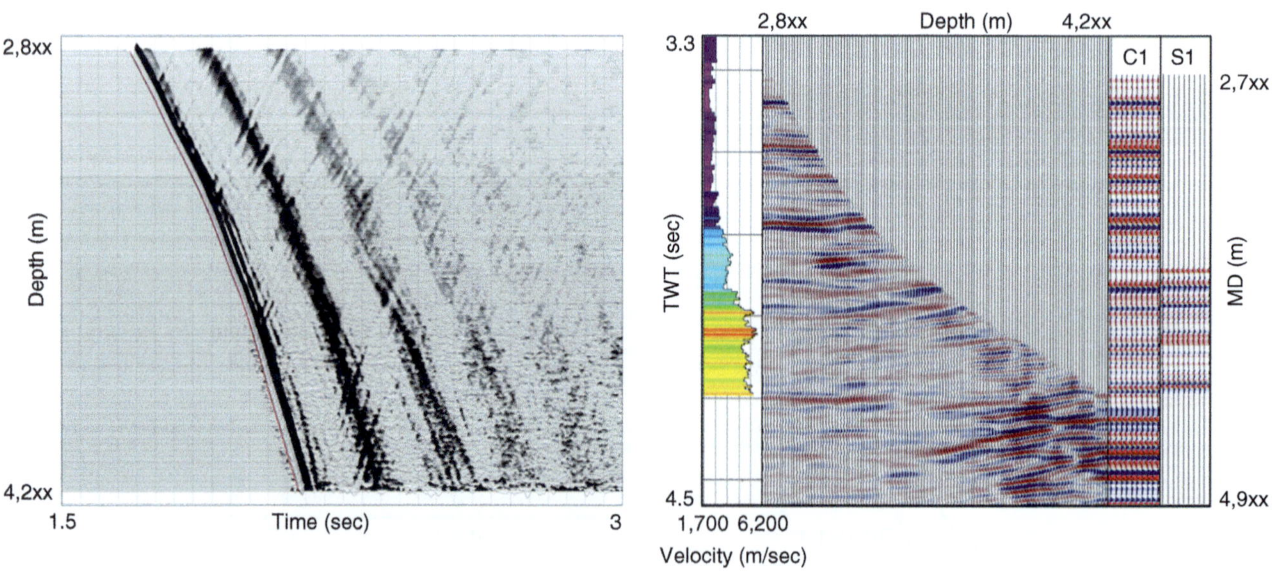

Figure 9.24 (Left) Stack data of the DAS acquisition in slant offshore well. (Right) Composite of VSP processing. From left to right: velocity profile (track 1), upgoing deconvolved wavefield (track 2), high-frequency corridor stack (C1), high-frequency synthetic seismogram (S1). Rosi et al. (2023) / Society of Petroleum Engineers (SPE).

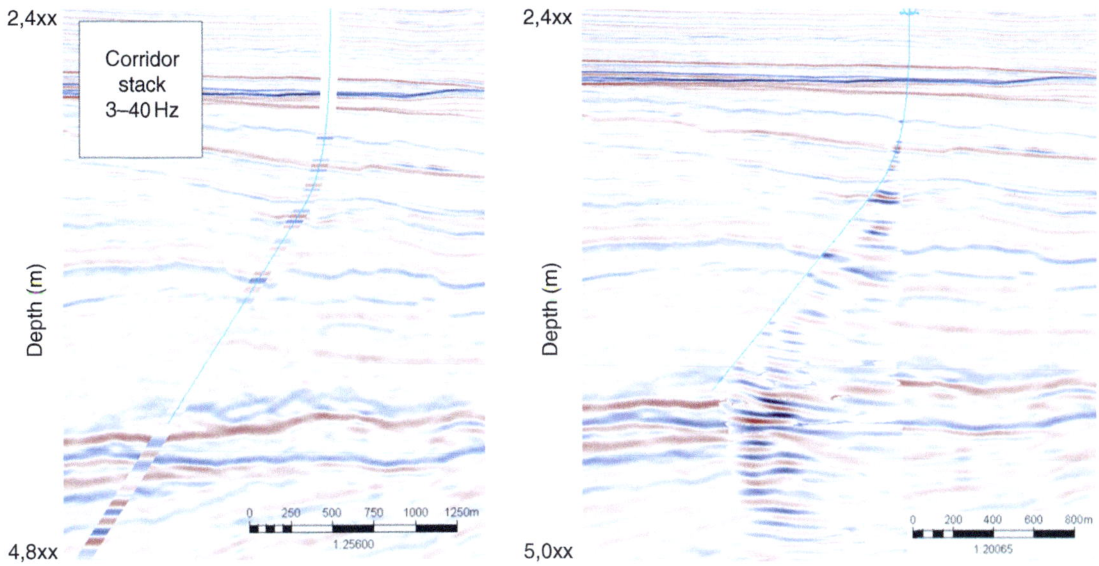

Figure 9.25 Corridor stack (left) and migrated image (right) superimposed to surface seismic for well-tie analysis. Rosi et al. (2023) / Society of Petroleum Engineers (SPE).

9.8. Q ANALYSIS FROM DAS ZERO-OFFSET DATA

During a zero-offset acquisition in an offshore deepwater well in Brazil, Q analysis was performed. This job also included additional VSP recorded with traditional geophone methods for comparison purposes. Similar analyses, as performed in Section 9.6 of this chapter, were done for this dataset. Full analysis and results are described in Rossi et al. (2023). The single spectral method (Leaney, 1999) was then used to estimate the Q factor, as shown in Figure 9.26 (Rossi et al., 2023; Sayed et al., 2021). The Q analysis was performed for both datasets and the results were similar across the same depth interval, 161 in the case of geophone and 155 for DAS.

9.9. OFFSHORE EXAMPLES

Three examples are discussed in this section. The first case study corresponds to an appraisal vertical well offshore Norway, reaching 1,775 m TD (Guerra et al., 2020). A high-quality rig-source VSP survey was required to deliver reliable time-depth-velocity and corridor stacks. The DAS VSP was recorded during the descent of the formation tester tool. A standard wireline VSP, planned as a contingency, was also deployed because the formation tester tool was held at some point in the wellbore. The customer expected that the slimmer geophone tool could negotiate the obstruction, but it could not. So, this well presents a comparison of DAS and geophone data. A $3 \times 250\,\text{in}^3$ G-gun cluster at 2,500 psi and 7 m depth controlled by an in-sea digital gun controller was used. The DAS acquisition parameters included GL = 20.4 m, trace spacing 5.1 m, and pulse frequency rate (PFR) = 4 kHz. The recording was done by allowing 20 m slack on the logging cable for improved acoustic coupling while recording 45 shots to form a stack. Good data quality is observed across the borehole except for the area below the casing shoe, which was muted (Figure 9.27). Note that DAS transit times are corrected, assuming a linear distribution of the cable slack in the well. The comparison of the transit times between corrected DAS and geophone is very good.

Smoothed traveltimes inversion of DAS data results in vertical velocities in overall good match with those of geophones (Figure 9.28). The usable frequency band of DAS strain data were 4–80 Hz, whereas geophone data exhibits 4–130 Hz. Note that no conversion of DAS to velocity was made here, which would boost the high frequencies. The difference is mostly attributed to acoustic coupling, where DAS relies on the slacking applied to the cable, whereas geophones are strongly anchored to the borehole wall.

The correlation between DAS and geophone data is of excellent quality, extending nearly 1 km below the deepest measurement in the borehole (e.g., lookahead). The correlation of the corridor stacks from both measurements ties in very well with the synthetic recording derived from logging-while-drilling (LWD) sonic and density logs (Figure 9.29).

Another example is presented in Horne et al. (2020). This is a nearly vertical appraisal well drilled deep-water in West Africa, where DAS seismic was simultaneously

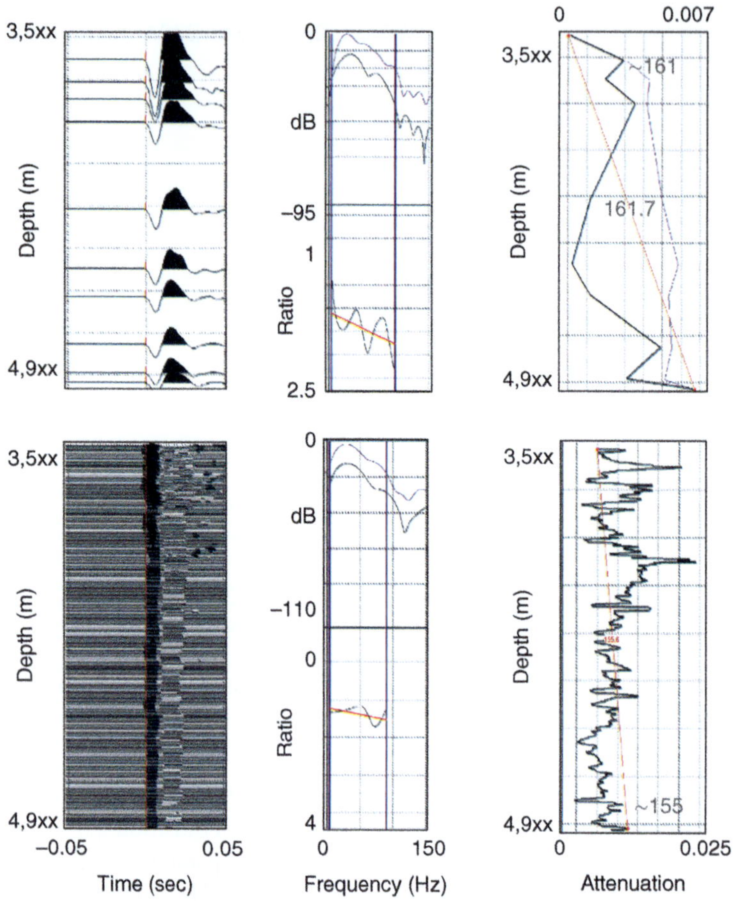

Figure 9.26 (Top) Geophone data subtracting TT (left), associated frequency spectrum (middle), and cumulative Q for depth interval. (Bottom) Same data for DAS. Adapted from Rossi et al. (2023)

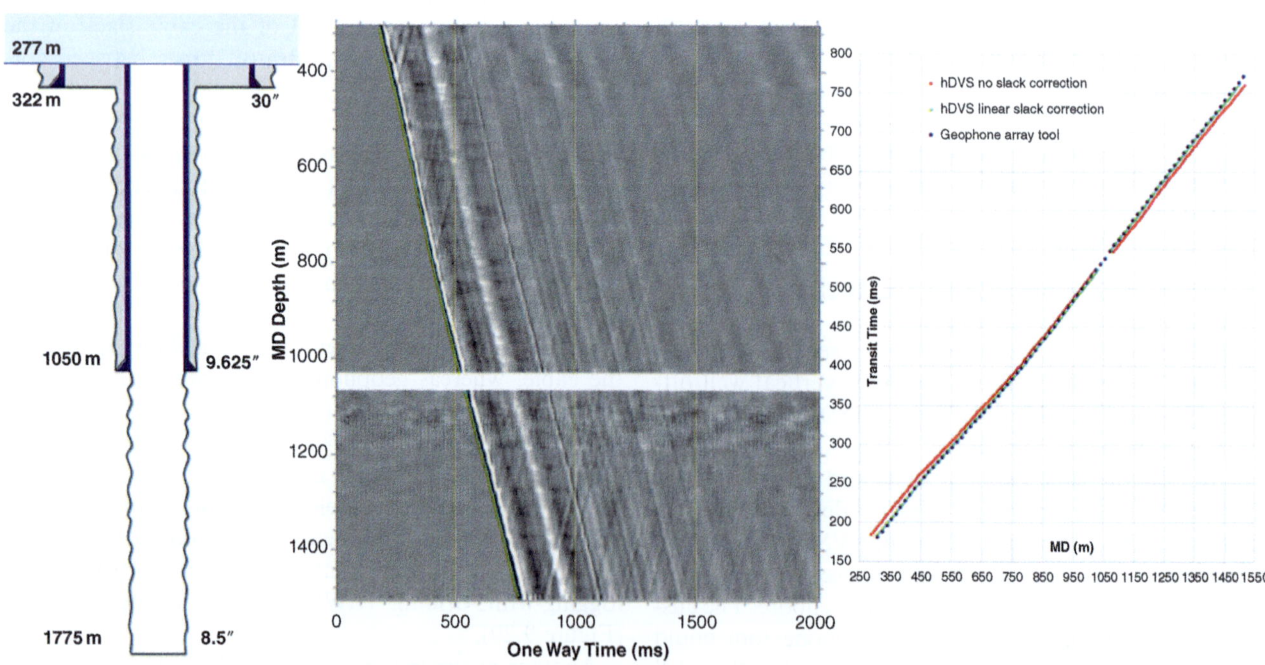

Figure 9.27 (Left) DAS offshore data recorded in open- and cased-hole section. (Middle) Good quality wavefield except area below casing affected by large washout. (Right) Transit times recorded with DAS corrected assuming linear slack distribution and with geophone tools. Guerra et al. (2020) / European Association of Geoscientists & Engineers.

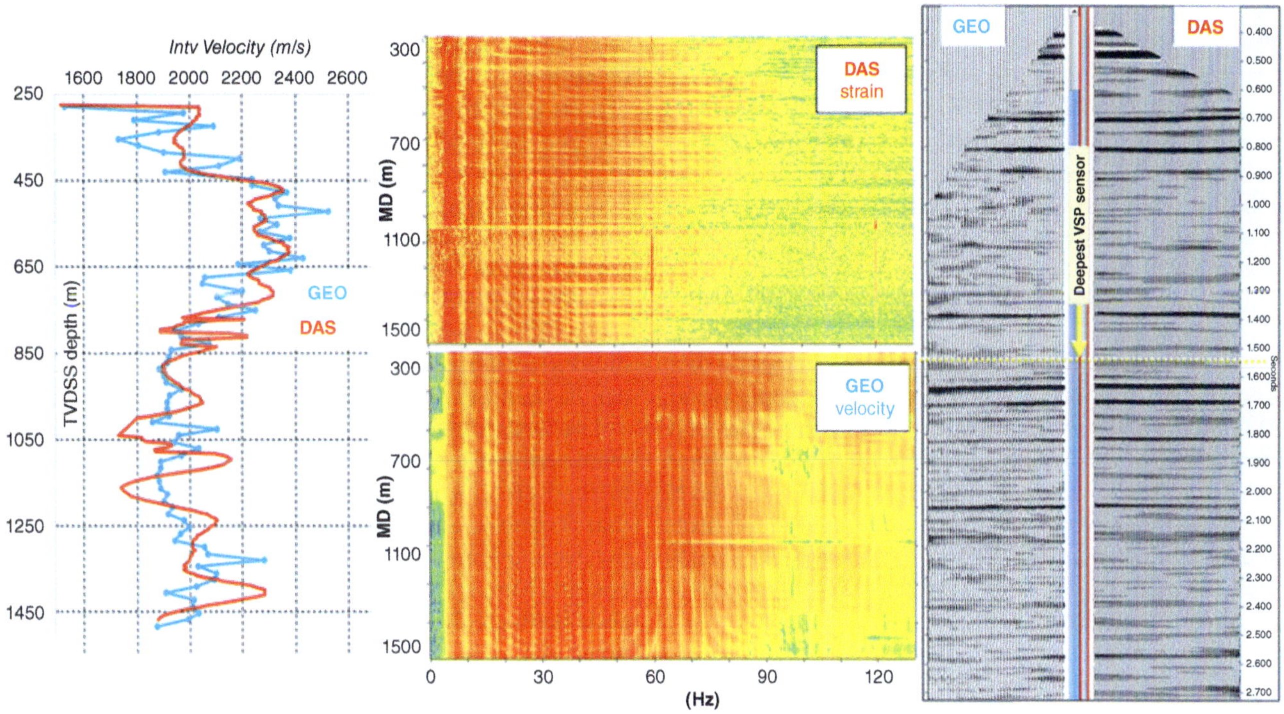

Figure 9.28 (Left) DAS and geophone interval velocities. (Middle) Frequency spectra of DAS and geophone. (Right) Transposed VSPs of deconvolved upgoing waves 4–80 Hz. Comparison is most accurate at the center of the display, and yellow dotted line indicates the deepest VSP measurement in the datasets. Guerra et al. (2020) / European Association of Geoscientists & Engineers.

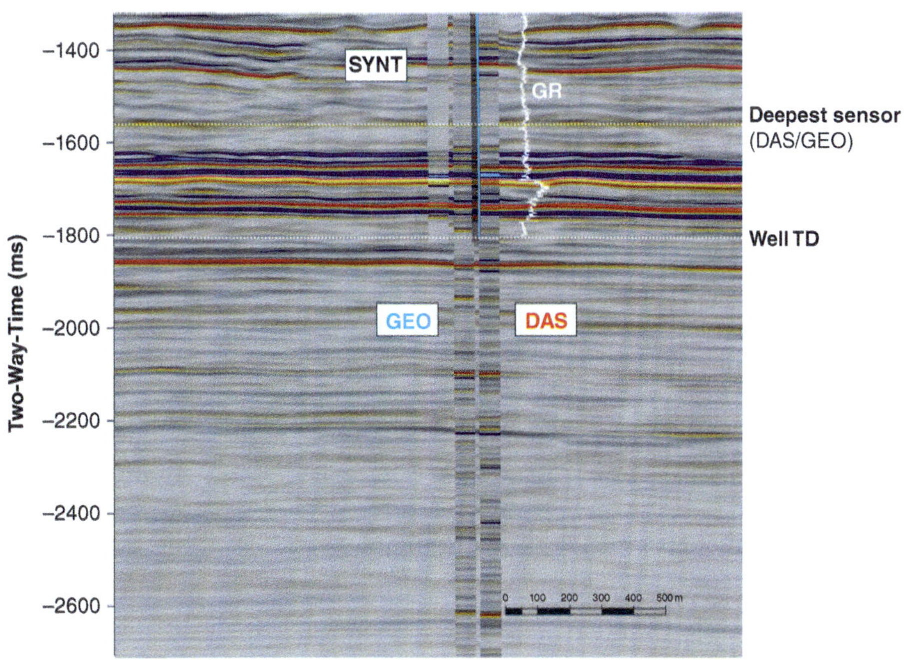

Figure 9.29 Seismic tie display showing corridor stacks of DAS and geophone data, as well as synthetic. Note excellent match of reflectivity in the lookahead portion marked well below the dotted yellow line of the deepest VSP measurement. Guerra et al. (2020) / European Association of Geoscientists & Engineers.

recorded with geophone data during a walkaway survey. The VSP program used a hybrid-replaceable conveying tool string, with 16 sections of the 3C geophone spaced at 15 m spacing and a GR tool. DAS VSP data were recorded while logging the geophone tool without additional downhole tools and saving rig time. Slacking the cable by 1.5 m allowed good coupling with the formation of over 600 m of deviated section (1–16°). Only eight shots were stacked during the DAS ZOVSP, whereas during the walkaway, a single shot per position was recorded (Figure 9.30).

A gun array of total 1,200 in^3 total volume operated at 2,000 psi and 7 m depth was towed from a supply vessel. Maximum offset in the walkaway was of the order of 5 km. A GL of 20 m with PFR of 40 Hz rendered good DAS data in the range of 4–80 Hz at the reservoir level, and as in the previous example, good quality lookahead VSP over more than 1 km below the deepest DAS measurement point, as confirmed with comparison with the geophone data (Figure 9.31).

A direct comparison between walkaway data recorded with DAS and geophone cannot be made as both systems recorded different-looking (complementary) depth intervals (Horne et al., 2020). A common receiver gather (CRG) of the shallowest vertical geophone in the tool is compared with a CRG of the DAS data 600 m above (Figure 9.32). Although the noise of the DAS data is higher, clear first breaks extend to large offsets, nearly 80° from the vertical. The impact of noise in DAS is minimal for processing, and the authors concluded that the benefit from DAS data is provided through the extended depth coverage; in this case, the DAS data spanned a depth range of 695 m (142 channels), whereas the wireline geophone system spanned less than a third of the depth range (225 m) with 16 channels. The value of such large direct arrival transit time datasets is considerable, as these can be used to constrain anisotropic velocity models.

The DAS transit times were used to derive local anisotropic parameters using the slowness curve construction method (Horne et al., 2008). The assumption here is that structural simplicity of the geological structure in the overburden is relatively flat, as seen in the surface seismic, so estimates with this method are reliable. The difference between the vertical and horizontal slownesses is indicative of a Thomsen's ε of around 0.35. Estimates of Thomsen's δ are around -0.05. These slowness curves can be used to compute local VTI anisotropy estimates at multiple depth levels (Figure 9.33).

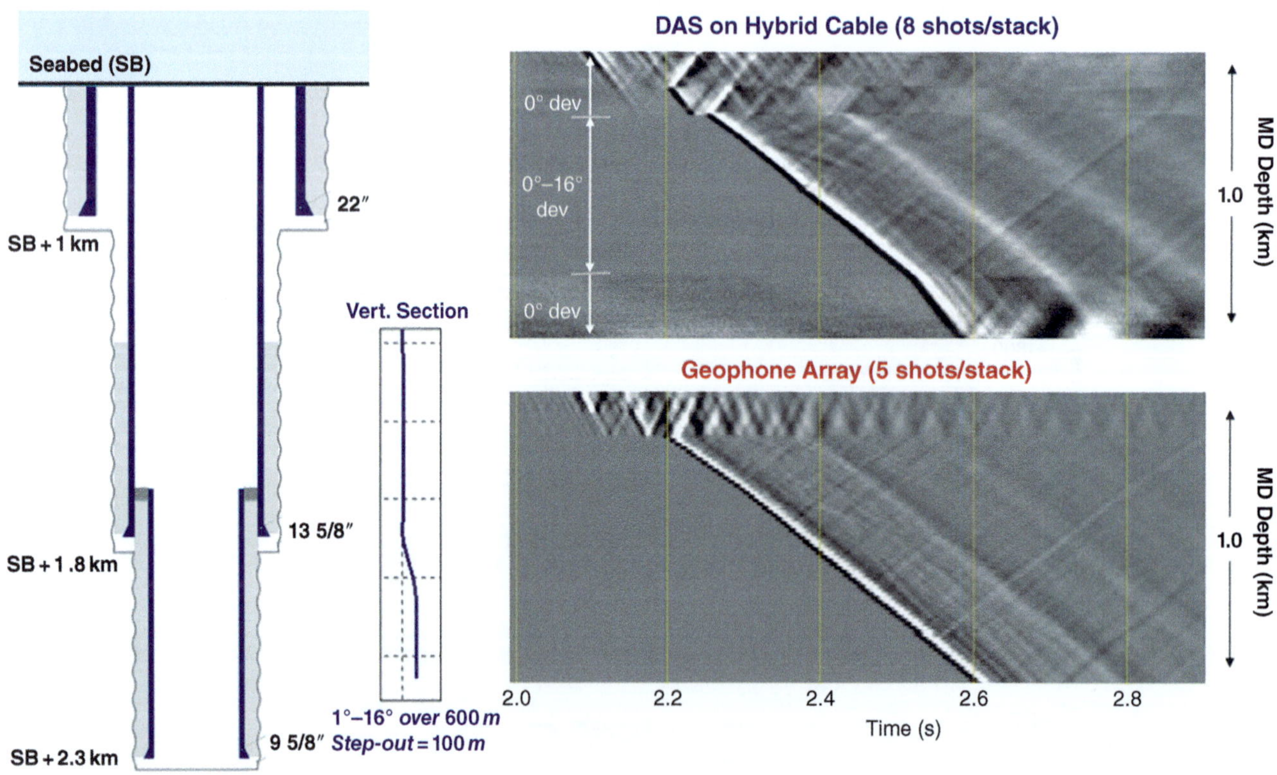

Figure 9.30 Well sketch and ZVSP data recorded by anchored geophones and DAS on hybrid heptacable. hDVS acquisition excellent data quality was recorded over 600 m intervals where the well deviation was greater than one degree. Schlumberger (2020) / SLB.

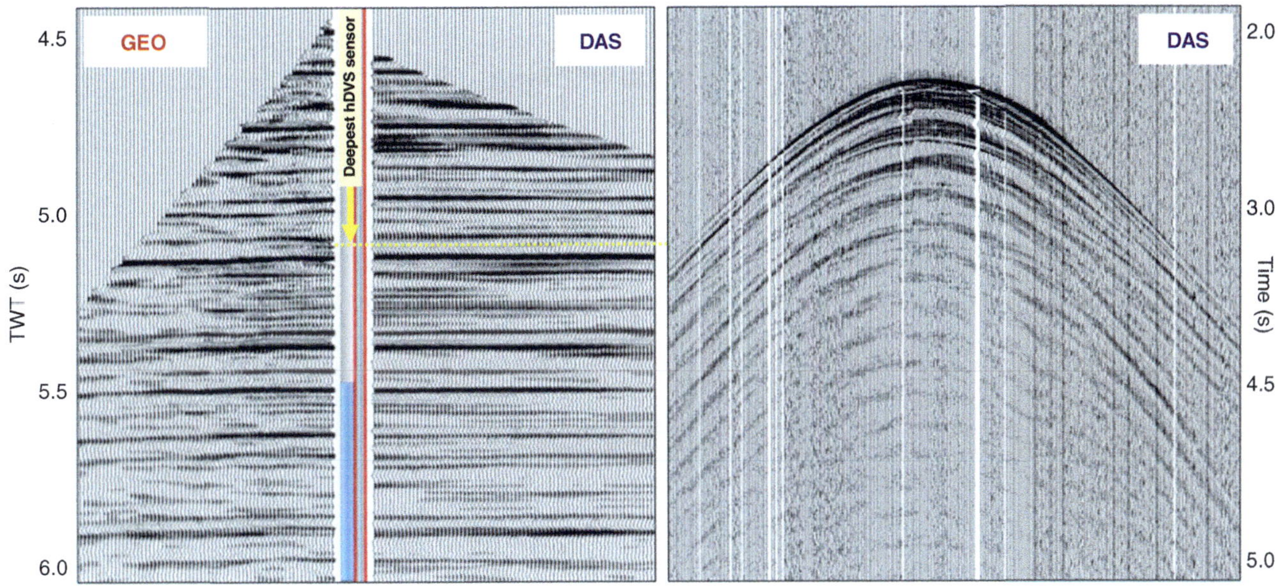

Figure 9.31 Transposed VSP display of geophone and DAS zero-offset VSP upgoing waves. Most accurate data is in the center, to correlate reflectors. DAS walkaway receiver gathers field raw shots. Asymmetric line with 4 km offset to the left of the apex and 5 km to the right, with 360 shots at 25 m with some gaps. Good SNR. Signature changes due to pressure drop. Schlumberger (2020) / SLB.

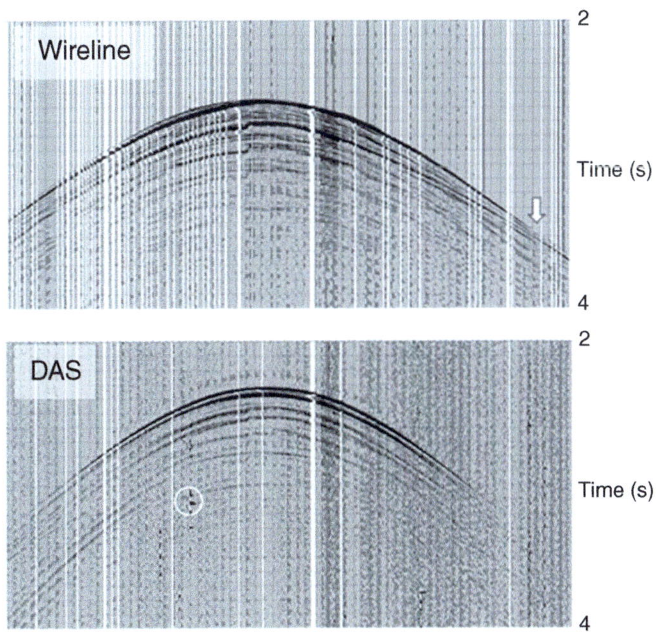

Figure 9.32 Common receiver gathers from the wireline-conveyed geophone (top) and DAS (bottom) systems. The arrow on the wireline plot indicates the turning ray offset. The circle on the DAS plot (bottom) highlights optical noise. White vertical bands correspond to gaps in shot points. Horne et al. (2020) / Schlumberger.

The third example is from an operator in the UK North Sea, where a ZOVSP was acquired in a 30° deviated well with TD at 3,352 m MD (Guerra et al., 2023). Thirty shots were stacked while the hybrid logging cable was slacked 5 ft. The operation was completed in nearly 10 min during the descent of a sonic and formation pressure tool combo. The source employed in this job was a triple-gun cluster of 750 in^3 total volume lowered at 5 m depth. DAS VSP traces

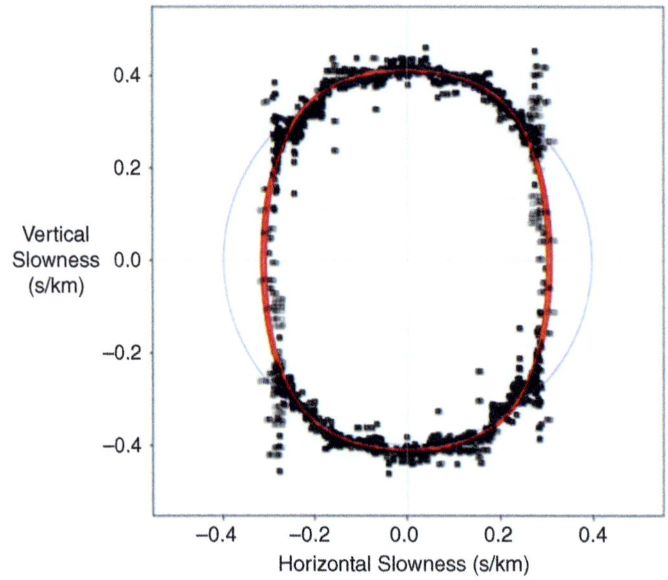

Figure 9.33 Phase slowness data (black dots) derived from first-break picks on DAS data. The blue circle represents isotropic behavior. The red line is the modeled phase slownesses for the best-fitting models derived from inversion. Slowness data in the lower half of the plot are duplicated from the upper half and are based on reciprocity. Horne et al. (2020) / Schlumberger.

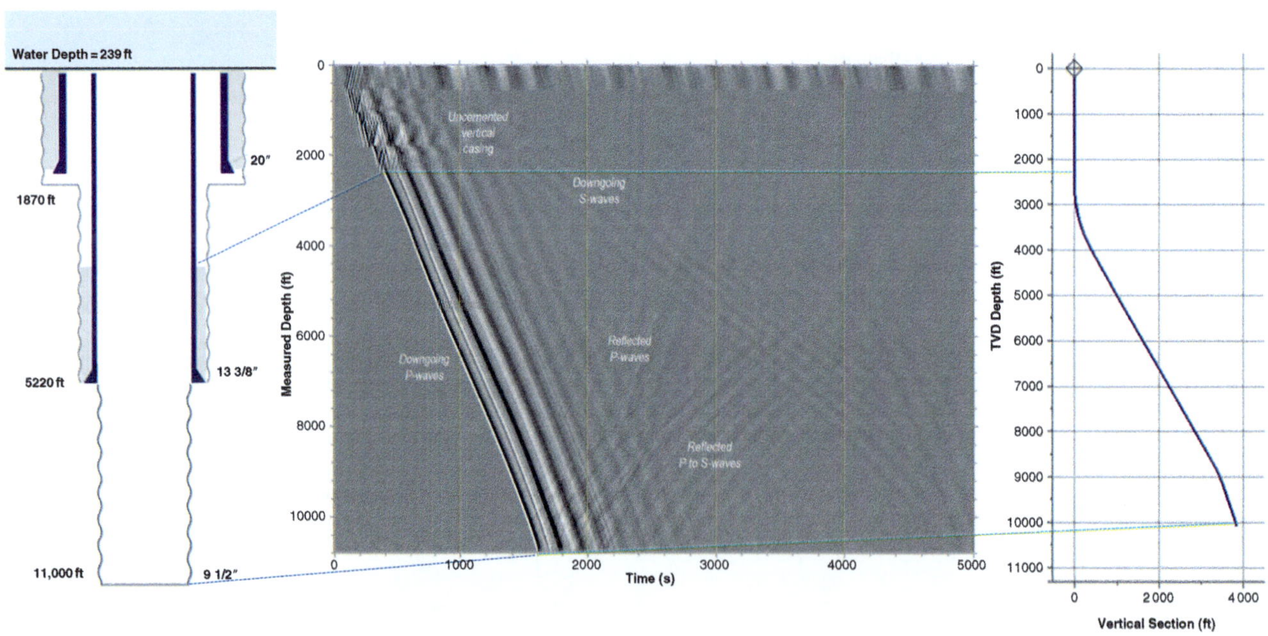

Figure 9.34 Raw field stack with excellent SNR, rich in P- and converted Sv-waves in central panel. Well construction detail (left) and deviation profile (right) are provided to visualize where data becomes noisy due to environmental conditions. Guerra et al. (2023) / European Association of Geoscientists & Engineers.

were output every 5.1 m, GL 30 m, PFR 10Khz, and light pulse width 6.6 m with 1 ms SR. Figures 9.34 and 9.35 show that the usable data is in the range of 4–64 Hz, and its excellent quality allowed full velocity analysis and corridor stacks. The wavefield becomes noisy in the uncemented vertical portion in the shallow section of the well; something similar would have been expected in a traditional vertical component geophone acquisition.

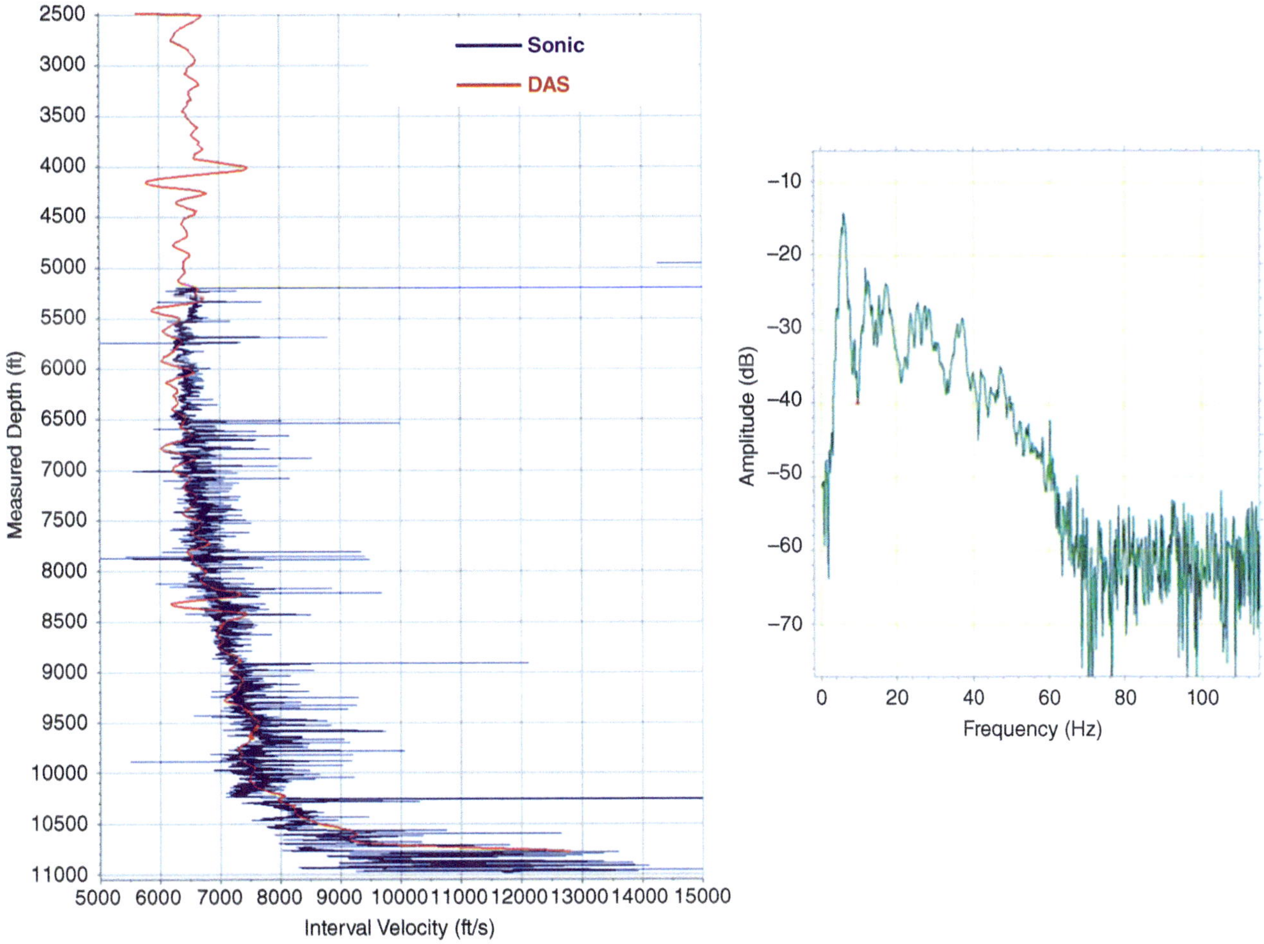

Figure 9.35 (Left) DAS VSP and sonic velocities comparison. (Right) Excellent SNR data over 4–64 Hz. Adapted from Guerra et al. (2023).

9.10. EFFICIENCY, HSE, AND COST CONSIDERATIONS

An important aspect to highlight is the value proposition of the DAS technology when it is run as part of the reservoir logging program. As could be seen through the examples, the quality of DAS data is very good for most applications and environments, and DAS renders similar and reliable data to its geophone counterpart. The fundamental difference is that the information is recorded in a fraction of the time employed for conventional techniques, at least in the ZO configuration, which is the bulk of VSP work in the industry. This has significant implications for rig time and hence for overall cost of the operation. Logistic aspects in offshore operations are also minimized as the seismic sources require the rig's crane to be tied up for seismic operations, which can be quite disruptive during conventional ZOVSP surveys (24 hr or more), as opposed to a few minutes with DAS.

Additional aspects, such as not requiring a fit-for-purpose seismic run, also help alleviate the risk of sticking. The number of shots required in typical DAS ZOVSP surveys is much lower, and the shots are concentrated in a shorter period than in the conventional approach. This could help in onshore environments to reduce the noise in general and exposure to communities nearby, as well as the nuisance of interrupting traffic and rig work during long hours. In jobs offshore, the reduction in sound exposure levels (SEL) is significant when fewer shots concentrated in a matter of minutes will create less disruption for fauna. In general, a reduction of the HSE exposure for personnel in the rig (i.e., high-pressure equipment, lifting operations, noise, etc.) is also realized with much shorter seismic logging operations.

A direct consequence of the lower time requirement of DAS operations is the reduction of CO_2 emissions and other greenhouse gases into the environment. Duque et al. (2023a) performed a comprehensive analysis of several DAS ZOVSP acquisitions using hybrid logging cable and compared these with both actual and theoretical conventional borehole seismic data acquisitions (i.e., estimated logging times to cover the interval of interest with the same resolution). This work concluded that DAS ZOVSP acquired with logging cables (for the case of one operator) helped with the reduction of more than 7 tons of CO_2 equivalent per job by eliminating on average over 15 hr of rig time (Figure 9.36).

9.11. SOME CHALLENGES AND LIMITATIONS OF THE TECHNOLOGY

Effective acoustic coupling for the hybrid wireline cable in the wellbore is possibly the main limitation of the technique. Data quality can be affected in places where the cable is not in good contact with the borehole walls. In wells with as little as 1° in deviation, gravity will force the cable to rest on the low side of the borehole, effectively

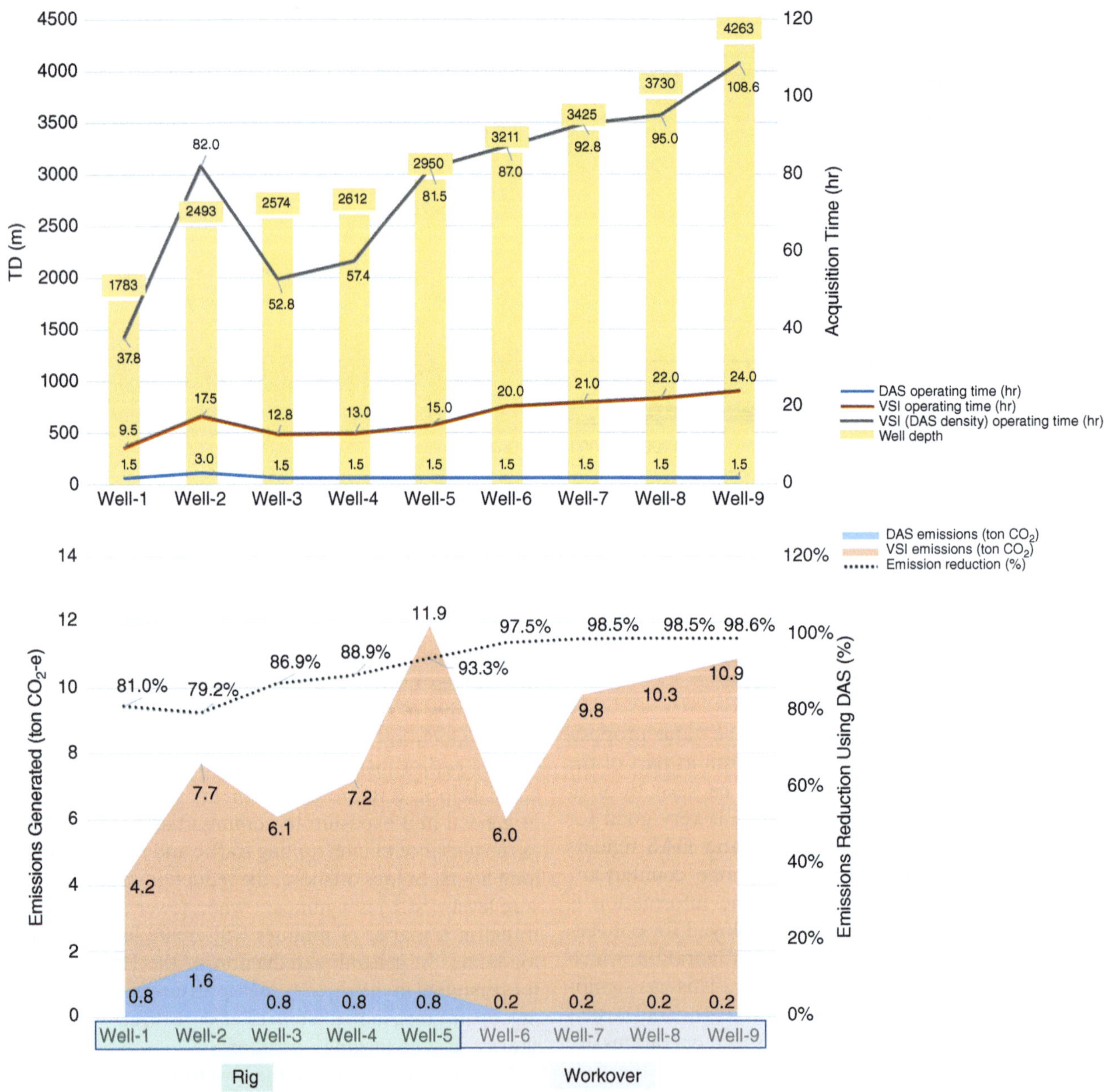

Figure 9.36 (Top) Rig time employed by ZOVSP operations DAS versus geophone. (Bottom) CO_2 emissions by type of land drilling unit when operating DAS vs geophone tools. Adapted from Duque et al. (2023a).

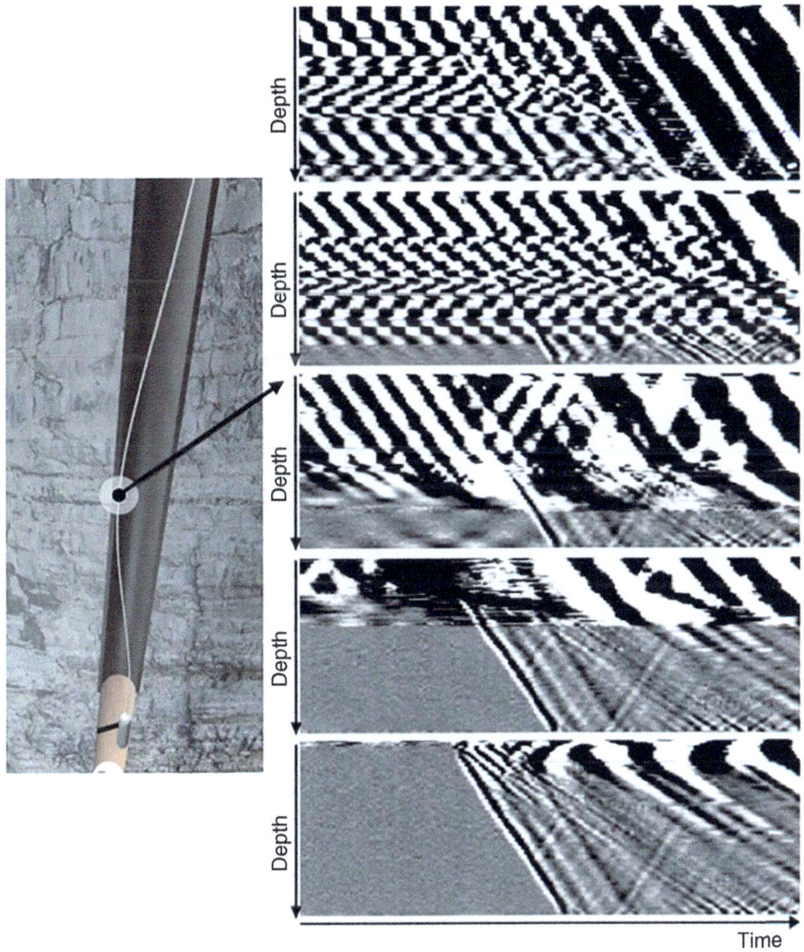

Figure 9.37 (Left) Schematic representation of cable de-tensioning to maximize cable coupling. Cable slack is exaggerated for visualization purposes. (Right) Seismic data vintages in the same well with varying values of cable tension from top full tension to bottom final tension. Frignet and Hartog (2014).

achieving acoustic coupling. In perfectly vertical wells, coupling cannot be anticipated. De-tensioning techniques are used to force the cable in that scenario to coil up along the borehole trajectory (Frignet & Hartog, 2014). Figure 9.37 shows an experiment in a vertical sallow well where different levels of de-tensioning are applied to the hybrid cable to improve data quality. This maneuver is typically accomplished by either resting on the calipers of the downhole tool string below the cable, sampling probes, packers, etc., or by sitting with the tool string on the bottom of the well.

In deep wells, intrinsic cable resonances could result in the need for de-tensioning the cable, regardless of the wellbore trajectory. This could be related to minute stress dislocations in the cable tension profile due to hole rugosity and internal cable strain. This translates into potential noise bursts while trying to acquire seismic. To overcome this issue, waiting a prudent time before seismic acquisition once the cable is in position, usually 10–20 min, allows the cable to be at rest. Such a process can be monitored by continuously looking at the cable strain profile (Figure 9.38). It is worth noting that the imperfect coupling may be more critical when the objective is imaging or dependent on reflections; however, it is less crucial when the objective is to obtain first-break or transit time information.

De-tensioning techniques introduce the need for depth correction, which could be a matter of concern. As shown in the example in Figure 9.27, the assumption of the linear distribution of the extra cable added works very well in the case of vertical wells. The same approach is used in a 30° deviated well, as shown in Figure 9.39, where the best DAS data were acquired with 15 ft of slack (Martinez et al., 2020). Consistent time picks between the DAS and 3C geophone data with less than 1 ms deviation in overlapping areas. A more quantifiable approach, using

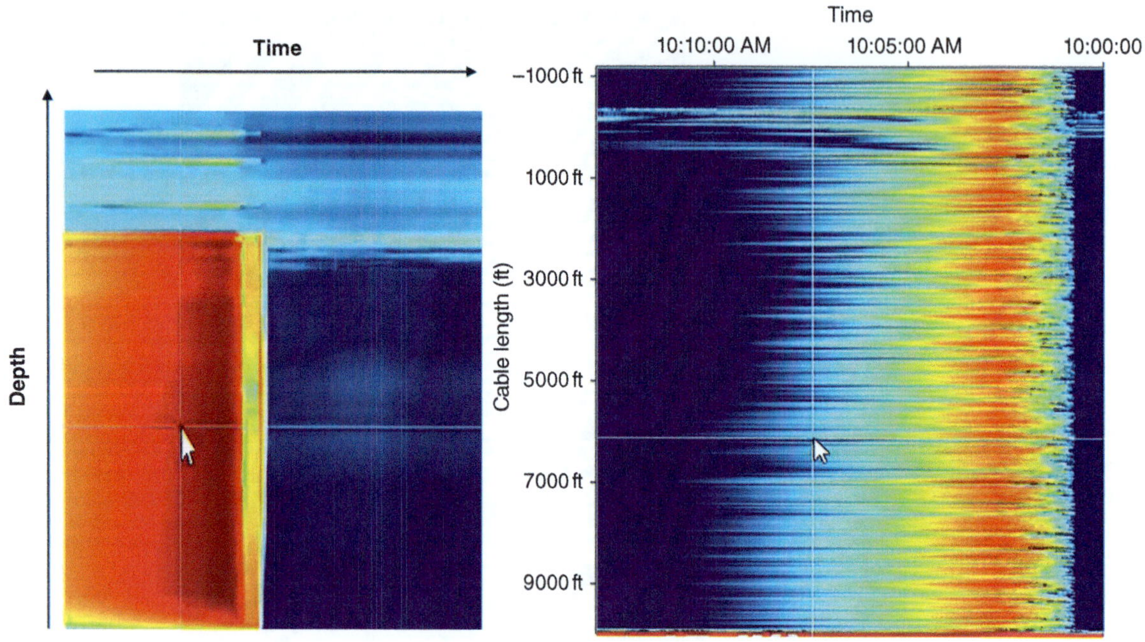

Figure 9.38 Low-frequency DAS monitoring of a cable downhole. Left indicates a cable that has come to a stop, but there is noise present along it. (Right) Strain in the same cable after a few minutes of rest, ready for seismic measurements. Adapted from Guerra and MacLeod (2024).

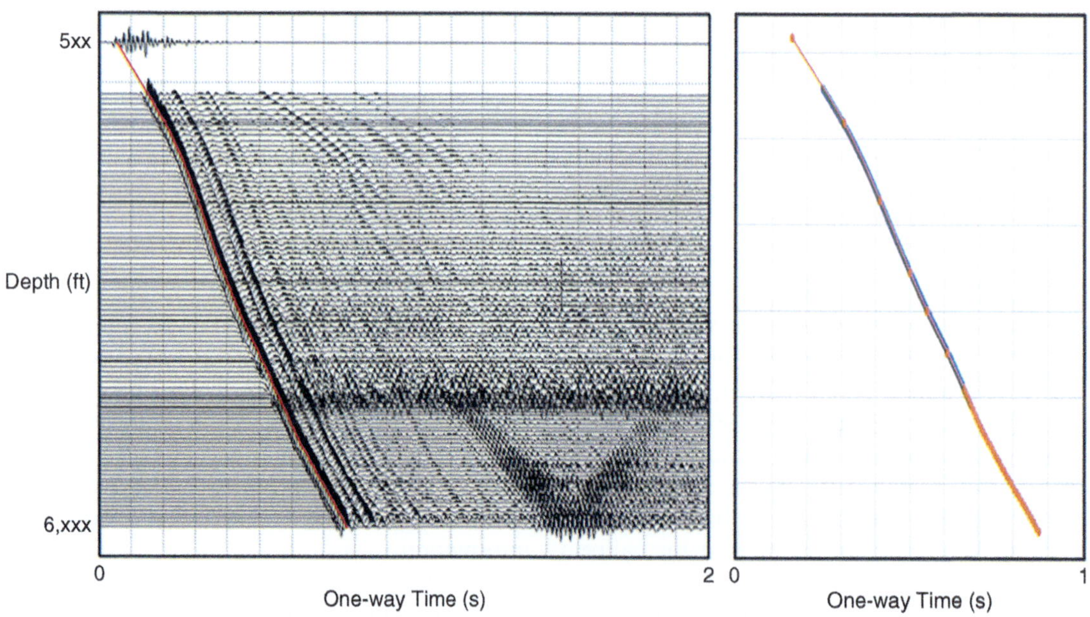

Figure 9.39 (Left) Merge of vertical component geophone stack recorded at TD and shallower DAS data. DAS data were decimated to 64 ft. to compare with the 50 ft interval of geophone data. (Right) Good overlay of transit time picks for both datasets. Blue curve denotes DAS transit times; orange curve is transit times from geophone data. Martinez et al. (2020) / Society of Exploration Geophysicists.

finite elements to model how the cable can accommodate the wellbore based on tension profile, well deviation, and varying hole size, would be desirable. However, for practical purposes, the linear correction of the slack is reasonable.

For high-complexity jobs, such as 3D VSP jobs, depending on the acquisition technique, datasets can be massive. In addition to proper data storage capacity, working on data can be cumbersome. For example, in large 3D VSP prospects where simultaneous sources with dithering are used, the downhole data are forced to be recorded continuously. Challenges with datasets are reported in Li et al. (2023). This challenging aspect is also applicable to any continuous type of recording, such as microseismic.

In other cases, the directivity of the measurement poses the need to complement DAS surveys with geophones, such as in the case of complex geology and/or specific objectives that require 3C measurements. For example, if the objective is HTI anisotropy, a hybrid approach of 3C geophone tool conveyed with fiber is most appropriate, as well as in the case of microseismic surveys where confident location of events is only possible with 3C geophone tools form one well monitoring location, or several well locations instrumented with 1C fiber. Even in this case, the presence of fiber optics serves as a potential survey optimization tool by allowing efficiency when measuring areas of need, such as 3C data combined with DAS, especially in cases where shallow portions of the data are needed for velocity calibration.

Current cable maximum safe working load (SWL) can also be a limiting factor in ultra-deep wells. The technology of hybrid heptacable is currently limited to 18,000 lb, which is already considered ultra-high-strength cable that addresses most wells typically encountered. Typically, in wells that require higher tension during open-hole logging, an effective way to record DAS on hybrid cable is to move the seismic acquisition to a cased-hole environment where the tension profile is lowered.

9.12. CONCLUSIONS

Most of the examples shown herein indicate that DAS VSP on hybrid cables, for certain applications of wellbore seismic acquisition, can replace conventional methods using geophone tools. The quality of DAS records is comparable to geophone-acquired data for many applications and environments in which the oil and gas industry operates. The cost–benefit characteristics of the DAS technique can position VSP as a standard measurement in exploration and appraisal campaigns. Canceling seismic jobs due to poor well results or budget limitations may be seen less often as DAS technology is adopted more widely. In many cases, as seen throughout this chapter, the benefits of DAS data acquired with hybrid wireline cables, especially in efficiency and fit-for-purpose data to cover objectives, outweigh some of the potential limitations of the technology.

Velocity profiles can be routinely obtained almost seamlessly with DAS using logging hybrid cables. DAS long aperture and fine sampling in depth enable detection of PS converted waves for extraction of S-wave velocities. Quality corridor stacks and VSP images from DAS conveyed by wireline have been generated in the examples described herein. Data-quality variations could be expected depending on well trajectory, total depth, and, more importantly, actual acoustic coupling achieved. After the collections of jobs shown in this chapter, it is demonstrated that VSP products that are most utilized in the workflows of the geophysical community can be obtained with DAS technology. This technology in hybrid cables has been successfully utilized in different environments: open-cased holes, subvertical to deviated well trajectories, shallow and deep wells, multiple tool combinations, and tension profiles. Also, the DAS in logging cable technique has been used for various seismic objectives, from classic velocity analysis to well-tie analysis and Q analysis, all the way to imaging and anisotropy calibration.

Furthermore, there are several advantages of using the DAS technology regarding the reduction of environmental footprint. CO_2 emission reductions are a direct consequence of less operating time needed to acquire DAS surveys, and there are associated risks that are also minimized, such as noise exposure, induced vibrations, and disruptions to other rig operations and community activities. All of these contribute to a more cost-effective seismic operation with reduced liability and, in general, greater sustainability in the future.

ACKNOWLEDGMENTS

The authors would like to thank SLB for the permission to publish this work. We also extend gratitude to Ecopetrol, Hocol, Petrosud, Alvopetro, GreatOil, Petrobras, OKEA, Petoro, Neptune Energy, and BP for granting permission to use their data. The authors would also like to acknowledge the contribution of various colleagues at SLB who worked on the processing of these datasets and to the operators who exchanged fruitful ideas. Also, the comments made by anonymous reviewers are greatly appreciated.

AVAILABILITY STATEMENT

Data associated with this research are confidential and cannot be released.

REFERENCES

Ali, S., Miranda, F., Spadavecchia, E., & Castaneda-Aguilar, R. (2021). *Processing of a large offshore 3DVSP DAS survey in a producing well.* Paper presented at the First International Meeting for Applied Geoscience & Energy - IMAGE, Denver.

Bakulin, A., Silvestrov, L., & Pevzner, R. (2019). *Surface seismic with DAS changes land acquisition.* Paper presented at the SPE Middle East Oil and Gas Show and Conference, Manama, Bahrain, SPE-194950-MS. doi:10.2118/194950-MS

Bostick, F. X., Purkis, D., Feherty, C., Rose, K., & Hitchcock, D. (2019a). *Disposable intervention tool for production logging.* Society of Petroleum Engineers.

Bostick, F. X., Purkis, D., Feherty, C., Rose, K., & Hitchcock, D. (2019b). *Disposable intervention tool for production logging.* Paper presented at the SPE/ICoTA Well Intervention Conference and Exhibition, The Woodlands, Texas, USA, SPE-194271-MS. doi:10.2118/194271-MS

Cherubini, A., Richard, S., Jestin, C., Calbris, G., & Lanticq, V. (2023). *Real-time downhole monitoring using DAS and DTS: a new technology for leak detection and well integrity.* Paper presented at the International Petroleum Technology Conference, Bangkok, Thailand, IPTC-23102-EA. doi:10.2523/IPTC-23102-EA

Danardatu, H., Gregersen, S., Altern, E., & Pellegrini, I. (2014). *Data acquisition and processing of carbon rod conveyed DTS and DAS in very long horizontal wells: first trial in North Sea Danish sector.* Paper presented at the SPE Annual Technical Conference and Exhibition, Amsterdam, The Netherlands, SPE-170663-MS. doi:10.2118/170663-MS

Duque, H., Duarte, A., Morales, D., Suarez, N., Useche, M., Martinez, A., et al. (2023a). Reducing environmental footprint of borehole seismic acquisition by using DAS on hybrid logging cable. Paper presented at SPE LACPEC, Trinidad & Tobago, SPE-213169-MS.

Duque, H., Morales, D., Useche, M., Martinez, A., & Sanchez, D. (2022). *DAS on hybrid logging cable: changing the status quo of borehole seismic acquisition.* Paper presented at the AAPG International Conference & Exhibition (ICE), Cartagena, Colombia.

Duque, H., Vanegas, J. W., Cardenas, D., Martinez, A., Singh, S., Mercado, S., Marin, D. (2023b). Repurposing multimode DTS fiber optic into high-quality DAS VSP facility for EOR monitoring. Paper presented at the EAGE Seventh Borehole Geophysics Workshop, Milan, Italy.

Ferla, M., Miranda, F., Malosi, A., Nutricato, G., Moriggi, S., Galli, G., et al. (2019). *Processing of pseudo 3D VSP acquired using permanent fiber optic technology deployed in an exploration well.* Paper presented at the 81st EAGE Conference & Exhibition, London, UK.

Ferla, M., Miranda, F., Nutricato, G., Galli, G., Moriggi, S., Malossi, A., et al. (2018). *Efficiency and innovation for borehole seismic acquisition in a well completed with permanent fiber: DAS technology for saving time and minimizing risk.* Paper presented at the Abu Dhabi International Petroleum Exhibition & Conference, Abu Dhabi, UAE, SPE-192616-MS.

Frignet, B. & Hartog, A. (2014). *Optical vertical seismic profile on wireline cable.* Paper presented at the SPWLA 55th Annual Logging Symposium.

Guerra, R., Bajwa, H., MacLeod, P., & Sanger, K. (2023). *Ultra-strength fibre optics logging system acquires DAS VSP in minutes, offshore UK.* Paper presented at the 84th EAGE Annual Conference & Exhibition, Vienna, Austria.

Guerra, R., Cuny, T., Kimura, T., Rufino, R., Danielsen, J., & Kamp, W. (2020). *hDVS/DAS VSP recorded while logging formation tester in vertical well, offshore Norway.* Paper presented at the 82nd EAGE Conference & Exhibition, Amsterdam, The Netherlands.

Guerra, R. & MacLeod, P. (2024). *A step forward in DAS VSP acquisition in vertical exploration wells: a North Sea case study.* Paper presented at the 3rd EAGE GeoTech Conference.

Hartog, A. H. (2017). *An introduction to distributed optical fibre sensors* (1st ed.). CRC Press. doi:10.1201/9781315119014-8

Hartog, A., Frignet, B., Mackie, D., & Clark, M. (2014). *Vertical* seismic optical profiling on wireline logging cable. *Geophysical Prospecting*, 693–701. doi:10.1111/1365-2478.12141

Horne, S., Armstrong, P., & Soulas, S. (2020). Deepwater DAS walkaway VSP analysis. Paper presented at the 90th SEG International Exposition and Annual Meeting. doi:10.1190/segam2020-3423715.1

Horne, S., Borland, W., Ali, S. Mercado, G., & Ikawa, H. (2008). *Anisotropic slowness inversion using 3D VSP data.* Paper presented at the SEG Technical Program. Expanded Abstracts, 2947–2951.

Hornman K., Kuvshinov B., Zwartjes P., & Franzen A. (2013). *Field trial of a broadside-sensitive distributed acoustic sensing cable for surface seismic.* Paper presented at the 75th EAGE Conference and Exhibition, London, U.K.

Khaitan, M. L., Sayed, A., & Martinez, A. (2021). *Distributed acoustic sensing vs. geophone accelerometer measurements.* Paper presented at the Sixth European Association of Geoscientists and Engineers Borehole Geophysics Workshop, Online. Extended Abstracts, BG16. doi:10.3997/2214-4609.2021613025

Kimura, T., Chen, Y., Kobayashi, Y., Xue, Z., & Adachi, K. (2019). *DAS VSP acquisition through coiled tubing fiber-optic cable.* Paper presented at the Fifth EAGE Workshop on Borehole Geophysics.

Kimura, T., Martinez, A., Santivanez, R., Cuny, T., Gmach, H., & Eliuk, N. (2017). *Borehole seismic acquisition using fiber optic technology: zero rig-time operation.* Paper presented at the SEG International Exposition and 87th Annual Meeting.

Kimura, T., Suzuki, K., Ali, S., Breninger, J., Tessman, J., Mahmoud, I.Z., & Hajjaj, S. (2018). *Hybrid 3D VSP using fiber-optic technology and a conventional borehole seismic array tool.* Paper presented at the 80th EAGE Conference & Exhibition, Copenhagen.

Kobayashi, Y., Uematsu, Y., Mochiji, S., & Xue, Z. (2020). A field experiment of walkaway distributed acoustic sensing vertical seismic profile in a deep and deviated onshore well in Japan using a fibre optic cable deployed inside coiled tubing. *Geophysical Prospecting, 68*, 501–520. doi:10.1111/1365-2478.12863

Lawrence, B. L., Uko, E. D., Eze, C. L., Chicozie, I., Tamuno-bereton-ari, I., & Umoren, N. A. (2021). Ground roll noise attenuation in 3d land seismic data in parts of Niger Delta, Nigeria. *Geological Behavior*, *5*(1), 04–06.

Le Calvez, J., Mizuno, T., Wilson, C., Bettinelli, P., & Fundytus, N. (2022). *Locating microseismic events and determining spatial uncertainty using 1C DAS fiber optic strain measurements or a combination of 1C (DAS) and 3C (geophones)*. Paper presented at the SPWLA 63rd Annual Logging Symposium, Stavanger, Norway, SPWLA-2022-0014. doi:10.30632/SPWLA-2022-0014

Leaney, S. (1999). *Walkaway Q inversion*. Paper presented at the 69th Annual International Meeting and Exposition, Houston, Texas. Expanded Abstracts, 1311–1314.

Li, F., Yu, G., Liu, H., Song, J., Zhang, Y., Chen, Y., et al. (2023). *The deblending processing of multi-well 3D DAS-VSP data from offshore Abu Dhabi*. Paper presented at the Seventh EAGE Borehole Geophysics Workshop.

Lizarralde, D., & Swift, S. (1999). Smooth inversion of VSP travel time data. *Geophysics*, *64*(3), 659–661.

Martinez, A. (2022). *DAS technology setting a new standard in borehole seismic acquisition*. Paper presented at the SEG Land & Shallow Water Acquisition Workshop, Dubai, UAE.

Martinez, A., Cuny, T., & Kimura, T. (2018). *Borehole seismic acquisition in every well and with zero additional rig time using hybrid optic-electric logging cable*. Paper presented at the Rio Oil & Gas Expo and Conference, Rio de Janeiro.

Martinez, A. & Jones, M. (2010). *Fundamentals of Borehole Seismic Technology*. Schlumberger book (www.slb.com). ISBN-13: 978-097885307-5, ISBN-10: 097885307-5.

Martinez, A., Useche, M., & Guerra, R. (2020). *Use of fiber optic acoustics to improve drilling efficiency and well placement*. Paper presented at the Offshore Technology Conference, OTC-30866-MS.

Martinez, A., Useche, M., Sayed, A., Duran, C., & Araujo, J. (2021). *DAS: an efficient and effective solution for VSP acquisition*. Paper presented at the First International Meeting for Applied Geoscience & Energy, Society of Exploration Geophysicists, Denver.

Mateeva, A., Lopez, J., Chalenski, D., Tatanova, M., Zwartjes, P., Yang, Z., et al. (2017). 4D DAS VSP as a tool for frequent seismic monitoring in deep water. *The Leading Edge*, *36*(12). doi:10.1190/tle36120995.1

Nakatsukasa, M., Ban, H., Kato, A., Shimoda, N., White D., & Nickel, E. (2018). *VSP acquisition with combined fiber-optic DAS and permanent seismic source ACROSS for permanent reservoir monitoring at the Aquistore site*. Paper presented at the 2018 SEG International Exposition and Annual Meeting, Anaheim, California, USA, SEG-2018-2995929. doi:10.1190/segam2018-2995929.1

Olofsson, B., & Martinez, A. (2017). Validation of DAS data integrity against standard geophones - DAS field test at Aquistore site. *The Leading Edge, Special Section: Fiber-optic Distributed Sensing*, 981–986. doi:10.1190/tle36120981.1

Omega-SLB Manual. (2022). Anomalous amplitude attenuation 27.2 (ANOMALOUS-AMP-ATTEN).

Rossi, T., Martinez, A., & Useche, M. (2023). *DAS VSP using hybrid wireline logging cable in deepwater wells*, Paper presented at the Offshore Technology Conference, Brazil, 23OTCB-P-751-OTC.

Sayed, A., Ali, S., & Stewart, R. R. (2020). *Distributed acoustic sensing (DAS) to velocity transform and its benefits*. Paper presented at the 90th Annual International Meeting, SEG. Expanded Abstracts, 3788–3792. doi:10.1190/segam2020-3424622.1

Sayed, A., Khaitan, M. L., Ali, S., & Martinez, A. (2021). *Characterizing anelastic attenuation using distributed acoustic sensing (DAS) data*. Paper presented at the 91st Annual International Meeting, Denver, Colorado, USA. Expanded Abstracts, 3505–3509. doi:10.1190/segam2021-3563438.1

Sayed, A., Khaitan, M. L., & Martinez, A. (2022). *Conditioning distributed acoustic sensing measurements for borehole seismic imaging*. Paper presented at the 2022 International Petroleum Technology Conference, 22IPTC-21944-MS.

Starr, J. & Jacobi, R. (2017). *Simultaneous use of microseismic monitoring and DAS-strain measurements to analyze the anatomy of a hydraulic stimulation in the Marcellus Shale*, Paper presented at the SEG International Exposition and Annual Meeting, Houston, Texas, SEG-2017-17559444.

Stewart, R. R., Huddleston, P. D., & Kan, T. K. (1984). Seismic versus sonic velocities—a vertical-seismic-profiling study. *Geophysics*, *49*(8), 1153–1168.

Useche, M., Martinez, A., Silva, M., Heredia, P., & Albarracin, L.F. (2020). *From checkshot to VSP using the same acquisition footprint enabled by fiber optics*. Paper presented at the 90th Annual International Meeting, SEG.

Varkey, J., Mydur, R., Sait, N., Wijnberg, W., Kunathikom, S., & Darpi, M. (2008). *Optical fiber cables for wellbore applications* (US Patent No. 7324730 B2). Washington, DC: U.S. Patent and Trademark Office.

10

Geothermal Reservoir Characterization Using Distributed Acoustic Sensing from Vertical Seismic Profiling in Six Geothermal Fields in Japan

Junzo Kasahara[1,2], Yoko Hasada[1,3], and Hitoshi Mikada[4]

ABSTRACT

Geophysical explorations of six geothermal fields in Japan were carried out using distributed fiber-optic seismic and temperature sensing in geothermal wells. Prior to the field tests, the usefulness of fiber-optic sensors for seismic study was evaluated using the distributed acoustic sensing (DAS) system in geothermal wells to obtain seismic shot records at each geothermal field. Four hundred eighty shot records cross-correlated with the source signatures were stacked. Most records showed P-wave first arrivals, PP reflections, and S-wave arrivals. The reflections and direct arrivals from stacked records were separated using frequency-wavenumber (F-K) filtering, and the reflections were migrated to obtain reflection images in the geothermal fields. Synthetic waveforms were also calculated to interpret the observations. Geothermal subsurface structures were obtained by integrating existing geological, geophysical, and drilling data. Our two case studies showed deep PP reflections reflected at unknown geothermal reservoirs. Assuming the study area and estimated geothermal subsurface model, the usefulness of full-waveform inversion (FWI) was examined, and the shape and depth were retrieved at a reasonable resolution. However, retrieval of the P- and S-wave velocity and density values depended on the source and receiver arrangement. Based on our research, the current approach could provide an advanced geothermal exploration method.

10.1. INTRODUCTION

Geothermal energy is a renewable energy source and emits less carbon dioxide than, nuclear solar, or wind energy. Especially since the disaster at the Fukushima Nuclear Power Plant caused by the 2011 Tohoku earthquake (Mw = 9.0) and colossal tsunami, the Japanese government has paid more attention to renewable energy sources other than nuclear power. Japan has the world's third largest geothermal resources because the Japanese islands are a volcanic archipelago. However, the development of geothermal energy in Japan was very slow before the earthquake because many geothermal areas are located near active volcanoes in designated national parks where development is severely restricted. Currently, there are 18 active geothermal power plants in Japan (Figure 10.1). The importance of renewable energy development was reevaluated after the earthquake, and the rules and regulations on geothermal energy development in national parks were relaxed in

[1]*Engineering Advancement Association of Japan, Tokyo, Japan*
[2]*Center for Integrated Research and Education of Natural Hazards, Shizuoka University, Shizuoka, Japan*
[3]*Daiwa Exploration and Consulting, Co., Ltd., Tokyo, Japan*
[4]*Department of Civil and Earth Resources Engineering, Kyoto University, Kyoto, Japan*

Distributed Acoustic Sensing in Borehole Geophysics, Geophysical Monograph 289, First Edition.
Edited by Yingping Li, Robert Mellors, and Ge Zhan.

DOI: 10.1002/9781394179275.ch10

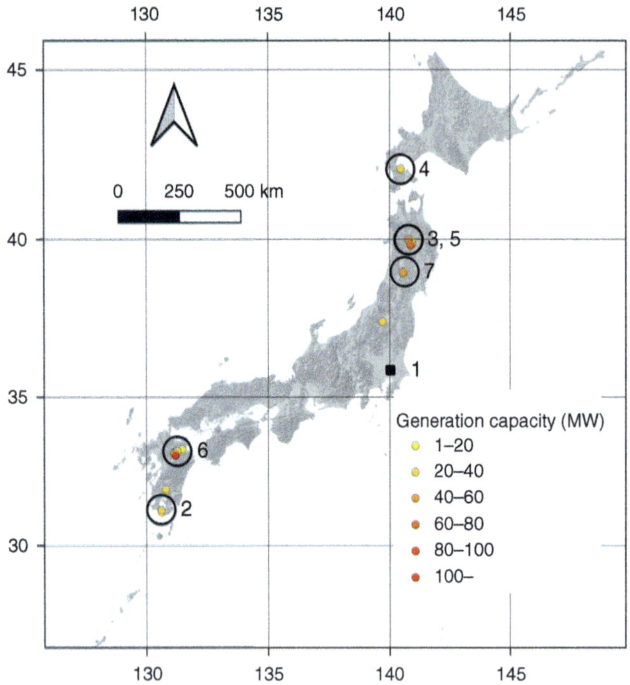

Figure 10.1 Geothermal power plants in Japan from the database of AIST (2009). The colors of the circles indicate the generation capacity of the power plants. Survey areas are also shown by numbers: (1) Abiko test site and (2–7) geothermal fields surveyed by distributed acoustic sensors (DAS) and distributed temperature sensors (DTS). 2: Medipolis; 3: Ohnuma; 4: Mori; 5: Sumikawa; 6: Takigami, and 7: Kijiyama geothermal fields.

2012. Unfortunately, the construction of new geothermal power plants is very costly, and the risk of exploration for identifying appropriate high-temperature geothermal reservoirs needs to be mitigated to accelerate the construction of such power plants in national parks.

The distribution of geothermal reservoirs seems to be controlled by numerous factors such as (a) fracture distribution, (b) the magnitude of permeability and permeable rock distribution, (c) faults, (d) the presence of seal layers, (e) upwelling of hot fluid circulation, (f) brittleness/ductility, (g) thermal conductivity of rocks, and (h) pressure. At present, several approaches, such as well drilling, magnetotelluric (MT) surveys, and gravity methods, have been used to explore geothermal fields. Seismic methods are generally considered the most reliable approach but are rarely used for geothermal exploration. Seismic exploration methods provide the Vp, Vs, density, and attenuation of seismic waves and images of seismic scattering intensities in the subsurface. We use seismic methods to infer the fluid distribution, density anomalies, fracture distribution, and many other anomalies in the distribution of physical properties in the subsurface. Geothermal fields suitable for constructing new power plants are mostly located in highly heterogeneous geological structures composed of massive igneous or metamorphic materials covered by volcaniclastics at shallow levels. Typical geological environments in geothermal fields usually attenuate the seismic waves from surface seismic sources due to the volcaniclastics near the surface, massive fractured materials, and fault systems in the subsurface. Considering this challenging environment in geothermal areas, we developed seismological tools to record seismic waves at deeper depths than before to reduce the effects of attenuation and facilitate exploration of the subsurface of geothermal fields. Our goal was to clarify the geothermal fluid and fracture distribution using seismic information. The target depth was deeper than that of ordinary geothermal reservoirs, namely supercritical water zones 3–4 km near or below the brittle-to-ductile transition zone. We used DAS for detecting seismic signals in high-temperature environments that electronics cannot withstand, and we installed distributed temperature sensors (DTS) for temperature measurements in the geothermal wells.

This chapter outlines our geophysical surveys rather than giving a detailed description of each case study. We present our method for obtaining accurate geothermal subsurface structures at several geothermal fields, showing some case studies and the results of a numerical approach using the full-waveform inversion (FWI) method. In addition, we briefly discuss the difficulty of using seismic waves to evaluate temperature.

Figure 10.1 shows the geothermal power plants used in this study.

10.2. SUPERCRITICAL WATER AS AN ENERGY SOURCE

Supercritical water is drawing the attention of the global geothermal community as an essential future renewable energy source for the world (e.g., Dobson et al., 2017; Reinsch et al., 2017) because it is expected to have a much higher geothermal production capacity due to its very high enthalpy (Dobson et al., 2017; Friðleifsson et al., 2007). The New Energy and Industrial Technology Development Organization (NEDO) in Japan is promoting the "Research and Development of Supercritical Geothermal Resources" project. As a part of this project, we have been studying new exploration methods for supercritical water located in much deeper and hotter zones than those of existing geothermal power plants.

Supercritical water can probably exist in plastic rock due to the high temperatures and can be trapped beneath cap rock. Although plastic rock could have low permeability for transporting supercritical water, Tsuchiya et al. (2016) proposed that supercritical water could be present beneath the silica cap rock deposited as a result of the subcritical to supercritical transition. Future drilling in the region could reveal the real nature of supercritical water deposition. If supercritical water is in the surrounding rocks at a moderate pressure, it could be detected as a high seismic reflective zone. The temperature and pressure of supercritical water are above the critical point of pure water at 374 °C and 22 MPa. In the Kakkonda geothermal field in Japan, scientific drilling of the WD-1a geothermal well was carried out in 1995. At a depth of 3,000 m, the temperature gradient changed from low to high, and at a depth of 3,729 m, the temperature exceeded 500 °C (Figure 10.2) (Muraoka et al., 1998). Although the temperature and pressure at the Kakkonda WD-1a well were above the critical point of water, no fluid was identified. The reason is considered that the high KCl and NaCl content (38%) raised the critical point of water to higher than 374 °C.

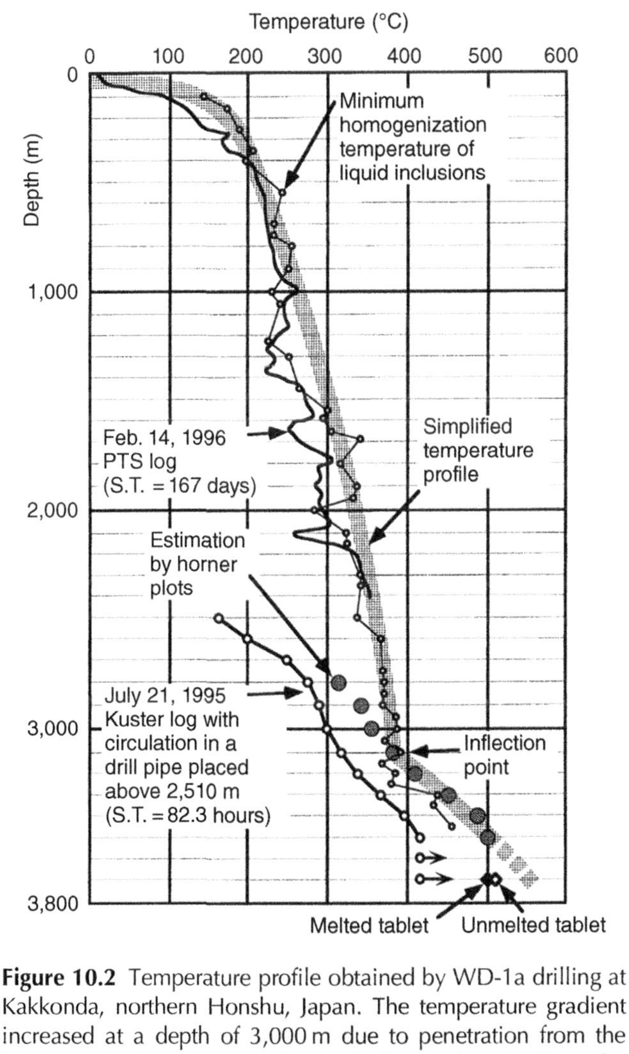

Figure 10.2 Temperature profile obtained by WD-1a drilling at Kakkonda, northern Honshu, Japan. The temperature gradient increased at a depth of 3,000 m due to penetration from the brittle to plastic regions. At depths shallower than 3,000 m, the rocks were in a brittle state and fractured. At depths deeper than 3,000 m, the rocks were in a plastic state and had low permeability. Muraoka et al. (1998) / with permission of Elsevier.

As an example of a successful case study on the drilling of a supercritical geothermal reservoir, the RN-15/IDDP-2 drilling at Reykjanes, Iceland, found supercritical water at a depth of 4,659 m because the temperature and pressure in the well were 426 °C and 34 MPa, respectively (Friðleifsson et al., 2017). Iceland mainly comprises a basaltic crust, whereas the Japanese islands have a mostly granitic upper crust. This difference in the type of rock affects the depth of an existing supercritical zone. Considering the geological differences between Iceland and Japan, we investigated supercritical geothermal reservoirs in Japan.

10.3. SENSITIVITY OF DIFFERENT DAS CABLE CONSTRUCTIONS

Recent advances in fiber-optic sensing technologies have enabled the use of optical fiber for seismic measurement: DAS (e.g., Hartog, 2017). There are three modes of scattering for incident light in optical fiber: Rayleigh, Brillouin, and Raman (see Hartog, 2017). DAS uses Rayleigh scattering of incident light by the strain or strain rate through the optical fibers. It is possible to measure the temperature distribution along the optical fibers using the same fibers with Raman scattering. We carried out geophysical exploration using DTS and DAS methods using optical fibers in the geothermal wells.

10.3.1. Comparison of Different Optical Fibers

To select the best fiber specifications for DAS measurement at the geothermal wells, we evaluated three different types of fiber: single-mode, multimode, and AcoustiSens optical fiber (Handerek et al., 2018; OFS Co., Ltd., 2021 GS8262). We examined both high (no attenuator) and low (10 dB attenuator) source laser intensities using optical fibers 100 m in length and placed nearly parallel to each other (Figure 10.3). The multimode fiber showed different waveforms from the other two types of fiber. The other two showed similar waveforms, but the noise floor of the single-mode fiber exhibited a higher noise level than the AcoustiSens fiber at frequencies greater than 10 Hz (Figure 10.4). In this study, the single-mode fiber was better than the multimode fiber. The AcoustiSens fiber was better than the single-mode fiber, but it cannot be used in geothermal circumstances because the temperature allowance of acoustic fiber is 40 °C, according to the datasheet.

Considering these study results, we decided to use single-mode fiber for geothermal DAS measurement. To resist high temperatures present in the geothermal wells, we used polyimide and carbon-coated optical fibers in dual stainless-steel armored tubes. These tubes provide a seal against hot water that can quickly penetrate the optical fiber, generating an "−OH" light absorption spectrum at a wavelength of 1,350 nm.

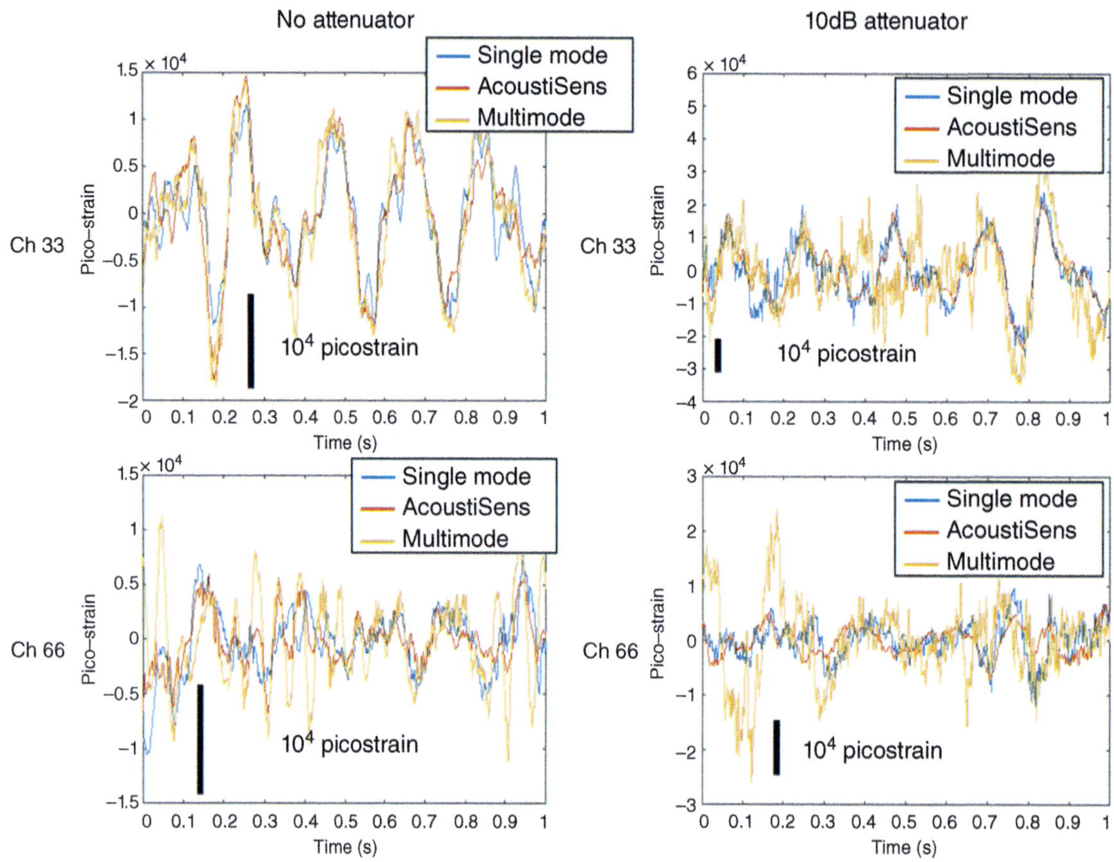

Figure 10.3 Comparison of three different fibers. Blue, single-mode; yellow, multimode; red, AcoustiSens fibers (OFS, GS82628). Each fiber is 100 m long and is placed nearly parallel to each other on the ground. The vertical axis is the amplitude in picostrain ($\times 10^4$), and the horizontal axis is the time in seconds. Ch 33: 33 m and Ch 66: 66 m from the end, respectively. Single-mode and AcoustiSens fiber with no attenuator showed similar ground motion waveforms. When the attenuated source intensity was used, the multimode fiber showed waveforms that were different from the other two types of fiber.

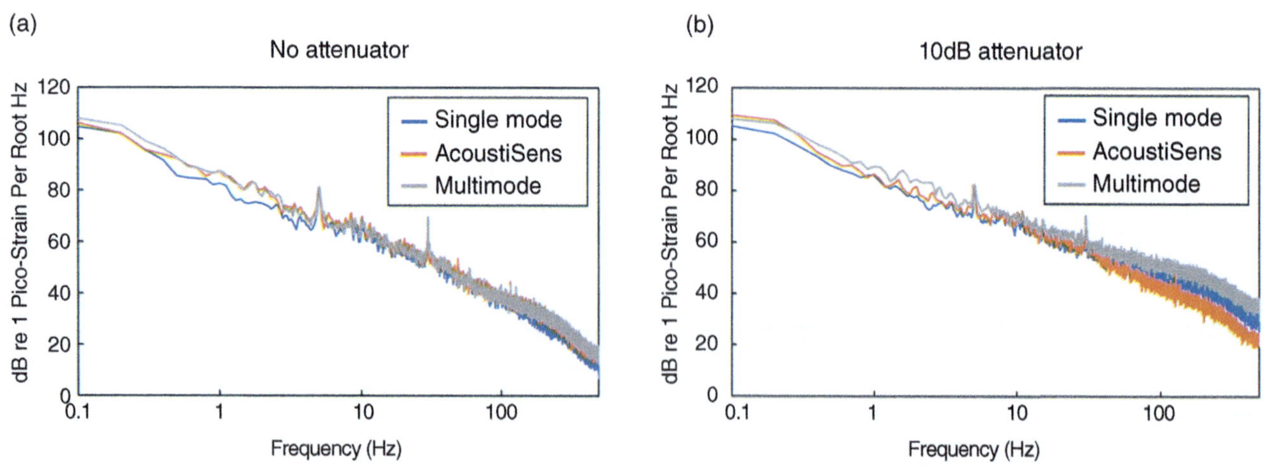

Figure 10.4 The noise floor of three different types of fiber in dB (for one picostrain/$\sqrt{\text{Hz}}$) versus the frequency (Hz) for (a) high (fiber without attenuator) and (b) low (10 dB attenuator) laser sources. When the 10 dB source attenuator was used, the single-mode and multimode fibers showed a higher noise level compared to that of the AcoustiSens fiber at ranges greater than several tens of Hz.

10.3.2. Field Test Using DAS and Seismometers

Kasahara et al. (2018a, 2018b) experimentally evaluated the quality of DAS seismic data at the Abiko area of the Central Research Institute of Electric Power Industry in 2017 (Figure 10.1). We used three fiber-optic cables that were 100 m long and buried 20 cm beneath the ground surface. Two of the single-mode optical fibers were in stainless-steel tubes and one in a vinyl tube. We also placed 101 4.5 Hz vertical geophones (sensitivity: 25 V/m/s) every 1 m. In addition, three-component seismometers (1 or 0.2 Hz) were used. Sledgehammer impacts and a MiniVib seismic source were used as the active sources. Figure 10.5 is a comparison of the three parallel

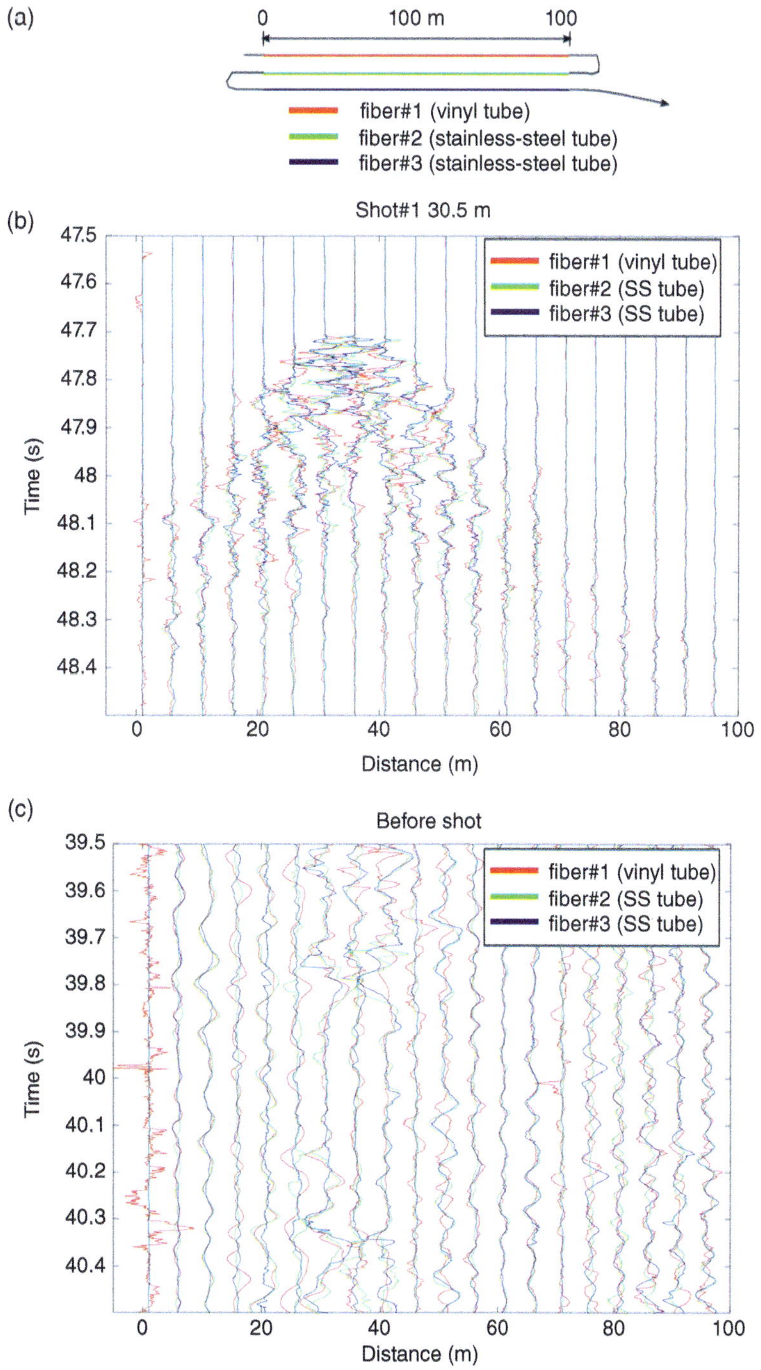

Figure 10.5 Comparison of seismic DAS signals acquired by three fibers (a), sledgehammer impact (b), and seismic tremors (c), respectively. The green and blue lines denote the fibers in stainless-steel tubes, and the red lines denote the fiber in a vinyl tube. Waveform discrepancies among the three fibers are due to ground coupling differences or sensitivity perturbation. Adapted from Kasahara et al. (2018a).

fibers in nearly the same places (Kasahara et al., 2018a, 2018b). Discrepancies can be seen in the waveforms of the three fibers due to the sledgehammer impact and ground noise. Even though two of the same single-mode fibers in the stainless-steel tubes show slightly different waveforms, the differences are due to ground coupling to the surrounding soil or sensitivity perturbation.

We calculated the strain rate waveforms using the horizontal components of seismometer records. The waveforms of a natural earthquake in 2017 acquired by seismometers and DAS were compared (Kasahara et al., 2018a, 2018b). The incident angle of this earthquake to the fiber is nearly perpendicular, and the two horizontal components of the seismometers are parallel and perpendicular to the fiber direction, respectively. A high-cut filter of 30 Hz and amplitude normalization were applied to the parallel components of 18 seismometers. We calculated the strain rate waveforms based on the differences in 17 adjacent pairs and averaged them all. The waveform comparison of the seismometer and DAS is shown in Figure 10.6. Both waveforms of the S-wave portion are similar, suggesting that the DAS could measure a physical quantity comparable to an ordinary seismometer (Hasada et al., 2018; Kasahara et al., 2018a, 2018b). Note that the amplitude units of DAS and strain rate seismometer records were not considered. Wang et al. (2018) supported our results using the ML 4.3 earthquake 150 km southeast of Brady Hot Springs, NV.

We found that the current DAS sensitivity was slightly less than that of the seismometers; however, it still provided very dense data, namely at intervals of a few meters.

10.4. GEOTHERMAL STUDIES USING DAS AND DTS

This section presents an outline of the data acquisition in the field and the data processing for the acquired records. The well deviations and temperature profiles obtained by the DTS method are summarized for six geothermal fields.

We used a set of active seismic sources at the surface and an optical fiber system in the wells to obtain the subsurface structure of the geothermal fields. Refracted and reflected seismic waves were measured by the DAS method using optical fibers in the geothermal wells. To understand the thermal circumstances in the target field, we also measured the temperature profile in the wells using the DTS method.

Temperature measurements were made every 1 m at intervals of an average of 15 min. The accuracy of the temperature measurements was approximately 0.1 °C. We measured the temperature profile before and after the DAS measurements. The DTS temperature was calibrated using the total length of the cable winch at the outside temperature. By carrying out the temperature measurements twice (at the beginning and the end) and referring to the

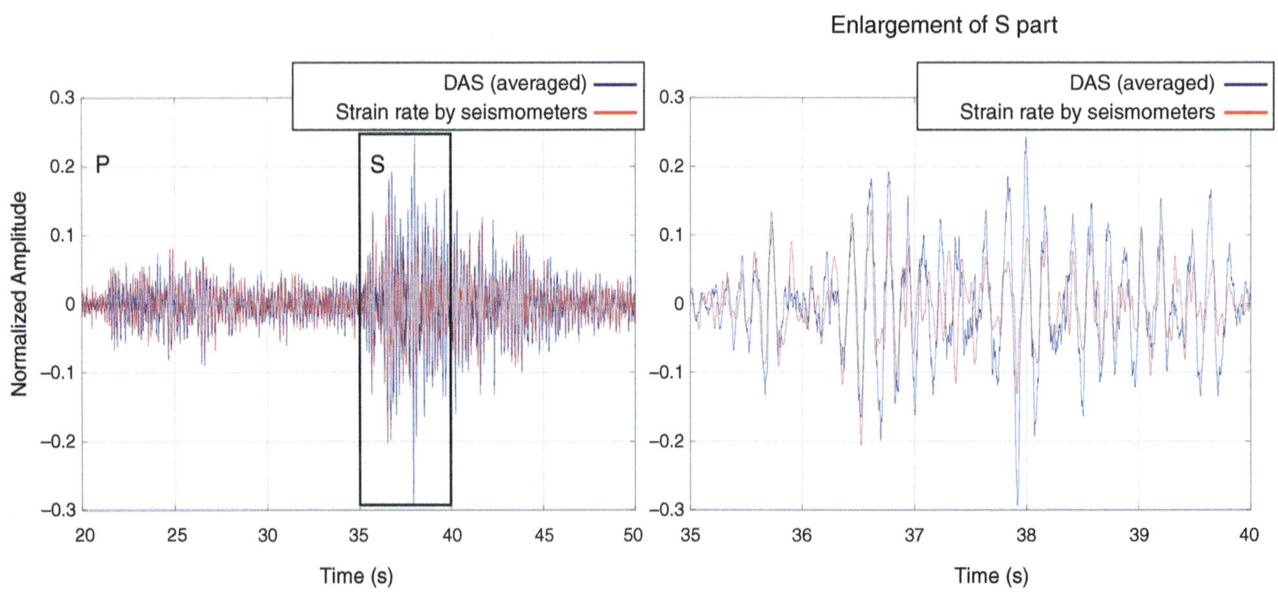

Figure 10.6 (Left) The M 3.1 Ibaraki earthquake occurred at 17:28 JST on 23 September 2017. P and S waves were recorded from 17:28:20 to 17:28:50. (Right) Enlarged waveforms of the rectangular box in the left. Blue lines, DAS; red lines, calculated strain rate using the seismometer component parallel to the fiber direction. The sensitivity of the seismometer was normalized, but both amplitudes were normalized. The vertical axis is the amplitude, and the horizontal axis is the time in seconds. Adapted from Hasada et al. (2018) and Kasahara et al. (2018a, 2018b).

previous measurements by the owner of the wells, we confirmed that the measurements were correct.

We switched the DTS to DAS measurements and continuously measured the seismic signals in the DAS mode during the shootings. In our previous surveys, we used three different interrogators: hDVS (Schlumberger Co., Ltd.) for the Medipolis and Ohnuma studies; ODH-4 (OptaSense Co., Ltd.) for the Mori, Sumikawa, Takigami, and Kijiyama studies; and N5200A (AP Sensing) for the Ohnuma study. We used an AP sensing interrogator for one shooting at Ohnuma. A comparison of the shooting DAS data obtained by the hDVS and N5200A revealed that the S/Ns of both instruments were similar after frequency response correction. We selected longer gauge lengths, such as 15 m. We used the IVI MiniVib for the Medipolis study and the IVI EnviroVibe for the others. The source signal comprised a 30 s sweep from 10 to 75 Hz and a 30 s sleep. We repeated the sweep-and-sleep cycle more than 480 times for each shooting location.

At first, we eliminated common noise for all traces and spike noise on each trace. After stacking the raw DAS records 480 times or more, we eliminated the source signature in the frequency domain. During the stacking, we inversely weighted each trace by its standard deviation over 40 s to suppress the outliers. We used the weight for stacking based on the inverse of the standard deviation of 40 s of DAS data.

F-K filtering was applied to the original DAS waveforms to obtain the reflection (upgoing) and direct (downgoing) arrivals. The *Vp* structure was estimated using the P-wave first arrivals on DAS and the surface seismic records. The traveltime inversion of the P-wave first arrival readings, reflection images interpretation, and forward modeling were used to compare the synthesized and observed waveforms. We calculated the migration images using the reflection (upgoing) waveforms obtained by the F-K filtering.

During 2018–2022, we carried out DTS and DAS measurements at six geothermal fields: Medipolis (Kasahara et al., 2020a, 2020b), Ohnuma (Kasahara et al., 2022a), Nigorikawa-Mori (Kasahara et al., 2022b), Sumikawa (Kasahara et al., 2022c), Takigami (in this monograph), and Kijiyama (Figure 10.1). Figure 10.7 shows the horizontal deviations of the wells and the measured temperature profiles in the six wells. The maximum depth of

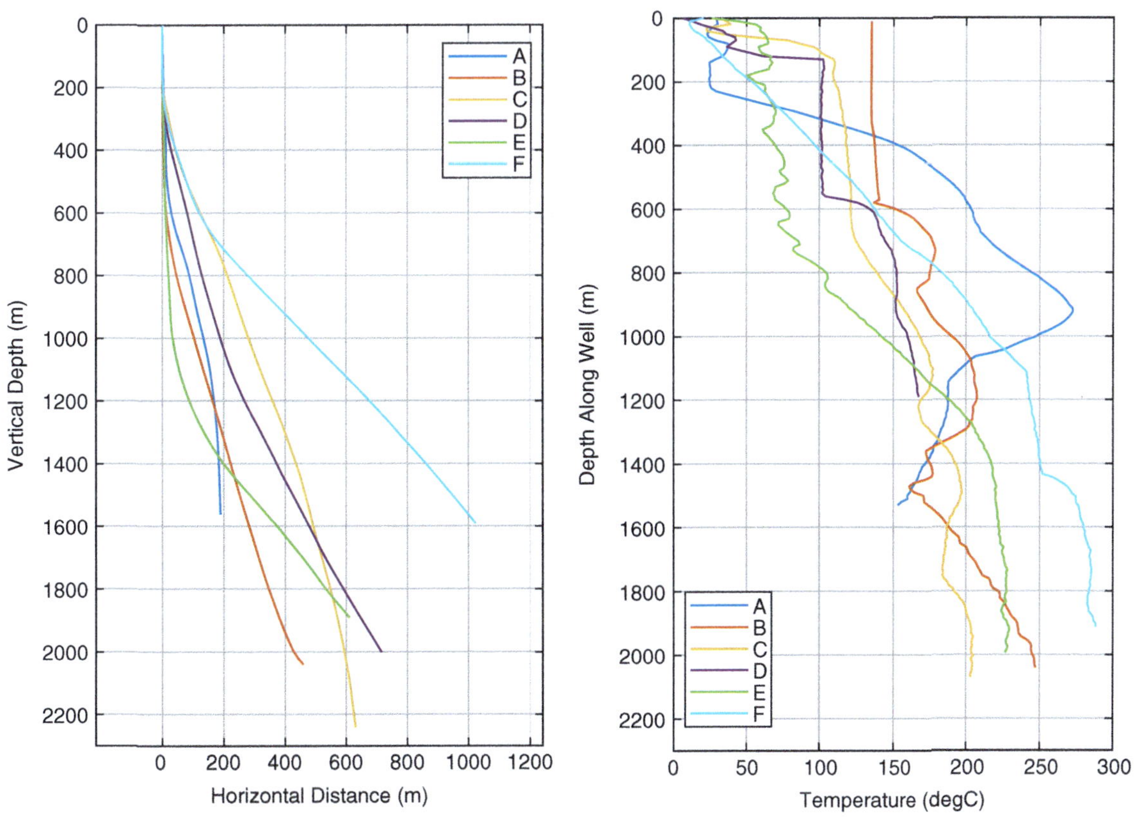

Figure 10.7 Horizontal deviations of wells and temperature profiles. (Left) Horizontal deviation profile of geothermal wells used in our surveys. (Right) Temperature-depth profiles of wells obtained by our studies. The vertical axis is the depth in meters.

optical fiber deployment was 2,200 m. The horizontal deviations differed among the six wells, but the signals obtained were of similar quality. The highest temperature reached 288 °C. The temperature gradients increased with the depth in four cases, but the temperature decreased from some depths in two cases. The depth profiles differed due to the presence of fractures, seal layers, water circulation, and production zones.

10.5. GEOPHYSICAL STUDIES AT MEDIPOLIS AND OHNUMA GEOTHERMAL FIELDS

This section presents part of the results of two case studies: the Medipolis and Ohnuma geothermal fields among the six locations (Figure 10.1). Additional details are provided in Kasahara et al. (2020a, 2020b) for Medipolis and in Kasahara et al. (2022b) for Ohnuma.

10.5.1. Case Study of Medipolis Geothermal Field

We carried out geophysical studies at the Medipolis geothermal field in 2018 and 2019 (Figure 10.1). In the 2018 study, we installed the optical fiber in the IK-4 well to a depth of 970 m. We observed several natural earthquakes and confirmed the validity of using optical fiber in the geothermal wells for the DTS and DAS measurements.

In the 2019 study, we used the same IK-4 well and installed the optical fiber to a depth of 1,500 m. We carried out the offset VSP study at five locations using the MiniVib as the seismic source. F-K filtering was applied to the DAS records for the seismic sources. The DAS record at 1,600 m offset distance is shown in Figure 10.8. The middle of Figure 10.8 shows the upgoing (reflection) arrivals and the bottom shows the downgoing (direct) arrivals. It was possible to identify the deep PP reflection and intense arrivals estimated as S-wave arrivals. Similar shallow and deep PP reflections were observed in the DAS records for the Mori (Kasahara et al., 2022a), Sumikawa (Kasahara et al., 2022c), and Takigami (in this monograph) geothermal fields.

We obtained the *Vp* distribution in the Medipolis field by the traveltime inversion (Kasahara et al., 2019b) shown in Figure 10.9.

We calculated the synthetic DAS waveforms to interpret the observed DAS records in the Medipolis geothermal field (Kasahara et al., 2019b). We also calculated the synthetic surface seismic records. Figure 10.10 shows the

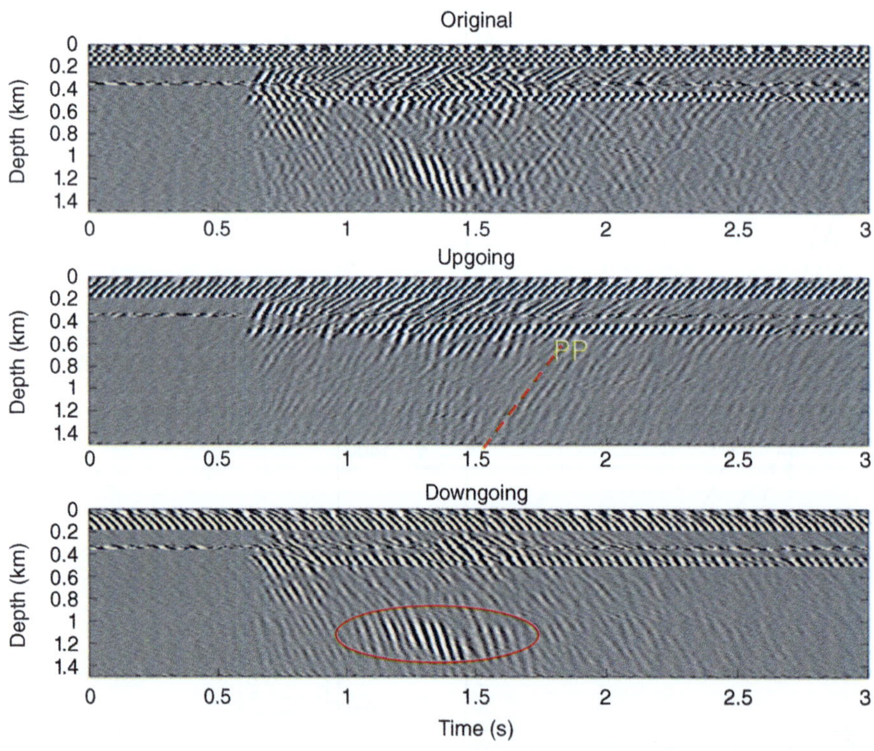

Figure 10.8 A set of stacked, source signature compensated DAS records at the Medipolis geothermal field. The three diagrams are the original, F-K filtering records of the upgoing and downgoing phases. The DAS records were obtained in the Medipolis geothermal IK-4 well at an offset distance of 1,600 m by the MiniVib seismic source. The vertical axis is the depth in the well, and the horizontal axis is the elapsed time in seconds. The broken red line shows the PP reflected phase from a depth of 3.6 km. The red ellipsoid is the S phase observed at a depth of 1,000–1,400 m.

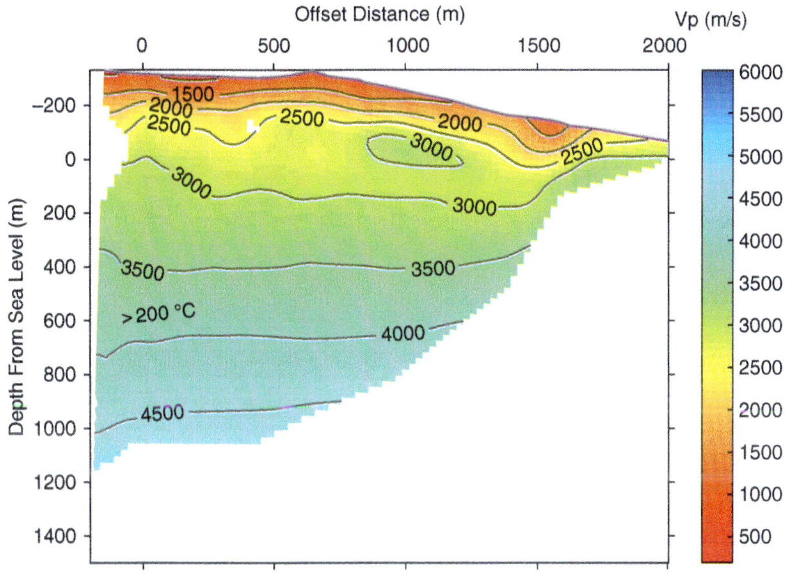

Figure 10.9 P-wave velocity distribution of Medipolis field estimated by 2D traveltime inversion The vertical axis is the depth from sea level in meters, and the horizontal axis is the offset from the wellhead in meters.

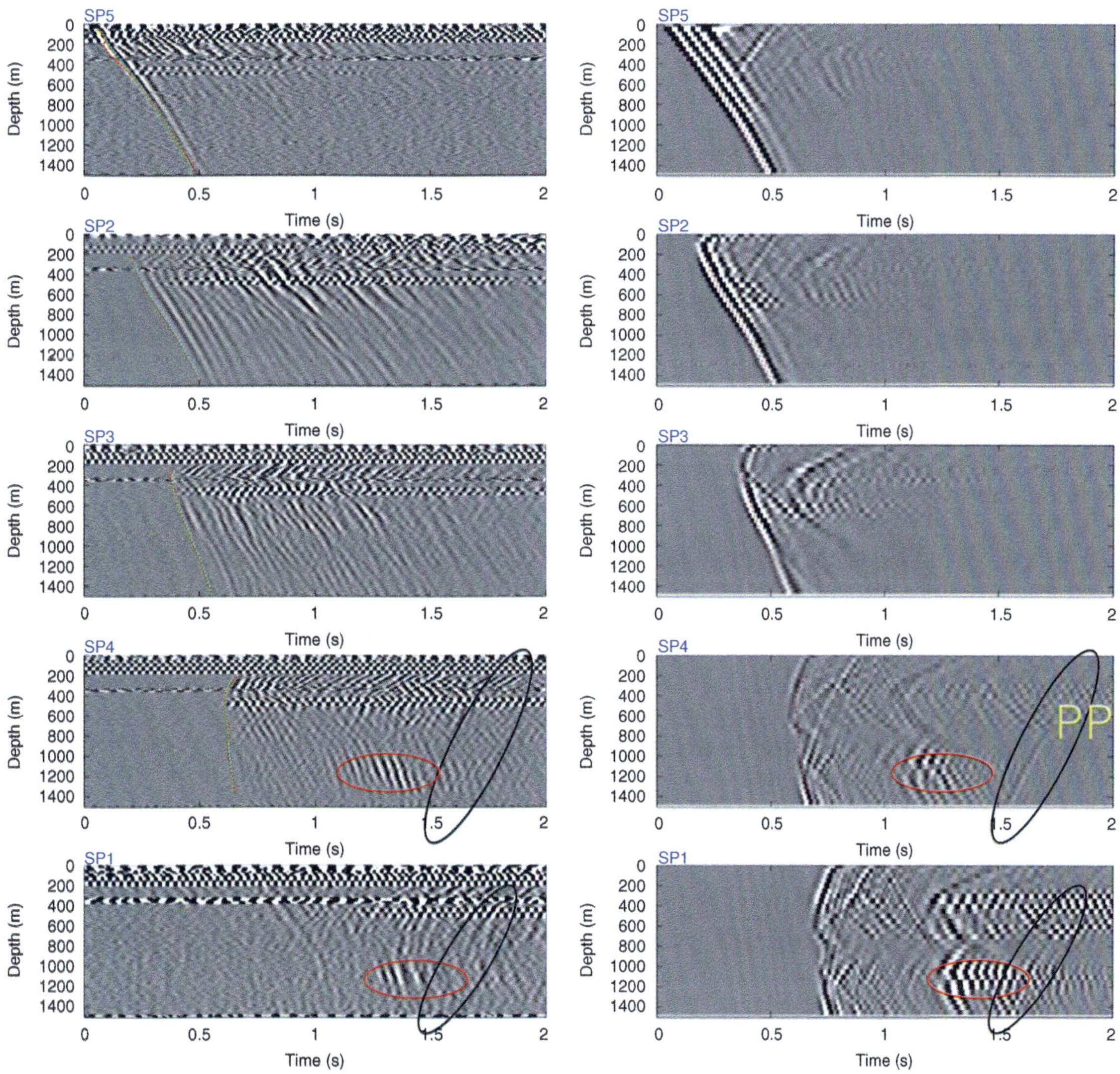

Figure 10.10 (Left) Observed DAS records excited by the MiniVib from 10 to 75 Hz. (Right) synthetic DAS records using the structural model for the Medipolis study (Figure 10.11). Black ellipsoids show the PP reflection from a depth of 3.6 km; red ellipsoids show the S phase. From top to bottom for the left and right sides: (SP5) wellhead, (SP2) 300 m, (SP3) 800 m, (SP4) 1,600 m, and (SP1) 2,000 m offsets. Kasahara et al. (2019) / with permission of the Stanford Geothermal Program.

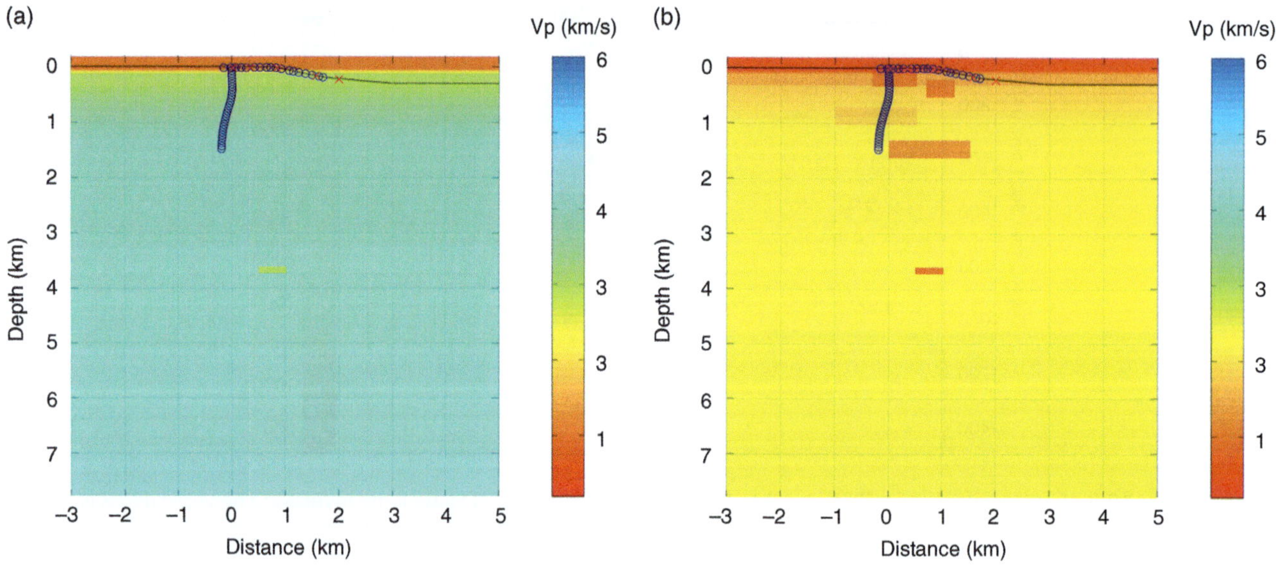

Figure 10.11 (a) *Vp* and (b) *Vs* models of the Medipolis geothermal field explaining the DAS records. The vertical and horizontal axes are the depth in kilometers and the horizontal distance in kilometers from the western side. Blue small circles indicate the IK-4 well profile and surface seismometers. The low *Vs* zone at a depth of 3.6 km is considered to be the cause of the deep PP reflection. Adapted from Kasahara et al. (2020a).

observed DAS records and the synthetic waveforms obtained from the structural model shown in Figure 10.11. In this comparison, we considered the P-wave first arrivals, the seismic reflection from a depth of 3.6 km at SP4, and the S phase at SP4 and SP1. The seismic reflection zone at a depth of 3.6 km can be interpreted as being in a supercritical water region because the temperature at a depth of 918 m was 272.8 °C, the pressure at 3.6 km was higher than 30 MPa, and the pressure and temperature conditions were above the critical point of pure water of 374 °C and 22 MPa (Kasahara et al., 2019b, 2019c).

Vp and *Vs* structure models explaining the DAS records were obtained by forward modeling (Figure 10.12) (Kasahara et al., 2020a). The low *Vs* zones at 3.6 km and 1.3–1.5 km are interpreted by the deep PP reflection and intense S phase.

We migrated the reflected arrivals of the DAS records and compared them to the *Vp*/*Vs* model obtained by the forward modeling. The high *Vp*/*Vs* zone around a depth of 3.6 km is identified in the migrated image, but the other zone at a depth of around 1,300 m is not evident in the migrated image by P reflections.

We identified the P-to-S conversion for natural earthquakes observed by the surface seismometers (Kasahara et al. 2020b). A depth of approximately 3.6 km was obtained using the time difference between the UD and NS-EW seismometers for the P-to-S conversion. We obtained $Vp/Vs > 2$ at the depth of 3.6 km, producing the P-to-S conversion. $Vp/Vs > 2$ suggests extremely high fluid contents (Winkler & Nur, 1982).

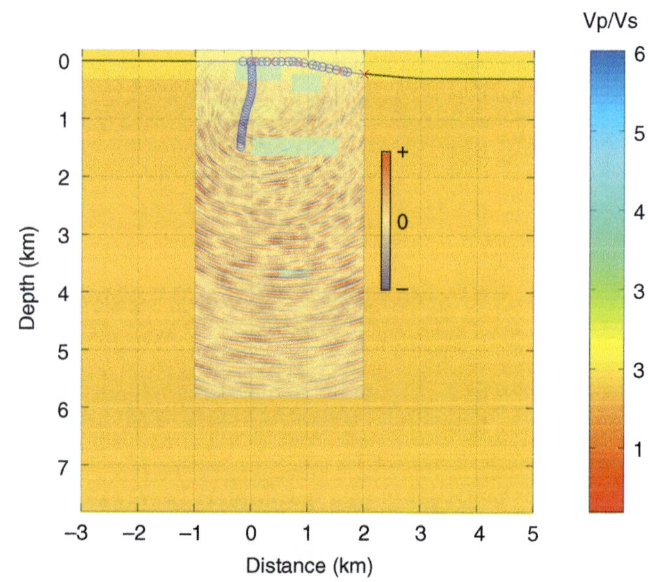

Figure 10.12 Migration image superimposed on the *Vp*/*Vs* model obtained by forward modeling.

10.5.2. Case Study of Ohnuma Geothermal Field

This section presents a geophysical case study carried out at the Ohnuma geothermal field in 2019 (see Figure 10.1). The details are described in Kasahara et al. (2022b). Optical fiber was installed to a depth of 2,000 m at the O-13R geothermal well, and the EnviroVibe seismic source was operated at nine locations. Figure 10.13 shows the results of the F-K filtering at the C4 seismic source

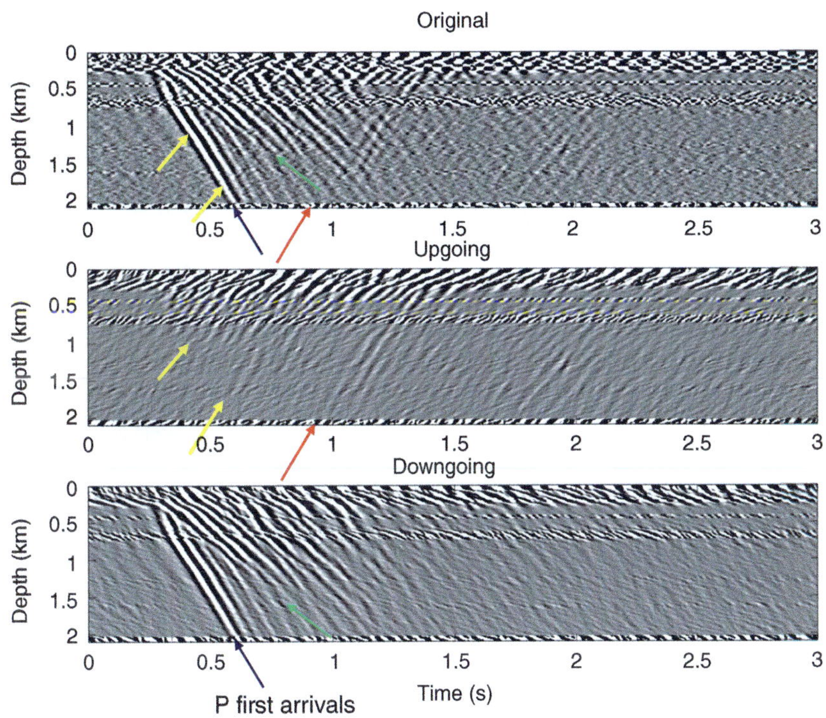

Figure 10.13 A set of stacked-source signature compensated DAS records obtained in the Ohnuma O-13R geothermal well by the C4 seismic source. The vertical axis is the depth in kilometers in the well, and the horizontal axis is the elapsed time in seconds. Black arrows are the P-wave first arrivals. Red arrows are PP reflections from a depth of 2.8–3.0 km. Yellow arrows are other shallow PP reflections. Green arrows are S-wave arrivals. Note that 2.8–3.0 km deep PP reflections are relatively intense compared to the P-wave first arrivals. Kasahara et al. (2022) / with permission of the Geothermal Research Society of Japan.

with the original DAS record that was obtained. We identified clear and intense deep PP reflections, as shown by the red arrows. Comparing the synthetic waveforms, migrations, and observed records, these PP reflections were interpreted as being generated at a depth of 2.8–3.0 km. Other shallow reflection phases (yellow arrows) were estimated to be reflected at a depth of 1.5–2.0 km. The S phase (green arrow) was also identified.

We migrated the reflection arrivals for the Ohnuma DAS records using the Kirchhoff migration method (Figure 10.14). The estimated background *Vp* structures in the study areas were used for the migration process. The migration image of Ohnuma (Figure 10.14) shows an intense reflection zone distributed from a depth of 2.8–3.0 km (Kasahara et al., 2022b). The 3D image is shown in Figure 10.15. The commercial production zone is 1.5–2.0 km in depth corresponding to the reflection zone at this depth range.

We calculated the synthetic waveforms to estimate the physical properties in an intense seismic reflection zone at a depth of 2.8–3.0 km (Kasahara et al., 2022b). Using the 3D model shown in Figure 10.16, we obtained nine synthetic DAS records for nine seismic locations. The two synthetic records for two sources are shown in Figure 10.17. Although the intensity of the synthetic PP reflections from a depth of 2.8–3.0 km is weak, the observed reflections were 0.2–0.3 times those of the observed P-wave first arrival amplitudes, as seen by the red arrows in Figure 10.17. We evaluated the density anomaly at this intense seismic reflection zone and obtained a value of 25% in contrast to that of the surrounding rocks. The 25% density anomaly might imply a fluid zone (Kasahara et al., 2022b).

10.5.3. Geothermal Importance of the Two Case Studies

The geophysical study at the Medipolis geothermal field obtained clear PP reflections from the DAS records at an offset distance of 1,600 m. The PP reflection arrivals at a depth of 3.6 km were interpreted as being caused by the low *Vs* zone where *Vp*/*Vs* was greater than 2.0. The PP reflection from 3.6 km is seen in the synthetic DAS records and the migration image. This low *Vs* zone could explain the observed P-to-S conversion. *Vp*/*Vs* > 2 suggests a high fluid zone and could be a supercritical water zone. We obtained the low *Vs* zone at a depth of 1.0–1.5 km and it is probably a production zone with fractures filled by fluid.

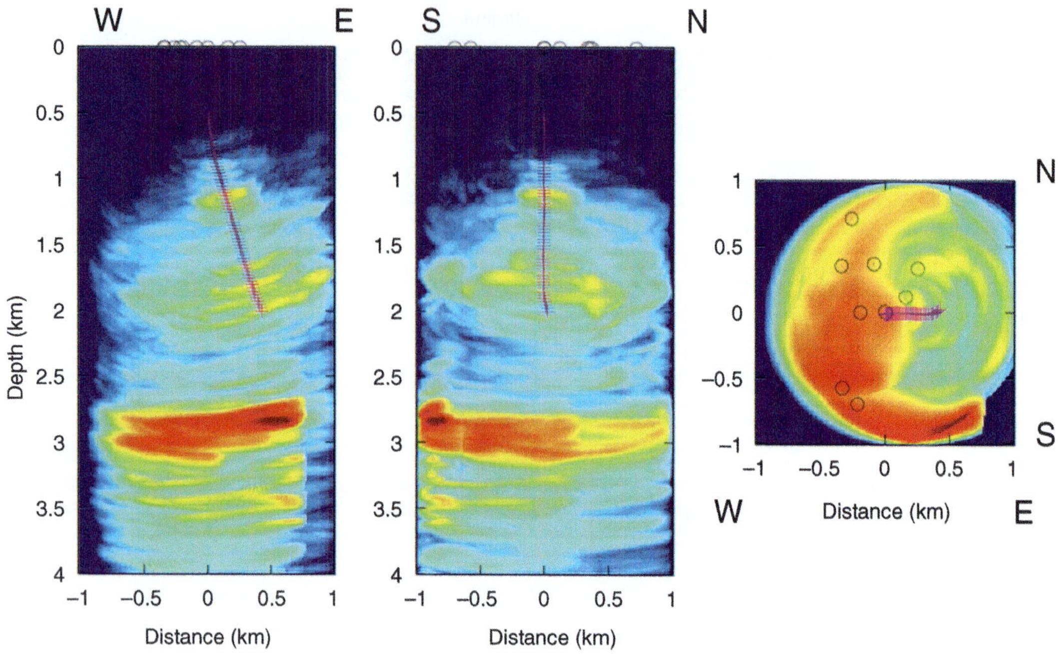

Figure 10.14 Migrated image of PP reflections at the Ohnuma geothermal field. (Left) E–W depth section, (center) N–S depth section, and (right) top view. Colors show the maximum intensity in the direction perpendicular to each diagram as high (red) to low (blue) intensity levels. The intense reflection region is concentrated at a depth of 2.8–3.0 km. The pink bar shows the O-13R well profile. The zone between 1.5 and 2.0 km is the production zone.

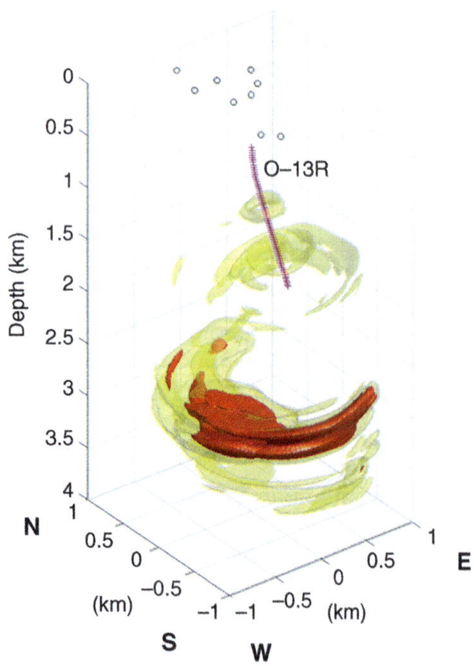

Figure 10.15 3D migrated image at the Ohnuma geothermal field. The yellow zones have high intensity, and the red zones have the highest intensity. The pink bar shows the O-13R well trace, and the black open circles are the source locations. Adapted from Kasahara et al. (2022b).

In the geophysical study at the Ohnuma geothermal field, we observed a significant reflection zone at a depth of 2.8–3.0 km through the DAS records and migrated image. This reflective zone was not previously identified. The amplitude of the reflected arrivals from 2.8 to 3.0 km suggested a 25% density anomaly, and this zone is possibly a new geothermal reservoir. If the 25% density anomaly at this depth is true, the PP reflections at 2.8–3.0 km could be caused by fluid in fractures. The reflections at a depth of 1–2 km correspond to a production zone.

10.6. DISCUSSION OF INFLUENCE OF TEMPERATURE ON *Vp* AND *Vs*

If the subsurface temperature distribution is known, it is possible to directly drill high-temperature zones for geothermal energy sources. Are seismic waves affected by temperature variations? Kern (1978), Kern and Richter (1981), and Kern et al. (1997) used numerous *Vp* and *Vs* laboratory measurements at high confining pressures and temperatures and found that *Vp* and *Vs* for most rocks except gneiss, granite, and quartzite decrease at 0.02–0.04 km/s/100 °C at a confining pressure of 600 MPa. *Vp* in granite, gneiss, and quartzite decreased rapidly at temperatures above 500 °C. *Vs* in amphibolite, gneiss, and granite showed a similar temperature

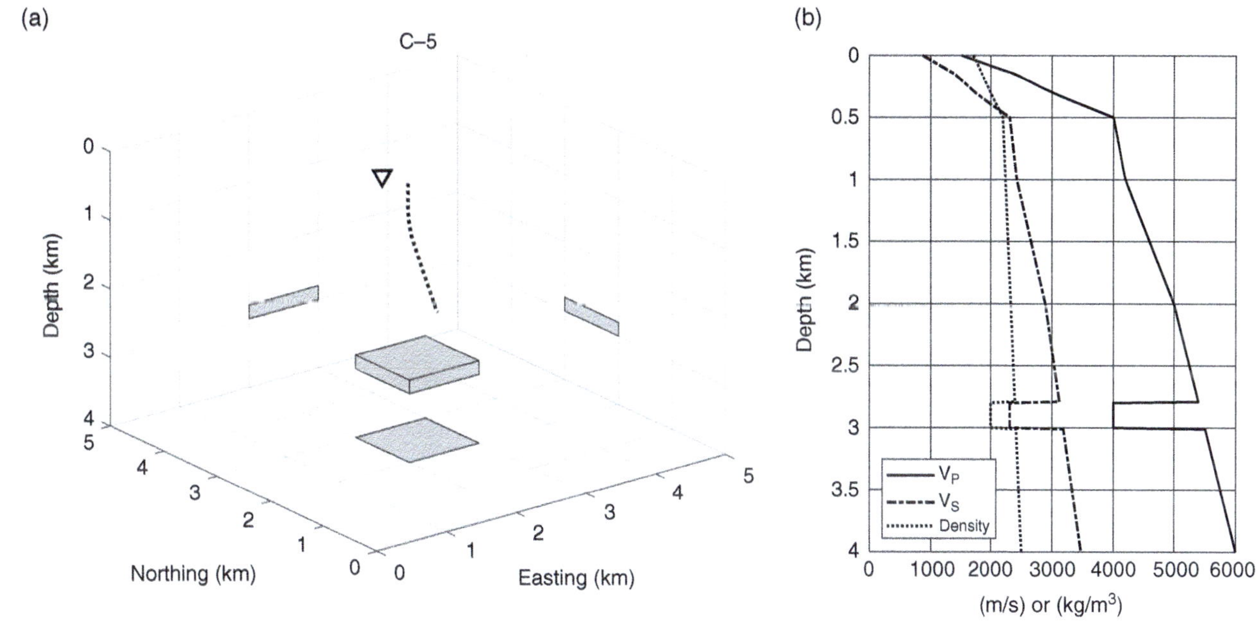

Figure 10.16 (a) 3D model used to calculate synthetic DAS waveforms for the C-5 source at the Ohnuma field. The inverted triangle and the dotted lines are the source and the DAS locations, respectively. The reflective region is placed at a depth of 2.8–3.0 km with a width of 1 km and a thickness of 200 m. (b) *Vp*, *Vs*, and density through the box zone at 2.8–3.0 km versus the depth.

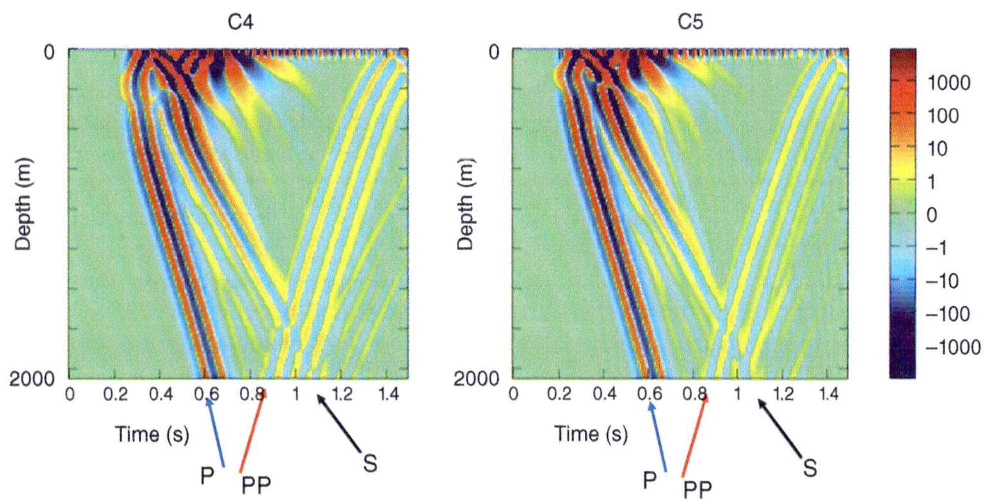

Figure 10.17 Synthetic DAS waveforms in the O-13R well for the C4 (left) and C5 (right) source locations. The vertical and horizontal axes are the depth in meters and the elapsed time in seconds. The color shows the amplitude level on the log scale on the right side of the diagrams. Blue arrows, first P-wave arrivals; black arrows, S; and red arrows, PP reflections from a depth of 2.8–3.0 km.

dependency. Kern et al. (1997) also evaluated *Qp* and *Qs* of serpentine and amphibolite under 600 MPa and 25–700 °C and found almost no changes in *Qp* and *Qs* with temperature variation. These experiments used serpentinite as rocks having hydrous minerals, but other major hydrous minerals such as clay and chloride minerals exist in the crust and show dehydration at some depth (Kasahara et al., 2003). It is reasonable to think that the dehydration of clay and chlorite could cause the changes in velocity, but unfortunately, there are no experimental results to confirm or refute this.

Because we obtained temperature profiles and DAS shooting records in the same wells, we could examine the temperature effects on seismic wave propagation. We compared the zero-offset DAS records and temperature profile of the IK-4 well in the Medipolis geothermal

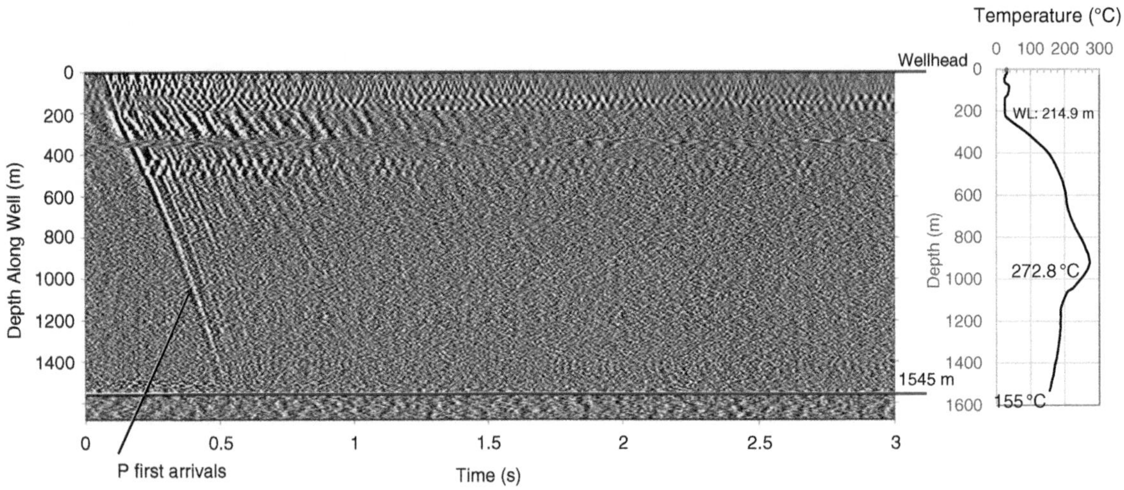

Figure 10.18 Zero-offset DAS records by a MiniVib shot (left) and temperature profile (right) of the IK-4 well in the Medipolis geothermal field. The temperature was 272.8 °C at a depth of 918 m. The P first arrivals could be identified. The vertical axis is the depth in meters. Adapted from Kasahara et al. (2019b).

field (Figure 10.18; Kasahara et al., 2019b). The temperature was 272.8 °C at a depth of 918 m and 155 °C at a depth of 1,530 m. Although the maximum temperature was at 918 m, no significant change in the P-wave first arrival amplitude and the interval *Vp* were found around this depth zone. Because *Vp* at this depth was 3–4 km/s (Figure 10.9), we excluded the presence of smectite clay. The dehydration of chlorite occurs at temperatures higher than 500 °C (Kasahara et al., 2003), and we can also exclude chlorite from the dehydration effect at 918 m. It is possible that the temperature dependence on *Vp* is not significant for the Medipolis field. Considering the results of the high-temperature and high-pressure experiments showing quite a small temperature dependence of *Vp*, *Vs*, and Q (Kern, 1978; Kern & Richter, 1981; Kern et al., 1997), our observation is consistent with the laboratory results. If a large velocity or amplitude change is observed with an increase in the depth in DAS, the velocity change can be interpreted as being caused by geological layer boundaries, the presence of fractures, or the presence of fluid rather than the temperature itself.

10.7. FWI APPROACH

If the physical properties of a geothermal field are known, it is possible to evaluate the geothermal structure and the nature of the geothermal reservoirs in this field.

10.7.1. Simulation Model

Seismic migration is a method frequently used in seismic reflection surveys. The FWI method has recently been applied to estimate high-resolution seismic velocity structures. Most FWI applications use a combination of forward wavefield extrapolation and reverse time migration based on the reciprocal principle of Green's function. Many studies have been conducted on FWI (e.g., Alkalifa, 2016; Schuster, 2017; Tarantola, 1984, 1986; Tromp et al., 2005; Virieus & Opeto, 2009). We evaluated the FWI method for geothermal seismic studies using the adjoint method proposed by Tromp et al. (2005). The sensitivity kernels for compressibility, rigidity, and density were calculated by backpropagation in the adjoint method. The iteration made it possible to minimize the residual waveforms between the observed ones and the synthetic ones of the modified model. Using this adjoint FWI method, we evaluated the possibility of detecting a subsurface reservoir even with the configuration of a few sources and dense receiver arrays.

Because a downhole seismic source can effectively transmit seismic waves for the geothermal study, we assumed a seismic source at a depth of 2 km in order to design such a device. For a borehole seismic source located at a depth of 2 km (Figure 10.19) at three different offsets from a target anomaly, we examined whether the target of a supercritical reservoir 1 km long and 200 m thick at a depth of 4 km could be reconstructed using FWI. The physical properties of the target were assumed to have a slightly different *Vp*, *Vs*, and density of $\Delta Vp = -5\%$, $\Delta Vs = -5\%$, and $\Delta\rho = -2\%$ compared to the surrounding medium.

10.7.2. Results of FWI Simulation

The reconstruction results from the simulation assuming three different epicenter locations from the borehole showed that the assumed zones were retrieved almost satisfactorily (Figure 10.20; Kasahara et al., 2019a). The

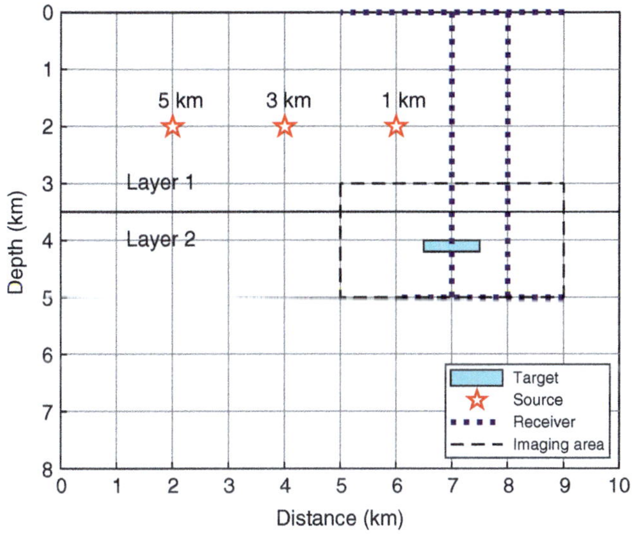

Layer 1	Vp: 3,500 m/s Vp/Vs: 1.7 ρ: 2,200 kg/m^3
Layer 2	Vp: 5,500 m/s Vp/Vs: 1.7 ρ: 2,500 kg/m^3
Target	ΔVp: –5% ΔVs: –5% Δρ: –2%

Figure 10.19 Two-dimensional numerical subsurface structural model for a rectangular reservoir filled with supercritical water at a depth of 4 km. The reservoir is assumed to be 1 km across and 0.2 km thick. As in the figure, two vertical and one horizontal boreholes equipped with seismic sensors are considered below the reservoir. Three buried seismic sources were selected at distances of 1, 3, and 5 km to the left vertical borehole. Surface seismic arrays were also deployed to test the detectability of changes in *Vp*, *Vs*, and density.

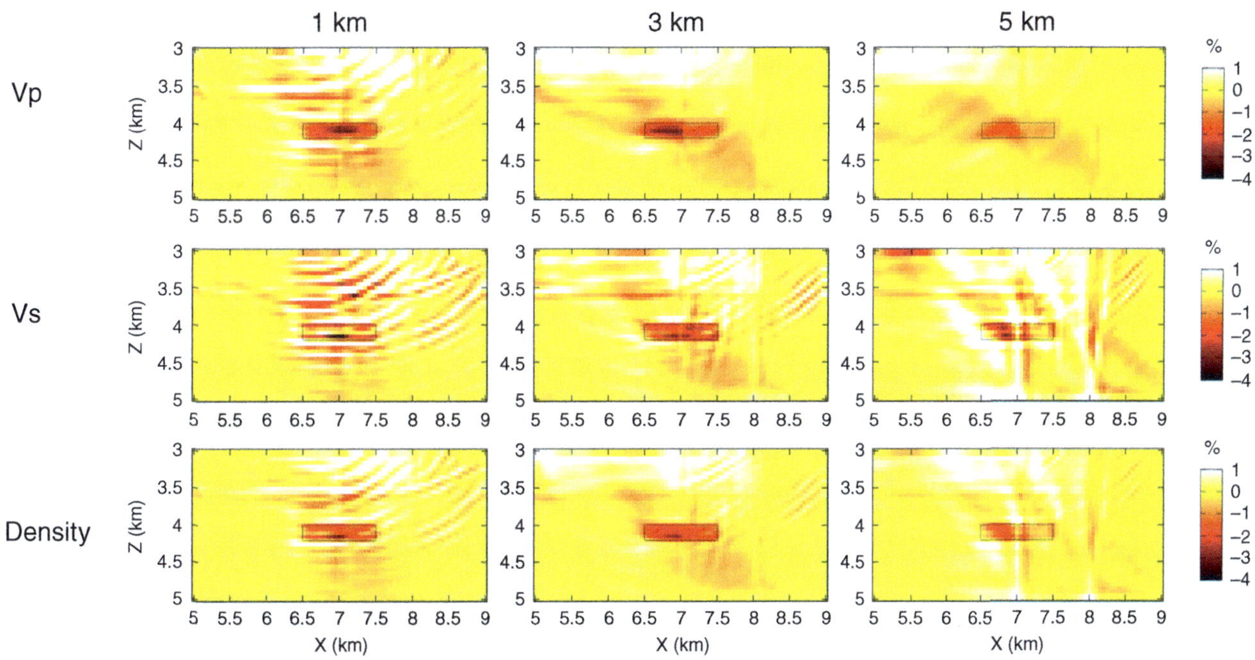

Figure 10.20 Subsurface anomalies in *Vp* (top), *Vs* (middle), and density (bottom), reconstructed from acquired seismic data for the seismic source locations with 1 (left), 3 (center), and 5 km (right) distance to the left vertical borehole shown in Figure 10.19. Retrieved values for each quantity are shown to the right of the diagrams.

results for the source location with an offset of 3 km showed the best reproducibility. It was clear from the simulations that the source location affects the results of the estimation of each physical property of the rectangular-shaped target. The results of several iterations showed that the *Vp*, *Vs*, and density values could be reproduced to show 4%, 4%, and 2% differences in the physical properties given to the model. In the case of the source at a

1 km offset, the *Vs* and density values were not obtained satisfactorily, whereas the *Vp* values were estimated reasonably well. Conversely, when the source was located 5 km from the target, the *Vs* and density values were obtained somewhat correctly, whereas the change in *Vp* was underestimated. Detailed analysis of estimating each physical property and its effectiveness in heterogeneous surroundings are future tasks.

10.8. INTEGRATION OF DAS SEISMIC RESULTS, DTS TEMPERATURE PROFILE, AND EXISTING DRILLING AND GEOLOGICAL AND GEOPHYSICAL DATA

In the Sumikawa geothermal field, we integrated the reflection intensity distribution, geothermal production, injection well profiles, and igneous intrusions (Kasahara et al., 2022c). The E–W section of the geological model of the Sumikawa geothermal field using well cuttings (Sakai et al., 1993) is shown in Figure 10.21. The N–S section was also given by the same authors. Two production wells reached granodiorite intrusion. The previous drilling of SA-1 in the Sumikawa field showed granodiorite at a depth of 1,725 m, and the southern well of N59-SN-5 reached granodiorite at a depth of 1,055–1,700 m (Ariki & Kato, 2010).

We conducted a field survey at the Sumikawa geothermal field in October 2021. An optical fiber system was installed in the SE-4 injection well to a depth of 1,200 m because part of the well deeper than 1,200 m was an uncased borehole. Nine seismic sources were positioned around the SE-4 well, and DAS records were obtained for nine seismic source excitations. We obtained the *Vp* depth distribution using the P first arrivals of the DAS and surface seismometers. The *Vp* distribution was compared to the geological model by Sakai et al. (1993) and geological descriptions of all wells (Mitsubishi Material Co., internal data), and we confirmed reasonable fitting among all available data.

Migrated images were generated using the upgoing arrivals. It was found that two high-reflection intensity regions were consistent with production and injection wells (Mitsubishi Material Co., internal data). The migration image between 1.8 and 2.2 km is shown in Figure 10.22. The production zone and injection zone are approximately 2 and 1.5 km deep, respectively. A 3D image of the reflection intensity, well trajectory, and granodiorite surface is shown in Figure 10.23. The reflections and production and injection distributions are nicely fitted.

The study results for the Takigami geothermal field (Kasahara et al., 2025) were also interpreted by integrating the existing data on the geology obtained by well

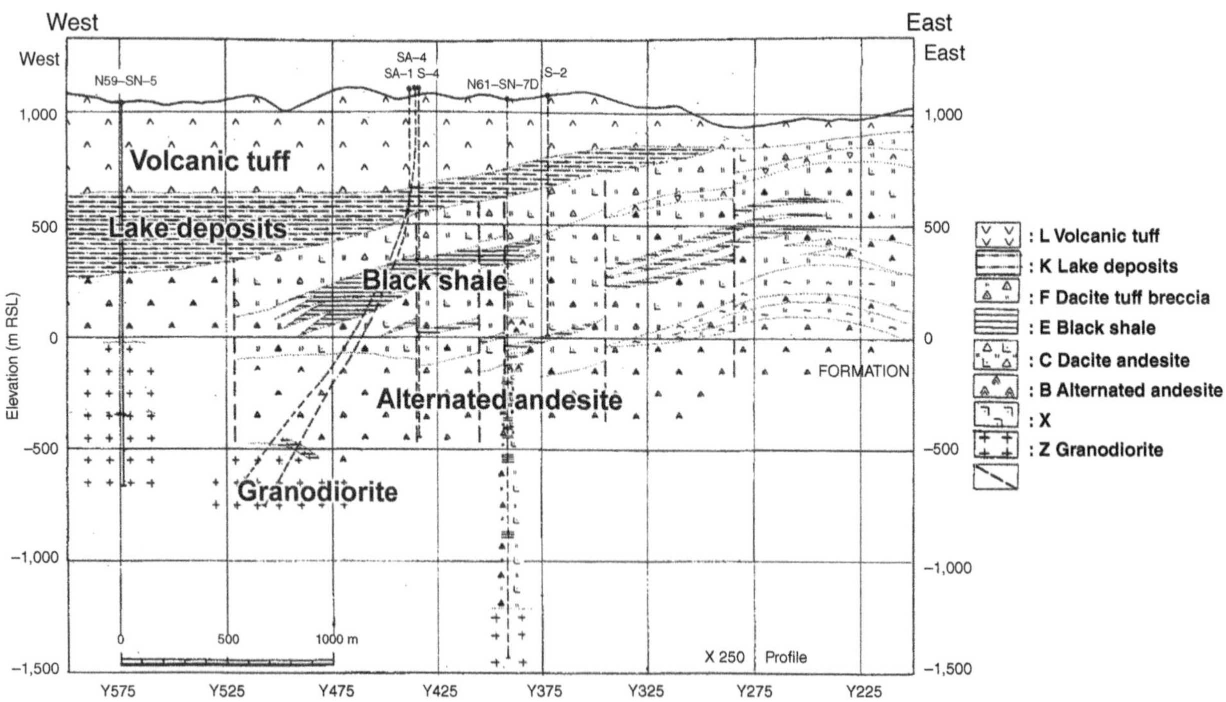

Figure 10.21 Geological model of the E–W cross-section in the Sumikawa geothermal field. Sakai et al. (1993) / with permission of the Society of Resource Geology.

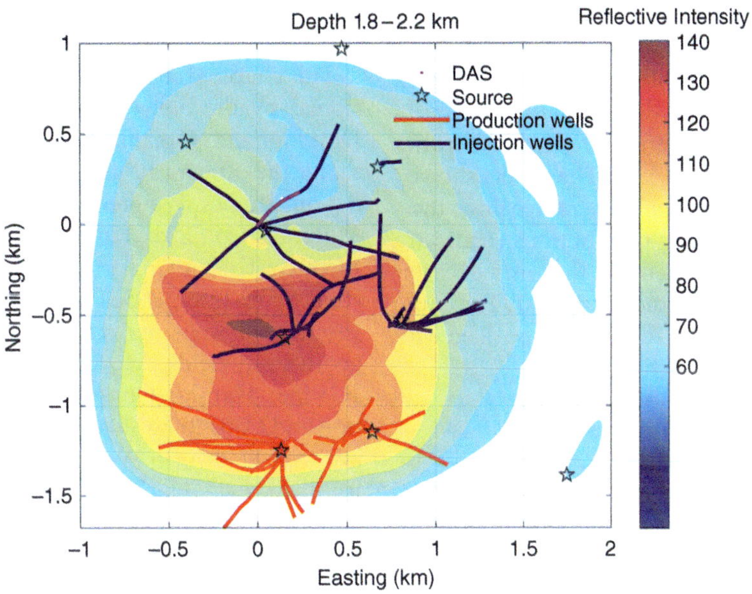

Figure 10.22 Results of migration image at a depth of 1.8 and 2.2 km. Blue lines are injection wells, and red lines are production wells. The stars are seismic source locations. Reflection intensity is shown on the right with arbitrary scales (red to blue).

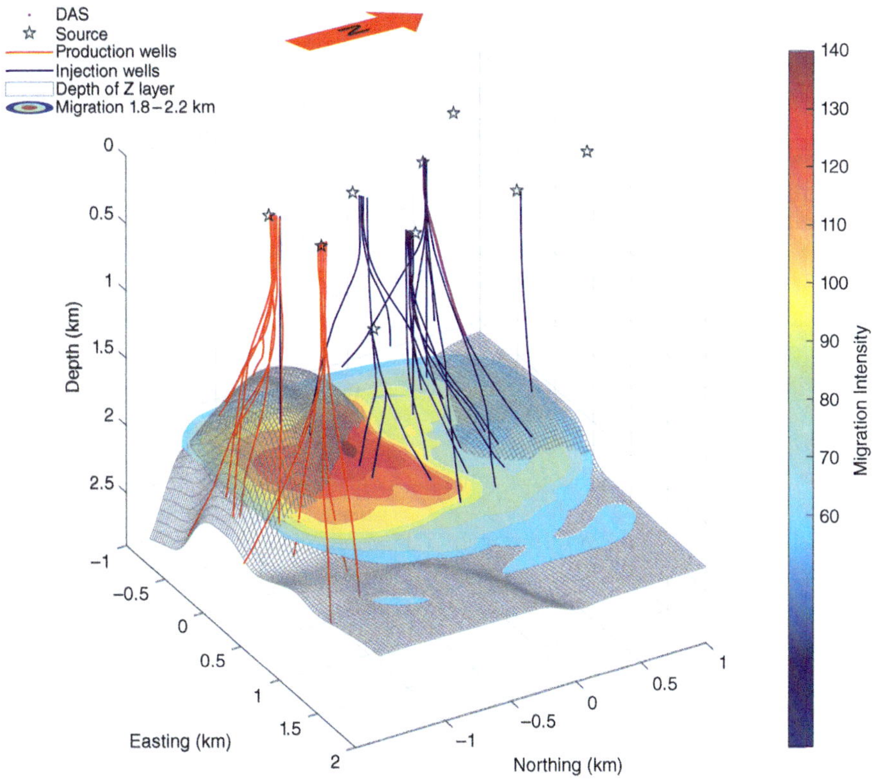

Figure 10.23 3D image of the Sumikawa study. Red lines are production wells, and blue lines are injection wells. The color surface is the reflection intensity distribution at a depth between 1.5 and 2 km. The gridding surface shows the top of the igneous intrusion. Adapted from Kasahara et al. (2022c).

drilling, gravity data, and past refraction studies. Details are described in Chapter 22 of this monograph. The results obtained at the Mori geothermal field (Kasahara et al., 2022a) were compared to the geothermal information collected by Hokkaido Electric Company, the owner of the Mori Geothermal Power Plant. The results showed satisfactory consistency in permeability, water flow, thermal field, and estimated density distributions (Kajiwara et al., 2022). It could be interpreted that the seismic scatter is related to geothermal fluid in fractured rocks. The cause of the high reflectivity could be a fractured zone filled with fluid in the production and injection zones.

10.9. DISCUSSION AND CONCLUSIONS

Geological and geothermal subsurface structures differ greatly from one place to another. The 3D geothermal reservoir distribution seems to be controlled by many factors, such as fracture distribution, magnitude of permeability and permeable rock distribution, faults, presence of seal layers, upwelling of hot fluid circulation, brittleness, thermal conductivity of rocks, and pressure. If the seismic reflectivity distribution is related to the distribution of vapor and fluid layers in fractured rocks, the seismic migration image will provide helpful information. The migrated image using PP reflections indicates reflectors beneath the geothermal area. Deeper reflectors suggest the possibility of future geothermal reservoirs.

Increasing the confidence level for understanding reservoir distribution, we made efforts to integrate the observed DAS records and existing data in the geothermal fields. The study in the Mori geothermal field is one of the best cases (Kasahara et al., 2022a). Sumikawa (Kasahara et al., 2022c) and Takigami (Kasahara et al., 2025) are other excellent cases.

We examined the usefulness of FWI and retrieved the shape and depth at a reasonable resolution, but the retrieval of *Vp*, *Vs*, and density depended on the source and receiver arrangement. However, field studies face limitations in real geothermal fields, making it difficult to achieve the ideal observation setting. The FWI approach could provide some clues for making observation plans considering the effect of the sources and receiver geometry.

Kern (1978), Kern and Richter (1981), and Kern et al. (1997) found that the temperature dependence of *Vp*, *Vs*, and Q of several types of rocks was not significantly large. Considering these laboratory experiments, it is considered that the changes in *Vp* or the amplitude with the depth could be interpreted as the presence of geological boundaries, fractures, or fluid rather than a temperature effect.

In conclusion, the seismic method in geothermal surveys is effective for obtaining reflector or scatterer distributions with the integration of other geological or geothermal information.

ACKNOWLEDGMENTS

This chapter is based on the results of a project commissioned by the New Energy and Industrial Technology Development Organization (NEDO JPNS21991). We thank the officers at NEDO for their active support during our studies. We greatly thank SNBL Ltd., Mitsubishi Materials Co., Ltd., Hokkaido Electric Power Co., Ltd., Idemitsu Kosan Co., Ltd., and Tohoku Sustainable and Renewable Energy Co., Ltd. for permission to use their geothermal wells and access their internal database on geothermal fields. We also thank WELMA Co., Ltd., JGI Inc., Hanshin Consulting Co., Ltd., Daiwa Exploration and Consulting Co., Ltd., and Oyo Co., Ltd. for helping with fieldwork. We thank two anonymous reviewers for their valuable comments. Their comments greatly improved the manuscript.

REFERENCES

AIST. (2009). Geothermal potential map in Japan GT-4 (CDROM).

Alkalifa, T. A. (2016). *Full waveform inversion in an anisotropic world* (Revised ed.). EAGE, 197pp

Ariki, K. & Kato, H. (2010). Reservoir management on the Sumikawa geothermal reservoir. *Chishitsu News, 665*, 0–27.

Dobson, P., Asanuma, H., Huenges, E., Poletto, F., Reisch, T., & Sanjuan, B. (2017). *Supercritical geothermal system – a review of past studies and ongoing research activities.* Paper presented at the 41st Workshop on Geothermal Reservoir Eng., Stanford Univ., Stanford, CA.

Friðleifsson, G. O., Albertsson, A., Stefansson, B., Gnnlangsson, E., & Adalsteinsson, H. (2007). *Deep unconventional geothermal resources: a major opportunity to harness new sources of sustainable energy. Paper presented at the* 20th World Energy Conference, Rome.

Friðleifsson, G. O., Elders, W. A., Zierenberg, R. A., Fánsson, A., Fowler, A. P. G., Weisenberger, T. B., et al. (2017). The Iceland Deep Drilling Project 4.5 km deep well, IDDP-2, in the seawater-recharged Reykjanes geothermal field in SW Iceland has successfully reached its supercritical target. *Scientific Drilling, 23*, 1–12. doi:10.5194/sd-23-1-2017

Handerek, V. A., Karimi, M., Nkansah, A., Yau, A., Westbrook, P. S., Feder, K. S., et al. (2018). *Improved optical power budget in distributed acoustic sensing using enhanced scattering optical fibre.* Paper presented at the 26th International Conference on Optical Fiber Sensors, Lausanne, Switzerland.

Hartog, A. H. (2017). *An introduction to distributed optical fiber sensors* (p. 440pp). CRC Press.

Hasada, Y., Kasahara, J., Kawashima, H., Yamauchi, Y., Sugimoto, Y., Yamaguchi, T., & Kubota, K. (2018). *Comparison of the records by optical fiber DAS (distributed acoustic sensor) and geophone using natural earthquakes.* Paper presented at the JPGU Annual Meeting.

Kajiwara, T., Kasahara, J., Moriya, Y., Nagaso, M., Maetou, K., & Ohnuma, H. (2022). *Comparison of DAS survey results with geological and geophysical models in the Mori geothermal field*. Paper presented at the Annual Meeting of the Geothermal Research Society of Japan.

Kasahara, J., Hasada, Y., & Kuzume, H. (2020b). *Possibility of high Vp/Vs zone in the geothermal field suggested by the P-to-S conversion*. Paper presented at the 44th Workshop on Geothermal Reservoir Engineering, Stanford University, Stanford, CA.

Kasahara, J., Hasada, Y., & Yamaguchi, T. (2019a). Seismic imaging of supercritical geothermal reservoir using full-waveform inversion method. Paper presented at the 44th Workshop on Geothermal Reservoir Engineering, Stanford University, Stanford, CA.

Kasahara, J., Hasada, Y., Furuya, S., Shimazaki, M., Mikada, H., Ohnuma, H., & Fujise, Y. (2025). Distributed acoustic sensing and distributed temperature sensing geophysical study in the Takigami geothermal field in northeastern Kyushu, Japan. In this monograph.

Kasahara, J., Hasada, Y., Kuzume, H., Fujise, Y., & Yamaguchi, T. (2019b). *Seismic feasibility study to identify supercritical geothermal reservoirs in a geothermal borehole using DTS and DAS. Paper presented at the* EAGE 2019 Annual Meeting, London.

Kasahara, J., Hasada, Y., Kuzume, H., Fujise, Y., Yamaguchi, T., & Mikada, H. (2019c). *Seismic time-lapse approach to monitor temporal changes in the supercritical water reservoir*. Paper presented at the 44th Workshop on Geothermal Reservoir Engineering, Stanford University, Stanford, CA, SGP-TR-24.

Kasahara, J., Hasada, Y., Kuzume, H., Mikada, H., & Fujise, Y. (2020a). *The second seismic study at the geothermal field in southern Kyushu, Japan using an optical fiber system and surface geophones*. Paper presented at the 44th Workshop on Geothermal Reservoir Engineering, Stanford University, Stanford, CA.

Kasahara, J., Hasada, Y., Kuzume, H., Mikada, H., & Fujise, Y. (2022b). The DAS geothermal study in the geothermal borehole of the Ohnuma geothermal power plant in Hachimantai, Honshu, Japan. *Journal of Geothermal Research Society of Japan*, *44*, 73–84.

Kasahara, J., Hasada, Y., Kuzume, H., Ohnuma, H., Mikada, H., & Fujise, Y. (2022a). A field experiment of temperature-tolerant distributed acoustic sensor in deep geothermal reservoir prospecting. *The Leading Edge*, *41*, 306–312.

Kasahara J., Y. Hasada, H. Ohnuma, H. Mikada, & Fujise, Y. (2022c). *Seismic study at the Sumikawa geothermal field in Akita Pref. Japan using a geothermal well*. Paper presented at the Society of Exploration l Geophysicists of Japan Spring Meeting.

Kasahara, J., Hasada, Y., Sugimoto, Y., Kawashima, H., Yamauchi, Y., Yamaguchi, T., & Kubota, K. (2018b). *Waveform comparison of DAS and geophone records using natural earthquakes*. Paper presented at EAGE2018.

Kasahara, J., Toriumi, M., & Kawamura, K. (Eds.) (2003). *The role of water in earthquake generation*. The University of Tokyo Press. ISBN4-13-06744-8, 392pp

Kasahara, J., Yamaguchi, T., Sugimoto, Y., Kawashima, H., Yamauchi, Y., Hasada, Y., & Kubota, K., (2018a). *Evaluation of the fiber-optic DAS as a seismic array sensor*, Paper presented at the SEGJ Spring Meeting.

Kern, H. (1978). The effect of high temperature and high confining pressure on compressional wave velocities I quartz-bearing and quartz-free igneous and metamorphic rocks. *Tectonophysics*, *44*, 185–203.

Kern, H., & Richter, A. (1981). Temperature derivatives of compressional and shear wave velocities in crust and mantle rocks a 6 kbar confining pressure. *Journal of Geophysics*, *49*, 47–56.

Kern, H., Liu, B., & Popp, T. (1997). Relationship between anisotropy of P and S wave velocities and anisotropy of attenuation in serpentine an amphibolite. *Journal of Geophysical Research: Solid Earth*, *102*(B2), 3051–3065.

Muraoka, H., Uchida, T., Sasada, M., Yagi, M., Akaku, K., Sasaki, M., et al. (1998). Deep geothermal resources survey program: igneous metamorphic and hydrothermal processes in a well encountering 500°C at 3729 m depth, Kakkonda, Japan. *Geothermics*, *27*, 507–534.

OFS, Co., Ltd. (2021). AcoustiSens optical fiber datasheet.

Reinsch, T., Dobson, P., Asanuma, H., Huenges, E., Poletto, F., & Sanjuan, B. (2017). Utilizing supercritical geothermal systems: a review of past ventures and ongoing research activities. *Geothermal Energy*, doi:10.1186/s40517-017-0075-y, *5*, 16.

Sakai, Y., Matsunaga, E., & Kubota, Y. (1993). Geothermal energy development in the Sumikawa field, Northeast Japan. *Resource Geology*, *43*(6), 409–425. (In Japanese)

Schuster, G. T. (2017). Seismic inversion. In *Investigations in geophysics*, No. 20. SEG.

Tarantola, A. (1984). Inversion of seismic reflection data in the acoustic approximation. *Geophysics*, *49*, 1259–1266.

Tarantola, A. (1986). A strategy for nonlinear inversion of seismic reflection data. *Geophysics*, *51*(10), 893–1903.

Tromp, J., Tape, C., & Liu, Q. (2005). Seismic tomography, adjoint methods, time reversal and banana-doughnut kernels. *Geophysical Journal International*, *160*(1), 195–216.

Tsuchiya, N., Yamada, R., & Uno, M. (2016). Supercritical geothermal reservoir revealed by a granite-porphyry system. *Geothermics*, *63*, 182–194.

Virieus, J., & Opeto, S. (2009). An overview of full-waveform inversion in exploration geophysics. *Geophysics*, *74*, WCC1–WCC26.

Wang, H. F., Zeng, X., Miller, D. E., Frantta, D., Fedighl, K. L., & Thurber, C. H. (2018). Ground motion response to an ML4.3 earthquake using co-located distributed acoustic sensing and seismometer array. *Geophysical Journal International*, *213*, 2020–2026. doi:10.1093/gji/ggy102

Winkler, K., & Nur, A. (1982). Seismic attenuation: effects of pore fluids and frictional-sliding. *Geophysics*, *47*, 1–15. doi:10.1190/1.1441276

11

Borehole and Surface Applications of Distributed Acoustic Sensing for Characterization of the Cryosphere and Glacial Environments

Alex M. Brisbourne[1], Thomas S. Hudson[2,3], and J. Michael Kendall[2]

ABSTRACT

Early borehole and surface distributed acoustic sensing (DAS) experiments have demonstrated the viability of the method in glacial settings and allowed new insights. The importance of good coupling and noise suppression are highlighted, as well as considerations for deployment and field methodology. The power of DAS lies in the high numbers of densely spaced channels that can be deployed with relatively low logistical cost. However, to exploit DAS fully, consideration must be made during experiment design for the method's inherent limitations, through enhanced source strength and frequency content, for example. Data from borehole deployments in Greenland and Antarctica have been used to derive englacial seismic velocity, attenuation, and reflectivity as well as subglacial structure. Due to the relatively simple velocity structure, glaciological settings provide useful environments for evaluating methodologies or analysis techniques, such as the application of artificial intelligence (AI) to analyze the huge data volumes. With fiber-optic cables already deployed in glaciological settings for temperature measurements, borehole DAS experiments are possible with a minimal additional logistical overhead. DAS will likely become an integral part of geophysical observations of the cryosphere, being of most value when used in conjunction with conventional methods such as surface geophone arrays or ice-penetrating radar.

11.1. INTRODUCTION

Despite the logistical challenges of working in such extreme environments, seismic methods have been used to investigate the glaciers and ice sheets since the late 1940s. These early experiments, partly driven by the geopolitics of the post-World War II era, addressed enormous gaps in knowledge of ice sheet thickness and revealed that the beds of both Greenland and Antarctica lay close to or below sea level. Robin (1958) presented results from the Norwegian-British-Swedish Antarctic Expedition of 1949–1951, and this was followed by a number of Antarctic experiments under the umbrella of the International Geophysical Year in 1957–1958 (Sullivan, 1957). Early large-scale refraction experiments revealed significant depth to the bed in West Antarctica, which would later lead to the "weak underbelly" moniker (Hughes, 1981) due to the implications for vulnerability to ice sheet collapse. Meanwhile, some of the first measurements of glacier ice and firn (the surface layer of snow transitioning to ice through burial) thickness using active seismic methods in the northern hemisphere were reported by Röthlisberger (1955) from the Penny Ice Cap, Baffin Island. It was not long before more advanced methods were employed. For example, Westphal (1965) measured attenuation in

[1]*British Antarctic Survey, Natural Environment Research Council, Cambridge, UK*
[2]*Department of Earth Sciences, University of Oxford, Oxford, UK*
[3]*Department of Earth Sciences, ETH Zurich, Zurich, Switzerland*

Distributed Acoustic Sensing in Borehole Geophysics, Geophysical Monograph 289, First Edition.
Edited by Yingping Li, Robert Mellors, and Ge Zhan.

DOI: 10.1002/9781394179275.ch11

temperate valley glacier ice using an explosive source and Robinson (1968) reported measurements of seismic anisotropy on the Ross Ice Shelf.

Active source seismic methods are now commonly applied to investigate a range of glaciological processes and properties: quantifying glacial and ice stream basal properties, processes, and materials with seismic reflectivity and amplitude variation with incidence angle (e.g., Anandakrishnan et al., 1998; Blankenship et al., 1986; Booth et al., 2015; Smith, 1997; Walter et al., 2013); measurement of englacial properties such as attenuation, which is closely related to englacial temperature (e.g., Peters et al., 2012); and quantifying strain and ice rheology by imaging ice crystal preferred orientation (termed "fabric") (e.g., Blankenship & Bentley, 1987; Brisbourne et al., 2019; Horgan et al., 2008, 2011). On ice shelves or over subglacial lakes, active seismic methods provide an unambiguous measure of bathymetry (e.g., Brisbourne et al., 2014; Johnson & Smith 1997; Nost, 2004; Rosier et al., 2018; Woodward et al., 2010) and can be used to infer lakebed or seabed properties (e.g., Brisbourne et al., 2023; Hofstede et al., 2021; Kristoffersen et al., 2014; Smith et al., 2018). Seismic refraction methods are also regularly used to investigate the near-surface firn layer (e.g., King & Jarvis, 2007; Kirchner & Bentley, 1990; Pearce et al., 2023).

Similarly, passive seismic methods can be used to monitor seismicity and also image the internal properties of ice. The first observations of "ice tremors" were made at Barter Island, Alaska, by Crary (1955). Since then, developments in instrumentation and logistics and concern for the ice sheet's potential contribution to global sea level have led to a surge in applications using naturally occurring seismic sources, ranging from the study of glacial earthquakes, stick–slip and basal motion, ice calving, ice rheology and fracture, tremor, and crevassing (see Table 1 of Podolskiy and Walter [2016] for a more exhaustive list of applications). With recent improvements in instrumentation, large icequake datasets recorded at the surface of fast-flowing glaciers and ice streams have been used to image properties such as ice fabric (e.g., Gajek et al., 2021; Kufner et al., 2023; Smith et al., 2017). In parallel with developments in conventional solid Earth seismic experiments, more recently developed methods such as noise interferometry are now also being applied in glaciology (e.g., Chaput et al., 2022; Zhou et al., 2022).

Recent developments in portable nodal seismometer technology are leading to unprecedented surface instrument density and large-N seismic arrays with hundreds or thousands of stations (e.g., Hansen & Schmandt, 2015; Lin et al., 2013). However, in remote cryosphere applications, logistical complexity still constrains instrument numbers and deployment duration. For example, results from a deployment of 98 three-component nodal stations on Argentière Glacier in the French Alps were reported by Gimbert et al. (2021). The infrastructure requirements for charging, configuring, and data harvesting of nodal systems may still limit array size or deployment duration in remote Antarctic experiments. Distributed acoustic sensing (DAS), using fiber-optic cable that serves as an array of sensors, therefore presents a potential step change in spatial coverage and resolution in glaciological experiments. The glaciological community is starting to exploit this with both surface and borehole experiments. Despite only recent deployment of DAS in glaciological settings, a handful of early studies have established the viability and potential benefits of the method in these difficult and inaccessible settings (e.g., Booth et al., 2020; Brisbourne et al., 2021; Fichtner et al., 2023a; Hudson et al., 2021; Walter et al., 2020). Not only is DAS providing enhanced insight into glaciological processes, but also the relatively simple velocity structure of the cryosphere in comparison to geological settings is proving useful for developing and verifying DAS data processing methods that can be adopted beyond glaciological applications. Here, we outline these findings, discuss the strengths and weaknesses of the DAS method, and make recommendations for the future use of DAS in glaciological settings.

11.2. BOREHOLE DRILLING METHODS

Borehole access to glaciers and ice streams remains challenging, although drilling is undertaken relatively frequently for both subglacial access and ice core extraction. Two methods are available. (a) Hot water drilling provides rapid access to the ice column and subglacial bed or underlying ocean cavity for instrument deployment or sampling (e.g., Anker et al., 2021; Makinson & Anker, 2014). In general, boreholes created using hot water drilling close rapidly due to either refreezing of water or a dynamic ice-flow regime and will likely result in good coupling between the ice and any fiber-optic cable deployed in the borehole. (b) Mechanical drilling is used to extract ice cores and again results in a suitable borehole (e.g., Mulvaney et al., 2021). Ice core drilling is logistically expensive and complex, and due to the time taken to recover the core to the surface, it is an inherently slow process. For example, the West Antarctic Ice Sheet (WAIS) Divide ice core took six summer seasons of drilling to reach 3,405 m depth. In general, deep ice core sites target regions of slow deformation of the ice column, and dynamic hole closure is therefore less rapid. Fluid used in the drilling processes keeps the holes open to aid drilling, core recovery, and instrument deployment. Upon completion, removal of drilling fluid then allows hole closure over a period of months or years.

Distributed temperature sensing (DTS) is already used in glaciology to derive the vertical temperature profile of the ice column following hot water drilling (e.g., Tyler et al., 2013) or ice core extraction (e.g., Mulvaney et al., 2021). As such, fiber-optic cables are already being deployed in cryosphere settings and may be used opportunistically for DAS (Brisbourne et al., 2021). In general, DTS measurements utilize multimode fiber-optic cables as the larger core diameter allows higher power sources to be used, thus increasing backscatter signal. However, multimode fiber-optic cables may result in lower-quality DAS data.

11.3. PREVIOUS BOREHOLE EXPERIMENTS WITH CONVENTIONAL INSTRUMENTATION

Previously, borehole seismic studies in ice have relied on conventional downhole geophones and hydrophones combined with surface sources and sensors. Diez et al. (2015) measured velocity anisotropy at an ice core site in East Antarctica with surface sources using a single-borehole geophone raised at 40 m intervals from a depth of 2,580 to 100 m. These data were combined with surface-deployed three-component geophones and a streamer. Gusmeroli et al. (2013) measured P-wave velocities in ice using four orthogonally oriented walkaway VSPs collected on Storglaciären, a polythermal glacier in northern Sweden. They used a hammer and plate source at the surface with a 12-channel (1 m spacing) hydrophone string down an 80 m borehole made with a hot water drill. In this case, the borehole was water filled. Lutz et al. (2020) also used hot water drilling to create a borehole on the Ross Ice Shelf in West Antarctica. Downhole sensors consisted of eight three-component 15 Hz seismometers frozen into the borehole from 80 to 325 m depth at 35 m intervals. A hammer source generated P and S waves, and anisotropic crystallographic preferred orientations were constrained using shear-wave splitting measurements.

11.4. BOREHOLE APPLICATIONS OF DAS

Booth et al. (2020) conducted the first downhole DAS experiment in a glaciological setting. They created a 1,043 m deep borehole at Qarassap Sermia (Store Glacier) in Greenland using a hot water drill. A reinforced cable containing two single-mode and four multimode fibers was deployed down the borehole for both DAS and DTS. Due to the ice dynamics and drilling method, the cable rapidly froze in to achieve excellent coupling. VSPs were then recorded using a hammer and plate source with walkaway sources over a range of azimuths (Figure 11.1 (a)). Both englacial and subglacial seismic properties were measured, including englacial seismic velocity, Poisson's ratio, and attenuation. An increased P-wave velocity at depth was interpreted as an indication of anisotropic ice, consistent with enhanced deformation toward the base of the ice column. Englacial attenuation values were linked to the temperature profile of the ice column, with $Q = 311 \pm 8$ measured over the majority of the ice column, consistent with previous conventional measurements made elsewhere, and $Q = 15 \pm 2$ in the lowermost 100 m, indicating warmer temperate ice toward the bed. A 20 m thick subglacial sediment layer was identified using subglacial reflections recorded along the downhole fiber. The success of this experiment was in large part attributable to the rapid freeze-in of the cable that resulted in excellent coupling to the ice.

Brisbourne et al. (2021) conducted a similar experiment at Skytrain Ice Rise in West Antarctica. There, a 600 m deep borehole was drilled previously to recover an ice core and remained open due to the presence of drilling fluid in the hole. A multimode fiber was deployed previously for the purposes of DTS. This same cable was subsequently used for a VSP DAS experiment. This deployment is markedly different from that of Booth et al. (2020) as good coupling to the borehole was achieved in the lower half of the borehole only, where drilling fluid was present. Surface noise disrupted seismic signals in the free-hanging shallower half of the cable (Figure 11.1(b)). Coupling in the shallower half of the cable was poor due to the cable hanging in air and not being directly coupled to the ice. Despite this, vertical P-wave velocities consistent with anisotropic ice were measured. A low Q value of 75 ± 12 (high attenuation) was measured, which is inconsistent with previous observations in ice or the observed temperature profile (Mulvaney et al., 2021) and likely reflects body wave estimates of Q being a combination of scattering and attenuation. Brisbourne et al. (2021) also observed P-wave energy reflected from the ice base as well as the surface multiple. Intriguingly, even though the fiber was suspended in fluid rather than frozen into the borehole, coupling was sufficient to measure both incident P-wave and converted S-wave energy. Glaciological experiments often require boreholes to remain open, and so this study evidences the value of DAS borehole experiments even if the fiber cannot be frozen in. In contrast to the results of Mad Zahir et al. (2023) in an industrial setting, Brisbourne et al. (2021) and Fichtner et al. (2023a) demonstrate that deploying fiber in fluid-filled boreholes in other environments (e.g., geothermal or water wells) can be useful.

Fichtner et al. (2023a) successfully deployed 1,500 m of fiber-optic cable at a deep ice core site in the onset region of the Northeast Greenland Ice Stream. In this borehole, drilling fluid was present at depths below ~70 m, and due to its density and viscosity, it presented challenges for

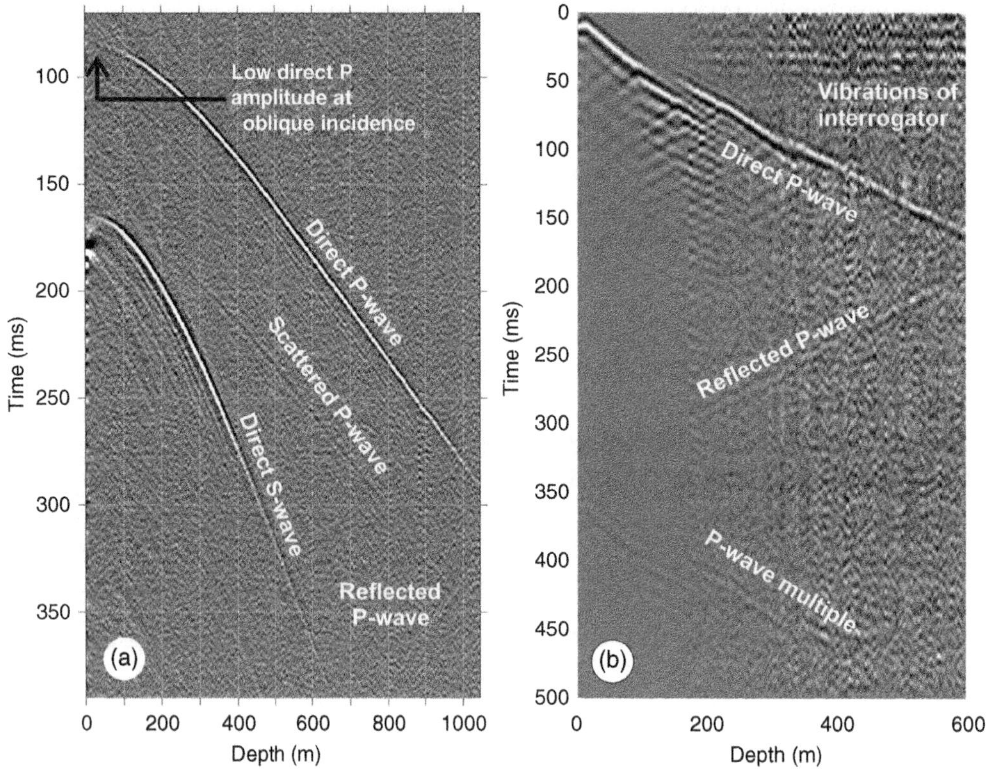

Figure 11.1 Example borehole DAS data recorded in glaciological settings. (a) 300 m offset VSP from Store Glacier, Greenland, with arrivals identified. Here the cable was frozen into the borehole, providing good coupling and low noise. Booth et al. (2020) / John Wiley & Sons / CC BY 4.0. (b) Borehole DAS data from Skytrain Ice Rise, Antarctica, with arrivals identified. Here the cable hung in the open borehole with fluid in the lower half. Brisbourne et al. (2021) / European Geosciences Union GmbH / CC BY 4.0. Vibrations of the power generator are observed at the interrogator unit and manifest as coincident arrivals across all DAS channels.

both deployment and retrieval of the cable. A series of active surface sources produced clear P- and S-wave signals, including internal reflections, along the length of the fiber. P- and S-wave velocities were derived with a depth-dependent vertical resolution of 20–70 m and uncertainties of the order 10 m/s. Variation in englacial reflectivity is consistent with a climate-induced grain size variation, and variation in seismic velocities hints at anisotropy. They determine that experiment resolution is not limited by the 2 m channel spacing but by the ~200 Hz upper frequency limit of the signal, poor coupling, and errors resulting from the use of the ray approximation in modeling. Instrumental rather than ambient noise controls the lower limit in amplitude of detectability of the signals.

These early experiments highlight the potential of the DAS method to obtain borehole seismic observations at unprecedented channel spacing and with a significantly lower logistical burden compared to conventional instrument deployments, especially where existing infrastructure can be utilized. However, Fichtner et al. (2023a) highlight that to achieve the full potential of DAS, additional logistical commitment may be necessary. Likewise, the study of Brisbourne et al. (2021) clearly demonstrates the potential shortcomings of utilizing pre-existing downhole fiber where coupling or noise issues may prohibit useful data recovery: Brisbourne et al. (2021) had access to multimode fiber only, which is less well suited to DAS than single-mode fiber.

11.5. SURFACE APPLICATIONS OF DAS

Castongia et al. (2017) reported on an early experiment testing DAS response on the frozen surface ice of Lake Mendota, Madison, Wisconsin, USA. Array geometry and coupling proved critical to data quality, and a tight-buffered cable more consistent for P-wave arrival picking than a loose-tube construction (typically used for DTS). Again, cable orientation with respect to the particle motion was highlighted as critical, and the effect of gauge length observed.

The first reported experiment with DAS on a glacier surface was by Walter et al. (2020), who deployed a triangular fiber array on Rhonegletscher, Switzerland. Measuring microseismic signals and ambient noise, they monitored glacier stick–slip, rockfalls, and crevassing. Identification of multiple reflected and refracted phases that would otherwise be missed with a conventional network clearly demonstrated the benefit of DAS's inherent multichannel acquisition (Figure 11.2(a)). They argue that the advantages of DAS outweigh the relatively low signal-to-noise ratio (SNR) of individual channels and recommend deployment prior to winter snowfall in alpine settings to improve coupling of the cable to the glacier ice to address this issue. As DAS interrogator technology develops, the SNR will inevitably become a more minor limitation in such studies.

The first surface DAS experiment in Antarctica (Hudson et al., 2021) evaluated a triangular and linear DAS fiber geometry against standard seismic network methods (Figure 11.2(b), (c)). A number of previous studies of microseismicity using geophones highlight Rutford Ice Stream as an excellent site for method evaluation (e.g., Kufner et al., 2021; Smith, 2006; Smith et al., 2015). Use of the triangular array removes the ambiguity in event location (Figure 11.2(b)). DAS outperformed a standard geophone network for source spectra analysis and full-waveform source-mechanism inversion. In addition, measurements of shear-wave splitting were made with the triangular array. Through stacking, the data were of sufficient quality to determine the slowness of the fast and the slow shear wave (Figure 11.2(c)), thereby enabling an assessment of the splitting parameters. The authors recommend a hybrid methodology in future experiments to benefit from the strengths of both DAS and conventional arrays. Butcher et al. (2021) used radon stacking methods to enhance detection of the microseismic events recorded by the linear array. Zhou et al. (2022) also used data from the same experiment to derive the S-wave velocity structure of the firn using ambient noise interferometry. Again, using the hybrid network, DAS interferograms were improved by cross-correlation with a collocated geophone, which removed coherent noise on the interrogator (Figure 11.1(b)), as well as selective stacking, where only the best cross-correlograms are used. Fichtner et al. (2023b) identified 15 wave propagation modes in 2,500 m of fiber deployed at ~50 cm depth orthogonal to an aircraft skiway in Greenland. Using an unconventional but high-amplitude source, a C-130 Hercules aircraft touching down, they derived an S-wave structure in the firn with meter-scale resolution and low uncertainties, and using an empirical relationship, derived a density profile consistent with the nearby EastGRIP ice core site.

Klaasen et al. (2021) demonstrated the logistical feasibility of DAS in a challenging volcano-glacial environment at around 2,000 m altitude in British Columbia, Canada. With 3 km of fiber spanning both a volcanic ridge and a glacier surface, a broad range of low-magnitude local seismicity including tremor and high-frequency earthquakes (5–45 Hz), associated with fluid movement within the geothermal reservoir of Mount Meager, was identified. Similarly, Fichtner et al. (2022) deployed 12.5 km of fiber across a subglacial lake within an active volcanic caldera at Grimsvötn, Iceland. Here, arguably, the most exciting finding was observing the resonance of the ice layer above the lake, which was only possible due to the low-frequency response and spatial sampling provided by DAS. In each case, the authors highlight the logistical feasibility of DAS in challenging environments.

Seismic methods have recently been demonstrated as viable for the monitoring of sea ice and the derivation of ice thickness and elastic properties (e.g., Moreau et al., 2020). Nziengui-Bâ et al. (2023) extended this capability with DAS, using a fiber cable deployed across the surface of a frozen lake in the French Alps. Ice thickness and Young's modulus values consistent with other field measurements were obtained. Early deployment that resulted in freeze-in of the cable produced significantly better coupling to the ice and an improved SNR, highlighting coupling as a critical consideration when designing experiments.

Paitz et al. (2023) use existing fiber-optic infrastructure to characterize ground deformation induced by avalanches at a test site in Switzerland. Strain rate signals registered up to 30 s prior to the arrival of the avalanche mass highlight the potential for early warning applications, and group and phase velocity analysis points to complexity in the internal structure of the avalanche.

11.6. OPTIMIZING THE APPLICATION OF DAS IN GLACIOLOGICAL SETTINGS

As well as DAS contributing to our understanding of cryosphere processes, deployments on glaciers are contributing to our understanding of DAS performance and testing new DAS data processing methods. In general, glacial ice presents a relatively homogeneous and well-constrained environment for evaluating seismic applications. The use of DAS in these environments therefore allows the evaluation of analytical methods without either the complexity of subsurface geology or the presence of manmade infrastructure dictating array configurations. Combined experiments with active source seismics in settings with naturally occurring seismicity have allowed new analytical methods to be developed and evaluated.

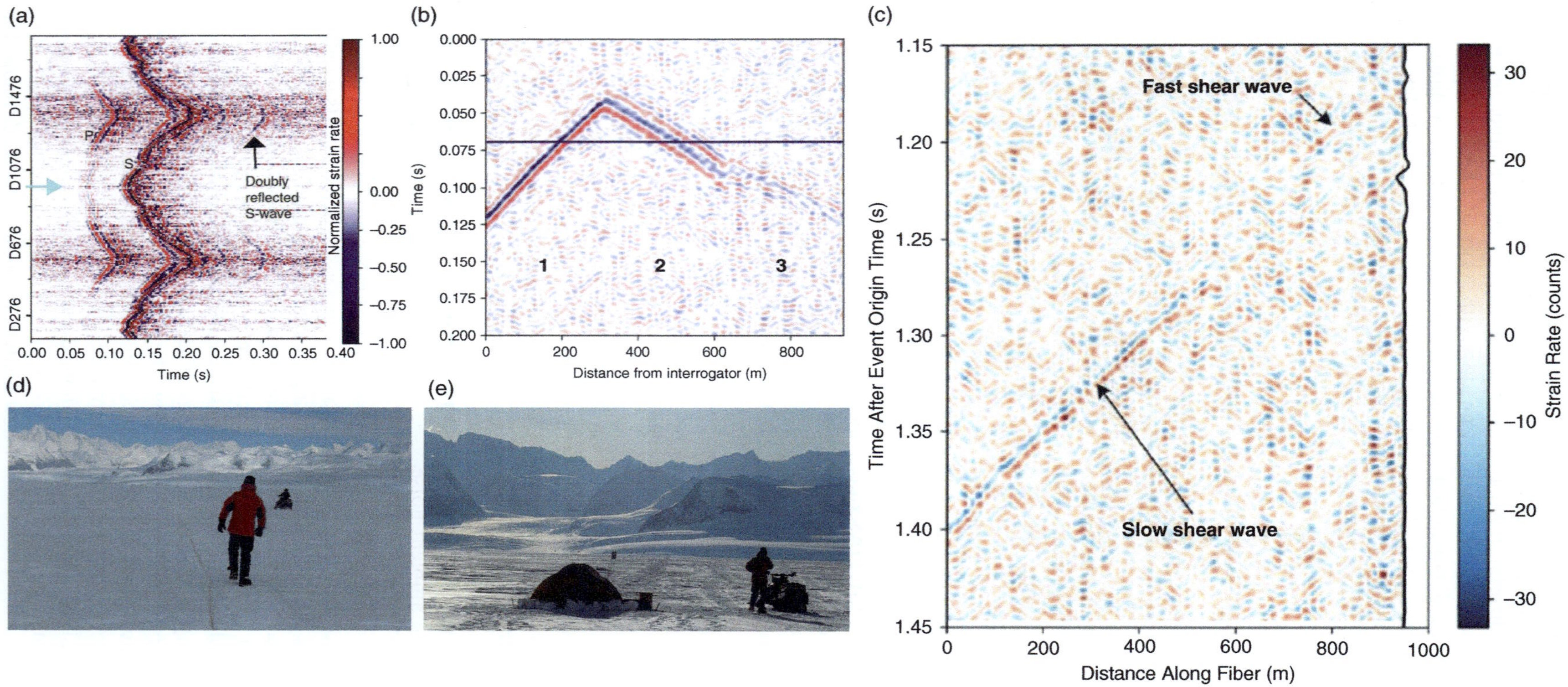

Figure 11.2 Surface DAS experiment examples from the Swiss Alps and Antarctica. (a) Full DAS record of a normalized stick–slip seismogram recorded at the surface of Rhonegletscher, Switzerland. The DAS laser is looped back at the cable end near Channel D988 along a second fiber core to provide two measurements at each location (cyan arrow). Higher channels sample similar cable locations to the ones below Channel D988, causing a symmetrical appearance. Walter et al. (2020) / Springer Nature / CC BY 4.0. (b) An example recording of an icequake from the bed of Rutford Ice Stream, Antarctica, recorded on a triangular array showing the S-wave arrivals. Clear linear moveout can be observed on the three linear segments of the array. (c) An S-wave arrival across the linear DAS fiber array showing shear-wave splitting. The north component of a geophone collocated at one end of the fiber is shown by the trace at ~950 m. Hudson et al. (2021) / John Wiley & Sons / CC BY 4.0. (d) In Antarctica, the fiber-optic cable is buried in snow at the surface to reduce noise and improve coupling. (e) The interrogator unit is deployed in a tent.

For example, Hudson et al. (2021) demonstrated the use of DAS to measure shear-wave splitting with a 2D surface-deployed cable configuration. In this case, a triangular fiber configuration is emulating a multicomponent sensor. They also used icequakes to test the performance of existing full-waveform earthquake source-mechanism inversion algorithms. The simple velocity structure of ice allowed them to verify that source inversion using DAS is possible, emphasizing the value of how dense spatial sampling of a small portion of the focal sphere can still yield well-constrained source mechanisms. Transferring this approach to more complex velocity structures is challenging, but glacier DAS deployments show that such inversions are feasible.

Lapins et al. (2023) used data from the same experiment to develop a method for suppressing strong random noise in DAS data. A weakly supervised machine learning method was used to suppress noise on the cable by using simultaneous recordings on a single strand of cable that extends to the end of the cable and then back to the start. They exploited a spliced fiber-optic cable, which allows two measurements across the extent of the cable at each location using the same interrogator. This denoising method is more efficient and orders of magnitude faster than more conventional filtering.

Paitz et al. (2020) use data from a DAS deployment at an alpine glacier to quantify the fiber response at high frequencies (up to 60 Hz). Quantifying DAS fiber frequency-dependent response and coupling is still an ongoing area of research, but it is needed for calculating the effective transfer function of the DAS recordings, something that is needed in calculating icequake magnitudes, for example. Another area of ongoing research in the DAS community is earthquake source inversion. Future glacier deployments will provide a fruitful environment for contributing insights in this research area.

These early studies in glaciological settings have highlighted both the potential of the method as well as the limitations and deployment requirements. The initial experiments (Hudson et al., 2021; Walter et al., 2020) used cables no greater than 1 km in length and so did not fully exploit one of the major advantages of DAS: sampling thousands of channels with a single interrogator unit. Subsequent experiments, with over 10 km of fiber, are proving successful (Fichtner et al., 2022). Walter et al. (2020) state that the major benefit of DAS lies in the spatial density of measurements available for relatively low logistical cost, and for them, this outweighs the relatively low signal-to-noise performance in surface deployments, especially when coupling can be maximized. Glaciological settings with surface snow cover, as opposed to exposed ice at the surface, lend themselves well to good coupling between the fiber and the surface. Burial of the fiber by snowfall or direct burial at deployment, for example, enhances coupling and reduces potential environmental noise such as wind or solar loading. To maximize the benefits of DAS, Fichtner and Hofstede (2022) introduced a basic algorithm for optimal experimental design, accounting for the cable length and locations of desirable targets and obstacles.

Likewise, both coupling and noise reduction are critical in borehole fiber deployments. In borehole settings, the best coupling is achieved by allowing sufficient time for the cable to freeze into the hole or by backfill. This has the additional benefit of reducing noise on the fiber, as has been an issue in situations where the cable hangs free. However, Fichtner et al. (2023a) demonstrate that adequate coupling and low noise acquisition are achievable in deep fluid-filled boreholes. Any methods that exploit amplitude or waveforms, and not just traveltimes, are reliant on consistent coupling along the length of the fiber.

Standard fiber-optic cables allow sampling of only one longitudinal component of strain rate. For borehole deployments, P-wave energy is sampled preferentially at near offsets in VSP experiments, for example. S-wave energy will be observed only when surface sources are offset from the borehole and particle motion becomes oblique to the fiber. For surface deployments of fiber addressing basal seismicity, for example, the steep velocity gradient in the near-surface firn turns seismic arrivals from depth toward vertical incidence and therefore precludes the measurement of P-wave energy for events directly beneath the array. An additional limitation of DAS results from the gauge length or the finite length of fiber over which the strain rate is measured. However, in general, this becomes an issue only where very fine-scale spatial sampling is required, such as for seismic refraction experiments, which are used to study the near-surface firn layer. The steep velocity gradient in the firn requires spatial sampling that is below the spatial resolution of the current generation of DAS interrogators. Nziengui-Bâ et al. (2023) highlight that a shorter gauge length results in a higher frequency content in comparable experiments. Fichtner et al. (2023a) highlight that rather than DAS channel spacing, the bandwidth of the waveforms and source amplitude, as well as the use of the ray approximation in modeling, are the primary factors currently limiting resolution.

Combining both conventional sensors and fiber-optic cables, hybrid seismic arrays exploit the potential of DAS while mitigating the single-component measurement limitations. Hybrid arrays incorporating DAS increase spatial coverage and spatial sample density, for example, while maintaining the benefits of conventional three-component measurements. In the case of seismic interferometry, independent measurements with nodal geophones remove correlated noise such as that inherent when using a single DAS interrogator for all fiber channels.

11.7. PRACTICAL CONSIDERATIONS

There are a number of considerations when acquiring DAS data in glacial environments. These include factors related to field deployment logistics, array geometry, power supply, data handling, and data processing.

In terms of field deployment, fiber-optic cable can be cumbersome to deploy, especially in remote environments. Fiber is normally shipped on cable drums, and depending on the nature of the cable armory, it can be challenging to unroll from and roll back onto the drum. Winch systems or mechanical winding mechanisms can be helpful.

Cable weight is variable, with bare cable weighing a few kg/km and heavily armored cable weighing in excess of 50 kg/km. The degree of armoring needed depends on the nature of the deployment. For Hudson et al. (2021), a reel with 1 km of fiber weighed around 50 kg, whereas for Klaasen et al. (2022), a reel of a similar weight held 4 km of cable. The fluid-filled borehole deployment reported by Fichtner et al. (2023a) highlights the requirement of weight on the fiber to ease deployment as well as the potential drag on the cable that may result from the high viscosity drilling fluid or friction at the wall of a nonvertical borehole.

Where a field team has only a single DAS interrogator, the cable must be deployed in a continuous array. Borehole data are generally much quieter than surface data. However, even shallow trenching of the fiber will improve the SNR in surface deployments (Figure 11.2(d)). Although repeat measurements with deployed fiber may be useful, the environmental impact of the permanent installation or nonrecovery of fiber must also be considered and may require a permit from the relevant authorities. In general, single-mode fiber provides better seismic data than multimode fiber. Cables are available with both single-mode and multimode fibers in a single bundle, allowing DAS (seismic), DTS (temperature), and distributed strain sensing (DSS; slow strain) recordings in any given setting.

Array design is particularly important (Williams et al., 2017). A linear array of fiber has an inherent 180° ambiguity in determining the direction of arrival of seismic energy. Ideally, an array will be large enough to cover a sufficient area of the subsurface and have a nonlinear geometry. For example, fiber in a borehole that extends in two directions across the surface of the ice would remove this ambiguity. Fiber in multiple boreholes, with interconnection at the surface to a single interrogator, would be ideal for imaging variations in depth of the ice column (Figure 11.3). Finally, due to the dense sampling of the seismic wavefield, DAS arrays can be useful in locating distal seismic signals using array stacking methods (e.g., Butcher et al., 2021; Hudson et al., 2023).

Another consideration is the installation of the DAS interrogator. Field setups require a reliable power source and usually a stable GPS signal. Interrogators that are not ruggedized or field-ready, or temperature insensitive, will also need protection from the elements (Figure 11.2(e)). Low-power, self-contained, and ruggedized interrogators with no requirement of peripheral devices are ideal. Where possible, the interrogator should be isolated from external sources of vibration and recording of the self-noise made. For the Rutford Ice Stream experiment (Hudson et al., 2021), power was provided using a small generator, where the petrol/gasoline generator needed careful attention to ensure a continual fuel supply. Furthermore, a power generator is a noise source, so ideally, the power cable from the generator to the interrogator needs to be hundreds of meters in length.

A further consideration is data format and storage. Where DAS data are not recorded in a commonly used seismic format, prior to analysis, data may require conversion or interfaces in common analysis tools for native format developed. Assuming 1 km of cable length, 1 m spatial sampling, and 8,000 Hz sampling, the data volume for 30 days can be over 40 TB. This requires large field-worthy data media and regular backups. The data recorded may be strain rate or strain and confined to recording in a direction parallel to the direction of the fiber. Strain rate is averaged over a fixed length known as the gauge length (normally somewhere between 1 and 10 m). This is not an issue with distant seismic signals but can be an issue in near-surface surveying and can lead to terracing effects in time in the recorded data. Furthermore, native strain measurements will introduce spatial integration noise that cannot be removed, whereas native strain rate measurements allow the user to avoid this issue where possible.

11.8. DISCUSSION AND CONCLUSIONS

The logistical benefits of a single fiber-optic cable and interrogator providing thousands of sensors at unprecedented density have huge potential. For example, instead of focusing on localized features, alpine studies can readily instrument the entire glacier to learn more about spatial variability in rate-determining processes. Potentially the most significant advance with DAS is in borehole deployments. Ice core extraction and hot water drill access for subglacial investigation and instrument deployment are well established and relatively common. The deployment of fiber is becoming routine for englacial temperature measurement, and therefore borehole

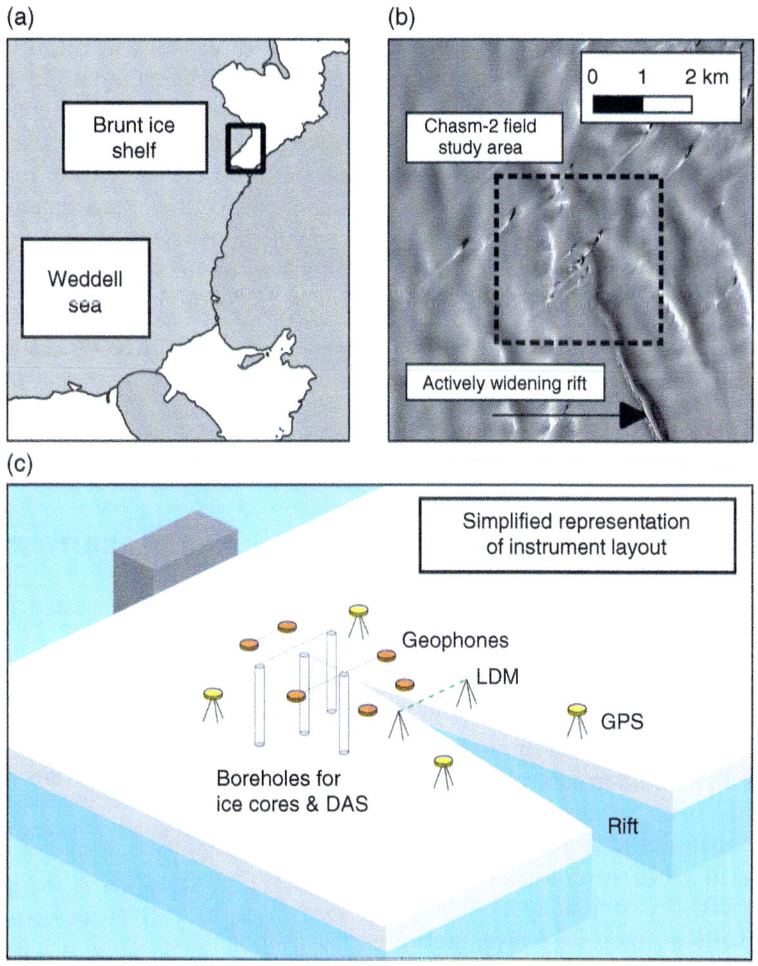

Figure 11.3 An example DAS deployment using multiple boreholes to investigate fracture processes at an ice shelf rift. (a) Location of the Brunt Ice Shelf in Antarctica (76°S; 025°W). (b) Landsat-8 imagery of the Brunt Ice Shelf with an actively widening rift labeled. (c) The experiment includes multiple boreholes with fiber deployed through the ice column, recording naturally occurring ice fractures using a single interrogator unit. DAS data will be supplemented with a three-component surface geophone array and interpreted using additional information from surface GPS measurements, ice-penetrating radar profiles, and strain monitoring at the surface with laser distance meters. Figure provided by Oliver Marsh, British Antarctic Survey.

DAS experiments are feasible with very little additional logistical overhead.

Early DAS experiments in glaciological settings have demonstrated the potential of this rapidly developing technology and allowed new insight. In borehole settings, the method enables observations that would be logistically prohibitive with conventional instrumentation. The technique has already led to new insight: the borehole experiments of Booth et al. (2020), Brisbourne et al. (2021), and Fichtner et al. (2023a) all infer the presence of seismic anisotropy in deeper ice; Walter et al. (2020) identified indirect seismic phases from basal stick–slip events that would otherwise have been missed; Fichtner et al. (2022) detected low-amplitude tremor on ice floating on a subglacial lake with application for subglacial volcanic monitoring; and Fichtner et al. (2023a) observed seismic reflectivity variation deep in the ice column that results from the climatic imprint on grain size.

For both surface and borehole deployments, coupling and noise suppression are critical. Efficient deployment methods are required to optimize data quality through measures such as predeployment of surface fiber before winter snowfall to ensure burial, advance deployment of downhole fiber to allow freeze-in, or reliable borehole-backfill or surface-burial methods for immediate utilization. Once good coupling is achieved, deployed fiber-optic cables may be revisited for both monitoring purposes and evaluation of temporal change in ice dynamics or physical properties, for example. Gauge length is currently a limitation for any method that requires submeter station spacing, such as seismic refraction for investigating near-surface firn properties with a steep vertical velocity gradient. To exploit the unprecedented channel spacing fully, source amplitude and frequency content must be considered.

Hybrid DAS deployments will likely become the standard, exploiting the three-component capability of conventional geophone sensors to aid interpretation of DAS data or reduce noise at the analysis stage. Helically wound fiber to measure strain in three dimensions may go some way to address this requirement. For event location, a sufficiently large aperture is required to constrain icequake hypocenters, but two-dimensional surface deployments combined with borehole fiber will improve depth constraints. A similar combination is required for studying englacial and subglacial properties, with englacial velocity anisotropy and attenuation potentially useful parameters for understanding present-day ice dynamics and establishing ice sheet history.

Initial experiments have in general evaluated and exploited existing conventional seismic processing and analysis methods. However, a feature of even moderate-scale DAS experiments is the size of the data volume produced, most significantly with passive monitoring. A likely fruitful route for analysis therefore lies with artificial intelligence. Booth et al. (2023) demonstrate a ~300-fold improvement in efficiency by combining a convolutional neural network with a frequency-wavenumber transformation. And as already mentioned, Lapins et al. (2023) developed a method for suppressing strong random noise in surface DAS data.

Looking forward, it is clear that DAS will play an integral role in seismic monitoring in glacial settings. Ideally, these rich datasets that sample the seismic wavefield in great detail, but often with poorer SNR than conventional seismic recordings, will complement and enhance other geophysical recordings such as ground-penetrating radar and GPS. An example of such an experiment to be conducted on the Brunt Ice Shelf in Antarctica is presented in Figure 11.3. Fracture processes at the tip of an active rift will be investigated. Ice cores will be extracted from four sites around the rift tip to allow laboratory measurement of crystal preferred orientation and fracture toughness. Fiber-optic cables will be deployed down the boreholes and backfilled with water to improve coupling. Downhole fiber will be linked across the surface in such a way that all four boreholes may be monitored concurrently with a single DAS interrogator. The DAS array will be supplemented by a surface deployment of conventional geophones and nodes, as well as dual-frequency GPS units and strain measurement with laser distance meters. Borehole DAS data will help constrain fracture timing, location, and mechanism to a degree that would otherwise not be possible. This will lead to a significant improvement in our understanding of fracture processes in ice and help constrain the future contributions of the ice sheet to sea level rise. It would be extremely difficult logistically and prohibitively expensive to conduct the downhole component of this experiment without DAS, which clearly demonstrates the transformative potential of this technology to glaciology.

ACKNOWLEDGMENTS

We are grateful to Emma Pearce, Coen Hofstede, and Andreas Fichtner for their insightful reviews that helped improve this manuscript.

AVAILABILITY STATEMENT

No new data were generated, analyzed, or presented in this manuscript.

REFERENCES

Anandakrishnan, S., Blankenship, D. D., Alley, R. B., & Stoffa, P. L. (1998). Influence of subglacial geology on the position of a West Antarctic ice stream from seismic observations. *Nature*, *394*, 62–65.

Anker, P. G. D., Makinson, K., Nicholls, K. W., & Smith, A. M. (2021). The BEAMISH hot water drill system and its use on the Rutford Ice Stream, Antarctica. *Annals of Glaciology*, *62* (85–86), 233–249. doi:10.1017/aog.2020.86

Blankenship, D. D., & Bentley, C. R. (1987). The crystalline fabric of polar ice sheets inferred from seismic anisotropy. *IAHS Publication*, *170*, 17–28.

Blankenship, D. D., Bentley, C. R., Rooney, S. T., & Alley, R. B. (1986). Seismic measurements reveal a saturated porous layer beneath an active Antarctic ice stream. *Nature*, *322* (6074), 54–57. doi:10.1038/322054a0

Booth, A. D., Emir, E., & Diez, A. (2015). Approximations to seismic AVA responses: validity and potential in glaciological applications. *Geophysics*, *81*(1), WA1–WA11. doi:10.1190/geo2015-0187.1

Booth, A. D., Christoffersen, P., Schoonman, C., Clarke, A., Hubbard, B., Law, R., et al. (2020). Distributed acoustic sensing of seismic properties in a borehole drilled on a fast-flowing Greenlandic outlet glacier. *Geophysical Research Letters*, *47*(13), e2020GL088148. doi:10.1029/2020gl088148

Booth, A. D., Christoffersen, P., Pretorius, A., Chapman, J., Hubbard, B., Smith, E. C., et al. (2023). Characterising sediment thickness beneath a Greenlandic outlet glacier using distributed acoustic sensing: preliminary observations and progress towards an efficient machine learning approach. *Annals of Glaciology*, *63*, 87, 79–89, 82. doi:10.1017/aog.2023.15

Brisbourne, A. M., Kendall, M., Kufner, S. K., Hudson, T. S., & Smith, A. M. (2021). Downhole distributed acoustic seismic profiling at Skytrain Ice Rise, West Antarctica. *The Cryosphere*, *15*(7), 3443–3458. doi:10.5194/tc-15-3443-2021

Brisbourne, A. M., Martín, C., Smith, A. M., Baird, A. F., Kendall, J. M., & Kingslake, J. (2019). Constraining recent ice flow history at Korff Ice Rise, West Antarctica, using radar and seismic measurements of ice fabric. *Journal of Geophysical Research: Earth Surface*, *124*(1), 175–194. doi:10.1029/2018jf004776

Brisbourne, A. M., Smith, A. M., King, E. C., Nicholls, K. W., Holland, P. R., & Makinson, K. (2014). Seabed topography beneath Larsen C Ice Shelf from seismic soundings. *The Cryosphere*, *8*(1), 1–13. doi:10.5194/tc-8-1-2014

Brisbourne, A. M., Smith, A. M., Rivera, A., Zamora, R., Napoleoni, F., Uribe, J. A., & Ortega, M. (2023). Bathymetry and bed conditions of Lago Subglacial CECs, West Antarctica. *Journal of Glaciology*, 1–10. doi:10.1017/jog.2023.38

Butcher, A., Hudson, T., Kendall, J., Kufner, S., Brisbourne, A., & Stork, A. (2021). Radon transform-based detection of microseismicity on DAS networks: a case study from Antarctica. *2021*(1), 1–4. doi:10.3997/2214-4609.202131039

Castongia, E., Wang, H. F., Lord, N., Fratta, D., Mondanos, M., & Chalari, A. (2017). An experimental investigation of distributed acoustic sensing (DAS) on lake ice. *Journal of Environmental and Engineering Geophysics*, *22*(2), 167–176. doi:10.2113/Jeeg22.2.167

Chaput, J., Aster, R., Karplus, M., & Nakata, N. (2022). Ambient high-frequency seismic surface waves in the firn column of central West Antarctica. *Journal of Glaciology*, *68*(270), 785–798. doi:10.1017/jog.2021.135

Crary, A. P. (1955). A brief study of ice tremors. *Bulletin of the Seismological Society of America*, *45*(1), 1–9. doi:10.1785/bssa0450010001

Diez, A., Eisen, O., Hofstede, C., Lambrecht, A., Mayer, C., Miller, H., et al. (2015). Seismic wave propagation in anisotropic ice - part 2: effects of crystal anisotropy in geophysical data. *The Cryosphere*, *9*(1), 385–398. doi:10.5194/tc-9-385-2015

Fichtner, A., & Hofstede, C. (2022). A simple algorithm for optimal design in distributed fibre-optic sensing. *Geophysical Journal International*, *233*(1), 229–233. doi:10.1093/gji/ggac458

Fichtner, A., Hofstede, C., Gebraad, L., Zunino, A., Zigone, D., & Eisen, O. (2023a). Borehole fibre-optic seismology inside the Northeast Greenland Ice Stream. Preprint, arXiv:2307.05976v1. doi:10.48550/arXiv.2307.05976

Fichtner, A., Hofstede, C., Kennett, B. L. N., Nymand, N. F., Lauritzen, M. L., Zigone, D., & Eisen, O. (2023b). Fiber-optic airplane seismology on the Northeast Greenland ice stream. *The Seismic Record*, *3*(2), 125–133. doi:10.1785/0320230004

Fichtner, A., Klaasen, S., Thrastarson, S., Çubuk-Sabuncu, Y., Paitz, P., & Jónsdóttir, K. (2022). Fiber-optic observation of volcanic tremor through floating ice sheet resonance. *The Seismic Record*, *2*(3), 148–155. doi:10.1785/0320220010

Gajek, W., Gräff, D., Hellmann, S., Rempel, A. W., & Walter, F. (2021). Diurnal expansion and contraction of englacial fracture networks revealed by seismic shear wave splitting. *Communications Earth & Environment*, *2*(1), 209. doi:10.1038/s43247-021-00279-4

Gimbert, F., et al. (2021). A multi-physics experiment with a temporary dense seismic array on the Argentière Glacier, French Alps: the RESOLVE Project. *Seismological Research Letters*, *92*(2A), 1185–1201. doi:10.1785/0220200280

Gusmeroli, A., Murray, T., Clark, R. A., Kulessa, B., & Jansson, P. (2013). Vertical seismic profiling of glaciers: appraising multi-phase mixing models. *Annals of Glaciology*, *54*(64), 115–123. doi:10.3189/2013AoG64A106

Hansen, S. M., & Schmandt, B. (2015). Automated detection and location of microseismicity at Mount St. Helens with a large-N geophone array. *Geophysical Research Letters*, *42*(18), 7390–7397. doi:10.1002/2015gl064848

Hofstede, C., et al. (2021). Evidence for a grounding line fan at the onset of a basal channel under the ice shelf of Support Force Glacier, Antarctica, revealed by reflection seismics. *The Cryosphere*, *15*(3), 1517–1535. doi:10.5194/tc-15-1517-2021

Horgan, H. J., Anandakrishnan, S., Alley, R. B., Burkett, P. G., & Peters, L. E. (2011). Englacial seismic reflectivity: imaging crystal-orientation fabric in West Antarctica. *Journal of Glaciology*, *57*(204), 639–650.

Horgan, H. J., Anandakrishnan, S., Alley, R. B., Peters, L. E., Tsoflias, G. P., Voigt, D. E., & Winberry, J. P. (2008). Complex fabric development revealed by englacial seismic reflectivity: Jakobshavn Isbrae, Greenland. *Geophysical Research Letters*, *35*(10). doi:10.1029/2008gl033712

Hudson, T. S., Baird, A. F., Kendall, J. M., Kufner, S. K., Brisbourne, A. M., Smith, A. M., et al. (2021). Distributed acoustic sensing (DAS) for natural microseismicity studies: a case study from Antarctica. *Journal of Geophysical Research: Solid Earth*, *126*(7), e2020JB021493. doi:10.1029/2020JB021493

Hudson, T. S., Brisbourne, A. M., Kufner, S. K., Kendall, J. M., & Smith, A. M. (2023). Array processing in cryoseismology. *EGUsphere*, *2023*, 1–18. doi:10.5194/egusphere-2023-657

Hughes, T. J. (1981). The weak underbelly of the West Antarctic ice sheet. *Journal of Glaciology*, *27*(97), 518–525. doi:10.3189/S002214300001159X

Johnson, M. R., & Smith, A. M. (1997). Seabed topography under the southern and western Ronne Ice Shelf, derived from seismic surveys. *Antartic Science*, *9*(2), 201–208.

King, E. C., & Jarvis, E. P. (2007). Use of shear waves to measure Poisson's ratio in polar firn. *Journal of Environmental & Engineering Geophysics*, *12*(1), 15–21.

Kirchner, J. F., & Bentley, C. R. (1990). RIGGS III: seismic short-refraction studies using an analytical curve-fitting technique. *Antartic Research Series*, *42*, 109–126.

Klaasen, S., Paitz, P., Lindner, N., Dettmer, J., & Fichtner, A. (2021). Distributed acoustic sensing in volcano-glacial environments—Mount Meager, British Columbia. *Journal of Geophysical Research: Solid Earth*, *126*(11), e2021JB022358. doi:10.1029/2021JB022358

Klaasen, S., Thrastarson, S., Fichtner, A., ubuk-Sabuncu, Y., & Jnsdttir, K. (2022). Sensing Iceland's most active volcano with a "buried hair". *EOS*, *103*. doi:10.1029/2022eo220007

Kristoffersen, Y., Hofstede, C., Diez, A., Blenkner, R., Lambrecht, A., Mayer, C., & Eisen, O. (2014). Reassembling Gondwana: a new high quality constraint from vibroseis exploration of the sub-ice shelf geology of the East Antarctic continental margin. *Journal of Geophysical Research: Solid Earth, 119*(12), 9171–9182. doi:10.1002/2014JB011479

Kufner, S. K., Brisbourne, A. M., Smith, A. M., Hudson, T. S., Murray, T., Schlegel, R., et al. (2021). Not all icequakes are created equal: basal icequakes suggest diverse bed deformation mechanisms at Rutford Ice Stream, West Antarctica. *Journal of Geophysical Research: Earth Surface, 126*(3), e2020JF006001. doi:10.1029/2020jf006001

Kufner, S.-K., Wookey, J., Brisbourne, A. M., Martín, C., Hudson, T. S., Kendall, J. M., & Smith, A. M. (2023). Strongly depth-dependent ice fabric in a fast-flowing Antarctic ice stream revealed with icequake observations. *Journal of Geophysical Research: Earth Surface, 128*(3), e2022JF006853. doi:10.1029/2022JF006853

Lapins, S., Butcher, A., Kendall, J.-M., Hudson, T. S., Stork, A. L., Werner, M. J., J. Gunning, et al. (2023). DAS-N2N: machine learning distributed acoustic sensing (DAS) signal denoising without clean data. Preprint. doi:10.48550/arXiv.2304.08120

Lin, F.-C., Li, D., Clayton, R. W., & Hollis, D. (2013). High-resolution 3D shallow crustal structure in Long Beach, California: application of ambient noise tomography on a dense seismic array. *Geophysics, 78*(4), Q45–Q56. doi:10.1190/geo2012-0453.1

Lutz, F., Eccles, J., Prior, D. J., Craw, L., Fan, S., Hulbe, C., et al. (2020). Constraining ice shelf anisotropy using shear wave splitting measurements from active-source borehole seismics. *Journal of Geophysical Research: Earth Surface, 125*(9), e2020JF005707. doi:10.1029/2020JF005707

Mad Zahir, M. H., Abdul Aziz, K. M., Ghazali, A. R., & Abdul Latiff, A. H. (2023). Effectiveness of fiber optic distributed acoustic sensing (DAS) in vertical seismic profiling (VSP) field survey. *Applied Sciences, 13*(8), 5002.

Makinson, K., & Anker, P. G. (2014). The BAS ice-shelf hot-water drill: design, methods and tools. *Annals of Glaciology, 55*(68), 44–52.

Moreau, L., et al. (2020). Sea ice thickness and elastic properties from the analysis of multimodal guided wave propagation measured with a passive seismic array. *Journal of Geophysical Research: Oceans, 125*(4), e2019JC015709. doi:10.1029/2019JC015709

Mulvaney, R., Rix, J., Polfrey, S., Grieman, M., Martìn, C., Nehrbass-Ahles, C., et al. (2021). Ice drilling on Skytrain Ice Rise and Sherman Island, Antarctica. *Annals of Glaciology, 62*, 85, 311–86, 323. doi:10.1017/aog.2021.7

Nost, O. A. (2004). Measurements of ice thickness and seabed topography under the Fimbul Ice Shelf, Dronning Maud Land, Antarctica. *Journal of Geophysical Research: Oceans, 109*(C10). doi:10.1029/2004jc002277

Nziengui-Bâ, D., Coutant, O., Moreau, L., & Boué, P. (2023). Measuring the thickness and Young's modulus of the ice pack with DAS, a test case on a frozen mountain lake. *Geophysical Journal International, 233*(2), 1166–1177. doi:10.1093/gji/ggac504

Paitz, P., Edme, P., Gräff, D., Walter, F., Doetsch, J., Chalari, A., et al. (2020). Empirical investigations of the instrument response for distributed acoustic sensing (DAS) across 17 octaves. *Bulletin of the Seismological Society of America, 111*(1), 1–10. doi:10.1785/0120200185

Paitz, P., Lindner, N., Edme, P., Huguenin, P., Hohl, M., Sovilla, B., et al. (2023). Phenomenology of avalanche recordings from distributed acoustic sensing. *Journal of Geophysical Research: Earth Surface, 128*(5), e2022JF007011. doi:10.1029/2022JF007011

Pearce, E., Booth, A. D., Rost, S., Sava, P., Konuk, T., Brisbourne, A., et al. (2023). Characterising ice slabs in firn using seismic full waveform inversion, a sensitivity study. *Journal of Glaciology, 69*(277), 1419–1433. doi:10.1017/jog.2023.30

Peters, L. E., Anandakrishnan, S., Alley, R. B., & Voigt, D. E. (2012). Seismic attenuation in glacial ice: a proxy for englacial temperature. *Journal of Geophysical Research: Earth Surface, 117*(F2), F02008. doi:10.1029/2011jf002201

Podolskiy, E. A., & Walter, F. (2016). Cryoseismology. *Reviews of Geophysics, 54*(4), 708–758. doi:10.1002/2016RG000526

Robin, G. d. Q. (1958). *Glaciology III: Seismic shooting and related investigations*. Norsk Polarinstitutt.

Robinson, E. S. (1968). Seismic wave propagation on a heterogeneous polar ice sheet. *Journal of Geophysical Research (1896–1977), 73*(2), 739–753. doi:10.1029/JB073i002p00739

Rosier, S. H. R., Hofstede, C., Brisbourne, A. M., Hattermann, T., Nicholls, K. W., Davis, P. E. D., et al. (2018). A new bathymetry for the southeastern Filchner-Ronne ice shelf: implications for modern oceanographic processes and glacial history. *Journal of Geophysical Research: Oceans, 123*(7), 4610–4623. doi:10.1029/2018JC013982

Röthlisberger, H. (1955). Studies in glacier physics on the penny ice cap, Baffin Island, 1953: part III: seismic sounding. *Journal of Glaciology, 2*, 539–552.

Smith, A. M. (1997). Basal conditions on Rutford Ice Stream, West Antarctica, from seismic observations. *Journal of Geophysical Research, 102*, 543–552.

Smith, A. M. (2006). Microearthquakes and subglacial conditions. *Geophysical Research Letters, 33*(24). doi:10.1029/2006gl028207

Smith, E. C., Baird, A. F., Kendall, J. M., Martín, C., White, R. S., Brisbourne, A. M., & Smith, A. M. (2017). Ice fabric in an Antarctic ice stream interpreted from seismic anisotropy. *Geophysical Research Letters, 44*(8), 3710–3718. doi:10.1002/2016gl072093

Smith, E. C., Smith, A. M., White, R. S., Brisbourne, A. M., & Pritchard, H. D. (2015). Mapping the ice-bed interface characteristics of Rutford Ice Stream, West Antarctica, using microseismicity. *Journal of Geophysical Research: Earth Surface, 120*(9), 1881–1894. doi:10.1002/2015JF003587

Smith, A. M., Woodward, J., Ross, N., Bentley, M. J., Hodgson, D. A., Siegert, M. J., & King, E. C. (2018). Evidence for the long-term sedimentary environment in an Antarctic subglacial lake. *Earth and Planetary Science Letters, 504*, 139–151. doi:10.1016/j.epsl.2018.10.011

Sullivan, W. (1957). *Quest for a continent* (p. 372pp). London: Secker and Warburg.

Tyler, S. W., Holland, D. M., Zagorodnov, V., Stern, A. A., Sladek, C., Kobs, S., et al. (2013). Using distributed temperature sensors to monitor an Antarctic ice shelf and sub-ice-shelf cavity. *Journal of Glaciology, 59*(215), 583–591. doi:10.3189/2013JoG12J207

Walter, F., Dalban Canassy, P., Husen, S., & Clinton, J. F. (2013). Deep icequakes: what happens at the base of Alpine glaciers? *Journal of Geophysical Research: Earth Surface*, *118* (3), 1720–1728.

Walter, F., Gräff, D., Lindner, F., Paitz, P., Köpfli, M., Chmiel, M., & Fichtner, A. (2020). Distributed acoustic sensing of microseismic sources and wave propagation in glaciated terrain. *Nature Communications*, *11*(1), 2436. doi:10.1038/s41467-020-15824-6

Westphal, J. A. (1965). In situ acoustic attenuation measurements in glacial ice. *Journal of Geophysical Research*, *70*(8), 1849–1853. doi:10.1029/JZ070i008p01849

Williams, A., Kendall, J. M., Clarke, A., & Verdon, J. (2017). *Challenges in locating microseismic events using distributed acoustic sensors*. Paper presented at the American Geophysical Union Fall Meeting, S33B-0873.

Woodward, J., Smith, A. M., Ross, N., Thoma, M., Corr, H. F. J., King, E. C., et al. (2010). Location for direct access to Subglacial Lake Ellsworth: an assessment of geophysical data and modelling. *Geophysical Research Letters*, *37*(11), L11501. doi:10.11029/12010GL042884

Zhou, W., Butcher, A., Brisbourne, A. M., Kufner, S. K., Kendall, J. M., & Stork, A. L. (2022). Seismic noise interferometry and distributed acoustic sensing (DAS): inverting for the firn layer S-velocity structure on Rutford ice stream, Antarctica. *Journal of Geophysical Research: Earth Surface*, *127*(12), e2022JF006917. doi:10.1029/2022jf006917

12

Stratigraphic Test Well (Hydrate-01) Distributed Acoustic Sensing 3D Vertical Seismic Profile Processing

Teck Kean Lim[1], Akira Fujimoto[2], Toshiaki Kobayashi[2], Sergei Tcherkashnev[3], and Michael Mondanos[4]

ABSTRACT

The emergence of distributed acoustic sensing (DAS) as a geophysical monitoring tool has provided a new opportunity to acquire large vertical seismic profiling (VSP) surveys with ease, as it offers the possibility of recording the acoustic signal over the entire fiber or well length with a single shot. This reduces acquisition time by removing the need to move the sensors, required for conventional point-sensor measurements. This chapter outlines the processing and fault interpretation performed for the largest known onshore DAS 3D VSP survey conducted at the gas hydrate research Stratigraphic Test Well Hydrate-01 on the North Slope of Alaska. The data were recorded on a single-mode fiber in a cable permanently cemented behind casing. Two vibroseis sources were used in the survey. The data were processed with a sequence designed to produce a precise and high-resolution P-wave image that included editing, redatum, bandpass filtering, denoise, upgoing/downgoing wavefield separation, deconvolution, and migration. Faults around the Hydrate-01 well were identified and interpreted using the 3D VSP volume and attributes. These faults were clearly observed in the 3D VSP volume but absent from the 3D surface seismic volume.

12.1. INTRODUCTION

The emergence of distributed acoustic sensing (DAS) technology, which uses a fiber-optic cable as a series of seismic receivers, has enabled large vertical seismic profiling (VSP) surveys to be conducted at an economical scale (Hartog, 2017). The cable is installed along the full length of the wellbore, which allows VSP data to be recorded throughout the well length with each shot. This can result in large volumes of data for processing that require specific workflows and computing power.

A data processing example of an onshore 3D DAS VSP survey is presented here. It is the largest known onshore 3D DAS VSP survey, with 1,701 shot points, and it was recorded in 2019 on the North Slope in Alaska. The objective of the survey was to obtain an accurate P-wave velocity cube, anisotropic parameter estimates, and high-resolution 3D VSP P-wave imaging to aid in seismic interpretation and production well planning and for use as a baseline survey for a potential future 4D project. The processing workflow was designed based on a single-component VSP processing workflow that includes data conditioning, wavefield separation, deconvolution, and migration using an elastic velocity model.

[1] *Consultant, Tokyo, Japan*

[2] *Technology & Research Center, Japan Organization for Metals and Energy Security, Chiba, Japan*

[3] *ASTO Geophysical Consulting, Perth, Australia*

[4] *Silixa Ltd., Elstree, UK*

Distributed Acoustic Sensing in Borehole Geophysics, Geophysical Monograph 289, First Edition.
Edited by Yingping Li, Robert Mellors, and Ge Zhan.

DOI: 10.1002/9781394179275.ch12

12.2. VSP METHODOLOGY AND DAS TECHNOLOGY

VSP is a borehole seismic method that measures velocity as a function of depth (Hardage, 1985). The measurements are made in a borehole using geophones, accelerometers, hydrophones, and DAS inside the wellbore, as well as a source at the surface, both near to and far from the well. VSP provides a time-to-depth link between surface seismic measurements recorded in time and the wireline logs (resistivity, pressure, saturation, nuclear, acoustic, density, etc.) recorded with respect to depth. VSP measurements include additional important information, such as direct measurements of seismic wavefields and characterization of elastic and anelastic rock properties and anisotropy parameters for the rock mass around the borehole. These measurements are used to validate and calibrate surface seismic processing and interpretation results.

DAS technology uses fiber-optic cables to provide strain-rate or strain measurements. The fiber is the sensing element, and the incidence of acoustic or seismic waves on the fiber induces strain changes. The data are recorded at the surface using an optoelectronic device called an interrogator. This broadband system allows the true acoustic strain signal to be detected over large distances with a large dynamic range and in harsh conditions. DAS has the sensitivity of a cosine squared function to incident P waves, whereas the geophone sensitivity is a cosine function. This means the sensitivity of DAS decreases more quickly than that of a geophone as the incidence angle increases. To compensate for this, typically more shots are required to improve DAS signal-to-noise ratio (SNR) at far offsets.

Figure 12.1 show the DAS sensitivity map for the Alaskan 3D VSP study. Due to the well geometry, the sensitivity map is tilted to the north–east direction. The left map shows the DAS sensitivity to the reflections from the top primary reservoir, which was 10–20% less sensitive than a geophone. To match geophone equivalent SNR, additional shots are required in the region of 10–20% extra stacks. The right figure shows the bin fold map of the reflected rays from the top primary unit.

12.3. DEPTH CALIBRATION AND DATA ACQUISITION

In March 2019, a large 3D VSP DAS survey was acquired on a deviated Stratigraphic Test Well: Hydrate-01 (STW), onshore Alaska, USA, using Silixa's intelligent distributed acoustic sensor (iDAS) as the interrogator on a standard single-mode (SM) fiber.

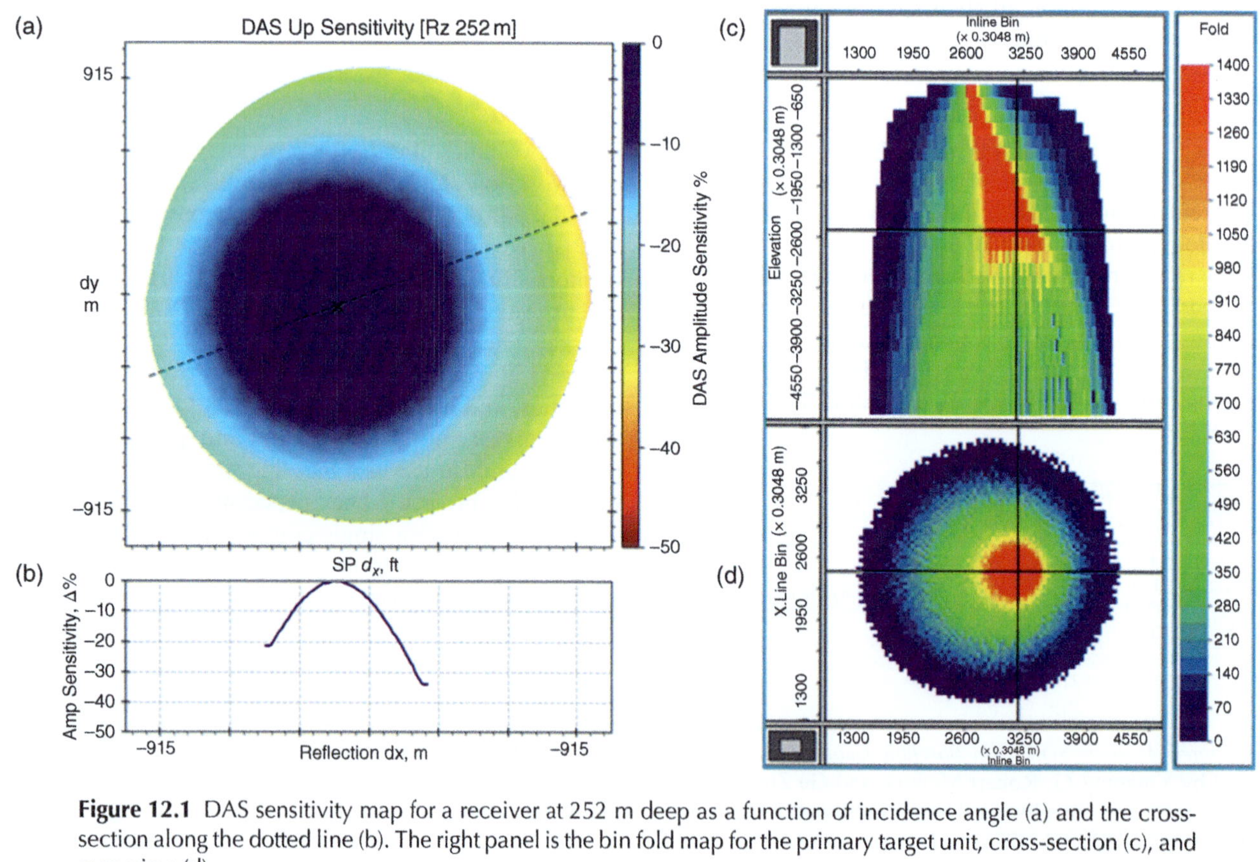

Figure 12.1 DAS sensitivity map for a receiver at 252 m deep as a function of incidence angle (a) and the cross-section along the dotted line (b). The right panel is the bin fold map for the primary target unit, cross-section (c), and map view (d).

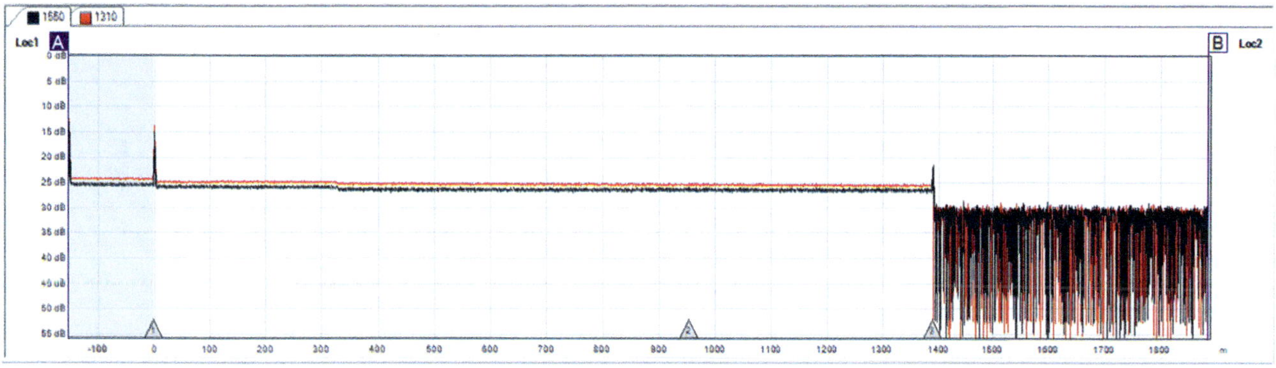

Figure 12.2 OTDR measurement on SM fiber (black line) in STW.

The optic cables were installed outside 5.5 in. casing in the STW well from 1,068.6 m measured depth (MD) up to the surface, and VSP data were acquired using the SM fiber with the lowest attenuation losses in the cable, based on an optical time domain reflectometer (OTDR) test, as shown in the Figure 12.2. Before the survey, so-called *tap tests* were performed. This is where a small signal is created by tapping lightly on the cable at a known location. This enables the channels (receivers) to be located in space. Based on the site condition, a tap test identified the location of the wellhead in distance along the fiber (fiber distance). Figure 12.3(a) shows the standard tap test at the cable exit from the wellhead outlet (WHO), thus identifying the location of the wellhead along the fiber. This was used as a reference point in the depth calibration of the fiber in the well. The recorded VSP interval velocity was compared to a sonic log to ensure a depth tie. As the DAS data were recorded in units of strain rate (nm/m/s), a conversion to geophone equivalent units (speed, m/s) was made prior to data processing using an algorithm developed by Silixa, which was based on spatial integration over a gauge length.

Two Inova AHV-IV Commander PLS-364 vibroseis units were used as the seismic energy source, which was controlled by the Seis Force-3 system using a radio connection. Before the survey, several tests were performed to determine the best vibrator parameters for acquisition. After evaluating DAS sensitivity and comparing the SNR and frequency content from the reservoir interval, a linear pilot sweep 20 s long with 250 ms starting and ending tapers and a frequency range of 2–200 Hz was used for the survey.

A spiral survey layout was used as it provided the best coverage around the wellbore. The radius of the spiral is approximately 3,200 ft, with the center at the midpoint between the wellhead and bottom hole location. Areas that were excluded from shooting are pipelines, roads, pingos, and a polar den, as shown on the daily production map (Figure 12.4). A total of 1,701 shot points (SPs) were recorded. The 3D VSP was acquired with 10 sweeps per SP for the near offsets, 15 sweeps per SP for the middle and far offsets, and 20 sweeps per SP for the far offsets in a segment of the area between approximately 45° and 135° from North. Table 12.1 shows the main acquisition parameters.

Real-time data quality control (QC) is performed to ensure the stacked gather has sufficient SNR, as shown in Figure 12.5. As more shots are taken at a location, stacking is performed immediately, and the SNR plot indicates the current SNR status. This ensures that an optimum number of shots are taken to obtain good data.

In addition to the tap tests performed during the fiber installation, the recorded DAS VSP interval velocity was compared to a logging while drilling (LWD) sonic log to ensure a depth tie. The left column in Figure 12.6 shows that the fiber-optic cable depth calibration matches the well logs (right column) on the main VSP seismic markers in the middle.

12.4. DATA QC

Borehole seismic data are acquired by using the dedicated surface and downhole equipment, as shown in Figure 12.7. The acquisition equipment must pass the internal hardware system and external user QC checks before and after a job to ensure that high-quality data are recorded. QC is an important step to assess borehole seismic data quality during processing. It helps to form an understanding of factors that impact data quality and hence contributes to more comprehensive and meaningful analysis during field acquisition and post-mortem analysis.

The QC results are focused on gaining detailed qualitative and quantitative inputs to inform a client of any data quality issues during the processing phase so that the survey can be processed to ensure the best data quality.

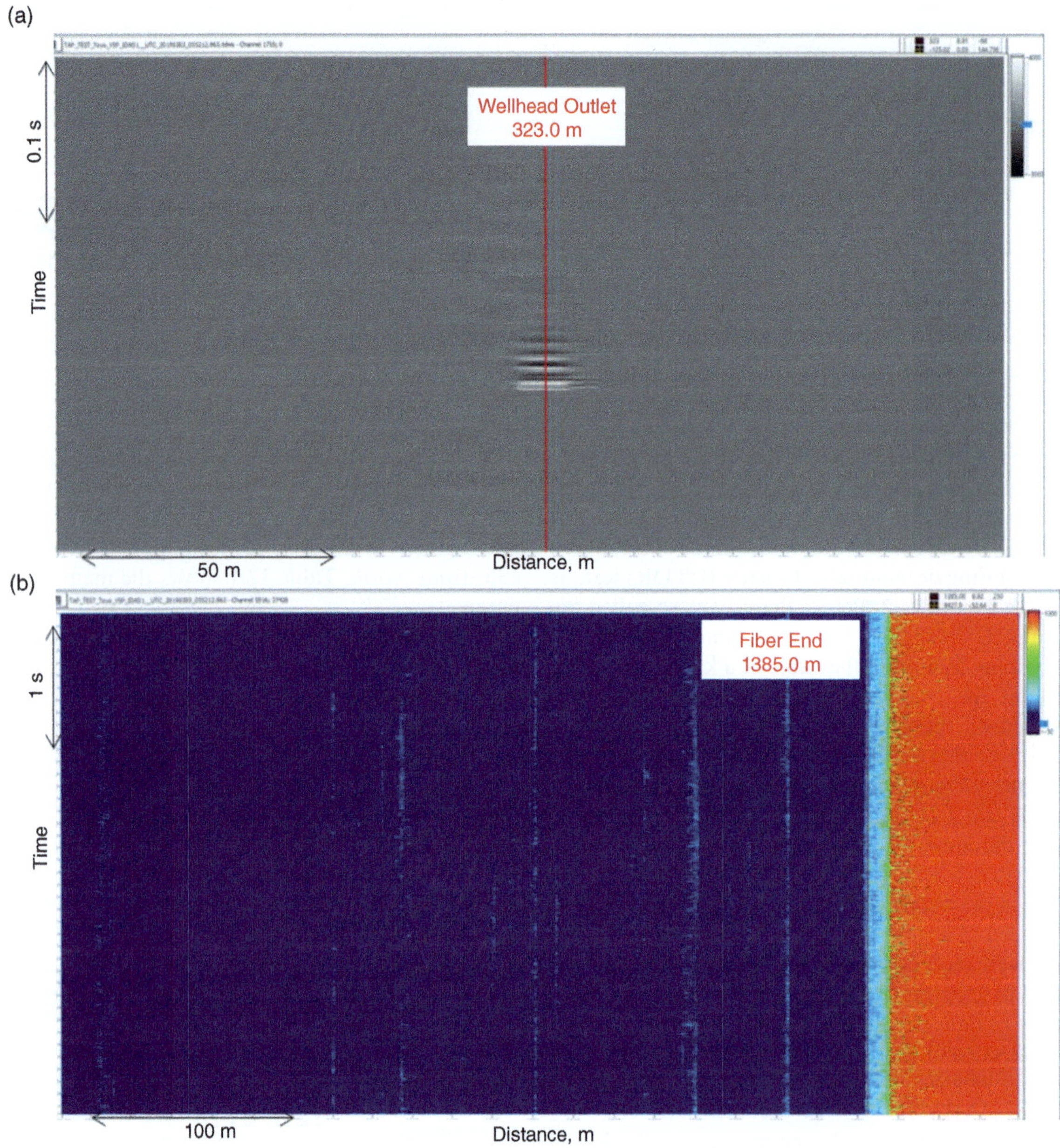

Figure 12.3 (a) Tap test at wellhead outlet. Tap location was at 323 m along the fiber and (b) 1,385 m mark at the end of the fiber.

The Hydrate-01 3D VSP DAS dataset was acquired with 10 sweeps per SP for the near offsets, 15 sweeps per SP for middle and far offsets, and 20 sweeps per SP for the far offsets.

Figure 12.8 shows an example of a strain gather recorded on DAS (top) and a geophone transformed gather (bottom) at a 300 ft offset source. Both datasets have well-defined first arrivals, and strong upgoing P events are clearly visible. There are some noisy traces on the geophone data below 914.4 m MD.

Figure 12.9 shows the strain-to-geophone converted data after stacking at each level at the mid and far offsets. Both datasets have well-defined first arrivals and strong upgoing P events clearly visible. Due to the well deviation, significant downgoing mode converted P-to-S wave arrivals are observed.

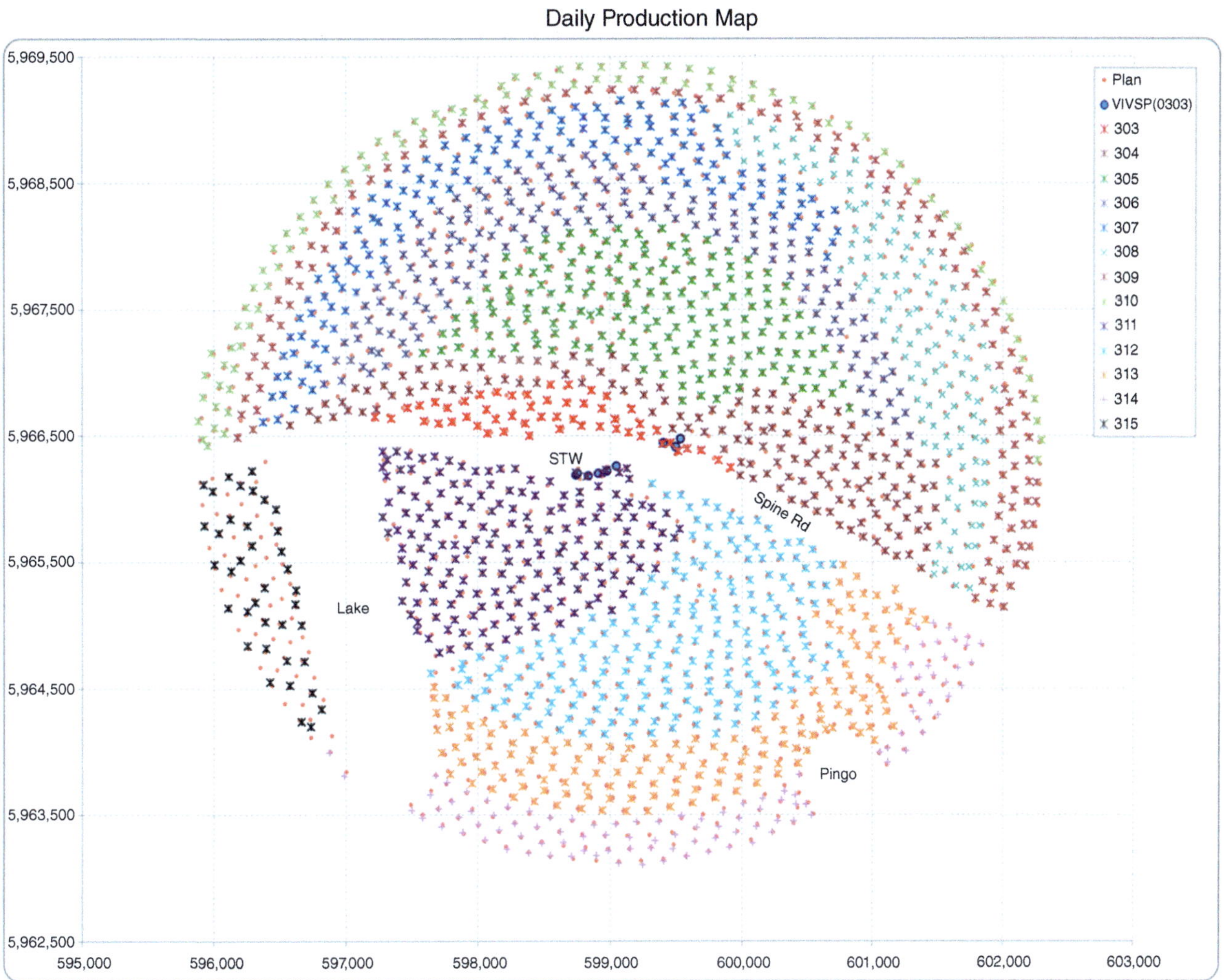

Figure 12.4 3D VSP spiral shooting daily production SPs. Dots represent the presurvey location, and cross marks indicate recorded SPs, color-coded by date (mdd). Voids are restricted areas.

Table 12.1 Hydrate-01 3D VSP acquisition parameters.

Well name	Hydrate-01 (STW: stratigraphic test well)
Location	TD at 1,084.5 m MD
Source type	Inova – AHV-IV Commander PLS-364
	Seis Force-3 Controller, varying SP locations
Source parameters	Correlated recording length: 3,000 ms
	Uncorrelated sweep length: 20,000 ms
	Sampling rate: 1 ms
	Linear sweep, frequency: 2–200 Hz
DAS recording unit	iDAS (Silixa)

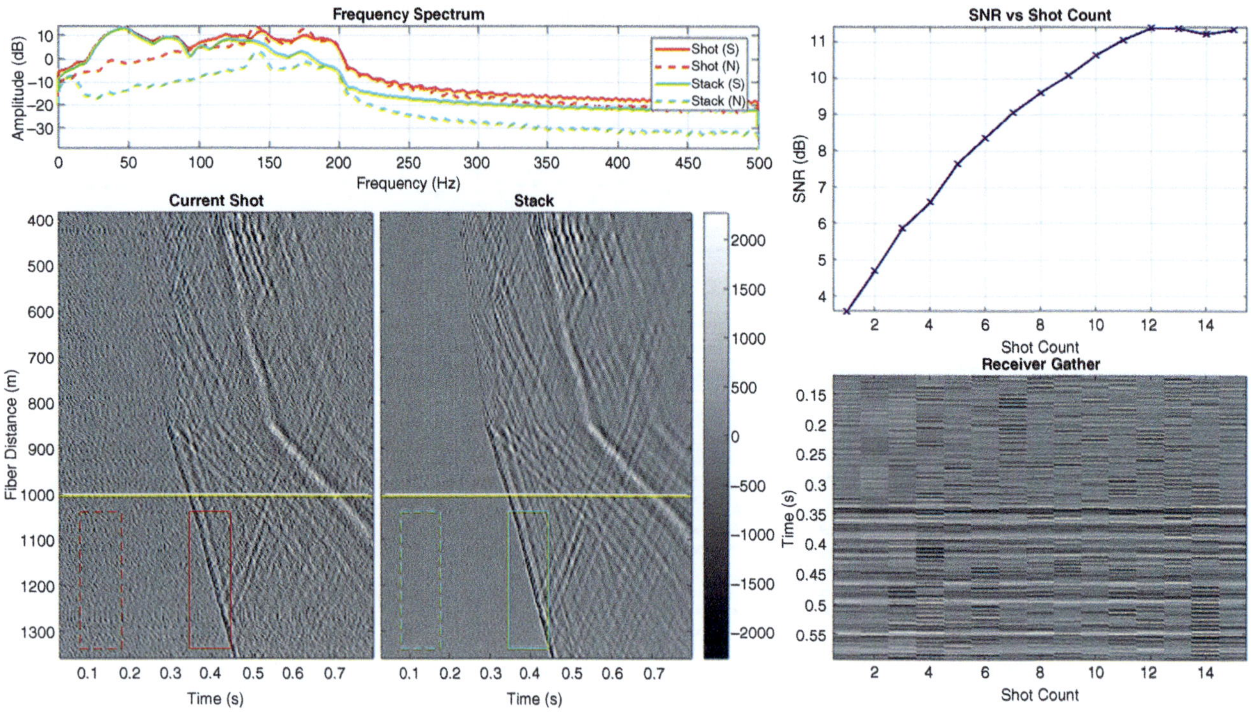

Figure 12.5 Example of data QC panels showing single (left) and stacked (right) waveform gathers for a source offset at 3,000 ft, along with the frequency spectrum (top left) and the SNR versus shot (stacking) count plot.

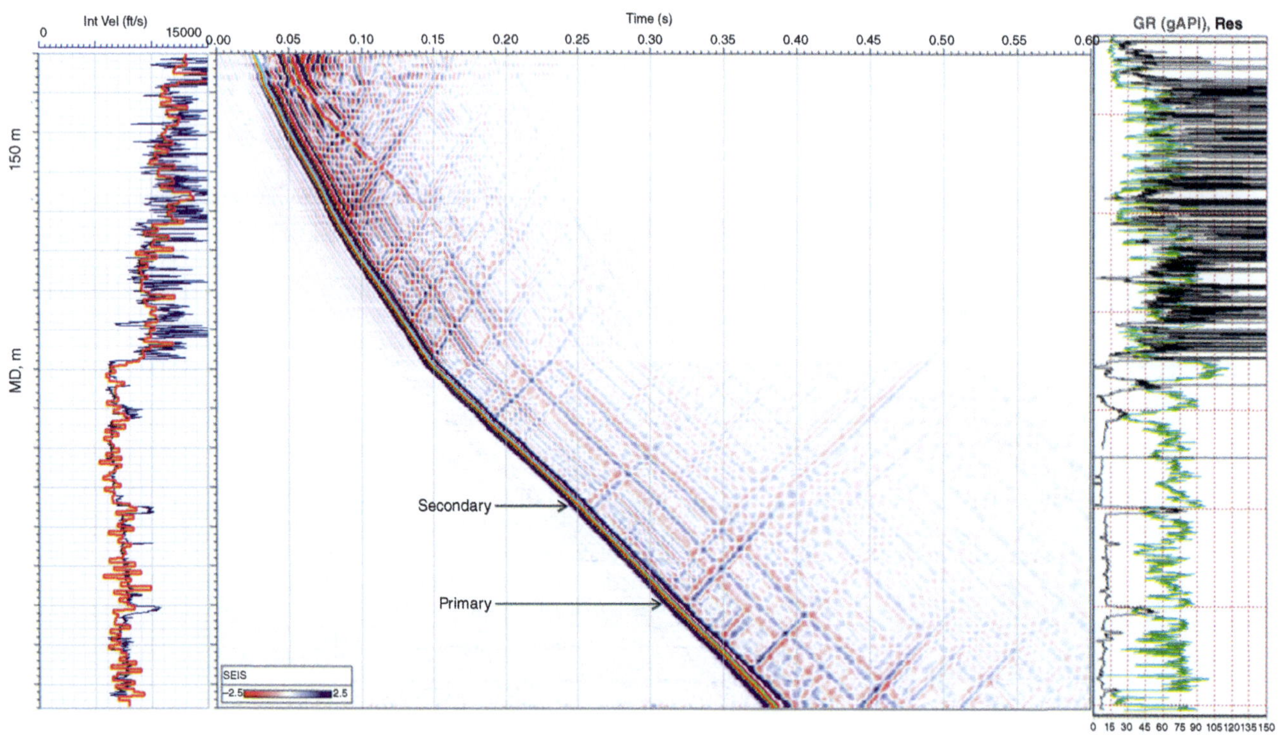

Figure 12.6 Fiber depth calibration (left) with VSP stacked gather and GR and resistivity logs (right).

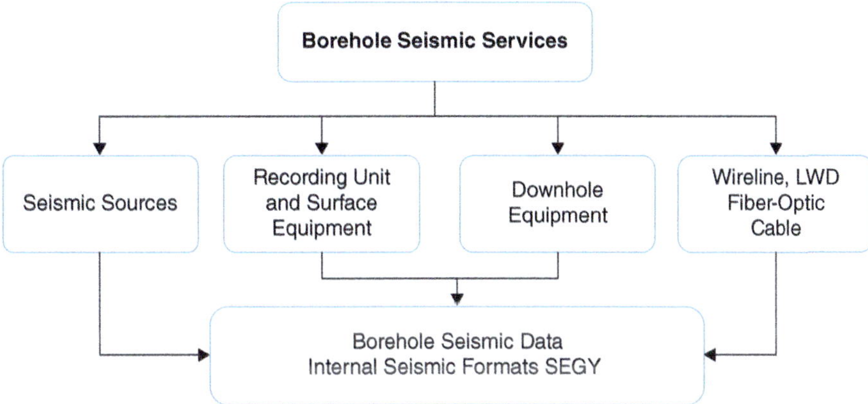

Figure 12.7 Borehole seismic acquisition components.

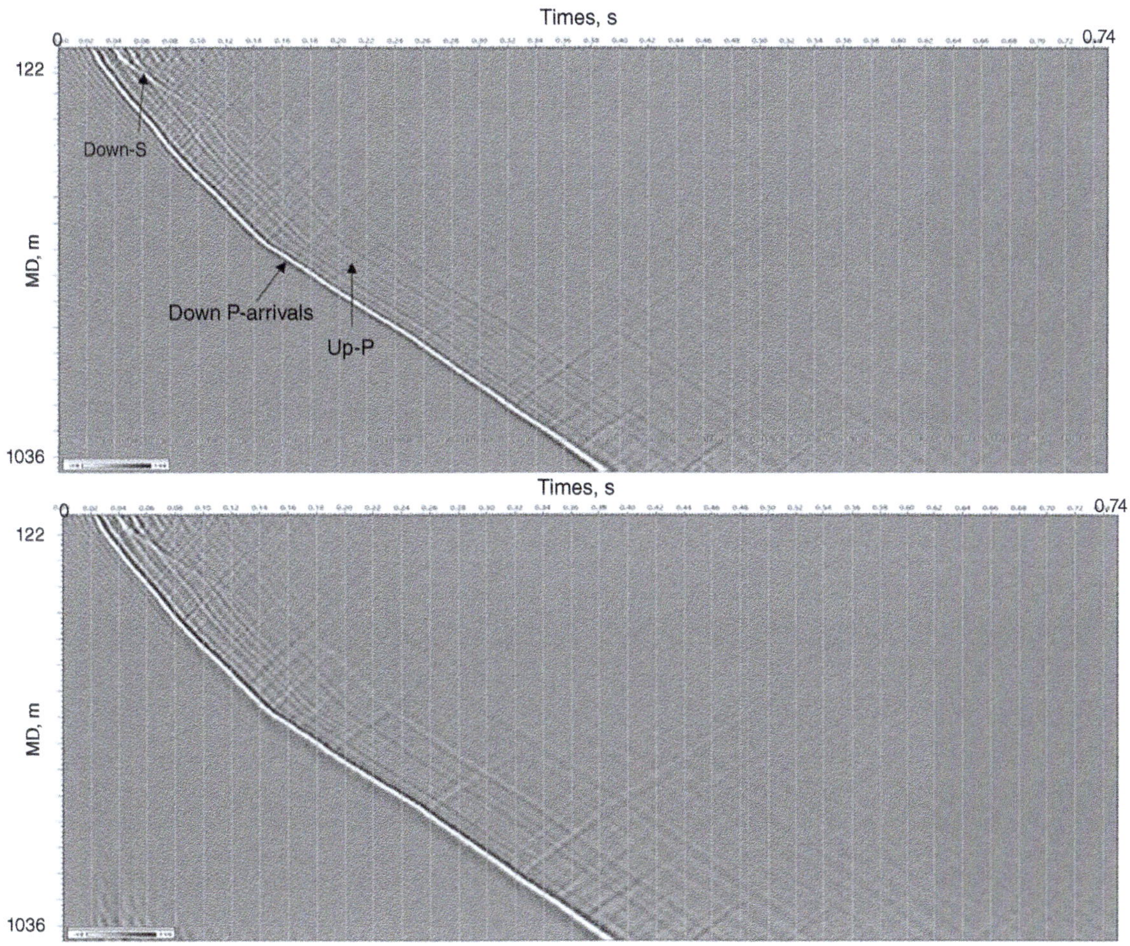

Figure 12.8 DAS data strain gather (top) and geophone transformed gather (bottom) at near offset. Sx = 300 ft.

The overall 3D VSP data quality is very good, with SNR of more than 40 dB and well-defined first arrivals, strong downgoing and upgoing P arrivals, and PS events clearly visible (Figures 12.8 and 12.9). SNR is high at near offsets and decreases to 40 dB at far offsets. Maximum signal frequency content is 200 Hz/0 dB at 0–300 ft and 170 Hz/–40 dB at 3,200 ft from the wellhead. The 3D VSP amplitude spectra in the northern and southern parts are shown in Figures 12.10 and 12.11.

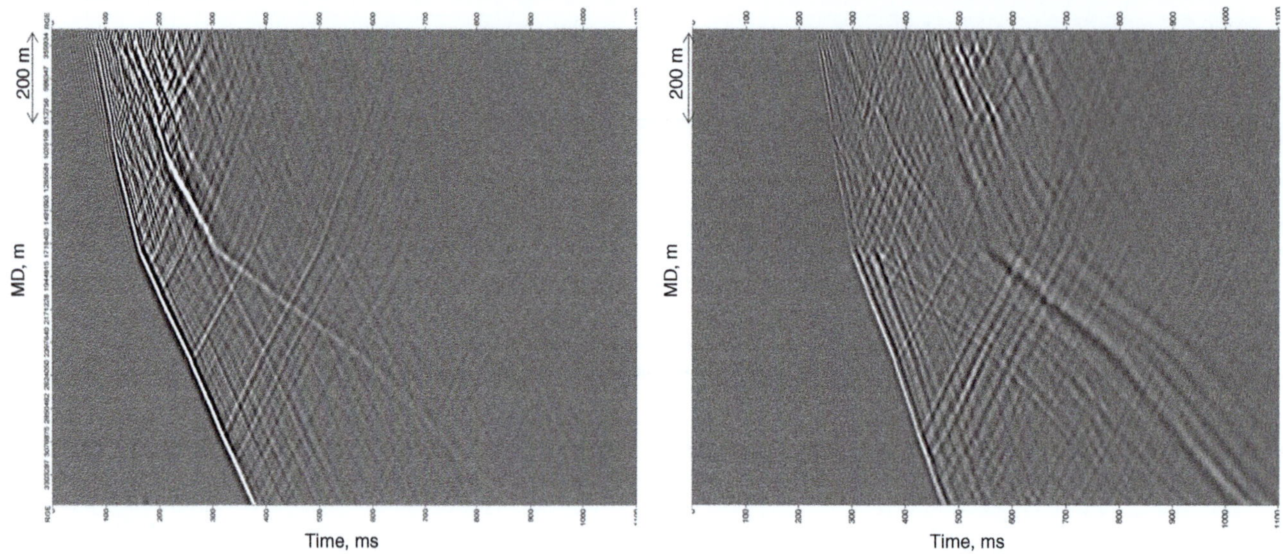

Figure 12.9 Geophone SP1118 (left) Sx = 1,500 ft, offset, and SP2682 (right), Sx = 3,200 ft, offset.

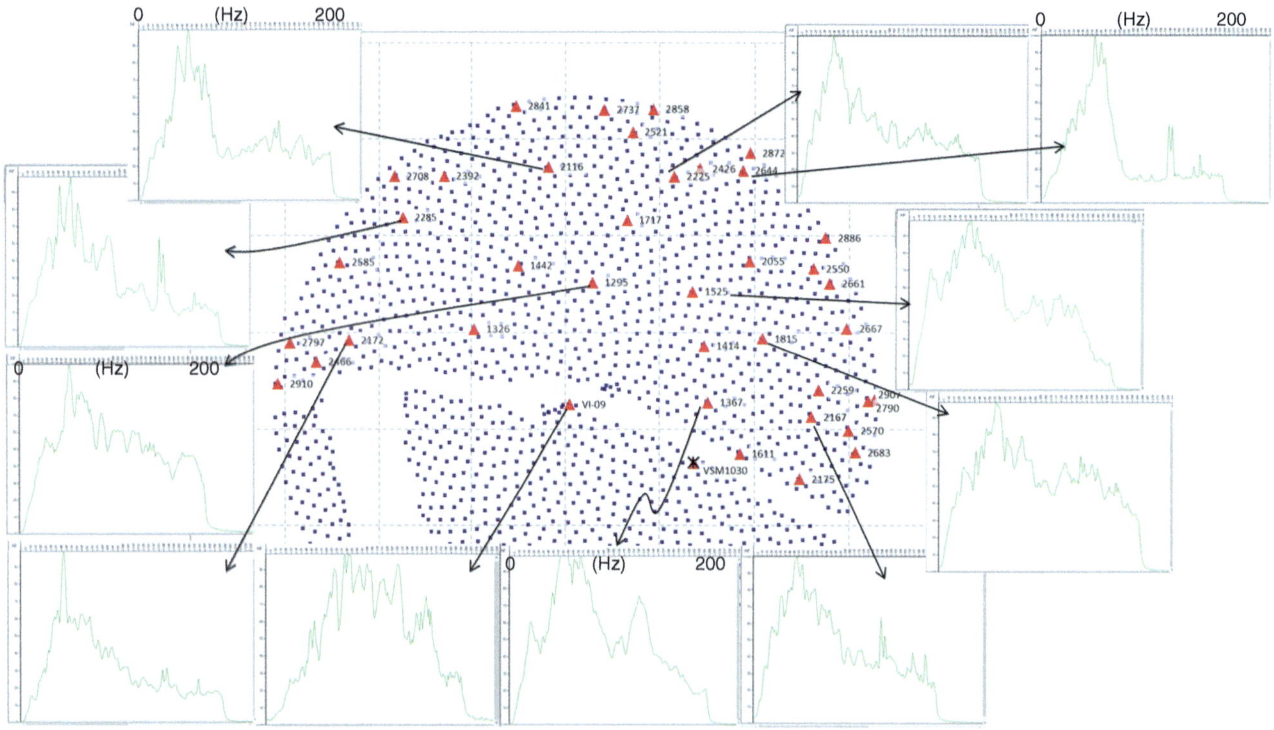

Figure 12.10 3D VSP amplitude spectrum (North site).

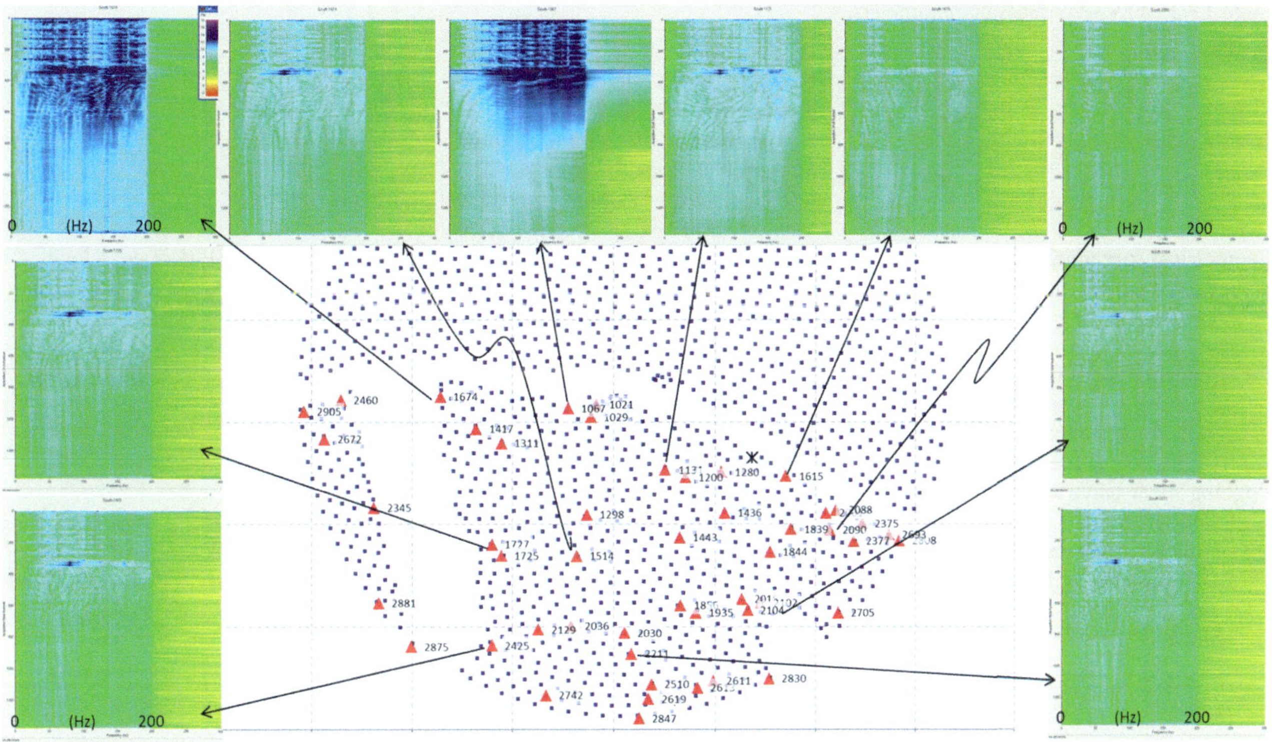

Figure 12.11 3D VSP amplitude spectrum (South site); same color legend (top-left insert) for all the spectrum plots.

12.5. PROCESSING

The purpose of the 3D VSP data processing is the separation of primary reflected P waves from regular and unwanted coherent energy and their transformation into a migrated image.

The processing parameters and processing sequence of the geophone equivalent data are shown in Table 12.2 and Figure 12.12.

12.5.1. Preprocessing

The first step of data processing is data editing and reconditioning the data.

Time Picking

The arrival time of each trace was computed on the trough (SEG normal polarity convention) of the downhole DAS geophone data. These picks are visually checked and interactively repicked. Visual and quantitative analysis of the first breaks of each shot shows that there are small lateral velocity variations and small azimuthal anisotropy. Six 2D walkaway lines were used for 3D travel-time inversion to simultaneously estimate tilted transverse isotropy (TTI) (Blias, 2012) anisotropic parameters for a single 3D depth velocity model (21 layers). The residual anisotropic transit times are small and could be related to the overburden velocity variations.

The break time is now the transit time from the source to the first compressional arrival.

Correction to Datum

A seismic reference datum (SRD) is the elevation point at which the traveltime is zero; mean sea level (MSL) is usually used as the SRD. This is set before further processing is conducted.

The 3D VSP traces must be corrected for elevation statics. Static corrections are needed as source depth varies per shot location (Figure 12.10). Elevation statics were computed using the replacement zone velocity equal to 10,000 ft/s. The 3D VSP data were shifted to MSL using elevation statics, and the transit times were updated accordingly.

Geometrical Spreading Gain

A model-based time-varying and offset-dependent gain have been applied to the 3D VSP data to compensate for the spherical divergence effect of the wavefront.

Table 12.2 3D VSP processing parameters.

Loading/editing	Data loading Navigation QC, editing of bad shots Median stacking
Preprocessing	Automated first break time picking Manual first break tuning Receiver geometry update for the well deviation Bandpass filtering (5–200 Hz bandwidth)
Geometrical spreading (GS) correction	Model-based geometrical spreading correction using the VSP-derived velocity
Wavefield separation (subtraction)	Estimation of optimum down P (163-level median filter) Subtraction down P FK filter to remove down S
Deterministic waveshaping deconvolution	Trace-by-trace deconvolution based on downgoing and applied to the upgoing wavefield. Down P amplitude normalization is applied. Zero phase, 5–200 Hz bandwidth Window = 0.2 s, 5% white noise
Upgoing shear wave attenuation	FK filter to remove upgoing shear waves in CSG
Model building	Model building from logs and vertical incidence VSP, VIVSP (1D model)
3D seismic velocity calibration	3D surface seismic velocity calibration by 3D VSP first break times tomography
VSPCDP mapping	3D VSP VSPCDP mapping on upgoing P wave after deconvolution using a 3D TTI media model
Kirchhoff migration	3D VSP Kirchhoff migration on upgoing P wave after deconvolution and UpS removal using a 3D TTI media model

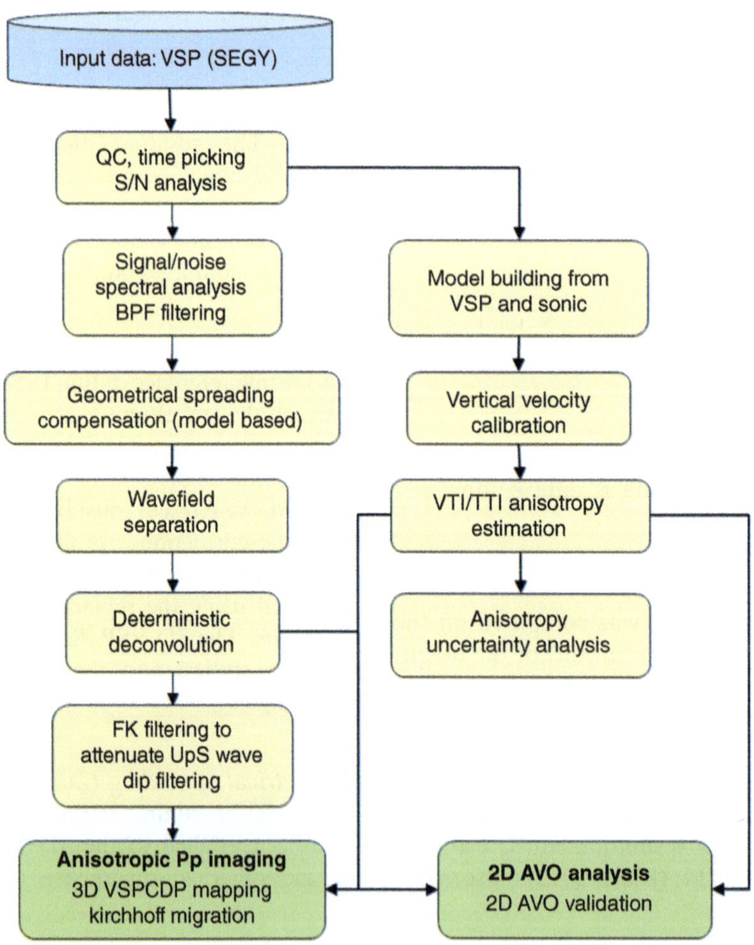

Figure 12.12 3D VSP processing workflow.

This geometrical spreading gain depends on the incident angle/ray parameter, shot-receiver position, velocity of the medium, and distance traveled along the ray path. A 3D VSP velocity model was used to compensate for the spherical divergence effect.

12.5.2. Wavefield Separation

The purpose of wavefield separation is to separate the upgoing and downgoing wavefields, as we are only interested in the upgoing component, which contains the reflection energy. Subtraction of downgoing P waves was performed using a median velocity filter (MVF) method, which involves three steps:

1. Align events using the first break header values.
2. Stack traces within a chosen window length using a median stack to estimate a downgoing wavefield.
3. Subtract the stacked downgoing wavefield from the total wavefield.

The window length depends on data quality and the output, the separated wavefields, and the conformity of events within the chosen window. A 163-trace median filter was used to estimate the downgoing wavefield after numerous tests. This wavefield is then subtracted from the total wavefield to give the residual upgoing wavefield. The residual wavefield is dominated by strong downgoing PS events, reflected compressional P events, and weak upgoing PS events (Figures 12.13 and 12.14).

Removing Downgoing PS Waves

Removing downgoing PS waves was performed in the wave-number domain using an FK filter. The wavefields before and after applying the FK filter are presented in Figure 12.15. Removal of the upgoing S-waves was performed in the wave-number domain using an FK filter after deconvolution (Figure 12.16).

Deconvolution

The trace-by-trace zero-phase deconvolution process shortens the seismic pulse and centers the amplitude peak on the reflector. It attenuates long periods and interbed multiples and collapses the multiple sequences of shear arrivals, diffractions, or out-of-plane reflections. The waveshaping deconvolution operator is double-sided and is designed on trace-by-trace openings of 100 ms before the first break, with a window length of 200 ms. The VSP decon is an inverse operator and therefore whitens the amplitude spectra from zero to the Nyquist frequency. A bandpass filter is applied to limit the output data to the signal range of the input data. The desired outputs were chosen to be zero phase with a bandwidth of 5–10–200–225 Hz. Once the design is made on the downgoing wavefield, it is applied to the downgoing and subtracted enhanced wavefield at the same level.

Numerous deconvolution tests, including predictive and spiking options, were performed to select the optimum deconvolution parameters.

Note that trace-by-trace VSP deconvolution will correct for all residual propagation effects by normalization

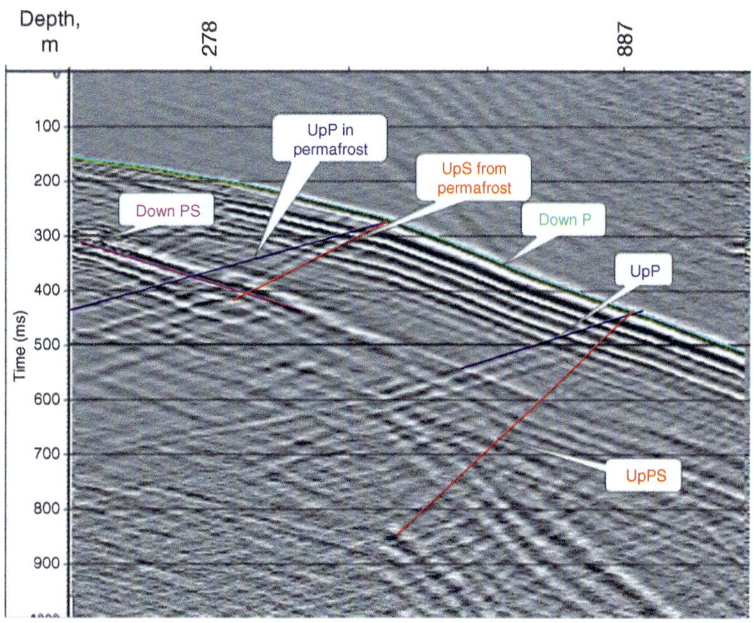

Figure 12.13 Total wavefield at SP 2,464, Sx = 2,423 ft in common shot gather (CSG).

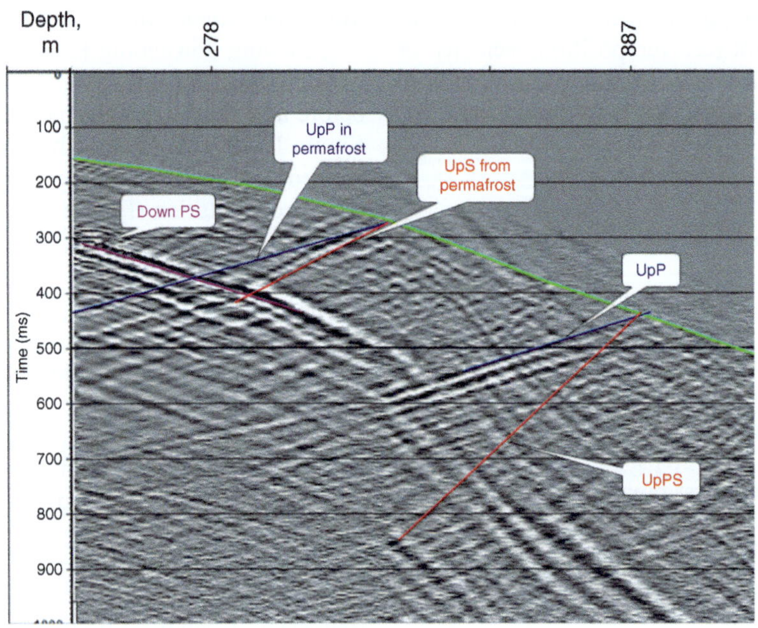

Figure 12.14 Residual wavefield at SP 2,464, Sx = 2,423 after downgoing P removal by median velocity filtering in CSG.

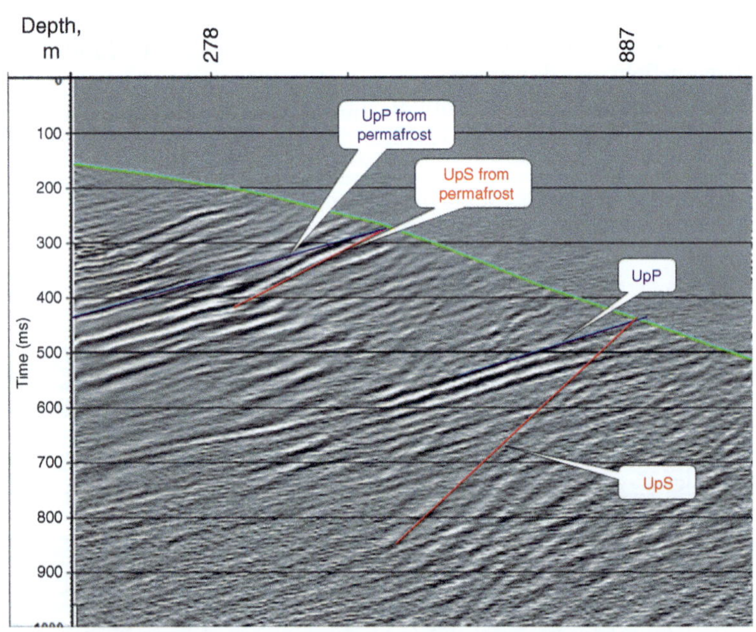

Figure 12.15 Residual wavefield at SP 2,464, Sx = 2,423 after downging P and downgoing S removal by FK filtering in CSG.

of downgoing amplitudes in the frequency domain. A VSP deconvolution operator normalization by downgoing amplitudes compensates for downgoing and upgoing wave amplitude variations from source to source, geometrical spreading, transmission losses, and *Qp*, which are included in the downgoing P-wave amplitudes. The deconvolved upgoing P amplitudes are the frequency band limited reflectivity series and can be compared directly with synthetic seismograms.

The predictive deconvolution involves the design of the prediction filters, which require only autocorrelation of the input series in which a repeating signal (e.g., primaries

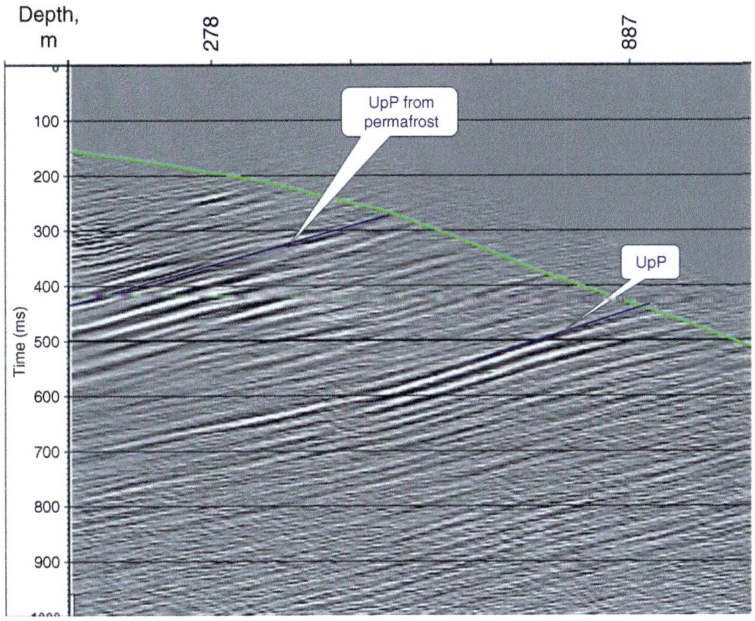

Figure 12.16 Residual wavefield at SP 2,464, Sx = 2,423 after downgoing P, downgoing S, and upgoing S removal by FK filtering in CSG.

and multiples) is shaped to one that does not repeat (primaries only). Predictive deconvolution suppresses multiple reflections and does nothing to the shape of the primary signal. Predictive and spiking deconvolution does not use any downgoing amplitude normalization and does not compensate for the upgoing amplitudes. Further amplitude correction is required after predictive and spiking deconvolution.

For QC purposes, a single walkaway line at the azimuth 70° along the Hydrate-01 well trajectory (Figures 12.17, 12.18, and 12.19) was extracted from the 3D VSP to test deconvolution parameters.

Figure 12.20 shows 2D WVSP (walkaway VSP) RMS measured amplitudes before (red) and after geometrical spreading compensation (green) and after deterministic VSP deconvolution (blue).

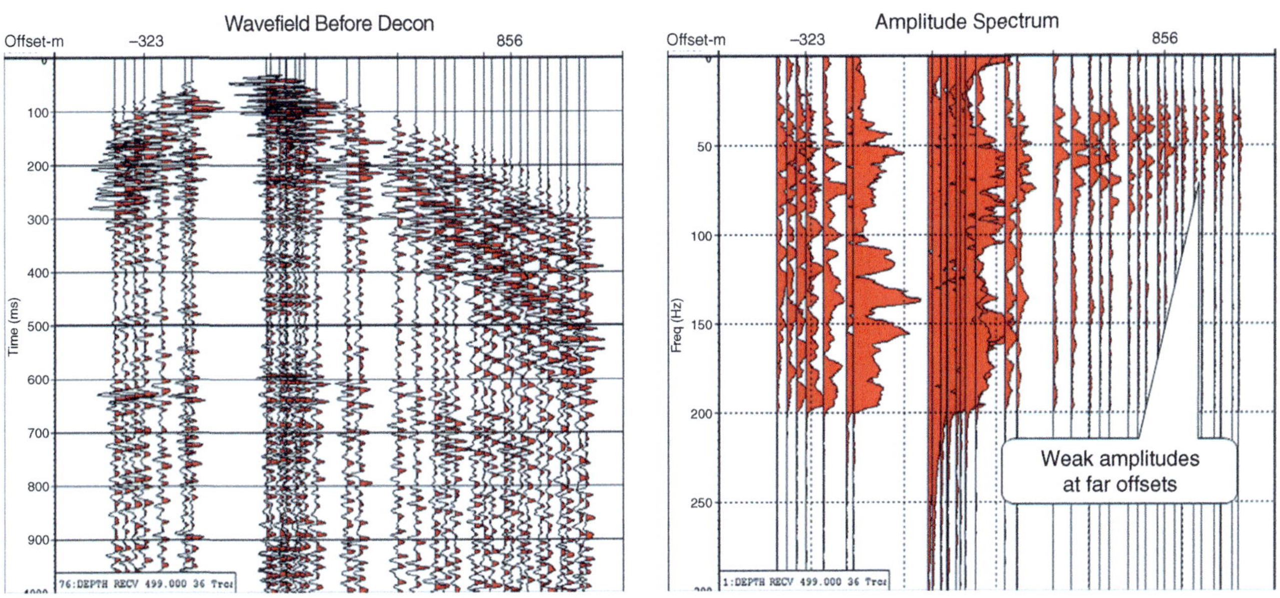

Figure 12.17 2D WVSP after down P, PS-wave removal, and amplitude spectrum.

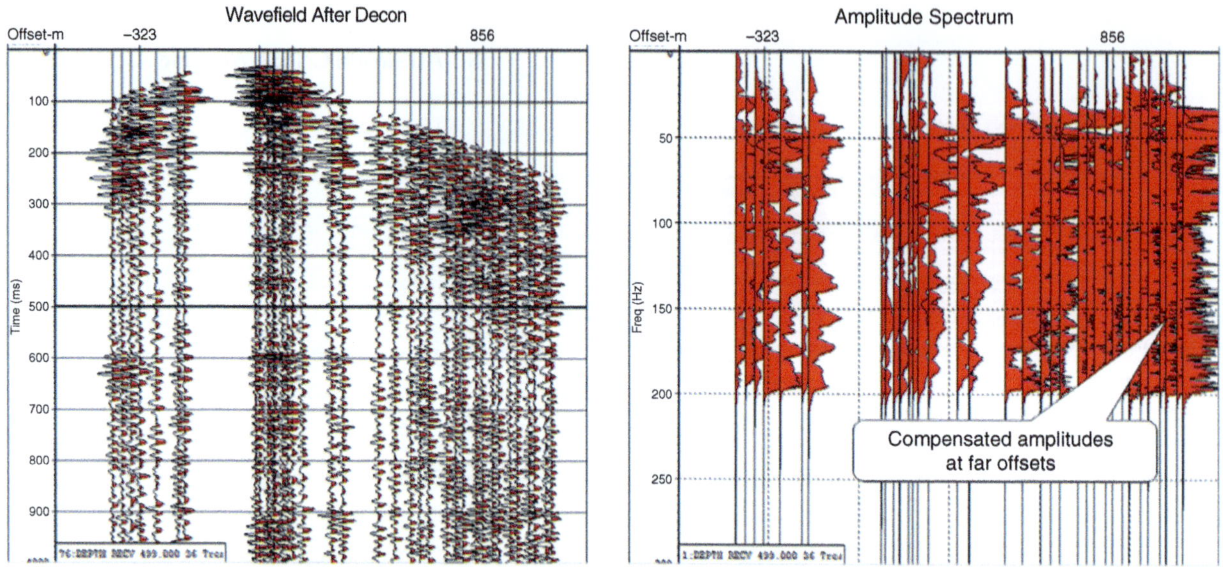

Figure 12.18 2D WVSP after down P, PS-wave removal, VSP decon, and amplitude spectrum.

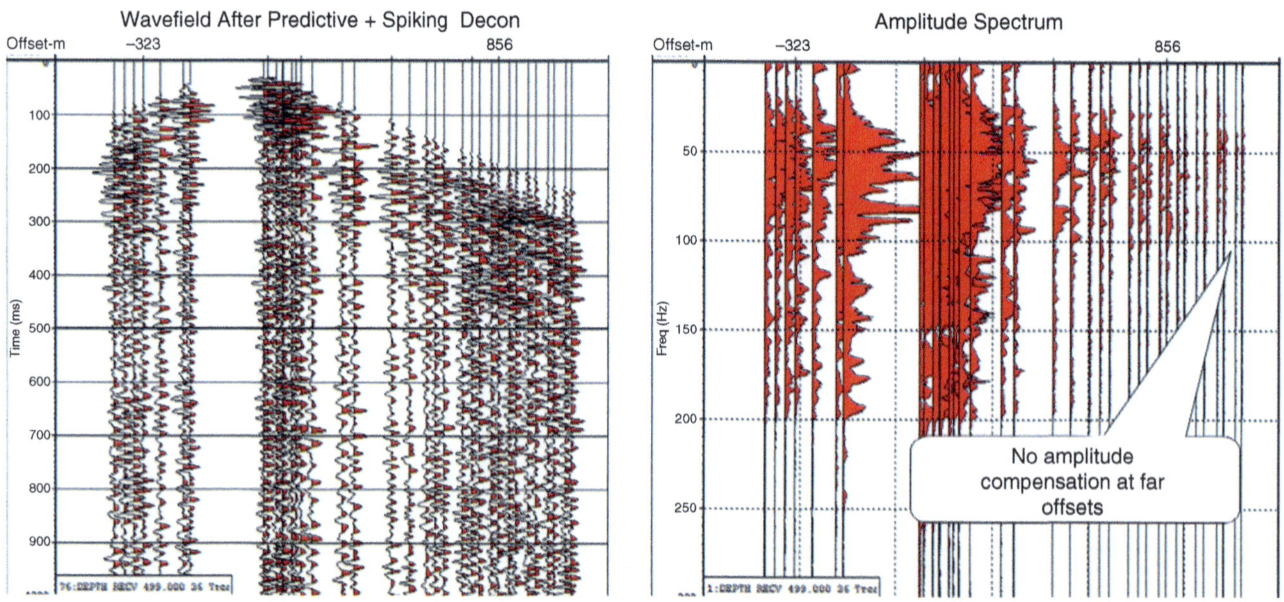

Figure 12.19 2D WVSP after down P, PS-wave removal, predictive and spiking deconvolution, and amplitude spectrum.

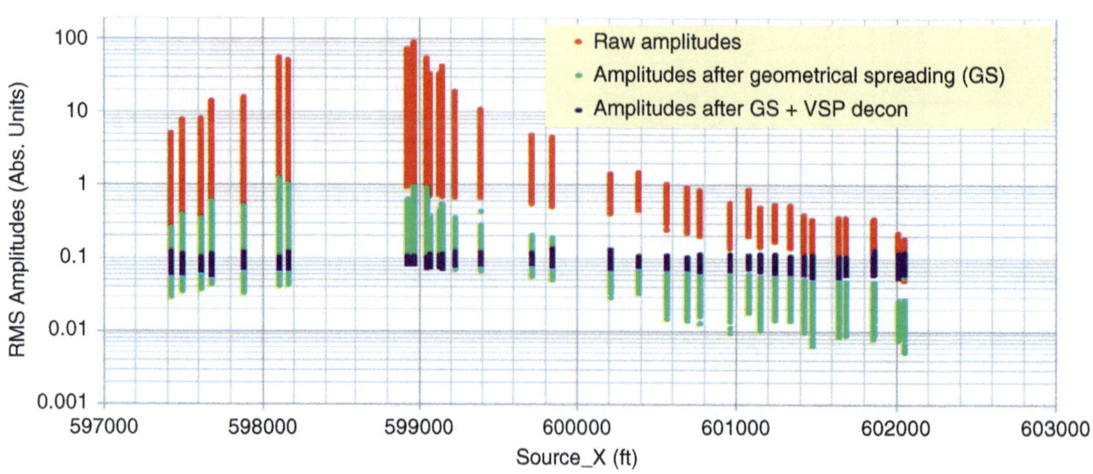

Figure 12.20 2D WVSP RMS measured (red) amplitudes after geometrical spreading compensation (green) and after VSP deconvolution (blue).

Mapping and Migration

VSP common depth point (CDP) mapping (Hardage, 1985) and Kirchhoff migration can image PP, PS, SP, or SS reflections from the depth-time domain to the offset-depth domain. Both processes are performed using a raytracing routine to first map the reflected signals and then transform the reflected signals from the depth-time domain (VSP domain) to the offset-depth domain.

VSP CDP mapping process is similar to the CDP normal moveout application and stacking. As a rule, the VSP CDP depth image is sharper than the Kirchhoff depth migration image. It does not smear coherent noise such as tube waves, guided waves, and converted waves as migration could, but it will not collapse diffractions.

VSP CDP mapping and Kirchhoff migration of the VSP data can provide an image around the borehole. It allows the mapping of upgoing reflections away from the borehole trajectory and the use of a mapped VSP image for interpretation. In addition, it helps in the evaluation of borehole seismic imaging capabilities. VSP CDP and Kirchhoff migration compute a fold that can be used to compensate for the amplitude difference due to stacking. A fold shows not how many traces are in each bin but how many samples are stacked in each sample of each image trace.

In this case, the deconvolved upgoing P wave, a 3D VSP TTI model, and the calibrated surface seismic velocity model were used to compute VSP CDP mapping and Kirchhoff migration image in depth. The anisotropic velocity model was used to convert a VSP CDP and a Kirchhoff migration depth image to a time domain, respectively.

Figure 12.21 show 3D VSP CDP mapping (aperture three bins) using a 3D TTI velocity model by 3D VSP time tomography at 30 ft receiver's spacing. There is evidence of a fault zone on a 3D WVSP VSPCDP image at 0.617 s

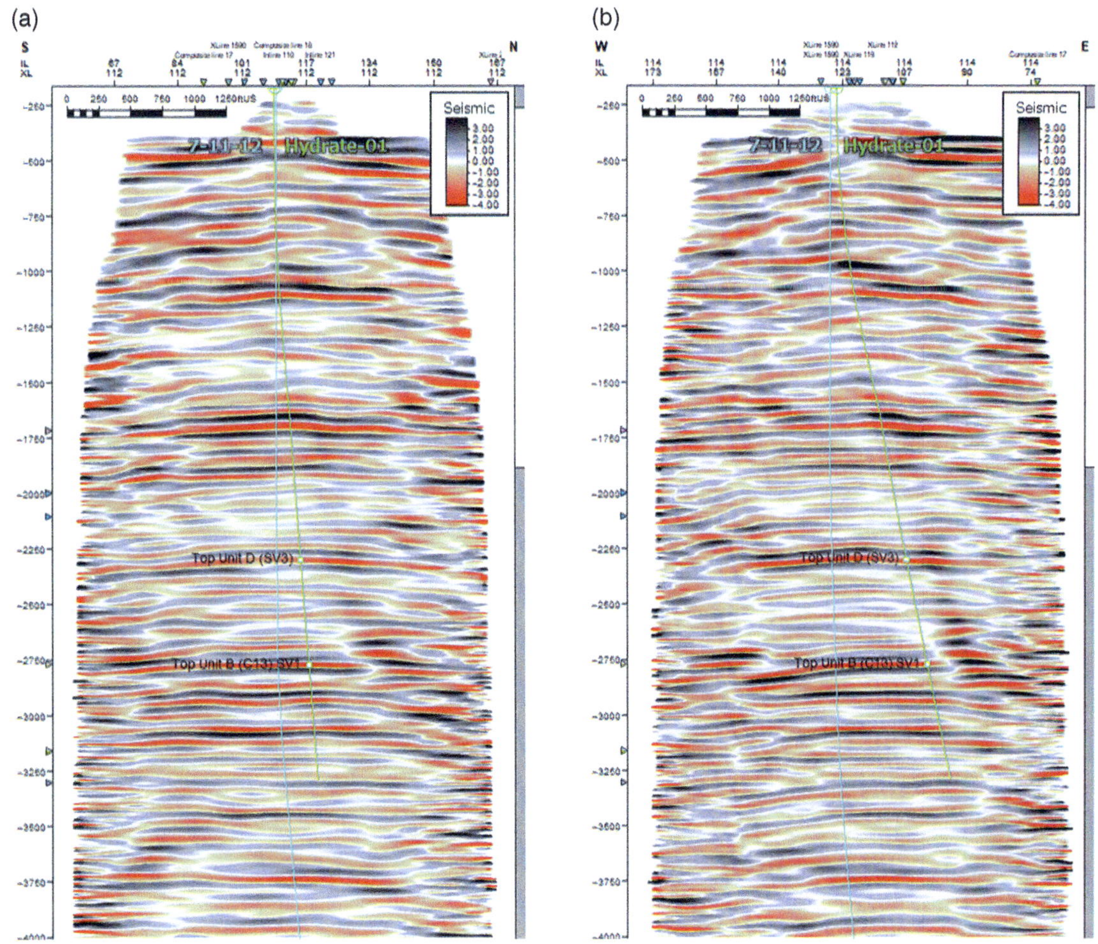

Figure 12.21 Example of anisotropic processed 3D VSP depth sections in North–South (a) and East–West (b) directions with the well trajectory projected. Solid line shows well trajectory (blue, 7-11-12 well; green, Hydrate-01 well), interpreted target horizons.

(two way time, MSL), 843.7 m (true vertical depth, MSL) approximately 150 ft from the well trajectory in the north–east direction.

12.6. STRUCTURAL INTERPRETATION

One of the key objectives of the 3D VSP survey was to improve the structural interpretation of the survey area (Lim et al., 2020). The first step in interpretation is to perform a well-tie analysis with the Hydrate-01 logs and a checkshot obtained from vertical incidence VSP (VI VSP). A key marker was picked on the peak seismic events that correspond to the top of the hydrate formations (Fujimoto et al., 2021). The high-resolution image from the 3D VSP allowed substructures to be delineated, as shown in the Figure 12.22 (a). In addition, the edge enhancement attribute cross-section was created using the 3D VSP volume to reveal more detailed fault distribution between the 7-11-12 (Lewis & Collett, 2013) and Hydrate-01 wells. This attribute emphasizes slight features of the data and improves the fault visualization, as shown in Figure 12.22(b) (Tamaki et al., 2022). Interpretation accuracy near the wellbore is very good due to the high quality of the seismic data. However, accuracy far from the wellbore decreases due to the poor signal-to-noise caused by the residual PS waves and the low seismic bin fold.

The variance attribute (Figure 12.23) was calculated using the 3D VSP volume to emphasize the discontinuity of the seismic events near Hydrate-01 and was estimated for the top of two hydrate-bearing horizons. These discontinuity patterns from the attributes indicate minor faults

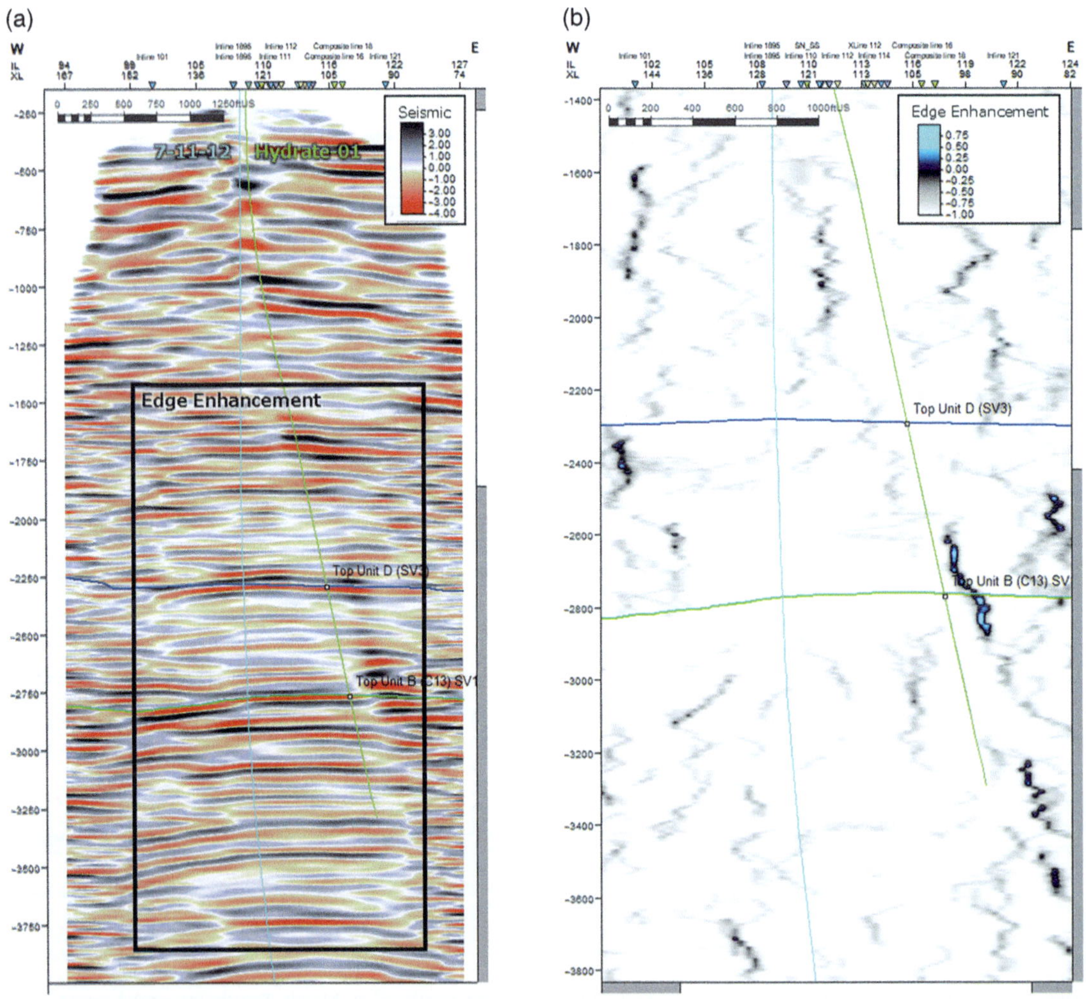

Figure 12.22 (a) Composite display of 3D VSP imaging along the well plane with Hydrate-01 wellpath. (b) Edge enhancement of the box in (a).

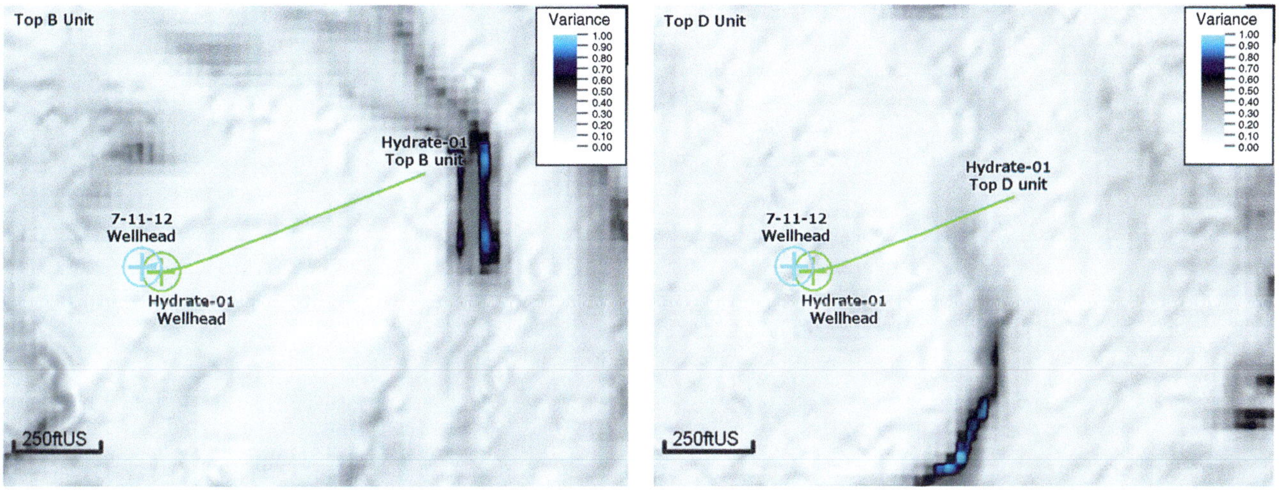

Figure 12.23 Variance attribute on primary (left) and secondary (right) hydrate horizons. Solid lines mean well trajectory from wellhead to the hydrate horizon (well markings: blue, 7-11-12; green, Hydrate-01).

that cannot be recognized by the 3D surface seismic. Therefore, subseismic scale faults are considered to be distributed in the vicinity of hydrate formations around the Hydrate-01 well.

12.7. CONCLUSION

Results from the largest known onshore DAS 3D VSP survey show the recorded DAS data were of high quality, and the processing result has revealed minor faults within the survey area, which cannot be imaged by the surface seismic data. This result is used to determine locations of planned monitoring and producer wells for the production test and will also be incorporated into a 3D geological model.

ACKNOWLEDGMENTS

The authors gratefully acknowledge Alaska MH Project owners the Ministry of Economy, Trade and Industry (METI), United States Department of Energy (DOE), National Energy Technology Laboratory (NETL), U.S. Geological Survey (USGS), and MH21-S consortium for permission to present this work.

The authors also would like to thank SAExploration for VSP shooting and Silixa for recording and converting the strain data. The data were processed and imaged by ASTO Geophysical Consulting. TOYO Engineering Corporation supervised and supported all the work.

Any use of trade, product, or firm names is for descriptive purposes only and does not imply endorsement by the U.S. Government or Japanese Government.

AVAILABILITY STATEMENT

The data presented in this manuscript are confidential and commercially restricted. They are not publicly available.

REFERENCES

Blias, E. (2012). Interval anisotropic parameter estimation from walkaway vertical seismic profiling data. *Geophysical Prospecting*, *60*, 1054–1067. doi:10.1111/j.1365-2478.2012.01108.x

Fujimoto, A., Lim, T., Tamaki, M., Kawaguchi, K., Kobayashi, T., Haines, S., et al. (2021). *DAS 3D VSP survey at stratigraphic test well (Hydrate-01)*. Paper presented at the 14th SEGJ International Symposium, Online. doi:10.1190/segj2021-006.1

Hardage, B. A. (1985). Vertical seismic profiling. *The Leading Edge*, *4*(11), 59. doi:10.1190/1.1487141

Hartog, A. (2017). *An introduction to distributed optical fiber sensors*. Series in fiber optic sensors. Boca Raton, FL: CRC Press. doi:10.1201/9781315119014

Lewis, K. A., & Collett, T. S. (2013). *Brookian sequence well log correlation sections and occurrence of gas hydrates, north-central North Slope, Alaska* (Scientific Investigations Report 2013–5050). US Geological Survey. Retrieved from http://pubs.usgs.gov/sir/2013/5050

Lim, T. K., Fujimoto, A., Kobayashi, T., & Mondanos, M. (2020). *DAS 3D VSP data acquisition for methane hydrate research*, Paper presented at the EAGE Workshop on Fiber Optic Sensing for Energy Applications in Asia Pacific. doi:10.3997/2214-4609.202070014

Tamaki, M., Fujimoto, A., Boswell, R., & Collett, T. (2022). Geological reservoir characterization of a gas hydrate prospect associate with the Hydrate-01 stratigraphic test well, Alaska North Slope. *Energy Fuels*, *36*(15), 8128–8149.

13

Potential of Seismic Attenuation for Exploring Complex Media: A Focus on Carbonate Rocks

Fateh Bouchaala, Seda Rouxel, and Bing Zhou

ABSTRACT

Considerable effort has been dedicated to the potential of seismic attenuation for investigating complex media, such as fractured reservoir zones. Several seismic attenuation studies have been conducted on carbonate rocks, but far fewer than in sandstones. Gray zones may be explained by the complex lithology and intricate porous network, which makes carbonate media particularly challenging to analyze in terms of seismic attenuation. Nevertheless, significant efforts have been devoted to shedding light on the sensitivity of seismic attenuation to petrophysical properties and its mechanism in carbonate rocks. The notable difference between sandstones and carbonate rocks was observed in the compressional to shear attenuation ratio $\frac{Q_P^{-1}}{Q_S^{-1}}$ and scattering magnitudes. In fully saturated media, $\frac{Q_P^{-1}}{Q_S^{-1}}$ exhibits magnitudes higher than one, whereas in partially saturated media, it exhibits magnitudes lower than one, contrary to siliciclastic media. It is worth noting that in carbonate rocks, the scattering mechanism is more significant than in siliciclastic media, where it is often considered insignificant. The squirt flow mechanism becomes more prominent at sonic frequencies, whereas global flow and frictional sliding may dominate at other frequencies. This highlights the complexity and frequency-dependent nature of attenuation mechanisms in carbonate rocks.

13.1. INTRODUCTION

Seismic wave attenuation largely contributes to the decay of wave amplitude during seismic propagation. Attenuation is often quantified by the dimensionless quality factor Q, which represents the ratio of stored energy to the energy lost in one cycle, $\Delta E (Q = 2\pi E/\Delta E)$. Hereafter in this chapter, the inverse of the quality factor Q^{-1} is referred to as *attenuation*.

Geoscientists and engineers have long been interested in attenuation because of its sensitivity to the physical properties of geological media: grain size (Hamilton, 1972; Pinson et al., 2008), clay content (Best & McCann, 1995; Bouchaala et al., 2019a), porosity (Bradley & Fort, 1966; Klimentos & McCann, 1990), and fluid saturation (Qi et al., 2017; Winkler & Nur, 1979). Furthermore, an accurate estimate of Q^{-1} helps compensate for the amplitudes of attenuated seismic waves, which can lead to improvements in seismic imaging (e.g., Harris & Zhu, 2014; Zhang et al., 2016). However, an accurate estimate of Q^{-1} can be challenging because of its frequency dependency, complex geological structures leading to complex attenuation patterns that are difficult to predict and model accurately, and model uncertainties due to simplified representations of complex subsurface conditions. Those challenges are largely due to the misunderstanding of attenuation mechanisms, which are unpredictable in complex media such as carbonate rocks. The latter host a significant portion of the world's oil reservoirs, and attenuation is a

Department of Earth Sciences, Khalifa University of Science and Technology, Abu Dhabi, UAE

Distributed Acoustic Sensing in Borehole Geophysics, Geophysical Monograph 289, First Edition.
Edited by Yingping Li, Robert Mellors, and Ge Zhan.

DOI: 10.1002/9781394179275.ch13

potential seismic parameter for exploring these media. Hence, seismic attenuation remains an ongoing and active area of research in carbonate rocks, with researchers continuing to address the mentioned challenges.

The complexity of carbonate media compared to sandstones comes from the complexity of their porosity structure, which is controlled by many factors: the initial depositional environment, cementation, dissolution, and biological processes. This leads to a complex and irregular distribution of porosity within the rock.

The objective of this chapter is to provide valuable insights into seismic wave attenuation in carbonate rocks by exploring the most relevant studies and incorporating new results obtained by the authors. Those studies use a multidisciplinary approach that combines seismic data analysis with geological and petrophysical information. Additionally, the use of advanced techniques, such as full-waveform inversion (FWI) and rock physics modeling, can help in better characterizing seismic attenuation in carbonate rocks and improving reservoir characterization. This chapter aims to enhance the understanding of attenuation mechanisms, investigate the impact of petrophysical properties on attenuation, and provide clarity in interpreting attenuation results. This study clarifies the challenges of attenuation studies in carbonate rocks, their causes, and the distinctions from sandstones. Furthermore, this study elucidates the connections between the effects of petrophysical parameters on seismic attenuation, its mechanisms, and frequency dependence. This clarification aids in the improved interpretation of attenuation, which in carbonate rocks typically is not straightforward and can occasionally be biased due to a misunderstanding of attenuation mechanisms and rock properties. For instance, interpretation of fluid saturation based on compressional to shear attenuation ratio $\frac{Q_P^{-1}}{Q_S^{-1}}$ is one of the most controversial seismic attenuation interpretations in carbonate rocks. In sandstones, it is widely accepted that the magnitude of $\frac{Q_P^{-1}}{Q_S^{-1}}$ is greater than one in partially saturated media and less than one in fully saturated media (Winkler & Nur, 1982). However, in carbonate rocks, there is no consensus regarding the magnitude of $\frac{Q_P^{-1}}{Q_S^{-1}}$ in saturated media, and this point will be addressed in this chapter.

13.2. ATTENUATION MECHANISMS

Attenuation is influenced by various frequency-dependent mechanisms, which can be broadly categorized into two main types: scattering and anelastic mechanisms. Scattering occurs due to the heterogeneities present in the medium, leading to small-scale variations in elastic properties. This generates low-amplitude reflected waves, causing the dissipation of energy from the primary waves. The losses caused by scattering are elastic in nature, as they involve the redistribution of energy without conversion into other forms of energy (Wu & Aki, 1988). Anelastic mechanisms, in contrast, involve the dissipation of energy through conversion, typically resulting in heat. These energy losses occur due to a variety of mechanisms within the medium (Figure 13.1). Wave-induced fluid flow (WIFF) is one of the most well-known anelastic mechanisms; it arises from frictional movement between pore fluids and solid grains induced by wave propagation (Bouchaala et al., 2019a; Müller et al., 2010). When a seismic wave passes through wet, porous rock, pressure gradients cause the fluid to move relative to the solid pore, creating internal friction until the pore pressure is equilibrated.

Several models have been proposed to explain the influence of fluid on seismic attenuation depending on the considered scale (Figure 13.1(a)) (Müller et al., 2010; Rubino & Holliger, 2013). At a macroscopic level, when the pressure gradients are of a magnitude similar to the wavelength, the WIFF is referred to as global flow. The global flow effect on seismic wave attenuation was first described by Biot with the elastic theory of poroelasticity (Biot, 1956). Rocks that are homogeneous, highly porous, and permeable are favorable for the occurrence of the global flow mechanism. It is commonly accepted that this mechanism is dominant at high frequencies above 0.1 MHz, which fall outside the seismic frequency range (Bourbié et al., 1992; Milani et al., 2016).

Conversely, at the microscopic scale, compliant pores such as intragrain cracks undergo more significant deformation compared to rigid intergranular pores during wave propagation, resulting in localized pressure gradients and high-speed frictional motion between the fluid and solid. This local-scale mechanism is known as the squirt flow mechanism and is believed to be particularly prominent at sonic and ultrasonic frequencies (Pride et al., 2004; Sun, 2021; Winkler, 1985).

In between, at a mesoscopic scale, the heterogeneous distribution of fluid and rock properties creates a compliance contrast, inducing a fluid pressure gradient when traversed by a seismic wave. Here, the mesoscopic scale is understood to cover dimensions from the largest pore size to the smallest wavelength, which, depending on the considered frequency, implies spatial dimensions ranging from centimeter to multimeter scale (Müller et al., 2010). Examples of mesoscopic heterogeneity include lateral lithological variation, bed layering, fractures, and patchy fluid saturation. Given this large range of action, it is believed that the mesoscopic WIFF has a significant influence on

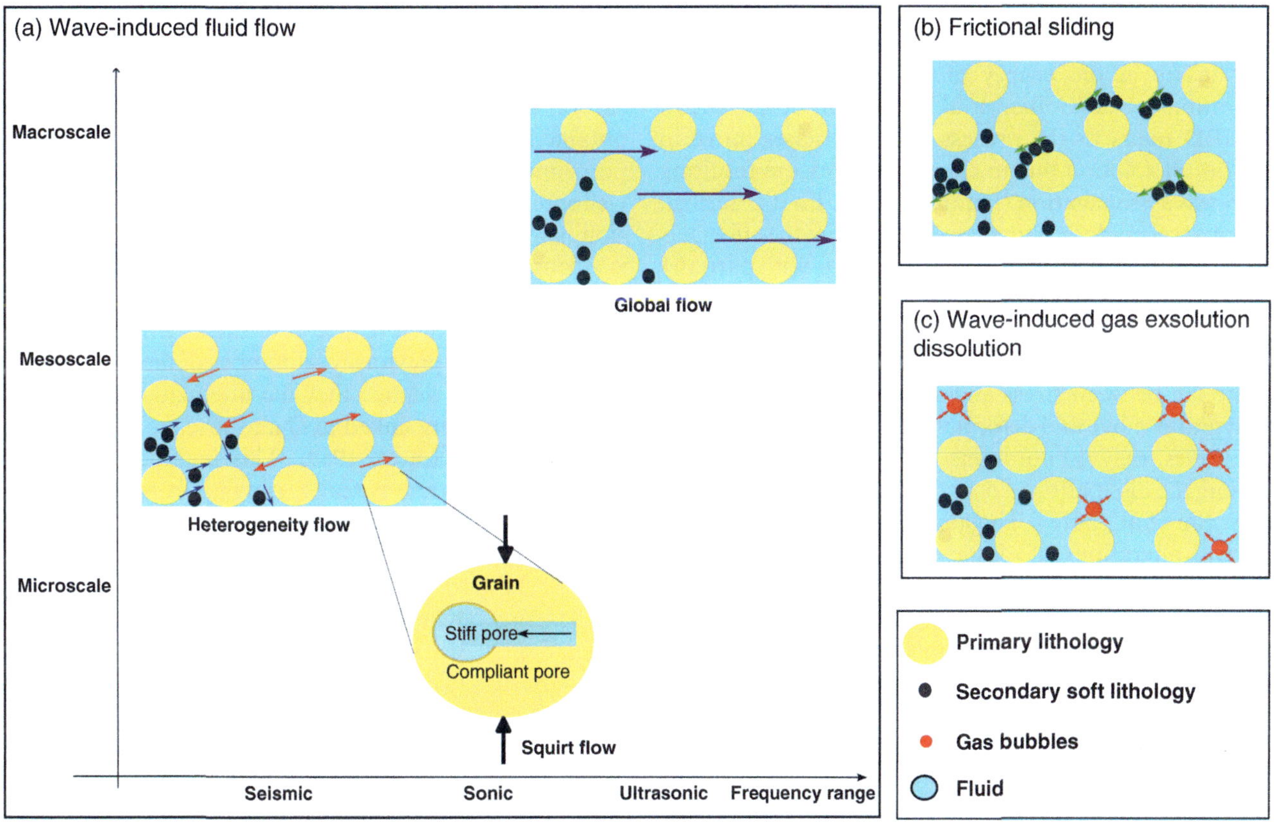

Figure 13.1 Illustration of main anelastic attenuation mechanisms occurring in geological media at different scales and frequency ranges.

seismic wave attenuation and at a large frequency range (Sun, 2021). For all scenarios, the viscous friction resulting from the relative fluid–solid motion causes energy dissipation, which results in seismic wave amplitude decay and dispersion (Rubino & Holliger, 2013). Thermal relaxation of pore fluids can also cause attenuation but with significant effects only at elevated temperatures (>150 °C) (Kjartansson, 1980).

Fluid-free anelastic mechanisms are also responsible for seismic attenuation and are grouped under the term *intrinsic matrix anelasticity*. It combines the effect of individual mineral matrix anelasticity, aggregated mineral anelasticity, and frictional sliding (Walsh, 1966). If mineral-related attenuation is extremely small, sliding manifests in geological environments exemplified by the wave-induced motion of solid grains at grain boundaries or narrow cracks (Mavko & Nur, 1979). The matrix anelasticity effect on attenuation has usually been considered negligible at the low-strain and high-confining regime, typical of oil and gas reservoirs and seismic surveys (Winkler & Nur 1982). However, Barton (2006) noticed that this mechanism could be nonnegligible in the near surface, less than a kilometer deep, where the confining pressure is much lower, particularly if the rock is considered at the rock mass scale. Kuteynikova et al. (2014) included the matrix anelasticity effect to validate the numerical mesoscopic fluid flow model at seismic frequency and surface conditions. When seismic waves pass through a medium with solid grains, the motion and interaction between these grains generate frictional forces, leading to energy losses. After conducting an experimental study on Solenhofen limestone samples, Peselnick and Outerbridge (1961) suggested the presence of shear internal friction at grain boundaries across a wide frequency range from 4 to 10^7 Hz. The strength of the relative motion of the grains is proportional to the elastic contrast between grains (Figure 13.1(b)) (Mavko & Nur, 1979). Guerin and Goldberg (2005) conducted a seismic wave attenuation study in gas hydrate-bearing sediments and suggested that the magnitude of Q_s^{-1} is dominated by the losses due to the frictional solid motion between solid grains (small deformation) and soft hydrates (large deformation). More recently, Bouchaala et al. (2019a), who estimated sonic attenuation from carbonate reservoir logging data, noticed a decrease in the magnitude of $\frac{Q_P^{-1}}{Q_S^{-1}}$ in clay-rich dense zones compared to the other zones. They explained it based on frictional solid motion between carbonate (hard) and clay (soft) grains.

Wave-induced gas exsolution dissolution (WIGED) is another attenuation mechanism that can cause significant losses in the energy of seismic waves for low-saturation rocks (Collier et al., 2006; Jin et al., 2023). WIGED refers to the process in which gases move in and out of solution due to changes in pressure resulting from wave propagation (Figure 13.1(c)). First described in global seismology and related to magma, the effect has since been described in laboratory settings with partially saturated sandstones at exploration seismic frequencies (Tisato et al., 2015). Seismic attenuation can also be attributed to the hysteresis phenomenon, which is due to the lag between stress and strain in the material. As solid grains move and deform in response to passing seismic waves, they do not immediately reach equilibrium, causing the dissipation of energy (e.g., Guyer & Johnson, 2009; Shi et al., 2018). To gain a deeper insight into the spectral alterations of elastic wave energy in partially saturated rock, Rozhko (2021) took into account the hysteresis exhibited by liquid within cracks, often called a liquid bridge. Rozhko explained the energy boost at lower frequencies by the nonlinear energy exchange between frequencies. Furthermore, the study noticed that attenuation due to the contact line is larger than fluid-flow-related viscous attenuation within the partially saturated crack. The contact line friction is related to the hysteresis phenomenon and occurs at the interface or boundary between two different phases, such as a solid–liquid interface or a solid–gas interface (e.g., Shi et al., 2018; Zhang, 2016).

13.3. CORRELATION BETWEEN SEISMIC ATTENUATION AND KEY PETROPHYSICAL PARAMETERS

Given the varied mechanisms controlling the attenuation of seismic waves, many characteristics of the rock mass can potentially influence seismic amplitude decay. Therefore, attenuation has been linked to several rock characteristics, and a focus was places on their link with fluid flow, porosity, permeability, fluid saturation, the presence of clay, fractures, and cracks. Among them, porosity characteristics and fluid saturation seem to have a strong influence on seismic attenuation (Ba et al., 2015; Winkler & Nur 1979). If most current studies are applied to hydrocarbon reservoirs, it is thought similar techniques could be used for CO_2 sequestration monitoring (Azuma et al., 2014), geothermal monitoring (Guo & Thurber, 2022), or underground aquifer levels (Lorenzo et al., 2013).

13.3.1. Porosity and Permeability

Heterogeneity and hydraulic properties of geological media depend significantly on porosity (Ø) and permeability (K). Hence, getting an estimate of these two petrophysical properties is essential for the characterization of porous and saturated media, such as hydrocarbon reservoirs and geological formations used for carbon storage. The values of Ø and K can be directly obtained in the laboratory from core samples using porosimeter and permeameter instruments, and from wellbores in the field using resistivity, neutron emission, and fluid saturation measurements. The fluid flow is strongly controlled by Ø and K, making attenuation largely influenced by Ø and K and hence a good alternative for a remote prediction of these two parameters.

As mentioned previously, a number of research studies have provided evidence that the attenuation of seismic waves is intricately linked to Ø and K (e.g., Klimentos & McCann, 1990; Lucet, 1989). This relationship can be assessed across various scales, ranging from seismic frequencies to ultrasonic frequencies. A positive correlation between Q^{-1} and Ø should be expected because the increase in porosity spurs fluid-related and scattering mechanisms (Klimentos & McCann, 1990; Prasad, 2003; Sang, 2016). In a seminal study that included 42 sandstones of varied provenance and covered a large range of porosities (2.9–35%), Klimentos and McCann (1990) found a quasi-linear relationship between Q^{-1} and Ø, with clay content acting as another parameter controlling Q^{-1} (Figure 13.2(a)). Sang (2016) conducted a similar study on carbonate samples and revealed more complex behavior (Figure 13.2(c)). Below 20% porosity, the correlation between Q^{-1} and Ø closely resembles that of sandstones, showing a linear trend. However, above this threshold, the data points exhibit increasing scatter, which may be attributed to differences in pore types (Sang, 2016). Based on the extensive work on attenuation in carbonate rocks conducted by the authors, the relationship in carbonate media is still unclear and ambiguous. The complexity of carbonate rocks primarily arises from the heterogeneities resulting from their intricate porosity network (Jouini et al., 2022) and lithology (Akbar et al., 1995).

The relationship between Q^{-1} and K is not straightforward, although it is intuitively expected that they should be positively correlated because the larger K is, the more fluid-related mechanisms are present (Prasad, 2003). However, Figure 13.2(b), (d) does not show a clear correlation between Q^{-1} and K for both sandstones and carbonates. In sandstones, a weak positive correlation can be observed when K is lower than 50 mD, before becoming negative for K higher than 50 mD. It is worth noting that most low-permeability samples have high clay content, whereas those with high permeability have low clay content (<4%), with the exception of two samples. Those two samples are highly permeable samples with low clay content and Q^{-1} magnitudes on par with low-permeability samples. Therefore, it appears that clay content exerts more control over attenuation magnitude than

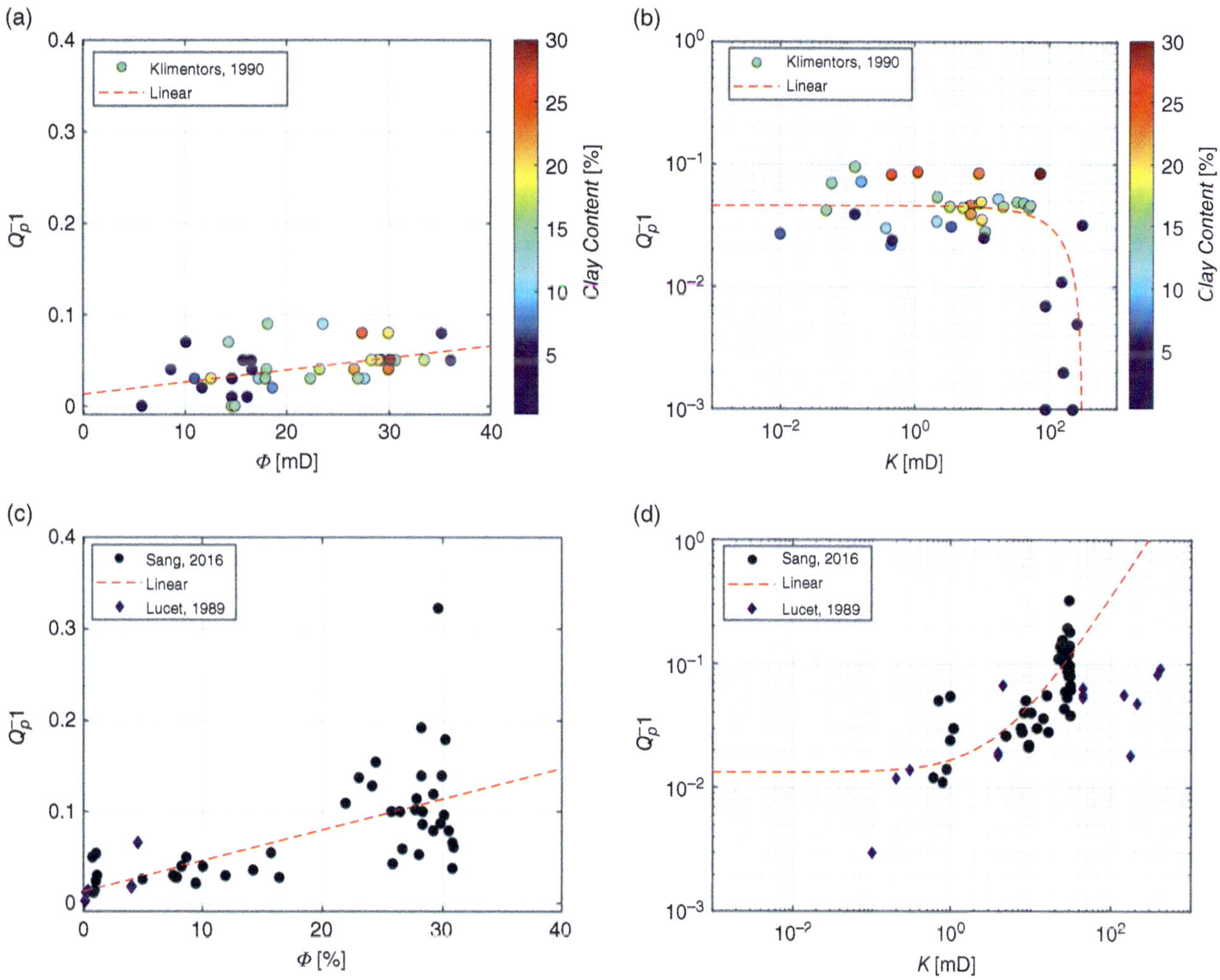

Figure 13.2 Cross-plots of (a) absorption versus porosity and (b) quality factor versus permeability on log/log scale, measured in a water-saturated sample. Adapted from Klimentos & McCann (1990). (c) Total attenuation versus porosity and (d) total attenuation versus permeability on log/log scale measure in water-saturated carbonates samples with linear fitting of Sang's data. Adapted from Lucet (1989) and Sang (2016).

permeability. This can be explained by the frictional solid motion mechanism resulting from elastic contrast and squirt flow due to fluid motion between sand and clay grains. Klimentos and McCann (1990) demonstrated that Q^{-1} magnitudes of highly permeable and free-clay samples fit with the Biot (1956) model, which predicts the global flow mechanism.

Sang (2016) showed a linear trend in the variation of Q^{-1} with K (Figure 13.2(d)) obtained in carbonate samples with low clay content.

Figure 13.3 shows a cross-plot of compressional attenuation Q_P^{-1} versus Ø with a K color bar for dry (Figure 13.3(a)) and water-saturated samples (Figure 13.3(b)). The attenuation was estimated in a laboratory at in situ pressure from carbonate core plugs from an oil reservoir located in Abu Dhabi, UAE. The cross-plots do not exhibit a clear trend under dry and saturation conditions. The tight samples display slight attenuation magnitude changes from dry to saturation conditions; however, significant changes can be noticed for porous and permeable samples (Figure 13.3). Furthermore, K does not show a clear correlation with Q_P^{-1}. Therefore, other parameters and factors may control the variation of Q_P^{-1} with these two parameters.

13.3.2. Fluid Saturation

Fluid saturation was linked early on to attenuation when Q^{-1} changes were first observed from dry to water-saturated samples. Toksöz et al. (1979), who experimentally measured Q_P^{-1} and Q_S^{-1} in sandstone

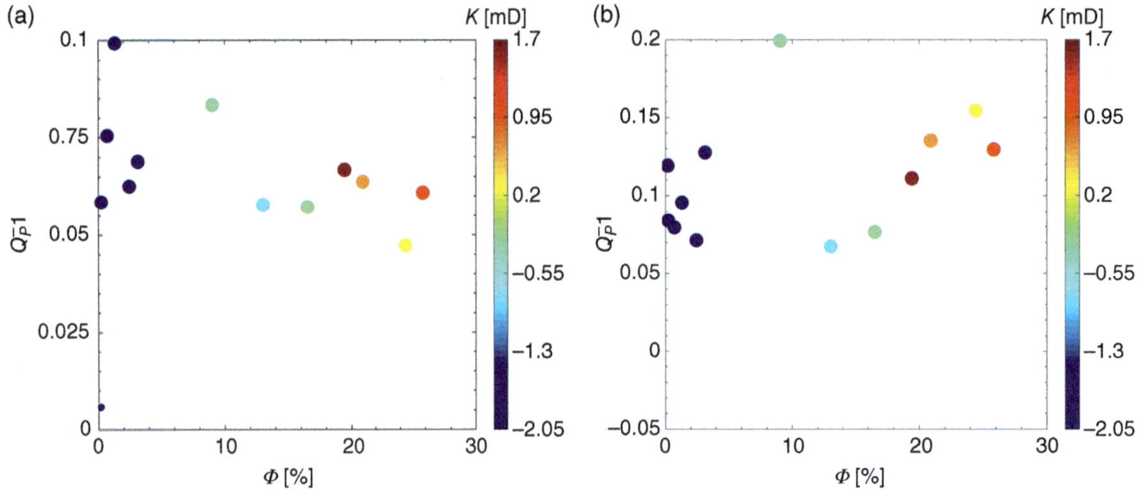

Figure 13.3 Cross-plots of total compressional attenuation versus porosity measured in (a) dry and (b) water-saturated samples. The permeability color bar is on a logarithmic scale.

and limestone samples, reported that they are frequency independent under dry and saturated conditions in the frequency range 0.1–1.0 MHz and they decrease with differential pressure. They also noticed that Q_P^{-1} is slightly higher than Q_S^{-1} under dry conditions and lower under saturation conditions. Whereas in dry intact rocks, the intrinsic attenuation is weak (Figure 13.4(a)), it increases sharply with fluid presence and peaks in fully saturated samples, and the effect is higher on S waves than P waves. Compressional and shear attenuation increase substantially when the rock is partially saturated, with Q_P^{-1} being affected about twice as much as Q_S^{-1} (Figure 13.4(b)). As total saturation is achieved, Q_S^{-1} becomes more important than Q_P^{-1} (Figure 13.4(c)). However, between those extrema, attenuation exhibits a nonlinear evolution with partial peak centered around 80% saturation only for P waves, whereas S-wave attenuation peaks sharply over 90% saturation (Walsh, 1995; Winkler & Nur, 1982). Therefore, the $\frac{Q_P^{-1}}{Q_S^{-1}}$ ratio is a good indicator of fluid-saturation degree in rocks, even better than the velocity ratio V_P/V_S.

Qi et al. (2017) estimated $\frac{Q_P^{-1}}{Q_S^{-1}}$ from sonic waveforms recorded in shale formations to identify gas-rich zones. Bouchaala et al. (2021) suggested that the $\frac{Q_P^{-1}}{Q_S^{-1}}$ ratio is mainly controlled by the degree of saturation in pore fluids as well as clay content. Conditions in a hydrocarbon reservoir are most realistically described by a mixed fluid partial saturation such as water plus gas or water plus oil,

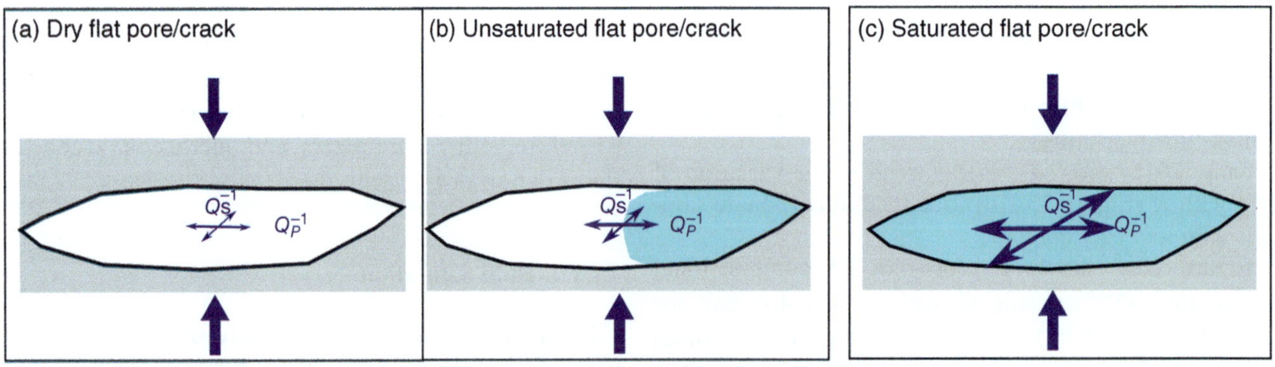

Figure 13.4 Attenuation increasing with increasing liquid saturation in a flat crack/pore: (a) dry conditions, (b) partially saturated conditions, and (c) fully saturated. Adapted from Winkler & Nur (1982).

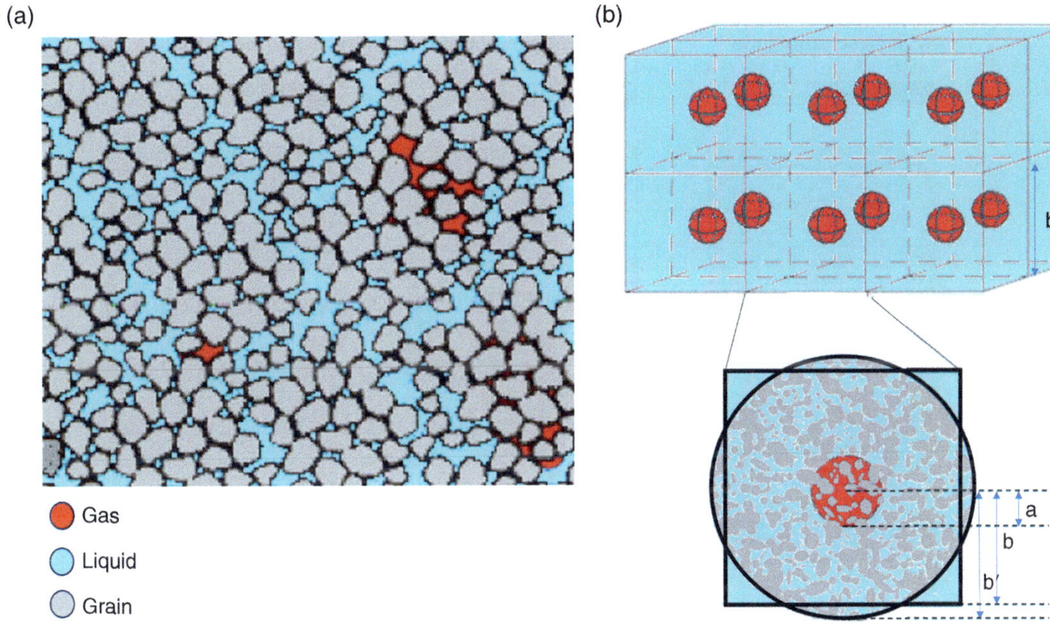

Figure 13.5 Representation of a patchy saturated model considering (a) gas bubbles intermingled between liquid and grain, and (b) modeling representing gas pockets by a series of spherical objects, each sphere surrounded by a liquid shell having a radius *b*. Adapted from Ba et al. (2015) and Vogelaar et al. (2010).

suggesting a combined effect on attenuation from the mixture referred to as *patchy saturation* (Figure 13.5(a)) (Chapman et al., 2021; Masson & Pride, 2011). White (1975) noticed a significant increase in the magnitude of Q_P^{-1} for partially saturated samples, whereas the magnitude of Q_S^{-1} was not affected as there is no shear modulus contrast between two nonviscous fluids. To explain this phenomenon, patchy saturation was considered a series of liquid elementary volumes in which partial saturation was modeled as a concentric spherical gas pocket of similar volume in which the inner pocket was filled with gas (Figure 13.5(b)). This model was later refined and made more complex by incorporating additional heterogeneities, such as double porosity, as suggested by Ba et al. (2015).

Various impacts of fluid saturation on seismic attenuation make it an attractive pathway to monitor reservoir production (Chapman et al., 2021). However, the dominance of attenuation modes in carbonate rocks is still ambiguous and controversial. Agersborg et al. (2008) and Adam et al. (2009) reported the dominance of Q_P^{-1} over Q_S^{-1} losses estimated in fully saturated carbonate samples at ultrasonic and sonic frequencies, respectively. Measurements by Best et al. (2007) of $\frac{Q_P^{-1}}{Q_S^{-1}}$ at ultrasonic frequencies in carbonate rock samples show mixed results, higher and lower than 1. The first author of this chapter has been significantly involved in many attenuation studies conducted in the Abu Dhabi subsurface, which has carbonate-dominated lithology; these have shown that $\frac{Q_P^{-1}}{Q_S^{-1}}$ is larger than 1 in saturated media, which is contradictory with clastic rocks. The results of a new study conducted in a reservoir in Abu Dhabi confirm the dominance of compressional over shear attenuation in fully saturated samples (Figure 13.6). These controversial results between carbonate and sandstone rocks may be explained by the difference in their textures and porosity structures, which are more complicated in carbonate rocks.

13.3.3. Fractures

Natural fractures, joints, and faults are important sources of rock heterogeneities and are ubiquitous over a wide range of length scales from micrometers to hundreds of kilometers. As they create conduits for fluids, as a result of the compressibility contrast between fracture and the host rock, flow can occur. Attenuation increases with fracture density (Peacock et al., 1994), which is mainly attributed to fracture-related WIFF. This mechanism can occur in two ways: either from fracture to fracture or from fracture to porous background (Figure 13.1) (Caspari et al., 2018).

Apart from density, the effect of fracture on seismic attenuation depends on several factors: size and

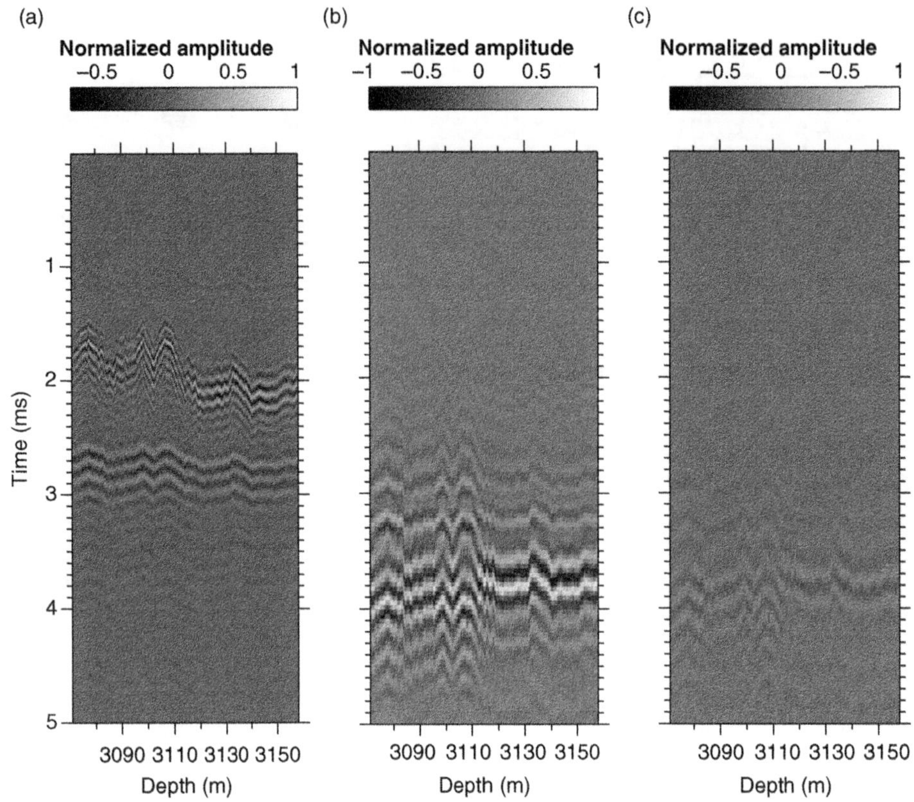

Figure 13.6 Compressional to shear total attenuation ratio measured in brine fully saturated carbonate samples from a Middle East reservoir.

connectivity of fractures within the rock. For example, highly interconnected fracture networks tend to cause more significant attenuation because they provide a better pathway for fluid flow. Quintal et al. (2013) observed that Q_S^{-1} variations could be linked to fracture connectivity. Based on the fact that the magnitude of attenuation is expected to be minimal at fracture strike direction, Bouchaala et al. (2019b) carried out a multiazimuthal analysis of seismic attenuation from walk-around vertical seismic profiling (VSP) data to determine the orientation of fractures. Furthermore, the scholars quantified the attenuation related to fracture-induced squirt flow, allowing them to distinguish between open and cemented fractures.

13.3.4. Lithology and Pore Structures

The literature has reported a large range of possible Q values associated with medium lithology; it is under 20 in some carbonates, 50 to 250 in sandstones, and several hundred for deep metamorphic and igneous rocks (Barton, 2006; Parra & Hackert, 2006). Although attenuation in sandstone reservoirs has historically been the most studied type of hydrocarbon reservoirs, recent years have witnessed a growing interest in carbonate rocks, because they hold a large amount of hydrocarbon resources. Compared to siliciclastic rocks (Figure 13.7(a)), carbonate rocks exhibit a highly heterogeneous, multiscale structure characterized by fractures and diverse lithological textures with a wide range of pore types (Figure 13.7(b)). The difference in porosity structure between carbonate rocks and sandstones can be attributed to two factors. First, the biological origin of carbonates results in specific growth frame pores and complex intraparticle pores (Ahr, 2011). Second, the chemical reactivity of carbonates makes them highly susceptible to diagenetic alterations, including dissolution and cementation, leading to significant mineralogical changes, particularly in cases of dolomitization, which, in turn, alters their pore structure (Eberli et al., 2003).

Carbonate textural complexity can create internal zones of varying elastic properties, which in turn affect fluid flow. Sang (2016) conducted an experimental seismic attenuation study on 47 core plug samples obtained from a Lower Cretaceous carbonate reservoir situated along a transition from a platform to a basin. These samples exhibit a wide spectrum of depositional facies encompassing those found in the platform interior, marginal areas, and slopes. Based on the porosity, permeability, and

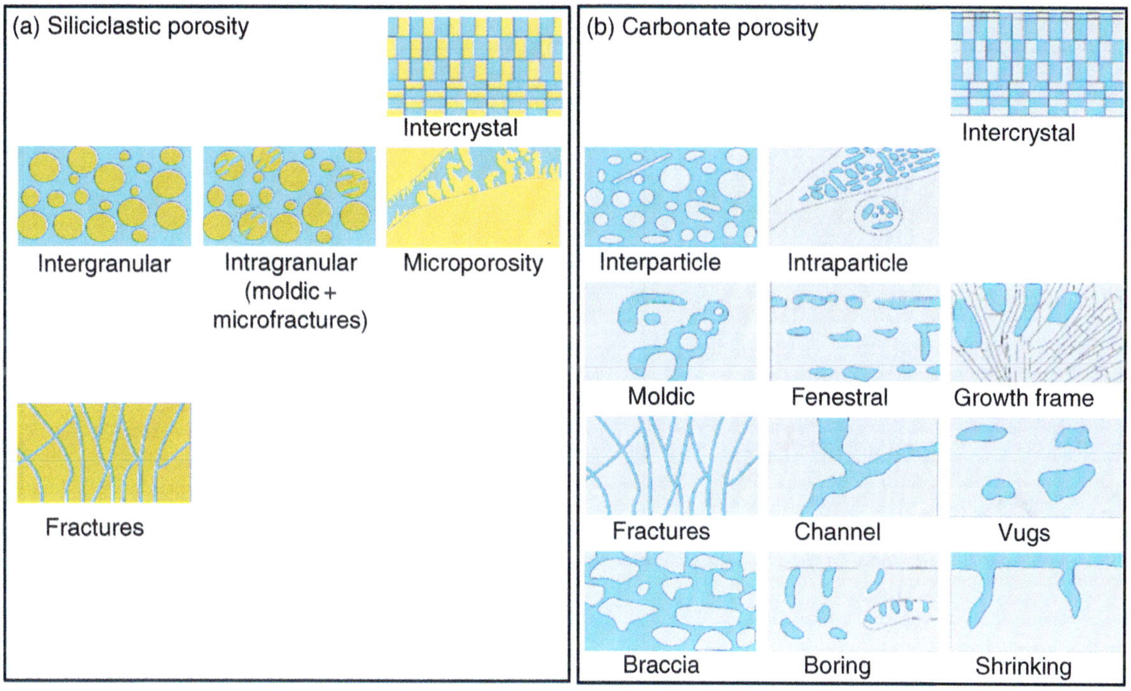

Figure 13.7 Typical porosity structure of (a) siliciclastic and (b) carbonate rocks. Adapted from Janjuhah et al. (2021) and Pittman (1979).

texture of the samples, Sang (2016) proposed four reservoir rock types (RRT): RRT-1 has a porosity less than 20% of micropore type and a permeability less than 1 mD, and its texture is dominated by mudstone, wackestone, and mud-dominated packstone. RRT-2 has a porosity higher than 20% of micropore type and a permeability in the range of 1–10 mD, and its texture is made of mud-dominated packstone. RRT-3 has a porosity higher than 20% of micropore-intraparticle type and a permeability in the range of 1–18 mD, and its texture is dominated by floatstone, wackestone, mud-dominated packstone, and boundstone. RRT-4 has a porosity higher than 20% of interparticle type and a permeability higher than 10 mD, and its texture is made of floatstone, packstone, and grainstone. The cross-plots displayed in Figure 13.8 show a slight increase of Q_P^{-1} and Q_S^{-1} versus porosity. Furthermore, RRT-1 and RRT-2 display the smallest attenuation magnitudes and are separated at a porosity of 20% corresponding to the cutoff porosity used for the rock typing. The RRT-4 samples have the largest attenuation, and those of RRT-3 type are complex with high and lower attenuation magnitudes, which was explained by the large-scale heterogeneous fabrics of the samples. Furthermore, carbonate attenuation properties can differ from clastics in the relationships between Q_P^{-1} and Q_S^{-1}, for example, when taking into account the nature of the saturating fluid (Adam et al., 2009).

The rise of unconventional exploration combined with the importance of seal characterization for storage applications has increased the interest in better understanding shale's seismic response. Shales exhibit low permeability and high clay content and are characterized by submicroscopic textures compared to conventional reservoir rocks, which all affect seismic attenuation (Deng et al., 2009). Ba et al. (2015) measured attenuation in siltstone samples and reported higher attenuation in dry than in fully water-saturated samples; the magnitude of Q_P^{-1} was found to decrease with water saturation, which is in disagreement with conventional reservoirs. Deng et al. (2009) found that attenuation in shales exhibits a strong anisotropic behavior related to grain and pore geometry. One study attempted to measure attenuation variation on a shale containing carbon dioxide (CO_2) (Rørheim et al., 2021). Success in quantifying attenuation and understanding its behavior helps in improving the quality of seismic data and imaging of salt basin borders, mining in salty environments, and monitoring of H_2 storage in salt formations.

13.3.5. Pressure and Temperature

Several studies have shown that intrinsic attenuation steadily decreases with depth due to an increase in confining pressure, causing progressive closure of microcracks. Prior studies showed that Q increases with temperature

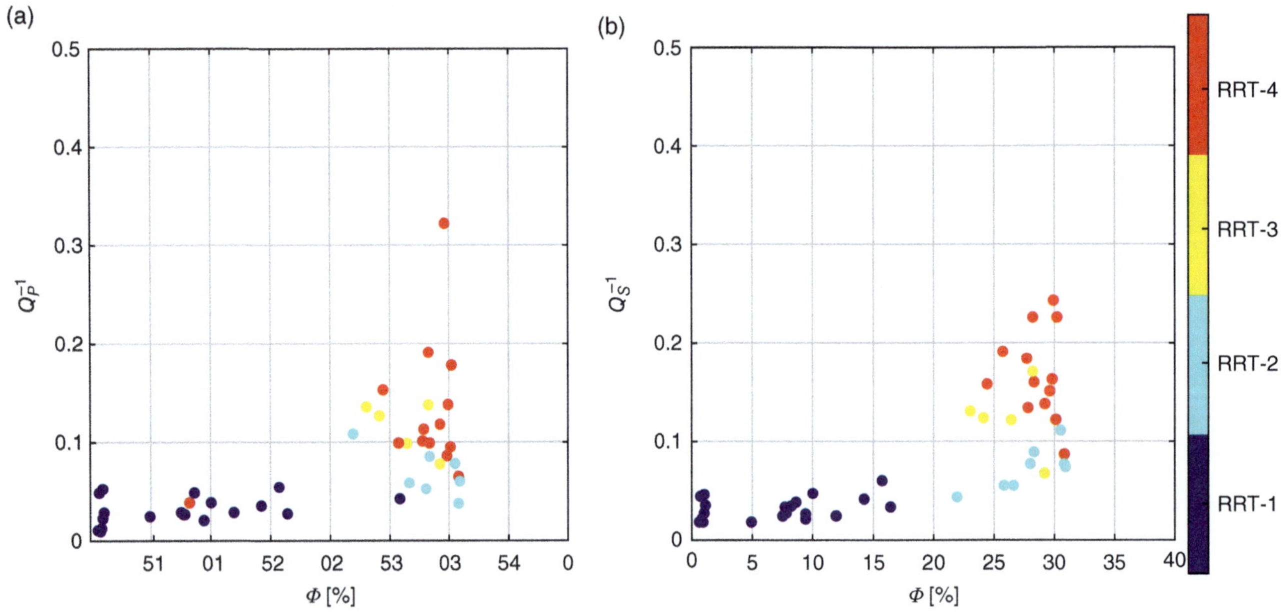

Figure 13.8 Total (a) compressional and (b) shear attenuation variations in carbonate rocks versus porosity and according to rock types. Sang (2016) / with permission of DuraSpace.

up to a peak, and then, beyond a certain temperature, it decreases rapidly (Jaya et al., 2010; Jones et al., 1980). Jaya et al. (2010) attributed this pattern to a lowering of the fluid viscosity, followed by the gas attenuation effect created by vapor bubbles, followed by thermal fracturing at the highest temperatures. Jones et al. (1980) also noted that at the transition phase, when the rock becomes partially saturated, the magnitude of Q_P^{-1} reaches its maximum. Theoretical development and numerical forward modeling may help in unraveling the effect of pressure and temperature on attenuation.

13.4. METHODOLOGY FOR ATTENUATION ESTIMATION

13.4.1. Laboratory Measurements

Several attenuation studies have been conducted in the laboratory under ultrasonic and subsequently under exploration frequencies. By conducting these laboratory experiments, researchers can estimate seismic attenuation in different materials, better understand the effect of petrophysical parameters on attenuation, gain insights into attenuation mechanisms, and validate theoretical models or numerical simulations. Experimental attenuation work is generally done by following this procedure:

1. Sample preparation: Rock samples are carefully selected and prepared to ensure they are representative of the target formation or material of interest. The samples are typically cylindrical pore plugs with diameters of a few centimeters (2.5–5 cm).
2. Poro-perma measurements: Various techniques can be employed to measure porosity, such as helium or gas pycnometry, water saturation, or mercury intrusion porosimetry. These methods involve measuring the volume of fluid or gas displaced by the rock sample. Permeability is determined by quantifying the flow of fluids through the sample under controlled conditions. Permeability testing can be conducted using techniques such as constant-head or falling-head permeametry, where fluids are forced to flow through the sample while measuring the flow rate and pressure drop. Other methods, like gas permeability testing, may be used for specific applications.
3. Saturation: The rock samples are saturated with the desired fluid to simulate the conditions of interest, such as reservoir fluids or groundwater. Saturation is achieved by immersing the samples in the fluid and allowing sufficient time for the fluid to fill the pore space.
4. Acoustic measurements: It is possible to perform the measurements at ultrasonic (MHz) and exploration frequencies (10–100 Hz). The ultrasonic setup typically consists of a transmitter and receiver–transducer pair that generates P and S waves separately. The transmitter generates ultrasonic waves, which are then transmitted through the rock sample. The receiver detects the transmitted waves and

records them as a signal in the time domain, which is compared to the ones recorded in a reference material (often aluminum) to measure attenuation, often by using the spectral ratio (McDonal et al., 1958) or centroid shift frequency (Quan & Harris, 1997) method.

Laboratory measurements of attenuation can also be conducted at sonic frequencies using acoustic resonant bar tests. In this technique, a uniformly cross-sectioned bar-shaped sample is vibrated by a piezoelectric transducer until it reaches the resonance frequency (f_R), at which point the transducer is stopped and the sample's vibration starts to exponentially decay (e.g., Coakley et al., 2003; Wang et al., 1997). The attenuation is calculated by using the formula

$$Q^{-1} = \frac{W}{2\pi f_R} \tag{13.1}$$

where W is the full-width half maximum of the resonance intensity spectrum (Figure 13.9).

Wyllie et al. (1962) utilized an acoustic resonant bar technique to verify Biot's (1956) theory on sandstone and limestone samples. They justified their choice of this technique due to the relative simplicity of the equipment and the ease with which vibrations could be generated. However, obtaining extended samples required for the tests is often a challenging task, especially when it comes to extracting them from the deep subsurface or from geological formations that are fragile and fractured. Furthermore, the resonance of the sample might be influenced by the resonance of the apparatus. Several efforts have been made to address those drawbacks, such as by Nakagawa (2011), who proposed the placement of small-length samples of two metal extended bars with attached transmitter–receiver. Due to the increased length and added mass, the resonance frequency of the entire experimental setup was notably reduced compared to the sample by itself.

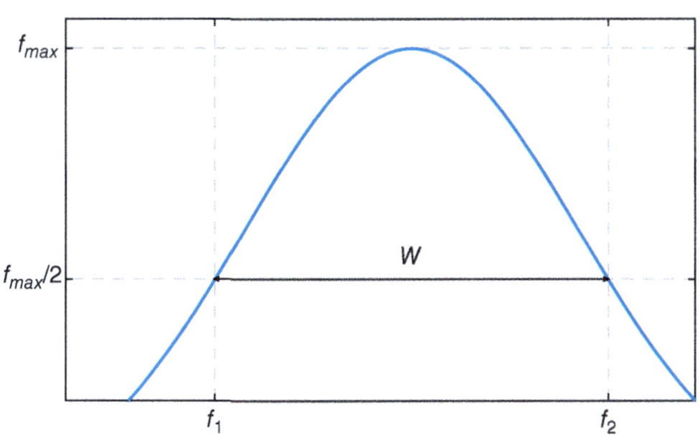

Figure 13.9 Description of full width at half maximum bandwidth for a resonance curve.

At exploration frequencies, the stress/strain phase difference method (Batzle et al., 2006) is used to estimate attenuation. In this case, the rock sample is subjected to a cyclic loading and unloading process using a testing apparatus. This apparatus applies a known stress or strain waveform to the sample while simultaneously measuring the corresponding stress and strain responses. The phase difference between stress and strain can be related to the attenuation of the material. By comparing the observed phase difference with a reference or calibration dataset, it is possible to estimate the attenuation of the rock sample.

13.4.2. Sonic Data

The routine recording of high-frequency waveforms in the kilohertz range is crucial in well logging due to their utility in measuring velocity and porosity logs. These logs are essential for evaluating the mechanical properties of rocks surrounding the borehole. Sonic waveforms are recorded with a sonic scanner tool, which consists of multiple transmitters and receivers that emit and detect high-frequency acoustic signals (Figure 13.10). Compressional and shear (dipole) (Figure 13.11) are generated by five transmitters integrated into the sonic scanner tool. The waveform acquisition is performed by gradually moving the sonic scanner tool from the bottom of the borehole toward the free surface, with a depth step of 0.1524 m, allowing a high-resolution record along the whole borehole. This practice enables a direct comparison between the estimated attenuation logs and other petrophysical logs, facilitating the exploration of potential relationships between them.

A dense record of sonic waveforms, given by the high-depth resolution of the well-logging acquisition, makes conventional methods developed for the estimate of attenuation from seismic data unsuitable for sonic data. To estimate Q^{-1} values from sonic logging data, Frazer et al. (1997) introduced a method based on the concept of median frequency shift (MFS) through the MFS-dependent item $\hat{\varnothing}(z)$:

$$\hat{\varnothing}(z) = Q^{-1}(z)\Delta t(z) + \hat{A} \tag{13.2}$$

where $\Delta t(z)$ is the source–receiver traveltime and $\hat{A}$ is a constant.

In equation (13.2), the constant calculation $\hat{A}$ depends on the choice of an arbitrary reference, which makes the magnitude of Q^{-1} directly dependent on this reference. Suzuki and Matsushima (2013) developed a modified MFS (MMFS) method to calculate absolute values of Q^{-1} not relative to an arbitrary reference, which makes the estimate of Q^{-1} from sonic data more robust and stable. An example of a Q^{-1} estimate by applying Suzuki and

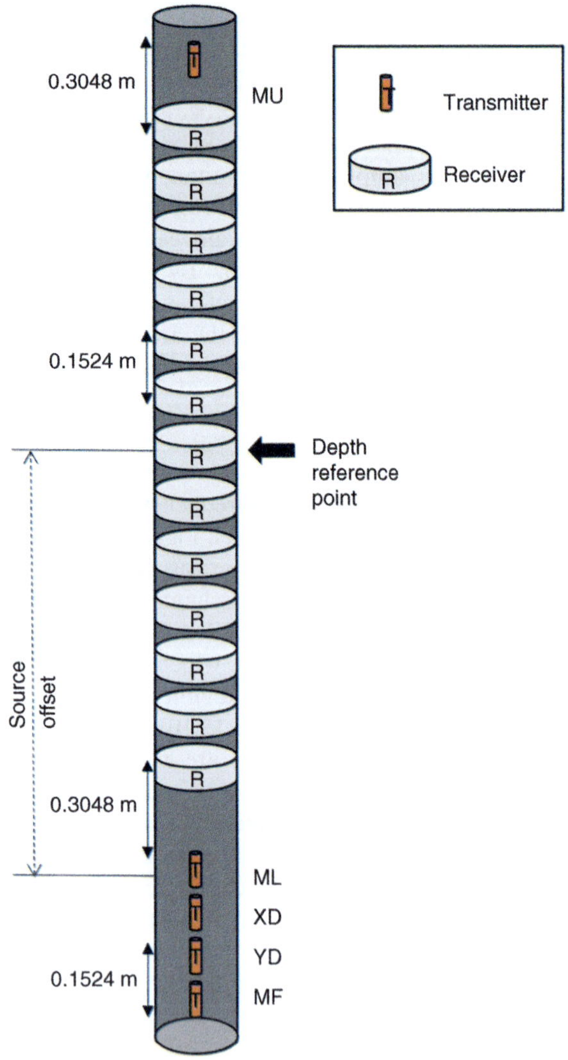

Figure 13.10 A sketch showing a sonic scanner tool used for the acquisition of sonic waveforms.

Matsushima (2013) on sonic data recorded in a carbonate reservoir is shown in Figure 13.12.

13.4.3. VSP Data

VSP data are considered the most suitable for attenuation studies as they enable sampling of the downgoing wavefield at various known depths. Indeed, the most-used seismic attenuation methods, namely, the spectral-ratio method (McDonal et al., 1958) and frequency shift (Quan & Harris, 1997) methods, are applicable for VSP data as they require waveforms recorded at different depths z_1 and z_2, as shown in equations (13.3) and (13.4)

$$Q^{-1} = -\frac{m}{\pi(t(z_2) - t(z_1))} \tag{13.3}$$

$$Q^{-1} = \frac{f_c(z_1) - f_c(z_2)}{\pi\sigma_1^2(t(z_2) - t(z_1))} \tag{13.4}$$

where m is the slope of the spectral ratio of the waveforms recorded at depths z_1 and z_2 versus frequency, f_c is the centroid frequency, and σ_1^2 is the variance.

Matsushima et al. (2016) adopted the MMFS for VSP data by adopting seismic interferometry technique in order to increase the resolution of attenuation profiles and their accuracy by taking into account the effect of both frequency and amplitude changes in attenuation estimate. The basic equation of the MMFS technique at depth z and time Δt from source to receiver is expressed as

$$\emptyset\,(z,\omega) = -Q^{-1}(z)\Delta t + L(\omega) \tag{13.5}$$

where $L(\omega)$ is a frequency-dependent term.

The MMFS method was applied to real zero-offset VSP (ZVSP) data acquired from an oilfield with carbonate-dominant lithology (Bouchaala et al., 2016b). The estimated attenuation profile obtained based on the MMFS method displayed comparable depth variations to those obtained by using spectral ratio and centroid frequency shift methods (Figure 13.13). However, the profiles derived from the interferometry method are of higher resolution and present slight discrepancies in magnitudes. These differences can be attributed to the improved sensitivity of interferometry methods to the upward-propagating waves in comparison to the other two methods.

As in Bouchaala et al. (2016b), compressional downgoing waves are often used for the estimate of seismic attenuation from VSP data. Most VSP attenuation studies have been performed on zero- or near-offset VSP data because they contain downgoing waves with a high signal-to-noise ratio (S/N). However, estimating attenuation from far- or mid-offset data is not common because of the decrease in S/N with offset increase. Some studies succeed in estimating attenuation by using P and S reflected waves extracted from no-zero-offset data (e.g., Bouchaala et al., 2021; Matsushima & Zhan, 2020). Combining P and S attenuation enhances geophysical interpretation in saturated media because, as explained earlier, the $\frac{Q_P^{-1}}{Q_S^{-1}}$ ratio has high sensitivity to fluid saturation. However, extraction of S waves is a laborious task and requires a lot of creativity in terms of signal processing and three-component (3C) data.

13.4.4. Surface Seismic Data

Surface seismic data refers to data collected using sources and receivers placed on the Earth's surface or just below the sea surface. Thanks to this acquisition

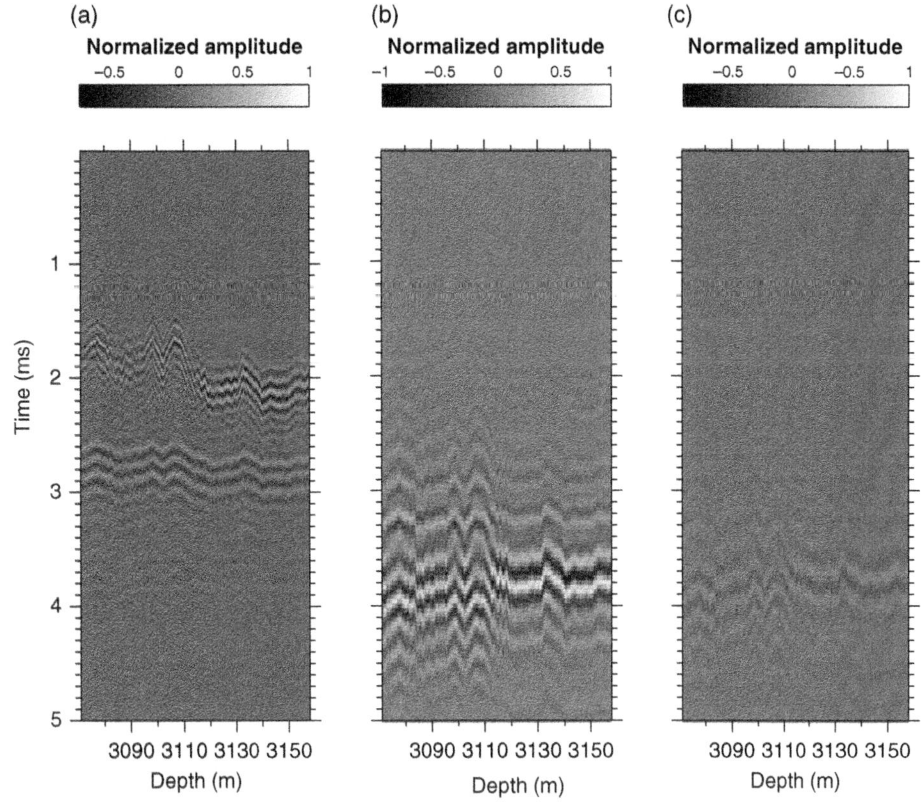

Figure 13.11 Sonic waveforms generated by (a) compressional MF, (b) inline XD, (c) offline XD sources, acquired by using the sonic scanner illustrated in Figure 13.1. Adapted from Bouchaala et al. (2016a).

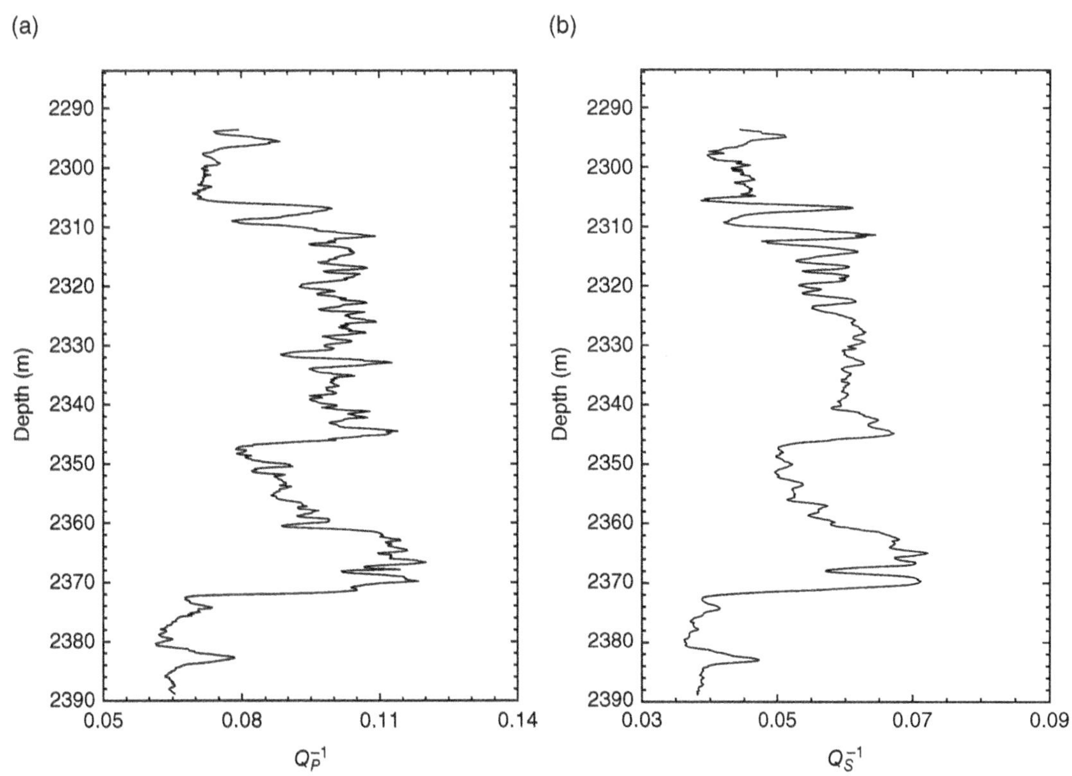

Figure 13.12 (a) Compressional and (b) shear attenuation measured from MF and XD inline sonic data shown in Figure 13.11 by using the MMFS method. Adapted from Suzuki & Matsushima, (2013).

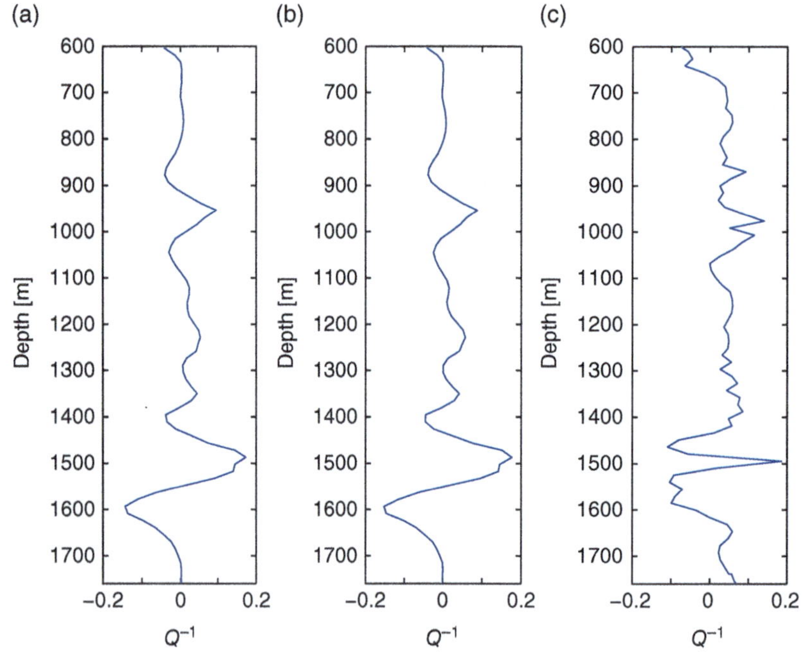

Figure 13.13 Compressional attenuation profiles estimated by applying (a) spectral ratio, (b) centroid frequency, and (c) MMFS methods on ZVSP data, acquired in a well crossing carbonate geological media. Bouchaala et al. (2016c) / with permission of EAGE.

geometry, seismic data are acquired with good lateral and vertical coverage. Hence, by utilizing this data, it becomes possible to track the lateral and vertical variation of Q^{-1}. However, the geometry of surface seismic data acquisition is not suitable for an accurate estimation of Q^{-1} because most attenuation methods work better for VSP geometry. Furthermore, at mid and large offsets, the S/N becomes insufficient for an accurate estimate of Q^{-1}. Therefore, it is common to combine multiple approaches and incorporate additional information from well data or other geophysical measurements to improve the accuracy of Q estimates. Dasgupta and Clark (1998) calculated the slope of spectral ratio-frequency curves at multiple common midpoint gathers, and then the intercept determined from the linear regression of slope versus squared offset variation was taken as Q^{-1} values at zero offset. However, the linearity of the slope versus squared offset is a strong assumption, which adds doubts about the accuracy of the Q^{-1} estimate. Bouchaala et al. (2022) proposed an innovative workflow to estimate attenuation from 3D ocean bottom cable seismic data by combining the centroid frequency shift method and Gabor transform (Gabor, 1946). The numerical parameters of the Gabor transform were optimized by using VSP waveforms as reference. The workflow was applied on a heterogeneous geological media to generate multiple two-dimensional Q^{-1} profiles, used for the interpretation of subsurface properties and delineating the reservoir zone (Figure 13.14). Unfortunately, the availability of VSP data, which is essential for this workflow, is not always guaranteed. Seismic forward modeling and inversion in viscoelastic media is another alternative for estimating Q^{-1} from 3D surface data. The viscoelasticity is introduced in the forward modeling of seismic wave propagation by adding a Q-dependent imaginary part to elastic parameters or to velocity. FWI using viscoelastic forward modeling has

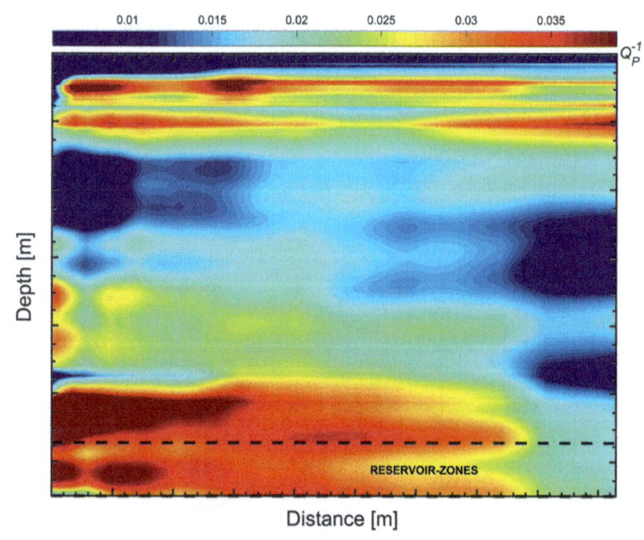

Figure 13.14 2D compressional attenuation profile estimated from surface seismic data. The dashed horizontal line denotes the top of reservoir's formation.

gained significant popularity and is being increasingly utilized for various applications: time-lapse studies for CO_2 injection and storage monitoring, geothermal exploration and monitoring, and even construction engineering. However, it is important to emphasize that seismic modeling and inversion often suffer from numerical instability and dispersion problems when they are dealing with complex media, in addition to the requirement of substantial numerical resources when dealing with complex media.

13.4.5. Distributed Acoustic System (DAS) Data

This technology converts standard optical fibers into an array of thousands of individual sensors for monitoring and detecting acoustic signals. DAS takes advantage of the optical fibers' sensitivity to strain and changes in temperature to enable continuous and distributed acoustic sensing along the entire length of the fiber. DAS technology has gained popularity in recent years due to its ability to leverage existing infrastructure and provide detailed, continuous monitoring over large areas. Its applications extend beyond geophysics and include security monitoring, traffic monitoring, environmental monitoring, and structural health monitoring.

The optical fiber of DAS can be deployed along a wellbore, allowing detailed depth imaging and analysis of subsurface rock formations. Therefore, comparing the amplitude or energy of the acoustic signal observed at different depths along the fiber, the Q^{-1} profile can be inferred for the analysis of the subsurface rock formations. The comparison of the acoustic signal can be done by using classical methods used for estimating attenuation from VSP data, namely spectral ratio and centroid frequency shift methods. Nevertheless, it is worth noting that estimating attenuation using DAS is not always straightforward and can be influenced by various factors. These factors include the nature of the medium through which the acoustic waves propagate, S/N, environmental conditions, and the specific setup of the DAS system. To obtain accurate and reliable attenuation estimates, it is often necessary to calibrate the DAS system, account for environmental influences, and consider other factors that may affect the acoustic signal's propagation characteristics. Additionally, combining DAS measurements with other conventional techniques, such as VSP or laboratory experiments, can provide complementary information and enhance the accuracy of Q^{-1} estimates.

13.4.6. Separation Between Scattering and Intrinsic Attenuation

The Q^{-1} profiles estimated from field data contain scattering and intrinsic attenuation effects, which are difficult to separate. Nevertheless, there are some techniques developed to estimate scattering $\left(Q_{sc}^{-1}\right)$ from well log data; then, based on equation (13.6) (Schoenberg, 1971), intrinsic attenuation $\left(Q_{in}^{-1}\right)$ is obtained by subtracting Q_{sc}^{-1} from Q^{-1}:

$$Q^{-1} = Q_{sc}^{-1} + Q_{in}^{-1} \tag{13.6}$$

Shapiro et al. (1994), modified the generalized O'Doherty–Anstey formula to take into account the additional complexity introduced by the finely layered structure of the medium. The scholars consider the effects of wave scattering, transmission, and reflection at each layer interface, as well as the multiple interactions between waves propagating through the layers. After undergoing mathematical development, Shapiro et al. (1994) proposed the following analytical expression for scattering at an angular frequency ω and incident angle θ

$$Q_{sc}^{-1}(\omega,\theta) = C(0)\frac{ka\cos^2\theta}{2\left(1 + k_0^2a^2\cos^2\theta\right)} \tag{13.7}$$

where $C(0)$ is a correlation function at zero lag calculated from velocity and density variations, a is the heterogeneity size at which the medium is assumed homogeneous, k is the wavenumber, and k_0 is the wavenumber calculated with mean velocity. Velocity and density values are obtained from well logs.

This technique for estimating Q_{sc}^{-1} is based on a stochastic model, not a physical one. Furthermore, it always provides positive magnitudes of Q_{sc}^{-1} (Figure 13.15(a)), which is not always physically true, as explained for example by Bouchaala et al. (2016b) and Mateeva (2003) who associated the negative values of Q_{sc}^{-1} with interference phenomenon and also negative impedance contrasts between fine layers.

An alternative way to estimate scattering is to apply conventional methods for estimating attenuation on synthetic seismograms generated by convolving reflectivity with a unit spike (Figure 13.15(b)). The reflectivity is calculated using ray theory and Z transforms, for example (Goupillaud, 1961). The impedance inputted in the model is calculated from sonic and density logs, generally obtained with a depth resolution of 0.15 m. However, the sonic logs need to be smoothed to upscale to VSP scale for a correct comparison of the calculated Q_{scat}^{-1} profile with Q^{-1} derived from zero-offset VSP field data. Bouchaala et al. (2016b) proposed a relationship between the wavelength and the length of the smoothing window, wl, for a correct upscaling of sonic logs:

$$wl = \frac{1}{2k_0\cos\theta} \tag{13.8}$$

Furthermore, it is necessary to apply a band-pass filter on the synthetic seismograms to estimate Q_{sc}^{-1} in the same

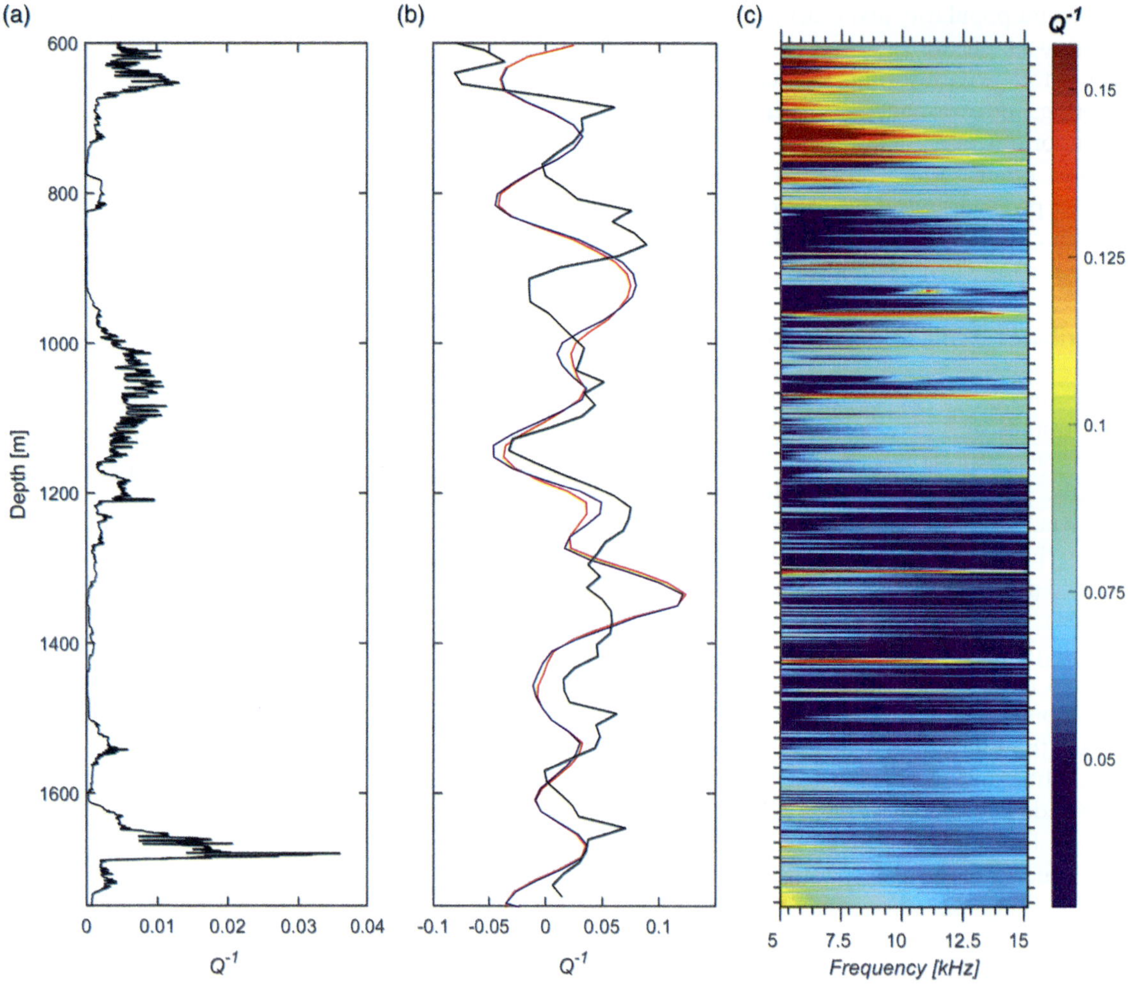

Figure 13.15 (a) Scattering attenuation estimated from a carbonate reservoir by applying (a) equation (13.7). Adapted from Shapiro et al. (1994). (b) Scattering attenuation estimated from a carbonate reservoir by applying spectral ratio (blue), centroid frequency shift (red), and MMFS methods on VSP synthetic waveforms. (c) Frequency-dependent scattering estimated by using equation (13.10). Adapted from Matsushima et al. (2017).

frequency range as Q^{-1} obtained from zero-offset VSP field data; hence, by using equation (13.6), Q_{in}^{-1} can be estimated in the same frequency range.

These techniques for estimating Q_{sc}^{-1} and isolating it from Q_{in}^{-1} assume that scattering and intrinsic attenuation are independent, but in reality, there is mutual interaction between these two parameters. Furthermore, they quantify the scattering mainly associated with heterogeneities caused by fine layering. To face these limitations and improve the Q_{sc}^{-1} estimate and its separation from intrinsic attenuation, Matsushima et al. (2017) developed a method based on the assumption that Q_{sc}^{-1} and Q_{in}^{-1} are frequency-dependent and -independent (Figure 13.15(c)). The intrinsic and scattering attenuation are described by the following equations

$$Q_{in}^{-1} = \frac{\left(-\hat{\text{Ø}}\,(z) + \hat{\text{Ø}}\,(\xi) + Q^{-1}(\xi)\Delta t(\xi)\right)}{\Delta t(z)} \tag{13.9}$$

$$Q_{sc}^{-1}(z,\omega) = \frac{\left(-\hat{\text{Ø}}_{F}(z,\omega) + \hat{\text{Ø}}_{F}(\xi,\omega) + Q_F^{-1}(\xi,\omega)\Delta t(\xi)\right)}{\Delta t(z)} \tag{13.10}$$

where $\hat{\text{Ø}}\,(z)$ and $\hat{\text{Ø}}_{F}(z,\omega)$ are frequency-independent and dependent attributes obtained by combining the mean-median concept (Frazer et al., 1997) at depth z, and ξ is an arbitrary depth reference at which $Q^{-1}(\xi)$ and $Q_F^{-1}(\xi,\omega)$ are known.

Scattering and intrinsic attenuation are valuable indicators of the degree of heterogeneity and fluid content of the seismogenic zones from which an earthquake occurs. The separation between scattering and intrinsic attenuation is often performed by fitting seismic coda. Warren (1972), fitted a diffusion equation (Aki, 1969) to volcanic coda recorded at a distance r, to describe the ratio of energy E to initial energy E_0 as follows

$$\frac{E}{E_0} = \frac{1}{(\pi kt)} exp\left(-\frac{r^2}{kt} - \omega t Q_s^{-1}\right) \tag{13.11}$$

where k is diffusivity, t is the traveltime, ω is the angular frequency, and Q_s^{-1} is the seismic attenuation.

Based on the assumption of the diffusive nature of seismic coda, Warren (1972) separately obtained scattering and intrinsic attenuation from analysis of different parts of the seismic record. Later, Hoshiba (1993) developed a robust technique called multiple lapse time window analysis (MLTWA) to estimate scattering and intrinsic attenuation separately from coda waves. MLTWA was used in many subsequent seismological studies (e.g., Rachman et al., 2017; van Laaten et al., 2022).

13.5. FREQUENCY DEPENDENCE OF ATTENUATION

The frequency dependence of seismic attenuation has important implications for seismic data processing and interpretation. It affects the resolution of seismic images and the accuracy of seismic velocity estimation. Therefore, understanding the frequency dependence of Q^{-1} is crucial for seismic exploration and earthquake studies.

Using different theoretical models, the frequency dependence of Q^{-1} can be described, such as the frequency-independent model, the power-law model, and the double-power-law model. In the frequency-independent model, Q^{-1} is assumed constant regardless of the frequency. It implies that the attenuation of seismic waves is the same at all frequencies. However, this model is a simplification and may not accurately represent the actual behavior of seismic waves in the subsurface. The power-law model states that Q^{-1} is proportional to a power of the frequency, f, meaning $Q^{-1} \propto f^{\alpha}$ (e.g., Beckwith et al., 2017), where α is a constant exponent that determines the rate at which Q^{-1} changes with frequency. This model suggests that higher-frequency waves experience greater attenuation compared to lower-frequency waves. The double-power-law model provides a more detailed description of the frequency dependence of the seismic quality factor (e.g., Carcione, 2007). It incorporates two power-law relationships to account for different frequency ranges. It consists of a low-frequency power-law region, where Q^{-1} is relatively constant or decreases with f, and a high-frequency power-law region, where Q^{-1} increases with f. This model better captures the complex behavior of seismic attenuation observed in the subsurface.

The most effective approach to examine the frequency dependence of Q^{-1} is to compare the magnitudes of Q^{-1} measured from real data collected at different frequency ranges. This is because changes in the magnitude of Q^{-1} caused by frequency within a particular frequency range may be within the range of estimation accuracy (Jones, 1986; Sams et al., 1997). Thus, we compiled attenuation measurements data estimated from different types of data collected at different frequency ranges in one carbonate reservoir: surface seismic (3–70 Hz) (Bouchaala et al., 2022), VSP (7–90 Hz, 5–85 Hz for P and S waves, respectively), sonic (kHz) (Bouchaala et al., 2021), and low (100 Hz) and high (MHz) frequency laboratory data recently collected in fully saturated carbonate samples. The plot of the results (Figure 13.16) revealed that the highest magnitude of attenuation was observed at sonic frequencies, which aligns with the findings of Sams et al. (1997), who compared compressional attenuation magnitudes estimated at different frequency ranges in fully saturated clastic rocks. The frequency dependence of Q^{-1} is generally attributed to pore fluids' ability to move during the wave passage in porous media (Bouchaala et al., 2021; Müller et al., 2010). Indeed, wave passage in rocks perturbs pore pressure, and returning to an equilibrium state depends on the relaxation time of fluids. As a result, when the relaxation time is greater than the wave period, the fluid will not have time to return to the equilibrium state, resulting in high attenuation, whereas in the inverse situation, lower attenuation should be observed. Furthermore, the magnitude of Q^{-1} depends

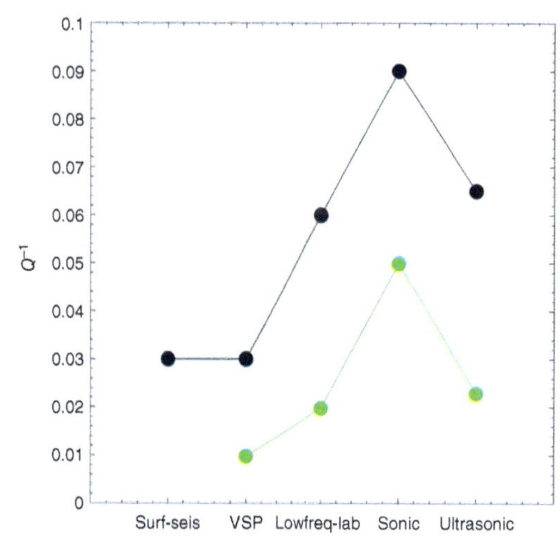

Figure 13.16 Average of seismic attenuation magnitudes calculated from compressional (black) and shear (green) waves acquired in a reservoir zone at different frequency ranges, from surface seismic to ultrasonic frequencies.

on attenuation mechanisms such as the squirt flow mechanism. The high elastic contrast between stiff carbonate grains and compliant fractures makes carbonate rocks favorable media for the squirt flow mechanism, which takes place between compliant (fracture) and stiff (intergranular) pores. This is one of the most attenuating mechanisms because it causes high-speed fluid motion, leading to high fluid–solid frictional losses at pore walls (Adelinet et al., 2010; Bouchaala et al., 2019a). Therefore, the high attenuation magnitude at sonic frequencies might be attributed to the dominance of the squirt flow mechanism. It is also important to mention that when the wavelength of a seismic wave is comparable to the size of the scattering heterogeneity, significant scattering can occur. On the other hand, if the wavelength is much larger or much smaller than the size of the heterogeneity, the wave tends to propagate through it without much effect of scattering (e.g., Matsushima et al., 2011). Sams et al. (1997) noticed a small amount of scattering at sonic frequencies and suggested that attenuation magnitude is dominated by a fluid-related mechanism, which may be the squirt flow mechanism. Bouchaala et al. (2016a) conducted an attenuation study in a carbonate reservoir and confirmed the nonsignificance of the scattering mechanism at sonic frequencies. Scholars have associated the presence of attenuation in the reservoir zone with the squirt flow mechanism. Furthermore, they explained the small magnitudes of sonic scattering because scattering magnitude depends on the wavelength and the size of heterogeneity, as was shown by Wu and Aki (1988). It is worth noting that the spatial sampling of 0.1524 m in the sonic logging data, as mentioned by Suzuki and Matsushima (2013), could potentially lead to an underestimation of scattering because such sampling might not be enough to recover the heterogeneities affecting sonic waveforms.

13.6. *Q*-COMPENSATION OF SEISMIC DATA

In carbonate media, it is common for seismic data to suffer from reduced signal amplitudes and poor signal-to-noise ratios due to high attenuation caused by the mechanisms explained earlier. Performing a Q^{-1} estimate followed by the *Q*-compensation procedure boosts the amplitudes of seismic reflection data, resulting in clearer and more distinct subsurface images. This is especially important when trying to identify carbonate reservoirs or other geological features. Harris and Zhu (2014), proposed a workflow to invert the Q^{-1} factor and restore prestack amplitude losses to enhance seismic images. They applied their workflow on cross-hole real data acquired in a carbonate mound located in west Texas, USA, to improve spatial resolution and continuity of seismic images. Liu et al. (2013) implemented a *Q*-compensation algorithm in a seismic processing workflow to increase the S/N of walkaway VSP data recorded in Ordovician carbonate rocks in the Tarim basin, western China. This led to a substantial improvement in seismic imaging and identifying a deep fractured oil reservoir, which was impossible with conventional processing due to low S/N. After deghosting surface seismic data acquired in a Malaysian offshore carbonate oilfield, Zhang et al. (2016) performed a *Q*-prestack depth migration (Q-PSDM) and succeeded in enhancing seismic imaging of the subsurface containing gas bodies and faults, which are an important source of absorption (Figure 13.17). Parts of this success come from the *Q*-tomography used to estimate the *Q* factor from PSDM data. Smith et al. (2019) proposed a processing workflow including *Q*-compensation to correct amplitudes and the arrival time of PSDM data, which was recorded in a carbonate onshore oilfield located in Abu Dhabi, UAE. The corrections converted seismic wavelets into zero-phase shape, which facilitates events interpretation and seismic tie analysis.

Although seismic *Q*-compensation techniques are valuable for improving seismic imaging, they are not without their challenges and limitations. *Q*-compensation often relies on models and assumptions about the subsurface properties and the *Q* factor itself. If these

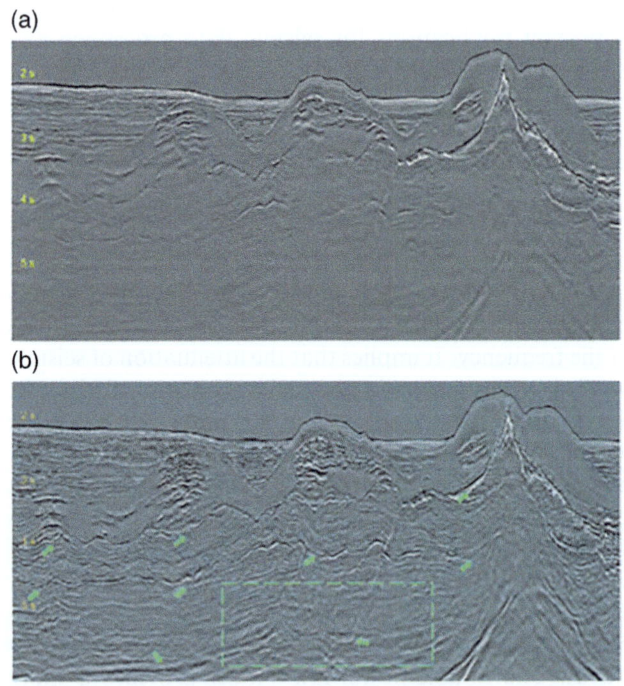

Figure 13.17 Time domain prestack depth migration (PSMD) with (a) deghosting and (b) *Q*-compensation of seismic data recorded in a Malaysian offshore carbonate oilfield. The enhancement of S/N in the zone covering gas pockets is indicated by green arrows, and the green box indicates the enhancement of fault imaging. Zhang et al. (2016).

assumptions are inaccurate, the compensation may introduce errors rather than improvements in the data. Accurate Q-compensation often requires well-log data to estimate subsurface Q values. Lack of well data or sparse well coverage can limit the accuracy of Q estimation and hence compensation of seismic amplitudes. Furthermore, zero-offset VSP data are the most suitable dataset for an accurate Q estimate, whereas the offset may have a negative impact on Q estimate accuracy. Therefore, this may lead to an over or under Q-compensation of seismic amplitudes. Another challenge is the geological complexity of subsurface formations, such as the presence of complex fractures or highly variable lithology, causing a low S/N ratio due to the high presence of noise, making an accurate Q estimate challenging. Such complexity implies significant variations in Q properties; hence, applying a uniform Q-compensation may not be appropriate. Recent studies have proposed stabilization schemes for accurate Q-compensation based on reverse time migration (RTM) (e.g., Li et al., 2023b; Wang et al., 2019b), implementation of an exponential window function to stabilize the time propagator of the compensated wave equation with decoupled fractional Laplacians (DFLs) in this exponentially divergent time propagator (Wang et al., 2019a), and the dictionary learning technique (Li et al., 2023a). Those recent techniques have not yet been tested on carbonate media and might potentially improve the Q-compensation of their seismic data.

13.7. CONCLUSIONS

This chapter sheds light on seismic attenuation with a focus on a complex media, namely carbonates. The complexity of carbonate rocks is attributed to their high heterogeneity, complex pore structure, and textures due to the processes and environments in which they form. This explains the presence of numerous fluid-related and free-fluid attenuation mechanisms, which are the cause of Q^{-1} frequency dependence and its indirect relationship with the petrophysical properties of media. Hence, interpreting attenuation results requires careful attention and deep investigation to avoid errors and extract valuable information about rock properties. Separation between the effects of attenuation mechanisms may substantially enhance the interpretation. Estimating attenuation at a wide range of frequencies may be useful for such separation, because each mechanism has a preferential frequency range. However, separation between scattering and the rest of the mechanisms, in which their effect is quantified by intrinsic attenuation, remains challenging in carbonate rocks. This is due to their interdependence and the modeling assumptions, which may not always hold true in highly heterogeneous media such as carbonates, potentially leading to inaccurate results. High-density DAS data offer high spatial resolution, meaning DAS can capture lateral variations in attenuation at very fine intervals along the fiber. However, the quality of the recorded data in carbonate media may be a challenge for an accurate estimate of attenuation. Therefore, innovative amplitude-preserved processing techniques and methods for estimating Q^{-1} should be developed to reduce uncertainties in estimating Q^{-1} from DAS data. Accurately estimating Q^{-1} will aid in effectively compensating DAS amplitudes, thus enhancing S/N, which is a current high priority in studies reliant on DAS data. Continued research and advancements in understanding seismic attenuation will contribute to our knowledge of Earth's subsurface, improve seismic hazard assessment, and enhance the design of structures in earthquake-prone regions.

ACKNOWLEDGMENTS

The authors express profound appreciation to Khalifa University of Science and Technology for providing a PhD funding award. We express profound appreciation to Abu Dhabi National Oil Company (ADNOC) for providing samples. We acknowledge the rock physics lab at Abu Dhabi Research Innovation Centre (ADRIC) for allowing us to use its facilities.

AVAILABILITY STATEMENT

Data associated with this research are confidential and cannot be released.

REFERENCES

Adam, L., Batzle, M., Lewallen, K., & van Wijk, K. (2009). Seismic wave attenuation in carbonates. *Journal of Geophysical Research: Solid Earth*, *114*(B6), 1–14. doi:10.1029/2008JB005890

Adelinet, M., Fortin, J., Guéguen, Y., Schubnel, A., & Geoffroy, L. (2010). Frequency and fluid effects on elastic properties of basalt: experimental investigations. *Geophysical Research Letters*, *37*(2), L02303. doi:10.1029/2009GL041660

Agersborg, R., Johansen, T. A., Jakobsen, M., Sothcott, J., & Best, A. (2008). Effects of fluids and dual-pore systems on pressure-dependent velocities and attenuations in carbonates. *Geophysics*, *73*(5), N35–N47.

Ahr, W. M. (2011). *Geology of carbonate reservoirs: The identification, description and characterization of hydrocarbon reservoirs in carbonate rocks*. John Wiley & Sons.

Akbar, M., Petricola, M., Watfa, M., Badri, M., Charara, M., Boyd, A., et al. (1995). Classic interpretation problems: evaluating carbonates. *Oilfield Review*, *7*(1), 38–57.

Aki, K. (1969). Analysis of the seismic coda of local earthquakes as scattered waves. *Journal of Geophysical Research, 74*(2), 615–631. doi:10.1029/JB074i002p00615

Azuma, H., Xue, Z., & Matsuoka, T. (2014). *Utilization of seismic attenuation in the monitoring of CO2 geological storage project*. Paper presented at Energy Procedia.

Ba, J., Carcione, J. M., & Sun, W. (2015). Seismic attenuation due to heterogeneities of rock fabric and fluid distribution. *Geophysical Journal International, 202*(3), 1843–1847.

Barton, N. (2006), *Rock quality, seismic velocity, attenuation and anisotropy. CRC Press.*

Batzle, M. L., Han, D.-H., & Hofmann, R. (2006). Fluid mobility and frequency-dependent seismic velocity—direct measurements. *Geophysics, 71*(1), N1–N9.

Beckwith, J., Clark, R., & Hodgson, L. (2017). Estimating frequency-dependent attenuation quality factor values from prestack surface seismic data. *Geophysics, 82*(1), O11–O22.

Best, A. I., & McCann, C. (1995). Seismic attenuation and pore-fluid viscosity in clay-rich reservoir sandstones. *Geophysics, 60* (5), 1386–1397.

Best, A. I., Sothcott, J., & McCann, C. (2007). A laboratory study of seismic velocity and attenuation anisotropy in near-surface sedimentary rocks. *Geophysical Prospecting, 55* (5), 609–625.

Biot, M. A. (1956). Theory of propagation of elastic waves in a fluid-saturated porous solid. II. Higher frequency range. *The Journal of the Acoustical Society of America, 28*(2), 179–191.

Bouchaala, F., Ali, M., & Matsushima, J. (2016a). Attenuation modes from vertical seismic profiling and sonic waveform in a carbonate reservoir, Abu Dhabi, United Arab Emirates. *Geophysical Prospecting, 64*(4), 1030–1047.

Bouchaala, F., Ali, M. Y., & Matsushima, J. (2016b). Estimation of seismic attenuation in carbonate rocks using three different methods: application on VSP data from Abu Dhabi oilfield. *Journal of Applied Geophysics, 129*, 79–91.

Bouchaala, F., Ali, M. Y., & Matsushima, J. (2019a). Attenuation study of a clay-rich dense zone in fractured carbonate reservoirs seismic attenuation of a dense zone. *Geophysics, 84*(3), B205–B216.

Bouchaala, F., Ali, M. Y., Matsushima, J., Bouzidi, Y., Takam Takougang, E. M., Mohamed, A. A., & Sultan, A. (2019b). Azimuthal investigation of compressional seismic-wave attenuation in a fractured reservoir. *Geophysics, 84*(6), B437–B446.

Bouchaala, F., Ali, M. Y., & Matsushima, J. (2021). Compressional and shear wave attenuations from walkway VSP and sonic data in an offshore Abu Dhabi oilfield. *Comptes Rendus. Géoscience, 353*(1), 337–354.

Bouchaala, F., Ali, M. Y., Matsushima, J., Bouzidi, Y., Jouini, M. S., Takougang, E. M., & Mohamed, A. A. (2022). Estimation of seismic wave attenuation from 3D seismic data: a case study of OBC data acquired in an offshore oilfield. *Energies, 15*(2), 534.

Bourbié, T., Coussy, O., Zinszner, B., & Junger, M. C. (1992). *Acoustics of porous media*. Acoustical Society of America.

Bradley, J. J., and A. N. Fort (1966), Internal friction in rocks, Handbook of Physical Constants, Geol. Soc. Amer. Mem., 97, 175–193.

Carcione, J. M. (2007). *Wave fields in real media: Wave propagation in anisotropic, anelastic, porous and electromagnetic media*. Elsevier.

Caspari, E., Novikov, M., Lisitsa, V., Barbosa, N. D., Quintal, B., Rubino, J. G., & Holliger, K. (2018). Attenuation mechanisms in fractured fluid-saturated porous rocks: a numerical modelling study. *Geophysical Prospecting, 67*(4), 935–955. doi:10.1111/1365-2478.12667

Chapman, S., Borgomano, J. V. M., Quintal, B., Benson, S. M., & Fortin, J. r. m. (2021). Seismic wave attenuation and dispersion due to partial fluid saturation: direct measurements and numerical simulations based on X-ray CT. *Journal of Geophysical Research: Solid Earth, 126*(4). doi:10.1029/2021jb021643

Coakley, K. J., Splett, J. D., Janezic, M. D., & Kaiser, R. F. (2003). Estimation of Q-factors and resonant frequencies. *IEEE Transactions on Microwave Theory and Techniques, 51*(3), 862–868.

Collier, L., Neuberg, J. W., Lensky, N., Lyakhovsky, V., & Navon, O. (2006). Attenuation in gas-charged magma. *Journal of Volcanology and Geothermal Research, 153*(1–2), 21–36.

Dasgupta, R., & Clark, R. A. (1998). Estimation of Q from surface seismic reflection data. *Geophysics, 63*(6), 2120–2128.

Deng, J., Wang, S., & Han, D.-h. (2009). The velocity and attenuation anisotropy of shale at ultrasonic frequency. *Journal of Geophysics and Engineering, 6*(3), 269–278. doi:10.1088/1742-2132/6/3/006

Eberli, G. P., Baechle, G. T., Anselmetti, F. S., & Incze, M. L. (2003). Factors controlling elastic properties in carbonate sediments and rocks. *The Leading Edge, 22*(7), 654–660.

Frazer, L. N., Sun, X., & Wilkens, R. H. (1997). Changes in attenuation with depth in an ocean carbonate section: Ocean Drilling Program sites 806 and 807, Ontong Java Plateau. *Journal of Geophysical Research: Solid Earth, 102*(B2), 2983–2997. doi:10.1029/96JB03189

Gabor, D. (1946). Theory of communication. Part 1: the analysis of information. *Journal of the Institution of Electrical Engineers-Part III: Radio and Communication Engineering, 93* (26), 429–441.

Goupillaud, P. (1961). An approach to inverse filtering of near surface effects from seismic records. *Geophysics, 26*(6), 754–760.

Guerin, G., & Goldberg, D. (2005). Modeling of acoustic wave dissipation in gas hydrate–bearing sediments. *Geochemistry, Geophysics, Geosystems, 6*(7), 1–21. doi:10.1029/2005GC000918

Guo, H., & Thurber, C. (2022). Temporal changes in seismic velocity and attenuation at the geysers geothermal field, California, from double-difference tomography. *Journal of Geophysical Research: Solid Earth, 127*(5). doi:10.1029/2021jb022938

Guyer, R. A., & Johnson, P. A. (2009). *Nonlinear mesoscopic elasticity: The complex behaviour of rocks, soil, concrete*. John Wiley & Sons.

Hamilton, E. L. (1972). Compressional-wave attenuation in marine sediments. *Geophysics, 37*(4), 620–646.

Harris, J. M. & Zhu, T. (2014). *Q-compensation for high resolution seismic imaging*. Paper presented at the 6th Simpósio Brasileiro de Geofísica.

Hoshiba, M. (1993). Separation of scattering attenuation and intrinsic absorption in Japan using the multiple lapse time window analysis of full seismogram envelope. *Journal of*

Geophysical Research: Solid Earth, *98*(B9), 15809–15824. doi:10.1029/93JB00347

Janjuhah, H. T., Kontakiotis, G., Wahid, A., Khan, D. M., Zarkogiannis, S. D., & Antonarakou, A. (2021). Integrated porosity classification and quantification scheme for enhanced carbonate reservoir quality: implications from the Miocene Malaysian carbonates. *Journal of Marine Science and Engineering*, *9*(12), 1410.

Jaya, M. S., Shapiro, S. A., Kristinsdóttir, L. H., Bruhn, D., Milsch, H., & Spangenberg, E. (2010). Temperature dependence of seismic properties in geothermal rocks at reservoir conditions. *Geothermics*, *39*, 115–123. doi:10.1016/j.geothermics.2009.12.002

Jin, Z., Liu, Z., Shi, Y., & Wang, W. (2023). A novel method to estimate Q from seismic data based on rock-physics modeling. *Geophysics*, *88*(2), MR55–MR63.

Jones, T. D. (1986). Pore fluids and frequency-dependent wave propagation in rocks. *Geophysics*, *51*(10), 1939–1953.

Jones, T., Murphy, W., & Nur, A. M. (1980). *Effects of temperature and saturation on the velocity and attenuation of seismic waves in rocks: applications to geothermal reservoir evaluation*. Paper presented at the Sixth Workshop on Geothermal Reservoir Engineering, Stanford University, Stanford, California.

Jouini, M. S., Bouchaala, F., Riahi, M. K., Sassi, M., Abderrahmane, H., & Hjouj, F. (2022). Multifractal analysis of reservoir rock samples using 3D X-ray micro computed tomography images. *IEEE Access*, *10*, 67898–67909.

Kjartansson, E. (1980). *Attenuation of seismic waves in rocks and applications in energy exploration* (PhD thesis), Stanford University.

Klimentos, T., & McCann, C. (1990). Relationships among compressional wave attenuation, porosity, clay content, and permeability in sandstones. *Geophysics*, *55*(8), 998–1014.

Kuteynikova, M., Tisato, N., Jänicke, R., & Quintal, B. (2014). Numerical modeling and laboratory measurements of seismic attenuation in partially saturated rock. *Geophysics*, *79*(2), L13–L20. doi:10.1190/geo2013-0020.1

Li, B., Lin, K., Wen, X., Tang, C., & Li, L. (2023a). Enhancement of the seismic data resolution through Q-compensated denoising based on dictionary learning. *IEEE Transactions on Geoscience and Remote Sensing, 61, 1–11*.

Li, S., Shi, Y., Wang, W., Wang, N., Song, L., & Wang, Y. (2023b). An explicit stable Q-compensated reverse time migration scheme for complex heterogeneous attenuation media. *Frontiers in Earth Science*, *11*, 1121648.

Liu, Y., Liang, X., Zhou, Y., Liu, S., Wang, X., & Li, Y. (2013). *Application of walkaway VSP for improving seismic imaging of fractured-cavernous carbonate reservoirs*. Paper presented at IPTC 2013: International Petroleum Technology Conference, European Association of Geoscientists & Engineers.

Lorenzo, J. M., Smolkin, D. E., White, C., Chollett, S. R., & Sun, T. (2013). Benchmark hydrogeophysical data from a physical seismic model. *Computers & Geosciences*, *50*, 44–51. doi:10.1016/j.cageo.2012.07.034

Lucet, N. (1989). *Vitesse et attenuation des ondes elastiques soniques et ultrasoniques dans les roches sous pression de confinement* (PhD thesis), Pierre and Marie Curie University, France.

Masson, Y. J., & Pride, S. R. (2011). Seismic attenuation due to patchy saturation. *Journal of Geophysical Research*, *116*(B3). doi:10.1029/2010jb007983

Mateeva, A. A. (2003). *Thin horizontal layering as a stratigraphic filter in absorption estimation and seismic deconvolution*. Colorado School of Mines.

Matsushima, J., & Zhan, L. (2020). S-wave attenuation estimation from walkaway vertical seismic profiling data in methane hydrate-bearing sediments at Nankai Trough, Japan. *Journal of Applied Geophysics*, *173*, 103931.

Matsushima, J., Ali, M. Y., & Bouchaala, F. (2016). Seismic attenuation estimation from zero-offset VSP data using seismic interferometry. *Geophysical Journal International*, *204*(2), 1288–1307.

Matsushima, J., Ali, M. Y., & Bouchaala, F. (2017). A novel method for separating intrinsic and scattering attenuation for zero-offset vertical seismic profiling data. *Geophysical Journal International*, *211*(3), 1655–1668.

Matsushima, J., Suzuki, M., Kato, Y., & Rokugawa, S. (2011). Estimation of ultrasonic scattering attenuation in partially frozen brines using magnetic resonance images. *Geophysics*, *76*(1), T13–T25.

Mavko, G. M., & Nur, A. (1979). Wave attenuation in partially saturated rocks. *Geophysics*, *44*(2), 161–178.

McDonal, F., Angona, F., Mills, R., Sengbush, R., Van Nostrand, R., & White, J. (1958). Attenuation of shear and compressional waves in Pierre shale. *Geophysics*, *23*(3), 421–439.

Milani, M., Monachesi, L., Sabbione, J. I., Rubino, J. G., & Holliger, K. (2016). A generalized effective anisotropic poroelastic model for periodically layered media accounting for both Biot's global and interlayer flows. *Geophysical Prospecting*, *64*(4), 1135–1148.

Müller, T. M., Gurevich, B., & Lebedev, M. (2010). Seismic wave attenuation and dispersion resulting from wave-induced flow in porous rocks—a review. *Geophysics*, *75*(5), 75A147–75A164.

Nakagawa, S. (2011). Split Hopkinson resonant bar test for sonic-frequency acoustic velocity and attenuation measurements of small, isotropic geological samples. *Review of Scientific Instruments*, *82*(4), 1–35.

Parra, J. O., & Hackert, C. L. (2006). Modeling and interpretation of Q logs in carbonate rock using a double porosity model and well logs. *Journal of Applied Geophysics*, *58*(3), 253–262. doi:10.1016/j.jappgeo.2005.07.003

Peacock, S., McCann, C., Sothcott, J., & Astin, T. R. (1994). Experimental measurements of seismic attenuation in microfractured sedimentary rock. *Geophysics*, *59*(9), 1342–1351. doi:10.1190/1.1443693

Peselnick, L., & Outerbridge, W. (1961). Internal friction in shear and shear modulus of Solenhofen limestone over a frequency range of 107 cycles per second. *Journal of Geophysical Research*, *66*(2), 581–588. doi:10.1029/JZ066i002p00581

Pinson, L. J., Henstock, T. J., Dix, J. K., & Bull, J. M. (2008). Estimating quality factor and mean grain size of sediments from high-resolution marine seismic data. *Geophysics*, *73*(4), G19–G28.

Pittman, E. D. (1979), Porosity diagenesis and productive capability of sandstone reservoirs. *American Association of Petroleum Geologists Bulletin*, *75*, 1279–1299.

Prasad, M. (2003). Velocity-permeability relations within hydraulic units. *Geophysics*, *68*(1), 108–117.

Pride, S. R., Berryman, J. G., & Harris, J. M. (2004). Seismic attenuation due to wave-induced flow. *Journal of Geophysical Research: Solid Earth*, *109*(B1). doi:10.1029/2003JB002639

Qi, Q., Müller, T. M., & Pervukhina, M. (2017). Sonic QP/QS ratio as diagnostic tool for shale gas saturation. *Geophysics*, *82*(3), MR97–MR103.

Quan, Y., & Harris, J. M. (1997). Seismic attenuation tomography using the frequency shift method. *Geophysics*, *62*(3), 895–905.

Quintal, B., Jänicke, R., & Steeb, H. (2013). *Effects of fracture connectivity on S-wave attenuation caused by wave-induced fluid flow*. Paper presented at the SEG International Exposition and Annual Meeting.

Rachman, A. N., Chung, T. W., & Chung, K.-H. (2017). Multiple lapse time window analysis using solely single events in South Korea. *Exploration Geophysics*, *48*(4), 504–511.

Rørheim, S., Bhuiyan, M. H., Bauer, A., & Cerasi, P. R. (2021). On the effect of CO_2 on seismic and ultrasonic properties: a novel shale experiment. *Energies*, *14*(16). doi:10.3390/en14165007

Rozhko, A. Y. (2021). On the spectral changes of seismic wave energy by a partially saturated crack due to the hysteresis of liquid bridges phenomenon. *Geophysics*, *86*(3), MR133–MR147.

Rubino, J. G., & Holliger, K. (2013). Research note: seismic attenuation due to wave-induced fluid flow at microscopic and mesoscopic scales. *Geophysical Prospecting*, *61*(4), 882–889.

Sams, M. S., Neep, J. P., Worthington, M. H., & King, M. S. (1997). The measurement of velocity dispersion and frequency-dependent intrinsic attenuation in sedimentary rocks. *Geophysics*, *62*(5), 1456–1464.

Sang, L. (2016). *Seismic attenuation in carbonate rocks and rock physics models of clay-bearing and organic-rich formations* (PhD thesis). Texas A&M.

Schoenberg, M. (1971). Transmission and reflection of plane waves at an elastic-viscoelastic interface. *Geophysical Journal International*, *25*(1–3), 35–47.

Shapiro, S. A., Zien, H., & Hubral, P. (1994). A generalized O'Doherty-Anstey formula for waves in finely layered media. *Geophysics*, *59*(11), 1750–1762.

Shi, Z., Zhang, Y., Liu, M., Hanaor, D. A., & Gan, Y. (2018). Dynamic contact angle hysteresis in liquid bridges. *Colloids and Surfaces A: Physicochemical and Engineering Aspects*, *555*, 365–371.

Smith, C., Vasilyev, P., Glushchenko, A., & Zarubov, D. (2019). *Q-compensation from near surface to reservoir and below: case study from onshore Abu Dhabi*. Paper presented at the Abu Dhabi International Petroleum Exhibition & Conference, OnePetro.

Sun, W. (2021). On the theory of Biot-patchy-squirt mechanism for wave propagation in partially saturated double-porosity medium. *Physics of Fluids*, *33*(7), 076603.

Suzuki, H., & Matsushima, J. (2013). Quantifying uncertainties in attenuation estimation at methane-hydrate-bearing zones using sonic waveform logs. *Geophysics*, *78*(5), D339–D353.

Tisato, N., Quintal, B., Chapman, S., Podladchikov, Y., & Burg, J.-P. (2015). Bubbles attenuate elastic waves at seismic frequencies: first experimental evidence. *Geophysical Research Letters*, *42*(10), 3880–3887. doi:10.1002/2015GL063538

Toksöz, M., Johnston, D. H., & Timur, A. (1979). Attenuation of seismic waves in dry and saturated rocks: I. Laboratory measurements. *Geophysics*, *44*(4), 681–690.

van Laaten, M., Eulenfeld, T., & Wegler, U. (2022). Comparison of multiple lapse time window analysis and Qopen to determine intrinsic and scattering attenuation. *Geophysical Journal International*, *228*(2), 913–926.

Vogelaar, B., Smeulders, D., & Harris, J. (2010). Exact expression for the effective acoustics of patchy-saturated rocks. *Geophysics*, *75*(4), N87–N96.

Walsh, J. B. (1966). Seismic wave attenuation in rock due to friction. *Journal of Geophysical Research (1896–1977)*, *71*(10), 2591–2599. doi:10.1029/JZ071i010p02591

Walsh, J. B. (1995). Seismic attenuation in partially saturated rock. *Journal of Geophysical Research: Solid Earth*, *100* (B8), 15407–15424. doi:10.1029/94jb03264

Wang, Y., Li, D., & Harris, J. M. (2019a). *A generalized stabilization scheme for seismic Q compensation*. Paper presented at the SEG International Exposition and Annual Meeting.

Wang, Y., Zhou, H., Zhao, X., Zhang, Q., & Chen, Y. (2019b). Q-compensated viscoelastic reverse time migration using mode-dependent adaptive stabilization scheme. *Geophysics*, *84*(4), S301–S315.

Wang, Z., Palaz, I., & Marfurt, K. (1997). Seismic properties of carbonate rocks. *Carbonate Seismology*, *6*, 29–52.

Warren, N. (1972). Q and structure. *The Moon*, *4*(3–4), 430–441.

White, J. E. (1975). Computed seismic speeds and attenuation in rocks with partial gas saturation. *Geophysics*, *40*(2), 224–232. doi:10.1190/1.1440520

Winkler, K. W. (1985). Dispersion analysis of velocity and attenuation in Berea sandstone. *Journal of Geophysical Research: Solid Earth*, *90*(B8), 6793–6800. doi:10.1029/JB090iB08p06793

Winkler, K., & Nur, A. (1979). Pore fluids and seismic attenuation in rocks. *Geophysical Research Letters*, *6*(1), 1–4. doi:10.1029/GL006i001p00001

Winkler, K. W., & Nur, A. (1982). Seismic attenuation: effects of pore fluids and frictional-sliding. *Geophysics*, *47*(1), 1–15.

Wu, R.-S., & Aki, K. (1988). Introduction: seismic wave scattering in three-dimensionally heterogeneous earth. In *Scattering and attenuations of seismic waves, part I* (pp. 1–6). Springer Nature, ISBN: 978-3-7643-2254-0

Wyllie, M., Gardner, G., & Gregory, A. (1962). Studies of elastic wave attenuation in porous media. *Geophysics*, *27*(5), 569–589.

Zhang, Y. (2016). *Static and dynamic behaviour of inter-granular liquid bridges: hysteresis of contact angle and capillary forces*. (PhD thesis). The University of Sydney, Australia

Zhang, Z., Jia, J., Fu, G., Zhang, H., Chow, D., Hung, B., et al. (2016). Fullband imaging. In *PESA News Magazine*.

14

Distributed Acoustic Sensing to Velocity Transform for Vertical Seismic Profiling Applications

Ali Sayed[1], Manish Lal Khaitan[2], Alejandro Martinez[1], and Robert R. Stewart[3]

ABSTRACT

Distributed acoustic sensing (DAS) or fiber-optical strain measurement is an effective and efficient method for vertical seismic profiling (VSP) applications. DAS-measured strain induces frequency and amplitude effects on VSP data, which must be corrected to compare with conventional velocity measurement. A DAS-to-velocity transform is developed here to convert DAS data to a corresponding velocity measurement. The transform consists of a gauge-length correction followed by a conversion of axial strain to velocity, through an integration step in depth and a differentiation step in time. Synthetic data are used to demonstrate the application of the DAS-to-velocity transform on VSP applications. The gauge-length correction applies appropriate gains to the wavenumbers and corrects significant errors in Q estimation. A 210% error in Q estimation with DAS data is reduced to 1% error after correcting the wavenumbers with the gauge-length correction. The DAS-to-velocity transform is also shown to remove dimming effects on borehole seismic images. Migrated images from DAS data show significant variation in amplitude along the image due to the variation in subsurface model properties, as compared to the benchmark migrated images generated by velocity data.

14.1. INTRODUCTION

Distributed acoustic sensing (DAS) or strain measurement using a fiber-optic cable has been commonly used for vertical seismic profiling (VSP) applications. DAS technology can provide dense sampling and long receiver arrays for VSP measurements. These features allow for large volumes of data to be acquired in a relatively short time and with less source effort (Martinez et al., 2021). Borland et al. (2016) demonstrate efficiency gains that DAS recording brings to VSP acquisition. Kimura et al. (2018) simultaneously acquired a 3D VSP survey over fiber and conventional arrays. DAS VSP data can be used to generate frequently used borehole seismic results such as time–depth relationships, corridor stacks, Q estimates, and VSP images.

DAS data also have some shortcomings. They have been demonstrated to have stronger directivity effects compared to the conventional geophone (Mateeva et al., 2014). The sensor response has a dependence on wavenumber (Bakku, 2015; Sayed et al., 2020). Several authors have studied DAS measurement, evaluated the response, and converted the data to the traditional velocity equivalent (Bona et al., 2017; Correa et al., 2017; Daley et al., 2016).

To better understand the sensor response and its impact on the acquired data, we revisit the DAS response and gauge-length effects, develop a DAS-to-velocity transform, and test it on synthetic VSP data to highlight its benefits. Most DAS systems are configured to acquire DAS

[1]*SLB, Houston, Texas, USA*

[2]*SLB, London, UK*

[3]*Department of Earth and Atmospheric Sciences, University of Houston, Houston, Texas, USA*

Distributed Acoustic Sensing in Borehole Geophysics, Geophysical Monograph 289, First Edition.
Edited by Yingping Li, Robert Mellors, and Ge Zhan.

DOI: 10.1002/9781394179275.ch14

strain data. In this chapter, we develop expressions to relate the raw DAS measurement to strain and velocity fields. The DAS-to-velocity transform comprises two steps: gauge-length correction to determine axial strain followed by the conversion of axial strain to velocity.

Synthetic data are generated using simple models to compare the DAS response against the conventional velocity geophone measurement. We then use synthetic data to demonstrate the effect of DAS on borehole seismic applications. The impact on frequency is demonstrated by a Q estimation study, and the impact on amplitude is demonstrated by a VSP imaging study.

14.2. GAUGE-LENGTH EFFECT AND CORRECTION

The DAS measurement determines strain across a finite length of the fiber at each synthesized sensor location. We follow Hartog (2017) and define the DAS strain measurement as

$$\varepsilon_{33}^{DAS}(z,t) = \frac{\int_{z-L/2}^{z+L/2} \varepsilon_{33}(z,t)}{L} \tag{14.1}$$

where ε_{33}^{DAS} is the DAS strain measurement, ε_{33} is the physical or actual axial strain, z is the depth along the fiber, and L is the gauge length. Equation (14.1) can be simplified in terms of displacement u_3 as follows

$$\varepsilon_{33}^{DAS}(z,t) = \frac{u_3\left(z+\frac{L}{2},t\right) - u_3\left(z-\frac{L}{2},t\right)}{L} \tag{14.2}$$

where $u_3(z,\ t) = Ae^{j(2\pi kz-\omega t)}$, k is the wavenumber along z, ω is the temporal frequency, and t is the time. Substituting u_3 in equation (14.2), we get

$$\varepsilon_{33}^{DAS}(z,t) = sinc_\pi(kL)\,\varepsilon_{33}(z,t) \tag{14.3}$$

where $sinc_\pi(x) = sin(\pi x)/(\pi x)$, is the normalized *sinc* function (Sayed et al., 2020, 2021, 2022). This result is similar to equation (3.5) of Bakku (2015). Bakulin et al. (2020) show that the acquisition pulse width also results in a similar wavenumber filtering effect as the gauge length. Dean et al. (2016) show that the effect of the acquisition pulse width is smaller than the gauge-length effect for pulse widths shorter than half the gauge length. We consider short acquisition pulse widths much smaller than half of the gauge lengths, and thus, pulse-width effects are small and neglected.

Equation (14.3) shows that DAS strain measurement is equal to the axial strain along the fiber scaled by a normalized *sinc* function. As gauge length, L, approaches zero, the *sinc* function approaches a value of 1, and DAS strain equals axial strain. The relationship in equation (14.3) holds true for strain-rate measurements as well. This can be derived by using velocity instead of displacement in equation (14.2). Figure 14.1 shows the scaling response of the normalized *sinc* function for wavenumbers, k, ranging from −0.2 to 0.2 cycles/m and a gauge length, L, of 15 m. The amplitude drops with an increase in wavenumber. The function has null points when kL is an integer value: that is, when $1/k$ (apparent wavelength along the fiber) is an integer multiple of L. The amplitude drop-off steepens, and null points occur at lower wavenumbers as gauge length, L, is increased. Further, the polarity of the scaling response changes at each null wavenumber. This results in destructive interference of the wavenumbers beyond the null wavenumber.

Axial strain, $\varepsilon_{33}(z, t)$, can be determined by dividing the DAS strain with the *sinc* function as given in equation (14.4). Other authors (Correa et al., 2017) have introduced a regularization term to the *sinc* function to avoid zero division. We replace the reciprocal of the *sinc* function with the equivalent *gamma* function by making use of Euler's reflection formula (Abramowitz & Stegun, 1965) to avoid this regularization step, as given in equation (14.5)

$$\varepsilon_{33} = \left[sin\,c_\pi(sL)\right]^{-1} \varepsilon_{33}^{DAS} \tag{14.4}$$

$$\varepsilon_{33} = \Gamma(1-sL)\Gamma(1+sL)\varepsilon_{33}^{DAS} \tag{14.5}$$

where Γ $[\Gamma(x) = \int_0^\infty t^{x-1}e^{-t}dt; \Gamma(1+n) = n!]$ is the *gamma* function and $\Gamma(1-sL)$ $\Gamma(1+sL)$ are the gauge-length correction factors.

The gauge-length correction factors are gains to the corresponding wavenumbers. Figure 14.1 displays the gauge-length correction factors (green) overlying the scaling factors. These correction factors can be clipped to avoid overamplification near the null wavenumbers. The maximum gain to be applied should be optimized for data quality. As an example, gauge-length correction factors limited to ±1 would only align the polarities of the wavenumbers within the bandwidth without applying any gain to individual wavenumbers. Tapers on either side of the signal bandwidth can also be used to avoid overamplification. The gauge length should be chosen to ensure that the signal bandwidth is within the first lobe of the *sinc* function to avoid drastically scaling down the higher wavenumber. Figure 14.1 shows that the gauge-length correction factors are in the range of 1–5 within the prominent part of the main lobe of the *sinc* function. Using regularization factors in the denominator to stabilize the correction at the notch wavenumbers results in the addition of errors within the useful part of the spectrum.

To gain a deeper understanding of gauge-length effects, displacement waveforms along z were generated with the wavenumber range of 0.01–0.11 cycles/m. Each wavenumber contributes equally to the summed displacement wavelet (Figure 14.2). When the gauge length is shorter

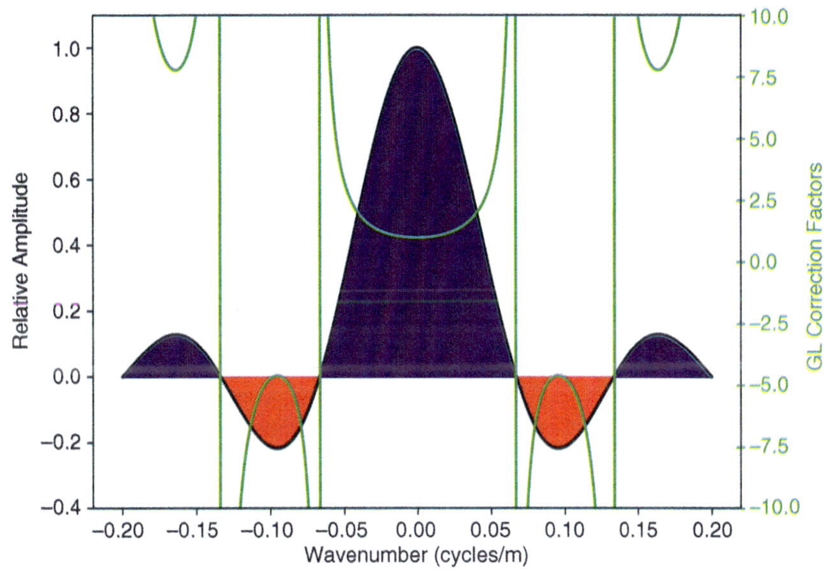

Figure 14.1 Amplitude response of the normalized *sinc* function shown as variable area curve against wavenumbers along the horizontal axis. The positive lobes of the *sinc* function are colored in blue, and negative lobes are in red. The gauge length used is 15 m. The gauge-length correction factors are shown in green.

than the smallest wavelength, the bandwidth is within the main lobe of the *sinc* function, and null wavenumbers are not observed. A 12 m gauge length, longer than the smallest wavelength, is chosen and applied to each wavenumber, as shown in Figure 14.2. Amplitudes near the null wavenumber are damped, and the waveforms at wavenumbers higher than the notch wavenumber are 180° out of phase with the wavenumbers lower than the notch wavenumber. These wavenumbers destructively interfere to produce a relatively broad wavelet. The resulting wavelet represents DAS strain with a gauge length of 12 m (Figure 14.2(b)). We corrected the DAS strain by applying appropriate gauge-length correction factors from equation 14.5 to each wavenumber. Gains are applied to each wavenumber to yield a sharper wavelet. The resultant wavelet represents axial strain (Figure 14.2(c)). The ramp in amplitudes toward higher wavenumbers is expected from the differentiation of u_3. The result obtained in Figure 14.2(c) is equivalent to having differentiated the waveforms in Figure 14.2(a), as opposed to applying a finite gauge-length operator. Integrating the waveforms in Figure 14.2(c) along the depth axis causes the waveforms to revert to the displacement demonstrated in Figure 14.2(d). The waveforms in Figure 14.2(d) are equivalent to the original displacement waveforms displayed in Figure 14.2(a), indicating a full recovery of amplitude and phase through the gauge-length correction and integration process.

Gauge-length correction factors can be considered a gain function applied in the wavenumber domain to raw DAS measurement. This conversion transforms the DAS strain measurement to axial strain. The gain function is clipped or tapered to a maximum value to avoid overamplification of noise. The minimum absolute value of the gain function is 1 at zero wavenumber. A maximum absolute value of 1 for the gain function corrects the wavenumber for the polarity reversals but not the amplitude. Comparing the waveforms without applying a gain function, with a gain function clipped at a value of 1, and with a gain function clipped at a value of 2 shows that the width of the wavelet is recovered with small gains and the larger gains recover larger amplitudes (Figure 14.3).

14.3. AXIAL STRAIN TO VELOCITY

Axial strain along the fiber axis is expressed as follows

$$\varepsilon_{33} = \frac{\partial u_3}{\partial z} \tag{14.6}$$

where u_3 is the particle displacement along the z axis (Bakku, 2015). Equation 14.6 can be rewritten in terms of particle velocity and apparent phase velocity as

$$\varepsilon_{33} = \frac{\partial u_3}{\partial z} = \frac{\partial t}{\partial z}\frac{\partial u_3}{\partial t} = \frac{v_3}{c_a} \tag{14.7}$$

where v_3 is the axial particle velocity and c_a is the apparent phase velocity.

The apparent phase velocity as observed by the fiber can be written as follows

$$c_a = \frac{c}{\cos(\theta)} \tag{14.8}$$

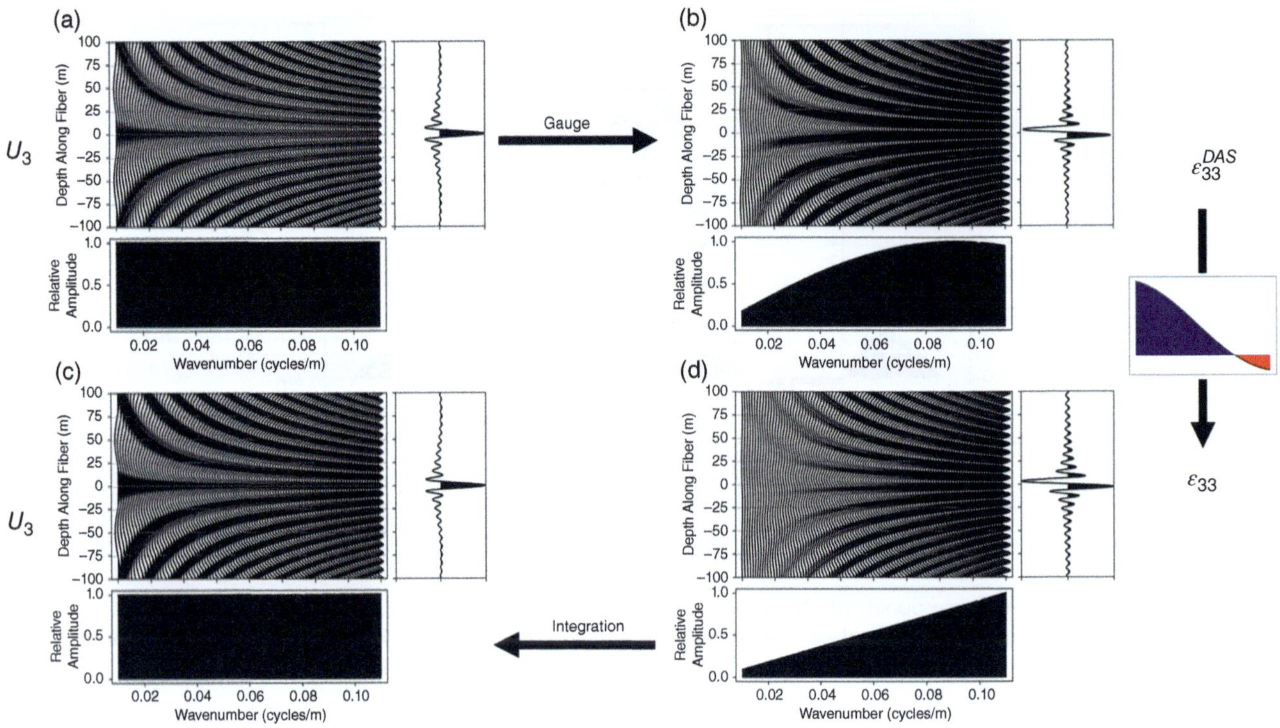

Figure 14.2 Demonstration of gauge-length effect and correction on monochromatic wavenumber waveforms where the gauge length is longer than the shortest wavelength. The stacked wavelet along the depth axis is displayed to the right of the waveforms, and the relative amplitude along the wavenumber axis is shown below the waveforms. (a) Wavenumber components of displacement u_3, (b) DAS measurement ε_{33}^{DAS} simulated by applying a 12 m gauge length, (c) axial strain ε_{33} determined by applying gauge-length correction factors to the DAS response, (d) displacement u_3 obtained on integrating the axial strain along depth.

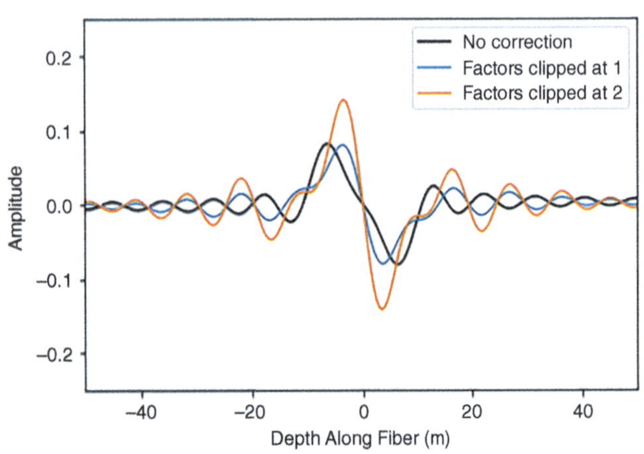

Figure 14.3 Plot of DAS waveform (black) with gauge-length corrected waveforms with correction factors clipped at an absolute value of 1 (blue) and 2 (orange). Gauge-length correction, limited to an absolute value of 1, recovers the bandwidth (width) of the wavelet. Further gain applied by the correction improves the signal amplitude.

where c is the phase velocity of the incident wave and θ is the wave propagation angle with respect to the fiber axis. The particle velocity field along the fiber can be expressed as

$$v_3 = \cos(\phi)v, \tag{14.9}$$

where v is the particle velocity and ϕ is the angle of the polarization vector with the fiber axis.

Substituting equations (14.8) and (14.9) into equation (14.7), we get

$$\varepsilon_{33} = \frac{\cos(\theta)}{c}\cos(\phi)v \tag{14.10}$$

Equation (14.10) represents a generalized relationship between axial strain and particle velocity for body waves. Assuming weak anisotropy, the polarization vector is approximately coincident with the propagation vector for P waves and approximately orthogonal for S waves. For P waves, equation (14.10) simplifies to

$$\varepsilon_{33} = \frac{\cos^2(\theta)}{c_p}v, \tag{14.11}$$

and for S waves it simplifies to

$$\varepsilon_{33} = \frac{\cos(\theta)}{c_s}\sin(\theta)v, \quad (14.12)$$

where c_p and c_s are the P- and S-wave phase velocities along the fiber axis, respectively. Equation (14.12) can be rewritten as

$$\varepsilon_{33} = \frac{\sin(2\theta)}{2c_s}v \quad (14.13)$$

Equations (14.11) and (14.13) show that the angular dependence of DAS amplitude is $\cos^2(\theta)/c_p$ and $\sin(2\theta)/2c_s$ for P and S waves, respectively.

The axial strain-to-velocity transformation can be achieved using multiple methodologies. Integrating equation 14.6 along the z axis yields particle displacement, u_3. Differentiating the particle displacement with respect to time results in the particle velocity component along the z axis, v_3. Alternatively, integration along the z axis and differentiation along the time axis can be performed in the f–k domain by scaling the data by ω/k factors to compute v_3. Equation (14.7) can be used to scale axial strain by corresponding apparent phase velocities ($1/p$ factors) in the *tau-p* domain to compute v_3.

A model-driven approach can be used to compute the $\cos(\theta)/c$ term in equation (14.10) to compute v_3. The corresponding correction to axial strain is $c/\cos(\theta)$. The isotropic assumption for polarization vector can be used to compute particle velocity, v. The corrections for P and S waves are $c_p/cos^2(\theta)$ and $2c_s/\sin(2\theta)$, respectively.

14.4. APPLICATION TO ZERO-OFFSET VSP SYNTHETIC

In this section, we test the proposed DAS-to-velocity transform using simulated zero-offset VSP (ZVSP) data. We recognize the challenges associated with real-world data, which often contain noise and complexities arising from unknown subsurface properties, and opt to test the transform with synthetic VSP data. We choose to use synthetic ZVSP data for controlled testing for a simple background model. This helps us identify subtle effects and variations. The synthetic data was generated using the finite-difference method, a numerical technique commonly employed in simulating wave propagation phenomena. By applying the transform to these synthetic data, we aim to demonstrate its effectiveness in converting DAS measurements to particle velocity measurements. This controlled experiment provides insights into the transform's performance and its potential for real-world VSP applications. Using synthetic data allows for a thorough evaluation of the method's accuracy and robustness before its deployment in analyzing field data, which often presents complexities not easily replicated in simulations.

A vertical-force source was simulated on the surface at 100 m offset from the well (Figure 14.4). A single flat interface is assumed within the subsurface model. The interface intersects the well at 1,000 m, and the receivers straddle the interface from 500 m to 1,500 m at 1 m receiver intervals. DAS waveforms in strain were synthesized from synthetic particle velocity using a 20 m gauge-length operator, followed by integration along the time axis. The synthesized DAS data were transformed to particle velocity by applying the DAS-to-velocity transform. Gauge-length correction factors, limited to a fivefold gain, are applied to DAS strain to compute axial strain. Axial strain is integrated along the depth axis and differentiated along the time axis to compute the DAS-transformed velocity (DtV).

Figure 14.5 illustrates three key wavefield representations: the zero-offset particle velocity synthetic (Figure 14.5(a)), the DAS synthetic waveforms with a 20 m gauge length (Figure 14.5(b)), and the DtV wavefield (Figure 14.5(c)). The DAS synthetic waveforms in Figure 14.5(b) exhibit consistent polarity for both upgoing and downgoing wavefields. In contrast, the DtV waveforms capture the polarity variations that are evident in the velocity synthetic seismogram. Specifically, the downgoing P and S events show opposite polarity compared to their upgoing counterparts (Figure 14.5 (c)). Although strain and velocity share a similar frequency bandwidth (other than the gauge-length effect), they exhibit opposite polarity in the downgoing events.

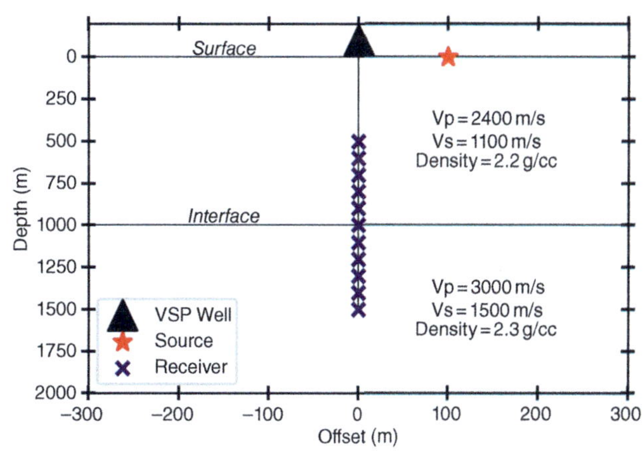

Figure 14.4 Survey simulation schematic for ZVSP synthetic. Receivers are decimated by a factor of 100 for display.

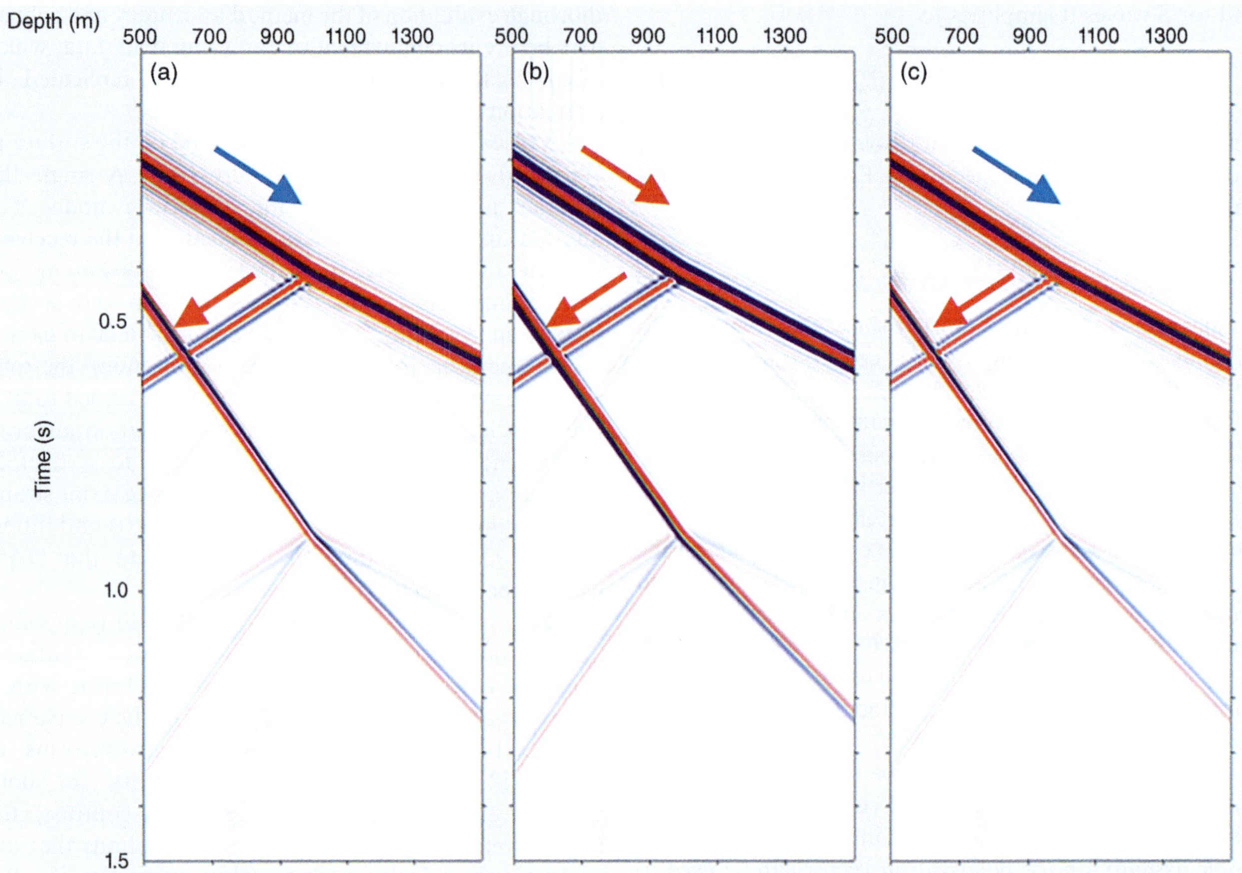

Figure 14.5 Finite difference zero-offset velocity synthetic (a), DAS strain generated by applying a 20 m gauge length (b), and DtV (c). Gauge-length application removes the directionality of the velocity measurement. However, the DAS-transformed velocity preserves the polarity of downgoing and upgoing waves as shown with the blue (positive) and red (negative) arrows. Adapted from Sayed et al. (2020).

These polarity relationships also depend on the polarity definition of the geophone sensor and the sign of the interface's reflection coefficient. It suggests that strain data can be effectively utilized for traveltime determination, provided that the correct phase is selected.

The gauge-length effect significantly diminishes amplitude at high wavenumbers. Slower shear waves, with their shorted wavelengths, fall into this higher wavenumber range and are thus heavily impacted by gauge-length scaling. Acting as a low-pass filter, the gauge-length operator suppresses these higher frequencies, but the DAS-to-velocity transformation works to reclaim the lost bandwidth (Figure 14.6). However, the recovery at the upper end of the spectrum is partial, as the maximum gain for gauge-length correction is restricted to a factor of five. To achieve optimal results, this maximum gain must be rigorously tested and fine-tuned for the specific dataset.

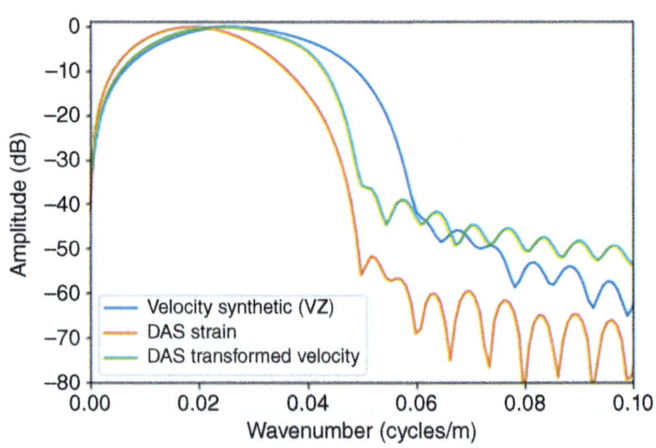

Figure 14.6 Wavenumber spectrum of velocity synthetic, DAS strain, and DtV for shear waves. The gauge-length correction factors are limited to a value of 5. Sayed et al. (2020) / with permission of the Society of Exploration Geophysicists.

14.5. Q ESTIMATION USING DAS

Seismic waves dissipate energy as they propagate through the subsurface because of geometrical spreading, anelastic attenuation, scattering, and transmission loss. Geometrical spreading refers to the loss of amplitude as the wavefront propagates along an expanding sphere. Anelastic attenuation represents the loss of energy due to several distinct mechanisms as the wave propagates through the subsurface (Batzle et al., 2005; Bouchaala et al., 2016; Muller et al., 2010). Scattering of the wavefield because of inhomogeneity of the subsurface also leads to amplitude decay (Matsushima et al., 2017; Wu & Aki, 1988). Transmission loss occurs because of energy partitioning at geological bed boundaries.

Anelastic attenuation is expressed by the seismic quality factor, Q. Q is commonly used in seismic processing and interpretation. An accurate estimate of Q can be used for compensating seismic data premigration or within the migration algorithm (Cavalca et al., 2013; Kristiansen et al., 2017) and supporting interpretation via lithology identification and as a direct hydrocarbon indicator (Klimentos, 1995).

Q is typically estimated by comparing energy levels of the seismic wavefront at different depths. The VSP geometry provides an ideal scenario for determining Q as receivers are positioned along the depth of the well. Several methods of estimating Q exist, with the spectral-ratio method being the most commonly used (Campbell et al., 2005; Leaney, 1999; Tonn, 1991).

We investigate measurement errors in estimating Q with the spectral-ratio method using synthetic DAS data. We use the method described by Leaney et al. (2019) to generate anelastic ray-trace synthetics. The method uses dynamic two-point ray tracing to supply the displacement polarization vector and phase slowness vector at the receiver. DAS strain is computed for any ray signature using a combination of the infinitesimal strain and gauge-length wavenumber filter. The synthetic data are used to demonstrate that applying the gauge-length correction factors can compensate for the error in Q estimation.

14.5.1. Spectral Ratio Method for Q Estimation

The spectral ratio method compares the spectra of the observed downgoing wavelet at two separate depths within the VSP data to estimate Q. The waveform amplitude propagating with this medium can be defined as follows

$$A_1(Z_1,f) = A_0(Z_0,f)\, e^{-\frac{\pi f}{QV}(Z_1 - Z_0)} \tag{14.14}$$

where Z represents the receiver depth, f is frequency, Q is the seismic quality factor, and V is the interval velocity between the two receivers. Taking the logarithm, we get

$$\ln\left[A_1(Z_1,f)/A_0(Z_0,f)\right] = -\pi f\,\frac{(T_1 - T_0)}{Q} \tag{14.15}$$

where T_1 and T_0 are the arrival times at the Z_1 and Z_0 receiver depths, respectively. If the logarithm of the spectral ratio (left-hand side of equation (14.15)) is crossplotted against frequency f, we can estimate Q by fitting a slope to the spectral ratio within the VSP bandwidth:

$$Q = -\frac{\pi(T_1 - T_0)}{\text{slope}} \tag{14.16}$$

In this method, the direct arrival is preserved and selected to generate spectra. Preprocessing steps such as median filtering, predictive deconvolution, and noise attenuation are conducted to isolate the direct arrival while suppressing reflections, converted waves, multiples, tubes waves, casing arrivals, and other noise. The slope of a linearly fit line to the computed spectral ratio of a pair of sensor waveforms is used to estimate Q. The method assumes that the source signature, receiver coupling, and stratigraphic filtering are unchanged between the two levels. It also assumes that Q is independent of frequency within the VSP bandwidth. Typically, an average Q along the well is computed by comparing the top and bottom geophones, and a depth-dependent Q is computed by using all possible receiver pairs (Campbell et al., 2005).

14.5.2. Synthetic Anelastic Data: DAS and Velocity

A two-layer model featuring a significant velocity contrast is employed to assess the impact of typical variations in local receiver velocities used for Q estimation. The selected velocities represent the extremes of the velocity range commonly observed in wells. Anisotropy was not considered in the model definition. The model's elastic properties are detailed in Table 14.1.

A vertical well is assumed. Sensors spaced at 1 m intervals from 500 to 1,500 m were simulated as vertical components of particle velocity and DAS cable along the well. The gauge length for DAS synthetic is 10 m. The source

Table 14.1 Elastic properties of the model used for generating synthetic data for Q estimation.

Layer no.	Z (m)	V_p (m/s)	V_s (m/s)	Density (g/cc)
1	0–1,000	2,000	1,000	2.0
2	1,000–2,000	4,000	2,000	2.5

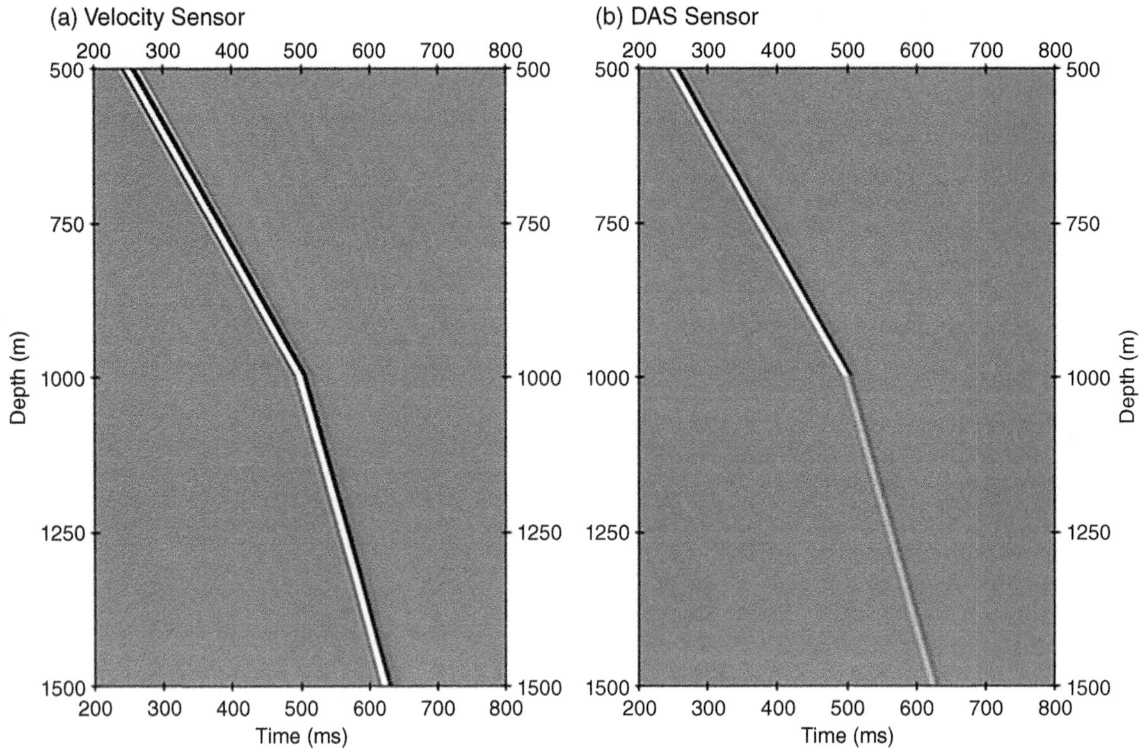

Figure 14.7 Synthetic VSP data for vertical velocity (a) and DAS (b) sensors along a vertical well and zero-offset source. The horizontal axis is observed time increasing to the right, and the vertical axis is sensor depth increasing downward. The interface between the faster and slower zones is at 1,000 m well depth. Note the weaker amplitude on the faster velocity event for DAS synthetic. Adapted from Sayed et al. (2021).

wavelet in displacement is of a trapezoidal frequency bandwidth of 1–10–60–120 Hz. The Q value is constant within the model at 100. Amplitude effects other than Q, such as geometrical spreading, transmission loss, and so on, are not modeled to isolate the Q effect on the synthetic waveforms.

Figure 14.7 presents the simulated velocity and DAS VSP data, with the DAS synthetic revealing amplitude variations correlated to the slope of the event. In Figure 14.8, the frequency spectra of the synthetic data are displayed. The velocity synthetic demonstrates a gradual decline in bandwidth along the well, attributed to Q. In contrast, the DAS spectra exhibit a discontinuity at the depth corresponding to the change in velocity, as the wavenumber filter imposed by the gauge length differentially impacts the frequency spectra of events traveling at varying velocities (Figure 14.1). The faster velocity zone generates a broader bandwidth compared to the DAS sensors within the slower velocity zone. The DAS measurement (DAS strain) corresponds to axial strain, scaled by a normalized *sinc* function in the wavenumber domain, as described in equation (14.3). This leads to fluctuations in wavenumber bandwidth as the event slope (apparent velocity) changes, which in turn causes variations in temporal frequencies corresponding to the changes in the event slope (apparent velocity).

14.5.3. *Q* Estimation: DAS and Velocity Sensor

To evaluate the gauge-length effect on estimating Q, we compute Q between sensors at 800 and 1,300 m from the two zones. Figure 14.9 shows the input spectra for the velocity synthetic and DAS synthetic and their corresponding spectral ratios. A linear fit across a frequency bandwidth of 10–100 Hz yields a Q estimate of 100 for the velocity synthetic. However, the same bandwidth yields a Q estimate for 310 for the DAS synthetic.

The gauge length applied within the DAS measurement has a detrimental effect on the spectral ratio due to variations in local velocity at the receivers. The DAS spectral ratios are nonlinear as the higher frequencies are scaled down differentially by the *sinc* function in the faster and slower velocity zones. The *sinc* function has a stronger effect on slower velocity events than the faster velocity event as the effect is wavenumber dependent. This results

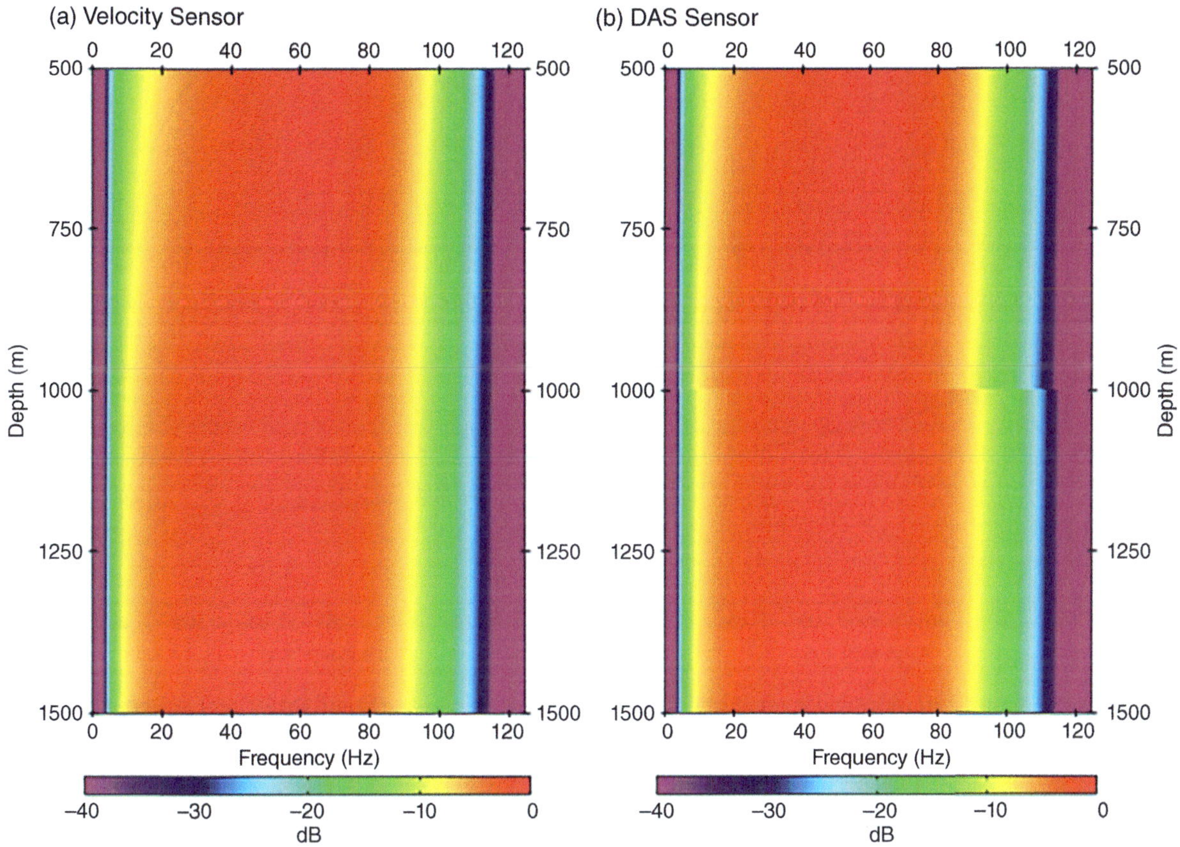

Figure 14.8 Trace normalized frequency spectra for the velocity (a) and DAS (b) synthetic data. The horizontal axis is frequency increasing to the right and the vertical axis is sensor depth increasing downward. Change in frequency bandwidth along the well is due to Q. In the case of the DAS synthetic, the drop-off on the higher end of the spectrum is earlier in the slower velocity zone and later in the faster velocity zone. Such discontinuity is not observed in the velocity synthetic data. Adapted from Sayed et al. (2021).

in an apparent increase in frequency with increasing velocity. Such an effect results in overestimation of the Q value. Intuitively, if the shallower receiver was within a faster velocity zone and the deeper receiver was within a slower velocity zone, the effect would be reversed. An inversion in velocity would show an apparent stronger decrease in frequency with depth resulting in an underestimation of the Q value.

14.5.4. Gauge-length Correction for *Q* Estimation

The gauge-length response was corrected by applying the multiplicative inverse of the *sinc* function, represented as an equivalent *gamma* function in the wavenumber domain (refer to the section on gauge-length effect and correction). These corrections on DAS strain data yield the true axial strain. Figure 14.10a illustrates the axial strain wavefield computed from the DAS synthetic data after applying the gauge-length correction. The corresponding input spectra and spectral ratio are shown in Figure 14.10(b), (c). The Q factor estimated from the axial strain between the receivers at 800 and 1,300 m is 101. Notably, gauge-length corrections significantly reduce the error in Q estimation, from 210% in DAS strain to just 1% in axial strain. The correction from DAS strain to axial strain compensates for variations in the wavenumber domain, subsequently correcting the temporal frequencies.

Gauge-length effects are particularly significant in DAS data, especially for Q estimation, and are exacerbated by stronger velocity contrasts. Additionally, higher-frequency content amplifies the impact of gauge length. Therefore, gauge-length corrections, whether applied independently or as part of the DAS-to-velocity transformation, are crucial for accurate processing and interpretation of DAS data.

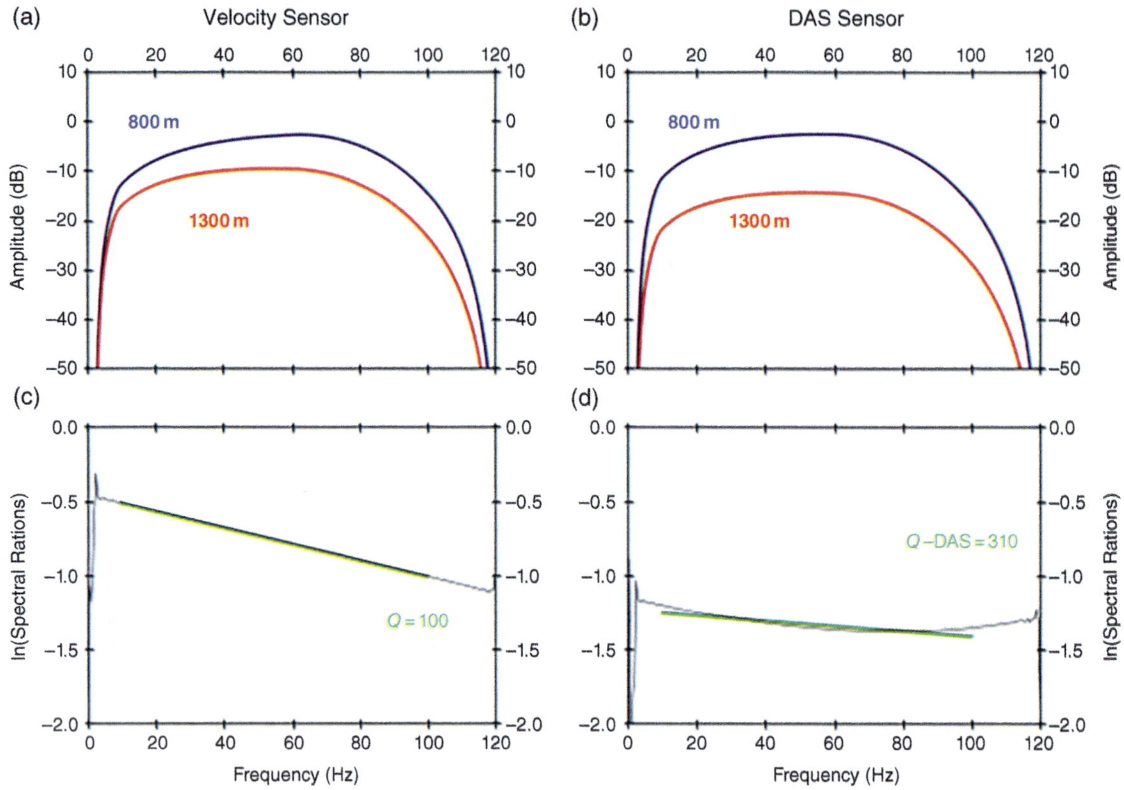

Figure 14.9 Input spectra (top) and spectral ratio (bottom) for the velocity (a and b) and DAS (c and d) synthetic data corresponding to receivers at 800 m and 1,300 m. Input spectra show spectra for 800 m (blue) and 1,300 m (red). The plot on the right shows the respective spectral ratio in black and the linear best-fit line in green. *Q* estimate for the velocity synthetic is 100 as used in the input model and 310 for the DAS synthetic. Note the nonlinear response of the DAS spectral ratio. Adapted from Sayed et al. (2021).

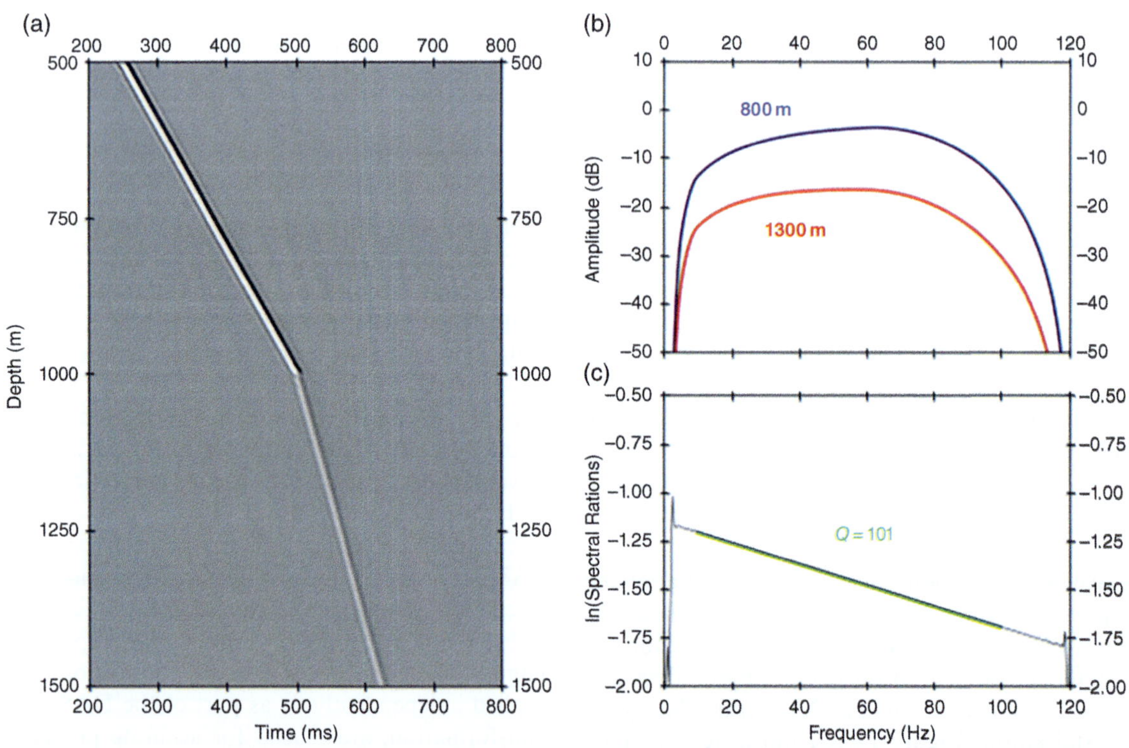

Figure 14.10 Axial strain wavefield computed from DAS synthetic after application of gauge-length correction (a). The vertical axis is sensor depth, and the horizontal axis is observed time. Input spectra (b) and spectral ratio (c) for axial strain. Spectra for receivers at 800 m (blue) and 1,300 m (red) are shown. The spectral ratio is plotted in black and the linear fit slope in green. *Q* estimate after gauge-length correction is 101, indicating a 1% residual error from the model value. Adapted from Sayed et al. (2021).

14.5.5. Sensitivity Analysis

We establish the sensitivity of Q estimation with DAS data by varying model and measurement parameters. While keeping the other model and measurement parameters constant, we vary the velocity at the deeper sensor and the gauge length. The velocity of the deeper zone is varied to understand the effect of contrasting velocities at the two sensors used for the analysis. The estimated Q shows larger errors for larger differences in local velocities at the sensors (Figure 14.11). An apparent compensative effect (larger Q value) is observed when the bottom sensor is in a faster medium, and an apparent attenuative effect (smaller Q value) is observed when the bottom sensor is in a slower medium (Figure 14.11(d)). In the special case when the two sensors have the same local velocity, the spectral ratio of DAS data correctly estimated the Q value (Figure 14.11(d)). This happens as the gauge-length effect on the two sensors is equivalent and the slope of the spectral ratio is not distorted. Another parameter that significantly affects Q estimation is gauge length. Varying the gauge length shows that a larger gauge length results in enormous errors and negative Q values (Figure 14.12). Larger gauge lengths have a stronger distortion on wavenumbers, resulting in an anomalously low bandwidth for shallower sensors at slower wave velocity.

14.6. APPLICATION TO WVSP SYNTHETIC

We generate a walkaway-VSP synthetic dataset to comprehensively investigate the influence of velocity on DAS imaging. A simple three-layer velocity model over half-space is built for this simulation. As salt environments are known for large velocity variations, velocities used in the three layers are representative of the suprasalt sediments, salt, and subsalt sediments, respectively (Table 14.2). Anisotropy and anelasticity effects are ignored in this study.

A vertical VSP well was assumed at the center of the model space. Velocity sensors were simulated along the depth axis from 250 to 1,500 m at 5 m intervals, spanning

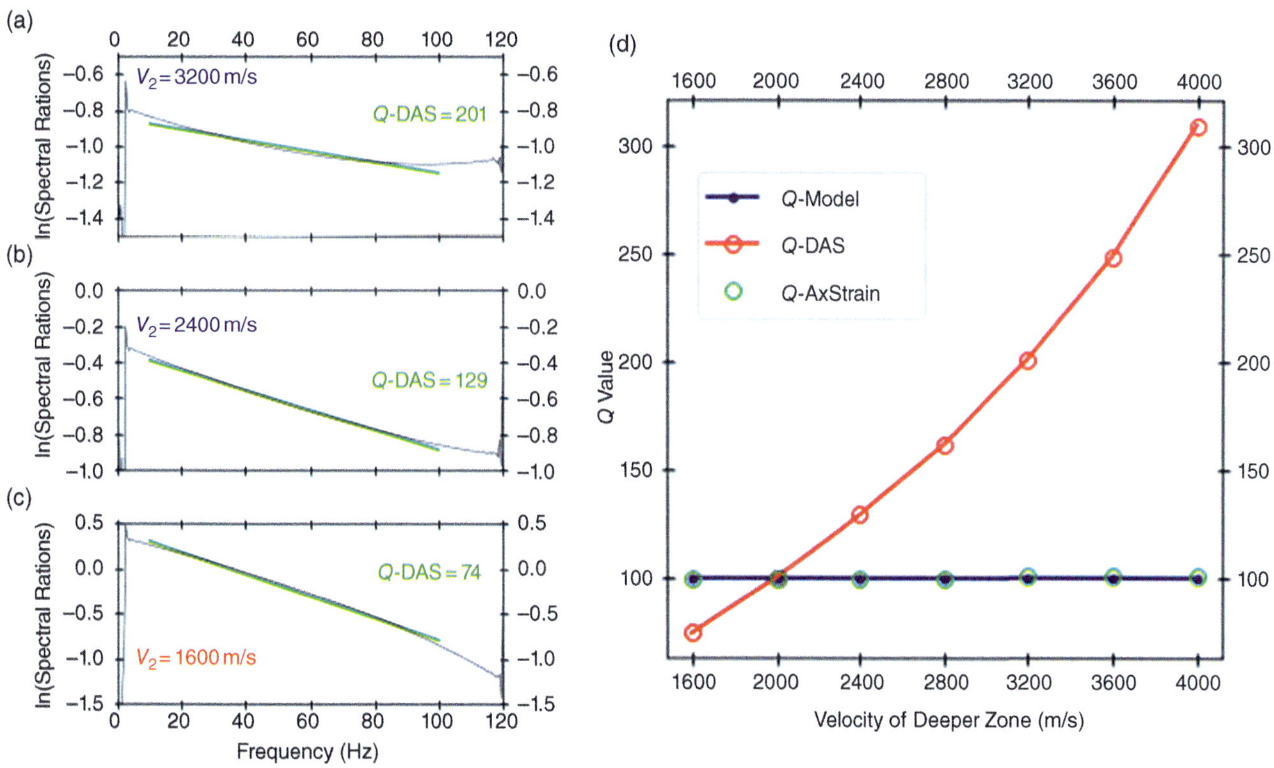

Figure 14.11 Sensitivity analysis for Q estimation of DAS synthetic to the velocity at the deeper sensor. Spectral ratios for three cases are shown: (a) V_2 = 3,200 m/s, (b) V_2 = 2,400 m/s, and (c) V_2 = 1,600 m/s. V_1 is constant at 2,000 m/s for all tests. The linear best-fit line and the estimated Q value are shown in green. The variation of results for Q measured on DAS (red) and Q measured on axial strain are plotted against the wave velocity at the deeper sensor (d). Faster velocities at the deeper sensor overestimate the Q value, whereas slower velocities underestimate the Q value. Q is accurately estimated when both sensors are located in a homogeneous velocity medium ($V_1 = V_2$ = 2,000 m/s).

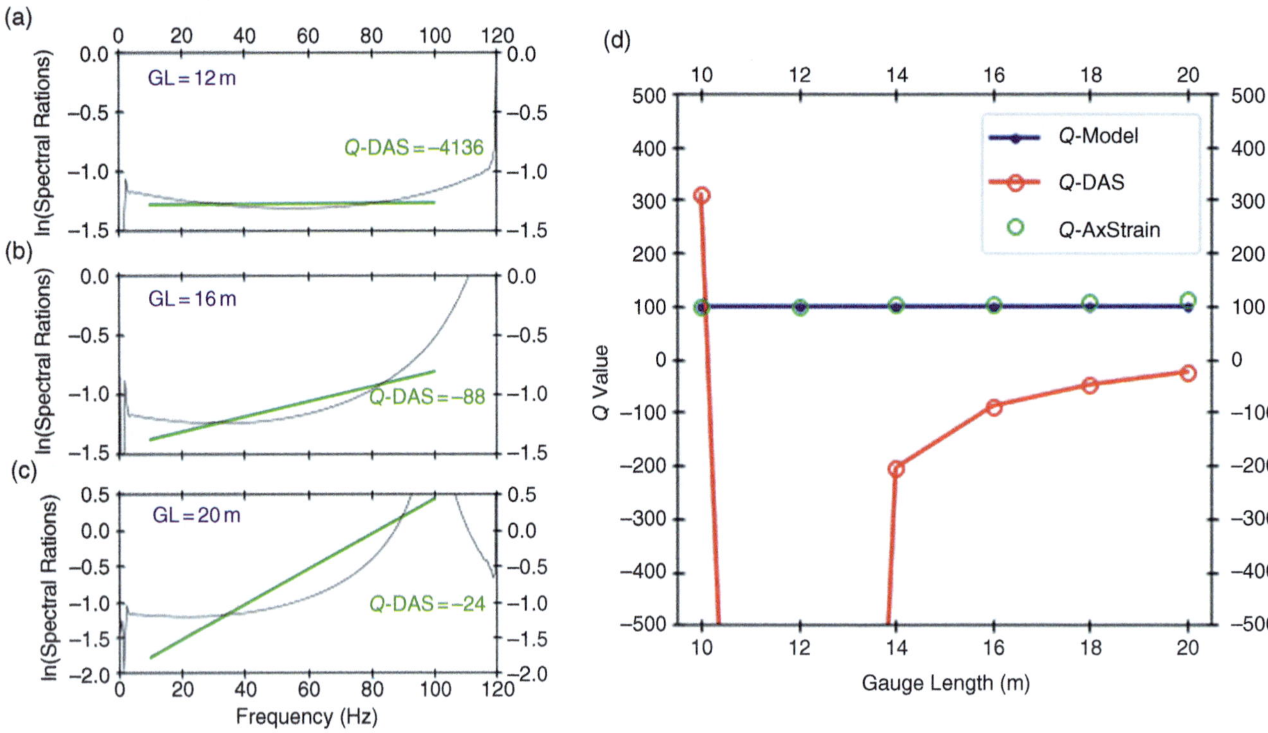

Figure 14.12 Sensitivity analysis for Q estimation of DAS synthetic to the gauge length. Spectral ratios for three cases are shown: (a) GL = 12 m, (b) GL = 16 m, and (c) GL = 20 m. The linear best-fit line and the estimated Q value are shown in green. The variation of results for Q measured on DAS (red) and Q measured on axial strain are plotted against the gauge length (d). Increasing gauge length causes severe distortion of the high frequencies, resulting in negative values of Q.

Table 14.2 Model properties for walkaway-VSP synthetic.

Layer	Depth range (m)	V_p (m/s)	V_s (m/s)	Density (g/cc)
Suprasalt sediments	0–500	2,500	1,200	2.2
Salt	500–1,000	4,480	2,500	2.16
Subsalt sediments	1,000–1,500	2,750	1,375	2.3
Half-space	1,500–	3,500	1,750	2.5

across the suprasalt sediments, salt, and subsalt sediments. Sources were simulated along the surface at 25 m intervals from −3,000 to +3,000 m surface offset from the well (Figure 14.13). Two-point ray tracing was used to simulate the velocity synthetic for this walkaway-VSP geometry (Leaney et al., 2019). A 1–10–80–120 Hz bandlimited source wavelet was used. The upgoing Pp reflection from the base of salt at 1,000 m and base of subsalt sediments at 1,500 m were modeled with a velocity sensor (Figure 14.14a). Amplitude effects such as geometrical spreading and transmission loss were not modeled to ensure that the response is free of these effects. The velocity sensor data was converted to DAS equivalent by integrating along the time axis and applying a gauge length of 20 m. The expected variation in amplitude was clearly observed for each layer in the DAS synthetic (Figure 14.14(b)).

14.6.1. DAS-to-Velocity Transform for WVSP

The DAS synthetic data are transformed to velocity using the steps described for gauge-length correction and axial strain to velocity conversion. The gauge-length corrections are applied in the wavenumber domain to compute axial strain. As the gauge-length corrections are applied as gains to wavenumbers, it also has a secondary amplitude effect. The gauge-length correction amplifies the amplitude differences between the fast and slow

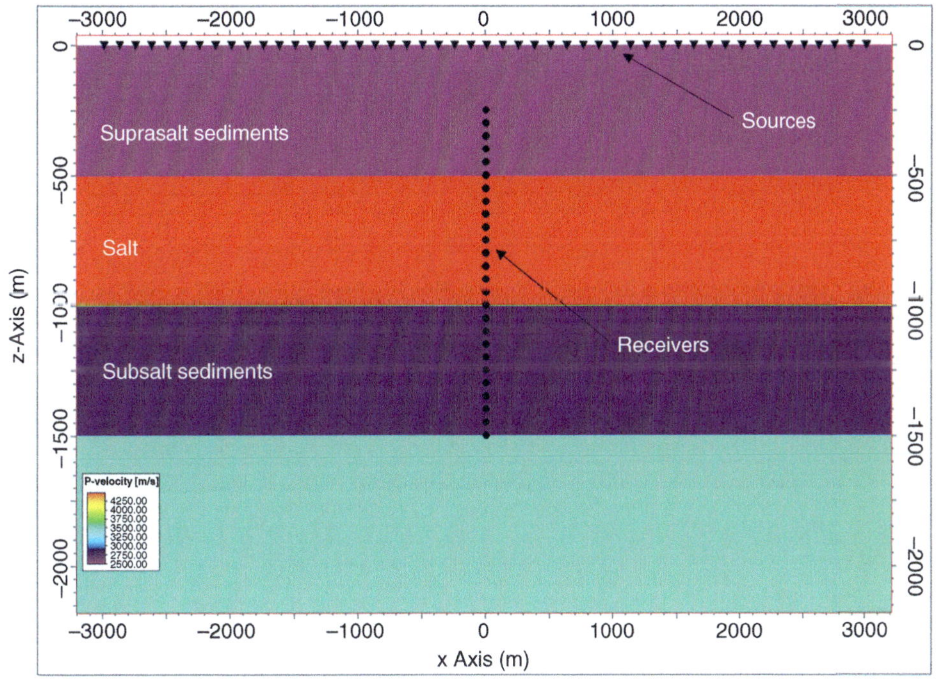

Figure 14.13 Subsurface velocity model overlain with simulated survey geometry. Source and receiver positions are decimated for this display by a factor of 5 and 10, respectively. Sayed et al. (2022) / with permission of OnePetro.

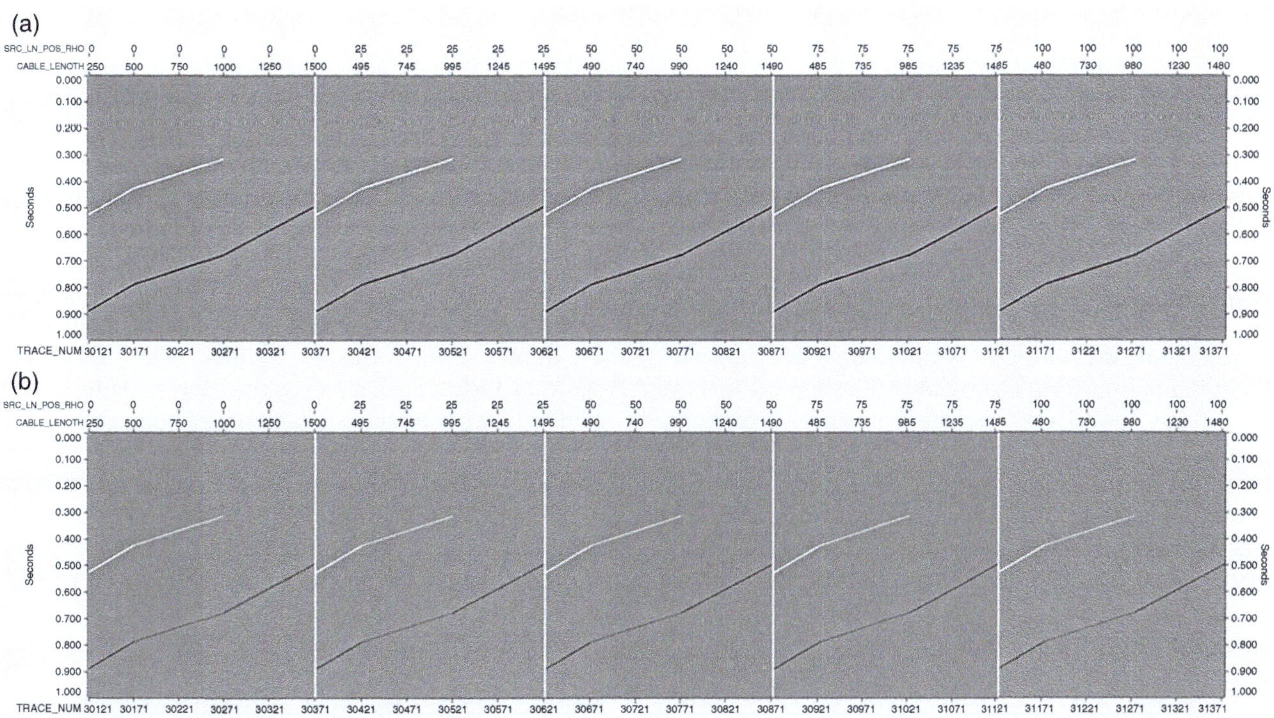

Figure 14.14 Simulated shot gathers for (a) velocity and (b) DAS sensors. Sensor depth increases to the right for each gather, and observed time is increasing downward. Note the variation in amplitude of DAS sensor synthetic data with change in velocity. Sayed et al. (2022) / with permission of OnePetro.

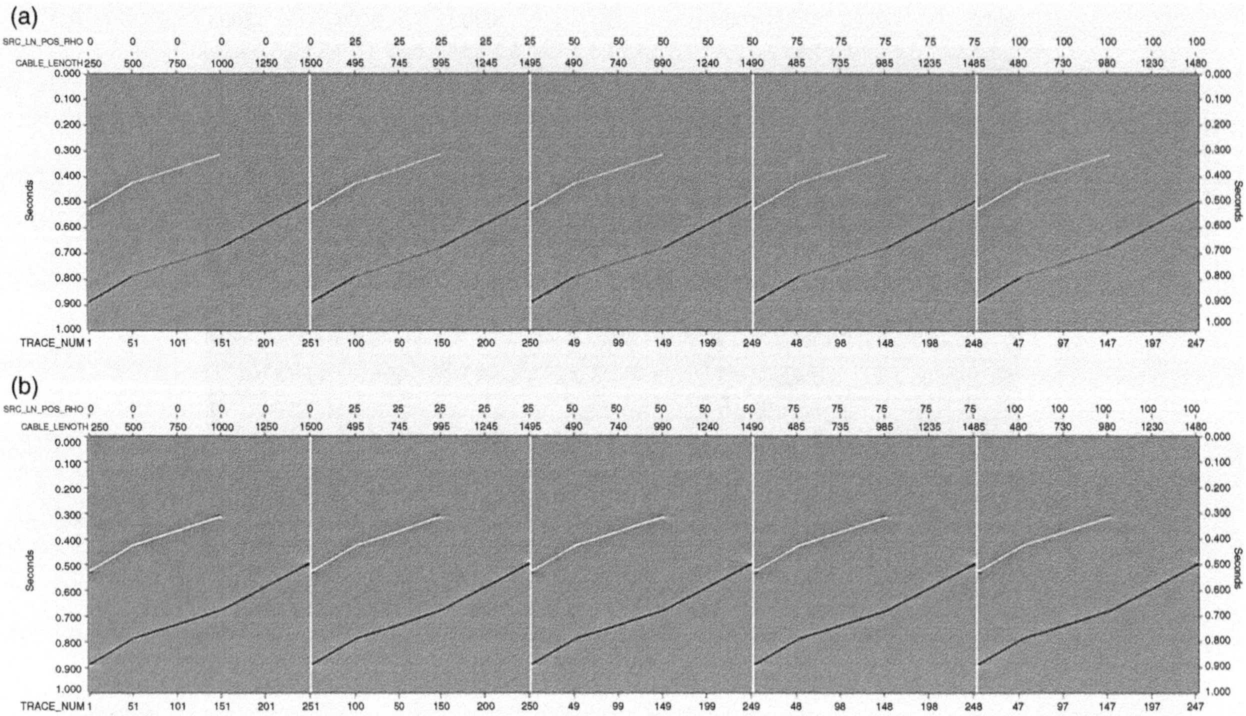

Figure 14.15 Shot gathers after applying the DAS-to-velocity transform. The gauge-length correction step produces axial strain (a), and *f-k* scaling step produces DtV data (b). Axial strain is corrected in the wavenumber domain and has secondary amplitude effects. DtV shot gathers display consistent amplitudes along the layers indicating a robust correction. Sayed et al. (2022) / with permission of OnePetro.

events on the shot gather (Figure 14.14(b) and 14.15(a)). Axial strain is scaled by ω/k factors in the *f-k* domain to compute the velocity response. The *f-k* scaling step compensates for the amplitude differences between axial strain and velocity data (Figure 14.15) to produce the DtV data.

14.6.2. Imaging

We migrate the computed synthetic data using a generalized radon transform algorithm (Miller et al., 1987). As the velocity model used for the synthetic is accurately known, we use a narrow-aperture migration to evaluate the common image point gathers. Such narrow-aperture migrations can be considered equivalent to traditional common depth point mapping in VSP imaging. The incident angle of the rays is limited to 60° from the interface normal. The receiver array was decimated to a 25 m interval to reduce computing time and cost.

We migrate the DAS synthetic with the conventionally accepted $1/\cos^2(\theta)$ correction (Mateeva et al., 2014). Large variations in amplitude and bandwidth corresponding to the changes in seismic velocity along the sensor array are observed on the common image point gathers (Figure 14.16(a)). The traces from sensors within the salt show weaker amplitude relative to the traces from sensors within the sediments. We also migrate the DAS synthetic with the $c_p/\cos^2(\theta)$ corrections as given in equation (14.11). Such a correction would be equivalent to applying amplitude-only compensation in the DAS-to-velocity transform without any gauge-length correction. The amplitude-only correction overcompensates the amplitude effect. This results in the traces from sensors within the salt having stronger amplitudes relative to the traces from sensors within the sediments (Figure 14.16(b)).

Similarly, we compute common image point gathers for a model-driven approach using axial strain with $c_p/\cos^2(\theta)$ corrections, a data-driven approach using vector migration of the DtV, and vector migration of the raw velocity synthetic. The three results (Figure 14.17) show common image point gathers with consistent amplitude and bandwidth across the sensors, validating the model-driven and data-driven approaches of conditioning DAS data for imaging.

We also analyze the final migrated images for four cases. Case 1, a reference image by migration of the velocity synthetic, Case 2, a conventional image by migrating DAS synthetic with $1/\cos^2(\theta)$ corrections, Case 3, a model-driven image by migrating axial strain with $c_p/\cos^2(\theta)$ corrections, and Case 4, a data-driven

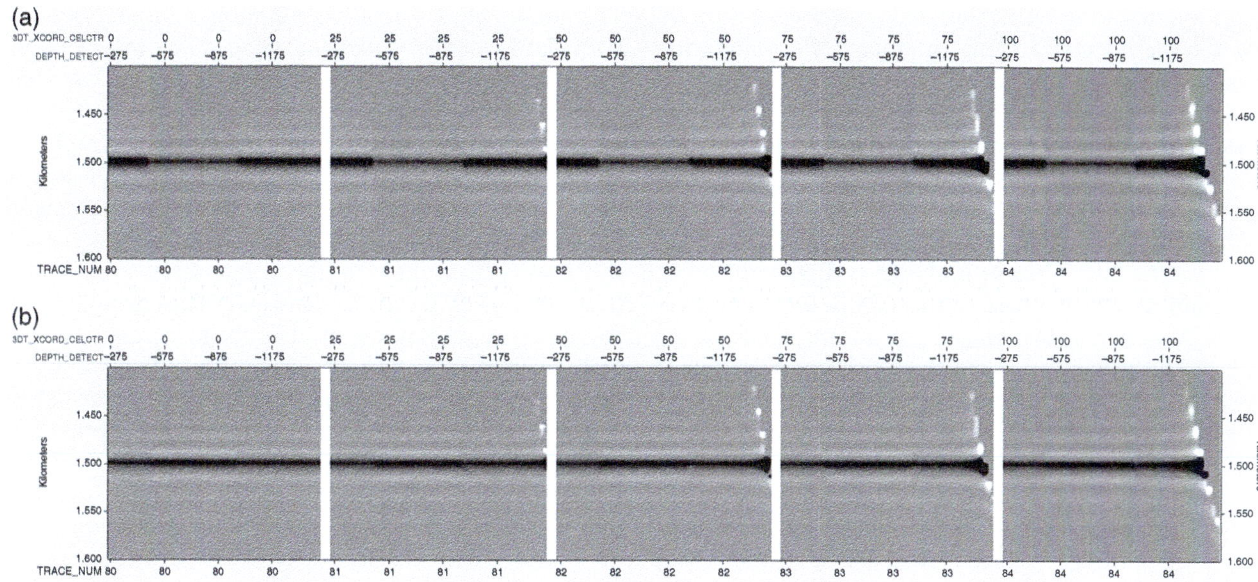

Figure 14.16 Common image point gathers for (a) DAS synthetic with $1/\cos^2(\theta)$ corrections and (b) DAS synthetic with $c_p/\cos^2(\theta)$ corrections. The base of the subsalt sediment layer is shown here. Sensor depth is increasing to the right for each gather, and vertical axis is model depth. Note the amplitude drop along the salt receivers in (a) and overcompensation in (b). Sayed et al. (2022) / with permission of OnePetro.

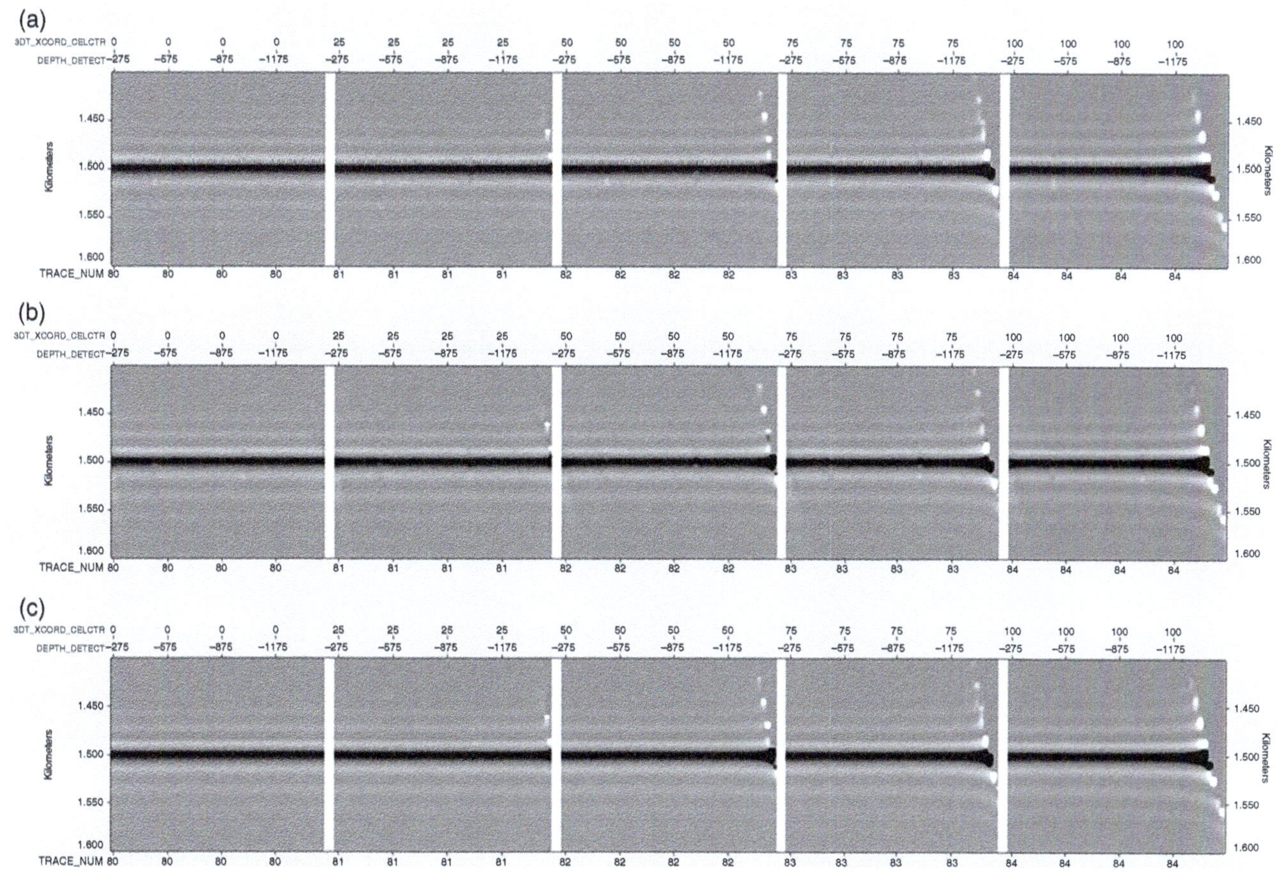

Figure 14.17 Common image point gathers for (a) axial strain after migration with $c_p/\cos^2()$ corrections, (b) DtV after migration, and (c) velocity synthetic after migration. The base of the subsalt sediment layer is shown here. Sensor depth is increasing to the right for each gather, and the vertical axis is model depth. Note the consistent amplitude on the three tests. Sayed et al. (2022) / with permission of OnePetro.

image by migrating DtV. Migrated images were normalized for the hit count along the image. Post-migration enhancements or AVO corrections are not applied on any image, and the velocity synthetic migration (Case 1) is used as the benchmark. The image comparisons use consistent survey geometry and migration parameters.

The reference image produced by migration of the velocity synthetic (Case 1) is shown in Figure 14.18(a). The amplitude of the base of the subsalt sediment layer is consistent across the image, and the amplitude of the base of the salt layer is mostly consistent with minor low-frequency smearing in the central half of the image. We use the velocity image as the reference for comparison with other images.

The amplitude of the bottom of the subsalt sediment layer is used to compare the results from each case. The Case 1 image, velocity synthetic migration with directivity correction, is used as the benchmark for comparison (Figure 14.18(a)). The DAS image with $1/\cos^2(\theta)$ corrections (Case 2) shows significant variation in amplitude across the image (Figure 14.18(b)). The image shows relative dimming in the center of both flanks of the walkaway image. The image points away from the well are progressively imaged by shallower sensors. This bias in capturing the reflections across the receiver array results in a variable amplitude along the image corresponding to the sensor depths that image it. Illumination near the well is dominated by receivers within each layer. As the image point shifts away from the well, the illumination contribution from shallower receivers increases. This results in the variation in amplitude of the image across the well with overamplification at near-offsets, dampening at mid-offsets, and overamplification at the far-offsets (Figure 14.18(b)). These variations are also captured on the common image point gathers in Figure 14.16(a). Results from the Case 3 image (Figure 14.18(c)) (model-based correction) and Case 4 image (Figure 14.18(d)) (data-driven correction) better match the Case 1 image (Figure 14.18(a)), benchmark velocity synthetic. The RMS amplitude extracted across the bottom of the subsalt sediment layer at 1,500 m depth is normalized and shown in Figure 14.19.

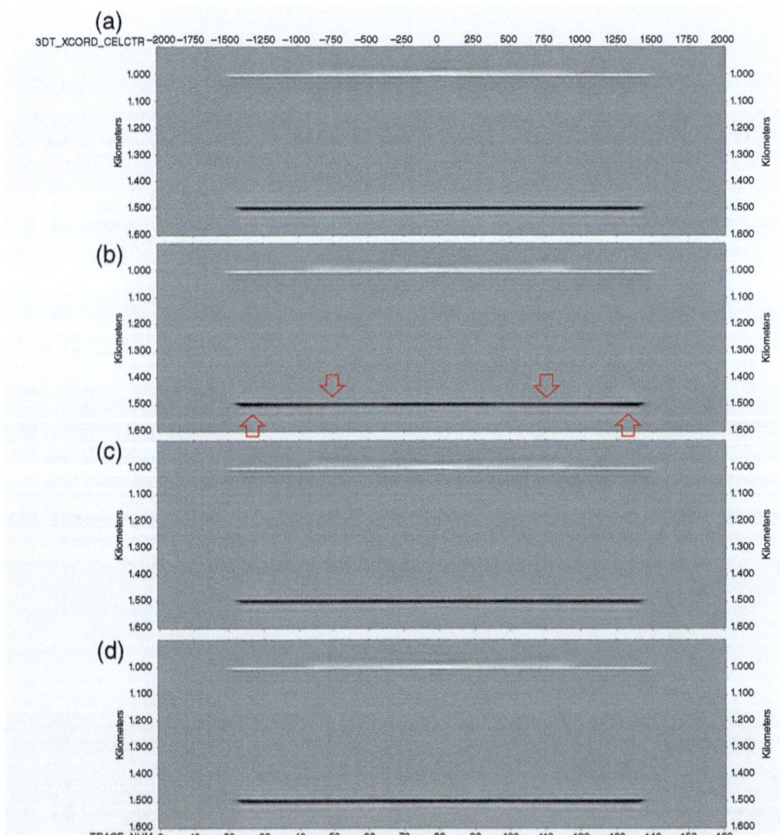

Figure 14.18 (a) Case 1, image produced by migrating the velocity synthetic; (b) Case 2, conventional image produced by migrating DAS synthetic with $1/\cos^2(\theta)$ corrections; (c) Case 3, image produced by migrating DAS synthetic with model-based corrections; (d) Case 4, image produced by migrating DAS synthetic with data-driven corrections. Horizontal axis is the surface offset from the well. Vertical axis is model depth. The red arrows highlight the dimming and overamplification of the conventional DAS image (b). Sayed et al. (2022) / with permission of OnePetro.

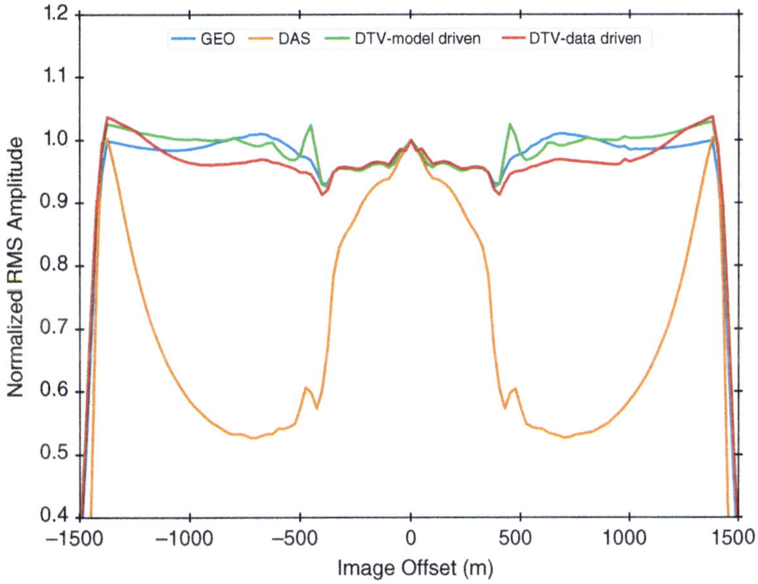

Figure 14.19 The normalized RMS amplitude across the bottom of the subsalt sediment layer at 1,500 m depth plotted against image offset for Case 1, image produced by migrating the velocity synthetic (blue); Case 2, conventional image produced by migrating DAS synthetic with $1/\cos^2(\theta)$ corrections (orange); Case 3, image produced by migrating DAS synthetic with model-based corrections (green); and Case 4, image produced by migrating DAS synthetic with data-driven corrections (red). Significant amplitude variations are observed in Case 2 without appropriate corrections applied.

The model-driven image and the data-driven image shown in Figure 14.18(c), (d) are consistent with the response observed in the reference image (Figure 14.18(a)), validating these approaches for DAS imaging.

14.7. CONCLUSIONS

We show that DAS strain measurement differs from the geophone particle velocity measurement in frequency and amplitude. DAS strain is scaled with a normalized *sinc* function in the wavenumber domain. Scaling in the wavenumber domain causes significant errors in Q estimation. Axial strain can be recovered by gaining DAS strain in the wavenumber domain by the corresponding gauge-length correction factors. We show that the gauge-length correction results in accurately estimating Q. The axial strain measurement has a scaling effect with apparent velocity. We correct for this effect by applying equivalent velocity gains to the axial strain data, which transforms the data to velocity. We show that axial strain has an amplitude dependence on the angle of propagation as well as the angle of polarization with respect to the fiber. In the isotropic homogeneous case, this is $\cos^2(\theta)$. The DAS-to-velocity transform recovers the amplitude and frequency effects observed on DAS data. We show the effects of using raw DAS, velocity-transformed DAS, and velocity synthetic seismograms on borehole seismic imaging for a simple case. The uncorrected DAS data show dimming effects away from the well. The DAS-transformed-velocity and velocity synthetic images are shown to be consistent.

ACKNOWLEDGMENTS

The authors would like to thank SLB for allocating resources to this study.

AVAILABILITY STATEMENT

Data associated with this research are confidential and cannot be released.

REFERENCES

Abramowitz, M., & Stegun, I. A. (Eds.) (1965). *Handbook of Mathematical Functions: With Formulas, Graphs, and Mathematical Tables* (0009-Revised edition). Dover Publications.

Bakku, S. K. (2015). *Fracture characterization from seismic measurements in a borehole* (PhD Thesis). Massachusetts Institute of Technology.

Bakulin, A., Silvestrov, I., & Pevzner, R. (2020). Surface seismics with DAS: an emerging alternative to modern point-sensor acquisition. *The Leading Edge*, *39*(11), 808–818. doi:10.1190/tle39110808.1

Batzle, M., Hofmann, R., Prasad, M., Kumar, G., Duranti, L., & Han, D. (2005). *Seismic attenuation: observations and mechanisms*. Paper presented at the 75th Annual International Meeting, SEG. Expanded Abstracts, 1565–1568. doi:10.1190/1.2147991

Bona, A., Dean, T., Correa, J., Pevzner, R., Tertyshnikov, K. V., & Van Zaanen, L. (2017). *Amplitude and phase response of DAS receivers* Paper presented at the 79th Conference and Exhibition, Paris, France. doi:10.3997/2214-4609.201701200

Borland, W., Hearn, R., & Sayed, A. (2016). *Improving the acquisition efficiency of acquiring borehole seismic data by recording optical distributed acoustic data on a wireline hybrid electrooptical cable*. Paper presented at the 86th Annual International Meeting, SEG. Expanded Abstracts, 800–804. doi:10.1190/segam2016-13951115.1

Bouchaala, F., Ali, M. Y., & Matsushima, J. (2016). *Attenuation modes from Vertical Seismic Profiling and sonic waveform in a carbonate reservoir, Abu Dhabi, United Arab Emirates. Geophysical Prospecting, 64*(4-Advances in Rock Physics), 1030–1047. doi:10.1111/1365-2478.12388.

Campbell, A., Fryer, A., & Wakeman, S. (2005). Vertical seismic profiles—More than just a corridor stack. *The Leading Edge, 24*(7), 694–697. doi:10.1190/1.1993259

Cavalca, M., Fletcher, R., & Riedel, M. (2013). *Q-compensation in complex media—ray-based and wavefield extrapolation approaches*. Paper presented at the 83rd Annual International Meeting, SEG. Expanded Abstracts, 3831–3835. doi:10.1190/segam2013-1130.1

Correa, J., Egorov, A., Tertyshnikov, K., Bona, A., Pevzner, R., Dean, T., et al. (2017). Analysis of signal to noise and directivity characteristics of DAS VSP at near and far offsets—A CO2CRC Otway Project data example. *The Leading Edge, 36*(12), 994a1–994a7. doi:10.1190/tle36120994a1.1

Daley, T. M., Miller, D. E., Dodds, K., Cook, P., & Freifeld, B. M. (2016). Field testing of modular borehole monitoring with simultaneous distributed acoustic sensing and geophone vertical seismic profiles at Citronelle, Alabama: field testing of MBM. *Geophysical Prospecting, 64*(5), 1318–1334. doi:10.1111/1365-2478.12324

Dean, T., Hartog, A., Cuny, T., & Englich, F. (2016). *The effects of pulse width on fibre-optic distributed vibration sensing data*. Paper presented at the 78th Conference and Exhibition, EAGE. Extended Abstracts, 1–5. doi:10.3997/2214-4609.201600684

Hartog, A. H. (2017). *An Introduction to Distributed Optical Fibre Sensors* (1st ed.). CRC Press. doi:10.1201/9781315119014-8

Kimura, T., Suzuki, K., Ali, S., Breninger, J., Tessman, J., Mahmoud, I. Z., & Hajjaj, S. (2018). *Hybrid 3D VSP using fiber-optic technology and a conventional borehole seismic array tool*. Paper presented at the 80th Conference and Exhibition, EAGE. Extended Abstracts, 1–5. doi:10.3997/2214-4609.201801499

Klimentos, T. (1995). Attenuation of P- and S-waves as a method of distinguishing gas and condensate from oil and water. *Geophysics, 60*(2), 447–458. doi:10.1190/1.1443782

Kristiansen, P., Abdelaziz, K., Saragoussi, E., Zaman, S. O., Zdraveva, O., Li, Q., & Chakraborty, S. (2017). *Accurate and efficient absorption compensation pre-imaging*. Paper presented at the 87th Annual International Meeting, SEG. Expanded Abstracts, 4572–4576. doi:10.1190/segam2017-17663231.1

Leaney, W. S. (1999). *Walkaway Q inversion*. Paper presented at the 69th Annual International Meeting, SEG. Expanded Abstracts, 1311–1314. doi:10.1190/1.1820751

Leaney, W. S., Mizuno, T., Velez, E., Cuny, T., & Perez, M. (2019). *Anisotropic model building, DAS simulation and imaging*. Paper presented at the 89th Annual International Meeting, SEG. Expanded Abstracts, 999–1003. doi:10.1190/segam2019-3215957.1

Martinez, A., Useche, M., Sayed, A., Duran, C., & Araujo, J. (2021). *DAS: an efficient and effective solution for VSP acquisition*. Paper presented at the First International Meeting for Applied Geoscience & Energy, SEG/AAPG. Expanded Abstracts, 3500–3504. doi:10.1190/segam2021-3581700.1

Mateeva, A., Lopez, J., Potters, H., Mestayer, J., Cox, B., Kiyashchenko, D., et al. (2014). Distributed acoustic sensing for reservoir monitoring with vertical seismic profiling. *Geophysical Prospecting, 62*, 679–692. doi:10.1111/1365-2478.12116

Matsushima, J., Ali, M. Y., & Bouchaala, F. (2017). A novel method for separating intrinsic and scattering attenuation for zero-offset vertical seismic profiling data. *Geophysical Journal International, 211*(3), 1655–1668. doi:10.1093/gji/ggx391

Miller, D., Oristaglio, M., & Beylkin, G. (1987). A new slant on seismic imaging: migration and integral geometry. *Geophysics, 52*(7), 943–964. doi:10.1190/1.1442364

Muller, T. M., Gurevich, B., & Lebedev, M. (2010). Seismic wave attenuation and dispersion resulting from wave-induced flow in porous rocks—A review. *Geophysics, 75*(5), 75A147–75A164. doi:10.1190/1.3463417

Sayed, A., Ali, S., & Stewart, R. R. (2020). *Distributed acoustic sensing (DAS) to velocity transform and its benefits*. Paper presented at the 90th Annual International Meeting, SEG. Expanded Abstracts, 3788–3792. doi:10.1190/segam2020-3424622.1

Sayed, A., Khaitan, M. L., Ali, S., & Martinez, A. (2021). *Characterizing anelastic attenuation using distributed acoustic sensing (DAS) data*. Paper presented at the First International Meeting for Applied Geoscience & Energy, SEG/AAPG. Expanded Abstracts, 3505–3509. doi:10.1190/segam2021-3563438.1

Sayed, A., Khaitan, M. L., & Martinez, A. (2022). *Conditioning distributed acoustic sensing measurements for borehole seismic imaging*. Paper presented at the International Petroleum Technology Conference. doi:10.2523/IPTC-21944-MS

Tonn, R. (1991). The determination of the seismic quality factor *Q* from VSP data: a comparison of different computation methods. *Geophysical Prospecting, 39*, 1–27.

Wu, R.-S., & Aki, K. (1988). Introduction: seismic wave scattering in three-dimensionally heterogeneous earth. In K. Aki & R.-S. Wu (Eds.), *Scattering and Attenuations of Seismic Waves, Part I* (pp. 1–6). Birkhäuser. doi:10.1007/978-3-0348-7722-0_1

15

Automatic Picking of Distributed Acoustic Sensing Vertical Seismic Profiling First-Arrival Traveltimes Using a Differential-Phase-Spectral Slope Method

David Li[1], Lianjie Huang[1], Yingcai Zheng[2], and Yingping Li[2,3]

ABSTRACT

Accurately picking first-arrival traveltimes is crucial for vertical seismic profiling (VSP) initial velocity model building, traveltime tomography, phase rotation, and wavefield separation. For conventional VSP surveys, which use 15–30 m spaced geophones to acquire data, accurate manual picking is feasible. By contrast, distributed acoustic sensing (DAS) vertical seismic profiling VSP uses densely spaced (1 m) downhole DAS channels to record data. These data contain thousands of channels, making manual picking tedious, time-consuming, and potentially cost prohibitive. We use a differential-phase-spectral slope method (DPSSM) to automatically pick first-arrival P-wave traveltimes on zero-offset DAS VSP data. DPSSM calculates the differential time between two similar wavelets by fitting a straight line to differential phase spectra in the frequency domain and calculating the slope, which is the differential time. We apply DPSSM to automatically pick 1,010 and 1,206 first-arrival P-wave traveltimes on two DAS VSP gathers. Using these picks, we derive interval and average velocity profiles for these wells. We compare these velocity profiles with a reverse DAS VSP survey and well logging data and find correlations between major velocity contrasts and formation boundaries. Our results show that DPSSM is effective for automatically picking first-arrival traveltimes in zero- and near-offset DAS VSP data.

15.1. INTRODUCTION

First-arrival picking of P and/or S waves is typically the first major step in basic seismic data processing for both earthquake seismology and exploration geophysics. Picking first-arrival times can be carried out manually, automatically, or as a combination of both. Manually picking first-arrival times is simple and straightforward but tedious and potentially operator dependent.

For similar waveforms, such as earthquake doublets/triplets and zero-offset vertical seismic profiling (VSP) data, cross-correlation either in the time domain (e.g., Frankel, 1982; Li et al., 1995, 1998; Pechmann & Kanamori, 1982; Thorbjarnadottir & Pechmann, 1987) or the frequency domain (e.g., Fremont & Malone, 1987; Ito, 1985; Li et al., 1995; Poupinet et al., 1984) has been used to quantitatively characterize the degree of similarity of seismic waveforms from a cluster of earthquakes close in space and time. Cross-correlation is also used to measure the differential arrival times of P and S waves in an accurate, objective, and consistent manner. However, conventional time-domain methods typically only enable arrival times to be picked, at best, to an accuracy of one sample interval in time. On the other hand, cross-spectral analysis can improve the timing precision by about an

[1]*Los Alamos National Laboratory, Los Alamos, New Mexico, USA*

[2]*Department of Earth and Atmospheric Sciences, University of Houston, Houston, Texas, USA*

[3]*BlueSkyDAS LLC, Sugar Land, Texas, USA*

Distributed Acoustic Sensing in Borehole Geophysics, Geophysical Monograph 289, First Edition.
Edited by Yingping Li, Robert Mellors, and Ge Zhan.

DOI: 10.1002/9781394179275.ch15

order of magnitude greater than the sampling interval (Poupinet et al., 1984). Both Poupinet et al. (1984) and Ito (1985) claimed that for a waveform with a sampling rate of 100 samples per second, their methods had a timing accuracy of about 1 ms, which is 10 times smaller than the sampling interval.

Li et al. (1995) applied the cross-correlation methods in both the time and frequency domains to an earthquake doublet in Southern Quebec, Canada. For seismic data with a sampling rate of 80 samples per second, conventional time-domain analysis results in a differential arrival time of 0.025 s, with a timing accuracy of 0.0125 s. Cross-spectral analysis of the P waveforms results in a differential arrival time between the two P waves of about 0.021 s and a timing resolution of about 1 ms. However, spectral analysis involves time-consuming calculations and fitting straight lines to the slopes of the phase spectrum. Li et al. (2003) first applied this differential-phase-spectral method based on the Fourier time-shift theorem to a land zero-offset VSP dataset acquired using an air gun seismic source. They employed this method to measure differential times between first-arrival P waves at different depth levels.

Distributed acoustic sensing (DAS) is a technology that uses a fiber-optic cable to measure seismic (strain) signals with a high spatial resolution in a wide frequency range. DAS can be used in high pressure and temperature environments. These properties make DAS an appealing tool in both earthquake and exploration seismological research (Li et al., 2021). In borehole seismology, a DAS observation system with fiber cables in a wellbore is usually deployed to acquire DAS VSP data (Correa et al., 2017; Daley et al., 2013, 2016; Harris et al., 2016; Li et al., 2015; Mateeva et al., 2013; Mestayer et al., 2011; Miller et al., 2016; Willis et al., 2016; Zhan et al., 2016). Compared with a conventional VSP survey that records data with downhole geophones at a receiver spacing of dozens of meters, a DAS VSP survey can have a channel/receiver spacing as low as 1 m or smaller, allowing for acquiring massive amounts of data at hundreds to thousands of receiver levels for many sources. Therefore, automatic first-arrival picking is indispensable as manual picking becomes infeasible.

In this chapter, we apply the differential-phase-spectral slope method (DPSSM) introduced by Li et al. (2003) to DAS VSP data acquired with the vibroseis sources at the Frontier Observatory for Research in Geothermal Energy (FORGE) site in Utah. Although DPSSM is a potentially powerful tool for first-arrival picking, this method has not been applied to DAS VSP data before. We apply DPSSM and successfully pick accurate first-arrival traveltimes of the down-going P waves on two zero-offset VSP (ZVSP) gathers recorded using fiber installed in two near vertical wells at the Utah FORGE site. The first is Well 78A-32, which contains 1,010 DAS channels. The second, Well 78B-32, is located approximately 85 m to the east of Well 78A-32, and it contains 1,206 DAS channels. The DAS channel spacing for this DAS VSP survey is 1 m. We show a map view and a cross-section of the survey area in Figure 15.1. Using the automatically picked first-arrival picks, we calculate interval and average P-wave velocity profiles for both wells. We compare the ZVSP interval velocity profiles with wireline logging data from both wells and find that the sharp velocity contrasts correlate well with the interfaces from both the gamma-ray and sonic logs. We also compare our velocity survey results with those of Lellouch et al. (2020) from a reverse DAS VSP survey using a perforating shot at depth as a seismic source. This application shows that DPSSM is an effective tool for automatic picking of first-arrival traveltimes on zero-offset DAS VSP data.

15.2. THE DIFFERENTIAL-PHASE-SPECTRAL METHOD AND A SYNTHETIC EXAMPLE

Li et al. (2003) introduced a DPSSM for velocity analysis. DPSSM calculates differential phase spectra and fits a straight line in the frequency domain. The slope of this fitted line is the differential time between two seismic waveforms with similar wavelets. For a time series of h(t), we can write its Fourier transform $H(f)$ as

$$H(f) = \int h(t)\ \exp\left(-i2\pi f t\right) \quad dt = |H(f)|\ \exp\left[-i\Phi(f)\right], \tag{15.1}$$

where f is the frequency, and $|H(f)|$ and $\Phi(f)$ are the amplitude and phase spectra, respectively. When $\Phi(f)$ is constrained to its principal value, either the interval $[-\pi, \pi]$ or $[0, 2\pi]$, it is called a wrapped phase that causes discontinuities of phase response. Otherwise, it is called the unwrapped phase, which is a continuous function. We use phase unwrapping to avoid phase discontinuity of phase spectra at $\pm\pi$ or $[0, 2\pi]$. The unwrapping phase is particularly important for our differential phase spectra method introduced in the following. Subtracting one unwrapped phase spectrum from another can avoid discontinuity of the differential phase spectrum. Many spectral analysis codes implement the unwrapping algorithm, and we can handily use those codes to calculate unwrapped phase spectra.

The Fourier transform of a time-shifted signal with a time lag Δt, $g(t) = h(t-\Delta t)$, is

$$G(f) = H(f)\ \exp(i2\pi f \Delta t). \tag{15.2}$$

The differential phase spectrum (DPS), $\Delta\Phi(f)$, of these two signals h(t) and g(t) is related to the time lag Δt by

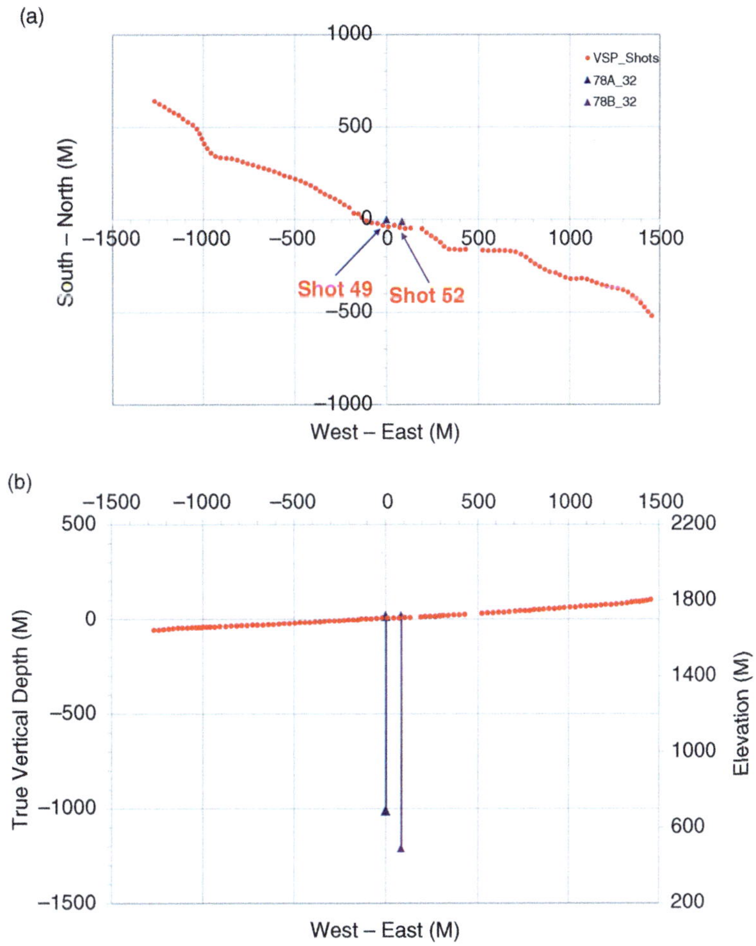

Figure 15.1 (a) Map view showing the geometry of the walkaway DAS VSP survey at the Utah FORGE site. The red dots show the seismic source locations. The blue and purple triangles are the locations of the wellheads for 78A-32 and 78B-32, respectively. All coordinates are with respect to the origin located at the wellhead 78A-32. Shots 49 and 52 are seismic sources of the DAS ZVSP survey. (b) A vertical cross-section shows VSP shots (red dots) and two fiber-optic cables installed in the Wells 78A-32 (blue line) and 78B-32 (purple line).

$$\Delta\Phi(f) = \Phi_G(f) - \Phi_H(f) = [-2\pi f(t-\Delta t)] - [-2\pi f t] = 2\pi f \Delta, \quad (15.3)$$

$$\text{or } \Delta\Phi(f)/(2\pi) = \Delta t f. \quad (15.4)$$

Equation (15.4) indicates that the slope in the plot of the differential phase spectra versus the frequency is the time lag Δt. Alternatively, the time lag Δt_{ij} or differential times T_{ij} between any pairs of waveforms can be calculated using

$$\Delta t_{ij} = T_{ij} = \Delta\Phi(f)/(2\pi f). \quad (15.5)$$

We illustrate this method using synthetic VSP data in Figure 15.2(a), (b). We construct a synthetic Trace 1 signal (Figure 15.2(a)) using a windowed wavelet from a field VSP dataset and zero padding. We time-shift Trace 1 by 51 ms to form Trace 2 and 155 ms to form Trace 3. We decrease the amplitudes of Traces 2 and 3 to simulate geometry spreading. By measuring the slopes of the differential phase spectra, we obtain the measured differential times of 51, 104, and 155 ms between Traces 1 and 2, Traces 2 and 3, and Traces 1 and 3, respectively (Figure 15.2(b)). These numbers are the exact time shifts among the traces.

15.3. ZERO-OFFSET DAS VSP DATASETS

The Utah FORGE site is a field laboratory used to test and develop enhanced geothermal system technologies (Moore et al., 2020). The Utah FORGE project conducted a 2D DAS-DAS survey (Nakata et al., 2023) on 23 April 2022 along a single walkaway line passing through both Wells 78A-32 and 78B-32. We show a

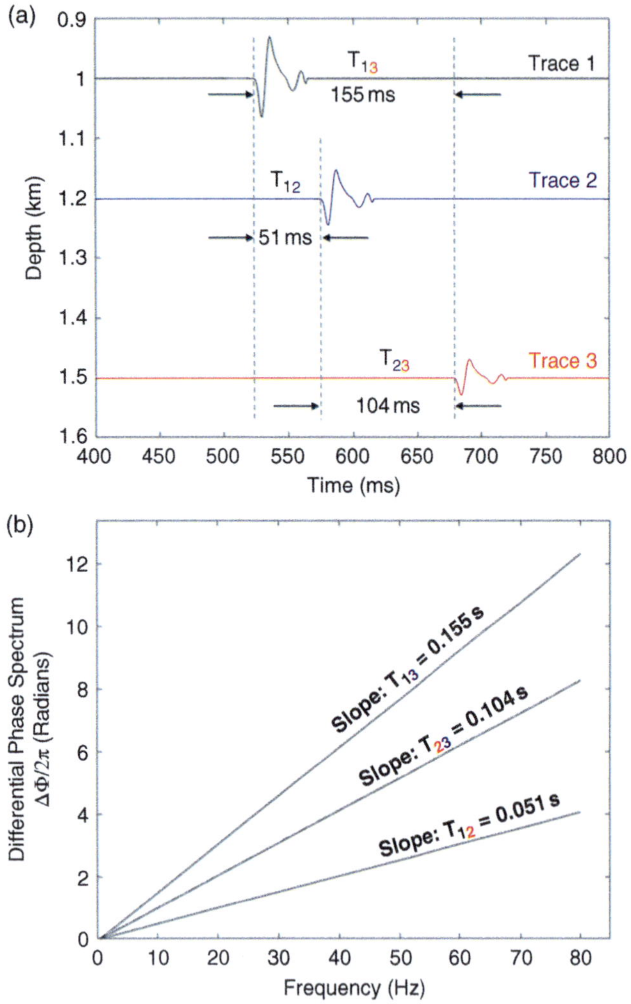

Figure 15.2 (a) Three synthetic traces with different time lags corresponding to three different receiver depths. Differential times between traces are labeled with T12, T23, and T13. (b) Differential phase spectra between three synthetic wavelets with different slopes.

map view and a cross-section of the walkaway VSP survey geometry in Figure 15.1. On the surface, 100 vibroseis sources were deployed along the survey line with a source spacing of about 30 m. The maximum source–receiver offset is about 1.5 km. At each source location, the vibroseis produced a 25 s sweep from 8 to 125 Hz. Three sweeps were performed at each source location, and signals were stacked to increase the signal-to-noise ratios (SNRs).

We use two different zero-offset gathers in our analysis. For Well 78A, we use the gather for shot 49, which has an offset of 35.8 m (closest to zero offset) from Well 78A-32, as the ZVSP shot gather. For Well 78B-32, we use the gather for shot 52, which has an offset of 37.1 m (the closest to zero offset) from Well 78B-32. We show the map view of these shots and the wells in Figure 15.1(a). Note that the source location for zero-offset VSP data acquisition is usually dozens of meters from the wellhead to avoid well damage.

Well 78A-32 does not have a wellbore trajectory survey and is assumed to be vertical. Well 78B-32 is near vertical with a maximum inclination of about 5°. To record DAS VSP data, a single fiber cable was installed in both Wells 78A-32 and 78B-32, respectively, by cementing the fiber cable behind the casing. The fiber cable has 1,010 channels in Well 78A-32 and 1,206 channels in Well 78B-32 (b). A Silixa iDAS observation system records the DAS data using a time sampling rate of 1 ms, a channel spacing of 1 m, and a gauge length of 10 m. The recording time duration of the DAS VSP data is 4 s. We show two zero-offset common-shot gathers for both Wells 78A-32 and 78B-32 in Figure 15.3(a), (b), respectively. We observe that deep portions of DAS ZVSP waveforms are noisy for both wells (Figure 15.3(a), (b)), but the first arrivals of P waves are relatively clean, with a few exceptions. For Well 78A-32, we observe distinct tube waves traveling down and up along the wellbore after hitting the bottom at about 0.8 s (Figure 15.3(a)). The distinct downgoing and upgoing tube wave in Well 78A-32 indicates that the vertical well assumption is reasonable.

As shown in the cross-section in Figure 15.1(b), the active sources of two zero-offset DAS VSPs are deployed at the surface. Thus the P-wave energy of the first arrivals arrive at shallower channels first and deeper channels later (Figure 15.3(a) and (b)). In contrast to this standard zero-offset DAS VSP survey geometry configuration, a reverse near-offset DAS VSP survey can use either a perforating shot or an induced earthquake as a seismic source at depths (Lellouch et al., 2020). For this type of survey, the P-wave first-arrival energy arrives at deeper channels first and at the shallower channels later if the seismic source is located beneath the cable depth range.

Our previous studies (Li et al., 1995, 2003) have demonstrated that DPSSM can be successfully applied to seismometer data from natural earthquakes (Li et al., 1995) and geophone data from a zero-offset VSP using an active source at the surface (Li et al., 2003). However, such a method has not yet been used to process DAS VSP data. Conventional VSP surveys with three-component (3C) geophones record particle velocity in a seismic-wave vector field, whereas DAS VSP measures the strain rate (or strain) tensor of the fiber cable caused by a passing seismic wave. DAS instrumentation manufacturers typically provide a coefficient or a formula to convert a DAS measurement to velocity to compare and calibrate with geophone data. The main difference between the conventional geophone VSP and a DAS VSP system is that the DAS system contains a far greater number of channels (fiber sensors). For a typical conventional VSP survey, a string array of 20–100 geophones with a depth interval of 15 to 30 m is used to collect data. For a general DAS VSP survey, there

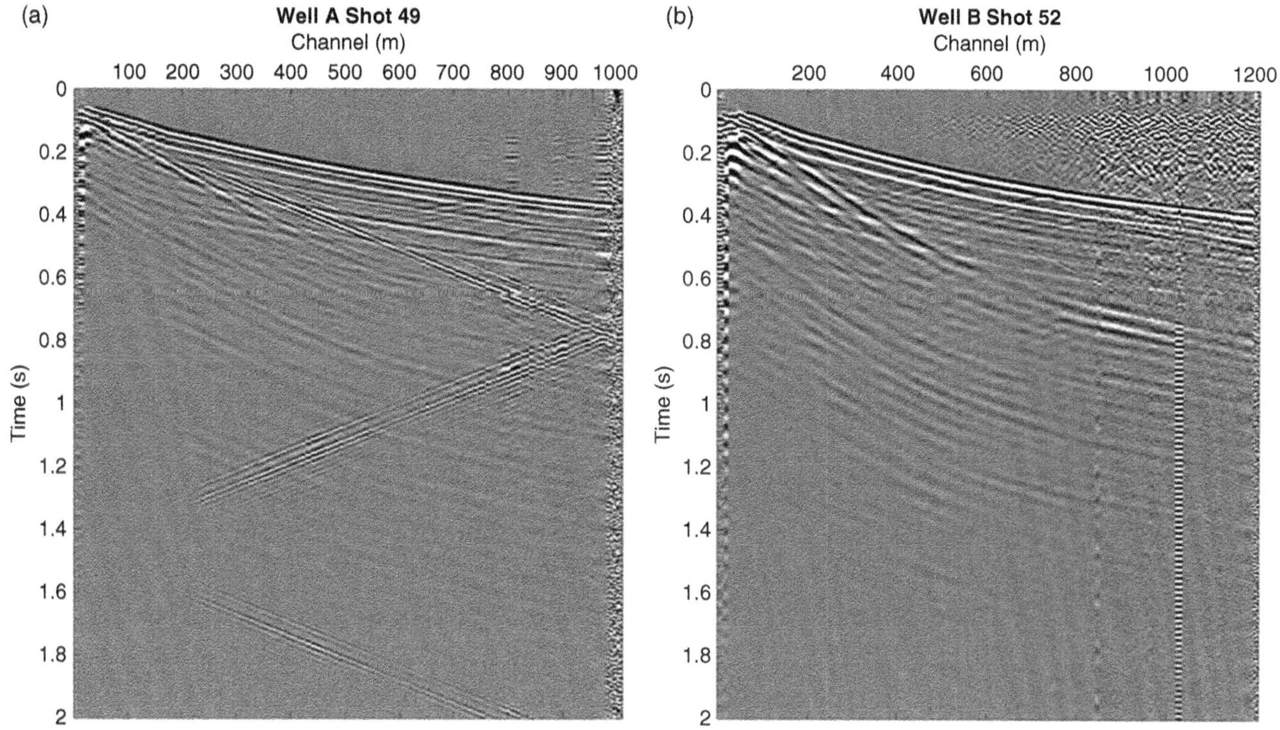

Figure 15.3 Two zero-offset DAS VSP datasets for sources at shots 49 and 52 in Figure 15.1 for data acquired in (a) Well 78A-32 and (b) Well 78B-32 at the Utah FORGE site.

could be thousands of channels with channel spacing as small as 1 m. For example, in Well 78A-32 at Utah FORGE, there are 12 geophones with a depth interval of 30 m and a 1-km-long DAS fiber cable with a sensor spacing of 1 m. DAS data are about 30 times as dense as geophone array data. Manually picking tens of first-arrival times is doable, but it is time-consuming to accurately and reliably pick thousands of DAS channels manually. Therefore, an automatic picking approach is essential to process large amounts of DAS VSP data with thousands of channels.

Another issue is the quality of DAS data. Typically, DAS data acquired using a single-mode common fiber cable are noisier than geophone data. One solution to increase the SNR of DAS systems is to use engineered fibers that have enhanced backscatters in the fiber. DAS measures strain rate or strain not at individual points but across a section of the fiber. The length at which the strain rate measurement is made across is called the gauge length (GL). The GL is an important parameter because it directly influences the SNR and sampling resolution of DAS data (Dean et al., 2017). Do those differences between DAS and geophone data affect accurately picking first-arrival times of P and S waves on a DAS record? Lellouch et al. (2020) quantitatively compared the DAS and geophone seismic records, trace by trace, of a microearthquake induced by fracture stimulations in Well 78A-32 at Utah FORGE. They found that the DAS and geophone recordings kinematically show a high degree of agreement. They also used the first-arrival times to derive the interval velocity profiles of P and S waves (Lellouch et al., 2020). It seems that the GL may have less effect on the waveform shape than the amplitude of waveforms, and it does not cause significant errors on picking first-arrival times of seismic waves on the DAS records. We examine the performance of DPSSM on zero-offset DAS VSP data acquired at the Utah FORGE site.

15.4. APPLYING DPSSM TO ZERO-OFFSET DAS VSP DATA FROM UTAH FORGE

15.4.1. DPSSM Analysis of DAS ZVSP Data

We show six representative ZVSP DAS traces windowed around the P-wave first arrivals acquired from Well 78A-32 in Figure 15.4. The depth interval between these traces is 20 m. Because of geometry spreading and attenuation, the amplitudes of the P waves typically decrease as the depth increases, with only the exception of trace 1. The amplitude of the P-wave first arrival of Trace 2 is comparable to that of Trace 1. However, the waveforms of the first-arrival P waves are similar to one another. Figure 15.5 shows that the FFT amplitude

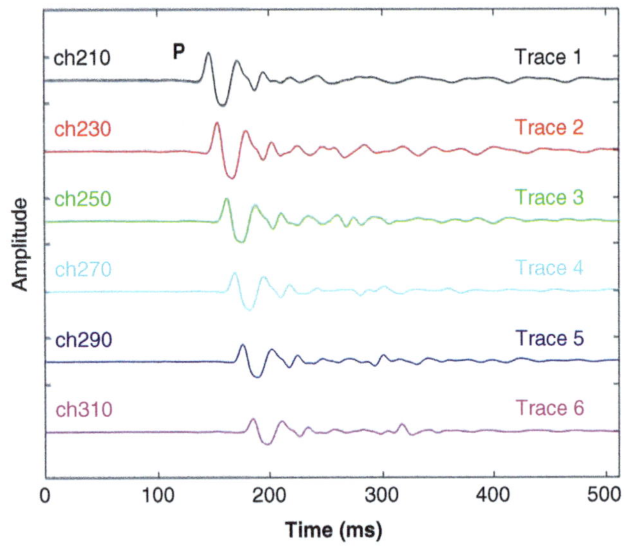

Figure 15.4 Six representative DAS VSP traces of 512 ms recorded in Well 78A-32. They are recorded 20 m apart in depth (labeled with channel numbers that are the same as those in Figure 15.3), and their P-wave first arrivals (P) are similar to one another.

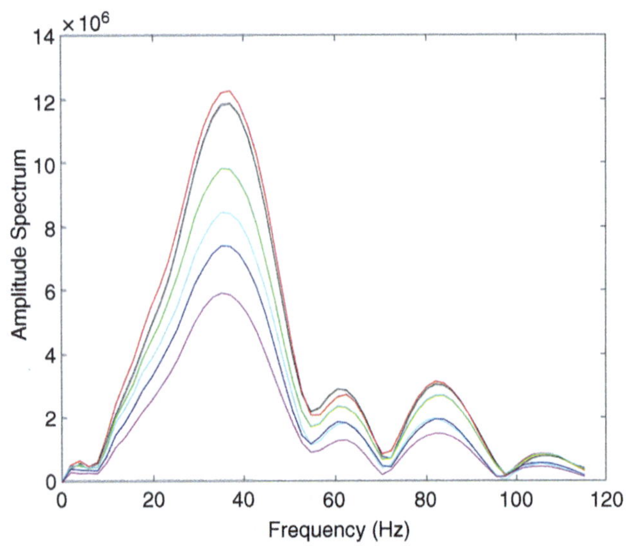

Figure 15.5 P-wave amplitude spectra of the six traces show in Figure 15.4. The dominant P-wave energy is distributed within a frequency band of ~10 to 50 Hz.

spectra of these six traces share a high degree of similarity. The spectra also indicate that the main energy of P waves concentrates in a frequency band from 10 to 50 Hz. The spectral analysis enables us to find the optimal frequency band to fit the DPS with a straight line.

We also note that the spectral amplitude of Trace 1 is slightly smaller than that of Trace 2 (Figure 15.5). Such an abnormal spectral amplitude may be attributed to different coupling conditions for the fiber sensors at different depths. However, we check the cement bond logging (CBL) data of Well 78A-32 and find no significant variations in CBL values around channels 210 to 230 (Figure 15.4). Alternatively, the amplitude differences may be caused by uneven distribution of backscatters at different portions of the fiber cable, resulting in stacked coherent backscatter signals with stronger or weaker amplitudes for different channels. If it is true, DAS amplitudes are affected not only by the fiber directivity but also by the uneven distribution of backscatters. An engineering fiber cable may help to solve the issue of unevenly distributed backscatters. The third possibility is that channel 230 is too close to an interface such that the amplitude of reflected seismic waves off the interface is superimposed on that of P-wave first arrival to produce larger amplitudes. Our method should be less sensitive to the abnormal amplitudes because we use DPS instead of the amplitude spectrum.

We calculate the differential phase spectra of six traces and fit a straight line in a frequency band of 11.7 to 48.8 Hz for each trace pair, with the first trace as a reference to obtain the slopes of the straight lines. The time lags between each of the six traces (T_{11}, T_{12}, T_{13}, T_{14}, T_{15}, and T_{16}) with the first trace are shown in Figure 15.6 (a)–(f). As expected, the slope of the auto-correlation case for Trace 1 (Figure 15.6(a)) is zero. The nonzero slopes of the cross-correlation phase spectra of Traces 2–6 with respect to Trace 1 (Figure 15.6(b)–(f)) are the differential P-wave arrival times between each of these traces and the first trace. The unit of the slope in Figure 15.6 is in seconds. High R^2 (coefficient determination) values ranging from 0.9973 to 0.9999 indicate an excellent linear regression. The fitting straight lines almost overlap with the calculated DPSs in Figure 15.6(d)–(f), implying nondispersive of the media. However, Figures 15.6(b), (c) display that calculated DPSs deviate very slightly from the fitting straight lines, indicating those formations are very weak dispersive media. The phase velocities vary with frequency in the dispersive media.

We plot differential times (T_{1j}, j = 2 to 6) between the j-th trace (j = 2 ~ 6) and the first trace with respect to a reference time zero in Figure 15.7. T_{1j} are the relative traveltimes between the first trace and the *j*th trace of the DAS sensor at different depths. Differential times T_{12}, T_{23}, T_{34}, T_{45}, and T_{56} can be used to calculate interval velocities, $V_{int} = (Z_i - Z_j) / (T_i - T_j)$, for the two DAS sensors (channels), *i* and *j*, at two different depths. For a zero-offset DAS VSP, the clocks of a source controller and DAS recording systems are synchronized to ensure time zero is the time to initiate a seismic source. Thus the first-arrival time of each trace in Figure 15.4 is the vertical one-way traveltime, t, from the seismic source to any DAS sensor at a different depth. We can measure the

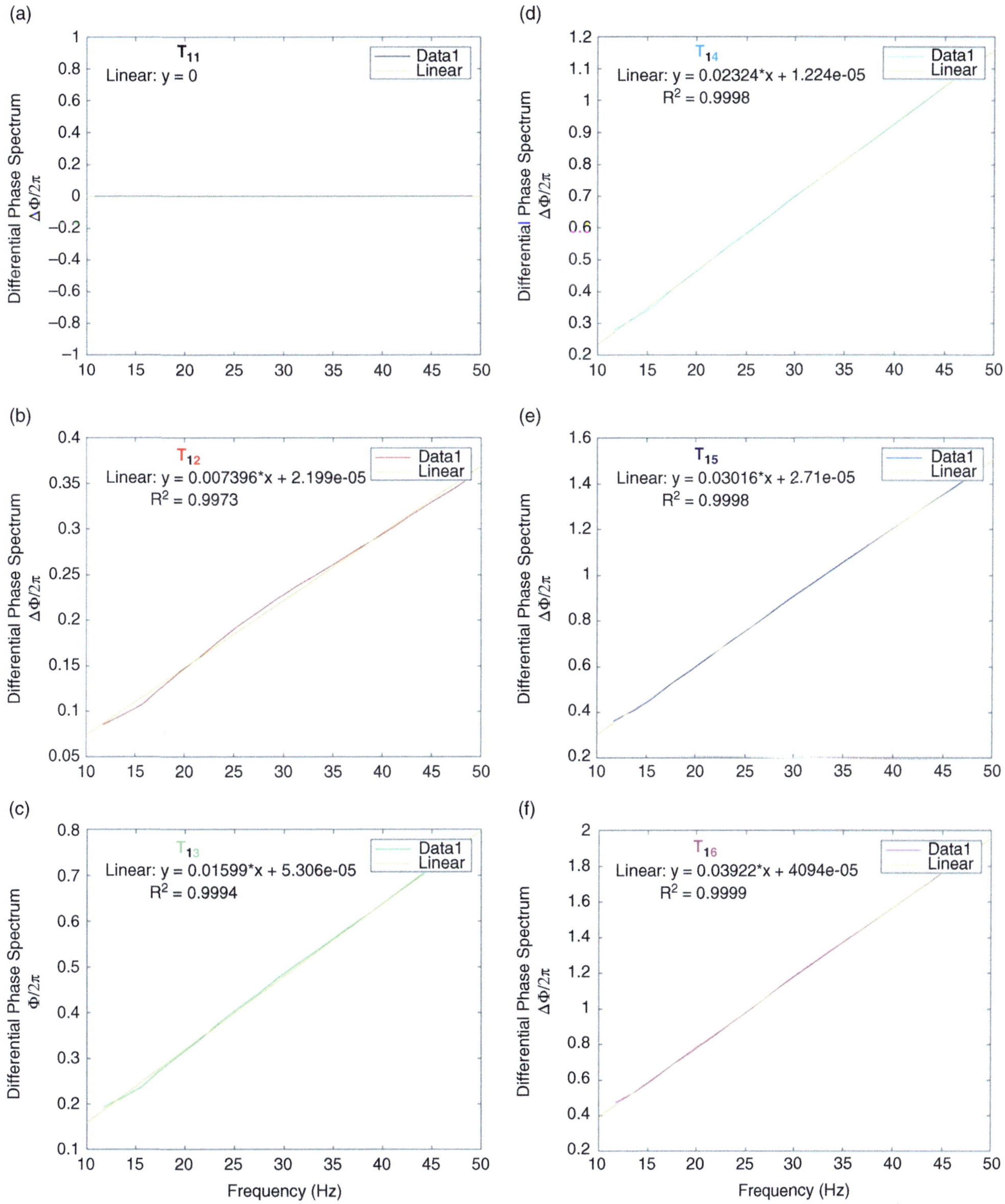

Figure 15.6 (a–f) Differential phase spectra of six trace pairs with different time lags and straight lines fitting of the differential phase spectra versus frequency. The slopes of the fitting lines are time lags Tij (Δt_{ij}) between traces at different depths. The slope of autocorrelation phase spectrum (a) of Trace 1 without a time lag is zero. (b–f) are cross-correlation phase spectra of the rest traces with respect to Trace 1.

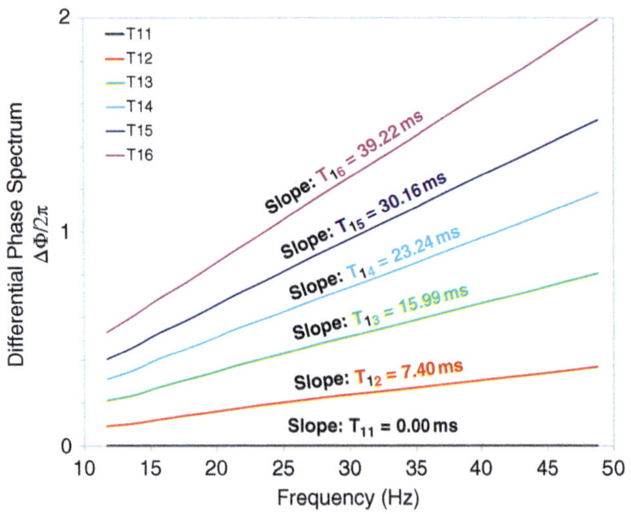

Figure 15.7 Calculated relative traveltimes of DAS VSP traces from slopes of differential phase spectra using the first trace as the reference. Interval velocities between DAS VSP traces at different depths can be derived from the time lags and depth intervals between two traces.

traveltime of trace 1 and then derive traveltime for the other traces using the differential traveltimes $T1j$. The average velocity between the surface and the depth Z is given by Z/t.

15.4.2. Results of Automatic Picks and P-Wave Velocity Analysis

Our DPSSM for analyzing DAS ZVSP data involves (a) manually picking the P-wave first-arrival time of the reference trace, (b) selecting a 36 ms time window of one-cycle P-wave wavelets, (c) deriving fast Fourier transform (FFT) amplitude spectra of P waves to determine the optimal frequency band for fitting straight lines to differential phase spectra, and (d) deriving differential phase spectra relative to that of a reference P-wavelet and fitting straight lines in the frequency domain to obtain the slopes. These slopes are the differential times between any trace with the reference trace.

We automate the DPSSM processing flow and apply it to two DAS ZVSP datasets acquired in Wells 78A-32 and 78B-32. For a common-shot gather with over 1,000 traces, our algorithm on a laptop takes about 20 s to calculate DPS, curve fit for the slope, and store the picking times of first arrivals. We show the automatic picks of first arrivals of P waves for Wells 78A-32 and 78B-32 in Figure 15.8(a), (b), respectively. For these two DAS ZVSP common-shot gathers, we select a single reference trace at channel 400 for both DAS ZVSP shot gathers as the reference for the local reference traces to obtain the traveltimes for all channels in the wells. However, for a DAS VSP survey with a fiber cable in a deep well

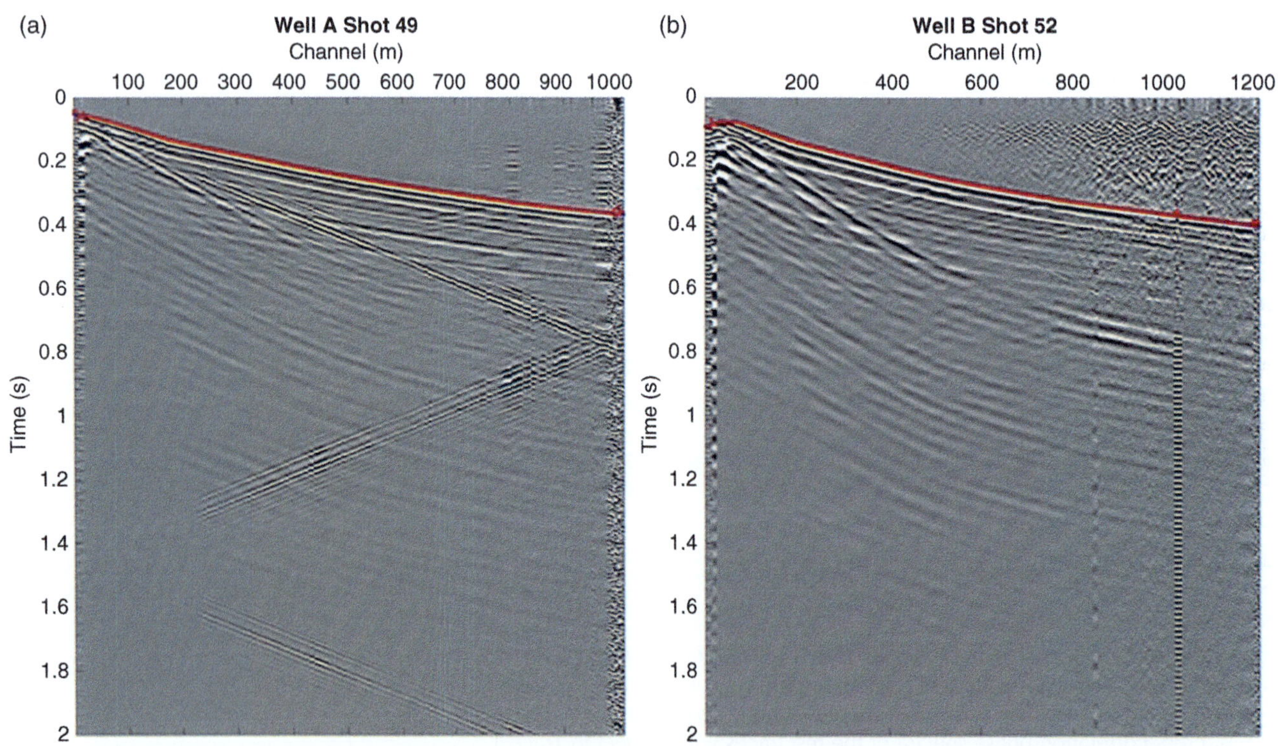

Figure 15.8 Automatically picked first-arrival times using differential-phase-spectral slopes of zero-offset DAS VSP common-shot gathers for (a) Well 78A-32 and (b) Well 78B-32. The red dots show the P-wave first-arrival picks. Note that the tube waves in (a) propagate from the surface to bottom of the well and bounce back and forth.

of 8 to 10 km depth, attenuation of formations may distort the waveforms of first-arrival P waves (lower amplitude and wider pulse width) significantly and result in decreasing waveform similarity. Thus, multiple reference traces are required. The fitting DPS plots and coefficient determination values (R^2) shown in Figure 15.6 serve as useful QC/QA tools to determine if and when multiple reference traces are required.

From Figure 15.8, we observe that the automatically picked first-arrival traveltimes are generally clean across the gathers. However, the picks for the very shallow and deepest channels in Well 78B-32 are worse than the others because of their poor SNR. The reason for this discrepancy might be that the DAS cable is not very well coupled at those depth ranges with the poor SNR. Those unreliable arrival picks would introduce errors for the velocity calculations, and we should exclude them from our velocity analysis. However, when DAS VSP data are acquired using fiber cables in a retrievable manner inside a well casing or tubing with fibers carried by a hybrid wireline cable (e.g., Henninges et al., 2021; Yu et al., 2016), the fiber sensors are not coupling with rock formations at all, resulting in cable reverberations (zigzag ringing noise) that can significantly distort seismic waveforms of first arrivals. Data denoising is needed before applying the DPSSM to such DAS VSP data. Chen et al. (2021) developed an algorithm to remove the cable resonances and recover the first-arrival waveforms reasonably well. Thus, the DPSSM is suitable for denoised DAS VSP data with similar first-arrival waveforms. Otherwise, this method is not reliable.

We use automatic picks of P-wave first arrivals to calculate average and interval velocities for Well 78A-32 and plot the P-wave velocity results in Figure 15.9. Because the ZVSP source offset is not exactly zero but 35.8 m, we need to perform the cosine correction for the small-incidence angles when calculating both the average velocity and the interval velocity, as we define in the previous section. The average velocity calculated using all automatically picked P-wave traveltimes is plotted in the thin blue line, and the interval velocity is plotted using the thick blue line. We use a depth interval of 10 m for interval velocity calculation because the 10 m GL of the DAS system reduces spatial resolution of the DAS ZVSP data. At each depth level with an interval of 10 m, we take an average interval velocity value over five adjacent sensors below and above this depth to smooth the profile and to account for the GL effect. We plot gamma-ray log data acquired in Well 78A-32 and P-wave sonic log data collected in the adjacent Well 58–32 for comparison. The gamma-ray log indicates changes in formations based on the radioactive content of different types of rocks. We observe that three major interfaces shown on the gamma-ray log coincide with those on the DAS interval velocity profile. The trend of the velocity from a sonic log is consistent with that of the interval velocity profile we derive using the DAS ZVSP data. However, we do not expect they could match each other in detail because the two datasets measure the velocity at different frequency bands, and the sonic log velocity is typically subject to the effects of the un-uniform borehole diameter and porous and fractured formations.

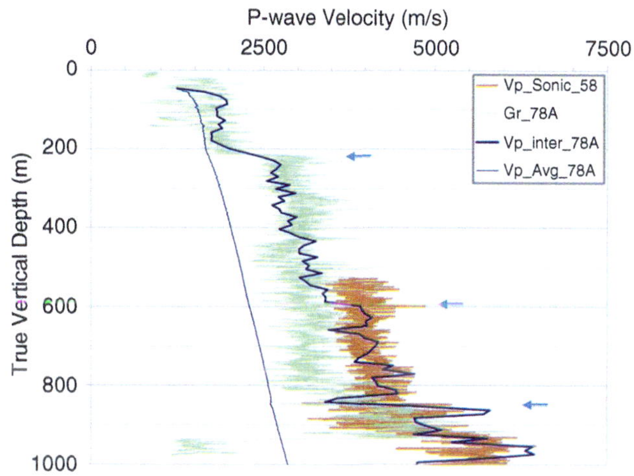

Figure 15.9 Interval (blue thick line) and average (blue thin line) P-wave velocity calculated with automatically picked first-arrival times of DAS ZVSP P waves for Well 78A-32, along with gamma-ray log data (green line) in Well 78A-32 and sonic P-wave velocity (orange line) from the adjacent Well 58-32.

The P-wave interval velocities increase from about 2,000 m/s at shallower depths above 200 m to 5,000 m/s at depths below 850 m in the granitic formations. The trend of P-wave interval velocities derived from the DAS ZVSP data in the monitor Well 78A-32 matches well with that of sonic P-wave velocities (orange curve) of the adjacent Well 58–32 at the depth range of 600 m to 1,000 m. We observe that the P-wave interval velocity profile has at least three major interfaces (blue arrows) with velocity jumps around depths at about 200, 600, and 850 m, which match the corresponding jumps shown in the gamma log (green curve).

Figure 15.10 shows the P-wave velocity-depth profiles of average (thin purple line) and interval (thick purple line) velocities for Well 78B-32 derived using DPSSM-picked first-arrival traveltimes of the DAS ZVSP data with the same parameters used for Well 78A-32 DAS ZVSP data. Unfortunately, no P-wave sonic log and gamma-ray log data are available for shallow depths, and we only plot the available log data at deep depths for comparison. We observe that the trend of interval velocity of sonic P waves in the granite formations is consistent with the P-wave velocity analysis with the DAS ZVSP data. The gamma-ray jumps in Well 78B-32 also

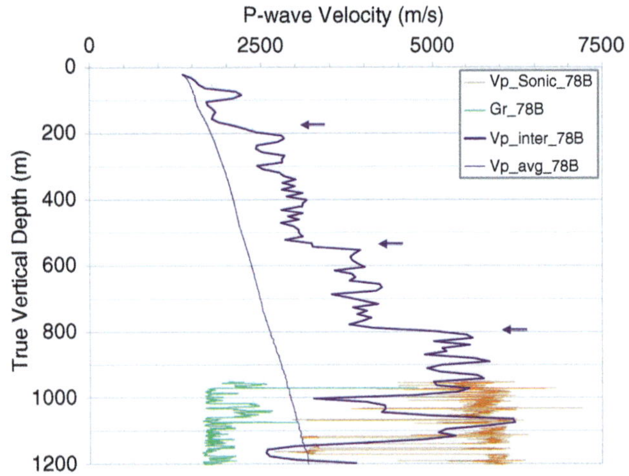

Figure 15.10 Interval (purple thick line) and average (purple thin line) velocity of P-wave calculated with automatically picked first-arrival times of DAS ZVSP P waves for Well 78B-32, along with gamma-ray log data (green line) and sonic P-wave velocity (orange line) in Well 78B-32.

correspond to interval velocity interfaces derived from the DAS ZVSP data.

The trend of P-wave interval velocity for Well 78B-32 is very similar to that of Well 78A-32, both exhibiting a similar velocity variation pattern (Figures 15.9 and 15.10). We also observe three major interfaces (purple arrows) with velocity jumps at depths of about 180, 560, and 800 m. Depths of interfaces with velocity jumps in Well 78B-32 are shallower than those in Well 78A-32, indicating that there are dips of formations along the west–east direction and the dip angle at the sediment–granite boundary, where P-wave velocities increase to about 5,000 m/s, is larger than those in the sedimentary rocks. The changes in dipping angles for different formations are as small as a few degrees but measurable.

We observe strong tube waves with large amplitudes (Figure 15.8(a)) in zero-offset and near-offset DAS VSP data acquired in Well 78A-32. The tube waves bounce back and forth between the surface and the well bottom several times in a DAS record of a 4 s length. From the move-out of arrival times for the first downgoing tube wave trains, we estimate the tube-wave velocity to be 1,432 m/s. We find apparent amplitude variations of the tube waves at a depth zone around channel 800. In the future, we would like to further study the possible correlations between the amplitude changes and variations of formation structure properties, such as fractures or wellbore washout. DAS ZVSP data with a source offset of 37.1 m acquired in Well 78B-32, which is only 85 m from Well 78A-32, does not contain any observable tube waves at all. We attribute that a maximum deviation angle of 5° in Well 78B-32 may hinder the generation of tube waves.

15.4.3. Comparison and Discussion

Using two perforating shots in treatment Well 58-32 and an induced earthquake as a seismic source, and 1 km long DAS fiber cable in a monitor Well 78A-32 as fiber receivers, Lellouch et al. (2020) derived P- and S-wave velocity profiles at the Utah FORGE with the DAS data acquired during a 2019 project to monitor fracture stimulations (Moore et al., 2020). The distance between the two wells is approximately 336 m. Calculating internal P- and S-wave velocities requires a cosine correction of the incidence angles from the sources to fiber receivers. One of the DAS ZVSP surveys for Well 78A-32 uses the same 1-km-long DAS fiber cable but a seismic source at the surface rather than using seismic sources at depth below the DAS fiber cable. Therefore, it is interesting to compare their data and velocity survey results with ours.

Figure 15.11(a) shows a DAS record for one of the perforating shots. In this shot, the first-arrival P-wave arrives at the deeper sensors first and at the shallower sensors later because the active seismic source is located below the DAS fiber cable. The derived P-wave velocity profile from DAS VSP, along with the sonic log velocity, is shown in Figure 15.11(b). Because the perforating shots generated few S waves, Lellouch et al. (2020) used the DAS record (Figure 15.11(c)) of a microearthquake induced by the fracture stimulation to obtain velocity profiles for both P and S waves (Figure 15.11(d)). Figure 15.11(c) indicates that both P and S waves arrive first at deeper sensors and later at the shallower sensors because the passive seismic source is located under the DAS fiber cable. This setup is in contrast with our DAS VSP data shown in Figure 15.8(a), where the first-arrival P-wave arrives at shallow sensors first and at deeper sensors later.

Figure 15.9 shows the P-wave average velocity profile (thin blue line). However, no average velocity profiles are available in Figure 15.11(b) or (d). In a standard DAS VSP survey, clocks of a seismic source controller and a DAS recording system are synchronized, and thus the source origin time is accurately known, enabling us to calculate the average velocity of P wave from the surface seismic source to sensors at any depth. Our P-wave interval velocity profile (thick blue line) shown in Figure 15.9 matches quite well with those (blue lines) derived by Lellouch et al. (2020) (Figure 15.11(b), (d)). Particularly, both results show three distinct velocity jumps in the velocity profiles at the formation boundaries.

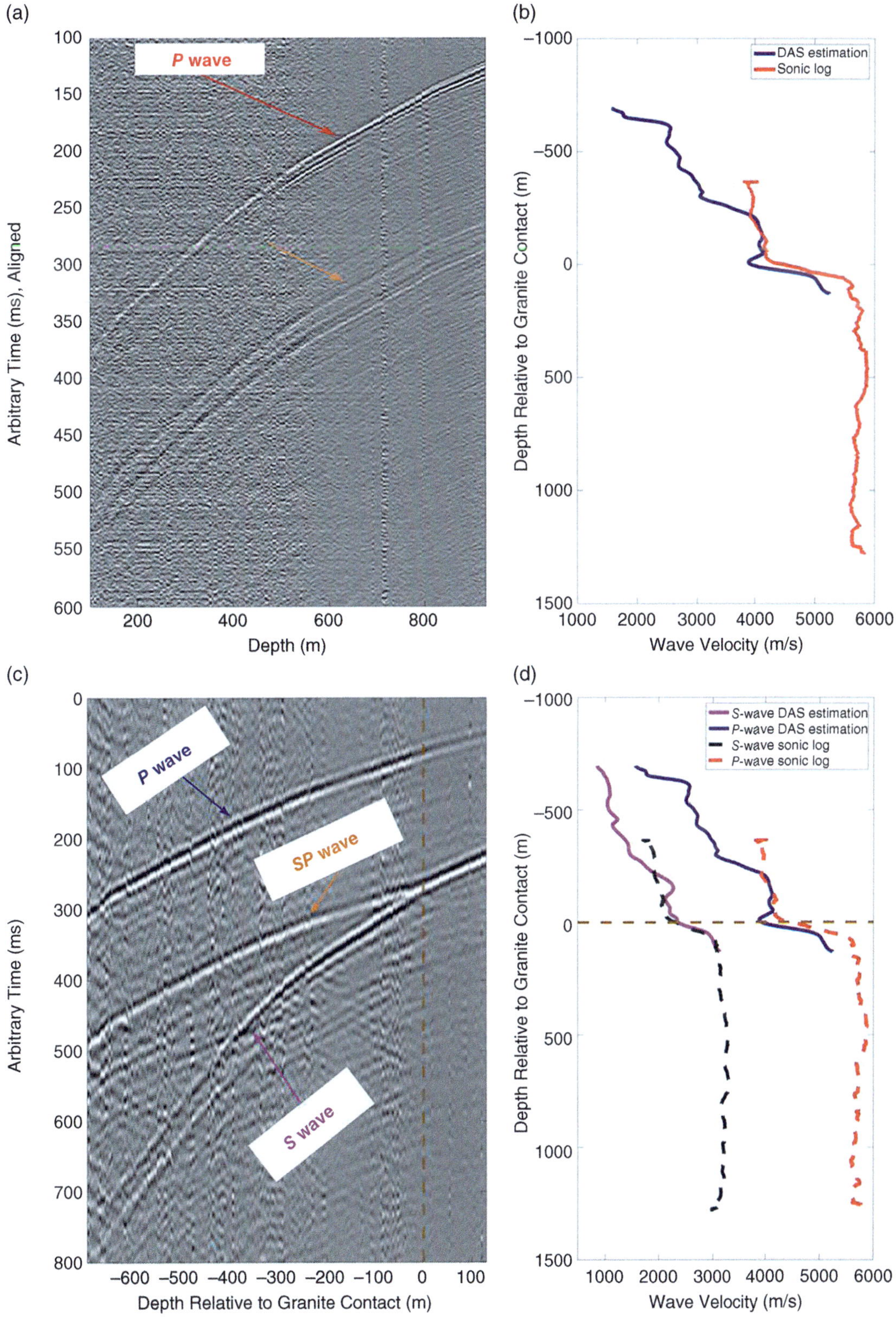

Figure 15.11 A DAS record of a perforate shot (a) and an induced microearthquake (c). (b) Interval P-wave velocity profile (blue line) and sonic log velocity (red line). (d) Interval P-wave velocity profile (blue line) interval S-wave velocity profile (magenta line) and sonic log velocities of P and S waves (dashed red and black lines). Note that vertical axes in (b) and (d) use a boundary between sediment and granitic rock as zero of the reference depth. Adapted from Lellouch et al. (2020).

The P-wave interval velocity profile in Figure 15.9 seems to have slightly more details and a relatively larger depth range compared to those in Figure 15.11(b), (d). For a depth range from 200 to 600 m (Figure 15.9), an average relative error between two interval velocity profiles by different authors is about 5%, and the error increases to about 8% for a depth range from 600 to 800 m. Consistency among the interval velocity profiles derived using both a standard and a reverse DAS VSP survey demonstrates the power of the automatic picking method based on the DPSSM and enhances our confidence for broader applications of this method.

Lellouch et al. (2020) used the advantage of an induced microearthquake radiating stronger S-wave energy to derive the S-wave interval velocity profile (magenta line in Figure 15.11(d)), whereas we have not applied the DPSSM to pick the direct S waves of the DAS VSP data yet. Figure 15.8(b) indicates S waves with smaller amplitudes converted at the surface because of the nonperfect vertical force of a vibrator source (Zhao et al., 2005). The converted S wave seems to be contaminated by the tube waves in Figure 15.8(a). Therefore, we should remove the tube waves by filtering before employing DPSSM to pick the S waves. For a zero-offset or near-offset DAS VSP survey, in addition to direct P waves, the automatic picking algorithm based on the DPSSM theoretically should be feasible for picking direct S waves and reflected P and S waves as long as they share the same or similar ray paths and preserve the waveform similarity. Large-amplitude S waves generated by induced microearthquakes in a catalog from Lellouch et al. (2020) along with some near-offset DAS VSP from the 2022 walkaway DAS VSP (Nakata et al., 2023) provide valuable datasets for us in the future to examine feasibility of our automatic picking method for picking S-wave arrival times and to investigate the maximum near-offset distance for the DPSSM to be reliable.

Nakata et al. (2023) used sonic log data in three wells to calibrate and update a 3D seismic velocity model and generated some preliminary migration images of the subsurface using some far-offset DAS VSP data at the Utah FORGE. The far-offset DAS VSP gathers they displayed indicate the refracted first arrival of P-wave arrives at deep sensors first and shallow sensors later, in contrast to that of the zero-offset DAS VSP data. Waveforms of those refracted seismic waves from far-offset sources are not similar to one another because they do not share the same or similar ray paths. We have applied our algorithm to one far-offset DAS VSP data and found the current version of the automatic picking method is not suitable for the far-offset DAS VSP data because of the lack of waveform similarity.

15.5. CONCLUSIONS

We have successfully employed a differential-phase-spectral slope method to automatically pick 1,010 and 1,206 first-arrival times on two zero-offset DAS VSP datasets for Well 78A-32 and Well 78B-32 at the Utah FORGE site. We have applied FFT to the DAS ZVSP waveform data and determined the differential traveltimes using the slope in the phase spectra between two traces in the frequency domain. We have built interval and average velocity profiles using the automatically picked first-arrival times and compared them with the velocity survey results in Well 78A-32 derived by Lellouch et al. (2020) using data from reverse DAS VSP surveys and wire-line sonic and gamma-ray logging data. Large velocity jumps in the interval velocity profiles correspond to formation changes shown in the gamma-ray logs. The interval P-wave velocity derived from our automatically picked traveltimes also shows similar velocities to the reverse DAS VSP results and the sonic P-wave log data. The presence of tube waves in Well 78A-32 but not in 78B-32 can be attributed to the slight deviation of Well 78B-32. Our results show that the DPSSM can automatically and accurately pick DAS ZVSP first-arrival traveltimes, which can be used for further velocity analysis. The method can also be applied to S-wave and converted and reflected wave data.

ACKNOWLEDGMENTS

This work was supported by the U.S. Department of Energy (DOE) through the Los Alamos National Laboratory (LANL), which is operated by Triad National Security, LLC for the National Nuclear Security Administration (NNSA) of U.S. DOE under Contract No. 89233218CNA000001, and through the University of Houston under Contract No. DE-EE0008764. The DAS VSP data used were obtained from the Utah FORGE Data Distribution (https://constantine.seis.utah.edu/datasets.html). We thank the coeditor and three anonymous reviewers for their careful review of the manuscript.

AVAILABILITY STATEMENT

The DAS data associated with this chapter are publicly available and can be downloaded from the Utah FORGE Data Distribution at https://constantine.seis.utah.edu/datasets.html. The VSP phase spectral picking code is available with LANL's approval to release.

REFERENCES

Chen, Y.-Z., Hu, G.-M., Wu, J.-J., Yu, G., Li, Y.-P., Huang, J.-H., et al. (2021). Denoising analysis and processing methods of distributed acoustic sensing (DAS) vertical seismic profiling (VSP) data, Chapter 7. In Y. Li, M. Karrenbach, & J. B. Ajo-Franklin (Eds.), *Distributed acoustic sensing in geophysics: Methods and applications* (pp. 93–99). AGU. doi:10.1002/9781119521808.ch7

Correa, J., Egorov, A., Tertyshnikov, K., Bona, A., Pevzner, R., Dean, T., et al. (2017). Analysis of signal to noise and directivity characteristics of DAS VSP at near and far offsets — A CO2CRC Otway Project data example. *The Leading Edge*, *36*, 994a1–994a7. doi:10.1190/tle36120994a1.1

Daley, T. M., Freifeld, B. M., Ajo-Franklin, J., Dou, S., Pevzner, R., Shulakova, V., et al. (2013). Field testing of fiber-optic distributed acoustic sensing (DAS) for subsurface seismic monitoring. *The Leading Edge*, *32*(6), 699–706.

Daley, T. M., Miller, D. E., Dodds, K., Cook, P., & Freifeld, B. M. (2016). Field testing of modular borehole monitoring with simultaneous distributed acoustic sensing and geophone vertical seismic profiles at Citronelle, Alabama. *Geophysical Prospecting*, *64*(5), 1318–1334.

Dean, T., Cuny, T., & Hartog, A. H. (2017). The effect of gauge length on axially incident P-waves measured using fibre optic distributed vibration sensing. *Geophysical Prospecting*, *65*(1), 184–193. doi:10.1111/1365-2478.12419

Frankel, A. (1982). Precursors to a magnitude 4.8 earthquake in the Virgin Islands: spatial clustering of small earthquakes, anomalous focal mechanisms and earthquake doublets. *Bulletin of the Seismological Society of America*, *72*(4), 1277–1294. doi:10.1785/BSSA0720041277

Fremont, M., & Malone, S. (1987). High precision relative locations of earthquakes at Mount St. Helens, Washington. *Journal of Geophysical Research*, *92*(B10), 10223–10233. doi:10.1029/JB092iB10p10223

Harris, K., White, D., Melanson, D., Samson, C., & Daley, T. M. (2016). Feasibility of time-lapse VSP monitoring at the aquistore CO_2 storage site using a distributed acoustic sensing system. *International Journal of Greenhouse Gas Control*, *50*, 248–260.

Henninges, J., Martuganova, E., Stiller, M., Norden, B., & Krawczyk, C. M. (2021). Wireline distributed acoustic sensing allows 4.2 km deep vertical seismic profiling of the Rotliegend 150 °C geothermal reservoir in the North German Basin. *Solid Earth*, *12*(2), 521–537. doi:10.5194/se-12-521-2021

Ito, A. (1985). High-resolution relative hypocenters of similar earthquakes by cross-spectral analysis method. *Journal of Physics of the Earth*, *33*(4), 279–294. doi:10.4294/jpe1952.33.279

Lellouch, A., Lindsey, N. J., Ellsworth, W. L., & Biondi, B. L. (2020). Comparison between distributed acoustic sensing and geophones: downhole microseismic monitoring of the FORGE geothermal experiment. *Seismological Research Letters*, *91*(6), 3256–3268. doi:10.1785/0220200149

Li, Y., Cheng, C. H., & Toksoz, M. N. (1998). Seismic monitoring of the growth of a hydraulic fracture zone at Fenton Hill, New Mexico. *Geophysics*, *63*(1), 120–131. doi:10.1190/1.1444304

Li, Y., Doherty, F., & Jackson, J. (2003). *VSP velocity analysis using a phase spectral method.* Paper presented at the 65th EAGE Conference & Exhibition, cp-6-00038. doi:10.3997/2214-4609-pdb.6.F41

Li, Y., Doll, C., & Toksoz, M. N. (1995). Source characterization and fault plane determination for Mblg = 1.2 to 4.4 earthquakes in the Charleviox seismic zone, Quebec, Canada. *Bulletin of the Seismological Society of America*, *85*(6), 1604–1621. doi:10.1785/BSSA0850061604

Li, Y., Karrenbach, M., & Ajo-Franklin, J. B. (2021). A literature review: distributed acoustic sensing (DAS) geophysical applications over the past 20 years. Chapter 17. In Y. Li, M. Karrenbach, & J. B. Ajo-Franklin (Eds.), *Distributed acoustic sensing in geophysics: Methods and applications* (pp. 229–291). AGU. doi:10.1002/9781119521808.ch17

Li, Y., Wu, H., Wong, W., Hewett, B., Liu, Z., Mateeva, A., and Lopez, J. (2015). *Velocity analysis and update with 3D DAS-VSP to improve borehole/surface seismic images.* Paper presented at the 85th SEG Annual International Meeting. Expanded Abstracts, 5285–5289.

Mateeva, A., Lopez, J., Mestayer, J., Wills, P., Cox, B., Kiyashchenko, D., et al. (2013). Distributed acoustic sensing for reservoir monitoring with VSP. *The Leading Edge*, *32*(10), 1278–1283.

Mestayer, J., Cox, B., Wills, P., Kiyashchenko, D., Lopez, J., Costello, M., et al. (2011). Field trials of distributed acoustic sensing for geophysical monitoring. Paper presented at the SEG Technical Program. Expanded Abstracts, 4253–4257.

Miller, D. E., Daley, T. M., White, D., Freifeld, B. M., Robertson, M., Cocker, J., & Craven, M. (2016). Simultaneous acquisition of distributed acoustic sensing VSP with multi-mode and single-mode fibre-optic cables and 3C-geophones at the aquistore CO_2 storage site. *CSEG Recorder*, *41*(6), 28–33.

Moore, J., McLennan, J., Pankow, K., Simmons, S., Podgorney, R., Wannamaker, P., et al. (2020). *The Utah Frontier Observatory for Research in Geothermal Energy (FORGE): a laboratory for characterizing, creating and sustaining enhanced geothermal systems.* Paper presented at the 45th Workshop on Geothermal Reservoir Engineering, Stanford University, CA, USA, SGP-TR-216.

Nakata, N., Vasco D. W., Shi, P., Bai, T., Lanza F., Dyer B., et al. (2023). *Elastic characterization at Utah FORGE: P-wave tomography and VSP subsurface imaging.* Paper presented at the 48th Workshop on Geothermal Reservoir Engineering, Stanford University, CA, USA, SGP-TR-224 https://pangea.stanford.edu/ERE/db/GeoConf/papers/SGW/2023/Nakata1.pdf

Pechmann, J. C., & Kanamori, H. (1982). Waveforms and spectra of preshocks and aftershocks of the 1979 Imperial Valley, California, earthquake: evidence of fault heterogeneity?

Journal of Geophysical Research, *87*, 10579–10597. doi:10.1029/JB087iB13p10579

Poupinet, G., Ellsworth, W. L., & Frechet, J. (1984). Monitoring velocity variations in the crust using earthquake doublets: an application to the calvears fault, California. *Journal of Geophysical Research*, *89*(B7), 5719–5731. doi:10.1029/JB089iB07p05719

Thorbjarnadottir, B. S., & Pechmann, J. C. (1987). Constraint on relative earthquake locations from cross-correlation of waveforms. *Bulletin of the Seismological Society of America*, *77*, 1626–1634.

Willis, M. E., Erdemir, C., Ellmauthaler, A., Barrios, O., & Barfoot, D. (2016). Comparing DAS and geophone zero-offset VSP data sets side-by-side. *CSEG Recorder*, *41*(6), 22–26.

Yu, G., Cai, Z., Chen, Y., Wang, X., Zhang, Q., & Li, Y. (2016). Walkaway VSP using multimode optical fibers in a hybrid wireline. *The Leading Edge*, *35*(7), 615–619. doi:10.1190/tle35070615.1

Zhan, G., Li, Q., Nahm, J., Kommedal, J., & Konkler, J. (2016). Field trials of distributed acoustic sensing for reservoir delineation with VSP. *CSEG Recorder*, *41*(6), 34–37.

Zhao, X., Zhou, R., Li, Y., Janak, P., & Dushman, D. (2005). *Shear waves from near-offset VSP survey and applications*. Paper presented at the SEG Technical Program. Expanded Abstracts, 2629–2632.

16

A Prospective Evaluation of Borehole Distributed Acoustic Sensing Technologies for Detecting and Locating Out-of-Network Microseismicity

Federica Lanza[1], Katinka B. Tuinstra[1], Antonio P. Rinaldi[1], Verónica Rodríguez Tribaldos[2,3], Chet Hopp[3], Falko Bethmann[4], Ben Dyer[4], Peter Meier[4], Francisco Serbeto[4], Marian Hertrich[1], Domenico Giardini[5], and Stefan Wiemer[1]

ABSTRACT

Downhole distributed acoustic sensing (DAS) has proven effective in monitoring microseismicity produced during hydraulic stimulations. It is also a promising tool for monitoring microseismicity occurring beyond the stimulation region. Downhole DAS arrays exhibit reduced sensitivity to surface noise compared to surface deployments, and they can remain in place undisturbed after operations cease, unlike downhole geophones. Identifying and locating microseismicity occurring at remote distances is crucial for reservoir monitoring, allowing us to assess stress conditions and the presence of pre-existing faults. Here, we evaluate the potential of downhole DAS in detecting and locating out-of-network microseismicity through two case studies: a field-scale investigation at the Frontier Observatory for Research in Geothermal Energy (FORGE) site in Utah and a mesoscale experiment in the Bedretto Underground Laboratory for Geosciences and Geoenergies (BULGG) in Switzerland. We demonstrate that DAS data have sufficient quality to detect and locate low-magnitude earthquakes ($M < 0$) occurring several hundred meters away. However, azimuthal ambiguity is unavoidable when only a single vertical borehole is available. We show that DAS signal characteristics and location capabilities perform equally at different scales, ranging from full-reservoir scale to mesoscale. Our evaluation supports the employment of longer-term experiments to explore downhole DAS suitability for passive microseismic monitoring.

16.1. INTRODUCTION

Several industry operations, from petroleum to geothermal applications, rely on locations of microseismic events to gain insight into underground processes. The goal of microseismic data processing is to transform continuous wavefield observations into precise and accurate estimates of event locations and magnitudes and other source characteristics. Knowledge of such parameters is key for surveillance and diagnostic characterization of induced seismicity both in the context of the development of the reservoir connectivity (e.g., reconstruction of spatiotemporal fracture development) and for safety protocols and risk governance (Chen et al., 2019; Grigoli et al., 2022; Van Der Baan et al., 2013). Typical microseismic monitoring systems use both a permanent network of sensors deployed at the surface and/or in shallow boreholes

[1]*Swiss Seismological Service, ETH Zurich, Zurich, Switzerland*
[2]*GFZ German Research Centre for Geosciences, Potsdam, Germany*
[3]*Energy Geosciences Division, Lawrence Berkeley National Laboratory, Berkeley, California, USA*
[4]*Geo-Energie Suisse, Zurich, Switzerland*
[5]*Institute of Geophysics, ETH Zurich, Zurich, Switzerland*

Distributed Acoustic Sensing in Borehole Geophysics, Geophysical Monograph 289, First Edition.
Edited by Yingping Li, Robert Mellors, and Ge Zhan.

DOI: 10.1002/9781394179275.ch16

and deep downhole geophone arrays as close as possible to the stimulation region (Eaton, 2018).

Recently, distributed acoustic sensing (DAS) technology has become available, turning fiber-optic cables (e.g., conventional telecommunication or built-for-purpose cables) into an ultra-dense network of strain sensing points that can sample the seismic wavefield continuously for tens of kilometers. Existing infrastructures such as telecommunication fiber-optic cables, can be used for distributed sensing. Thus, DAS technology has been increasingly used as a promising and cost-effective subsurface monitoring tool for a wide range of applications across many research fields and environments: from urban seismology (Dou et al., 2017; Song et al., 2022; Spica et al., 2020; Yuan et al., 2020) to glacial studies (Booth et al., 2020; Brisbourne et al., 2021; Fichtner et al., 2023; Hudson et al., 2021), volcanology (Currenti et al., 2021; Jousset et al., 2022; Klaasen et al., 2021; Nishimura et al., 2021), and regional and local earthquake monitoring as well as structural imaging (Biondi et al., 2021a; Biondi et al., 2021b; Jousset et al., 2018; Karrenbach et al., 2019; Li et al., 2021; Lindsey et al., 2017). Borehole fiber-optic seismology where cables are installed along the casing or behind the tubing of wells has also gained attention with successful applications for reservoir and fracture imaging (Lellouch et al., 2019b; Martuganova et al., 2022; Staněk et al., 2022; Titov et al., 2022) and microseismic monitoring for different industrial operations, including oil and gas exploration (Jiang et al., 2016; Shultz & Simmons, 2019; Zhan & Nahm, 2020), CO_2 sequestration monitoring (Correa et al., 2019; Daley et al., 2013; Götz et al., 2018; Humphries et al., 2015; Wilson et al., 2021), hydraulic stimulation in unconventional gas exploitation (Bakku, 2015; Karrenbach et al., 2017; Liu et al., 2022; Luo et al., 2021), and enhanced geothermal system (EGS) sites (Lellouch et al., 2020; Lellouch et al., 2021).

In particular, for geothermal systems, DAS technology offers an advantage over traditional seismic sensors for microseismic monitoring because it does not have the same temperature limitations as standard electronic equipment. The sensing fiber can therefore be installed at high-temperature reservoir conditions. The maximum temperature at which fibers can maintain satisfactory performance depends on the materials used for fiber coating; typically, these values range between 300 and 400 °C (Mellors et al., 2018). This increased durability and resistance to corrosion of fiber-optic cables (Reinsch et al., 2015) is also beneficial to other applications where pressure, salinity, or high acidity are elevated. We note, however, that fiber-optic cable deployments in borehole settings are still challenging and are not devoid of practical issues. These challenges include safety concerns during installation, cable breakdowns, management of lengthy cables, and environmental conditions such as temperature and humidity, which may degrade the integrity of interrogator units. When deployments are successful, seismic DAS cables installed at target depth reservoirs can provide ideal conditions to investigate the behavior of the subsurface during stimulation and utilization. Consequently, the distance between the detecting sensor and the microseismic events can be minimized, thus maximizing the detection capability. In addition, DAS has the potential for recording many more microseismic events than conventional surface arrays or sparse downhole geophone arrays due to its dense spatial coverage over long distances (typical DAS acquisition samples the wavefield at meter or submeter spacing and sampling frequencies of 1 kHz or higher; Lellouch et al., 2020), which can also be used to image subsurface structures at high-resolution (Lellouch et al., 2019a; Staněk et al., 2022). On the other hand, as DAS records the strain or strain rate only along the fiber axial direction, a primary limitation is its cylindrical symmetry, which creates azimuthal ambiguity, especially in purely vertical boreholes and single-borehole monitoring configurations (Karrenbach et al., 2017; Lellouch & Biondi, 2021).

Many studies have tackled the problem of earthquake detection and location with DAS. For surface earthquake monitoring studies, DAS technology demonstrated promising outcomes when traditional detection and location algorithms were employed (e.g., Jousset et al., 2018). However, early applications of conventional detection and location algorithms to DAS data in borehole settings were relatively unsuccessful and underperforming, yielding fewer detections and greater location errors compared to conventional geophones (Hull et al., 2017; Webster et al., 2016). Thus, to fully exploit the high spatial and temporal sampling of DAS data, as well as to deal with the large data volumes (up to several TB per day) of inherently noisier recordings, new processing techniques have been proposed across several different fields of application. For example, the PoroTomo DAS array at Brady Hot Springs geothermal field in Nevada has provided a rich DAS dataset and testbed for template matching techniques (Li & Zhan, 2018), evaluation of beamforming potential in DAS data (van den Ende & Ampuero, 2021), as well as subsurface mapping and seismic event monitoring (Piana Agostinetti et al., 2022). Template matching and beamforming have also been used for earthquake aftershock detection (Li et al., 2021), small earthquake detection at regional distances (Nayak & Ajo-Franklin, 2021), and identifying and locating volcanic events (Jousset et al., 2022). In recent years, machine learning models are being adapted to and trained with continuous data recorded on DAS arrays for earthquake detection (Huot et al., 2018a; Huot & Biondi, 2020), earthquake classification (Currenti et al., 2023), seismic arrival-time picking (Zhu et al., 2023), as well as traffic and infrastructure monitoring (Huot et al., 2017; Huot et al., 2018b; Huot & Biondi, 2018; Martin et al., 2018).

Specifically in the field of microseismic monitoring with downhole DAS, significant advances have been made toward enhancing and/or tailoring detection and location algorithms to conform with DAS data (Lellouch & Biondi, 2021). In their work, Lellouch et al. (2020), using a single vertical fiber, proposed a beamforming approach targeted at event detection and event location without azimuthal information. Machine learning models for microseismic event detection have also shown promising progress, as evidenced by successful applications (e.g., Binder & Chakraborty, 2019; Stork et al., 2020). More recently, deep learning approaches have also emerged, demonstrating significantly enhanced accuracy and outcomes (Huot et al., 2022). Furthermore, Porras et al. (2024) proposed a fast semblance-based seismic event detection method in which waveform coherence is used as the detector threshold, overcoming the high self-noise of DAS records by exploring all the channels at once. The near-real-time performance of the proposed detector has the potential to unlock valuable information during underground operations (e.g., hydraulic fracturing, fluid injection, circulation, and extraction). Tuinstra et al. (2024) tackle the location ambiguity of a single vertical DAS cable by using distance geometry solvers to retrieve the shape and relative location of clustered microseismicity. The method is adapted from the single-station location algorithm for clustered earthquakes (eartHquake locAtion via Distance gEometry Solvers [HADES]) described in Grigoli et al. (2021), and it was used to locate off-shore clustered seismicity (Cesca et al., 2021).

Downhole DAS deployments have proven effective in monitoring microseismic events generated during hydraulic stimulation operations and in the proximity of the well being stimulated. However, it has been shown that they can also detect distant earthquakes and often outperform a surface-based regional seismic network by an order of magnitude (Lellouch et al., 2021). This is due to their deeper deployment, which allows bypassing the highly dissipative heterogeneous shallow layers of the Earth and lower sensitivity to surface noise sources such as anthropogenic activities and wind. The inherent lower signal-to-noise ratio (SNR) of individual DAS channel recordings, especially at far-offsets due to the narrower directivity patterns (Correa et al., 2017), can be overcome by beamforming and then, once time delays have been properly corrected for, stacking all the available channels. The ability to detect and locate distant microseismicity is paramount to reservoir monitoring both during and after the end of operations, as it allows us to assess the geomechanical conditions, seismogenic processes, and potential for reactivation of pre-existing, favorably oriented faults that could affect the stimulation region.

In this chapter, we evaluate the potential of DAS technology in out-of-network microseismicity contexts. Here we define seismic events as "out-of-network" when they exhibit an azimuthal gap exceeding 180°, where the gap is defined as the largest angle between two recording stations as seen from the epicenter. We present two case studies in which DAS cables were installed in boreholes and recorded distant seismicity at both the full-reservoir scale and mesoscale, respectively. We analyze the signal characteristics and DAS-based locations of a seismicity cluster recorded ~3 km east of the Frontier Observatory for Research in Geothermal Energy (FORGE) site, Utah, during the 2019 pilot stimulation. Next, we apply the same methods to a seismicity cluster recorded in 2021 ~ 900 m southeast of the Bedretto Underground Laboratory for Geosciences and Geoenergies in Switzerland (BULGG). Leveraging the different scales, we compare performance in terms of data quality, SNR, detectability, and applicability of the HADES-R location methodology (Tuinstra et al., 2024) for the analysis of out-of-network microseismicity in borehole settings.

16.2. METHODS

16.2.1. Preprocessing of DAS Signal

DAS recordings at each sensing point have low sensitivity and generally lower SNRs than downhole geophones. This has been observed in several studies (e.g., Correa et al., 2017; Lellouch et al., 2020), where a direct comparison with colocated geophones highlighted the inherent noisier data quality of DAS data. It is thus usually recommended that a preprocessing step, aimed at denoising the DAS raw data, is performed before further processing. It has been shown that the removal of the mean and linear trends of each trace of the array (linear and mean filtering step) is essential, as it removes most of the interrogator imprint (laser drift, reference loop vibration, etc.). This can be followed by a bandpass filter in the frequency range of interest, where the energy signal is generally concentrated. Further coherent noise can be removed by applying a frequency-wavenumber (FK) filter. FK filters are often used in active seismic exploration (Duncan & Beresford, 1994), and both traditional and adaptive methods have been applied to DAS data (e.g., Isken et al., 2022). In the following examples, a traditional FK filter is applied along the $k = 0$ wavenumber to remove noise generated by the interrogator, and, for the FORGE datasets, only the upgoing part of the wavefield was preserved to also remove downgoing free-surface reflections.

16.2.2. HADES-R Location Algorithm

The one-component nature of a vertically deployed DAS cable can lead to difficulties when locating events. When only a single cable is available, there is no information on the azimuth of the seismic event unless prior information is provided (Karrenbach et al., 2019; Lellouch et al., 2021). To overcome some of these challenges, Tuinstra et al. (2024) put forward a new methodology to locate clustered microseismic events with a fiber-optic cable deployed downhole in nonideal network configurations. The method builds on the work of Grigoli et al. (2021), which proposed an efficient method based on distance geometry problem (DGP) solvers for locating clustered seismicity with sparse surface networks of one or two stations (HADES). The method assumes that the events to be located belong to the same cluster and that this cluster is smaller than the distance between the receivers and the cluster. This is a condition that is easily satisfied in an out-of-network setting.

The new version of HADES (HADES-R) described in Tuinstra et al. (2024) can reconstruct the shape and relative location of a seismicity cluster using P- and S-wave arrival times and one a priori located event. To find event locations in a relative frame, HADES-R uses the inter-event distance between two events computed from the double difference in arrival times of S and P phases (t_s and t_p) and measured at a seismic station (equation (16.1)):

$$\left\|r^{s}_{e1e2}\right\| = k_v\left|\left(t^{e1}_s - t^{e1}_p\right) - \left(t^{e2}_s - t^{e2}_p\right)\right| \qquad (16.1)$$

where $\boldsymbol{r}^s$ is the inter-event distance between event $e1$ and event $e2$, and $k_v = \frac{v_p v_s}{(v_p - v_s)}$.

In Grigoli et al. (2021) a minimum of four master events with known locations was required to find the absolute location and rotation of the events in the cluster. Any rotational error of the cluster was minimized in the horizontal (latitude–longitude) plane, whereas rotational errors in depth remained unresolved. The main advantage of this method lies in the ability to provide the correct shape of clustered seismicity, particularly in ill-posed event-receiver configurations. In Tuinstra et al. (2024), the master event dependency is reduced from four to one, and multiple random combinations of DAS channels are considered instead of the original two station condition. Once a master event is selected (e.g., a perf shot, the largest event in the sequence), it represents the start of a coordinate system, and three other reference events are then located relative to it to construct a relative, arbitrarily oriented coordinate system. It is not necessary to know the absolute location of the reference events. However, an estimate of the width of the reference frame is required.

As the cluster's initial location is derived from arbitrary-located reference events, a rotation step is necessary to determine the cluster's absolute location. The rotation of the cluster is found by using quaternions, which can describe rotations around arbitrary axes in higher-dimensional space, specifically using a four-dimensional representation (Shoemake, 1985). The cluster rotation procedure consists of searching for the correct distribution of observed arrival times with respect to distance to the receivers (e.g., the DAS cable). To this end, the master event serves as the rotational origin of the seismic cluster, and three rotation optimization axes are defined: (a) a vertical axis from the origin pointing upward to the surface, (b) an orthogonal axis from the origin pointing to one selected station on the horizontally projected plane, and (c) the cross product of the first two axes. The choice of station that governs axis (b) can be interchanged. The optimal rotation angle for each of the three axes is then the one that maximizes the rectilinearity of the absolute distance between the receiver and the events with respect to the differential traveltime t_s–t_p at each receiver. Ideally, the distribution of t_s–t_p is linearly increasing with increasing distance to the receivers. We caution, however, that in some angular situations, the solution to the rotation problem may be nonunique: that is, multiple rotations of the cluster will result in equivalent distance-traveltime distributions. This can be the case of single vertical boreholes, where lateral rotations could result in equivalent traveltimes. Despite the challenges posed by these unfavorable network configurations, the cluster's overall shape remains intact. Although the method does not fully resolve the azimuthal ambiguity in such cases, the abundance of channel pairs along the length of the DAS cable enables the application of an iterative bootstrap approach and thus allows us to retrieve location uncertainties. For each channel pair, we can apply HADES-R followed by a rotation optimization. The ensemble of solutions from all the channel pairs can aid in constraining uncertainties and identifying possible biases or ambiguities in azimuthal orientation. Ensemble solutions are typically shown as two-dimensional histograms of the output of multiple iterations of HADES-R and the associated rotation step conducted over multiple channel pairs.

As the input data for the method consists of P- and S-wave arrival times, Tuinstra et al. (2024) also propose a semiautomated procedure that allows us to pick efficiently the large volume of data usually recorded by the DAS systems. (Figure 16.1(a)–(c)). P- and S-wave arrivals are initially manually identified on traces with strong SNRs. Subsequently, a user-defined search range is selected encompassing the initial arrivals. Typically, the search area for the S-phase arrival is broader than that used for P-phase arrivals. Finally, the P and S master traces, generated by stacking the manually selected

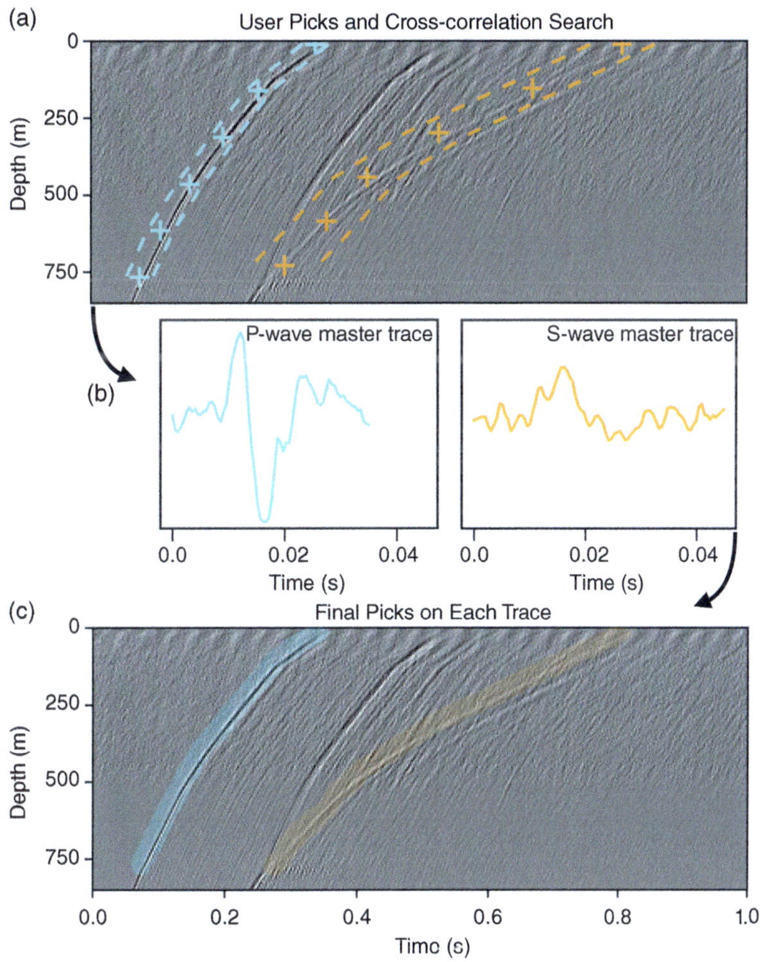

Figure 16.1 Semiautomated workflow for picking P- and S-wave arrivals for DAS data. (a) Manual picks for P- and S-wave arrivals (crosses) and user-defined search range (dotted lines). (b) Master trace derived from stacking manually picked traces over the predefined search range. (c) Final automatic picks for all traces after cross-correlation. Adapted from Tuinstra et al. (2024).

arrivals, are cross-correlated within the predefined search range. This process automatically identifies arrivals in all traces. This is just one of the methods to retrieve P and S picks. Several methods exist in the literature, from semi-automated to semisupervised machine learning methods, to address the phase-picking task for the ultra-dense spatial sampling of DAS data (e.g., Zhu et al., 2023).

16.3. CASE STUDIES

16.3.1. UTAH FORGE EGS Test Site

The U.S. Department of Energy FORGE is a dedicated international underground field laboratory in Utah. Its purpose is to develop, test, and accelerate breakthroughs in EGS in a controlled environment. FORGE is located in an unpopulated area on the western flank of the Mineral Mountains in the vicinity of Milford in Beaver County, Utah, a small community of 1,400 people (Figure 16.2; Moore et al., 2020; Wannamaker et al., 2020; Pankow et al., 2020), and close to the active Blundell geothermal plant in the Roosevelt Hot Springs (Gwynn et al., 2016) within Utah's Renewable Energy corridor. FORGE sits on alluvial fan deposits of 300–1,000 m in thickness, with the target EGS reservoir to be created in the underlying basement granitoid rocks where temperatures have been estimated to exceed 190 °C at less than 3 km depth. The site is in continuous development, with new wells being drilled and new instrumentation being added.

The first pilot stimulation took place in 2019 between 21 April and 3 May. At the time, in addition to stations of the Utah Regional Seismic Network (URSN), surface monitoring consisted of five surface broadband stations

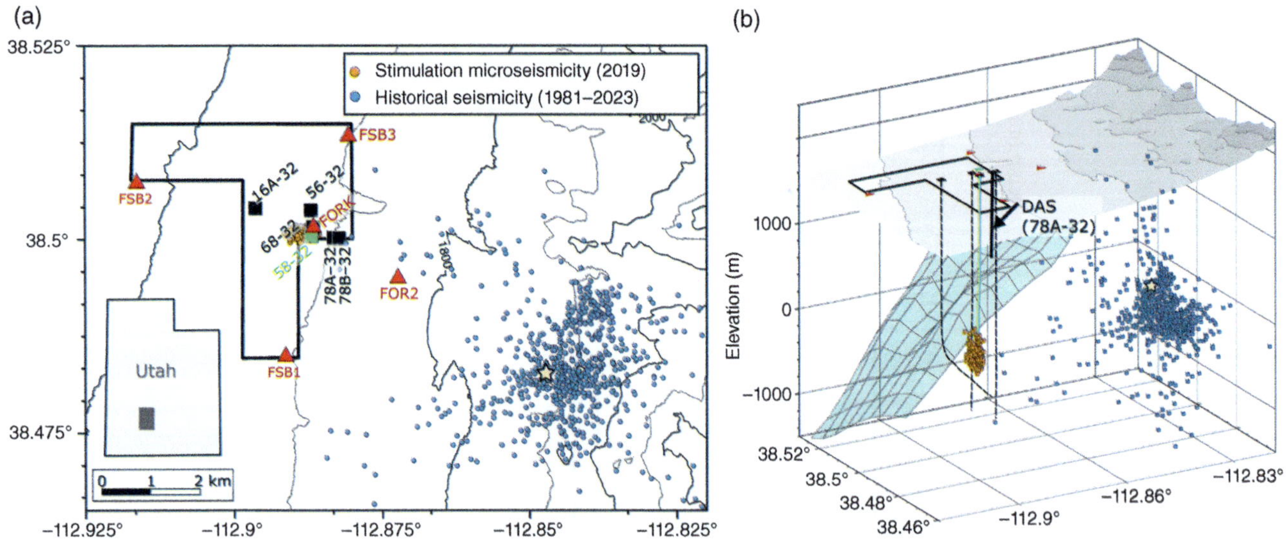

Figure 16.2 (a) Map of the FORGE geothermal test site in Utah, USA. Historical seismicity for the period between January 1981 and March 2023, cataloged by the University of Utah Seismograph Stations, is shown with solid blue circles (this represents what hereafter is referred to as a "southeast cluster"). The microseismicity recorded during the 2019 FORGE pilot stimulation is shown with solid orange circles. The green square illustrates the stimulation well (58-32) used during the pilot stimulation in 2019. All wells drilled to date (78A-32, 78B-32, 68-32, 56-32, and 16A-32) are indicated by black squares, and the seismic stations of the seismic network currently active are shown with red triangles. (b) 3D rendition of the site highlighting the dipping granitoid layer (dark green surface), the pilot injection Well 58-32 (light green line), and the monitoring Well 78-32 instrumented with DAS (solid black line). The yellow star indicates the event shown in Figure 16.3, which was recorded by both DAS and the surface stations.

(red triangles in Figure 16.2); three accelerometers, one placed in a shallow well (68-32); and several temporary high-density nodal arrays in various configurations covering the footprint of the FORGE site. Additional details about the instrumentation in the surrounding vicinity can be found in previous studies (e.g., Pankow et al., 2020; Potter, 2017) as well as on the IRIS website (https://ds.iris.edu/mda/UU/). Among the wells drilled at FORGE, Well 58-32 was used for the hydraulic stimulation, and Well 78-32, located about 400 m east-southeast of 58-32, was used for deep seismic monitoring (Figure 16.2). Well 78-32 was equipped with an industry-grade 12-level geophone string deployed by Schlumberger and an optical fiber in a metal tube cemented behind the casing. The geophones were positioned at depths of 645–980 m, with 30.5 m spacing between each sensor, and they were recording only during the stimulation. The installed fiber cable included two multimode (MM) fibers, two single-mode (SM) fibers, and a specialized Silixa Constellation fiber (Diaz-Meza et al., 2023). Fibers were looped in U bends and protected in bottom hole assembly (BHA) housing. The cables were interrogated using the Carina system (Naldrett et al., 2020). Correa et al. (2017) demonstrated that this engineered fiber yields a substantial 20 dB improvement in the optical SNR with respect to recordings using standard fibers. DAS data were acquired with a 1 m channel spacing using a 10 m gauge length and a sampling rate of 2 kHz for a 1,000-m-long cable. Additionally, before being written to disk, the laser sampling rate underwent a 16-fold internal stacking process (Lellouch et al., 2021). This process enables acquiring native data at very high sampling rates to achieve high SNR while avoiding extremely large data volumes by reducing the output frequency.

During the 2019 stimulation, in addition to the near-well microseismicity originated by the stimulated fractures, the DAS fiber installed in the monitoring Well 78-32 recorded a distant cluster of 82 earthquakes located 3–5 and 8–10 km southeast of the stimulation Well 58-32 (Lellouch et al., 2021). Of these, only one event (Figure 16.3) was detected by the routine processing of the regional surface network and reported in the University of Utah Seismograph Stations (UUSS) catalog. This activity is likely not directly associated with the stimulation but rather belongs to a cluster of seismicity well aligned with the historical seismicity in the area, and which is possibly connected to the Mineral Mountains swarms and natural hydrothermal fluid-flow processes (Lellouch et al., 2021; Zandt et al., 1982). In the next sections, we explore the signal characteristics of the proximal

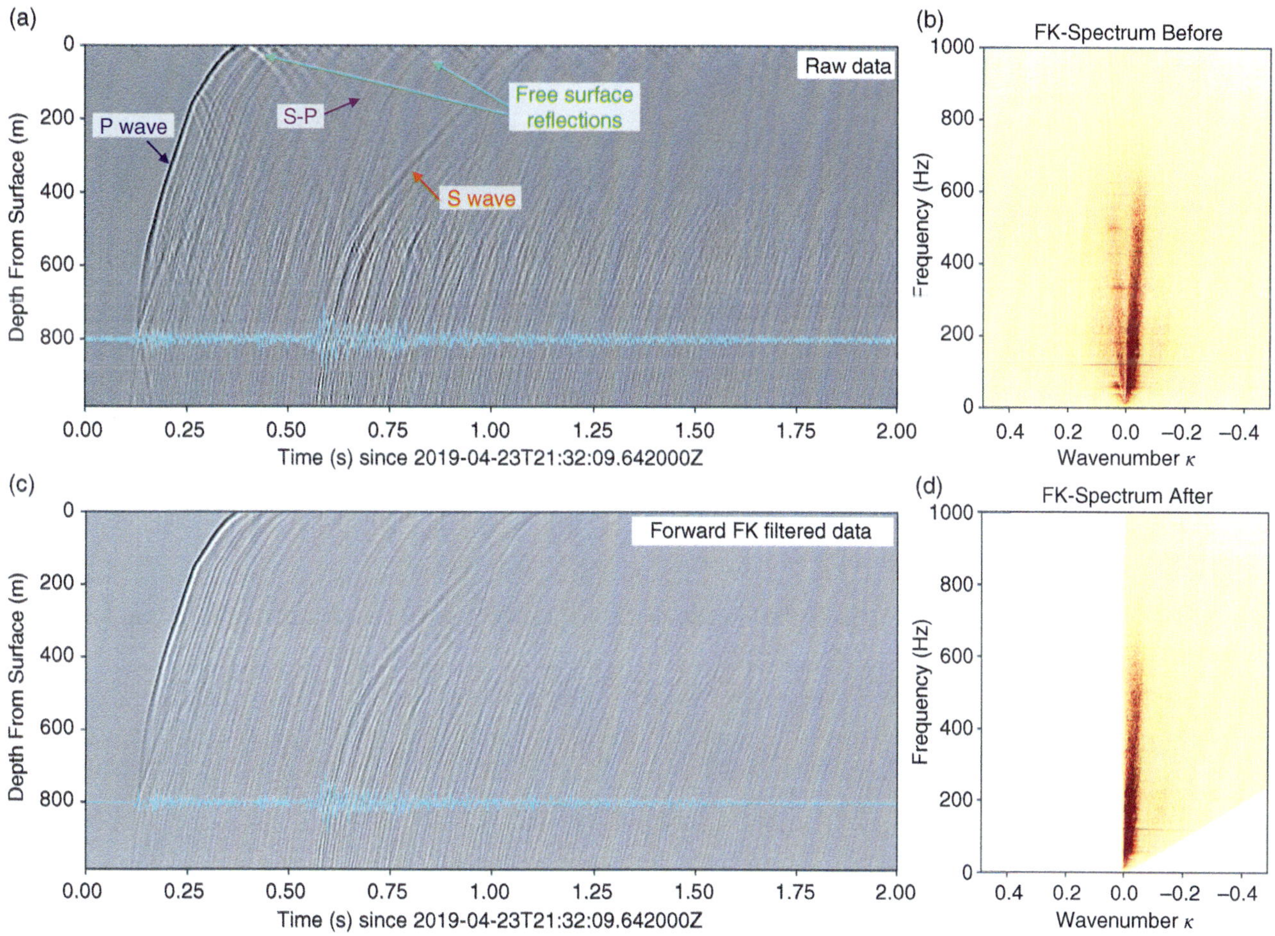

Figure 16.3 Event recorded by the DAS cable installed in Well 78A-32 on 23 April 2019 21:32:09; $M_L = 0.91$. The earthquake was also recorded by the surface array and cataloged by the UUSS. Many different phases can be observed, some of which are labeled with arrows in both the raw data (a) and after a forward FK filter is applied (c): direct P waves (blue), direct S waves (red), their conversion to a P wave at the granite contact (purple), and free-surface reflections (green). In (a) and (c), we also show the waveform for channel 800 of the fiber cable (cyan line). (b) and (d) FK spectrum before and after applying a forward FK filter, respectively.

subcluster (3–5 km) as recorded by the DAS system and test the DAS location capabilities in challenging out-of-network conditions.

Figures 16.3 and 16.4 present illustrative examples of 2 s DAS records corresponding to two earthquakes with magnitudes M_L 0.91 and M_L –0.8, respectively. These earthquakes are part of the detection catalog of Lellouch et al. (2021) and belong to the proximal cluster (3–5 km) located ~3 km to the southeast of the stimulation region, hereafter referred to as the "southeast cluster" (Lellouch et al., 2021). Here, we report the magnitude as calculated by Lellouch et al. (2021) using the methodology described in Lellouch et al. (2020). For both earthquakes, it is clear that DAS offers a significant advantage through its quasi-continuous spatial and temporal sampling of the seismic wavefield, where we can recognize distinct arrival patterns as a function of depth along the fiber. Figures 16.3(c) and 16.4(c) show the data for both events after denoising, with their corresponding FK spectra. For the weaker event (Figure 16.4(a)), the energy signal of the raw data is not clearly visible either in the time domain or in the FK spectrum. However, after denoising, the coherent $k = 0$ noise is attenuated as well as the high-frequency random noise, allowing the spatial coherency of the filtered wavefield to be visible in both time and frequency domains (Figure 16.4(c)).

Spectrograms obtained for the two events also exhibit distinct features between the higher and lower magnitude events, which can be directly attributed to the noise levels in each record. The higher-magnitude event displays clear energy peaks corresponding to the P- and S-phase arrivals (Figure 16.5(a)). On the other hand, in the case

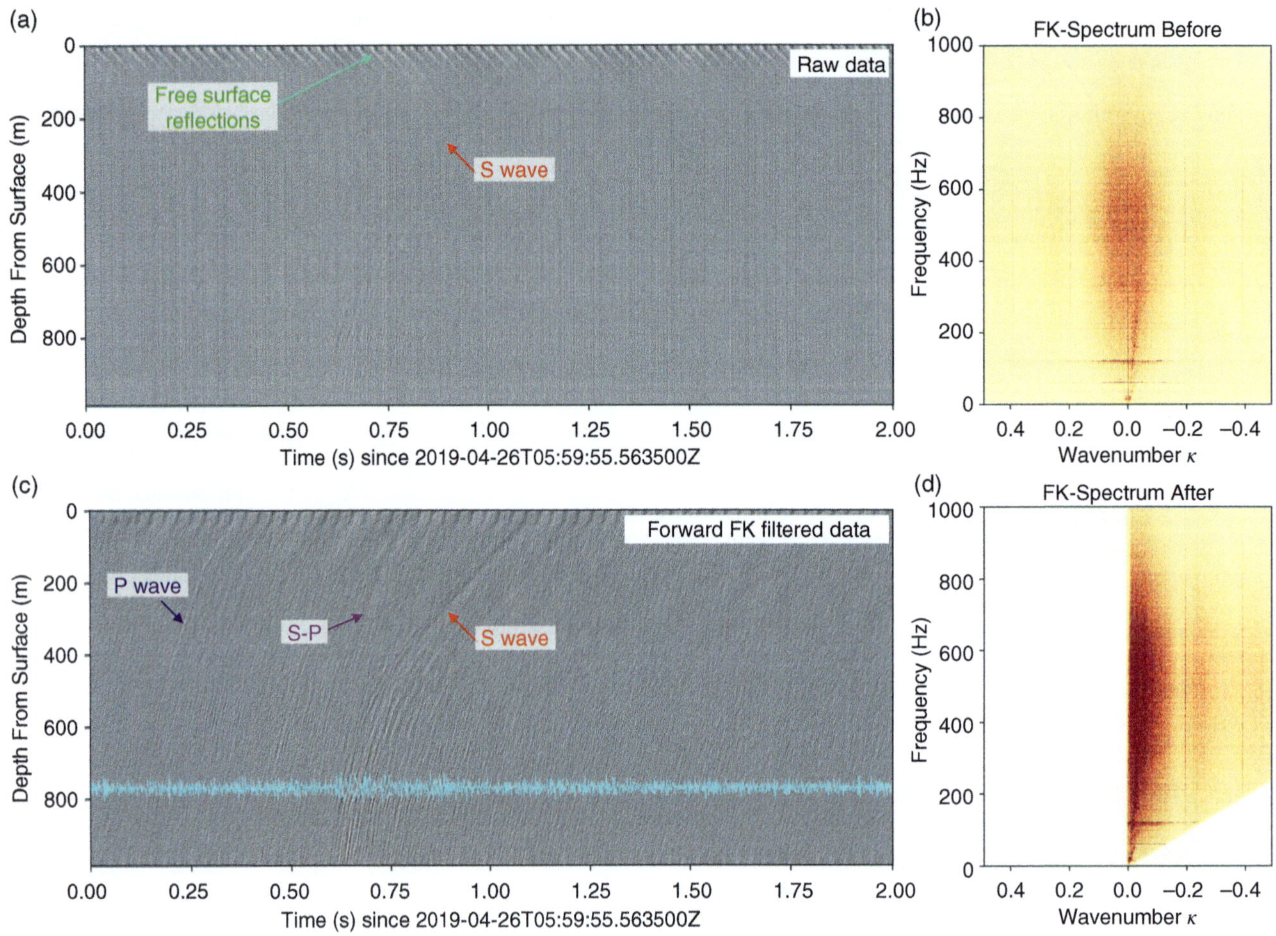

Figure 16.4 Event recorded by the DAS cable installed in Well 78A-32 on 26 April 2019 05:59:55; $M_L = -0.8$. Note how the event phases are hardly visible in the raw data (a), whereas they can be better identified after a forward FK filter is applied (c): direct P waves (blue), direct S waves (red), their conversion to a P wave at the granite contact (purple), and free-surface reflections (green). In (c), we also show the waveform for channel 770 of the fiber cable (cyan line). (b) and (d) FK spectrum before and after applying a forward FK filter, respectively.

of the low-magnitude event, although the SNR varies between the individual channels, for most channels we cannot distinguish P or S phases unless a filter is applied (Figure 16.5(b)).

DAS-Based Locations

Figure 16.6 shows the DAS-based locations and their uncertainties obtained by applying HADES-R to the southeast seismicity cluster located ~3 km from the stimulation patch. Of the 82 earthquakes cataloged by Lellouch et al. (2021), 39 had clearly visible P- and S-phase arrivals that could be picked on the lower 626 m section of the DAS cable, up to a depth of 877 m below the surface. To serve as the master event, we selected a well-recorded earthquake that had been previously located using the surface network, as documented by Lellouch et al. (2021). Similarly to Lellouch et al. (2021), we fix the depth of the master event at 1 km, as the initial location derived with solely surface stations had large depth uncertainties of ~3 km. We then repeat the HADES-R location procedure for 100 random channel pairs along the DAS cable and retrieve the distribution of the location uncertainty. We show this in Figure 16.6, where the frequency of earthquake locations found at a specific spatial coordinate by HADES-R over 100 iterations is indicated by the two-dimensional histogram. We observe a certain degree of rotational uncertainty around the master event, with multiple possible solutions for which the rectilinearity is maximized. However, there is a clear tendency for the events to cluster to the southeast of the master event (yellow solid circle in Figure 16.6). The location density distribution agrees with the DAS-derived locations of Lellouch et al. (2021), with the distinction that we are able to obtain a better constraint on depths

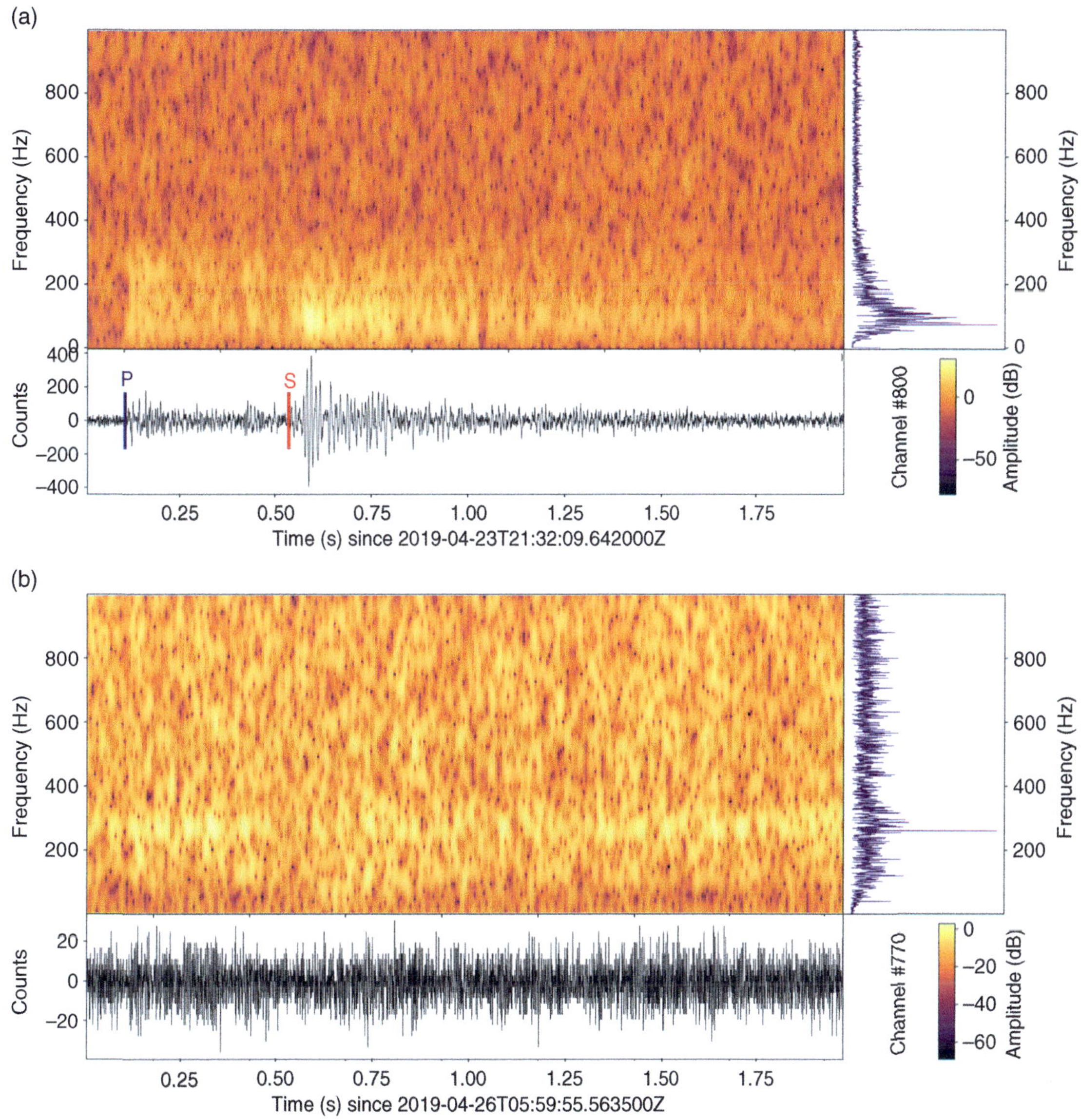

Figure 16.5 Spectrograms for a single channel of the fiber-optic cable installed in Well 78A-32. (a) Event 2019-04-23 21:32:09; $M_L = 0.91$; DAS channel 800. The bottom panel shows the raw seismogram recorded by DAS channel 800, and the right panel indicates the corresponding amplitude spectrum. (b) Event 2019-04-26 05:59:55; $M_L = -0.8$. DAS channel 770. The bottom panel shows the raw seismogram recorded by DAS channel 770, and the right panel indicates the corresponding amplitude spectrum. Adapted from Lellouch et al. (2021).

and a smaller uncertainty density cloud. Both location sets from Tuinstra et al. (2024) and Lellouch et al. (2021) show a strong correlation with the historical seismicity underlying the Mineral Mountains.

16.3.2. Bedretto Underground Rock Laboratory

Whereas FORGE offers a testbed for experiments at the full-reservoir scale, the Bedretto Underground Laboratory for Geosciences and Geoenergies ("BedrettoLab" hereafter) provides the ideal environment for controlled mesoscale in situ experiments to further improve our understanding of hydro-seismo-mechanical processes taking place in fractured crystalline rock masses (Ma et al., 2022).

The BedrettoLab is located in the Bedretto Tunnel in the Swiss Central Alps, near the Gotthard Pass (Figure 16.7a). With a length of 5,218 m, the tunnel connects the Furka Base Tunnel in the northwest to the Bedretto Valley in the southeast and was originally excavated to transport muck as part of the Furka Base Tunnel construction logistics (Keller & Schneider, 1982). Since 2018,

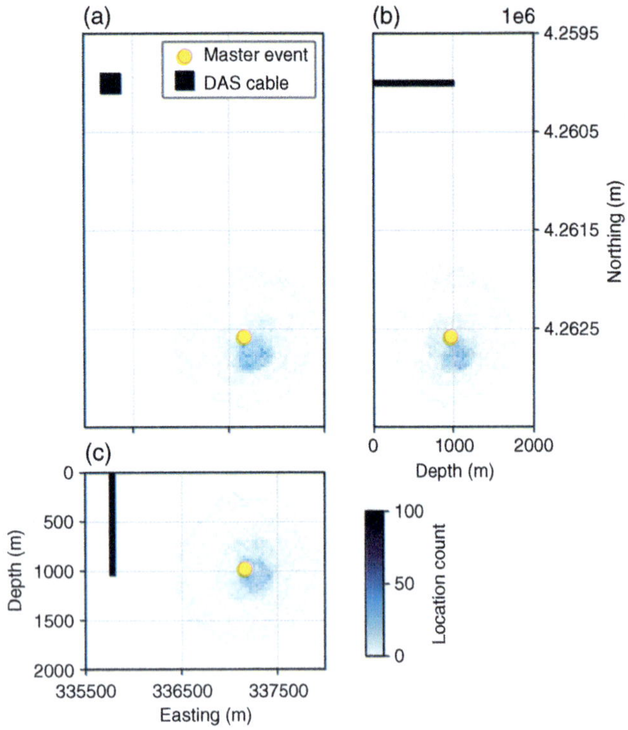

Figure 16.6 HADES-R derived locations of the southeast seismicity cluster recorded by a vertical downhole DAS during the 2019 FORGE pilot stimulation in the horizontal (a), longitude-depth (b), and latitude-depth plane (c). The master event is indicated by the yellow solid circle, and the uncertainty cloud is shown as a two-dimensional histogram (bluish colors). Adapted from Tuinstra et al. (2024).

the Bedretto Tunnel has been made available by its owner, Matterhorn Gotthard Bahn (MGB), to ETH Zurich to perform experimental research focusing on the behavior of the deep underground in various domains, including geothermal energy and earthquake physics.

Since then, several boreholes along the tunnel have been excavated and instrumented for detailed geophysical characterization of the experimental site. The goal is to provide a better understanding of the potential for geothermal energy by researching hydro-fracturing via various injection strategies in a mesoscale, controlled environment close to the injection point (Plenkers et al., 2023). Several injection activities have been conducted at the BedrettoLab, with the first operations in the Fall of 2020 (Hertrich et al., 2021) and more occurring in 2022/2023 (Broeker et al., 2023).

Here, we focus on a seismicity cluster that was observed in May 2021 about 900 m from the instrumented laboratory, toward the Bedretto Valley (Figure 16.7). At the time of occurrence of the cluster, the seismic network inside the tunnel was composed of the following: three seismic stations along the tunnel itself; an eight-level three-component (3C) digital geophone chain in borehole ST1; an Avalon single 3C analog geophone with downhole amplifier in borehole ST1; and a single 3C IMS analog geophone in one of the monitoring boreholes. Borehole ST1, which is slightly deviated, features a multi-packer system (14 intervals in total) for a total length of about 400 m, with a hybrid cable installed through the packer system. This hybrid cable, manufactured by Solifos and referred to as BRUsens Heatable, features distributed fiber-optic sensing with an actively heated element. The fibers are installed in a metal tube (FIMT) with two SM and two MM loose fibers. Active heating can be achieved using copper conductors. The MM fibers are looped at the end of the boreholes and were used for monitoring of distributed temperature (DTS), and the SM fibers are both single ended, both used for injection monitoring with DAS and long-term monitoring with distributed strain sensing (DSS) interrogators.

The dataset employed here was recorded with an iDASv2 that recorded continuously in the period 24 April to 10 May. DAS data were acquired with a 1 m channel spacing using a 10 m gauge length and a sampling rate of 1 kHz, with 20-fold internal stacking. The seismicity cluster contains 17 events detected in the period between 24 April and 10 May that were identified in the DAS dataset. Figure 16.7 shows the location of a subset of 11 events. These events were additionally detected and located using the geophone array, which was, however, continuously recording only from 2 to 9 May by Geo-Energie Suisse (GES). For its location estimates, GES used P- and S-wave arrival times, as well as hodogram azimuths from the 3C eight-level geophone chain and L2 minimization approach over a search grid of 1 m. Despite the fact that Vp and Vs velocities were not optimized, the direction of the events was constrained by hodograms analysis and resulted in an uncertainty of +/−2°. The relative location resolution was estimated to be between 80 and 90 m in depth and between 20 and 30 m in the horizontal plane. Of the recorded events, the largest magnitude $M_L = 0$ was clearly detected by the routine processing of the regional surface network and reported in the Swiss Seismological Service (SED) catalog (Figure 16.8).

The cluster of events observed in the DAS occurred before and after some experiments conducted by GES featuring injections in borehole ST1, which generated little to no seismicity in the instrumented reservoir. Similarly to our first case study at the FORGE site, we refrain from interpreting the origin of the cluster recorded during the 24 April to 10 May period and focus here on the location of events and analysis of the seismic recordings of the fiber optics cable installed in borehole ST1.

Figure 16.8 illustrates an example of a 0.6 s DAS record for the largest event with magnitude $M_L = 0$, as reported by the SED catalog. We caution here that the magnitude

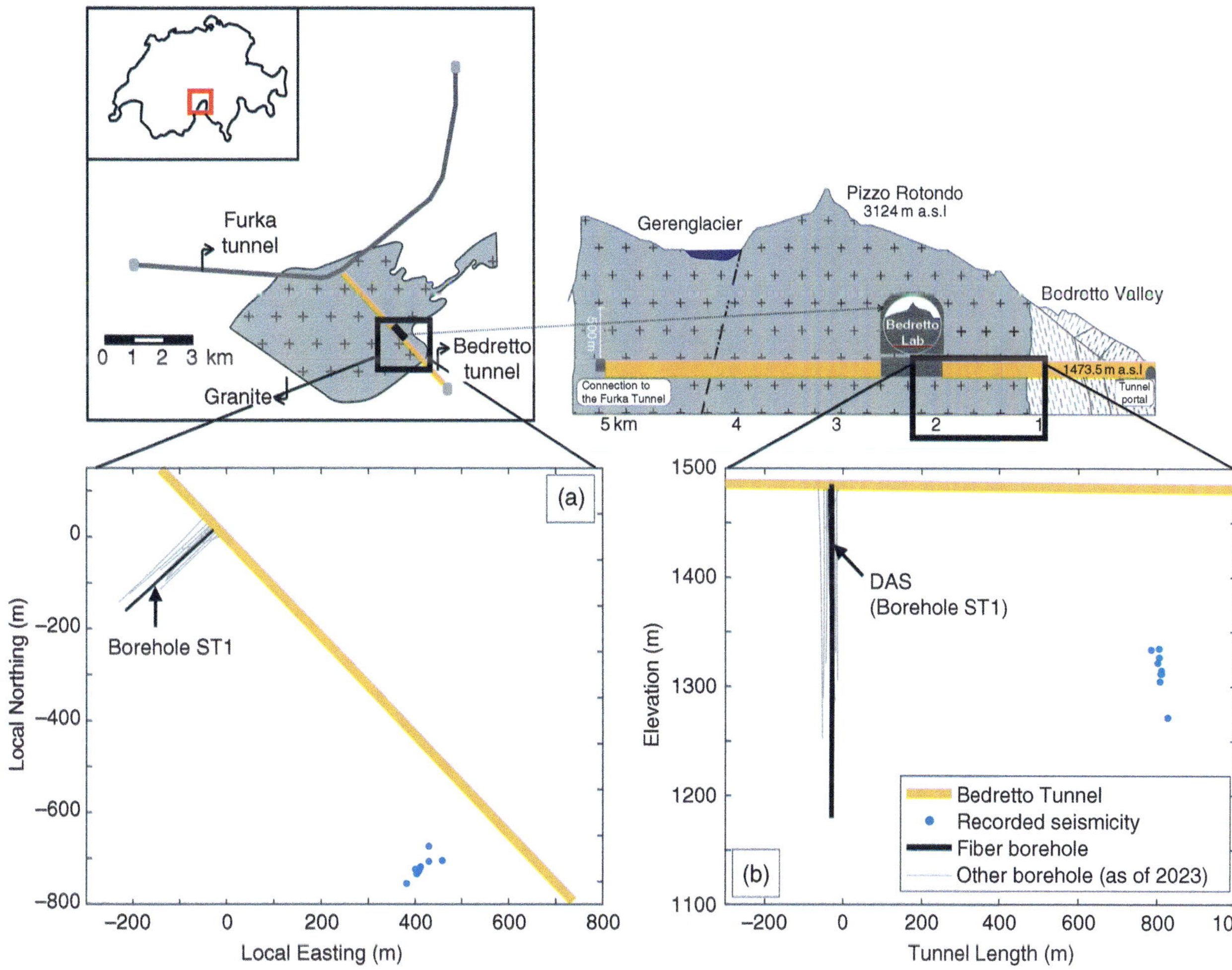

Figure 16.7 Overview map of the BedrettoLab, in Switzerland, in plane view (a) and cross-section (b). The Bedretto Tunnel is shown by the orange line, and all the boreholes drilled as of 2023 are shown with gray lines. The thick black line indicates borehole ST1 instrumented with DAS.

calculated by the national network may be overestimated. In its estimation, GES followed a time-domain approach over a window of 20 ms from the P- and S-wave first arrivals picked on the geophone chains and thus may be better calibrated with respect to that of the national network. For the same event shown in Figure 16.8, GES calculated a $M_w = 0.27$. As we have seen at the FORGE site, we can apply FK filters to increase the SNRs of small, weak events (Figure 16.8(c)–(e)). In this case, we also observe a strong scattering effect of the first arriving phases, which could be because the fiber passes through the packer systems installed in the same borehole ST1. Figure 16.8(c)–(e) show various filters that were tested to attenuate these scattering effects. With a bidirectional filter that allows only high apparent velocities, the scattering noise is much attenuated but cannot be removed completely. When we apply a one-directional filter, it seems to remove some broadside energy that could be linked to reflections within the packer systems, although we may also lose some signal information. For P- and S-phase picking tasks, we opted for the more standard bidirectional FK filter as we can still observe quite strong P- and S-wave phases (Figure 16.8(c)), also clearly visible in the spectrogram (Figure 16.9).

DAS-Based Locations

Similar to the southeast cluster at FORGE, we applied HADES-R to the distant seismicity cluster recorded between 24 April and 10 May 2021. On the DAS record, of the 17 events detected, 14 were suitable to pick both the P and S phases and thus were used in the HADES-R workflow. The master event was selected among the events located using the geophone chains by GES.

Figure 16.10 shows the DAS-based locations obtained by applying HADES-R to the cluster. In this case, we repeat the location procedure for 150 random channel

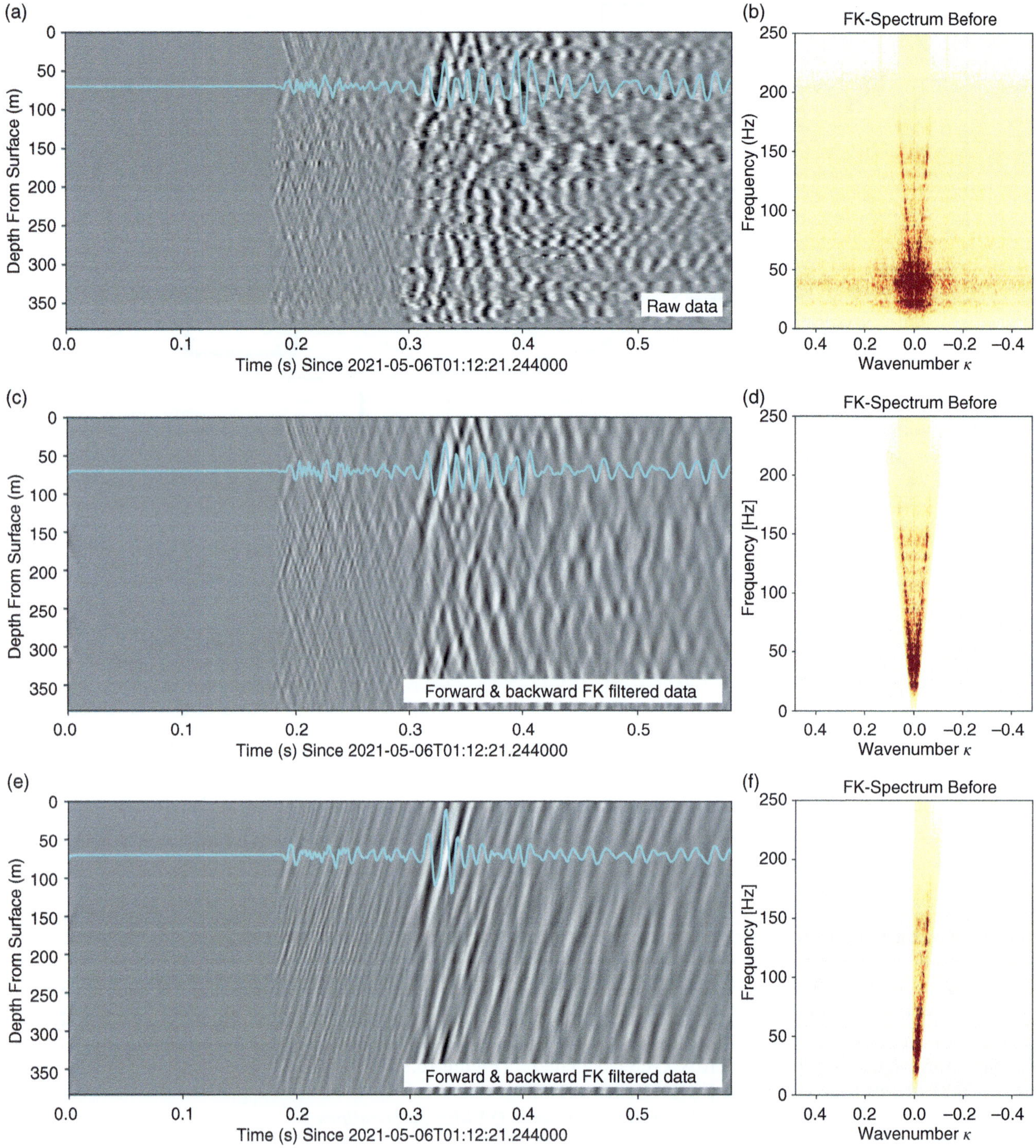

Figure 16.8 Event recorded by the DAS cable installed in borehole ST1 on 6 May 2021 01:12:21; $M_L = 0.0$ from the SED catalog solution ($M_w = 0.27$ as calculated by GES). Note the strong scattering effects due to the fiber passing through the packer system. (a) Raw data; (c) filtered data following a forward and backward bidirectional FK filter; (e) filtered data following a forward and backward one-directional FK filter; in (a), (c), and (e), we also show the waveforms for channel 70 of the fiber cable (cyan line). (b), (d), (f) FK spectrum before and after applying forward and backward FK filters, respectively.

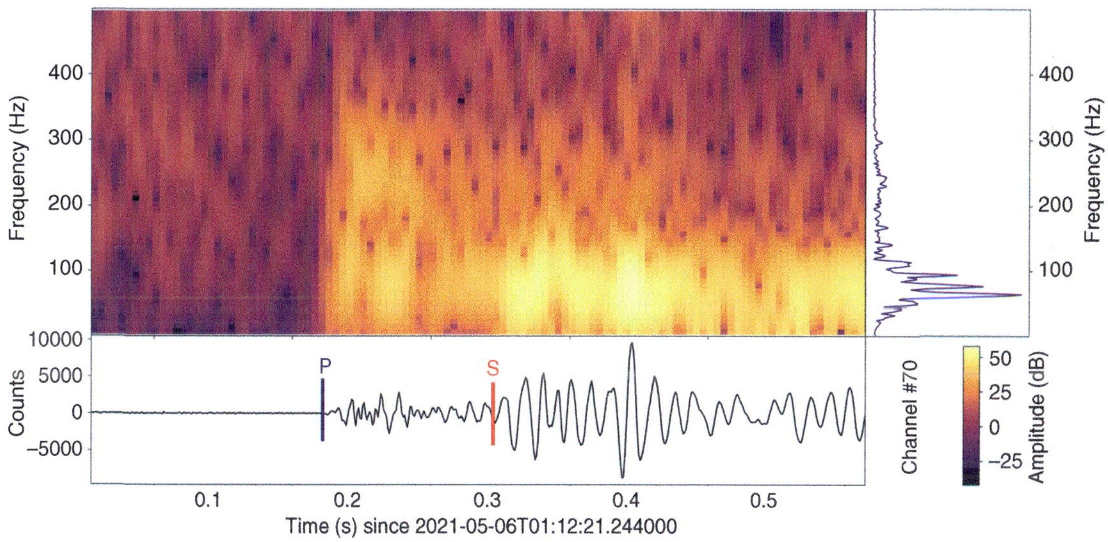

Figure 16.9 Spectrograms for channel 70 along the fiber. The bottom panel shows the raw seismogram recorded by DAS channel 70, and the right panel indicates the corresponding amplitude spectrum.

pairs along the DAS cable and retrieve the distribution of the uncertainties. The location density cloud agrees quite well with the geophone-derived locations of the GES catalog. The depth uncertainties observed are also consistent with those reported by GES, which similarly ranged in the hundreds of meters. Here, the presence of a nonvertical monitoring well, and thus a better angular configuration between stations and events, significantly reduces the ambiguity in the azimuthal direction, leading to a reduction in nonuniqueness associated with the rotations performed to find the absolute location of the cluster from its relative initial position. As evident from the two-dimensional histograms, a distinct preferred orientation of the cluster in the NE–SW direction, rather than a circular pattern, becomes apparent.

16.4. DISCUSSION

The two showcased examples at FORGE and BedrettoLab demonstrate the capability of downhole DAS technologies to effectively detect and locate low-magnitude earthquakes and microseismic events, even at distances of several meters to kilometers from the respective monitoring wells where the DAS systems are deployed. It also shows that DAS technologies are well suited for passive monitoring in both the far and near fields across different scales, ranging from low-magnitude events ($-4.0 < M < 0.0$) detected in an underground field laboratory to hydraulic fracturing-induced microseismicity ($-2.0 < M < 0.5$), to local earthquakes ($M > 1.0$).

Detectability capabilities of DAS systems are somewhat hindered by the inherently noisier nature of single channels of DAS data with respect to conventional downhole geophone arrays (Correa et al., 2017). However, we have shown that traditional preprocessing filtering routines and FK filters can already perform quite well. We see this clearly at the FORGE site, where the wavefield phases become visible in both the frequency and time domains after the application of FK filters (Figure 16.4 (c)). Yet certain types of noise cannot always be eliminated through conventional filtering techniques. This is observed in the BedrettoLab dataset, where we are unable to separate the signal from the scattering. We argue that such an effect could be produced by the presence of packer systems in the well and the fiber-optic cable feeding through the packers (more details about the well can be found in Plenkers et al., 2023). For these cases, more advanced techniques may be needed, such as deep learning methods (e.g., Jousset et al., 2022; Schwarz, 2019; van den Ende et al., 2021) or adaptive filtering (e.g., Isken et al., 2022). Although the performance of downhole geophones is undoubtedly superior to single-channel downhole DAS in terms of recovered SNR, if we compare the DAS records with surface stations, it becomes apparent that the deeper installation of downhole DAS renders it unaffected by both anthropogenic noise and wind noise, as well as by the dissipative heterogeneous shallow layers of the Earth (Hutt et al., 2017; Lellouch et al., 2021). DAS deeper deployments at reservoir depths are also possible due to the DAS robustness to high temperatures and pressure. As DAS cables can be installed permanently and in a

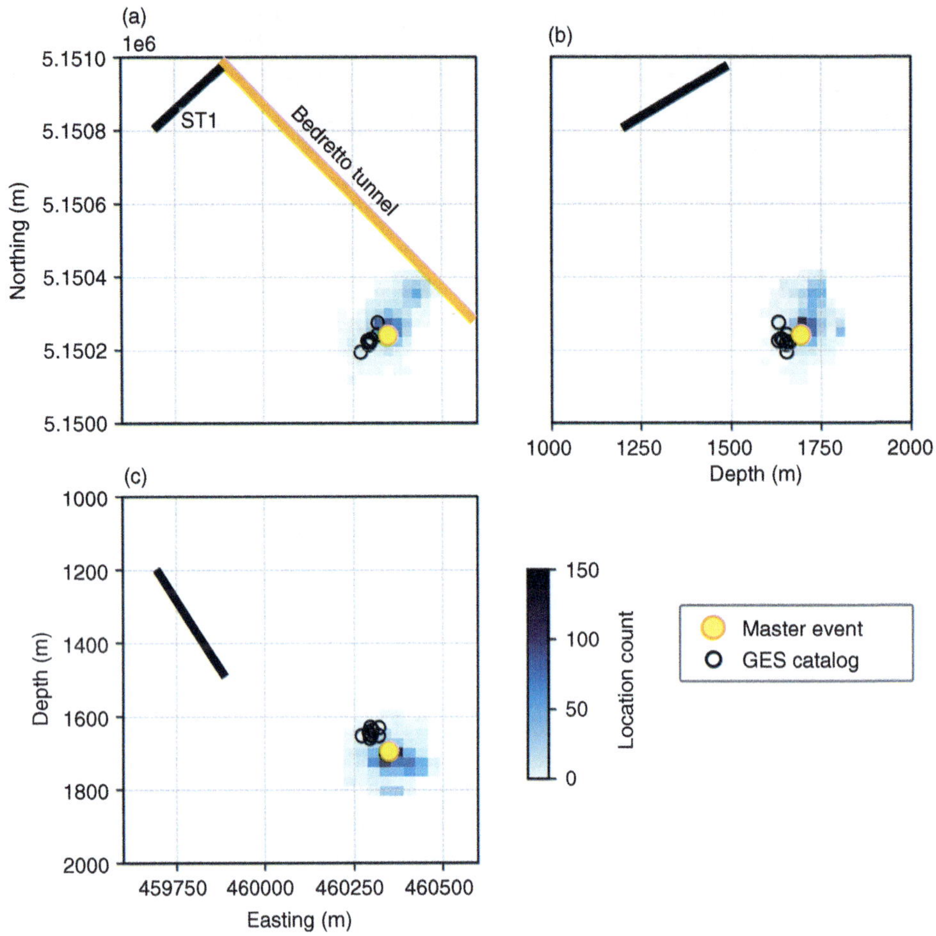

Figure 16.10 HADES-R derived locations of the southeast seismicity cluster recorded by the downhole DAS array in borehole ST1 in the horizontal (a), longitude-depth (b), and latitude-depth plane (c). The master event is indicated by the yellow solid circle, and the uncertainty cloud is shown as a two-dimensional histogram (bluish colors). Open black circles indicate the locations derived by GES using colocated geophone chains.

cost-effective way in boreholes (both monitoring and stimulation ones), they offer a promising alternative for long-term passive monitoring across a wider area beyond the immediate vicinity of the stimulation patch. At FORGE, Lellouch et al. (2021) were able to improve the magnitude completeness of 0.5–1.0, well into negative magnitudes (as also observed in BedrettoLab), even though the well was situated 3–4 km from the nearest seismically active region.

In settings where geophones might experience mechanical issues due to elevated temperature conditions or where surface noise obstructs the identification of deep, low-magnitude events, event localization using DAS-only network configurations has gained traction in microseismic monitoring applications. In terms of DAS technology's locatability potential, the inherent azimuthal ambiguity of a single vertical downhole DAS cable poses a limitation on the directional resolvability of DAS-based locations. However, we have shown that the HADES-R algorithm can retrieve the general shape of a seismic cluster if the distance between the receivers and the cluster centroid is much larger than the overall size of the cluster and at least one event in the cluster has a well-known absolute location. These assumptions are often valid for the illustrated cases of out-of-network seismicity.

Additionally, HADES-R can deliver information on location uncertainties by leveraging the DAS channel abundance through an iterative approach in which the cluster orientation uncertainty cloud is computed. In extremely ill-posed network configurations (e.g., single vertical well) and where clustered seismicity occurs out-of-network, Tuinstra et al. (2024) demonstrated that by incorporating a subset of events that are also visible on surface stations, a better estimate of the cluster shape and network aperture can be achieved. Similarly, incorporating 3C geophones can be effective in reducing the azimuthal ambiguity as information derived from hodogram analysis can provide crucial a priori azimuthal constraints. This highlights how additional instruments can improve location estimates. This approach of merging dynamic strain sensing and seismic data has also been tested and demonstrated by Flóvenz et al. (2022) and Binder and Abatchev (2021). With the Bedretto case study, we also show that even a slightly deviated well could improve the azimuthal ambiguity and provide a reliable DAS-based only location with minimal prior information and no assistance from a surface network.

DAS-based locations provide a more accurate depth estimation compared to the surface array. Thus, network geometry is also important and strongly affects the location capability of DAS. A generalization requires specific forward simulations, and very often, given the cost to drill boreholes, their geometry is driven by other factors (e.g., to maximize energy exploitation). From a seismological point of view, the drilling of a horizontal or inclined well enhances the azimuthal gap, and it is always favored over a vertical borehole.

Although this is also true for any other type of borehole array, fiber-optic cables provide several advantages with respect to conventional sensor arrays, as detailed previously. Hence, we believe that downhole DAS arrays, even when deployed in a single vertical well, can serve as a valuable complement to existing local networks, whether in boreholes or at the surface.

One further outcome from our evaluation is the ability of DAS to perform equally across scales. DAS signal characteristics and location capabilities are tested at the FORGE site (full-scale reservoir) and BedrettoLab (mesoscale reservoir), with results showing overall good performance. This has favorable impacts on the usage of downhole DAS in vertical or in deviated wells, at close or far distances, for high- or low-magnitude events. Nonetheless, we must emphasize that acquiring more extensive data through longer experiments is crucial to obtain a more robust statistical understanding of DAS capability to complement conventional earthquake monitoring network arrays.

16.5. CONCLUDING REMARKS

DAS offers numerous operational advantages for microseismic monitoring: it can be permanently installed in extreme environments (e.g., high pressure and temperatures), can remain in place undisturbed behind casing, and can interrogate long distances with high-density spatial and temporal coverage. Here, we have shown two case studies at the mesoscale and full-scale reservoir to evaluate the potential of downhole DAS to detect and locate out-of-network microseismicity. Our results at FORGE and BedrettoLab indicate that with adequate preprocessing, downhole DAS data quality allows the detection of very low-magnitude earthquakes several meters or kilometers from the well. HADES-R is able to retrieve the shape of out-of-network clustered seismicity across scales, produce more accurate depths, and provide uncertainty distribution. To avoid azimuthal ambiguity, even a slight deviation of the wells can greatly aid in the recovery of the correct rotation for obtaining robust locations. Our evaluation supports the employment of longer term experiments to keep exploring downhole DAS suitability for continuous passive microseismic monitoring.

ACKNOWLEDGMENTS

This work is supported by the De-Risking Enhanced Geothermal Energy Projects (Innovation for DEEPs). DEEP is subsidized through the Cofund GEOTHERMICA by the Swiss Federal Office of Energy (SFOE), which is supported by the European Union's HORIZON 2020 program for research, technological development, and demonstration under Grant Agreement Number 731117.

Analysis support with respect to the HADES-R location algorithm was provided by Francesco Grigoli, who played a key role in the theoretical development of the methodology.

AVAILABILITY STATEMENT

Both geophone data and DAS data at FORGE are openly available for download at the Geothermal Data Repository (GDR) website (https://gdr.openei.org/forge). DAS data acquired in May 2021 at the BedrettoLab from borehole ST1 are available upon request.

REFERENCES

Bakku, S. K. (2015). *Fracture characterization from seismic measurements in a borehole*, (PhD thesis). Massachusetts Institute of Technology.

Binder, G. & Abatchev, Z. (2021). *Joint microseismic event location with surface geophones and downhole DAS at the FORGE geothermal site*, Paper presented at the First International Meeting for Applied Geoscience & Energy, Society of Exploration Geophysicists.

Binder, G. & Chakraborty, D. (2019). *Detecting microseismic events in downhole distributed acoustic sensing data using convolutional neural networks*, Paper presented at the SEG Technical Program. Expanded Abstracts, 4864–4868.

Biondi, B. L., Clapp, R. G., Yuan, S., & Huot, F. (2021a). *Scaling up to citywide dark-fiber seismic arrays: lessons from five years of the Stanford DAS array project*, Paper presented at the First International Meeting for Applied Geoscience and Energy, Society of Exploration Geophysicists.

Biondi, B. L., Yuan, S., Martin, E. R., Huot, F., & Clapp, R. G. (2021b). *Using telecommunication fiber infrastructure for earthquake monitoring and near-surface characterization*. In Y. Li, M. Karrenbach, & J. B. Ajo-Franklin (Eds.), *Distributed acoustic sensing in geophysics: methods and applications* (pp. 131–148). AGU. doi:10.1002/9781119521808.ch10

Booth, A. D., Christoffersen, P., Schoonman, C., Clarke, A., Hubbard, B., Law, R., et al. (2020). Distributed acoustic sensing of seismic properties in a borehole drilled on a fast-flowing greenlandic outlet glacier. *Geophysical Research Letters*, *47*, e2020GL088148. doi:10.1029/2020GL088148

Brisbourne, A. M., Kendall, M., Kufner, S.-K., Hudson, T. S., & Smith, A. M. (2021). Downhole distributed acoustic seismic profiling at Skytrain Ice Rise, West Antarctica. *The Cryosphere*, *15*, 3443–3458. doi:10.5194/tc-15-3443-2021

Broeker, K., Ma, X., Gholizadeh, N., Rinaldi, A. P., Hertrich, M., Obermann, A., et al. (2023). *Hydro-geomechanical observations during multistage hydraulic stimulation at the Bedretto Underground Laboratory, Switzerland*. Paper presented at the 57th U.S. Rock Mechanics/Geomechanics Symposium.

Chen, Y., & Huang, L. & the EGS Collab team(2019). Optimal design of 3D borehole seismic arrays for microearthquake monitoring in anisotropic media during stimulations in the EGS collab project. *Geothermics*, *79*, 61–66.

Cesca, S., Stich, D., Grigoli, F., et al. (2021). Seismicity at the Castor gas reservoir driven by pore pressure diffusion and asperities loading. *Nature Communications*, *12*, 4783. doi:10.1038/s41467-021-24949-1

Correa, J., Egorov, A., Tertyshnikov, K., Bona, A., Pevzner, R., Dean, T., et al. (2017). Analysis of signal to noise and directivity characteristics of DAS VSP at near and far offsets - a CO_2CRC Otway project data example. *The Leading Edge*, *36*(12), 994a1–994a7. doi:10.1190/tle36120994a1.1

Correa, J., Pevzner, R., Bona, A., Tertyshnikov, K., Freifeld, B., Robertson, M., & Daley, T. (2019). 3D vertical seismic profile acquired with distributed acoustic sensing on tubing installation: a case study from the CO_2CRC Otway Project. *Interpretation*, *7*, SA11–SA19. doi:10.1190/INT-2018-0086.1

Currenti, G., Jousset, P., Napoli, R., Krawczyk, C., & Weber, M. (2021). On the comparison of strain measurements from fibre optics with a dense seismometer array at Etna volcano (Italy). *Solid Earth*, *12*, 993–1003. doi:10.5194/se-12-993-2021

Currenti, G., Allegra, M., Cannavò, F., et al. (2023). Distributed dynamic strain sensing of very long period and long period events on telecom fiber-optic cables at Vulcano, Italy. *Scientific Reports*, *13*, 4641. doi:10.1038/s41598-023-31779-2

Daley, T. M., Freifeld, B. M., Ajo-Franklin, J., Dou, S., Pevzner, R., Shulakova, V., et al. (2013). Field testing of fiber-optic distributed acoustic sensing (DAS) for subsurface seismic monitoring. *The Leading Edge*, *32*, 699–706.

Diaz-Meza, S., Jousset, P., Currenti, G., Wollin, C., Krawczyk, C., Clarke, A., & Chalari, A. (2023). On the comparison of records from standard and engineered fiber optic cables at Etna volcano (Italy). *Sensors*, *23*, 3735. doi:10.3390/s23073735

Duncan, G., & Beresford, G. (1994). Slowness adaptive *f-k* filtering of prestack seismic data. *Geophysics*, *59*(1), 140–147.

Dou, S., Lindsey, N., Wagner, A. M., Daley, T. M., Freifeld, B., Robertson, M., et al. (2017). Distributed acoustic sensing for seismic monitoring of the near surface: a traffic-noise interferometry case study. *Scientific Reports*, *7*, 11620. doi:10.1038/s41598-017-11986-4

Eaton, D. W. (2018). *Downhole Microseismic Processing* (pp. 158–189). Cambridge University Press.

Fichtner, A., Hofstede, C., Gebraad, L., Zunino, A., Zigone, D., & Eisen, O. (2023). Borehole fibre-optic seismology inside the Northeast Greenland Ice Stream. *Geophysical Journal International*, *235*(3), 2430–2441. doi:10.1093/gji/ggad344

Flóvenz, Ó. G., Wang, R., Hersir, G. P., et al. (2022). Cyclical geothermal unrest as a precursor to Iceland's 2021 Fagradalsfjall eruption. *Nature Geoscience*, *15*, 397–404. doi:10.1038/s41561-022-00930-5

Götz, J., Lüth, S., Henninges, J., & Reinsch, T. (2018). Vertical seismic profiling using a daisy-chained deployment of fibre-optic cables in four wells simultaneously – case study at the Ketzin carbon dioxide storage site. *Geophysical Prospecting*, *66*, 1201–1214. doi:10.1111/1365-2478.12638

Grigoli, F., Clinton, J. F., Diehl, T., Kaestli, P., Scarabello, L., Agustsdottir, T., et al. (2022). Monitoring microseismicity of the Hengill geothermal field in Iceland. *Scientific Data*, *9*(1), 1–11.

Grigoli, F., Ellsworth, W. L., Zhang, M., Mousavi, M., Cesca, S., Satriano, C., et al. (2021). Relative earthquake location procedure for clustered seismicity with a single station. *Geophysical Journal International*, *225*(1), 608–626.

Gwynn, M., Allis, R., Hardwick, C., Hill, J., & Moore, J. (2016). A new look at the thermal regime around Roosevelt Hot Springs. *Utah, Geothermal Resources Council Transactions*, *40*, 551–558.

Hertrich, M., Brixel, B., Broeker, K., et al. (2021). *Characterization, hydraulic stimulation, and fluid circulation experiments in the Bedretto Underground Laboratory for Geosciences and Geoenergies*. Paper presented at the 55th U.S. Rock Mechanics/Geomechanics Symposium, Virtual.

Hudson, T. S., Baird, A. F., Kendall, J. M., Kufner, S. K., Brisbourne, A. M., Smith, A. M., et al. (2021). Distributed acoustic sensing (DAS) for natural microseismicity studies: a case study from Antarctica. *Journal of Geophysical Research. Solid Earth*, *126*, e2020JB021493. doi:10.1029/2020JB021493

Hull, R. A., Meek, R., Bello, H., & Miller, D. (2017). *Case history of DAS fiber-based microseismic and strain data, monitoring horizontal hydraulic stimulations using various tools to highlight physical deformation processes (Part A)*, Paper presented at the SPE/AAPG/SEG Unconventional Resources Technology Conference, Austin, TX, USA.

Humphries, M., Vidal, J., & de Dios, J. (2015). *VSP monitoring for CO_2 migration tracking in fractured rock massifs*. Paper presented at the 77th EAGE Conference and Exhibition no. 1, Madrid, Spain. doi:10.3997/2214-4609.201412672

Huot, F. & Biondi, B. L. (2018). *Machine learning algorithms for automated seismic ambient noise processing applied to DAS acquisition*. Paper presented at the SEG Technical Program. Expanded Abstracts, 5501–5505.

Huot, F., & Biondi, B. L. (2020). *Detecting earthquakes through telecom fiber using a convolutional neural network*. Paper presented at the SEG Technical Program. Expanded Abstracts, 3452–3456.

Huot, F., Biondi, B. L., & Beroza, G. (2018a). *Jump-starting neural network training for seismic problems*. Paper presented at the SEG Technical Program. Expanded Abstracts, 2191–2195.

Huot, F., Lellouch, A., Given, P., Luo, B., Clapp, R. G., Nemeth, T., et al. (2022). Detection and characterization of microseismic events from fiber-optic DAS data using deep learning. *Seismological Research Letters*, *93*(5), 2543–2553. doi:10.1785/0220220037

Huot, F., Ma, Y., Cieplicki, R., Martin, E. R., & Biondi, B. L. (2017). *Automatic noise exploration in urban areas*. Paper presented at the SEG Technical Program. Expanded Abstracts, 5027–5032.

Huot, F., Martin, E. R., & Biondi, B. L. (2018b). *Automated ambient noise processing applied to fiber optic seismic acquisition (DAS)*. Paper presented at the SEG Technical Program. Expanded Abstracts, 4688–4692.

Hutt, C. R., Ringler, A. T., & Gee, L. S. (2017). Broadband seismic noise attenuation versus depth at the Albuquerque seismological laboratory. *Bulletin of the Seismological Society of America*, *107*(3), 1402–1412. doi:10.1785/0120160187

Isken, M. P., Vasyura-Bathke, H., Dahm, T., & Heimann, S. (2022). De-noising distributed acoustic sensing data using an adaptive frequency–wavenumber filter. *Geophysical Journal International*, *231*(2), 944–949. doi:10.1093/gji/ggac229

Jiang, T., Zhan, G., Hance, T., Sugianto, S., Soulas, S., & Kjos, E. (2016). *Valhall dual-well 3D DAS VSP field trial and imaging for active wells*. Paper presented at the 86th SEG Annual Meeting, Dallas, Texas, USA. Expanded Abstracts, 5582–5586. doi:10.1190/segam2016-13871754.1

Jousset, P., Currenti, G., Schwarz, B., Chalari, A., Tilmann, F., Reinsch, T., et al. (2022). Fibre optic distributed acoustic sensing of volcanic events. *Nature Communications*, *13*, 1753. doi:10.1038/s41467-022-29184-w

Jousset, P., Reinsch, T., Ryberg, T., et al. (2018). Dynamic strain determination using fibre-optic cables allows imaging of seismological and structural features. *Nature Communications*, *9*, 2509. doi:10.1038/s41467-018-04860-y

Karrenbach, M., Cole, S., Ridge, A., Boone, K., Kahn, D., Rich, J., et al. (2019). Fiber-optic distributed acoustic sensing of microseismicity, strain and temperature during hydraulic fracturing. *Geophysics*, *84*, D11–D23.

Karrenbach, M., Kahn, D., Cole, S., Ridge, A., Boone, K., Rich, J., et al. (2017). Hydraulic-fracturing-induced strain and microseismic using in situ distributed fiber-optic sensing. *The Leading Edge*, *36*, 837–844.

Keller, F., & Schneider, T. R. (1982). Geologie und Geotechnik. *Schweizer Ing. Archit*, *24*, 512–520.

Klaasen, S., Paitz, P., Lindner, N., Dettmer, J., & Fichtner, A. (2021). Distributed Acoustic Sensing in Volcano-Glacial Environments – Mount Meager, British Columbia. *Journal of Geophysical Research. Solid Earth*, *126*, e2021JB022358. doi:10.1029/2021JB022358

Lellouch, A., & Biondi, B. L. (2021). Seismic applications of downhole DAS. *Sensors*, *21*(9), 2897. doi:10.3390/s21092897

Lellouch, A., Horne, S., Meadows, M. A., Farris, S., Nemeth, T., & Biondi, B. L. (2019a). DAS observations and modeling of perforation induced guided waves in a shale reservoir. *The Leading Edge*, *38*, 858–864.

Lellouch, A., Lindsey, N. J., Ellsworth, W. L., & Biondi, B. L. (2020). Comparison between distributed acoustic sensing and geophones: downhole microseismic monitoring of the FORGE geothermal experiment. *Seismological Research Letters*, *91*, 3256–3268.

Lellouch, A., Schultz, R., Lindsey, N. J., Biondi, B. L., & Ellsworth, B. L. (2021). Low-magnitude seismicity with a downhole distributed acoustic sensing array-examples from the FORGE geothermal experiment. *Journal of Geophysical Research*, *126*, e2020JB020462. doi:10.1029/2020JB020462

Lellouch, A., Yuan, S., Spica, Z., Biondi, B., & Ellsworth, W. L. (2019b). Seismic velocity estimation using passive downhole distributed acoustic sensing records: examples from the San Andreas Fault Observatory at Depth. *Journal of Geophysical Research. Solid Earth*, *124*, 6931–6948. doi:10.1029/2019JB017533

Li, Z., Shen, Z., Yang, Y., Williams, E., Wang, X., & Zhan, Z. (2021). Rapid response to the 2019 Ridgecrest earthquake with distributed acoustic sensing. *AGU Advances*, *2*, e2021AV000395. doi:10.1029/ 2021AV000395

Li, Z., & Zhan, Z. (2018). Pushing the limit of earthquake detection with distributed acoustic sensing and template matching: a case study at the Brady geothermal field. *Geophysical Journal International*, *215*, 1583–1593. doi:10.1093/gji/ggy359

Lindsey, N. J., Martin, E. R., Dreger, D. S., Freifeld, B., Cole, S., James, S. R., et al. (2017). Fiber-optic network observations of earthquake wavefields. *Geophysical Research Letters*, *44*, 11–792.

Liu, Y., Huff, O., Luo, B., Jin, G., & Simmons, J. (2022). Convolutional neural network-based classification of microseismic events originating in a stimulated reservoir from distributed acoustic sensing data. *Geophysical Prospecting*, *70*(5), 904–920.

Luo, B., Lellouch, A., Jin, G., Biondi, B., & Simmons, J. (2021). Seismic inversion of shale reservoir properties using microseismic-induced guided waves recorded by distributed acoustic sensing (DAS). *Geophysics*, *86*(4), 1–58.

Ma, X., Hertrich, M., Amann, F., et al. (2022). Multi-disciplinary characterizations of the BedrettoLab – a new underground geoscience research facility. *Solid Earth*, *13*, 301–322. doi:10.5194/se-13-301-2022

Martin, E. R., Huot, F., Ma, Y., Cieplicki, R., Cole, S., Karrenbach, M., & Biondi, B. L. (2018). A seismic shift in scalable acquisition demands new processing: fiber-optic seismic signal retrieval in urban areas with unsupervised learning for coherent noise removal. *IEEE Signal Processing Magazine*, *35*, 31–40.

Martuganova, E., Stiller, M., Norden, B., Henninges, J., & Krawczyk, C. M. (2022). 3D deep geothermal reservoir imaging with wireline distributed acoustic sensing in two boreholes. *Solid Earth*, *13*, 1291–1307. doi:10.5194/se-13-1291-2022

Mellors, R. J., Sherman, C., Ryerson, R., Morris, J., Messerly, M., Yu, C., & Allen, G. (2018). *Modeling potential EGS signals from a distributed fiber optic sensor deployed in a borehole. Paper presented at the 43rd Workshop on Geothermal Reservoir Engineering*. CA: Palo Alto.

Moore, J., McLennan, J., Pankow, K., Simmons, S. Podgorney, R., Wannamaker, P., et al. (2020). *The Utah Frontier Observatory for Research in Geothermal Energy (FORGE): a laboratory for characterizing, creating and sustaining enhanced geothermal systems*. Paper presented at the 45th Workshop on Geothermal Reservoir Engineering, Stanford University.

Naldrett, G., Parker, T., Shatalin, S., & Mondanos, M. (2020). High-resolution Carina distributed acoustic fibre- optic sensor for permanent reservoir monitoring and extending the reach into subsea fields. *First Break*, *38*, 71–76.

Nayak, A., & Ajo-Franklin, J. (2021). The imperial valley dark fiber team; distributed acoustic sensing using dark fiber for array detection of regional earthquakes. *Seismological Research Letters*, *92*(4), 2441–2452. doi:10.1785/0220200416

Nishimura, T., Emoto, K., Nakahara, H., Miura, S., Yamamoto, M., Sugimura, S., et al. (2021). Source location of volcanic earthquakes and subsurface characterization using fiber-optic cable and distributed acoustic sensing system. *Scientific Reports*, *11*, 6319. doi:10.1038/s41598-021-85621-8

Pankow, K., Mesimeri, M., Mclennan, J., Wannamaker, P., & Moore, J. (2020). *Seismic monitoring at the Utah frontier observatory for research in geothermal energy*. Paper presented at the 45th Workshop on Geothermal Reservoir Engineering, Stanford University.

Piana Agostinetti, N., Villa, A., & Saccorotti, G. (2022). Distributed acoustic sensing as a tool for subsurface mapping and seismic event monitoring: a proof of concept. *Solid Earth*, *13*, 449–468. doi:10.5194/se-13-449-2022

Plenkers, K., Reinicke, A., Obermann, A., Gholizadeh Doonechaly, N., Krietsch, H., Fechner, T., et al. (2023). Multi-disciplinary monitoring networks for mesoscale underground experiments: advances in the Bedretto reservoir project. *Sensors*, *23*, 3315. doi:10.3390/s23063315

Porras, J., Pecci, D., Bocchini, G. M., De Solda, M., Tuinstra, K. B., Lanza, F., et al. (2024). A semblance-based microseismic event detector for DAS data. *Geophysical Journal International*, *236*(3), 1716–1727. doi:10.1093/gji/ggae016

Potter, S.A. (2017). *Characterizing background seismicity in the region surrounding Milford, Utah* (Masters thesis). Utah: University of Utah.

Reinsch, T., Henninges, J., Götz, J., Jousset, P., Bruhn, D., & Lüth, S. (2015). *Distributed acoustic sensing technology for seismic exploration in magmatic geothermal areas*. Paper presented at the World Geothermal Congress, Melbourne, Australia. Retrieved from https://gfzpublic.gfz-potsdam.de/rest/items/item_1182900_4/component/file_1182899/content

Schwarz, B. (2019). Coherent wavefield subtraction for diffraction separation. *Geophysics*, *84*, 157–168. doi:10.1190/geo2018-0368.1

Shoemake, K. (1985). *Animating rotation with quaternion curves*. Paper presented at the 12th Annual Conference on Computer Graphics and Interactive Techniques.

Shultz, W. & Simmons, J. (2019). *3D DAS VSP in unconventionals: a case study*. Paper presented at the 89th SEG Annual Meeting. Expanded Abstracts, 979–983. doi:10.1190/segam2019-3214518.1

Song, Z., Zeng, X., Chi, B., Bao, F., & Osotuyi, A. G. (2022). Using the three station interferometry method to improve urban DAS ambient noise tomography. *Frontiers in Earth Science*, *10*, 952410. doi:10.3389/feart.2022.952410

Spica, Z. J., Perton, M., Martin, E. R., Beroza, G. C., & Biondi, B. (2020). Urban Seismic Site Characterization by Fiber-Optic Seismology. *Journal of Geophysical Research. Solid Earth*, *125*, e2019JB018656. doi:10.1029/2019JB018656

Staněk, F., Jin, G., & Simmons, J. (2022). Fracture imaging using DAS-recorded microseismic events. *Frontiers in Earth Science*, *10*, 907749. doi:10.3389/feart.2022.907749

Stork, A. L., Baird, A. F., Horne, S. A., Naldrett, G., Lapins, S., Kendall, J.-M., et al. (2020). Application of machine learning to microseismic event detection in distributed acoustic sensing data. *Geophysics*, *85*, KS149–KS160.

Titov, A., Jin, G., Binder, G., & Tura, A. (2022). Distributed acoustic sensing time-lapse vertical seismic profiling during zipper-fracturing operations: observations, modeling, and interpretation. *Geophysics*, *87*(6), B329–B336.

Tuinstra, K., Grigoli, F., Lanza, F., Fichtner, A., & Wiemer, S. (2024). Locating clustered seismicity using distance geometry solvers: applications for sparse and single-borehole DAS networks. *Geophysical Journal International*, *238*(2), 661–680. doi:10.1093/gji/ggae168

van den Ende, M. P. A., & Ampuero, J.-P. (2021). Evaluating seismic beamforming capabilities of distributed acoustic sensing arrays. *Solid Earth*, *12*, 915–934. doi:10.5194/se-12-915-2021

van den Ende, M. P., Lior, I., Ampuero, J. -P., Sladen, A., Ferrari, A., & Richard, C. (2021). A self-supervised deep learning approach for blind denoising and waveform coherence enhancement in distributed acoustic sensing data. *IEEE Transactions on Neural Networks and Learning Systems*, *34* (7), 3371–3384. doi:10.1109/TNNLS.2021.3132832

Van Der Baan, M., Eaton, D., & Dusseault, M. (2013). *Microseismic monitoring developments in hydraulic fracture stimulation*. Paper presented at the ISRM International Conference for Effective and Sustainable Hydraulic Fracturing.

Wannamaker, P., Simmons, S., Miller, J., Hardwick, C., Erickson, B., Bowman, S., et al. (2020). *Geophysical activities over the Utah FORGE site at the outset of project phase 3*. Paper presented at the 45th Workshop on Geothermal Reservoir Engineering, Stanford University.

Webster, P., Molenaar, M., & Perkins, C. (2016). *DAS microseismic fiber-optic locating DAS microseismic events and errors* (dissertation). Stanford. Retrieved from https://search.proquest.com/openview/8d477d31956825c13eb091f6f14f2c35/1?pq-origsite=gscholar&cbl=18750&diss=y.

Wilson, G. A., Willis, M. E., & Ellmauthaler, A. (2021). Evaluating 3D and 4D DAS VSP image quality of subsea carbon storage. *The Leading Edge*, *40*, 261–266. doi:10.1190/tle40040261.1

Yuan, S., Lellouch, A., Clapp, R. G., & Biondi, B. (2020). Near-surface characterization using a roadside distributed acoustic sensing array. *The Leading Edge*, *39*, 646–653. doi:10.1190/tle39090646.1

Zandt, G., McPherson, L., Schaff, S., & Olsen, S. (1982). *Seismic baseline and induction studies: Roosevelt Hot Springs, Utah and Raft River Idaho* (DOE final technical report DOE/ID/01821-T1). Earth Science Laboratory/University of Utah Research Institute.

Zhan, G. & Nahm, J. (2020). *Multi-well 3D DAS VSPs: illumination and imaging beyond the wellbores.* Paper presented at the 90th SEG Annual Meeting, Online. Expanded Abstracts, 3798–3802. doi:10.1190/segam2020-3426032.1

Zhu, W., Biondi, E., Li, J., et al. (2023). Seismic arrival-time picking on distributed acoustic sensing data using semi-supervised learning. *Nature Communications*, *14*, 8192. doi:10.1038/s41467-023-43355-3

Part III

Borehole Distributed Acoustic Sensing Imaging and Inversion

17

Distributed Acoustic Sensing Vertical Seismic Profiling P-P and P-S Wave Imaging Method: A Fast Ray-Based Approach Through a Multitemplate Fast-Advancement Algorithm

Yuanzhong Chen[1,2], Jingjing Zong[3], Shujun Xia[4], Fei Wang[4], Hao Zhang[5], and Guangmin Hu[3]

ABSTRACT

This chapter explores a fast ray-based vertical seismic profiling (VSP) imaging method and presents its application in distributed acoustic sensing VSP (DAS VSP) P-wave and converted-wave imaging. We use a multitemplate fast-advancement algorithm (MSFM) to obtain the traveltime table for each shot–receiver pair, which is then summed for each shot–receiver pair before being sorted at each depth. Given the number of common depth points (CDPs), the traveltime table is sorted by traveltime. The field records are projected to the image domain via the sorted traveltime table. The projected wave field data is then stacked while the fold of projections at each position is calculated to obtain the average value of each position. The data and traveltime table are then converted one by one and stacked to form an imaging profile. By allowing a flexible migration aperture, this method combines the advantages of both the VSP CDP transform and migration. It focuses imaging near the reflection path, reduces migration artifacts, and addresses the problem of uneven coverage imaging amplitude by accounting for the fold of coverages during the imaging process. Using the grid ray-tracing imaging method, satisfactory DAS VSP P-wave and converted-wave images are obtained in both numerical and field data examples.

17.1. INTRODUCTION

Vertical seismic profiling (VSP) is advantageous in providing a high signal-to-noise ratio (SNR) and high-resolution seismic signal in comparison with surface seismic surveys. In addition, it provides wider illumination for steeply dipping structures through the borehole receiver arrangement. Hence, VSPs can help identify fine structures such as small faults and micro fractures near wells. In particular, it is a powerful technology in current seismic exploration to solve problems such as fine imaging of underground structures and fine description of oil and gas reservoirs. With the rapid development of a large array of downhole instruments, the current downhole receiver capacity has now reached 100–400 levels (Müller et al., 2010; Paulsson, 2006). In recent years, the development of distributed acoustic sensing (DAS) has allowed for denser borehole sampling and more convenient joint VSP and 3D surface seismic acquisition. Meanwhile, new challenges have appeared in denoising, processing, and interpretation with the rapid evolution of the DAS VSP technique.

The main processing steps of VSP generally include noise removal, shot point static correction, velocity

[1]*School of Information and Communication Engineering, University of Electronic Science and Technology of China, Chengdu, China*

[2]*Optical Science and Technology (Chengdu) Ltd., Chengdu, China*

[3]*School of Resources and Environment, University of Electronic Science and Technology of China, Chengdu, China*

[4]*BGP, Inc., China National Petroleum Co., Chengdu, China*

[5]*Institute of Geophysics, Shanghai Petroleum Corporation Ltd., Shanghai, China*

Distributed Acoustic Sensing in Borehole Geophysics, Geophysical Monograph 289, First Edition.
Edited by Yingping Li, Robert Mellors, and Ge Zhan.

DOI: 10.1002/9781394179275.ch17

analysis, wave field separation, and VSP imaging. VSP image processing is an evolving and challenging technology. The main problems include the following: (a) currently, the acoustic imaging condition remains dominant, which requires appropriate VSP wave field decomposition; (b) VSP surveys present highly uneven and asymmetric illumination in comparison to the surface seismic surveys; and (c) the signal from the underground reflection surface has wide azimuth ranges. The processing procedures for walkaway VSP and 3D VSP are more complicated and less standardized compared to zero-offset VSP or offset VSP, and many practices are still being explored (Kuzmiski et al., 2008).

Establishing an appropriate velocity model that satisfies VSP imaging is one of the main difficulties in VSP processing because VSP imaging is more sensitive to velocity errors (Blias & Hughes, 2015). Two types of velocity models are commonly applied, including the velocity model built from the surface seismic survey conducted in the same region of the VSP survey and the velocity model built from the VSP survey (via, e.g., traveltime tomography inversion). Currently, in addition to various migration methods (including ray-based Kirchhoff migration wave-equation migration) aimed at imaging steeply dipping structures and salt flanks abroad, another trend is the development from isotropic post-stack time imaging to anisotropic prestack depth imaging, especially in complex areas (Etgen et al., 2009; Huang et al., 2022). Anisotropic prestack depth imaging technology improves the focus of seismic energy and the positional accuracy of geological structure imaging and is adopted In this study (Huang and Greenhalgh, 2019).

DAS technology is an emerging technology that turns a fiber-optic cable into a densely distributed strain sensor array. DAS has the advantages of low cost, high density, and highly repeatable observations and has been widely used in the field of borehole seismic exploration (Chen et al., 2023). Zhang et al. (2020) studied the Q value estimation of DAS VSP. Horne et al. (2020) studied the anisotropic parameter analysis of DAS walkaway VSP. Mizuno and Ali (2021) studied generalized radon transform migration imaging of DAS VSP. Wilson et al. (2021) used time-shifted 3D DAS VSP imaging to evaluate the quality of carbon dioxide storage.

This chapter mainly explores and studies DAS VSP P-wave and converted-wave imaging methods. With the goal of addressing issues such as VSP asymmetric observation and uneven coverage, a VSP imaging method based on grid ray tracing is examined. This method combines the advantages of VSP common depth point (CDP) transform and migration, focusing imaging around the reflection path, reducing migration artifacts, and accounting for the coverage fold during the imaging process to solve the problem of uneven coverage in imaging amplitude. The accuracy of the method is shown in both numerical simulated data and a field survey. Using the grid ray-tracing imaging method, satisfactory DAS VSP P-wave and converted-wave images are obtained.

17.2. GRID RAY-TRACING METHOD

Based on the characteristics of VSP upgoing and downgoing waves, grid ray tracing is performed from both the source side and receiver side, respectively, to calculate traveltime tables. This chapter proposes a fast ray-tracing method that involves two processes. First, a multitemplate fast-advancing algorithm (MSFM) is used to calculate the wavefront propagation time of each grid node in a known velocity field from the source point. Then, the fastest descent method is used to trace the ray path from the receiving point to the source point along the fastest descent direction of the traveltime gradient. Compared to the traditional fast marching method (FMM) and its improved algorithms, the MSFM uses two templates to calculate adjacent point travel, taking into account information in the horizontal, vertical, and diagonal directions. This greatly improves the computational accuracy and efficiency.

MSFM is an improved algorithm based on FMM (Sethian, 1996) and the high precision fast propulsion method (HAFMM) (Sethian, 1999). Both the FMM and HAFMM algorithms ignore the information provided by nodes on the diagonal, resulting in substantial numerical errors in the diagonal direction. The MSFM algorithm combines multiple templates and directional derivatives on the basis of FMM and HAFMM algorithms, thereby improving the accuracy of solving functional equations.

Figure 17.1 shows two templates used in the MSFM algorithm, where template 1 covers the nearest neighbor point and template 2 covers the neighbor point in the diagonal direction. Let $r_1 = [r_{11}, r_{12}]^T$ and $r_2 = [r_{21}, r_{22}]^T$ represent unit vectors along the $\overrightarrow{p_2p_1}$ and $\overrightarrow{q_2q_1}$ directions, respectively, and U_1 and U_2 represent directional derivatives along r_1 and r_2, respectively. Then U_1 and U_2 can be expressed as

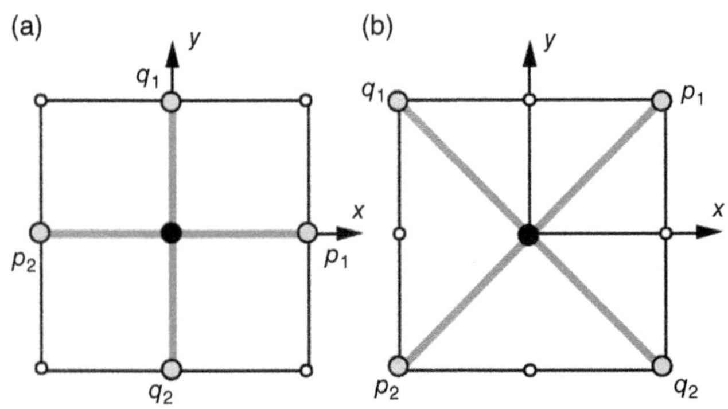

Figure 17.1 Two templates used in the MSFM algorithm (a) template 1; (b) template 2.

$$U_1 = \boldsymbol{r}_1 \nabla T(x,y) = r_{11} T_x + r_{12} T_y \quad (17.1)$$

$$U_2 = \boldsymbol{r}_2 \nabla T(x,y) = r_{21} T_x + r_{22} T_y \quad (17.2)$$

where $T_x = \frac{\partial T}{\partial x}$ and $T_y = \frac{\partial T}{\partial y}$ represent the partial derivatives of T with respect to x and y, respectively. Equations (17.1) and (17.2) can be written in the form of matrix multiplication:

$$\begin{bmatrix} U_1 \\ U_2 \end{bmatrix} = \begin{bmatrix} r_{11} & r_{12} \\ r_{21} & r_{22} \end{bmatrix} \begin{bmatrix} T_x \\ T_y \end{bmatrix} \quad (17.3)$$

Namely,

$$\boldsymbol{U} = \boldsymbol{R} \nabla T(x,y) \quad (17.4)$$

Owing to $|\nabla T(x,y)|^2 = [\nabla T(x,y)]^{\boldsymbol{T}} \nabla T(x,y)$, equation (17.4) can be transformed into

$$|\nabla T(x,y)|^2 = \boldsymbol{U}^T \left(\boldsymbol{R}\boldsymbol{R}^T\right)^{-1} \boldsymbol{U} = S^2(x,y) \quad (17.5)$$

Make $\Delta x = \Delta y = h$; $(\boldsymbol{R}\boldsymbol{R}^T)^{-1} = \boldsymbol{I}$, substituted into equation (17.5), is

$$U_1^2 + U_2^2 = S^2(x,y) \quad (17.6)$$

To improve accuracy, MSFM approximates U_1 and U_2 using a second-order difference. The second-order difference approximation of the directional derivatives U_1 and U_2 is

$$U_v = max\left[\frac{3}{2\Delta d_v}\left(T_{i,j} - T_v\right), 0\right] \qquad v = 1, 2 \quad (17.7)$$

where $\Delta d_v \Delta d_v$ represents the distance between the corresponding node and the current node when T_v is the minimum value. For template 1, substitute equation (17.7) into equation (17.6), with

$$max\left[\frac{3}{2h}\left(T_{i,j} - T_1\right), 0\right]^2 + max\left[\frac{3}{2h}\left(T_{i,j} - T_2\right), 0\right]^2 = S_{i,j}^2 \quad (17.8)$$

Similarly, for template 2, equation (17.7) is substituted into equation (17.6), with

$$max\left[\frac{3}{2\sqrt{2}h}\left(T_{i,j} - T_1\right), 0\right]^2 + max\left[\frac{3}{2\sqrt{2}h}\left(T_{i,j} - T_2\right), 0\right]^2 = S_{i,j}^2 \quad (17.9)$$

The two quadratic equations shown in equations (17.8) and (17.9) produce two solutions. Because the value of $T(x, y)$ is unknown, it is necessary to check whether the calculated $T(x, y)$ is larger than the neighboring T_1 and T_2. Only solutions larger than T_1 and T_2 are acceptable. A good method is to use the following upwind conditions before solving the eikonal equation:

$$|T_1 - T_2| < f(\Delta x, \Delta y) S(x,y) \quad (17.10)$$

For template 1 and template 2, $f(\Delta x, \Delta y)$ is $2\,min\,(\Delta x, \Delta y)$ and $2\sqrt{\Delta^2 x + \Delta^2 y}$, respectively.

After deriving the local traveltime algorithm for MSFM, it is necessary to calculate the wavefront propagation time from the source point to each grid node: that is, the wavefront expansion process. Similar to the FMM and HAFMM algorithms, the MSFM algorithm needs to select a node with the minimum wavefront propagation time from the narrow band near the current wavefront during each wavefront advance and then use this point as a secondary source point to expand the wavefront outward. As shown in Figure 17.2, the MSFM algorithm divides all grid nodes into three types of points during the wavefront expansion:

1. Near point (M = 2): Indicates a node where the wavefront propagation time has been calculated and is no longer updated
2. Narrow band point (M = 1): Indicates a node where the wavefront propagation time has been calculated but may be updated
3. Far point (M = 0): Indicates a node whose wavefront propagation time has not been calculated

The process of wavefront expansion of the MSFM algorithm can be summarized as follows

1. Mark the source point as a near point, mark all other nodes as far points, and assign a sufficiently large initial wavefront propagation time. Use equations (8)

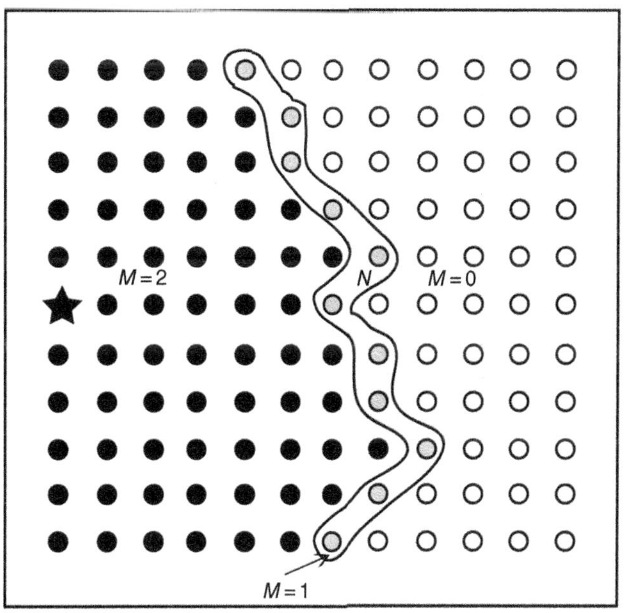

Figure 17.2 Schematic diagram of MSFM two-dimensional wavefront expansion, where M = 2 represents the proximal point, M = 1 represents the narrowband point, M = 0 represents the distal point, and N represents the wavefront narrowband.

and (9) to calculate the wavefront propagation time of adjacent points at the source point, and select the solution that meets the upwind condition in equation (10), marking it as a narrowband point.

2. Find the node with the minimum wavefront propagation time in the narrow band, and mark it as a near point.
3. Find the points marked as narrowband points or distant points in the adjacent points of this node.
4. Update the wavefront propagation time of these points using equations (8) and (9), and upwind condition equation (10) (if the updated value is smaller than the previous value).
5. Repeat steps 2–4 until there is no space in the narrow band.

Figure 17.3 shows the calculation of VSP traveltime using the MSFM. The true velocity model is displayed in Figure 17.3(a). Figure 17.3(b) and 17.3(c) show the traveltime tables calculated from the source side and receiver side, respectively. A summation of these two timetables is shown in Figure 17.3(d), which is used for data-trace projection. For instance, the trace recorded at the example shot (red star) and receiver (purple block) block can be mapped by the sorted timetable marked in the purple polygon. The wider window around the minimum traveltime corresponds to a larger migration aperture.

17.3. VSP IMAGING BASED ON GRID RAY TRACING

The process of the proposed VSP imaging method based on grid ray tracing is demonstrated in Figure 17.4. An MSFM is used to calculate the travel timetable for each shot and receiver pair. Further, the two travel timetables are summed and sorted by depth. Given the number of grids (migration aperture), the wave field data is projected at the corresponding position according to the traveltime

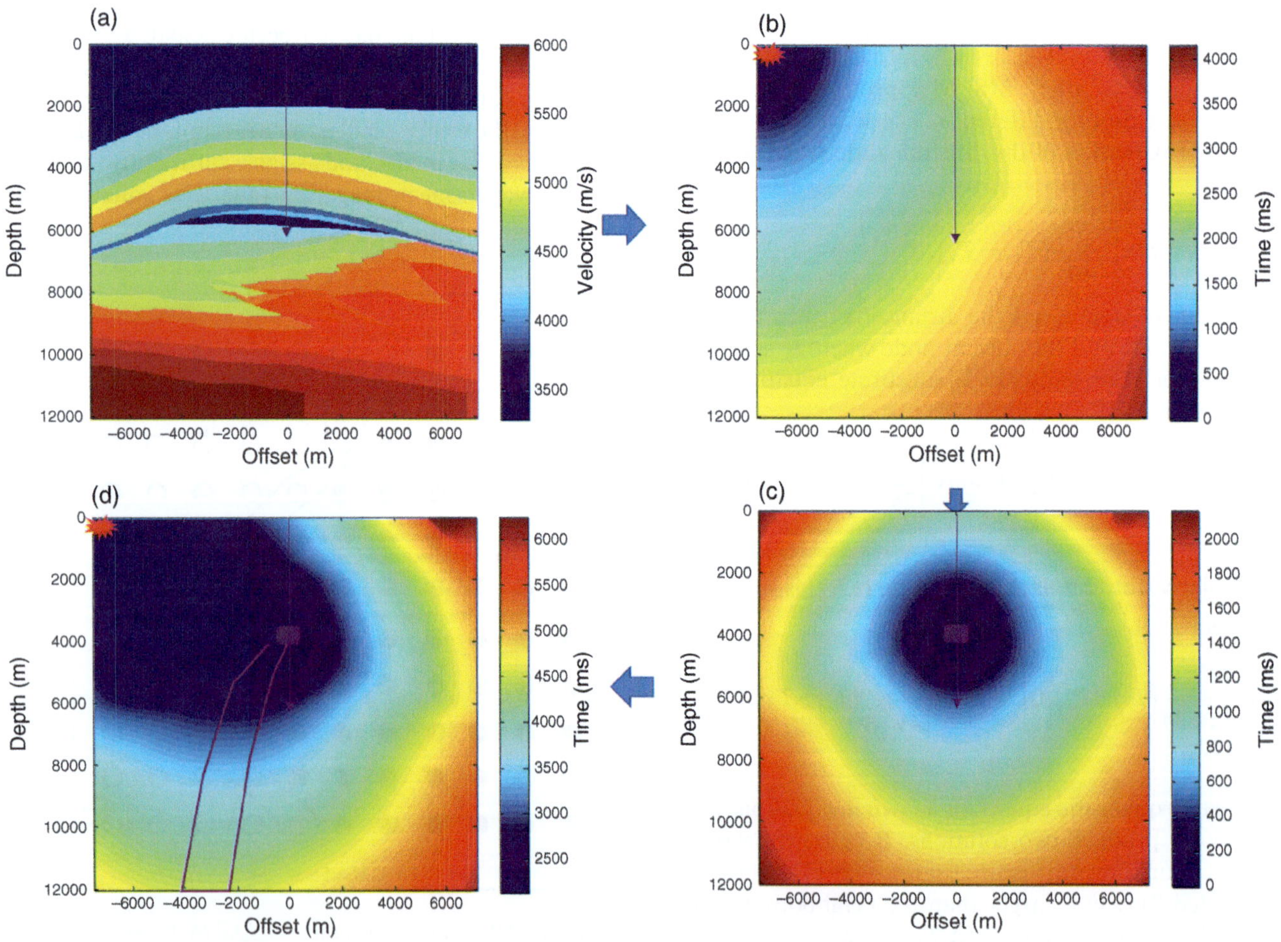

Figure 17.3 The process of using MSFM for VSP traveltime calculation. (a) True velocity model; (b) traveltime tables on the source side; (c) traveltime tables on the receiver side; (d) summation of the traveltime tables in (b) and (c). The purple polygon marks the areas with smaller traveltimes sorted below the depth of the chosen receiver point; the number of imaging grids determines the width of purple polygons.

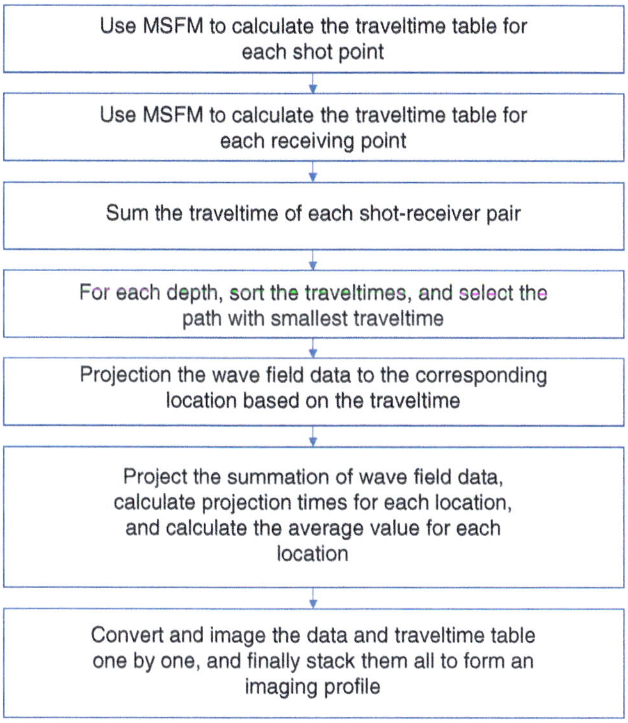

Figure 17.4 VSP imaging process flow based on grid ray tracing.

and stacked. We repeat the projection for each location, followed by calculating the coverage fold to average the seismic amplitude anomaly due to uneven coverage. Finally, we stack them all to form an imaging profile. This process is applicable for both P-wave and converted-wave imaging. This method combines the advantages of CDP conversion and migration, focusing imaging near the reflection path, reducing migration artifacts, and calculating the number of coverage times during the imaging process, solving the problem of abnormal imaging amplitude due to uneven coverage.

Figure 17.5 shows an imaging profile obtained from shot pair data using grid ray-tracing imaging. Figure 17.5(a) shows shot pair VSP reflected wave data; Figure 17.5(b) shows the traveltime table corresponding to the 200 grids with the shortest traveltimes (each depth, traveltime sorted from smallest to largest, with the top 200 points in time value); Figure 17.5(c) shows an imaging section.

17.4. NUMERICAL DATA TESTING

For the velocity model shown in Figure 17.3(a), we generate numerical walkaway VSP data and perform grid ray-tracing imaging and one-way wave imaging comparisons. Figure 17.6(a) shows the reflected wave data of a walkaway VSP common receiver point. Figure 17.6(b) shows a one-way wave migration imaging section. Figure 17.6(c) shows a grid ray-tracing imaging profile. The comparison between the two imaging methods shows that the major imaged structures obtained by the two methods are consistent, with a higher SNR and better profile quality than the one-way wave migration results. The one-way wave migration wellhead energy is strong, and the energy away from the well reflection axis becomes weak. Due to the consideration of coverage times in grid ray-tracing imaging, the lateral image quality is more uniform, and the imaging results are of higher fidelity.

Figure 17.7 shows the migration imaging section after stacking the walkaway VSP velocity model and all depth data, and Figure 17.7(a) shows the velocity model.

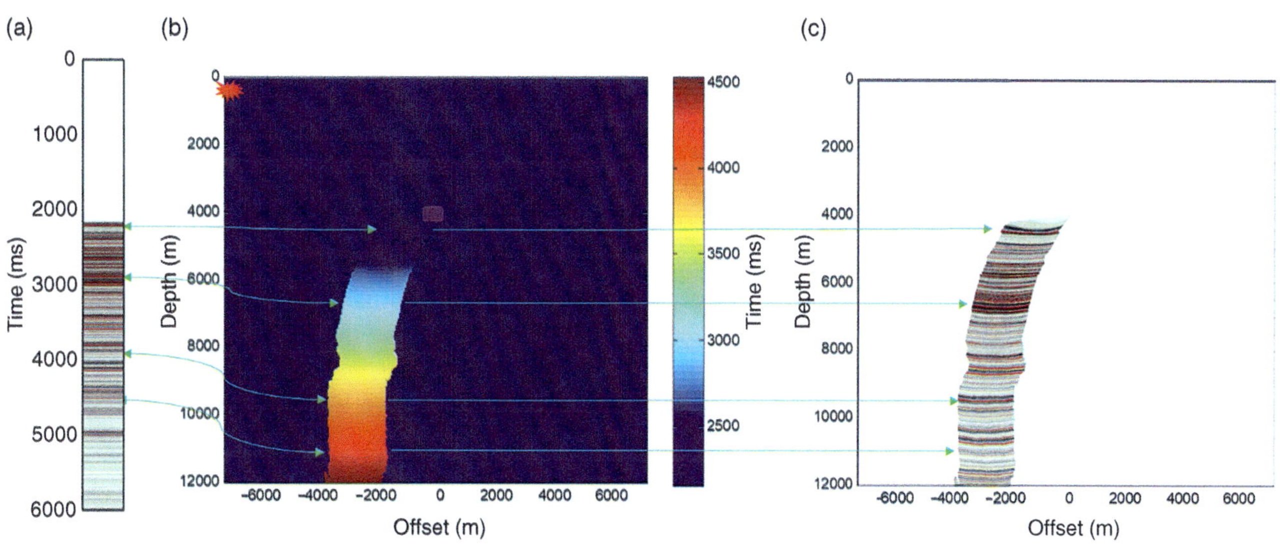

Figure 17.5 VSP imaging process based on grid ray tracing. (a) VSP reflected wave data; (b) traveltime table corresponding to 200 grids with the smallest traveltime; (c) imaging profile.

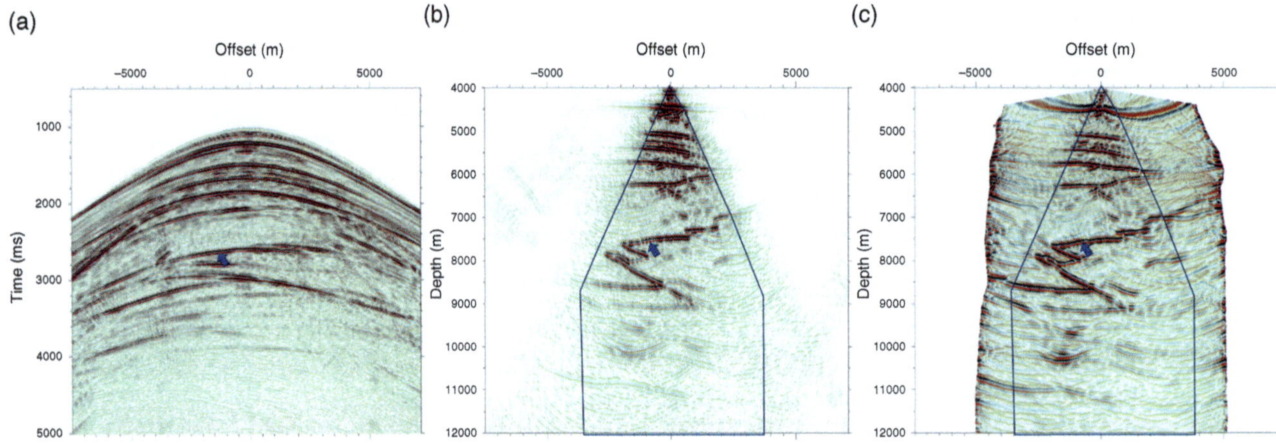

Figure 17.6 Images of the numerical walkway VSP survey. (a) Simulated data sorted and displayed by common receiver point; (b) one-way wave-equation migration imaging result; (c) the grid ray-tracing imaging profile.

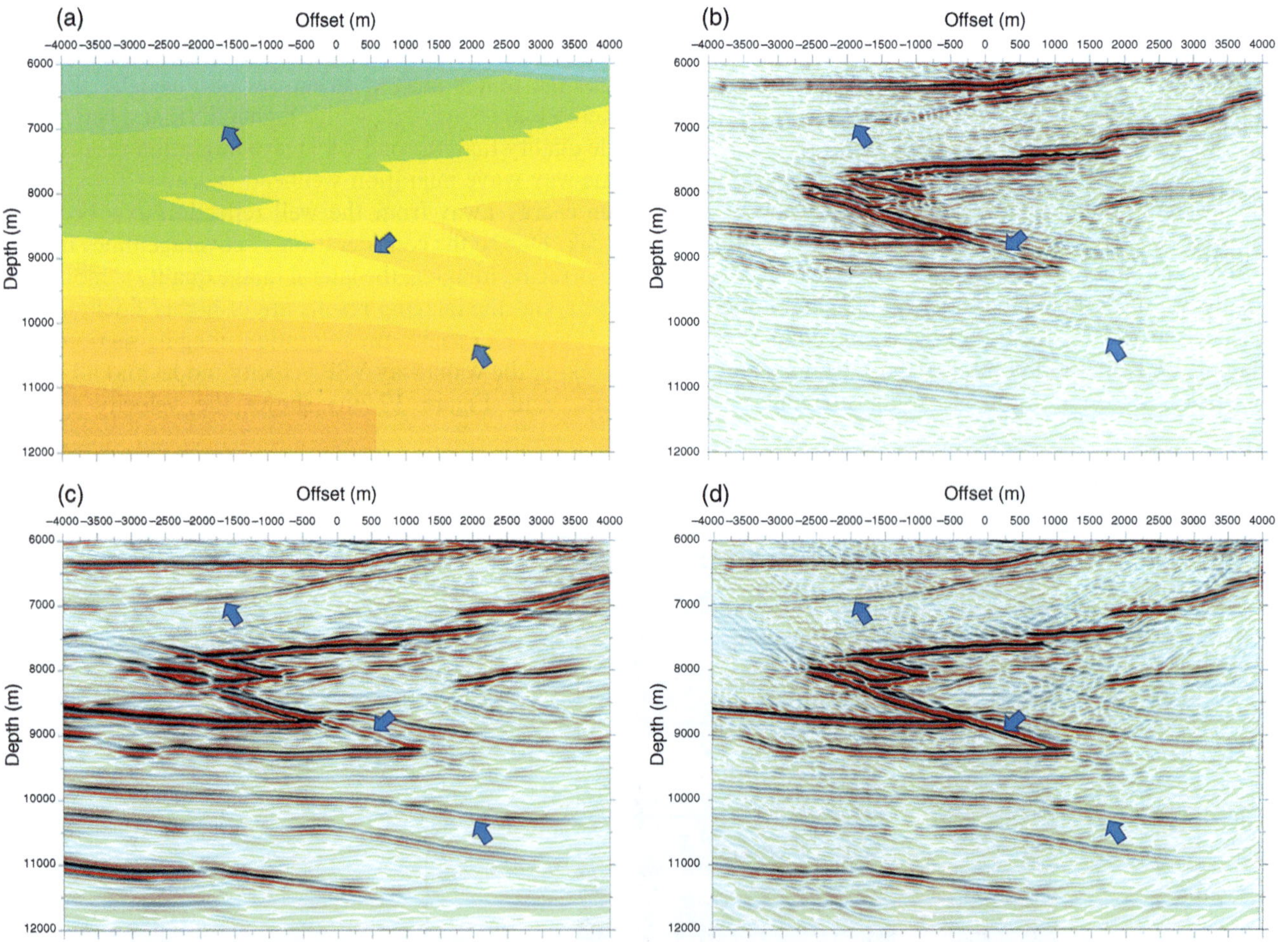

Figure 17.7 (a) True velocity model; (b) image profile obtained by the one-way wave migration; (c) image profile obtained by the proposed grid ray-tracing imaging method with 200 grids applied; (d) image profile obtained by the proposed grid ray-tracing imaging method with all grids applied. A large number of grids provides better imaging for complex structures, and a small number of grids provides better imaging for simple structures.

Figure 17.7(b) shows one-way wave migration imaging. Figure 17.7(c) shows grid ray tracing imaging, with 200 grids selected for each depth during imaging; Figure 17.7(d) shows full grid ray-tracing imaging. It can be seen from the figure that when the underground structure is simple, the imaging profile obtained by small grid imaging has fewer arc offsets, and the imaging SNR is higher. Large grid imaging can capture underground structural features under complex structural conditions.

17.5. FIELD DATA PROCESSING

The data of this study are located in the Pinghu Oil and Gas Field in the East China Sea, which is excited by an air gun source and received by DAS. The converted-wave imaging process extracts a shot line in the direction of the well trajectory for testing, as shown in Figure 17.8. The maximum offset is 4,190 m. The shot point distance is 50 m. The total number of shots is 148. The measured depth of the optical cable is 1,357 m. The maximum offset is –1,533 m, and the DAS receiver channel distance is 2 m. Figure 17.8 shows the upgoing converted-wave ray and polarization direction. From the figure, it can be seen that the polarization of the upgoing converted wave in the direction of the well trajectory is basically perpendicular to the optical cable, and the well trajectory is in the opposite direction. The upgoing converted wave is basically parallel to the optical cable. The polarization direction of the converted S wave at #1 is vertical to the optical cable, so the converted S wave is not developed (Figure 17.9, #1); the polarization direction of the converted S wave at #4 is parallel to the optical cable, so the converted S wave is developed (Figure 17.9, #4). Figure 17.9 shows the near-offset DAS VSP single shot (shot #3 in Figure 17.8) recording and the forward traveltime curves of the upgoing P wave and converted wave. Figure 17.9(b) shows the traveltime curves of the upgoing P wave for 1,000 m, 1,500 m, 2,000 m, 2,500 m, 3,000 m, and 3,500 m. Figure 17.9(c) shows the traveling time curve of the forward converted wave. From the figure, it can be seen that the energy of the upgoing P wave of a near-offset single shot is developed, and some depth-converted shear wave energy is stronger than that of the P wave.

Figure 17.10 shows the single shot records of DAS VSP with different well source distances, corresponding to the five shot locations shown in Figure 17.8. From the analysis of the reflected wave energy on the map, it can be seen that the reflected P waves above #1 are mainly reflected, the upgoing P waves and upward converted shear waves at #2 are both developed, and the converted shear waves above #3, #4, and #5 are mainly reflected. Figure 17.11 shows the separated upgoing P-wave record, and Figure 17.12 shows the separated upgoing converted S-wave record. The wave field separation method uses median filtering and frequency-wavenumber (FK) filtering, where FK filtering is performed on the upgoing P-wave normal moveout (NMO) record, and the separation result has relatively high fidelity.

Figure 17.13(a), (b) shows the DAS walkaway VSP P- and converted-wave imaging profiles obtained by the grid ray-tracing imaging method, respectively. The imaging results indicate that the upgoing converted-wave imaging range is smaller than that of the upgoing P wave. Both

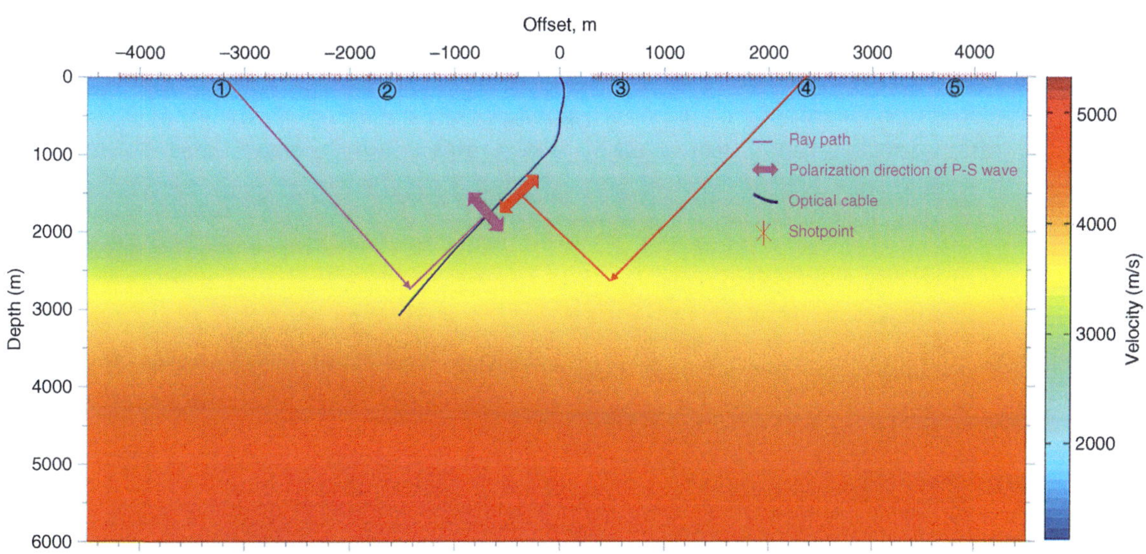

Figure 17.8 DAS walkaway VSP observation system.

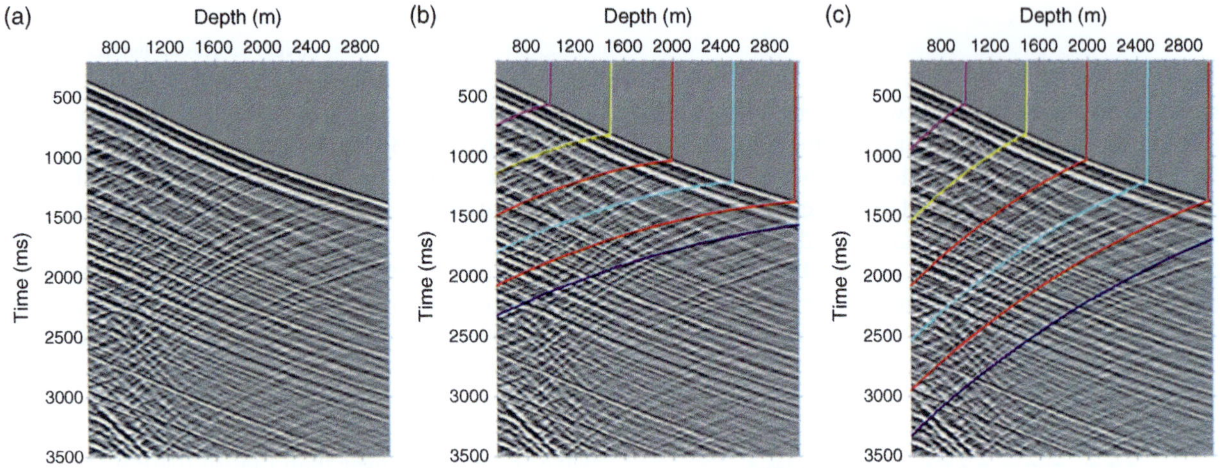

Figure 17.9 DAS VSP near-offset single shot recording. (a) Full wave field recording; (b) upgoing P-wave traveltime at different depths; (c) upgoing converted shear wave traveltime at different depths. Six lines from left to right represent the traveltime of P-P and P-S waves at depths of 1,000 m, 1,500 m, 2,000 m, 2,500 m, 3,000 m, and 3,500 m.

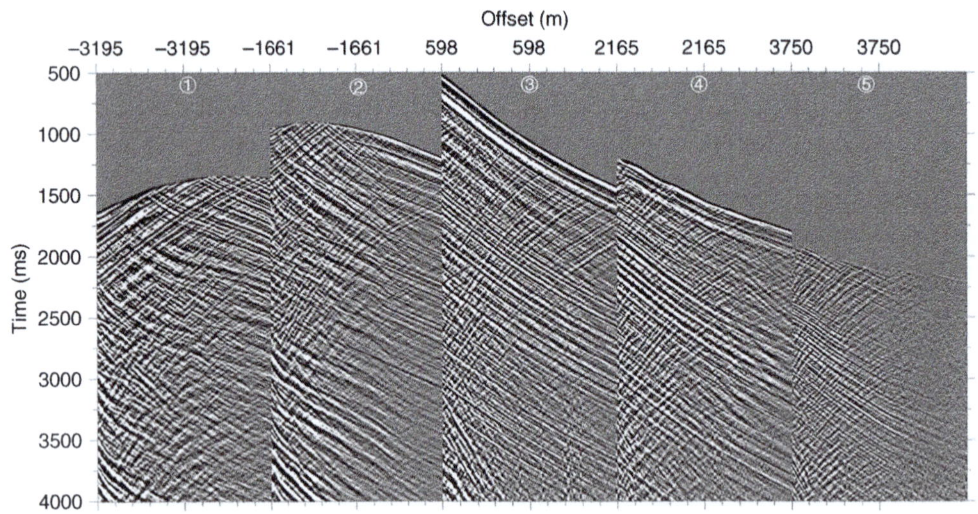

Figure 17.10 DAS VSP single shot records with different offsets.

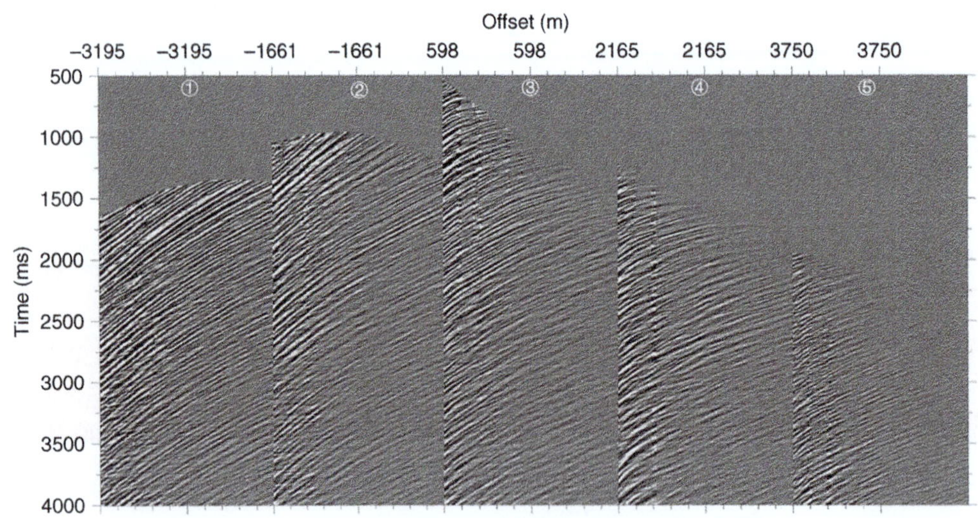

Figure 17.11 Upgoing P-wave record obtained after wave field separation.

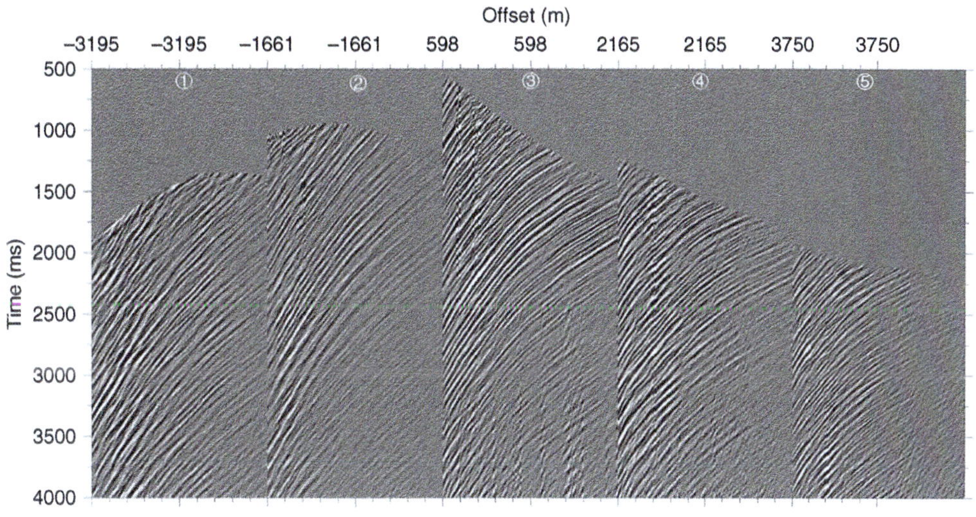

Figure 17.12 Upgoing converted shear wave record obtained after wave field separation.

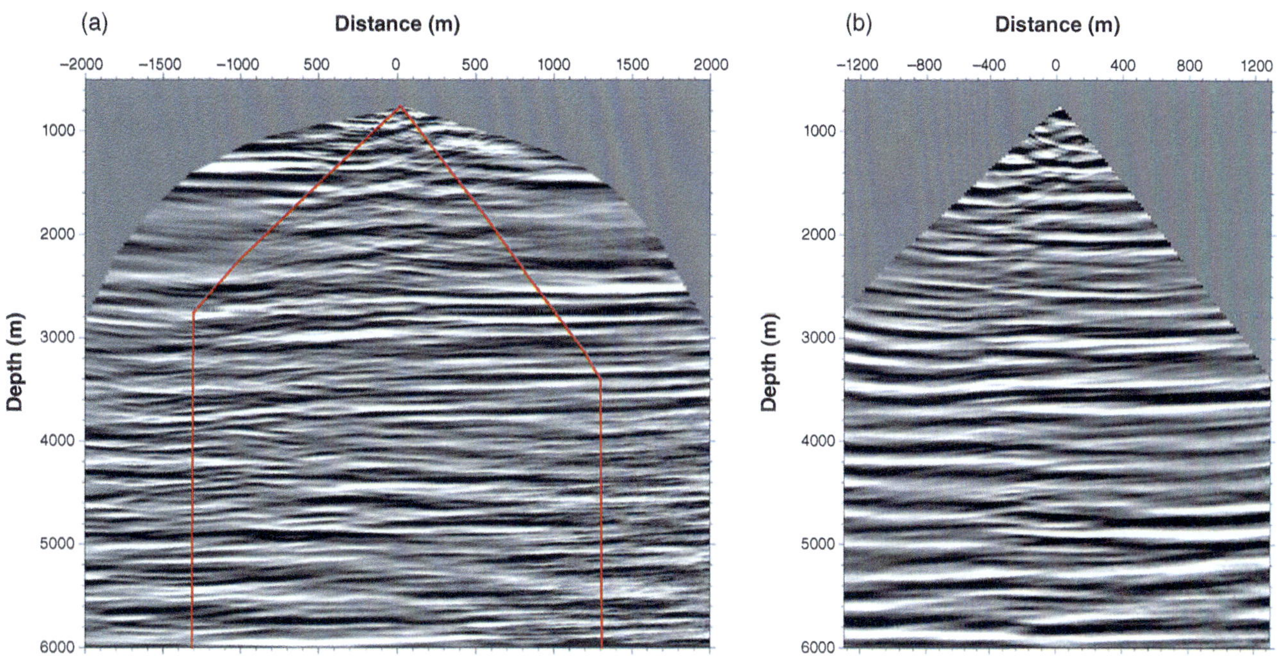

Figure 17.13 DAS walkaway VSP imaging profile obtained based on grid ray tracing imaging method. (a) Upgoing P-wave imaging; (b) upgoing converted-wave imaging.

images have considerable correspondence in the dominant reflectors. Moreover, deeper in the section, the SNR of converted-wave imaging is higher than the P-wave results, which proves that DAS VSP performs well in converted-wave imaging. The current results demonstrate the capability of the proposed grid-based ray-tracing imaging method to image both P waves and converted waves.

17.6. CONCLUSION

This chapter studies a VSP imaging method based on grid ray tracing, which offers several advantages, including the following:

1. More focused imaging around the reflectors significantly reduces migration artifacts, which are common under a VSP configuration.

2. By calculating the fold coverage during the imaging process, the proposed method allows for a straightforward solution to solve the problem of abnormal imaging amplitude due to the uneven coverage.
3. The flexibility in choosing the imaging aperture is suitable for imaging structures with variable complexity.
4. The calculation efficiency of the proposed method is higher than that of wave-equation migration.

Herein, the proposed VSP imaging method based on grid ray tracing combines the advantages of VSP CDP transform and migration, which can achieve VSP imaging profiles under complex structural conditions in a cost-efficiency mode, as demonstrated in the numerical example. The field data example further shows the effectiveness of the proposed method in imaging both P and S wavefields.

ACKNOWLEDGMENTS

The authors wish to thank Shanghai Petroleum and Natural Gas Co., Ltd. for the research project funding and permission to publish the data, as well as BGP management for the support and guidance during the data acquisition and processing phases of this project.

REFERENCES

Blias, E. & Hughes, B.2015). *3D VSP imaging: some general problems.* Paper presented at the 85th Annual International Meeting. Expanded Abstracts, 5630–5634.

Chen, Y., Zong, J., Liu, C., Cao, Z., Duan, P., Li, J., & Hu, G. (2023). Offshore subsurface characterization enabled by fiber-optic distributed acoustic sensing (DAS): an East China Sea 3D VSP survey example. *Frontiers in Earth Science, 11*, 1033456.

Etgen, J., Gray, S. H., & Zhang, Y. (2009). An overview of depth imaging in exploration geophysics. *Geophysics, 74*(6), WCA5–WCA17.

Huang, X., & Greenhalgh, S. (2019). Traveltime approximation for strongly anisotropic media using the homotopy analysis method. *Geophysical Journal International, 216*(3), 1648–1664.

Huang, X., Greenhalgh, S., Han, L., & Liu, X. (2022). Generalized effective Biot theory and seismic wave propagation in anisotropic, poroviscoelastic media. Journal of Geophysical Research: Solid. *Earth, 127*(3), e2021JB023590.

Horne, S., Armstrong, P., Soulas, S., et al. (2020). *Deepwater DAS walkaway VSP analysis.* Paper presented at the SEG Technical Program. Expanded Abstracts, 495–498.

Kuzmiski, R., Charters, R., & Galbraith, M. (2008). Processing considerations for 3D VSP. *First Break, 26*(7). doi:10.3997/1365-2397.26.1289.28424

Mizuno, T. & Ali, S. (2021). *Benchmarking DAS VSP migration methods with geophone VSP migration: Case study for the generalized Radon transform migration.* Paper presented at the SEG Technical Program. Expanded Abstracts, 422–425.

Müller, K. W., Soroka, W. L., Paulsson, B. N., Marmash, S., Al Baloushi, M., & Al Jeelani, O. (2010). 3D VSP technology now a standard high-resolution reservoir-imaging technique: part 1, acquisition and processing. *The Leading Edge, 29*(6), 686–697.

Paulsson, B. N. (2006). *Development of a 400 level 3C clamped downhole seismic receiver array for 3D borehole seismic imaging of gas reservoirs.* Paulsson Geophysical Services Inc.

Sethian, J. A. (1996). A fast marching level set method for monotonically advancing fronts. *Proceedings of the National Academy of Sciences of the United States of America, 93*(4), 1591–1595.

Sethian, J. A. (1999). *Level sets methods and fast marching methods* (2nd ed.). Cambridge University Press.

Wilson, G. A., Willis, M. E., & Ellmauthaler, A. (2021). Evaluating 3D and 4D DAS VSP image quality of subsea carbon storage. *The Leading Edge, 40*, 261–265.

Zhang, M., Zhao, X., Glenn A., et al. (2020). *Q estimation from geophone and DAS VSPs.* Paper presented at the SEG Technical Program. Expanded Abstracts, 475–478.

18

Distributed Acoustic Sensing Vertical Seismic Profiling Imaging Workflow: Transitioning from the Geophone Era

Takashi Mizuno[1], Shujaat Ali[2], Pierre Bettinelli[3], and Joel Le Calvez[1]

ABSTRACT

This chapter covers two topics regarding DAS imaging problems: survey design for DAS VSP and computation of subsurface images to obtain results comparable to geophones. Automation of survey design is useful for a DAS array to help explore the wide range of possibilities efficiently and quantify the expected data quality. We define the survey design as an inversion problem, and the number of shots and their distributions are estimated automatically in the method. Regarding image computation, we discuss case studies of the necessity for corrections to DAS data to obtain comparable images with geophones. We confirm that correcting the DAS response yields a result similar to the three-component (3C) geophone migration result, especially in deviated wells. Furthermore, dense shot spacing is a more critical factor for imaging than geophone data. Migration artifacts are observed in DAS images for a sparse shot configuration, whereas 3C geophone migration performs robustly because 3C data cancel artifacts from undesired wave modes in the data.

18.1. INTRODUCTION

Distributed acoustic sensing (DAS) has wide applications in seismology, including exploration seismology. Due to the fewer logistical constraints raised by DAS deployment compared to geophone array deployment, and also recent advances for robust acquisition setups, vertical seismic profiling (VSP) is one of the areas where the application of DAS is emerging. A fundamental question concerns the best use of DAS. We have long-established workflows for processing geophone-acquired data, but they require updating when applied to DAS-acquired data. Table 18.1 summarizes DAS-based and geophone-based acquisition. A DAS system has a long aperture that benefits from the distributed sensing extending from the surface to the bottom of the monitoring well. In contrast, a geophone array has a smaller aperture because of the far fewer receivers in the system. The polarization of the seismic wavefield cannot be determined from DAS data because the measurement is limited to a single, axial component, and it can easily be measured by three-component (3C) geophone data. A DAS system measures strain, which is the spatial derivative of the displacement wavefield along the borehole, whereas the geophone measurement is a point measurement of particle velocity. Excluding the fiber-specific noise source, the signal-to-noise ratio (SNR) of a DAS system tends to be lower than that of a geophone array. Those differences should be considered in survey design and processing. In this chapter, we review the changes in imaging workflow from a geophone array to a DAS system.

[1]*SLB, Sugar Land, Texas, USA*
[2]*SLB, Houston, Texas, USA*
[3]*SLB, Kuala Lumpur, Malaysia*

Distributed Acoustic Sensing in Borehole Geophysics, Geophysical Monograph 289, First Edition.
Edited by Yingping Li, Robert Mellors, and Ge Zhan.

DOI: 10.1002/9781394179275.ch18

Table 18.1 Comparison of specification of geophone and DAS arrays.

	Geophone array	DAS
Aperture	1,170 m (40 levels with 30 m spacing) 2,000 m (100 levels with 20 m spacing)	From wellhead to total depth
Number of components	3	1
Measurement	Particle velocity/acceleration at the receiver	Optical phase difference between two measurement points separated by gauge length (L_G)
Frequency/wavenumber response	Flat response over given frequency band	Gauge length-related wavenumber (frequency and apparent phase velocity) response
Source of noise	Environment noise Electric noise	Environment noise, standing waves (ringing noise), optical noise

The objective of survey design is to determine the best source and receiver configurations to illuminate the subsurface target for specific objectives. In the conventional geophone-acquired dataset workflow, survey design is completed by experienced geophysicists relying on a qualitative analysis of attributes from ray tracing using several different source and receiver geometries. In the case of a DAS-acquired dataset, the workflow stays the same, but the number of receivers increases dramatically by two to three orders of magnitude compared to the case of a geophone-acquired dataset. Also, the survey design requires additional considerations: a DAS array is a single-component (1C) array, and a geophone array is most often a 3C array. Furthermore, the strain seismogram is sensitive to the apparent velocity local to the receiver. This is the so-called cosine square effect for P waves: one cosine comes from the projection of 3C to 1C, and another is apparent slowness projected to the axis of the fiber-optic cable. Thus even if a ray hits the image area, it does not mean the data quality is sufficient. In addition to typical considerations to qualify an illumination point for a geophone VSP survey, such as incidence angle at the image point, the DAS-specific landing angle relative to the fiber must also be considered. Forward modeling should carry on considering such DAS response to evaluate data quality.

Considering the increase in number of receivers, and further complicated response of a DAS system, automation of the survey design process is important when working with a DAS system. We prototyped an algorithm optimizing a DAS VSP survey for subsurface imaging. The algorithm enables the qualification of the survey design for a DAS-based survey in a semiautomated approach, making the entire process less human-dependent than previous approaches.

When it comes to processing, we need to revise the computation to correct for the nature of DAS-acquired data. The nature of DAS-acquired data is important to consider in several areas of seismology, including the migration problem. The P-wave amplitudes change more in a DAS dataset than in a 1C geophone dataset for a mid-offset source due to the cosine versus cosine squared factor (e.g., Mateeva et al., 2012; Willis et al., 2016). In this chapter, we discuss the DAS VSP migration problem using generalized Radon transform (GRT) migration. We use simulated DAS data and geophone data to benchmark DAS migration results with geophone results. We briefly review the application of the method to small-scale real data.

18.2. SURVEY DESIGN FOR DAS VSP MIGRATION

Although the theory of migration assumes several wave propagation modes, including reflection, diffraction, and scattering, we often only assume reflection for survey design of imaging problems. One of the main purposes of the survey design is to obtain source and receiver positions through illumination study: that is, to check if the reflection ray hits the target image area within the desired range of illumination attributes. Traditionally, an experienced geophysicist performs ray tracing with various source–receiver geometries for an imaging target (horizon) and then analyzes the attributes qualitatively to identify the source and receiver geometry used for the survey.

Although such a human-intensive approach can be applied to DAS, automation is more desirable for DAS compared to the geophone case. In the case of DAS VSP, we assume the entire borehole is the sensor. Therefore, the number of receivers increases significantly compared to a geophone-acquired dataset. Also, the DAS sensor has only a single component (parallel to the axis of the fiber array: so, in most cases, the axis of the well), and recorded amplitude is a function of the incident angle to the borehole. Therefore, data quality is considered in addition to the illumination problem in the case of DAS. Due to the complexity associated with the shape of the borehole, and the long aperture, the problem is best solved by computation rather than an interactive human-driven process.

For automation, we need to formulate the survey design as an optimization problem that can be solved by computer. Liner et al. (1998) implement the objective function based on the common rules of seismic survey design (bin size, number of folds, etc.). They optimize the objective function to solve those parameters using local and global searches. Alvarez et al. (2004) and Monsegny (2017) formulate the problem considering the 3D velocity model and apply different types of optimizations that are appropriate for the problem where multiple minima exist. We use different types of considerations and an optimization technique, and quantify the data quality.

Figure 18.1 shows the algorithm used in the present study. First, we obtain the shot position for each receiver–image point combination. To estimate shot points, reverse ray tracing is employed: we compute the ray from the receiver to every image point and then compute the corresponding reflection ray to estimate the source position on the surface. We call those shot positions "ideal shot positions" because they should exactly hit the target image points. The total number of ideal shots becomes $N_{is} = N_r \times N_i$, where N_r and N_i are the number of receivers and image points, respectively. In the case of DAS-acquired data, N_r becomes several hundred because of the long aperture and dense sampling along the fiber-optic cable. In most cases, the first step leads to oversampled shot clusters, as many receiver–image point combinations yield nearly identical shot locations. It is obvious that this number of shots is not necessary because the total number of image points by N_{is} shots are $N_{is} \times N_r = (N_r \times N_i) \times N_r$, which is much higher than N_i. We estimate the optimum shot positions from the ideal shot positions. The algorithm employs clustering analysis for these ideal shot positions to infer optimum shot positions as follows. In this algorithm, the algorithm uses the center of the clusters as optimum shot positions. The computation of optimum shot positions starts with a small number of clusters: that is, a small number of shots. Once the candidate shot position is estimated, the algorithm computes the ray from the source to the receiver and confirms illumination. This iteration continues until the ray hits the entire target image zone within the given threshold.

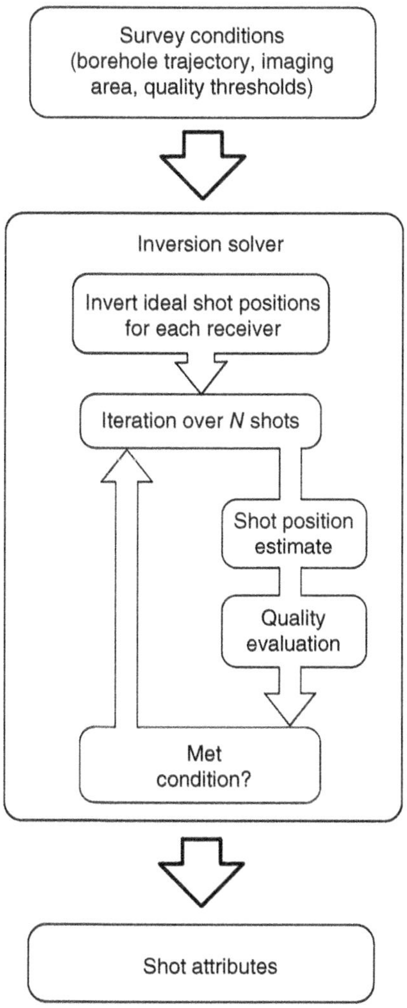

Figure 18.1 Computation flow chart in automated DAS survey design. The borehole trajectory, imaging area, and quality thresholds are given. Once ideal shot positions are computed, optimum shot positions are computed based on ideal shot positions. The number of shots is increased until the quality meets the conditions required.

Figure 18.2 shows an example of the algorithmic survey design. DAS cable is deployed into an S-shaped borehole, and the receiver points are distributed from 100 to 3,000 m along the borehole with 10 m spacing. The target image area is east of the well, 5,000 m deep in true vertical depth, and extends 1,000 m in the east and north directions. A homogeneous velocity model is assumed in this case study to demonstrate the concept of the method. Figure 18.2(a) shows the ideal shot positions, which are computed for each receiver–image point combination. The total number of ideal shot positions is about 340,000. The optimized shot points are computed from the ideal shot positions using the algorithm in Figure 18.1. About 1,000 shot points are populated as optimum shot points, which cover the image area (Figure 18.2(b)).

Using the optimum shots estimated in the previous step, the data quality is computed. We define the quality measure for the DAS data. Because the data SNR is used in most cases, we employ SNR in the quality measure. We define the following as an equation for SNR of DAS data:

$$SNR = \frac{D_{DAS}}{A_N}, \tag{18.1}$$

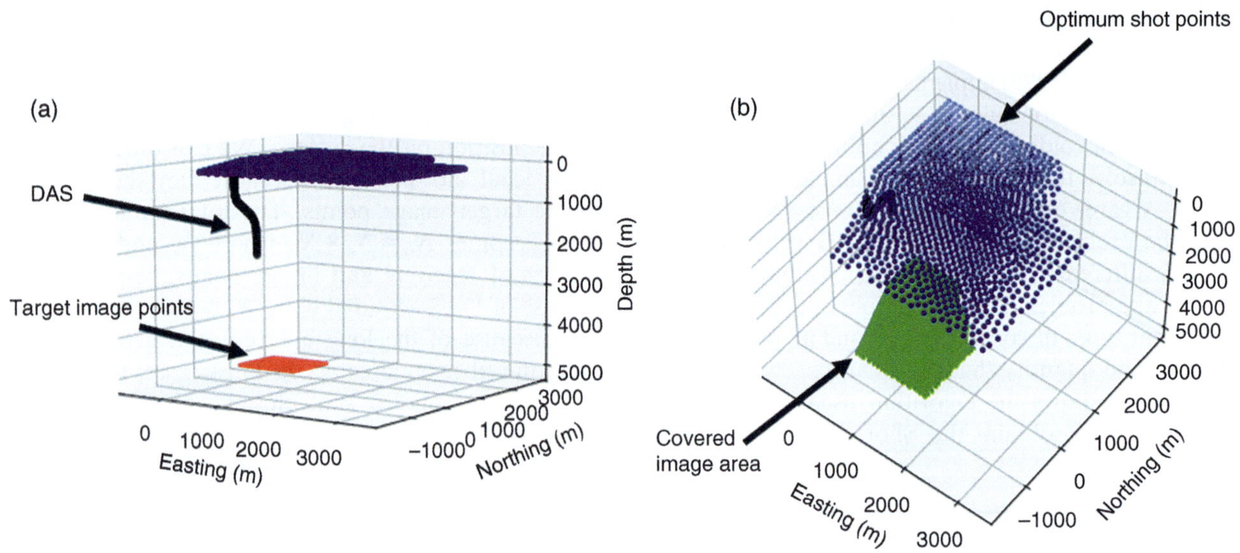

Figure 18.2 Example of automated 3D DAS VSP imaging survey design for a deviated well. The total number of receivers is about 1,000. (a) Distribution of target image points (red) and ideal shot points (blue) computed from DAS receiver points (10 m spacing) (black). The total number of shot points is about 340,000. (b) Optimum shot points for the survey. Green is the image point for the shot points. The green points overlap the target image points in (a).

where D_{DAS} is the amplitude of the DAS data and A_N is the noise amplitude. We can formulate the amplitude of DAS data as follows:

$$D_{DAS} = S \times P \times R. \quad (18.2)$$

where S is the amplitude at the source in particle velocity, P is the amplitude decay during the propagation of seismic wave to the receiver from the source, and R is the receiver response. To compute SNR, we need precise knowledge of all the parameters in equations (18.1) and (18.2), which is impractical. Therefore, we define the quality measure as the ratio to the best case: how much data SNR there will be compared to our best observation. The best observation is when DAS is in the vertical well and the source is located just above the measurement point (Figure 18.3), where the total ray path is shortest and the apparent velocity is smallest (amplitude is maximum). This also means the quality measure is defined as the ratio to zero-offset reflection data acquired at a vertical well. Assuming the PP reflection problem, the SNR for the best and actual cases are

$$SNR_{ref} = \frac{D_{refDAS}}{A_{refN}} = \frac{S \times P_{ref} \times \frac{1}{V_{refA}}}{A_{refN}}, \quad (18.3)$$

$$SNR_{act} = \frac{D_{actDAS}}{A_{actN}} = \frac{\sqrt{E} \times S \times P_{act} \times \frac{\cos\theta}{V_{actA}}}{A_{actN}}, \quad (18.4)$$

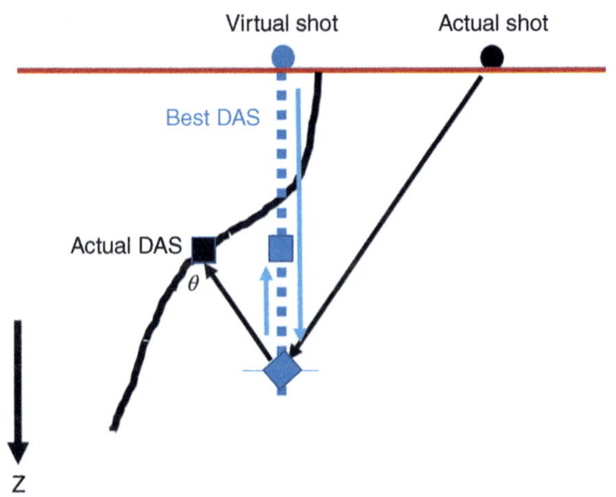

Figure 18.3 Relation between the real and virtual reference DAS measurement to define quality measure. Virtual DAS measurement assumes vertical well deployment with zero-offset source.

where θ is the ray incident angle to each measurement point in the borehole. E is the effort factor, which is the number of sweeps in the case of a vibration source. $\cos\theta$ in equation (18.4) expresses the projection of the three-component amplitude of particle velocity to the borehole. Then the inverse of apparent velocity (V_A) is multiplied to compute the DAS response from particle velocity as the first-order approximation of the DAS measurement (when the wavelength is a few times longer than the gauge

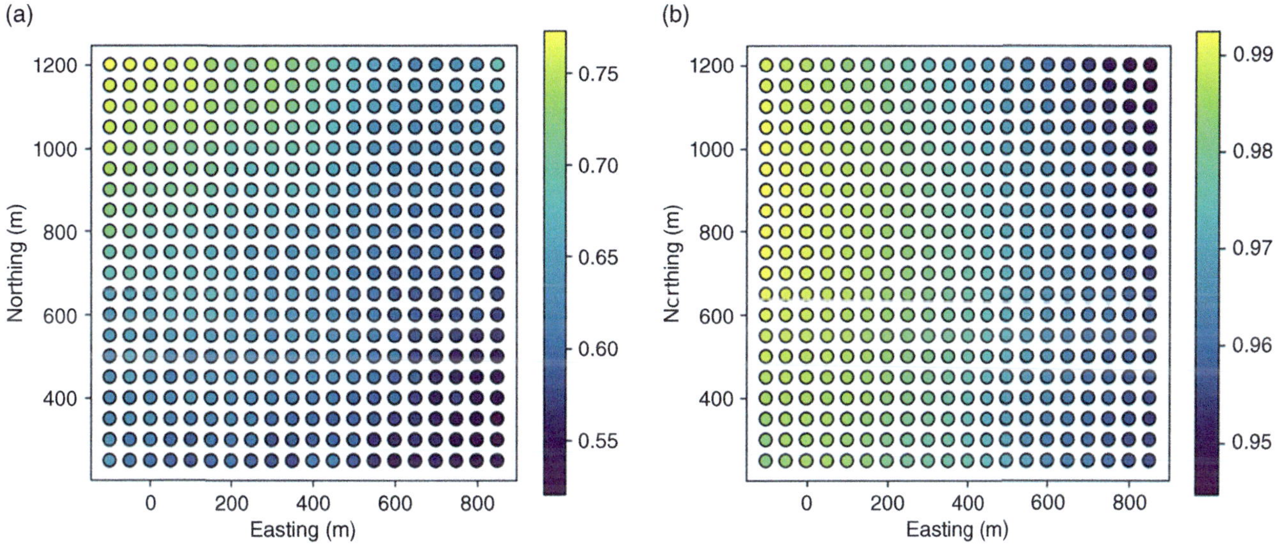

Figure 18.4 Data quality measure for each image point for the survey in Figure 18.2. (a) DAS and (b) geophone.

length) (Mizuno et al., 2019). By taking the ratio of equations (18.3) and (18.4), we define the quality measure O in terms of SNR:

$$O_{SNR} = \frac{SNR_{act}}{SNR_{ref}} = \cos\theta^2 \times \frac{P_{act}}{P_{ref}} \times \sqrt{E}, \tag{18.5}$$

where we assume (a) the noise amplitude is the same between the virtual reference and the real configuration (assuming noise amplitude is solely coming from the measurement system, not the well condition) and (b) the radiation pattern of the source is isotropic. Equation (18.5) indicates the quality measure is computed by the source geometry, DAS measurement point, and effort factor.

Figure 18.4(a) shows the data quality measure for the DAS survey in Figure 18.2, and Figure 18.4(b) is for the geophone case as a reference. Because this is a homogenous velocity case, the coefficients in equation (18.5) can be estimated solely from the geometry of shots, receivers, and well trajectory. The quality measure changes in space in both cases. However, the DAS case shows large variation compared to the geophone case. This is mainly because of the measurement differences, as summarized in Table 18.1. In this example, the SNR of the DAS data changes between 0.55 and 0.75, and the SNR of the geophone data ranges from 0.95 to 0.99. We anticipate that lack of data quality control (QC) will be seen as inconsistency in the quality of the image because of the variation of SNR in the DAS case. This is the difference from the geophone case in which the variation of data quality is less significant. To obtain more consistent data quality, we need to consider increasing the number of stacks for far-offset shots or doubling the source effort, which was not required in the geophone case. Also, real-time data QC is important to observe the mitigation works as expected during the acquisition. The algorithm could work with any type of ray-tracing engine to fully replace conventional workflow in day-to-day operations. We introduce a case study for a 1D velocity model in the latter part of this chapter.

18.3. GENERALIZED RADON TRANSFORM AND ITS APPLICATION TO DAS

The main purpose of imaging is twofold: to obtain the subsurface structure for interpretation purposes and to obtain true amplitude reflectivity to infer formation properties. GRT migration is an extension of Kirchhoff migration (e.g., Dillon, 1990). This method assumes seismic wave scattering at the image point due to the velocity model variation, and formulation is applicable to a flexible source-and-receiver configuration. This brings the applicability of GRT migration to VSP configurations. It was initially formulated as a scalar wavefield imaging problem (Miller et al., 1987). Jackson et al. (1991) and Takahashi (1995) extended it to multicomponent problems. Dillon (1990) compared the difference of GRT compared to Kirchhoff and identified the better performance of true amplitude recovery in GRT in the VSP case. We found its application in global seismology (e.g., Wang et al., 2006) and in the oil and gas industry (e.g., Ali et al., 2019). We discuss its extension to DAS.

GRT migration can be formulated as per the following equation:

$$m_x = \int dW \cdot u[r, s, t = \tau(r, x, s)] \qquad (18.6)$$

where m is the image amplitude, u is the data amplitude, r is the receiver, x is the scattering point, s is the source location, and τ is the traveltime of the scattered waves. This is similar to the form of Kirchhoff migration: stacking amplitudes of each trace for a given traveltime computed from each receiver, source, and scattering point pair and applying a migration weight. dW is a weighing factor. However, the details of GRT are different from Kirchhoff migration. GRT is formulated based on the plane wave scattering at the image point, and it incorporates Jacobian, obliquity, and geometrical spreading corrections and dip weight in the following form (e.g., Miller et al., 1987):

$$\int dW = \frac{1}{\pi^2 c_0{}^3(x_0)} \int d^2\xi(r, x_0, s) \cdot \left| \cos^3\beta(r, x_0, s) \right| \cdot \frac{1}{A(r, x_0, s)} \cdot W_{dip}(x_0) \qquad (18.7)$$

The first term on the right side is the Jacobian weighting, the second term is obliquity compensation, the third is geometrical spreading compensation, and the last is the dip filter. Jacobian weighting is for the correction of uneven energy flux density at the scattering point, and it contributes to a stable image even for the limited illuminations that can be found in VSP in general. Those weights are defined to compensate for the limitation in terms of the receiver and source distribution in space. 3D ray tracing is used to compute the ray path and ray parameters for all shots–image point–receiver paths to compute weights.

Regarding the application of the migration to DAS, we often see two approaches depending on assumptions made for DAS data. The first approach assumes DAS data are scalar data like hydrophone data and so a scalar migration workflow is applied to the data. Considering the availability of computation resources and objectives, it can be a practical option. The other approach assumes DAS data are vector strain data. In this case, we need to apply an image-domain correction as the additional weight, compensating for the DAS P-wave data amplitude response. The DAS response can be explained by two factors: the projection of 3C polarization to a single component (parallel to the borehole axis), and the DAS response as strain, which is approximated by the normalization of particle velocity by the apparent velocity. This assumption is valid when the apparent wavelength is a few times longer than the gauge length (e.g., Mizuno et al., 2019). Assuming polarization angle ~ group angle ~ phase vector, the following equation can be delivered as the approximated form for image-domain correction for DAS data in a GRT migration:

$$W_{DAS}(r, x, s) = 1/\left(SL(\cos\theta(r, x))^2\right), \qquad (18.8a)$$

where θ is the angle between a group velocity vector at the receiver and a well deviation vector at the landing point of the ray to the well, and SL is the formation slowness of the target phase at the depth of each receiver. If receivers are in a homogeneous area, equation (18.8b) can be employed in its simplified form:

$$W_{DAS}(r, x, s) = 1/(\cos\theta(r, x))^2 \qquad (18.8b)$$

We compare the DAS migration images described earlier with images from the geophone data in two examples. We refer to Jackson et al. (1991) for the weighting factor for 3C geophone migration. We need to note that weighting is changed by source and receiver geometry and also the local dip of the borehole. Therefore, the weighting can be changed with the dip of the well. In the following section, we test two different borehole geometries: a vertical and a deviated well. We review the synthetic data case to verify the method and then review briefly the real data case to discuss the significance of the method in practical aspects.

The first example is the SEAM-1 VSP synthetic (SEG Advanced Modeling Corporation, 2013). Sources are located along the east–west line (walkaway), and receivers are deployed in a vertical well. Because the SEAM-1 VSP synthetic is an accelerometer dataset, we simulate DAS data (Figure 18.5) from the accelerometer data using a gauge length of 15 m (Mizuno et al., 2021). In this case, apparent wavelength is long enough compared to the gauge length. This simulation demonstrates the typical nature of DAS data: amplitude is different from the 1C geophone because the amplitude of DAS data is also scaled by apparent slowness. The S wave tends to remain in a high amplitude because its apparent slowness is larger than the P wave. Figure 18.6 shows the geometry of receivers and the velocity model. We used receivers at depths ranging from 1,000 to 1,550 m and shots within 2,000 m of the wellhead, assuming a small scale of VSP imaging survey. We apply GRT migration to both DAS and 3C geophone datasets. To control image quality, we use incident rays between 20° from the specular reflections. Figure 18.7 shows the comparison of 3C geophone migration, 1C geophone migration, DAS image-domain correction migration, and DAS scalar migration. At the area below the receiver array (the area inside the dotted rectangle in Figure 18.7), those four images are comparable in terms of amplitude, polarity, and image extent. Because illumination below the receivers occurs with rays that have

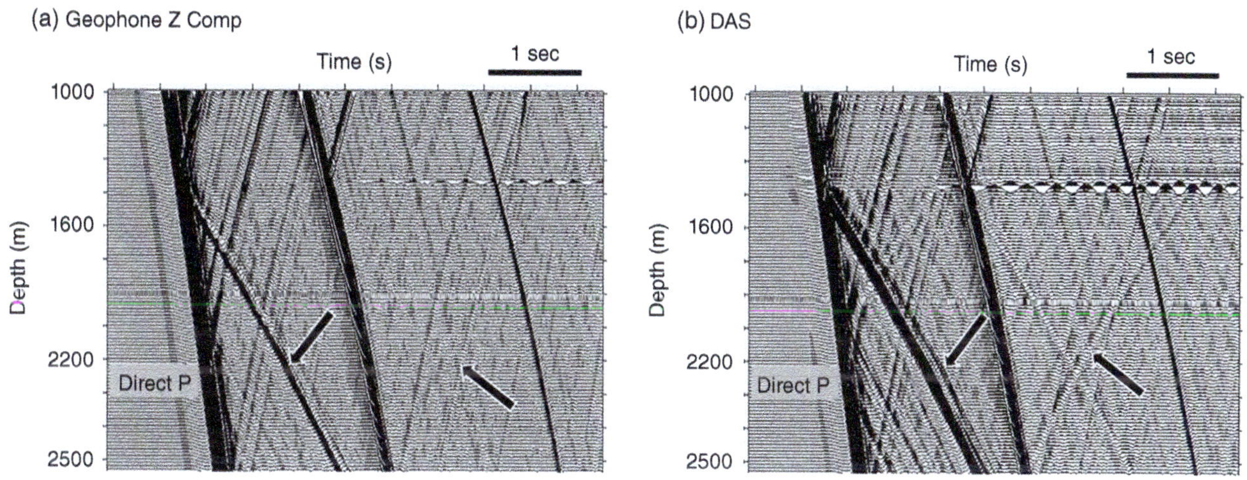

Figure 18.5 The example of common shot gathers of geophone synthetic (a) and DAS synthetic (b) used in the study for SEAM-1 dataset (trace normalized). PS converted phases (marked by arrows) are clearly seen in the DAS synthetic compared to the vertical geophone component. The receiver spacing in the geophone array is 15 m. The gauge length of DAS synthetic is 15 m.

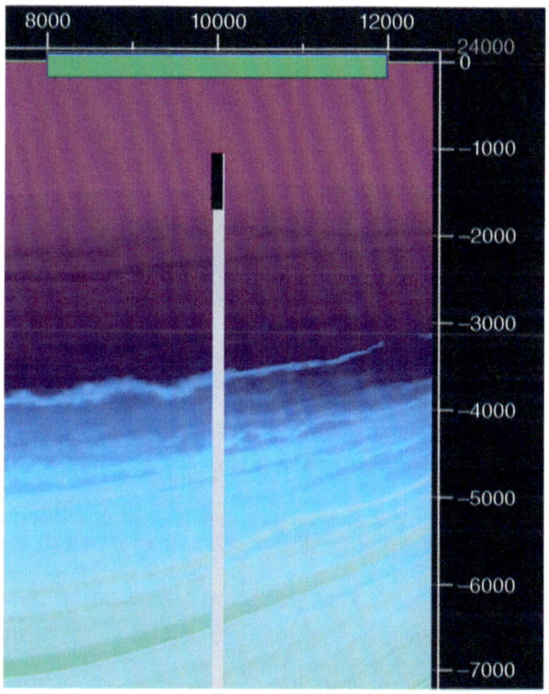

Figure 18.6 Velocity model and survey geometry. Receivers are located in the depth range of 1,000 to 1,600 m (blank line). Source offsets are about 2,000 m from the wellhead (green line on the surface).

a small angle relative to receiver orientation, the contribution of the 1C correction (equations (18.8a) and (18.8b) for DAS) is not significant. Even if the DAS image-domain correction is not available for resource and other constraints, we anticipate that scalar migration achieves the objective of the imaging project for a vertical DAS array. The images elucidate structures that are seen in the velocity model (Figure 18.6). However, they also include some events that do not correspond to the velocity model. This is because GRT migration assumes primary reflections/diffraction/scattering for specific scattering modes (PP or PS), and the dataset includes conversion modes, multipathing, and multiples. We also see the migration artifact due to the limitation of 1C migration. For the area near the receiver depths, some increase in amplitudes with image point offset is visible in all 1C migrations in Figure 18.7. This increase is due to the instability of the 1C correction exacerbated by a lack of illumination. Such an issue is not seen in 3C migration as a weighted sum of the 3C wavefield provides a more stable amplitude than the scaling of 1C data. We will review the other example of the limitation of 1C migration later in this chapter.

As mentioned earlier, weight is a function of shot and receiver point geometry and also well trajectory local to the receiver point. As a second example, we looked at the case when the well is highly deviated. We employ a ray-based synthetic where we could simulate primary reflections and control conversion modes (Leaney et al., 2019). Figure 18.8(a) shows the geometry of the survey. To test the deviated well case, the well has a dip of 45° toward the east, and the maximum depth of the receiver is about 1.5 km. The velocity model is a layer-cake model that has three layers. The maximum shot offset is set to 5 km, and the shot spacing is 20 m. The dominant frequency of the source wavelet is 60 Hz. Isotropic radiation of the P wave is assumed at the source. The dataset contains both PP and PS reflections (Figure 18.8(b)). Figure 18.9 (a) shows the shot point geometry and expected data

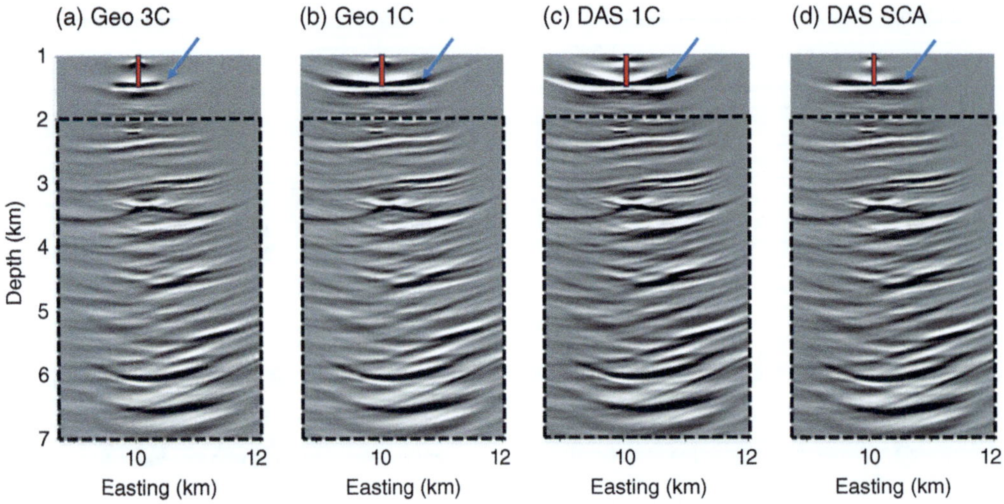

Figure 18.7 PP migration image by GRT migration for SEAM-1 synthetic data case. (a) 3C geophone migration, (b) 1C geophone migration, (c) DAS migration with image-domain correction, (d) DAS migration as scalar wavefield. The red bar indicates the location of the vertical DAS receiver array. The blue arrow points to the area where we see differences between the images. The black dotted rectangle outline indicates the area where a consistent image is obtained among the four methods.

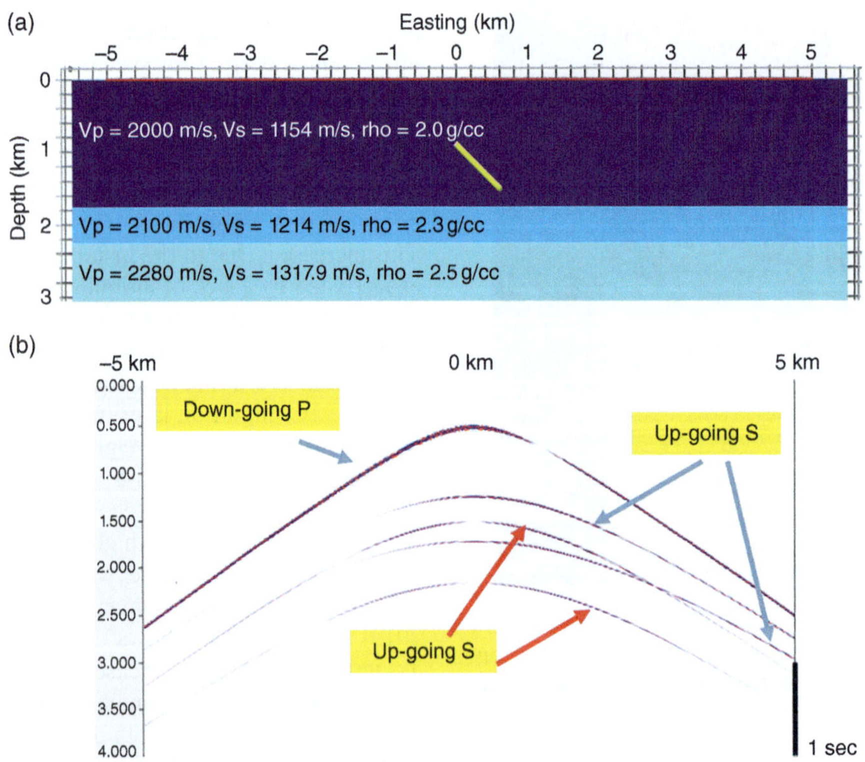

Figure 18.8 (a) Cross-sectional view (from the south) of the velocity model, sources (red), and receivers (yellow) for the walkaway synthetic dataset in the ray-based synthetic study. (b) The common shot gathers at receiver number 9 (depth 1,440 m).

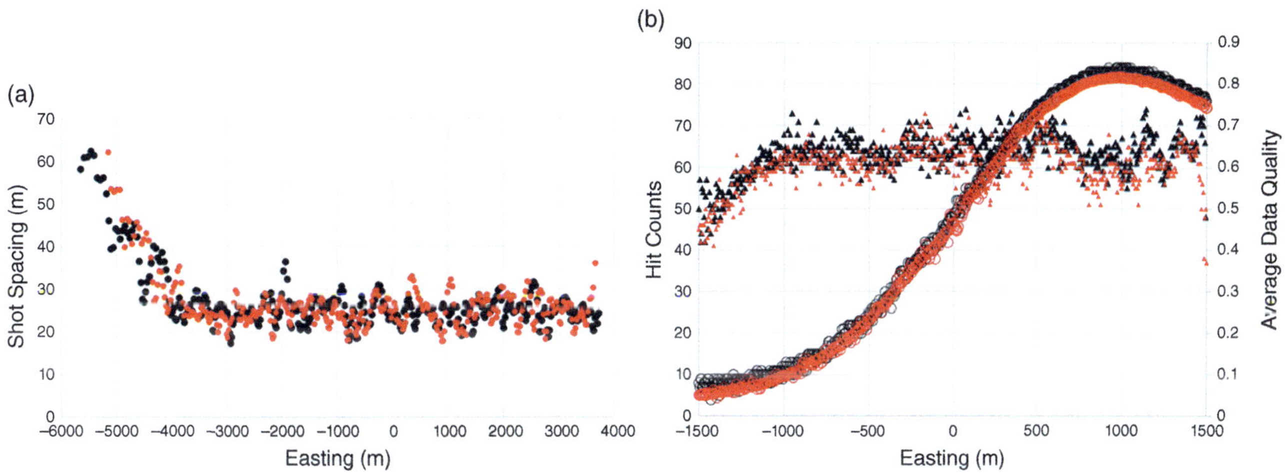

Figure 18.9 Results of the algorithmic survey design for walkaway imaging with deviated well example. Shot position and data quality are computed assuming the imaging for a 3-km-long section at a depth of 2.8 km beneath the receivers. The image bin size is set at 5 m. Two computation results are presented: black, homogeneous velocity model; red, 1D velocity model (compare to Figure 18.8(a)). (a) Shot locations and shot spacing. The shot spacing is estimated at 20–30 m for the 10-km-long walkaway line. (b) Number of stacks (triangles) and associated data quality (circles) at each image point along the walkaway line. Data quality is generally poor on the west side of the walkaway because of high angle incidence to the DAS receivers.

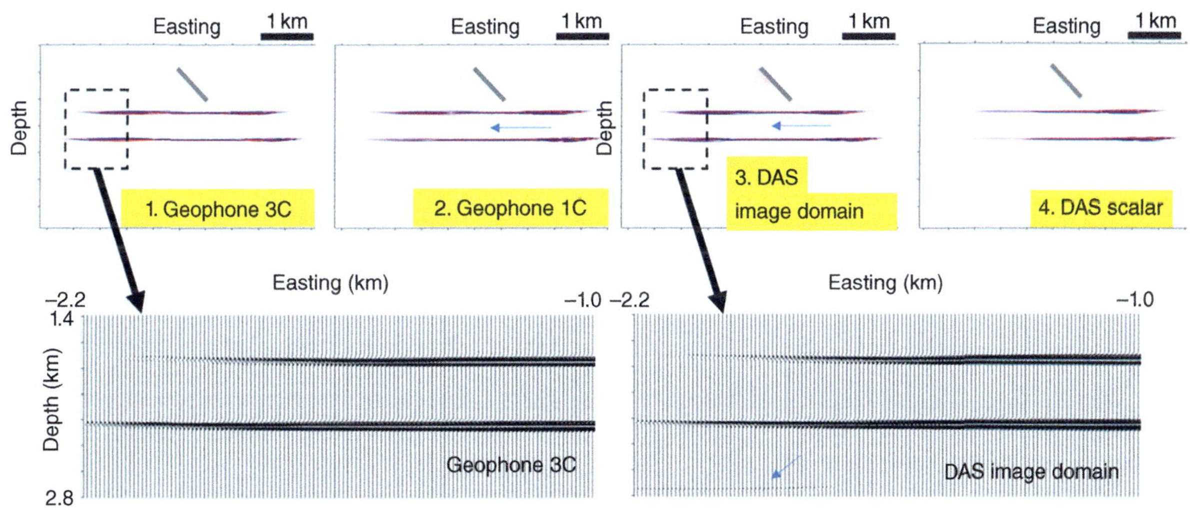

Figure 18.10 GRT PP migration image for the ray-based synthetic data. From left, 3C geophone migration, 1C geophone migration, DAS image-domain correction migration, and DAS scalar migration. The DAS image-domain correction provides a result that is comparable to 3C geophone images, although there is an artifact, as marked by the blue arrow.

quality derived from the algorithmic survey design introduced in the earlier section. We present two cases: (a) homogenous velocity model and (b) actual 1D velocity model. Those two computations fundamentally provide an almost comparable nature of shot positions and expected data quality. The total length of walkaway is estimated to be about 10 km. The shot spacing is estimated to be about 20–30 m, indicating that the configuration used in the study is an optimum setup.

We compute the PP image from PP- and PS-reflection data. Figure 18.10 compares four cases: 3C geophone, 1C geophone, DAS image-domain correction migration, and DAS scalar migration. We used the rays 5° from specular reflection for the imaging because the synthetic data contain only reflections. The open-angle limit is set at 180° to construct an image using a wide range of offsets. DAS scalar migration exhibits illumination issues for the reflectors west of the array, as expected because the amplitude of the

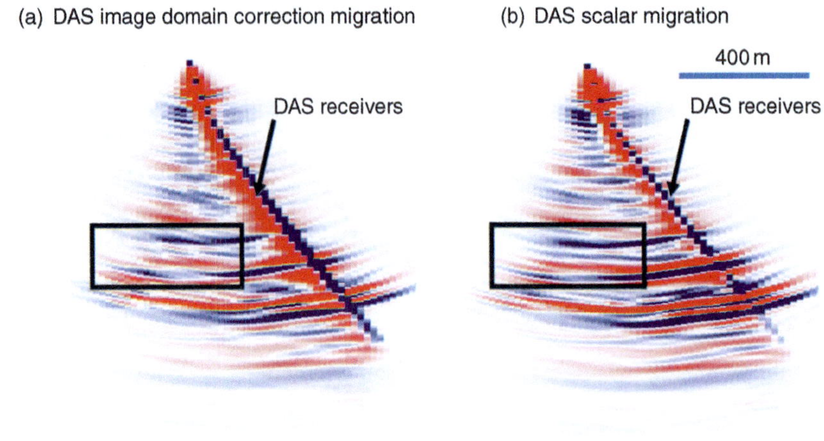

Figure 18.11 Comparison of DAS image-domain correction migration (a) and DAS scalar migration (b). The black rectangle outline indicates the area where we see the differences between the two images.

PP reflection is weak ($\cos^2\theta \to 0$) for that area (Figure 18.9(b)). Both 1C geophone and DAS image-domain correction show an image comparable to that of the 3C geophone migration. Although the necessity for DAS image-domain correction was not obvious in the SEAM-1 case where the well is vertical, this correction is important to address the illumination issue in the deviated well case. Although the DAS image-domain corrected image obtains an image fairly comparable to that of the 3C geophone, one may note that there is a slight imaging artifact seen only in the DAS image-domain correction. We will discuss this in the next section.

To discuss the criticality of the image-domain correction in the real data, Figure 18.11 shows the comparison of the GRT migration for the real walk-above dataset acquired in offshore Indonesia (Sakina et al., 2023). The well is J-shaped, and the maximum hole deviation is about 40°. Both scalar and DAS image-domain correction migration provide similar major features. However, we see differences away from the well, where the amplitude of DAS data is expected to be weak (Figure 18.11). This can be understood as a similar phenomenon observed with the synthetic test example in Figure 18.10. DAS image-domain correction migration maximizes the information extracted from the DAS shot gathers and is recommended for the DAS imaging workflow, particularly for deviated well cases.

18.4. LIMITATION OF DAS GRT MIGRATION

Because the migration artifact is significant in incomplete data, we simulate a challenging acquisition configuration (Figure 18.12): use the same deviated DAS array and layer boundaries in the ray-based synthetic, but increase shot spacing (100 m) and increase the velocity contrast between layers to enhance the migration artifact problem associated with contamination of PS mode. Figure 18.13 compares the GRT migration images. The artifact is not seen in the 3C geophone data. However, we observe the migration artifact about 2,000 m west of the receiver array in the 1C geophone and DAS image-domain correction migration. As seen in both 1C geophone and DAS, this artifact is not associated with the DAS and geophone systems but rather is a result of 1C versus 3C measurement. Haldorsen (2020) noted that the spatial aliasing of the wavefield is significant in the 1C sensor compared to the 3C sensor and that polarization from the 3C sensor reduces the spatial aliasing issue. Such artifacts can be characterized in the image due to shot spacing and wavelength (λ) relation. In this present case, the artifact is due to the conversion of the S wave at the layer boundary. Using the S-wave wavelength, shot spacing is increased from 1λ to 5λ. Although no imaging artifact is seen in the 5λ shot spacing case in 3C VSP, the artifact is significant in single-component measurement cases, including DAS. To further improve the DAS image, technology development for P- and S-wave field separation is needed for single-vector component data including DAS.

18.5. SUMMARY

This section describes the transition of the acquisition to processing workflow with the introduction of DAS into subsurface imaging in exploration seismology. We note the following changes between the DAS-based workflow and the geophone-based workflow:

1. One of the success factors for migration with DAS is obtaining data of reasonable quality. Therefore, the survey design is a critically important step. By defining the new quality measure and implementing the survey design problem as an inversion problem, we

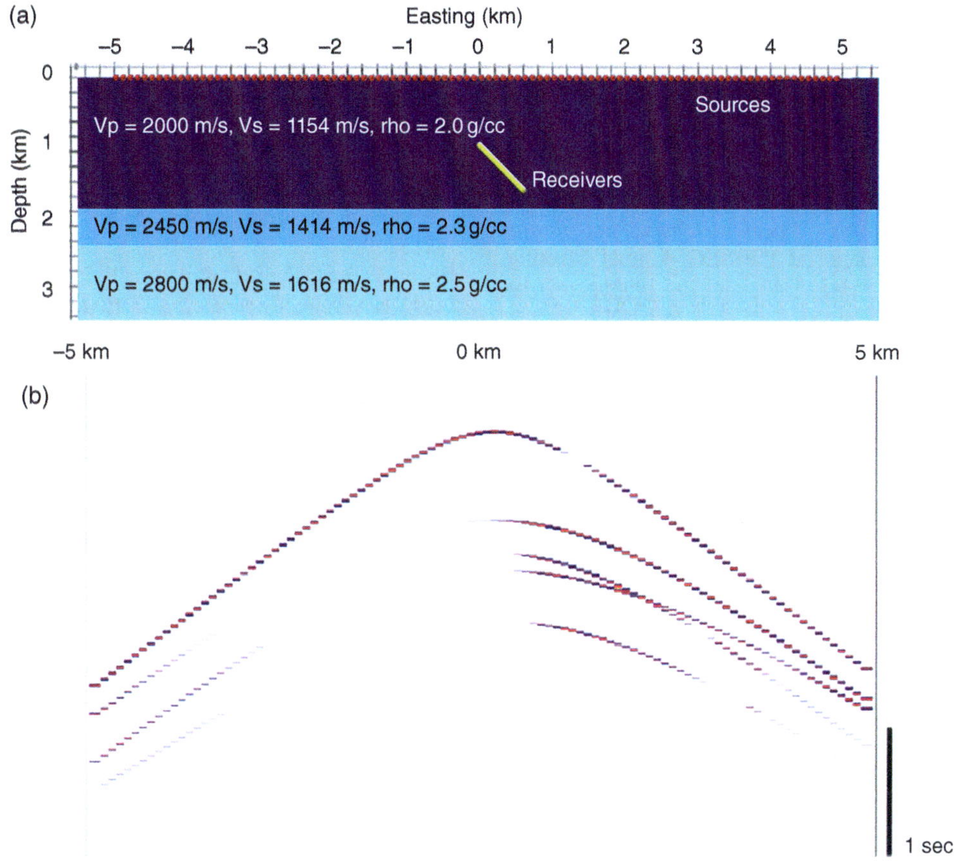

Figure 18.12 (a) Cross-sectional view (from the south) of the velocity model, sources (red), and receivers (yellow) for the walkaway synthetic dataset for the spatial aliasing test. The receiver position and walkaway length are the same as in Figure 18.6. (b) Common receiver gathers at receiver number 9 (depth 1,440 m).

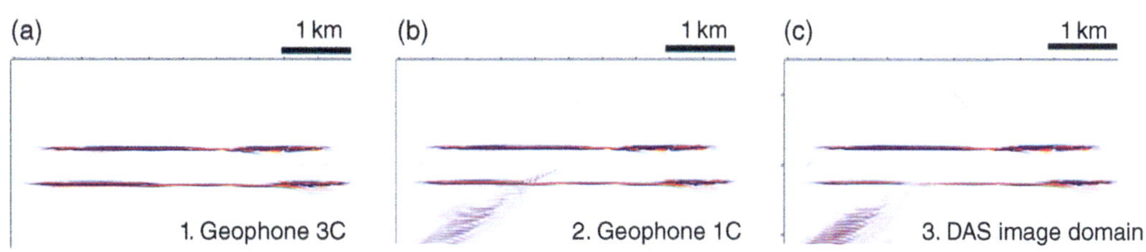

Figure 18.13 GRT PP migration image for configuration in Figure 18.11. (a) 3C geophone migration, (b) 1C geophone migration, and (c) DAS image-domain correction migration. Both 1C geophone and DAS image-domain correction present the migration artifact on the west (left) side of the receivers.

could design the optimum survey and measure the quality of data automatically. This enables the comparison with the geophone case in terms of data quality. Using the measure, we could adjust acquisition by increasing stacks or increasing source effort to obtain quality data.

2. In terms of the migration computation engine, minimal changes are required to obtain comparable results between 3C geophone and 1C DAS. Correction of DAS data becomes most significant in the case of a deviated well, but not in the vertical well case.
3. The remaining challenge for DAS migration stems from the limitation of 1C measurement. It can result in imaging artifacts in cases where spatial aliasing of the wavefield is occurring. Forward modeling is a useful tool to assess the risk and further optimize the survey prior to the project.

AVAILABILITY STATEMENT

No data are available for this chapter.

REFERENCES

Ali, S., Zhu, J., & Sayed, A. (2019). *Integrational anisotropic earth-model building using vertical and horizontal-vibrator VSP data*. Paper presented at the SEG Annual Meeting. Expanded Abstracts, 5345–5349. doi:10.1190/segam2019-3216768.1

Alvarez, G., Pereyra, V., & Carcione, L. (2004). *Model-based 3D seismic survey design as an optimization problem*. Paper presented at the SEG Annual Meeting. Expanded Abstracts, 63–66. doi:10.1190/1.1845276

Dillon, P. B. (1990). A comparison between Kirchhoff and GRT migration on VSP data. *Geophysical Prospecting*, *38*, 757–770. doi:10.1111/j.1365-2478.1990.tb01873.x

Haldorsen, J. (2020). *Spatial aliasing*. Paper presented at the SEG Annual Meeting. Expanded Abstracts, 3823–3827. doi:10.1190/segam2020-3421471.1

Jackson, G. M., Mason, I. M., & Lee, D. (1991). Multicomponent common-receiver gather migration of single-level walk-away seismic profiles. *Geophysical Prospecting*, *39*, 1015–1029. doi:10.1111/j.1365-2478.1991.tb00357.x

Leaney, S., Mizuno, T., Velez, E., Cuny, T., & Perez, M. (2019). *Integrated anisotropic model building, DAS simulation and imaging*. Paper presented at the SEG Annual Meeting. Expanded Abstracts, 999–1003. doi:10.1190/segam2019-3215957.1

Liner, C. L., Gobeli, R., & Underwood, W. D. (1998). *3D seismic survey design as an optimization Problem*. Paper presented at the SEG Annual Meeting. Expanded Abstracts, 117–120. doi:10.1190/1.1820103

Mateeva, A., Mestayer, J., Cox, B., Kiyashchenko, D., Wills, P., Lopez, J. et al., (2012). *Advances in distributed acoustic sensing (DAS) for VSP*. Paper presented at the SEG Annual Meeting. doi:10.1190/segam2012-0739.1

Miller, D., Oristaglio, M., & Beylkin, M. (1987). A new slant on seismic imaging: migration and integral geometry. *Geophysics*, *52*, 943–964. doi:10.1190/1.1442364

Mizuno, T., Leaney, S., Le Calvez, J., Naseer, F., & Lal Khaitan, M. (2019). *The significance of gauge length in particle velocity estimation from DAS data: VSP and microseismic cases*. Paper presented at the SEG Annual Meeting. Expanded Abstracts, 4869–4873. doi:10.1190/segam2019-3215250.1

Mizuno T., Le Calvez, J., & Bettinelli, P. (2021). *Development of DAS geophysics by simulating DAS data using geophone data*. Paper presented at the SEG AGU Advances in Distributed Sensing for Geophysics Workshop.

Monsegny, J. (2017). *Bi-objective optimization for seismic survey design*. Paper presented at the SEG Annual Meeting. Expanded Abstracts, 274–278. doi:10.1190/segam2017-17632413.1

Sakina, M., Rahmani, D. L, Prasetia, F. T., Majesta, C., & Miranda, F. (2023). *First implementation of hybrid fiber optic cable in Indonesia – carbon reduction through leading-edge technology*. Paper presented at the Indonesian Petroleum Association Forty-Seventh Annual Convention & Exhibition, IPA23-G-17.

SEG Advanced Modeling Corporation. (2013). SEAM open data. Retrieved from https://seg.org/News-Resources/Research-Data/Open-Data

Takahashi, T. (1995). Prestack migration using arrival angle information. *Geophysics*, *60*, 154–163. doi:10.1190/1.1443742

Wang, P., de Hoop, M. V., van der Hilst, R. D., Ma, P., & Tenorio, L. (2006). Imaging of structure at and near the core mantle boundary using a generalized radon transform: 1. Construction of image gathers. *Journal of Geophysical Research*, *111*, B12304. doi:10.1029/2005jb004241

Willis, M., Barfoot, D., Ellmauthaler, A., Wu, X., Barrios, O., Erdemir, C., et al. (2016). Quantitative quality of distributed acoustic sensing vertical seismic profile data. *The Leading Edge*, *35*, 605–609. doi:10.1190/tle35070605.1

19

Seismic Imaging of Reservoir Fractures Using Distributed Acoustic Sensing Data

Brian N. Fuller

ABSTRACT

Time-lapse seismic data were recorded in the horizontal leg of an oil and gas well using fiber-optic distributed acoustic sensing (DAS) "receivers" and surface vibroseis energy sources. The data were used to generate reservoir reflection and diffraction images. The first time-lapse dataset was recorded before hydraulic fracture treatment. The second dataset was recorded 4 weeks after hydraulic fracture treatment. The 4-week time interval between fracture stimulation and the second seismic survey included flowback and initial production. The primary goal of this study was to investigate the validity of modeling results that predict diffractions are generated when seismic waves interact with fractures that have a width on the scale of a millimeter. Time-lapse reflection images were generated from the DAS data and used to identify where some natural and hydraulically induced reservoir fractures existed that are clearly larger than millimeter scale. Diffraction imaging was used to test the thin-fracture diffraction hypothesis. To the author's knowledge, this is the first attempt to test the thin-fracture diffraction hypothesis using real seismic data. The analysis indicates that diffractions from hydraulically stimulated fractures likely occur, but more work on the topic is warranted to gain more confidence and certainty in that conclusion.

19.1. INTRODUCTION

The purpose of this chapter is to investigate the hypothesis that reservoir fractures with fracture width on the millimeter scale generate diffraction wavefields when an incident seismic wave moves through the fracture. Leviant et al. (2019) developed a grid-characteristic scheme to model the seismic response of thin fractures in 2D and 3D models. Their modeling results show that both P-P and P-S reflected and diffracted wavefields from fractures of the same fracture width as those produced by hydraulic fracture stimulation. If in fact diffractions are generated from seismic waves encountering hydraulicly stimulated fractures, then diffraction wavefields may provide a means to directly map the location, height, length, and possibly fracture width of hydraulicly stimulated fractures. Time-lapse fracture maps would provide a direct measurement of the effectiveness of fracture stimulation methods and the longevity of fractures through the production history of a well. Modeling by Leviant et al. (2019) indicates that dominant seismic wavefield frequencies near only 20 Hz should generate measurable diffractions.

Testing the diffraction hypothesis requires recording seismic wavefields where thin fractures are known to exist. Distributed acoustic sensing (DAS) fiber deployed in the horizontal leg of a well, and therefore within a stimulated reservoir, might provide a nearly ideal seismic recording geometry for recording diffractions generated within a stimulated reservoir. The DAS fiber is obviously near hydraulically stimulated fractures, so if DAS recording is sufficiently sensitive, it should provide among the best commonly available opportunities to record diffractions.

Sterling Seismic & Reservoir Services, Littleton, Colorado, USA

Distributed Acoustic Sensing in Borehole Geophysics, Geophysical Monograph 289, First Edition.
Edited by Yingping Li, Robert Mellors, and Ge Zhan.

DOI: 10.1002/9781394179275.ch19

Deployment of fiber-optic DAS cables in boreholes has provided revolutionary capabilities for borehole seismic data acquisition. Traditional three-component sensor arrays are typically limited to sensor spacing greater than 15 ft (5 m) and fewer than 100 three-component sensors on a single wireline. The term "sensor" in the case of DAS does not imply that a DAS fiber has specific seismic sensor units within it. A DAS fiber is continuous, but the interrogator unit at the surface divides the DAS fiber response to seismic wavefields and assigns spatial strain rates to uniformly spaced data channels along the fiber in the borehole. With that understood, the term "sensor" is used in the context of DAS data acquisition for the remainder of the chapter without quotation marks.

DAS sensors are effectively only single-component sensors with sensitivity along the direction of the fiber. DAS fibers have the advantage of being closely spaced (on the order of a meter to a few meters) and can simultaneously occupy the entire length of a borehole, including the horizontal leg of the borehole. Three-component seismometers provide particle motion in all three spatial dimensions and full-wavefield amplitude fidelity; however, DAS has the economic advantage of simultaneously recording high-quality data over the entire length of the borehole. Additionally, fiber-optic equipment can be used for multiple purposes, including time-variant temperature analysis, low-frequency stress/strain analysis, and borehole fluid flow analysis. Fenta et al. (2021) present a recent application of DAS data and a significant list of references on the topic.

Diffractions play an integral part in surface seismology because diffractions occur when wavefields meet discontinuities, such as faults in the subsurface. Diffracted wavefields are recorded by seismometers at the surface of the Earth or at sea and contribute to fault images through the process of seismic migration (Pedersen & Marcy, 2022; Tyiasning et al., 2016). A key difference between diffraction imaging in seismic reflection imaging and the diffraction imaging done in this chapter is that the diffractions processed in this work are specifically used for imaging after removal of reflection data from common-source gathers so that only the diffractions are preserved and used to image diffracting points in the reservoir.

The dataset used for this study was a time-lapse seismic survey recorded in both the vertical and horizontal legs of a borehole using DAS recording and vibroseis sources at the surface of the Earth. The baseline survey was recorded prior to hydraulic fracture stimulation, and the second survey was recorded 4 weeks after stimulation. Flowback and initial production operations were carried out during the 4-week interval between stimulation and the second seismic survey. The dataset was used to generate pre- and post-stimulation reflection images over the horizontal leg of the borehole (Fuller et al., 2020) and was also used to generate pre- and post-stimulation diffraction images over the horizontal leg of the borehole. The reflection images were used to identify where stimulation and production-related fractures are likely to have occurred. The diffraction images were used to investigate the theoretical hypothesis that diffractions can be generated by even millimeter-scale fractures (Leviant et al., 2019).

19.2. TIME-LAPSE DATA ACQUISITION AND SURVEY GEOMETRY

The spatial measurement system used in this chapter is the English system of feet to remain consistent with well logs and DAS deployment information. Spatial measurement system consistency throughout data processing helps avoid data processing mistakes. Thus, all plots with depth as an axis are labeled in feet. All spatial specifications in the text are given in feet, followed by meters in parentheses.

Figure 19.1 shows a cross-sectional sketch of the survey geometry used in this project. The vertical leg of the

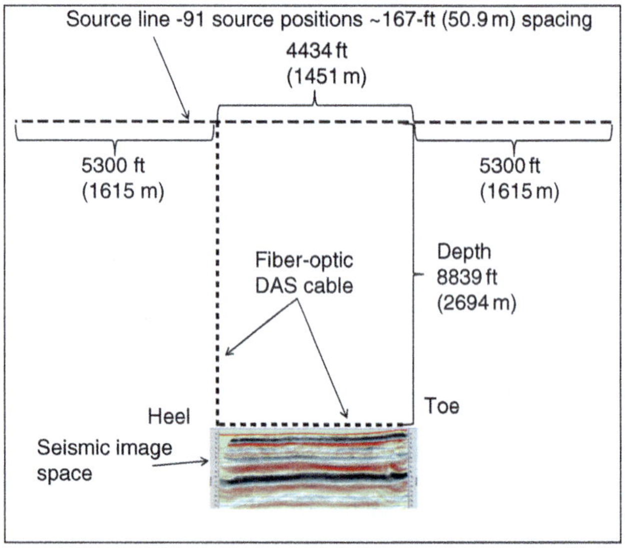

Figure 19.1 Field site geometry sketch. This sketch shows the source and receiver field geometry for the DAS data used in this chapter. The wellbore extended vertically to a depth of 8,839 ft (2,694 m) and then turned horizontal for a distance of 4,434 ft (1,451 m) in a toe-up orientation. The DAS cable was cemented behind the casing in the wellbore. The seismic source was vibroseis, and 91 roughly colinear source points at the surface were used in processing. The reflection and diffraction images shown in this chapter were all below the horizontal part of the well and within the lateral bounds of the horizontal well. A reflection image is shown in this figure to demonstrate the vertical and lateral extent of the imaging as being below the horizontal borehole and limited laterally to the horizontal leg of the wellbore.

borehole extends to a depth of 8,839 ft (2,694 m). The borehole then extends laterally in a toe-up orientation, with the fiber extending 4,434 ft (1,351 m) horizontally in the lateral section of the borehole.

The DAS fiber was permanently installed and cemented behind the casing, providing receiver repeatability in time-lapse surveys. A total of 91 vibroseis source points located at the surface of the Earth were used for this survey. Figure 19.2 shows a map of the source points and a projection of the wellbore on the surface of the Earth.

The surface seismic source points extended approximately 5,300 ft (1,615 m) off each end of the wellbore lateral section. The nominal source point interval was 167 ft (50.9 m) along a roughly north–south trend (north-0.58° east). The vibroseis sweep parameters were 2–120 Hz, 18 s in length, with a 0.8 s cosine taper at the beginning of the sweep and 0.3 s cosine taper at the end of the sweep. Three sweeps were taken at each source point. The recording channel interval provided by the DAS interrogator was 13.12 ft (4.0 m). Based on the discussions earlier regarding the frequency content of modeling in Leviant et al. (2019), the vibroseis sweep here should be sufficient to generate diffractions from fractures that are on the scale of a millimeter in width.

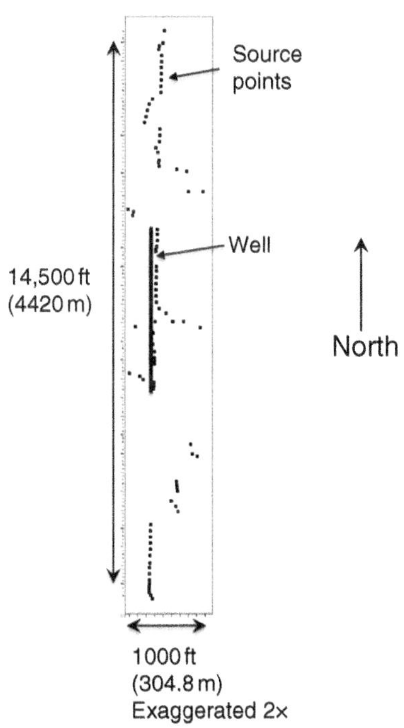

Figure 19.2 Map view of well and seismic source points. A projection of the wellbore is shown as an annotated solid dark line. The entry point of the well is at the south end of the well. The source points are shown as small dots. There were 91 source points distributed in a roughly north–south orientation. Source points were not in a straight line due to surface facilities and surface features.

The baseline survey was recorded prior to fracture stimulation of the wellbore, which was a "plug-and-perf" completion. The second seismic survey was recorded 4 weeks after stimulation. The flowback procedure was completed and the well was put into production in the time interval between stimulation and the second seismic survey. Repeated shots used in time-lapse data processing were restricted to source points that were repeated within 10 ft (3.3 m) of the baseline survey surface position. Repeated source positions can be important for source repeatability of the seismic source signature (Kragh & Christie, 2002). The DAS fiber was cemented in place behind the casing, and there was no evidence in the data that the receiver response changed between the first and second surveys.

Data quality in the zone of interest was generally good, with high-quality first arrivals, high-quality reflections, and good repeatability of the source signature. A notable exception to the good data quality is for DAS receivers that were vertically beneath or at high ray-path incidence angles. The angle θ is measured between the ray path direction and the direction of the horizontal DAS cable at the point of intersection. DAS fibers generally measure the strain rate along the length of the fiber, and as a result, the recorded signal amplitude for P waves is proportional to $\cos^2(\theta)$, and S-wave amplitudes are proportional to $\sin(2\theta)$ (Mateeva et al., 2014; Booth et al. 2020). The P-wave first arrival was quite weak for source points directly above the DAS receivers, with the value of $\cos^2(\theta)$ being zero. In practice, there is generally a low-amplitude P-wave signal even for P waves at normal incidence, although the signal amplitude is quite low relative to signal amplitudes recorded with higher incidence angles. A discussion of amplitudes leads naturally to a discussion of signal-to-noise ratio (SNR).

SNR values of the time-lapse datasets were estimated in the following way. The signal amplitude was estimated by measuring the root mean squared (RMS) amplitude in a 10 ms window around the P-wave first arrival. The noise amplitude was estimated by measuring the RMS amplitude within a 10 ms window centered at 100 ms before the first arrival. The ratio of signal estimate/noise estimate was computed to arrive at a value of SNR for each data trace. The median SNR value of all traces within each common-source record was computed to provide an estimate of SNR for each common-source gather. Figure 19.3 shows a plot of SNR value computed for each common-source gather in the survey plotted against the Y coordinate of the source point location. The source line was oriented generally north–south, so the source position Y coordinate is unique to each source location. In Figure 19.3, the gray squares represent the SNR values for source points in the first survey (the baseline survey), and black circles represent the SNR values for the second survey.

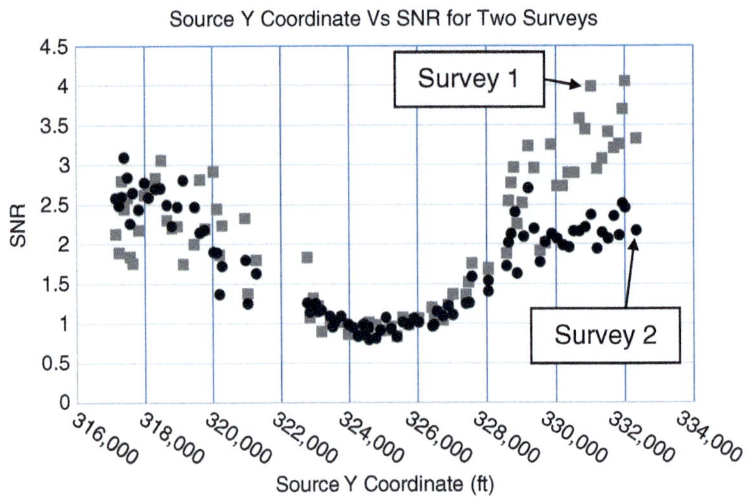

Figure 19.3 Source Y coordinate versus SNR for the two surveys. Gray squares show SNR for common-source gathers in the first survey, and black circles show SNR for common-source gathers in the second survey.

Figure 19.3 shows that SNR values are higher for the first survey than for the second survey in the northernmost 3,500 ft of the source line (from source coordinate 322,500 northward). The average difference between SNR values for the two surveys is the neighborhood of 40%.

The data show that the noise level was low for the northernmost shots in the first survey and then increased approximately 40% for the remainder of the survey. The noise level was uniform for the entirety of the second survey. Perhaps field operations such as nearby hydraulic stimulation or drilling were at a low when the northern shots were acquired for the first survey. The center third of the data points (from source Y coordinate 322,800 to 328,500) have low SNR values for both surveys. The low SNR values are most certainly due to the low response of DAS fiber to P-wave wavefronts encountering the DAS cable at high angles of incidence, as discussed earlier. Source points in the southern third of the survey (from source Y coordinate 317,000 to 321,200) show a relatively consistent SNR between the surveys, although the first survey clearly has more SNR scatter than the second survey. Nothing in the data acquisition field notes gives a reason for the higher scatter of SNR values in the first survey, but the scatter in background noise was greater during the acquisition of these common-source gathers.

The distinct U-shaped pattern of the SNR values in Figure 19.3 is expected. Source points directly above the horizontal DAS fiber (middle third of source points) have lower signal amplitude than source points that are at either end of the source line with higher P-wave incidence angles and therefore higher first arrival amplitudes.

Figure 19.4 shows time-lapse common-source gathers for the source Y coordinate 318304.3. The horizontal distance from the nearest receiver is 4,282.6 ft (1,305.3) m. Figure 19.4(a) is from the baseline survey, and Figure 19.4(b) is data from the same source point but from the second survey. The similarity between the common-source gathers is evident, but for these common-source gathers, there is a slightly higher SNR in the gather from the second survey (Figure 19.4(b)) than in the first survey (Figure 19.4(a)). The SNR value for Figure 19.4(a) is 2.85 and for Figure 19.4(b) is 3.12. The source position for records in Figure 19.4 is at the south end of the survey among common-source gathers with relatively consistent SNR values between the surveys, as seen in Figure 19.3.

The data shown in Figure 19.4 is in its original measured form of strain rate. For the work presented here, there was no purpose in converting the data to simulate voltage output from velocity-sensitive geophones or accelerometers. It was therefore left in the form of the data delivered from the data acquisition system.

19.3. REFLECTION DATA PROCESSING

19.3.1. Initial Data Processing

The initial data processing steps that account for time-lapse factors are given here:

1. Load data to disk from SEG-Y files.
2. Assign source and receiver coordinates.
3. Remove noisy traces, including corresponding traces from other time-lapse datasets.
4. Remove interrogator noise.
5. Cross-correlate with vibroseis sweep.
6. Sum multiple sweeps.
7. Surface consistent amplitude analysis and application.

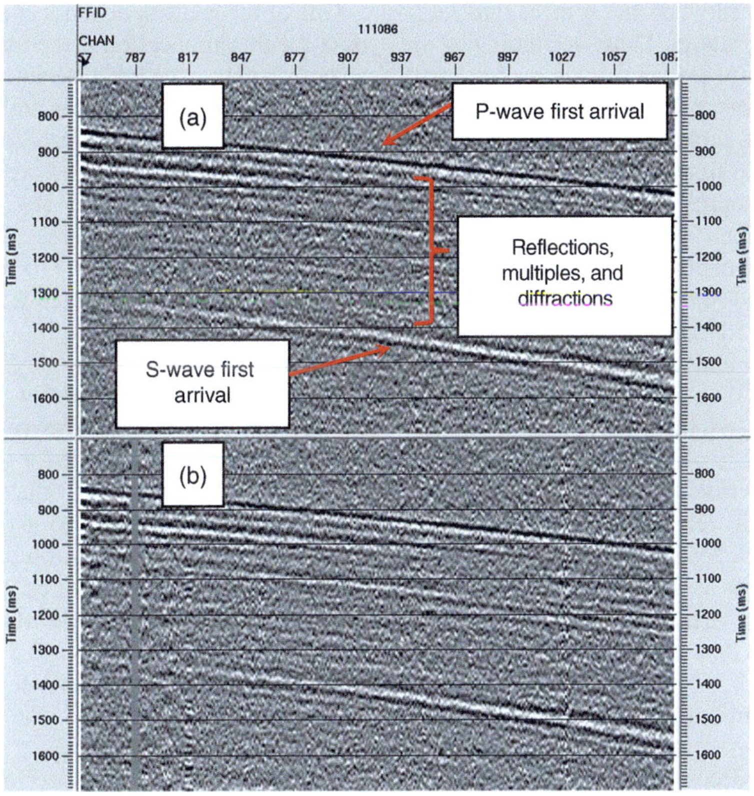

Figure 19.4 Common-source gathers from the same source point location. The P-wave first arrival, S-wave first arrival, and the time interval in which reflections, multiples, and diffractions (if they exist) are labeled.

8. Derive and apply minimum phase equivalent filter.
9. Surface consistent deconvolution.
10. Surface consistent amplitude analysis and application.
11. Pick first arrivals.
12. Remove direct and reflected S waves and S-wave multiples with spatial filters. This step is identical to the removal of P waves and P-wave reflections and multiples above only using S-wave velocities.

At this point in the processing flow, virtual source points are created on the DAS cable. The process of virtual source creation is discussed next.

19.3.2. Virtual Source Creation

The purpose of reflection processing in this chapter is to generate a reservoir reflection image below the DAS cable to identify potential fracture locations. Although it is possible to also generate a reservoir reflection image above the DAS fiber, it is beyond the scope of this chapter to do so and would distract from the central discussion point of diffraction imaging.

Standard time-domain seismic reflection imaging with seismic sources and receivers at the surface of the Earth makes extensive use of the common midpoint assumption in which the reflection point at depth is midway between the source and receiver. The common-midpoint assumption requires that the seismic sources and receivers have somewhere close to the same elevation, a condition that is deliberately not met in the case of borehole seismic data where the receivers are in the borehole and sources are at the surface. The approach used to mitigate the disparity in source and receiver elevations in this dataset was to generate virtual source points on the DAS cable, effectively moving the seismic sources from the surface of the Earth to the DAS cable in the horizontal borehole. Virtual source imaging has been effectively used in other borehole seismic applications (Bakulin et al., 2007), and those concepts influenced the virtual source software that was developed for this project. All of the virtual source software was developed and deployed on a commercially and commonly available seismic data processing system. The virtual source operation applied to the data presented in this chapter resulted in a new virtual source point at every DAS receiver location. So for every DAS receiver in the subsurface, a virtual source point was also created coincident with the DAS receiver position.

There are two significant benefits provided by the virtual source approach in this project. First, standard time-domain seismic reflection imaging methods and technologies can be applied when the sources and receivers are at the same elevation. Those technologies are highly developed, and software for them is readily available. The second advantage is that the creation of virtual source points on the DAS cable only requires a combination of the easily obtained first arrival times at the DAS cable and the common-source records. Contrast this to a prestack depth migration (PSDM) approach in which the interval-velocity-with-depth field is needed between the surface sources and downhole receivers. Derivation of the interval-velocity-with-depth profile can be laborious and error-prone with limited data.

19.3.3. Reflection Imaging with Virtual Source Data

The virtual source data discussed earlier was input to a standard 2D time-domain reflection processing flow in the commercially available seismic data processing package. The processing flow is as follows:

1. Assign geometry to virtual source data.
2. Move data to a flat datum 5 ft (1.5 m) above the toe of the wellbore.
3. Surface consistent amplitude analysis and application.
4. Surface consistent deconvolution (applied this second time to compensate for suppression of frequencies due to small errors that occur in first arrival times).
5. Surface consistent amplitude analysis and application.
6. Sort to CMP (common midpoint) gathers.
7. Generate super gathers from groups of five adjacent CMP gathers.
8. Perform NMO (normal moveout) velocity analysis.
9. Apply NMO velocities.
10. Residual statics (as expected, statics shifts were virtually zero).
11. Stack NMO-corrected CMP gathers.
12. Smooth NMO velocity field.
13. Post-stack time-domain migration.
14. Stretch time migration to depth using P-wave velocity (16,388 ft/s, 4995 m/s).

Some processes in a normal surface 2D processing flow were unnecessary in this processing flow. For example, picking first arrivals was not done as the virtual source method does not produce a first arrival. Nor was there a near-surface statics problem to resolve with refraction statics, so picking first arrivals was not needed in any case. Just as there was not a refraction statics problem to solve, there was also not a residual statics problem to resolve. Residual statics computations were tested, but as expected, they yielded essentially zero-valued statics. Regardless of their value, the residual statics step is left in the processing flow.

One of the greatest benefits of producing virtual source data for this application is the use of time-domain NMO velocity analysis to produce stacking velocities for optimal reflection image stacking and for a migration velocity profile. An NMO velocity profile can be picked for every common midpoint if desired, although typically picking velocities at some interval of CMPs is sufficient, as it was with this dataset, with velocities picked at an interval of about 100 CMPs (660 ft, 201 m). One practical benefit of the virtual source approach is that the entire data processing flow can be executed by a skilled data processor in one to two normal working days.

Figure 19.5(a) shows the P-wave reflection profile for the prestimulation seismic survey. Figure 19.5(b) shows the P-wave reflection profile for the post-stimulation seismic survey. The red curve at the top of both images is the wellbore location relative to a flat datum elevation 5 ft (1.5 m) above the toe elevation of the borehole. The vertical axis of the plots in Figure 19.5 is two-way time in ms relative to the flat datum elevation. Trace spacing along the horizontal axis is 6.56 ft (2.0 m). The frequency content of the images includes frequencies from 8–110 Hz. The sweep frequency of the vibroseis source was 2–120 Hz.

Some time-lapse effects are readily apparent, such as the change in reflection coefficient pointed to by the black arrow in Figure 19.5(b). A change in P-wave interval velocity up to 5% in the upper 100 ft (30.4 m) was measured, although it varied to as small as having no change in other places.

Often, in time-lapse imaging examples, a time-lapse subtraction is used to show the difference between "before and after" effects. In this case, however, the interval velocity in the uppermost rock interval changed due to fracture stimulation and production. Changes in interval velocity naturally resulted in changes in reflection times between the two surveys, so subtracting one image from another would have resulted in reflections being subtracted from noncorresponding reflectors. With better interval-velocity-with-depth information, the images could be converted to depth before subtraction and may be a subject for future research.

As noted earlier, reflection imaging in this chapter is only discussed as a basis for the diffraction imaging part of the work, which is the primary subject of interest in this chapter. Discussion of the time-lapse reflection images is therefore continued later in the chapter as they relate to diffraction imaging.

19.3.4. Thin-Fracture Diffraction Modeling

Seismologists have known for decades that faults and large fractures act as discontinuities in the subsurface and generate diffraction energy from seismic waves. In

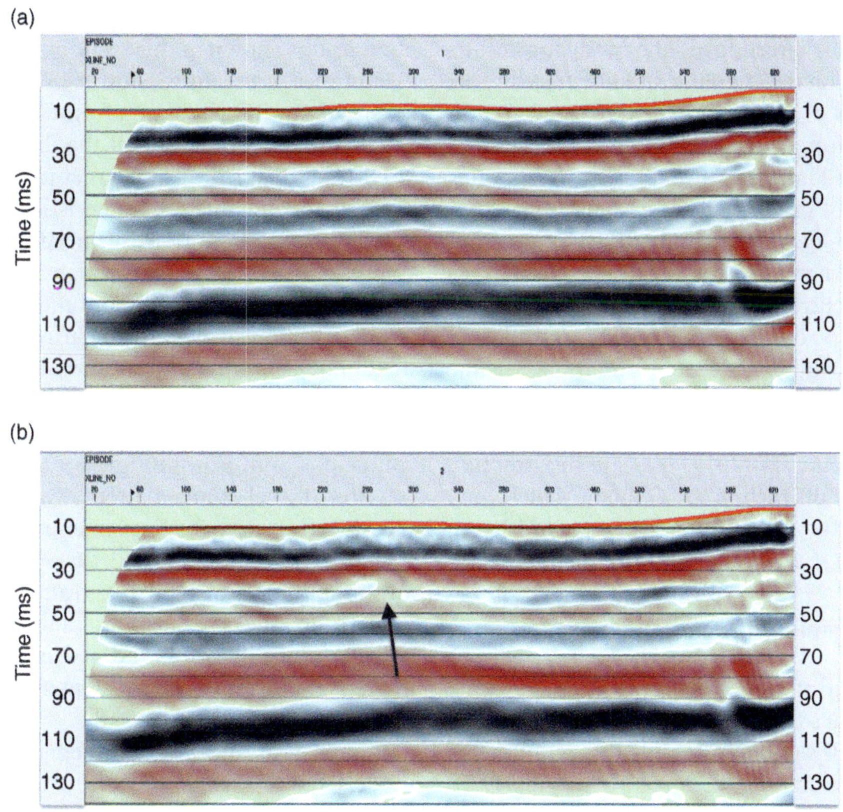

Figure 19.5 Pre- and post-stimulation P-wave reflection profiles. The time-domain virtual source processing methods described in the text were used to generate the 2D prestimulation P-wave reflection profile (a) and post-stimulation P-wave reflection profile (b). There are multiple locations where the time-lapse images are different from one another. Analysis later in the chapter is focused primarily on the uppermost 30 ms of the reflection images where stimulation and production are known to have occurred.

fact, diffraction energy recorded in seismic data provides important energy for seismic migration programs to sum into images of faults. The topic of the relationship between diffractions, faulting, and seismic imaging is extensively discussed in geophysical literature, for example in Yilmaz (1987). Diffractions associated with imaging faults in reflection seismic processing are generally attributed to velocity discontinuities where rocks of different velocities and densities have been juxtaposed by faulting. An SEG Geophysical Developments publication by Leviant et al. (2019), however, investigated the theoretical diffraction response of even millimeter-width fractures where there is no displacement along the fracture. The source of diffractions is apparently the presence of the mechanical discontinuity of the rock due to the fracture, and the fracture is filled with either gas or fluid. Leviant et al. (2019) computed dozens of synthetic seismograms for different field scenarios based on their grid-characteristic theory. They also documented their results with seismic plots and amplitude tables.

Leviant et al. (2019) motivated the investigation that is presented in this chapter toward the end of potentially being able to directly measure reservoir fractures associated with hydraulic fracture stimulation. If diffractions associated with hydraulic fracture stimulation occur in real seismic data, then DAS data may be well-suited to the task of recording diffractions for the following reasons:

- DAS fiber is often in the same reservoir that has been hydraulically stimulated. Thus the seismic sensors are near, if not collocated with, the fractures that are the source of diffractions. In this environment, even low-amplitude events might be recorded because the sensors are near the source of diffraction energy.
- DAS data is often recorded with a data channel spacing between a meter and a few meters, so data recording is spatially dense relative to seismic wavelengths.

As noted earlier, DAS recording is almost entirely sensitive to the strain rate in the direction of elongation and contraction of the DAS cable. The capability of DAS systems to record diffractions will be dependent on the orientation of the DAS fiber relative to the particle motion generated by the seismic wavefield. Modeling results by Leviant et al. (2019) predict that both P-P and P-S diffracted waves will exhibit particle motion in both the vertical and horizontal direction. If the modeling results are true, then the diffraction signal should be measurable by a horizontal DAS cable that is sensitive to motion in the horizontal direction.

Direct detection and imaging of thin reservoir fractures would be of significant value to reservoir engineers. Comparison of fracture images before and after fracture stimulation would show where new fractures have appeared as a direct result of fracture stimulation and give engineers quantitative estimates of the efficiency of stimulation parameters. Subsequent time-lapse diffraction seismic surveys would show where fractures close, open, or remain open in response to production. Diffraction modeling results by Leviant et al. (2019) indicate that fracture width plays a role in diffraction amplitudes, which may provide an avenue for quantifying fracture width, an important parameter in reservoir fluid production.

19.3.5. Borehole Seismic Diffraction Kinematics

A discussion of diffraction kinematics or travel time characteristics provides the necessary background before the presentation of diffraction processing discussions. The data presented in this chapter were recorded using DAS fiber in only the horizontal leg of a single well. The modeling results shown in this section show just the computed event times of the diffraction events without regard to the sensitivity of the direction of motion in the DAS cable or the theoretical amplitude of the diffraction events.

Diffractions from thin fractures, if they exist, will exhibit travel time characteristics in common-source records. Some important properties of diffraction travel times are discussed here in which a diffractor is generally modeled as an acoustic discontinuity that is small relative to the seismic wavelength. A diffractor in a homogenous medium essentially scatters seismic energy in a spherical volume. Importantly, the scattering diffractor behaves identically in terms of travel times regardless of the seismic source position. For multiple source points, the same diffraction travel time pattern will be repeated when recorded on nearby seismic recording instruments. Leviant et al. (2019) noted that diffraction wave phase and amplitude can be a function of the source position relative to the fracture orientation, but travel times between the diffractor and stationary recording instruments should remain consistent regardless of the phase of the diffracted wave.

Figure 19.6 shows computed diffraction travel time models for a surface seismic source and diffractors at

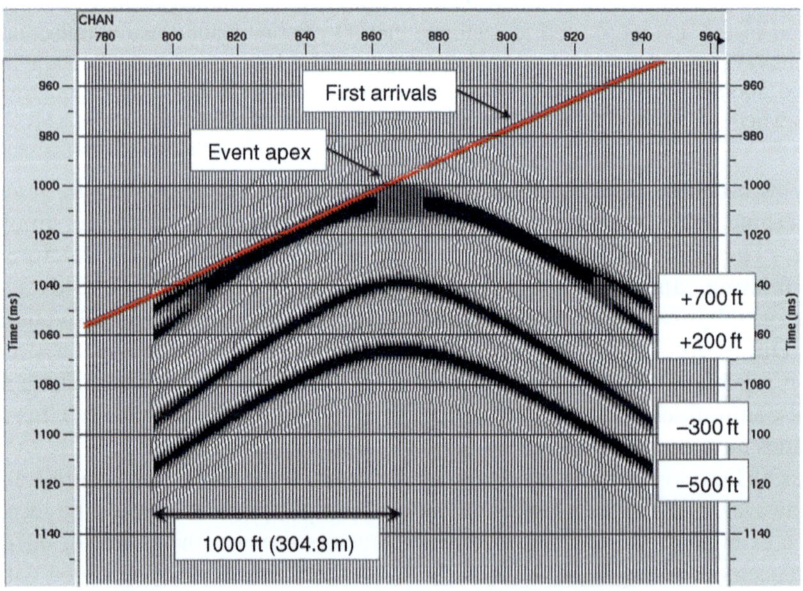

Figure 19.6 Modeled diffraction travel times. Diffractions were modeled using the source and receiver geometry of the dataset in this chapter. Diffractors were placed 700 ft (213 m) and 200 ft (61.9 m) above the DAS cable, and another two diffractors were placed 300 ft (91.4 m) and 500 ft (152 m) below the DAS cable. All diffractors were colinear in a vertical line that also included the DAS cable. A modeled seismic source was offset 5,000 ft (1,524 m) off the north end of the DAS cable and colinear with the DAS receiver cable. The key result is that colinear sources, receivers, and diffractors generate diffractions originating above the DAS cable that strongly overlap and provide no depth resolution. Diffractions that originate below the DAS cable however are well-separated in the time domain and provide good depth resolution. Other source–receiver geometries can provide depth resolution for diffractors above the DAS cable.

different depths relative to the DAS cable. The record shown is one of the common-source records from the dataset, but the real data amplitudes were set to zero for the modeling procedure to display just the model results. The seismic source is about 5,000 ft (1,524 m) off the north (right) end of the DAS cable. The first arrival times are annotated. For this simple model the diffractors are all in the same vertical line that intersects the horizontal DAS cable. Annotations on diffraction events in Figure 19.6 show the distance in feet that the diffractor is above or below the DAS cable. Diffractors above the DAS cable are marked with a "+" and diffractors below the DAS cable are marked with a "−". Note that the diffraction travel time models shown earlier do not incorporate the amplitude sensitivity of DAS data to the wavefront. The purpose of the models was solely to investigate diffraction travel times, so the angular dependency of amplitudes was not included in the model output.

The two diffractors above the DAS cable are 700 ft (213.4 m) above and 200 ft (60.9 m) above the DAS cable, respectively. The two diffraction seismic events from diffractors above the DAS cable overlay one another almost entirely within 800 ft (243.8 m) of the event apex despite the diffractors being separated by 500 vertical feet (152.4 m). The diffraction wave interference shown occurs because the diffraction wavefield from each diffractor is produced when the downgoing wavefront encounters the respective diffractors. The time delay between the generation of the diffraction events results in arrival times at the seismic receivers being very similar. The general lesson is that diffraction events provide relatively poor vertical resolution when the diffractors are colinear in a vertical line above a horizontal DAS cable. Depth resolution improves with the horizontal distance of receivers from the event apex. In Figure 19.6, for example, diffraction events begin to separate from one another at a distance of about 800 ft (243.8 m), thus providing vertical resolution given that the diffraction events propagate far enough with sufficient amplitude to be recorded.

Depth resolution for diffractors below the DAS cable is the opposite of diffractions above the DAS cable. Figure 19.6 shows diffraction events below the DAS cable with "-" symbols next to the vertical distance from the cable. In this case, the two diffractors are 300 ft (91.4 m) and 500 ft (152 m) below the DAS cable, respectively. The diffraction events from different depths are well-separated in time because the downgoing wave from the seismic source first encounters the DAS cable at the apex receiver, followed by interaction with the diffractor at a depth of 300 ft (91.4 m) and then the diffractor at a depth of 500 ft (152 m). By the time the downgoing wave interacts with the diffractor at 500 ft (152 m), the diffraction event from the diffractor at 300 ft (91.4 m) is already well on its way to being recorded by the receivers. Diffraction events originating from below the DAS cable are analogous to two-way reflection times in reflection seismology though with obviously different travel time curves. The lesson from this part of the modeling is that diffraction events originating from below the DAS cable provide good depth resolution for diffractors even when the diffractors are vertically colinear. Source geometries that are different from the one shown earlier can provide vertical diffractor resolution above the borehole, and this topic is discussed next.

19.3.6. A Diffraction Imaging Program

A diffraction summation computer program was written to test the concept of generating 3D diffraction images based on geometries of the kind modeled in Figure 19.6. The general case of diffractor geometry with respect to sources and receivers is that both the surface seismic sources and diffractors are offset some horizontal distance perpendicular to the direction of the DAS cable, and the program accommodates the general case as well as the specific case in which the sources, receivers, and diffractors all lay in the same vertical plane.

The diffraction imaging program takes the following input:

- Common-source gathers with known source, receiver, and elevation coordinates
- Known first-arrival times for each common-source gather
- A velocity field that can include both P-wave and S-wave velocities simultaneously

Input parameters to the program include a grid node spacing for each of the X, Y, and vertical directions. For each grid node in the volume and each common-source gather, the seismic wavefield is summed over the estimated diffraction arrival times at the DAS cable.

Output from the diffraction program is an image in the depth domain with summed amplitudes along predicted diffraction curves. If a diffractor is on or near a grid node, then the output image has a larger amplitude than a grid node for which no diffractor is nearby. The program was tested on synthetic data with modeled diffraction signals as an input. The output image contained the expected result that located the modeled diffractors.

The diffraction imaging program described here is strongly similar to a Kirchoff-type depth migration program, which is ubiquitous in the seismic industry (Yilmaz, 1987).

As discussed earlier, the seismic wave amplitudes recorded by the DAS fiber are sensitive to the ray path arrival angle, θ, relative to the DAS fiber. As a result, P-wave arrival amplitudes are observed to be quite small at normal incidence and increase with increasing source–receiver offset. Conversely, recorded S-wave amplitudes are proportional to sin(2θ), and therefore, S-waves recorded by the DAS cable will be largest on receivers at a small horizontal offset from the diffractor and generally decrease with a horizontal offset from the seismic energy source.

19.3.7. Diffraction Seismic Data Preprocessing

Much of the seismic data processing for diffraction analysis was completed in the seismic reflection processing flow. The data processing flow for *diffraction* processing branched away from the *reflection* processing flow after the second application of surface consistent amplitude in the reflection processing flow. First-arrival information from reflection processing is also used in diffraction processing algorithms.

Modeling by Leviant et al. (2019) predicted that diffracted waves would have amplitudes about two orders of magnitude lower than the incident wave amplitude, with the P-S diffracted wave having a higher amplitude than the P-P diffracted wave. Consistent with that result is the fact that there are no diffraction events in the common-source records that are readily visible in the way that reflections are often visible in VSPs and surface seismic common-source records. Synthetic events were added to the data with an amplitude that was two orders of magnitude less than the incident first arrival P wave, and indeed, those synthetic events were not visible. Although no entire diffraction events are visible in common-source gathers, there is a small number of partial events in some common-source gathers that fit the expected moveout of diffractions. Those events may or may not be diffractions, but they do fit the predicted moveout.

Given the relatively low expected diffraction amplitudes, it was prudent to apply signal processing methods that optimize the desired diffraction signal relative to the nondiffraction signal while minimizing harm to the desired diffraction signal. Observable "noise" or interference relative to diffractions is the first arrivals and reflection/multiple wavefields that are identified in Figure 19.7. Figure 19.7 is a common-source gather after the application of surface consistent amplitude and deconvolution processing discussed earlier.

The reflections, multiples, and first arrivals exhibit a consistent moveout over many data channels, which makes them easy targets for spatial filtering algorithms. Various spatial filtering approaches were tested to suppress the reflection/multiple parts of the wavefield while minimizing suppression of events with diffraction moveout.

Spatial filtering tests were done after adding synthetic diffraction events to the common-source gathers, followed by the application of the spatial filtering algorithms. Filtering algorithms included an FK filter, median filter, trace mixing, and an adaptation of a noise projection and subtraction method (Fuller, 1999) which is referred to in this chapter as a "dot-product filter".

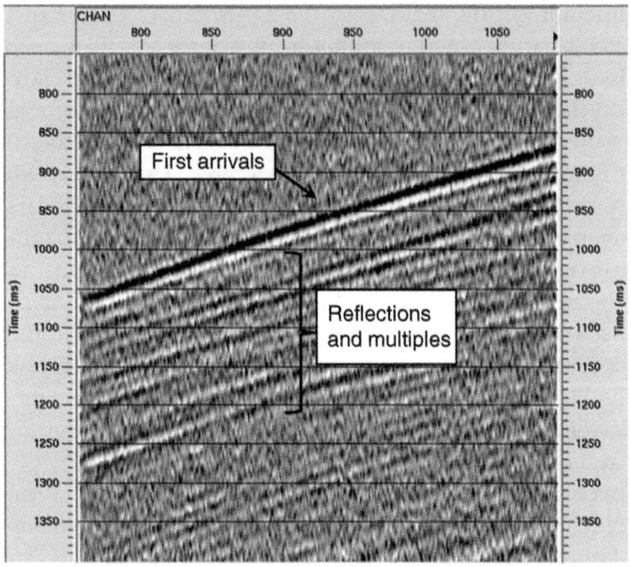

Figure 19.7 A processed common-source gather. The common-source gather in this figure is shown after the application of surface consistent amplitude and deconvolution. Data in this form were further processed by flattening first arrivals to a constant time, which also flattened the reflections and multiples. A dot-product filter discussed in the text was applied to remove the first arrivals, reflections, and multiples while preserving diffractions because diffractions have a different moveout or slope in time–space than reflections and first breaks.

From among the spatial filtering algorithms, the one that best reduced noise while preserving the desired signal was the last one listed previously, the noise projection and subtraction method (Fuller, 1999). A complete discussion of the algorithm is outside the scope of this chapter and would distract from the purpose of the chapter; however, a brief description of the algorithm follows. An estimate of the noise (first arrivals, multiples, and reflections) was generated by shifting the data traces so that the first arrivals were at a constant time. This time shift also approximately aligned the multiples and reflections over 20–30 adjacent traces. The data traces were then smoothed over 30 traces with a running average, shifted to the original first break time position, and output as a noise estimate. The original data was then input to the spatial filter along with the noise estimate. Corresponding data trace segments of 75 ms were then extracted from the noise estimate and the original data. The noise estimate vector was normalized, and the dot product between the normalized noise and the original data was computed. The normalized noise estimate vector was multiplied by the dot-product value, which provided an estimate of the noise content in the original data. The scaled noise vector was then subtracted from the original data and the trace time segment was output. All common-source gathers were treated via this algorithm.

The original algorithm cited in Fuller (1999) was somewhat different in that the signal part was estimated via eigenvector analysis, and then the signal part was preserved via eigenvector reconstruction without reconstructing the noise part of the original data.

19.3.8. Diffraction Imaging

The diffraction summing/imaging software discussed in section 19.3.6 was applied to the data using all 91 shot records. Pre- and post-stimulation 2D diffraction images were generated for both the P-P wavefield and P-S wavefields. The images were spatially coincident with the seismic reflection images shown earlier.

The velocity field used to sum diffractions was derived from limited velocity sources. At the heel end of the horizontal borehole there was a zero-offset VSP provided by the DAS fiber in the vertical part of the borehole and the nearest-offset shot to the wellhead. The SNR of the zero-offset VSP at the reservoir depth was poor, so picking reliable first breaks within the reservoir on the zero-offset VSP was impossible. As a result, extraction of an interval velocity value within the reservoir was not possible from the zero-offset VSP. The lower signal quality was probably in part due to the DAS fiber being in the well curve inside the reservoir and so was at a poor angle to be sensitive to the vertically traveling wavefield. The higher-precision source of velocity information was from microseismic events generated by stimulation of an adjacent well and recorded by the DAS fiber, which is the subject of this chapter. The microseismic events provided high-quality linear P-wave and S-wave events that extended the entire length of the horizontal leg of the well.

The P-wave velocity at the reservoir level was measured to be 16,388 f/s (4,995 m/s), and the S-wave velocity was 9,911 ft/s (3,020 ms/s). The Vp/Vs ratio of the reservoir rock was 1.65 in the horizontal direction. The velocities derived from the microseismic events varied less than 1% over the length of the horizontal borehole so the diffraction imaging was computed with essentially constant velocities. There were no reliable interval velocity data available to provide vertical velocity information above or below the horizontal borehole. At the current stage of this technology, the author has not developed an objective and repeatable velocity analysis tool that uses the data itself. Thus, velocities from microseismic event data were relied upon as an objective source of velocity information.

Four diffraction images were generated: pre- and post-stimulation P-P images (Figures 19.8(a) and 19.8(b)) and pre- and post-stimulation P-S images (Figures 19.9(a) and 19.9(b)). The trace position, trace spacing, and depth sample interval are identical for all four diffraction images. The trace spacing and depth sample interval of the diffraction images is 20 ft (6.1 ft). Additionally, the two-dimensional diffraction images overlay the location of the 2D reflection images in Figure 19.5 so that the diffraction and reflection images could be compared.

The diffraction images are shown in depth relative to thc top of the diffraction image at an elevation of −7,640 ft (−2,329 m) AMSL (above mean sea level), and the borehole location is plotted as a red curve near the top of the images. The full images were originally computed from a datum elevation of −7,200 ft (−2,195 m), which is 440 ft (134 m) above the top of the image shown in Figures 19.8 and 19.9. As stated earlier, however, the vertical resolution of an image is poor above a DAS cable if the surface seismic sources are largely colinear with the DAS cable. The colinearity condition is met in this dataset, and the diffraction images are largely vertical streaks above the DAS cable. Although the vertical streaks above the wellbore may be associated with diffractions, the diffractor locations are unknown in the vertical dimension, and therefore they add little to the discussion. The images were therefore shifted to focus on the part of the images near and below the DAS cable.

19.3.9. Comparison of P-Wave and S-Wave Diffraction Images

If diffractions are generated when the downgoing wavefield interacts with fractures, it might be expected that the P-P and P-S diffraction images will have some similarities. The prestimulation and post-stimulation images should provide an estimate of where fractures were generated in the reservoir as a result of stimulation, proppant emplacement, and production. In general, the post-stimulation P-wave and S-wave diffraction images do not share many common diffractor features. The generally low level of similarity between the P-P and P-S diffraction images may be an indication that the signal-to-noise ratio for either one or both of the P-P and P-S diffractions in the data are low and therefore there is interference that makes both images noisy, and thus they do not look very much alike.

One location in which the P-P and P-S diffraction images are similar is a post-stimulation feature that is surrounded by an oval in both of the post-stimulation images in Figures 19.8(b) and 19.9(b) respectively. The oval superimposed on Figures 19.8(b) and 19.9(b) surrounds an event trending upward and to the right in the same location for both the post-stimulation P-P and P-S diffraction images. Because this event did not appear in the prestimulation diffraction images, it is likely that stimulation

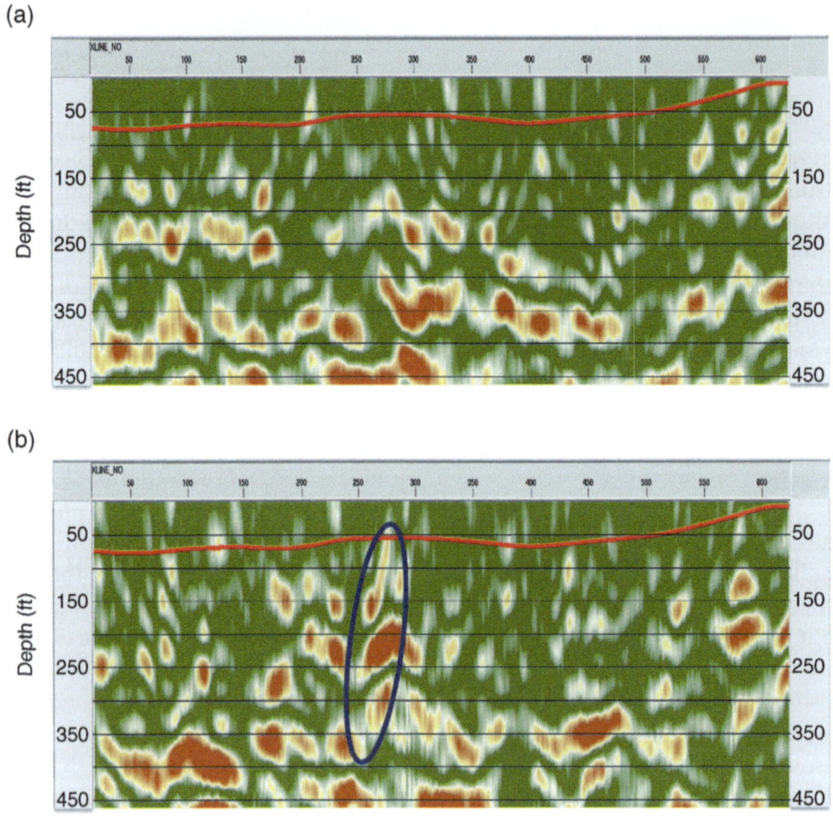

Figure 19.8 P-wave diffraction images. The diffraction imaging program discussed in section 19.3.6 was applied to the 91 common-source records. (a) Prestimulation P-wave diffraction image; (b) post-stimulation P-wave diffraction image. The oval in (b) discussed in the text as it is a location similar to the S-wave diffraction image.

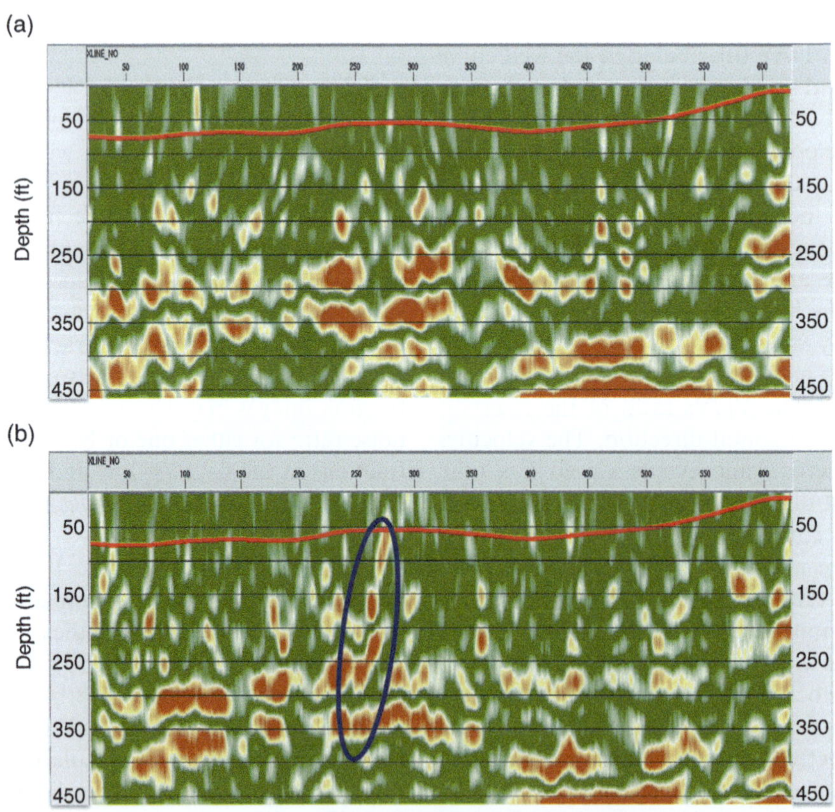

Figure 19.9 P-S wave diffraction images. The diffraction imaging program discussed in section 19.3.6 was applied to the 91 common-source records. (a) Prestimulation S-wave diffraction image; (b) post-stimulation S-wave diffraction image. The oval in (b) is discussed in the text as it is a location similar to the P-wave diffraction image.

and/or production caused a change in velocity or a fracture that is generating diffractions. This event indicates that there is a diffractor or set of diffractors common between the two datasets and provides evidence that diffractions from fractures do exist.

Another possible cause of the dissimilarity between the P-P and P-S diffraction images has to do with potential phase reversals in P-P diffractions. Seismic models produced by Leviant et al. (2019) show that vertical fractures with aspect ratios in the range seen in oil and gas fields produce both P-P and P-S diffractions. The models also indicate that a phase reversal can occur in P-P diffractions on opposite sides of diffracting fracture when the fracture aspect ratio is on the scale of those seen in oil and gas fields. Phase reversals in real data are difficult to observe when diffraction events are not readily observable in common-source records. It is therefore not obvious how to handle the phase reversals in processing. If phase reversals also occur in real data, then summing the reversed and nonreversed diffraction traces will suppress the diffraction signal. Diffraction phase reversals may be a source of the dissimilarity between Figures 19.8 and 19.9.

Based on the lower likelihood of phase reversals in the P-S diffraction images, the choice was made to rely on the P-S diffraction images for interpretation. The interpretive work is presented in the following section.

19.3.10. Comparison of Diffraction and Reflection Images

As noted in the Introduction section, the seismic reflection images were used to identify where stimulation and production-related fractures are likely to have occurred. Millimeter-scale fractures will not be detected by the reflection images but the effects of fracture stimulation and production can be seen in amplitude variations of the reflection amplitudes. Attention to the reflection images for this chapter is therefore focused on the approximately 100 ft (30.48 m) immediately below the horizontal well that was stimulated. Fracture density is most likely greater adjacent to the stimulated borehole than farther away from the borehole because the closely spaced borehole perforations are within that rock mass. Analysis of the reflection images also shows that local interval velocity changes as a result of stimulation and/or production range up to 5% for P waves and 10% for S waves in the 100 ft (30.5 m) immediately adjacent to the borehole. It is therefore likely that hydraulically stimulated fractures penetrated the reservoir from which fluids were produced. Stress changes due to pressure drawdown and changes in pore contents both lead to velocity changes (Sammonds et al., 1989).

Figure 19.10(a), (b) shows, respectively, the pre- and post-stimulation P-wave reflection images converted from the time domain to the depth domain using the constant

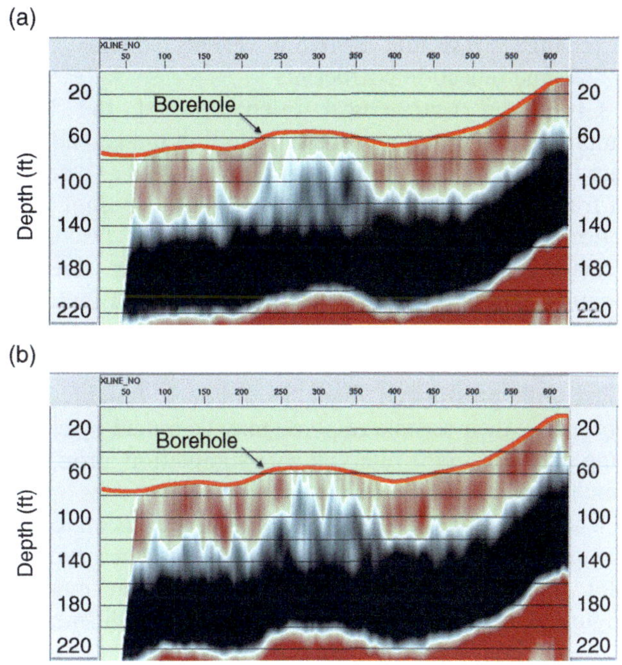

Figure 19.10 Pre- and post-stimulation P-wave reflection images stretched to the depth domain. The P-wave reflection images were stretched to the depth domain and shown here as the prestimulation image (a) and post-stimulation image (b). The well location is shown as a red curve across the top of the nonzero data samples and annotated as "borehole." Interpretation emphasis is placed on the 100 ft (30.5 m) immediately below the horizontal well because that reflection interval is closest to the wellbore perforations and was therefore fractured by the stimulation. Red values have positive amplitude, and black values have negative amplitude.

velocity of 16,388 ft/s (4,995 m/s) used for P-wave diffraction imaging. The image amplitudes are scaled to emphasize the uppermost reflector to show where amplitude discontinuities occur in the reflecting interval. The red curve near the top of the image is the horizontal wellbore location.

Figure 19.10(a) clearly exhibits prestimulation amplitude discontinuities in the reflecting interval immediately adjacent to the borehole. Figure 19.10(b), the post-stimulation reflection image, shows a higher spatial frequency of amplitude discontinuities with higher-amplitude contrasts than the prestimulation image. Interestingly, many of the low-amplitude discontinuities seen in Figure 19.10a are replaced by amplitude highs and vice versa in Figure 19.10(b). The physical difference in the reservoir between the prestimulation and post-stimulation reflection images are

- Introduction of new fractures in the reservoir
- Introduction of proppant in the reservoir to hold some fractures open

- Production-related pressure drawdown that can change stress and therefore velocity, density, and seismic reflection coefficients

Given the near-vertical orientation of the low-amplitude discontinuities in the reflection image and the fact that the amplitude highs and lows moved as a result of fracture stimulation, it is assumed that low-amplitude reflection discontinuities are associated with fractures. Fractures obviously do not move laterally due to fracture stimulation. Pre-existing fractures and faults in Figure 19.10(a) could, however, close in response to fracture stimulation when adjacent fractures are opened and filled with pressurized fluid. So although fractures do not move, their seismic response may be muted by changing pressures in the reservoir with stimulation, proppant emplacement, and production.

The discussions that follow include an overlay on top of the reflection images in which the overlay is a new type of plot that is computed from the diffraction images. Figure 19.11(a) shows the post-stimulation P-S diffraction image. Figure 19.11(b) was derived from the diffraction image and is an outline of the diffraction events such that all amplitudes within the black outline have an amplitude greater than a specific amplitude threshold value. In the following discussion, a diffraction event outline of the post-stimulation P-S diffraction image is superimposed over the seismic data so that the diffractions and reflections can be viewed simultaneously.

Figure 19.12 shows the post-stimulation reflection image converted to depth using the constant P-wave diffraction imaging velocity of 16,388 f/s (4,995 m/s). The seismic reflection image is scaled to emphasize amplitude variations in the uppermost reflection and is also overlayed with the post-stimulation P-S diffraction outline image discussed in Figure 19.11. Figure 19.12 also shows an annotated set of picks that follow the uppermost reflection. A plot above the seismic data shows a black curve that is the seismic amplitude at each of the reflection picks in the seismic data. Black down-pointing arrows indicate where locally low seismic amplitudes are identified.

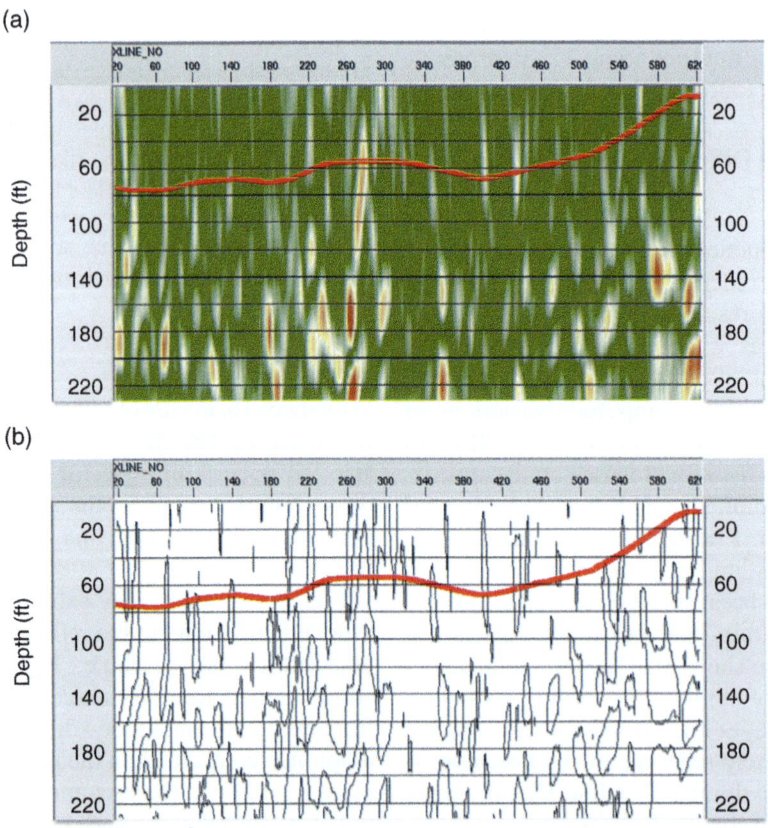

Figure 19.11 A diffraction outline image. (a) P-S post-stimulation diffraction image. (b) Plot generated from the P-S diffraction image in (a), which is the outline of the diffraction events shown in (a). Data points inside the closed outlines have a diffraction amplitude greater than a threshold amplitude value. The smallest diffraction amplitudes were discarded to minimize clutter in the plots. The red curve at the top of both images is the borehole location.

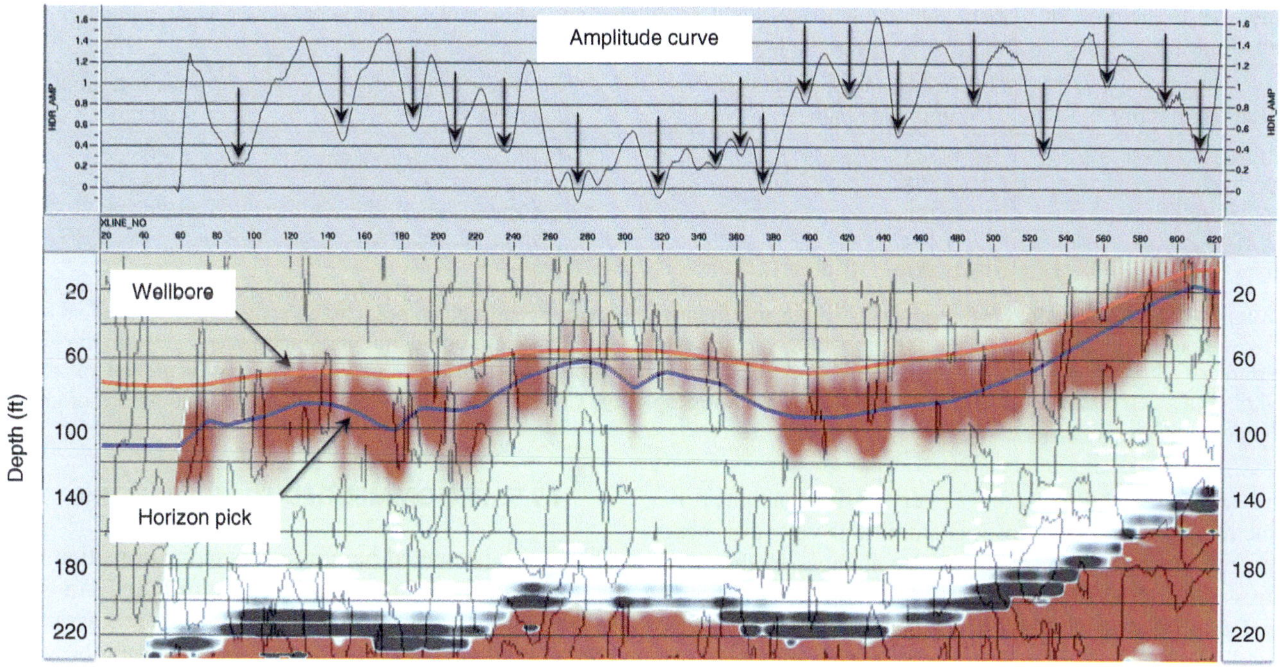

Figure 19.12 The diffraction outline image is overlayed on the scaled reflection image that is displayed to emphasize breaks in the reflector amplitude immediately adjacent to the wellbore (annotated). A horizon pick (annotated) along the reflection next to the wellbore was used to extract reflector amplitudes, which are plotted in the curve above the seismic plot and annotated "Amplitude curve." The downward-pointing black arrows mark local seismic reflection amplitude lows in the amplitude curve.

Visually, at least, there appears to be a correlation between local low seismic reflection amplitudes and diffraction image amplitude peaks in Figure 19.12. A more analytical analysis of the correlation between diffraction image highs and reflection amplitude lows was done, however, by computing the probability that a diffraction image peak amplitude is within a given distance of a local reflection amplitude low. Table 19.1 shows the computed probability that a high amplitude diffraction image event is within the given distance (column 1) of a local reflection amplitude low along the picked horizon. The table also shows the probability that a random location in the image along the horizon pick is within the same distance of a seismic reflection amplitude low.

In all cases tested, the probability that a diffraction image peak is near a low seismic reflection amplitude is between 1.77 and 1.80 times the probability that a random location along the picked horizon is near a low seismic reflection amplitude.

The probability values shown in Table 19.1 were computed by first measuring the horizontal distance between peak values in the diffraction image along the picked horizon in Figure 19.12 and the *nearest* reflection amplitude low along the same picked horizon. Recall that the reflection amplitude lows along the picked horizon are indicated by the downward arrows in Figure 19.12. The mean and standard deviation of distances between reflection amplitude lows and diffraction image peaks were then computed. The probability of a diffraction amplitude peak being within a given distance of a reflection amplitude low was computed by assuming that the distance measurement values earlier were Gaussian distributed and then computing twice the area under the normalized Gaussian probability density function from zero distance to the distance in column 1 in Table 19.1 (Peebles, 1980). The probability values for random locations were done in the same way as for diffraction image highs, except that the random locations were every data trace in Figure 19.12 along the picked horizon.

In Figure 19.12, the diffraction amplitude highs occur near and sometimes on seismic reflection amplitude lows, but diffraction highs also occur directly on top of reflection amplitude highs in some places. Despite the statistical evidence in Table 19.1, the correlation between diffraction amplitude highs and reflection amplitude lows is not perfect. There are several possible explanations for the lack of correlation in places. One is that multiple factors affect seismic reflection amplitudes including that energy from outside the image plane is contributing to the reflection amplitudes. Another is that although a fracture may have penetrated the picked horizon, resulting in a high diffraction image amplitude, the fracture did not disrupt the reflection amplitude on the picked horizon. The picked horizon may have remained as a continuous reflector

Table 19.1 The probability that a diffraction image is high is within a given horizontal distance of a low seismic reflection amplitude.

Distance in ft (m)	Probability of diffraction amplitude high	Probability of random location	Ratio column 2 to column 3
20 (6.09)	0.243	0.135	1.80
25 (7.62)	0.296	0.166	1.78
30 (9.14)	0.354	0.197	1.79
35 (10.6)	0.404	0.228	1.77
40 (12.2)	0.458	0.258	1.77

Note: Column 1 is the horizontal distance in feet (meters in parentheses) between a given point in the image and an identified local low seismic amplitude. Column 2 is the probability that the center of a high amplitude diffraction image event is within the given distance (Column 1) of a local low seismic amplitude. Column 3 is the probability that a random location along the picked horizon is within the given distance (Column 1) of a local low seismic amplitude. Column 4 is the ratio of Column 2 to Column 3 (Column 2 divided by Column 3).

either because the fracture was very narrow or that production from the fracture was small and velocities near the reflector were not disrupted by changes in the rock stress field. It may also be that the correlations shown in Table 19.1 are a statistical accident and there is no diffraction energy actually shown in the earlier figures. This final option assumes that the modeling results of Leviant et al. (2019) are significantly in error.

19.4. DISCUSSION AND CONCLUSIONS

The work in this chapter and the values presented in Table 19.1 show that P-S diffraction image maxima are about 1.8 times more likely to occur within a given distance of a relatively low reflection amplitude than if a point in the image were randomly selected. This result does not stand alone as proof that the diffraction events hypothesized by Leviant et al. (2019) either exist or that the DAS data has captured the diffractions. Neither do the results discount the possibility that the diffraction images are measuring diffraction energy. The higher probability of a diffraction image event being located near a low seismic amplitude in fact supports that the diffractions exist. Additional evidence supporting that the DAS data has captured at least some diffractions comes from the similar event shape in Figures 19.8(b) and 19.9(b) that show a similarly shaped diffraction event in both the P-P diffraction image and the P–S diffraction image.

Additional albeit lower-confidence evidence also exists that diffractions exist. Low-amplitude events that rise above the background noise in common-source gathers fit the expected curvature of diffraction events, but they are only partial events that do not stand out like fully formed synthetic events shown in Figure 19.6. It is possible that in a different field location with higher SNR or with more sensitive recording equipment, diffraction events would be visible in common-source records, thus providing definitive proof that the diffractions have been recorded.

19.5. RECOMMENDED FURTHER WORK

The work presented here provides indications that diffractions from thin fractures are present in the DAS data but does not prove it. Further work is therefore warranted to either confirm or dismiss the hypothesis given the large potential benefits for reservoir analysis if diffractions from thin fractures exist and can be used to directly image fractures. Work of this type should be extended to work sites where detailed core analysis and logs from reservoirs can definitively identify the presence of reservoir fractures in the borehole from which the core was retrieved. The Hydraulic Fracture Test Site (HFTS) projects (Gale et al., 2019) would be well-suited to this type of analysis because most of the data types listed earlier are generally collected with new projects, provided surface seismic sources are included in the project.

The seismic modeling presented by Leviant et al. (2019) was useful for multiple purposes in the work presented in this chapter. The modeling provided theoretical evidence that thin fractures might be able to produce diffractions and that those diffractions would be produced at common seismic frequencies. As stated, further work in determining if thin-fracture diffractions exist should be done on specific reservoirs with known fracture widths and estimates of fracture height and length. In that research environment, it would be useful to compute seismic models of the expected seismic diffraction response of the known fracture geometry to assist in determining if diffractions are in fact produced and if the seismic modeling provides realistic results. The modeling step obviously requires that the software be available either directly to the researchers or via another research or commercial service. Currently, there does not appear to be a strong market for diffraction modeling via the grid-characteristic method of Leviant et al. (2019), so it may be difficult to find access to the software.

If 3D thin-fracture diffraction modeling software were to become available, there would be many fracture environments to which it could be applied. The software should be used to produce particle motion direction results for P-P and P-S diffracted waves for varying wavefront arrival angles relative to the fracture orientation, fracture fill material, fracture rugosity, and seismic wave frequency bands. Particle motion direction is particularly important in DAS recording, given the sensitivity of particle motion with respect to the DAS fiber direction. Modeling would provide estimates of which waves (P-P or P-S) to use in diffraction imaging for different positions of diffractors relative to the DAS fiber or fibers in adjacent boreholes.

ACKNOWLEDGMENTS

I am grateful to David Langton with Devon Energy for his advice and attention in development of the reflection and diffraction data processing technology presented in this chapter. I am also grateful to S. Rutt Bridges for conceiving of, funding, and playing the lead role in development of the ProMAX seismic data processing software package. ProMAX has provided an efficient avenue over the last three decades for thousands of developers, including myself, to express new geophysical ideas in a professional-level seismic data processing system.

AVAILABILITY STATEMENT

The data and software used in this chapter are proprietary to private industrial organizations and are not available to the public.

REFERENCES

Bakulin, A., Mateeva, A., Mehta, K., Jorgensen, P., Ferrandis, J., Sinha Herhold, I., & Lopez, J. (2007). Virtual source applications to imaging and reservoir monitoring. *The Leading Edge*, *26*, 732–740.

Booth, A. D., Christoffersen, P., Schoonman, C., Clarke, A., Hubbard, B., Law, R., et al. (2020). Distributed acoustic sensing of seismic properties in a borehole drilled on a fast-flowing Greenlandic outlet glacier. *Geophysical Research Letters*, *47*, e2020GL088148. doi:10.1029/2020GL088148

Fenta, M. C., Potter, D. K., & Szanyi, J. (2021). Fibre optic methods of prospecting: a comprehensive and modern branch of geophysics. *Surveys in Geophysics*, *42*, 551–584.

Fuller, B. N. (1999). Principal component methods for suppressing noise and detecting subtle reflection character variations. In R. L. Kirlin & W. J. Done (Eds.), *Covariance Analysis for Seismic Signal Processing, Geophysical Developments* (Vol. *8*, (Chapter 11, pp. 227–240). Tulsa, OK: Society of Exploration Geophysicists.

Fuller, B. N., Bratton, T., & Langton, D. (2020). *Time lapse seismic imaging with a horizontal DAS cable for reservoir surveillance*. Paper presented at the Rocky Mountain 3D Seismic Symposium, Denver, CO (virtual presentation).

Gale, J. F. W., Elliott, S. J., & Laubach, S. E. (2019). *Direct observation of hydraulic fractures in the Hydraulic Fracture Test Site (HFTS) Slant Core, Reagan, Co., Midland Basin, Texas: insights into distribution, morphology, intensity and interaction with natural discontinuities*. Paper presented at New Directions in Geosciences for Unconventional Resources, Banff, Canada. https://www.cspg.org/common/Uploaded%20files/pdfs/documents/conference_website/gussow/archives/2019/Session%203/5-Gale.pdf

Kragh, E., & Christie, P. (2002). Seismic repeatability, normalized rms, and predictability. *The Leading Edge*, *21*(7), 640–647.

Leviant, V., Kvasov, I., & Petrov, I. (2019). *Numerical modeling of seismic responses from fractured reservoirs by the grid-characteristic method. Geophysical developments no. 17*. Society of Exploration Geophysicists.

Mateeva, A., Lopez, J., Potters, H., Mestayer, J., Cox, B., Kiyashchenko, D., et al. (2014). Distributed acoustic sensing for reservoir monitoring with vertical seismic profiling. *Geophysical Prospecting*, *62*(4), 679–692. doi:10.1111/1365-2478.12116

Pedersen, O., & Marcy, F. (2022). Diffraction Imaging: a valuable complement to reflection imaging for sand injectite interpretation in the Norwegian North Sea. *The Leading Edge*, *41*(4), 247–252.

Peebles, Z. P. (1980). *Probability, random variables, and random signal principles*. McGraw-Hill Series in Electrical Engineering. McGraw-Hill Book Company.

Sammonds, P. R., Ayling, M. R., Meredith, P. G., Murrell, S. A. F., & Jones, C. (1989). *A laboratory investigation of acoustic emission and elastic wave velocity changes during rock failure under triaxial stresses in Rock at Great Depth*, Paper presented at the ISRM International Symposium.

Tyiasning, S., Merzlikin, D., Cooke, D., & Fomel, S. (2016). A comparison of diffraction imaging to incoherence and curvature. *The Leading Edge*, *35*(1), 86–89.

Yilmaz, O. (1987). *Seismic data processing. Investigations in geophysics no. 2*. Society of Exploration Geophysicists.

20

Multiwell Distributed Acoustic Sensing 3D Vertical Seismic Profile Under Flowing Conditions: Toward Solving the Processing and Imaging Challenges

Shujaat Ali[1], Francesco Miranda[2], Spadavecchia Emmanuel[2], and Olga Podgornova[3]

ABSTRACT

Borehole seismic data acquisition using distributed acoustic sensing (DAS), particularly through permanent installations, has recently attracted a lot of interest especially for large 3D vertical seismic profile (VSP) surveys. DAS surveys offer large receiver apertures and mostly allow VSP acquisition without disrupting well operations or production, thereby significantly reducing costs. In multiwell DAS imaging, it is not uncommon for migration images to be impacted by strong illumination artifacts and poor stacking due to misalignment of images from different wells. The illumination artifacts arise from laterally sparse receiver geometry and are further exacerbated in cases when only part of the DAS receiver is usable for imaging. We provide a detailed processing and imaging workflow to address multiwell DAS 3D VSP-related challenges by applying image domain receiver directivity correction, single and multiwell image alignment prior to stacking, least-squares migration in image domain (LSM-ID), and post-image migration swing noise reduction to obtain high-resolution 3D image volume. It is also noted that proper denoise of the DAS data acquired in producing conditions, especially when fiber is deployed over tubing, is key in obtaining high-resolution subsurface images.

20.1. INTRODUCTION

Multiwell imaging with a DAS sensor imposes several challenges to obtaining high-resolution images. Multiwell DAS VSP images are commonly affected by uneven illumination and misalignment of reflections between wells (Yurikov et al., 2021); fiber directivity in deviated wells, such as in the present case, also impacts the amplitudes and must be compensated for. Image misalignment between the wells is primarily due to the imperfections of the velocity model used for migration and the differences in incidence angles from different wells. Receiver directivity is corrected by applying a ray-based correction to individual receiver migration images (Ali et al., 2021). Image misalignment between the receivers and wells is resolved by an image-based trim alignment. Finally, an image domain least-squares reverse time migration (RTM) LSM-ID process is used to address the variable illumination due to receiver geometry. LSM-ID reduces the illumination artifacts and improves the coherency of reflection events by mitigating the variable nature of the 3D image wavelet.

In the present case, a significant amount of coherent noise related to ringing caused by poor fiber/tubing coupling and production flow is observed in the data. Multi-domain noise attenuation schemes are used to improve the signal-to-noise ratio of the recorded data.

Initially, the available isotropic velocity model from surface seismic pre-stack depth migration (PSDM) processing is updated with VSP direct-arrival traveltimes

[1] *SLB, Houston, Texas, USA*
[2] *ENI Natural Resources, Milan, Italy*
[3] *SLB, Cambridge, Massachusetts, USA*

Distributed Acoustic Sensing in Borehole Geophysics, Geophysical Monograph 289, First Edition.
Edited by Yingping Li, Robert Mellors, and Ge Zhan.

DOI: 10.1002/9781394179275.ch20

and reflection VSP-common image point (VSP-CIP) tomography. Tomography inversion is done with a tilted transversely isotropic (TTI) background. Tilt axis information is derived from surface seismic dip field and anisotropy parameters, and P-wave velocities are inverted by tomography.

20.2. DATA ACQUISITION

DAS fibers deployed as part of the completion strings (strapped to the outside of tubing) on four wells in an offshore field are used to record a 3D VSP survey with ongoing production, using a source grid with more than 10,000 shots. The data are acquired in a grid pattern with shot spacing of 50 m and line spacing of 125 m. Figure 20.1 shows the source and receiver locations. The colored dots are receivers, and empty circles are sources.

Imaging targets of interest are truncating sedimentary features and complex fault geometries; therefore, high-resolution imaging with 3D VSPs is desired to improve the details that are poorly resolved by existing surface seismic.

The quality of the acquired data is quite variable within and across wells. Well deviation generally improves fiber coupling, resulting in a higher signal-to-noise ratio than observed in the vertical section of the well. As an example, for the data shown on the left side of Figure 20.2, the well is nearly vertical up to 610 m measured depth (MD). From 610 to 717 m, the well builds up a maximum deviation angle of 4.8° and gradually drops to near zero at 950 m. From 1,100 m depth, the well deviation starts to build up again to reach a maximum deviation of 40° around 1,700 m. Below this depth, the well deviation is between 40 and 34°. The DAS data shows remarkable correspondence between the well deviation angle and the presence of casing arrivals. Casing arrivals are seen both in intervals above 600 m and near vertical intervals around 900–1,100 m. Production noise is present on all wells; however, the amplitude of this noise varies over time as well as over receiver depths.

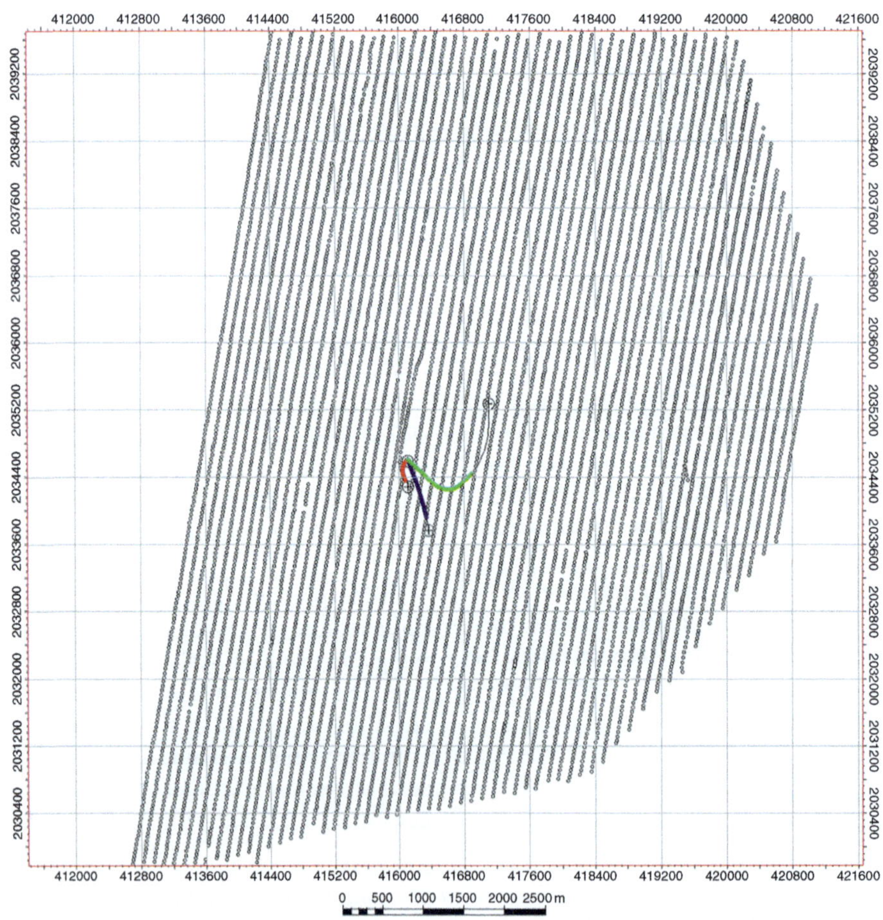

Figure 20.1 Map of source and receiver geometry. Horizontal and vertical axes are easting and northing in meters.

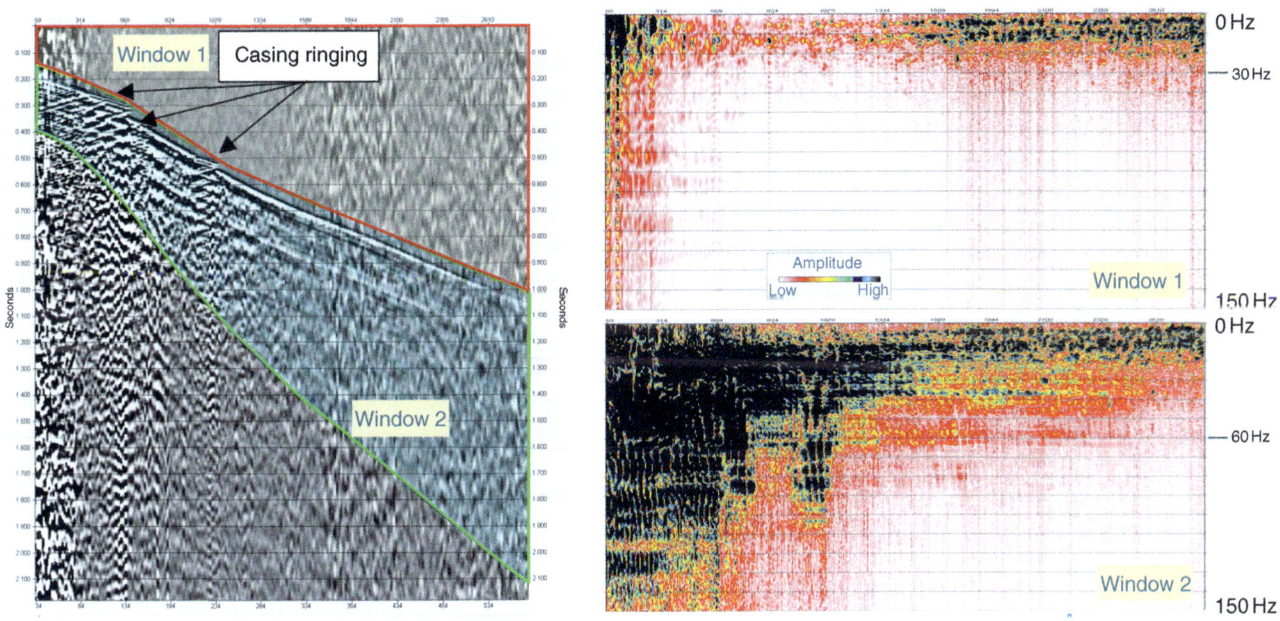

Figure 20.2 Near-offset shot gather (left) and amplitude spectra of signal and noise windows (top and bottom right, respectively). Trace index shown above the data is measured depth from Kelly bushing (KB).

20.3. DATA PROCESSING

20.3.1. Shot Domain Denoising and Processing

Production-related noise was dominant at lower frequencies (up to 35 Hz) and was spread over a range of velocities in the shot domain. This is evident from the analysis of the near-offset shot record shown in Figure 20.2. Amplitude spectra are computed in time widows of roughly equal lengths before and after first arrivals. Production-related background noise is more dominant in deeper parts of the well, whereas in the vertical part of the well, casing ringing-related noise is dominant.

Tau-p and frequency-space infinite impulse response (FXIIR) dip filters were used to remove very low-velocity components of this noise. A tau-p domain denoise was also done to attenuate constant velocity noise by algorithmically identifying and suppressing the anomalous amplitudes in this domain.

After applying the ray-based directional deghosting step, a shot-by-shot source signature deconvolution was applied to debubble and zero-phase the signal. The far-field signature for this operation was computed from the near-field hydrophone for each shot.

To suppress the repetitive ringing noise caused by poor fiber/tubing coupling, a predictive deconvolution operator was used. Figure 20.3 compares one such gather after shot domain processing. After denoising, a downgoing wavefield is removed in the shot domain by a tau-p domain filter.

A shot domain scaling by 1/p values was applied in the tau-p domain. This step effectively provides amplitude-only DAS to velocity scaling. For example, a common receiver gather before and after scaling is compared in Figure 20.4, where it can be seen that scaling by 1/p improves amplitude balancing on either side of the minimum offset location.

20.3.2. Receiver Domain Denoise and Interpolation

Shot domain denoised and strain-to-velocity converted DAS data are normal moveout (NMO) corrected by ray tracing through a 1D velocity function extracted from the isotropic PSDM velocity model at the wellhead location. The NMO-corrected data are then sorted in the receiver domain. When such data are analyzed along or orthogonal to the shooting direction, several parabolic events are identified with their origin at the zero-offset location. A 3D parabolic Radon transform (3D PRT) is used to model these events. The coherent noise modeled through this process is further isolated by curvelet transform before adaptively subtracting it from the data. The parameters for this 3D PRT transform are carefully selected for signal protection and are varied with receiver depth. For example, an NMO-corrected common receiver gather from a near well location line, before and after 3D PRT denoise, is shown in Figure 20.5. The data after this step are regularized to a 50 m × 50 m grid through a 4D matching pursuit Fourier interpolation (4D MPFI) algorithm. This step further attenuates the production noise that, because of its time-variant nature, remains incoherent in

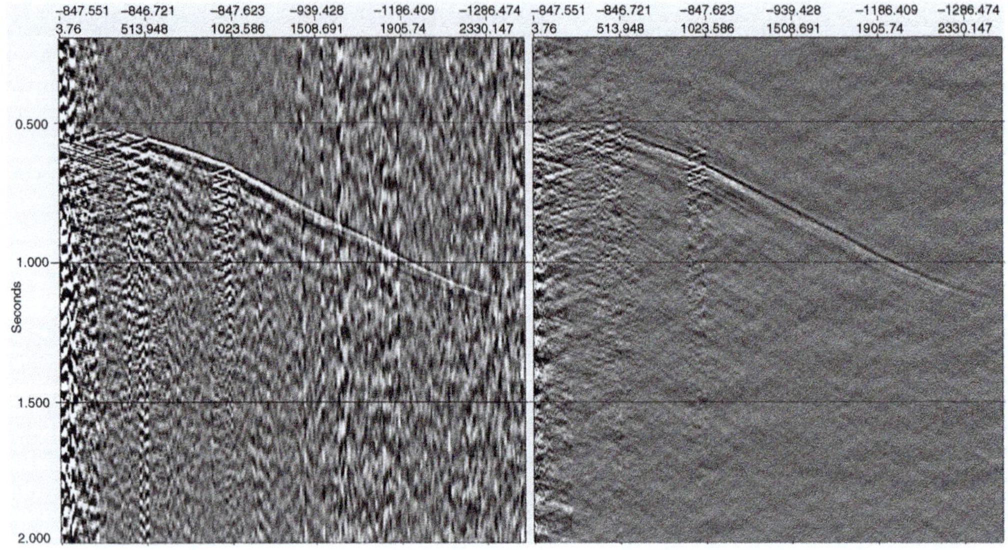

Figure 20.3 Near-offset common source gather before (left) and after (right) source domain processing. Top annotations indicate source-receiver offset (top) and receiver depth (bottom).

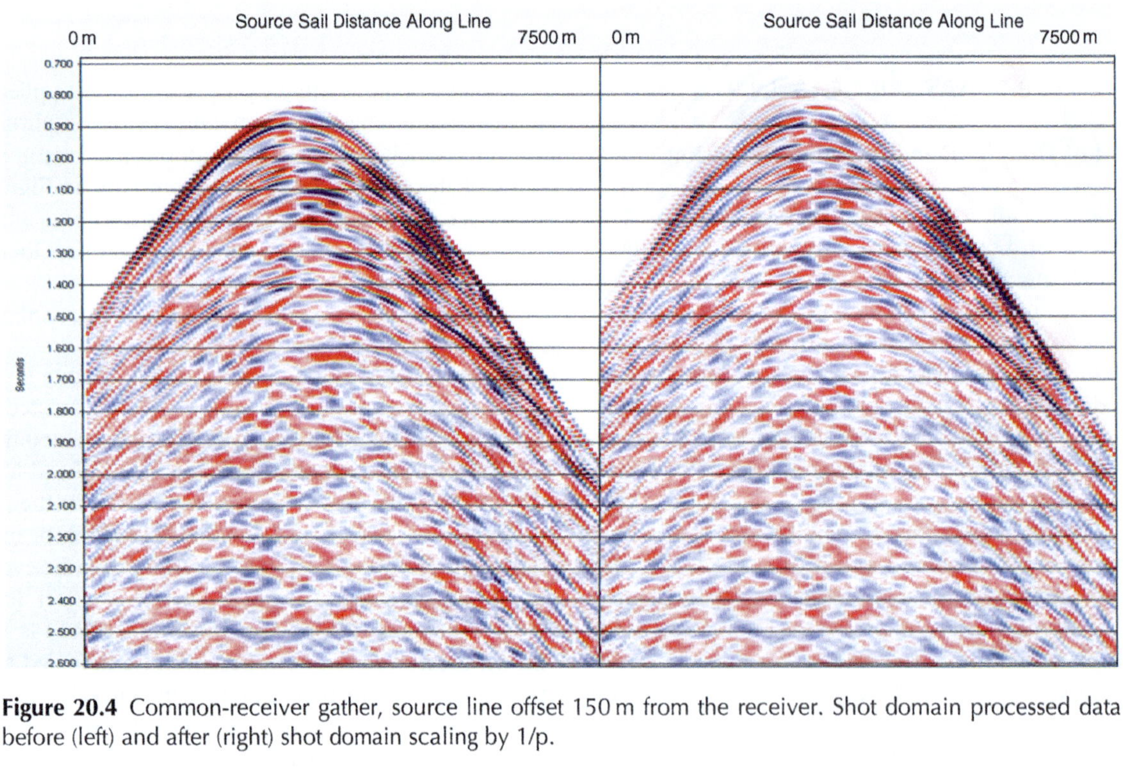

Figure 20.4 Common-receiver gather, source line offset 150 m from the receiver. Shot domain processed data before (left) and after (right) shot domain scaling by 1/p.

the receiver domain. Following this step, NMO correction is removed. Examples of common shot and common receiver gathers of raw and processed data after this step are compared in Figure 20.6.

As a final spectral shaping step, a bandwidth extension algorithm is used to enhance amplitudes both on the low and high ends of the input spectrum.

20.4. VELOCITY MODEL CALIBRATION

The velocity model is calibrated using the 3D VSP direct-arrival times and VSP-CIP gathers with reflection data through a tomography workflow.

The initial velocity is an isotropic model from PSDM processing of marine seismic data. Overall, such initial

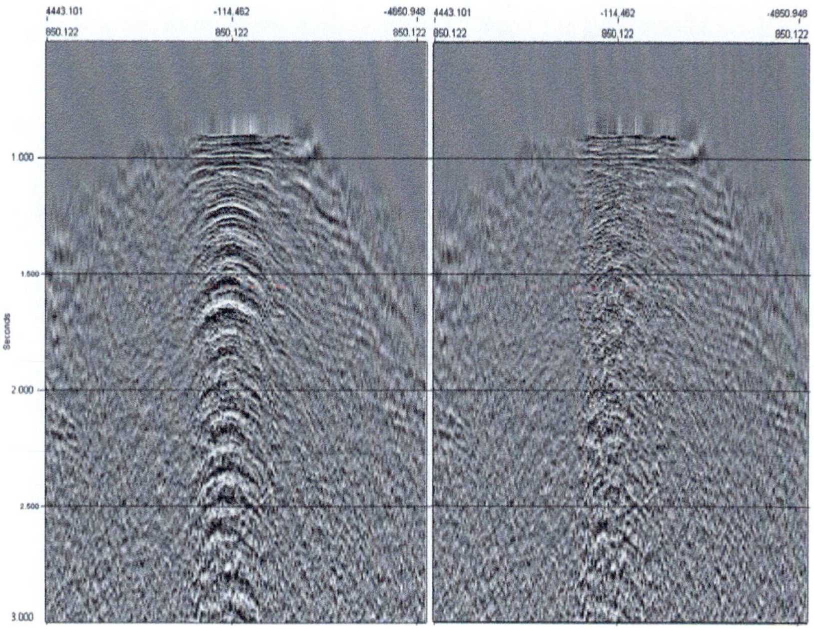

Figure 20.5 NMO-corrected common receiver gather from a near well line, before (left) and after (right) 3D PRT noise estimation and removal processing. Top labels indicate source-receiver offset (top) and receiver depth (bottom).

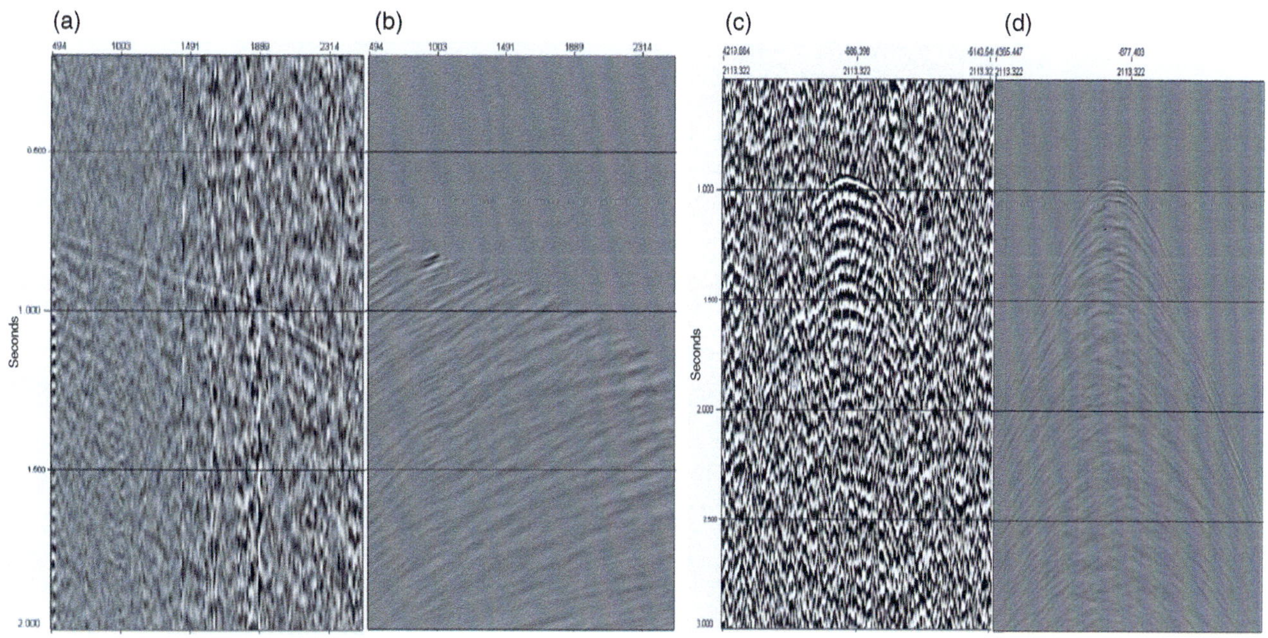

Figure 20.6 Example mid-offset shot record raw (a) and processed upgoing wavefields (b). Top labels indicate receiver depth (m). Receiver gather from a near well line, raw (c) and processed upgoing wavefield (d). Top labels indicate source-receiver offset (top) and receiver depth (bottom).

velocity reasonably explains the VSP times and provides adequately flat VSP-CIP gathers. However, some local differences between the VSP times and the PSDM model are observable; in addition, CIP gathers at the target level have moderate moveouts. As an example, Figure 20.7 shows residual times of one receiver with initial and calibrated velocity models. The calibrated model significantly reduces the traveltime residuals. Large positive residuals in the calibrated model (extreme south end) are caused by salt arrivals, which cannot be modeled in the sediment-only model used in the study.

VSP-CIP gathers are produced by migrating individual receivers and sorting by receiver depth for each image point. A set of VSP-CIP gathers from initial and

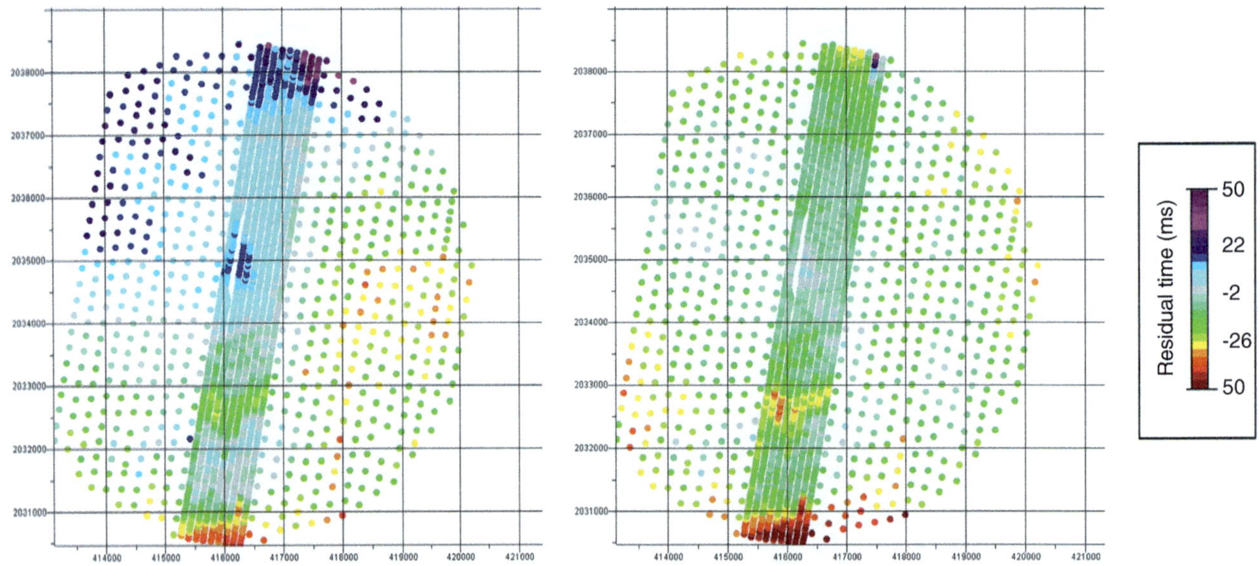

Figure 20.7 Map of travel time residuals from initial (left) and final model (right) for a single receiver. Minimization of residual times is seen at near offsets as well as at long offsets in the north direction. Some increase in residual traveltimes is seen on the south side, mainly due to some ray paths through the salt (salt is not included in the velocity model).

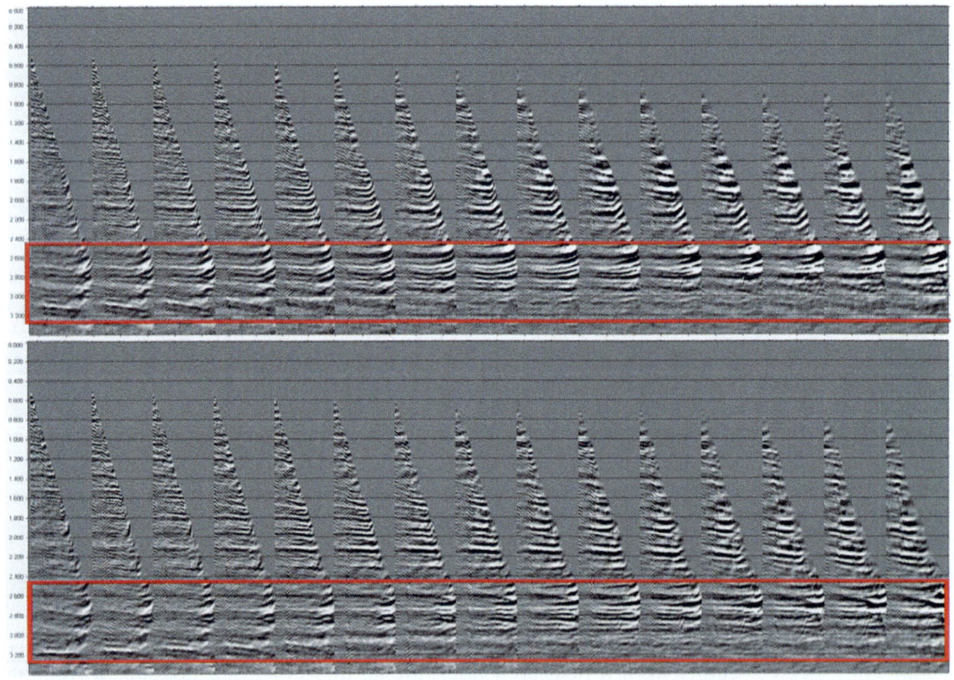

Figure 20.8 VSP-CIP gathers from initial (top) and calibrated (bottom) velocity model. The target depth interval is highlighted by the red rectangle.

calibrated models is shown in Figure 20.8. It is evident from this figure that the calibrated model step improves the reflection alignment in the target interval, highlighted in red. Figure 20.9 compares the initial and calibrated model image from a single well. In addition to resolution improvements in the target interval, a significant uplift in structural clarity is observed, such as in the area highlighted by the rectangle.

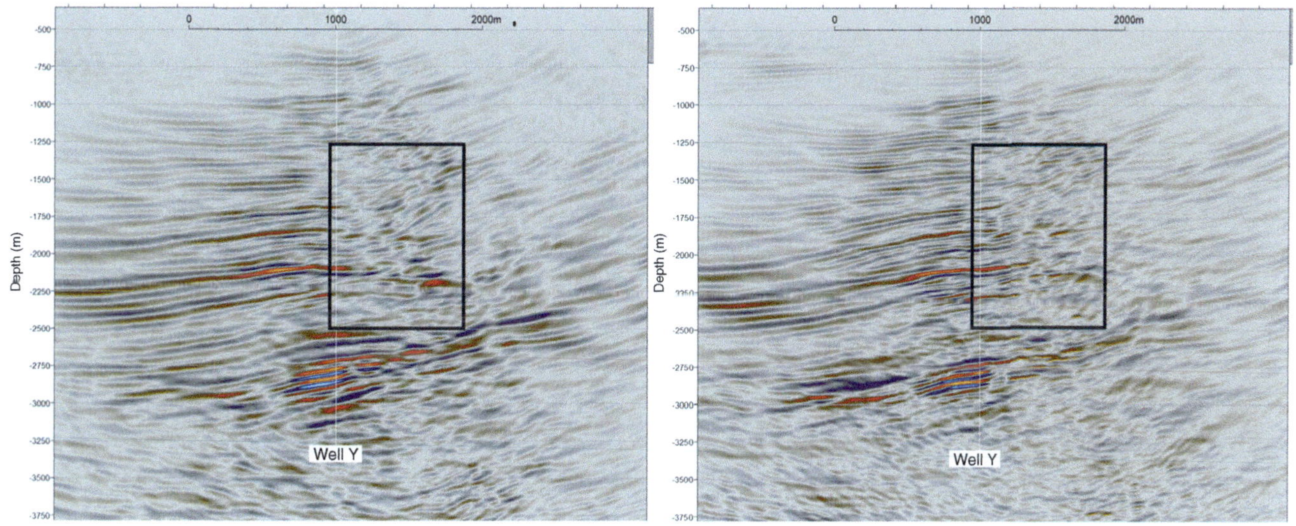

Figure 20.9 Migration section from single well data. Left: migrated with initial model. Right: migrated with calibrated model. The area highlighted by the black square shows significant improvement in resolution of the faulted region.

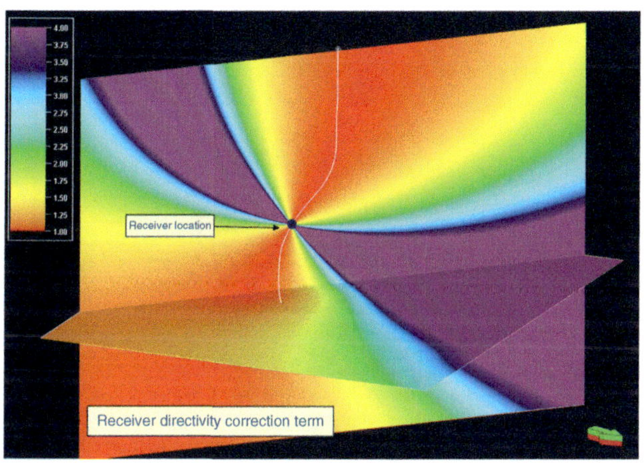

Figure 20.10 Receiver directivity correction vertical and horizontal slices for a receiver in deviated section of the well. Colors indicate correction term, scaled between 1 and 4.

20.5. IMAGING AND POSTMIGRATION ENHANCEMENT

The processed data are migrated through least-squares RTM. The mathematical formulations of finite-difference modeling for the DAS sensors are given in Appendix 7.

In RTM, a small group of consecutive DAS receivers are migrated together. Each group consists of 10 receivers (~50-m span). These receiver group image volumes are individually scaled by applying the corresponding local fiber directivity correction volume. This correction term is computed by taking the inner product of receiver slowness vectors (from 3D ray tracing) and fiber direction vector (from well trajectory) at the center of the receiver subgroup span. An example of a receiver directivity correction factor is shown in Figure 20.10. Image cross-sections before and after correction are shown in Figure 20.11. The directivity correction display is scaled between 1 and 3.

Receiver image volumes are scanned for data quality, and only good-quality receivers from all the wells are selected for stacking. An interpretation-guided selective stacking (IGSS) process is applied to the receiver image volumes, where each image volume is decomposed into several 3D dip components. These dip components are weighted using a background dip field created from surface seismic volume and restacked. This reduces migration smiles while retaining spatial resolution. Receiver group image volumes from each well are corrected for misalignment of reflection events; the correction is obtained by sequentially aligning the shallow to deep receiver images within each well by using a 3D dynamic warping algorithm. This step provides the shift field needed to align consecutive receivers. This shift field is smoothed in the CIP domain before applying to the individual receiver group images without IGSS processing. A stack of unaligned and aligned images are compared in Figure 20.12. A significant improvement in reflector continuity amplitude consistency is observed after the alignment step. After individual, well-aligned stacked images are generated, another alignment step is applied following the same process, and the stack of the all-well images is formed.

20.5.1. LSM-ID

Multiwell DAS 3D VSPs generally comprise locally dense but overall sparse receivers. This uneven receiver geometry, coupled with complex subsurface structures, often results in uneven illumination of subsurface structures and spatially variable image wavelets. In addition,

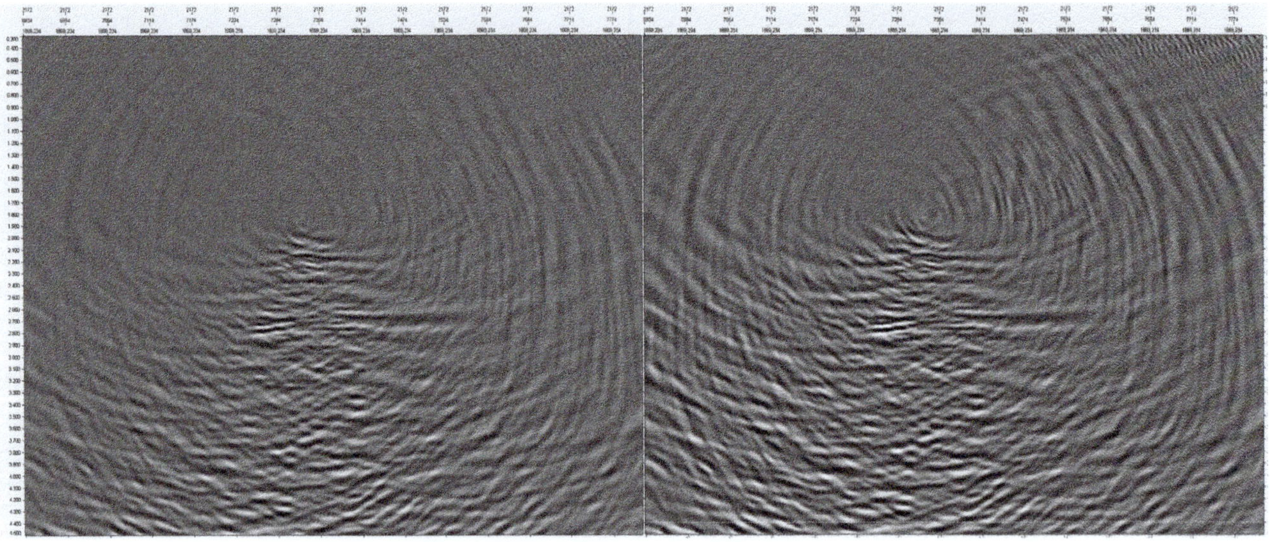

Figure 20.11 Receiver gather RTM vertical cross-section before (left) and after (right) receiver directivity correction.

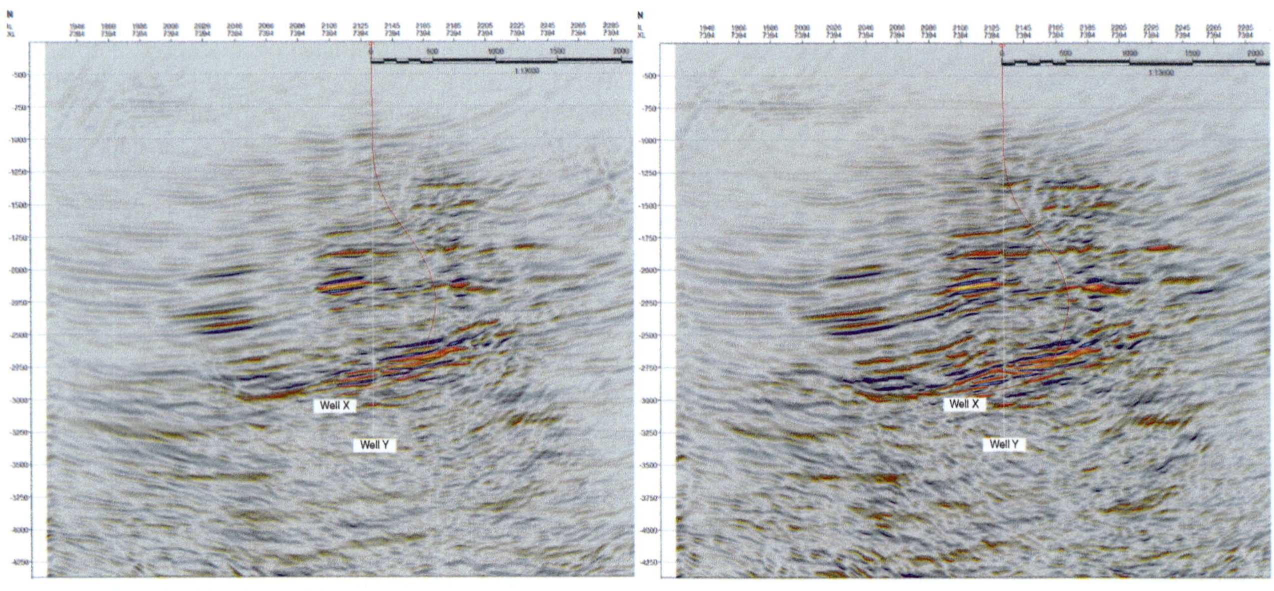

Figure 20.12 Migration section from all wells, stacked after receiver directivity correction. (Left) before alignment; (right) after alignment. Dip-guided selective stacking has been applied to both volumes.

migrated image quality could be further degraded by lack of resolution caused by wavelet stretch, attenuation, and suboptimal deghosting (Fletcher et al., 2016). Because the VSP illumination drops off laterally away from receivers, the 3D image wavelet is also spatially variable. This variability of the seismic wavelet is evident in point spread functions (PSF) shown in Figure 20.13. The migration images before and after LSM-ID are shown in Figure 20.14. LSM-ID is seen to significantly reduce illumination artifacts in the image, most notably when comparing amplitudes along depth. Noticeable improvements in resolution are also seen, and fault-related truncations are better resolved in the LSM-ID image.

The LSM-ID image of 3D VSP is compared to surface seismic in Figure 20.15 (cross-section) and Figure 20.16 (depth slices). The 3D VSP image shows significant improvement in vertical and lateral resolution compared to surface seismic.

20.6. CONCLUSIONS

A DAS 3D VSP survey recorded in very challenging data acquisition conditions is handled through a rigorous processing sequence to yield a high-resolution, relative-amplitude friendly subsurface image volume. We

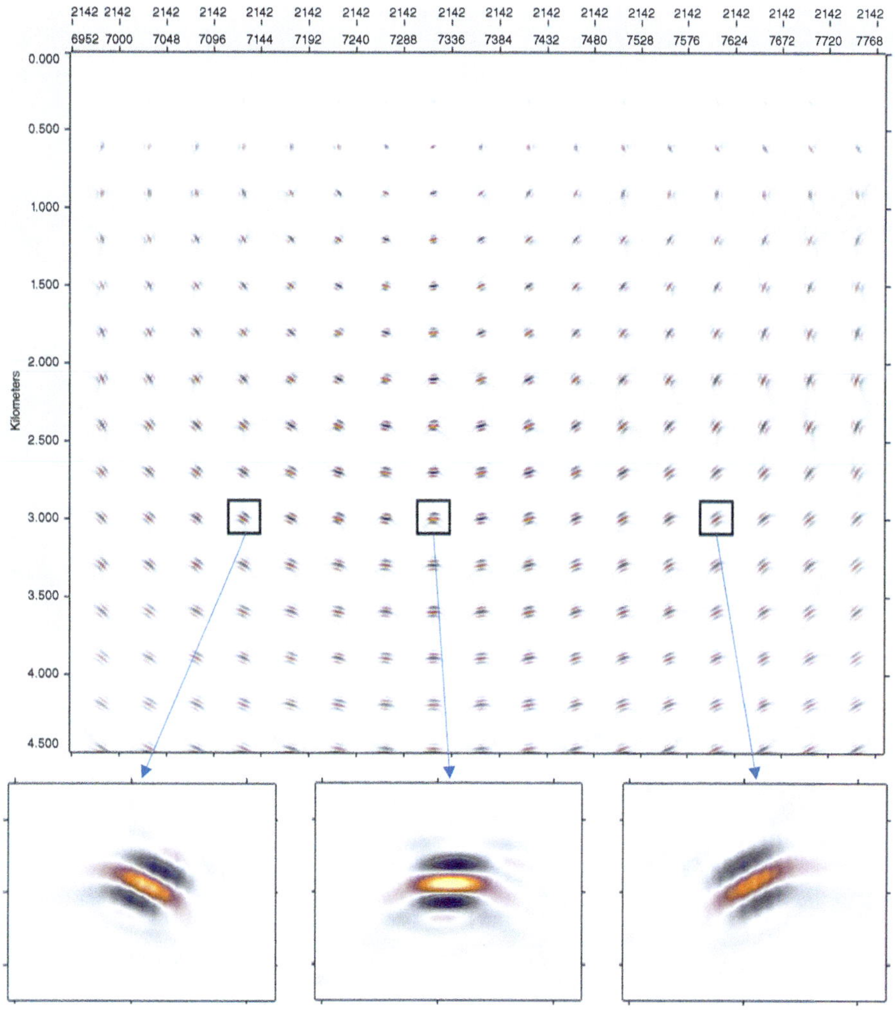

Figure 20.13 PSF functions from combined contributions of the multiwell DAS data. Zoomed-in view of the PSF functions (bottom row) shows location-dependent variability of the seismic wavelet.

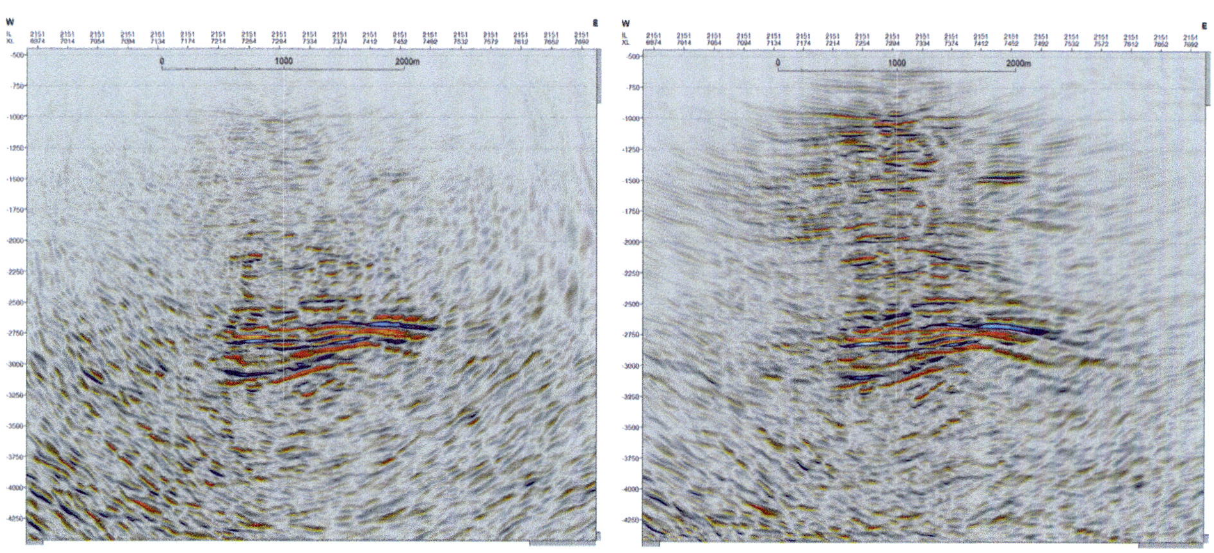

Figure 20.14 Aligned stacked data before (left) and after LSM-ID (right). Some migration smiles have been removed by an IGSS process on the LSM-ID output.

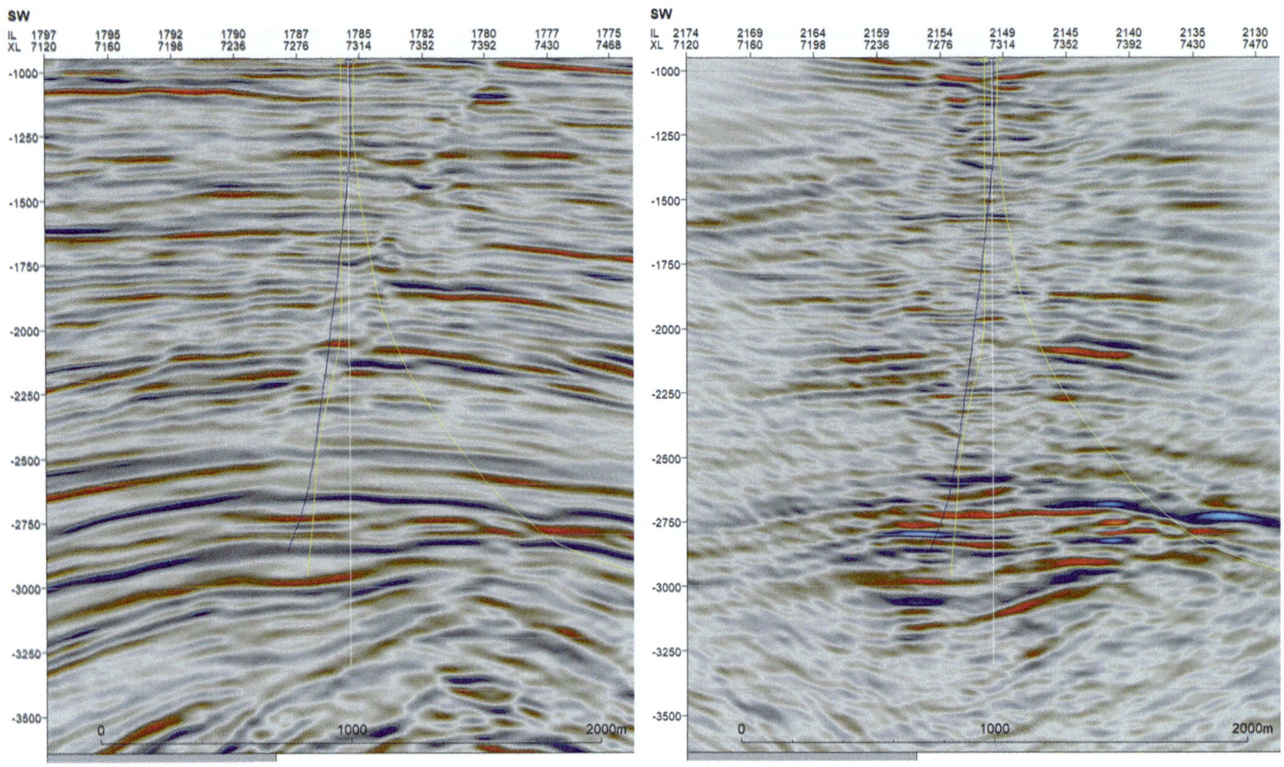

Figure 20.15 Surface seismic (left) and multiwell DAS 3D VSP image after LSM-ID (right).

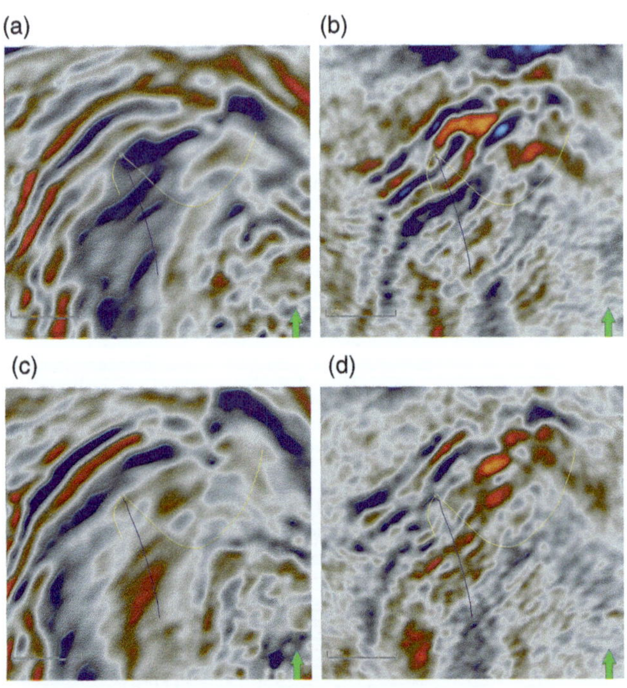

Figure 20.16 Surface seismic (a, b) and 3D VSP (c, d) depth slice comparisons at depths close to the target interval.

demonstrated that through careful processing, DAS data acquired in a high-noise, producing well environment can yield excellent-quality images with improved resolution in comparison to surface seismic. The ability to handle the high flow noise is key for successful seismic monitoring projects using DAS data.

Commonly observed challenges to multiwell DAS 3D VSP imaging, such as receiver directivity, image misalignment between different receivers, variable illumination, and spatially varying 3D image wavelet, are effectively addressed by the proposed processing scheme. LSM-ID has effectively removed a very strong survey footprint, both in terms of amplitudes and 3D variability of the wavelet, resulting in a high-resolution image.

ACKNOWLEDGMENTS

The authors would like to thank ENI and SLB for permission to publish the results of this study.

AVAILABILITY STATEMENT

Data used in this study are confidential and cannot be shared.

APPENDIX: DAS MODELING AND IMAGING

A.1. Modeling of DAS

For a fiber perfectly coupled with the well, DAS data at the receiver $\mathbf{x}_{rcv}$ give the strain of the elastic formation taken along the fiber and integrated over a gauge window

(Daley et al., 2015; Dean et al., 2015; Frignet & Hartog, 2014; Mateeva et al., 2014)

$$d_{DAS}(\mathbf{x}_{rcv}) = \int\limits_{\tau \in [\mathbf{x}_{rcv} - G/2,\, \mathbf{x}_{rcv} + G/2]} \varepsilon_{\tau\tau} d\tau. \tag{20.1}$$

Here $\varepsilon_{\tau\tau}$ is the strain along the fiber, parameter τ parametrizes the cable, integration is done along the (curvilinear) cable, and G is the gauge length. We model strain by solving elastic wave equations

$$\begin{aligned} \rho \partial_{tt} \mathbf{u} &= \nabla \cdot \boldsymbol{\sigma} + \mathbf{f}, \\ \boldsymbol{\sigma} &= \mathbf{C}(\mathbf{m}) : \boldsymbol{\varepsilon}(\mathbf{u}) + \mathbf{g}, \end{aligned} \tag{20.2}$$

where $\mathbf{u}$ is a displacement vector, $\boldsymbol{\sigma}$ is a stress tensor, ρ is bulk density, $\mathbf{C}(\mathbf{m})$ is a stiffness tensor, $\mathbf{m}$ is the vector of the medium properties, $\mathbf{f}$ and $\mathbf{g}$ are source terms, and strain $\boldsymbol{\varepsilon}$ is

$$\boldsymbol{\varepsilon}(\mathbf{u}) = \frac{1}{2}\left(\nabla \mathbf{u} + \nabla \mathbf{u}^T\right). \tag{20.3}$$

The strain taken along the fiber with the tangential direction $\boldsymbol{\tau}$ is

$$\varepsilon_{\tau\tau} = \boldsymbol{\tau}^T \boldsymbol{\varepsilon} \boldsymbol{\tau} \tag{20.4}$$

A vector $\mathbf{d}_{DAS}$ collecting all DAS data over all receivers can be written as

$$\mathbf{d}_{DAS} = \mathbf{S} \boldsymbol{\varepsilon}_{\tau\tau} \tag{20.5}$$

where $\mathbf{S}$ is an approximation of the spatial integration in (20.1), and $\boldsymbol{\varepsilon}_{\tau\tau}$ is a vector collecting all strains along the cable.

A.2. RTM for DAS

To derive RTM formulation, we consider elastic full waveform inversion (FWI) in the time domain (Tarantola, 2005) with the least-squares data misfit functional

$$J(\mathbf{m}) = \frac{1}{2}(\mathbf{d}(\mathbf{m}) - \mathbf{d}_{obs})^T (\mathbf{d}(\mathbf{m}) - \mathbf{d}_{obs}). \tag{20.6}$$

Here $\mathbf{d}_{obs}$ are observed data, $\mathbf{m}$ is medium properties, and synthetic $\mathbf{d}(\mathbf{m})$ are computed with equation (20.1).

The misfit $J(\mathbf{m})$ is minimized over the medium parameters $\mathbf{m}$. For DAS data, the gradient is essentially the same as for conventional measurements (Tarantola, 2005), which is a cross-correlation in time of the forward-propagated wavefield $\mathbf{u}$ from the source and backward-propagated residuals $\mathbf{w}$ from the receivers

$$\partial_{\mathbf{m}} J = -\int_0^T \left(\partial_{\mathbf{m}} \rho \partial_{tt} \mathbf{u} \cdot \mathbf{w} + \partial_{\mathbf{m}} \mathbf{C} : \boldsymbol{\varepsilon}(\mathbf{u}) : \boldsymbol{\varepsilon}(\mathbf{w})\right) dt. \tag{20.7}$$

where vector $\mathbf{w}$ is backward-propagated residuals injected from the receivers. For conventional measurements of velocities/accelerations, the backward-propagated field $\mathbf{w}$ is injected as a force. For integrated strain measurements, the injected residuals are additionally multiplied by the transposed averaging matrix $\mathbf{S}$ and then injected as a moment tensor source term

$$\mathbf{g} = \mathbf{S}^T(\mathbf{d}(\mathbf{m}, \mathbf{q}) - \mathbf{d}_{obs}) \mathbf{M} \delta(\mathbf{x} - \mathbf{x}_{rcv}) \tag{20.8}$$

where $\mathbf{M} = \boldsymbol{\tau}^T \boldsymbol{\tau}$ with $\boldsymbol{\tau}$ being a well deviation direction as before. In the Cartesian coordinate system (x, y, z) with $\boldsymbol{\tau} = (\tau_x, \tau_y, \tau_z)$

$$\mathbf{M} = \begin{pmatrix} \tau_x \\ \tau_y \\ \tau_z \end{pmatrix} \begin{pmatrix} \tau_x & \tau_y & \tau_z \end{pmatrix} = \begin{pmatrix} \tau_x^2 & \tau_x\tau_y & \tau_x\tau_z \\ \tau_x\tau_y & \tau_y^2 & \tau_y\tau_z \\ \tau_x\tau_z & \tau_y\tau_z & \tau_z^2 \end{pmatrix}. \tag{20.9}$$

REFERENCES

Ali, S., Miranda F., Emmanuel S., & Aguilar, R. C. (2021). *Processing of a large offshore 3DVSP DAS survey in a producing well.* Paper presented at the SEG Technical Program. Expanded Abstracts, 427–431. doi:10.1190/segam2021-3594397.1

Daley, T. M., Miller, D. E., Dodds, K., Cook, P., & Freifeld, B. M. (2015). Field testing of modular borehole monitoring with simultaneous distributed acoustic sensing and geophone vertical seismic profiles at Citronelle, Alabama. *Geophysical Prospecting*, *26*, 1318–1334. doi:10.1111/1365-2478.12324

Dean, T., Cuny, T., & Hartog, A. (2015). *Determination of the optimum gauge length for borehole seismic surveys acquired using distributed vibration sensing.* Paper presented at 77th EAGE Conference and Exhibition, Madrid. doi:10.3997/2214-4609.201412740

Fletcher, R. P., Nichols, D., Bloor, R., & Coates, R. T. (2016). Least-squares migration — Data domain versus image domain using point spread functions. *The Leading Edge*, *35*, 157–162. doi:10.1190/tle35020157.1

Frignet, B. G. & Hartog, A. H. (2014). *Optical vertical seismic profile on wireline cable.* Paper presented at SPWLA 55th Annual Logging Symposium, Abu Dhabi, UAE.

Mateeva, A., Lopez, J., Potters, H., Mestayer, J., Cox, B., Kiyashchenko, D., et al. (2014). Distributed acoustic sensing for reservoir monitoring with vertical seismic profiling. *Geophysical Prospecting*, *62*, 679–692. doi:10.1111/1365-2478.12116

Tarantola, A. (2005). *Inverse problem theory and methods for model parameter estimation.* Society for Industrial and Applied Mathematics. doi:10.1137/1.9780898717921

Yurikov, A., Tertyshnikov, K., Isaenkov, R., Sidenko, E., Glubokovskikh, S., Barraclough, P., et al. (2021). Multi-well 3D DAS VSP imaging with engineered fibres: CO2CRC Otway project case study. *Geophysics*, *86*, D241–D248. doi:10.1190/geo2020-0670.1

21

Enhanced Geothermal Reservoir Characterization Using Wireline Distributed Acoustic Sensing

Evgeniia Martuganova[1,2], Ben Norden[2], Guido Blöcher[2], and Charlotte M. Krawczyk[2,3]

ABSTRACT

Successful across-scale reservoir characterization is a prerequisite for sustainable utilization of the subsurface. Here, we discuss aspects of distributed acoustic sensing (DAS) vertical seismic profiling (VSP) acquisition in Germany and provide an example of data integration for reservoir characterization at the geothermal research platform Groß Schönebeck, aiming to utilize Permian and Permo-Carboniferous reservoir sections at depths below 4 km. At this site, DAS VSP wireline data obtained in two boreholes were processed successfully for 3D interpretation. The data and their interpretation serve as input for a fundamental update of the structural and property characterization of the reservoir zones, relying on the integration of conventional surface 3D seismics and joint core-log analysis. The new model provides a much more detailed characterization of the target reservoirs, allowing for the first time a geological-based (facies-driven) parameterization of the subsurface, considering hydraulic and thermal properties of the reservoir-forming rocks.

21.1. INTRODUCTION

One of the European Union (EU)'s key development objectives is to become the first climate-neutral continent by 2050. Alongside various proposed actions, such as cutting greenhouse gas emissions and developing carbon capture and storage (CCS) facilities, the expansion of renewable energy sources is an essential part of the EU's decarbonization efforts (The European Commission, 2018). Among the currently available renewable resources, geothermal energy, unlike solar and wind, has the advantage of not being dependent on weather variations and does not require a large amount of land, which becomes especially important in urban areas (Roadmap for Deep Geothermal Energy for Germany, 2022). By 2050, the geothermal sector is expected to achieve 2,570 TWh (The European Commission, 2021).

Currently, Germany has in total approximately 190 geothermal installations operational for direct utilization of geothermal energy with an available geothermal capacity equal to 406.9 MWth with a geothermal heat production of 6,183.7 TJ (Weber et al., 2022). Most of the installations are realized in the shallow subsurface. In February 2023, 42 deep geothermal energy projects were in operation with a total capacity of 417 MWth (Bundesverband Geothermie, 2023). Most of the deep geothermal district heating and power plants are situated in the Molasse Basin in Southern Germany, owing to the favorable geological conditions showing high productivity and low salinity of the geothermal fluid originating from a karstic aquifer system (Figure 21.1(a)). Deep geothermal installations are more challenging to apply in the Upper Rhine Graben and the North German Basin. Here,

[1]*Department of Geoscience & Engineering, Delft University of Technology, Delft, The Netherlands*

[2]*Helmholtz Centre Potsdam, GFZ German Research Centre for Geosciences, Potsdam, Germany*

[3]*Department of Applied Geophysics, Technische Universität Berlin, Berlin, Germany*

Distributed Acoustic Sensing in Borehole Geophysics, Geophysical Monograph 289, First Edition.
Edited by Yingping Li, Robert Mellors, and Ge Zhan.

DOI: 10.1002/9781394179275.ch21

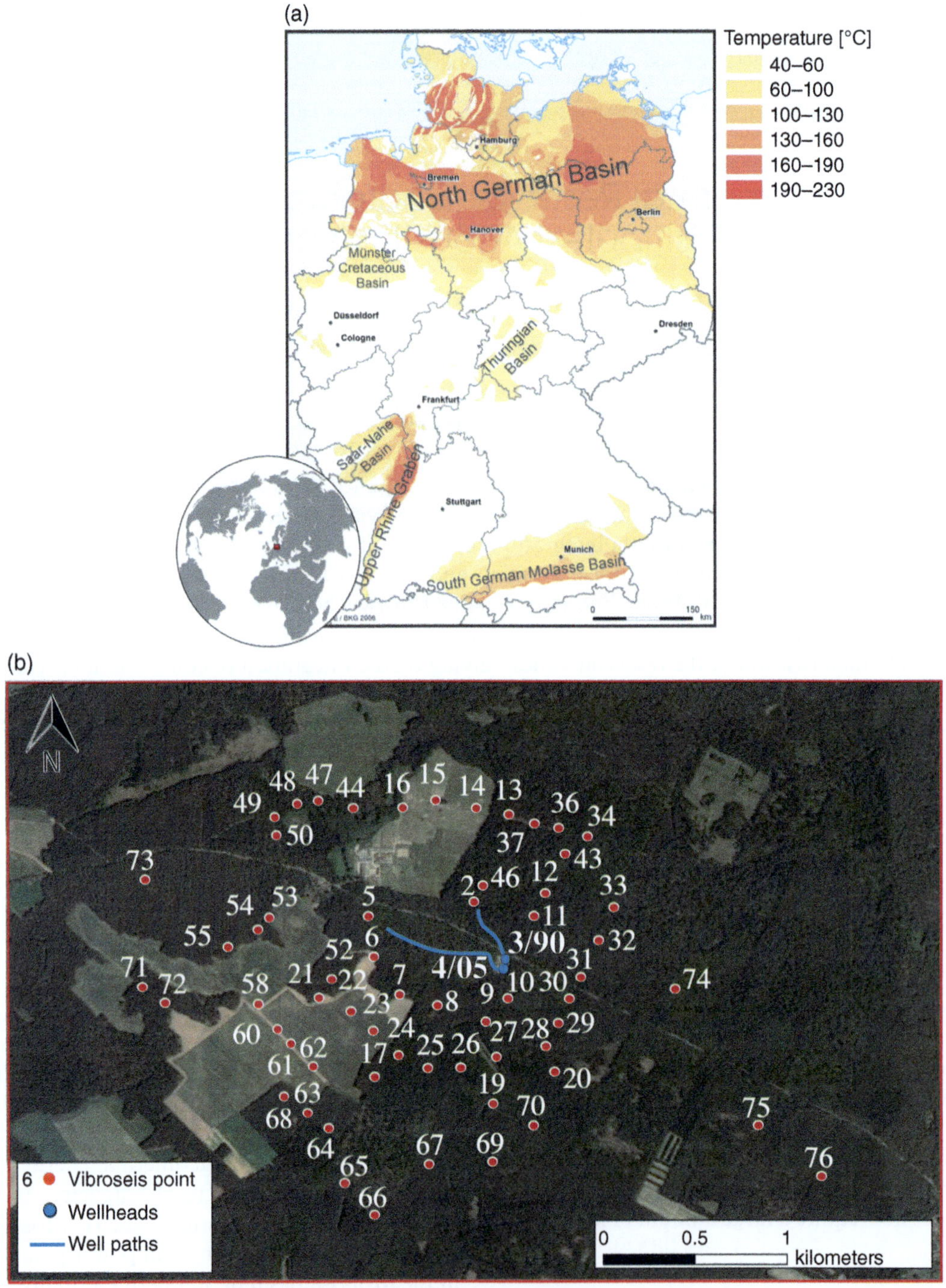

Figure 21.1 Location of the Groß Schönebeck geothermal site. (a) German hydrothermal resources regions with associated temperature ranges. Adapted from Suchi et al. (2014). (b) Base map of the Groß Schönebeck geothermal site with two boreholes (blue circles) in the center and 61 VSP source points (red dots) arranged in a spiral pattern around.

the more complex geological setting requires a customized exploration approach to account for possible risks in the geothermal exploitation of deep reservoirs. Thereby, the geothermal market in Germany has a significant potential for expansion, aiming to cover over 300 TWh/yr (Roadmap for Deep Geothermal Energy for Germany, 2022). To further stimulate the demand for geothermal energy, the Federal Ministry for Economic Affairs research funding has a significant focus on continuously developing exploration techniques, which will allow for minimizing the risk of failure to find a suitable reservoir and lower high upfront investment costs for geothermal projects (Federal Ministry for Economic Affairs and Climate Action of Germany, 2014).

The application of distributed acoustic sensing (DAS) measuring techniques to acquire seismic data reduces acquisition costs. It provides high-resolution data at considerable depths, which is especially important for deep geothermal reservoirs due to the high drilling costs. The installation of fiber-optic cables and the usage of DAS acquisition methods in Germany are quite limited and have more of a research focus. However, further technological advancement may allow the implementation of this data acquisition method into a standard geothermal exploration program.

Besides the application of DAS for deep geothermal systems, the technology is also applied in so-called aquifer thermal energy storage (ATES) systems. In total, 56% of the primary energy consumption in Germany is utilized for heating and cooling purposes (Federal Ministry for Economic Affairs and Climate Action of Germany, Berlin, Germany, 2019), whereas often, heat (or cooling) is only temporarily needed. Therefore, besides direct uses of geothermal energy, applying ATES systems in the shallow subsurface is increasingly becoming popular as seasonal energy storage for district heating or cooling. The limited deployment of such systems in Germany is caused partially due to the fear of potential interference with shallow freshwater aquifers used for drinking water supplies (Frick et al., 2022). Thus, licensing authorities and developers have shifted their focus toward deeper and saline aquifers, demanding a profound characterization of the target aquifers and their overburden. Moreover, such subsurface installations need to be close to the user. Conventional exploration using 2D and 3D geophysical surveys is limited in urban areas. In Berlin, an ongoing research project focuses on strategies for developing adjusted exploration and utilization concepts for safe, reliable, and efficient integration of ATES systems in the local distribution network.

The research project GeoFern investigates potential Mesozoic storage targets for seasonal heat storage in Berlin-Adlershof. Within this framework, a research well was drilled to study the geological sequence and its storage suitability, and an innovative seismic exploration approach using DAS technology was conducted. The latter included an acquisition using a so-called dark fiber position along a major road. Here, the data were analyzed using multichannel analysis of surface waves with the coherency-based enhancement of wavefields (Ehsaninezhad et al., 2024). In addition, vertical seismic profiling data recorded using fiber-optical cable inside a shallow borehole behind the casing and weight drop as a source are planned to be used for further investigations, such as velocity profile characterization and imaging.

The growing demand for deep geothermal energy utilization poses new challenges not only for exploration campaigns but also for monitoring geothermal sites. For instance, installing permanent monitoring systems in geothermal production wells equipped with an electrical submersible pump is technically demanding because the monitoring method should refrain from intervening with the production process. Equipment of the borehole with fiber-optic cable can provide a solution for permanent monitoring.

The first attempt of using a fiber-optic monitoring system along the whole length of the production well was tested in the Molasse at the geothermal site Schäftlarnstraße in Munich, Germany. The production borehole was equipped with a fiber-optical cable, installed inside the well using the mounting method on the rigid pump sucker rod, which acts as a centralizer and helps to guide flexible cable and prevent it from coiling, which is especially beneficial in deviated boreholes (Lipus et al., 2022). The installed cable comprises two DAS fibers, two distributed temperature sensing fibers, and a single fiber for a down-hole fiber-optic pressure/temperature gauge that was attached to the sucker rods, reaching the measured depth of 3691 m (Schölderle et al., 2021). The data collected in a post-drilling period (during borehole tests and geothermal operational activities) were used to characterize hydraulically active zones in the karstic Upper Jurassic reservoir (Schölderle et al., 2021). The DAS data provided confirmation that the temperature drop that occurred during the injection of cold water was responsible for the contraction/expansion moves of the sucker rod/fiber cable setup (Lipus et al., 2022). In addition, the data recorded during the beginning of the injection test agree well with the injection profile data (Schölderle et al., 2021). Overall, the fiber-optical sensing setup performed well during the tests, proving the possibility of further investigation to detect microseismicity events and provide information on well integrity.

In the case of repurposing accessible drilled (legacy) boreholes for geothermal characterization of the subsurface, the application of the wireline DAS technique can provide a convenient and cost-efficient way to acquire additional data necessary for successful exploration. In the following, we will focus on this type of DAS application based on the Groß Schönebeck case study.

21.2. THE GEOTHERMAL RESEARCH PLATFORM GROß SCHÖNEBECK

The in situ geothermal laboratory Groß Schönebeck aims to develop and test concepts for heat extraction from deep, naturally low-permeable rock units below the Permian Zechstein salt. The site is located approximately 50 km NE of Berlin in the natural reserve Schorfheide, Brandenburg, Germany, at the southern margin of the North German Basin (NGB) (Figure 21.1(a)). Two research wells encountered sedimentary Permian Rotliegend and Permo-Carboniferous volcanic rocks, providing access to water-bearing horizons at depths between 3.9 and 4.3 km at temperatures of 150 °C but with low productivity (Zimmermann et al., 2010). A viable exploitation of the deeply stored geothermal energy might be possible by applying engineering measures to allow sufficient heat extraction. At the Groß Schönebeck site, research is focusing on possibilities of reservoir stimulation to support the further development of strategies for the successful implementation of so-called enhanced geothermal systems (EGSs; e.g., Tester et al. (2006); Olasolo et al. (2016)). Altogether, seven stimulation treatments were carried out between 2002 and 2007 in different depth intervals (Blöcher et al., 2016).

However, successful geothermal utilization of the target horizons has not been possible so far. One of the boreholes of the research platform, the E GrSk 3/90 well, was initially drilled for gas exploration and reopened and deepened for geothermal production in 2001. The second well, the Gt GrSk 4/05 borehole, was drilled in 2006 for geothermal production from a geothermal doublet. After first promising stimulation treatments showing an increase in productivity by more than a factor of 4 (Zimmermann et al., 2011), the further site development resulted in shutdown periods for the geothermal production. From 2011 to 2013, in total, 139 hydraulic experiments were conducted at the site (Blöcher et al., 2016), trying to establish an enduring geothermal heat production. However, the experiments document a sharp reduction of productivity, possibly caused by the accumulation of minerals and copper precipitation (scales) clogging the production well, forming a wellbore skin (Regenspurg et al., 2015). Other reasons for the productivity decrease were discussed by Blöcher et al. (2016), as a single process will not account for all well-test results (e.g., two-phase flow due, free gas, and possible compartmentalization of the target horizons by fault zones).

The influence of fault zones on the hydraulic behavior of the reservoir has been investigated in several numerical studies. A direct pressure response between the two wells could be traced back to a poroelastic behavior (Jacquey et al., 2016), and microseismic events measured during a stimulation treatment were outside the previously interpreted fault zones (Blöcher et al., 2018). The conclusion can be drawn here that the previously interpreted fault zones are not hydraulically active or even nonexistent.

To allow well-founded further site development, further geophysical exploration campaigns were initiated in 2017, covering a 3D surface seismic survey and a DAS vertical seismic profiling (VSP) seismic surveying. The resulting data form the basis for the presented case study.

21.3. DATA INTEGRATION FOR GEOLOGICAL AND PETROPHYSICAL CHARACTERIZATION

21.3.1. 3D Seismic and Wireline DAS Acquisition Campaign

A 3D surface seismic campaign was conducted in February–March 2017 (Krawczyk et al., 2019; Stiller et al., 2018). Until then, the deeper geology of the site was interpreted from sparsely distributed legacy 2D seismic lines from former hydrocarbon exploration. The 3D data should therefore deepen the structural understanding of the reservoir section, locate possible faults and fractures, and help to understand the general geological setting better. The 3D surface seismic survey covers an area of 8 km by 8 km, focusing on target reservoir depth from 4 to 4.3 km. The reservoir section is overlain by more than 1 km of thick Zechstein salt, significantly reducing the frequency range and, therefore, the resolution at the target depths, and providing a challenging geological setup for the seismic campaign. The 3D seismics allows for the first time a consistent geological interpretation and model parameterization of the site (Norden et al., 2023).

The 3D surface seismic campaign was accompanied by a VSP experiment using wireline DAS to bridge resolution gaps from the borehole to the field scale. The VSP method benefits from the shorter travel times and, therefore, can provide higher-resolution data. Thus, to improve the resolution of the seismic imaging results around the boreholes, a complementary VSP experiment was conducted before the main 3D surface seismic campaign. Furthermore, the application of fiber-optical cable for seismic data measurements enabled data acquisition at elevated temperatures of 150° during four acquisition days (Henninges et al., 2021). The temperature at the Groß Schönebeck geothermal site is within the temperature limit of geophone's applications, and moreover, geophones can withstand these temperatures only for a limited amount of time (Kästner et al., 2020). Thus, wireline DAS VSP allowed for significant time and cost savings for the survey that was conducted.

The dataset was recorded in February 2017 within four acquisition days: one day of start-up testing and three acquisition days. VSP data were acquired from two 4.3-km-deep wells: E GrSk 3/90 (maximum inclination 7.2°) and Gt GrSk 4/05 (maximum inclination 49°), using a hDVS (Heterodyne Distributed Vibration Sensing, Schlumberger) interrogator system connected to a wireline cable. The spatial spacing along the cable was processed to 5 m. A summary of receiver parameters is presented in Table 21.1.

Table 21.1 Receiver acquisition parameters.

Parameter	E GrSk 3/90	Gt GrSk 4/05
Interrogator type	Schlumberger hDVS unit	Schlumberger hDVS unit
Type of wireline cable	GFZ–Rochester (Henninges et al., 2011)	Schlumberger–NOVA-F
Type of fiber	Single-mode	Single-mode
Slack on the cable	1 m	1 m
Recorded type of data	Strain	Strain
Depth interval receiver channels	Ground level – 4,251 m MD	Ground level – 3,716 m MD
Recorded number of vibroseis points	61	18
Spatial sampling along borehole	5 m	5 m
Gauge length	40 m	20 m
Sampling interval	2 ms	2 ms
Trace length (after correlation)	4 s	4 s
Polarity convention	European/EAGE normal	European/EAGE normal

Table 21.2 Source acquisition parameters.

Parameter	Value
Seismic source	Four vibrator trucks: Mertz M12 Hemi 48, peak force 200 kN (45,100 lbf) each
Number of source points	61
Offset	188–2,036 m
Sweep (near offsets)	10–112 Hz, linear up, 36 s, 360 ms taper
Sweep (far offsets)	10–96 Hz, linear up, 36 s, 360 ms taper
Average vertical stacking fold	16 per source location

The 61 vibroseis points (VPs) were positioned in a spiral configuration with offsets varying between 188 and 2,036 m (Figure 21.1(b)). Source acquisition parameters are summarized in Table 21.2. In borehole E GrSk 3/90, a total of 61 VPs were recorded, with a maximum depth of 4,251 m measured depth (MD). Due to cable failure, the number of VPs and maximum measured depth recorded using the cable placed in Gt GrSk 4/05 is limited to 18 VPs and 3,716 m MD, respectively. More information and possible explanations of the incident can be found in Henninges et al. (2021).

21.3.2. Coupling Noise in Wireline DAS VSP Measurements

In wireline installations, the measuring fiber-optical cable is freely suspended inside the borehole. This allows for cost-effective data acquisition in already-drilled boreholes, but it comes at the expense of reduced data quality due to coupling conditions that are difficult to control and are therefore suboptimal. Wireline DAS data are often affected by specific coupling noise, which we observed in our data. Raw strain data have a pronounced pattern of high-amplitude areas (Figure 21.2(a)), which, after correlation with the pilot sweep, creates a distinct zigzag pattern (Figure 21.2(b)). This characteristic noise is typical for wireline DAS recordings and was described in the literature by Hartog et al. (2014), Dean et al. (2015), Yu et al. (2016), and Martuganova et al. (2021).

In Martuganova et al. (2021), the authors propose that this noise represents a standing wave phenomenon generated on poorly coupled depth intervals of the fiber-optical cable (Figure 21.3(a)) by a vibroseis signal. A classical displacement formula can be utilized to compute a synthetic model for the first four harmonics for fixed-end string case and thus clarify the nature of zigzag patterns found in the data:

$$U = 2A \sin \frac{n\pi x}{L} \cos \omega t, \qquad (21.1)$$

where A is the displacement amplitude, n is an integer mode number, x is a spatial coordinate, L is the length of the string, ω is an angular frequency, and t is time.

Figure 21.3(b) shows that the displacement for fixed cable ends is theoretically described for the fundamental mode and three overtones. The output of the hDVS interrogator unit is strain. To convert data from displacement to strain, the data were differentiated over spatial coordinates (Figure 21.3(c)). The positions of nodes and antinodes will be interchanged due to differentiation, and therefore, the standing wave for closed-end strain recording will look like displacement for the free-ends case. In the case of an odd number of the harmonic, the strain at the ends of the string is out of phase, and for an even number, it is in phase (Figure 21.3(c)). The transition of the mode from odd to even will change the phase of the one end of the string. This disproportion in phases at the string ends will create the zigzag pattern after correlation with different up and down velocities (Figure 21.3(d)).

21.3.3. 3D DAS VSP Data Processing

The processing flow for the 3D DAS VSP dataset consists of typical elements for 3D VSP data processing (Table 21.3), with necessary adjustments for the specifics of the Groß Schönebeck survey, such as the elimination of the coupling noise described earlier. We applied the

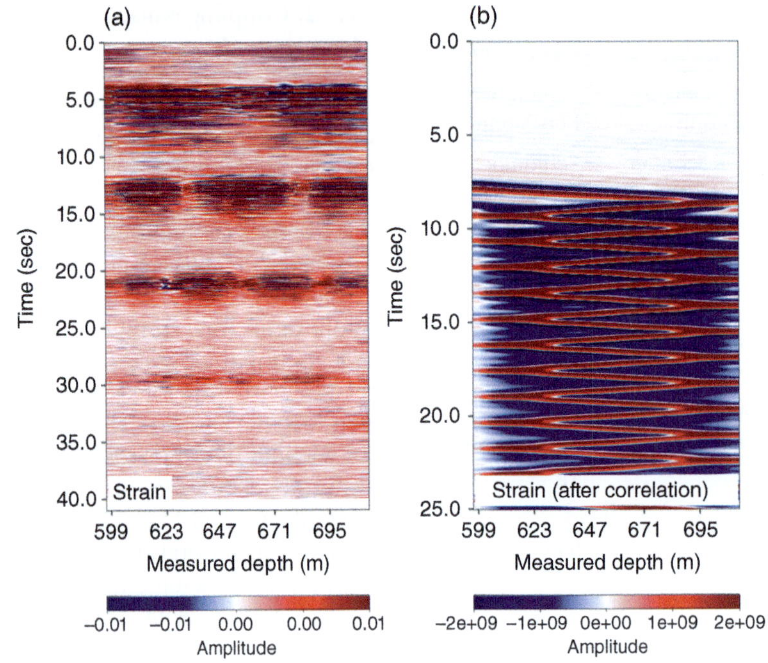

Figure 21.2 (a) An example of field strain data recording for the depth interval 597–711 m. (b) Field strain data recording after correlation.

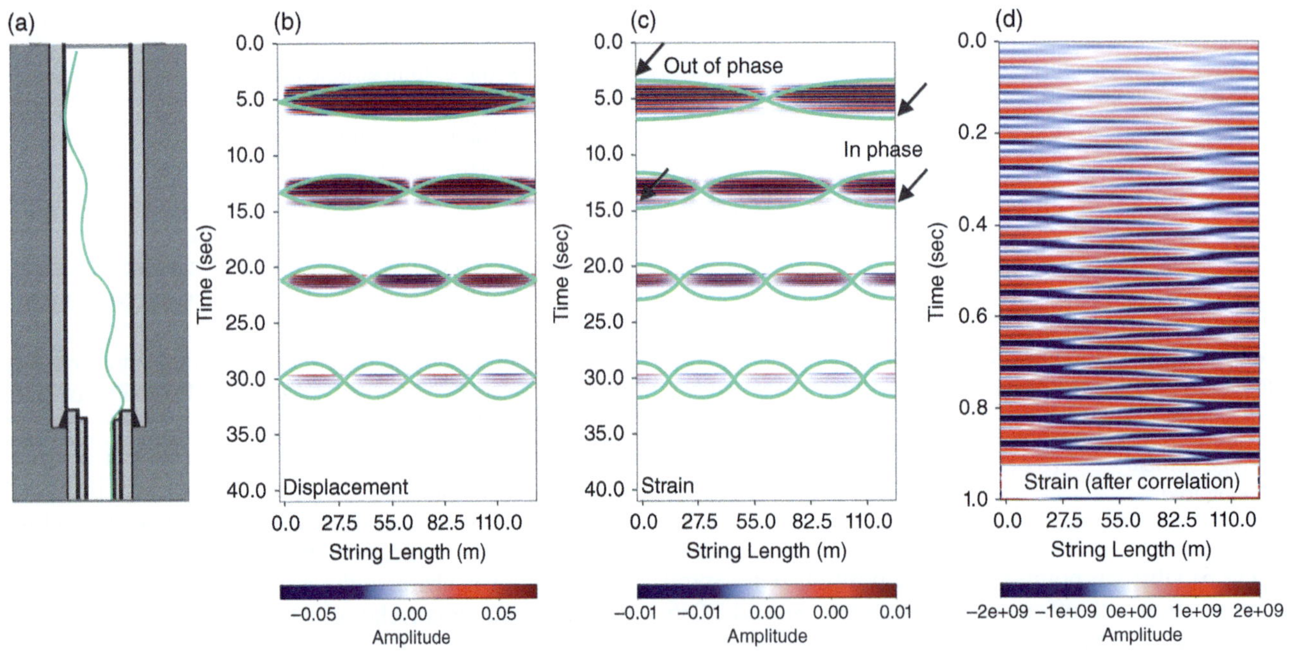

Figure 21.3 An explanation of the noise pattern generation during the wireline survey. (a) A sketch representing wireline cable positioning inside the casing. (b) Modeled displacement for fixed-end cable vibrations for one the main tone and three overtones; the bold green line shows theoretical cable movement for the corresponding frequency mode. (c) Modeled strain; the bold green line shows theoretical cable movement for the corresponding frequency mode. (d) Modeled strain after correlation with a sweep signal.

matching pursuit decomposition (MPD) method (Martuganova et al., 2021) first to decompose a raw stacked seismic trace (Figure 21.4(a)) into elementary Gabor waveforms (an example of elementary Gabor functions is shown in Figure 21.4(b)). Through detailed analysis of elementary functions parameters used for signal decomposition, such as frequency, phase, amplitude, and temporal position, the part of the signal associated

Table 21.3 3D DAS VSP data processing flow for wells E GrSk 3/90 and Gt GrSk 4/05.

Processing step	Description and parameters
Geometry input	Source and receiver coordinates, depths
Trimmed mean stack	Suppression of impulsive noise
Ringing noise suppression	MPD with Gabor atoms
Correlation with pilot sweep	10–112 Hz, 10–96 Hz, 36 s, 360 ms taper
Conversion to strain rate	Time derivative
First-arrival time picking	The peak of the direct downgoing P-wave
Amplitude corrections	Spherical divergence compensation ($t^{1.5}$) and lateral equalization
Coherency enhancement	Moderate wavefield sharpening by tau-p method
Velocity model building	Migration velocities from 3D surface seismics and sonic log from E GrSk 3/90 as initial velocity model, checked and optimized by ray tracing
Ray tracing	Mapping of the reflectivity for all source receiver pairs
Wavefield separation	Subtraction of downgoing P-wave field (median filter) by a nine-trace median operator
Deterministic deconvolution	Waveshaping zero-phasing of the upgoing wavefield, removal of multiples
Polarity reversal	180° phase shift to match polarity convention of conventional geophone data
3D imaging	3D Kirchhoff migration with restricted aperture of 12°

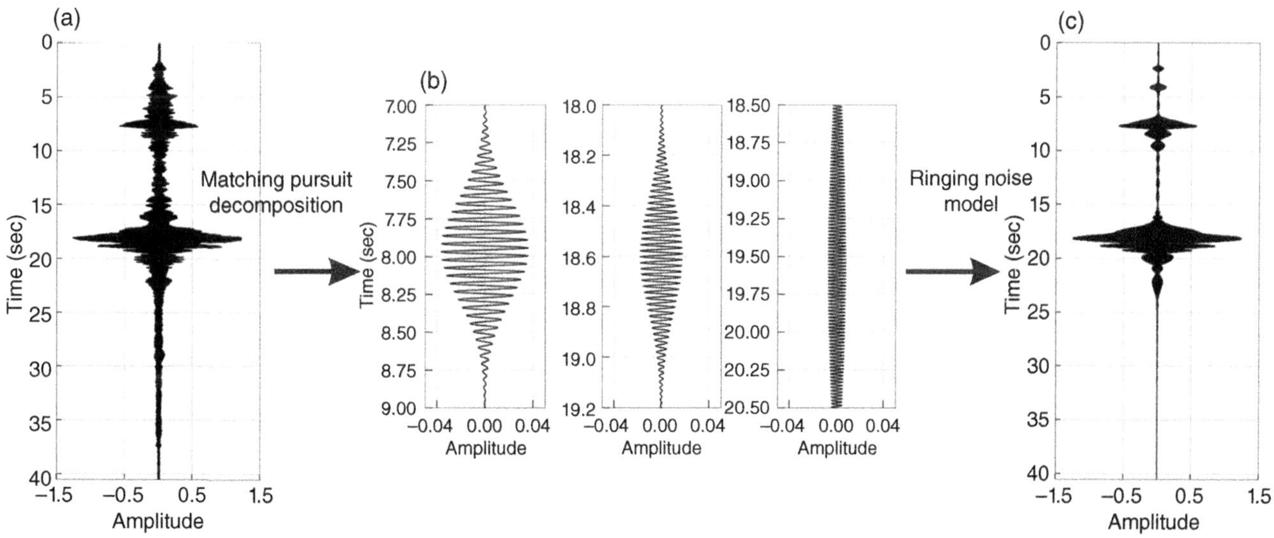

Figure 21.4 An explanation of the matching pursuit algorithm. (a) Raw stacked strain data before the decomposition process. (b) Examples of Gabor elementary functions used to fit the original seismic trace. (c) A coupling noise model, which was generated using selected Gabor atoms.

with cable slapping and ringing was identified and used to create a model of the noise (Figure 21.4(c)). Compared to the original data with standard preprocessing applied (stacking, correlation with pilot sweep, and differentiation in time domain) shown in Figure 21.5(a), we observe that the amplitude and phase of the coupling noise (Figure 21.5(b)) is accurately reconstructed. The created noise model was subtracted from the original raw strain data, and the same preprocessing flow was applied to the seismic section shown in Figure 21.5(c).

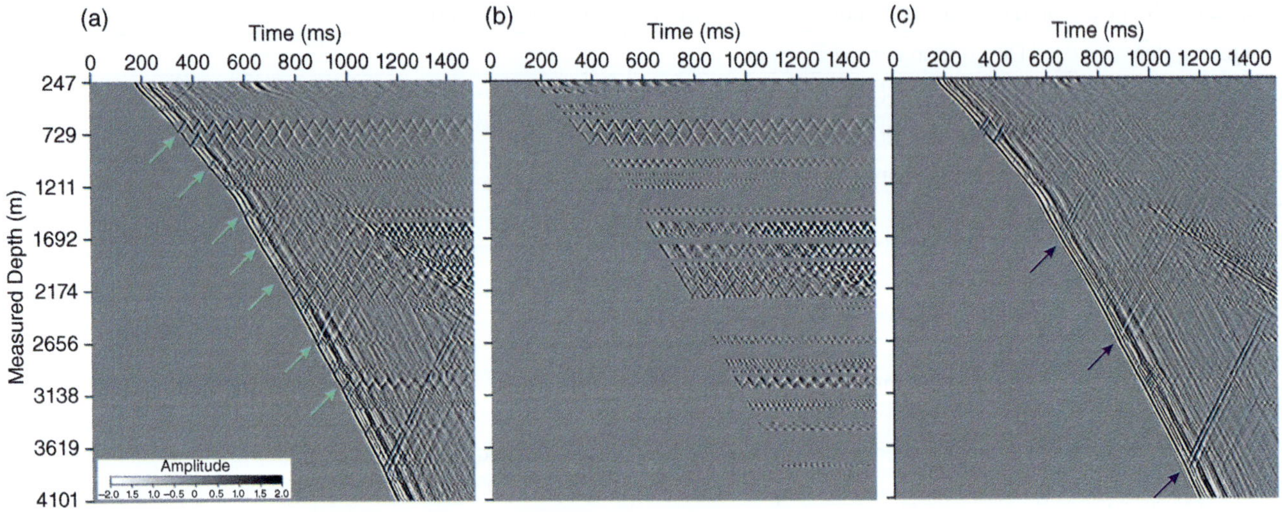

Figure 21.5 Common-source gather acquired at source position 10 (Figure 21.1(b)). (a) Before the denoising; (b) the coupling noise model, created using MPD decomposition method; (c) common-shot gather after the noise model was subtracted. Light green arrows in (a) highlight depth intervals affected by the coupling noise; blue arrows in (c) show improved traceability of the reflections in VSP data.

After vertical trimmed mean stacking and removal of the typical wireline acquisition coupling noise, the VSP data preparation consisted of geometry input, correlation with the pilot sweep, and subsequent differentiation in the time domain to convert data into strain rate and achieve acceleration-like characteristics. Further processing included first-arrival picking, spherical divergence, lateral amplitude equalizations, and moderate wavefield sharpening using the tau-p method. To obtain a 3D seismic image in the depth domain, first we constructed a preliminary velocity model using velocities from 3D surface seismics and sonic log from well E GrSk 3/90, which was checked and adjusted using the ray-tracing method. The following processing step included wavefield separation by subtracting the downgoing wavefield using a median filter, followed by deterministic deconvolution. The 3D DAS VSP image was computed using the commercial algorithm (VSProwessX) 3D Kirchhoff migration algorithm with a restricted aperture of 12°. A more detailed description of 3D DAS VSP data processing can be found in Martuganova et al. (2022a).

21.3.4. Geological Modeling and Petrophysical Model Parameterization

The geological and petrophysical characterization could rely on different data types ranging from the laboratory scale to the borehole and across borehole scales, including the reservoir level. Information on the lithology and stratigraphy and the overall geological setting is provided by the Groß Schönebeck site's boreholes and neighboring wells. Petrophysical data are accessible from the E GrSk 3/90 borehole. Starting from a seismic-well tie interpretation (Krawczyk et al., 2019), seismic horizons were mapped and lithologically interpreted (Norden et al., 2023). The well control allows linking the seismically mapped Rotliegend succession to lithotypes and sedimentary facies patterns. To better constrain the internal structure pattern of the sedimentary Rotliegend, a wavelet transform-based seismic facies classification using neural network clustering (Bauer et al., 2020) was applied to resolve information stored in the signal attributes calculated along the target horizon. Based on the constructed structural model, a reservoir model was set up for further site development (Norden et al., 2023). The model comprises the Permo-Carboniferous volcanic rock section and the sedimentary Rotliegend. In the first step, the structural units and facies types were modeled using a stochastic approach. Then, the respective units were parameterized with petrophysical properties (e.g., total and effective porosity, permeability, density, bulk thermal properties; see also Figure 21.6), considering each facies type with its respective property distribution characteristics. The Groß Schönebeck dataset provided by Norden et al. (2022) forms the basis for this second step in model building. It comprises an extensive evaluation of legacy data and newly acquired measurements and interpretations for core-log integration.

For the facies-dependent petrophysical parameterization, the total porosity distribution as the most robust allocatable property was modeled based on the respective geostatistical input derived from the core-log integration. For the sedimentary Rotliegend, the Elbe reservoir sandstone's porosity distribution was modeled using the

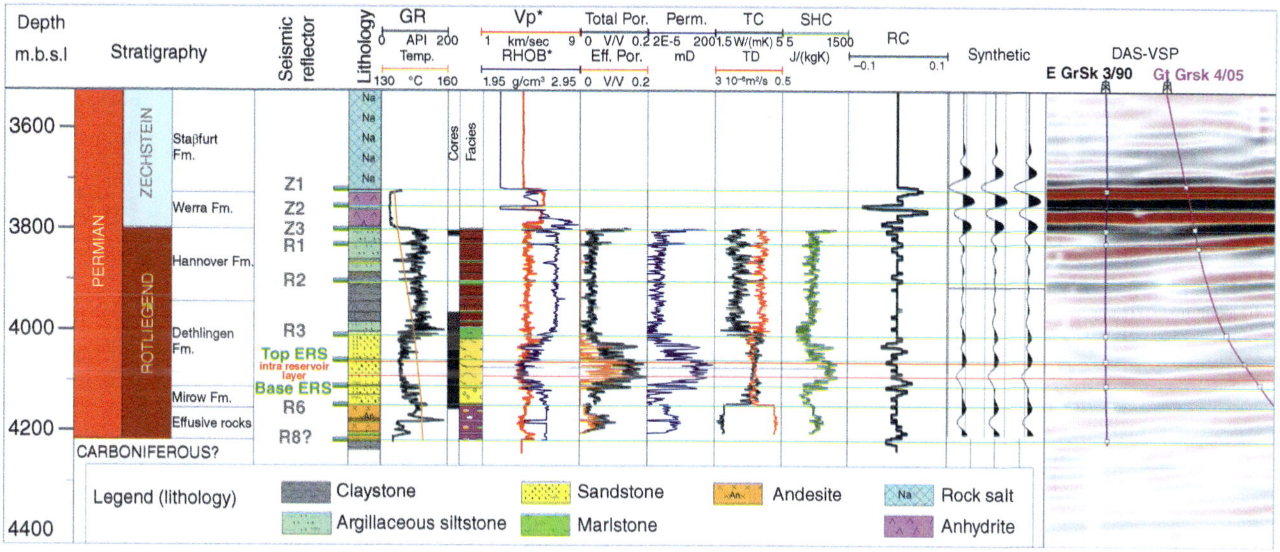

Figure 21.6 DAS VSP seismic cross-section through the boreholes plotted together with well-log related data from the E GrSk 3/90 borehole (GR, gamma ray; Vp, sonic velocity; RHOB, bulk density), petrophysical properties from core-log integration (Total Por., total porosity; Eff. Por., effective porosity; Perm., permeability; TC, thermal conductivity; TD, thermal diffusivity; SHC, specific heat capacity), lithology, stratigraphy, reflector coefficient (RC), and synthetics.

seismic facies analysis of Bauer et al. (2020) and the observations of the DAS VSP analysis of Martuganova et al. (2022a) as trend data. For the sedimentary units, effective porosity was modeled in relation to total porosity using collocated co-kriging and bulk density of the rocks, and the shale content was assigned to the grid cells. They were simulated based on the distribution of total porosity (for bulk density) and effective porosity (for the shale content from GR) using the determined geostatistical input parameter. Permeability was calculated on total porosity. In situ thermal conductivity and diffusivity were estimated using the density and shale content distributions based on formulas provided by Fuchs et al. (2015) for well-log interpretation. The specific heat capacity was calculated from the evaluated distributions of thermal conductivity, thermal diffusivity, and density. The approach for the parameterization of the Permo-Carboniferous volcanic rocks was similar. Here, density, effective porosity of interconnected pores, and permeability were simulated using the geostatistical input data and the collocated co-kriging with the total porosity distribution. For details, see Norden et al. (2022) and Norden et al. (2023).

21.4. RESULTS

The geothermal research platform Groß Schönebeck has focused on two possible targets for geothermal exploration so far: the low-porous Permo-Carboniferous volcanic rocks and the sandstones of the Lower Elbe Subgroup (Dethlingen formation). The geophysical surveys performed at the site should help to understand the reservoir setting and clarify possible structural obstacles that may hinder geothermal utilization. In the following subsections, we characterize these subsalt geothermal targets based on the 3D surface and the 3D DAS VSP seismic cubes, available core information, well logs, laboratory data, and the results of the stochastically parameterized facies-dependent petrophysical property distributions.

21.4.1. Permo-Carboniferous Volcanic Rocks

The top of Permo-Carboniferous volcanic rocks was encountered in the E GrSk 3/90 borehole at a depth of 4216 m TVDSS (Reflector R8, Figure 21.6). This reflector is traceable within the 3D surface and the 3D DAS VSP seismic data. In the 3D DAS VSP cube, this boundary is characterized by a change of the seismic reflection pattern: from a more or less horizontal to a more inclined layering (red line, Figure 21.7), with numerous pinch-outs below. Martuganova et al. (2022a) refer to this feature as the Lower Rotliegend unconformity. According to Guterch et al. (2010), Lower Rotliegend has significant time gaps in sedimentation; therefore, the mapped unconformity horizon could indicate a difference in layering caused by erosion.

Volcanic rock cores from the E GrSk 3/90 borehole are characterized by several layered lava beds and tuffs with a single bed thickness of 1 to over 3 m; the mean flow

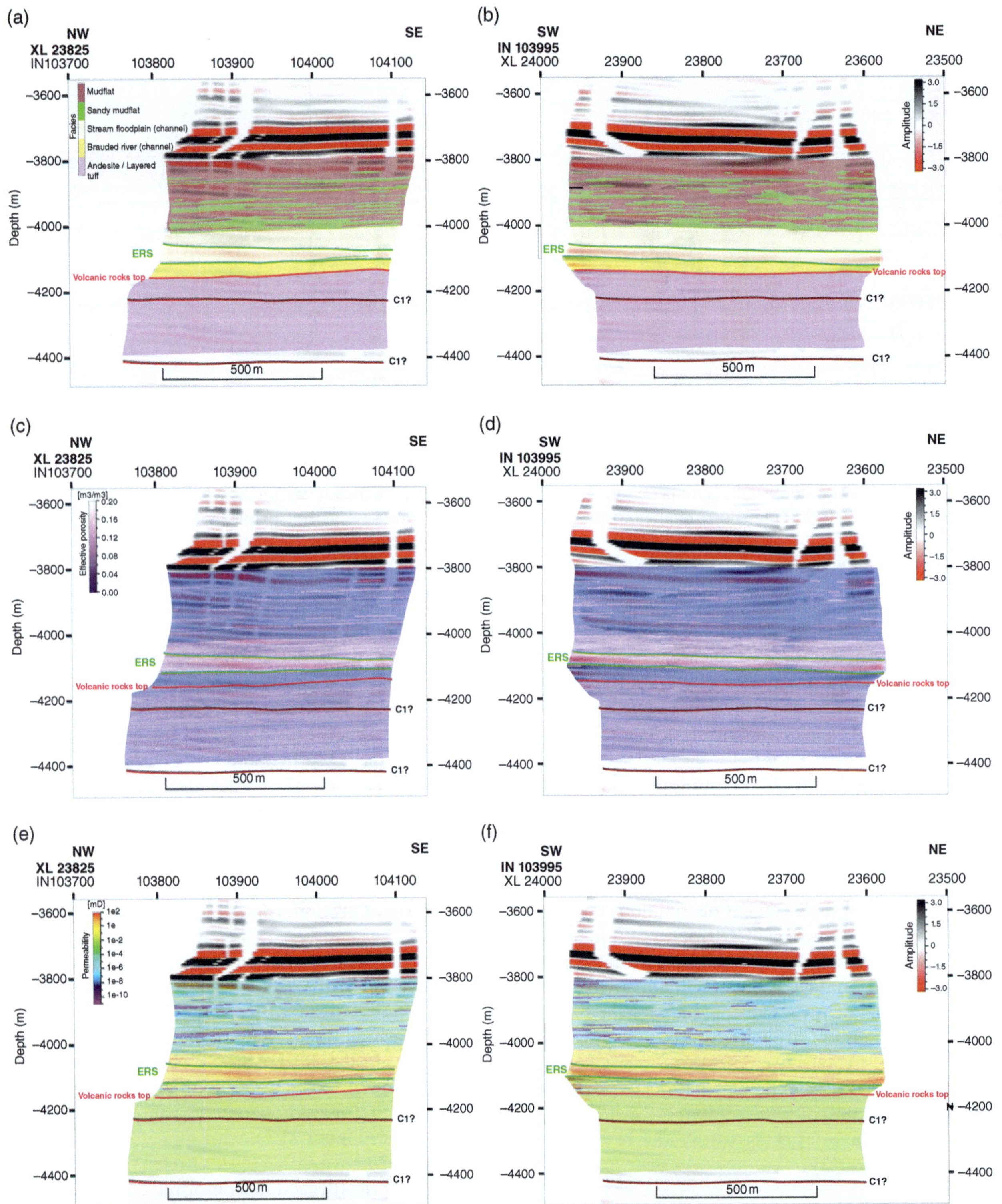

Figure 21.7 Seismic lines extracted from the 3D DAS VSP cube with facies (a, b) and petrophysical parameterization (c–f). Green lines show the borders of the Elbe basis reservoir sandstone layer, red lines show the top of the volcanic rocks, and burgundy lines annotated C1? highlight possible interpretations for the base of Carboniferous.

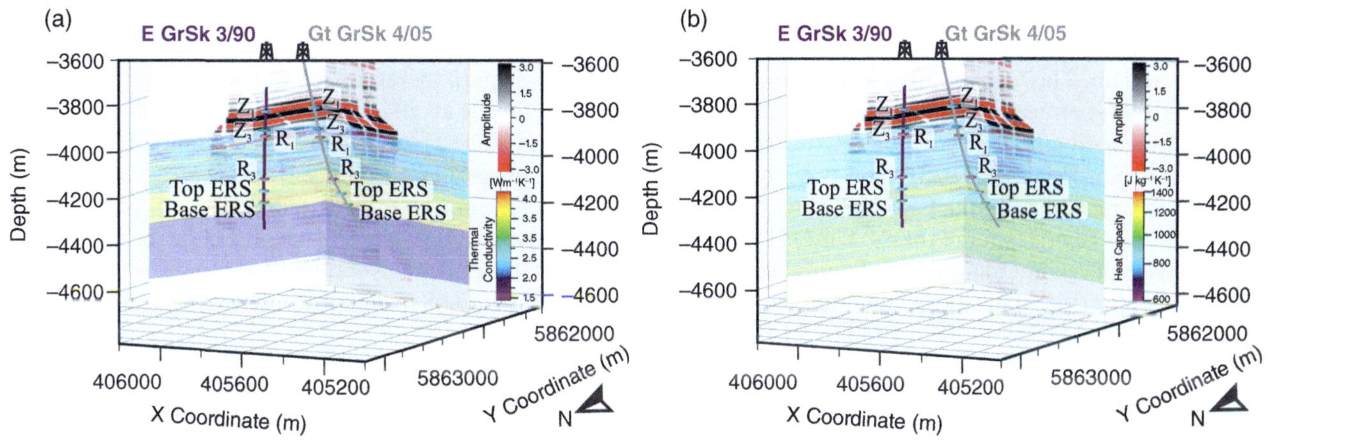

Figure 21.8 The 3D visualization of the thermophysical properties illustrates the expected variability of property distribution in the vicinity of the boreholes. 3D DAS VSP cube overlaid by (a) thermal conductivity and (b) heat capacity.

direction of the andesitic lava indicated from image logs show north to north–northwest direction (Norden et al., 2023). Thus, the internal structure of the volcanic sequence is below seismic resolution. Here, the image log of the E GrSk 3/90 provides the main input for the geological model. Based on gamma-ray (GR) and sonic logs and to some extent on laboratory measurements on cores (density, porosity, permeability, thermal conductivity), the modeled volcanic sequence was parameterized (Figures 21.7 and 21.8). Effective porosity shows mean values of about 2% and 4% for tuffitic and andesitic layers and mean matrix permeability of 9.9 E–18 m^2. Mean values of the volcanic sequence for the thermal conductivity amount to 1.9 $Wm^{-1}K^{-1}$ (Figure 21.8(a)), the mean thermal diffusivity to 0.69 $10^{-6}/m^2/s^{-1}$, and the specific heat capacity to 1025 $Jkg^{-1}K^{-1}$(Figure 21.8(b)).

Unfortunately, the top of the sedimentary Carboniferous underlying the volcanic succession could not be resolved with certainty from either seismic or borehole data (Norden et al., 2023). The thickness of the volcanic may amount to a 70 m or about 200 m thick sequence (burgundy lines with annotations C1? in Figure 21.7). Therefore, because we have only limited data from one borehole, the shown property distribution within the volcanic sequence represents a first-order estimation. This will be used to design hydraulic stimulations and estimate the geothermal potential of this sequence.

21.4.2. Sedimentary Rotliegend (Including Elbe Reservoir Sandstone)

Sedimentary Rotliegend succession is located in the depth interval between 3,800 and 4,216 m comprising Mirow, Dethlingen, and Hannover Formations (Figure 21.6). In this lithostratigraphic unit, four different sedimentary facies types were considered in the modeling: the mudflat playa, the sandy mudflat, the ephemeral stream floodplain, and the braided river facies (Figures 21.7(a), (b)). The latter facies type dominates the Mirow Formation, overlaying the volcanic succession. According to Holl et al. (2005), most of the coarse-grained bed-load-dominated and conglomeratic deposits of this formation are interpreted as multistoried channel sediments of a braided plain fluvial system with a paleocurrent direction toward NNE. The conglomerates, consisting of cemented volcanic rocks, show a porosity similar to that of volcanic rocks and low matrix permeability (Norden et al., 2023). However, the mean thermal conductivity of this unit is higher than that of the volcanic rocks and amounts to about 3.7 $Wm^{-1}K^{-1}$ (Figure 21.8(a)).

According to the core information from the E GrSk 3/90 and Gt GrSk 4/05 wells, a more porous sandstone is present in the Dethlingen Formation above the Mirow Formation (Bauer et al., 2020; Norden et al., 2023). This so-called Elbe reservoir sandstone (ERS) was encountered in the depth interval 4,060–4,100 m at the E GrSk 3/90 borehole. On well logs, this interval is observed as a low P-wave velocity zone due to the increased porosity of this section (Figure 21.6). The ERS unit in the geological model is based on the depth of the mapped horizons of the top and base ERS from the 3D surface seismic cube. These horizons were correlated to the drilled well tops and adjusted considering the information from the thickness distribution pattern extracted from the wavelet analysis, providing trend information on possible thickness changes below seismic resolution. In the vicinity of the boreholes, the picked horizons from the 3D DAS VSP cube were integrated into the structural model (Norden et al., 2023). In Figure 21.7, the borders of this unit are shown using green lines. The ERS represents sediments that were initially formed in an aeolian environment and later redeposited by fluvial processes of the ephemeral stream floodplain. The rocks of these facies show the most robust porosity–permeability relationship because they represent quartz-dominated sandstones with only minor clay content (Norden et al., 2023). However, the derived in situ fluid permeability for the ERS unit ranges from less than 1 mD (1 E–15 m^2) to more than 100 mD (1 E–13 m^2) with an overall geometric mean of only some mD (Figure 21.7(e), (f)). The mean thermal conductivity amounts to about 3.6 $Wm^{-1}K^{-1}$ (Figure 21.8(b)). The joint interpretation of the geophysical data did not provide evidence of structural compartmentalization of the reservoir; the range of variation is covered by the expected geological heterogeneity (Norden et al., 2023).

The sediments of the Upper Dethlingen and Hannover Formations are dominated by rocks of the mudflat and sandy mudflat environment. They show a more complex facies distribution pattern. Effective porosities are low (in the order of 0–3% (Figure 21.7(c), (d)), and due to the higher clay content compared to the ERS, the thermal conductivity of these formations is lower and in the order of 3.2 $Wm^{-1}K^{-1}$ for the sandy mudflat and 2.2 $Wm^{-1}K^{-1}$ for the mudflat facies (Figure 21.8(a)).

21.5. DISCUSSION

Although the seismic data could not provide evidence for structural boundaries (block or fault compartments), the facies and parameter modeling results show that the ERS slightly reduced reservoir properties along the Gt GrSk 4/05 borehole. They could reflect the natural heterogeneity of the reservoir. Based on the new reservoir characterization, detailed information on the principle geometry in three dimensions is given. Former models did apply a simple layer-cake characterization approach without consideration of lateral variability. The more realistic characterization of the presented model allows better planning of future measures. For example, future studies can investigate, in detail, how the natural heterogeneity will influence the long-term behavior of the reservoir.

As stated in Norden et al. (2023), a further enhancement of the presented model parameterization could be achieved by inversion of the seismic data to better guide the distribution of petrophysical parameters. The inversion will also be limited by the scale and resolution of the geophysical (seismic) data. In complex geological environments and cases when understanding heterogeneous reservoir sections is crucial for further geothermal development, VSP and DAS VSP surveys, in particular, can help significantly improve the understanding of the subsurface because of the broader frequency range and dense depth coverage. In Groß Schönebeck case study, the 3D DAS VSP cube has better resolution (vertical resolution up to 16 m (Martuganova et al., 2022a)) than 3D surface seismics; therefore, it will be beneficial to use it as an input for the inversion and as a result resolve a detailed geological facies pattern and the accompanying petrophysical properties on a finer scale.

In recent years, after the wireline 3D DAS VSP measuring campaign conducted at Groß Schönebeck in 2017, DAS interrogators have evolved dramatically. The first generation of DAS measuring units mainly utilized coherent light sources (Lindsey et al., 2020). Their pulse length restricts their spatial resolution, and their energy limits the dynamic range of the measurement. Most recent advancements in DAS data measurements, such as the usage of engineered fibers (Sun et al., 2022), multiple stable lasers, and the chirped compression technique (Waagaard et al., 2021), allow interrogating much longer fibers with significantly reduced measurement noise. Thus, acquiring another dataset with the newer-generation interrogator unit and installing engineered fibers will improve the signal-to-noise ratio, which can further enhance the resolution of 3D DAS VSP data and extract even smaller-scale geological details from the reservoir section.

The future perspective of the in situ laboratory of Groß Schönebeck envisages heat supply in rural regions. In contrast to the previous usage concept, which focused on heat and electricity, geological units with lower temperatures can be developed here. It is, therefore, also possible to drill into the shallower Mesozoic units of the Muschelkalk (shell limestone) and Buntsandstein formations. Furthermore, the EGS development of the Rotliegend (Permian) can be refrained from, and hydrothermal development of the previously mentioned units can take place. The focus is not on reusing the existing wells but rather on drilling a new doublet system to provide the heat demanded by the surrounding villages.

21.6. CONCLUDING REMARKS

This chapter integrates and summarizes the available geophysical and geological data for the geothermal in situ research platform Groß Schönebeck. A more accurate understanding of the Rotliegend reservoir was achieved by incorporating the existing reconnaissance data with the interpretations made using new 3D seismic survey, DAS VSP data, and borehole data. The created reservoir model allows an understanding of the site on various scales, permitting more realistic planning for further development of the geothermal site. The successful 3D DAS VSP application at the Groß Schönebeck may have a critical impact on the advancement and implementation of modern and effective geothermal exploration practices in the NGB and areas with similar geological settings, taking advantage of existing (reconnaissance) boreholes for DAS VSP wireline acquisition.

ACKNOWLEDGMENTS

Funding for the present work was provided by the German Federal Ministry of Economic Affairs and Energy (grant no. 0324065) as well as the European Commission Horizon 2020 Framework Programme (DESTRESS (grant no. 691728) and EPOS IP (grant no. 676564)).

AVAILABILITY STATEMENT

3D DAS VSP data used for this study are published at GFZ Data Services (Martuganova et al., 2022b) and are accessible at https://doi.org/10.5880/GFZ.4.8.2022.014. The parameterized reservoir model, main geological horizons, core, and well-log data are also available via the GFZ repository, https://doi.org/10.5880/GFZ.4.8.2022.013 (Norden et al., 2022). The 3D surface seismic datasets are available on request. No new data were generated, analyzed, or presented in this chapter.

REFERENCES

Bauer, K., Norden, B., Ivanova, A., Stiller, M., & Krawczyk, C. M. (2020). Wavelet transform-based seismic facies classification and modelling: application to a geothermal target horizon in the NE German Basin. *Geophysical Prospecting*, *68* (2), 466–482. Retrieved from https://onlinelibrary.wiley.com/doi/abs/10.1111/1365-2478.12853

Blöcher, G., Cacace, M., Jacquey, A. B., Zang, A., Heidbach, O., Hofmann, H., et al. (2018). Evaluating micro-seismic events triggered by reservoir operations at the geothermal site of Groß Schönebeck (Germany). *Rock Mechanics and Rock Engineering*, *51*(10), 3265–3279. doi:10.1007/s00603-018-1521-2

Blöcher, G., Reinsch, T., Henninges, J., Milsch, H., Regenspurg, S., Kummerow, J., et al. (2016). Hydraulic history and current state of the deep geothermal reservoir Groß Schönebeck. *Geothermics*, *63*, 27–43. doi:10.1016/j.geothermics.2015.07.008

Bundesverband Geothermie. (2023). Tiefe Geothermie in Deutschland 2023/24. Retrieved from https://www.geothermie.de/bibliothek/downloads.html

Dean, T., Constantinou, A., Cuny, T., Frignet, B., Hartog, A., Kimura, T., et al. (2015). *Vertical seismic profiles: now just another log?* Paper presented at the 85th SEG Annual Meeting. Expanded Abstracts, 5544–5548. doi:10.1190/segam2015-5804007.1

Ehsaninezhad, L., Wollin, C., Rodríguez Tribaldos, V., Schwarz, B., & Krawczyk, C. M. (2024). Urban subsurface exploration improved by denoising of virtual shot gathers from distributed acoustic sensing ambient noise. *Geophysical Journal International*, *237*(3), 1751–1764. doi:10.1093/gji/ggae134

The European Commission. (2018). *A european strategic long-term vision for a prosperous, modern, competitive and climate neutral economy*. Brussels (Belgium): Publications Office of the European Union. Retrieved from https://eur-lex.europa.eu/legal-content/EN/TXT/?uri=CELEX:52018DC0773

The European Commission. (2021). *Report from the commission to the European Parliament and the Council. Progress on competitiveness of clean energy technologies* (part 4/5). Brussels (Belgium): Publications Office of the European Union.

Federal Ministry for Economic Affairs and Climate Action of Germany. (2014). *Bekanntmachung Forschungsförderung im 6. Energieforschungsprogramm "Forschung für eine umweltschonende, zuverlässige und bezahlbare Energieversorgung"*. Retrieved from https://www.bmwk.de/Redaktion/EN/Artikel/Energy/research-priorities-deep-geothermal-energy.html

Federal Ministry for Economic Affairs and Climate Action of Germany, Berlin, Germany. (2019). *Energiedaten: Bundesministerium für Wirtschaft und Klimaschutz der Bundesrepublik Deutschland (BMWi)*. Retrieved from https://www.bmwk.de/Redaktion/EN/Artikel/Energy/research-priorities-deep-geothermal-energy.html

Frick, M., Kranz, S., Norden, B., Bruhn, D., & Fuchs, S. (2022). Geothermal resources and ATES potential of Mesozoic Reservoirs in the North German basin. *Energies*, *15* (6). Retrieved from https://www.mdpi.com/1996-1073/15/6/1980

Fuchs, S., Balling, N., & Förster, A. (2015). Calculation of thermal conductivity, thermal diffusivity and specific heat capacity of sedimentary rocks using petrophysical well logs. *Geophysical Journal International*, *203*(3), 1977–2000. doi:10.1093/gji/ggv403

Guterch, A., Wybraniec, S., Grad, M., Chadwick, A., Krawczyk, C., Ziegler, P., et al. (2010). In J. C. Doornenbal & A. G. Stevenson (Eds.), *Crustal structure and structural framework, in: Petroleum geological atlas of the southern Permian Basin area* (pp. 11–23). Houten, The Netherlands: EAGE Publications.

Hartog, A., Frignet, B., Mackie, D., & Clark, M. (2014). Vertical seismic optical profiling on wireline logging cable. *Geophysical Prospecting*, *62*(4), 693–701. doi:10.1111/1365-2478.12141

Henninges, J., Baumann, G., Brandt, W., Cunow, C., Poser, M., Schrötter, J., & Huenges, E. (2011). *A novel hybrid wireline logging system for downhole monitoring of fluid injection and production in deep reservoirs [Conference Proceedings]*. Paper presented at the 73rd EAGE Conference & Exhibition. Retrieved from doi:10.3997/2214-4609.20149727

Henninges, J., Martuganova, E., Stiller, M., Norden, B., & Krawczyk, C. M. (2021). Wireline distributed acoustic sensing allows 4.2 km deep vertical seismic profiling of the Rotliegend 150°c geothermal reservoir in the North German Basin. *Solid Earth*, *12*(2), 521–537. doi:10.5194/se-12-521-2021

Holl, H. G., Moeck, I., & Schandelmeier, H. (2005). *Characterization of the tectonosedimentary evolution of a geothermal reservoir—implications for exploitation (Southern Permian Basin, NE Germany)*. Paper presented at the World Geothermal Congress. Retrieved from https://www.geothermal-energy.org/pdf/IGAstandard/WGC/2005/0614.pdf

Jacquey, A. B., Cacace, M., Blöcher, G., Watanabe, N., Huenges, E., & Scheck-Wenderoth, M. (2016). Thermo-poroelastic numerical modelling for enhanced geothermal system performance: case study of the Groß Schönebeck reservoir. *Tectonophysics, 684*, 119–130. doi:10.1016/j.tecto.2015.12.020

Kästner, F., Giese, R., Planke, S., Millett, J. M., & Flóvenz, Ó. G. (2020). Seismic imaging in the Krafla high-temperature geothermal field, NE Iceland, using zero- and far-offset vertical seismic profiling (VSP) data. *Journal of Volcanology and Geothermal Research, 391*, 106315. doi:10.1016/j.jvolgeores.2018.02.016

Krawczyk, C. M., Stiller, M., Bauer, K., Norden, B., Henninges, J., Ivanova, A., & Huenges, E. (2019). 3-D seismic exploration across the deep geothermal research platform Groß Schönebeck north of Berlin/Germany. *Geothermal Energy, 7*(1), 15. doi:10.1186/s40517-019-0131-x

Lindsey, N. J., Rademacher, H., & Ajo-Franklin, J. B. (2020). On the broadband instrument response of fiber-optic DAS arrays. *Journal of Geophysical Research: Solid Earth, 125* (2), e2019JB018145. doi:10.1029/2019JB018145

Lipus, M. P., Schölderle, F., Reinsch, T., Wollin, C., Krawczyk, C., Pfrang, D., & Zosseder, K. (2022). Dynamic motion monitoring of a 3.6 km long steel rod in a borehole during cold-water injection with distributed fiber-optic sensing. *Solid Earth, 13*(1), 161–176. doi:10.5194/se-13-161-2022

Martuganova, E., Stiller, M., Bauer, K., Henninges, J., & Krawczyk, C. M. (2021). Cable reverberations during wireline distributed acoustic sensing measurements: their nature and methods for elimination. *Geophysical Prospecting, 69*(5), 1034–1054. doi:10.1111/1365-2478.13090

Martuganova, E., Stiller, M., Norden, B., Henninges, J., & Krawczyk, C. M. (2022a). 3D deep geothermal reservoir imaging with wireline distributed acoustic sensing in two boreholes. *Solid Earth, 13*(8), 1291–1307. doi:10.5194/se-13-1291-2022

Martuganova, E., Stiller, M. N., Henninges, B., & Krawczyk, C. M. (2022b). *3D DAS-VSP Data from the Groß Schönebeck Site, Germany, February 2017*. GFZ Data Services. doi:10.5880/GFZ.4.8.2022.014

Norden, B., Bauer, K., & Krawczyk, C. M. (2022). *Input and resulting structural and parameterized subsurface data for a geological model of the geothermal research platform Groß Schönebeck (North German Basin)*. GFZ Data Services. doi:10.5880/GFZ.4.8.2022.013

Norden, B., Bauer, K., & Krawczyk, C. M. (2023). From pilot knowledge via integrated reservoir characterization to utilization perspectives of deep geothermal reservoirs: the 3D Model of Groß Schönebeck (North German Basin). *Geothermal Energy, 11*(1), 1. doi:10.1186/s40517-022-00242-2

Olasolo, P., Juárez, M., Morales, M., D'Amico, S., & Liarte, I. (2016). Enhanced geothermal systems (EGS): a review. *Renewable and Sustainable Energy Reviews, 56*, 133–144. doi:10.1016/j.rser.2015.11.031

Regenspurg, S., Feldbusch, E., Byrne, J., Deon, F., Driba, D. L., Henninges, J., et al. (2015). Mineral precipitation during production of geothermal fluid from a Permian Rotliegend reservoir. *Geothermics, 54*, 122–135. doi:10.1016/j.geothermics.2015.01.003

Roadmap for deep geothermal energy for Germany. (2022). Retrieved from doi:10.24406/publica-248

Schölderle, F., Lipus, M., Pfrang, D., Reinsch, T., Haberer, S., Einsiedl, F., & Zosseder, K. (2021). Monitoring cold water injections for reservoir characterization using a permanent fiber optic installation in a geothermal production well in the Southern German Molasse Basin. *Geothermal Energy, 9*(1), 21. doi:10.1186/s40517-021-00204-0

Stiller, M., Krawczyk, C. M., Bauer, K., Henninges, J., Norden, B., Huenges, E., & Spalek, A. (2018). 3D Seismik am Geothermiestandort Groß Schönebeck. *BBR - Fachmagazin für Brunnen- und Leitungsbau, 1*, 84–91. ((last access: 1 April 2022))

Suchi, E., Dittmann, J., Knopf, S., Müller, C., & Schulz, R. (2014). *Geothermieatlas zur darstellung möglicher nutzungskonkurrenzen zwischen CO2-Einlagerung (CCS) und tiefer geothermie in Deutschland*. Retrieved from https://www.geotis.de/homepage/sitecontent/info/publication_data/final_reports/final_reports_data/Endbericht_Geothermie_Atlas.pdf

Sun, Y., Li, H., Fan, C., Yan, B., Chen, J., Yan, Z., & Sun, Q. (2022). Review of a specialty fiber for distributed acoustic sensing technology. *Photonics, 9*(5). doi:10.3390/photonics9050277

Tester, J., Anderson, B., Batchelor, A., Blackwell, D., DiPippo, R., Drake, E., et al. (2006). *The future of geothermal energy: impact of enhanced geothermal systems (EGS) on the United States in the 21st century* (Technical Report No. 0-615-13438-6). Massachusetts Institute of Technology.

Waagaard, O. H., Rønnekleiv, E., Haukanes, A., Stabo-Eeg, F., Thingbø, D., Forbord, S., et al. (2021). Real-time low noise distributed acoustic sensing in 171 km low loss fiber. *OSA Continuum, 4*(2), 688–701. doi:10.1364/OSAC.408761

Weber, J., Born, H., Pester, S., Schifflechner, C., & Moeck, I. (2022). *Geothermal energy use in Germany, country update 2019–2021*. Paper presented at the European Geothermal Congress 2022. doi:10.24406/publica-592

Yu, G., Cai, Z., Chen, Y., Wang, X., Zhang, Q., Li, Y., et al. (2016). Walkaway VSP using multimode optical fibers in a hybrid wireline. *The Leading Edge, 35*(7), 615–619. doi:10.1190/tle35070615.1

Zimmermann, G., Blöcher, G., Reinicke, A., & Brandt, W. (2011). Rock specific hydraulic fracturing and matrix acidizing to enhance a geothermal system — concepts and field results. *Tectonophysics, 503*(1), 146–154. doi:10.1016/j.tecto.2010.09.026

Zimmermann, G., Moeck, I., & Blöcher, G. (2010). Cyclic waterfrac stimulation to develop an enhanced geothermal system (EGS)—conceptual design and experimental results. *Geothermics, 39*(1), 59–69. Retrieved from doi:10.1016/j.geothermics.2009.10.003

22

Distributed Acoustic Sensing and Distributed Temperature Sensing Geophysical Study in the Takigami Geothermal Field in Northeastern Kyushu, Japan

Junzo Kasahara[1,2], Yoko Hasada[1,3], Shigetsugu Furuya[4], Masahiro Shimazaki[4], Hitoshi Mikada[5], Hiroshi Ohnuma[1], and Yoshihiro Fujise[6]

ABSTRACT

A distributed acoustic sensing (DAS) seismic study and distributed temperature sensing (DTS) temperature measurement were carried out in 2022 at the Takigami geothermal field in Kokonoe town, northeastern Kyushu, Japan. Optical fiber cable was deployed down to a depth of 2,000 m in the TP-2 geothermal well with 12 seismic sources in the vicinity of the Takigami field. A comparison of observed and theoretical travel times suggested the existence of a distinct vertical fault with an N–S strike, concordant with the regional geology, as inferred by the E–W P-wave velocity (*Vp*) profile. The migrated reflection phases obtained by F-K filtering were calculated for some reflection images of the deeper part. Three significant reflected (higher-amplitude) zones were identified at depths of 1, 1.5, and 2–3 km, respectively. DAS and surface geophone records captured the S-to-P converted phases when natural earthquakes occurred during the survey. The conversion depths were estimated at around 1 and 2.5 km, coinciding with the depth of the major reflected zones.

22.1. INTRODUCTION

The detection of ground vibrations has traditionally relied on moving-coil-type seismometers (or geophones). We use velocity, acceleration, or displacement-type seismometers. Around 2010, optical fiber sensing technology called distributed acoustic sensing (DAS) (Hartog, 2017) was used in borehole seismic studies (Barberan et al., 2012; Bettinelli & Frignet, 2015; Daley et al., 2013; Frignet & Hartog, 2014; Götz et al., 2015; Hartog et al., 2014; Johannessen et al., 2012; Madsen et al., 2013; Mateeva et al., 2013, 2014; Miller et al., 2012, 2016; Mellors et al., 2018; Varkey et al., 2008). Because optical fiber can be converted to multiple sensors for the total length, DAS has the advantage of extremely dense sensing. In applying the DAS technology to a vertical seismic profile (VSP), the seismic data can be acquired over the optical fiber in the borehole with designated high-density sampling spacing at one shot.

The fiber-optic methodology and technology are advantageous for application to geothermal exploration and production, particularly for wider and higher temperature ranges than the previous survey technology up to 300 °C without magnetic degradation or damage to electrical circuits. The covered ranges are almost enough for normal geothermal fields and the exploration targets. However,

[1]*Engineering Advancement Association of Japan, Tokyo, Japan*
[2]*Center for Integrated Research and Education of Natural Hazards, Shizuoka University, Shizuoka, Japan*
[3]*Daiwa Exploration and Consulting Co., Ltd., Tokyo, Japan*
[4]*Idemitsu Kosan Co., Ltd., Tokyo, Japan*
[5]*Department of Civil and Earth Resources Engineering, Kyoto University, Kyoto, Japan*
[6]*WELMA Co., Ltd., Fukuoka, Japan*

Distributed Acoustic Sensing in Borehole Geophysics, Geophysical Monograph 289, First Edition.
Edited by Yingping Li, Robert Mellors, and Ge Zhan.

DOI: 10.1002/9781394179275.ch22

application under conditions over 300 °C is still difficult and development of the new technology is still underway due to hydrogen darkening caused by OH absorption peaks in the optical fiber, which prevents continuous long-term observation (Kaizer et al., 1973; Hartog, 2017; Matsuda et al., 2023a, 2023b). Because hydrogen atoms or molecules are smaller than those of other materials, hydrogen in hot water penetrates optical fiber coated by metals such as stainless steel, copper, or gold. The hydrogen in optical fiber creates H or OH absorption peaks around 1,250, 1,350, 1,400, 1,600, and 1,700 nm, and these peaks created by even 1 ppm of hydrogen cause darkening of Raman scattering for DTS measurements, especially at temperatures greater than 300 °C, and the lobe of the 1,600 nm absorption peak also affects the DAS measurement at 1,550 nm for longer than a few weeks and a high temperature close to 300 °C (Matsuda et al., 2023a, 2023b).

Because of the advantages of temperature resistance, some DAS case studies were recently conducted for geothermal regions (e.g., Chalari et al., 2019; Chang & Nakata, 2022; Feigl & Parker, 2019; Schölderle, et al., 2021). Since 2018, we have carried out DAS and distributed temperature sensing (DTS) surveys for geothermal evaluation by deploying optical fiber systems in the wells of several geothermal fields in Japan (Kasahara et al., 2020a, 2020b, 2022a, 2022b, 2022c). The summary of our past studies is presented in this monograph (Kasahara et al., 2025). The primary purpose of the studies was to develop an efficient and optimized technology that would allow the survey to provide better subsurface information for selecting high-potential geothermal targets.

In our new survey in 2022 at the Takigami field, we applied the technology for geothermal concepts and settings such as rocks, temperature, depth, and other types that were different from the previous applications, in order to build up our knowledge, skills, and versatility following the previous experiments for the technology for everyday use. This chapter summarizes the results of the application, focusing on the subsurface velocity modeling and the migration images, as well as the utilization of natural seismic sources such as earthquakes to understand the subsurface structure. We believe that the results provide a good example of the application of DAS and DTS technology to obtain details of the subsurface structure in geothermal fields for commercial use. In this study, the temperature profile in a borehole was first obtained by DTS survey, and the location of the cap rock layer was identified from a comparison with the assumed geological structure and the temperature gradient. Next, a stratigraphic model was used to explain the borehole DAS system's observed arrival times for the signals from the active seismic source at the surface. Then, the depths of the layer interfaces with a large impedance contrast could explain the characteristic seismic reflection waves that appeared in the observation record. Finally, the implied depth of the discontinuity that generates the S-to-P conversion waves was estimated from natural seismic observations. We were able to show how seismic observations in a high-temperature wellbore help clarify the subsurface structure of geothermal areas.

22.2. GEOLOGICAL BACKGROUND OF THE SURVEY AREA AND OUTLINE OF THE SURVEY

The Takigami geothermal field, which started generating electricity in 1996, is geologically located north of the Beppu-Shimabara graben with an E–W strike, representing the dominant N–S extensional force in this area. The Mount Aso volcano, about 20 km southwest of Takigami, is located in the middle of this graben, and the 2016 Kumamoto Earthquake (Mw 7.3) occurred at the southwestern end of this graben.

The geology of the geothermal survey area was studied based on the regional geology, subsurface structure, borehole data, resistivity measurements, and gravity field data from late 1980 to the 1990s (Hoshizumi et al., 1988; Motomatsu, 1999; NEDO, 1990; Takenaka et al., 1995). A distinct vertical fault with an N–S strike that subdivides the geothermal system is inferred by the gravity distribution reported by Takenaka et al. (Figure 22.1). The borehole geological data in the Takigami field (Idemitsu Kosan internal data) also supports the existence of an N–S fault. The inferred location of the fault is about 1 km east of the TP-2 wellhead, which is one of the geothermal wells close to the Takigami Geothermal Power Plant and was selected for the 2022 DAS and DTS survey.

The geological cross-section in the Takigami field is shown in Figure 22.2. The key layers (from shallow to deep) are the Noine-dake volcanic rocks, the Ajibaru formation, the Takigami formation, and the Mizuwake andesite formation. The upper part of the Takigami formation is the sealing layer containing smectite and illite clays. Based on the well-to-well correlation in the Takigami field, the vertical displacement associated with the N–S fault is about 1,000 m, as shown in Figure 22.2. The thicknesses of the Ajibaru and Takigami formations are thinner toward the east, and the depth of the Takigami formation is shallower in the east.

We carried out the DAS and DTS survey from 15 May to 1 June 2022, deploying the optical fiber system down to a depth of 2,000 m in the TP-2 geothermal well. The locations of the 12 seismic sources are shown in Figure 22.3. After setting up the survey tools, we measured the temperature profile and seismic signals under the following conditions and parameters: utilizing an OptaSense Interrogator unit, 20 kHz ping rate, 1 kHz sampling, 16.4 m gauge length, and data acquisition at intervals of 1 m. We operated the IVI EnviroVibe system at each source for 8 h with a 30 s sweep from 10–75 Hz and 30 s of sleep and stacked the data for better images. In addition to the seismic measurement along the well, we used

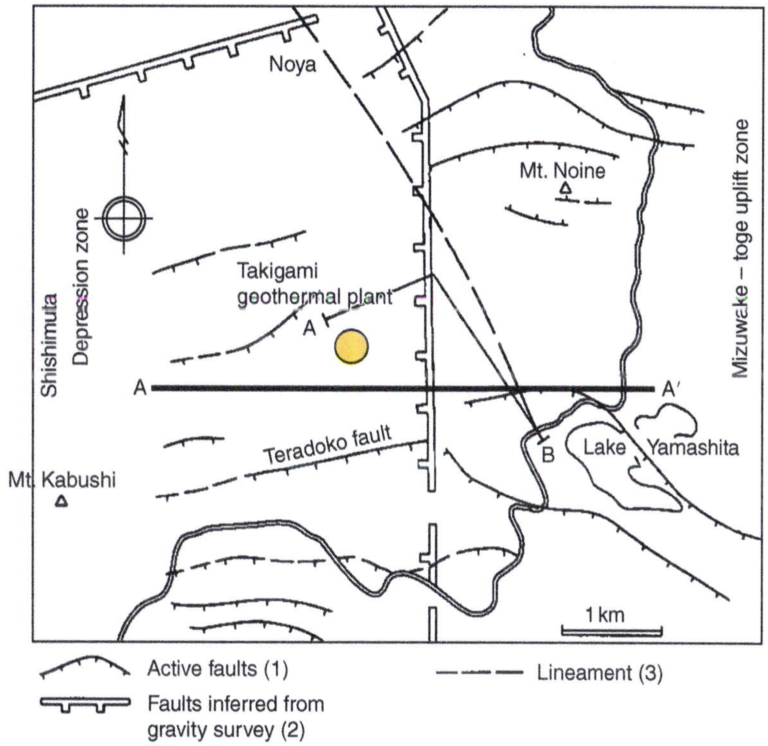

Figure 22.1 N–S fault inferred by gravity data. The geological cross-section A–A′ is shown in Figure 22.2. The yellow circle shows the location of the Takigami Geothermal Power Plant and the TP-2 wellhead. Adapted from Takenaka et al. (1995).

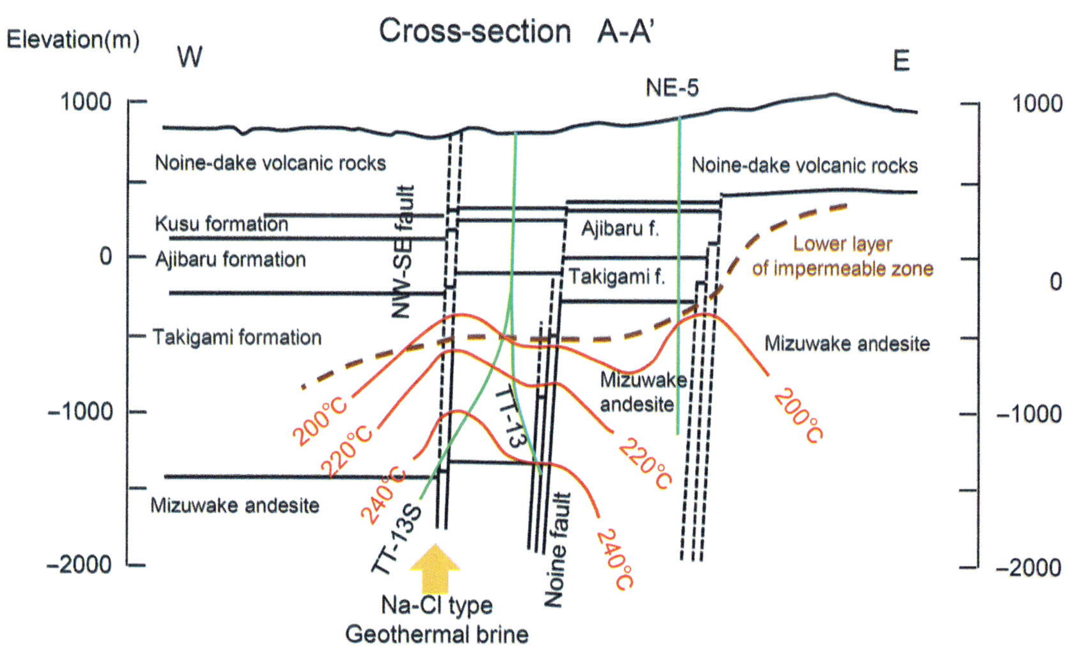

Figure 22.2 E–W geological cross-section. The four key layers (from shallow to deep) are the Noine-dake volcanic rocks, Ajibaru formation, Takigami formation, and Mizuwake andesite formation. Adapted from Idemitsu Kosan (2022).

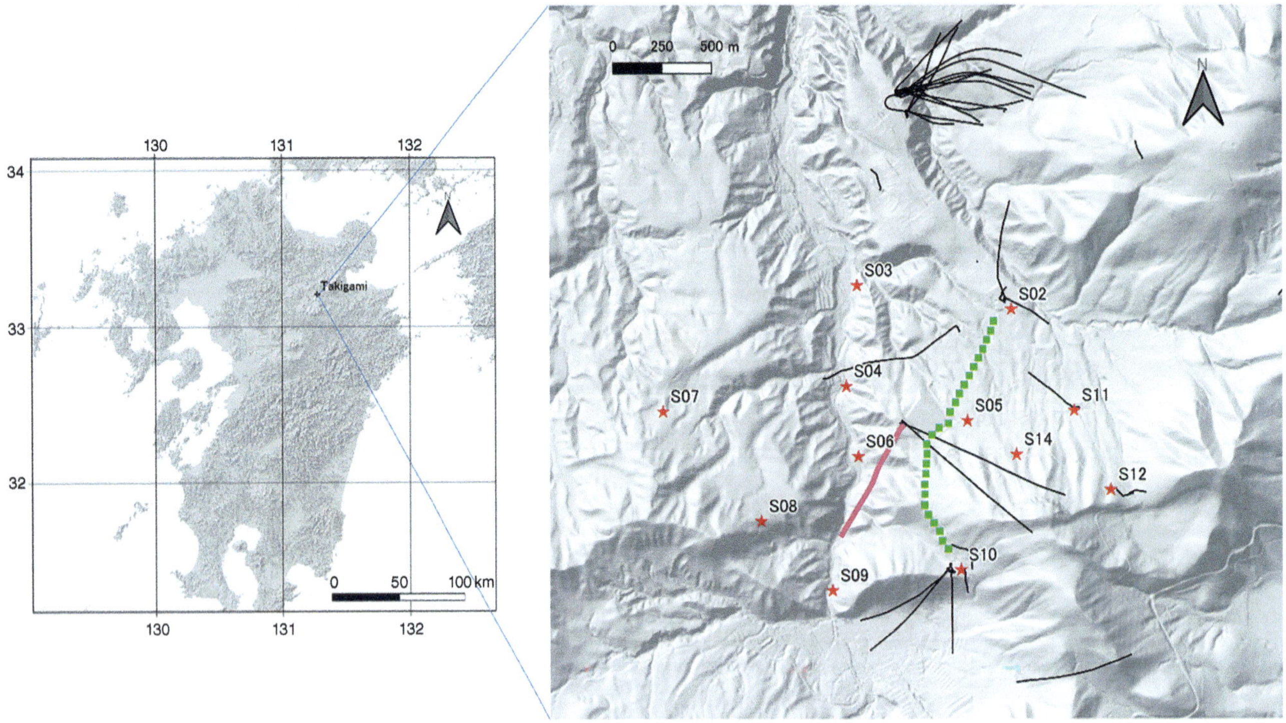

Figure 22.3 Location map of TP-2 geothermal well (pink line), seismic sources (red stars), and surface geophones (green squares). Black lines are other wells. The hill shade map is adapted from a topographic map from the Geospatial Information Authority of Japan.

three-component seismometers (1 or 2 Hz) at the ground surface, as shown in Figure 22.3. The seismometers were set along narrow roads within the Takigami geothermal field, which had limited access to the public and reduced artificial noise such as traffic.

22.3. RESULTS OF TEMPERATURE MEASUREMENTS

The temperature profile of the TP-2 well derived from DTS is shown in Figure 22.4. The temperature above 800 m is nearly flat, suggesting that meteoric water circulation occurs at depths shallower than 800 m. The temperature gradually increases from 800 to 1,400 m, ranging between 100 °C and 200 °C, implying that the dehydration process of hydrous smectite clay forms less-hydrous illite at 100 °C to 190 °C. The smectite absorbs a considerable amount of water in the dual electric layer, and this causes low permeability, which is a critical factor of the sealing layer (Kawamura & Ichikawa, 2003). The bottom of the TP-2 well reached the Mizuwake andesite formation, and the temperature was 223 °C at 2,000 m. The recorded temperature profile is concordant with the previous surveys conducted by Idemitsu Kosan.

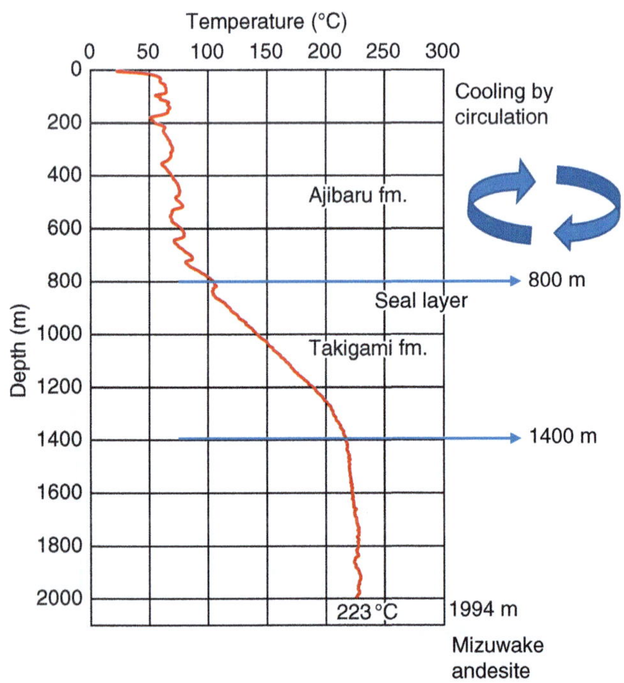

Figure 22.4 Temperature profile obtained by DTS measurement. The vertical axis is the depth of the well in meters. The horizontal axis is the temperature in °C.

22.4. DAS RECORDS, SYNTHETIC DAS RECORDS, AND VELOCITY ANALYSIS

We obtained 12 sets of DAS records representing each seismic source. The summary of the records is shown in Figure 22.5. The first process of the analysis is the determination of the velocity structure.

Two existing *Vp* data sets were referred to when constructing the velocity structure in the survey area. One was a refraction analysis done in 1979 (Ministry of International Trade and Industry report, 1980), and the other was the check shot data from a well near the TP-2 well (1998 Idemitsu Kosan internal data). Using the results of previous seismic studies mentioned previously, the geological descriptions from the Takigami wells (2022 Idemitsu Kosan internal data), and the zero-offset DAS records, we constructed the starting model (m000) with a distinct N–S strike fault assumed to be located 800 m east of the TP-2 wellhead. The usefulness of the model (Figure 22.6) was examined by synthetic waveforms along the TP-2 well in 3D space (Figure 22.7). The vertical displacement of the lower layer (Mizuwake andesite) was 1 km. The dots in the same figure show the observed P-wave first arrivals.

After numeral examination of the *Vp* models, we obtained the best *Vp* profile (m023) where the N–S fault location was moved further west from the initial assumed location to match the observed P-wave first arrivals (Figure 22.8). This was determined by examining the S11 and S12 locations after we recognized a significant travel-time misfit between the assumed model and the observation.

We calculated the synthetic DAS waveform using m023, as shown in Figure 22.8, and compared it to observed P first arrivals (Figure 22.9). Larger displays of the synthetic waveforms calculated by the m023 model at S02, S08, S09, and S12 are shown in Figure 22.10.

The best location of the N S fault after examination is shown in Figure 22.11. The location is 600 m east from the TP-2 wellhead, compared with the initial location of 800 m east from TP-2.

22.5. REFLECTED PHASES AND MIGRATED IMAGE

We separated the DAS waveforms into the downgoing and upgoing phases using F-K filtering. Although we tested the median filter for the separation of downgoing and upgoing arrivals (Hardage, 1983; Yilmaz, 2001), we could not obtain better results than those by F-K filtering. The downgoing phases correspond to direct arrivals, and the upgoing phases correspond to the reflected phases. Representative DAS records at S02, S08, S09, and S12 source locations are shown in Figure 22.12. Numerous

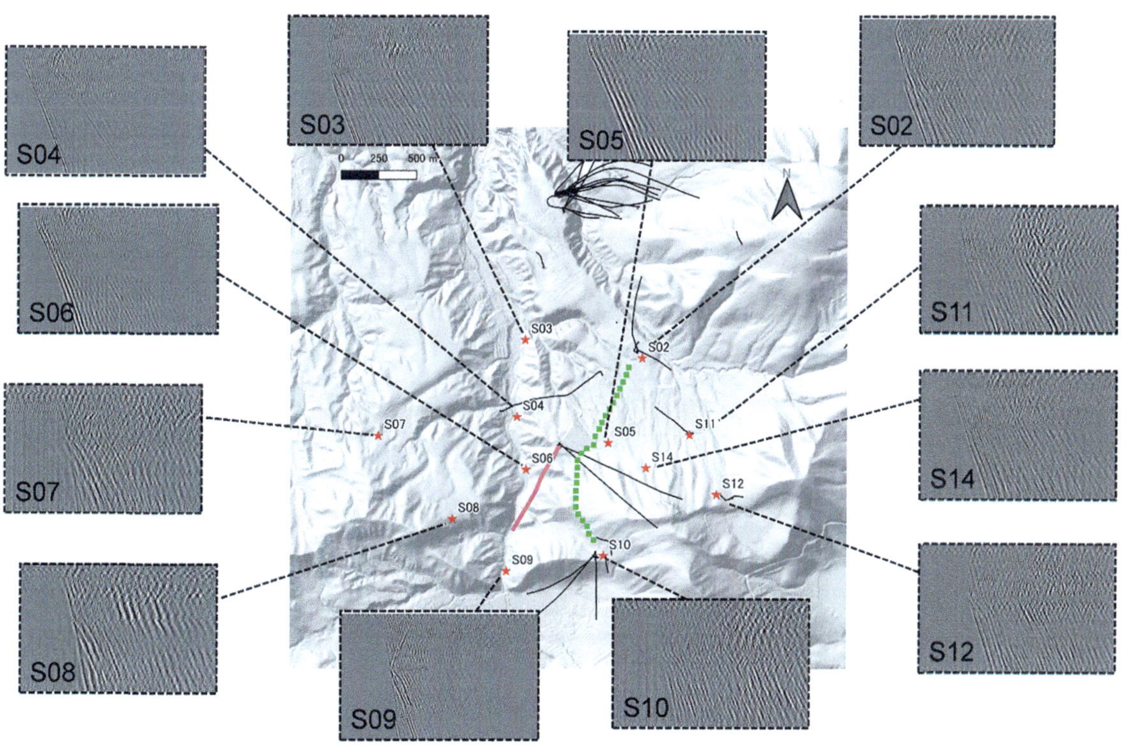

Figure 22.5 Seismic source location map with the related DAS records. The vertical and horizontal axes of each DAS diagram are the depth between 0 and 2,000 m and the elapsed time between 0 and 2 s, respectively. The hill shade map is adapted from a topographic map from the Geospatial Information Authority of Japan.

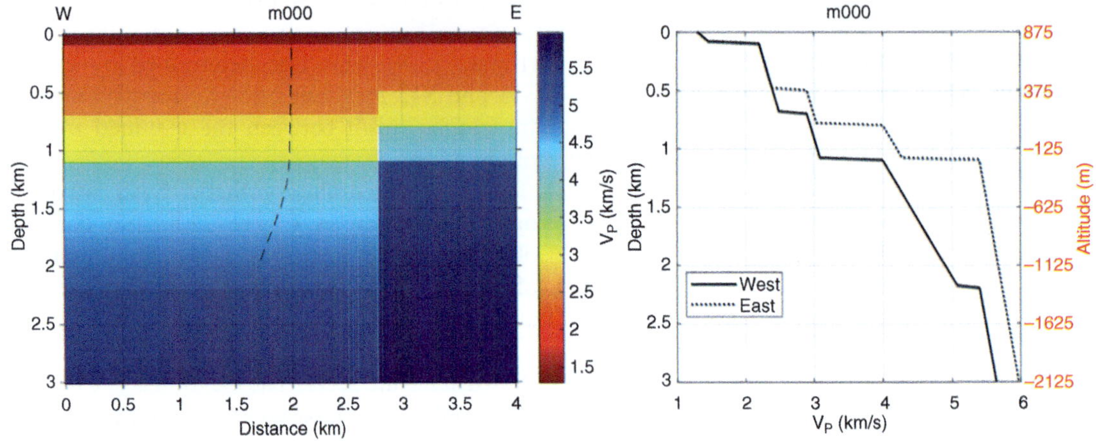

Figure 22.6 Starting model (m000) for examining the fitting to the observed DAS records. The vertical axis is the depth. The horizontal axis of the diagram on the right is the *Vp* in kilometers per second, and the *Vp* profile across the N–S fault is shown in the diagram on the left. Based on previous studies, the location of the N–S fault was assumed to be 800 m east of the TP-2 wellhead. The TP-2 well profile is shown by the broken black line.

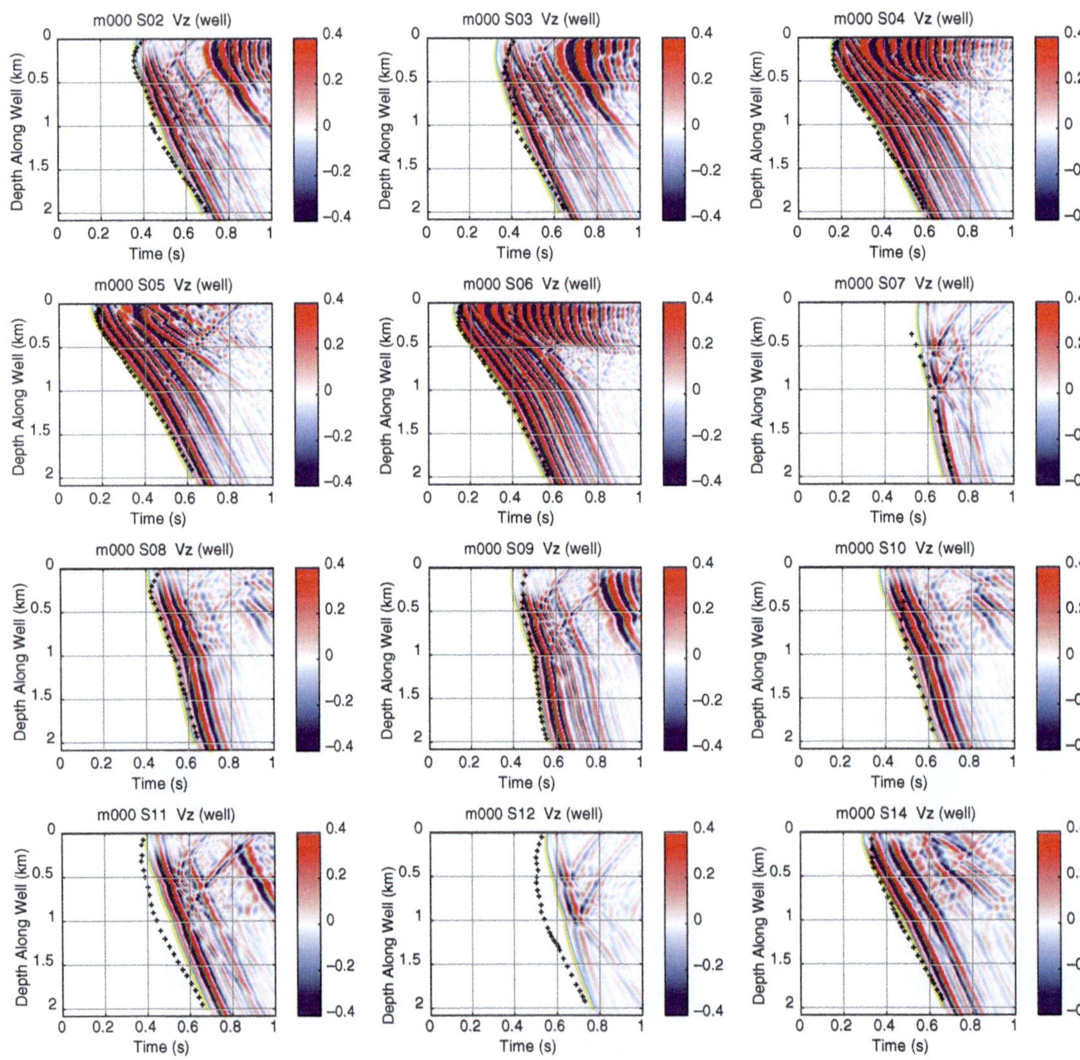

Figure 22.7 Synthetic DAS waveforms at 12 source locations calculated by the m000 model. Dots show the observed P-wave first arrivals. The vertical and horizontal axes of each diagram are the depth of the TP-2 well and the travel time, respectively. The amplitude scale is shown on the right side of each diagram. S11 and S12 show significant travel time discrepancies.

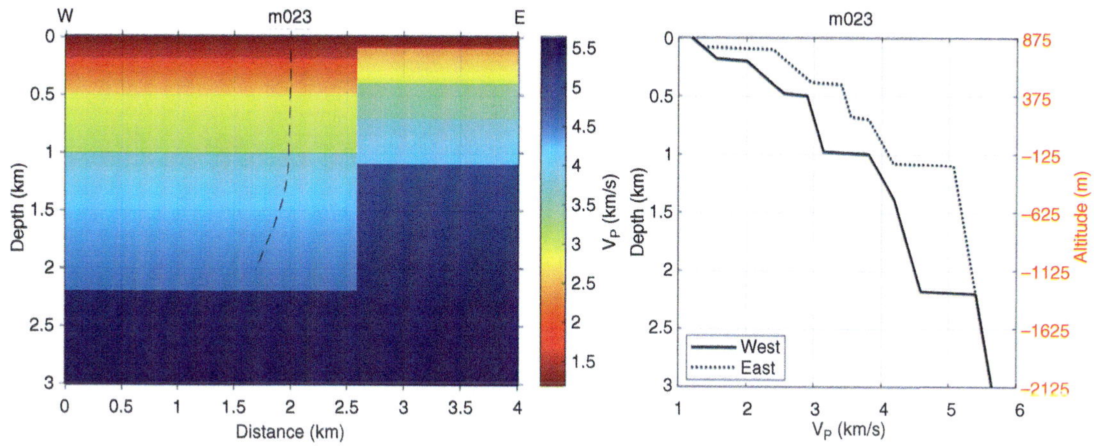

Figure 22.8 *Vp* depth model (m023) along the E–W profile across the TP-2 wellhead, which best matches the observed DAS data. The N–S vertical fault is modeled at around 600 m east from the TP-2 wellhead location. The broken black line shows the TP-2 well trace. The boundary of the Ajibaru and Takigami formations is at a depth of about 1 km. On the right is the east (dotted) and west (solid) depth profile, and on the left is the velocity model. The horizontal axis is the distance in kilometers from the west.

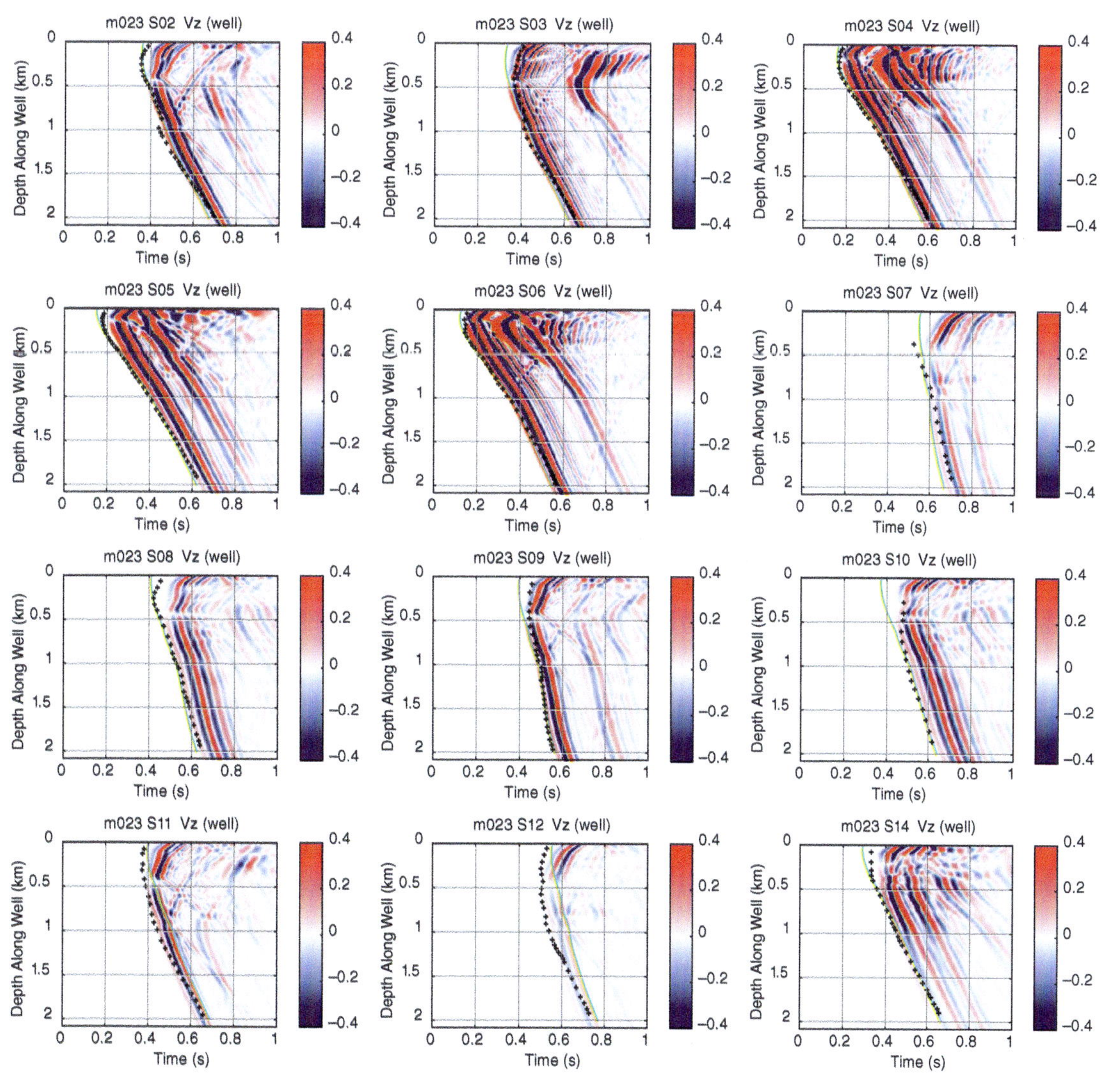

Figure 22.9 Synthetic DAS waveforms at 12 source locations calculated by the m023 model, as discussed earlier. Dots show the observed P-wave first arrivals. The vertical and horizontal axes of each diagram are the depth of the TP-2 well and the travel time, respectively. The amplitude scale is shown on the right side of each diagram.

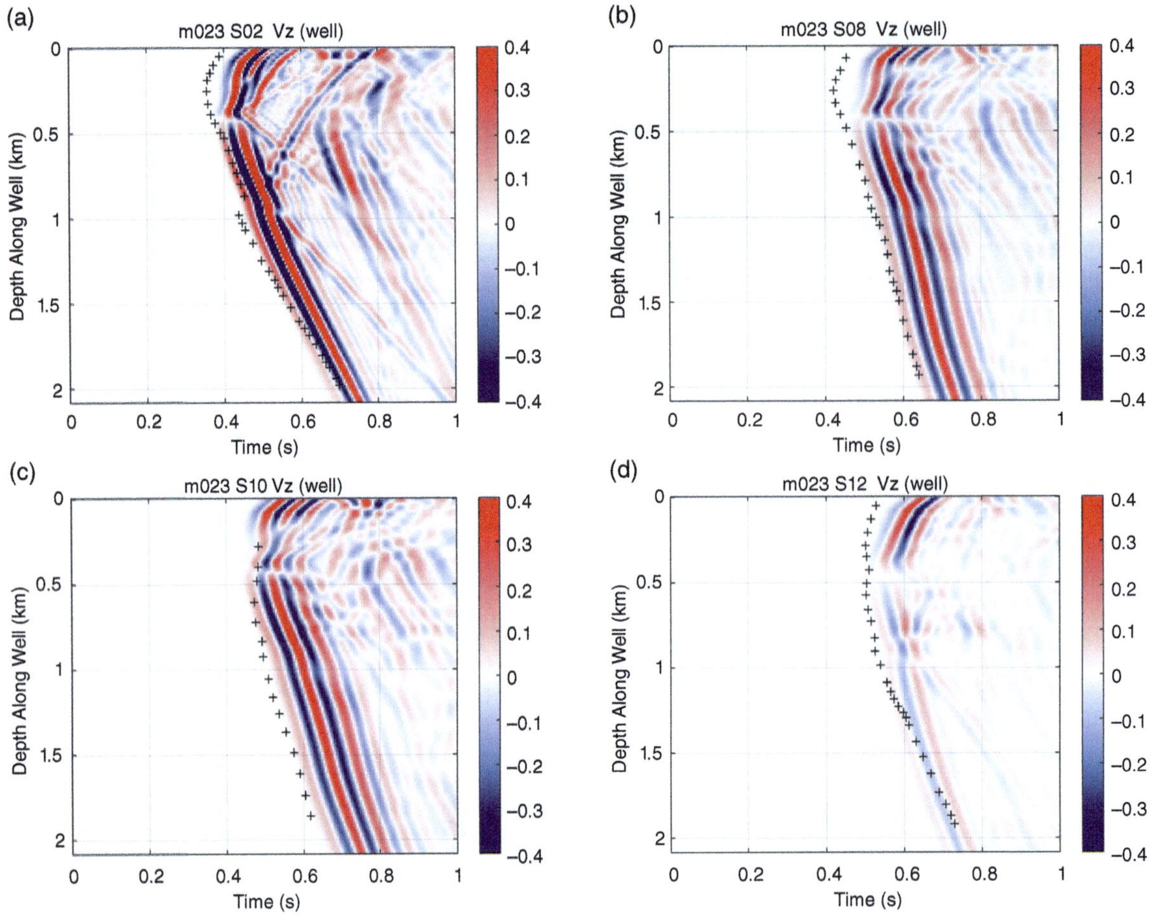

Figure 22.10 Synthetic DAS waveforms at (a) S02, (b) S08, (c) S09, and (d) S12. The vertical and horizontal axes of each diagram are the depth of the TP-2 well and the travel time, respectively. (+), observed P-wave first arrivals.

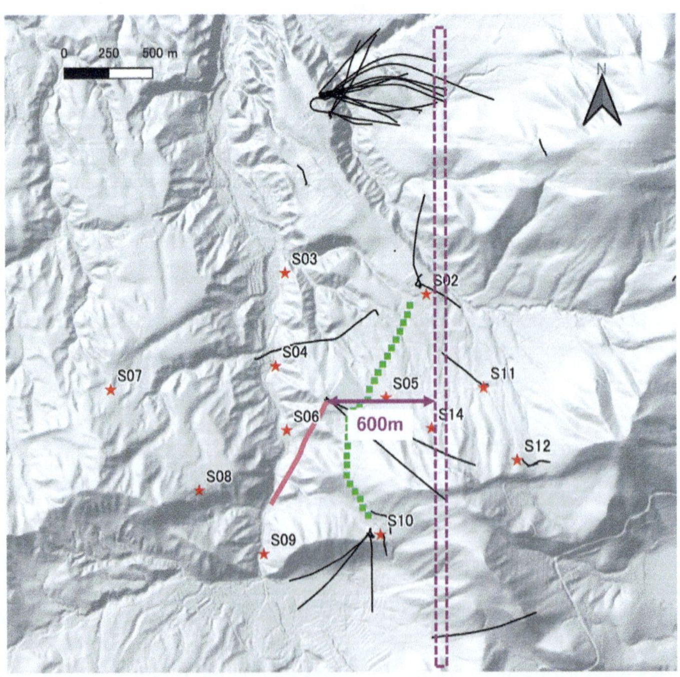

Figure 22.11 Inferred final location of the N–S fault (pink rectangular box) 600 m from the TP-2 wellhead toward E (pink arrow). Lines show the trajectory of the wells in the Takigami geothermal field. The red stars are the seismic source locations, and the green squares are surface geophones.

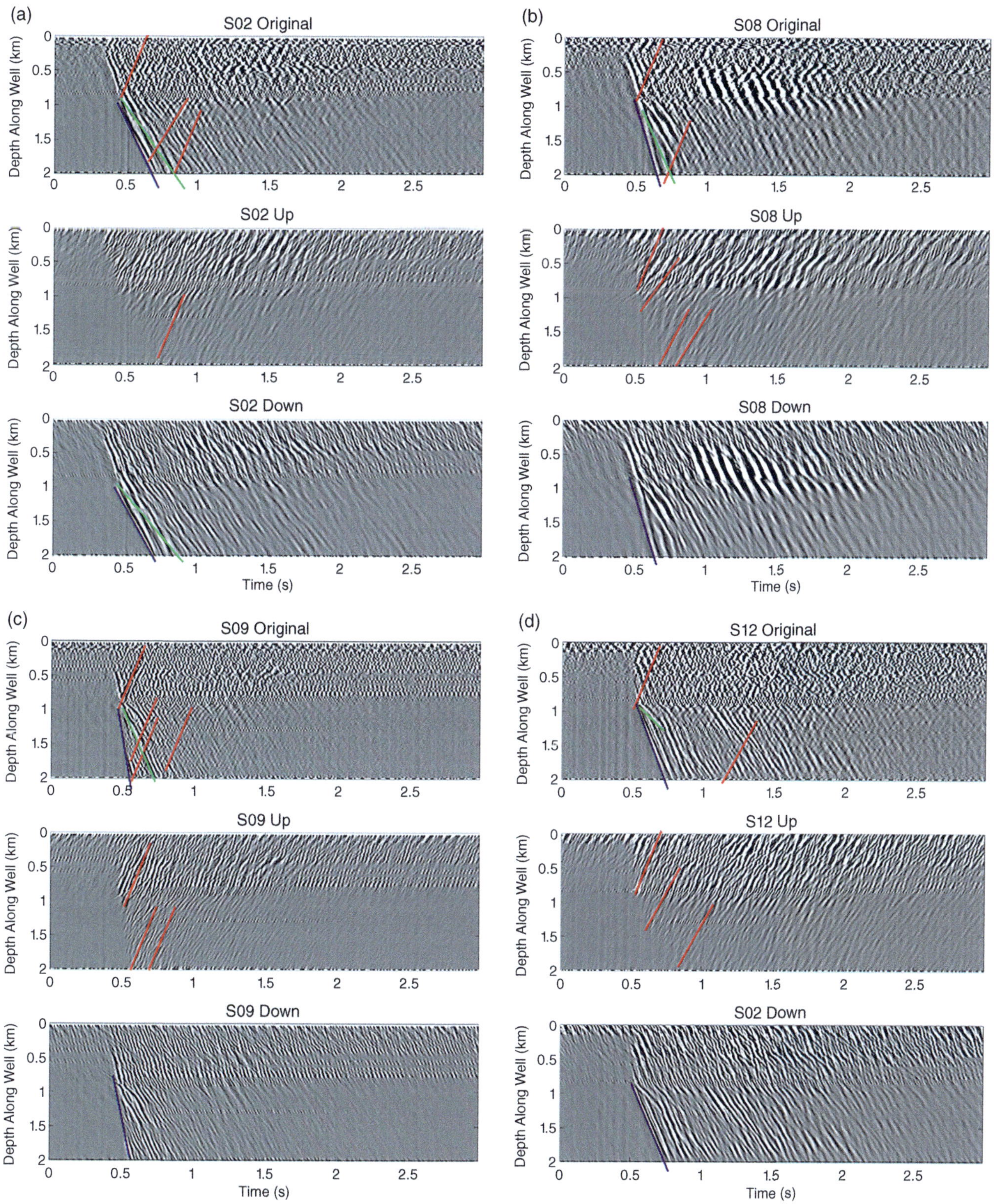

Figure 22.12 Four sets of observed DAS waveforms at (a) S02, (b) S08, (c) S09, and (d) S12. Blue lines, P-wave first arrivals; red lines, reflected P phases; green lines, S phases. Middle and bottom of each set are DAS waveforms for upgoing and downgoing arrivals.

reflectors were identified in the original and upgoing panels. It was also found that reflected P waves and converted downgoing S waves were generated at a depth of about 1 km. This means that there was a significant change in the physical condition of the formation at that depth, and the depth coincides with the top sealing layer of the Takigami formation, which is discussed in Section 22.3.

We finally migrated 11 upgoing record sections (excluding S12) to obtain reflection images using the Kirchhoff method and the *Vp* structure of the western part obtained in this study. Although we examined reverse time migration, the migrated image obtained by this method did not have better coherency than the results obtained by the Kirchhoff migration. The 3D Kirchhoff migration for each source position was calculated. The envelopes of migrated traces were added for the 11 source positions to obtain the final results. We extracted five migration images from the 3D migration image cube volume for sections containing the source(s) and the wellhead of TP-2 (Figures 22.13 and 22.14). Because only 11 sources were used for the migration images, there are smile-like and poor images, especially for a cross-section far from the source(s).

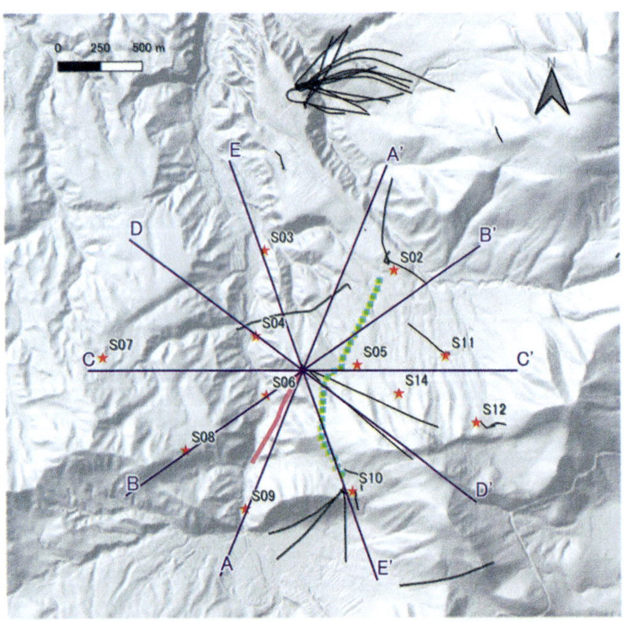

Figure 22.13 Location map for migration cross sections crossing the sources and the TP-2 wellhead (blue lines). Red stars are the locations of seismic sources S02 to S14. We did not use S12 for the migration. The center is the location of the TP-2 wellhead. The thick reddish-purple line is the trajectory of the TP-2 borehole. Green squares are surface seismometers, but were not used for the migration. Black lines are other wells in this field. The hill shade map is adapted from a topographic map from the Geospatial Information Authority of Japan.

Three high-amplitude anomalous zones were identified at depths of 1, 1.5, and 2–3 km in all cross sections. We imaged the zone at a depth deeper than TP-2 by reflections identified in the DAS reflection records.

The 3D image in Figure 22.15 has better areal distribution. The most reflective zone at a depth of 1 km is shown in red. The other reflective zones at 1.5 and 2–3 km are shown in yellow.

22.6. S-P CONVERSIONS

Compressional and shear seismic waves convert each other, such as P to S and S to P, at interfaces with significant impedance contrasts. Because geothermal reservoirs seem to contain fluid in a fracture system having a significant impedance contrast to surrounding tight rocks, the depth and location of the conversions could provide good evidence of geothermal reservoirs. For this reason, we checked the possibility of phase conversion in the DAS and surface seismic records. We observed several earthquakes during the survey period (Figure 22.16). Based on the DAS records, we identified the occurrence of S-to-P converted arrivals for some earthquakes. An example is shown in Figure 22.17. In addition to this observation, other natural earthquake observations by DAS and surface seismometer records are described in this monograph (Hasada et al., 2025; Kasahara et al., 2025). Two kinds of S-to-P converted phases were recognized: one at a depth of 2.5 km and the other at a depth of 1 km.

Hasada et al. (2025) discuss more details about P-to-S and S-to-P conversion and the general utilization of converted waves in geothermal fields.

22.7. DISCUSSION AND CONCLUSIONS

The following observations and conclusions are based on the results of the DAS and DTS survey carried out in 2022 at the TP-2 geothermal well in the Takigami geothermal field.

22.7.1. Temperature Profile (DTS)

The obtained temperature profile is the same as the previous measurements obtained by Idemitsu Kosan. This means that the DTS survey validates and reproduces the necessary information using DTS technology to deploy optical fiber cable in the well. The survey results show essential geological features of the dehydration process of smectite clay forming smectite–illite at 100 °C to 190 °C at a depth between 800 and 1,400 m, implying

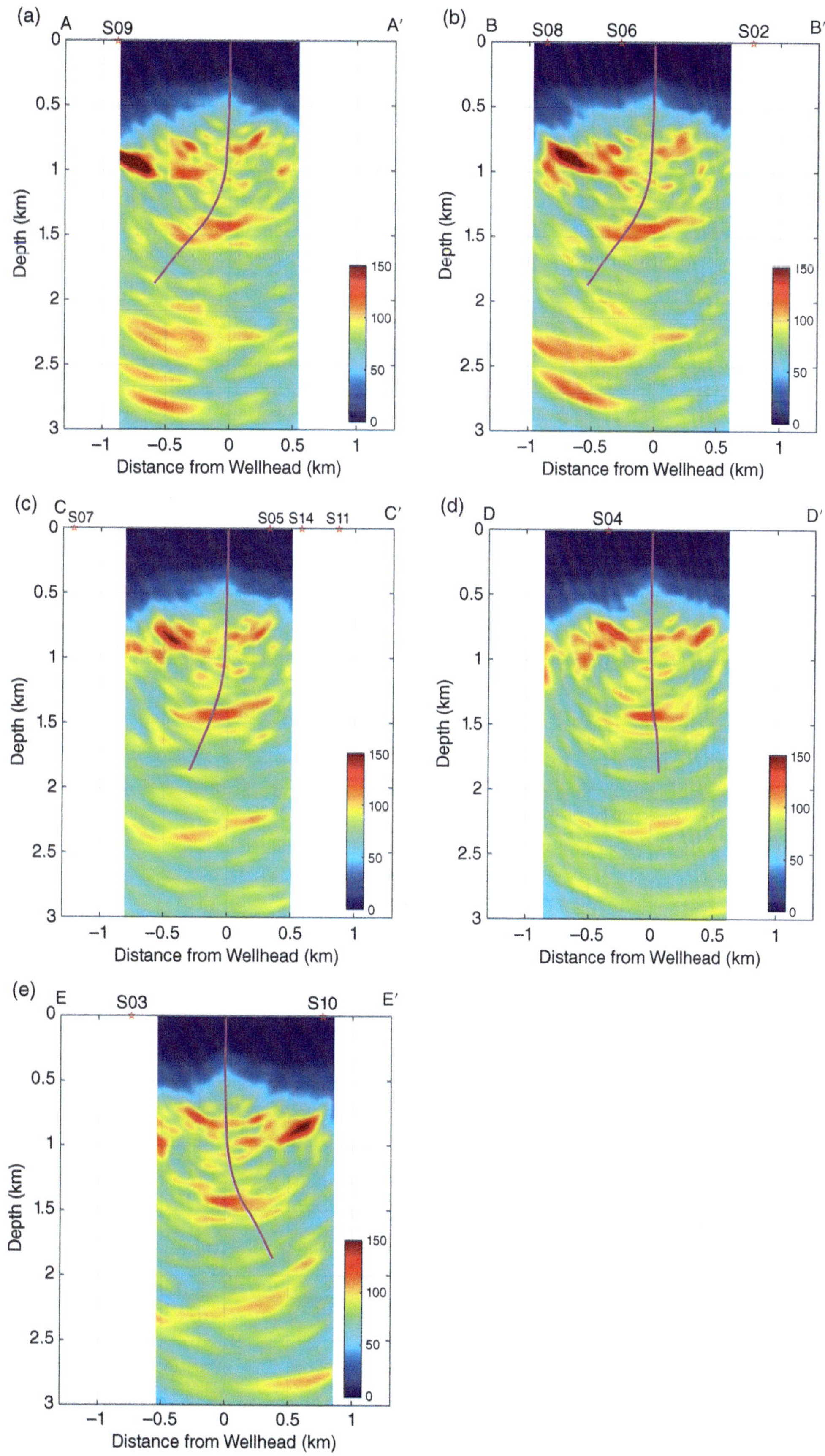

Figure 22.14 The results of migration along the lines A–A′ to E–E′ including the sources in Figure 22.13. The pink line is the TP-2 borehole trajectory. Each cross-section shows the sources along each line. The colors show the absolute amplitude level of the migration from blue (low) to red (high). The unit is arbitrary.

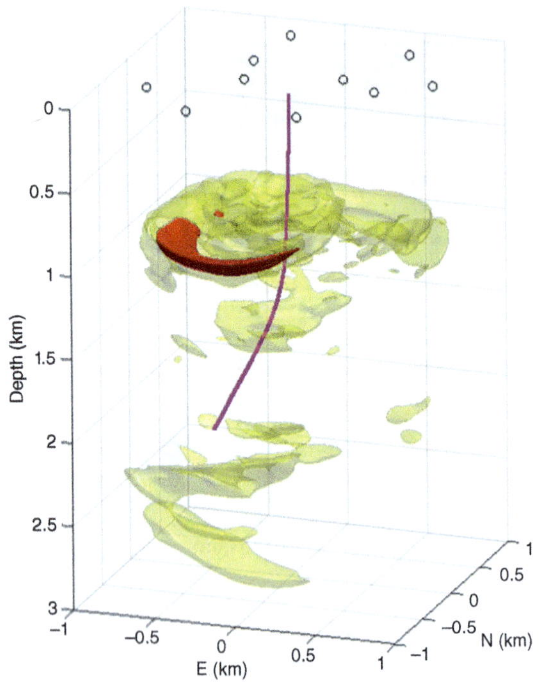

Figure 22.15 The 3D image of the migration results as P reflections. The pink line shows the TP-2 trajectory. The most intense reflection is at a depth of about 1 km. The 1.5 km depth reflective zone is near the well. The 2–3 km reflections exist southwest of the TP-2 bottomhole location. The colors show the migration amplitude intensity from higher (red) to lower (yellow).

the existence of the sealing layer of the Takigami geothermal field.

22.7.2. DAS Analysis

After utilizing all the measured and calculated data derived from the DAS and surface seismometer records, we obtained the final *Vp* structure with the N–S vertical fault located 600 m east of the TP-2 wellhead. The vertical displacement across the N–S fault is approximately 1 km.

We prepared migration images using reflected phases obtained by F-K filtering of DAS waveforms. Because 11 seismic sources were used, the results of migration had smiles and poor resolution at locations far from the sources and the TP-2 trajectory. However, we identified three major reflection zones. The shallow one exists at a depth of about 1 km, almost over the survey area, whereas the deeper ones were located at depths of about 1.5 and 2–3 km around the well and at the southwest area of the TP-2 wellhead, respectively. Due to the limited number of sources for the migration imaging, the results are ambiguous for the horizontal direction, but the vertical depth of the intense reflection zone was more accurate because we applied the *Vp* subsurface distribution using DAS and surface seismometer arrivals.

22.7.3. Interpretation

The shallower reflection zone is probably the bottom of the Ajibaru formation and/or the boundary of the Ajibaru

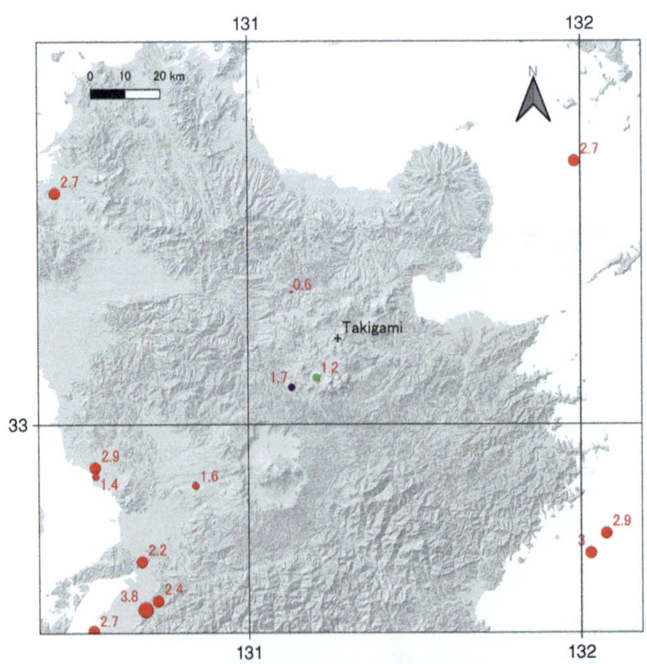

Figure 22.16 Fourteen natural earthquakes occurred during the observation period. Each epicenter is shown as a red circle. The solid blue circle is the Aso region M1.7 earthquake shown in Figure 22.17. The black + symbol is the location of the Takigami field.

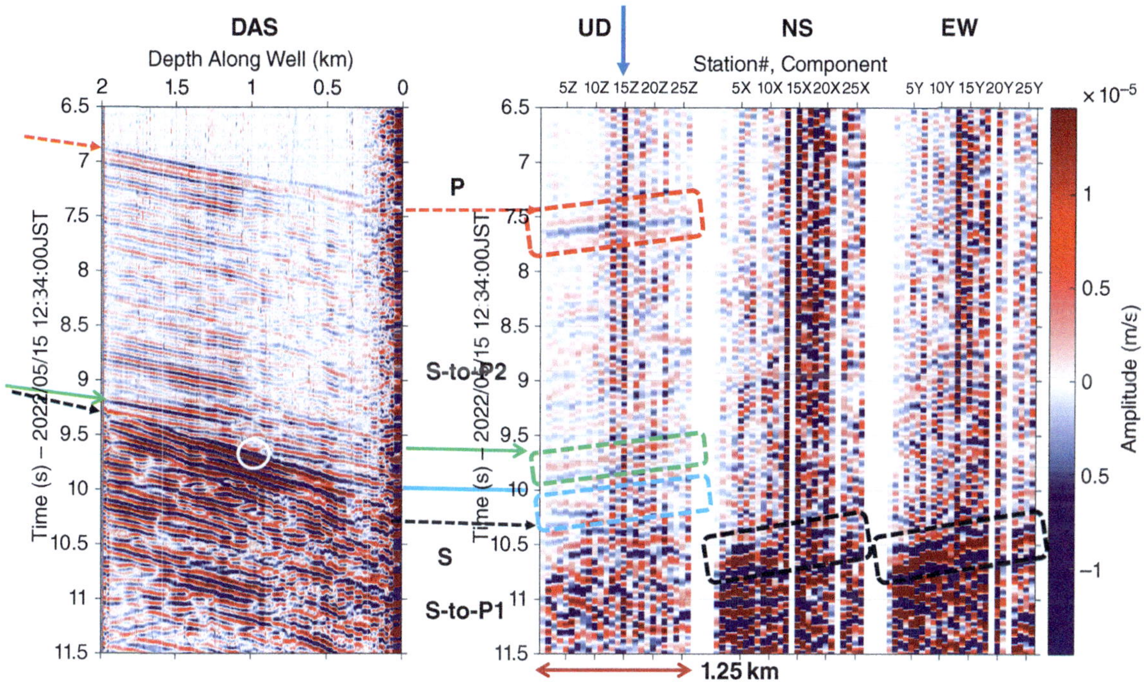

Figure 22.17 An example of a DAS record showing the S-to-P phase at the Aso region M1.7 earthquake 20 km from the field (Figure 22.16). (Left) DAS records. DAS records show S-wave arrival and S-to-P2 converted arrivals at 10.5 and 9.6 s at the wellhead, respectively, corresponding to the formation boundary at a depth of 2.5 km. The gradient of S-to-P2 is the same as the direct P phase. The S gradient corresponds to *Vp*/*Vs* = 1.75. Another S-to-P1 conversion event can be identified at a depth of 1 km. (Right) Surface seismometer records of up-down (UD), E–W, and N–S. Each component has 26 traces with 50 m separations. The locations of the surface seismometers are shown in Figure 22.3. Green and light blue rectangles on the UD component are S-to-P arrivals, and the black rectangles on the E–W and N–S components are S arrivals. The white circle shows the conversion at a depth of 1 km.

and Takigami formations, and the meteoric water circulation within the Ajibaru formation causes the flat temperature profile. The intense reflection at around 1,000 m suggests higher fluid saturation at this depth. Time-lapse gravity research indicated a temporal gravity change around the TP-2 wellhead (Nishijima et al., 2000), suggesting that higher fluid saturation by migration data at this depth may be the reason for the gravity change. The deeper zone at 2–3 km may be a new geothermal reservoir. The S-to-P conversion coincides with these depths, suggesting higher saturation of the liquid layer.

ACKNOWLEDGMENTS

The present research was supported by the NEDO Supercritical Development Project (JPNS21991). We sincerely thank the officers of NEDO for their active and continuous support. We also express our gratitude to Idemitsu Kosan Co., Ltd. and Idemitsu Oita Geothermal Co., Ltd. for kindly allowing us to use their TP-2 well and related geothermal data in the geothermal field. We also thank WELMA, Oyo Geomechanics Co., Ltd., and Hanshin Consultants Co., Ltd. for field support. We express our thanks to two anonymous reviewers. Their comments helped to greatly improve our manuscript.

REFERENCES

Barberan, C., Allanic, C., Avila, D., Hy-Billiot, J., Hartog, A., Frignet, B., & Lees, G. (2012). Multi-offset seismic acquisition using optical fiber behind tubing. Paper presented at the 74th EAGE Conference and Exhibition incorporating EUROPEC 2012, Copenhagen, Denmark. doi:10.3997/2214-4609.20148798

Bettinelli, P., & Frignet, B. 2015, Using optical fiber seismic acquisition for well-to-seismic tie at the Ketzin pilot site (CO_2 storage) site. Paper presented at the 2015 SEG Annual Meeting, New Orleans, LA, SEG-5846235.

Chalari, A., Mondanos, M., Coleman, T., Farhadiroushan, M., & Stork, A. (2019). Seismic methods for geothermal reservoir characterization and monitoring using fiber optic distributed acoustic and temperature sensor. Paper presented at the EAGE/BVG/FKPE Joint Workshop on Borehole Geophysics and Geothermal Energy.

Chang, H., & Nakata, N. (2022). Investigation of time-lapse changes with DAS borehole data at the brady geothermal field using deconvolution interferometry. *Remote Sensing, 14*(1), 185. doi:10.3390/rs14010185

Daley, T. M., Freifeld, B. M., Ajo-Franklin, J., Dou, S., Pevzner, R., Shulakova, V., et al. (2013). Field testing of fiber-optic distributed acoustic sensing (DAS) for subsurface seismic monitoring. *The Leading Edge, 32*(6), 699–706. doi:10.1190/tle32060699.1

Feigl, K. L. & Parker, L. M. (2019). Porotomo final technical report: poroelastic tomography by adjoint inverse modeling of data from seismology, geodesy, and hydrology (Technical Report). University of Wisconsin. doi:10.2172/1499141

Frignet, B.G. & Hartog, A. H. (2014). Optical vertical seismic profile on wireline cable. Paper presented at the SPWLA 55th Annual logging Symposium, Abu Dhabi, UAE.

Götz, J., Lüth, S., Henninges, J., & Reinsch, T. (2015). Using a fiber optic cable distributed acoustic sensor for vertical seismic profiling at the Ketzin CO_2 storage site, Paper presented at the 77th EAGE Conference and Exhibition, Madrid, Spain, Th P2 13.

Hardage, B. A. (1983). *Vertical seismic profiling, Part A., Principles* (p. 182). Amsterdam Geophysical Press.

Hartog, A., Frignet, B., Mackie, D., & Clark, M. (2014). Vertical seismic optical profiling o wireline logging cable. *Geophysical Prospecting, 62*, 693–701.

Hartog, A. H. (2017). *An introduction to distributed optical fiber sensors* (440pp). CRC Press.

Hasada, Y., Kasahara, J., Mikada, H., & Ohnuma, H. (2025). Investigation of distributed acoustic sensing characteristics by complementary observations of distributed acoustic sensing in geothermal wells and by surface seismometers, in this monograph.

Hoshizumi, H., Ono, K., Mimura, K., & Noda, T. (1988). *Beppu geological map* (55(522.6) (084.32 M 50) (083)). Geological Survey of Japan.

Idemitsu Kosan (2022). Unpublished raw data.

Johannessen, K., Drakeley, B., & Farhadiroushan, M. (2012). *Distributed acoustic sensing-a new way of listening to your well/reservoir*. Paper presented at the SPE Intelligent Energy International held in Utrecht, The Netherlands, SPE 149602.

Kasahara, J., Harada, Y., & Kuzume, H. (2020a). *Possibility of high Vp/Vs zone in the geothermal filed suggested by the P-to-S conversion*. Paper presented at the 45th Workshop on Geothermal Reservoir Engineering, Stanford University, Stanford, CA.

Kasahara, J., Harada, Y., Kuzume, H., Mikada, H., & Fujise, Y. (2020b). *The second seismic study at the geothermal field in southern Kyushu, Japan using an optical fiber system and surface geophones*. Paper presented at the 45th Workshop on Geothermal Reservoir Engineering, Stanford University, Stanford, CA.

Kasahara, J., Hasada, Y., Kuzume, H., Ohnuma, H., Mikada, H., & Fujise, Y. (2022a). A field experiment of temperature-tolerant distributed acoustic sensor in deep geothermal reservoir prospecting. *The Leading Edge, 41*, 331–337.

Kasahara, J., Hasada, Y., Ohnuma, H., Mikada, H., & Fujise, Y. (2022b). *Seismic study at the Sumikawa geothermal field in Akita Pref. Japan using a geothermal well*. Paper presented at the Society of Exploration Geophysicists, Japan, Spring Meeting.

Kasahara, J., Hasada, Y., Kuzume, H., Mikada, H., & Fujise, Y. (2022c). DAS-Seismic study in the Ohnuma geothermal borehole in Honshu, Japan. *Journal of the Geothermal Research Society of Japan, 44*, 73–84.

Kasahara, J., Hasada, Y. & Mikada, H. (2025). *Geothermal reservoir characterization using distributed acoustic sensing from vertical seismic profiling in six geothermal fields in Japan*, in this monograph.

Kawamura, Y., & Ichikawa, Y. (2003). Permeability of clay and deformation mechanics. In J. Kasahara, M. Toriumi, & K. Kawamura (Eds.), *The role of water in earthquake generation* (392pp). University of Tokyo Press.

Keizer, P., Tynes, A. R., Astle, H. W., Pearson, A. D., French, W. G., Jaeger, R. E., & Cherin, A. H. (1973). Spectral losses of unclad vitreous silica and soda-lime-silicate fiber. *Journal of the Optical Society of America, 63*(9), 1141–1148.

Madsen, K. N., Dummong, S., Kisrki, A., Pedersen, A. S., Finfer, D., Gillies, A., & Travis, P. (2013). *Simultaneous multiwell VSP in the North Sea using distributed acoustic sensing*. Paper presented at the 75th EAGE Conference & Exhibition Incorporating SPE EUROPEC, London, UK.

Mateeva, A., Mestayer, J., Yang, Z., Lopez, J., Wills, P., Roy, J., & Bown, T. (2013). *Dual-well 3D VSP in deepwater made possible by DAS*. Paper presented at the Annual Meeting of Society of Exploration Geophysicists, Houston, TX, USA, DEG, 0667.

Mateeva, A., Lopez, J., Potters, H., Mestayer, J., Cox, B., Kiyashchenko, D., et al. (2014). Distributed acoustic sensing for reservoir monitoring with vertical seismic profiling. *Geophysical Prospecting, 62*, 679–692.

Matsuda, T., Kitaoka, S., Nakahira, K., Itou, T., Katou, T., & Kasahara, J. (2023a). *In-situ evaluation of light-transmitting loss of optical fibers under hydrothermal conditions*, abstract of LST68-P02, 1-3. Tokyo 2023.

Matusda, T., Kitaoka, S., Nakahira, K., Kato, T., & Kasahara, J. (2023b). Evaluation of in-situ optical loss of polyimide-coated optical fiber under hydrothermal environments. *Optical Fiber Technology, 82*, 103582.

Mellors, R., Sherma, C. T., Ryerson, R., Morris, J., Messerly, J., Yu, C., et al. (2018). *Modeling potential EGS signals from a distributed fiber optic sensor deployed in a borehole*. Paper presented at the Stanford Geothermal Workshop.

Miller, D., Parker, T., Kashikar, S., Todorov, M., & Bostic, T. (2012). *Vertical seismic profiling using a fiber optic cable as a distributed acoustic sensor*. Paper presented at the 74th EAGE Conference & Exhibition Incorporating SPE EURPOPEC, Copenhagen, Denmark.

Miller, D., Daley, T. M., White, D., Freifeld, B. M., Robertson, M., Cocker, J., & Craven, M. (2016). Simultaneous acquisition of distributed acoustic sensing VSP with multi-mode and single mode fiber optic cables and 3C geophones at the Aquistore CO_2 storage site. *CSEG Recorder, 41*, 28–33.

Ministry of International Trade and Industry. (1980). *Environmental report of the ultra-deep geothermal power plant in 1979.*

Motomatsu, T. (1999). Geothermal development in Takigami area. *Journal of the Geothermal Research Society of Japan, 21*, 267–275.

NEDO (New Energy and Industrial Technology Development Organization). (1990). *Geothermal development promotion* (Report B-4). Kuenohirayama District.

Nishijima, J., Shimosako, N., Motoyama, T., Fujimitsu, Y., Ehara, S., Ikeda, K., et al. (2000). Gravity monitoring of geothermal reservoirs in the Takigami geothermal field, Central Kyushu, Japan-gravity changes caused by production and reinjection of geothermal fluids. *Journal of the Geothermal Research Society of Japan, 11*, 117–129.

Schölderle, F., Lipus, M., Pfrang, D., Reinsch, T., Haberer, S., Einsiedl, F., & Zosseder, K. (2021). Monitoring cold water injections for reservoir characterization using a permanent fiber optic installation in a geothermal production well in the Southern German Molasse Basin. *Geothermal Energy, 9* (1), 1–36. doi:10.1186/s40517-021-00204-0

Takenaka, T., Goto, H., Yamamoto, Y., & Furuya, S. (1995). Exploration and development of the Takigami geothermal system, Kyushu, Japan. *Resource Geology, 45*, 361–376.

Varkey, J., Mydur, R., Sait, N., Wijnberg, W., Kunathikom, S., & Darpi, M. (2008). *Optical fiber cables for wellbore applications.* U.S. Patent No. 7324730 B2. Washington, DC: U.S. Patent and Trademark Office.

Yilmaz, O. (2001). *Seismic data analysis* (2027pp). Investigation in geophysics, No. 10. Society of Exploration Geophysicists.

23

Multiazimuth Elastic Full-Waveform Inversion of Fiber-Optic and Accelerometer Vertical Seismic Profile Data: A Field Example from the Carbon Management Canada Newell County Facility, Alberta, Canada

Xiaohui Cai[1], Kristopher A. Innanen[1], Mattew V. Eaid[2], Scott D. Keating[3], Qi Hu[4], and Donald C. Lawton[1,5]

ABSTRACT

Distributed acoustic sensing (DAS) is a rapidly developing technology enabling the recording of seismic data using fiber-optic cables. Intensive efforts have been devoted to optimizing the application of seismic processing, imaging, and inversion methods to DAS data. We examine the response of an elastic full-waveform inversion (FWI) approach, combining DAS and accelerometer vertical seismic profile (VSP) data. The problem is formulated by combining strain and displacement components in one objective function. Accelerometer data are proportional to particle acceleration, whereas DAS data are proportional to strain rate/strain along the fiber axis; thus, both datasets require conversion to displacement and strain. To prepare the DAS VSP field data for inversion, we develop a depth registration method based on cross-correlation scanning and use first-break picking as quality control to obtain a robust DAS depth for each trace. An effective source scheme is incorporated into the VSP FWI to address complex near-surface wave propagation. Application of the FWI approach to 2D two-azimuth walkaway VSP datasets acquired at Newell County, Alberta, reveals horizontal layering consistent with the site's known geology and limited azimuthal variations. Reverse-time migration imaging further corroborates the inversion results.

23.1. INTRODUCTION

Full-waveform inversion (FWI) is now regularly used for producing and assessing images of structures under complex geological conditions (Alkhalifah & Plessix, 2014; Eaid et al., 2020; Hu et al., 2023; Keating & Innanen, 2019; Lee et al., 2010; Pan et al., 2019; Pratt, 2004; Vigh et al., 2014; Virieux & Operto, 2009; Zhang & Alkhalifah, 2022). FWI makes use of the waveform (amplitude and phase) information contained in seismic wave data (Tarantola, 1984), gradually approximating plausible high-resolution models by matching field and simulation data and admitting a wide range of regularization types and error/noise estimates. With recent rapid increases in computational power and novel computational architectures, FWI is increasingly found outside theoretical studies and in application to field data. Single-parameter velocity inversion based on the scalar-acoustic approximation is in most common use.

[1]*Department of Earth, Energy, and Environment, University of Calgary, Calgary, Alberta, Canada*
[2]*Chevron, Houston, Texas, USA*
[3]*Department of Earth and Planetary Sciences, ETH Zurich, Zurich, Switzerland*
[4]*Department of Energy Science and Engineering, Stanford University, Stanford, California, USA*
[5]*Carbon Management Canada, Calgary, Alberta, Canada*

Distributed Acoustic Sensing in Borehole Geophysics, Geophysical Monograph 289, First Edition.
Edited by Yingping Li, Robert Mellors, and Ge Zhan.

DOI: 10.1002/9781394179275.ch23

Elastic FWI, although more physically researched, remains in limited use, especially for land data, being more complex algorithmically and involving more complex amplitude and phase matching, and introducing the independent determination of multiple simultaneous parameter classes (Aragao & Sava, 2020; Brittan & Jones, 2019; Brossier et al., 2009; Eaid et al., 2022; Hu et al., 2021; Keating & Innanen, 2020; Métivier et al., 2017; Singh et al., 2018). Two key challenges in the land data application are the construction of a reasonable initial model and the suppression of cross-talk between different elastic parameters. The nonlinearity and nonuniqueness of the three-parameter inversion problem are strong, aggravating the likelihood of becoming trapped in local minima. The workflow or algorithmic structure design for FWI of land data must somehow accommodate these two issues.

Distributed acoustic sensing (DAS) is a low-cost acquisition method, and it is the general trend to incorporate DAS data into traditional FWI research. DAS is an increasingly mature seismic signal acquisition technology suitable for vertical seismic profile (VSP) acquisition. DAS systems rely on fiber-optic technology, using Rayleigh scattering to sense local changes in the fiber caused by external fluctuations in strain (Posey et al., 2000) and then acquire data by spatially and temporally registering the seismic variations along the fiber. DAS measures are limited to sensing strain or strain rate along the fiber, meaning that the technology is sensitive only to strain components of the wavefield aligned tangentially with the fiber (Eaid et al., 2020; Kuvshinov, 2016). In VSP acquisition settings (if not elsewhere), the DAS data have the following advantages compared to the traditional geophone or accelerometer data. First, the fiber is less expensive to deploy and is resistant to corrosion, high temperature, and high pressure; moreover, it can be installed quasi-permanently in the well to acquire seismic data or monitor subsurface fluids, which represents potentially large savings in labor and material resources. Second, fiber-optic systems are simpler and more efficient than other sensors because they require little manual intervention and are entirely controlled by the instrumentation system once installed. Third, fiber-optic systems provide increased coverage or aperture within instrumented wells. Conventional geophones or accelerometers need to be moved along the well during acquisition, which is time-consuming and leads to spatial aliasing; in contrast, the fiber instantaneously samples the whole length of the well with generally small sampling intervals, leading to signals with very low spatial aliasing. The resolution of the data obtained by fiber-optic depends on the pulse width emitted by the DAS system, but within these constraints, in principle, a high resolution can be reached by varying the pulse width (Daley et al., 2016; Mateeva et al., 2014; Spikes et al., 2019).

In 2018 (see Hall et al., 2019) and again in 2022 (see Innanen et al., 2022), researchers from the University of Calgary CREWES project (Consortium for Research in Elastic Wave Exploration Seismology) and various external partners collaborated with Carbon Management Canada (CMC) at their Newell County Facility in Alberta, Canada, to acquire baseline and monitoring multiazimuth and multioffset VSP datasets near a point of injection of CO_2. In the study reported here, we combine three-component accelerometer VSP data and single-component DAS VSP data acquired during the baseline survey to facilitate a validation test of frequency-domain elastic FWI. First, the advantages of DAS data are dense sampling and low-cost (Egorov et al., 2018). Second, the VSP data usually contain more transmission waveform than the surface data, which helps to reconstruct the low wavenumber information (Egorov et al., 2017). Finally, well-log data are generally available in a VSP survey, which can be important prior information for model parameters and reduce the uncertainty of the inversion. Our particular purpose in this chapter is to invoke repeated instances of 2D FWI to assess the velocities in the field and compare the inversion results across two azimuths within the full dataset. To achieve this purpose, we need strategies to cope with the complexities inherent in the field data, and various tests of the robustness of the FWI approach based on the accelerometer-DAS VSP data. Therefore, we make use of a single-parameter compressional velocity (V_p) parameterization based on well-log data analysis and use an effective source scheme that mitigates the demand on accurate source characteristics and near-surface model information (Keating et al., 2021).

23.2. DATA AND PROCESSING

Generally, the goal of processing field data for FWI is to prepare data that matches the simulation data, and minimal processing is sufficient. We extract shots from line 4 (blue dots in Figure 23.1) and line 1 (red dots in Figure 23.1) of the CMC Newell County Facility field data, called Snowflake, and line 4 is chosen because the injection and observation wells are located on this line, which provides a better basis for future research on time-lapse data monitoring of CO_2. As shown in Figure 23.1, shots from line 4 are located from northeast to southwest, and shots from line 1 are from east to west. Observation well 2 is located at the center of the shot points, and the injection well is located at the northeast of observation well 2 at a distance of 20 m. Line 4 has

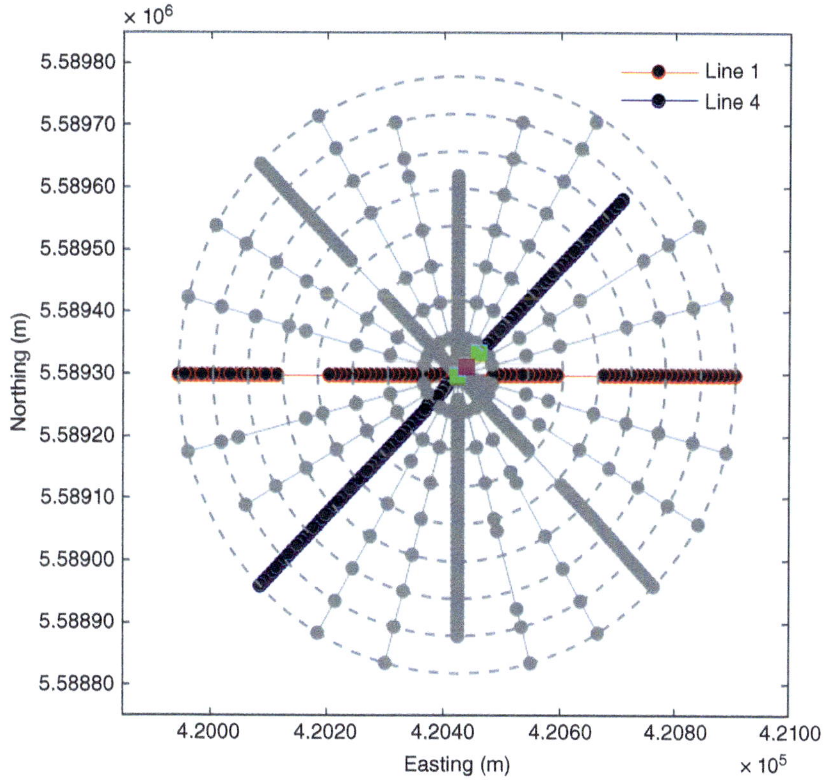

Figure 23.1 Shot geometry of the 3D walkaway-walkaround VSP data in CMC Newell County Facility field data, Alberta, Canada. These blue and red circles represent the shot point locations of line 4 and line 1, respectively. The two green squares are the locations of the two observation wells (observation well 2 is at the center of the shot points). The magenta square is the location of the injection well.

a maximum offset of 390 m on the northeast and 480 m on the southwest of observation well 2, and line 1 has a maximum offset of 480 m on both sides of observation well 2. Each shot is assigned a five-digit point number; the first two digits are the line number, and the last three are the shot number. The shot on the northeast edge of line 4 is number 110, and the 82 shots almost increase for 10 m shot spacing, and the shot on the east edge of line 1 is number 101, and the 77 shots also increase for approximately 10 m. The accelerometer and DAS datasets require similar but different processing steps (Figure 23.2).

23.2.1. Accelerometer Data Processing

The trace spacing for the accelerometer receivers is changing from 1 m (over the interval of 0–266.4 m) to 2 m (from 266.6 to 324.2 m) due to a hardware problem during acquisition. Most of the signal occurs in the first 500 ms of the dataset, so line 4 and line 1 data are windowed to the first 500 ms, and they have the same data processing workflow; we use line 4 data as an example to show the specific data processing.

The accelerometer field data consists of the three orthogonal components of particle acceleration associated with the propagating wavefield. However, the horizontal components are orientated in an unknown and arbitrary direction that varies for each receiver. To ensure that the horizontal data for each receiver has a consistent azimuth, a coordinate rotation needs to be applied to each receiver. The maximum horizontal (called Hmax) component is more coherent than the H1 and H2 components and consists of clear downgoing P-wave and S-wave energy and reflections. Therefore, we use the QC Hodogram Display by Vista to rotate the H1 and H2 components to the Hmax and minimal horizontal components, and then implement QC rotation angles and adjust the polarity traces one by one. Figure 23.3 shows the accelerometer data after coordinate rotations and polarity correction for three shots (near, medium, and far offset distances) on line 4. Dead traces are replaced through interpolation, and a low-pass filter (25 Hz) is applied to

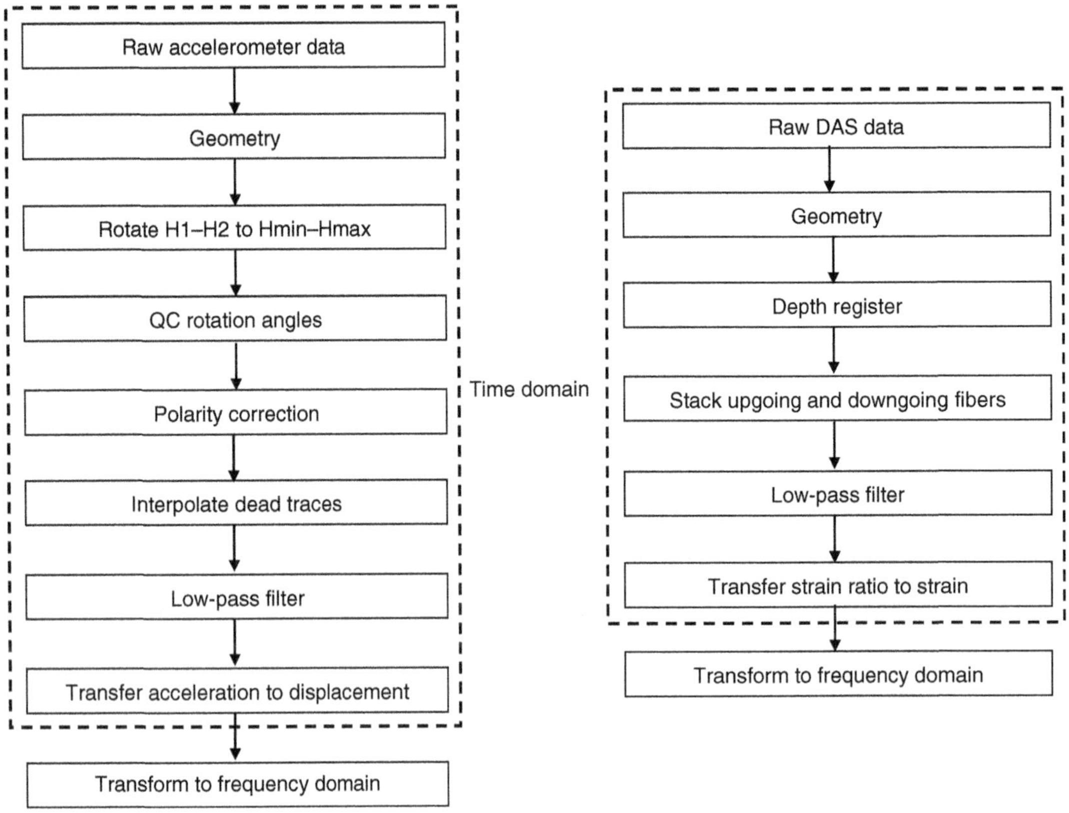

Figure 23.2 Data processing workflow of the accelerometer VSP data (left) and DAS VSP data (right).

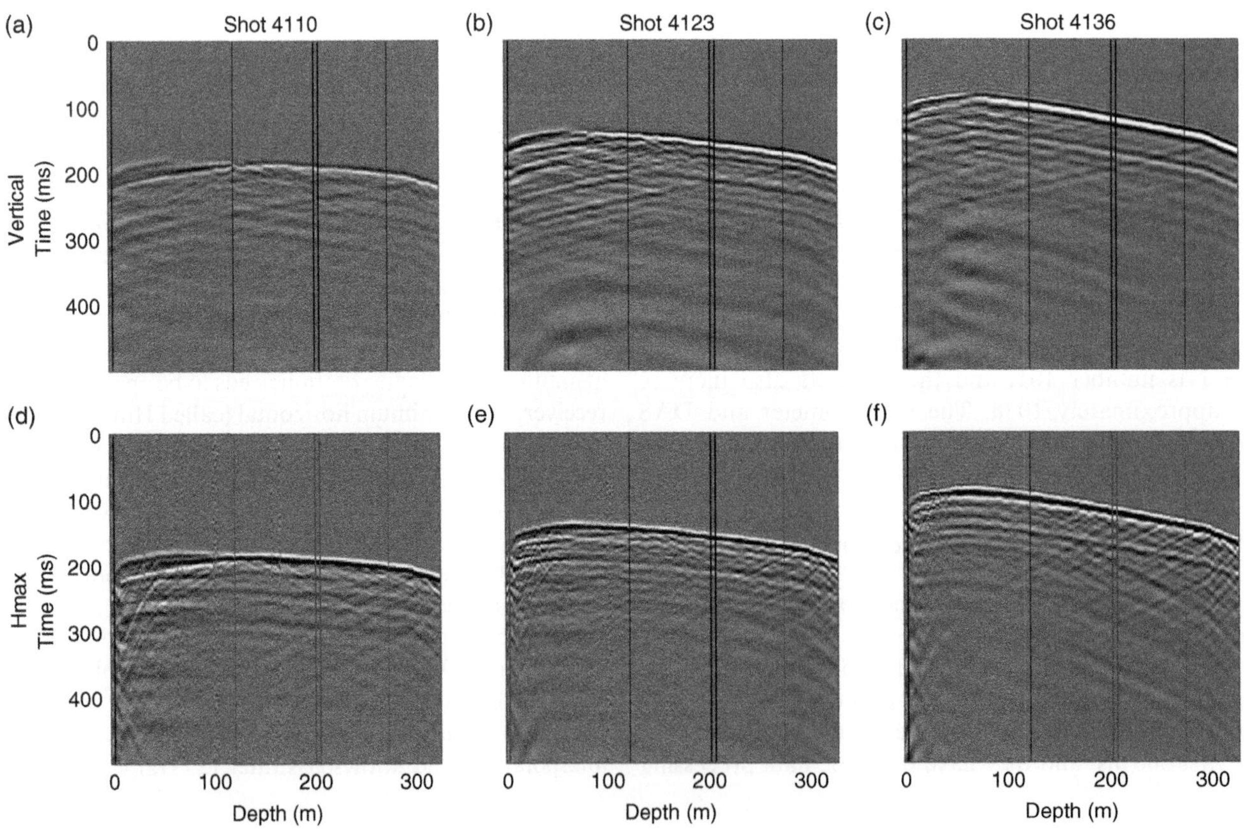

Figure 23.3 Accelerometer data after coordinate rotations and polarity correction on line 4. The top row (a–c) and bottom row (d–f) plot the vertical and Hmax components, respectively.

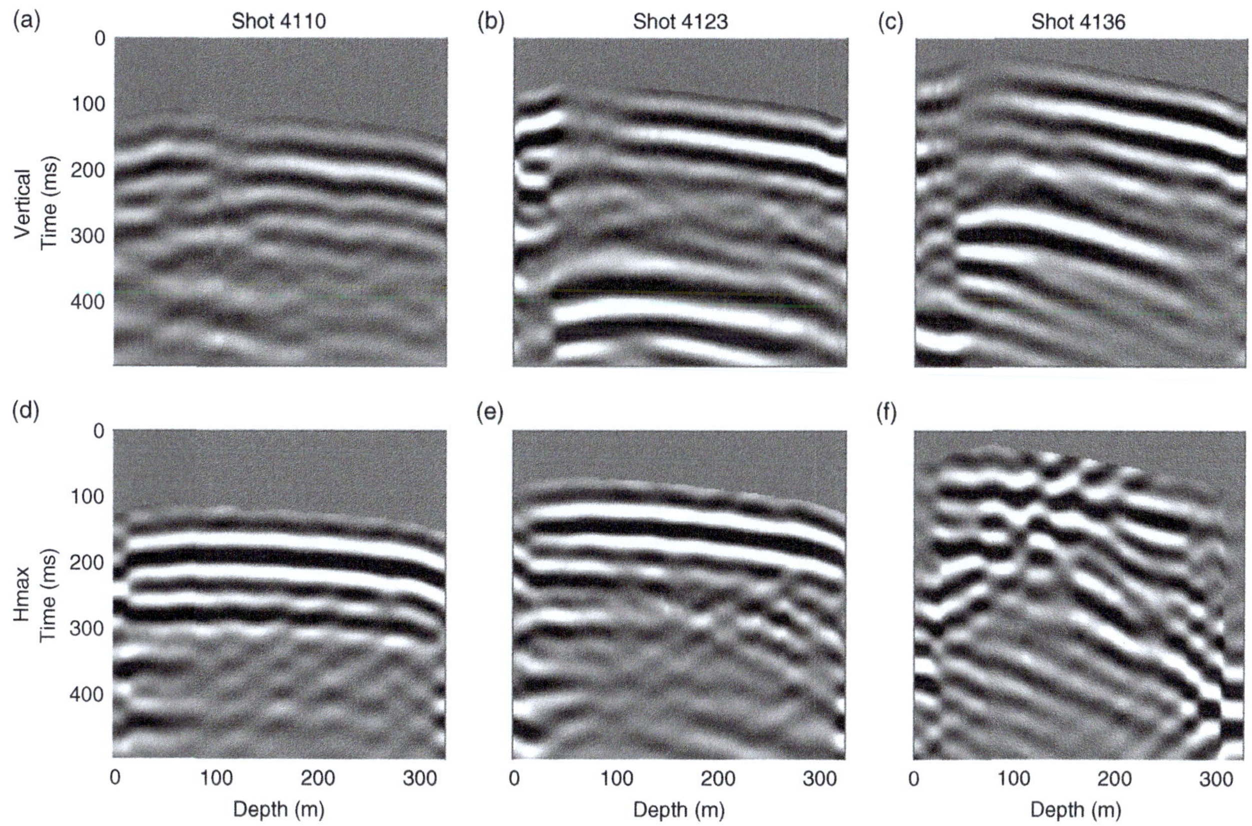

Figure 23.4 Accelerometer data after interpolation and low-pass filter (25 Hz). The top row (a–c) and bottom row (d–f) plot the vertical and Hmax components, respectively.

the vertical and Hmax components of the accelerometer data. The utilized 25 Hz here is prepared for the following FWI. Figure 23.4 presents both two components of accelerometer data after interpolation and low-pass filter (25 Hz) for the previous three shots. Finally, acceleration is transformed into displacement via −180° phase transformation.

23.2.2. DAS Data Processing

DAS measures strain rate along the fiber, a characteristic influenced by the vertical orientation of the well and its integration into a continuous loop, as elucidated by Hall et al. (2019), resulting in both downgoing and upgoing vertical regions (Figure 23.5). It is unnecessary to complete coordinate rotations, polarity correction, and interpolation for the DAS data. However, it is necessary to register depth because the trace spacing for the DAS data is unknown. From the raw data, we have analyzed that the noise here is dominated by Gaussian noise, which is less detrimental to FWI than coherent noise (Lines, 2014). To prepare for data for FWI, we apply a 25 Hz low-pass filter in the raw data (shown in Figure 23.6) and stack the upgoing and downgoing regions after depth registration. Finally, the −90° phase transformation is applied to the DAS data to achieve the strain component.

Depth Registration

We develop a depth registration scheme of DAS data based on the cross-correlation scanning scheme, as shown schematically in Figure 23.7. Channel 1 (C1) is the data with known depth, and Channel 2 (C2) is the data whose depth needs to be estimated. The scheme has two requirements: a fixed sampling interval on C2 and simultaneous acquisition of C1 and C2. Three steps occur. First, on C2, we set the start depth, the depth interval, and the time difference (T) between the two channels, outputting the depth of each trace on C2. Then, we locate the nearest trace in C2 to each trace in C1 and sum the cross-correlations of these traces. Finally, we obtain a contour map by scanning those three parameters. Figure 23.8(a), (c), (e) shows the contour map obtained from line 4. The estimated start depth, depth interval, and T are

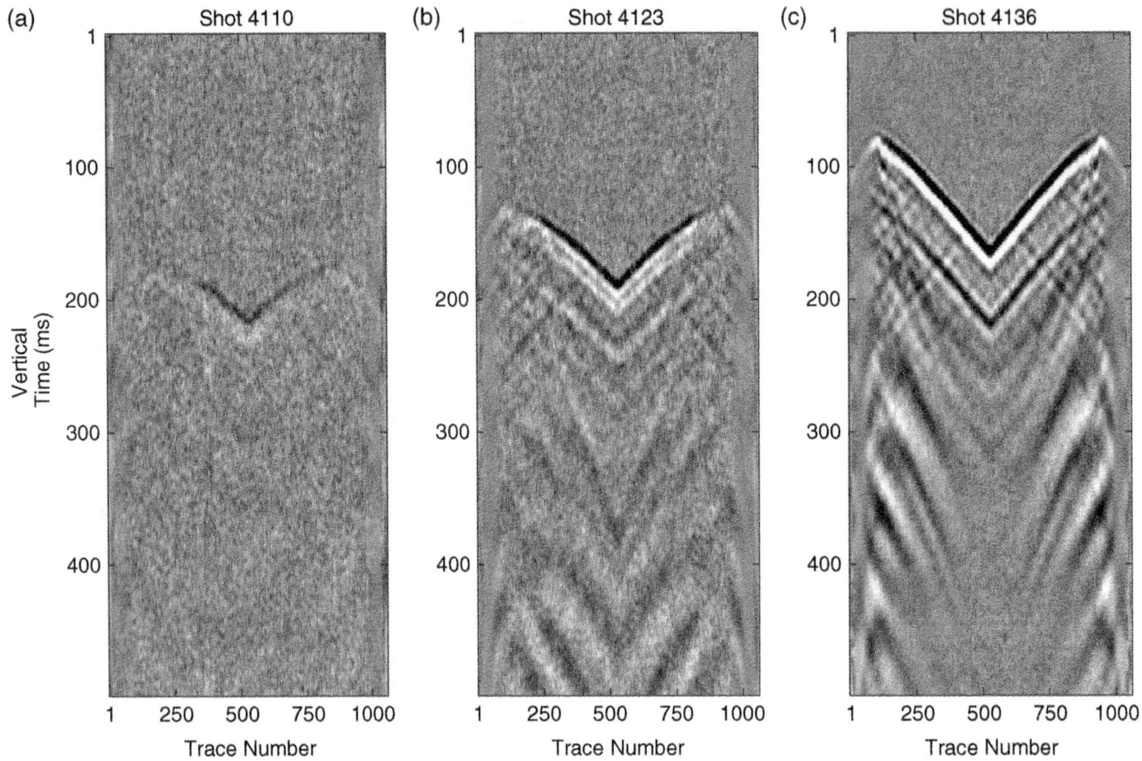

Figure 23.5 Raw DAS data.

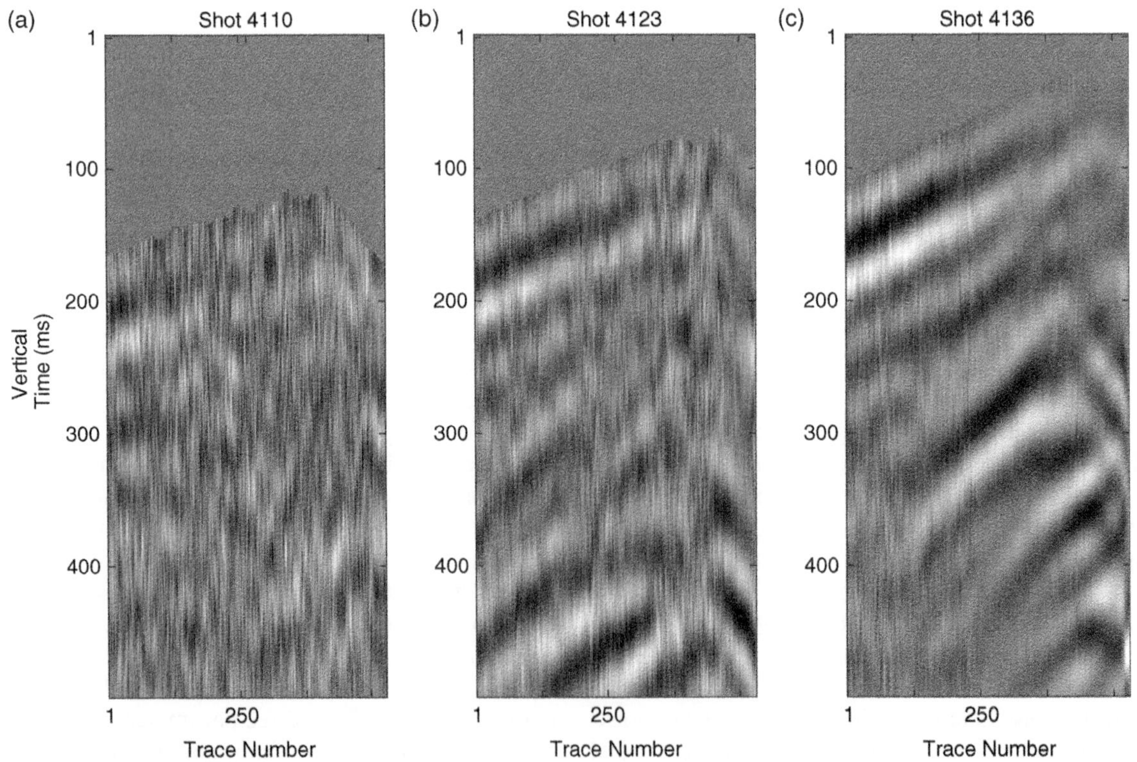

Figure 23.6 DAS data after 25 Hz low-pass filter and stack upgoing and downgoing fibers.

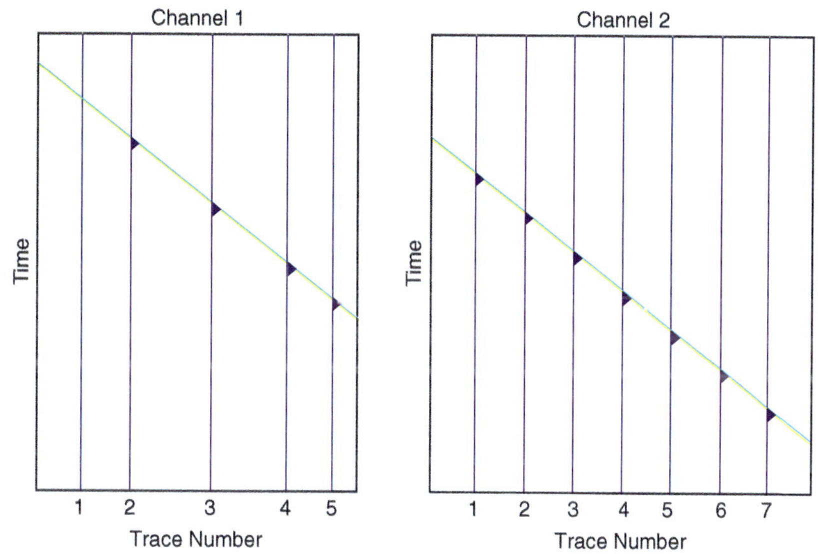

Figure 23.7 Schematic diagram of the developed depth registration scheme based on cross-correlation scanning.

determined by picking the maximum value in the contour map. Table 23.1 shows the maximum values of the cross-correlation with different T and their corresponding start depth and depth interval. The values of the maximum cross-correlation are relatively close when T is −2, 0, and 2 ms. For quality control, we use first-break matching between the accelerometer and DAS data on line 4 (shown in Figure 23.8(b), (d), (f)), and we obtain the best fit with T = 2 ms. Through this process, we register the DAS data in depth, from 0 m with an interval of 0.698 m, as shown in Table 23.1. The total calculation time takes about 5 min based on the Intel Core i7-7800X CPU processor @ 3.50 GHz.

23.3. INVERSION METHODOLOGY

23.3.1. 2D Isotropic Frequency-Domain FWI

We use codes originally developed for multiparameter elastic FWI by Keating and Innanen (2020). These are based on the isotropic displacement elastic wave equation (Pratt, 1990)

$$\mathbf{A}(\mathbf{m})\mathbf{u} = \mathbf{f} \tag{23.1}$$

where $\mathbf{A}$ is the impedance matrix, $\mathbf{m}$ is the model parameter, $\mathbf{u}$ is the displacement wavefield vector, and $\mathbf{f}$ is the source vector. Here, we apply the fourth-order finite-difference staggered grid scheme to solve equation (23.1), and a perfectly matched layer (PML) method (Berenger, 1994) is applied to suppress artificial boundary reflections.

The objective function is based on the L2 norm of the difference between field data $\mathbf{d}_{\text{obs}}$ and simulated data $\mathbf{d}_{\text{syn}}$

$$\min \phi(\mathbf{m}) = \frac{1}{2}\left\|\mathbf{d}_{\text{obs}} - \mathbf{d}_{\text{syn}}\right\|_2^2 \tag{23.2}$$

where $\mathbf{d}_{\text{syn}} = \mathbf{Ru}$ and $\mathbf{R}$ is the receiver matrix. Generally, $\mathbf{R}$ samples the vertical and horizontal displacement components at the location of the receivers. To incorporate the DAS data in the FWI, we follow the approach of Eaid et al. (2020), augmenting the $\mathbf{R}$ matrix to include additional rows to compute the tangential strain field from displacements. The way to simulate the combined accelerometer and DAS data can be written as

$$\begin{bmatrix} d_x^1 \\ \vdots \\ d_x^{Na} \\ d_z^1 \\ \vdots \\ d_z^{Na} \\ \tau_z^1 \\ \vdots \\ \tau_z^{Nd} \end{bmatrix} = \begin{bmatrix} 1 & & & & & \\ & \ddots & & & & \\ & & 1 & & & \\ 1 & & & & & \\ & \ddots & & & & \\ & & & 1 & & \\ w_1 & w_2 \cdots w_3 & w_4 & & & \\ & & \ddots & & & \\ & & & w_1 & w_2 \cdots w_3 & w_4 \end{bmatrix} \begin{bmatrix} u_x^1 \\ u_z^1 \\ u_x^2 \\ u_z^2 \\ \vdots \\ u_x^N \\ u_z^N \end{bmatrix} \tag{23.3}$$

where $\mathbf{u} = \begin{bmatrix} u_x^1 & u_z^1 & u_x^2 & u_z^2 & \cdots & u_x^N & u_z^N \end{bmatrix}^T$, d_x, and d_z are horizontal and vertical displacement, respectively, Na and Nd are accelerometer and DAS receiver numbers, respectively, τ_z is vertical strain, and $\tau_z = \partial u_z / \partial z$.

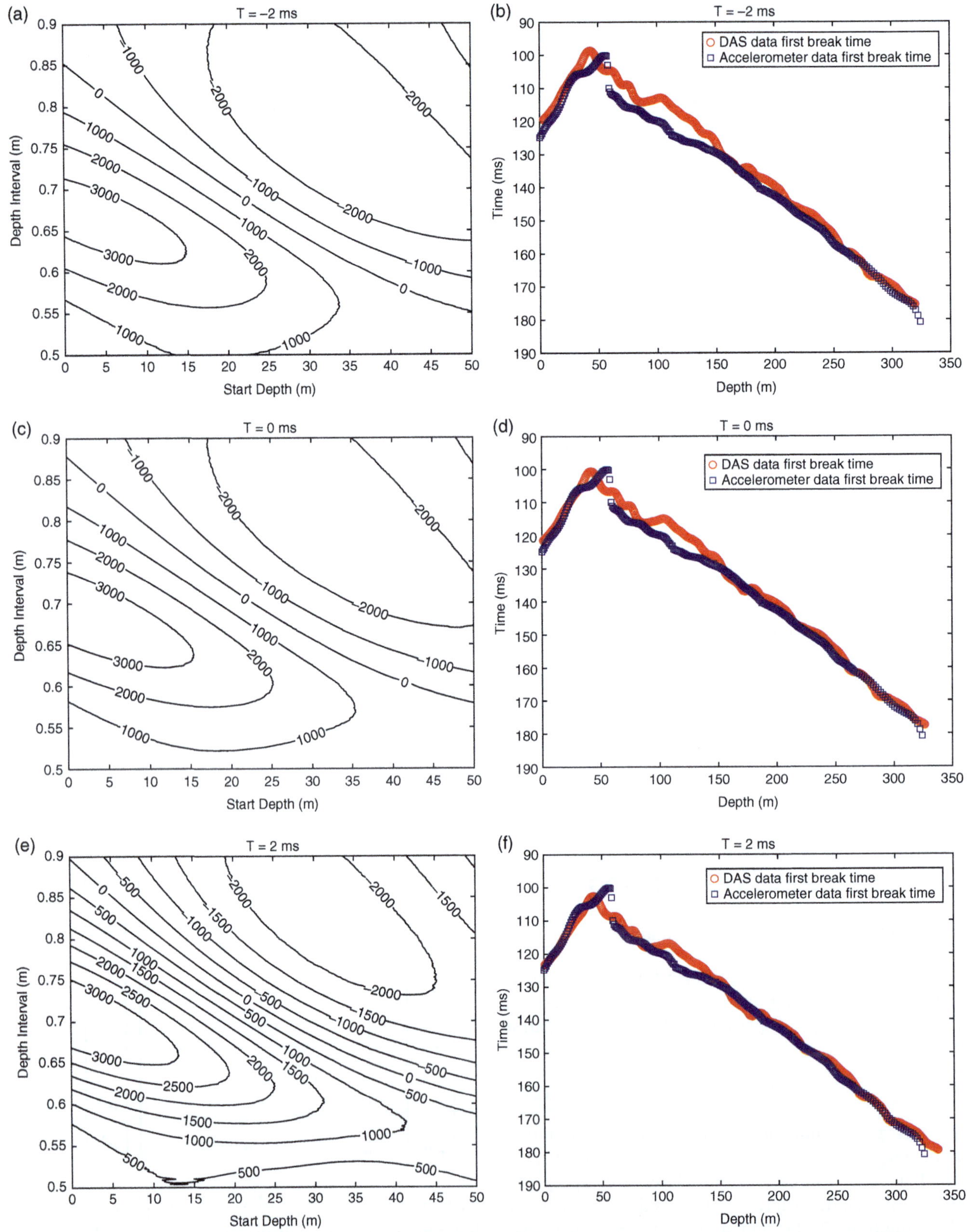

Figure 23.8 The contour map obtained from the upgoing fiber on line 4 with (a) T = −2 ms, (c) T = 0 ms, (e) T = 2 ms, and the first-break matching with DAS data and accelerometer data when the T is (b) −2 ms, (d) 0 ms, and (f) 2 ms.

Table 23.1 The maximum values of the cross-correlation with different T and their corresponding start depth and depth interval.

T (ms)	The maximum value of the cross-correlation	Start depth (m)	Depth interval (m)
−10	1,664.467	8	0.574
−8	2,177.812	8	0.591
−6	2,706.502	6	0.616
−4	3,191.605	3	0.643
−2	3,546.747	3	0.659
0	3,688.065	1	0.678
2	3,559.106	0	0.698
4	3,162.45	0	0.716
6	2,612.309	0	0.74
8	2,534.529	38	0.634
10	2,927.834	38	0.651

The near surface is highly heterogeneous and largely unknown, as even density and velocity values derived from well logs are missing at depths above 223 m. FWI strongly depends on the initial velocity-density model: a poor initial model can significantly compromise the convergence of inversion. To address the challenge, we use an effective source approach (Keating et al., 2021; Eaid et al., 2023) to take into account incomplete near-surface model information, source characteristics, and wavefield physics. The effective source-based FWI optimization problem is formulated as

$$\min \phi(\mathbf{m}^*, \mathbf{f}^*) = \frac{1}{2}\|\mathbf{d}^*_{\text{obs}} - \mathbf{R}^*\mathbf{u}^*\|_2^2 \text{ subject to } \mathbf{A}(\mathbf{m}^*)\mathbf{u}^* = \mathbf{f}^* \tag{23.4}$$

where each of the $*$ variables is only allowed to take values below the effective source depth z^* (blue stars in Figure 23.9). Equation (23.4) represents a FWI problem similar to the conventional one (equations 23.1 and 23.2) but defines the problem within a smaller model domain, and it represents the inversion models, including an unknown model, $\mathbf{m}^*$, and an unknown source term, $\mathbf{f}^*$. As a result, $\mathbf{f}^*$ will embody a collection of multiple line sources situated around the depth z^*. The method removes the near-surface part of the model from the inversion by introducing an unknown source $\mathbf{f}^*$ at the depth that characterizes the source wavefield after propagation through the near-surface (see Figure 23.9). The effective line source $\mathbf{f}^*$ is placed at a depth that best explains the data. In this approach, we do not need to solve model parameters in the near surface: that is, in regions whose role is largely to obstruct a coherent waveform from reaching the zone of interest.

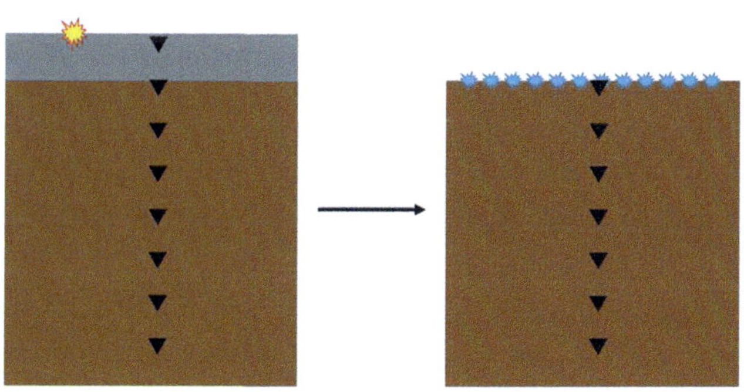

Figure 23.9 Schematic depiction of the effective source approach for VSP FWI.

23.3.2. Parameterization

Choosing appropriate parameterization is critical in elastic FWI (Köhn et al., 2012; Pan et al., 2018). The velocity-density parameterization (compressional velocity V_P, shear velocity V_S, and density ρ) is suitable when performing FWI of VSP data because those transmitted arrivals are dominant in terms of energy (Roberts et al., 2008). In the field, the well-log data were interpreted by the soft-sand model and Gassmann's equations (Hu et al., 2024) to provide P-wave sonic, S-wave sonic, and density logs from the surface to 223 m of the injection well, and that information from 224 m to the bottom of the injection well is acquired by well logs. Figure 23.10 shows the regression relationship between V_p, V_s, and ρ. Based on the strong correlation between V_p, V_s, and ρ, we use a model parameterization based on V_p only to implement the 2D elastic frequency-domain FWI, which incorporates prior information of the model into the inversion and helps reduce cross-talk by only allowing updates in one parameter. The relationships in Figure 23.10 are used to obtain the V_s and ρ models from V_P.

23.4. RESULTS

We develop a starting model based on smoothed V_p, V_s, and ρ logs with a sampling interval of 1 m over a 100 m window (Figure 23.11). The model size is 840 m in the x direction and 350 m in the z direction, and the space interval we adopted is 5 m.

To mitigate the nonlinearity of the inverse problem, a multiscale strategy (Bunks et al., 1995) is applied. The data are inverted using frequencies within the interval of 10–25 Hz. The seismic signal-to-noise ratio was considered acceptable at a minimum frequency of 10 Hz, although opting for lower frequencies might be more advantageous. Within this dataset, the main frequencies significantly surpass the threshold we investigated, which was constrained due to the computational costs tied to adopting a smaller finite-difference grid necessary for higher frequency. We consider five frequency bands, each containing six evenly spaced frequencies from 10 Hz to a maximum frequency; the maximum frequency is 12 Hz for the first band and increases to 25 Hz for the last

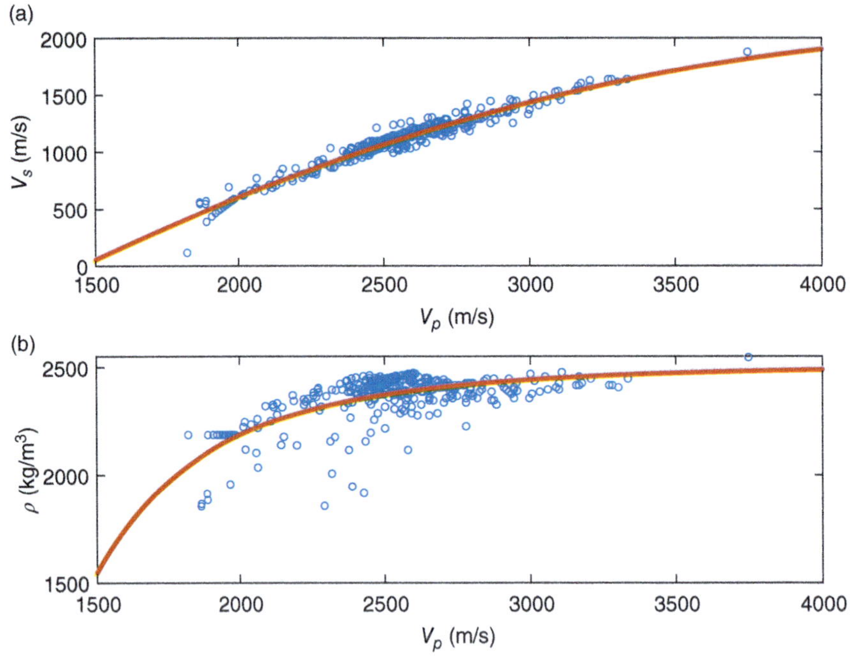

Figure 23.10 (a) The relationship between V_p and V_s, and (b) the relationship between V_p and ρ. The solid red lines indicate the trend lines.

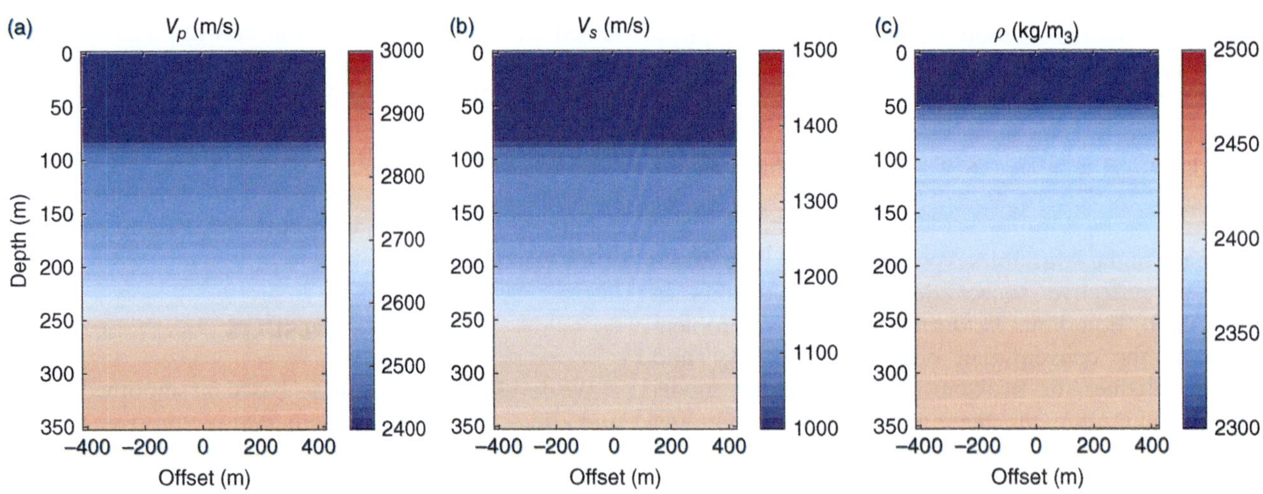

Figure 23.11 Initial V_p, V_s, and ρ models for line 1 and line 4.

band. We use the limited-memory Broyden–Fletcher–Goldfarb–Shanno (L-BFGS) optimization method and run 10 iterations for each band. The inversion is performed in two steps: the source-only $\mathbf{f}^*$ inversion and the simultaneous source $\mathbf{f}^*$ and model $\mathbf{m}^*$ inversion. Both the ricker wavelet and transformed minimum phase wavelet are used as the initial wavelet, and the inversion experimentations prove that the initial wavelet has a minor effect on the inversion because the source-only $\mathbf{f}^*$ inversion will obtain the wavelet close to the field data waveform. Through some experimentations, we select an effective source depth of 40 m. After source-only updates, we recalculate the amplitude scaling terms and proceed to the inversion. In pursuit of elucidating the merits inherent in the combined DAS-accelerometer data, we conducted inversion using only accelerometer data, only DAS data, and the combined DAS-accelerometer data, and then the inversion models and the corresponding well-log model profiles are summarized in Figure 23.12 and Figure 23.13a. We observe that the results from the three types of data are comparable, and they all demonstrate significant improvements from the initial model. The

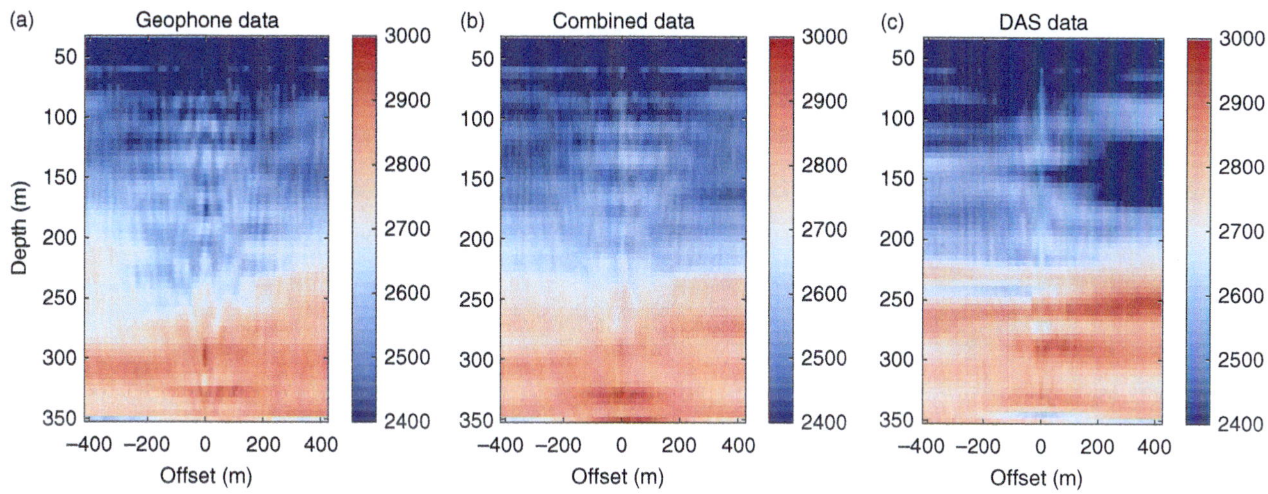

Figure 23.12 Inverted V_p models for line 4 by the accelerometer data, the combined DAS-accelerometer data, and the DAS data, respectively.

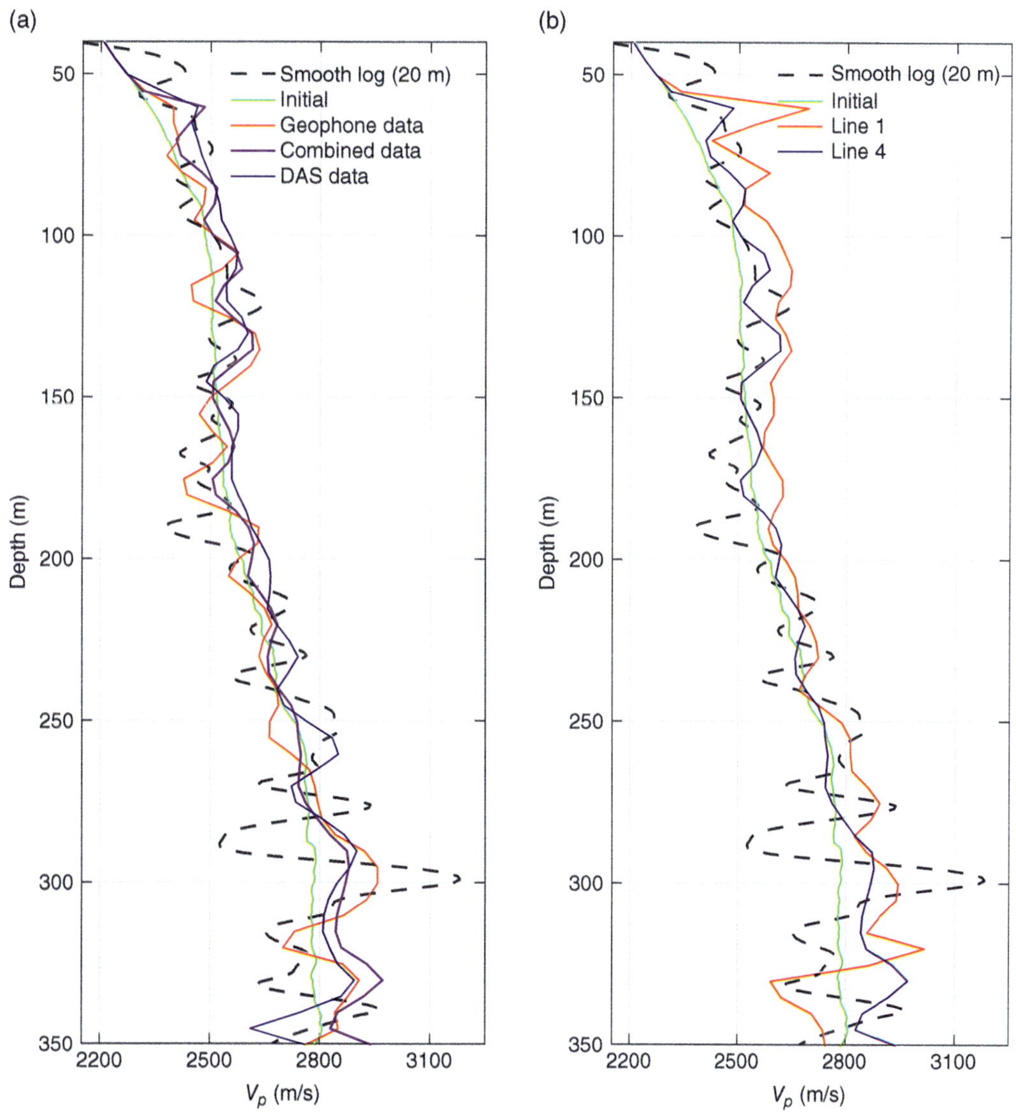

Figure 23.13 A comparison of the inverted V_p model with the well-log data for line 4. (a) The dashed black line denotes the smoothed well log, the green line denotes the initial model, and the red, purple, and blue lines denote the inverted V_p models from the accelerometer data, the combined DAS-accelerometer data, and the DAS data, respectively. (b) The inverted models from line 1 (red line) and line 4 (blue line) are compared.

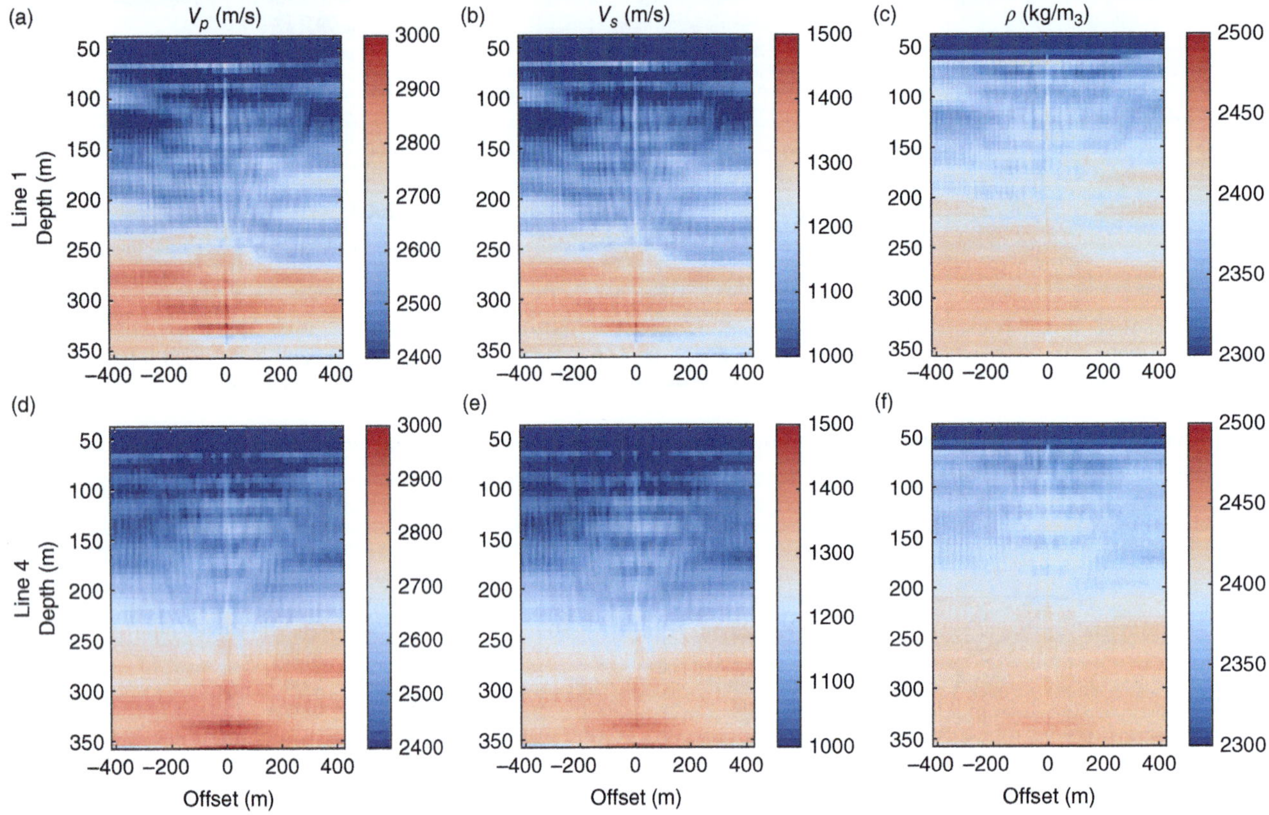

Figure 23.14 Inverted models of V_p, V_s, and ρ with combined DAS-accelerometer data for line 1 (a–c) and line 4 (d–f).

inverted model based on accelerometer data provides a reasonable level of detail, which is beneficial to its three-component acquisition advantage, but it exhibits relatively weak lateral continuity in the vicinity of the observation well. On the other hand, the inverted model based on DAS data is more correlated with the whole well-log, which is beneficial to the data's continuity, but the target layer amplitude with the well-log data is relatively poor. Finally, the inversion based on the combined DAS-accelerometer data balances the strengths and drawbacks of the above individual data.

Figure 23.14 plots the inverted models of V_p, V_s, and ρ of both line 1 and line 4 based on the combined DAS-accelerometer data. Notably, we observe that the overall V_p, V_s, and ρ of line 1 and line 4 are highly similar, exhibiting only minor differences: The two-azimuth inversion results are dominated by layered models, and velocities and densities at 300 and 330 m are obviously higher than that at other depths, but the lateral span of line 1 is slightly wider than that of line 4. Further, a comparison between the inverted V_p model (at 5 m offset) and the V_p well-log data (Figure 23.13(b)) shows that the inversion results for both line 1 and line 4 partially reconstruct the velocity structure. To ascertain the accuracy of the inversion procedure, we initially assessed the error curve for the objective function throughout the inversion process (Figure 23.15). Figure 23.15(a) illustrates the frequency values encompassing the five frequency bands employed in the inversion procedure, and Figure 23.15(b) delineates the errors of the objective function across these same frequency bands throughout the inversion process. The figure reveals that higher frequency values correspond to a larger initial error in the objective function. Moreover, with the inversion process, the error of the objective function in each frequency band is gradually decreasing. Subsequently, the frequency-domain data derived from the inversion outcomes underwent an inverse Fourier transformation to yield corresponding time-domain simulated data, which was then merged into the field data for comparison (Figure 23.16). Figure 23.16 demonstrates that the modeled data match closely with the field data. It can be seen from Figure 23.16(c) that because the modeled data has a high degree of fitting between the effective signal of the field data and does not reconstruct the noise of the field data, which also indicates that Gaussian noise influence on FWI is limited.

The elastic reverse-time migration (RTM) is a robust seismic imaging technology for the P wave and S wave, and is a sufficiently mature methodology to validate FWI results. We carry out an elastic RTM (Cai et al.,

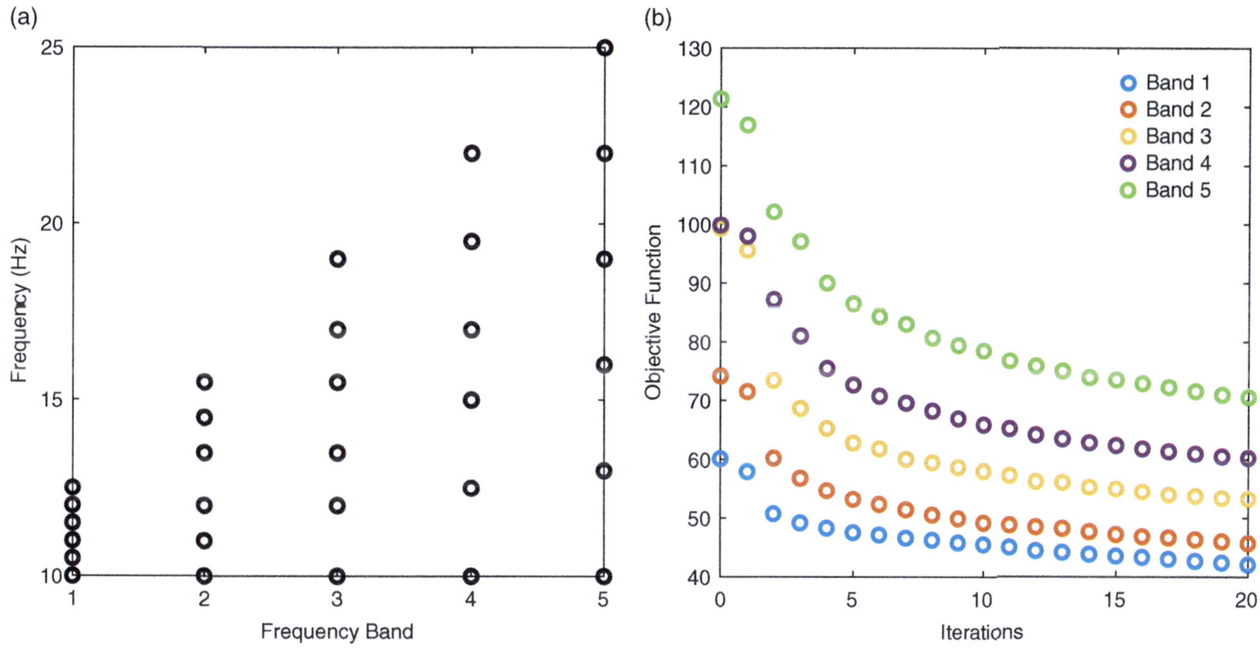

Figure 23.15 Converge properties. (a) Frequencies and (b) objective function errors for different frequency bands.

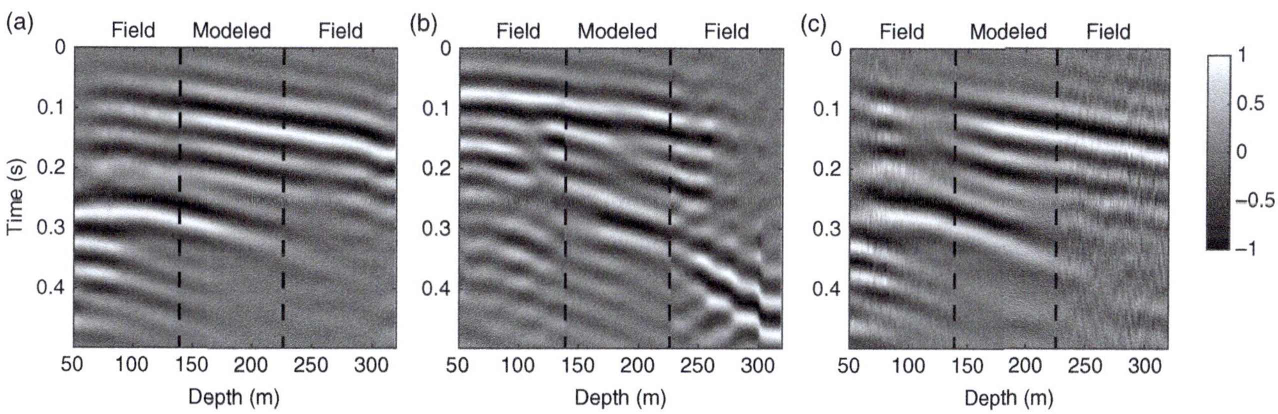

Figure 23.16 A mixture of field and modeled data in the time domain for a shot at 120 m offset. (a) The vertical component of acceleration; (b) the horizontal component of acceleration; (c) the DAS data.

2015) procedure along the same lines as the field data. The accelerometer data has a processing workflow similar to that in Figure 23.2, and the difference is that the data are not applied to the low-frequency filter and are separated into upgoing and downgoing field data by median filtering and FK filtering methods. Finally, the upgoing field data is used in the time-domain elastic RTM. Here, we implement a fourth-order finite-difference staggered grid scheme within the forward and backward propagations, as well as use PML to suppress artificial boundary reflections and apply the Helmholtz equation (Morse & Feshbach, 1954) to separate the P wave and S wave. The forward modeling wavelet is obtained by summing the flattened downgoing wavefield. The velocity model is the same as the FWI initial model (Figure 23.11), but the space interval we use in RTM is 2 m. As mentioned earlier, the near-surface velocity in this area has significant lateral variations, and the RTM imaging is as sensitive as the inversion to the initial velocity. Therefore, the imaging results will be compromised at shallow depths and far offsets. In Figure 23.17, the PP and PS imaging results of line 1 and line 4 are summarized. The PP imaging result shows that the subsurface layers are reasonably flat, and there is a strong reflective interface around 300 m. Moreover, these results reveal the relatively limited difference between the two azimuths because the two PP imaging results are close but limitedly different. PS imaging is of relatively poor quality, likely originating from

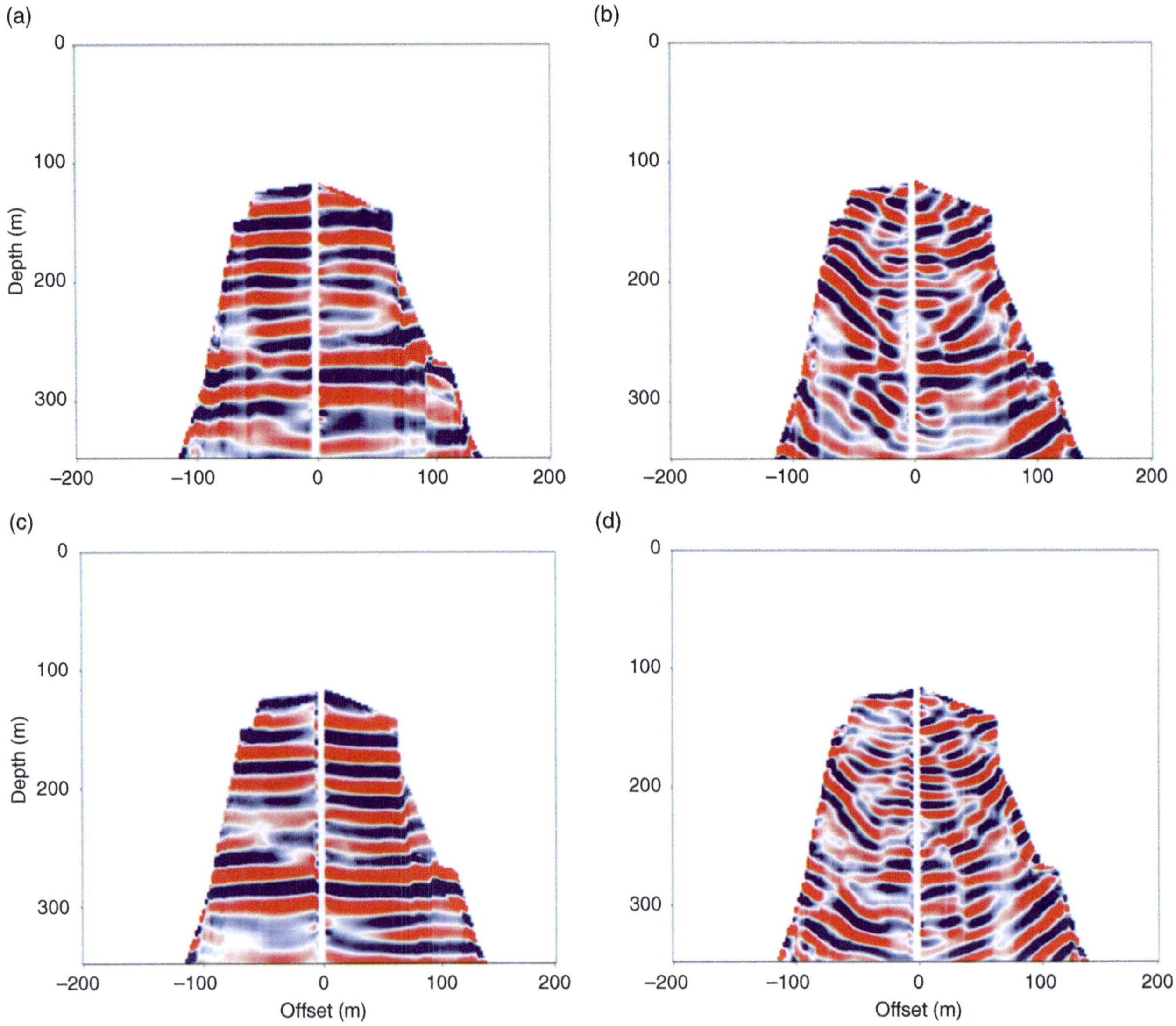

Figure 23.17 Elastic RTM imaging results: (a) PP and (b) PS imaging for line 1; (c) PP and (d) PS imaging for line 4.

two aspects: one is that the near-surface is highly variable in the x direction, and the other is that the predicted S-wave velocity data is not accurate enough at shallow depths (0–223 m). The predicted P-velocity seems reasonable through the two-azimuth PP RTM tests; however, the S-velocity is overestimated. The reflection interfaces exhibited in the PP images are consistent with the FWI models, still verifying the validity of the inversion result.

23.5. DISCUSSION

We consider these results to be very positive and point to the possibility of practical, relatively low-cost FWI monitoring of target areas in which (for instance) CO_2 containment failure requires diagnosis. However, plainly, much development and research remain necessary. The depth registration method presented in this work sets to fixed positions geophone or accelerometer and DAS data acquisition points. Of course, this method is not applicable to cases where only one of the DAS or geophone/accelerometer data is available. In addition to the apparent complementarity of the two data types for FWI, this may be considered a good reason to supplement DAS data with (perhaps sparsely deployed) standard geophones. The RTM imaging results indicate the negative impact of near-surface complexity; therefore, the effective source approach introduced into the FWI could mitigate the impact of the near-surface on the inversion results. The single-parameter V_p inversion might help reduce the cross-talk between different parameter classes and improve convergence. However, it also represents a severe loss of elastic medium property information in the result, which is essential for reservoir property FWI characterization. In addition, the assumption made for this parameterization, namely that the model variables such as velocity and density are perfectly correlated, can cause inversion errors if they violate the real subsurface condition. Finally, 3D FWI would represent a more complete method for generating inversions, especially for exploring lateral differences along azimuths. This naturally is a more computationally expensive proposition, but it would more fully realize the potential of full-waveform methods, and we consider it to be an important next step to maximize resolution and reliability.

23.6. CONCLUSIONS

Accelerometers or geophones are conventional seismic acquisition tools, whereas DAS is an evolving seismic acquisition tool with great potential. Currently, seismic data acquired using both geophone/accelerometer and DAS fibers are increasing, so an inversion method for the combined DAS-accelerometer datasets is important. First, we develop the depth registration scheme based on cross-correlation scans of the two datasets and use first-breaking matching to implement the quality control, leading to reliable depth estimates for each channel of DAS data. Then, we propose an elastic FWI method based on the DAS-accelerometer data that suppresses cross-talk in the inversion, and the inversions of two lines of the walkaway VSP data verify the robustness of the proposed FWI methods. Finally, the RTM imaging using the same accelerometer data is consistent with the FWI results, verifying our inversion scheme's effectiveness.

ACKNOWLEDGMENTS

We thank the sponsors of CREWES for their continued support. This work was funded by CREWES industrial sponsors and NSERC (Natural Science and Engineering Research Council of Canada) through the grant CRDPJ 543578-19. The authors thank Carbon Management Canada, the CaMI.FRS Joint Industry Project, and the University of Calgary Global Research Initiative for access to the site and the data used in the chapter.

AVAILABILITY STATEMENT

The data used in this study are not available for public release.

REFERENCES

Alkhalifah, T., & Plessix, R. É. (2014). A recipe for practical full-waveform inversion in anisotropic media: an analytical parameter resolution study. *Geophysics*, *79*(3), R91–R101. doi:10.1190/geo2013-0366.1

Aragao, O., & Sava, P. (2020). Elastic full-waveform inversion with probabilistic petrophysical model constraints. *Geophysics*, *85*(2), R101–R111. doi:10.1190/geo2019-0285.1

Berenger, J. P. (1994). A perfectly matched layer for the absorption of electro-magnetic waves. *Journal of Computational Physics*, *114*, 185–200. doi:10.1006/jcph.1994.1159

Brittan, J., & Jones, I. (2019). FWI evolution-from a monolith to a toolkit. *The Leading Edge*, *38*(3), 179–184. doi:10.1190/tle38030179.1

Brossier, R., Operto, S., & Virieux, J. (2009). Seismic imaging of complex on shore structures by 2D elastic frequency-domain full-waveform inversion. *Geophysics*, *74*(6), WCC105–WCC118. doi:10.1190/1.3215771

Bunks, C., Saleck, F. M., Zaleski, S., & Chavent, G. (1995). Multiscale seismic waveform inversion. *Geophysics*, *60*(5), 1457–1473. doi:10.1190/1.1443880

Cai, X., Liu, Y., & Ren, Z. (2015). *Elastic reverse-time migration based on the optimal staggered-grid finite-difference scheme*. Paper presented at the SEG International Exposition and Annual Meeting, SEG-2015-5906925. doi:10.1190/segam2015-5906925.1

Daley, T. M., Miller, D. E., Dodds, K., Cook, P., & Freifeld, B. M. (2016). Field testing of modular borehole monitoring with simultaneous distributed acoustic sensing and geophone vertical seismic profiles at Citronelle, Alabama. *Geophysical Prospecting*, *64*(5), 1318–1334. doi:10.1111/1365-2478.12324

Eaid, M. V., Keating, S., & Innanen, K. A. (2020). Multiparameter seismic elastic full-waveform inversion with combined geophone and shaped fiber-optic cable data. *Geophysics*, *85*(6), R537–R552. doi:10.1190/geo2020-0170.1

Eaid, M. V., Keating, S., & Innanen, K. A. (2022). *Combined elastic FWI of accelerometer and DAS VSP data from a CO2 sequestration test site in Newell County, Alberta*. Paper presented at the SEG/AAPG International Meeting for Applied Geoscience & Energy, SEG-2022-3746014. doi:10.1190/image2022-3746014.1

Eaid, M. V., Keating, S. D., Innanen, K. A., Macquet, M., & Lawton, D. (2023). Field assessment of elastic full-waveform inversion of combined accelerometer and distributed acoustic sensing data in a vertical seismic profile configuration. *Geophysics*, *88*(6), WC163–WC180. doi:10.1190/geo2023-0066.1

Egorov, A., Correa, J., Bóna, A., et al. (2018). Elastic full-waveform inversion of vertical seismic profile data acquired with distributed acoustic sensors. *Geophysics*, *83*(3), R273–R281. doi:10.1190/geo2017-0718.1

Egorov, A., Pevzner, R., Bóna, A., et al. (2017). Time-lapse full waveform inversion of vertical seismic profile data: workflow and application to the CO2CRC Otway project. *Geophysical Research Letters*, *44*(14), 7211–7218. doi:10.1002/2017GL074122

Hall, K. W., Bertram, K. L., Bertram, M., Innanen, K. A., & Lawton, D. C. (2019). *Simultaneous accelerometer and optical fibre multi-azimuth walk-away VSP experiment: Newell County, Alberta, Canada*. Paper presented at the SEG International Exposition and Annual Meeting, SEG-2019-3216606. doi:10.1190/segam2019-3216606.1

Hu, Q., Eaid, M. V., Innanen, K. A., Keating, S. D., & Cai, X. (2024). Estimation of rock physics properties via full waveform inversion of vertical seismic profile data recorded by accelerometer and fibre-optic sensors. *Geophysical Journal International*, *237*(2), 805–817. doi:10.1093/gji/ggae074

Hu, Q., Grana, D., & Innanen, K. A. (2023). Feasibility of seismic time-lapse monitoring of CO_2 with rock physics parametrized full waveform inversion. *Geophysical Journal International*, *233*(1), 402–419. doi:10.1093/gji/ggac462

Hu, Q., Keating, S., Innanen, K. A., & Chen, H. (2021). Direct updating of rock-physics properties using elastic full-waveform inversion FWI for rock physics properties. *Geophysics*, *86*(3), MR117–MR132. doi:10.1190/geo2020-0199.1

Innanen, K. A., Hall, K. W., & Lawton, D. C. (2022). *A time-lapse multi-offset, multi-azimuth VSP acquired as a candidate for low-cost monitoring of CO_2 injection and storage* (Research Report 34(29)), CREWES.

Keating, S., Eaid, M. V., & Innanen, K. A. (2021). *Effective sources: removing the near surface from the VSP FWI problem* (Research Report 33(26)), CREWES.

Keating, S., & Innanen, K. A. (2019). Parameter crosstalk and modeling errors in viscoacoustic seismic full-waveform inversion. *Geophysics*, *84*(4), R641–R653. doi:10.1190/geo2018-0410.1

Keating, S., & Innanen, K. A. (2020). Parameter crosstalk and leakage between spatially separated unknowns in viscoelastic full-waveform inversion. *Geophysics*, *85*(4), R397–R408. doi:10.1190/geo2019-0370.1

Köhn, D., De Nil, D., Kurzmann, A., et al. (2012). On the influence of model parametrization in elastic full waveform tomography. *Geophysical Journal International*, *191*(1), 325–345. doi:10.1111/j.1365-246X.2012.05633.x

Kuvshinov, B. N. (2016). Interaction of helically wound fibre-optic cables with plane seismic waves. *Geophysical Prospecting*, *64*(3), 671–688. doi:10.1111/1365-2478.12303

Lee, H. Y., Koo, J. M., Min, D. J., et al. (2010). Frequency-domain elastic full waveform inversion for VTI media. *Geophysical Journal International*, *183*(2), 884–904. doi:10.1111/j.1365-246X.2010.04767.x

Lines, L. (2014). *FWI and the "noise" quandary* (Research Report 26(49), 1–6), CREWES.

Mateeva, A., Jorge, L., Hans, P., et al. (2014). Distributed acoustic sensing for reservoir monitoring with vertical seismic profiling. *Geophysical Prospecting*, *62*(4), 679–692. doi:10.1111/1365-2478.12116

Métivier, L., Brossier, R., Operto, S., & Virieux, J. (2017). Full waveform inversion and the truncated Newton method. *SIAM Review*, *59*(1), 153–195. doi:10.1137/16M1093239

Morse, P. M., & Feshbach, H. (1954). Methods of theoretical physics. *American Journal of Physics*, *22*(6), 410–413. doi:10.1119/1.1933765

Pan, W., Geng, Y., & Innanen, K. A. (2018). Interparameter trade-off quantification and reduction in isotropic-elastic full-waveform inversion: synthetic experiments and hussar land data set application. *Geophysical Journal International*, *213*(2), 1305–1333. doi:10.1093/gji/ggy037

Pan, W., Innanen, K. A., Geng, Y., et al. (2019). Interparameter trade-off quantification for isotropic-elastic full-waveform inversion with various model parameterizations. *Geophysics*, *84*(2), R185–R206. doi:10.1190/geo2017-0832.1

Posey, R., Johnson, G., & Vohra, S. (2000). Strain sensing based on coherent Rayleigh scattering in an optical fiber. *Electronics Letters*, *36*, 1688–1689. doi:10.1049/el:20001200

Pratt, R. G. (1990). Frequency-domain elastic wave modeling by finite differences: a tool for crosshole seismic imaging. *Geophysics*, *55*(5), 626–632. doi:10.1190/1.1442874

Pratt, R. G. (2004). *Velocity models from frequency domain waveform tomography–Past, Present and Future*. Paper presented at the 66th EAGE Conference and Exhibition-Workshops, 133. doi:10.3997/2214-4609.201405674

Roberts, M. A., Singh, S., & Hornby, B. E. (2008). Investigation into the use of 2D elastic waveform inversion from look-ahead walk-away VSP surveys. *Geophysical Prospecting*, *56*(6), 883–895. doi:10.1111/j.1365-2478.2007.00689.x

Singh, S., Tsvankin, I., & Naeini, E. Z. (2018). Bayesian framework for elastic full-waveform inversion with facies information. *The Leading Edge*, *37*(12), 924–931. doi:10.1190/tle37120924.1

Spikes, K. T., Tisato, N., Hess, T. E., et al. (2019). Comparison of geophone and surface-deployed distributed acoustic sensing seismic data. *Geophysics*, *84*(2), A25–A29. doi:10.1190/geo2018-0528.1

Tarantola, A. (1984). Inversion of seismic reflection data in the acoustic approximation. *Geophysics*, *49*(8), 1259–1266. doi:10.1190/1.1441754

Vigh, D., Jiao, K., Watts, D., & Sun, D. (2014). Elastic full-waveform inversion application using multicomponent measurements of seismic data collection. *Geophysics*, *79*(2), R63–R77. doi:10.1190/geo2013-0055.1

Virieux, J., & Operto, S. (2009). An overview of full-waveform inversion in exploration geophysics. *Geophysics*, *74*(6), WCC1–WCC26. doi:10.1190/1.3238367

Zhang, Z., & Alkhalifah, T. (2022). Regularized elastic full-waveform inversion using deep learning. In S. Bhattacharya & H. Di (Eds.), *Advances in subsurface data analytics* (pp. 219–250). Elsevier. doi:10.1016/B978-0-12-822295-9.00009-1

Part IV
Borehole Distributed Acoustic Sensing Monitoring

24

Seismological Applications of Downhole Distributed Acoustic Sensing

Ariel Lellouch

ABSTRACT

Whereas the analysis of earthquakes recorded over surface fibers is relatively common, downhole fibers have so far been scarcely utilized for earthquake seismology studies. In this short review chapter, I discuss the potential, capabilities, and limitations of downhole distributed acoustic sensing (DAS) in a broad seismological context. The spatial continuity of the DAS measurement is especially important for microseismic monitoring and the study of local earthquakes, whereas for regional and teleseismic earthquakes, the lower noise levels at depth may play a significant role. Whereas the higher noise levels in DAS compared to traditional sensors may be mitigated through various array-based approaches, the directional nature of the measurement remains a primary limitation in any source parameter estimation. I suggest various approaches to mitigate the shortcomings of DAS and integrate it into groundbreaking seismological studies.

24.1. INTRODUCTION

In traditional seismological applications, downhole sensors are known to provide high-quality data that can help study small-scale earthquakes and microseismic events. Deploying downhole monitoring arrays is common in reservoir monitoring applications (Maxwell, 2014), often in conjunction with surface receivers. It has been shown that downhole receivers significantly improve the depth estimation of microseismic events (Chambers et al., 2010; Eisner et al., 2010) and can help recover their focal mechanism (Willacy et al., 2019). In earthquake seismology, downhole tools are key to studying low-magnitude earthquakes (Abercrombie, 1995; Hofstetter et al., 2020; Ide et al., 2003), despite needing careful treatment and potentially suffering from adverse effects due to destructive interference between upgoing and downgoing waves, reflected at the surface and potentially other interfaces (Shearer & Orcutt, 1987). Whereas shallow downhole sensors are used as part of the global seismic network, they remain relatively rare compared to surface stations due to the operational challenges of deploying and maintaining borehole instruments.

In distributed acoustic sensing (DAS) applications, the situation is somewhat similar. A significant amount of seismological studies using DAS rely on existing fiber-optic infrastructure installed close to the surface on land or at sea (Ajo-Franklin et al., 2019; Lindsey et al., 2017, 2019). The practical benefits of utilizing existing infrastructure, whether through dark fibers or during cable downtime, are especially prominent for academic disciplines that often lack the funding for cable deployment. In marine environments (e.g., Williams et al., 2019), cable deployment is even more prohibitive in terms of effort and cost. Operating with existing, horizontal near-surface cables imposes several limitations. First, cable coupling in existing installations is generally suboptimal for seismic sensing. Usually, fiber deployment was conducted taking completely different considerations

School of the Environment and Earth Sciences, Tel Aviv University, Tel Aviv, Israel

Distributed Acoustic Sensing in Borehole Geophysics, Geophysical Monograph 289, First Edition.
Edited by Yingping Li, Robert Mellors, and Ge Zhan.

DOI: 10.1002/9781394179275.ch24

into account. Thus, cables are often laid in conduits and can be above ground or underground depending on the landscape, trenching depth varies, in-fill material and quality can be inconsistent, etc. As a consequence, signal quality is generally lower than what is offered by DAS systems, and it varies considerably along the cable. The precise layout of the fiber is also unknown, and mapping its exact spatial coordinates can be time-consuming (Biondi et al., 2022; Martin et al., 2017) due to inaccurate maps, cable spools, and hard-to-access locations. The orientation of the fiber can also significantly vary, and it needs to be accurately mapped. Even with correct mapping, DAS recording over unfavorable sensing directions can be detrimental to the recorded seismological data (Kennett, 2022). Purposedly deployed fibers can mitigate and even overcome these limitations. There are examples of purposedly deployed fiber-optic cables that were used for seismological studies (Klaasen et al., 2021; Li & Zhan, 2018), but the aperture of these arrays is often limited due to budgetary and logistical constraints.

Regardless of the fiber installation, the near-surface nature of the measurement suffers from inherent seismic noise in the form of surface waves, cultural noise, and natural (wind-induced) noise (Carter et al., 1991). Ambient noise studies show that most of the seismic noise onshore propagates as surface (Rayleigh) waves (Shapiro et al., 2005), which are strongest at the surface level. In marine environments, cables laid on the sea bottom observe mostly Scholte waves (Cheng et al., 2021), which travel along the solid–liquid contact. Like onshore surface waves, Scholte waves are most prominent close to the cable depth. Although these properties make DAS arrays useful for ambient noise interferometry, signals from distinct seismological sources, such as earthquakes or microseismic events, are often masked.

24.1.1. Advantages of Downhole DAS Installations

Downhole fibers have significant advantages compared to near-surface installations, especially those utilizing dark fiber. Noise levels due to the ambient seismic field quickly decrease with depth, thus allowing for quiet recording even in anthropogenically active areas (Lellouch et al., 2020), which are often highly problematic for recording over horizontal fibers. Optimal coupling with the well can be achieved through installation behind casing. Although it commands significant planning and installation efforts, it guarantees consistent coupling along the entire well trajectory. It is also possible to have the fiber return to the surface, thus doubling the amount of acquired data and potentially helping constrain the accurate location of individual fiber DAS channels.

Installation at depth also offers significant benefits when recording subsurface seismic sources. First, the geometrical distance to the fiber is shorter, resulting in weaker geometrical spreading. Although this effect is negligible in global seismology, it is crucial for microseismic monitoring. Because the area of seismic activity is generally predictable, the recording array can be optimized so that the source-receiver distance is as low as a few hundred meters. For regional-scale earthquakes, the benefit of approaching the source depends on the location of the source and the extent of the DAS array. In addition, the shallow subsection is known to be highly heterogeneous and dissipative. Dissipation can be intrinsic (anelasticity) or extrinsic (scattering). By recording below the depth of the shallow subsection, wavefronts from deep sources undergo much weaker attenuation and reach the receivers as distinct seismic waves, facilitating data analysis.

From a practical perspective, installing downhole fiber for long-term monitoring is easier than deploying geophones, mostly because the fiber itself is a passive element. Given an appropriate coating, fibers can withstand high temperatures and pressure in areas that are potentially inaccessible to traditional seismometers (Cai et al., 2019). They can remain in place for years or decades, although their quality might degrade due to chemical processes occurring in long-term exposure to high temperatures.

Finally, a primary advantage of the DAS measurement is its extremely wide frequency response (Paitz et al., 2021), spanning from mHz to kHz and further. Traditional sensors implemented mechanically have a much narrower, fit-for-purpose frequency response. The ability to record all types of seismic events with a single instrument is highly beneficial.

24.1.2. Limitations of Downhole DAS Installations

The primary limitation of DAS is the directional nature of its measurement, which is parallel to the orientation of the fiber. Traditional downhole instruments, on the other hand, generally have at least triaxial measurements. For surface DAS, this limitation can be somewhat mitigated by recording over fibers with complex trajectories, but it is impossible to do so in downhole installations. Because the measurement direction follows the borehole trajectory, it will generally be along the vertical direction, but other directions are possible in deviated wells.

A secondary downside of DAS is the measurement of strain or strain rate over a certain gauge length. The conversion between DAS measurements and particle velocity/acceleration has been extensively studied. Particle velocity can be obtained from strain measurement through multiplication by the local phase velocity. This conversion may be performed in either the F-K or time domain (Daley et al., 2016; Lindsey et al., 2020; Lior et al., 2021; Wang et al., 2018), but it requires careful treatment to avoid instabilities and numerical errors. The gauge length (Dean et al., 2016) can also potentially limit high-frequency DAS measurements if the apparent recorded wavelength is on the order of, or shorter, than the gauge length, which is typically on the order of 10 m.

In addition to these fundamental limitations, DAS data suffer from certain technical weaknesses. The impulse response of the measurement is not as well understood as for traditional sensors, mostly due to complicated coupling between the fiber, borehole, and surrounding media. Colocated seismometers can greatly help calibrate DAS response as well as obtain absolute particle motion values from DAS measurements. Besides that, DAS measurements might saturate if the induced strain (or strain rate) within the gauge length is above a certain value, causing the phase unwrapping algorithm to fail (Mellors et al., 2021; Willis, 2022). If we utilize a strain formulation for a gauge length of 10 m and typical optical parameters (Lindsey et al., 2020), strain values above ~35 nano strain will cause phase unwrapping to fail using a naïve implementation. Depending on the magnitude and distance of the recorded earthquake, this can be a limiting factor without specialized processing. Finally, accurate geospatial positioning of downhole DAS channels is challenging, and permanent DAS observatories will likely require active surveys for calibration.

24.1.3. Downhole DAS Seismological Studies

Although there are many studies documenting local, regional, and teleseismic earthquakes over surface fibers (e.g., Ajo-Franklin et al., 2019; Li & Zhan, 2018; Lindsey et al., 2017, 2019; Wang et al., 2018; Yu et al., 2019), there are relatively few peer-reviewed publications documenting earthquakes recorded over a downhole DAS array. This is potentially due to the relative difficulty of obtaining access to existing downhole fiber, especially for academic groups. Downhole fiber deployment is even more challenging and is often achievable only by commercial entities (Lellouch & Biondi, 2021). Nonetheless, several open downhole DAS datasets could be used for earthquake seismology studies (Spica et al., 2023). Some of them (Lellouch et al., 2019a, 2020) have already been studied in the context of microseismic analysis or local earthquakes, but dedicated reprocessing could potentially allow for the analysis of regional and teleseismic events.

In this chapter, I introduce several seismological applications of downhole DAS recordings. Although, as of today, they remain a minority among academically driven DAS publications, downhole fiber deployments hold vast potential to enhance capabilities in most of the fields they are being gradually introduced to, despite their limitations.

24.2. EARTHQUAKE SEISMOLOGY

24.2.1. Local Earthquakes (D < 100 km)

In the seismological context, the spatial continuity of DAS measurements is most useful at the local scale because the array's aperture can be significant compared to the distance from the source. In addition, high-frequency events can be recorded without spatial aliasing. For example, assuming a central frequency of 50 Hz and P-wave velocity of 4,000 m/s, the spatial wavelength is 800 m—much shorter than the representative distance between stations in conventional seismological networks. Such frequencies were observed by Lellouch et al. (2019a, 2019b) for local earthquakes close to the San Andreas Fault Observatory at Depth (SAFOD) borehole. This allows for the coherent tracking of the different arrival phases, as can be seen in Figure 24.1.

The spatial continuity of DAS measurements allows for the automatic detection of low-magnitude events. Although individual DAS channels are noisier than traditional sensors, the large number of measurements can compensate for the lower signal-to-noise ratio. Different approaches have been employed for removing noise in DAS signals based on the assumption that the signals should be spatially coherent with reasonable traveltime differences between nearby channels (e.g., Chen et al., 2022, 2023; Dong & Li, 2021; Tian et al., 2022a, 2022b; Yang et al., 2023). Some of these studies employ signal processing approaches, whereas others utilize machine learning capabilities to differentiate between noise and signal. These studies show a superior signal-to-noise ratio compared to basic filtering steps of DAS data, such as common-mode suppression and frequency filtering.

A straightforward approach to event detection is applying an automatic picking method to individual DAS channels after denoising, and only accepting events that are statistically coherent along the different channels. Picking can be conducted with conventional signal processing approaches, such as short-term average / long-term average (STA/LTA), or using machine learning approaches (Mousavi et al., 2020; Zhu & Beroza, 2018). An alternative approach is to directly train a neural network for detection, based on DAS images instead of individual channels (Huot et al., 2022; Wamriew et al., 2021). However, because the location, magnitude, and focal mechanism of the earthquakes strongly vary between events, ensuring a complete and representative training dataset is a major challenge. Another class of approach is to detect events based on predicted traveltime arrivals along the DAS array. Given knowledge of the velocity structure, traveltime arrivals along a vertical DAS array can be predicted as a function of the incidence angle (Lellouch et al., 2019a, 2020). By scanning over all possible angles and

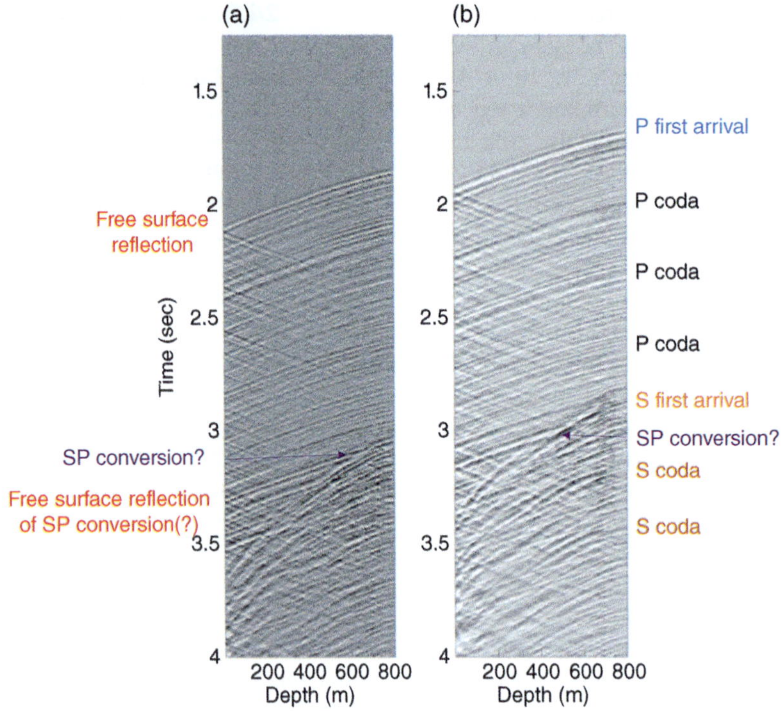

Figure 24.1 Two local earthquakes of magnitude (a) M = 1.33 and (b) M = 2.46 recorded by a downhole DAS array at SAFOD. Distances from the array are approximately 11.5 km for both events. Thanks to the spatial continuity of DAS recording, many different seismic phases can be tracked. Lellouch, Yuan, Spica, et al. (2019) / with permission of John Wiley & Sons.

applying a certain coherency threshold, low-magnitude earthquakes can be recorded. The application of this approach to 10 days of recorded DAS data in the FORGE site resulted in magnitude completeness better by 0.5 to 1 order of magnitude compared to a conventional surface array (Lellouch et al., 2021).

Event location using downhole fiber suffers from the directional nature of the DAS measurement. The depth and horizontal distance from the array can theoretically be retrieved for a single vertical borehole, but there is no azimuthal information on the source location (Lellouch et al., 2019b). A multiwell recording is a natural solution to the location problem, but array geometry design is a crucial step of the process. For example, a dual-well DAS recording in the FORGE site, which is seismically active, has been publicly released (University of Utah Seismograph Stations, 2022). Two recording DAS wells are at a depth of approximately 1 km and located approximately 70 m apart. Although there are currently no published location results based on DAS data only, initial tests indicate that the acquisition geometry is unfavorable for event location due to the short distance between the wells.

Magnitude and focal mechanism estimation using DAS are still very much in their infancy. Generally speaking, the majority of the magnitude estimation methods have been developed for particle motion or acceleration. Converting DAS records to particle motion can thus be useful in applying conventional magnitude estimation methods based on the recorded spectrum (Lior et al., 2023). Another approach is to use strain (or strain rate) data directly, based on heuristical formulations developed with strain meters (Lellouch et al., 2020). Estimating the focal mechanism is significantly more challenging, and only partial solutions have been proposed for both surface and downhole DAS (Hudson et al., 2021; Luo et al., 2021; Vera Rodriguez & Wuestefeld, 2020). Having only one direction of measurement strongly limits the ability to resolve the full focal mechanism, and more research about the suitability of DAS for this task is needed. Notwithstanding, along certain planes, the spatial coverage of DAS may potentially allow for a very accurate estimation of the focal mechanism projection thanks to our ability to continuously track the wavefront amplitude at different angles.

24.2.2. Regional Earthquakes (100 < D < 1,000 km)

Regional earthquakes have been shown to be recorded over downhole DAS arrays (Kobayashi et al., 2020; Pevzner et al., 2020; Shashkin et al., 2022). However, they were not the direct focus of a seismological study. In some studies, earthquakes were used to detect changes in subsurface conditions. By assuming vertically propagating wavefronts, they used different earthquakes to detect changes in propagation through the array and link them, for example, to carbon dioxide injection. (Shashkin et al., 2022) showed that using a multiwell DAS recording, the event location is possible through traveltime picking combined with horizontal slowness estimation, but that the constraint on the horizontal location is much weaker than the distance from the recording arrays. For five regional events that were compared to cataloged locations, the location differences were between 2 and 14 km. An example of a regional earthquake recorded over downhole DAS is in Figure 24.2.

Especially at the lower distance range of the regional scale, and depending on the particular acquisition geometry, downhole DAS arrays can reach beamforming capabilities. Studies with horizontal fiber (Lindsey et al., 2017, 2019; van den Ende & Ampuero, 2021) show the benefits and limitations of utilizing DAS arrays for beamforming at regional distances. However, in these cases, the beamforming was along the horizontal plane. For a vertical downhole DAS array, beamforming can ideally resolve the vertical angle of incidence, measured in relation to the axis of the fiber, yet the azimuthal symmetry remains. In addition, as discussed by Pevzner et al. (2020), the angle of incidence may often be close to vertical, or parallel to the DAS array, due to the subsurface velocity structure inducing diving waves. In this case, the beamforming capabilities of the DAS array will be very limited.

24.2.3. Teleseismic Earthquakes (D > 1,000 km)

Recent publications have documented teleseismic events recorded over downhole fiber (Glubokovskikh et al., 2021; Shulakova et al., 2022; Sidenko et al., 2022) in a frequency range of 0.1 to 1 Hz. In the context of global seismology, the broadband nature of DAS allows for straightforward temporal filtering to the frequency range of interest. However, the aperture of a downhole array is insignificant compared to the representative distance to the source. Therefore, the focusing power of the DAS array will be highly limited. Instead, it can be effectively treated as a point, or point-like, sensor. The seismic frequencies that we expect are very low, and multiple nearby DAS channels can be summed to improve the signal-to-noise ratio (Yu et al., 2019). Denoising methods that take into account the spatial coherency of the DAS measurements and potentially correct for different arrival times along the array can also be considered (Atterholt et al., 2022; Chen et al., 2022) to maximize the signal-to-noise ratio.

Generally, seismic noise levels decrease with depth as noise propagates mostly in the form of surface waves (Shapiro et al., 2005), as can be observed in downhole DAS arrays (Lellouch et al., 2021). Therefore, summation over different depth ranges of the DAS array could enhance or destroy surface waves, thus facilitating either their analysis or removal, mostly to study late body wave phases that may be obscured by the surface waves (Kennett & Engdahl, 1991). Such a selective summation can also help mitigate the imprint of near-surface

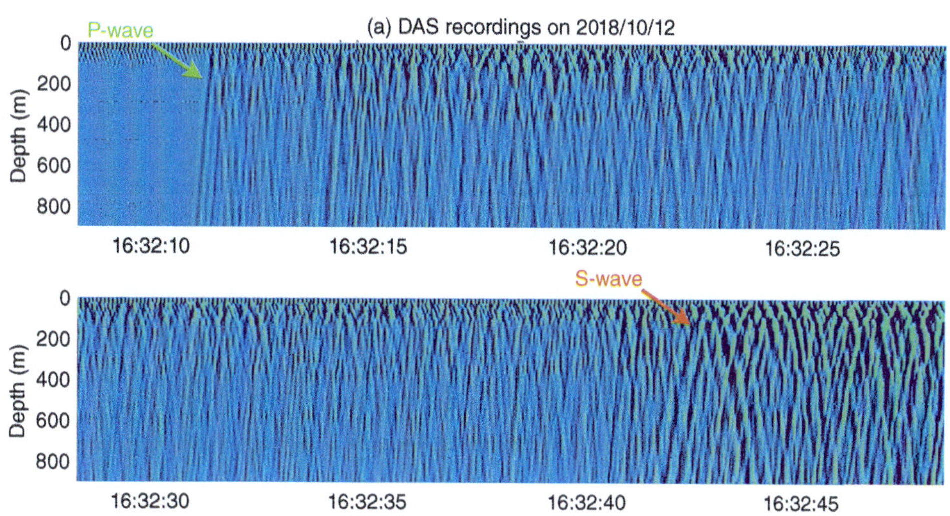

Figure 24.2 Regional earthquake (M_L 4.6) at a distance of approximately 285 km recorded over a vertical DAS array. P- and S- wave arrivals are visible, as well as higher noise levels in the shallow (<100 m) part of the well. Pevzner et al. (2020) / GeoScienceWorld.

scattering, often observed in horizontal DAS deployments (Li et al., 2021; Lindsey et al., 2019; Yu et al., 2019), because it propagates mostly as surface waves.

However, DAS directivity may be especially limiting for distant earthquakes. We expect primarily upgoing waves, turning a vertical downhole DAS array into an ideal P-wave sensor. However, S waves should in theory have no imprint on the DAS measurement if propagation is exactly vertical. In practice, S waves are recorded even for near-vertical propagation (Lellouch et al., 2019a), but they are not as dominant as in traditional seismometers. In addition, the azimuthal sensitivity of DAS to S waves significantly differs from traditional seismometers due to the strain radiation patterns. Recording over deviated boreholes may significantly improve S-wave sensitivity, but ensuing data processing must be conducted carefully to avoid unwanted crosstalk between the vertical and quasi-horizontal parts of the array.

24.3. MICROSEISMIC MONITORING

Downhole DAS deployments play an increasing role in microseismic monitoring. Because the locations of microseismic events are approximately known, DAS arrays can be deployed in proximity to the microseismic sources. This allows for high-resolution recording over a spatially limited area, thus fully utilizing the potential of DAS. In conjunction with the ability to record DAS data over wells with other purposes than monitoring, these benefits have led to a significant increase in the usage of downhole fiber for microseismic monitoring. Whereas monitoring hydrocarbon reservoirs has been a strong incentive for DAS deployments, it is also important in enhanced geothermal systems (EGSs) and carbon capture, utilization, and storage (CCUS).

Downhole DAS recording was first recognized as a potential tool for monitoring about a decade ago (Webster et al., 2013). As DAS fibers were deployed in deviated wells targeting unconventional hydrocarbons, events were primarily recorded over deep horizontal fiber sections. The stronger events could also be recorded along the deviated and vertical parts of the borehole, as can be seen in Figure 24.3.

Initial attempts at microseismic event detection with DAS showed very little success, missing more than 90% of events detected by traditional downhole sensors (Hull et al., 2017; Molteni et al., 2017; Webster et al., 2016). A possible cause for these unsatisfactory results was that DAS channels were primarily analyzed individually. Due to their lower signal-to-noise ratio compared to downhole sensors, traditional detection methods underperformed. For example, by utilizing an array-based beamforming approach, Lellouch et al. (2020) significantly improved the catalog generated by individual DAS channel processing (Mondanos & Coleman, 2019).

Machine learning approaches, treating DAS data as 2D images, have been applied with relative success.

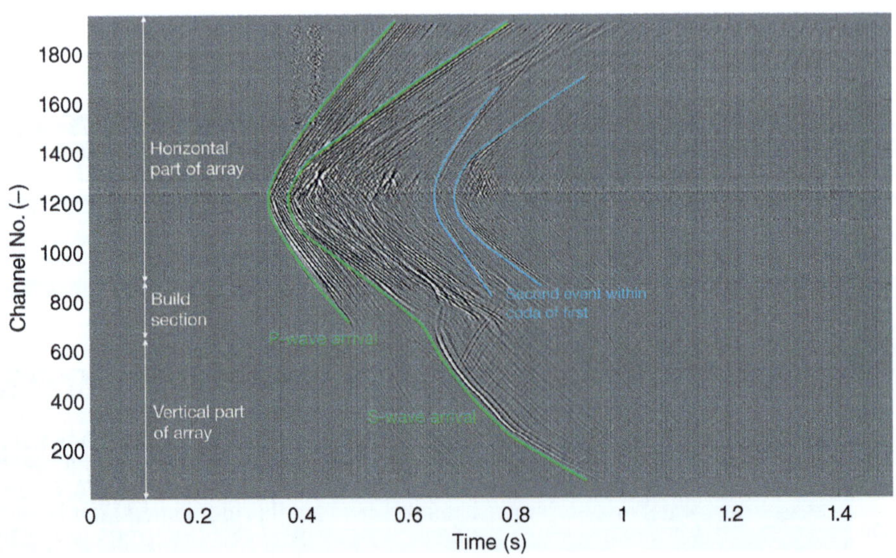

Figure 24.3 Example microseismic event recorded by a DAS array. The moveout of the P- and S-wave arrivals (green) can be observed on the horizontal and vertical parts of the array. A second event, marked in blue, is also visible within the coda of the first. Verdon et al. (2020) / GeoScienceWorld.

Binder and Tura (2020) used a homogenous elastic medium for forward modeling of microseismic events and trained a network for event detection. Stork et al. (2020) expanded their forward modeling to anisotropic with constant elastic properties and used the pretrained state-of-the-art YOLOv3 for image object detection. Wamriew et al. (2021) generated training data using 1D anisotropic representations of the subsurface and ray tracing. All studies added real-life noise taken from DAS records to their training datasets. Another approach, taken by Huot et al. (2022), was to utilize a curated dataset of about 7,000 microseismic events to train an optimized neural network. By doing so, they were able to detect over 100,000 microseismic events, exceeding human detection capabilities and that of a surface monitoring array.

The limitations in event location due to the directional measurement of DAS were quickly recognized (Karrenbach et al., 2017; Webster et al., 2016). Although the location along the axis of the fiber could be pinpointed, the distance from the array was more challenging to estimate. Naturally, the azimuth measured in relation to the horizontal axis could not be resolved for a horizontal well. Karrenbach et al. (2019) showed that using traveltime picking over horizontal, deviated, and vertical parts of the well, event location to some extent was somewhat possible. To eliminate the remaining location uncertainties, some assumptions derived from the production schedule were required. Multiwell recording would further resolve location uncertainties up to the point of fully eliminating them. However, certain geological conditions may allow for event location using a single horizontal fiber. If a fiber is deployed inside a low-velocity reservoir, and the event originates inside or close to the reservoir, distinct guided waves will be excited. Thus, the existence of guided waves in the DAS data indicates the source origin inside the reservoir—thus partially resolving the azimuthal location uncertainty (Huff et al., 2020; Liu et al., 2022). Utilizing this property, and neglecting the width of the reservoir, an absolute location of events originating inside the low-velocity reservoir is possible (Lellouch et al., 2022). Through the analysis of dispersive propagation velocity, the event distance from the DAS array can be constrained. Left–right symmetry is resolved by the production schedule. Given that many reservoirs have lower velocities than their surroundings, this ostensibly niche approach may prove highly beneficial.

A particularity of microseismic focal mechanism estimation is that near- and intermediate-field terms can be recorded and used to constrain the inversion in addition to the commonly used far-field term. Whereas the mentioned studies have shown a manual focal plane estimation for field data, I am not aware of any systematic inversion approach that has been published.

24.4. OVERCOMING THE LIMITATIONS OF DAS

Throughout this chapter, I have discussed the two main limitations of DAS: lower signal-to-noise ratio at each measurement point, and the directional nature of the measurement. Whereas the former can be resolved by different multichannel analysis approaches, the latter imposes fundamental limitations on event location, magnitude estimation, and focal mechanism reconstruction. There are several principal approaches to handling these difficulties, and they can be combined.

First, deploying fiber over complex downhole acquisition geometries can resolve some of the symmetry planes arising from the DAS measurement. For example, a circular symmetry in event location arises in any vertical DAS array. Two downhole arrays, theoretically, reduce this symmetry to only two possible points: the true location and a spurious mirror point. Three downhole arrays yield a complete location. Recording over deviated wells is highly beneficial because there is a combination of vertical and horizontal recording, eliminating symmetry planes. The exact effect of the array geometry can be modeled before deployment to assess DAS capabilities. Such modeling should also include a sensitivity analysis as given noise and velocity model errors will always be present.

A second approach is to record a combination of surface and downhole fiber. Whereas downhole deviated wells are better suited to this task, surface arrays may be useful in better resolving source parameters. Nonetheless, they are expected to operate in worse signal-to-noise conditions and may not be usable for all events. In multiwell recording, connecting the wells with surface fiber is a natural choice to allow recording with a single interrogator. The effort required to correctly install the surface fiber and ensure good coupling is negligible compared to downhole operations, and it is thus recommended to maximize surface sensing capabilities.

The deployment of helical fibers may also be considered (Kuvshinov, 2016; Lim Chen Ning & Sava, 2018) to alleviate the directionality limitation. Helical fiber has been shown to yield good agreement with downhole accelerometers in the recording of controlled explosions (Mellors et al., 2021), but a substantial benefit over straight fiber has yet to be proven experimentally. In addition, the analysis of data from helical fibers is more challenging and requires highly accurate information about the cable design and sensing locations.

Finally, joint analysis with traditional three-component seismometers can help resolve the X-Y location and certain components of the moment tensor. Existing surface-based networks and borehole instruments are relatively easy to integrate into a processing workflow after ensuring temporal synchronization between the two systems, ideally through GPS. Deploying dedicated instruments, colocated with the DAS fiber, can also prove highly beneficial. Colocated instruments may also help calibrate the DAS response, which is often difficult to obtain due to its dependence on coupling. They can also be used in converting DAS signals to particle motion equivalent, which cannot be done without reference points (Wang et al., 2018). This is especially important for moment magnitude and focal mechanism estimation, as the vast majority of formulations have been made for particle motion or acceleration data. Although strain (or strain rate) derivations exist for different moment tensors (Luo et al., 2021; Vera Rodriguez & Wuestefeld, 2020), they are not commonly used.

24.5. FUTURE USES AND POTENTIAL

There are several domains in which downhole DAS could yield significant progress. First, DAS instrumentation of a deviated well crossing an active strike-slip fault may yield an unprecedented recording of the fault behavior. By recording on both sides of the fault, low-magnitude earthquakes could be detected and analyzed. These low-magnitude events are key to understanding the fault's behavior and, potentially, detecting precursors to major events. A detailed study of fault zone guided waves (Ellsworth & Malin, 2011) could also be made possible through such a deployment. For example, the San Andreas, Dead Sea, and North Anatolian faults could be considered targets, with the depth of the seismogenic zone guiding the well trajectory design.

In microseismic monitoring scenarios, DAS has been shown to record low-frequency strain and pressure fronts that precede the propagating fractures (Jin & Roy, 2017). Downhole deployment near active faults could allow for the detection of strain/stress changes induced by fault activity, potentially preceding earthquakes. In conjunction with the low-magnitude earthquakes described above, a better understanding of the geomechanical conditions and behavior of the fault could be achieved.

Downhole DAS could be used in marine boreholes to record high-resolution data close to subduction zones. By measuring the seismic wavefield below the seafloor, earthquakes with lower magnitudes can be detected. In addition, the S waves will not be attenuated, potentially improving location and focal mechanism solutions.

For all applications, installed downhole DAS will allow for subsurface characterization in addition to passive monitoring, assuming that seismic sources can be operated (VSP-type processing). Alternatively, ambient noise can be used. More accurate subsurface knowledge will be useful in improving event location and characterization.

REFERENCES

Abercrombie, R. E. (1995). Earthquake source scaling relationships from −1 to 5 M_L using seismograms recorded at 2.5-km depth. *Journal of Geophysical Research: Solid Earth, 100* (B12), 24015–24036. doi:10.1029/95JB02397

Ajo-Franklin, J. B., Dou, S., Lindsey, N. J., Monga, I., Tracy, C., Robertson, M., et al. (2019). Distributed acoustic sensing using dark fiber for near-surface characterization and broadband seismic event detection. *Scientific Reports, 9*(1), 1–14. doi:10.1038/s41598-018-36675-8

Atterholt, J., Zhan, Z., Shen, Z., & Li, Z. (2022). A unified wavefield-partitioning approach for distributed acoustic sensing. *Geophysical Journal International, 228*(2), 1410–1418. doi:10.1093/gji/ggab407

Binder, G., & Tura, A. (2020). Convolutional neural networks for automated microseismic detection in downhole distributed acoustic sensing data and comparison to a surface geophone array. *Geophysical Prospecting, 68*(9), 2770–2782. doi:10.1111/1365-2478.13027

Biondi, E., Wang, X., Williams, E. F., & Zhan, Z. (2022). Geolocalization of large-scale DAS channels using a GPS-tracked moving vehicle. *Seismological Research Letters, 94*(1), 318–330. doi:10.1785/0220220169

Cai, Z., Wang, S., Liu, W., Li, F., Wang, C., Ma, L., & Li, Q. (2019). *Application of walkaway-VSP based on joint observation by DAS and geophones in the Tarim Basin, northwest China*. Paper presented at the SEG International Exposition and Annual Meeting, 974–978. doi:10.1190/segam2019-3214449.1

Carter, J. A., Barstow, N., Pomeroy, P. W., Chael, E. P., & Leahy, P. J. (1991). High-frequency seismic noise as a function of depth. *Bulletin of the Seismological Society of America, 81* (4), 1101–1114. doi:10.1785/BSSA0810041101

Chambers, K., Kendall, J. M., Brandsberg-Dahl, S., & Rueda, J. (2010). Testing the ability of surface arrays to monitor microseismic activity. *Geophysical Prospecting, 58*(June), 821–830. doi:10.1111/j.1365-2478.2010.00893.x

Chen, W., Oboué, Y. A. S. I., & Chen, Y. (2023). Retrieving useful signals from highly corrupted erratic noise using robust residual dictionary learning. *Geophysics, 88*(1), WA55–WA64. doi:10.1190/geo2022-0168.1

Chen, Y., Savvaidis, A., Fomel, S., Chen, Y., Saad, O. M., Wang, H., et al. (2022). Denoising of distributed acoustic sensing seismic data using an integrated framework. *Seismological Research Letters, 94*(1), 457–472. doi:10.1785/0220220117

Cheng, F., Chi, B., Lindsey, N. J., Dawe, T. C., & Ajo-Franklin, J. B. (2021). Utilizing distributed acoustic sensing and ocean bottom fiber optic cables for submarine structural characterization. *Scientific Reports, 11*(1), 1–14. doi:10.1038/s41598-021-84845-y

Daley, T. M., Miller, D. E., Dodds, K., Cook, P., & Freifeld, B. M. (2016). Field testing of modular borehole monitoring with simultaneous distributed acoustic sensing and geophone vertical seismic profiles at Citronelle, Alabama. *Geophysical Prospecting, 64*(5), 1318–1334. doi:10.1111/1365-2478.12324

Dean, T., Cuny, T., & Hartog, A. H. (2016). The effect of gauge length on axially incident P-waves measured using fibre optic distributed vibration sensing. *Geophysical Prospecting*, *65*(1), 184–193. doi:10.1111/1365-2478.12419

Dong, X., & Li, Y. (2021). Denoising the optical fiber seismic data by using convolutional adversarial network based on loss balance. *IEEE Transactions on Geoscience and Remote Sensing*, *59*(12), 10544–10554. doi:10.1109/TGRS.2020.3036065

Eisner, L., Hulsey, B. J., Duncan, P., Jurick, D., Werner, H., & Keller, W. (2010). Comparison of surface and borehole locations of induced seismicity. *Geophysical Prospecting*, *58*(5), 809–820. doi:10.1111/j.1365-2478.2010.00867.x

Ellsworth, W. L., & Malin, P. E. (2011). Deep rock damage in the San Andreas fault revealed by P- and S-type fault-zone-guided waves. *Geological Society Special Publication*, *359* (1), 39–53. doi:10.1144/SP359.3

Glubokovskikh, S., Pevzner, R., Sidenko, E., Tertyshnikov, K., Gurevich, B., Shatalin, S., et al. (2021). Downhole distributed acoustic sensing provides insights into the structure of short-period ocean-generated seismic wavefield. *Journal of Geophysical Research: Solid Earth*, *126*(12), e2020JB021463. doi:10.1029/2020JB021463

Hofstetter, R., Malin, P., & Ben-Avraham, Z. (2020). Seismic observations of microearthquakes from the masada deep borehole. *Seismological Research Letters*, *91*(4), 2298–2309. doi:10.1785/0220190391

Hudson, T. S., Baird, A. F., Kendall, J. M., Kufner, S. K., Brisbourne, A. M., Smith, A. M., et al. (2021). Distributed acoustic sensing (DAS) for natural microseismicity studies: a case study from Antarctica. *Journal of Geophysical Research: Solid Earth*, *126*(7), e2020JB021493. doi:10.1029/2020JB021493

Huff, O., Lellouch, A., Luo, B., Jin, G., & Biondi, B. (2020). Validating the origin of microseismic events in target reservoir using guided waves recorded by DAS. *The Leading Edge*, *39* (11), 776–784. doi:10.1190/tle39110776.1

Hull, R. A., Meek, R., Bello, H., & Miller, D. (2017). *Case history of DAS fiber-based microseismic and strain data, monitoring horizontal hydraulic stimulations using various tools to highlight physical deformation processes (Part A)*. Paper presented at the SPE/AAPG/SEG Unconventional Resources Technology Conference, 3050–3062. doi:10.15530/urtec-2017-2695282

Huot, F., Lellouch, A., Given, P., Luo, B., Clapp, R. G., Nemeth, T., et al. (2022). Detection and characterization of microseismic events from fiber-optic DAS data using deep learning. *Seismological Research Letters*, *93*(5), 2543–2553. doi:10.1785/0220220037

Ide, S., Beroza, G. C., Prejean, S. G., & Ellsworth, W. L. (2003). Apparent break in earthquake scaling due to path and site effects on deep borehole recordings: small earthquake scaling in long valley. *Journal of Geophysical Research: Solid Earth*, *108*(B5). doi:10.1029/2001JB001617

Jin, G., & Roy, B. (2017). Hydraulic-fracture geometry characterization using low-frequency das signal. *Leading Edge*, *36* (12), 975–980. doi:10.1190/tle36120975.1

Karrenbach, M., Cole, S., Ridge, A., Boone, K., Kahn, D., Rich, J., et al. (2019). Fiber-optic distributed acoustic sensing of microseismicity, strain and temperature during hydraulic fracturing. *Geophysics*, *84*(1), D11–D23. doi:10.1190/geo2017-0396.1

Karrenbach, M., Kahn, D., Cole, S., Ridge, A., Boone, K., Rich, J., et al. (2017). Hydraulic-fracturing-induced strain and microseismic using in situ distributed fiber-optic sensing. *The Leading Edge*, *36*(10), 837–844. doi:10.1190/tle36100837.1

Kennett, B. L. N. (2022). The seismic wavefield as seen by distributed acoustic sensing arrays: local, regional and teleseismic sources. *Proceedings of the Royal Society A: Mathematical, Physical and Engineering Sciences*, *478* (2258), 20210812. doi:10.1098/rspa.2021.0812

Kennett, B. L. N., & Engdahl, E. R. (1991). Traveltimes for global earthquake location and phase identification. *Geophysical Journal International*, *105*(2), 429–465. doi:10.1111/j.1365-246X.1991.tb06724.x

Klaasen, S., Paitz, P., Lindner, N., Dettmer, J., & Fichtner, A. (2021). Distributed acoustic sensing in volcano-glacial environments—mount meager, British Columbia. *Journal of Geophysical Research: Solid Earth*, *126*(11). doi:10.1029/2021JB022358

Kobayashi, Y., Uematsu, Y., Mochiji, S., & Xue, Z. (2020). A field experiment of walkaway distributed acoustic sensing vertical seismic profile in a deep and deviated onshore well in Japan using a fibre optic cable deployed inside coiled tubing. *Geophysical Prospecting*, *68*(2), 501–520. doi:10.1111/1365-2478.12863

Kuvshinov, B. N. (2016). Interaction of helically wound fibre-optic cables with plane seismic waves. *Geophysical Prospecting*, *64*(3), 671–688. doi:10.1111/1365-2478.12303

Lellouch, A., & Biondi, B. L. (2021). Seismic applications of downhole das. *Sensors*, *21*(9). doi:10.3390/s21092897

Lellouch, A., Lindsey, N. J., Ellsworth, W. L., & Biondi, B. L. (2020). Comparison between distributed acoustic sensing and geophones: downhole microseismic monitoring of the FORGE geothermal experiment. *Seismological Research Letters*, *91*(6), 3256–3268. doi:10.1785/0220200149

Lellouch, A., Schultz, R., Lindsey, N. J., Biondi, B. L., & Ellsworth, W. L. (2021). Low-magnitude seismicity with a downhole distributed acoustic sensing array—examples from the FORGE geothermal experiment. *Journal of Geophysical Research: Solid Earth*, *126*(1), e2020JB020462. doi:10.1029/2020JB020462

Lellouch, A., Yuan, S., Ellsworth, W. L., & Biondi, B. (2019a). Velocity-based earthquake detection using downhole distributed acoustic sensing—examples from the San Andreas fault observatory at depth. *Bulletin of the Seismological Society of America*, *109*(6), 2491–2500. doi:10.1785/0120190176

Lellouch, A., Yuan, S., Spica, Z., Biondi, B., & Ellsworth, W. L. (2019b). Seismic velocity estimation using passive downhole distributed acoustic sensing records: examples from the San Andreas fault observatory at depth. *Journal of Geophysical Research: Solid Earth*, *124*(7), 6931–6948. doi:10.1029/2019JB017533

Lellouch, A., Luo, B., Huot, F., Clapp, R. G., Given, P., Biondi, E., et al. (2022). Microseismic analysis over a single horizontal distributed acoustic sensing fiber using guided waves. *Geophysics*, *87*(3), KS83–KS95. doi:10.1190/geo2021-0418.1

Li, Z., & Zhan, Z. (2018). Pushing the limit of earthquake detection with distributed acoustic sensing and template matching: a case study at the Brady geothermal field. *Geophysical Journal International*, *215*(3), 1583–1593. doi:10.1093/gji/ggy359

Li, Z., Shen, Z., Yang, Y., Williams, E., Wang, X., & Zhan, Z. (2021). Rapid response to the 2019 ridgecrest earthquake with distributed acoustic sensing. *AGU Advances*, *2*(2), e2021AV000395. doi:10.1029/2021AV000395

Lim Chen Ning, I., & Sava, P. (2018). High-resolution multi-component distributed acoustic sensing. *Geophysical Prospecting*, *66*(6), 1111–1122. doi:10.1111/1365-2478.12634

Lindsey, N. J., Dawe, C. T., & Ajo-Franklin, J. B. (2019). Illuminating seafloor faults and ocean dynamics with dark fiber distributed acoustic sensing. *Science*, *366*(6469), 1103–1107. doi:10.1126/science.aay5881

Lindsey, N. J., Martin, E. R., Dreger, D. S., Freifeld, B., Cole, S., James, S. R., et al. (2017). Fiber-optic network observations of earthquake wavefields. *Geophysical Research Letters*, *44*(11), 792–799. doi:10.1002/2017GL075722

Lindsey, N. J., Rademacher, H., & Ajo-Franklin, J. B. (2020). On the broadband instrument response of fiber-optic DAS arrays. *Journal of Geophysical Research: Solid Earth*, *125* (2), 1–16. doi:10.1029/2019JB018145

Lior, I., Rivet, D., Ampuero, J.-P., Sladen, A., Barrientos, S., Sánchez-Olavarría, R., et al. (2023). Magnitude estimation and ground motion prediction to harness fiber optic distributed acoustic sensing for earthquake early warning. *Scientific Reports*, *13*(1), 1. doi:10.1038/s41598-023-27444-3

Lior, I., Sladen, A., Rivet, D., Ampuero, J., Hello, Y., Becerril, C., et al. (2021). On the detection capabilities of underwater DAS. *Journal of Geophysical Research: Solid Earth*, *126*, 1–20. doi:10.1029/2020jb020925

Liu, Y., Huff, O., Luo, B., Jin, G., & Simmons, J. (2022). Convolutional neural network-based classification of microseismic events originating in a stimulated reservoir from distributed acoustic sensing data. *Geophysical Prospecting*, *70*(5), 904–920. doi:10.1111/1365-2478.13199

Luo, B., Jin, G., & Stanek, F. (2021). Near-field strain in distributed acoustic sensing-based microseismic observation. *Geophysics*, *86*(5), P49–P60. doi:10.1190/geo2021-0031.1

Martin, E. R., Castillo, C. M., Cole, S., Sawasdee, P. S., Yuan, S., Clapp, R., et al. (2017). Seismic monitoring leveraging existing telecom infrastructure at the SDASA: active, passive, and ambient-noise analysis. *The Leading Edge*, *36*(12), 1025–1031. doi:10.1190/tle36121025.1

Maxwell, S. (2014). *Microseismic imaging of hydraulic fracturing: Improved engineering of unconventional shale reservoirs*. SEG Distinguished Instructor Series No. 17. Society of Exploration Geophysicists. doi:10.1190/1.9781560803164

Mellors, R. J., Abbott, R., Steedman, D., Podrasky, D., & Pitarka, A. (2021). Modeling subsurface explosions recorded on a distributed fiber optic sensor. *Journal of Geophysical Research: Solid Earth*, *126*(12), e2021JB022690. doi:10.1029/2021JB022690

Molteni, D., Williams, M. J., & Wilson, C. (2017). Detecting microseismicity using distributed vibration. *First Break*, *35* (4), 51–55. doi:10.3997/1365-2397.35.4.87841

Mondanos, M., & Coleman, T. (2019). Application of distributed fibre-optic sensing to geothermal reservoir characterization and monitoring. *First Break*, *37*(7), 51–56. doi:10.3997/1365-2397.n0040

Mousavi, S. M., Ellsworth, W. L., Zhu, W., Chuang, L. Y., & Beroza, G. C. (2020). Earthquake transformer—An attentive deep-learning model for simultaneous earthquake detection and phase picking. *Nature Communications*, *11*(1), 1. doi:10.1038/s41467-020-17591-w

Paitz, P., Edme, P., Gräff, D., Walter, F., Doetsch, J., Chalari, A., et al. (2021). Empirical investigations of the instrument response for distributed acoustic sensing (Das) across 17 octaves. *Bulletin of the Seismological Society of America*, *111*(1), 1–10. doi:10.1785/0120200185

Pevzner, R., Gurevich, B., Pirogova, A., Tertyshnikov, K., & Glubokovskikh, S. (2020). Repeat well logging using earthquake wave amplitudes measured by distributed acoustic sensors. *Leading Edge*, *39*(7), 513–517. doi:10.1190/tle39070513.1

Shapiro, N. M., Campillo, M., Stehly, L., & Ritzwoller, M. H. (2005). High-resolution surface-wave tomography from ambient seismic noise. *Science*, *307*(5715), 1615–1618. doi:10.1126/science.1108339

Shashkin, P., Gurevich, B., Yavuz, S., Glubokovskikh, S., & Pevzner, R. (2022). Monitoring injected CO_2 using earthquake waves measured by downhole fibre-optic sensors: CO2CRC Otway stage 3 case study. *Sensors*, *22*(20), 20. doi:10.3390/s22207863

Shearer, P. M., & Orcutt, J. A. (1987). Surface and near-surface effects on seismic waves—theory and borehole seismometer results. *Bulletin of the Seismological Society of America*, *77* (4), 1168–1196. doi:10.1785/BSSA0770041168

Shulakova, V., Tertyshnikov, K., Pevzner, R., Kovalyshen, Y., & Gurevich, B. (2022). Ambient seismic noise in an urban environment: case study using downhole distributed acoustic sensors at the Curtin University campus in Perth, Western Australia. *Exploration Geophysics*, *53*, 620–633. doi:10.1080/08123985.2021.2021802

Sidenko, E., Tertyshnikov, K., Gurevich, B., Isaenkov, R., Ricard, L. P., Sharma, S., et al. (2022). Distributed fiber-optic sensing transforms an abandoned well into a permanent geophysical monitoring array: a case study from Australian South West. *The Leading Edge*, *41*(2), 140–148. doi:10.1190/tle41020140.1

Spica, Z. J., Ajo-Franklin, J., Beroza, G. C., Biondi, B., Cheng, F., Gaite, B., et al. (2023). PubDAS: A PUBlic distributed acoustic sensing datasets repository for geosciences. *Seismological Research Letters*, *94*(2A), 983–998. doi:10.1785/0220220279

Stork, A. L., Baird, A. F., Horne, S. A., Naldrett, G., Lapins, S., Kendall, J. M., et al. (2020). Application of machine learning to microseismic event detection in distributed acoustic sensing data. *Geophysics*, *85*(5), KS149–KS160. doi:10.1190/geo2019-0774.1

Tian, Y., Lin, T., Li, Y., & Wu, N. (2022a). Efficient SPSNet for downhole weak DAS signals recovery. *IEEE Geoscience and Remote Sensing Letters*, *19*, 1–5. doi:10.1109/LGRS.2022.3226759

Tian, Y., Sui, J., Li, Y., Wu, N., & Shao, D. (2022b). A novel iterative PA-MRNet: multiple noise suppression and weak signals recovery for downhole DAS data. *IEEE Transactions on Geoscience and Remote Sensing*, *60*, 1–14. doi:10.1109/TGRS.2022.3170635

University of Utah Seismograph Stations. (2022). Utah FORGE DAS seismic data 2022. Retrieved from https://gdr.openei.org/submissions/1393

van den Ende, M. P. A., & Ampuero, J.-P. (2021). Evaluating seismic beamforming capabilities of distributed acoustic sensing arrays. *Solid Earth*, *12*(4), 915–934. doi:10.5194/se-12-915-2021

Vera Rodriguez, I., & Wuestefeld, A. (2020). Strain microseismics: radiation patterns, synthetics, and moment tensor resolvability with distributed acoustic sensing in isotropic media. *Geophysics*, *85*(3), KS101–KS114. doi:10.1190/geo2019-0373.1

Verdon, J. P., Horne, S. A., Clarke, A., Stork, A. L., Baird, A. F., & Kendall, J.-M. (2020). Microseismic monitoring using a fibre-optic distributed acoustic sensor (DAS) array. *Geophysics*, *85*(3), KS89–KS99. doi:10.1190/geo2019-0752.1

Wamriew, D., Pevzner, R., Maltsev, E., & Pissarenko, D. (2021). Deep neural networks for detection and location of microseismic events and velocity model inversion from microseismic data acquired by distributed acoustic sensing array. *Sensors*, *21*(19), 6627. doi:10.3390/s21196627

Wang, H. F., Zeng, X., Miller, D. E., Fratta, D., Feigl, K. L., Thurber, C. H., & Mellors, R. J. (2018). Ground motion response to an ML 4.3 earthquake using co-located distributed acoustic sensing and seismometer arrays. *Geophysical Journal International*, *213*(3), 2020–2036. doi:10.1093/GJI/GGY102

Webster, P., Molenaar, M., & Perkins, C. (2016). DAS microseismic. *CSEG Recorder*, *41*(6), 38–39.

Webster, P., Wall, J., Perkins, C., & Molenaar, M. (2013). *Micro-seismic detection using distributed acoustic sensing*. Paper presented at the Society of Exploration Geophysicists International Exposition and 83rd Annual Meeting, 2459–2463. doi:10.1190/segam2013-0182.1

Willacy, C., Van Dedem, E., Minisini, S., Li, J., Blokland, J. W., Das, I., & Droujinine, A. (2019). Full-waveform event location and moment tensor inversion for induced seismicity. *Geophysics*, *84*(2), KS39–KS57. doi:10.1190/geo2018-0212.1

Williams, E. F., Fernández-Ruiz, M. R., Magalhaes, R., Vanthillo, R., Zhan, Z., González-Herráez, M., & Martins, H. F. (2019). Distributed sensing of microseisms and teleseisms with submarine dark fibers. *Nature Communications*, *10*(1), 1–11. doi:10.1038/s41467-019-13262-7

Willis, M. E. (2022). *Distributed acoustic sensing for seismic measurements – what geophysicists and engineers need to know*. Society of Exploration Geophysicists. doi:10.1190/1.9781560803850

Yang, L., Fomel, S., Wang, S., Chen, X., Chen, W., Saad, O. M., & Chen, Y. (2023). Denoising of distributed acoustic sensing data using supervised deep learning. *Geophysics*, *88*(1), WA91–WA104. doi:10.1190/geo2022-0138.1

Yu, C., Zhan, Z., Lindsey, N. J., Ajo-Franklin, J. B., & Robertson, M. (2019). The potential of DAS in teleseismic studies: insights from the goldstone experiment. *Geophysical Research Letters*, *46*(3), 1320–1328. doi:10.1029/2018GL081195

Zhu, W., & Beroza, G. C. (2018). PhaseNet: a deep-neural-network-based seismic arrival-time picking method. *Geophysical Journal International*, *216*(1), 261–273. doi:10.1093/gji/ggy423

25

Investigation of Distributed Acoustic Sensing Characteristics by Complementary Observations in Geothermal Wells and by Surface Seismometers

Yoko Hasada[1,2], Junzo Kasahara[1,3], Hitoshi Mikada[4], and Hiroshi Ohnuma[1]

ABSTRACT

We evaluated the optical fiber distributed acoustic sensing (DAS) system to obtain the subsurface structure of geothermal fields by seismic measurements in geothermal wells, through complementary observation with surface seismometers. Deploying the optical fiber in existing geothermal wells, we obtained good coupling to the casing and the area surrounding the well for receiving seismic arrivals. Many natural earthquakes were recorded in different geothermal wells and fields. The lowest frequency of DAS waveforms for a large earthquake was 0.25 Hz. By comparing DAS and surface seismometer records, we identified P-to-S and S-to-P conversion phases. In DAS records, we observed the conversion by the intersection of the S and S-to-P phases with different apparent velocities. For the surface seismometers, we used the time differences between the vertical and horizontal components to estimate the depth of the conversion zone. Observation of the converted waves by DAS is useful for estimating seismic impedance contrasts in the subsurface of geothermal fields, which could be related to fluid contents of geothermal reservoirs.

25.1. INTRODUCTION

Seismometers (or geophones) have long been used as conventional seismic sensors to detect ground vibrations and different types of seismometers are used to measure the velocity, acceleration, or displacement. Over a decade ago, the use of optical fiber distributed acoustic sensing (DAS) (Hartog, 2017) became widespread. The advantage of this technology is its extremely dense sensing. During measurements, we can choose a gauge length of roughly 15 m and focus on the objective geothermal zone by backpropagating the observed seismic signals toward the scattered zones. Because DAS is a dense sensing array, we have a dense source array for backpropagation. Backpropagation enables reverse time migration, waveform inversion, and beam forming. However, we are still determining the difference in characteristics between seismometers and DAS. This chapter provides information on this subject and also discusses the P-to-S and S-to-P seismic phase conversions in the observed seismograms.

[1]*Engineering Advancement Association of Japan, Tokyo, Japan*

[2]*Daiwa Exploration and Consulting Co., Ltd., Tokyo, Japan*

[3]*Center for Integrated Research and Education of Natural Hazards, Shizuoka University, Shizuoka, Japan*

[4]*Department of Civil and Earth Resources Engineering, Kyoto University, Kyoto, Japan*

25.2. COMPARISONS OF DAS AND SEISMOMETERS ON THE GROUND SURFACE

In 2017, we carried out an evaluation test to compare DAS and seismometers on the ground at the Abiko site of the Central Research Institute of Electric Power Industry (Hasada et al., 2018; Kasahara et al., 2018). We used

Distributed Acoustic Sensing in Borehole Geophysics, Geophysical Monograph 289, First Edition.
Edited by Yingping Li, Robert Mellors, and Ge Zhan.

DOI: 10.1002/9781394179275.ch25

the Schlumberger heterodyne distributed vibration sensing (hDVS) system as the interrogator equipment. The DAS records were obtained with a spacing of 1 m, but the gauge length was 4 m. The DAS used three 100-m-long single-mode optical fibers parallel to each other and buried 20 cm in the ground. Vertical geophones (4.5 Hz) with 1 m spacing and three-component (3C) seismometers (1 or 0.2 Hz) with 5–10 m spacing were deployed along the fibers (Figure 25.1(a)).

DAS and seismometers (geophones) measure the strain or strain rate and particle motions (acceleration, velocity, or displacement), respectively. Figure 25.1(b) shows the results of DAS and the 4.5 Hz vertical geophone recorded after a sledgehammer strike. The DAS records showed weak P-wave and precise S-wave arrivals because the hammer input caused an up-down (UD) motion, and DAS was sensitive to the horizontal strain or strain rate in the direction of the fiber elongation. Although DAS detected S-waves, the UD geophones were not sensitive to horizontal motion. The DAS with a 4 m gauge length was not sensitive to the surface waves. The sensitivity of the 4.5 Hz geophone was 25 V/m/s.

We compared DAS records and those of the 3C seismometers having a sensitivity of 400 V/m/s for an M = 3.1 earthquake in northern Ibaraki (Kasahara et al., 2018), shown in Figure 25.2. Table 25.1 lists the earthquakes in Japan mentioned in this chapter. The epicentral distance was 112 km from Abiko in the N25°E direction,

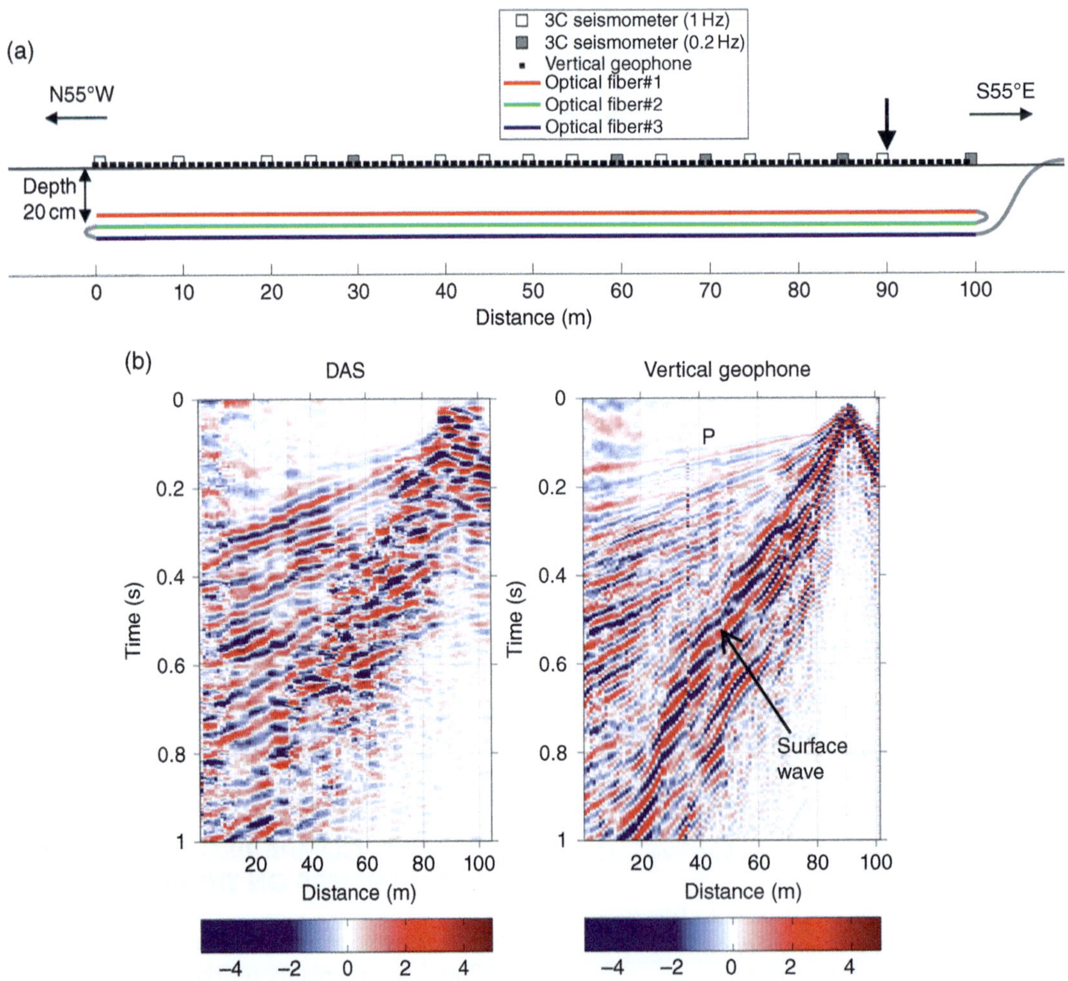

Figure 25.1 (a) Schematic diagram of the observation layout. The horizontal axis is the distance along the survey line. Small black squares and large white/gray squares denote the positions of the vertical geophones and 3C seismometers, respectively. The three optical fibers were buried at a depth of 20 cm. The down arrow is the source location. (b) A comparison of DAS (left) and 4.5 Hz geophone (right) records of the sledgehammer impact at a 90 m distance along the 100 m long line. The vertical axis is the travel time in seconds, and the horizontal axis is the distance from the end in meters. The color shows the normalized amplitude. Kasahara et al. (2018) / with permission of the Society of Exploration Geophysicists.

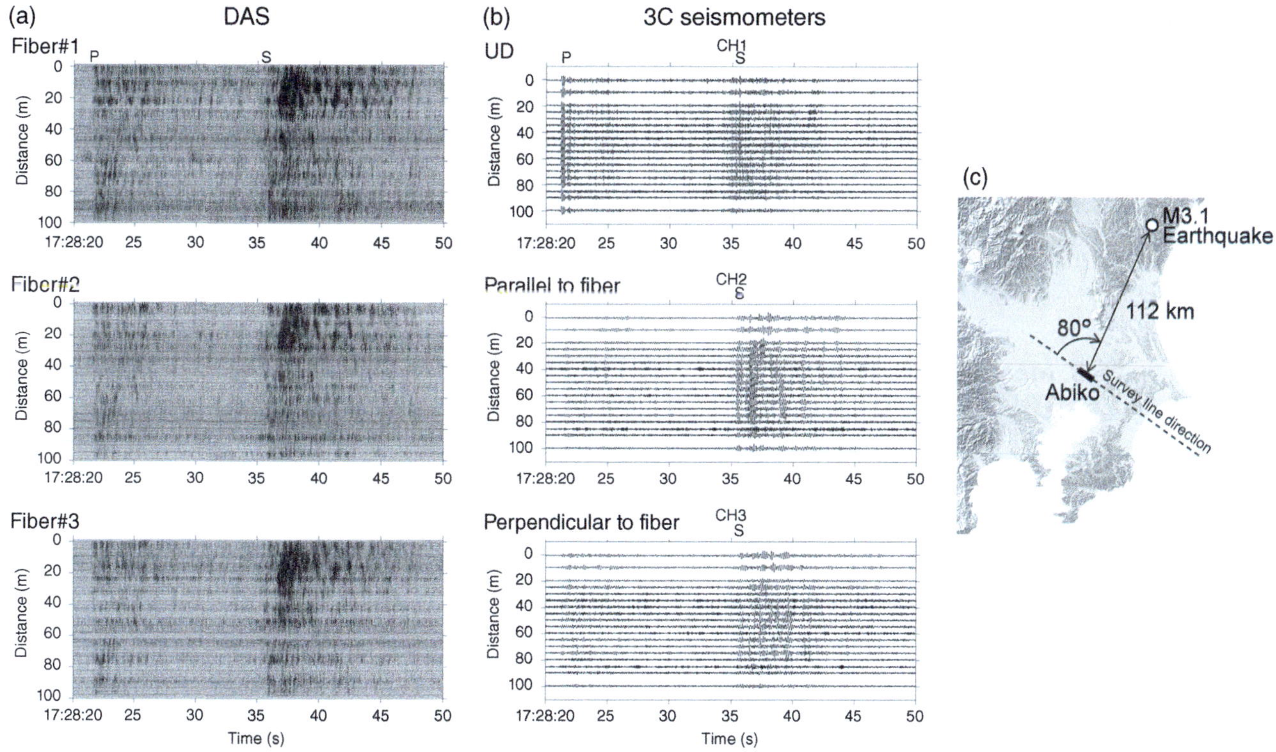

Figure 25.2 Comparison of the records of the three-optical-fiber DAS (a) and 3C 5 s seismometer (b) for the M = 3.1 earthquake in northern Ibaraki. Kasahara et al. (2018) / with permission of the Society of Exploration Geophysicists. (c) The locations of the Abiko site near Tokyo and the epicenter. The origin time was 2017/09/23 17:28:02.2. The epicentral distance was 112 km. The direction of the fiber and the earthquake were N55°W and N25°E, respectively. Adapted the Geospatial Information Authority of Japan (GSI).

Table 25.1 List of earthquakes.

Site	Interrogator	Epicenter	Magnitude	Epicentral distance (km)	Focal depth (km)	Origin time (JST)
Abiko	Schlumberger hDVS	Northern Ibaraki Pref.	3.1	112	9	2017/9/23 17:28
Medipolis	Schlumberger hDVS	Near Tanegashima Island	5.2	123	100	2018/11/21 4:10
Sumikawa	OptaSense ODH-4	East. Off Aomori Pref.	5.4	250	38	2021/10/19 21:36
		(microearthquake)	–	3	3	2021/10/19 7:11
Takigami	OptaSense ODH-4	A: NE Kumamoto Pref.	1.7	19	7	2022/5/15 12:34
		B: Western Oita Pref.	1.2	12	6	2022/5/13 12:43
		C: Hyuganada Region	3	92	42	2022/5/18 23:44
		D: NW Kumamoto Pref.	2.9	78	10	2022/5/31 7:42

and the seismic waves were expected to arrive roughly perpendicular to the survey line. The DAS seismograms obtained by the three parallel fibers closely resembled each other. The DAS records showed the P and S arrivals because the optical fibers were sensitive to horizontal motion along the fiber length. The UD component of the seismometers detected the P phase. The horizontal components parallel to the fibers detected the S arrivals. The sensitivities of the three fibers were approximately the same and enough to record the arrival phases detected by the seismometers. The results of this test encouraged us to use optical fiber DAS for seismic measurements.

The particle velocity recorded by the seismometers was converted to the strain rate, confirming the resemblance between the waveforms from the seismometers and DAS for the S parts of the earthquake event (Hasada et al., 2018; Kasahara et al., 2019; Kasahara et al., 2025, chapter 10). Wang et al. (2018) also compared DAS and seismometer records using test data from Brady Hot Springs, Nevada, in 2016, and obtained results similar to ours.

25.3. MEASUREMENTS IN GEOTHERMAL WELLS

25.3.1. First Experiment at the Medipolis Geothermal Field

In the past decade, many researchers have used DAS for vertical seismic profiling (VSP) studies (e.g., Johannessen et al., 2012; Madsen et al., 2013; Mateeva et al., 2014; Miller et al., 2012, 2016). The traditional VSP seismic measurements use 3C geophones with arms attached to the wall. On the other hand, we used pre-existing wells for DAS measurements, and the optical fibers were laid down along the well casing without any cementing, so we wondered whether the seismic coupling to the surrounding ground would be sufficient for the VSP study.

As the first objective of the geothermal investigations, we deployed optical fibers by simply dropping them into pre-existing geothermal wells in an attempt to observe natural earthquakes. In 2018, we used the IK-4 well at the Medipolis geothermal field in the southern part of Kyushu, Japan. The DAS records of the M = 5.2 earthquake near Tanegashima (Table 25.1, origin time: 21 November 2018, 04:10 JST) are shown in Figure 25.3 (Kasahara et al., 2018). The bottom of DAS measurements was at a depth of 977 m. We could see clear P-wave first arrivals from the bottom to the wellhead. Strange noises were detected at a depth of 350 m, and between 350 and 450 m, some kick-like signals were identified. The P-wave first arrivals were propagated from the 977 m depth to the wellhead at an apparent velocity of 3.5 km/s. We confirmed good coupling to the casing even though we only deployed the fiber system in a mostly vertical geothermal well. This observation strongly suggests that DAS measurements in a pre-existing well provide a reasonable signal-to-noise ratio.

Figure 25.4 shows an enlargement of the P-wave first arrivals of the same M = 5.2 earthquake near Tanegashima. Distributed temperature sensing (DTS) detected the highest temperature of 264 °C at a depth of 914 m.

As the second objective of the study, we compared DAS and seismometer records in several geothermal studies. During DAS seismic survey at the Medipolis geothermal field in 2018, we used 20 sets of 3C 1 Hz surface seismometers near the IK-4 well in addition to DAS. DAS and surface seismometer records for the M = 5.2 earthquake near Tanegashima are shown in Figure 25.5 (Kasahara et al., 2018). For this earthquake, the UD component of the seismometers shows the P-wave arrivals,

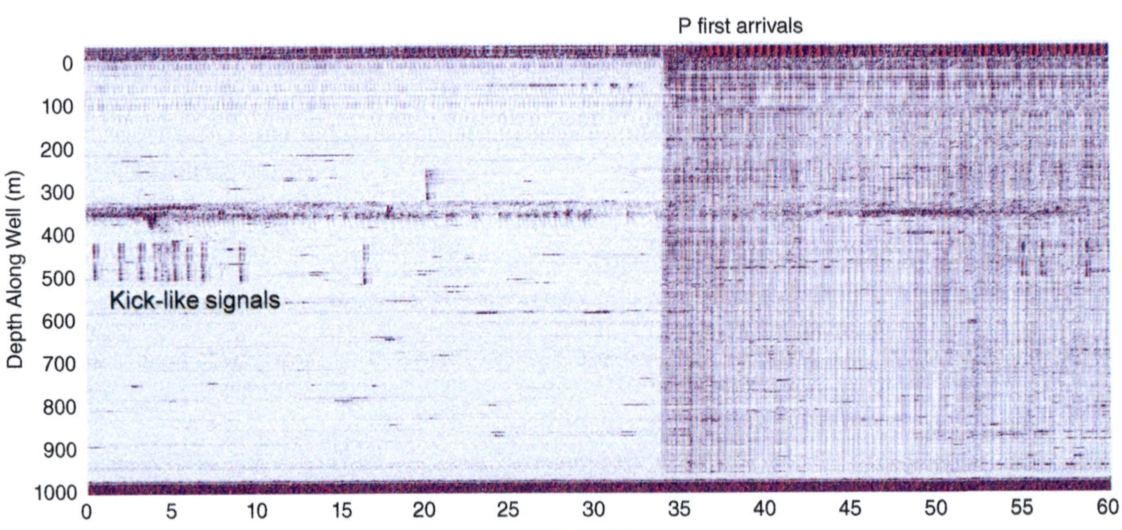

Figure 25.3 DAS records of the M = 5.2 earthquake near Tanegashima (21 November 2018, 04:10 JST). The focal depth and epicentral distance were 100 and 123 km, respectively. The vertical and horizontal axes are the depth in meters along the well and the elapsed time, respectively.

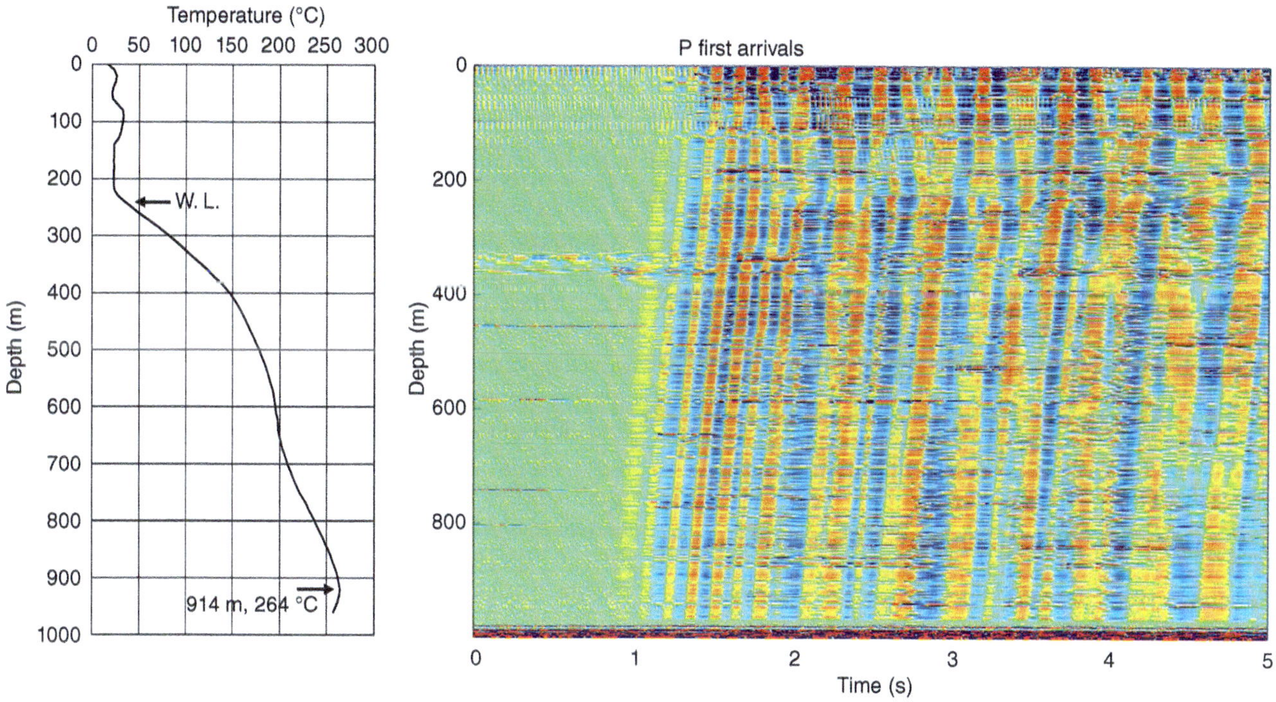

Figure 25.4 (Right) Enlargement of DAS records of the M = 5.2 earthquake near Tanegashima (21 November 2018, 04:10 JST) (Adapted from Kasahara et al., 2019). (Left) Temperature profile. The vertical axis is the depth in meters. The well above 240 m in depth was filled with air, and the well below 240 m was filled with hot water. The highest temperature was 264 °C, at a depth of 914 m.

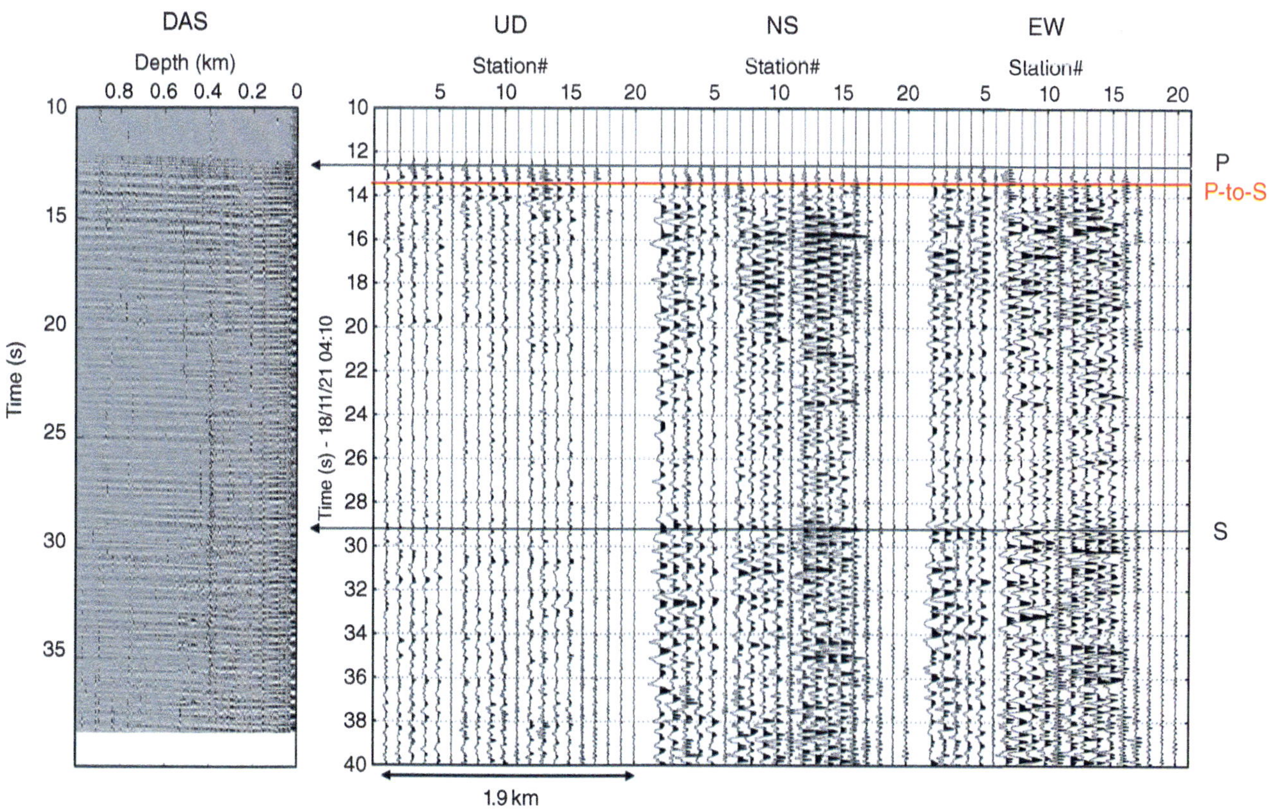

Figure 25.5 Comparison of DAS (left) and surface seismometer (right) records. The records for twenty 3C 1 Hz surface seismometers with 100 m spacing are shown for the UD, NS, and EW components. Clear P-to-S conversion phases arrived on the NS and EW components with larger amplitudes than the P-wave first arrivals on the UD components.

and DAS records at the wellhead fit these arrivals. The horizontal components of the surface seismometers show the P-to-S conversion arrivals, which appear 0.8 s after the P-wave first arrivals. The S-wave arrivals came roughly 16 s later than the P-wave arrivals. The P-to-S conversion arrivals have larger amplitudes than the P-wave first arrivals on the UD component. The simulation study suggested that the P-to-S phase can be explained as being generated at a depth of 3.6 km (Kasahara et al., 2019).

The study conducted in 2019 using the same IK-4 well and an active seismic source showed that DAS records at a 1.6 km offset distance had a PP reflection from a 3.6 km depth estimated as a zone with $Vp/Vs > 2$ (Kasahara et al., 2019). The zone at the 3.6 km depth is the location of the P-to-S conversion source and possibly includes some supercritical water in the rocks.

25.3.2. Examples of Earthquake Records at the Sumikawa Geothermal Field

In 2021, we carried out a DAS seismic study at the Sumikawa geothermal field located in Akita prefecture, northern Honshu, Japan (Kasahara et al., 2022b). We used the SE-4 injection well, cased to a depth of 1,200 m. We used OptaSense ODH-4 interrogator equipment with a 16 m gauge length and 25 kHz ping rates.

We have yet to learn how DAS technology can measure the low-frequency components in geothermal wells. It would be very useful if it could measure low-frequency components such as broadband seismometers. Figure 25.6 is an example of DAS records from the SE-4 geothermal well, where the 2021 M = 5.4 natural earthquake was detected at an epicentral distance of 250 km and focal depth of 38 km (Table 25.1). We compared the frequency spectra of 60 s just before the earthquake (Figure 25.6(a)) and during the earthquake (Figure 25.6(b)). Significant spectral amplitudes at 0.25 Hz can be identified in the DAS records in Figure 25.6(b), and the P-wave arrivals can be seen in Figure 25.6(c). Based on this observation, we can expect to detect seismic signals >0.25 Hz by DAS. Fichtner et al. (2022) observed a volcanic tremor with a peak frequency of 0.22 Hz using DAS deployed on a floating ice sheet, which behaved as a resonator. Their study and our observation in the geothermal well indicate the

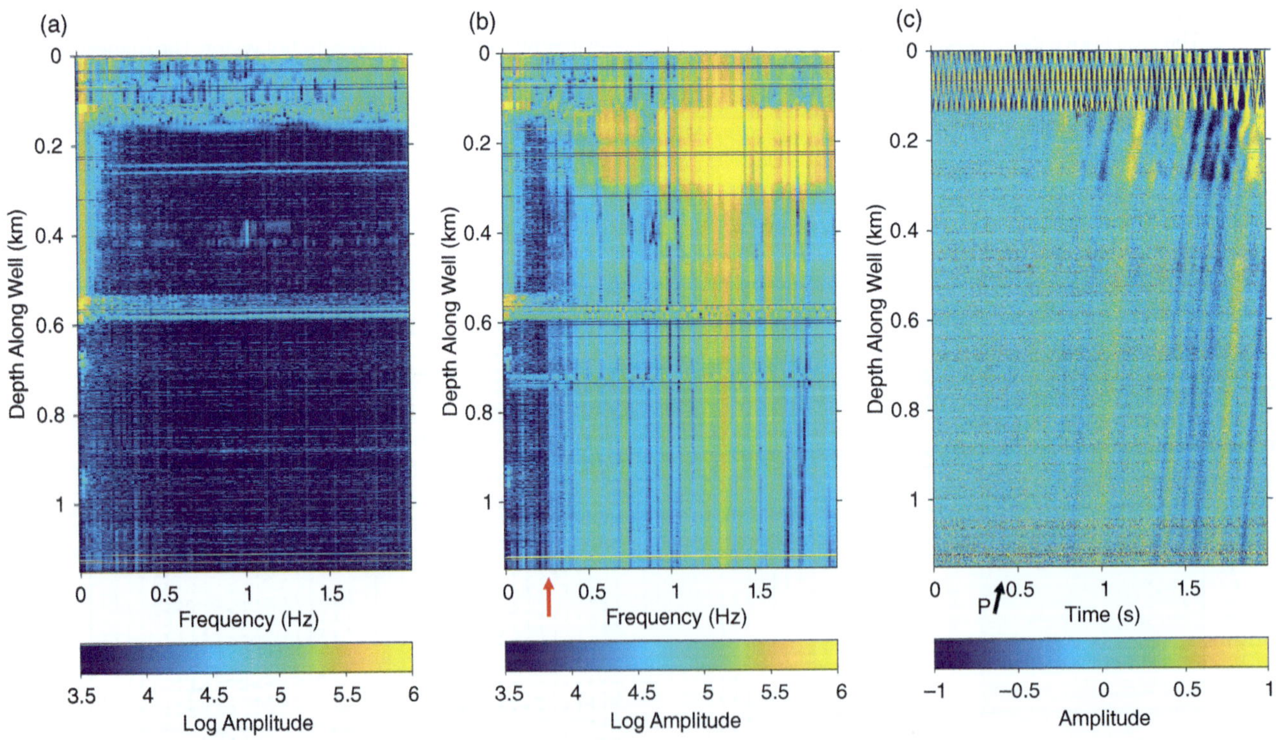

Figure 25.6 DAS records of the M = 5.4 earthquake in 2021 observed in the SE-4 geothermal well at the Sumikawa geothermal field. (a) Amplitude spectra of 60 s DAS records before the earthquake. (b) Amplitude spectra of 60 s DAS records containing the earthquake. (c) P-wave first arrival portion (black arrow) of the earthquake. Horizontal axes of (a–c) are the frequency in Hz and the time in seconds, respectively. The vertical axis is the depth in kilometers. The depth of the well is approximately 1.2 km. DAS records for the strain rate showed the spectral amplitude at a frequency > 0.25 Hz (red arrow).

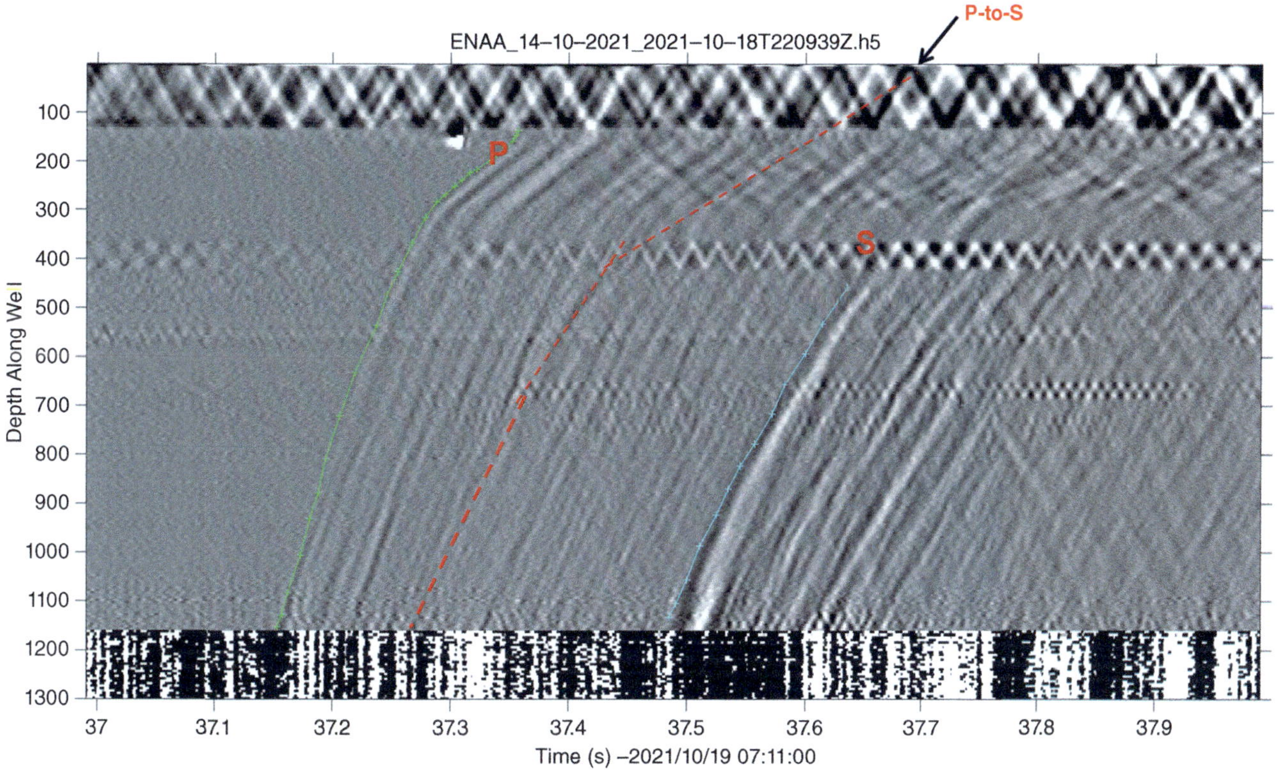

Figure 25.7 DAS record of the microearthquake event on 10 October 2021 07:11:36.378. The vertical and horizontal axes are the depth in meters and the elapsed time in seconds, respectively. We can identify the phases of the P (green line) and S (light blue line). The broken red line is an explanation of the P-to-S conversion phase.

capability of DAS to detect low-frequency events in volcanic or geothermal areas.

Figure 25.7 is the case of a microearthquake for which the S-P time was approximately 0.4 s, suggesting that the hypocenter was near the well (Table 25.1). Based on this observation, the P-to-S conversion depth could be 2 km. The *Vp* decreased at a depth above 300 m. The surface seismometer records in Figure 25.8 indicate clear P-to-S arrivals at stations #24 to #26.

25.3.3. S-to-P Conversion Observed at the Takigami Geothermal Field

In May 2022, we carried out a DAS seismic study in the TP-2 geothermal well at Takigami, located in Kokonoe in northeastern Kyushu, Japan. We installed the optical fiber system to a depth of 2,000 m in the TP-2 well, which has a heading of S45°W. We used ODH-4 OptaSense interrogator equipment with 16 m gauge length and 25 kHz ping rates in the well and 26 sets of 3C 1 Hz ground surface seismometers on the 1.25-km-long survey line near the well. We compared DAS and surface seismometer records for four earthquakes (Table 25.1, A to D in Figure 25.9) during the Takigami survey. The structural analysis of the Takigami DAS seismic study is described in a separate chapter in this monograph (Kasahara et al., 2025).

The first example is earthquake A (M = 1.7) in Figure 25.10. In this earthquake record, we can identify two S-to-P arrivals: one was converted at deeper than 2 km, and the other was converted at 1 km, shown by a white circle. Based on the detailed analysis of the Takigami DAS data, we found that 1 km is a distinct structural interface and 2.5 km is a reflective zone, possibly a hydrothermal reservoir (Kasahara et al., 2025).

Figure 25.11 shows earthquake B (13 May 202212:43:50 JST. M = 1.2, focal depth of 6 km). P-wave first arrivals could be more precise using the UD component of the surface seismometers, but they are clear on the DAS records. The S-to-P1 arrivals on UD are earlier than the S-wave arrivals on the NS and EW components, and these phases are initiated at a depth of 1 km.

Earthquake C is shown in Figure 25.12. The S part of the DAS records (Figure 25.12(b)) shows similar intense arrivals after the S-wave first arrivals. In addition, weak S-to-P arrivals could be identified.

The case of earthquake D is simple (Figure 25.13). The S-to-P arrivals appear to be weak.

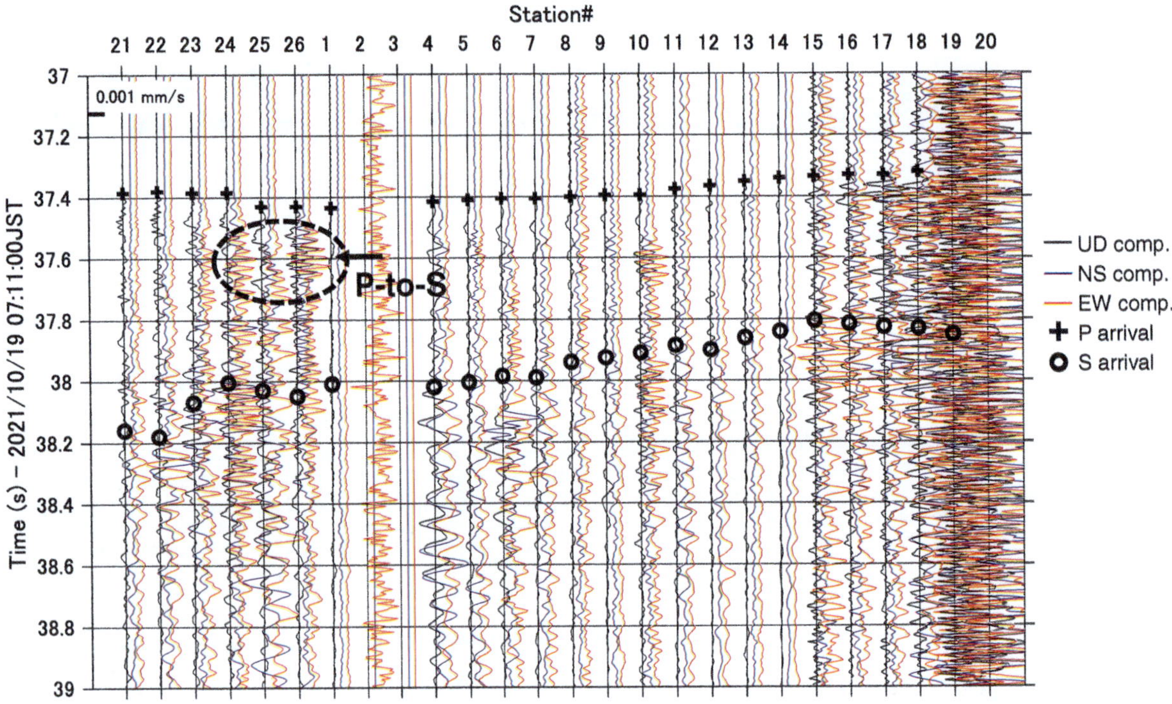

Figure 25.8 Twenty-six traces of 3C 1 Hz surface seismometer records for the microearthquake event on 10 October 2021 07:11:36.378. The horizontal components (NS in blue and EW in red) show the P-to-S conversion phase (indicated by the dashed ellipsoid and black arrow) approximately 0.18 s after the P-wave first arrivals (EW components of stations #24, #25, and #26). UD components are indicated by black lines.

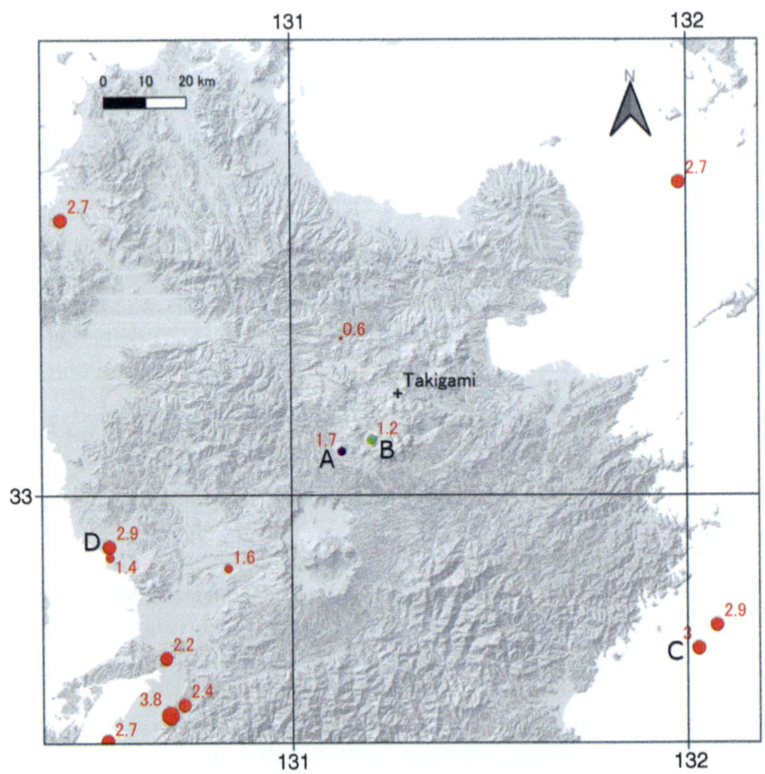

Figure 25.9 Location map of the TP-2 well (+) at the Takigami geothermal field and natural earthquake events (solid circles) with numbers denoting the magnitude. Epicenters for waveform comparison are indicated by A (M = 1.7), B (M = 1.2), C (M = 3.0), and D (M = 2.9) earthquakes. Map adapted from GSI.

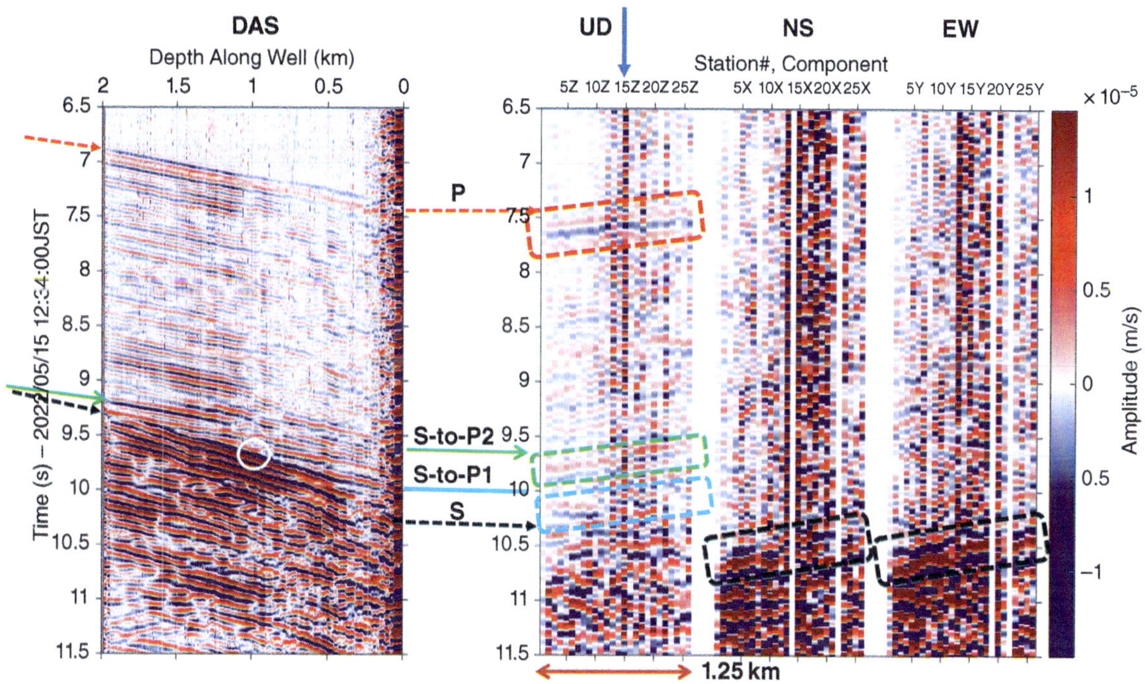

Figure 25.10 DAS records (left) and seismometer records (right) of earthquake A, 15 May 2022. M = 1.7, focal depth of 7 km. The horizontal axis of the left and right diagrams are the depth in kilometers and the station number, respectively. The DAS wellhead is close to the #15 surface seismometer station (blue down arrow). Dashed rectangles are P-wave first arrivals (red) and S-wave arrivals (black). Green dashed rectangles denote the S-to-P2 conversion at a depth of approximately 2.5 km. Blue dashed rectangles denote the S-to-P1 conversion at a depth of 1 km, with the white circle showing the conversion depth.

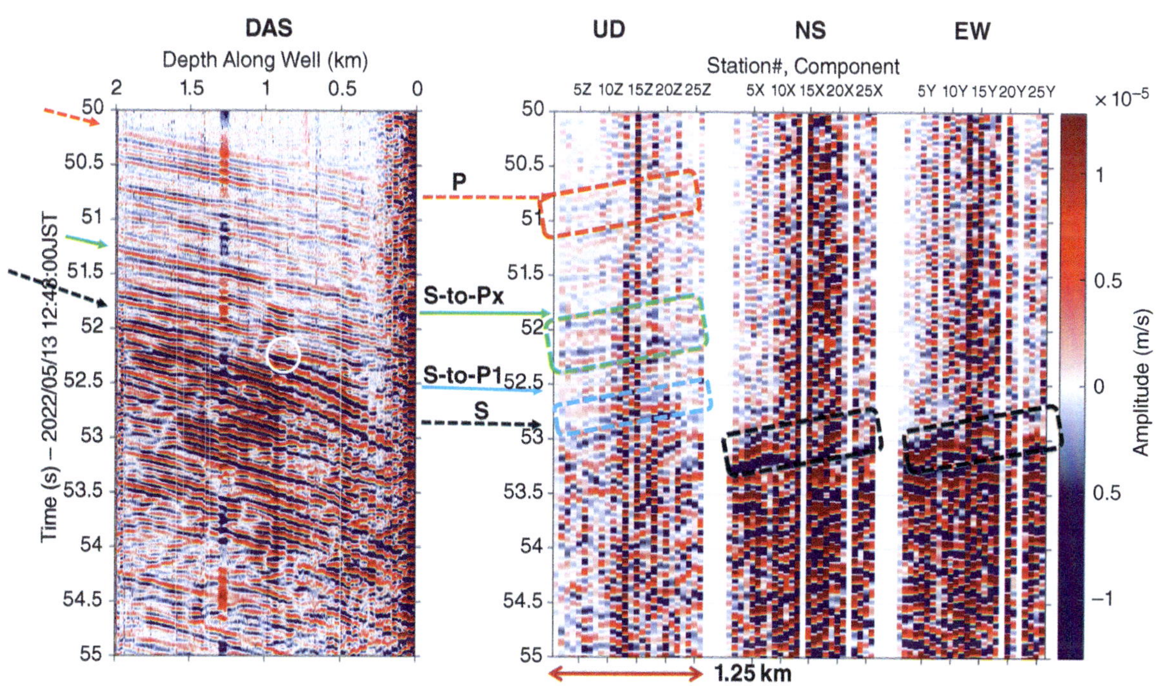

Figure 25.11 DAS (left) and surface seismometer (right) records of earthquake B; 13 May 2022 12:43:50 JST. M = 1.2 with a 6 km focal depth and 12 km focal distance. Black dashed rectangles are the S phase on the NS and EW components. The P-wave (red) does not have a large amplitude on the UD component. The green dashed S-to-Px shows the S-to-P converted at a depth deeper than 2 km, and the blue dashed rectangle shows the S-to-P conversion at a depth of 1 km. The S-wave arrived at approximately 53 s on the NS and EW components. The S-to-P1 could be at 52.6 s.

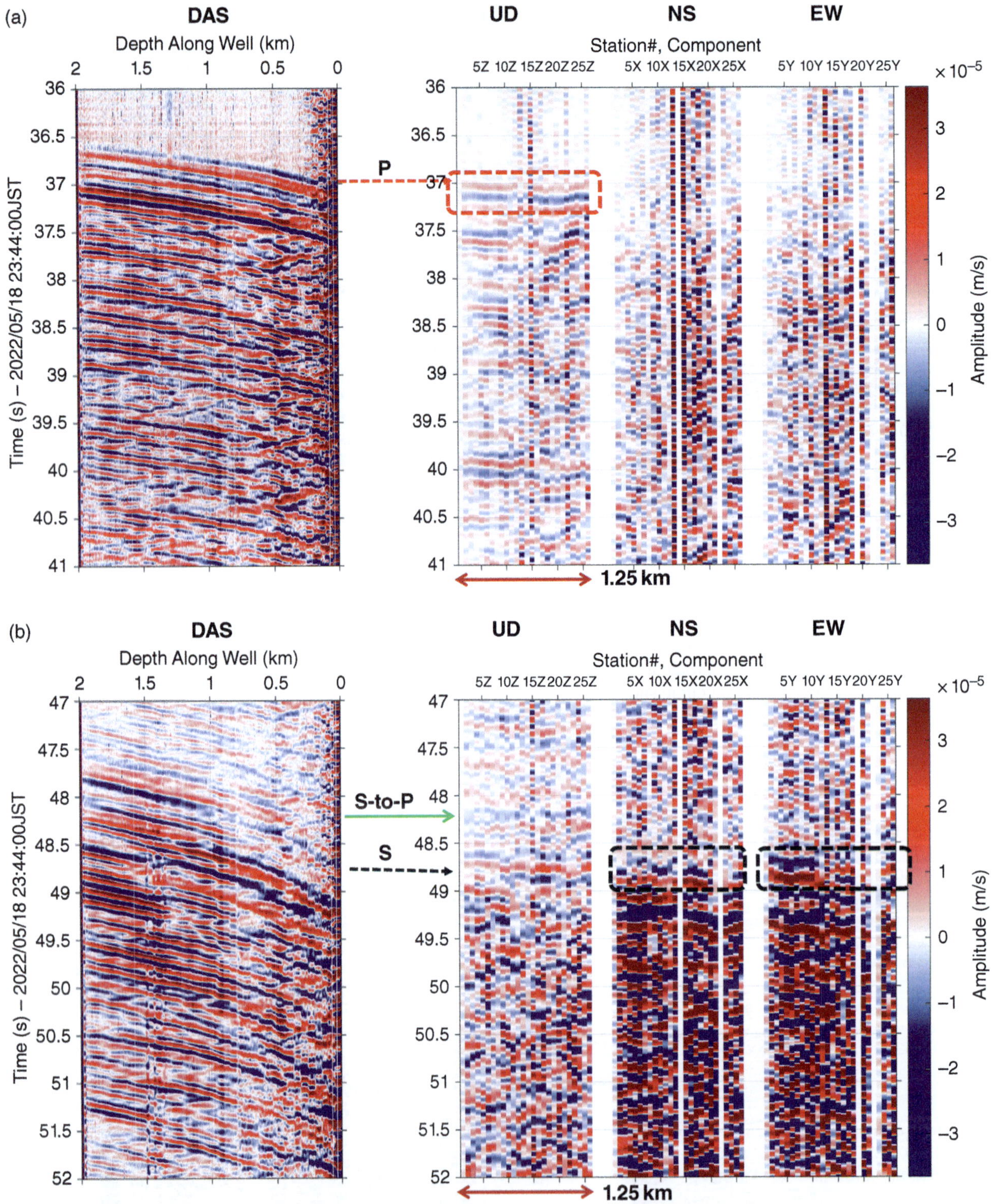

Figure 25.12 (a) P-wave and (b) S-wave parts of DAS (left) and seismometer (right) records of earthquake C; 18 May 2022 23:44:00 JST, M = 3.2 with a focal depth of 42 km. The red dashed rectangle in (a) is the P-wave arrival on the UD component, and the black dashed rectangles in (b) denote the S-wave phase on the NS and EW components with slightly different arrival times, suggesting S-wave anisotropy.

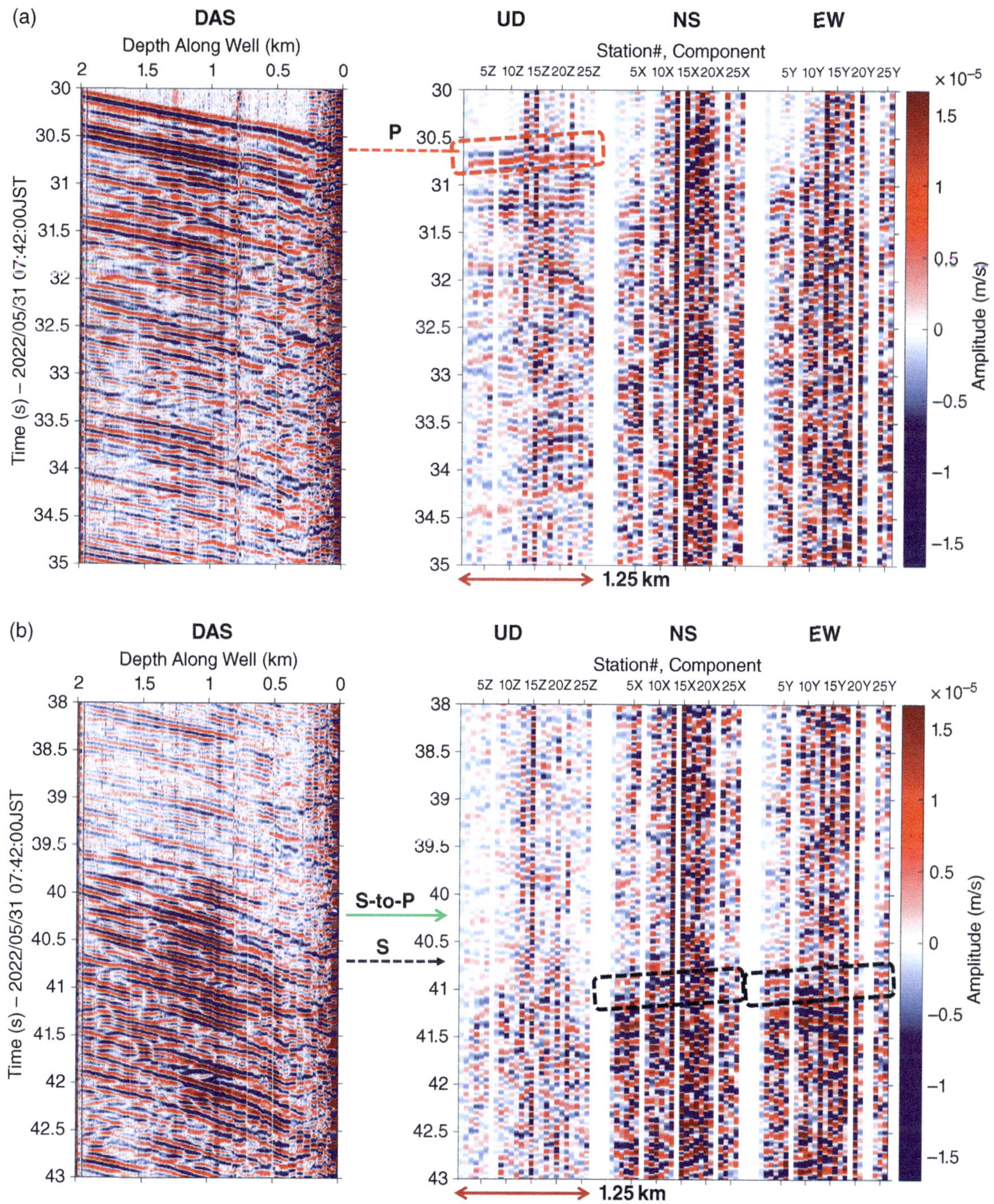

Figure 25.13 (a) P-wave and (b) S-wave parts of DAS (left) and seismometer (right) records of earthquake D; 31 May 2022 07:42:00 JST. M = 2.9 with a focal depth of 10 km. The red dashed rectangle in (a) is the P-phase on the UD component, and the black dashed rectangles in (b) denote the S phase on the NS and EW components.

25.4. DISCUSSION OF P-TO-S AND S-TO-P CONVERSIONS

We observed P-to-S conversion and S-to-P conversion arrivals on the natural earthquake records. The surface seismometers showed different arrival times on the UD and horizontal components. Using the time differences between P and P-to-S or between S-to-P and S, we estimated the depth of the conversion zone for the case at Medipolis. Kasahara et al. (2019) discussed that the conversion depth estimated by the natural earthquake and the reflection depth estimated by the MiniVib seismic source were at the same depth of 3.6 km. Kasahara et al. (2020) calculated synthetic waveforms to examine the P-to-S conversion in the Medipolis observation shown in Figure 25.5 using the model shown in Figure 25.14. The synthetic waveforms without and with a slanted reservoir are shown in Figure 25.15. When the reservoir is at a depth of 4 km with $Vp/Vs = 3$, the horizontal component shows the P-to-S conversion arrivals. The waveforms at an 8.8 km distance above the reservoir are shown. The time difference between P and P-to-S is approximately 0.8 s, which was the time difference in the Medipolis case. One possible cause of the generated P-to-S phase conversions is the fluid saturation zone, which could be a supercritical water state (Kasahara et al., 2019, 2020, 2022a).

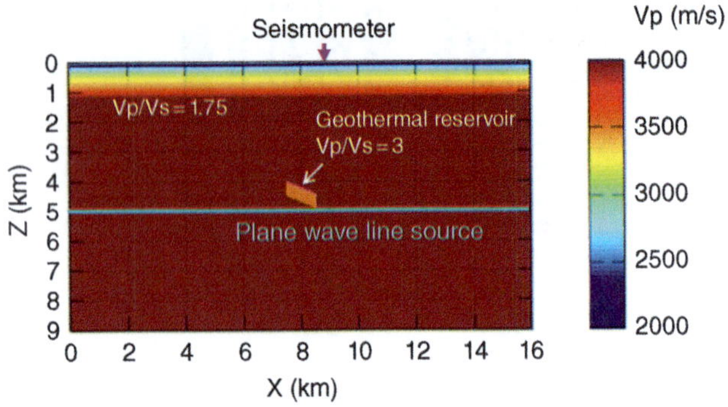

Figure 25.14 Test velocity model for the P-to-S conversion. A slanted reservoir is placed at a depth of 4 km with *Vp/Vs* = 3, and the surface seismograms are calculated for an incident plane wave from a depth of 5 km. Adapted from Kasahara et al. (2020).

We observed the conversion by the intersection of the S and S-to-P phases on DAS records obtained at the Takigami geothermal field. The conversion zone could be related to seismic impedance contrasts between the zone and the surrounding materials.

The P-to-S or S-to-P conversion of seismic waves generally indicates a change in the physical properties. Observing converted waves provides information about fluid-related fractures, such as in plate boundaries or active volcanoes. Geothermal reservoirs with high fluid saturation are expected to cause converted waves and the extent of conversion depends on the physical state of the reservoir. Observation of converted waves, as well as reflected waves, can be useful for determining the reservoir distribution, depth, and temporal changes related to the migration of reservoirs.

Full-wave inversion (FWI) or backpropagation techniques using converted waves are expected to be powerful tools for understanding geothermal structures despite the various difficulties still to be overcome.

25.5. CONCLUSIONS

Through test observations on the ground at the Abiko site, we confirmed that DAS could measure P- and S-waves, although the amplitude depends on the arrival

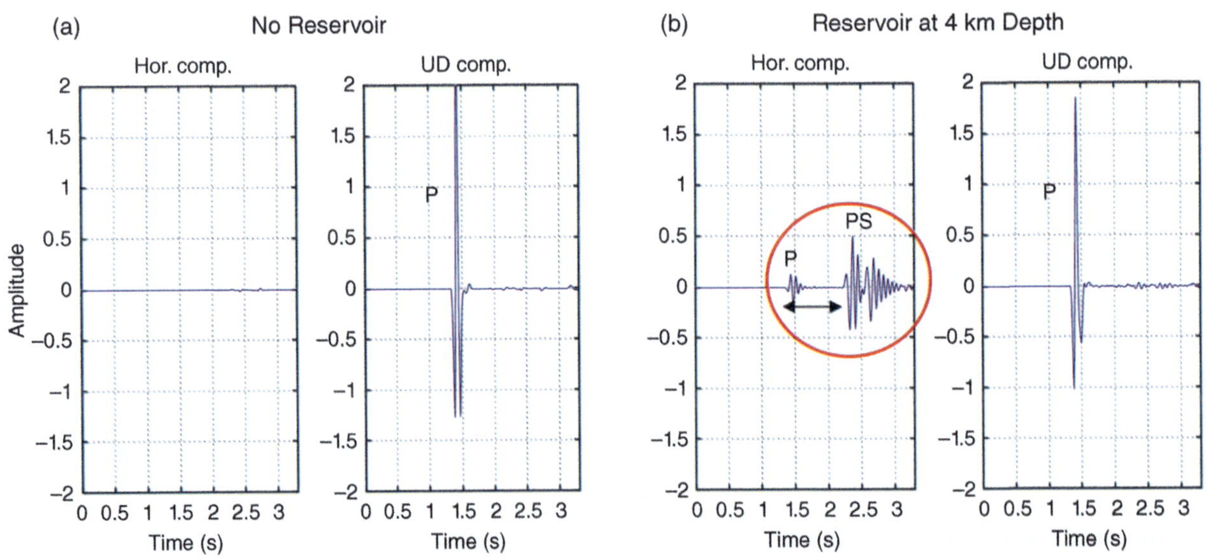

Figure 25.15 Synthetic seismic records of a seismometer at X = 8.8 km in Figure 25.14 without (a) or with (b) a reservoir at 4 km. The left and right diagrams of each set are the horizontal and vertical components, respectively. The vertical and horizontal axes of the diagrams are the amplitude and the time in seconds, respectively. Adapted from Kasahara et al. (2020).

direction and incidence angles. The detectability of seismic signals was comparable to that of a 1 Hz seismometer.

Deploying the optical fiber in existing geothermal wells, we obtained good coupling to the casing and the area surrounding the well for receiving seismic arrivals even without cementing the optical fiber. In the case of earthquakes for which the hypocenters are not close, the focal depths are not very shallow, and the incident S-waves are nearly perpendicular to the well, DAS could detect the S-wave phase. For active seismic sources, we observed S-wave arrivals for some cases (e.g., Kasahara et al., 2025).

We examined natural earthquake records in different geothermal wells and fields. P-to-S and S-to-P conversion arrivals were identified on some natural earthquake records of both DAS and surface seismometers. Because the conversion zones could be related to seismic impedance contrasts between the zone and the surrounding materials, one possibility for the generation of phase conversions identified at some geothermal fields is the fluid saturation zone, which could be a supercritical water state (Kasahara et al., 2019, 2022a). Observing converted waves by DAS in geothermal wells is useful for geothermal investigation.

ACKNOWLEDGMENTS

This chapter is based on the results obtained from a project (project number: JPNS21991) commissioned by the New Energy and Industrial Technology Development Organization (NEDO). We thank the officers at NEDO for their active support during our studies. We greatly thank SNBL Ltd., Geothermal Co., Mitsubishi Materials Co., Ltd., and Idemitsu Kosan Co., Ltd. for permission to use their geothermal wells and their informal database of geothermal fields. We also thank WELMA Co., Ltd. for helping with fieldwork. We thank Prof. Tuncay Taymaz and an anonymous reviewer for their helpful comments.

REFERENCES

Fichtner, A., Klaasen, S., Thrastarson, S., Çubuk-Sabuncu, Y., Paitz, P., & Jónsdóttir, K. (2022). Fiber-optic observation of volcanic tremor through floating ice sheet resonance. *The Seismic Record*, *2*(3), 148–155. doi:10.1785/0320220010

Hartog, A. H. (2017). *An introduction to distributed optical fiber sensors* (p. 440). CRC Press.

Hasada, Y., Kasahara, J., Kawashima, H., Yamauchi, Y., Sugimoto, Y., Yamaguchi, T., & Kubota, K. (2018). *Comparison of the records by optical fiber DAS (Distributed Acoustic Sensor) and geophone using natural earthquakes*. Paper presented at the JPGU 2018 Annual Meeting.

Johannessen, K., Drakeley, B., & Farhadiroushan, M. (2012). *Distributed acoustic sensing-a new way of listening to your well/reservoir*. Paper presented at SPE Intelligent Energy International, Utrecht, The Netherlands, SPE 149602.

Kasahara, J., Hasada, Y., Kawashima, H., Sugimoto, Y., Yamauchi, Y., Yamaguchi, T., & Kubota, K. (2018). *Comparison of DAS (distributed acoustic sensor) and seismometer measurements to evaluate physical quantities in the field*. Paper presented at the SEG 2018 Annual Meeting.

Kasahara, J., Hasada, Y., & Kuzume, H. (2019). *A possible geothermal source at around 4 km depth estimated by the seismic observation in Ibusuki geothermal area*. Paper presented at the JPGU 2019 Annual Meeting.

Kasahara, J., Hasada, Y., & Kuzume, H. (2020). *Possibility of high Vp/Vs zone in the geothermal field suggested by the P-to-S conversion*. Paper presented at the 44th Workshop on Geothermal Reservoir Engineering, Stanford University, Stanford, CA.

Kasahara, J., Hasada, Y., Kuzume, H., Mikada, H., & Fujise, Y. (2022a). The DAS geothermal study in the geothermal borehole of the Ohnuma geothermal power plant in Hachimantai, Honshu, Japan. *Jour: The Geothermal Research Society of Japan*, *44*, 73–84.

Kasahara J., Hasada, Y., Ohnuma, H., Mikada, H., & Fujise, Y. (2022b). *Seismic study at the Sumikawa geothermal field in Akita Pref. Japan using a geothermal well*. Paper presented at the Society of Exploration Geophysicists of Japan Spring Meeting.

Kasahara, J., Hasada, Y., Furuya, S., Shimazaki, M., Mikada, H., Ohnuma, H., and Fujise, Y., (2025). DAS and DTS geophysical study in the Takigami geothermal field in northeastern Kyushu, Japan. In this monograph.

Madsen, N, K., Dümmong, S., Kritski, A., Pedersen, L. S., Finfer, D., Gillies, A., & Travis, P. (2013). *Simultaneous multiwell VSP in the North Sea using distributed acoustic sensing*. Paper presented at the 75th EAGE Conference & Exhibition Incorporating SPE EUROPEC, London, UK.

Mateeva, A., Lopez, J., Potters, H., Mestayer, J., Cox, B., Kiyashchenko, D., et al. (2014). Distributed acoustic sensing for reservoir monitoring with vertical seismic profiling. *Geophysical Prospecting*, *62*, 679–692.

Miller, D., Daley, T. M., White, D., Freifeld, B. M., Robertson, M., Cocker, J., & Craven, M. (2016). Simultaneous acquisition of distributed acoustic sensing VSP with multi-mode and single-mode fibre-optic cables and 3C-geophones at the Aquistore CO_2 storage site. *CSEG Recorder*, *41*, 28–33.

Miller, D., Parker, T., Kashikar, S., Todorov, M., & Bostic, T. (2012). *Vertical seismic profiling using a fibre-optic cable as a distributed acoustic sensor*. Paper presented at the 74th EAGE Conference & Exhibition Incorporating SPE EUROPEC, Copenhagen, Denmark.

Wang, H. F., Zeng, X., Miller, D. E., Fratta, D., Feigl, K. L., Thurber, C. H., & Mellors, R. J. (2018). Ground motion response to an ML 4.3 earthquake using co-located distributed acoustic sensing and seismometer arrays. *Geophysical Journal International*, *213*, 2020–2036. doi:10.1093/gji/ggy102

26

Spatiotemporal Monitoring of a Frequent-Slip Fault Zone Using Downhole Distributed Acoustic Sensing at the MiDAS Project

Hsin-Hua Huang[1,2], Kou-Fong Ma[1,3,4], En-Shih Wu[5], Yun-Ze Cheng[1], Chin-Jen Lin[1], Chin-Shang Ku[1], Po-Li Su[1], and MiDAS Working Group

ABSTRACT

The occurrence of seismic or aseismic slips on a fault is mainly controlled by fault zone structures and properties. However, high-resolution in situ observations are challenging for active faults with uncertain locations or weathered outcrops. The Milun Fault, which repeatedly ruptured during the 1951 and 2018 Hualien earthquakes in Taiwan, offers a unique opportunity to investigate active fault zones using distributed acoustic sensing (DAS). DAS translates the phase shifts of scattering echoes in optical fibers into dynamic strain measurements, enabling continuous, high-resolution monitoring across fault zones. The Milun Fault Drilling and All-inclusive Sensing (MiDAS) project, initiated in late 2021, drilled two holes on either side of the fault and reached the fault zone at a depth of approximately 500 m in hanging wall. The high-density sampling of downhole fibers enables us to identify the 20-meter-thick Milun Fault zone and other hidden faults at various depths. An amplitude-based method that utilizes strain-rate data properties is proposed to map subsurface velocity structures over time, with the derived velocity profiles consistent with logging data. This method shows great potential for detailed temporal structural monitoring in broad applications such as energy exploitation, groundwater management, and geohazard monitoring.

26.1. INTRODUCTION

26.1.1. 2018 Hualien Earthquake and Frequently Slip Milun Fault

On 6 February 2018, an M_L6.4 earthquake occurring at the Hualien offshore area in eastern Taiwan produced a surface rupture of the Milun Fault and extensive damage around Hualien City, including the collapse of large hotels and breakage of bridges (Huang et al., 2019; Huang & Huang, 2018; Lee et al., 2018). The Hualien area is particularly a tectonically complex region at the structural transition from waning collision to subduction (Chen et al., 2014; Huang et al., 2012; Jian et al., 2022; Yen et al., 2019), where the Milun Fault connects the Longitudinal Valley Fault, a plate boundary suture zone system, to the south and the offshore Ryukyu subduction zone system to the east (Shyu et al., 2016; Figure 26.1(a)). Based on the historical surface rupture survey (Shyu et al., 2007), the Milun Fault has ruptured during the Hualien earthquakes of 1951 and 2018, giving an equivalent value of a short recurrence time considered in the Taiwan Earthquake Model (Chan et al., 2020). To investigate its frequently slip properties, the Milun Fault Drilling

[1]*Institute of Earth Sciences, Academia Sinica, Taipei, Taiwan*
[2]*Department of Geosciences, National Taiwan University, Taipei, Taiwan*
[3]*Department of Earth Sciences, National Central University, Taoyuan, Taiwan*
[4]*Earthquake-Disaster & Risk Evaluation and Management Center, Taoyuan, Taiwan*
[5]*Institute of Oceanography, National Taiwan University, Taipei, Taiwan*

Distributed Acoustic Sensing in Borehole Geophysics, Geophysical Monograph 289, First Edition.
Edited by Yingping Li, Robert Mellors, and Ge Zhan.

DOI: 10.1002/9781394179275.ch26

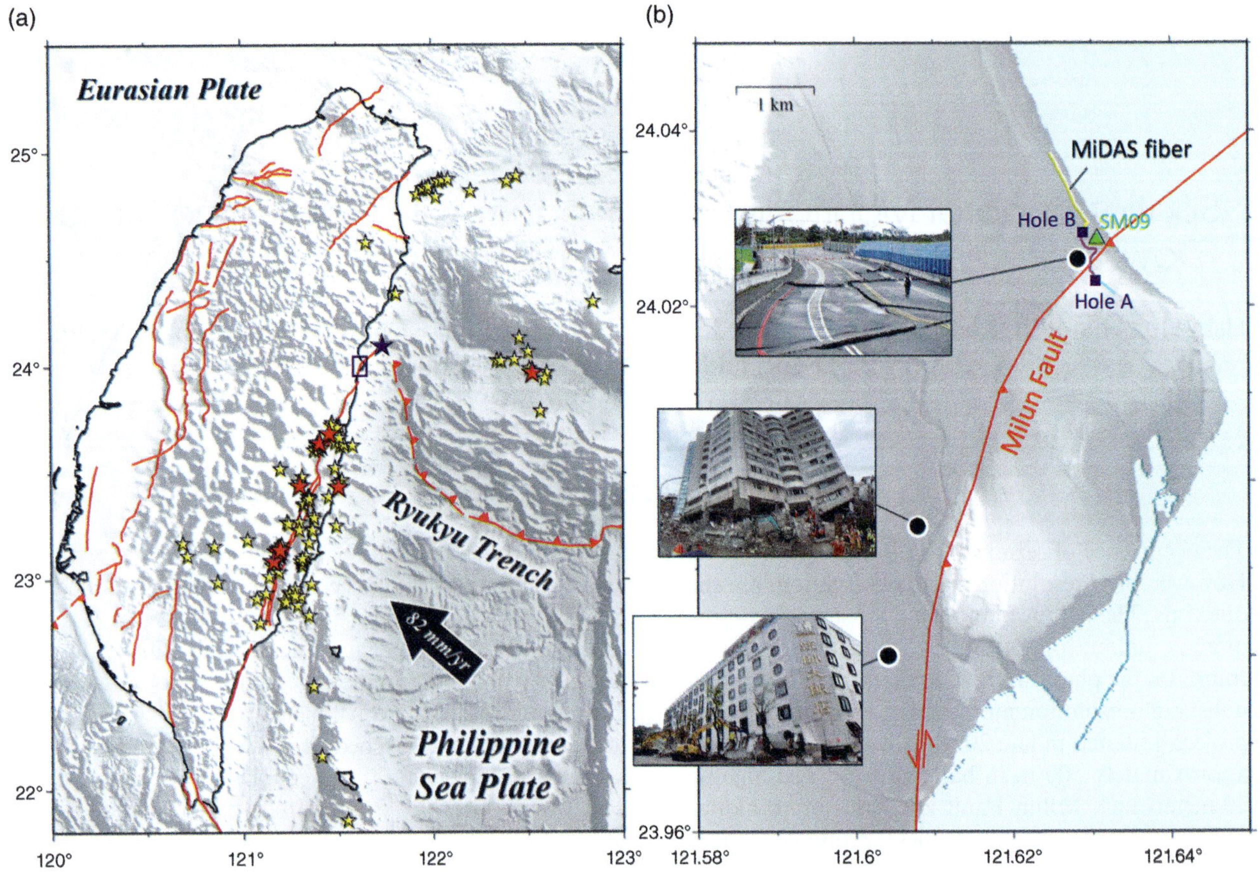

Figure 26.1 Taiwan regional tectonics and MiDAS project site. (a) Tectonic map showing Taiwan as a plate suture between the Eurasian Plate and the Philippine Sea Plate. Red lines indicate active faults. Blue box indicates the Hualien area struck by the 2018 M_L6.4 Hualien earthquake (blue star). Yellow stars are collected earthquakes for analysis during the MiDAS operation time from January 2022 to January 2023; the red stars are referenced later in the chapter, in Figure 26.6. (b) The Milun Fault and photos of damage caused by the Hualien earthquake. The blue squares and colored lines are MiDAS drilling holes and fiber routes. Green triangle indicates a nearby geophone station SM09.

and All-inclusive Sensing project (MiDAS) was launched in late 2021 (Huang et al., 2022; Ma et al., 2022), which aims to drill through the Milun Fault zone and serves as a borehole observatory toward an in-depth understanding of earthquake dynamics, triggers, and possible precursors in a complex multifault system.

26.1.2. MiDAS Project and 3D Fiber Network

The goal of the MiDAS project is to drill and core the Milun Fault using two boreholes, one drilled into the hanging wall (Hole A) and one drilled into the footwall (Hole B). Based on a seismic profile survey (Figure 26.1(b)), the project expects to intersect the fault zone in Hole A at a depth of approximately 500 m. The project also includes a variety of onsite multidisciplinary observations such as conventional wireline logging, temperature profiling, a groundwater and geochemistry monitoring system, and a borehole seismic array (Ma et al., 2022). For seismic monitoring, one of the key components is to introduce the distributed acoustic sensing (DAS) technique and install fibers into the boreholes directly through the active fault zone to obtain high-resolution in situ seismic observations to study the fault zone characteristics and earthquake dynamics. Surface fibers are also trenched and buried to connect two downhole fibers in Hole A and B to form a 3D fiber sensor route (Figure 26.2). The entire fiber network was completed in the summer of 2022, and altogether it includes two borehole segments and three surface segments, where surface Segment 2 crosses the surface trace of the Milun Fault and uses commercial fiber from the Chunghwa Telecom Company (CHT).

The DAS employed is an emerging new technique that can turn a fiber cable into a linear seismic sensor array by

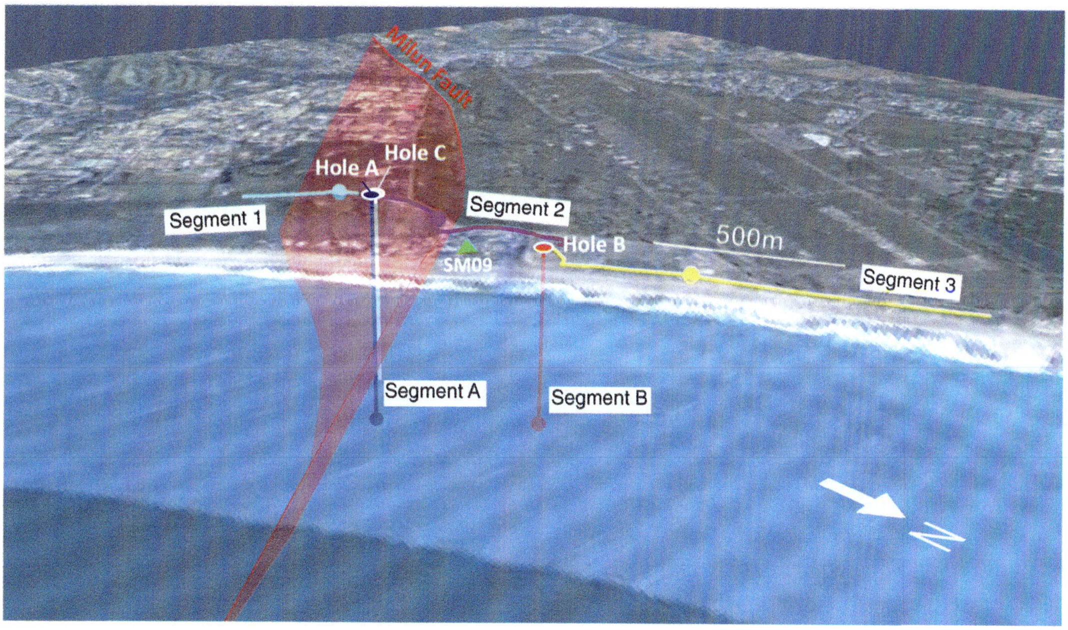

Figure 26.2 3D perspective of MiDAS fiber array. Colored lines indicate different fiber segments deployed subsequently. Shaded red plane represents the schematic subsurface Milun Fault plane. Green triangle indicates the geophone station SM09. Hole C is an auxiliary hole for coring and does not have fibers installed. Colored dots indicate five representative channels of each segment selected for analysis later in the chapter (see Figures 26.7 and 26.9).

sending a pulse of laser light down the fiber and analyzing the echoes scattering back from the intrinsic defects along the fiber. A variety of scattering behaviors are sensitive to different external factors (e.g., temperature and strain) that influence the fiber and cause changes in the amplitude or phase of laser light echoes (Zhan, 2019). Distributed strain sensing (DSS) uses Brillouin scattering; distributed temperature sensing (DTS) uses Raman or Brillouin scattering; and in particular, DAS uses Rayleigh backscattering to infer the longitudinal dynamic strain or strain rate over a gauge length of every few meters along the fiber. Such unprecedented meter-scale resolution thus creates new opportunities and has shown great promise in a wide variety of applications, including studies of fault zones (Jousset et al., 2018), site effect (Spica et al., 2020), volcanoes (Jousset et al., 2022), glaciers (Walter et al., 2020), ocean dynamics (Lindsey et al., 2019), groundwater monitoring (Tribaldos & Ajo-Franklin, 2021), thunder events (Zhu & Stensrud, 2019), and aftershock activity (Li et al., 2021). The ease of installation and maintenance of fibers in harsh environments (e.g., high temperatures) also renders DAS a powerful alternative tool in recent oil/gas, CO_2 sequestration, and geothermal exploration (Daley et al., 2016). However, compared to the recent blossom of surface DAS experiments carried out spanning various tectonic settings and scales, borehole DAS experiments have still been quite limited (Correa et al., 2017; Glubokovskikh et al., 2021; Lellouch et al., 2019; Li et al., 2022).

26.1.3. Unique Opportunities to Monitor Ground Motions Along Depth Across the Fault Zone

The unique setting of the MiDAS project, which penetrates the active fault zone, provides rare opportunities to study seismic responses to both strong (e.g., earthquakes) and weak motions (e.g., microseism) along depth and across fault zones. An increase in seismic rate has been observed after the 2018 $M_L6.4$ earthquake, with a series of M5.5+ earthquakes occurring in the northern Longitudinal Valley (Huang & Wang, 2022). This seismic crisis shifted to the middle and southern Longitudinal Valley in 2022 and led to the occurrence of the 22 March $M_L6.7$ earthquake, 20 June $M_L6.0$ earthquake, and destructive 18 September $M_L6.9$ Chihshang earthquakes in 2022, which were all recorded by the MiDAS fiber network and provided valuable near-distance strong motion DAS data to investigate.

In this chapter, we will first compare the seismic characteristics between the DAS and seismometer data for the 2022 Longitudinal Valley earthquake sequence with one near geophone SM09 (Figure 26.1(b)). We then analyze the spectral characteristics of DAS data, explore the amplification effect across the Milun Fault zone, and validate and discuss the potential of using strain-rate root-mean-square (RMS) amplitude ratios for temporal monitoring of subsurface structures.

26.2. DATA ACQUISITION AND METHODS

26.2.1. Fiber Experiment and DAS Data Processing

The MiDAS fiber network consists of five segments that include two borehole segments, A and B, with Hole A (700 m deep) and Hole B (500 m deep) and three surface segments 1, 2, and 3. Segment 1 is along the Haian Road and is connected to Hole A; Segment 3 is along the coast and connected to Hole B; and Segment 2 connects segments 1 and 3 crosses the Milun Fault trace (Figure 26.2). Segment 2 is using dark fiber in the CHT commercial cables because it is in a residential region. The entire deployment of fibers was conducted in several stages, as indicated in Figure 26.3. The data recording started immediately after the installation of Segment A on 31 December 2021. Although the recording was interrupted a few times by sequential deployments of different fiber segments, it is still ongoing. Up to completion, the total length of fibers gradually reached about 7.5 km up to completion. Detailed metadata is available at https://www.elab.earth.sinica.edu.tw/midas-metadata. We use one interrogator at Hole B, so we connect different segments of the fiber to each other, as demonstrated by the waterfall plot in Figure 26.4.

The interrogator is iDAS3 from Silixa. It has a fixed gauge length of 10 m. The sampling rate and spatial interval are set at 1,000 sps and 4 m, respectively, which results in 1,875 channels and produces 300 GB of data per day. All the data is downsampled to 100 sps onsite and transmitted back to the Academia Sinica Grid Computing Centre for real-time archiving and analysis.

26.2.2. Logging, Coring, and the Milun Fault Zone

The well logging was conducted right after the completion of drilling at Hole A and B in January and June, respectively. Figure 26.5(b)–(f) shows the logging results of gamma-ray, resistivity, spontaneous potential (SP), caliper width, temperature, and P-wave velocity (Vp). The gamma ray that mainly reflects the lithological properties shows anomalies at a depth of 490–540 m. Anomalies of resistivity and SP are also observed at similar depths of around 520–540 m. A lithological column is constructed by the retrieved cores in a depth range of 440–540 m in Figure 26.5(a). Above 490 and below 540 m, there is Milun Conglomerate. At 490–520 m is black schist, and at 520–540 m, there is gray gouge, which are identified as the damage zone and Milun Fault zone, respectively.

26.2.3. Monitoring Methods

Different from an inertial seismometer recording velocity or acceleration, DAS records the axial strain or strain rate along the fiber. The relationship between strain rate and acceleration is

$$\dot{\varepsilon} = \frac{\ddot{u}}{c^a} = \frac{\cos\theta}{c}\ddot{u} = \frac{\cos\theta}{c} S * G(\theta) * A(\theta) \qquad (26.1)$$

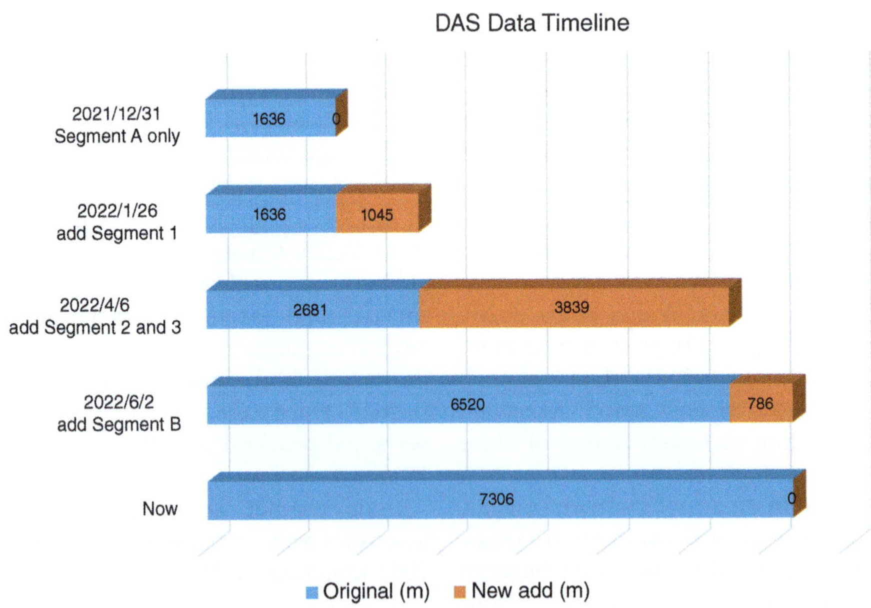

Figure 26.3 DAS data timeline with progressive fiber deployments. The total fiber length grows from 1,636 m at Hole A to 7,306 m at the end.

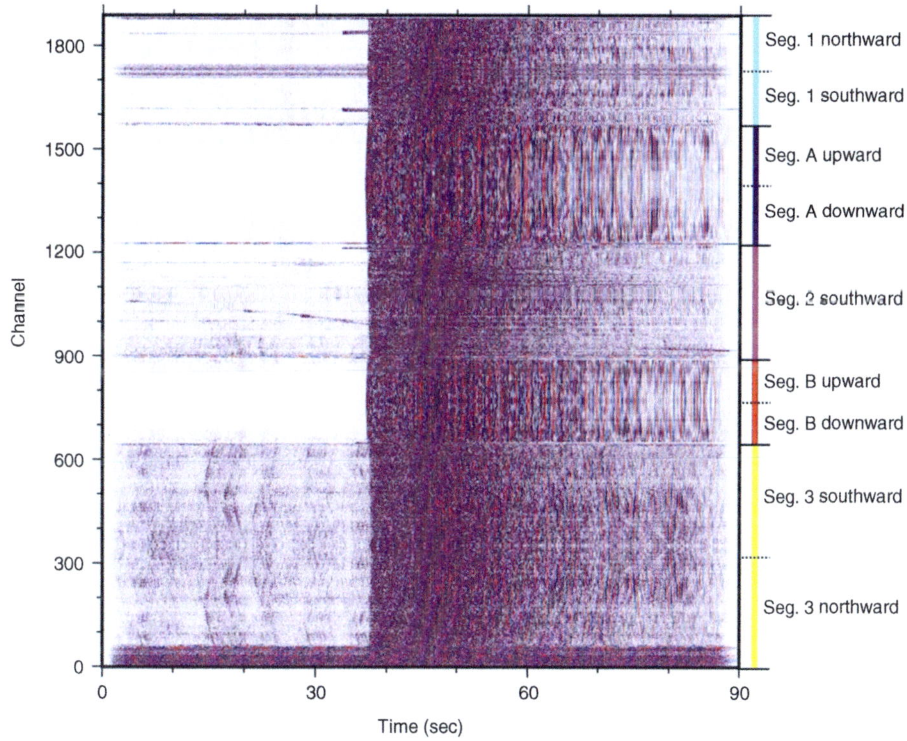

Figure 26.4 Waterfall plot of one event data recorded by the final MiDAS fiber array. The event is an earthquake on 21 January 2023. The colored lines on the right represent different fiber segments in Figure 26.2.

where $\dot{\varepsilon}$ is strain rate, $\ddot{u}$ is acceleration, and c^a and c are apparent and true phase velocity depending on the incident angle θ of seismic waves. Because this variation is mainly on the strain-rate amplitude and not the phase, phase-based analyses such as earthquake traveltime estimation (Lellouch et al., 2019) and ambient noise cross-correlation (Dou et al., 2017) can be applied as done for seismometers in the literature. The amplitude-based analyses that require the conversion from strain/strain rate to velocity/acceleration are, however, more complicated and less quantified. Although unbiased amplitude conversion is under vigorous development (Lindsey et al., 2020; Lior et al., 2021; Wang et al., 2018), we explore the direct use of strain-rate amplitude to infer and monitor subsurface structure, inspired by the work of Li et al. (2022).

Assume $\ddot{u}$ is the convolution of source (S), path (G), and site-specific amplification terms (A) in equation (26.1). Because we only consider the amplitude here, G will mainly represent the geometric spreading term that depends on the wave types and their incident angle. Assuming a far-field plane wave crossing through the entire fiber array, taking the relative strain rate between two channels cancels the source and path effect and the incident angle dependency of phase velocity as

$$\frac{\dot{\varepsilon}_1}{\dot{\varepsilon}_2} = \frac{c_2 A_1(\theta)}{c_1 A_2(\theta)} = \frac{c_2'}{c_1'} \quad (26.2)$$

where we further define this ratio as effective velocity contrast c_2'/c_1' including the amplification effect A. Although A may differ by different incident angles and frequency bands of seismic waves due to site-specific amplification and possible wave interference, we could largely minimize its effect by using long-wavelength (low-frequency) signals. This would approximate the effective velocity contrast to the true velocity contrast. We will validate later that this approximation could be extremely useful for monitoring purposes, which mainly focus on the temporal variations rather than absolute values in seismic velocity and appreciate the property of strain measurement itself.

Recently, Li et al. (2022) found a high correlation between the RMS of strain-rate amplitudes and subsurface elastic properties obtained from logging data by using quiet and carefully selected noise data within 48 hr. But how this correlation holds and whether it works for longer time periods under different ground motions conditions remain untested. Following our assumptions, as c is an invariant local property, at least within a short time period, taking the RMS of strain-rate amplitudes to equation (26.1) yields

$$RMS(\dot{\varepsilon}) = \frac{1}{c} RMS(\ddot{u}\cos\theta) \quad (26.3)$$

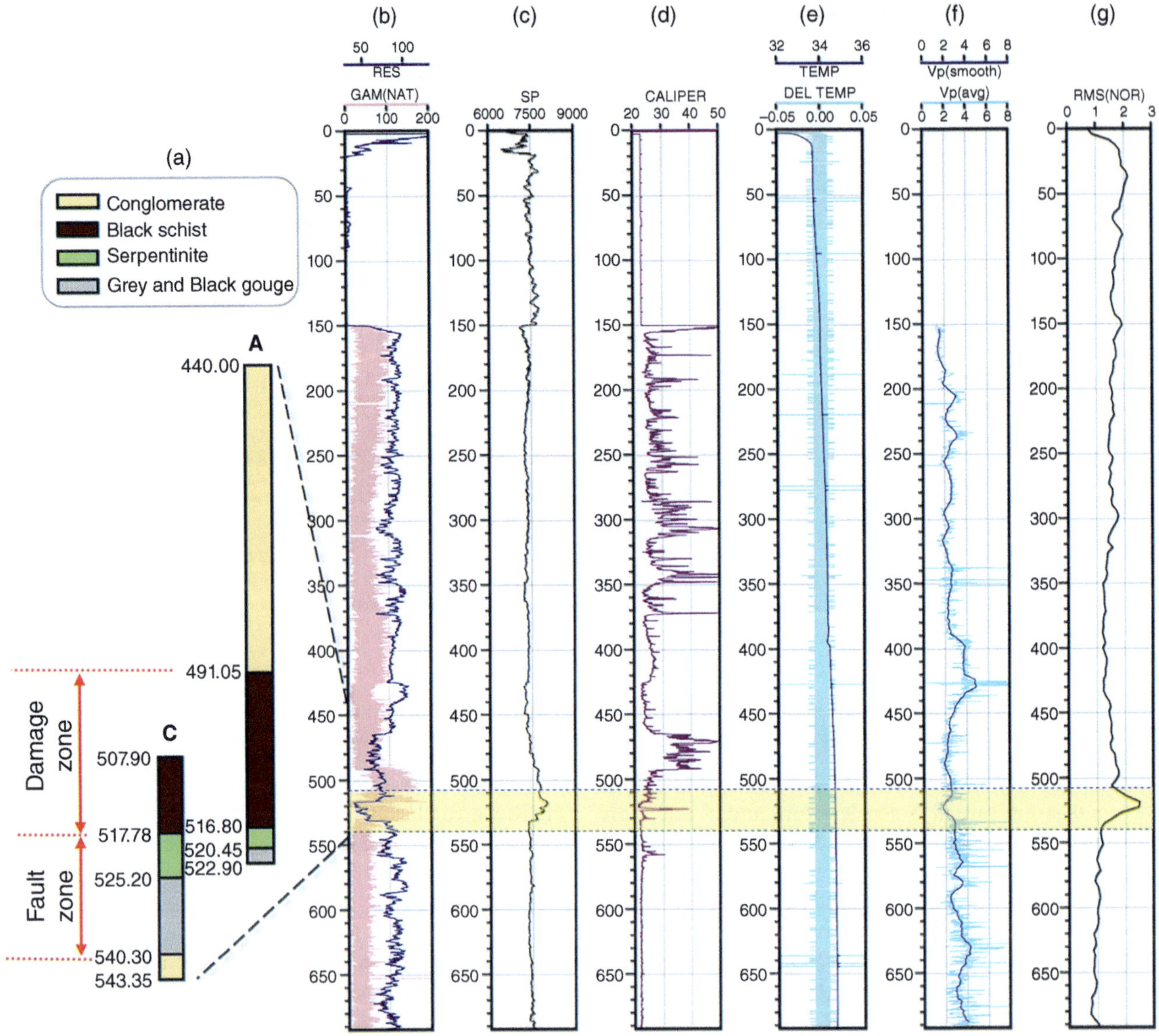

Figure 26.5 Seismic profile of DAS amplitude RMS ratios with coring and logging profiles for Hole A. Wireline logging at Hole A for (b) gamma ray and resistivity, (c) spontaneous potential, (d) caliper, (e) temperature, and (f) P-wave velocity. The measurements of 0–150 m are influenced by the well casing. A lithological column based on coring at Hole A and close to Hole C (15 m from Hole A) at depths of 440–543 m is indicated in (a). A DAS strain-rate RMS amplitude ratio profile for an hour of data is derived in (g). The Milun Fault zone is indicated by the shaded yellow zone.

and therefore equation (26.2) also holds for the ratio between strain-rate RMS amplitudes at different channels as

$$\frac{RMS(\dot{\varepsilon}_1)}{RMS(\dot{\varepsilon}_2)} = \frac{c_2'}{c_1'} \tag{26.4}$$

which could explain the findings in Li et al. (2022).

Essentially, when local amplification effects are negligible if the seismic wavefield $\ddot{u}$ is recorded by the entire fiber array within a time window (regardless of phase shifts), then the strain-rate amplitude ratio between different channels will mainly reflect the ratio of elastic property (i.e., phase velocity in equation 26.4). One major difference from Li et al. (2022) is that we look at the RMS amplitude ratio rather than the RMS amplitude itself, which is important to eliminate the source effect and incident angle dependency of phase velocity (equations 26.1 and 26.2). Figure 26.5(g) shows the strain-rate RMS amplitude ratio profile of Hole A using 1 hr DAS data at UTC time 7–8 p.m. (3–4 a.m. Taiwan time) on

20 January 2023. The ratios are taken between each channel to the bottom channel, which is presumed quietest and less influenced by the surface activity. It reveals consistent peaks at a depth of 520–540 m with the logging and coring results.

26.3. STRONG MOTION MONITORING

26.3.1. The 2022 Longitudinal Valley Earthquake Sequence and Saturation Issue

As indicated in Figure 26.1(a), a sequence of M6+ earthquakes that occurred in the middle and southern Longitudinal Valleys with extensive aftershocks in 2022 were well recorded by the MiDAS fiber array. Earthquakes in the Ilan offshore area to the north and the Ryukyu trench to the east are also collected for later analysis. Figure 26.6 shows a number of selected earthquakes (red stars in Figure 26.1a) with increasing local magnitude (M_L) of 4–7 on representative channels of the vertical borehole Segment A and horizontal surface Segment 1, in comparison with the vertical and the horizontal components rotated to the direction of Segment 1 (~N148E) of geophone SM09. The overall shape and duration of seismograms are comparable to each other. Using the earthquake on 19 September 2022 as an example, the zoom-in comparison also shows high similarity between the DAS and geophone signals (Figure 26.7). However, when earthquake magnitudes are greater than 6, an amplitude saturation phenomenon with clipping waveforms is observed on the DAS recording of surface Segment 1 but not downhole Segment A. Such saturation can also be seen clearly in the spectrogram in Figure 26.8, where a wide-band strong energy appears when waveforms are clipped.

If we scale these events with their local magnitude M_L and hypocentral distance R as

$$M_L = logA - log\,A_0(R) \tag{26.5}$$

where A is the peak amplitude of DAS seismograms (bandpasses in 0.1–50 Hz) with a local attenuation correction factor A_0 as

$$\log A_0(R) = -0.00716R - logR - 0.39 \tag{26.6}$$

for shallow earthquakes (focal depth < 35 km) within 80 km (Shin, 1993), we can see that DAS data holds good linearity until the saturation threshold is met (Figure 26.9 (a)), which is ~$3.6 * 10^{-4}$ 1/s with 1,000 sps sampling rate and 10 m gauge length for a Silixa iDAS3 unit. The higher amplitude of the surface-segment records reflects both the noisier surface environment (compared to the quiet downhole environment) and the site amplification effect. The relative scattering distribution, particularly seen in Segment 2, is possibly due to poor coupling at some parts of the CHT commercial fibers. The same analysis for geophone SM09 obtains a linear trend similar to those of surface fiber segments but higher than those of downhole segments (Figure 26.9(b)). The similar slopes between the surface and downhole segments and the geophone suggest that maybe only a static correction is needed to include DAS data in magnitude estimation in routine earthquake monitoring (Lellouch et al., 2020) and also the earthquake early warning system in which ocean cables could play a crucial role for offshore megathrust large earthquakes (Lior et al., 2023).

Ajo-Franklin et al. (2022) have reported the DAS saturation issue in the Imperial Valley Dark Fiber Project for $M_L > 4$ earthquakes. This difference in saturation magnitude could arise from different event distances and DAS settings. As the saturation occurring in DAS is due to incorrect phase unwrapping, shortening the gauge length or increasing the sampling rate could potentially mitigate the saturation problem in principle (personal communication with Silixa). However, doing so will also increase the data volume drastically, challenge real-time data archiving and transmission, and reduce the signal-to-noise ratio and sensing range (Dean et al., 2017). Hence, to include existing telecom and ocean-bottom cables for large earthquake hazard mitigation in practice without saturation, developments of managing and processing a large volume of DAS data in real time, such as rapid algorithms and slim data format, will be central to demands (Lindsey & Martin, 2021).

26.3.2. Spectral Characteristics Along Depth and Across the Fault Zone

To explore the spectral characteristics along depth and across the fault zone, Figure 26.10(a) shows the relative power spectral density (PSD) diagram for Hole A and Hole B using an 80 s time window of earthquakes on 25 March 2022 and 18 September 2022, respectively. The PSD at each frequency k (Pk) is calculated by

$$P_k = \frac{2\Delta t}{N}|\dot{\varepsilon}(f_k)|^2 \tag{26.7}$$

where Δt and N are the sample interval (0.001 s) and the number of samples within analyzed time windows (McNamara & Buland, 2004). The calculated PSD at each channel is then divided by that at the bottom channel of two holes to obtain the relative PSD for enhancing the amplification effect relative to the quietest channel. First of all, the most compelling feature is a clear exponential decay in frequency from the surface to a depth of 300–400 m. This likely results from surface noise sources,

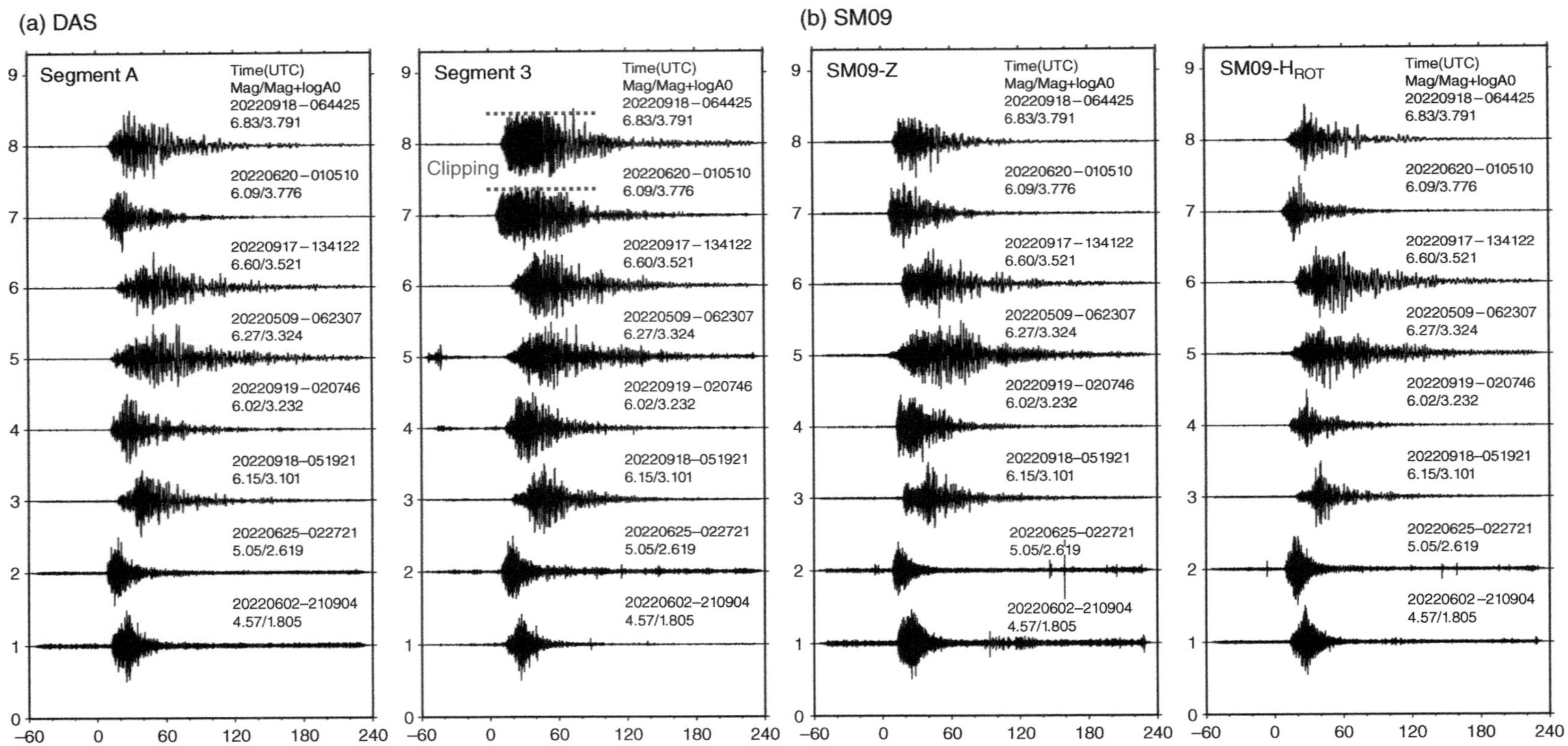

Figure 26.6 Waveform comparison for the selected earthquakes (red stars in Figure 26.1a) between DAS strain-rate data at representative channels of fiber Segment A (vertical) and 3 (horizontal) and acceleration data at geophone SM09. The horizontal component of SM09 is rotated to the orientation of Segment 1. The pink dotted lines indicate the clipping phenomena observed in DAS data.

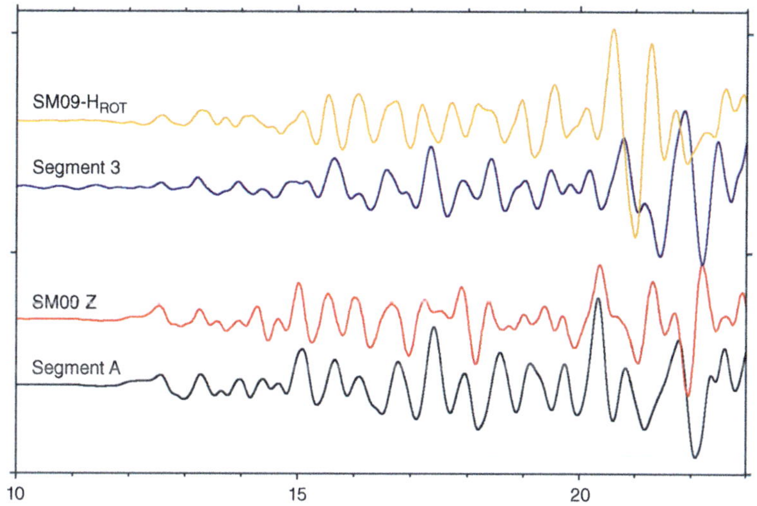

Figure 26.7 Zoom-in waveform comparison between DAS strain-rate data and SM09 acceleration data for the 19 September 2022 earthquake. Both vertical and horizontal waveforms show a high similarity.

such as cultural and traffic activity, where the higher frequency of energy attenuates faster than the lower frequency. This diagram also shows that the cultural noises are mostly restricted within the top 400 m and above 1 Hz.

Second, a clear amplification spanning a wide frequency band is observed at depths of 520–540 m, corresponding to the Milun Fault zone identified by logging and coring data (Figure 26.5). In addition, weaker amplifications are also seen at depths of around 170 and 320 m, which may suggest secondary fault structures developed in the thick conglomerate strata above the major fault zone (blue arrows). More interestingly, this fault-zone amplification at depths of 520–540 m appears not only during strong shaking of earthquakes but also in normal quiet times (i.e., ambient noise). Figure 26.10(b) shows the same relative PSD diagram for 1 hr. To avoid cultural signals during the day, we look at 3–4 a.m. local time (7–8 p.m. UTC) on 27 January 2023. The amplification is seen at depths similar to using earthquake data and spans a wide frequency band. Because the Milun Fault zone identified by logging and coring is only 20 m thick, the sensitivity of seismic waves to this zone should be diminished toward lower frequencies. Thus, this near-constant wide-band amplification feature is likely dominated by the local elastic properties of the medium as derived in section 26.2.3 (equation 26.2), although true amplification due to the low-velocity fault zones and interference of trapped waves could exist to some extent in higher frequency bands.

As a comparison, Hole B, located on the footwall of the fault, shows no significant amplification for both strong and weak motions in the relative PSD diagram, except for a slight amplification at a depth of ~270 m, which may need further investigation (Figure 26.10(c), (d)).

26.3.3. RMS Amplitude Ratio Method Validation on Earthquake Data

To examine whether the amplitude variations (i.e., strain-rate amplitude in equation 26.1) mainly reflect the local elastic properties along depth and its temporal stability, we calculate the strain-rate RMS amplitude ratios (equation 26.4) between each channel and the bottom channel during a local earthquake on 19 September 2022 (Figure 26.11(a)). A moving window of 40 s with a 20 s step and a relatively low frequency band of 0.1–1 Hz are used. We recall from equation (26.2) that because for the strain-type data, the amplitude ratio reflects a combinational effect of the medium velocity and site amplification, the purpose of using a low-frequency band is to minimize the site amplification effect caused by the shallow sediments and narrow fault damage zone so that the derived amplitude ratio could approximate the medium velocity ratio. (Figure 26.10). The results show that the derived RMS amplitude ratio profiles are almost invariant over time and agree well with the logging slowness profile ($1/Vp$), except for a slight deviation during the initial P-wave train. This deviation may result from wave–structure interaction during strong motions, but the high consistency in the later times implies that when the latter coda phases reflect or scatter in space, as long as they are recorded by the entire fiber array within the analyzed window, equations (26.2) and (26.4) may still roughly hold. The use of RMS amplitude ratio (equation 26.4) rather than amplitude ratio (equation 26.2) is also important to obtain stable amplitude ratios within analyzed windows.

Another test is on the recent M_W7.8 Turkey earthquake on 6 February 2023, which shows well-recorded

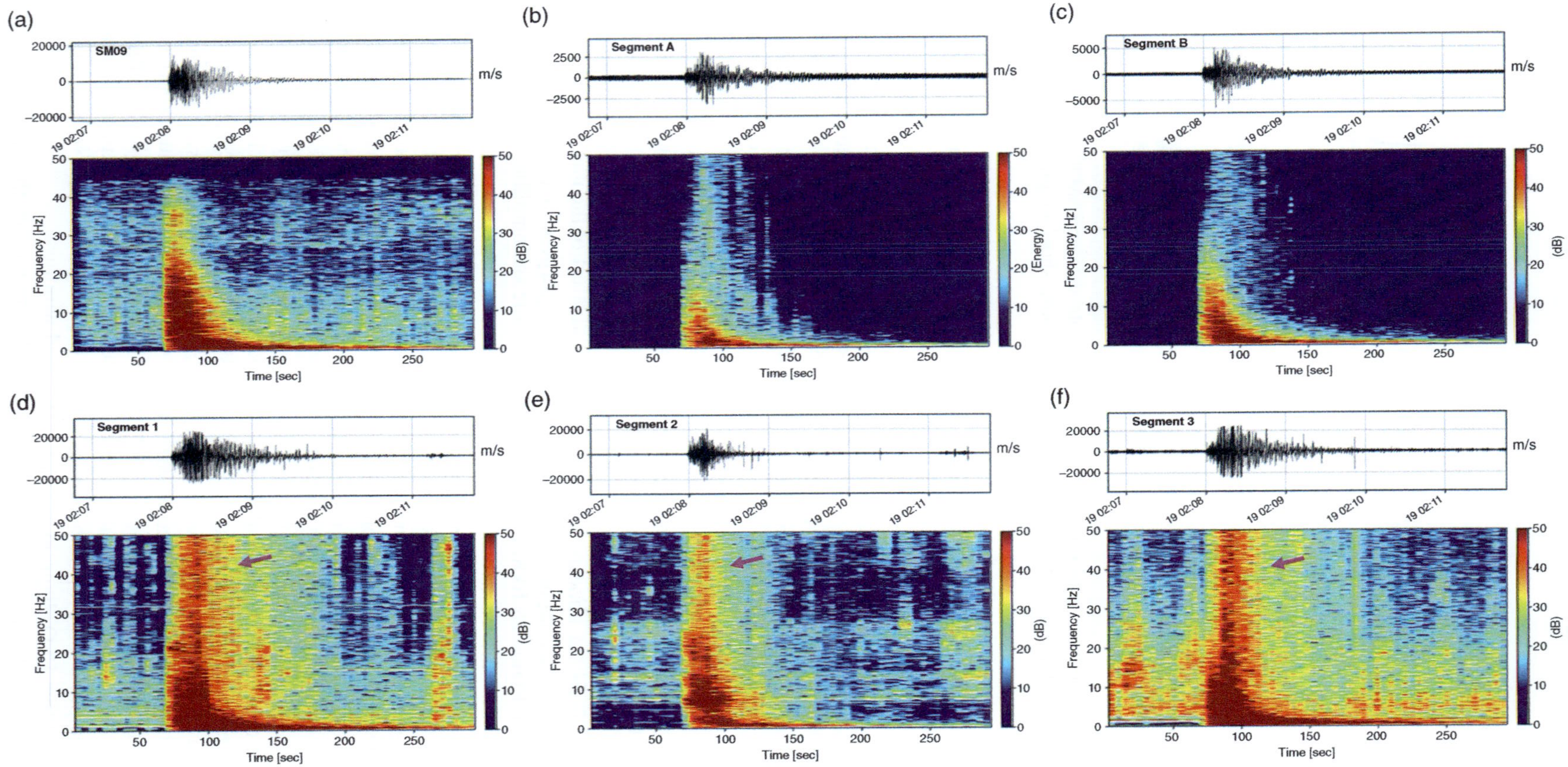

Figure 26.8 Spectrogram comparison for the 19 September 2022 earthquake between (a) SM09 and (b–f) representative channels of five fiber segments. The location of representative channels refers to Figure 26.2. (b, c) and (d–f) are downhole and surface segments, respectively. Pink arrows indicate the saturation effect at the horizontal fiber segments (d–f).

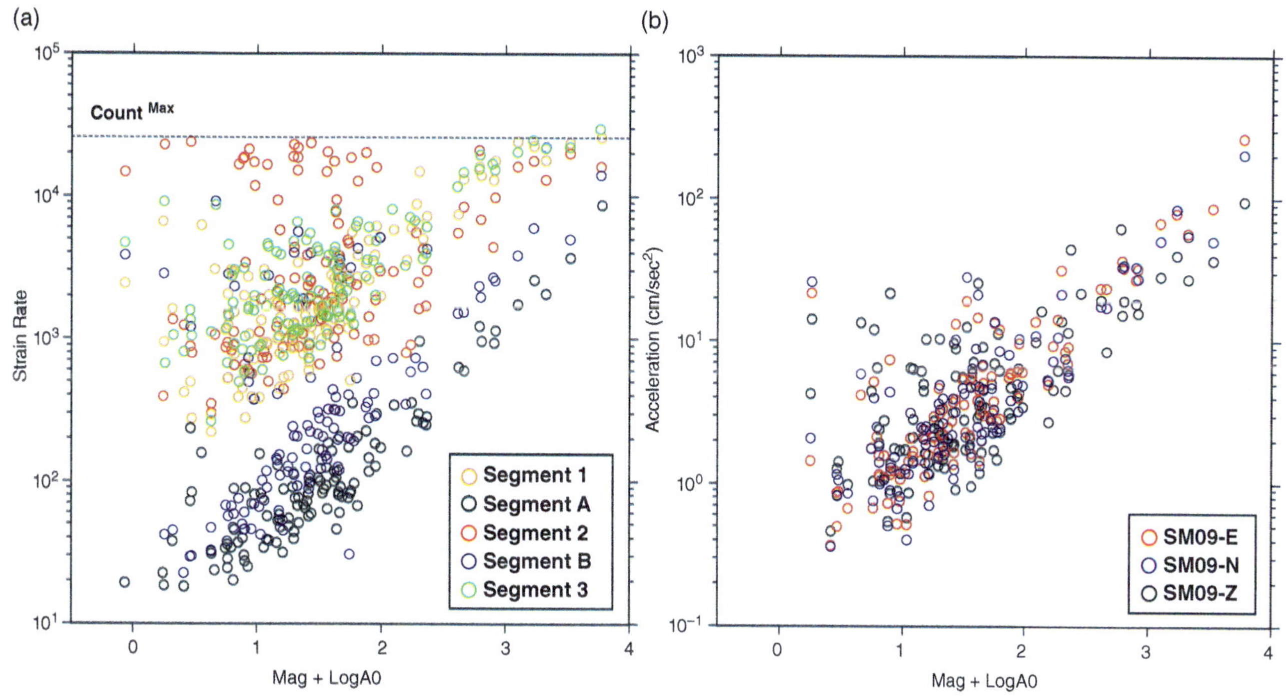

Figure 26.9 Peak ground motion attenuation relationship for the collected earthquakes (yellow stars in Figure 26.1a) between (a) DAS strain-rate data and (b) SM09 acceleration data. The analysis is conducted with the representative channels of five fiber segments as indicated by different colored circles in (a). The dotted black line denotes the instrumental upper limit of the strain rate. Results for the three components of SM09 are also shown in different colors in (b).

and separated P-wave and Rayleigh wave trains (Figure 26.11(b)). The relatively narrower azimuthal sensitivity of DAS naturally attenuates most S-wave energy in the waveforms. With a moving window of 400 s with a 200 s interval, the results show consistent and stable profiles of the RMS amplitude ratio, which is again a good agreement with the slowness structure with depth. We also observe that the RMS amplitude ratio (slowness) profiles remain very similar during the passage of Rayleigh wave trains (Figure 26.11(b)), where the Rayleigh wave should mostly be sensitive to the S-wave velocity (*Vs*) structure. Because the slowness profile constructed by the RMS amplitude ratio is relative (equation 26.4), the consistency between the P-wave and S-wave slowness profile may imply near-constant *Vp*/*Vs* ratios within the depth range of Hole A.

26.4. WEAK MOTION MONITORING

26.4.1. RMS Amplitude Ratio Method on Ambient Noise Data

We have demonstrated that when the downhole fiber array records a common wavefield from a single source such as earthquakes, a simple calculation of RMS amplitude ratio can reveal the local slowness structure along the depth regardless of incoming angles. Now we explore the possibility of extending this method to use ambient noise for continuous monitoring of subsurface structures. The success of seismic interferometry has proven that the microseism prevails in intermediate frequency bands of around 1–20 s. Our working hypothesis to test is whether the microseism signals could constantly provide the Rayleigh wave wavefields to be recorded by the entire downhole fiber so that an analysis similar to that done for earthquakes (Figure 26.11(b)) can be applied to continuous ambient noise data. The dominance of Rayleigh wave signals in ambient noise data has been proposed by Glubokovskikh et al. (2021) in the CO2CRC Otway Project site in Australia, where the authors found that vertical strain amplitude variations follow the Rayleigh wave motions in response to variations in rock stiffness with depth. Here, using the frequency band of 0.1–1 Hz to focus on secondary microseism signals and avoid busy cultural noise (>1 Hz) in the region, we test this idea for temporal monitoring purposes.

26.4.2. Temporal Monitoring of Subsurface Structures

We apply the amplitude RMS ratio method on Hole A for a time period of 18–25 January 2023 (Figure 26.12). As a comparison, in the results of the

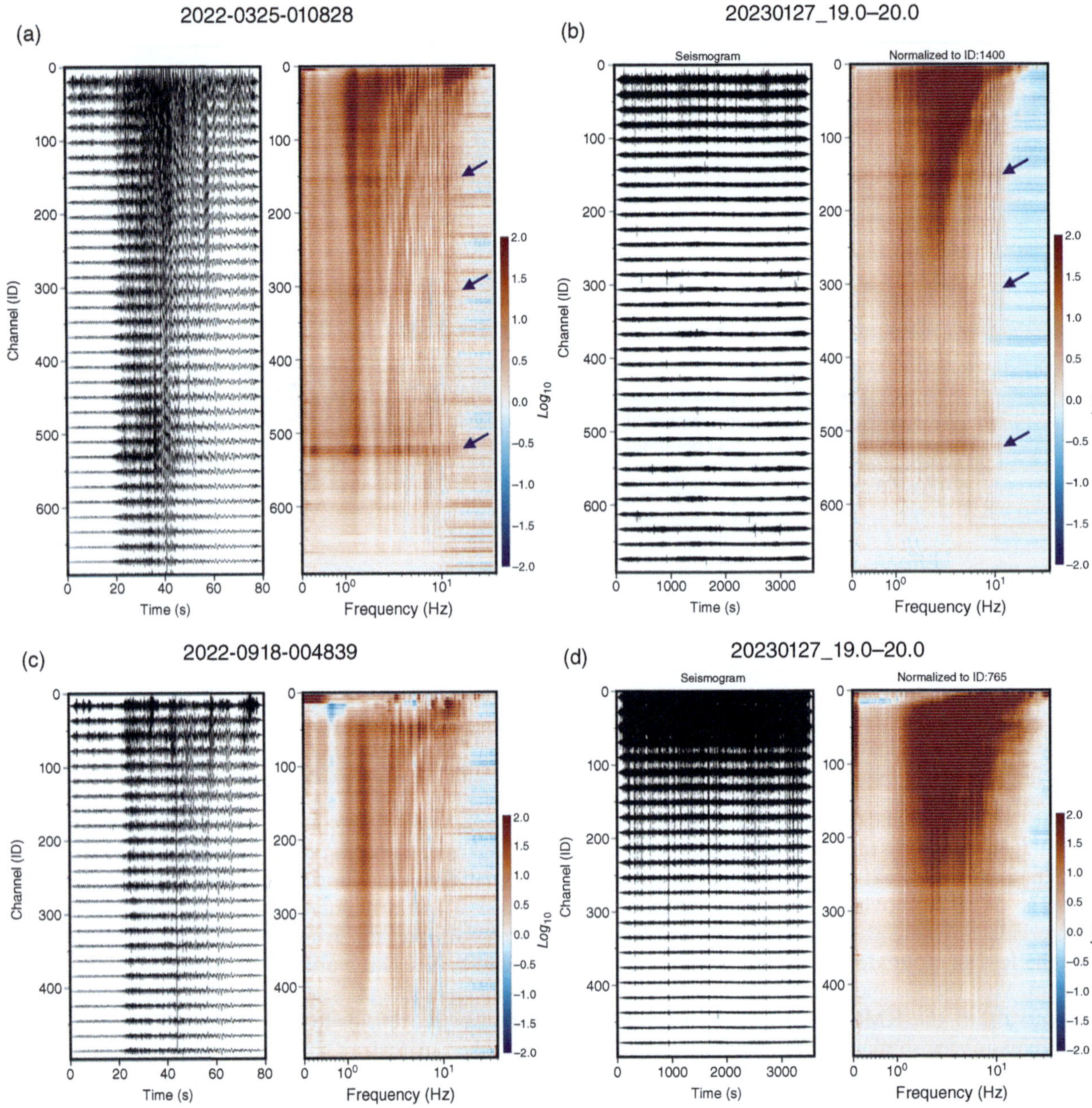

Figure 26.10 Depth-dependent spectral characteristics during strong motions (e.g., earthquakes) and weak motions at Hole A and Hole B. Left and right panels in (a–d) show DAS strain-rate waveforms and relative PSD diagram. (a) and (b) are results of Hole A for an earthquake on 25 March 2022 with an 80 s window and 1 hr ambient noise data on 27 January 2023 midnight Taiwan time, respectively. (c) and (d) are results of Hole B for an earthquake on 18 September 2022 with an 80 s window and 1 hr ambient noise data on 27 January 2023 midnight Taiwan time, respectively. Blue arrows indicate the depths of strain-rate amplification.

1–30 Hz frequency band, strong daily cultural noise and variations can be observed within the top approximately 100–150 m. Because our method requires the entire fiber array to record more or less the same wavefields (section 26.2.3), this frequency band is therefore less suitable to use (Figure 26.12(b)). In contrast, the results of the 0.1–1 Hz frequency band display much greater stability and high consistency between profiles and with the *Vp* logging profile (left panel), showing a high slowness zone (low-velocity zone) at the major fault zone of 520–540 m and around depths of 170 and 320 m for possible hidden fault structures (blue arrows, Figures 26.10 and 26.12(a)). Although the RMS amplitude ratio profiles in 1–30 Hz also show some temporally continuous features at depths

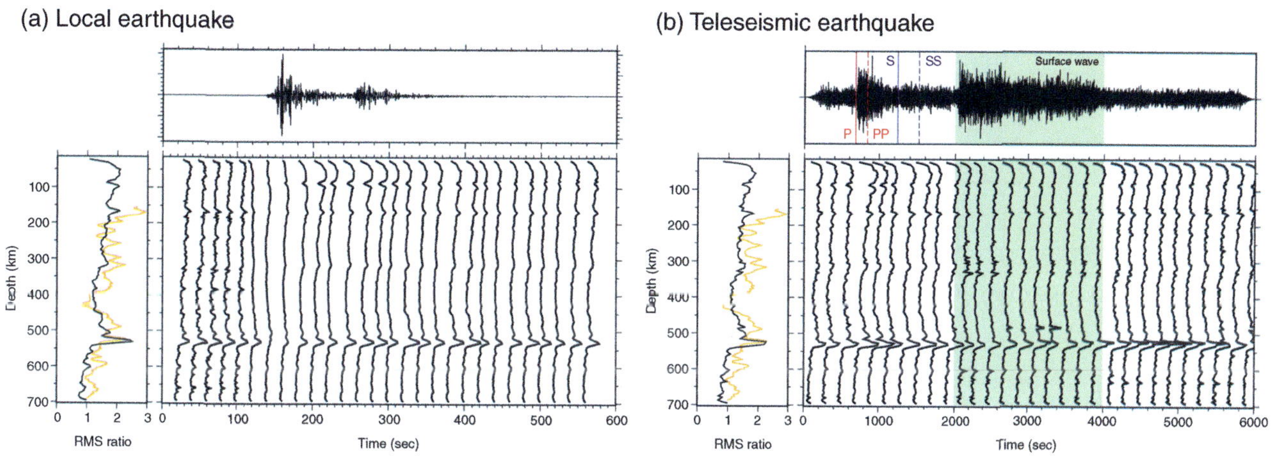

Figure 26.11 Temporal RMS amplitude ratio profiles at Hole A during (a) an M_L6.7 local earthquake on 22 March 2022 and (b) the M_W7.8 Turkey teleseismic earthquake on 6 February 2023. Upper panel shows the DAS strain-rate waveform, where the red and blue vertical lines and green shaded zones in (b) indicate the predicted P- and S-related arrivals and surface wave train. Lower panel shows the 20 s and 200 s moving-window calculation of amplitude RMS ratios for local and teleseismic events, respectively. Left panel shows the comparison between the mean amplitude RMS ratio profile (black) and logging P-wave slowness (1/velocity) profile (orange).

of 320, 490, and 610 m, they are not consistent with the logging *Vp* profile and may represent not only the velocity structure but also local amplification due to wave interference and may therefore be less stable for monitoring purposes. These complex wave–wave or wave–structure interactions could be largely reduced by considering only long-period waves such as in 0.1–1 Hz, which is mostly dominated by microseism Rayleigh waves. The daily precipitation from a nearby weather station and the daily peak ground motions calculated from SM09 are also displayed for comparison. We find a plausible correlation between the slowness variations at 0–200 m depths and a rainfall event on 24 January 2023. However, because the total amount of rainfall is low (with a peak below 5 mm), more data will be required for the validation.

Figure 26.13 shows the temporal RMS amplitude ratio analysis for Hole B during the same time period. The results of 0.1–1 Hz again outperform those of 1–30 Hz and agree well with those derived from the teleseismic event and the logging *Vp* profile (red and orange lines, left panel). Although the results of 1–30 Hz are relatively stable for Hole B compared to Hole A, the upper part of the profile is still dominated by the surface noise sources and significantly varies with time. The similarity of the lower part of Hole B between the two frequency bands possibly results from less wave–wave or wave-structure interactions without strong structural heterogeneity (e.g., fault zones) as in Hole A. Instead of the high slowness (low velocity) zones, we find two relatively low slowness (high velocity) zones at depths of 150 and 270 m (blue arrows). They may indicate variations in geological formations but require further investigations.

26.5. CONCLUSIONS

The MiDAS project that drills through and monitors the frequently slip Milun Fault with downhole fiber-optic sensing technique (DAS) provides unique opportunities to probe subsurface fault zones at high resolution in both space and time. With continuous and high-quality DAS data, we demonstrate that earthquake signals in terms of spectral characteristics and amplitude recorded by DAS are compatible or at least adaptable with geophone data for typical seismological applications such as routine earthquake monitoring and earthquake early warning, although a saturation phenomenon appears for earthquakes with M_L greater than 6.0. We note that this saturation issue could potentially be mitigated by increasing the sampling frequency or shortening the gauge length, but at the cost of increases in data volume and signal-to-noise ratios.

Moreover, although the DAS strain/strain-rate data share many similarities with seismometer velocity/acceleration data, the strain data that measure the differential displacements in space inherently possess local elastic properties compared to point-wise particle motion data. We therefore derive and examine a direct use of strain-rate amplitude rather than phase information to infer subsurface velocity structures. We show that under certain circumstances, such as in long frequency bands

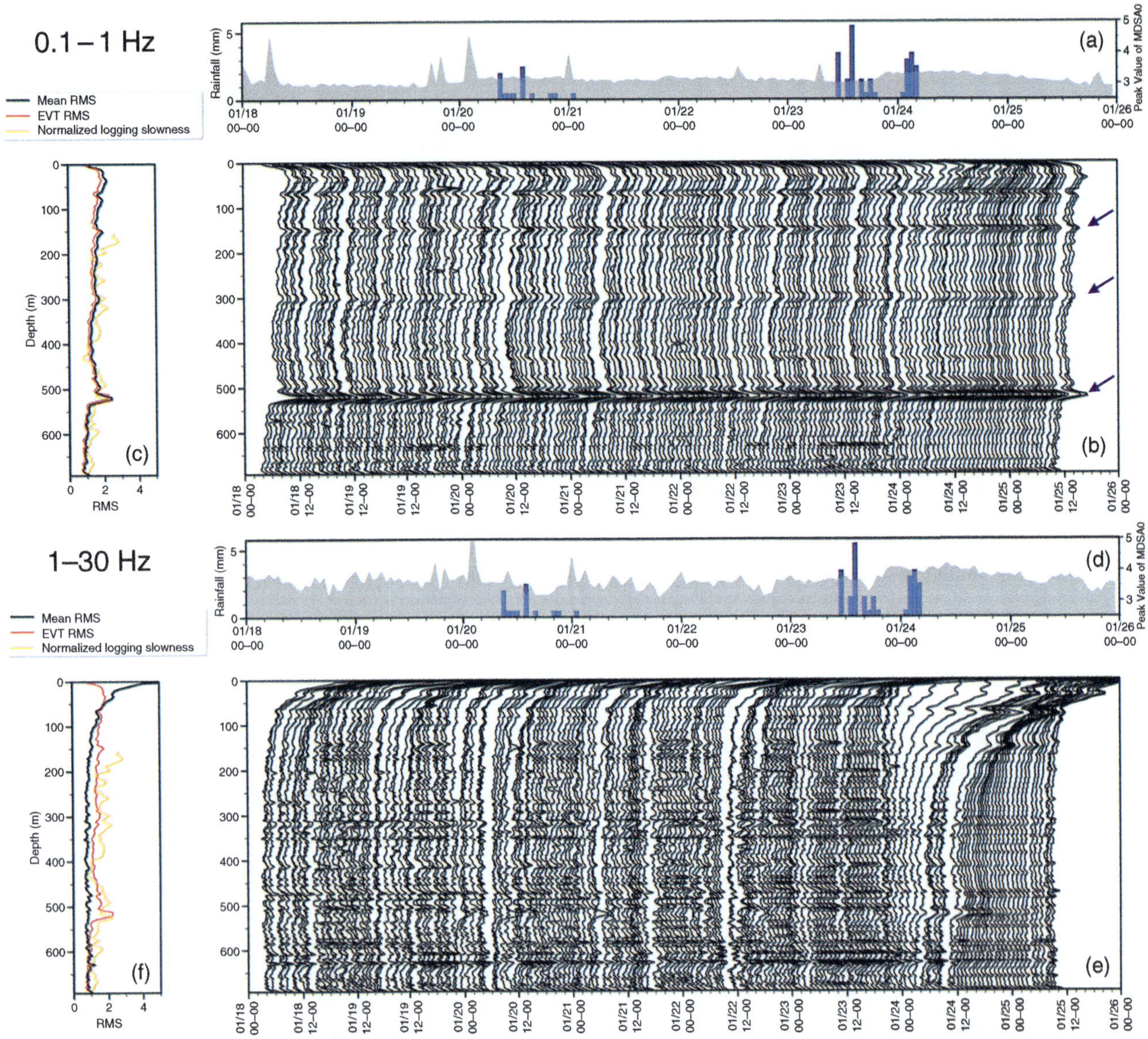

Figure 26.12 Long-term temporal RMS amplitude ratio profiles at Hole A from 18–25 January 2023 in (a–c) 0.1–1 Hz and (d–f) 1–30 Hz frequency bands. (a) and (d) show the hourly precipitation in blue bars and the hourly peak ground motions in gray. (b) and (e) show the hourly moving-window calculation of RMS amplitude ratios in corresponding frequency bands. (c) and (f) show the comparison between the mean RMS amplitude ratio profile (black), mean RMS amplitude ratio profile derived from earthquakes in Figure 26.11 (red) and logging P-wave slowness (1/velocity) profile (orange). Blue arrows indicate the low velocity zones or rigidity contrasts.

(e.g., 0.1–1 Hz), the proposed RMS amplitude ratio method could quickly and rightly capture variations in elastic properties (i.e., slowness) with depth and agree well with the logging P-wave velocity profile for both strong and weak motions. More examinations of the method in different conditions and tectonic settings are still required. Because borehole fiber experiments exist widely in the oil/gas industry and may start to increase in green energy fields, this method may provide a new means for high-resolution and continuous spatiotemporal monitoring of subsurface medium changes, in addition to common micro-seismicity monitoring. Further applications in, but not limited to, groundwater, infrastructure health, and geohazards monitoring are highly anticipated in the near future.

ACKNOWLEDGMENTS

We are grateful for the helpful discussion with the Silixa consultant team for borehole fiber installation and DAS

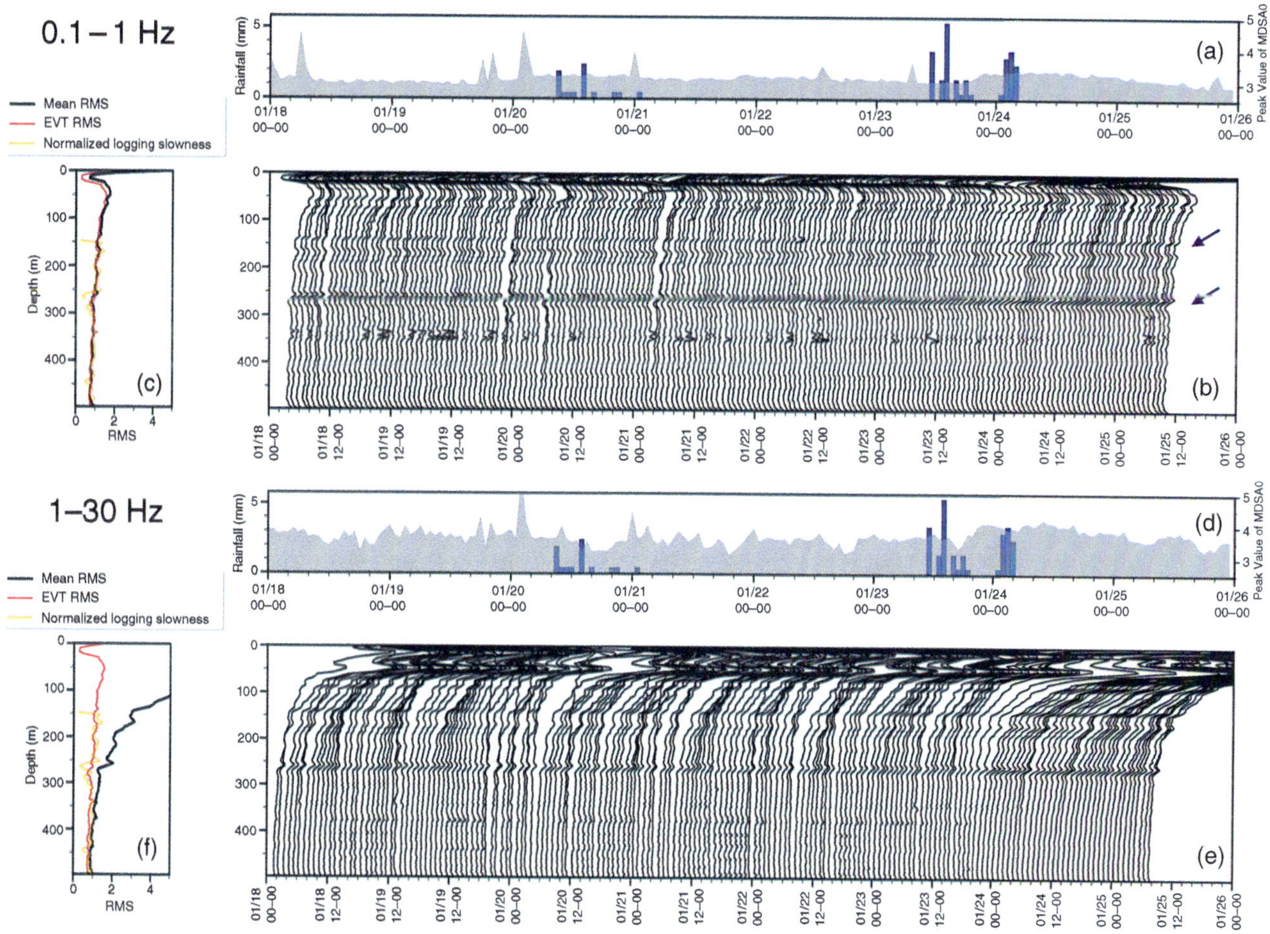

Figure 26.13 Long-term temporal RMS amplitude ratio profiles at Hole B from 18–25 January 2023 in (a–c) 0.1–1 Hz and (d–f) 1–30 Hz frequency bands. (a) and (d) show the hourly precipitation in blue bars and the hourly peak ground motions in gray. (b) and (e) show the hourly moving-window calculation of RMS amplitude ratios in corresponding frequency bands. (c) and (f) show the comparison between the mean RMS amplitude ratio profile (black), mean RMS amplitude ratio profile derived from earthquakes in Figure 26.11 (red), and logging P-wave slowness (1/velocity) profile (orange). Blue arrows indicate the low-velocity zones or rigidity contrasts.

strain data processing and to Kyle Smith for improving the manuscript. We thank the Industrial Technology Research Institute for assisting with borehole logging. We also thank two anonymous reviewers for their constructive comments that improved the work. This work was supported by the Academia Sinica grants AS-GC-110-07 and AS-CDA-111-M01. The MiDAS working group includes C.-Y. Wang, W.-Y. Chang, L.-W. Kuo, Y.-Y. Lin, C.-H. Kuo, C.-C. Fu, S.-J. Wang, S. von Specht, and C.-R. Lin.

AVAILABILITY STATEMENT

The DAS data analyzed in the study are made public via the Data Management Center of the Institute of Earth Sciences, Academia Sinica (IESDMC) (https://dmc.earth.sinica.edu.tw/Contributor/HuangHH/Huang2023/index.html), including 122 event-based data for the ground motion attenuation analysis and one-weak continuous data (18–25 January 2023) for the temporal monitoring analysis.

REFERENCES

Ajo-Franklin, J., Tribaldos, V. R., Nayak, A., Cheng, F., Mellors, R., Chi, B., et al. (2022). The Imperial Valley dark Fiber project: toward seismic studies using DAS and telecom infrastructure for geothermal applications. *Seismological Research Letters*, *93*, 2906–2919. doi:10.1785/0220220072

Chan, C.-H., Ma, K.-F., Shyu, J. B. H., Lee, Y.-T., Wang, Y.-J., Gao, J.-C., et al. (2020). Probabilistic seismic hazard assessment for Taiwan: TEM PSHA2020. *Earthquake Spectra*, *1–23*. doi:10.1177/8755293020951587

Chen, C.-Y., Lee, J.-C., Chen, Y.-G., & Chen, R.-F. (2014). Campaigned GPS on present-day crustal deformation in northernmost Longitudinal Valley preliminary results, Hualien Taiwan. *Terrestrial, Atmospheric and Oceanic Sciences, 25*, 337–357.

Correa, J., Egorov, A., Tertyshnikov, K., Bona, A., Pevzner, R., Dean, T., et al. (2017). Analysis of signal to noise and directivity characteristics of DAS VSP at near and far offsets — a CO2CRC Otway project data example. *The Leading Edge, 994a1*. doi:10.1190/tle36120994a1.1

Daley, T., Miller, D., Dodds, K., Cook, P., & Freifeld, B. (2016). Field testing of modular borehole monitoring with simultaneous distributed acoustic sensing and geophone vertical seismic profiles at Citronelle, Alabama. *Geophysical Prospecting, 64*, 1318–1334.

Dean, T., Cuny, T., & Hartog, A. H. (2017). The effect of gauge length on axially incident P-waves measured using fibre optic distributed vibration sensing. *Geophysical Prospecting, 65*, 184–193.

Dou, S., Lindsey, N., Wagner, A. M., Daley, T. M., Freifeld, B., Robertson, M., et al. (2017). Distributed acoustic sensing for seismic monitoring of the near surface: a traffic-noise interferometry case study. *Scientific Reports, 7*(1), 11620. doi:10.1038/s41598-017-11986-4

Glubokovskikh, S., Pevzner, R., Sidenko, E., Tertyshnikov, K., Gurevich, B., Shatalin, S., et al. (2021). Downhole distributed acoustic sensing provides insights into the structure of short-period ocean-generated seismic wavefield. *Journal of Geophysical Research: Solid Earth, 126*, e2020JB021463. doi:10.1029/2020JB021463

Huang, H.-H., Ma, K.-F., Lin, C.-J., Ku, C.-S., Kuo, L.-W., Wu, E.-S., & Cheng, Y.-Z. (2022). *Probing an active fault zone through fibers: the Milun fault drilling and all-inclusive sensing (MiDAS) project*. Paper presented at the AGU Fall Meeting.

Huang, H.-H., Shyu, J. B. H., Wu, Y.-M., Chang, C.-H., & Chen, Y.-G. (2012). Seismotectonics of northeastern Taiwan: structural characteristics of a transitional area from Wanning collision to subduction and post-collisional extension. *Journal of Geophysical Research, 117*, B01313. doi:10.1029/2011JB008852

Huang, H.-H., & Wang, Y. (2022). Seismogenic structure beneath the northern Longitudinal Valley revealed by the 2018-2021 Hualien earthquake sequences. *Terrestrial, Atmospheric and Oceanic Sciences, 33*, 17. doi:10.1007/s44195-022-00017-z

Huang, M.-H., & Huang, H.-H. (2018). The complexity of the 2018 Mw6.4 Hualien earthquake in East Taiwan. *Geophysical Research Letters, 45*, 13–249. doi:10.1029/2018GL080821

Huang, S.-Y., Yen, J.-Y., Wu, B.-L., Yen, I.-C., & Chuang, R. Y. (2019). Investigating the Milun Fault: the coseismic surface rupture zone of the 2018/02/06 M_L 6.2 Hualien earthquake, Taiwan. *Terrestrial, Atmospheric & Oceanic Sciences, 30*(3), 311–335.

Jian, P.-R., Liang, W.-T., & Kuo, B.-Y. (2022). Three-dimensional stress model of the collision-subduction junction east of Taiwan: implications for the decoupling of the Luzon arc during subduction. *Journal of Geophysical Research: Solid Earth, 127*, e2022JB024054. doi:10.1029/2022JB024054

Jousset, P., Currenti, G., Schwarz, B., Chalari, A., Tilmann, F., Reinsch, T., et al. (2022). Fibre optic distributed acoustic sensing of volcanic events. *Nature Communications, 13*, 1753.

Jousset, P., Reinsch, T., Ryberg, T., Blanck, H., Clarke, A., Aghayev, R., et al. (2018). Dynamic strain determination using fiberoptic cables allows imaging of seismological and structural features. *Nature Communications, 9*, 2509.

Lee, S.-J., Lin, T.-C., Liu, T.-Y., & Wong, T.-P. (2018). Fault-to-fault jumping rupture of the 2018 mw 6.4 Hualien earthquake in eastern Taiwan. *Seismological Research Letters, 90* (1), 30–39. doi:10.1785/0220180182

Lellouch, A., Lindsey, N. J., Ellsworth, W. L., & Biondi, B. L. (2020). Comparison between distributed acoustic sensing and geophones: downhole microseismic monitoring of the FORGE geothermal experiment. *Seismological Research Letters, 91*, 3256–3268. doi:10.1785/0220200149

Lellouch, A., Yuan, S., Ellsworth, W. L., & Biondi, B. (2019). Velocity-based earthquake detection using downhole distributed acoustic sensing—examples from the San Andreas fault Observatory at Depth. *Bulletin of the Seismological Society of America, 109*, 2491–2500.

Li, D., Huang, L., Zheng, Y., Li, Y., Wannamaker, P., & Moore, J. (2022). Feasibility of source-free DAS logging for next-generation borehole imaging. *Scientific Reports, 12*, 11910.

Li, Z., Shen, Z., Yang, Y., Williams, E., Wang, X., & Zhan, Z. (2021). Rapid response to the 2019 Ridgecrest earthquake with distributed acoustic sensing. *AGU Advances, 2*, e2021AV000395. doi:10.1029/2021AV000395

Lindsey, N. J., Dawe, T. C., & Ajo-Franklin, J. B. (2019). Illuminating seafloor faults and ocean dynamics with dark fiber distributed acoustic sensing. *Science, 366*, 1103–1107. doi:10.1126/science.aay5881

Lindsey, N. J., & Martin, E. R. (2021). Fiber-optic seismology. *Annual Review of Earth and Planetary Sciences, 49*, 309–336.

Lindsey, N. J., Rademacher, H., & Ajo-Franklin, J. B. (2020). On the broadband instrument response of fiber-optic DAS arrays. *Journal of Geophysical Research: Solid Earth, 125*, e2019JB018145. doi:10.1029/2019JB018145

Lior, I., Rivet, D., Ampuero, J. P., Sladen, A., Barrientos, S., Sánchez-Olavarría, R., et al. (2023). Magnitude estimation and ground motion prediction to harness fiber optic distributed acoustic sensing for earthquake early warning. *Scientific Reports, 13*(1), 424.

Lior, I., Sladen, A., Mercerat, D., Ampuero, J., Rivet, D., & Sambolian, S. (2021). Strain to ground motion conversion of distributed acoustic sensing data for earthquake magnitude and stress drop determination. *Solid Earth, 12*, 1421–1442.

Ma, K.-F., Lin, C.-J., Ku, C.-S., & Huang, H.-H. (2022). *Taiwan Milun fault drilling and all-inclusive sensing (MiDAS) project: downhole optical fiber through frequent slip active fault zone*. Paper presented at the Seismological Society of America Annual Meeting.

McNamara, D. E., & Buland, R. P. (2004). Ambient noise levels in the continental United States. *Bulletin of the Seismological Society of America, 94*(4), 1517–1527.

Shin, T.-C. (1993). The calculation of local magnitude from the simulated Wood-Anderson seismograms of the short-period seismograms in the Taiwan area. *Terrestrial, Atmospheric and Oceanic Sciences*, *4*(2), 155–170.

Shyu, J. B. H., Chung, L.-H., Chen, Y.-G., Lee, J.-C., & Sieh, K. (2007). Re-evaluation of the surface ruptures of the November 1951 earthquake series in eastern Taiwan, and its neotectonic implications. *Journal of Asian Earth Sciences*, *31*(3), 317–331. doi:10.1016/j.jseaes.2006.07.018

Shyu, J. B. H., Chuang, Y.-R., Chen, Y.-L., Lee, Y.-R., & Cheng, C.-T. (2016). A new on-land seismogenic structure source database from the Taiwan earthquake model (TEM) project for seismic hazard analysis of Taiwan. *Terrestrial, Atmospheric and Oceanic Sciences*, *27*(3), 311. doi:10.3319/tao.2015.11.27.02(tem)

Spica, Z., Perton, M., Martin, E., Beroza, G., & Biondi, B. (2020). Urban seismic site characterization by fiber-optic seismology. *Journal of Geophysical Research - Solid Earth*, *125*, e2019JB018656.

Tribaldos, V. R., & Ajo-Franklin, J. B. (2021). Aquifer monitoring using ambient seismic noise recorded with distributed acoustic sensing (DAS) deployed on dark fiber. *Journal of Geophysical Research: Solid Earth*, *126*, e2020JB021004. doi:10.1029/2020JB021004

Walter, F., Gräff, D., Lindner, F., Paitz, P., Köpfli, M., Chmiel, M., & Fichtner, A. (2020). Distributed acoustic sensing of microseismic sources and wave propagation in glaciated terrain. *Nature Communications*, *11*, 2436.

Wang, H. F., Zeng, X., Miller, D. E., Fratta, D., Feigl, K. L., Thurber, C. H., & Mellors, R. J. (2018). Ground motion response to an M_L 4.3 earthquake using co-located distributed acoustic sensing and seismometer arrays. *Geophysical Journal International*, *213*, 2020–2036.

Yen, J.-Y., Lu, C.-H., Dorsey, R. J., Kuo-Chen, H., Chang, C.-P., Wang, C.-C., et al. (2019). Insights into Seismogenic deformation during the 2018 Hualien, Taiwan, earthquake sequence from InSAR, GPS, and Modeling. *Seismological Research Letters*, *90*(1). doi:10.1785/0220180228

Zhan, Z. (2019). Distributed acoustic sensing turns Fiber-optic cables into sensitive seismic antennas. *Seismological Research Letters*, *91*, 1–15. doi:10.1785/0220190112

Zhu, T., & Stensrud, D. J. (2019). Characterizing thunder-induced ground motions using Fiber-optic distributed acoustic sensing Array. *Journal of Geophysical Research: Atmospheres*, *124*, 12,810–12,823. doi:10.1029/2019JD031453

27

Monitoring Granitic and Sedimentary Rocks at the Utah FORGE Site Using Borehole Distributed Acoustic Sensing Ambient Noise

David Li[1], Lianjie Huang[1], Yingcai Zheng[2], and Yingping Li[2,3]

ABSTRACT

Distributed acoustic sensing (DAS) has great potential for subsurface monitoring in high-temperature, high-pressure environments such as enhanced geothermal systems (EGSs). A 1 km DAS fiber was installed in an observation well (78-32) at the Frontier Observation for Research in Geothermal Energy (FORGE) site in Utah to continuously monitor induced seismicity during fracture stimulation in an injection well (58-32). We analyze 48 hours of continuously recorded borehole DAS ambient noise data to infer elastic properties of geologic formations. We calculate root-mean-squares (RMS) amplitudes of ambient noise for each DAS channel to construct noise RMS depth profiles. The DAS ambient noise profiles show noise peaks at some depths and quiet zones at others. We compare the DAS RMS profiles with wireline logging data and find apparent correlations between major RMS peaks and highly fractured low-velocity layers bounded by the structural interfaces. We apply time-lapse and time-frequency analysis and find that the noise RMS amplitudes vary with time for quiet zones. We also perform spectral analysis and find that the noisy channels contain both blue and white noises. Our results show that borehole DAS ambient noise can be used as a source-free logging tool to monitor temporal changes in subsurface rock properties.

27.1. INTRODUCTION

The development of long-term borehole monitoring tools is crucial for monitoring the development of an enhanced geothermal system (EGSs). However, the hostile environment of an EGS makes acquiring long-term monitoring data difficult, if not impossible. DAS systems have become popular in acquiring seismic data in high-temperature and high-pressure environments (Li et al., 2021) because the sensor relies on a fiber-optic cable, which can function properly under the hostile conditions of an EGS. By contrast, conventional geophones and sonic well-logging tools contain sensitive electronics that can be damaged in high-temperature environments such as EGS reservoirs where the temperature is usually higher than 150 °C. However, because the electronic components of a DAS system are in an interrogator box, the sensitive electronic components can be placed at a safe distance from the heat source. In this chapter, we use passive borehole DAS ambient noise to study and monitor temporal changes in granite and sedimentary rocks at the Utah FORGE site (Figure 27.1).

DAS ambient noise has been used to study shallow S-wave velocity structures in earthquake seismology and to locate underground water tables using DAS fibers deployed along the surface (Ajo-Franklin et al., 2015; Ajo-Franklin et al., 2019; Dou et al., 2017; Martin & Biondi, 2017; Tribaldos et al., 2021; Yang et al., 2022;

[1]*Los Alamos National Laboratory, Los Alamos, New Mexico, USA*

[2]*Department of Earth and Atmospheric Sciences, University of Houston, Houston, Texas, USA*

[3]*BlueSkyDAS LLC, Sugar Land, Texas, USA*

Distributed Acoustic Sensing in Borehole Geophysics, Geophysical Monograph 289, First Edition.
Edited by Yingping Li, Robert Mellors, and Ge Zhan.

DOI: 10.1002/9781394179275.ch27

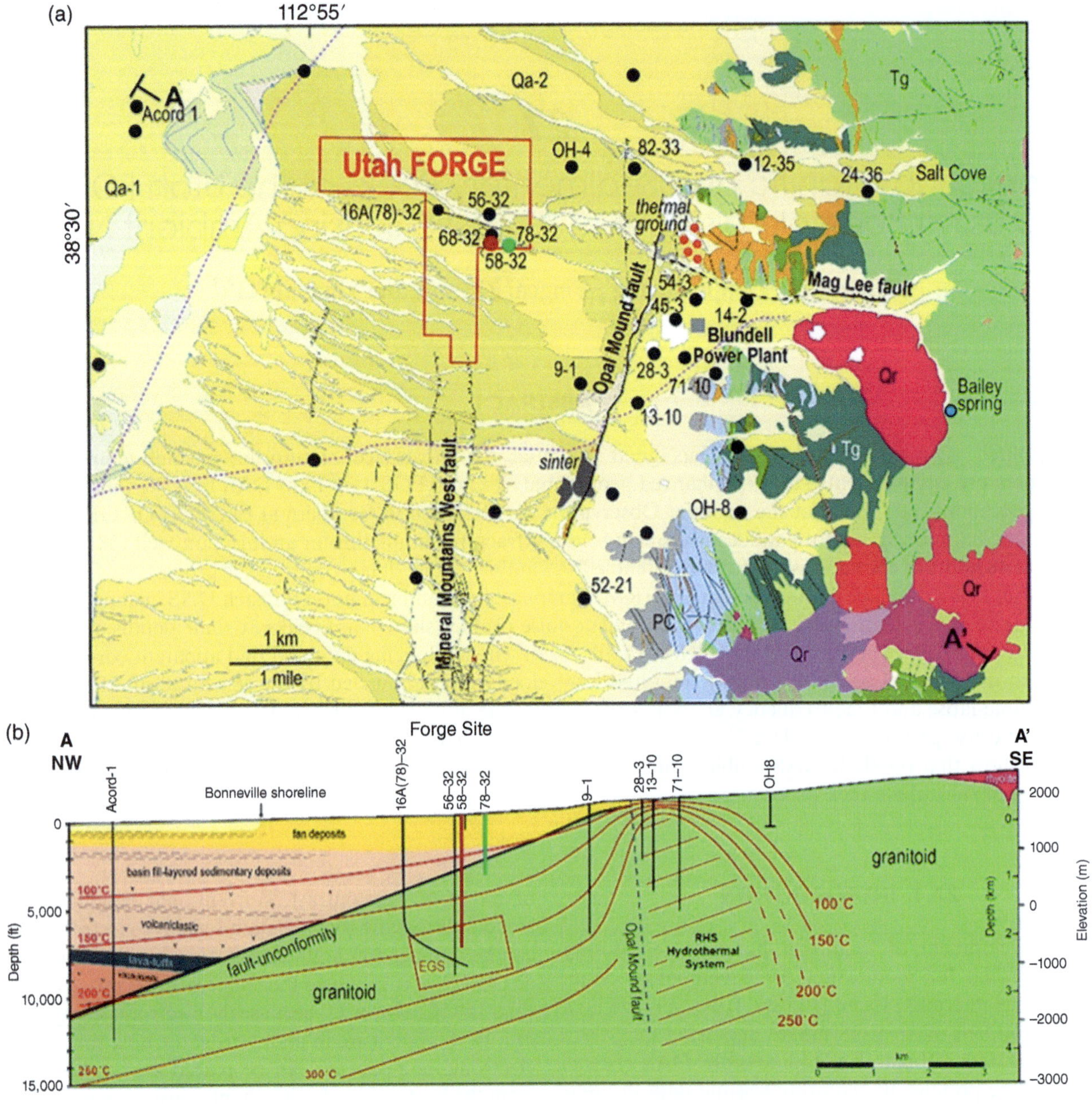

Figure 27.1 Geological setting. (a) Map view and (b) cross-section of the Utah FORGE geothermal site. The treatment well (58-32) was drilled to the EGS reservoir with a temperature of ~200 °C. Shallow Well 78-32 serves as an observation well using 1 km DAS fiber with 1 m sensor spacing to monitor treatment well activities. Li et al. (2022) / Springer Nature / CC BY 4.0.

Zeng et al., 2017, 2021). In boreholes, DAS is usually deployed to acquire active seismic data, such as DAS vertical seismic profiling (Correa et al., 2017; Daley et al., 2013, 2016; Harris et al., 2016; Li et al., 2015; Mateeva et al., 2013; Mestayer et al., 2011; Miller et al., 2016; Willis et al., 2016; Zhan et al., 2016). There are also a few studies on borehole DAS ambient noise (Chang & Nakata, 2022; Li et al., 2022). Li et al. (2022) conducted a feasibility study using borehole DAS ambient noise as a source-free logging tool. They found that areas of high ambient noise can be correlated with mechanical properties inferred from wireline well-logging data. In this chapter, we apply that principle to monitor temporal changes of mechanical properties in granite and sedimentary rocks at the Utah FORGE site.

We first give a brief background of the geology at the Utah FORGE site and discuss the parameters of the DAS observation system that collected the borehole

DAS ambient noise data. We then discuss how we process the raw borehole DAS ambient noise data to remove other types of noise and isolate the ambient noise. We construct the DAS Ambient Noise RMS Amplitude Profile (DANRAP), which quantifies the level of ambient noise on each channel in the monitoring well. We first obtain the DANRAPs for borehole DAS ambient noise data acquired in granite and shallow sedimentary rocks. The DANRAPs of both granite and sedimentary rocks show depths with high RMS amplitude noise peaks and other depths that are quiet. We compare the DANRAPs in granite and sedimentary rocks to wireline well-logging data and find that the peaks in the DANRAPs correlate with rock formations that are in low-velocity layers (LVLs) bound by formation boundaries and have low Poisson's ratio, low densities, and high porosities. We suggest that these properties are correlated with fracture zones and that trapped waves in the fracture zones may be the source of the observed noise peaks. We perform amplitude spectra analysis and find that the noise is characterized by blue noise below 500 Hz and white noise above 500 Hz. We also find that the noise RMS amplitudes of quiet channels vary with time, showing daily variations of high and low amplitudes. Our results show that borehole DAS ambient noise can be used to detect fractured media and monitor temporal variations in sedimentary and granitic rocks.

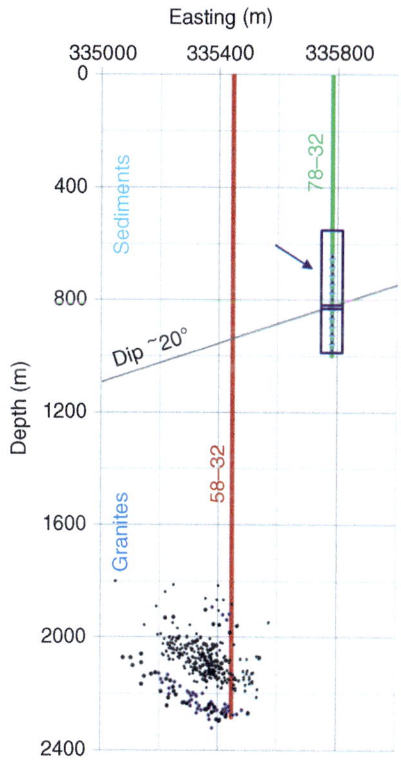

Figure 27.2 A zoomed view of treatment Well 58-32 and monitoring Well 78-32 together with induced microseismic events in black dots. The blue boxes show the depth range of sedimentary and granite rocks for this study. The triangles in the box are 3C geophones. Li et al. (2022) / Springer Nature / CC BY 4.0.

27.2. DAS OBSERVATION SYSTEM AT THE UTAH FORGE SITE

The Utah FORGE site is in Beaver County, Utah, about 100 miles south of Salt Lake City, UT. It is developed as a field laboratory for advancing EGS technologies. The geology of the area includes sedimentary rocks on top of a granitic basement where EGS reservoirs are created. We show a geologic map and a cross-section of the Utah FORGE site provided by the Utah Geologic Survey (Kirby et al., 2019) in Figure 27.1. The two main wells relevant to this study are Wells 58-32 and 78-32. Well 58-32 is the injection well that was drilled vertically to a measured depth of 2297 m. Well 78-32 is the monitoring well. This well is located approximately 336 m east of the injection well and was drilled near vertically to a depth of approximately 1,000 m.

A DAS system at the Utah FORGE site was installed in Well 78-32. The fiber-optic cable of the DAS system is 1 km long and is cemented behind the casing. The fiber-optic cable is connected to a Silixa iDAS interrogator box. The gauge length is 10 m, the channel spacing is 1 m, and the temporal sampling rate is 0.5 ms. This DAS system continuously recorded microseismic data during three stimulation cycles from 19 April to 3 May 2019, as part of Phase 2C for the FORGE project. Additionally, 12 three-component (3C) geophones are also installed in Well 78-32. We show a cross-section (Figure 27.2) showing in detail the locations of Wells 78-32 and 58-32 and the detected microseismic events. In this chapter, we process and analyze borehole DAS ambient noise recorded from 29 April to 1 May 2019, between two stimulation cycles to avoid noise caused by stimulation activities.

27.3. CLASSIFYING AND PROCESSING DAS AMBIENT NOISE DATA

Before analyzing borehole DAS ambient noise, we first classify what DAS ambient noise is and separate it from other types of noise. We observe four distinct types of noises in the data. The first is ambient noise. This noise occurs on all channels of the DAS data, with larger amplitudes on some channels and smaller on others, qualitatively indicating depth variations of noise amplitudes. We refer to ambient noise as Type A noise. We show a figure of the DAS ambient noise acquired in granite in Figure 27.3(a).

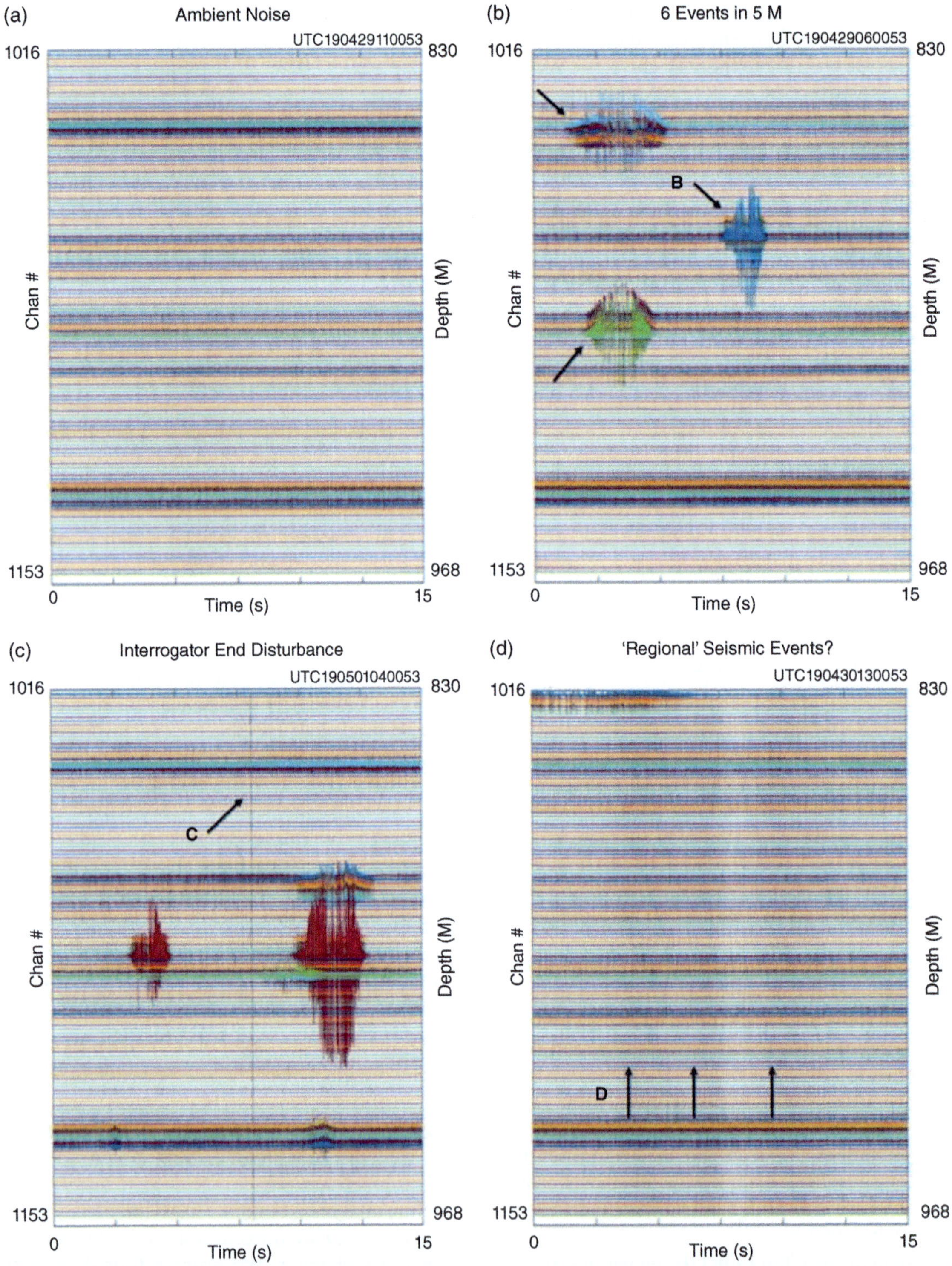

Figure 27.3 (a) 15 s time records of borehole DAS ambient noise from the Utah FORGE site. (b) Type B noise, six events occurring in a 5 m interval. (c) Type C noise, IED. (d) Type D noise, possibly a regional seismic event.

The second type of noise in the borehole DAS data is six events occurring in a 5 m spacing. We refer to this type of noise as Type B noise. We show the Type B noise in Figure 27.3(b). The Type B noise is peculiar as it always occurs randomly across a six-channel set. We show a zoomed view of the Type B noise in Figure 27.4(a). From the figure, we observe that the Type B noise waveforms on the six channels in 5 m are similar to one another and that

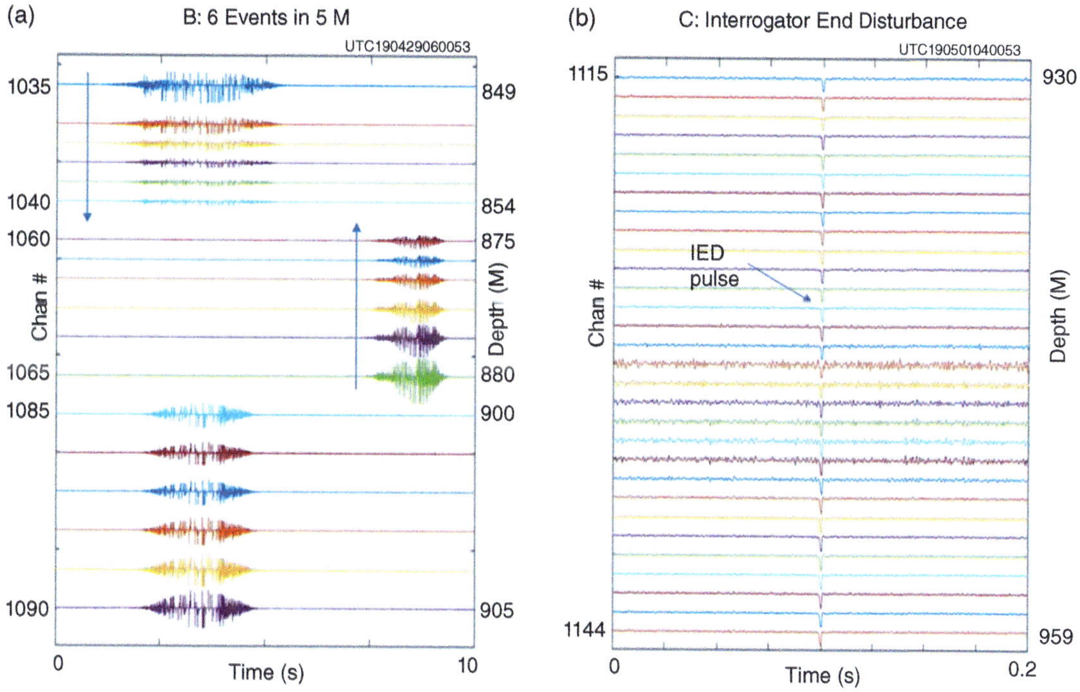

Figure 27.4 Zoomed-in view of (a) Type B noise: six events in a 5 m interval and (b) Type C noise: IED with a simple negative pulse. Note that in each group of Type B noise, amplitudes either decrease or increase as the channel depth increases.

the amplitude of this noise increases either in the upgoing or the downgoing direction. We do not know the source of this noise that occurs often in the data.

The third type of noise is the interrogator end disturbance (IED). We call the IED noise as Type C noise. We show an example of Type C noise in Figure 27.3(c). This noise occurs when the interrogator box collecting and processing the data is disturbed by some outside force. As such, this noise occurs across all channels in the DAS data. We show a zoomed view of the Type C noise in Figure 27.4(b). This zoomed-in graph shows that the Type C noise is a simple pulse that occurs across all channels.

Finally, the last type of noise is a low-frequency rumble that occurs on one of the 15 s time records. We refer to this low-frequency noise as Type D noise and show an example of this noise in Figure 27.3(d). The source of this noise might be from a regional seismic event, but we do not find an earthquake in the catalog associated with this event. Fortunately, this type of noise only occurs once in the 48 h window of the data we use.

To analyze the borehole DAS ambient noise data, we must process the data to remove all other types of noise except the ambient noise. We apply a processing approach outlined in Figure 27.5 to remove Types B, C, and D noises while preserving Type A noise. Figure 27.5(a) shows a 15 s time record of DAS data recorded in granite with Type A, B, and C noises. To remove the Type B noise, we select the time windows when this noise occurs and mute the traces with Type B noise. Figure 27.5(b) shows the data after muting the Type B noise. After removing the Type B noise from the data, we apply a despiking method outlined in Figure 27.6 to remove the Type C noise. To remove the IED noises, we first stack all the traces together to obtain the IED spikes; then we subtract the IED spikes from each trace to remove the IED spikes while preserving the ambient noise. This despiking method works well to remove the noisy traces from both quiet traces (Figure 27.6(b)) and noisy traces (Figure 27.6(c)). Figure 27.5(c) shows the data after removing both Type B and Type C noises. To remove the Type D noise, we simply exclude the data file with the Type D noise from our analysis. After removing the Type B, C, and D noises from the data, we obtain RMS noise amplitude profiles to quantify the ambient noise level versus depth.

27.4. DAS AMBIENT NOISE RMS AMPLITUDE PROFILE

After removing Type B, C, and D noises from the data, we quantify the ambient noise data on each channel of DAS data by finding the RMS amplitudes on each channel using

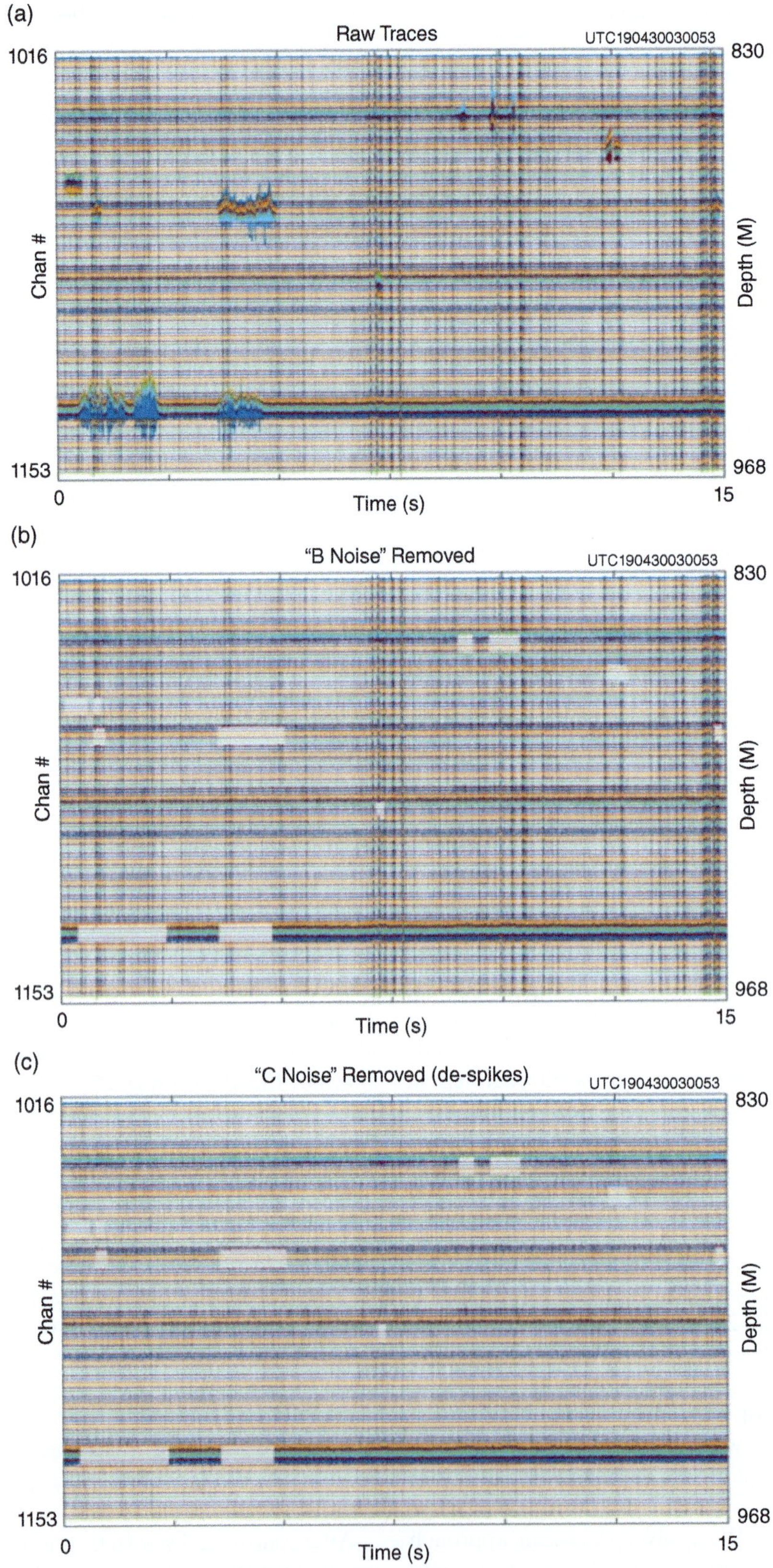

Figure 27.5 Method showing processing step to isolate ambient noise and remove other types of noise. (a) Raw traces showing data with Type A, B, and C noises. (b) Data after removing Type B noise by muting. (c) Data after removing Type C noise using a despiking method.

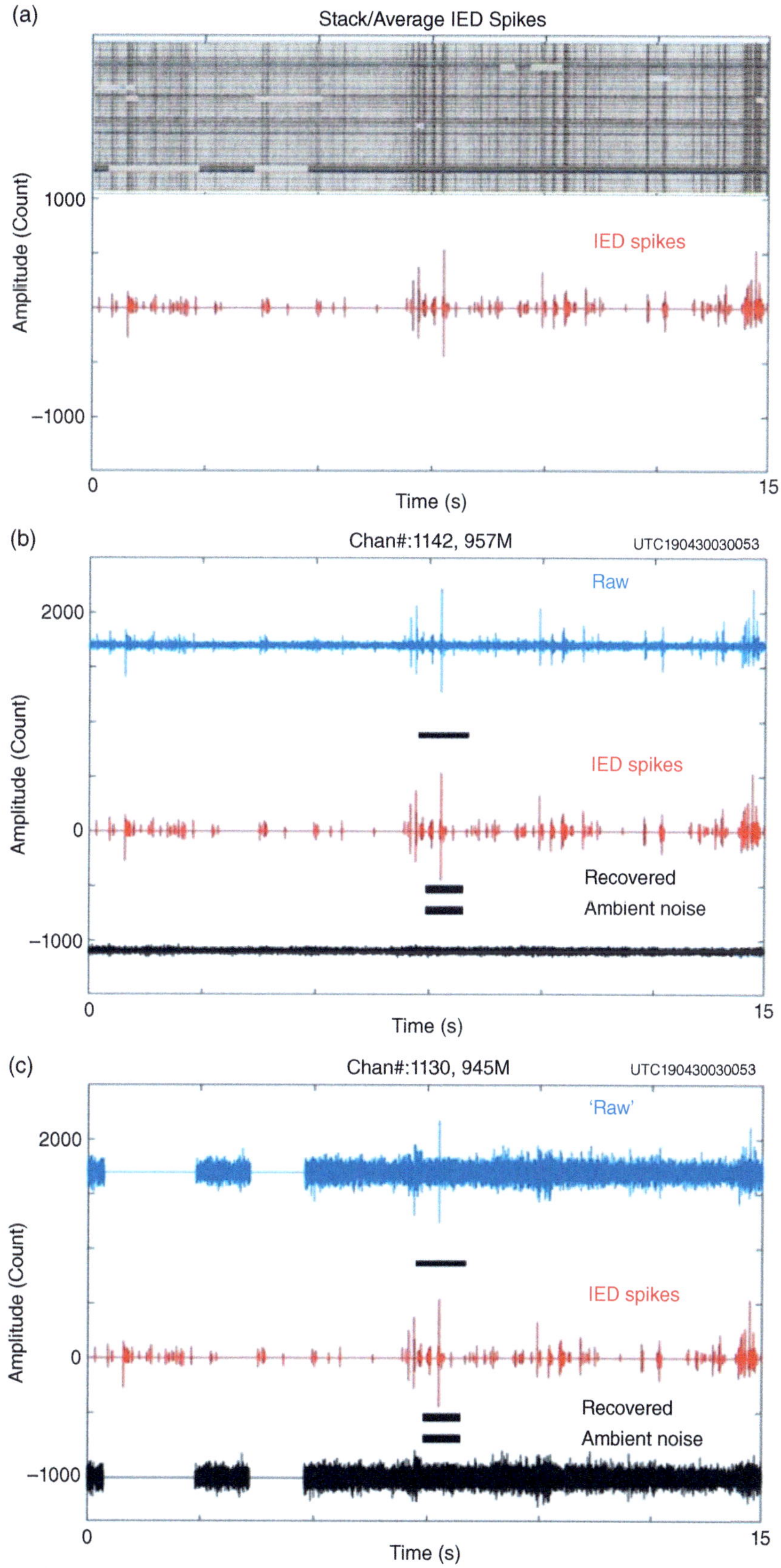

Figure 27.6 Despiking method to remove IED spikes. (a) Stacked IED spikes. (b) The trace of IED spikes is subtracted from the raw noise trace to preserve the ambient noise of a relatively quiet channel (1142) at a depth of 957 m. (c) Preserved useful ambient noise trace of a relatively noisy channel (1130) at a depth of 945 m.

$$N_{rms} = \sqrt{\left(\frac{1}{n}\sum_i y_i^2\right)},$$

where N_{rms} is the noise RMS amplitude, n is the total number of time samples, and y_i is the amplitude at each time sample i. We calculate the noise RMS amplitude on each channel of the DAS data to construct a DANRAP (Figure 27.7(b)). The DANRAP shows how the level of ambient noise varies with depth. Some depths have large noise peaks, whereas the other depths are quiet. In this section, we first calculate and analyze DANRAPs from the DAS data recorded mainly in granite (830–969 m) and then calculate and analyze the DANRAPs from the data recorded mainly in sedimentary rocks (550–850 m).

27.4.1. DAS Ambient Noise Depth Profile in Granite

We show a 15 s seismogram of the DAS ambient noise collected in the granitic rock (830–968 m) along with the calculated noise RMS amplitude profile in Figure 27.7. Figure 27.7(b) shows that the noise is stronger on some channels and weaker on others. We plot 49 noise RMS amplitude profiles recorded in the 48 h period in the granitic rock in Figure 27.8(a). We observe that the 49 DANRAPs in the granitic rock share a similar shape over the entire 48 h period. We show 49 superimposed DANRAPs along with the average of those DANRAPs in Figure 27.8(b). The average DANRAP has the same shape as the individual DANRAPs. From the averaged DANRAP in the granitic rock, we find five distinctive noise peaks and label them GN-1, GN-2, GN-3, GN-4, and GN-5.

27.4.2. DAS Ambient Noise Depth Profile in Sedimentary Rocks

We show a 15 s seismogram of the DAS ambient noise collected in the sedimentary rocks (550–850 m) along with the calculated noise RMS amplitude profile in Figure 27.9. The figure shows that the noise is stronger on some channels and weaker on the others. We plot 49 noise RMS amplitude profiles recorded in the 48 h period in the sedimentary rocks in Figure 27.10(a). As with the DANRAPs in granite, we observe that the 49 DANRAPs in the sedimentary rocks have a similar shape over the entire 48 h period. We show 49 superimposed DANRAPs along with the average of those DANRAPs in Figure 27.10(b). The average DANRAP maintains a similar shape as the individual DANRAPs. From the averaged DANRAP of the sedimentary rocks, we find seven major noise peaks with large amplitudes and seven minor noise peaks with smaller amplitudes on the

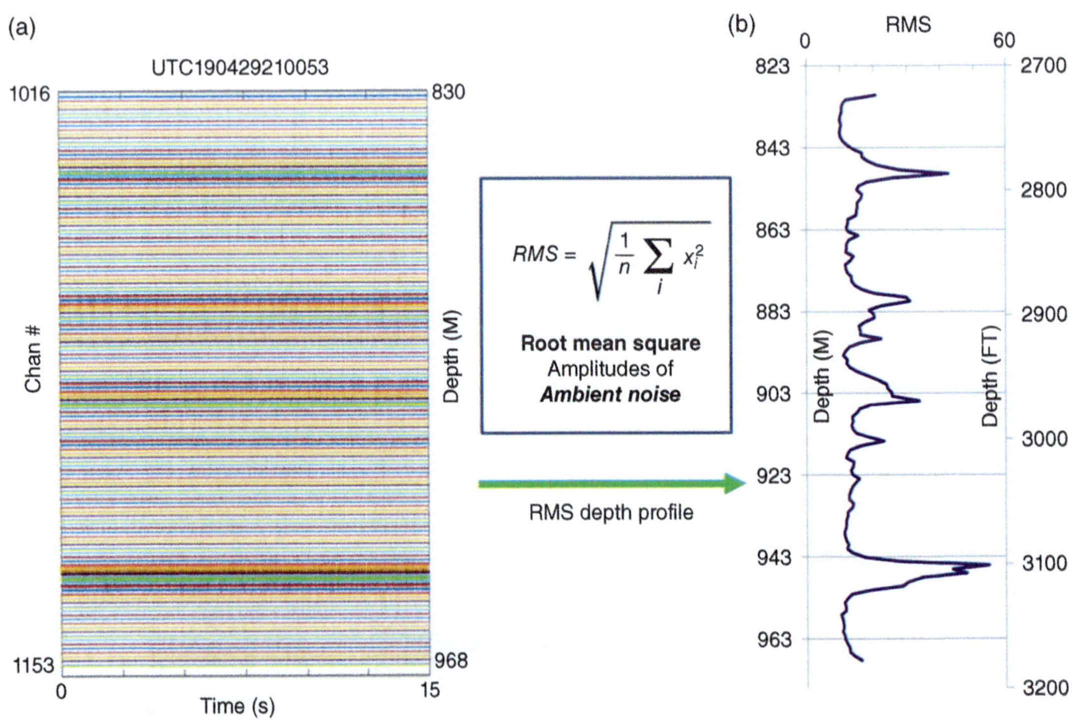

Figure 27.7 (a) 15 s seismogram showing channels of DAS ambient noise waveform data recorded in granitic rock (measured depth 830–968 m). (b) DANRAP calculated using the data in (a).

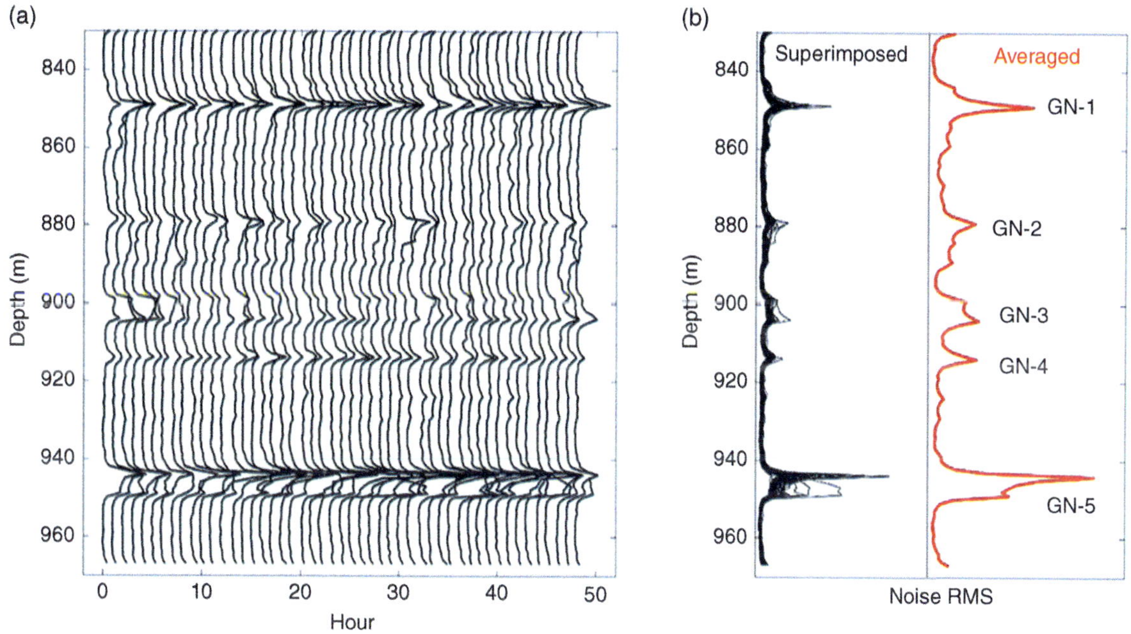

Figure 27.8 (a) 48 h of noise RMS amplitude profiles of the DAS ambient noise data collected in granitic rock (830–968 m). (b) Superimposed DANRAP of the 48 h on the left and averaged DANRAP. The DANRAPs all have a similar shape with five distinctive noise peaks labeled GN-1, GN-2, GN-3, GN-4, and GN-5.

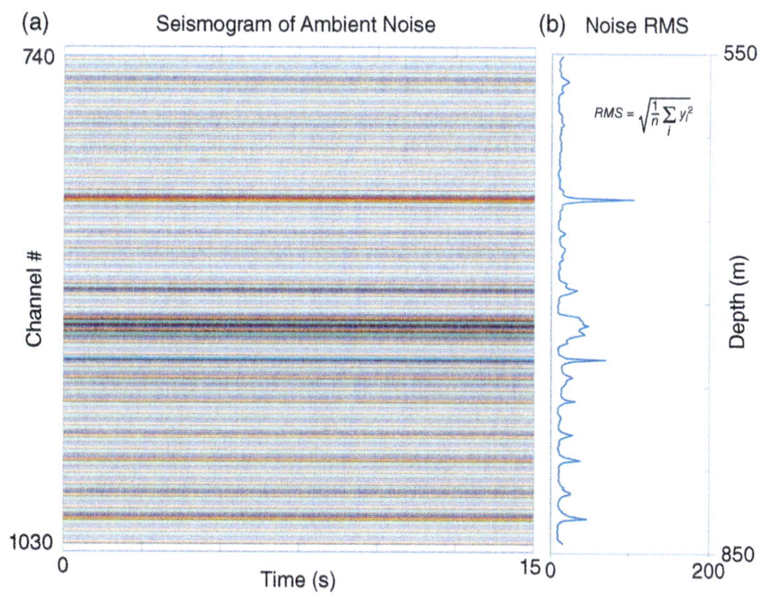

Figure 27.9 (a) 15 s traces of DAS ambient noise data recorded in sedimentary rocks (550–850 m). (b) Noise RMS amplitude profile calculated from the traces in (a).

DANRAP from the sedimentary rocks. We label the major noise peaks SNL-1 to SNL-7 and the minor noise peaks sns-1 to sns-7. To determine a potential cause for these noise peaks, we compare the DANRAPs in the granite and sedimentary formations with wireline well-logging data acquired in monitoring Well 58-32.

27.5. CORRELATING NOISE RMS AMPLITUDE DEPTH PROFILES TO WELL-LOGGING DATA

To understand why some channels have a higher level of ambient noise than other channels, we compare our DANRAPs in the granitic and sedimentary rocks with

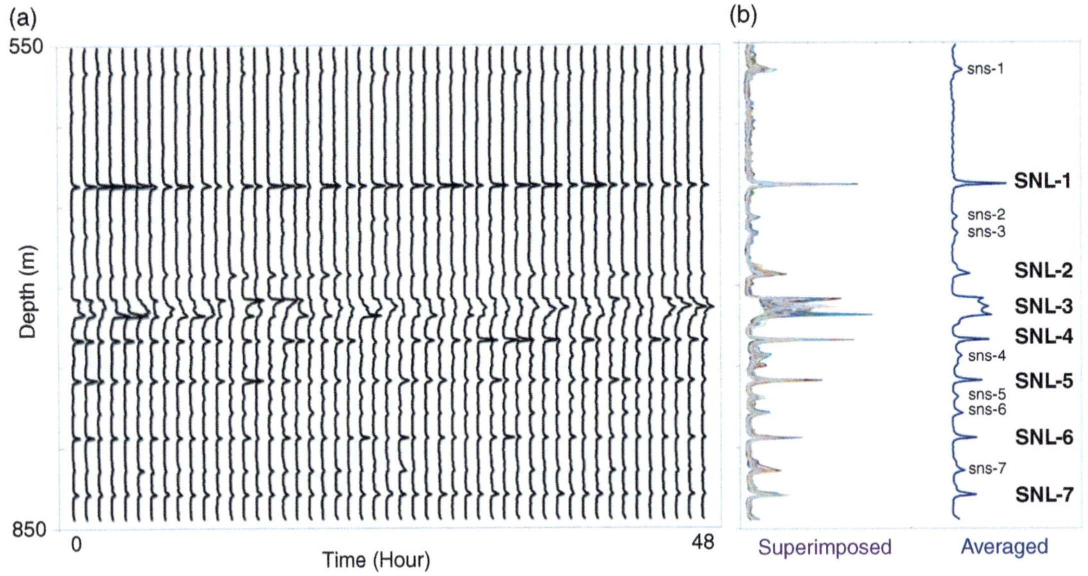

Figure 27.10 (a) 48 h of noise RMS amplitude profiles of the DAS ambient noise data collected in sedimentary rocks (550–850 m). (b) Superimposed DANRAP of the 48 h on the left and averaged DANRAP. The DANRAPs all have a similar shape, with seven major peaks and seven minor peaks. We label the seven major peaks SNL-1, SNL-2, SNL-3, SNL-4, SNL-5, SNL-6, and SNL-7 and the seven minor peaks sns-1, sns-2, sns-3, sns-4, sns-5, sns-6, and sns-7.

wireline well-logging data acquired in both observations, Well 78-32 and injection Well 58-32. Initially, we suspect that a bad cement job is the cause of the noise peaks. Thus, we compare the DANRAP in both the granite rock and sedimentary rocks with the cement bond log (CBL) dataset from Well 78-32. This dataset was collected after well completion and includes sonic waveform data (FWS), casing collar locator (CCL) data, bond index log (BI), and cement bond log amplitude (CBLA). We show a figure comparing the averaged DANRAP (blue line) from granite with the CCL (gray curve), BI (light blue curve), and CBLA (red curve) in Figure 27.11. Higher values on the CBLA log indicate a poor cement or casing job. However, the high values marked in red circles in Figure 27.11 do not correlate with the noise RMS peaks on the DANRAP. We compare the DANRAP with the CCL, BI, and CBLA from the sedimentary rocks shown in Figure 27.12 and find no correlation between the CBL log and the DANRAP, suggesting that a poor well-completion job is not the source of the ambient noise. We then correlate the DANRAP with additional logs in monitoring Well 58-32 to further seek the source of the noise.

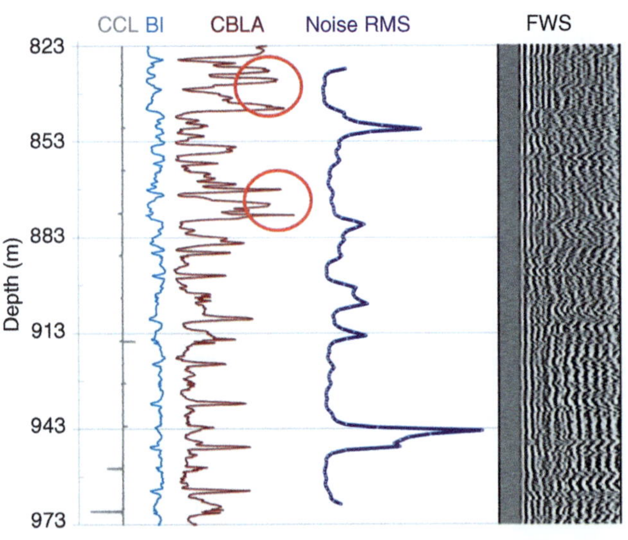

Figure 27.11 Comparing DANRAP in granite with cased collar log (gray line), bond index (light blue line), and CBLA. The red circles show high values in the CBLA, which usually indicate a poor cement job. However, these high values do not correlate with the peaks on the DANRAP, suggesting that the source of the ambient noise is not from a poor well completion job.

27.5.1. Correlating RMS Amplitude Depth Profiles in Granite With Wireline Well-Logging Data

After ruling out defects in well completion as the source of the noise, we correlate the DANRAP in the granitic rock with additional wireline well-logging data acquired in injection Well 58-32. Well 58-32 is located about 300 m from monitoring Well 78-32. We expect that geology does not change much between the two

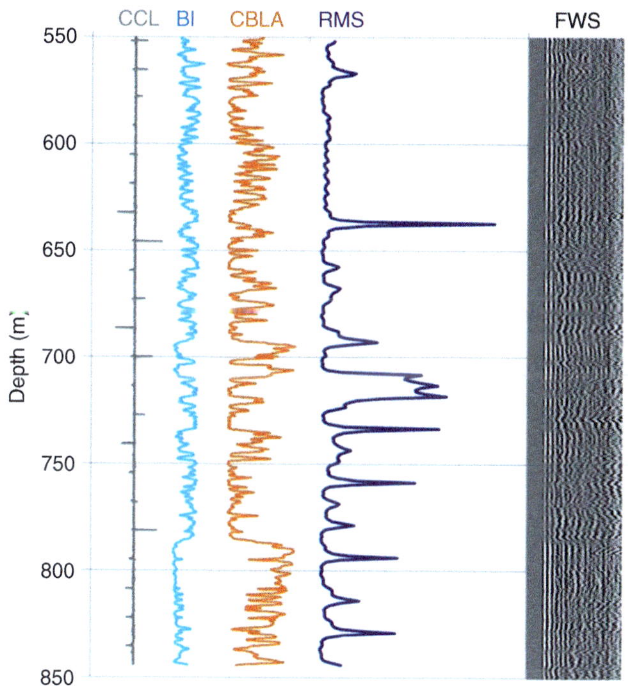

Figure 27.12 Noise RMS profile compared with the cased-hole CBL data in Well 78-32. Ambient noise RMS depth profile (blue curve) plotted with CCL, bond index (BI), and CBL amplitude. Large values in the CBL amplitude curve typically indicate poor-quality casing/cementing. No correlation observed between RMS peaks and large CBL amplitude values.

wells. However, to account for the approximately 20° dip in the geologic formations at this location, we correct the depths of the well-logging data by shifting all depths up by 125 m. We compare the averaged DANRAP in granite with the reflection coefficient (RC), density (Dn), Vp/Vs ratio (Vp/s), Poisson's ratio (Pr), S-wave impedance (Bp), P-wave impedance (Ap), S-wave velocity (Vs), P-wave velocity (Vp), density porosity (Dn%), sonic derived porosity (Sp%), thermal neutron porosity (Tp%), Well 58-32 gamma ray (Gr), and Well 78-32 gamma ray (Gr-78) logs in Figure 27.13. From the figure, we first immediately observe that peak GN-1 correlates almost perfectly with a peak in the gamma-ray log at about 850 m. We also observe that the locations of the DANRAP peaks correlate well with lower velocity layers bounded by sharp interfaces in the Vp log. These interfaces are marked by vertical lines on the Vp and DANRAP profiles in Figure 27.13. This apparent correlation is consistent for all five noise peaks on the DANRAP, and we find that the exact peak location matches exactly the trough locations in the Vp log. Notably, the apparent gradient at GN-2 on the DANRAP correlates with a gradient in the Vp log at the same depth. This gradient is marked by the bronze dotted line on both the DANRAP and the Vp log.

For further analysis, we show a more detailed comparison of the DANRAP in granite with the P-wave velocity (Vp), P-wave velocity squared (Vp2), Poisson's ratio (Pr),

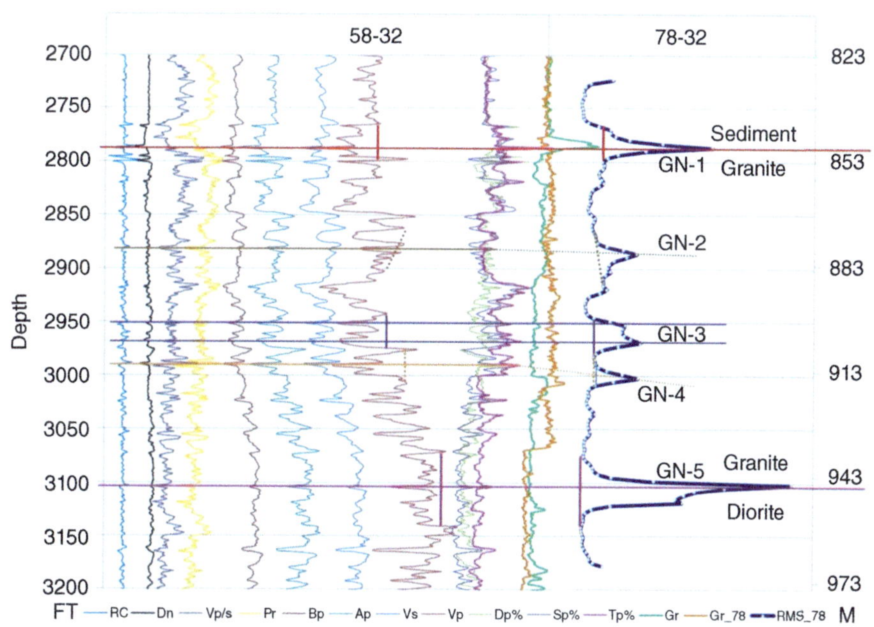

Figure 27.13 Comparing DANRAP in granite rock to depth-adjusted wireline logs acquired in monitoring Well 58-32. The major noise peaks correspond to the LVL regions.

and neutron porosity (TP%) logs in Figure 27.14. From this detailed view of the logs, we observe that the two major peaks on the DANRAP, GN-1 and GN-5, correlate with low Vp velocity, low Poisson's ratio, and high porosity. For GN-1, we find that the RMS noise peak correlates perfectly with an LVL at 843–853 m. This LVL is marked by the upper red box in Figure 27.14. Within this red box, we find a sharp trough on the Poisson's ratio log showing a low Poisson's ratio, and a sharp peak on the porosity log showing high porosity. Peak GN-5 is marked by the lower red box in Figure 27.14. This peak is also correlated with an LVL at 833–853 m and a low Poisson's ratio.

The red vertical dotted line in the Poisson's ratio log shows a Poisson's ratio value of 2/7 (0.286) that indicates a transition zone between the ductile and brittle regimes derived from empirical rock mechanic studies (Cao & Li, 2016; Christensen et al., 2018; Lewandowski et al., 2005; Paugh, 1954; Rice & Thomson, 1974). Lower Poisson ratios (0.1–0.25) indicate that the rock is brittle and easier to fracture, whereas higher Poisson ratios (0.35–0.45) mean the rock is ductile and harder to fracture (Belyadi et al., 2019). The Poisson's ratios near GN-1 and GN-5 are less than 0.25, which means the rocks are easy to fracture in those two areas. The high porosity values at those depths also suggest that fractures are present in those areas because in granite, porosity can usually be attributed to faulting, fracturing, or damage zones (Stanek & Geraud, 2019). Therefore, we suggest that the source for this high-amplitude DAS ambient noise at these depths is caused by trapped guided waves in an LVL caused by fracturing. These trapped guided waves in the LVL are comparable to large-amplitude surface waves trapped in a fault zone in earthquake seismology. These large-scale trapped waves have been recorded by both seismometer arrays (Ben-Zion, 1998; Ben-Zion & Aki, 1990; Li et al., 1990; Qiu et al., 2020; Thakur et al., 2020) and DAS arrays (Jousset et al., 2018; Pevzner et al., 2020; Yang et al., 2022). Different from seismometers that record particle velocity, DAS systems record strain rate changes that might make them more sensitive to lower velocity changes (Daley et al., 2013; Daley et al., 2016; Mateeva et al., 2013; Mateeva & Zwartjes, 2017; Pevzner et al., 2020).

The green box in Figure 27.14 marks an LVL between 859 and 868 m. However, this LVL is not associated with a low Poisson's ratio or high porosity. The Poisson's ratio in this area is about 0.35, which is in the ductile zone, and the rock at this depth is hard to fracture. As such, we do not observe a peak in the DANRAP in that depth range. The ability of our borehole DAS ambient noise analysis to detect fracture zones near the borehole may be significant for developing nonconventional energy sources by finding the best locations to stimulate.

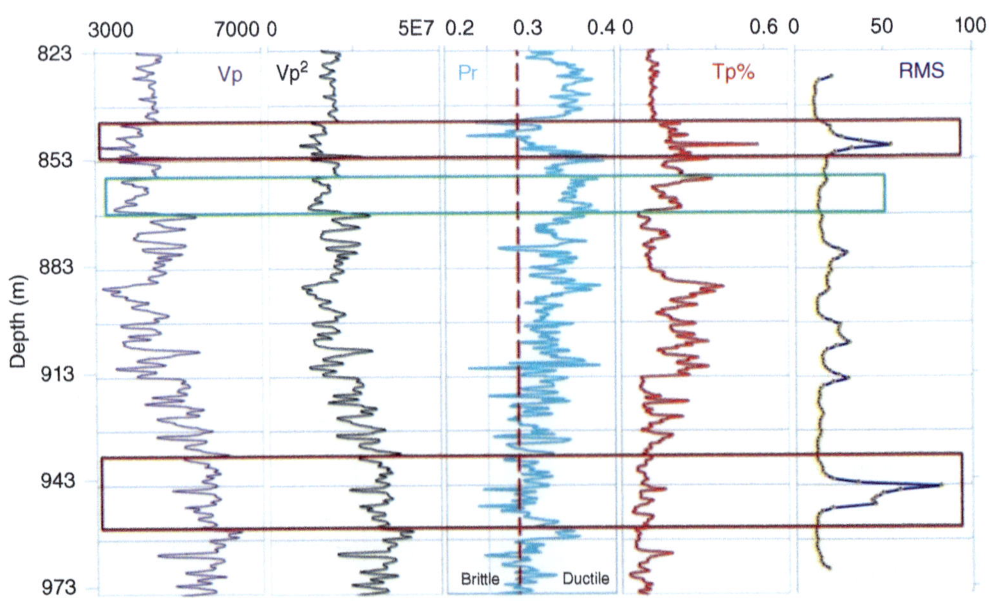

Figure 27.14 Comparison of DANRAP in granite with P-wave velocity (Vp), P-wave velocity squared, Poisson's ratio, thermal porosity (Tp%), and the DAS amplitude RMS noise profile (blue curve). The red box on the top marks an LVL around GN-1, and the red box at the bottom marks an LVL around GN-5. These red boxes mark LVL with low Poisson's ratio, high porosities, and high RMS amplitudes. The green box marks an LVL that is not associated with a low Poisson's ratio or high porosity. As such, there is no RMS noise peak at this depth range.

27.5.2. Correlating RMS Amplitude Depth Profiles in Sedimentary Rocks

We next compare the DANRAP in the sedimentary rocks with wireline logging data collected in Well 58-32. Figure 27.15 shows the DANRAP in the sedimentary formations plotted alongside the RC, S-wave impedance (Zs), P-wave impedance (Zp), Vp/Vs ratio (Vp/Vs), Poisson's ratio (Pr), density (Dn), S-wave velocity (Vs), P-wave velocity, (Vp), porosity (%Sp), and gamma ray

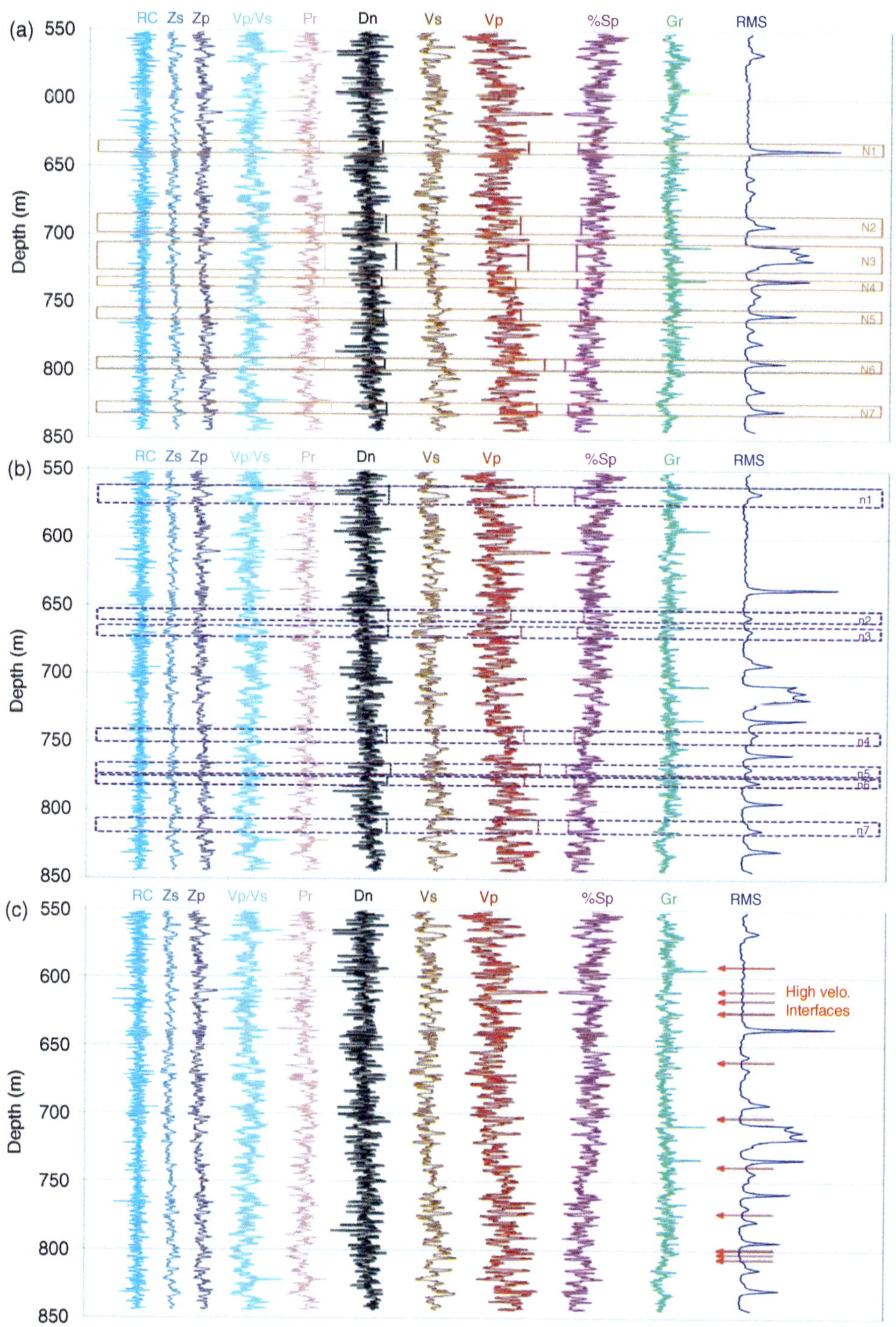

Figure 27.15 Noise RMS depth profile compared with Well 58–32 logging data after depth adjustment. Large and small noise RMS peaks of noisy zones, respectively, in (a) and (b) correlate well with layers of low velocity, low density, low Poisson ratio, and high porosity, bounded by sharp/gradually varying interfaces. (c) High-velocity and low-porosity interfaces correspond to Q-zones with small RMS values. DAS noise RMS profile can be used to identify subsurface major interfaces (fractures or faults) of low-velocity, high-porosity, and brittle zones.

(Gr) log of Well 58-32. To correct for the averaged 19° dip in the sedimentary formations, estimated by best matching gamma-ray logs in both Wells 58-32 and 78-32, we adjust the depths of the log by bringing them up 118 m. We show the areas around the major peaks N1-N7 in the sedimentary rocks in Figure 27.15(a) in orange boxes. The layers with anomalous logging data are bound by sharp interfaces or gradual ones. These noise peaks are bounded by low P-wave velocity layers bound by interfaces, low density, low Poisson's ratio, and high porosity.

We show the areas around minor amplitude peaks n1–n7 (lowercase *n*) in the sedimentary rocks in dotted purple boxes in Figure 27.15(b). As with the major peaks, the depths around the minor peaks correlate with areas on the log with low P-wave velocity, low density, or high porosity. These results also show that the amplitudes of DAS ambient noise data in the sedimentary rocks may also be caused by trapped waves in LVLs caused by fracturing.

The red arrows in Figure 27.15(c) point to high-velocity interfaces found in the Vp log. The RMS amplitude log does not show noise peaks at these areas. Because there are no peaks at high velocity boundaries, waves cannot be trapped in those layers to cause larger amplitude noise at those depths. The DAS noise RMS profile can be used to identify subsurface major interfaces (fractures or faults) of low-velocity, high-porosity, and brittle zones.

27.6. SPECTRAL FEATURES OF BOREHOLE DAS AMBIENT NOISE

In this section, we apply fast Fourier transform (FFT) to perform spectral analysis of the quiet and noisy DAS channels. To obtain the spectra of trapped noise, we calculate the amplitude spectra of the quiet-zone (Q-zone) traces, average the amplitude spectra across all quiet traces to create a quiet amplitude spectra model, and then subtract this quiet trace amplitude spectra model from the amplitude spectra of the noisy traces to isolate the amplitude spectra of the source of the noise. We show the averaged DANRAP in the sedimentary layers in Figure 27.16(a). On the DANRAP, we show 16 Q-zones that have weak amplitudes of RMS noise. We label these Q-zones Q1–Q16.

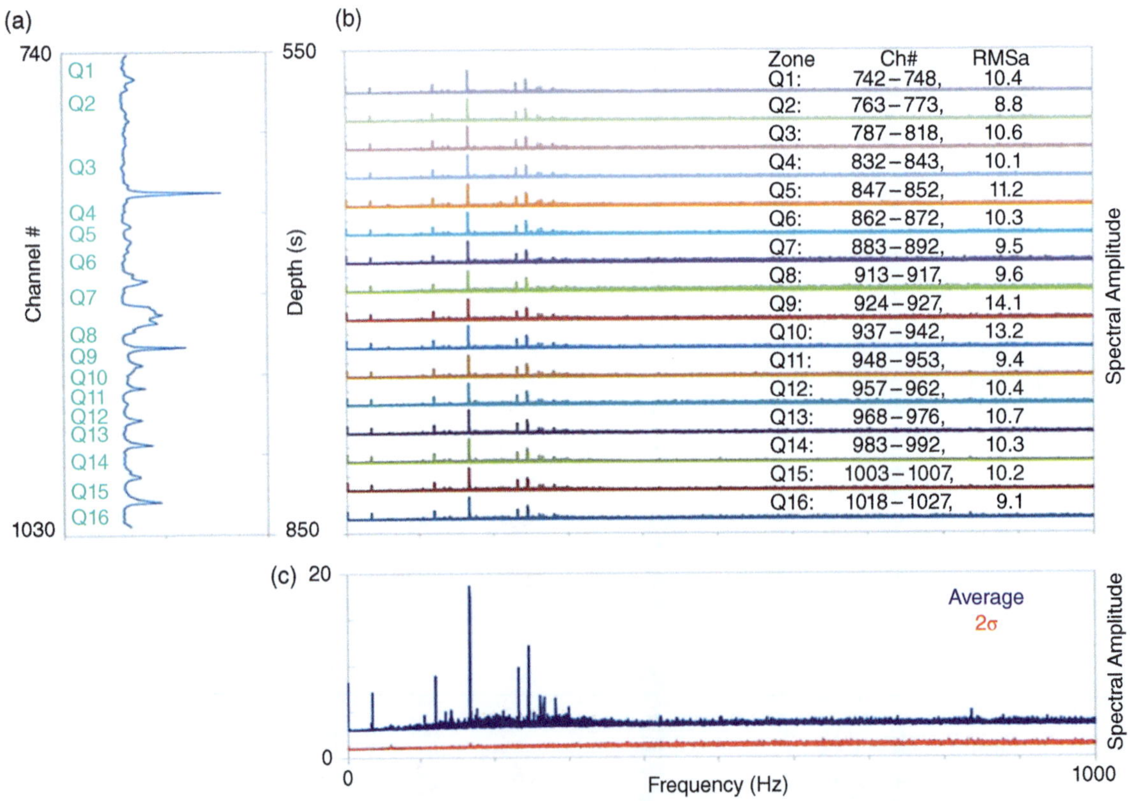

Figure 27.16 (a) DANRAP of sedimentary rocks with 16 Q-zones labeled. (b) Averaged spectra in the 16 Q-zones show similar patterns. (c) An averaged spectrum of 150 noise traces in Q1–Q16 to construct a quiet spectral model and its two standard deviations.

We show the averaged spectra of 150 noise traces for each of the 16 Q-zones in Figure 27.16(b). We show the zone number, the channels that we use to calculate the amplitude spectra, and the noise RMS amplitude in Figure 27.16(b). The figure shows that the amplitude spectra at each of the Q-zones look similar to one another with distinctive spectral peaks at frequencies lower than 300 Hz. We calculate the average of all the amplitude spectra (150 of them) in the 16 Q-zones in Figure 27.16(c). The blue curve shows the averaged amplitude, and the red curve depicts the two standard deviations as a function of frequency.

After calculating the amplitude spectra for the Q-zone traces, we apply FFT to the noisy traces to obtain amplitude spectra for the noisy zones. We compare the amplitude spectra of the quiet traces with those of the noisy traces in Figure 27.17. Figure 27.17(a) displays the amplitude spectra of three quiet channels (937, 1026, and 767), showing that they are similar to one another with similar amplitudes and distinct spectral peaks at certain frequencies <300 Hz. We also depict the amplitude spectra of three noisy channels (825, 920, and 1015) in Figure 27.17(b). The spectral shapes of these noisy traces are similar to one another but have different amplitudes. We compare the amplitude spectra of the noisiest channel (channel 825) to the averaged Q-zone amplitude spectra model (Qavg) in Figure 27.17(c). Finally, we display the ambient noise data of the trapped noise after subtracting Qavg in Figure 27.17(d). The amplitude spectra of the trapped noise indicate that it is blue noise at frequencies from 0 to 500 Hz and white noise at frequencies from 500 to 1,000 Hz. White noise has a flat frequency spectrum, whereas blue noise spectral amplitudes increase with increasing frequency over a finite frequency range.

We show the amplitude spectra of the seven major noise zones in sedimentary rocks before and after removing the Qavg in Figure 27.18(a), (b), respectively. Although the spectral amplitudes vary for the seven major noise zones, the general features/shapes of the spectra do not.

Finally, we calculate the amplitude spectra of the Type B noise and compare them with the amplitude spectra of the trapped ambient noise. We show traces containing Type B noise in a Q-zone in Figure 27.19(a) and traces containing Type B noise in a noisy zone in Figure 27.19(b). We display the amplitude spectra of Type B noise from quiet and noisy traces in Figure 27.19(c), (d), respectively. The amplitude spectra of the Type B noise differ from those of the ambient noise. The cause of the Type B noise, or why it always occurs in groups of six channels across 5 m depths, is unclear. The source for this Type B noise might not be a natural one. We remove the Type B noise from the data before creating DANRAPs.

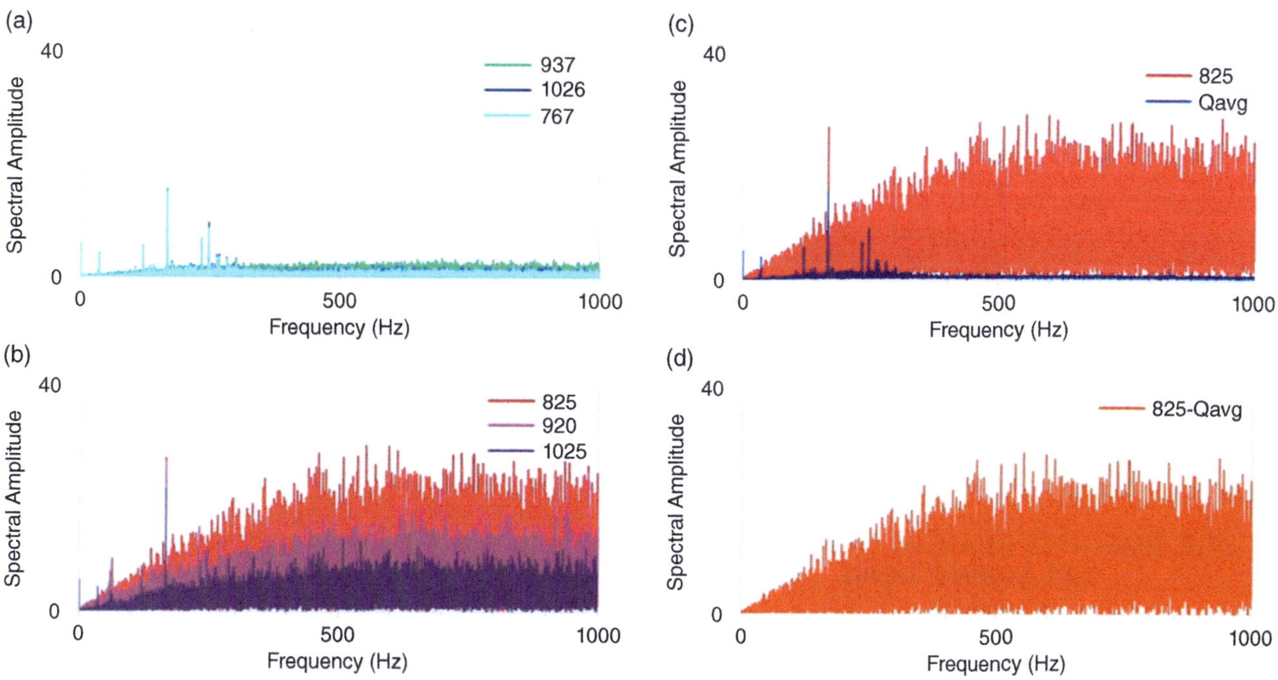

Figure 27.17 Amplitude spectra of (a) three quiet traces and (b) three noisy traces. (c) A spectrum of the noisiest trace (channel 825) versus the Q-zone spectral model. (d) Subtracting the averaged Q-zone spectrum (Qavg) from the noisiest spectrum to generate a trapped noise spectrum that shows features of blue noise (f < 500 Hz) and white noise (f > 500 Hz).

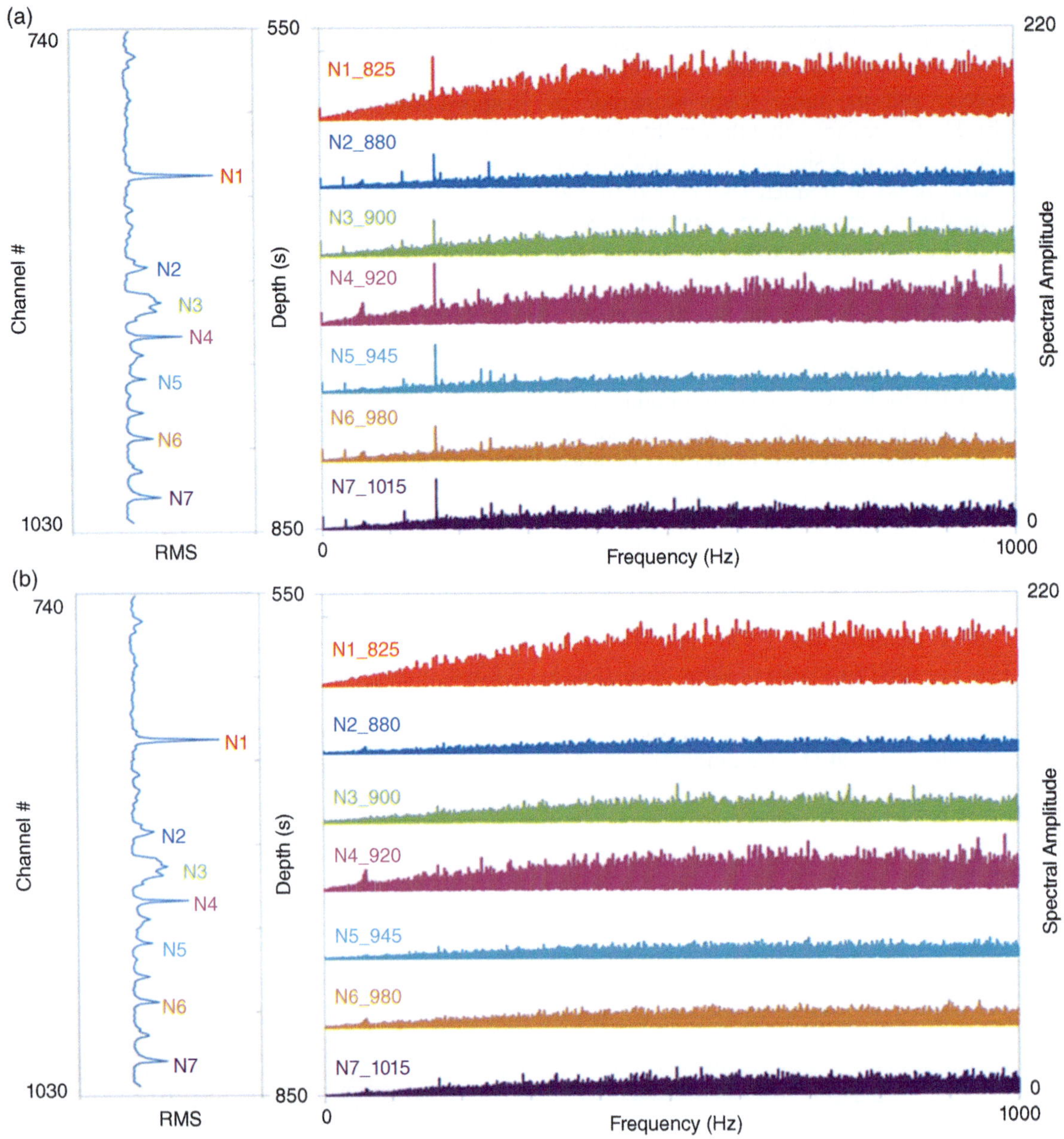

Figure 27.18 (a) Amplitude spectra of seven traces with peak RMS values in the seven noisiest zones. (b) Recalculated amplitude spectra of the same seven traces by removing the amplitude spectra of the Q-zone spectral model. Although spectral amplitudes of trapped noise energy vary, spectral features of blue–white noise remain unchanged.

27.7. TIME-LAPSE CHANGES OF BOREHOLE DAS AMBIENT NOISE

We analyze time-lapse changes in the DAS ambient noise data. Figure 27.20(a) shows averaged noise RMS amplitudes of the 16 Q-zones from Figure 27.16 over a period of 48 h in the top panel. In the bottom panel, we depict the superimposed noise RMS amplitude variations and the averaged noise RMS amplitude variation over time (black line). The daily variation of noise RMS for the Q-zones show a daily change of low-to-high-to-low-to high in the 48 h of data we analyze. At each time step, the average RMS is for a 15 s trace recorded in that hour. We display the temporal variation curves of the noise RMS amplitude for the seven major noise peaks in the sedimentary rocks in Figure 27.20(b). The top panel shows seven individual curves of noise RMS amplitudes versus time, and the bottom panel shows the superimposed

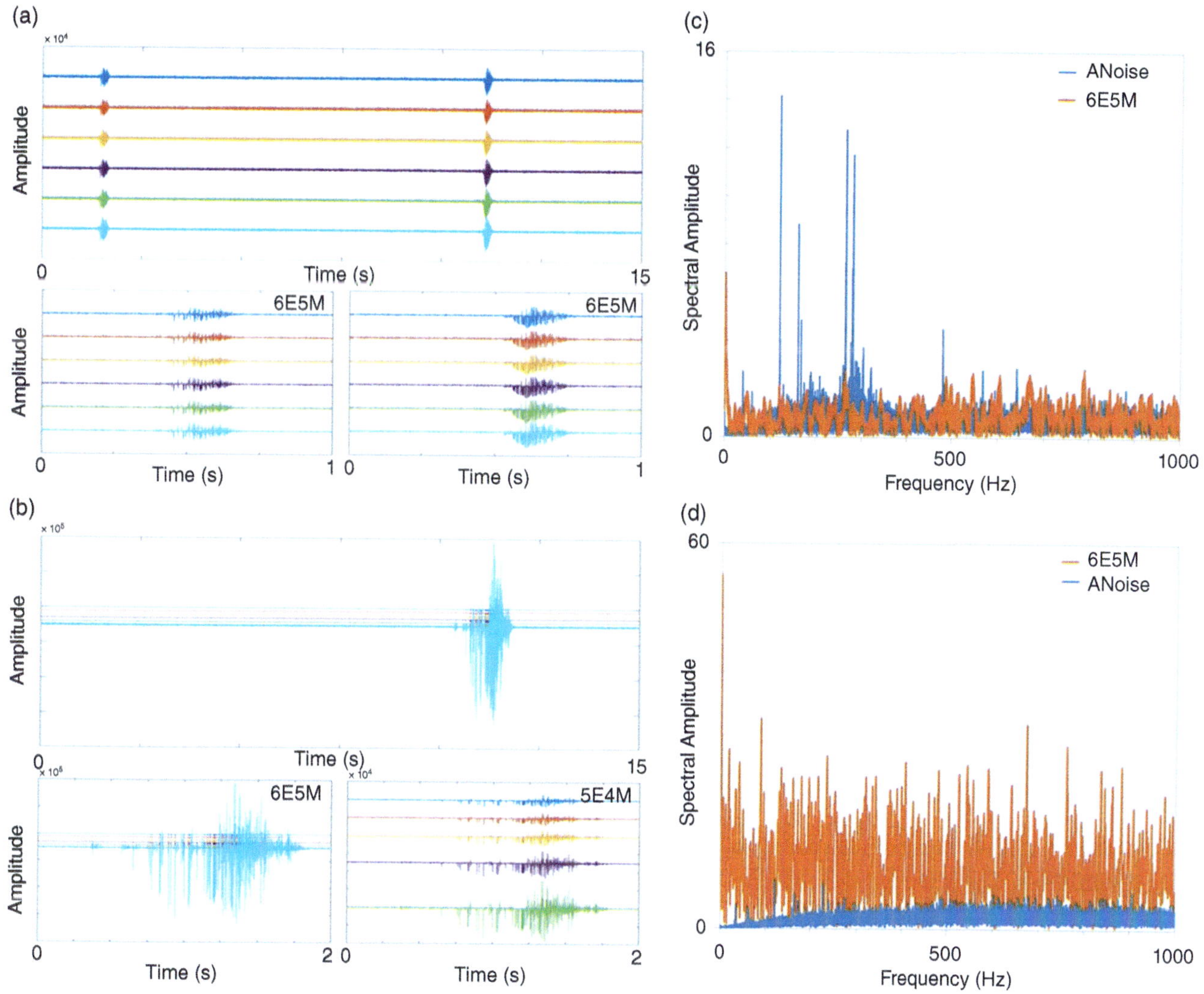

Figure 27.19 DAS traces of Type B noise (six events within 5 m, 6E5M) within (a) a Q-zone and (b) a noisy zone. (c) and (d) Average spectral amplitudes of ambient noise (light blue) and 6E5M (orange) for six quiet and six noisy traces shown in (a) and (b). We remove those Type B noises, 6E5M (might be from non-natural origins), in our data processing.

and averaged (thick black line) of these seven noise RMS amplitudes versus time curves. There is no discernable pattern of temporal changes in the RMS amplitudes of the noisy data.

We then perform a time-frequency analysis on the quietest channel (channel 767) and the noisiest channel (channel 825) (Figure 27.21). We plot our time-frequency analysis (TFA) of the two channels in Figure 27.21; it shows a similar result to those in Figure 27.20. From the TFA of the quiet traces, we observe daily variations in the approximately 100–300 Hz range. This variation is not seen in the TFA of the noisy channel.

We plot the average amplitude spectra of channel 825 over a 48 h period alongside the averaged amplitude spectra of channel 767 in Figure 27.22(a). We show the amplitude spectra of the noisy channel after subtracting the amplitude spectra of the quiet channel in Figure 27.22(b). The black line is the mean value of the amplitude spectra for 48 h on channel 825. The spectra of the trapped noise energy in an LVL show a blue–white noise combination. The corner frequency of this noise is about 460 Hz. These results show that borehole DAS ambient noise can be used to detect daily variations in the data.

27.8. CONCLUSIONS

We have processed and analyzed 48 h of borehole DAS ambient noise data collected at the Utah FORGE site in both the overlying sedimentary rocks and the granitic

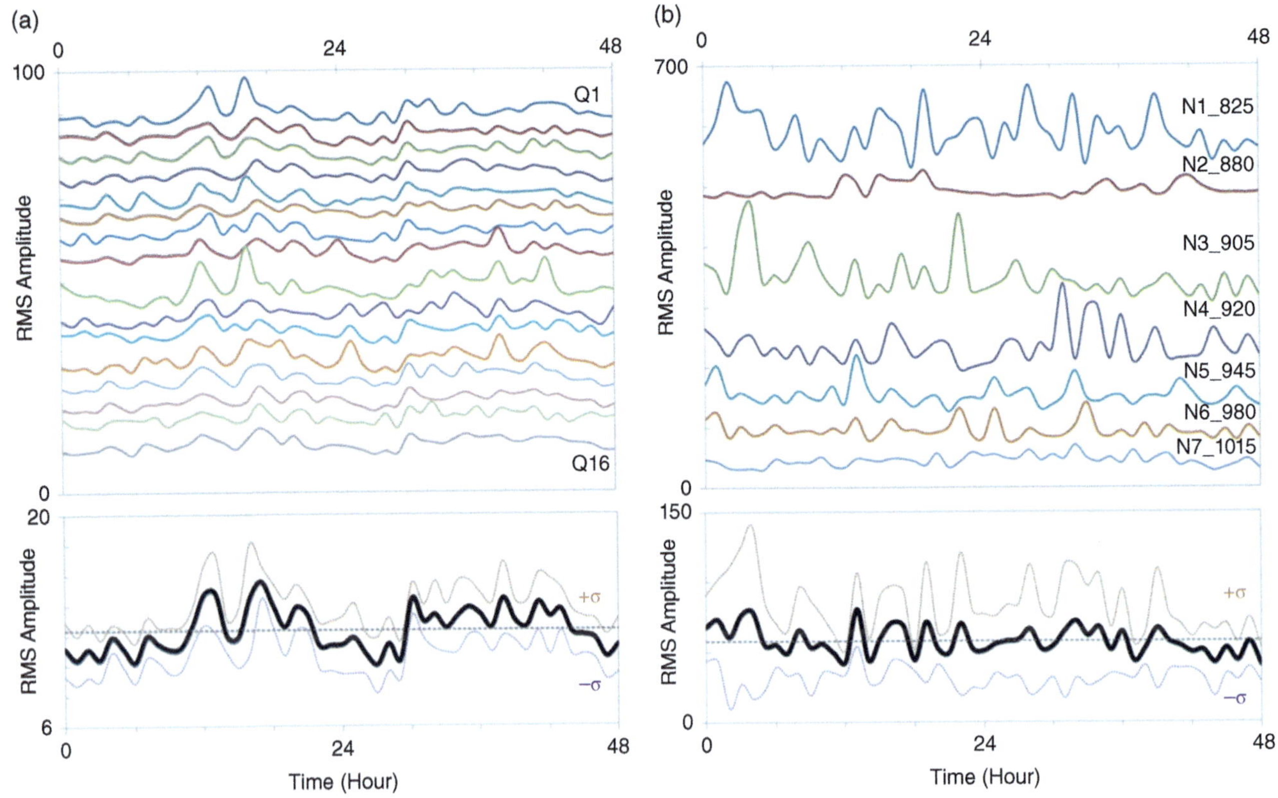

Figure 27.20 (a) Average noise RMS amplitudes of 16 Q-zones in a 48 h period along with an averaged curve in black over 16 zones, which show an apparent daily variation pattern of RMS amplitudes. (b) RMS amplitudes of seven noisy zones along with an averaged curve in black of seven noisy traces with large variations of standard deviations.

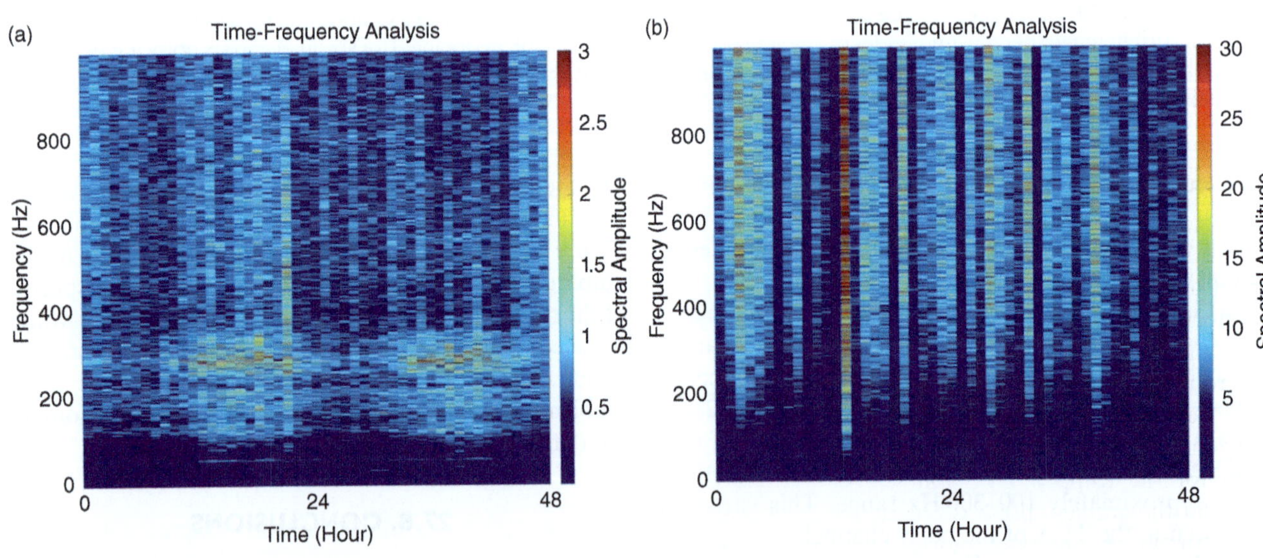

Figure 27.21 TFA over a 48 h time duration for (a) the quietest trace (channel 767) and (b) the noisiest trace (channel 825). Both figures show that spectral amplitudes increase with frequency (blue noise). We also observe the quiet and noisy daily variation patterns in (a), similar to that in Figure 27.20.

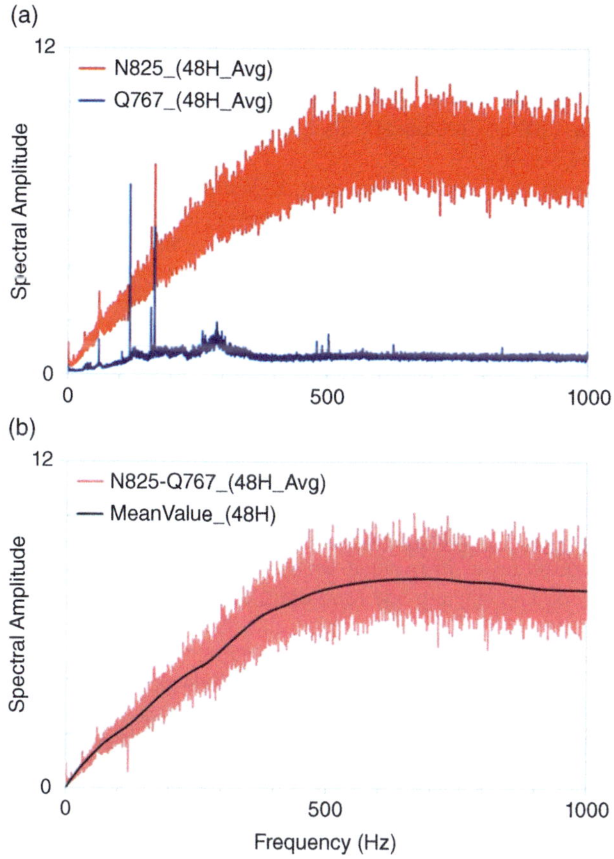

Figure 27.22 (a) Averaged spectral amplitudes over a 48 h period of the noisiest trace (red, channel 825) and the quietest trace (blue, channel 767). (b) Averaged spectral amplitude of the noisiest trace 825 after subtracting that of the quietest trace 767. The black line is the mean value. This spectrum of the trapped noise energy in a low-velocity zone shows a spectral feature of the blue-white noise with a corner frequency of about 460 Hz.

basement to obtain DAS ambient noise RMS amplitude profiles. These noise RMS depth profiles show distinct noise peaks at certain depths and Q-zones at other depths. We have compared our DAS ambient noise RMS amplitude profiles with well-logging data and found apparent correlations of peaks in the RMS profile with logs of low velocity, low Poisson's ratio, and high porosity, suggesting that trapped waves in the LVLs caused by fractures result in the high-amplitude ambient noise at certain depths. We have performed spectral analysis of the Q-zone traces and the noisy traces to obtain amplitude spectra for the trapped LVL ambient noise. We have found that the ambient noise is blue noise at frequencies lower than 500 Hz and white noise at frequencies higher than 500 Hz. Our time-lapse analysis shows daily variations of higher and lower amplitudes of noise on the quiet traces, but we do not find this variation in the noisy traces. Our results show that borehole DAS ambient noise can be used as a source-free logging tool to detect for fractures and monitor associated time-lapse changes in the subsurface.

ACKNOWLEDGMENTS

This work was supported by the U.S. Department of Energy (DOE) through Los Alamos National Laboratory (LANL), which is operated by Triad National Security, LLC for the National Nuclear Security Administration (NNSA) of the U.S. DOE under Contract No. 89233218CNA000001, and through the University of Houston under Contract No. DE-EE0008764. We thank Thomas Coleman and Mahmoud Farhadiroushan for discussions of the spikes of the DAS IED and the DAS system installed at the Utah FORGE site. The authors appreciate volume editor G. Zhan and two anonymous reviewers for their constructive comments and suggestions. The DAS ambient noise data used were obtained from the DOE Geothermal Data Repository (https://gdr.openei.org/).

AVAILABILITY STATEMENT

The DAS data associated with this chapter are public and can be readily downloaded from the DOE Geothermal Data Repository at https://gdr.openei.org/submissions/1185. The ambient noise analysis code is available on request with LANL's approval to release.

REFERENCES

Ajo-Franklin, J. B., Dou, S., Lindsey, N. J., Monga, I., Tracy, C., Robertson, M., et al. (2019). Distributed acoustic sensing using dark fiber for near-surface characterization and broadband seismic event detection. *Scientific Reports*, *9*(1), 1328.

Ajo-Franklin, J., Lindsey, N., Dou, S., Daley, T.M., Freifeld, B., Martin, E.R., et al. (2015). A field test of distributed acoustic sensing for ambient noise recording. Paper presented at the SEG Annual Meeting.

Belyadi, H., Fathi, E., & Belyadi, F. (2019). *Hydraulic fracturing in unconventional reservoirs: theories, operations, and economic analysis*. Gulf Professional Publishing.

Ben-Zion, Y. (1998). Properties of seismic fault zone waves and their utility for imaging low-velocity structures. *Journal of Geophysical Research: Solid Earth*, *103*(B6), 12567–12585.

Ben-Zion, Y., & Aki, K. (1990). Seismic radiation from an SH line source in a laterally heterogeneous planar fault zone. *Bulletin of the Seismological Society of America*, *80*(4), 971–994.

Cao, J., & Li, F. G. (2016). Critical Poisson's ratio between toughness and brittleness. *Philosophical Magazine Letters*, *96*, 425–431. doi:10.1080/09500839.2016.1243264

Chang, H., & Nakata, N. (2022). Investigation of time-lapse changes with DAS borehole data at the Brady geothermal field using deconvolution interferometry. *Remote Sensing*, *14*(1), 185.

Christensen, R. M., Li, Z., & Gao, H. (2018). An evaluation of the failure modes transition and the Christensen ductile/brittle failure theory using molecular dynamics. *Proceedings of the Royal Society A*, *474*, 20180361.

Correa, J., Egorov, A., Tertyshnikov, K., Bona, A., Pevzner, R., Dean, T., et al. (2017). Analysis of signal to noise and directivity characteristics of DAS VSP at near and far offsets — A CO2CRC Otway Project data example. *The Leading Edge*, *36*, 994a1–994a7. doi:10.1190/tle36120994a1.1

Daley, T. M., Freifeld, B. M., Ajo-Franklin, J., Dou, S., Pevzner, R., Shulakova, V., et al. (2013). Field testing of fiber-optic distributed acoustic sensing (DAS) for subsurface seismic monitoring. *The Leading Edge*, *32*(6), 699–706.

Daley, T. M., Miller, D. E., Dodds, K., Cook, P., & Freifeld, B. M. (2016). Field testing of modular borehole monitoring with simultaneous distributed acoustic sensing and geophone vertical seismic profiles at Citronelle, Alabama. *Geophysical Prospecting*, *64*(5), 1318–1334.

Dou, S., Lindsey, N., Wagner, A. M., Daley, T. M., Freifeld, B., Robertson, M., et al. (2017). Distributed acoustic sensing for seismic monitoring of the near surface: a traffic-noise interferometry case study. *Scientific Reports*, *7*(1), 11620.

Harris, K., White, D., Melanson, D., Samson, C., & Daley, T. M. (2016). Feasibility of time-lapse VSP monitoring at the Aquistore CO_2 storage site using a distributed acoustic sensing system. *International Journal of Greenhouse Gas Control*, *50*, 248–260.

Jousset, P., Reinsch, T., Ryberg, T., Blanck, H., Clarke, A., Aghayev, R., et al. (2018). Dynamic strain determination using fibre-optic cables allows imaging of seismological and structural features. *Nature Communications*, *9*(1), 2509.

Kirby, S. M., Simmons, S., Inkenbrandt, P. C., & Smith, S. (2019). Groundwater hydrogeology and geochemistry of the Utah FORGE site and vicinity. Utah Geological Survey Miscellaneous Publication 169-E; two plates, scale 1:24,000.

Lewandowski, J. J., Wang, W. H., & Greer, A. L. (2005). Intrinsic plasticity or brittleness of metallic glasses. *Philosophical Magazine Letters*, *85*, 77–87. doi:10.1080/09500830500080474

Li, D., Huang, L., Zheng, Y., Li, Y., Wannamaker, P., & Moore, J. (2022). Feasibility of source-free DAS logging for next-generation borehole imaging. *Scientific Reports*, *12* (1), 11910.

Li, Y. G., Leary, P., Aki, K., & Malin, P. (1990). Seismic trapped modes in the Oroville and San Andreas fault zones. *Science*, *249*(4970), 763–766.

Li, Y., Karrenbach, M., & Ajo-Franklin, J. B. (2021). A literature review: distributed acoustic sensing (DAS) geophysical applications over the past 20 years. In Y. Li, M. Karrenbach, & J. B. Ajo-Franklin (Eds.), *Distributed Acoustic Sensing in Geophysics: Methods and Applications* (pp. 229–291). AGU. doi:10.1002/9781119521808.ch17

Li, Y., Wu, H., Wong, W., Hewett, B., Liu, Z., Mateeva, A., & Lopez, J. (2015). Velocity analysis and update with 3D DAS-VSP to improve borehole/surface seismic images. Paper presented at the 85th SEG Annual International Meeting. Expanded Abstracts, 5285–5289.

Martin, E. R. & Biondi, B. L. (2017). *Ambient noise interferometry across two-dimensional DAS arrays*. Paper presented at the SEG Technical Program. Expanded Abstracts, 2642–2646.

Mateeva, A., Lopez, J., Mestayer, J., Wills, P., Cox, B., Kiyashchenko, D., et al. (2013). Distributed acoustic sensing for reservoir monitoring with VSP. *The Leading Edge*, *32*(10), 1278–1283.

Mateeva, A. & P.M. Zwartjes (2017). Depth calibration of DAS VSP channels: a new data-driven method. Paper presented at the 79th EAGE Conference and Exhibition.

Mestayer, J., Cox, B., Wills, P., Kiyashchenko, D., Lopez, J., Costello, M., et al. (2011). *Field trials of distributed acoustic sensing for geophysical monitoring*. Paper presented at the SEG Technical Program. Expanded Abstracts, 4253–4257.

Miller, D. E., Daley, T. M., White, D., Freifeld, B. M., Robertson, M., Cocker, J., & Craven, M. (2016). Simultaneous acquisition of distributed acoustic sensing VSP with multi-mode and single-mode fibre-optic cables and 3C-geophones at the Aquistore CO_2 storage site. *CSEG Recorder*, *41*(6), 28–33.

Paugh, S. F. (1954). Relations between the elastic moduli and the plastic properties of polycrystalline pure metals. *Philosophical Magazine*, *45*, 823–843. doi:10.1080/14786440808520496

Pevzner, R., Gurevich, B., Pirogova, A., Tertyshnikov, K., & Glubokovskikh, S. (2020). Repeat well logging using earthquake wave amplitudes measured by distributed acoustic sensors. *The Leading Edge*, *39*, 513–517. doi:10.1190/tle39070513.1

Qiu, H., Allam, A. A., Lin, F. C., & Ben-Zion, Y. (2020). Analysis of fault zone resonance modes recorded by a dense seismic array across the San Jacinto fault zone at Blackburn Saddle. *Journal of Geophysical Research: Solid Earth*, *125*(10), e2020JB019756.

Rice, J. R., & Thomson, R. (1974). Ductile versus brittle behavior of crystals. *Philosophical Magazine*, *29*, 73–97. doi:10.1080/14786437408213555

Stanek, M., & Geraud, Y. (2019). Granite microporosity changes due to fracturing and alteration: Secondary mineral phases as proxies for porosity and permeability estimation. *Solid Earth*, *10*, 251–274. doi:10.5194/se-10-251-2019

Thakur, P., Huang, Y., & Kaneko, Y. (2020). Effects of low-velocity fault damage zones on long-term earthquake behaviors on mature strike-slip faults. *Journal of Geophysical Research: Solid Earth*, *125*(8), e2020JB019587.

Tribaldos, V. R., Ajo-Franklin, J. B., Dou, S., Lindsey, N. J., Ulrich, C., Robertson, M., et al. (2021). Surface wave imaging using distributed acoustic sensing deployed on dark fiber: moving beyond high-frequency noise. In Y. Li, M. Karrenbach, & J. B. Ajo-Franklin (Eds.), *Distributed Acoustic Sensing in Geophysics: Methods and Applications* (pp. 197–212). American Geophysical Union.

Willis, M. E., Erdemir, C., Ellmauthaler, A., Barrios, O., & Barfoot, D. (2016). Comparing DAS and Geophone Zero-Offset VSP Data Sets Side-By-Side. *CSEG Recorder*, *41*(6), 22–26.

Yang, Y., Atterholt, J. W., Shen, Z., Muir, J. B., Williams, E. F., & Zhan, Z. (2022). Sub-kilometer correlation between near-surface structure and ground motion measured with distributed acoustic sensing. *Geophysical Research Letters*, *49*, e2021GL096503. doi:10.1029/2021GL096503

Zeng, X., Lancelle, C., Thurber, C., Fratta, D., Wang, H., Lord, N., et al. (2017). Properties of noise cross-correlation functions obtained from a distributed acoustic sensing array at Garner Valley, California. *Bulletin of the Seismological Society of America*, *107*(2), 603–610.

Zeng, X., Thurber, C. H., Wang, H. F., Fratta, D., & Feigl, K. L. (2021). High-resolution shallow structure at brady hot springs using ambient noise tomography (ANT) on a trenched distributed acoustic sensing (DAS) array. In Y. Li, M. Karrenbach, & J. B. Ajo-Franklin (Eds.), *Distributed Acoustic Sensing in Geophysics: Methods and Applications* (pp. 101–110). American Geophysical Union.

Zhan, G., Li, Q., Nahm, J., Kommedal, J., & Konkler, J. (2016). Field trials of distributed acoustic sensing for reservoir delineation with VSP. *CSEG Recorder*, *41*(6), 34–37.

28

Enhancing Unconventional Reservoir Development Through Low-Frequency Distributed Acoustic Sensing Technology

Yanrui Ning and Ge Jin

ABSTRACT

Low-frequency distributed acoustic sensing (LF-DAS) is a recent technology that utilizes fiber-optic cables as a distributed sensor array to monitor strain changes in the subsurface with high spatial and temporal resolution. This review starts with an introduction of LF-DAS and its advantages for subsurface imaging, followed by discussions on deployment methods of sensing cables, qualitative interpretations of LF-DAS data for various oil-field applications, geomechanics modeling, and quantitative investigation. We also delve into the emerging use of Rayleigh frequency shift (RFS)-based distributed fiber-optic strain sensing (DSS) technology for strain monitoring during production. This chapter highlights the advancements made in the use of LF-DAS technology and its value in enhancing our understanding of subsurface processes in oil-field applications and beyond. It also emphasizes the ongoing research efforts and prospects for LF-DAS, making it a valuable reference for researchers and practitioners in the field of geosciences.

28.1. INTRODUCTION

Distributed acoustic sensing (DAS) technology has been recently applied in the oil industry for hydraulic fracture characterization (e.g., Jin & Roy, 2017; Webster et al., 2013a), microseismic monitoring (Huot et al., 2022; Karrenbach et al., 2019; Verdon et al., 2020; Webster et al., 2013b), vertical seismic profiling (VSP) (e.g., Byerley et al., 2018; Kiyashchenko et al., 2020), injection fluid/proppant allocation (e.g., Crickmore et al., 2015; Molenaar et al., 2012; Sakaida et al., 2022), production logging (e.g., Fitzel et al., 2015; Jin et al., 2019; Ouyang & Belanger, 2006), and early gas detection (Sharma et al., 2020). It involves using an optical fiber deployed along the wellbore to guide laser energy into the reservoir. The heterogeneities within the fiber core scatter a small portion of the laser energy back to the surface-sensing unit, which is also called an interrogator unit (IU). The strain change or strain rate along the fiber is measured using the phase of Rayleigh backscatters. By sampling the sensing fiber at a high temporal rate (5,000–10,000 Hz) with a spatial resolution between 1 and 10 m, DAS provides critical spatial and temporal measurements for detecting near-wellbore changes (Jin & Roy, 2017). This review study primarily focuses on the low-frequency DAS (LF-DAS) method, which involves significantly downsampling raw DAS data after a low-pass filter with a corner frequency of 0.5 Hz or lower. The downsampling process increases the signal-to-noise ratio and preserves the strain-rate polarity, enabling the identification of thermal variation and/or mechanic strain perturbation.

LF-DAS offers versatile surveillance capabilities through both horizontal and vertical well deployments. In horizontal wells, it enables detailed characterization of fracture propagation and fluid flow along the entire wellbore length, providing insights into fracture initiation, growth patterns, and network connectivity. In vertical

Department of Geophysics, Colorado School of Mines, Golden, Colorado, USA

Distributed Acoustic Sensing in Borehole Geophysics, Geophysical Monograph 289, First Edition.
Edited by Yingping Li, Robert Mellors, and Ge Zhan.

DOI: 10.1002/9781394179275.ch28

wells, LF-DAS monitors fracture height growth, assesses containment within the target reservoir interval, and identifies preferential flow paths. A hybrid monitoring approach combining horizontal and vertical well deployments allows for simultaneous monitoring in different directions, enhancing fracture network characterization and reservoir performance assessment. Fiber-optic cables serve as distributed sensors, capturing high-resolution acoustic signals. These recorded data encompass amplitude, frequency, and arrival time information, which can be further processed to extract crucial insights regarding fracture growth, fluid movement, stress variations, and overall reservoir behavior.

LF-DAS has proven to be a powerful tool in characterizing hydraulic fracture geometry through field case studies and geomechanical modeling. Recent field studies have demonstrated the potential of LF-DAS data to characterize hydraulic fracture geometry in detail, including constraining fracture length (Liang et al., 2022; Zhu & Jin, 2022a), estimating fracture density (Ichikawa et al., 2019, 2020; Ugueto et al., 2021a), identifying cross-well communication (Ning et al., 2022a), and detecting fracture hits (Andrade & Baan, 2021; Jin et al., 2019; Liu et al., 2021a). Corresponding geomechanical models have been developed to confirm and support the interpretation of LF-DAS signatures (Sherman et al., 2019; Wang et al., 2022; Zhang et al., 2020a). Furthermore, the potential of LF-DAS to identify gas leakage (Sharma et al., 2020) and predict downhole pressure (Ekechukwu & Sharma, 2021) has also been explored. In addition to the oil-field application, Sidenko et al. (2022) applied LF-DAS to monitor downhole pressure during the carbon sequestration process. Becker et al. (2018) also suggest the application of LF-DAS in geothermal projects to measure hydraulic connectivity between wells.

In addition to the qualitative interpretation of LF-DAS signals, quantitative interpretation has been used to investigate temperature effects (Leggett et al., 2022), estimate fracture width (Liu et al., 2021b, 2021c; Ramos Gurjao et al., 2022), and assess fracture height (Liu et al., 2022a). More recently, the Rayleigh frequency shift (RFS)-based distributed fiber-optic strain sensing (DSS) technology has gained interest as a new type of distributed strain sensing system due to its higher sensitivity ($<1\ \mu\varepsilon$) and spatial resolution (0.2 m) than the traditional LF-DAS (Jin et al., 2021). DSS has been applied to evaluate perforation efficiency (Jin et al., 2021), identify productive fractures (Liang et al., 2022), and measure drainage rock volume (Dhuldhoya & Friehauf, 2022).

This review chapter aims to provide a comprehensive overview of the current state of the art in LF-DAS and distributed strain sensing applications in oil fields. It will discuss fiber cable deployment methods, as well as qualitative and quantitative interpretation of LF-DAS signals, with a particular focus on its application in multiple unconventional reservoirs across North America. In addition, geomechanical modeling will be discussed to demonstrate and extend the application of LF-DAS technology, followed by a review of recent applications of DSS. The objective of this review is to identify the potential of LF-DAS in unconventional reservoirs, as well as to highlight areas for future research.

28.2. FIBER CABLE DEPLOYMENT

LF-DAS targets to measure strain perturbation in the formation. The measurements depend on the mechanical coupling between sensing fiber and formation rocks, which are highly dependent on the fiber deployment methods. Currently, there are three major types of deployment methods: permanent installation, wireline conveyance, and disposable fiber deployment. Consideration of fiber coupling, cost, and project complexity is crucial for diagnostic objectives before acquiring LF-DAS data, as the method with better coupling conditions and data qualities typically incurs higher cost and complexity to install or deploy the sensor. Table 28.1 summarizes the differences between the three types of deployments.

Sensing fiber deployment methods significantly impact data quality, especially for LF-DAS measurements. The major reason is due to various mechanical coupling conditions between the sensing fiber and the formation rocks. Leblanc et al. (2022) conducted a detailed study to evaluate the coupling conditions of different types of deployment methods. Permanent installed optical cables have the best mechanical coupling with the formation, provided by cement bonding during completion. However, minor slippage can still be observed at the fracture-hit locations (Jin & Roy, 2017; Richter et al., 2019; Ugueto et al., 2023), likely due to decoupling between the cement and the formation. Cement bonding with the cable and the formation may also impact coupling and data quality. Wireline-deployed sensing fibers rely on gravity-induced friction force between cable and casing to provide mechanical coupling, which is dependent on cable density and surface friction coefficient. If the spatial gradient of formation strain exceeds the friction limitation, decoupling and cable slippage within the casing will happen. Also, cable slippage at the wireline and casing interface exhibits a stick–slip behavior due to the significant difference between the dynamic friction coefficient (μ_d) and the static friction coefficient (μ_s). This behavior arises from the threshold force required to overcome the static friction and initiate motion, followed by a lower force necessary to maintain motion due to dynamic friction. This behavior can be quantified by defining the net force (F_{net}) acting on the cable coupling in the direction of motion, which

Table 28.1 Comparison of three major types of fiber deployment methods: permanent installation, wireline conveyance, and disposable deployment.

	Permanent installation	Wireline conveyance	Disposable deployment
Placement	The fiber is situated inside a protective housing consisting of an inner metal tube(s) filled with protective gels and a hardened plastic/metal jacket.	The braided wireline contains insulation material, a conductive braided copper line and a metal tube to house sensing fibers. The metal tube is also filled with protective gels.	The bare fiber lies on the bottom of the casing after deployment. The fiber features microcoating for protection, along with a coating of grease to improve coupling. The fiber can be discarded after measurement.
Installation	Clamped outside well casing during well completion and cemented in place	Pumped down or deployed using borehole tractors	Pumped down and deployed through specially designed tools
	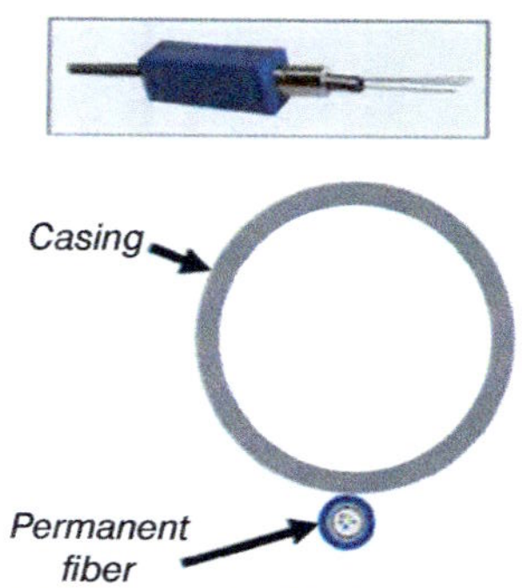 (Ugueto et al., 2023)	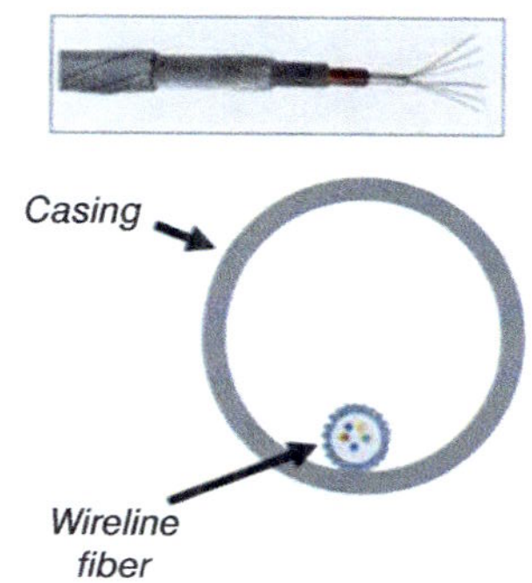 (Ugueto et al., 2023)	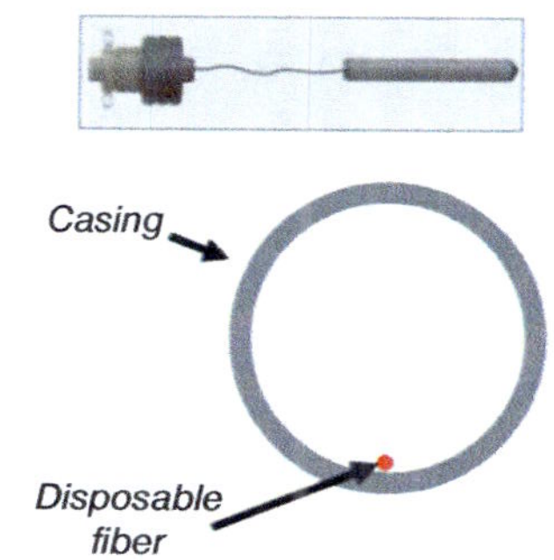 (Ugueto et al., 2023)
Cost	High initial cost, but lower cost for later acquisitions	Median cost	At least 40% lower than wireline fiber (Mantell et al., 2022)
Lifetime	More than 10 years	2–3 weeks	2–3 weeks; typically single-use, replaced for each well
Data quality	Minor slippage and decoupling	Spike noise generated by cable slippage	Higher noise levels due to fiber attenuation can have severe coupling issues in some cases
Coupling mechanism	Cement bonding	Gravity-induced frictional forces between the wireline cable and the casing (Leblanc et al., 2022)	Grease applied to the fiber
Field application examples	HFTS-2 in Permian Basin (Ugueto et al., 2019; Zhang et al., 2021)	Canadian Dip-in DAS project (Eaton et al., 2022; Wang et al., 2022)	DJ Basin (Barhaug et al., 2022; Ning & Jin, 2023a, 2023b); Marcellus Shale (Mantell et al., 2022)

is $F_{net} = T - F_f$, where F_f is the frictional force. The frictional force varies depending on the state of the cable. When the cable is at rest or moving slowly, the frictional force is equal to the static friction force, F_s, given by the equation $F_s = \mu_s * N$, where μ_s is the static friction coefficient and N is the normal force between the surfaces in contact. However, when the cable is slipping, the frictional force is equal to the dynamic friction force, F_d, given by $F_d = \mu_d * N$, where μ_d is the dynamic friction coefficient. When the tension force, T, gradually increases, the cable coupling will eventually reach a point where the force applied exceeds the static friction force, F_s. At this point, the cable will start to slip, and the frictional force will reduce to the level of the dynamic friction force, F_d.

As shown in Figure 28.1(a), (b), LF-DAS signals measured using wireline follow a pattern similar to those measured using permanently installed fiber but are contaminated by the vertical spike noise due to decoupling of the cable and the cable slippage. In most cases, this type of noise can be largely removed through data processing. Sensing fiber deployed by low-cost disposable tools (Higginson et al., 2017) relies on the grease applied on the fiber surface to provide mechanical coupling. It works sufficiently well in some cases (e.g., LeBlanc et al., 2022; Mantell et al., 2022; Ning & Jin, 2023), but coupling performance is not guaranteed in all cases (Figure 28.1(d)). Poor coupling can spread and blur the strain signal spatially and makes data interpretation challenging if not impossible. In addition, the sensing fiber is exposed to the harsh borehole environment, leading to stronger optical attenuation, which induces a higher noise level in the acquired data, as illustrated in Figure 28.1(c).

28.3. QUALITATIVE INTERPRETATION AND APPLICATION

LF-DAS has various applications in and beyond the oil industry, but most current studies in this field have focused on characterizing hydraulic fractures created during stimulation for unconventional reservoirs.

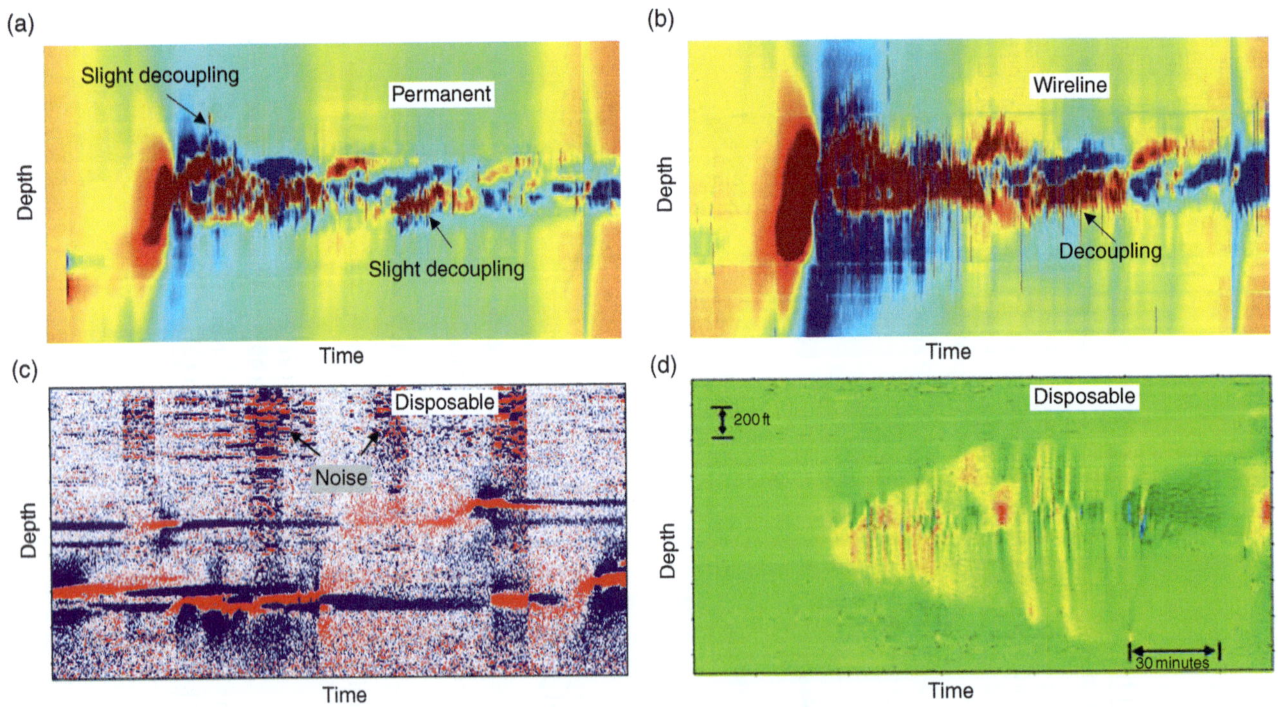

Figure 28.1 Qualitative comparison of cross-well strain induced by fracturing stimulation, measured using fibers deployed by various methods. Warm and cold colors represent strain extending and compressing, respectively. (a) Permanent installed optical cables have the best mechanical coupling with the formation, leading to the highest data quality. However, limited fiber slippage can still be observed at the fracture-hit locations. Richter et al. (2019) / with permission of EAGE. (b) This wireline-deployed sensing fiber is inside the same casing as the permanently installed fiber in (a). Frequent vertical striations indicate that there is a greater degree of decoupling compared to permanent fibers. Richter et al. (2019) / with permission of EAGE. (c) Signals recorded by fiber with disposable deployment with good coupling are interpretable, although there is noise present. Ning and Jin (2023) / with permission of the Society of Exploration Geophysicists. (d) An example of LF-DAS patterns recorded using the fiber deployed by disposable tools with poor coupling. Ugueto et al. (2023) / with permission of OnePetro.

Meanwhile, LF-DAS measurements can also be applied to prevent frac hits, identify gas leakage, support production logging, and predict and monitor downhole pressure. In this section, we will discuss these applications in detail.

28.3.1. Hydraulic Fracture Characterization in Unconventional Reservoirs

During fracturing stimulation operation, the propagation of hydraulic fractures can induce strain changes in surrounding formation rocks. The strain change can be measured by the fiber installed in offset monitoring wells, which can be either horizontal or vertical wells. This type of measurement is often referred to as cross-well measurements.

Figure 28.2 represents a typical cross-well LF-DAS strain-rate signal recorded in a horizontal monitor well during the hydraulic fracturing operation. In this case, the sensing fiber is cemented outside the casing. When a stage is stimulated (as represented by the red slurry rate curve in Figure 28.2), the newly created hydraulic fractures intersect with the monitor well, which results in fiber extension at the intersection, as shown by the red color in the LF-DAS signals. A heart-shaped extending pattern can usually be observed at the tip of fractures, which is marked "new fracture opening" in Figure 28.2. This extensional signal is associated with the processing zone near the fracture tip. On both sides of the opening fracture, a compressional signal can be observed, which is presented by the blue color. The compressional signal is also known as stress shadow. After pumping stops, most of the strain-rate measurements changed polarity immediately. This phenomenon is called "polarity flip/reversal" (Ning et al., 2022b), indicating that the newly created fractures are still hydraulically connected to the injector and are closing due to leakoff.

The geometry of hydraulic fractures can be determined through visualizing and interpreting the LF-DAS signals recorded during stimulation. These parameters play a critical role in designing unconventional reservoir completions for decreased cost and higher efficiency. The length and height of fractures can be used to optimize well spacing and stimulation job size; the density can be applied to evaluate cluster efficiency and design cluster spacing for each stimulation stage; the azimuth shows the orientation of fracture networks that impact drilling directions. In the following, we summarize the application of LF-DAS in characterizing hydraulic fractures and optimizing completion in unconventional fields across North America.

Eagle Ford

In 2015, ConocoPhillips conducted its first fully instrumented four-well pilot in the Eagle Ford play (Raterman et al., 2018, 2019, 2020), with the goal of quantifying the impact of completion design on interference and production at a fixed well spacing. The pilot used a staggered 80-acre high-low configuration, deployed optical fibers on each well to measure temperature and acoustics, and tested multiple completion designs. As illustrated in Figure 28.3(a), the pilot had four horizontal wells (B2, B3, B4, B5) arranged in a two-layer pattern, with 660 ft spacing within a layer and 330 ft interlayer lateral spacing.

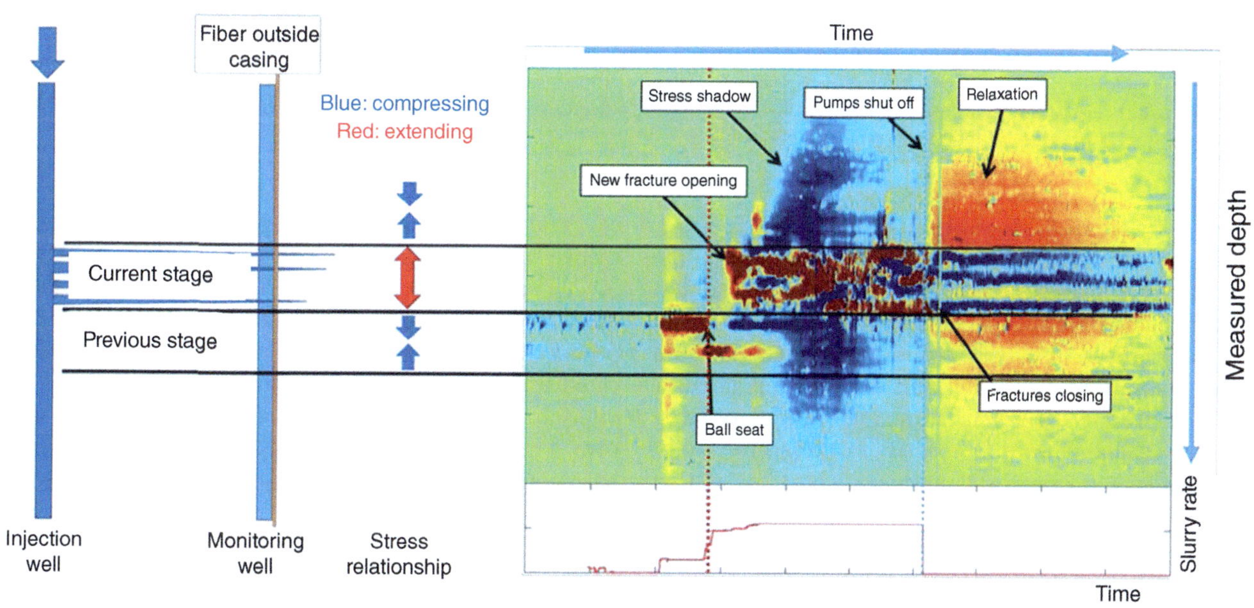

Figure 28.2 Typical LF-DAS response at an offset horizontal well during hydraulic fracturing operation. Jin and Roy (2017) / with permission of the Society of Exploration Geophysicists.

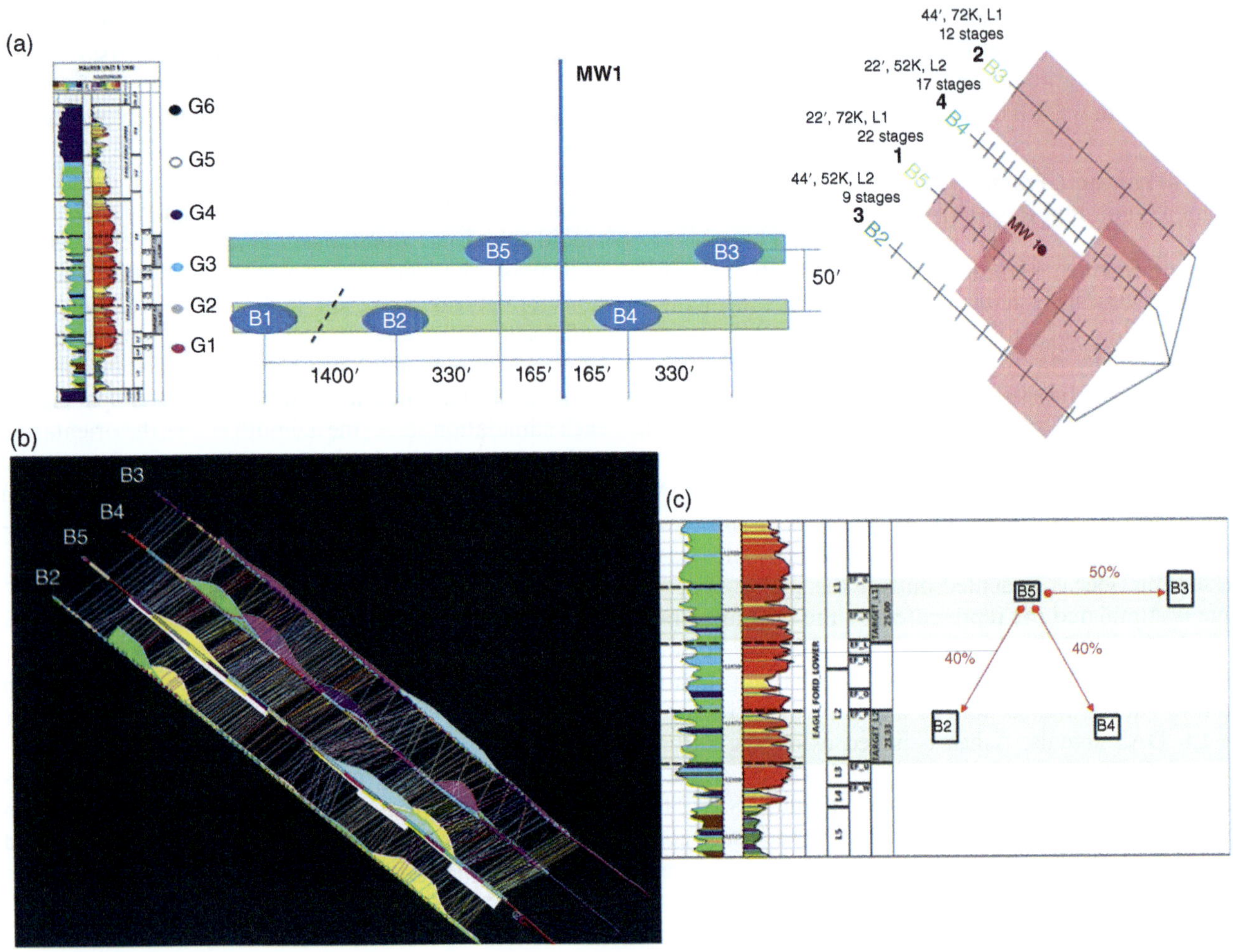

Figure 28.3 (a) The pilot had four horizontal wells (B2, B3, B4, B5) arranged in a two-layer pattern, with 660 ft spacing within a layer and 330 ft interlayer lateral spacing; a vertical monitor well (MW1) was also drilled midway between B5 and B4 for pressure measurement and microseismic acquisition. (b) When B5 was stimulated, the strain signals were recorded by the three offset wells B2, B3, and B4. This interwell fracture connection map shows the fracture hits between B5 and the three offset wells. (c) The percentage of total B5 perf clusters initiating a frac hit at each offset is marked in the cross-section view. The zone between B5 and B3 showed a relatively higher proportion (50%) of hits.

B2 and B4 were in the lower third of Lower Eagle Ford, with B2 offset from B1 by 1,400 ft. B3 and B5 were in the upper third of Lower Eagle Ford, and there was a 50 ft vertical spacing between well layers. A vertical monitor well (MW1) was also drilled midway between B5 and B4 for pressure measurement and microseismic acquisition. When one well was stimulated, the cross-well LF-DAS signals were recorded in adjacent wells. Figure 28.3(b), (c) show the map view and cross-section view of the fracture connection among each well pair. It was found that the number of events did not increase in proportion to the size of the treatment, which ranged from 52,000 to 72,000 to 100,000 lbs/cluster. LF-DAS and monitor well pressure data also confirm that many fractures extend distances greater than the interwell spacing of 1,000 ft. Cross-well DAS indicates that the conduits themselves are highly localized between wells. Integrated in-well and cross-well DAS, DTS, and pressure data were used to create a reservoir model that has history-matched production and interwell interference. The model indicated that drainage remained patchy and localized, and interference between wells was significant but not competitive. The completion and spacing design can be improved based on these insights.

Permian Basin

The primary application of LF-DAS in the Permian Basin is the project named Hydraulic Fracturing Test Site 2 (HFTS-2). It is a hydraulic fracturing research project in the Permian Delaware Basin, which employs multiple

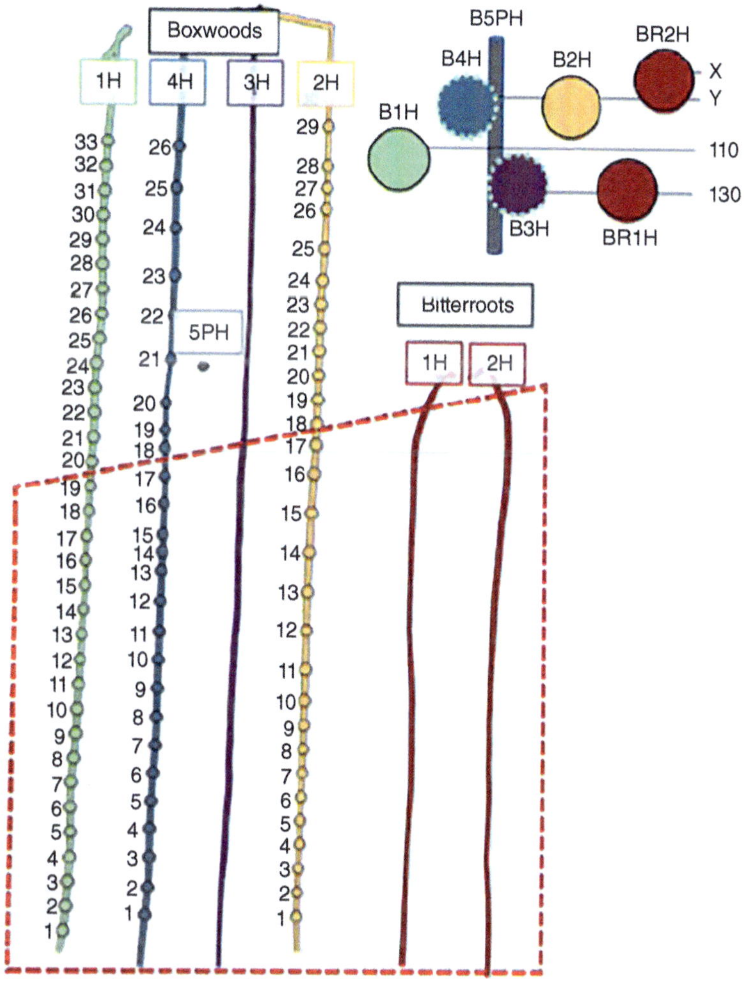

Figure 28.4 Map view and cross-section view of the HFTS-2 project, which includes two parent wells (Bitterroot wells) and five new drills (Boxwood wells). The red polygon marks, based on fracture azimuth, where the expected impact of the Bitterroots would be located for Boxwood stages. Haffener et al. (2022) / with permission of OnePetro.

monitoring techniques, including fiber-optics sensing, microseismic, and pressure/temperature gauges, to improve the understanding of the hydraulic fracturing process. It is a United States Department of Energy (DOE)-funded project with extensive industry and academia participation (Ciezobka, 2021).

The field site of HFTS-2 is presented in Figure 28.4. The two parent wells, Bitterroot 1H and 2H (BR1H and BR2H) are located in the east of the acreage. There are also four new producing wells, Boxwood 1H-4H (B1H–B4H). Vertical well B5PH is a dedicated monitor well. Permanent fibers were deployed along B3H, B4H, and B5PH to monitor cross-well strain during the stimulation of B1H and B2H. When B4H was stimulated, the LF-DAS data were acquired along B3H and B5PH (Zhang et al., 2021). There are multiple publications discussing the LF-DAS applications in this HFTS-2. Following is a summary of three of them.

Zhang et al. (2021) compare the LF-DAS signals recorded by the vertical monitor well B5PH (Figure 28.5(a)) and the horizontal monitor well B3H (Figure 28.5(c)). Figure 28.5(a)–(c) presents the LF-DAS and microseismic measurements during the stimulation of Stage 20 in Well B4H. The similar trend between the pressure change rate and the strain rate (Figure 28.5(a)) reveals that the fiber-optic and pressure measurements can mutually validate each other. As displayed in Figure 28.5(a), the LF-DAS signals detected in the vertical monitoring well provide clear evidence of upward fracture propagation, which is further corroborated by microseismic data (Figure 28.5(b)). This phenomenon can be attributed to the shallow zone's low-pressure gradient. In addition, signatures of different fracture shapes are observed. Figure 28.5(d) shows the LF-DAS signals recorded by the vertical well B5PH when Stage 21 of the horizontal well B2H was stimulated. Two types of

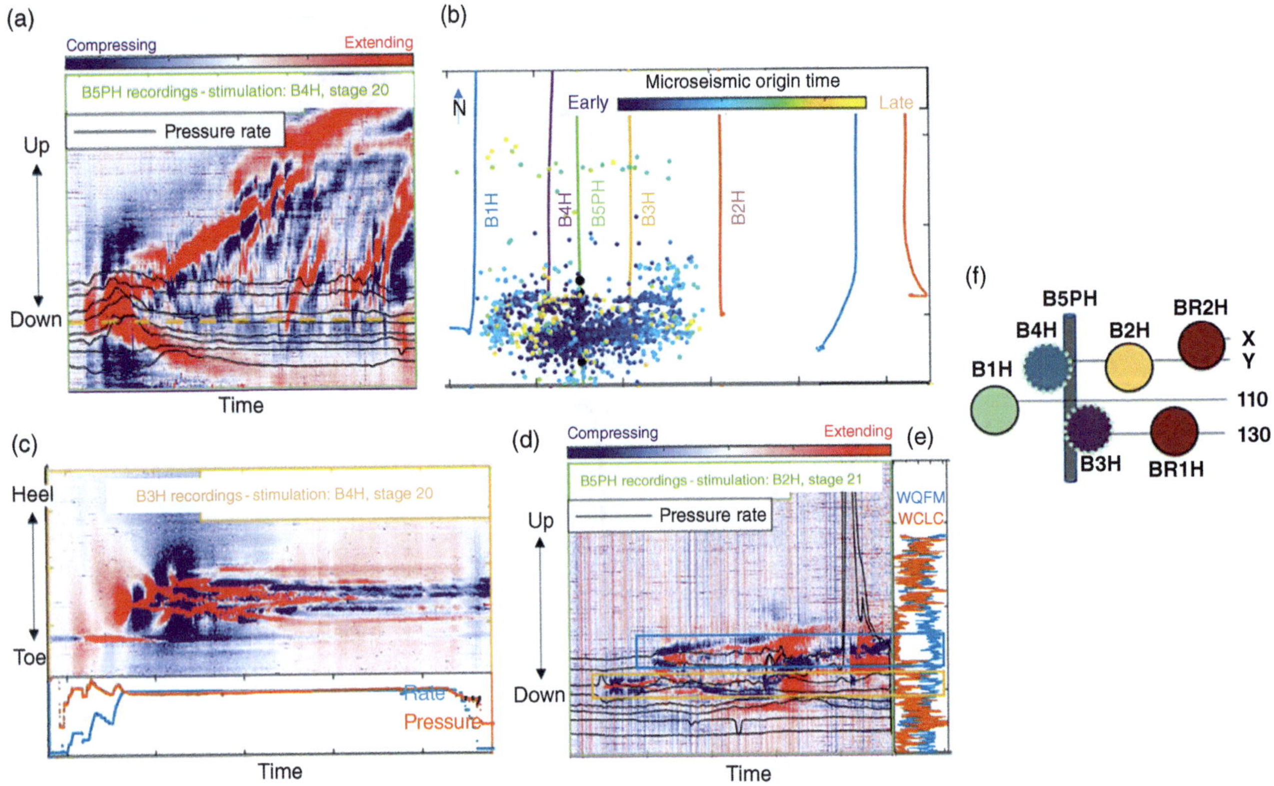

Figure 28.5 (a–c) Observations during the stimulation of Stage 20 in Well B4H. (a) LF-DAS recordings from vertical monitor well B5PH, where the pressure change rate measured by the P/T gauges (black curves) exhibits a trend similar to the LF-DAS measured strain rate. The upward growth of the hydraulic fracture is more prominent than the downward growth. (b) Cross-section view of the wells, with colored dots representing the microseismic events. (c) LF-DAS recordings from the offset parallel monitor well B3H, where multiple fractures can be identified from the measurements. (d) Two types of fracture shapes observed when B5PH records the stimulation of Stage 21 of well B2H: layered fractures in the yellow rectangle, and elliptically shaped fracture in the blue rectangle. (e) Dry weight of quartz–feldspar–mica (WQFM) and dry weight of calcite (WCLC) along the B5PH. The layered fractures in the yellow rectangle correspond to a vertically heterogeneous well log interval, whereas the *elliptic*-shaped fracture in the blue rectangle is located in a relatively homogenous layer well log interval. Zhang et al. (2021) / with permission of OnePetro. (f) Cross-section view of the wells, presented for reference. Haffener et al. (2022) / with permission of OnePetro.

fractures, layered fractures (yellow rectangle) and elliptic-shaped fracture (blue rectangle) are identified. The integration of LF-DAS signals with well logs (Figure 28.5(e)) demonstrates that the elliptical-shaped fracture corresponds to an interval where the well log is vertically homogeneous, whereas the layered fracture is associated with a vertically heterogeneous well log interval. It is possible that in the heterogeneous sections of the reservoir, the fractures encounter stress barriers and/or layered material properties, which limit vertical fracture growth. It was concluded that the integration of a comprehensive dataset provides insightful information for understanding hydraulic fracture propagation.

In the LF-DAS study conducted by Ugueto et al. (2021a), it was discovered that the hydraulic fractures in HFTS-2 are primarily vertical and parallel and exhibit asymmetry both vertically and horizontally, which is in line with the interpretation from Zhang et al. (2021). Figure 28.6 presents the plan view of the LF-DAS interpreted azimuth map during the stimulation of Well B1H (a), B2H (b), and B4H (c). The fractures show a uniform azimuth aligned with the maximum horizontal stress (SHmax). In Figure 28.6, red and blue polygons illustrate the areas affected and unaffected by the depletion of parent wells, respectively. For the stimulation of all three wells, more fractures are observed in the stages within the blue polygons than in the red polygons, confirming that depletion and pre-existing fractures affect fracture geometry. During the stimulation of B1H (Figure 28.6(a)), more fractures were observed in B4H

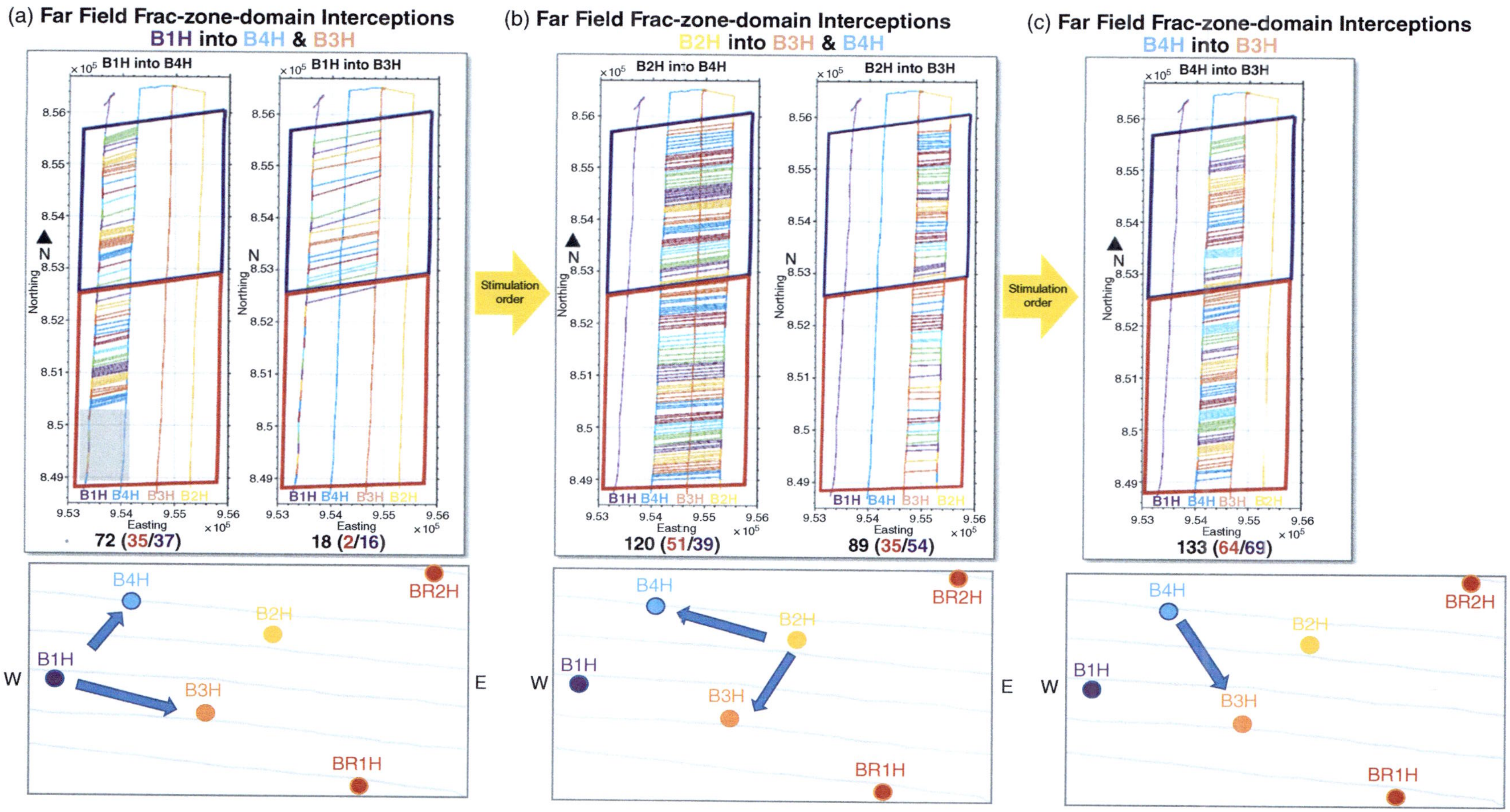

Figure 28.6 Plan view of the LF-DAS interpreted azimuth map during the stimulation of Wells B1H (a), B2H (b), and B4H (c). Red polygons represent the area affected by the depletion of parent wells, and blue polygons represent the nonaffected area. The number of fractures identified within the red and blue polygons is presented below each fracture connection map, where red and blue values represent the number of fractures within the respective polygons (e.g., the fracture connection map between B1H and B4H in (a) indicates that 35 fractures were identified in the red region, and 37 fractures were identified in the blue region). For the stimulation of all three wells, more fractures are observed in the stages within the blue polygons than in the red polygons, confirming that depletion and pre-existing fractures affect fracture geometry. Consistent azimuth is observed from all hydraulic fractures of the three stimulation wells. Ugueto et al. (2021a) / OnePetro.

than in B3H, mainly due to the relative well spacing of B4H (~660 ft) and B3H (~1,320 ft).

Haffener et al. (2022) demonstrate an interesting comparison between LF-DAS and microseismic and identified different fracture propagation speeds for new and pre-existing fractures. The reactivation signatures on LF-DAS show a rapid arrival with no tensional front and weak compressional lobes. The reactivation speed is likely tied to reservoir stress and pressure within the reactivated fractures. Specifically, B1H reactivated closed, potentially depleted parent fractures at ~25 ft/min, and B2H reactivated recharged parent fractures and new child fractures with minimal leakoff at ~50 ft/min.

The perforation gun pumpdown measurement showed the fastest rate of reactivation at the B4H monitor of about 100 ft/min. Furthermore, it was observed that B2H reactivated fractures from B1H at the B4H monitor, but no reactivation was perceived at the B3H monitor. This observation implies that active clusters, which intersect and reactivate earlier fractures, are unable to reexpand into regions of higher stress and, instead, propagate along lower stress pathways, such as laterally and upward. The study also suggests that LF-DAS signatures provide a powerful tool for monitoring and understanding fracture propagation in unconventional reservoirs, providing valuable insights for hydraulic fracturing design and optimization.

Bakken

In 2022, Cipolla et al. conducted a case study in the Middle Bakken to assess the cluster efficiency by integrating near-wellbore (NWB) measurements obtained through high-frequency DAS, sealed wellbore pressure monitoring (SWPM), and cross-well strain obtained from LF-DAS. The results indicated that stages with a more uniform distribution based on NWB measurements demonstrated higher volume to first responses (VFRs), as determined by cross-well strain and SWPM analysis.

Liang et al. (2022) optimized well spacing and evaluated cluster efficiency in XTO's pilot test by integrating data acquired from cemented fiber and pressure gauges that were installed in three wells—one single Middle Bakken producer (P1) and two monitor wells (M1 and M2) strategically designed around it. As shown in Figure 28.7(a), the two monitor wells' trajectories intersected hydraulic fractures of the producer at varying distances from the producer. The monitor well M1 is in the Middle Bakken. The wellbores of M1 and P1 are closer near the toe (200 ft) and further at the heel (1,000 ft). The monitor well M2 is a slant well drilled toe-up in the Three Forks and directly below the producer. The distance between M2 and P1 ranges from 50 ft near the toe and 130 ft near the heel. The LF-DAS measurements were used to identify the number of fractures at varying distances within Middle Bakken and from Middle Bakken to Three Forks. The interpretation of the LF-DAS data on a 6-cluster stage shows four distinct fracture hits (Figure 28.7(b)). The different fracture arrival time demonstrates that the speed of fracture propagation from the same stage can be highly nonuniform. Figure 28.7(b) also indicates fracture propagation in the Middle Bakken formation may not be continuous. Despite heart-shaped strain-rate patterns (green dots) appearing on the waterfall plot, suggesting that fractures were approaching the monitor well, these patterns did not immediately transform into the compression-tension-compression pattern associated with fracture intersection with the fiber. This implies that fractures were propagating in height or a different direction at some point during treatment, despite all six clusters taking a significant amount of fluid based on fiber monitoring in the producer. One of the pausing fractures did eventually continue to propagate toward M1 after a nearly 2 hr pause. Triangulation of perforation and frac-hit locations revealed an average frac azimuth of N55°E across all monitored stages, consistent with other operators' observations in the basin (Cipolla et al., 2022).

Completion of the producer from toe to heel led to an expected decrease in the number of identified frac hits as the distance between the stage and monitor well M1 increased. However, Figure 28.8(a) demonstrates that the number of frac hits to monitor well M1 remained almost constant over a certain distance from P1 before declining with increasing distances. The final stage, located approximately ~1,500 ft from M1, exhibited 2–3 frac hits to the in-plane monitor well. Differently, all stages showed around 6 frac hits to monitor well M2 (Figure 28.8(c)), which suggests that fractures in the Middle Bakken formation penetrated the Lower Bakken Shale and propagated into the first and second benches of Three Forks, even though the vertical distance to M2 increases from toe to heel of P1. The LF-DAS results (Figure 28.8(a), (c)) are consistent with later pressure gauge measurements of drainage (Figure 28.8(b), (d)). Figure 28.8(b) shows that the pressure decline curves recorded by the pressure gauges installed in well M1 can be separated into two groups: one group (marked by the red box) drained faster, and the other group (marked by the blue box) drained less, consistent with the stimulation measurement in Figure 28.8(a). Stages with higher stimulation efficiency (marked by the red box in Figure 28.8(a)) resulted in faster depletion, with each perforation growing a fracture hit to the monitor

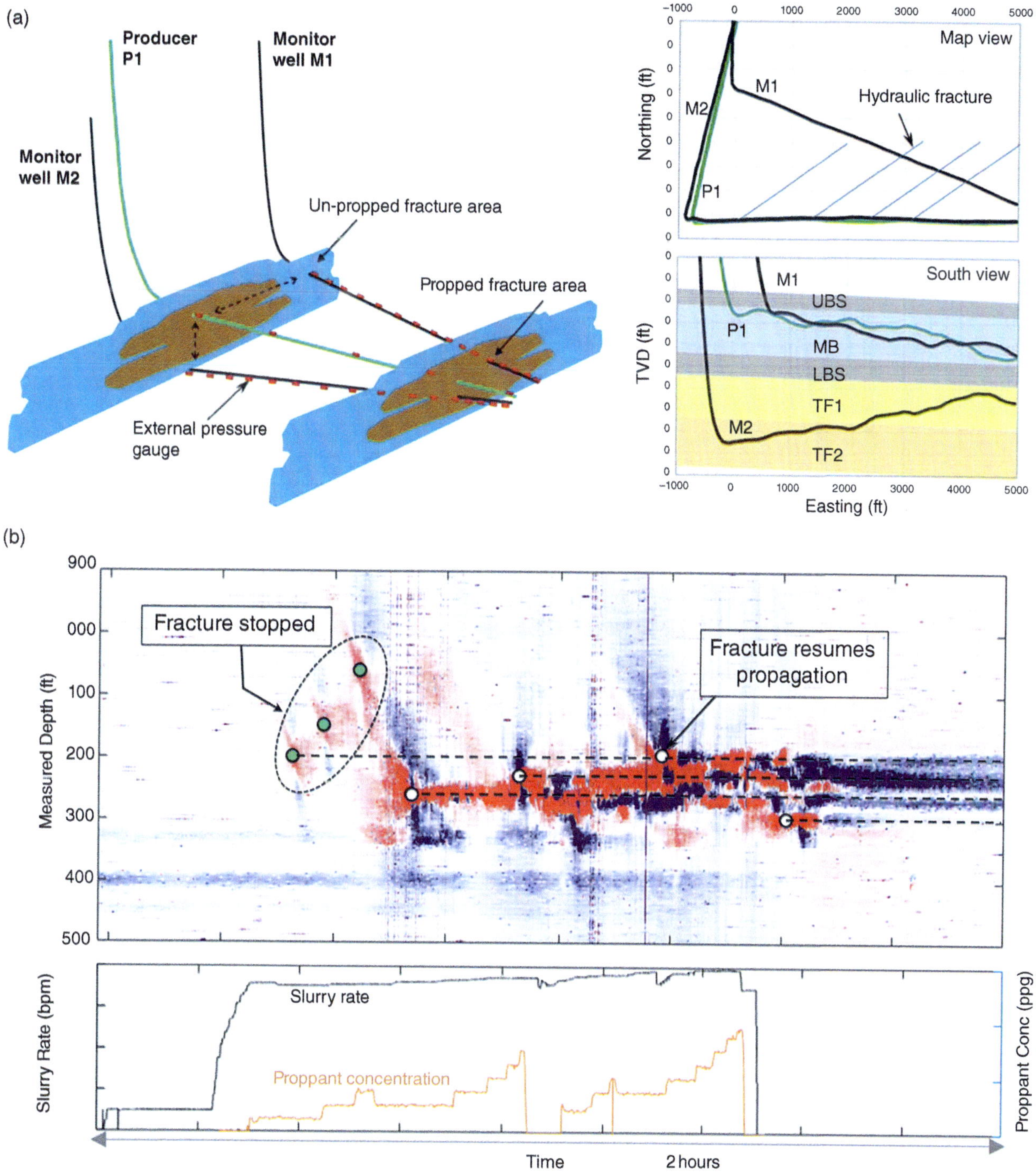

Figure 28.7 (a) Well setup in Liang et al. (2022) study. Well P1 is the producer, and Wells M1 and M2 are the monitor wells. In the south view, the formation names from top to bottom are UBS (Upper Bakken Shale), MB (Middle Bakken), LBS (Lower Bakken Shale), TF1 (Three Forks first bench), and TF2 (Three Forks second bench). (b) In the LF-DAS signals observed from Well M1, four fracture hits (white dots) were identified when a six-cluster stage of Well P1 was stimulated. Within this 2 hr stimulation window, the propagation speed from the same stage can be highly nonuniform. Liang et al. (2022) / reprinted with permission from URTeC, whose permission is required for further use.

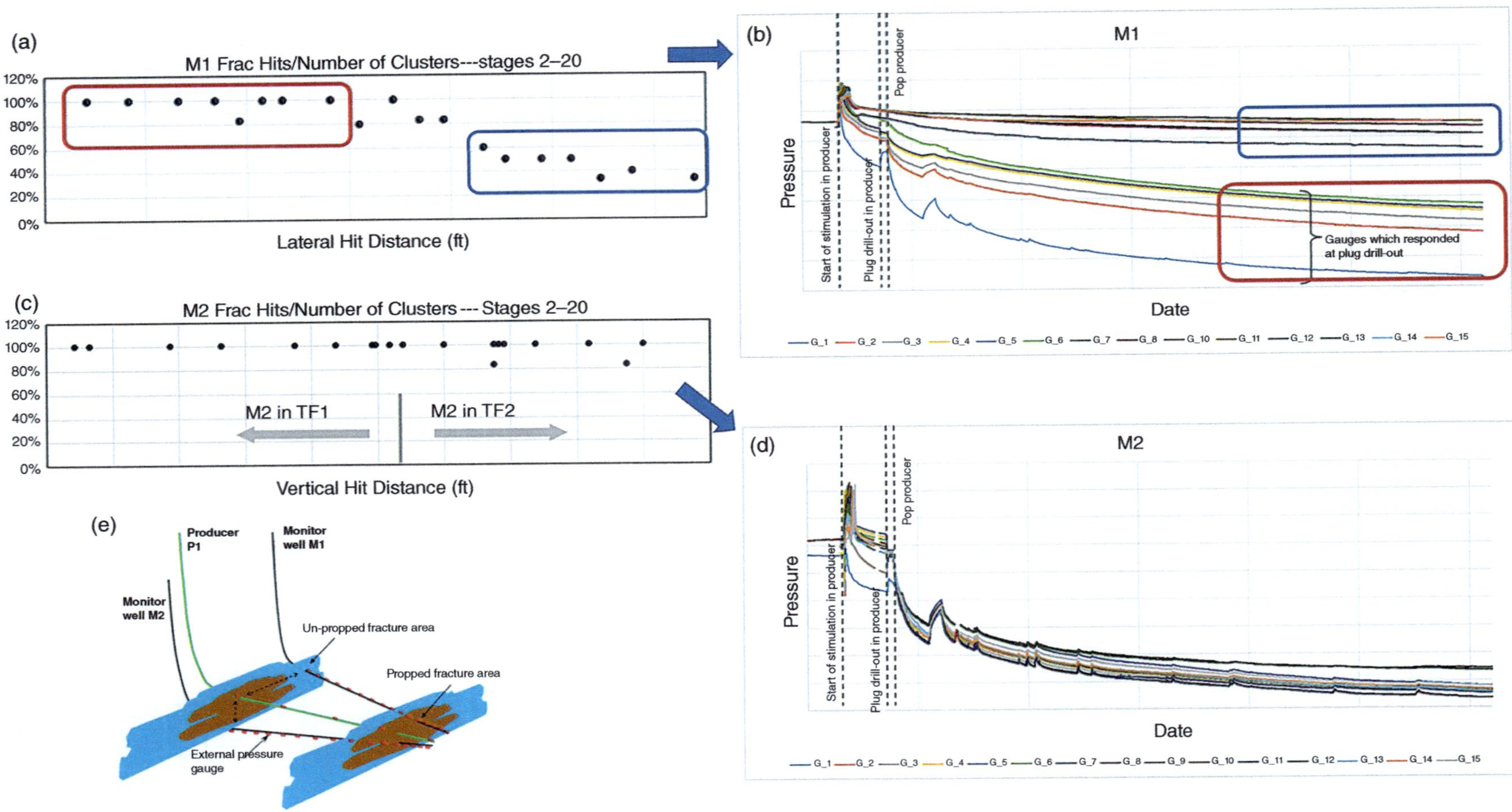

Figure 28.8 (a) More fracture hits are observed in the toe of well M1 and fewer are observed in the heel section of well M1. (b) The pressure decline curves recorded by the pressure gauges installed in well M1 can be separated into two groups: one group (marked by the red box) drained faster, and the other group (marked by the blue box) drained less. This is consistent with the LF-DAS measured results in (a). (c) All stages of P1 showed around six frac hits to monitor well M2, although the vertical distance to M2 increases from the toe to the heel of P1. (d) M2 pressure response of external pressure gauge array over time. External gauges along M2 show a tight grouping of the gauge responses. The pressure-declining observations at the M2 indicate good drainage of the Three Forks formation from the Middle Bakken well. (e) The relative location of the three wells: P1, M1, and M2.

well, and stages with lower stimulation efficiency (marked by the blue boxes) corresponded to regions with less depletion. These observations were imported into the integrated fracture and reservoir models and resulted in significant capex reduction by allowing larger well spacing and reduced freshwater usage in XTO's Bakken acreage.

Denver-Julesburg Basin

In the Denver-Julesburg (DJ) Basin Colorado, two fiber-optic research projects have been published: the Chalk Bluff project operated by HighPoint Resources (now acquired by Civitas) and the Postle project launched by Great Western Petroleum (first acquired by PDC Energy, later by Chevron). Studies on LF-DAS data in the two projects have been published and will be discussed in detail in the following section.

Chalk Bluff Project in the Northern DJ Basin. The Chalk Bluff study area is located in the northern DJ Basin, close to the Wyoming-Colorado state line. According to LF-DAS data from two fiber wells in the Chalk Bluff study area, Zhu and Jin (2021, 2022) analyzed the strain measurements when four wells in Niobrara and Codell were zipper fractured (Figure 28.9(a)). The fracture-hit interpretation from cross-well LF-DAS data provided insights into fracture geometry and propagation across and within Niobrara and Codell formations. Results showed significant differences in fracture propagation between the formations, with longer and faster fractures in Codell than in Niobrara, with a fracture propagation velocity of 13 ft/min within Niobrara and 24 ft/min within Codell. These differences were attributed to higher natural fracture density and potentially lower stress anisotropy in Niobrara. Different fracture orientations were also observed between the two formations and inconsistent orientations within Niobrara (Figure 28.9(b)). Strong cross-formation fracture connections were observed, with different up- and down-going fracture propagation velocities. The fracture propagation velocity is about 15 ft/min from Codell to Niobrara and 8 ft/min from Niobrara to Codell. These observations improved the understanding of complex fracture geometry in the DJ Basin.

Based on that, Ning et al. (2022b) related the LF-DAS signals with the daily production data from 23 horizontal wells in the Chalk Bluff project. This study demonstrates that the strong well interference during production correlates with a strong fracture connection identified from LF-DAS, and interwell fracture connection can potentially divert up to 80% of a shut-in well's production to an offset well in the adjacent formation.

Postle Project in the DJ Basin. The Postle study area, located in the Wattenberg field of the DJ Basin, aims to reduce well completion costs while maintaining production by examining the feasibility of breaking down more clusters over increased stage lengths. In Q2 2020, 13 child wells were drilled, and hydraulic fracturing operations began in November 2020 (Figure 28.9(c)). The fracturing of Zipper Group 1, consisting of seven wells, was monitored by a permanent fiber, a wireline fiber, and a disposable fiber located in different formation intervals, and Zipper Group 2, consisting of six wells, was monitored by the permanent fiber in the Niobrara B formation only. A tremendous amount of data was collected during the simultaneous fracturing operation of Zipper Group 1, with strain signals from multiple wells interfering with each other, making the dataset complex to interpret (Barhaug et al., 2022). To address the challenges, Ning et al. (2022c) and Ning & Jin (2023) shared the best practice of visualizing and interpreting the LF-DAS data, which provides confidence in the investigation of well communication during stimulation. The disposable fiber proved to be a viable fiber acquisition method, with data quality just as good as the acquisition by wireline and permanent fibers. Barhaug et al. (2022) pointed out that the disposable fiber experienced signal loss at the end of the fiber, which can be fixed by deploying an additional single-use fiber within 5 h for a 2-mile-long lateral. The velocities for all fractures in Zipper Group 1 were estimated to be between 10–20 ft/min, similar to what Zhu and Jin (2022a) estimated for the fracture propagation velocity in the Chalk Bluff project. Additionally, the azimuth map generated from the Postle project showed varying fracture azimuth within Niobrara wells (Figure 28.9(d)), which is consistent with observations from the Chalk Bluff study area (Figure 28.9(b)).

Midland Basin

Hull et al. (2019) conducted a case study at Pioneer Natural Resources in the Midland Basin, where a fiber-optic system was installed outside the casing of a vertical well to monitor the hydraulic stimulation process of two adjacent horizontal wells using the zipper-fracture method (Figure 28.10(a)). The study found that LF-DAS strain data showed the presence of downward fracture growth below the area where most microseismic events were detected (Figure 28.10(b)). The results also showed that the strain signals took around 15 min after the stimulation to propagate to the adjacent well over hundreds of feet of distance, demonstrating the importance of monitoring hydraulic fracturing operations in real time to better understand fracture growth and propagation.

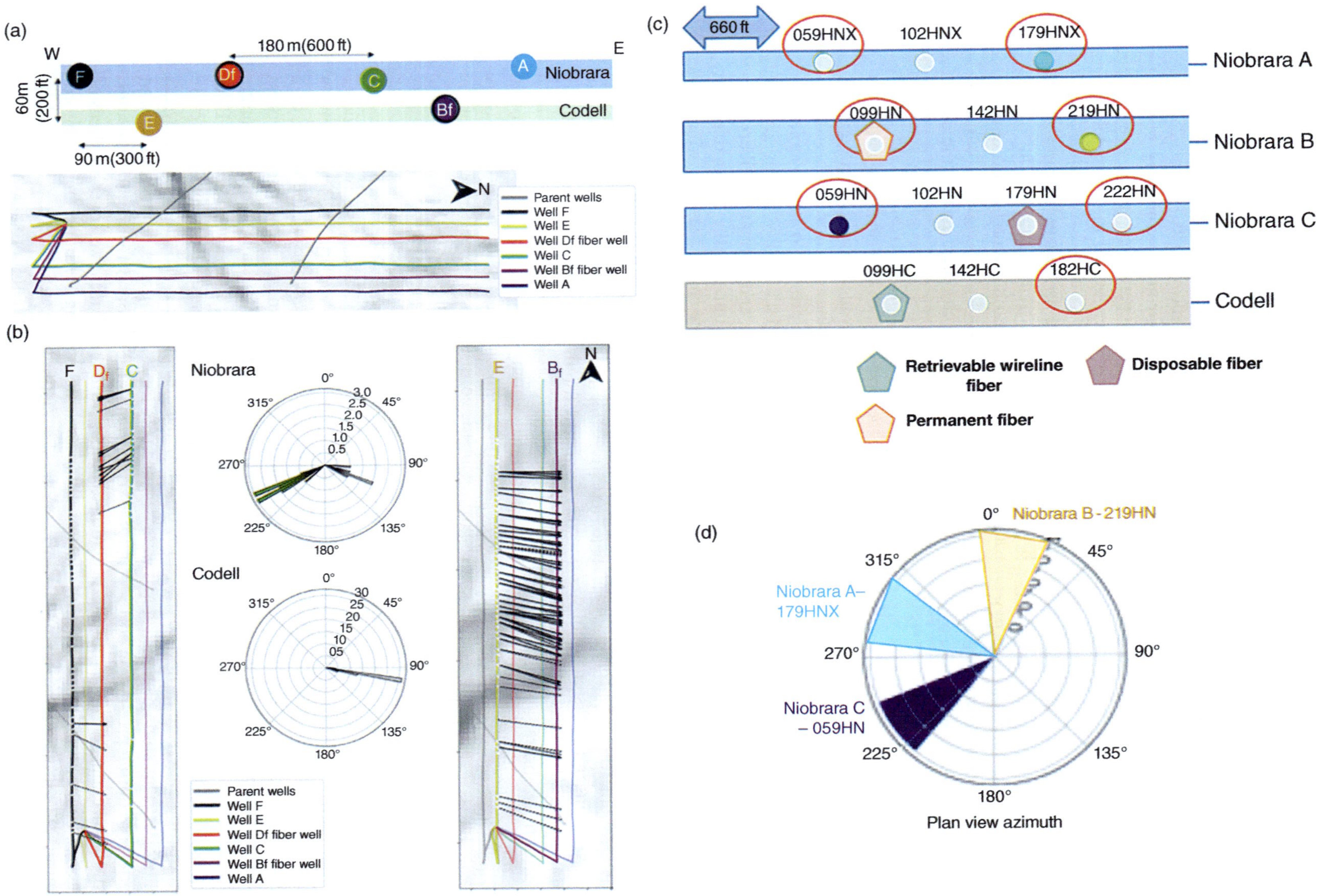

Figure 28.9 Overview of Chalk Bluff project (a, b) and Postle project (c, d) in the DJ Basin. (a) Cross-section and map view of well locations in the Chalk Bluff project. Two fiber wells, Bf and Df, monitored the stimulation process of the offset four wells (A, C, E, and F). In the map view, the parent well locations are shown in gray, with the seismic coherence attribute of Niobrara plotted in the background. Zhu and Jin (2022a) / with permission of OnePetro. (b) Map view of the fracture-hit connections (solid arrows) and approaching-fracture connections (dashed arrows) within the Niobrara formation (left) and within the Codell formation (right). The seismic coherence attributes of the two formations are plotted in the background respectively; a rose diagram of the fracture azimuth for each formation is shown in the middle. Arrows originate from the stage center to the observed fracture-hit locations. White dots indicate the stages with usable LF-DAS data. Zhu and Jin (2022a) / with permission of OnePetro. (c) Cross-section view of the Postle Study area. The wireline fiber and disposable fiber monitored the stimulation of seven wells in the Zipper Group 1, as marked by the red circles. Ning et al. (2022c) / with permission of the Society of Exploration Geophysicists. (d) Different fracture azimuths are observed from three Niobrara wells. Barhaug et al. (2022) / with permission of OnePetro.

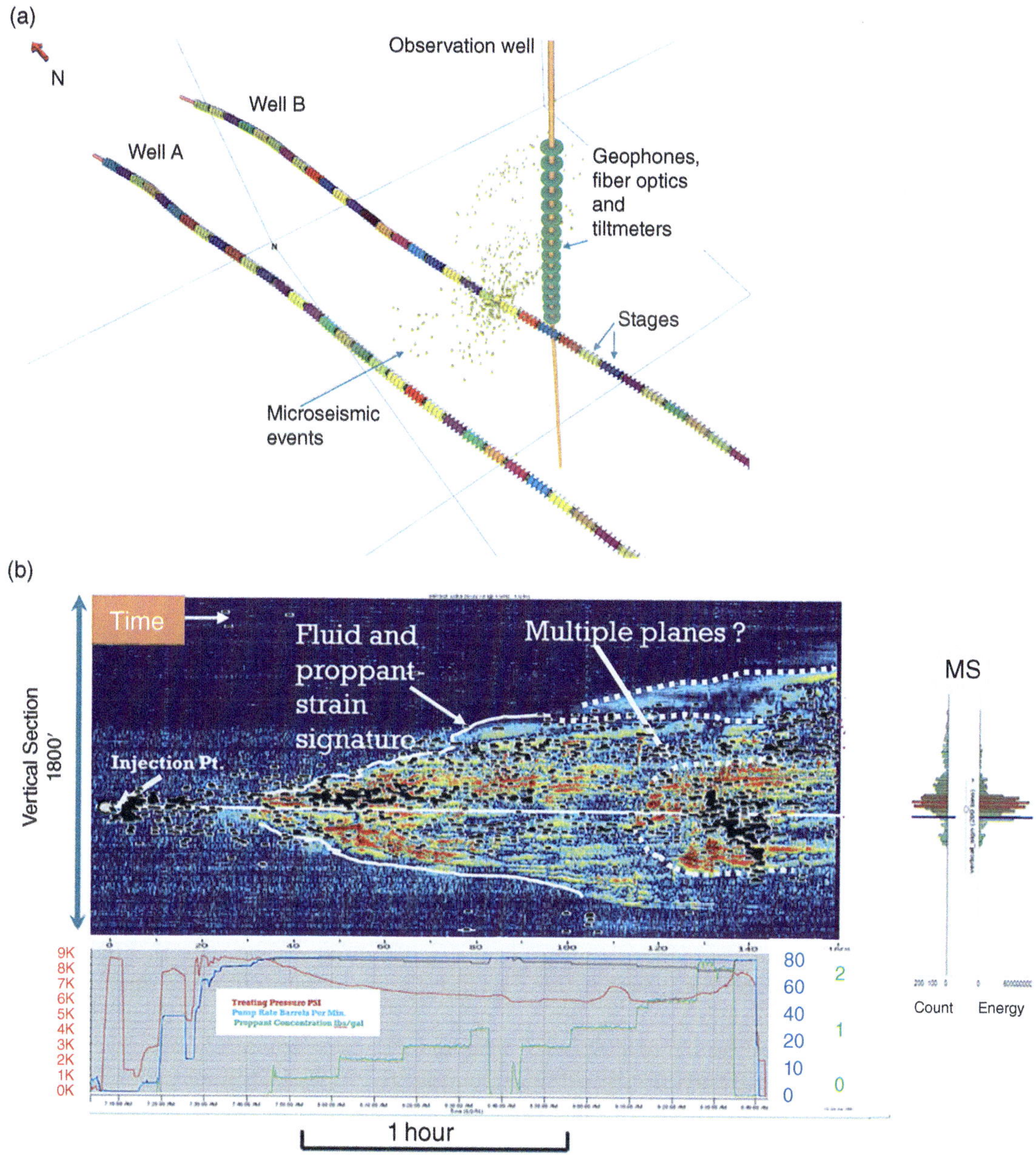

Figure 28.10 (a) Two horizontal wells (A and B) in the Midland Basin were hydraulically stimulated using a zipper sequence. The microseismic data for one stage are shown. (b) Both strain-rate signatures and microseismic events are utilized to monitor hydraulic fracture growth, with LF-DAS strain indicating some downward growth below where most microseismic events occur. The strain-rate signatures captured from a vertical monitor well illustrate the hydraulic fracture growth over time. The black dots represent the depths of microseismic events detected by geophone data. A histogram of microseismic events is also presented on the far right, with counts and energy displayed as a function of depth. Hull et al. (2019) / with permission of the Society of Exploration Geophysicists.

Marcellus Shale

In 2022, Mantell et al. presented the application of a disposable fiber-optic sensor in the Marcellus Shale for Chesapeake Energy. The single treatment well and single observation well project identified 36 cross-well events, including 29 fracture hits, 5 bypassing fractures, and 2 reactivation events. The fracture azimuths were very consistent between stages at N72°E, typical for this part of the Marcellus Shale. The P90 geometry plots show that the fracture half-length ranges from 761 to 1,051 ft, depending on the location of the stage, proximity to the fault, and asymmetry of the east–west hydraulic fracture wing growth. This project demonstrates the feasibility of using disposable fiber optics to effectively understand hydraulic fracture propagation, with little concern for downhole impact, at a low cost.

Montney

Ugueto et al. (2019) presented a LF-DAS case study in the Groundbirch Montney in Canada by Shell and OptaSense. The study demonstrated the application of DAS technology in wells with different completion systems, including plug-and-perforating (PnP), open hole packer sleeves (OHPS), ball-activated cemented single point entry sleeves (Ba-cSPES), and coil-tubing activated cemented single point entry sleeves (CTa-cSPES). Strain signals were measured in offset wells during stimulation, and most stages produced far-field strain gradient responses. The strain responses agreed with and confirmed the dominant planar fracture geometry, with fractures propagating perpendicular to the minimum stress. However, unexpected off-azimuth events were observed, corresponding with stage isolation defects in the stimulated well, leading to inefficient resource operation. The technology also identified opportunities for improvement in future completion programs by detecting restimulation and poor stage isolation.

Ichikawa et al. (2019, 2020) utilized DAS data during a hydraulic fracturing operation in the Montney tight gas area. Four horizontal wells were parallel with an average well spacing of 200 m (Figure 28.11(a)). Well C monitored the zipper fracturing operation of Wells A and D in 2017. Low-frequency components were extracted, followed by the generation of strain-rate maps (Figure 28.11(b)), cumulative strain maps (Figure 28.11(c)), and fracture azimuth map (Figure 28.11(d)). The maps showed pressure changes near the fiber-optic cable and detected the position and time of fracture hits from adjacent wells, providing an estimate of fracture propagation direction and speed. It is concluded that the main fracture propagation direction is estimated at approximately 37°–39°, which is consistent with the maximum principal stress trend near the target area.

Ortega Perez (2022) investigated another study area in the Montney Shale. Both treatment and monitor wells were drilled in the direction of minimum principal stress in the middle Member of the Montney Formation, with a well spacing of 150 m. The 38-stage stimulation process was monitored by the offset monitor well. The hydraulic fractures in this treatment propagated mostly perpendicular to the well, and the fracture propagation direction is consistent with the observation of Ichikawa et al. (2020) (Figure 28.11(d)). The hydraulic fractures in this study area had an average velocity of 21.13 m/min. When compared to the toe stages, the fractures took less time to close in the heel stages, indicating the dispersion of pressures and fluids into the formation or the effect of cumulative stress shadows.

The Canadian Dip-in DAS (CanDiD-1) project was investigated by Wang et al. (2022a, 2022b) and Eaton et al. (2022) to evaluate the feasibility of using retrievable fiber for fracture monitoring. This project is one of the first in Canada to use a temporary optical fiber deployment for monitoring HF operations. Wang et al. (2022a) report that although the signal-to-noise ratio of high-frequency signals was lower than that of permanently installed fibers, advanced preprocessing techniques were used to improve data quality. They developed a machine learning-assisted workflow to detect microseismic events and determine fracture-driven interactions, and the integration of high- and low-frequency signals provided valuable information about fracture geometry and local stress orientations. Eaton et al. (2022) also analyzed the LF-DAS recordings from six horizontal wells and determined fracture azimuths. They observed a systematic variation in azimuth with depth, which is interpreted as a depth-dependent rotation in the maximum horizontal stress direction. These findings demonstrate the potential for dip-in DAS systems to eliminate the need for permanent optical fibers in mapping fracture geometry and suggest that this technology can provide important insights into the dynamics of HF operations.

Anadarko Basin

In Anadarko Basin, two case studies regarding LF-DAS application were presented by Devon Energy. Haustveit et al. (2020) used LF-DAS stain measurements to validate a cheaper pressure-monitoring technique, SWPM (Figure 28.12). This technique installed a surface pressure gauge on the offset monitor well that has a sealed casing. During the hydraulic fracturing operation of one well located in the Anadarko Basin, the sealed wellbore pressure and the cross-well strain using a fiber-optic cable were recorded on the same monitor well. The first fracture arrival observed from the strain-rate plot occurred at the same time as the pressure inflection read on the derivative plot of the sealed wellbore pressure, demonstrating the consistency of the SWPM technique and the LF-DAS strain measurements.

In their 2022 field case study, Haffener et al. investigated the hydraulic fracture treatment of the A, B, and C wells in the Anadarko Basin targeting the Meramec formation using LF-DAS technology. Figure 28.13 analyzed the LF-DAS signals during the stimulation of a stage in Well A with an eight-cluster design. The analysis revealed that only two new fractures were created (as indicated by green cross symbols), and the remaining fractures were the reactivation of previously created hydraulic fractures (red plus symbols). LF-DAS measurements showed two noticeable differences between reactivated and new fractures. Reactivation signals displayed a narrow zone of tension, and newly created fractures showed heart-shaped

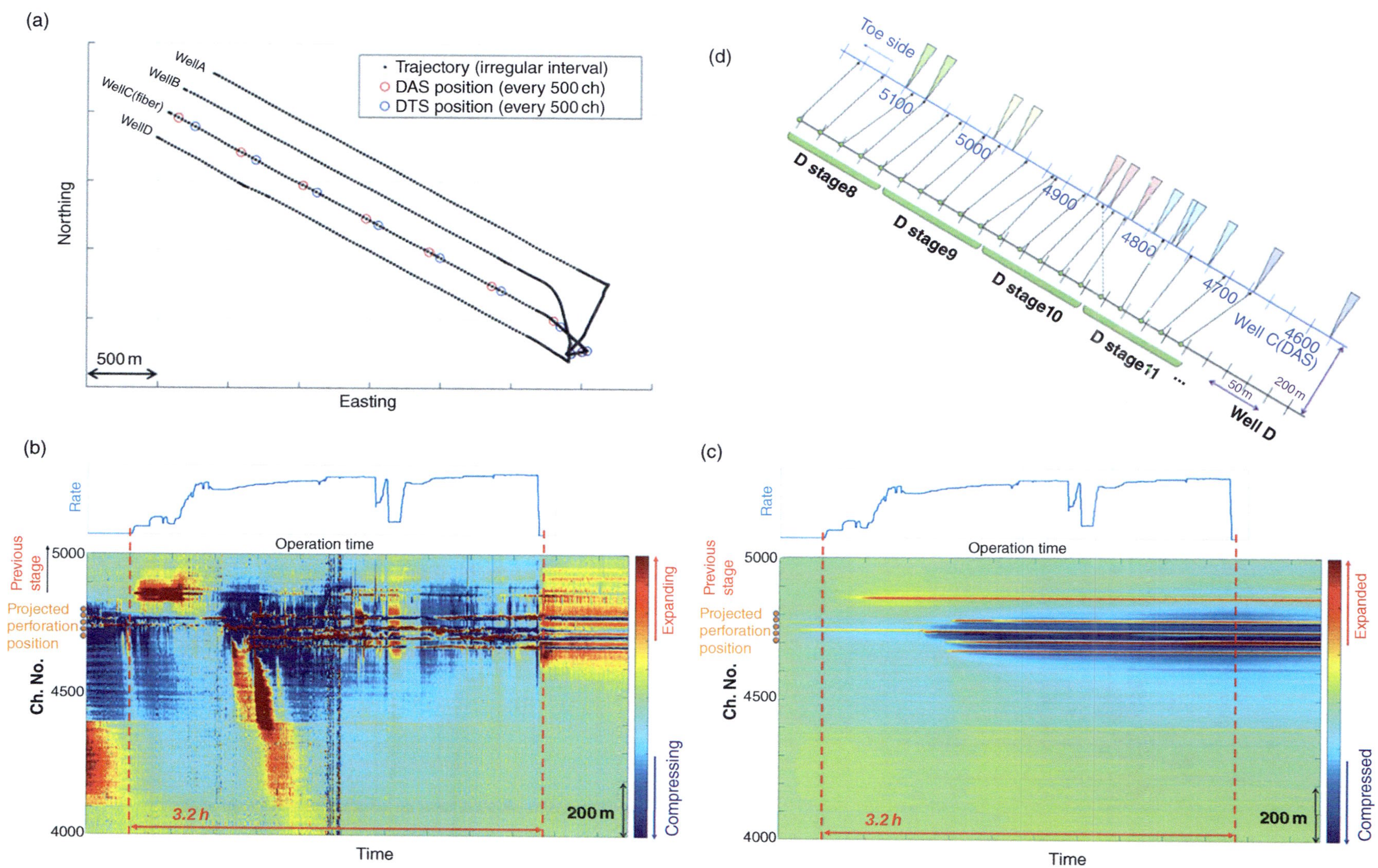

Figure 28.11 (a) Position of operation and observation wells in the Montney tight gas area (Ichikawa et al. (2020) / with permission of the Society of Exploration Geophysicists.). Four horizontal wells were parallel with an average well spacing of 200 m. Well C monitored the zipper fracturing operation of Well A and D in 2017 first and then monitored the stimulation of Well B. (b) Strain-rate map of the LF-DAS data with an injection rate curve at the top. (c) Cumulative strain map. The expanding points shown in (b) and (c) were highly correlated to the projected perforation position. (d) Fracture azimuth map between Well D and fiber well C. The main fracture propagation direction is about 37–39°, consistent with the maximum principal stress trend near the target area. Ichikawa et al. (2020) / with permission of the Society of Exploration Geophysicists.

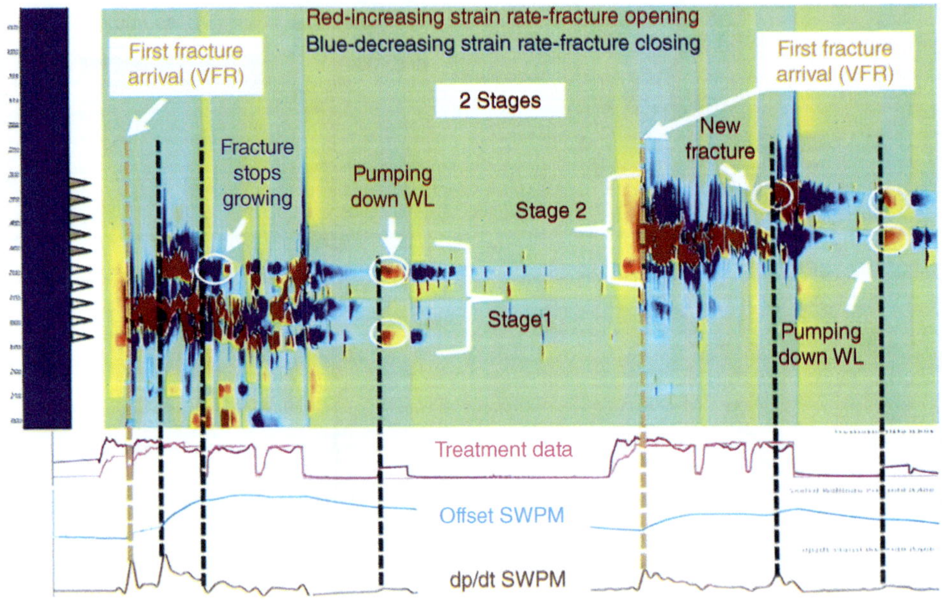

Figure 28.12 Two stages were stimulated within 4 hours. For these two stages, cross-well strain (top track) is compared to treatment data (pink and red curves represent pressure and pump rates) and sealed wellbore pressure responses (blue and brown curves at the bottom). Left side marks perforation clusters (five for each stage) as triangles, which are then colored by stage. Timing of VFR (brown vertical dash curve) for each stage is noted by dashed gold lines, and pumping down of wireline is noted on this figure. Haustveit et al. (2020) / with permission of OnePetro.

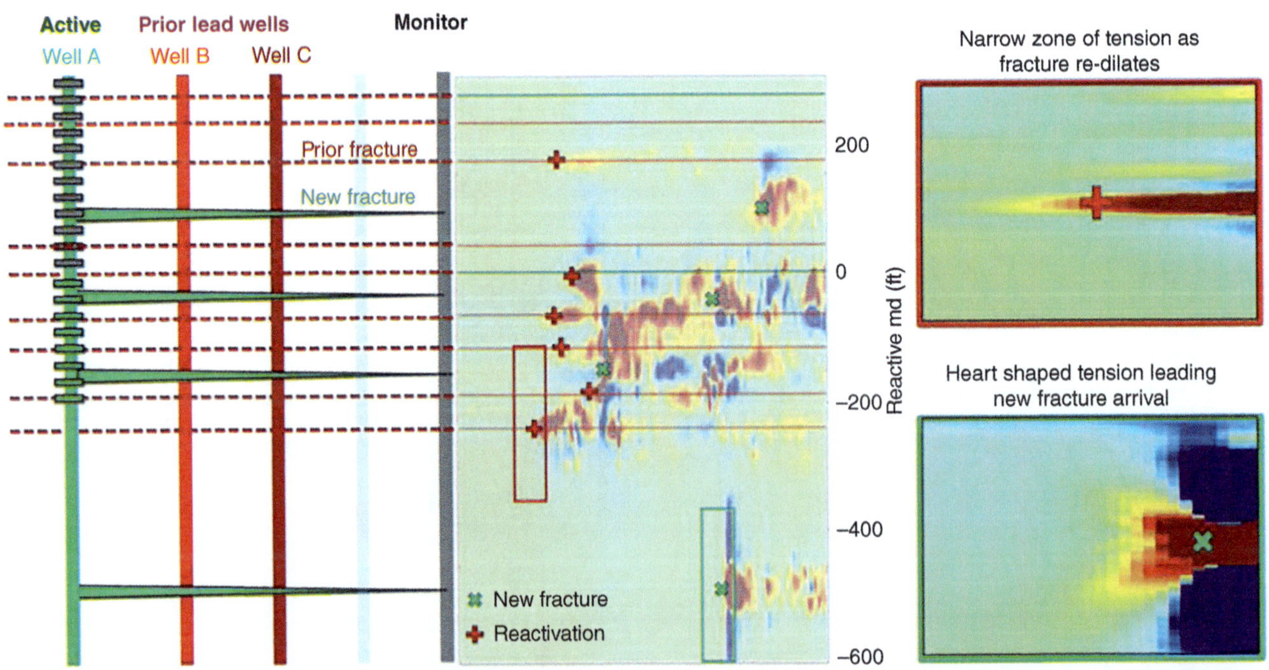

Figure 28.13 (Left) Zipper fracturing treatment of Wells A, B, and C was monitored by the offset gray well. Red horizontal dash curves present pre-existing fractures created by Wells B and C; green horizontal lines represent the new fractures created by the active stage of Well A. (Middle) LF-DAS arrivals were observed during the stimulation operation. Pre-existing fracture intersections are marked with horizontal lines. Strain responses were separated as reactivation (red plus symbols) and new fractures (green cross symbols). (Right) The reactivation signal shows a narrow zone of tension, and the new fracture arrival shows heart-shaped tension. Haffener et al. (2022) / with permission of OnePetro.

tension. Reactivation could also be distinguished from new fractures by a faster arrival time. The authors concluded that LF-DAS is preferred over SWPM for understanding reactivation. However, once the faster rate and distinct response of SWPM in a formation are understood, dual-instrumented SWPM and fiber-optic wells can serve as a cost-effective method to determine whether reactivation is taking place.

28.3.2. Summary

Table 28.2 presents a comparison of hydraulic fracture half-length, fracture azimuth, and fracture propagation speed in various basins, including the Eagle Ford, Permian, Bakken, DJ Basin, Marcellus Shale, and Montney formations. This table was generated based on the publications discussed earlier, which have investigated these parameters. The data reveals that most basins have an estimated fracture half-length larger than 1,000 ft, except for the Niobrara formation in the DJ Basin. Additionally, the fracture propagation velocity varies significantly across basins, with a range of 10–69.32 ft/min. It is noteworthy that results can differ greatly, even within the same basin, due to variations in geology and completion sequencing effects. Thus, the data presented in Table 28.2 is expected to be a valuable reference for planning future fracturing operations in different basins.

28.4. FRAC-HIT DETECTION

Detecting and preventing fracture hits is critical in hydraulic fracturing operations to avoid production loss and damage to adjacent wells (Ajani & Kelkar, 2012; Zhu & Jin, 2021). LF-DAS data are an essential tool for real-time monitoring of the reservoir during hydraulic fracturing completion to avoid fracture hits at offset wells and optimize production.

Jin et al. (2019) proposed a machine learning algorithm to automate fracture-hit identification in LF-DAS data (Figure 28.14). The study used a simple neural network model, which provided a fracture-hit probability estimation based on eight features extracted from the data. The workflow was applied in an unconventional case study, and the results showed that a smaller cluster spacing completion design resulted in denser hydraulic fractures in the reservoir.

Liu et al. (2021a) investigated the use of LF-DAS data to detect fracture hits and characterize fracture geometry in unconventional reservoirs during multistage hydraulic fracturing treatments. The study used a fracture propagation model to simulate simultaneous multiple-fracture propagation and the corresponding strain and strain-rate responses along an offset monitor well. The proposed guideline for fracture-hit detection is based on the quantitative analysis of strain/strain-rate responses and

Table 28.2 A comparison of hydraulic fracture half-length, azimuth, and propagation speed across different basins, as reported in the previously discussed studies.

Publications	Basin/formation	Longest distance a frac hit is observed (estimate fracture half-length)	Hydraulic fracture azimuth	Hydraulic frac propagation speed
Raterman et al., (2018, 2019, 2020)	Eagle Ford	Exceed 305 m (1,000 ft)	—	—
Ugueto et al. (2021a), Tan et al. (2020)	Permian	Approximately 402 m (1,320 ft)	~N80°E	9 m/min (30 ft/min)
Liang et al. (2022) and Cipolla et al. (2022)	Bakken	Larger than 457 m (1,500 ft)	N55°E	—
Zhu and Jin (2022), Ning et al. (2022b), Barhaug et al. (2022)	DJ Basin – Niobrara	198 m (650 ft)	N20°E, SE around 110°, SW around 240°, NW around 290°	3–6 m/min (10–20 ft/min)
	DJ Basin – Codell	387 m (1,270 ft)	SE around 100°	7 m/min (24 ft/min)
Mantell et al. (2022)	Marcellus Shale	320 m (1,051 ft)	N72°E	—
Ugueto et al. (2019), Ichikawa et al. (2019, 2020), Ortega Perez (2022)	Montney	400 m (1,312 ft)	N37°E, N60°E	21.13 m/min (69.32 ft/min)

Note: Variations may exist in other study areas within the same basin.

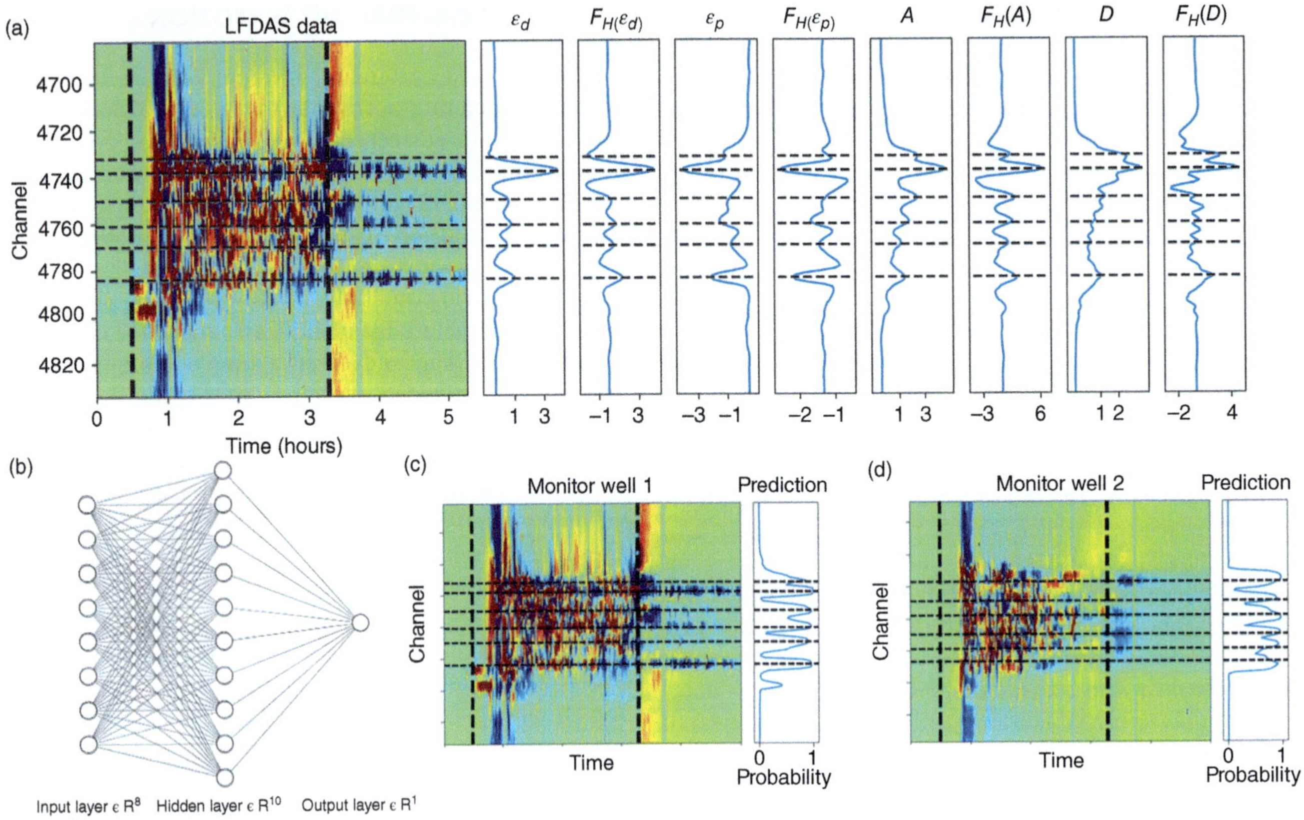

Figure 28.14 (a) LF-DAS signal and features. The color map shows the extension (red) and contraction (blue) of the fiber. Horizontal dashed lines indicate the locations of manually picked fracture-hit. Thick vertical dashed lines are the beginning and end of the pumping period. Eight features were used in the model: ε_d(accumulate strain during pumping), $F_H(\varepsilon_d)$ (high-pass filter with a corner wavelength of 30 channels (30 m)), A (root mean square of LF-DAS data during pumping), and $F_H(A)$ are calculated using data between the two dashed lines (0.5–3.2 h). ε_p (accumulate strain post pumping) and $F_H(\varepsilon_p)$ are calculated from the data after the second dashed line (3.2 h). D (difference between the current channel and nearby channels) and $F_H(D)$ are calculated using the entire data frame. (b) The architecture of a selected neural network model. (c–d) Model prediction on the same stimulation stage at two different monitor wells. (c) The data of monitor well 1 are in the training dataset; (d) the data of monitor well 2 are in the test dataset. Jin et al. (2019) / with permission of the Society of Exploration Geophysicists.

includes the maximum strain rate, the summation of strain rates, and the summation of strain-rate amplitudes. The field examples showed the effectiveness of the proposed guideline for fracture-hit detection with multicluster completion, with the fractures propagating nearly perpendicular to the horizontal wellbore and a "heel-biased" fracture pattern being observed.

Andrade and Baan (2021) introduced a computer vision workflow to automate the detection of frac hits in LF-DAS data. The workflow uses the Oriented FAST and rotated BRIEF (ORB) object detection algorithm to match key points between two or more images of the same object and subtract the frac-hit strain signature using image transformations and warping. The subtraction process helps to detect multiple frac hits at the same stage and understand their underlying strain response. The workflow is applied to a real LF-DAS dataset from Western Canada, and the results demonstrate the effectiveness of the proposed methodology for automating the detection of fracture hits.

Based on the HFTS-2 dataset, Chen Ning et al. (2021) proposed an early warning system to identify precursory events leading to a fracture hit by utilizing image-based algorithms on LF-DAS signals, including template matching and a convolutional neural network (CNN). Zhu et al. (2022) established a workflow for the processing and classification of LF-DAS signals and utilized machine learning techniques to detect fracture-hit precursors at an early stage. Among the machine learning models evaluated, the random forest classifier demonstrated the highest performance. The workflow was applied to the HFTS-2 dataset, achieving an accuracy of 93% in detecting fracture hits.

Chen et al. (2022) developed a machine learning-based workflow to identify and locate fracture-hit events in LF-DAS data for real-time event detection. The workflow starts with building a single fracture propagation model to generate labeled synthetic strain-rate responses, which are used to train a custom CNN model. The results showed near-perfect predictions for both event classification and localization, proving the feasibility of using CNN for real-time event detection from fiber-optic sensing data. The study also compared edge detection techniques in recognizing fracture-hit event patterns and found them to be plausible but less robust compared to CNN models. This research highlights the need for CNN applications in image-based real-time fiber-optic sensing event detection.

28.5. GEOMECHANICS MODELING

The development of geomechanical models is essential for the accurate interpretation of LF-DAS data. To determine the stress and strain distribution around hydraulic fractures, domain discretization methods, such as finite-element and finite-difference methods, have been commonly used (Lecampion, 2009). However, these methods can be computationally expensive due to the need to discretize the entire domain. An alternative approach is the use of boundary element methods, such as the displacement discontinuity method (DDM). DDM is a highly efficient and straightforward method for calculating stress and strain fields around hydraulic fractures (Crouch, 1976; Shou et al., 1997; Wu & Olson, 2015). The following section will discuss publications on LF-DAS-based geomechanical modeling, which is mainly employed to quantitatively characterize hydraulic fracture geometry.

28.5.1. Models Based on Recordings From Horizontal Monitoring Wells

Zhang et al. (2020a, 2020b) developed a DDM to model and interpret the LF-DAS recorded from a horizontal monitoring well (Figure 28.15). Modeling results explain heart-shaped extending patterns before a fracture hit, polarity flips during stimulation due to fracture interaction, and V-shape patterns when a fracture bypasses the monitoring well from above or below without intersecting. The modeling results are further confirmed by the field observations of HFTS-2 data.

Liu et al. (2020) also adopted the DDM approach to analyze fracture propagation and its impact on rock deformation and strain-rate variations. Results from five synthetic cases with single fracture propagation and a field example with four clusters are analyzed. The strain-rate waterfall plots of LF-DAS measurements can be divided into an extending heart-shaped region before the fracture hits the monitor well and a compressing region with two wings after the fracture hits. The size and shape of the observable signatures are influenced by fracture width, height, and height growth. The field example analysis indicates that the distinct signatures in field data convey valuable information on fracture characteristics, including fracture-hit time, propagation speed, and perforation efficiency.

Chen et al. (2021) extend a multiple-planar 3D fracture model to consider strain and strain-rate responses in a horizontal monitor well during hydraulic fracturing treatments of a parallel offset horizontal well. The model was validated against analytical plane-strain solutions and used to analyze the strain and strain-rate evolution of the fiber in the observation well for both single and multi-fracture cases. The findings suggest that the extending and compressing strain-rate patterns during hydraulic fracturing stimulation proposed by Ugueto et al. (2019) can be reproduced. The study also found that fracture hits can be interpreted based on strain-rate evolution. The delay of strain-rate reversal after pumping ends may reflect the continuing growth of hydraulic fractures after shut-in.

Gurjao et al. (2021) explored the impact of fluid, rock, and natural fractures on hydraulic fracture propagation by employing forward modeling simulations using both analytical and numerical solutions. The authors propose the integration of synthetic and field data to identify field fracture scenarios using machine learning classification techniques and estimate the magnitude of DSS/DAS using regression models. The results show that the LF-DAS data captures classic characteristics such as the elastic effect of a fracture approaching the fiber, a heart-shaped pattern, stress shadow, and fracture hits. Additionally, the results indicate that a single hydraulic fracture in tensile mode propagates faster under conditions of high Young's modulus, high injection rate, and low fluid viscosity. Numerical modeling of hydraulic and natural fractures interaction reveals complexities in DSS/DAS time histories and the impact of natural fracture geometric characteristics on strain and strain-rate results.

Tang and Zhu (2022) presented an integrated methodology for simulating both rock deformation-based and fiber-optic measurement-based strain rates during hydraulic fracturing. The approach is designed for monitoring a well equipped with DAS in the vicinity of a fracturing well. The methodology was applied to synthetic examples to examine the connection between strain-rate

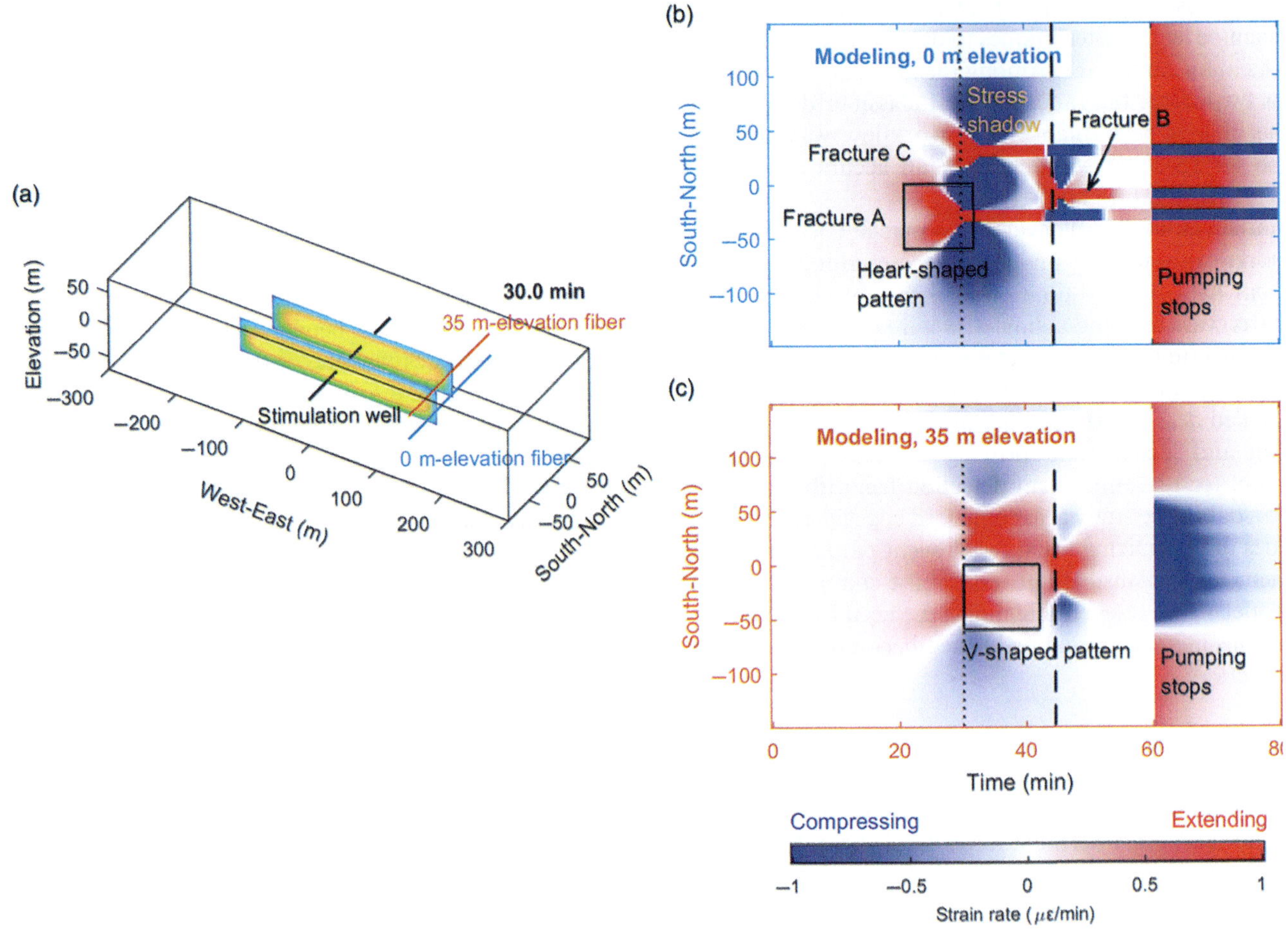

Figure 28.15 (a) Fracture shape and aperture distribution at 30.0 min after the stimulation starts. The stimulation well is in the middle, and the strain rate was observed by fibers with elevations of 0 and 35 m. (b) LF-DAS reveals heart-shaped patterns that precede fracture hits. These patterns are distinct and noticeable before the occurrence of fractures A and C. However, the heart-shaped pattern preceding fracture B is indistinct due to the interference of the stress shadow generated by fractures A and C. During hydraulic stimulation, a polarity flip of the strain rate is observed, which can be attributed to fracture interaction. Notably, this flip occurs during pumping and not after the pumping stops. (c) When a DAS does not intersect but is located above or below a propagating fracture, a V-shaped pattern can be observed, which can differentiate between intersecting and nonintersecting cases. Zhang et al. (2020a) / with permission of the Society of Exploration Geophysicists.

responses and fracture propagation during the injection. Results show that the strain-rate response to a fracture propagating toward the DAS-equipped monitoring well exhibits a cone-shaped front before intercepting the monitoring well and a strip-shaped front after passing it. The slope of the cone shape is inversely proportional to the injection rate within a certain range (10–25 bbl/min in this study) and can be used to estimate fluid distribution during the injection. The cone-shaped front tip in this study is actually consistent with the heart-shaped signature mentioned in other studies.

28.5.2. Models Based on LF-DAS Signals from Vertical Monitoring Wells

To better understand LF-DAS signals when a vertical offset well monitors the stimulation of a horizontal well, Sherman et al. (2019) developed idealized models to simulate synthetic DAS measurements and to estimate fracture location, geometry, and extent through a finite-element (FE) approach. By using a multiphysics code including rock physics, fluid flow, and elastic-wave propagation to generate synthetic DAS measurements from simple models of hydraulic fracturing, the relationship between the synthetic DAS signals and fracture attributes such as height, width, and aperture was studied. The results show that synthetic DAS measurements can be used to interpret field DAS measurements and improve the design of future sensor deployments.

Zhang et al. (2020a) developed a DDM method to model the cross-well strain field around propagating fractures. For vertical monitoring well, modeling confirms that LF-DAS responses show different characters when a fracture is near to (10 m) and far (50 m) from a vertical monitoring well (Figure 28.16(c), (e)), which correspond to field data from HFTS-2 in Figure 28.16(f), (h).

Tan et al. (2020) explored the interpretation of LF-DAS signals in multiple scenarios by utilizing a fracture simulator and FE solver. The strain-rate simulation results based on vertical monitoring wells show that hydraulic fractures continue to propagate upwards even after pumping has stopped. Also, the fluid movement is only upward after pumping has stopped and horizontal movement along the fracture has stopped. The continued upward propagation is believed to be driven by buoyancy, as the pressure within the fracture has a hydrostatic pressure gradient and the stress gradient in the formation is lithostatic. Additionally, the study found that with proppant transport in the simulation, the single fracture had more net pressure changes along the hydraulic fracture, resulting in a wider fracture near the wellbore. This resulted in an additional compression zone close to the wellbore, showing that the horizontal strain is sensitive to changes in net pressure along the fracture.

Bourne et al. (2021) used the deconvolution method to observe the evolution of hydraulic fracture heights and aperture distributions during stimulation and depressurization. They developed a computationally efficient electrostatic kernel model for fracture opening in a linear elastic medium. The application of this model to the HFTS-2 data showed that the ultimate fracture heights were typically higher than indicated by microseismic monitoring and that fractures continued growing for hours after injections ceased. The model rapidly simulates the time-depth distribution of strains induced by hydraulic fractures and infers the time-depth distribution of fracture apertures. The results of the model applied to the HFTS-2 data indicate that the shallowest reported microseismic depth is not a reliable indicator of the ultimate fracture height, and fractures can extend significantly above the shallowest event depth, with the possibility of upward growth during fluid pressure relaxation.

Srinivasan et al. (2022a, 2022b) estimated fracture height by building a physical rock deformation model to simulate strain-rate signals in vertical monitoring wells during hydraulic fracturing. A workflow is proposed to determine fracture height at different fiber-to-fracture distances, which is followed by sensitivity analysis to understand the impact of various factors on the strain-rate signals. The sum of the absolute strain rate along the fiber length is also calculated as a 1D feature to determine the fracture passing time. Results from the field data from HFTS-2 are expected to provide better interpretations of LF-DAS signals from vertical monitoring wells.

Wang et al. (2022) used field data from HFTS-2 to investigate hydraulic fracture height growth. By combining LF-DAS strain data from a vertical pilot hole with advanced geomechanical modeling, they observed excessive upward height growth (over 1,000 ft) and limited downward growth during pumping, as well as considerable additional upward height growth after well shut-in. The study also found that the strain-rate patterns resulting from single fractures showed different characteristics due to various fracture-fiber offsets. The HFTS-2 LF-DAS strain data revealed complex patterns when the stage was close to the monitoring vertical fiber, which was attributed to the simultaneous propagation of multiple fractures at different growth rates. The advanced geomechanical modeling supported the observed asymmetric upward and downward fracture height growth and showed distinct patterns for different fiber-to-fracture distances.

28.6. QUANTITATIVE INTERPRETATION

With the support of geomechanical models, LF-DAS data can be quantitatively inverted and interpreted. This section explores the quantitative interpretations of LF-DAS to estimate fracture width, assess fracture height, evaluate well integrity, and investigate the temperature impact on strain measurements.

28.6.1. Fracture Width Estimation

The width of hydraulic fractures influences the proppant transportation and fracture permeability. Liu et al. (2021b) proposed an inversion algorithm to estimate fracture width near the monitor well using LF-DAS data, which is interpreted as linear-scaled strain variations and monitored during hydraulic fracturing treatments. The algorithm uses the least-squares method and Markov-chain Monte Carlo (MCMC) inversions to relate the strains monitored by LF-DAS to the fracture widths through a Green function. The inversion results, which are obtained using the least-squares method and confirmed by MCMC simulation, show that the LF-DAS strain data have a dominant sensitivity to the fracture width near the monitor well (Figure 28.17(c)). The algorithm successfully inverts the width evolution near the monitor well as a function of injection time and provides accurate information to calibrate hydraulic fracturing

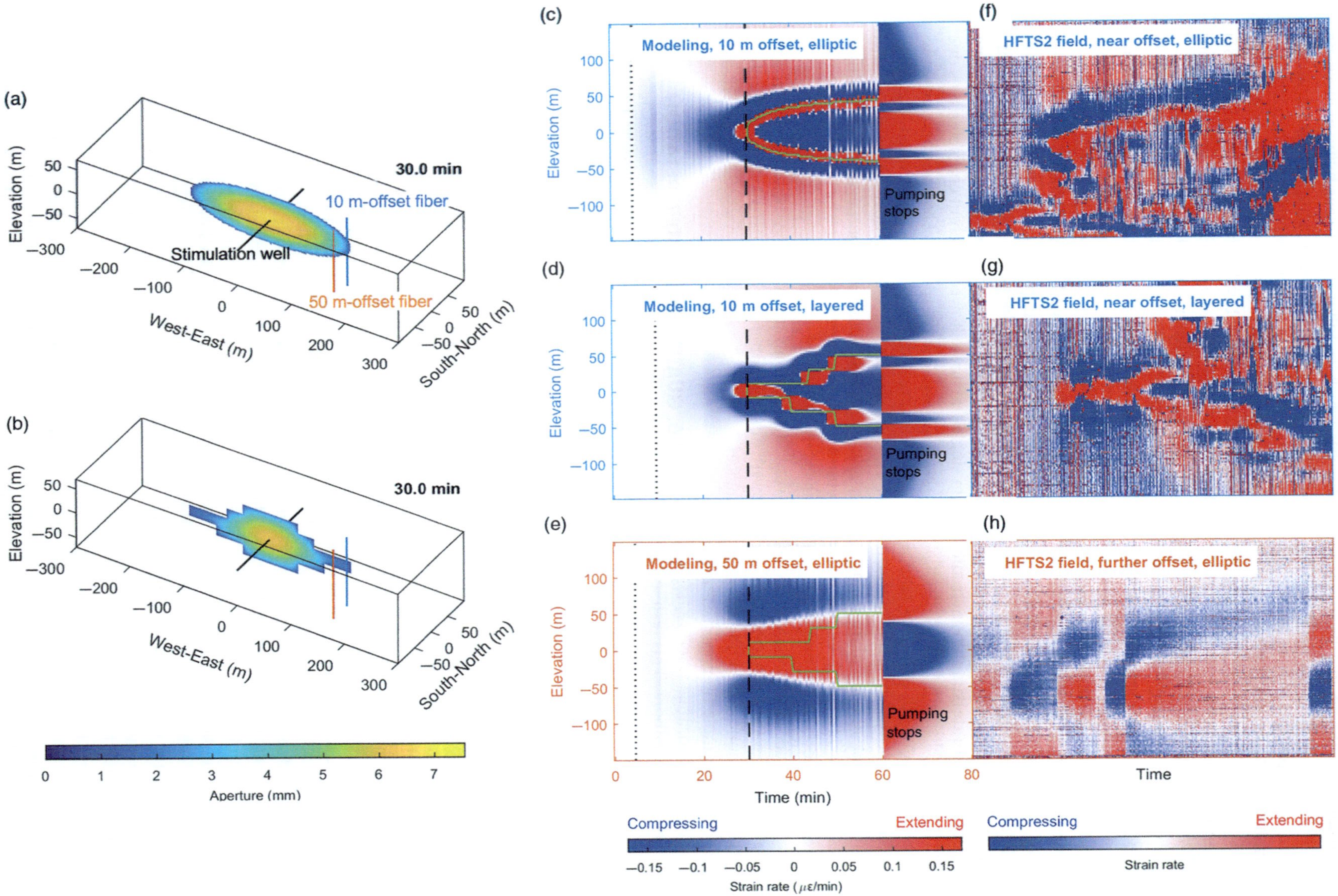

Figure 28.16 (a) Propagation of elliptic fractures at 30 min after stimulation starts. The stimulation well is a horizontal well located in the middle, and the stimulation is monitored by two vertical fibers. (b) Propagation of the layered fractures. (c)–(e) Modeled fiber-optic strain-rate responses to the elliptic and layered fractures in a 10 m-offset vertical well and a 50 m-offset vertical well. The black dashed lines at 30.0 min correspond to the time of (a) and (b). The green lines represent the outlines of fractures (f, g, and h) The corresponding field LF-DAS observed patterns from HFTS-2. Zhang et al. (2020a) / with permission of the Society of Exploration Geophysicists.

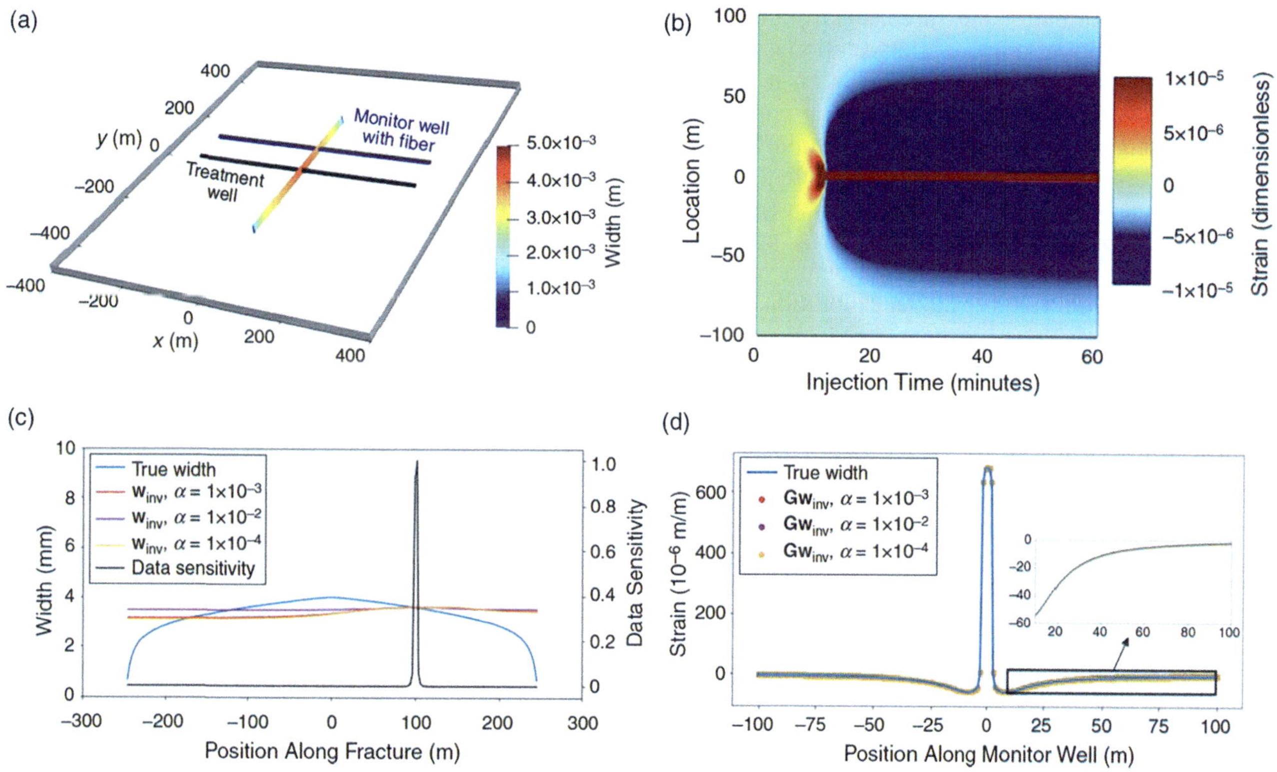

Figure 28.17 (a) Schematic of the well pair consisting of a treatment well and a monitor well, and the fracture geometry at the end of injection. The well spacing is 100 m. (b) Waterfall plot of strain measured along the monitor well during fracture propagation. (c–d) Comparisons of true widths and inverted widths (c) and true strains and calculated strains (d) for the smoothness-regularized least-squares method with three different smoothness-weighting coefficients α. (c) The inverted width is equal to the true width at the fracture-hit location (i.e., 100 m from the perforation), where the data sensitivity also shows the maximum value. (d) The magnitude of α does not have a significant effect on the inversion results. It indicates that strain data measured by LF-DAS along the monitor well has a maximum constraint on fracture geometry at the fracture-hit location. Liu et al. (2021b) / with permission of OnePetro.

models, improve completion parameters, and possibly characterize the whole fracture geometry. Based on this study, Liu et al. (2021c) extended the algorithm to handle multiple fractures, address uncertainties, and add a time-dependent constraint for better fracture width estimation. A synthetic case study and a field example are presented, showing consistent results between inverted fracture width and LF-DAS measured strain data.

Ramos Gurjao et al. (2022) developed a machine learning algorithm to estimate the magnitude of far-field LF-DAS response using spatial–temporal inputs. The study shows that a discontinuity in displacement results following a fracture hit may indicate fracture width. Tangent displacement data were generated through the DDM and fed into an artificial neural network (ANN) trained with position and time inputs, resulting in accurate predictions with negligible error. The findings demonstrate the potential of machine learning algorithms in characterizing created fracture systems beyond the near-wellbore region.

Hu et al. (2023) estimated the fracture length and width distribution over time by using a hydraulic fracture geometry inversion model that combines LF-DAS data, high-frequency DAS data, and injection rate data. The model utilizes a flow-solid coupling algorithm that relates the Green function and flow distribution functions and uses the Picard method to solve the algorithm. The study also explores the impact of the distance between the fiber and the injection wellbore, spatial sampling spacing, and fracture spacing on the computational stability of the algorithm. The results suggest that the inversion model is able to accurately and stably estimate multifracture geometry of different scales by combining DAS data, injection rate, and the coupling of the fluid distribution function and the Green function.

Moradi et al. (2022) proposed a geomechanical modeling method for estimating hydraulic fracture width that combines LF-DAS and downhole tiltmeter fracture mapping (DTFM) data. By jointly inverting near-field and far-field slow deformation measurements from these systems, the ambiguity of fracture interpretation is reduced, leading to higher confidence and resolution in fracture opening estimates. The joint data inversion approach is validated using synthetic data, with uncertainty quantified. The presented joint inversion tool offers a single answer from a combined dataset rather than generating different solutions that are not necessarily consistent, which provides more confidence in fracture geometry characterization and fracture model calibration.

28.6.2. Fracture Height Assessment

Liu et al. (2022a) explored the effect of fracture height on LF-DAS inversion results using a numerical simulation with a four-cluster completion design and then developed a method for estimating fracture height based on the inversion results. The results showed that the inverted widths were not sensitive to assumed fracture height, and the error between calculated and true strain decreased as the height approaching the true value. The authors applied their inversion algorithm to two field examples, where the results demonstrated the robustness and accuracy of the algorithm as the estimated fracture height matched well with other field measurements.

28.6.3. Temperature Impact Investigation

All strain measurements are sensitive to temperature changes due to thermal expansions. As a result, the effects of temperature changes on LF-DAS measurements for hydraulic fractures need to be investigated.

Leggett et al. (2022) developed a model relating strain, temperature, and optical phase shift of the LF-DAS signal, which provides insights into the relative strength of temperature and strain effects and quantifies uncertainties in differential temperature-gradient sensors. A workflow was presented to simulate the LF-DAS response accounting for both strain and temperature effects, including a comparison with a published field case to enhance the interpretation of waterfall plots. The results suggest that before, during, and immediately following a fracture, the strain drives the LF-DAS signal, but at later times, the temperature component can equal or exceed the strain component. The temperature effect was found to be only limited to the fracture hit locations. A sensing configuration was proposed to identify the events when fractures with fluids intersect the observation well.

28.7. STRAIN SENSING DURING PRODUCTION

LF-DAS has been used as a highly sensitive DSS method to investigate strain changes at offset monitor wells during hydraulic fracturing stimulation, with limited attention paid to strain changes during production periods. As an emerging technology, DSS measurement based on RFS is gaining interest as a new type of distributed sensing system with higher sensitivity (<1 με) and spatial resolution (0.2 m) compared to traditional LF-DAS measurements (Jin et al., 2021). This technology has been applied to evaluate perforation efficiency, identify productive fractures, and understand near-wellbore fracture properties, as well as their impact on stimulation and production, as summarized in the following studies.

Jin et al. (2021) conducted a field study using RFS-based DSS technology as part of the HFTS-2 project (Figure 28.18). The study aimed to characterize near-wellbore fractures and perforation cluster efficiency during production. The results showed that the DSS system can be used to evaluate perforation efficiency and identify productive fractures through the observation of extensional strain changes around perforation clusters. A novel diagnostic plot was also presented that showed different patterns of strain change at perforation locations compared to wellbore pressure. The study has the potential to improve the understanding of near-wellbore hydraulic fractures and their relationships with stimulation and production from unconventional oil and gas wells.

Ugueto et al. (2021b) utilized DSS-RFS strain monitoring to assess the characteristics of near-wellbore fractures and cluster productivity for unconventional wells. This technique provided accurate data with high temporal sampling rates and spatial resolution during steady flow, shut-in, and reopening operations. The strain-change variations observed indicated changes in near-wellbore fracture aperture, and the shape and magnitude of strain-change peaks were associated with the geometry of conductive near-wellbore fracture zones, as depicted in Figure 28.19. In Figure 28.19(a), the strain changes for six and nine perforation cluster (PC) stages were compared, demonstrating a significant correlation between the frac-zone domains' characteristics at each cluster. Figure 28.19(b) displays the relationship between geometry and strain changes for both 9–10 PC and 6 PC stages, revealing significant variations in strain change magnitude and geometry between the two stages. The authors suggested that the signals collected during stable flow provided the best indication of relative production and could be obtained without shutting in the well, allowing for more frequent time-lapse data collection.

Liang et al. (2022) applied DSS during a well interference test in the Bakken and found meaningful strain changes in the producing well and monitor well during a shut-in period. The extensional strain peaks at the cluster locations in the producing well indicated that most fractures were conductive, and the peak locations in the

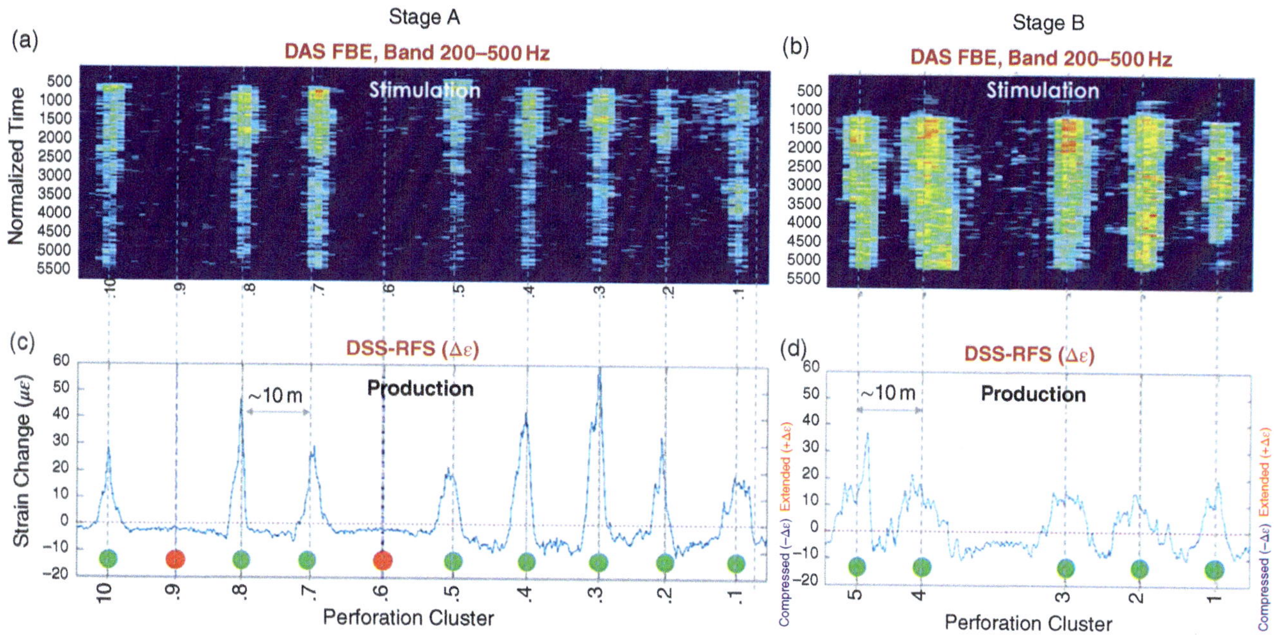

Figure 28.18 Comparison between DSS-RFS measurement (c, d) with DAS acoustic intensity measurement (a,b) during stimulation. (a–b) Waterfall plots for DAS acoustic intensity at 200–500 Hz frequency band, where FBE represents frequency bands extracted. The warmer color bar indicates a higher noise level during the injection. (c, d) Strain change during the shut-in. The left (a, c) and right (b, d) panels are from two different stages of the instrumented well. In (c, d), the dashed lines highlight the perforation locations. The red horizontal dashed line marks the level of zero strain. The red dot and green dot illustrate inefficient and efficient perforation clusters, respectively. Adapted from Jin et al. (2019).

monitor well coincided with frac-hit locations, where the extensional strain peaks represented conductive fractures at the far field. However, the study showed that only about 50% of frac-hit locations exhibited strain changes, suggesting that some fractures may have experienced conductivity loss. This test confirmed the potential of fiber monitoring in detecting conductive fracture distributions at the far-field.

Dhuldhoya and Friehauf (2022) applied DSS to characterize drained reservoir volume (DRV). The first application, in-well production strain (IWPS), involved measuring strain changes at producers induced by pressure transients during well operations, which demonstrated a relationship between fracture stage length, uniformity during stimulation, and productivity. The second application involved measuring DSS data on drill-through monitoring well during an interference test, providing insight into identifying and mapping effective fracture dimensions horizontally and laterally. The use of DSS measurements has shown promising results in enhancing the understanding of fracture features and their impact on stimulation and production in unconventional oil and gas wells.

Liu et al. (2022b) conducted a study to improve understanding of RFS-based DSS measurements and their relation to near-wellbore fracture properties. The study applied a coupled flow and geomechanics model to simulate the observed DSS signals and analyzed the strain responses. The results showed that the strain responses captured the observed signatures of field DSS measurements, and the strain-pressure diagnostic plot is path dependent. The study provides valuable insights for near-wellbore fracture characterization and completion optimization.

28.8. DISCUSSION

LF-DAS technology holds significant potential for monitoring in unconventional reservoir development. However, several challenges and limitations need to be addressed to fully exploit its capabilities and ensure accurate and reliable results.

One of the primary challenges is related to the signal-to-noise ratio. LF-DAS signals can be affected by various noise sources, including instrument noise, environmental noise, and cross-talk from nearby monitoring wells. Efforts should be made to improve signal quality through advanced signal processing techniques, noise reduction algorithms, and optimized sensor placement strategies.

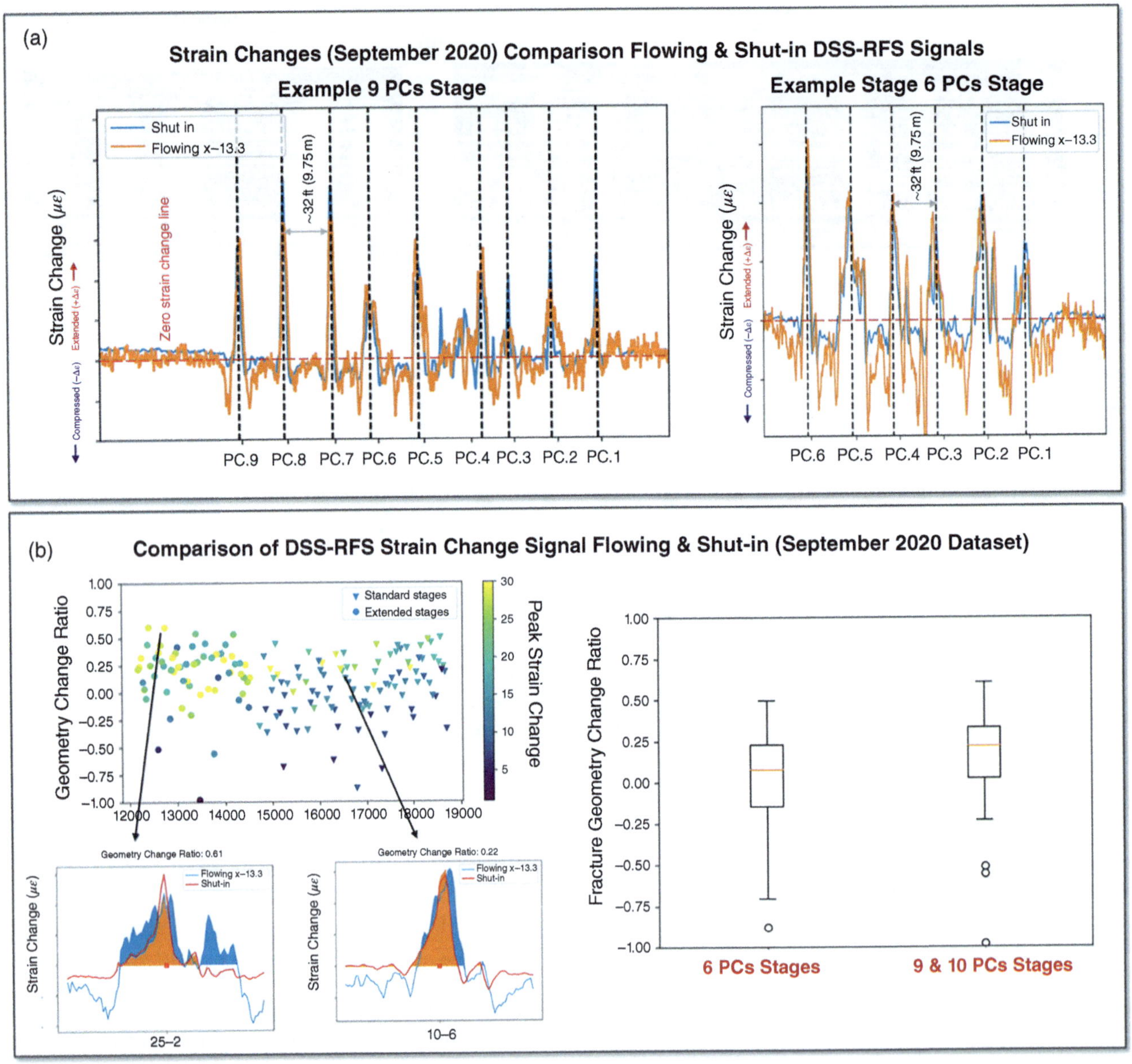

Figure 28.19 (a) Comparison of DSS-RFS strain-change profiles for six and nine PC stages during flowing (orange curve) and shut-in (blue curve) operations. The location and shape of the positive strain-change peaks are similar, demonstrating a significant correlation between the frac-zone domains' characteristics at each cluster. (b) The relationship between geometry and strain change for 9–10 PC and 6 PC stages, where significant differences in strain change magnitude and geometry were observed between the two stages.

Another challenge lies in the data processing and interpretation of LF-DAS measurements. The large volume of data generated by LF-DAS systems requires efficient algorithms for real-time analysis and interpretation. Developing robust and automated data processing workflows will enable timely decision-making and enhance the effectiveness of fracture monitoring and reservoir characterization. Additionally, poor data quality recorded by disposable fiber cables poses a significant challenge for LF-DAS applications. These cables, although cost-effective, may suffer from higher levels of signal degradation and lower sensitivity compared to permanent fiber installations. Addressing this challenge requires the development of improved disposable fiber cables with enhanced signal integrity and sensitivity, enabling more reliable data acquisition during hydraulic fracturing operations.

The limitations of LF-DAS technology include its spatial and temporal resolution. The ability to accurately resolve small-scale fractures and capture transient flow behavior remains a challenge. Further research is required to enhance the resolution capabilities of LF-DAS, both spatially and temporally, to capture fine-scale features and dynamic processes in the subsurface.

In terms of future directions, several avenues hold promise for advancing LF-DAS technology in unconventional reservoirs. Integrated modeling approaches that combine LF-DAS measurements with reservoir simulation can provide a more comprehensive understanding of fracture propagation and reservoir performance. Additionally, advancements in machine learning and artificial intelligence techniques can aid in the automated interpretation of LF-DAS data, leading to real-time fracture monitoring and decision support systems. Furthermore, the application of LF-DAS technology can be expanded to other areas of earth science, such as geothermal reservoir monitoring, carbon sequestration leakage detection, and induced seismicity monitoring. Exploring these applications and leveraging the versatility of LF-DAS technology will contribute to its wider adoption and foster new research opportunities.

In conclusion, although LF-DAS technology has demonstrated its potential in monitoring and surveillance of unconventional reservoirs, it is crucial to address challenges related to signal quality, data processing, and resolution limitations. By overcoming these obstacles and exploring future directions, LF-DAS can become a powerful tool for real-time fracture monitoring, reservoir characterization, and optimization of unconventional reservoir development. Moreover, extending the applications of LF-DAS to other areas of earth science will open new frontiers for research and development in the field.

28.9. CONCLUSIONS

LF-DAS and other types of DSS measurements have emerged as valuable technologies for oil-field applications. Their high spatial and temporal resolution enable detailed measurements of subsurface strain changes, facilitating both qualitative and quantitative interpretation of hydraulic fracturing, fracture monitoring, geomechanics modeling, and downhole pressure monitoring. Although this review has primarily focused on the use of DSS in optimizing hydraulic fracturing operations and ensuring well performance in unconventional reservoirs, recent studies have extended its application to carbon sequestration for monitoring pressure and to geothermal projects for measuring periodic hydraulic responses. This review study provides valuable insights into the broader application of LF-DAS in the oil and gas industry, as well as other subsurface projects.

REFERENCES

Ajani, A. & Kelkar, M. (2012). *Interference study in shale plays.* Paper presented at the SPE Hydraulic Fracturing Technology Conference.

Andrade, J. N. & van der Baan, M. (2021). *Automatic fracture hit detection in low-frequency distributed acoustic sensing using a computer vision workflow.* Paper presented at the First International Meeting for Applied Geoscience & Energy.

Barhaug, J., Bussey, J., Schaeffer, B., Shemeta, J., Lawrence, M., Tran, J., & Stark, P. (2022). *Testing XLE for cost savings in the DJ Basin: a fiber optic case study.* Paper presented at the SPE Hydraulic Fracturing Technology Conference and Exhibition, SPE, The Woodlands, Texas, USA. doi:10.2118/209155-MS

Becker, M. W., Ciervo, C., & Coleman, T. (2018). *A slimhole approach to measuring distributed hydromechanical strain in fractured geothermal reservoirs.* Paper presented at the 43rd Workshop on Geothermal Reservoir Engineering Stanford University, Palo Alto, CA, USA.

Bourne, S., Hindriks, K., Savitski, A. A., Ugueto, G. A., & Wojtaszek, M. (2021). *Inference of induced fracture geometries using fiber-optic distributed strain sensing in hydraulic fracture test site 2.* Paper presented at the Unconventional Resources Technology Conference, American Association of Petroleum Geologists, Houston, Texas, USA. doi:10.15530/urtec-2021-5472

Byerley, G., Monk, D., Aaron, P., & Yates, M. (2018). Time-lapse seismic monitoring of individual hydraulic frac stages using a downhole DAS array. *The Leading Edge, 37*(11), 802–810.

Chen, M., Guo, T. K., Qu, Z. Q., Zhou, T. & Li, N. (2021). *Numerical simulation of strain and strain rate evolution of optical fiber in the observation well during multi-cluster fracturing.* Paper presented at the 55th US Rock Mechanics/Geomechanics Symposium.

Chen, M., Tang, J., Zhu, D., & Hill, A. (2022). Classification and Localization of Fracture-Hit Events in Low-Frequency Distributed Acoustic Sensing Strain Rate with Convolutional Neural Networks. *SPE Journal, 27*(02), 1341–1353.

Chen Ning, I. L., Nemeth, T., Zhang, Z., Tan, Y., Bevc, D., Bartel, D., et al. (2021). *Image-based fracture hit warning system with distributed acoustic sensing.* Paper presented at the SEG/AAPG/SEPM First International Meeting for Applied Geoscience & Energy.

Ciezobka, J. (2021). *Overview of hydraulic fracturing test site 2 in the Permian Delaware Basin (HFTS-2).* Paper presented at the Unconventional Resources Technology Conference.

Cipolla, C., McKimmy, M., Hari-Roy, S., Wolters, J., Haffener, J., Haustveit, K., & Almasoodi, M. (2022). *Evaluating stimulation effectiveness with permanent optical fiber and sealed wellbore pressure monitoring: Bakken case study.* Paper presented at the SPE Hydraulic Fracturing Technology Conference and Exhibition, SPE, The Woodlands, Texas, USA. doi:10.2118/209129-MS

Crickmore, R., Werdeg, Z., Laing, C. & Molenaar, M. (2015). *Monitoring hydraulic fracturing operations using fiber-optic distributed acoustic sensing.* Paper presented at the SPE/AAPG/SEG Unconventional Resources Technology Conference.

Crouch, S. L. (1976). Solution of plane elasticity problems by the displacement discontinuity method. I. Infinite body solution. *International Journal for Numerical Methods in Engineering, 10*, 301–343. doi:10.1002/nme.1620100206

Dhuldhoya, K. & Friehauf, K. (2022). *Applications of distributed strain sensing via Rayleigh frequency shift: illuminating near-well and far-field fracture characteristics*. Paper presented at the 10th Unconventional Resources Technology Conference. doi:10.15530/urtec-2022-3721749

Eaton, D.W., Ma, Y., Wang, C., & MacDougall, K. (2022). *Effectiveness of dip-in DAS observations for low-frequency strain and microseismic analysis: The CanDiD experiment*. Paper presented at the Second International Meeting for Applied Geoscience & Energy. doi:10.1190/image2022-3745367.1

Ekechukwu, G. K., & Sharma, J. (2021). Well-scale demonstration of distributed pressure sensing using fiber-optic DAS and DTS. *Scientific Reports, 11*, 12505. doi:10.1038/s41598-021-91916-7

Fitzel, S., Sekar, B. K., Alvarez, D., & Gulewicz, D. (2015). *Gas injection EOR optimization using fiberoptic logging with DTS and DAS for remedial work*. Paper presented at the SPE/CSUR Unconventional Resources Conference, 175891-M.

Gurjao, K. G. R., Gildin, E., Gibson, R., & Everett, M. (2021). *Modeling of distributed strain sensing (DSS) and distributed acoustic sensing (DAS) incorporating hydraulic and natural fractures interaction*. Paper presented at the 9th Unconventional Resources Technology Conference. doi:10.15530/urtec-2021-5414

Haffener, J., Haustveit, K., & Ingle, T. (2022). *Did we break new rock? Utilizing diagnostics to differentiate new fracture creation vs old fracture reactivation: a Meramec and Wolfcamp study*. Paper presented at the SPE Hydraulic Fracturing Technology Conference and Exhibition, The Woodlands, Texas, USA. doi:10.2118/209123-MS

Haustveit, K., Elliott, B., Haffener, J., Ketter, C., O'Brien, J., Almasoodi, M., et al. (2020). *Monitoring the Pulse of a well through sealed wellbore pressure monitoring, a breakthrough diagnostic with a multi-basin case study*. Paper presented at the SPE Hydraulic Fracturing Technology Conference and Exhibition, The Woodlands, Texas, USA. doi:10.2118/199731-MS

Higginson, P., Purkis, D., & Webster, M. (2017). *Advancing the application of downhole fibre optics with a novel low cost disposable deployment method*. Paper presented at the SPE Offshore Europe Conference & Exhibition, Aberdeen, United Kingdom. doi:10.2118/186124-MS

Hu, X., Tu, Z., Zhou, F., Zhao, Y., Li, X., Zhou, Q., & Gou, X. (2023). A hydraulic fracture geometry inversion model based on distributed-acoustic-sensing data. *SPE Journal*, 1–17. doi:10.2118/214306-PA

Hull, R., Meek, R., Bello, H., Woller, K., & Wagner, J. (2019). Monitoring horizontal well hydraulic stimulations and geomechanical deformation processes in the unconventional shales of the Midland Basin using fiber-based time-lapse VSPs, microseismic, and strain data. *The Leading Edge, 38*, 130–137. doi:10.1190/tle38020130.1

Huot, F., Lellouch, A., Given, P., Luo, B., Clapp, R. G., Nemeth, T., et al. (2022). Detection and characterization of microseismic events from fiber-optic DAS data using deep learning. *Seismological Society of America, 93*(5), 2543–2553.

Ichikawa, M., Kurosawa, I., Uchida, S., Kato, A., Ito, Y., Takagi, S., et al. (2019). *Case study of hydraulic fracture monitoring using low-frequency components of DAS data*. Paper presented at the SEG Technical Program. Expanded Abstracts, 948–952. doi:10.1190/segam2019-3214251.1

Ichikawa, M., Uchida, S., Katou, M., Kurosawa, I., Tamura, K., Kato, A., et al. (2020). Case study of hydraulic fracture monitoring using multiwell integrated analysis based on low-frequency DAS data. *The Leading Edge, 39*, 794–800. doi:10.1190/tle39110794.1

Jin, G., Mendoza, K., Roy, B., & Buswell, D. G. (2019). Machine learning-based fracture-hit detection algorithm using LFDAS signal. *The Leading Edge, 38*, 520–524. doi:10.1190/tle38070520.1

Jin, G., & Roy, B. (2017). Hydraulic-fracture geometry characterization using low-frequency DAS signal. *The Leading Edge, 36*, 975–980. doi:10.1190/tle36120975.1

Jin, G., Ugueto, G., Wojtaszek, M., Guzik, A., Jurick, D., & Kishida, K. (2021). Novel near-wellbore fracture diagnosis for unconventional wells using high-resolution distributed strain sensing during production. *SPE Journal, 26*, 3255–3264. doi:10.2118/205394-PA

Karrenbach, M., Cole, S., Ridge, A., Boone, K., Kahn, D., Rich, J., et al. (2019). Fiber-optic distributed acoustic sensing of microseismicity, strain and temperature during hydraulic fracturing. *Geophysics, 84*(1), D11–D23.

Kiyashchenko, D., Mateeva, A., Duan, Y., Johnson, D., Pugh, J., Geisslinger, A., & Lopez, J. (2020). Frequent 4D monitoring with DAS 3D VSP in deep water to reveal injected water-sweep dynamics. *The Leading Edge, 39*(7), 471–479.

LeBlanc, M., Suh, K., Machovoe, S., Byrd, D., Jaaskelainen, M., Bland, H., et al. (2022).Theory and Practice of a Flexible Fiber Optic Cable in a Horizontal Well Used for Cross-Well and MicroseismicHydraulic Fracture Monitoring. Paper presented at the SPE Hydraulic Fracturing Technology Conference andExhibition, The Woodlands, Texas, USA, February 2022. 10.2118/209121-MS.

Lecampion, B. (2009). An extended finite element method for hydraulic fracture problems. *Communications in Numerical Methods in Engineering, 25*, 121–133. doi:10.1002/cnm.1111

Leggett, S. E., Zhu, D., & Hill, A. D. (2022). Thermal effects on far-field distributed acoustic strain-rate sensors. *SPE Journal, 27*(02), 1036–1048.

Liang, Y., Meier, H., Srinivasan, K., Tahir, H., Ortega, A., Amalokwu, K., et al. (2022). Accelerating development optimization in the Bakken using an integrated fracture diagnostic pilot. Paper presented at the 10th Unconventional Resources Technology Conference. 10.15530/urtec-2022-3719696.

Liu, Y., Jin, G., Wu, K., & Moridis, G. (2021b). Hydraulic-fracture-width inversion using low-frequency distributed-acoustic-sensing strain data—Part I: algorithm and sensitivity analysis. *SPE Journal, 26*(01), 359–371.

Liu, Y., Jin, G., Wu, K., & Moridis, G. (2021c). Hydraulic-fracture-width inversion using low-frequency distributed-acoustic-sensing strain data Part II: extension for multifracture and field application. *SPE Journal, 26*(05), 2703–2715.

Liu, Y., Jin, G., & Wu, K. (2022b). New insights on characteristics of the near-wellbore fractured zone from simulated high-resolution distributed strain sensing data. *SPE Reservoir Evaluation & Engineering, 25*, 99–112. doi:10.2118/208587-PA

Liu, Y., Jin, G., Wu, K., & Moridis, G. (2022a). Quantitative hydraulic-fracture-geometry characterization with low-frequency distributed-acoustic-sensing strain data: fracture-height sensitivity and field applications. *SPE Production & Operations, 37*, 159–168. doi:10.2118/204158-PA

Liu, Y., Wu, K., Jin, G., & Moridis, G. (2020). Rock deformation and strain-rate characterization during hydraulic fracturing treatments: insights for interpretation of low-frequency distributed acoustic-sensing signals. *SPE Journal, 25*, 2251–2264. doi:10.2118/202482-PA

Liu, Y., Wu, K., Jin, G., Moridis, G., Kerr, E., Scofield, R., & Johnson, A. (2021a). Fracture-hit detection using lf-das signals measured during multifracture propagation in unconventional reservoirs. *SPE Reservoir Evaluation & Engineering, 24*, 523–535. doi:10.2118/204457-PA

Mantell, M. E., Mercer, A., Jackson, J. B., Murphy, D. J., Conaway, J., Machovoe, S. R., et al. (2022). *Application of disposable fiber technology to evaluate far-field communication and fracture performance in the Marcellus Shale*. Paper presented at the SPE Hydraulic Fracturing Technology Conference and Exhibition, SPE, The Woodlands, Texas, USA. doi:10.2118/209148-MS

Molenaar, M. M., Hill, D. J., Webster, P., Fidan, E., & Birch, B. (2012). First downhole application of distributed acoustic sensing for hydraulic-fracturing monitoring and diagnostics. *SPE Drilling & Completion, 27*(01), 32–38.

Moradi, P., Angus, D., & Sharma, M. (2022). *Hydraulic fracture width inversion using simultaneous downhole tiltmeter and distributed acoustic sensing data*. Paper presented at the SPE Hydraulic Fracturing Technology Conference and Exhibition, SPE, The Woodlands, Texas, USA. doi:10.2118/209149-MS

Ning, Y., & Jin, G. (2023a). Challenges and best practices in interpreting crosswell strain signals to monitor multicrew zipper-fracturing operations. *Interpretation, 11*(2), SB21–SB32.

Ning, Y., Jin, G., Zhu, X., & Tura, A. (2022a). *Integration of production communication and hydraulic fracture connection in unconventional reservoirs*. Paper presented at the 10th Unconventional Resources Technology Conference. doi:10.15530/urtec-2022-3722164

Ning, Y., Jin, G., Zhu, X., & Tura, A. (2022b). Well interference and fracture geometry investigation using production and low-frequency distributed acoustic sensing data in an unconventional reservoir. *SPE Reservoir Evaluation & Engineering, 25*, 509–519. doi:10.2118/209788-PA

Ning, Y., Mjehovich, J., Zhu, X., & Jin, G. (2022c). *Low-frequency DAS interpretation upon multi-crew zipper fracturing configuration*. Paper presented at the Second International Meeting for Applied Geoscience & Energy. doi:10.1190/image2022-3738168.1

Ning, Y., & Jin, G. (2023b). Challenges and best practices in interpreting cross-well strain signals to monitor multi-crew zipper fracturing operations. *Interpretation, 11*(2), 1–43. doi:10.1190/int-(2022)-0092.1

Ortega Perez, A. K. (2022). *Hydraulic fracture monitoring: integrated analysis of DAS, pumping information, microseismicity and PKN modelling* (Master's Thesis). University of Alberta

Ouyang, L. B., & Belanger, D. (2006). Flow profiling by distributed temperature sensor (DTS) system—expectation and reality. *SPE Production & Operations, 21*(02), 269–281.

Ramos Gurjao, K. G., Gildin, E., Gibson, R., & Everett, M. (2022). Investigation of strain fields generated by hydraulic fracturing with analytical and numerical modeling of fiber optic response. *SPE Reservoir Evaluation & Engineering, 25*, 367–379. doi:10.2118/206049-PA

Raterman, K. T., Farrell, H. E., Mora, O. S., Janssen, A. L., Gomez, G. A., Busetti, S., et al. (2018). Sampling a stimulated rock volume: An eagle ford example. *SPE Reservoir Evaluation & Engineering, 21*, 927–941. SPE-191375-PA (November 2018). doi:10.2118/191375-PA

Raterman, K., Liu, Y., Roy, B., Friehauf, K., Thompson, B., & Janssen, A. (2020). *Analysis of a multi-well Eagle Ford Pilot*. Paper presented at the 8th Unconventional Resources Technology Conference. doi:10.15530/urtec-2020-2570

Raterman, K., Liu, Y. & Warren, L. (2019). *Analysis of a drained rock volume: an Eagle Ford example*. Paper presented at the Unconventional Resources Conference, Denver, Colorado, URTeC-2019-263. doi:10.15530/urtec-2019-263

Richter, P., Parker, T., Woerpel, C., Wu, Y., Rufino, R., & Farhadiroushan, M. (2019). Hydraulic fracture monitoring and optimization in unconventional completions using a high-resolution engineered fibre-optic Distributed Acoustic Sensor. *First Break, 37*, 63–68. doi:10.3997/1365-2397.n0021

Sakaida, S., Pakhotina, I., Zhu, D., & Hill, A.D. (2022). Evaluating effects of completion design on fracturing stimulation efficiency based on DAS and DTS interpretation. Paper presented at the SPE Hydraulic Fracturing Technology Conference and Exhibition. doi:10.2118/209167-MS

Sharma, J., Cuny, T., Ogunsanwo, O., & Santos, O. (2020). Low-frequency distributed acoustic sensing for early gas detection in a wellbore. *IEEE Sensors Journal, 21*, 6158–6169. doi:10.1109/JSEN.2020.3038738

Sherman, C., Mellors, R., Morris, J., & Ryerson, F. (2019). Geomechanical modeling of distributed fiber-optic sensor measurements. *Interpretation, 7*, SA21–SA27. doi:10.1190/INT-2018-0063.1

Shou, K.-J., Siebrits, E., & Crouch, S. L. (1997). A higher order displacement discontinuity method for three-dimensional elastostatic problems. *International Journal of Rock Mechanics and Mining Sciences, 34*, 317–322. doi:10.1016/S0148-9062(96)00052-6

Sidenko, E., Tertyshnikov, K., Gurevich, B., & Pevzner, R. (2022). DAS signature of reservoir pressure changes caused by a CO_2 injection: experience from the CO_2CRC Otway Project. *International Journal of Greenhouse Gas Control, 119*, 103735. doi:10.1016/j.ijggc.2022.103735

Srinivasan, A., Liu, Y., Wu, K., Jin, G., & Moridis, G. (2022a). *Geomechanical modeling of fracture-induced vertical strain measured by distributed fiber optic strain sensing. Paper presented at the*10th Unconventional Resources Technology Conference. doi:10.15530/urtec-2022-3722188

Srinivasan, A., Liu, Y., Wu, K., Jin, G., & Moridis, G. J. (2022b). *Analysis of strain responses in vertical monitoring wells for low-frequency distributed acoustic sensing*. Paper presented at the 56th U.S. Rock Mechanics/Geomechanics Symposium, ARMA, Santa Fe, New Mexico, USA, ARMA-2022-0263. doi:10.56952/ARMA-2022-0263

Tan, Y., Wang, S., Rijken, P., Hughes, K., Lim Chen Ning, I., Zhang, Z., & Fang, Z. (2020). Geomechanical template for DAS fiber strain patterns during hydraulic fracturing. Paper presented at the SPE Annual Technical Conference and Exhibition, D031S038R002.

Tang, J. & Zhu, D. (2022). *Characterize fracture development through strain rate measurements by distributed acoustic sensor DAS*. Paper presented at the SPE International Hydraulic Fracturing Technology Conference & Exhibition, Muscat, Oman, D011S004R001. doi:10.2118/205267-MS

Ugueto, G. A., Wojtaszek, M., Huckabee, P. T., Savitski, A. A., Guzik, A., Jin, G., et al. (2021a). *An integrated view of hydraulic induced fracture geometry in hydraulic fracture test site 2*. Paper presented at the Unconventional Resources Technology Conference, American Association of Petroleum Geologists, Houston, Texas, USA. doi:10.15530/urtec-2021-5396

Ugueto, G. A., Todea, F., Daredia, T., Wojtaszek, M., Huckabee, P. T., Reynolds, A., et al. (2019). *Can you feel the strain? DAS strain fronts for fracture geometry in the BC Montney, Groundbirch*. Paper presented at the SPE Annual Technical Conference and Exhibition, D021S029R005. doi:10.2118/195943-MS

Ugueto, G. A., Wojtaszek, M., Mondal, S., Guzik, A., Jurick, D., & Jin, G. (2021b). *New fracture diagnostic tool for unconventionals: high-resolution distributed strain sensing via rayleigh frequency shift during production in hydraulic fracture test 2*. Paper presented at the Unconventional Resources Technology Conference, American Association of Petroleum Geologists, Houston, Texas, USA. doi:10.15530/urtec-2021-5408

Ugueto, G., Wu, K., Jin, G., Zhang, Z., Haffener, J., Mojtaba, S., et al. (2023). A catalogue of fiber optics strain-rate fracture driven interactions. Paper presented at the SPE Hydraulic Fracturing Technology Conference and Exhibition, The Woodlands, Texas, USA, D021S003R001. doi:10.2118/212370-MS

Verdon, J. P., Horne, S. A., Clarke, A., Stork, A. L., Baird, A. F., & Kendall, J. M. (2020). Microseismic monitoring using a fiber-optic distributed acoustic sensor array Microseismic monitoring using a DAS array. *Geophysics*, *85*(3), KS89–KS99.

Wang, C., Eaton, D. W., & Ma, Y. (2022b). *Numerical modeling of low-frequency distributed acoustic sensing signals for mixed-mode fracture activation*. Paper presented at the Second International Meeting for Applied Geoscience & Energy, Society of Exploration Geophysicists and American Association of Petroleum Geologists, Houston, Texas. doi:10.1190/image2022-3749919.1

Wang, C., Ma, Y., Eaton, D., MacDougall, K., & Furlong, C. (2022a). *Hydraulic fracture monitoring using distributed acoustic sensing: a case study using dip-in fiber*. Paper presented at the 10th Unconventional Resources Technology Conference. doi:10.15530/urtec-2022-3723692

Wang, J., Tan, Y., Rijken, M., Liu, X., Singh, A., & Li, Y. (2022). Observations and modeling of fiber optic strain on hydraulic fracture height growth in Hydraulic Fracturing Test Site 2 (HFTS-2). *SPE Journal*, *27*, 1109–1122. doi:10.2118/209216-PA

Webster, P., B. Cox, & M. Molenaar (2013a). *Developments in diagnostic tools for hydraulic fracture geometry analysis*. Paper presented at the Unconventional Resources Technology Conference. doi:10.1190/urtec2013-025

Webster, P., Wall, J., Perkins, C., & Molenaar, M. (2013b). Micro-seismic detection using distributed acoustic sensing. In *SEG technical program expanded abstracts 2013* (pp. 2459–2463). Society of Exploration Geophysicists.

Wu, K., & Olson, J. E. (2015). A simplified three-dimensional displacement discontinuity method for multiple fracture simulations. *International Journal of Fracture*, *193*, 191–204. doi:10.1007/s10704-015-0023-4

Zhang, Z., DiSiena, J., Bevc, D., Ning, I. L. C., Tan, Y., Swafford, L., et al. (2021). *Hydraulic fracture characterization by integrating multidisciplinary data from the Hydraulic Fracturing Test Site 2 (HFTS-2)*. Paper presented at the Unconventional Resources Technology Conference, American Association of Petroleum Geologists, Houston, Texas, USA. doi:10.15530/urtec-2021-5039

Zhang, Z., Fang, Z., Stefani, J., DiSiena, J., Bevc, D., Lim Chen Ning, I., et al. (2020a). Modeling of fiber-optic strain responses to hydraulic fracturing. *Geophysics*, *85*, A45–A50. doi:10.1190/geo2020-0083.1

Zhang, Z., Fang, Z., Stefani, J., DiSiena, J., Bevc, D., Ning, I. L. C., et al. (2020b). *Fiber optic strain monitoring of hydraulic stimulation: geomechanical modeling and sensitivity analysis*. Paper presented at the Unconventional Resources Technology Conference, American Association of Petroleum Geologists, Online. doi:10.15530/urtec-2020-2648

Zhu, X. & Jin, G. (2021). *Analysis of cross-well fracture hits in DJ Basin, Colorado using low-frequency DAS data*. Paper presented at the First International Meeting for Applied Geoscience & Energy, Society of Exploration Geophysicists, Denver, CO and virtual. doi:10.1190/segam2021-3582658.1

Zhu, X., & Jin, G. (2022). Hydraulic fracture propagation in Denver-Julesburg basin constrained by cross-well distributed strain measurements. *SPE Journal*, *27*, 3446–3454. doi:10.2118/210582-PA

Zhu, X., Jin, G., & Hammack, R. (2022). *Hydraulic fracture-hit detection system using low-frequency DAS data*. Paper presented at the Second International Meeting for Applied Geoscience & Energy, Society of Exploration Geophysicists and American Association of Petroleum Geologists, Houston, Texas. doi:10.1190/image2022-3750851.1

29

Recent Advances in the Use of Temporary Optical Fiber Deployment for Downhole Hydraulic Fracture Monitoring

Yuanyuan Ma[1,2], Chaoyi Wang[1], Vikram Maji[1,3], Rebecca O. Salvage[1,4], Kelly MacDougall[1], and David Eaton[1]

ABSTRACT

Distributed acoustic sensing (DAS) has transformed hydraulic fracture monitoring in recent years. We report two of the first projects in Canada, called the Canadian Dip-in DAS (CanDiD) projects, in which a temporary optical fiber was deployed to monitor hydraulic fracturing operations. The main goal of CanDiD is to evaluate the effectiveness of a retrievable optical fiber for frac monitoring based on the analysis of both microseismic and low-frequency DAS signals. The DAS recordings from zipper-frac completions in horizontal wells show clear signatures of crosswell strain associated with fracture-driven interactions (FDIs). These signals enable fracture azimuth to be determined, indicative of the maximum horizontal stress (SH_{max}) direction. Using a machine learning–based approach, microseismic events were detected and processed, although it was challenging to obtain process hypocenters from a single fiber. Numerous coherent noise events, which we interpret as high-frequency waves that propagate along the wireline due to fiber slip, initiate in close proximity to the FDIs. During another hydraulic fracturing program in western Canada, low-frequency DAS signals from the CanDiD-2 project provide evidence for fracture initiation, reactivation, and termination. The results of these investigations demonstrate the utility of temporary DAS deployments to provide insights about fracture geometry and stress orientations.

29.1. INTRODUCTION

In recent years, permanent fiber-optic cables have been increasingly deployed in boreholes to monitor hydraulic fracturing (HF) programs. These systems include distributed acoustic sensing (DAS), which uses an interrogator to emit coherent laser pulses into an optical fiber and to measure the phase change of back-scattered light (Hartog, 2017). This information can be converted into strain or strain rate using interferometry. Compared to traditional geophone systems, fiber-optic sensing allows continuous monitoring of the entire wellbore with high spatial resolution (spatial sampling as small as ~1 m), avoiding the necessity to reposition the toolstring during vertical seismic profiling or HF process (Hartog, 2020; Lindsey & Martin, 2021; Lumens, 2014). Moreover, fiber-optic sensors are unaffected by electromagnetic interference and can be deployed in harsh and corrosive environments, making them well-suited for downhole monitoring of fluid injection. When assisted with advanced processing techniques, fiber-optic sensing enables operators to make real-time adjustments to optimize the well completion process and reduce the risk of

[1]*Department of Earth, Energy and Environment, University of Calgary, Calgary, Canada*

[2]*Department of Earth, Environmental, and Planetary Sciences, Rice University, Houston, Texas, USA*

[3]*Department of Earth Science, University of Bergen, Bergen, Norway*

[4]*Natural Resources Canada, Geological Survey of Canada, Calgary, Canada*

Distributed Acoustic Sensing in Borehole Geophysics, Geophysical Monograph 289, First Edition.
Edited by Yingping Li, Robert Mellors, and Ge Zhan.

DOI: 10.1002/9781394179275.ch29

accidents or environmental damage, thereby expanding the use of this technology (Molenaar et al., 2012).

Unconventional applications of DAS have primarily centered around HF efficiency determinations from cross-well strain and DAS microseismic monitoring. The low-frequency bands (<1 Hz) of DAS (LF-DAS) measurement are sensitive to mechanical strains caused by HF propagation, therefore making it suitable for monitoring the growth and geometry of hydraulic fractures (Becker et al., 2017). LF-DAS provides direct observations of fracture opening (extension) and closing (compression), enabling the correlation of fractures initiated at the treatment well with fracture-driven interactions (FDIs) at the monitoring fiber (Jin and Roy, 2017; Ugueto et al., 2023; Zhang et al., 2020). In addition to FDI analysis using fiber in horizontal wells, LF-DAS data from vertical monitoring wells can provide a direct tool to determine fracture height and offer valuable insights into vertical fracture growth (J. Wang et al., 2022). Another significant application in the realm of HF involves the deployment of a fiber-optic cable within a horizontal well to detect microseismic events (Karrenbach et al., 2019; Lellouch et al., 2020a). Benefiting from high spatial resolution and broadband frequency, DAS recordings offer rich microseismic wavefields in addition to the detection of a large number of microseismic events. For example, DAS data can contain near-field strain response (Luo et al., 2021), guided waves (Lellouch et al., 2020b), and reflected waves (Ma et al., 2023b; Staněk et al., 2022), allowing for detailed characterization of wave propagation and subsurface imaging. In addition to DAS fiber systems, another fiber-optic diagnostic method, distributed strain sensing based on Rayleigh frequency shift (DSS-RFS), has shown effectiveness in quantitative characteristics of near-wellbore fracture networks (Jin et al., 2021; Ugueto et al., 2021).

The fiber-optic cables used for DAS are typically permanently installed by cementing behind the well casing (Boone et al., 2015). This installation method is advantageous inasmuch as it provides excellent coupling between the cable and the surrounding medium. It also enables treatment-well monitoring during HF operations. If properly deployed, a permanent DAS cable does not interfere with production, so it can be used as a long-term monitoring device for multiple purposes (Daley et al., 2013; Karrenbach et al., 2017; Mateeva et al., 2013; Ugueto et al., 2019; Zhang et al., 2021). However, once compromised by mislocated perforation shots or excessive casing deformation, cemented fiber is almost impossible to repair or replace. The high cost and time associated with permanent is thus a deterrent to more widespread use of this technology.

Recently, the development of a flexible optical fiber cable, either as a wireline or a disposable fiber, has been used to monitor well interactions (Eaton et al., 2022; Henninges et al., 2021; LeBlanc et al., 2022; Shahri et al., 2021; C. Wang et al., 2022). Richter et al. (2019) presented pioneering work directly comparing permanent fiber and wireline fiber strain measurements. The advantages of a temporary wireline DAS system include flexibility, ease of repair, and generally lower deployment costs. However, the coupling between the cable and the surrounding medium is primarily reliant on gravity and friction (LeBlanc et al., 2022), which may render it less sensitive compared with cemented fiber. Figure 29.1 shows different deployment schematics for cement-in DAS fiber and wireline DAS fiber (C. Wang et al., 2022).

This chapter presents case studies from the Canadian Dip-in DAS (CanDiD) experiment using a wireline-deployed DAS system that operated during HF programs in western Canada. Two different wireline DAS systems were deployed in lateral wells at two field sites to monitor hydraulic treatments in horizontal wells drilled from a single pad, as described in the following section. The DAS observations have been processed and interpreted for both microseismic monitoring and LF-DAS analysis of FDIs (or "frac hits"), as detailed in the following sections. These analyses enable a quantitative evaluation of the deployed wireline system's performance. Additionally, waves caused by wireline slip are described and documented; the point of

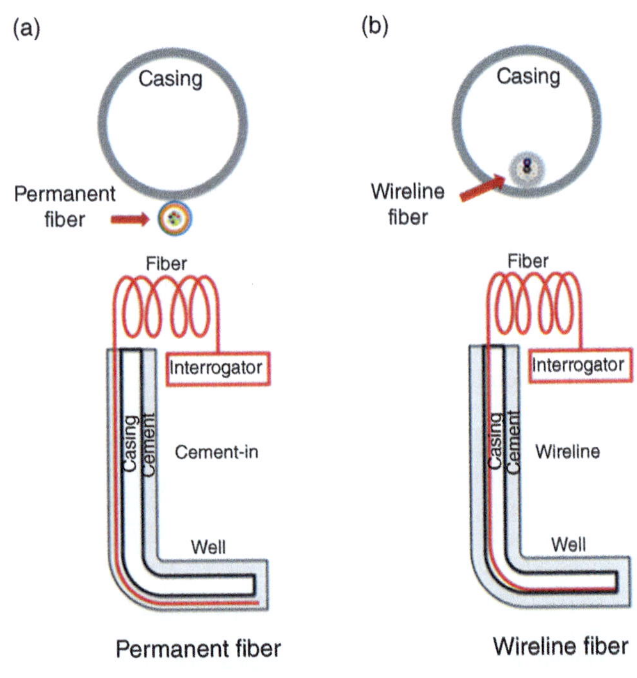

Figure 29.1 Schematic for (a) permanent DAS fiber cemented behind well casing, and (b) wireline DAS fiber, which is deployed inside the well casing. For temporary deployment, coupling between the fiber and the surrounding medium is provided by gravity and friction. Wang et al. (2022) / with permission of the Society of Exploration Geophysicists.

origin of these waves appears to pinpoint locations of high strain rate associated with FDIs.

29.2. PROJECT OVERVIEW

The CanDiD projects involved wireline deployment of an optical fiber in a lateral well to record DAS observations. The programs were carried out through the University of Calgary as part of the Microseismic Industry Consortium, in partnership with oil and gas operators as well as wireline and DAS service providers. Two project locations were used, with different wireline conveyance methods and different types of optical fibers (Figure 29.2). The primary objective of the CanDiD projects is to evaluate the effectiveness of a dip-in optical fiber for hydraulic fracture monitoring based on analysis of both microseismic and LF-DAS signals.

To illustrate the complexity of these operations, Figure 29.3 shows a simplified schematic of the equipment required to perform a multiwell completion operation. The addition of dip-in distributed fiber optical sensing (DFOS) acquisition to such an operation involves even more equipment on the well pad, such as an interrogator, a conveyance, and coil-tubing units. Depending on the number of wells and completion order for HF operations, other required equipment may include a crane (displayed in Figure 29.3 to the right side of the vertical line of red dots representing the well heads), which is used to suspend the DAS wireline above the surface at the monitoring well so that completion operations can move efficiently from one well head to the next. Additional equipment is required when using the option of a pump-down conveyance of the dip-in DAS fiber.

29.2.1. CanDiD-1

The first phase of the CanDiD project (CanDiD-1) took place in January 2021 during hydraulic fracturing operations in the Montney Formation (C. Wang et al., 2022). A total of nine horizontal wells were drilled in a stacked arrangement at a single well pad (Figure 29.4a). Wells A, D, and G were drilled into the upper part of the upper Montney member, whereas wells B, E, H, and wells 0, C, and F were located in the middle and lower parts of the upper member, respectively (Figure 29.4(b)). Adjacent wells are separated by a horizontal distance of 300 m and a vertical distance of 65–85 m. Zipper-frac treatments were performed sequentially in sets of three wells, consisting of the A, B, and 0 wells, then C, D, and H wells, and finally E, F, and G wells. In CanDiD-1, a standard single-mode fiber was deployed using a wireline system with a tractor (Figure 29.2(a), (c)) in the E well before the start

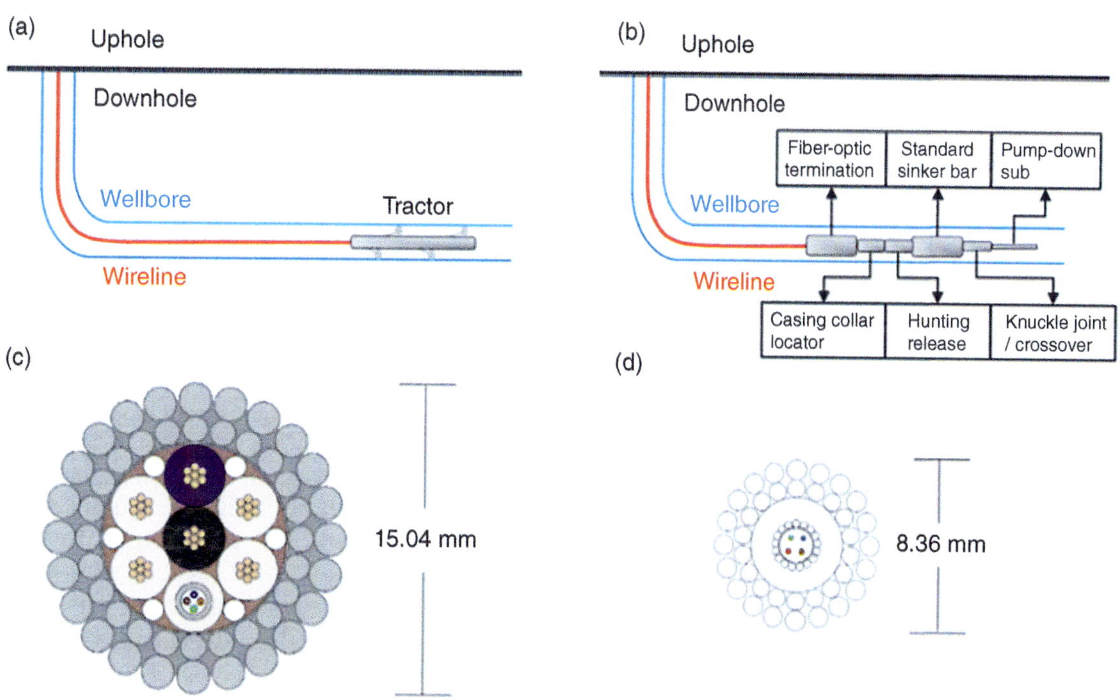

Figure 29.2 Comparison of (a) the tractor conveyance system used for CanDiD-1 with (b) the pump-down conveyance used for CanDiD-2. Cross sections of (c) the wireline with multimode fiber used for CanDid-1 and (d) the engineered fiber used for CanDiD-2. In each case, the optical fibers are represented by four small colored circles.

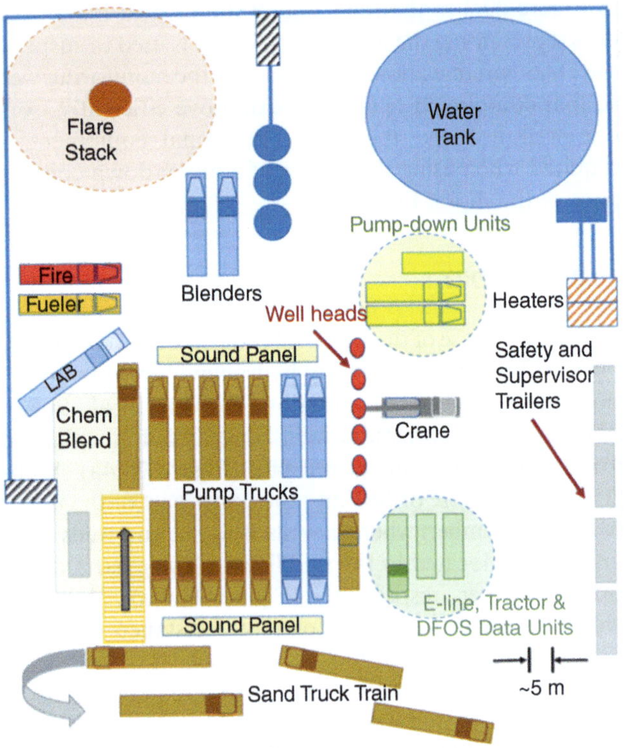

Figure 29.3 Generalized sketch of the well pad layouts for the CanDiD projects. Within the complex operational environments, a fiber-optic cable was deployed in one well with the recording system (E-line and DAS interrogator) located within control units at the surface.

of the HF treatment in the A well. For this recording system, the DAS data were output in units of strain. The wireline was removed before treatments in the E well, which recorded approximately 1 week of continuous data during the HF operations (Figure 29.4(c)). The monitoring well was subsequently hydraulically fractured after the removal of the fiber.

29.2.2. CanDiD-2

The second phase of the CanDiD projects (CanDiD-2) took place in June 2022. In this program, an engineered fiber (Figure 29.2(b), (d)) was deployed into a horizontal monitoring well Figure 29.5(a), (c), (M-2) located centrally within a stack of 15 horizontal wells that underwent zipper-frac operations in the Montney Formation. Using pumping equipment available on site, the wireline was pumped down into the horizontal monitoring well to a depth of over 4,900 m. Over the ~14 day monitoring period, the closest six wells (2-2, 2-3, 2-11, 2-12, 2-13, and 214) produced low-frequency strain signals captured by DAS. Raw DAS data were continuously sampled at 1 kHz, with a spatial resolution of 1 m and a gauge length of 10 m, using a single-mode fiber. The data was output as strain rate rather than strain. Continuous distributed temperature sensing (DTS) measurements were also taken with a spatial resolution of 0.5 m using a multimode fiber within the same monitoring well. Over 220 treatment stages were performed on the six closest wells to the monitoring fiber over ~14 days of data acquisition (Figure 29.5(b)). Treatment stages were not necessarily evenly spaced in time; instead, multiple stages in multiple wells were active at the same time.

29.3. CROSSWELL STRAIN MONITORING OF HYDRAULIC FRACTURING

29.3.1. Crosswell Low-Frequency DAS Monitoring

Understanding the geometry of the fractures is critical for designing successful unconventional well completions. LF-DAS is sensitive to strain perturbations caused by fracture propagation. Figure 29.6 shows the full acquisition of crosswell LF-DAS strain rate signals for the CanDiD-1 project, which includes stages from A, 0, B, D, C, and H wells.

A typical HF stage includes fractures growing from multiple closely spaced perforation clusters. Therefore, the LF-DAS signal for an HF stage can be complex. There are numerous factors that can affect the growth of fractures. These include competition between closely spaced fractures, which can result in uneven growth due to the stress shadow effect retarding the growth of some fractures (He et al., 2017; Morrill & Miskimins, 2012); lithology and pre-existing faults, which can modify or impede fracture propagation (Dehghan et al., 2015; Zhou et al., 2008); and screen-outs, where proppants clog the fracture, negatively impacting growth (Garagash et al., 2019). To address these issues, our case study presents a workflow that accomplishes three tasks: identifying treatment stages, analyzing FDI, and creating a fracture connection map.

29.3.2. Stage Identification: CanDiD-1

HF stages were identified by matching the timing of the LF-DAS data with the treatment pressure data. Figure 29.7 shows an example of a single treatment stage (C. Wang et al., 2022). When a hydraulic fracture is approaching a DAS fiber, the stress at the fracture tip deforms the rock mass ahead, creating a distinctive strain field that can be recognized on the waterfall plot. In general, when a hydraulic fracture approaches, it displays an extensional strain in the shape of a heart, which is surrounded by compressional strain. The extensional strain is the result of the fracture opening, and the compressional strain in the surrounding rock is due to the compression caused by the opening of the fracture. The signature evolves to a narrow band of extensional strain enclosed

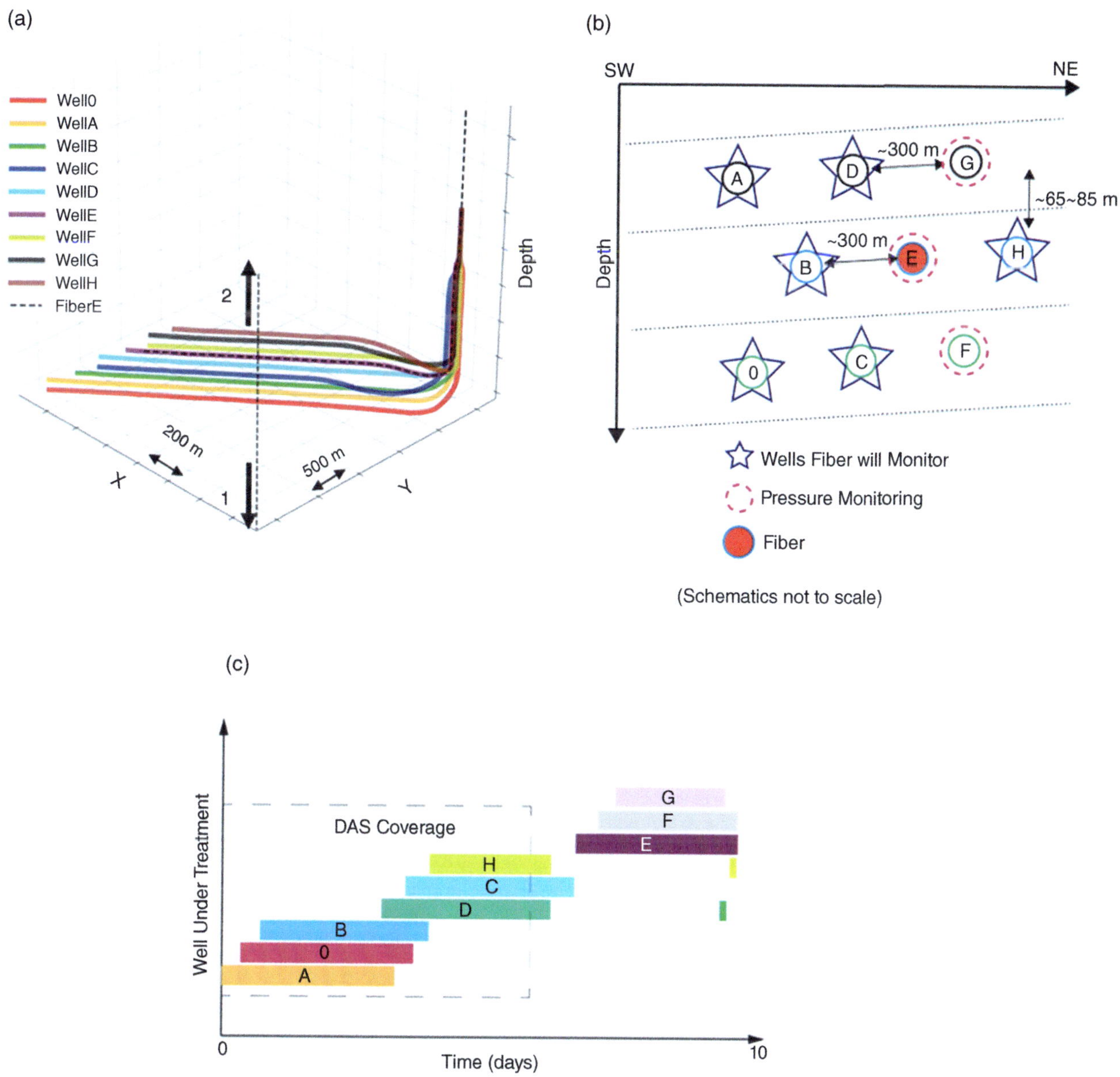

Figure 29.4 Well and fiber setup for CanDiD-1. (a) 3D view of the nine wells drilled into the Montney Formation. (b) Gun-barrel view of the horizontal part of the wells. (c) Treatment timeline and DAS coverage. The DAS monitoring duration lasted 1 week and covered most of the treatment stages for the A, 0, B, D, C, and H wells. Wang et al. (2022) / with permission of the Society of Exploration Geophysicists.

by compressional strains as the fracture continues to propagate. This signature represents an FDI. When the fracture starts to close, typically immediately after the time of pump shut-in (dashed black line in Figure 29.7), a reversal of the extensional and compressional strain field is apparent due to the relaxation of the rock mass. A time delay from the breakdown pressure to the initial FDI reflects the time that it takes for the fractures to propagate through the medium between the perforation clusters and the DAS fiber. Treatment wells with a larger physical distance from the fiber experience a longer time delay between the breakdown pressure and the first FDI.

For CanDiD-1, most of the stimulation stages covered by the DAS monitoring duration can be clearly identified. However, few FDIs were identified from A well treatment stages (the first stimulation well), possibly due to the large physical distance between A well and the DAS fiber. The apparent lack of FDIs could also be a consequence of the fractures having limited downward growth, potentially failing to reach the detection range of the fiber. Some early

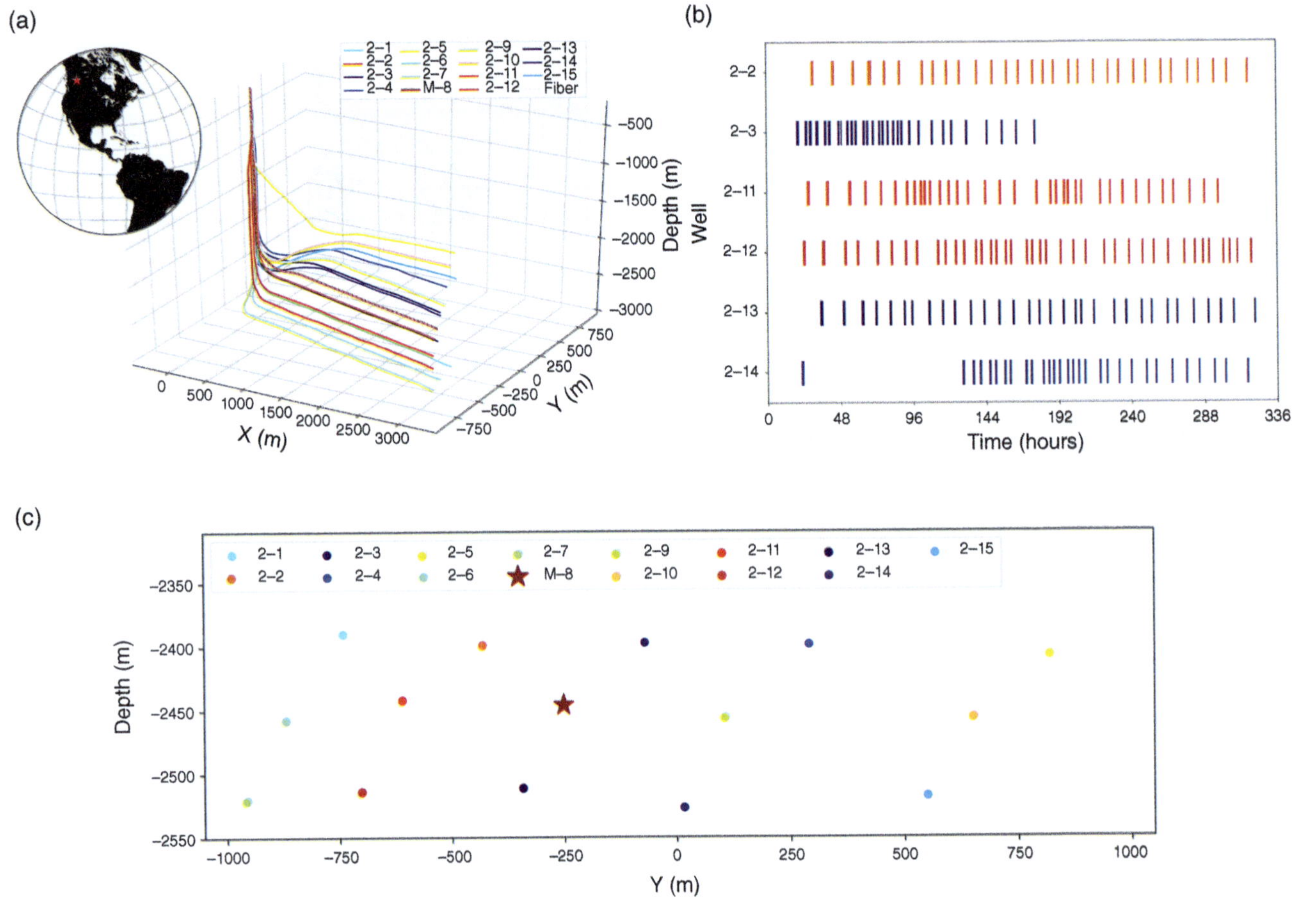

Figure 29.5 Well and fiber setup for CanDiD-2. (a) 3D view of the 15-well pad targeting the Montney Formation. (b) Treatment timeline for the closest six wells to the monitoring well (M-8), which were captured by the fiber. (c) Gun-barrel view showing the landing depth of each horizontal well. The monitoring well is indicated by the star symbol.

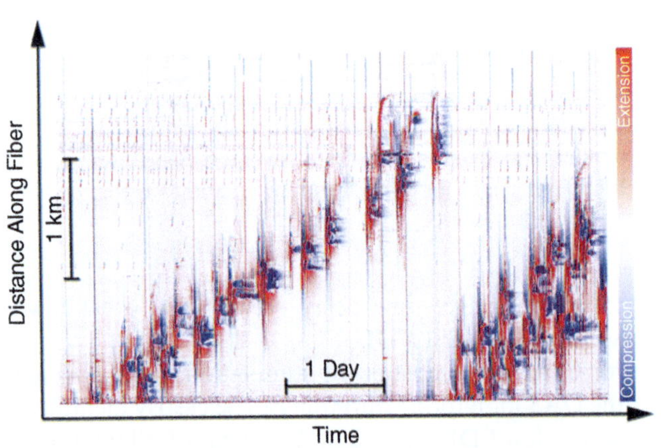

Figure 29.6 Waterfall plot showing CanDiD-1 LF-DAS response during the full acquisition period. Individual compression/extension signals represent FDIs from the treatment well. The staggered locations of the FDIs reflect the geometry of the zipper-frac operation.

treatment stages of 0 well and B well were not captured by the DAS fiber because these stages were located outside of the measuring aperture of the DAS fiber, as shown in Figure 29.4. Heel stages of C, D, and H wells were not recorded due to the removal of the DAS fiber.

29.3.3. Analysis of FDIs

A treatment stage typically includes multiple perforations, which lead to multiple fractures propagating at different speeds, reaching the DAS fiber at different times. Multiple FDIs complicate the LF-DAS signal, making the identification of individual FDIs challenging. A recent study found that the sum of the absolute values of measured strain or strain rate along the time axis is sensitive to FDI, whose local maxima suggest possible individual FDIs (Liu et al., 2020). We implemented a similar method to identify possible FDIs. In addition to identifying the local maxima of the summed absolute strain for each channel of DAS data along the time axis, we also utilize the signature

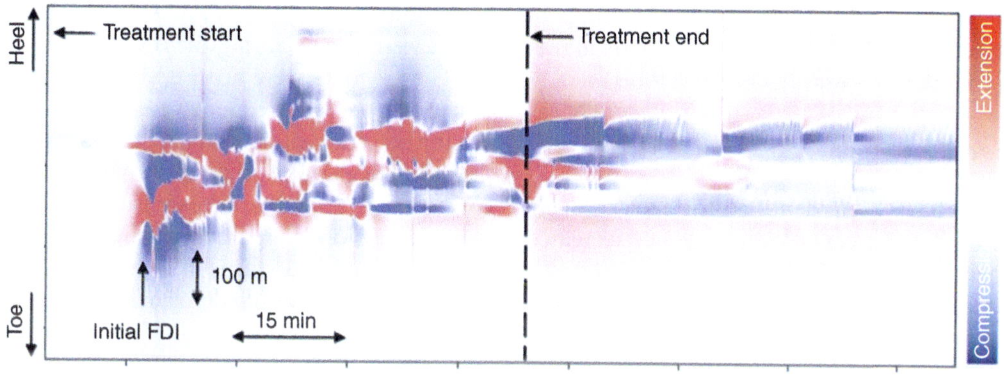

Figure 29.7 Waterfall plot for CanDiD-1 B well stage 10, demonstrating the LF-DAS expression of FDIs for a single treatment stage. The fracture closure caused by pump shut-in exhibited a polarity reversal of the strain field on the waterfall plot.

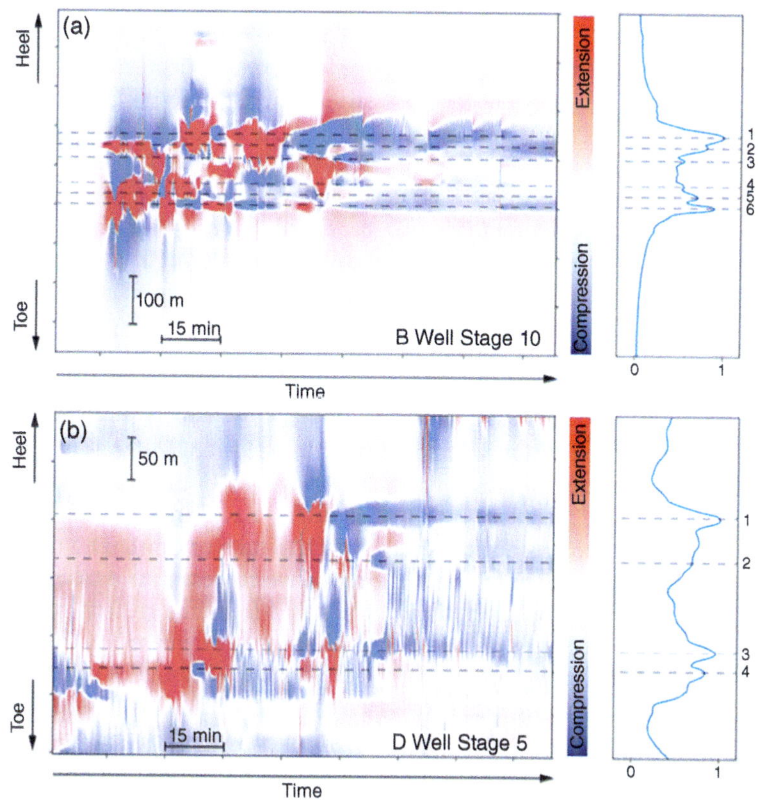

Figure 29.8 CanDiD-1: FDI analysis. (a) B well treatment stage 10 with six possible FDIs, (b) D well treatment stage 5 with four possible FDIs. Wang et al. (2022) / with permission of the Society of Exploration Geophysicists.

of fracture closure following the cessation of pumping to verify the picks. Figure 29.8 shows two examples of FDI analysis in this study (C. Wang et al., 2022): B well stage 10 (Figure 29.8(a)) and D well stage 5 (Figure 29.8(b)). The normalized sum of absolute strain for each DAS channel is plotted along the waterfall plots. The sum is calculated over the time periods shown in the waterfall plot, which includes the start to the end of pumping in the stage. The local maxima are picked to reflect possible FDIs. However, not all local maxima are picked as FDIs (e.g., Figure 29.8(b) between fracture 1 and 2) due to the consideration that no distinct extensional strain pattern is observed before the pump shut-in for initial FDI, and no fracture closure signature is present after the pump shut-in. This semi-quantitative approach is applied to all the analyzed stages. Advanced methods with higher accuracy and confidence are still being developed.

29.3.4. Fracture Connection Map

Once the FDIs have been identified for each detected treatment stage, a fracture connection map can be created for the treatment well by linking the locations of FDIs recorded on the DAS fiber with the nearest perforation clusters. This map provides valuable information on fracture length, perforation efficacy, and the orientation of maximum horizontal stress. In this study, the wells are presented in a top-down view (Figure 29.9), with green markers separating treatment stages (C. Wang et al., 2022). Interpreted fractures are displayed as solid black lines, and red circles indicate probable perforation clusters that failed to produce an FDI on the DAS fiber. The absence of an FDI may suggest that the perforation cluster was ineffective in producing a fracture or that fracture growth was asymmetrical such that fractures primarily grew in the direction away from the fiber. Figure 29.9(a) illustrates the interpreted fracture connection map for well D, which is located in the upper part of the Upper Montney formation. The connection map indicates that approximately half of the perforation clusters in each stage produced FDIs on the DAS fiber, providing insight into the heights of the hydraulic fracture. The fracture connection map for well B, located in the middle part of the formation, is shown in Figure 29.9(b), indicating excellent perforation effectiveness. Most treatment stages in well B have only one or two perforation clusters that did not produce an FDI, with the number of unconnected clusters decreasing from toe stages to heel stages. The fracture connection map for well 0, drilled into the lower part of the formation, is depicted in Figure 29.9(c), with only a few FDIs detected per treatment stage due to the large distance from the DAS fiber. This observation suggests that the treatment stages in well 0 can produce large fractures that reach 600 m in half-length, potentially indicating that treatments in other wells can produce fractures with similar geometry at similar frequencies.

29.3.5. CanDiD-2 LF-DAS Response

A similar analysis was applied for the CanDiD-2 project. The LF-DAS signals from the treatment stages of the HF operations in a treatment well were captured by the DAS fiber deployed in the monitoring well (M-8). In this case, the treatment wells (11 and 12) in the western part of the monitoring well were far apart, limiting the responses to reach the monitoring well. However, several treatment wells (03 and 14) located on the eastern side of the pad operations exhibit FDIs along the entire length of the well during fluid injection operation. Figure 29.10(a) illustrates the LF-DAS strain rate waterfall plot during the operation in treatment well 03. The raw data were processed through a low-pass filter to capture the low-frequency component of the DAS signal for detection of occurrences of slow slip events. As in the previous example, the initiation of a hydraulic fracture consists of a heart-shaped extensional region surrounded by compression, suggesting tensile mode crack opening. All the operational stages in Figure 29.10(a) exhibit similar signatures of tensile deformation (red: extension) surrounded by compression (blue: compression) regions on either side. With time, the extending region experiences compression as HF stages finish, allowing the previously opened crack to close gradually. In every instance, fluid injection induces additional compressive pressure on the previously operational stages, as reflected in the lower part of the operational stages while absent in the upper section.

The LF-DAS signals show examples of interpreted reactivation of fracture networks developed during the previous stages from the crosswell comparison of DAS strain where the activities were separated by time interval but occurred at equivalent depth ranges (Figure 29.10(b)). The first visible FDI of the treatment operation in well 13 was observed in stage 7 and was reactivated in stage 8. The dynamic subsurface behavior of a specific stage covering representative examples from both the eastern and western sections around the monitoring well are illustrated in Figure 29.10(c), (d). One of the stages (well 3, stage 26) contained an anomalous signal that is currently being investigated to determine if it represents a slow slip on a fracture or bedding surface. A similar mode of deformation occurred during stage 29.

Applying the same processing workflow that was developed for the CanDiD-1 project, we picked FDIs from LF-DAS data of CanDiD-2 to characterize fracture geometry. Two representative cases—one from the western and the other from the eastern part of the monitoring well—are discussed here based on the capacity of fracture corridor development (Figure 29.11). The treatment well 3, located on the eastern side, witnessed effective development of the fracture network during the operation of all 30 stages along the horizontal aspect of well 3, whereas treatment well 13 recorded limited responses of fracture network development out of the entire operational length of treatment stages.

29.4. DAS MICROSEISMIC OBSERVATION

29.4.1. DAS Microseismic Data Processing

This section introduces the processing workflow and results from DAS microseismic data. To analyze microseismic signals and tackle the big-data challenge, we have developed key elements of a processing workflow that is

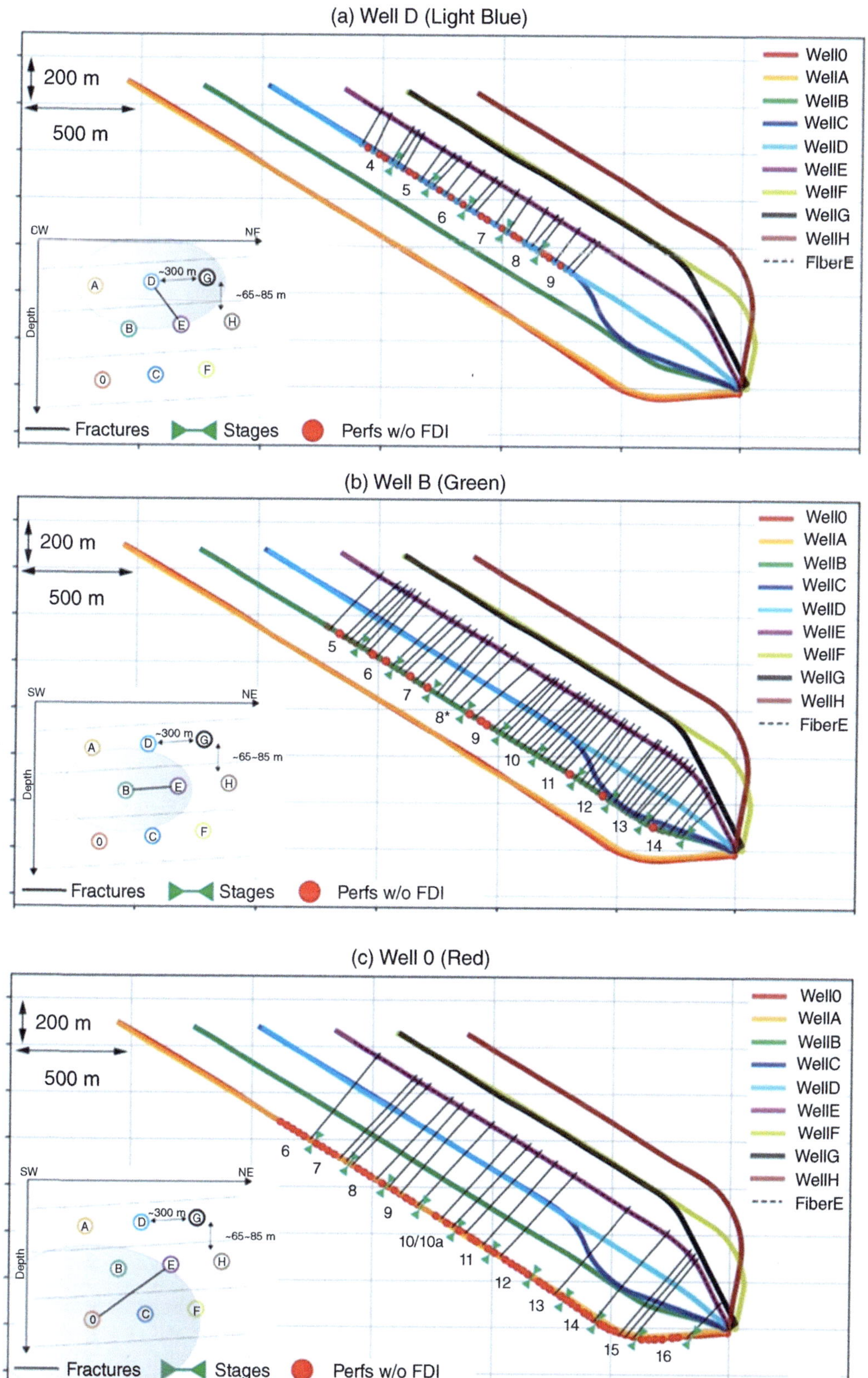

Figure 29.9 CanDiD-1: Fracture connection map for (a) D well, (b) B well, and (c) 0 well. The insets in each connection map show estimated fracture geometry in the gun-barrel view. Wang et al. (2022) / with permission of the Society of Exploration Geophysicists.

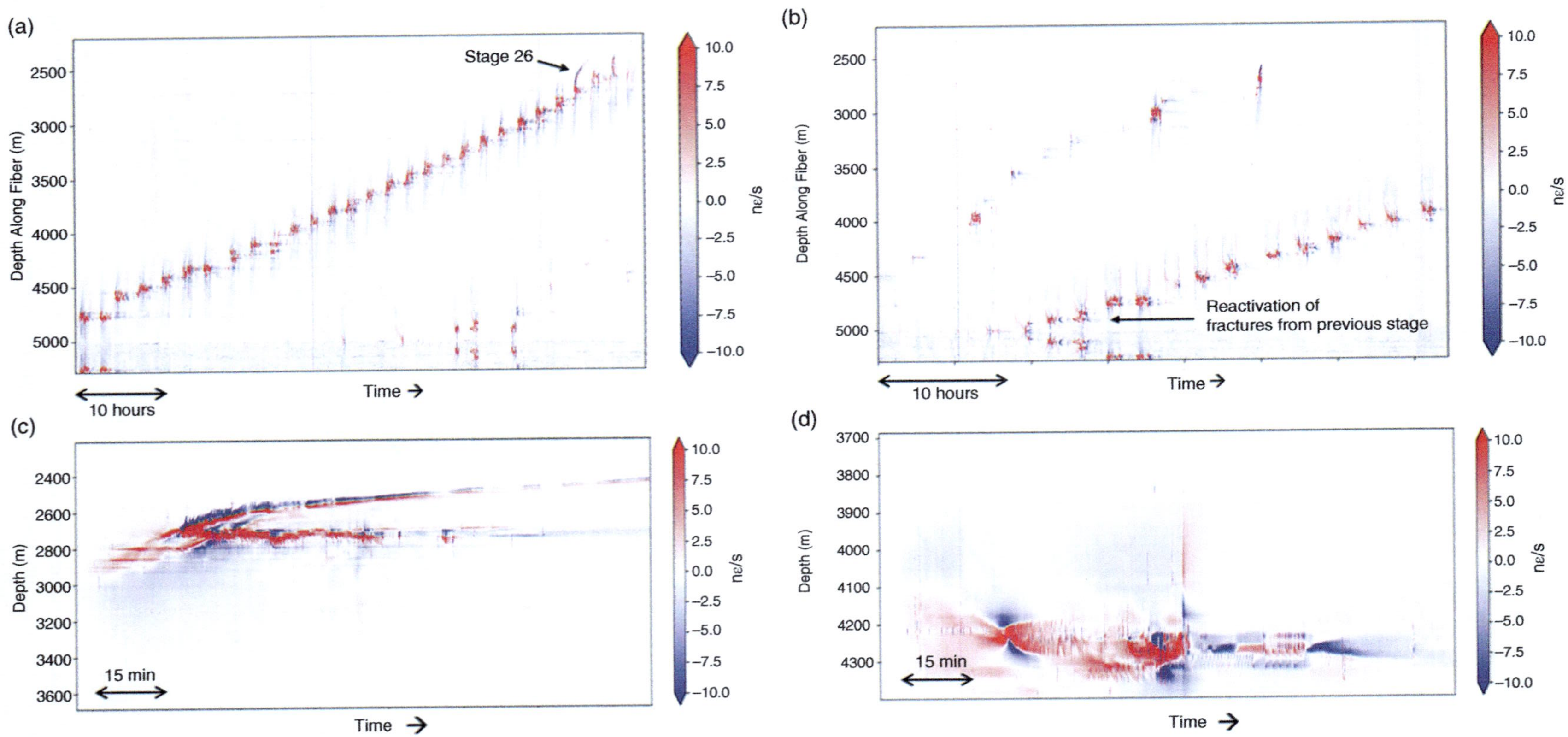

Figure 29.10 CanDiD-2 LF-DAS response full acquisition for treatment operations in (a) Well 03 and (b) Well 13. LF-DAS strain rate data from (c) Well 03, stage 26 and (d) Well 13, stage 14.

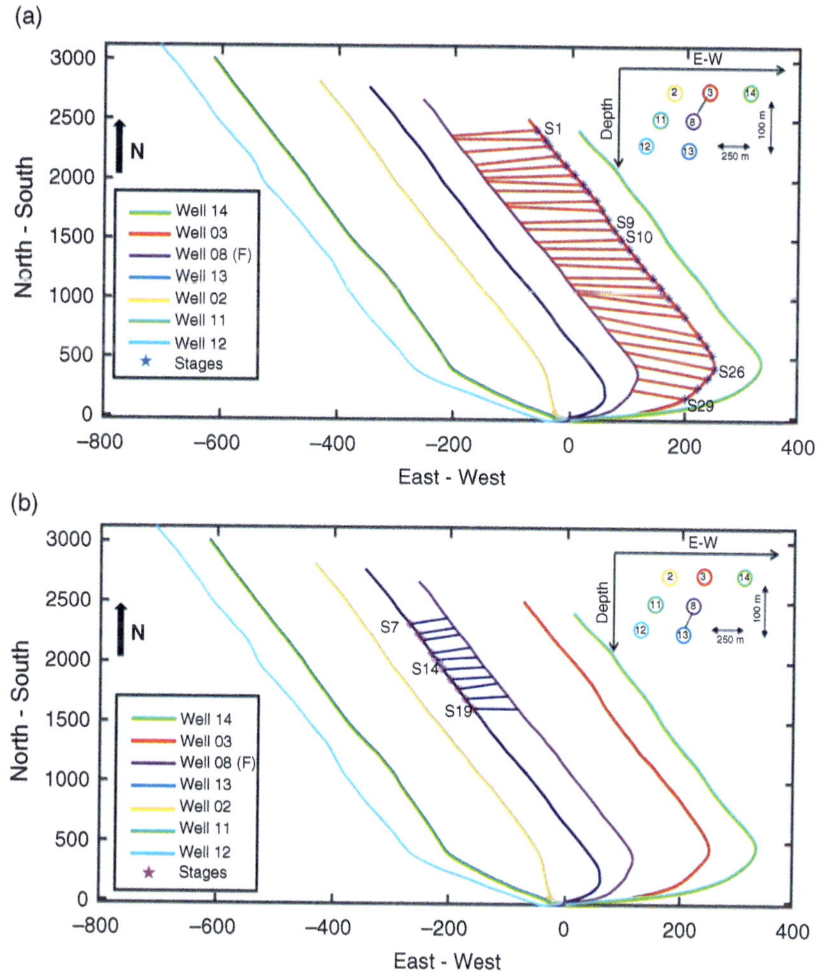

Figure 29.11 CanDiD-2: fracture connection maps for (a) Well 03 and (b) Well 13.

assisted by machine learning techniques. Briefly, we first remove bias in the background (DC) amplitude level and spurious data spikes and then bandpass filter the data between 15 and 150 Hz. This frequency band was chosen to avoid low-frequency noise and facilitate event detection based on the frequency band with the highest signal-to-noise ratio (SNR). In addition, we subtract the median value (using all channels) at each time sample of the DAS recordings to attenuate the system noise. We treat event detection as a task of 2D binary image classification and build a convolutional neural network (CNN) to address it. To pick arrival times, we used an image-segmentation approach with a simplified U-Net architecture (Ronneberger et al., 2015) to pick both P- and S-arrival times automatically (Ma et al., 2020; Zhu & Beroza, 2018). The machine learning workflow was built using Keras (Chollet, 2015) and TensorFlow (Abadi et al., 2015).

Figure 29.12 shows the overall processing workflow for multifiber DAS microseismic data using machine learning, including preprocessing, event detection using CNN, arrival time picking using U-Net, and 3D hypocenter location using conventional ray tracing and grid search (Ma et al., 2023a). Machine learning techniques can offer many advantages for automated DAS microseismic processing, including low computation cost and minimal human intervention requirements.

29.4.2. DAS Microseismic Analysis

Figure 29.13(a) shows a representative microseismic example with clear P- and S-wave arrivals from both horizontal and vertical sections of the well, although dip-in fiber data contain stronger noise than permanently deployed fiber systems (C. Wang et al., 2022). In addition to microseismic signatures, we also observed a significant number of coherent signals with linear moveout and strong amplitude, as indicated in Figure 29.13(b). The coherent noise events have an apparent velocity of approximately 950 m/s, which is too slow to be conventional tube waves. We refer to these events as "fiber waves" that possibly

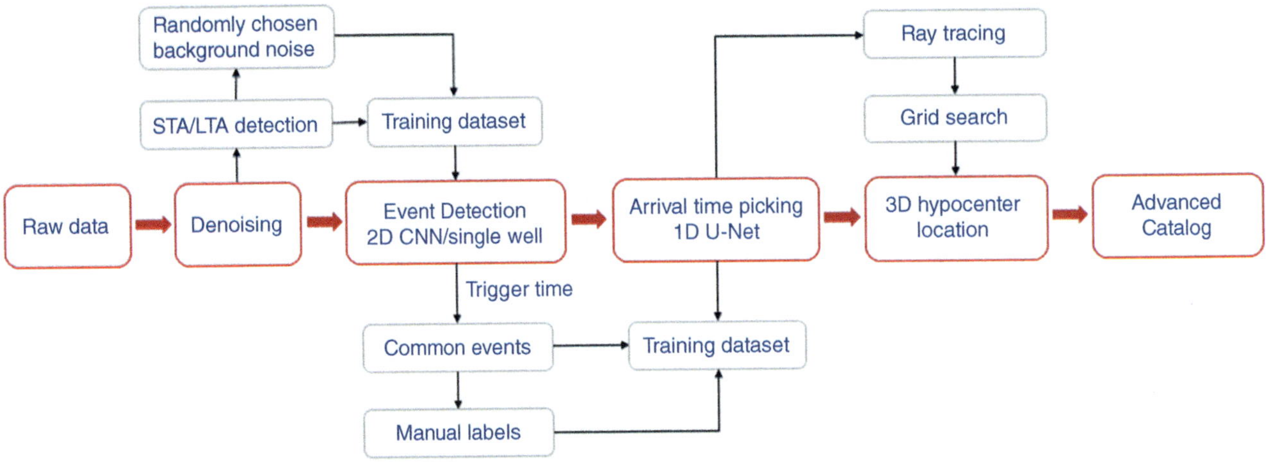

Figure 29.12 The machine learning-assisted processing workflow for DAS microseismic data. The primary workflow is highlighted in red. Ma et al. (2023) / with permission of Society of Exploration Geophysicists.

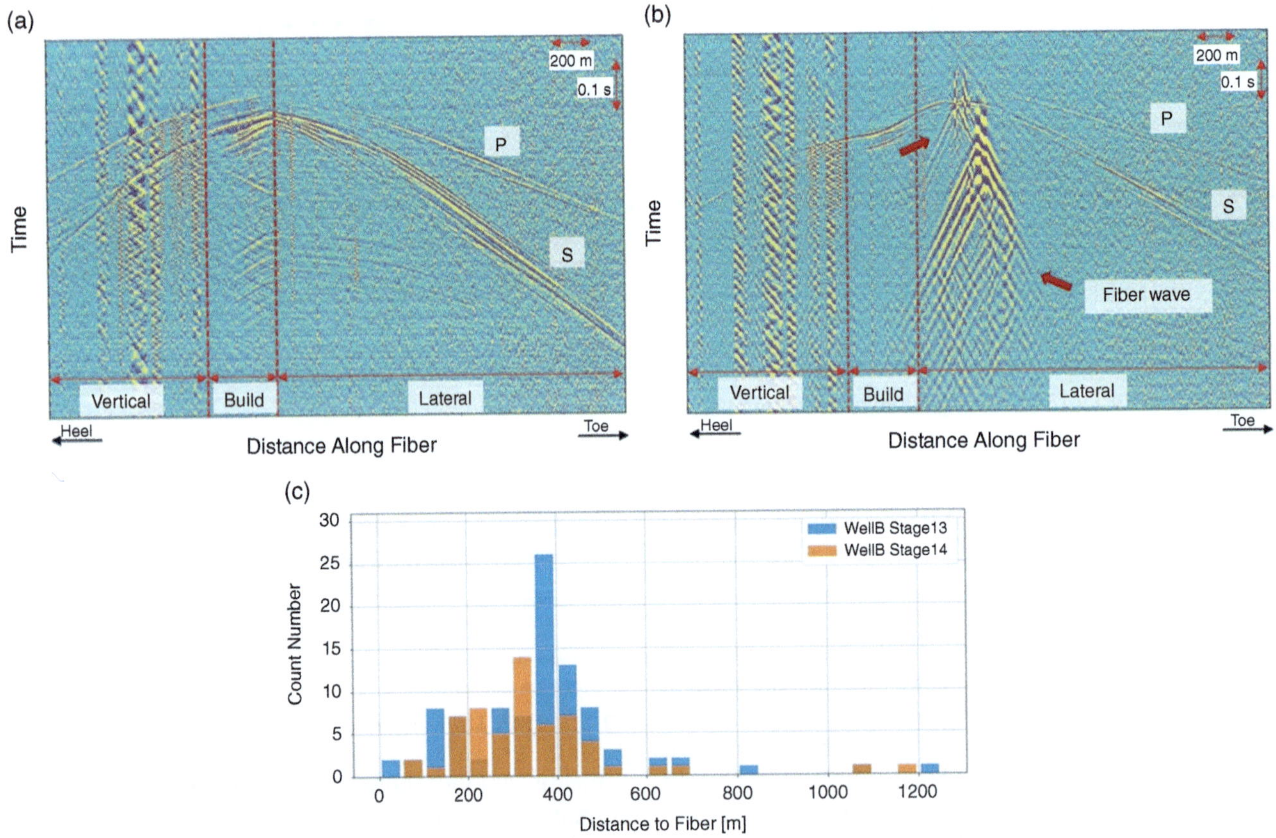

Figure 29.13 CanDiD-1: (a) Representative waveform examples of microseismic events with P and S arrivals over the entire cable length, as well as (b) coherent noise events characterized by linear moveout. (c) Histogram of the distance of microseismic hypocenters from the monitoring fiber in Well B, stages 13–14. Wang et al. (2022) / with permission of the Society of Exploration Geophysicists.

propagate along the wireline when the wireline slips in the wellbore. Interestingly, these events appear to have a close spatiotemporal association with FDIs, as discussed in the next section.

Applying the machine learning workflow to B well treatment stages, detection results are shown in Table 29.1 (C. Wang et al., 2022). Although dip-in fiber may be less sensitive to microseismic signals compared

Table 29.1 CanDiD-1: wireline fiber detection results of B well treatment stages. Fiber waves refer to waves with linear moveout (Figure 29.13).

Stage number	5	6	7	8	9	10	11	12	13	14
Pump time (hrs)	1.43	1.21	1.21	0.18	1.25	1.05	1.08	1.15	1.20	1.22
Microseismic	0	0	8	3	8	18	21	24	159	160
Fiber wave	242	251	137	152	93	72	158	173	449	216

Wang et al. (2022) / with permission of the Society of Exploration Geophysicists.

Table 29.2 Cemented fiber detection results elsewhere in the Montney Formation, using the same machine learning workflow.

Stage number	1	2	3
Pump time (hrs)	3.90	3.68	4.03
DAS microseismic	129	146	311

Wang et al. (2022) / with permission of the Society of Exploration Geophysicists.

to a cemented fiber system elsewhere in the Montney Formation, as shown in Table 29.2 (C. Wang et al., 2022), likely due to higher noise amplitude and poorer fiber coupling with the casing, it nevertheless captured comparable microseismic events counts in stages that are relatively close to the monitoring fiber (stages 13–14). For the wireline-deployed fiber, we see continuous strong background noise and coherent noise from the vertical section due to poor coupling and cable ringing (Figure 29.13(a), (b)). However, the presence of those noise bands does not prevent the machine learning detection approach because we input mainly horizontal sections of the fiber into the network (more than 2 km).

Because there is only a single DAS fiber available in the CanDiD-1 project, microseismic hypocenters in 3D cannot be uniquely located. However, we applied 2D raytracing and a grid search to estimate the distance from detected events to the monitor fiber in B well stages 13–14. (Figure 29.13(c)). Most detected microseismic events are located within a distance of 500 m from the fiber, which is consistent with the fracture connection map from LF-DAS (Figure 29.9(b)). Conversely, a few detected events appear to be significantly more distant, being located at least 1,200 m away from the fiber. This observation raises the possibility that larger structural features were activated at greater distances from the well.

29.5. FIBER WAVES AND FIBER COUPLING

In addition to microseismicity, we observed a large number of coherent waves with linear moveout during the CanDiD-1 project. Two typical examples of a fiber wave from CanDiD-1 are shown in Figure 29.14(a), (b). The wave exhibits linear moveout (apparent velocity) along two main branches that meet at the apex, characterized by strong amplitude. Both examples display reflections from both sides of the apex. Although these noise events are similar in appearance to tube waves (acoustic waves in water with a velocity of ~1,500 m/s), the velocity of these waves (<1,000 m/s) precludes this interpretation (Figure 29.14(c)). Unlike tube waves, the observed fiber waves show evidence of different characteristics, including some modes showing polarity flipping across the apex (Figure 29.14(a)) and others appearing that are not reversed (Figure 29.14(b)). Some examples of tube waves with a velocity of 1,500 m/s were observed, as shown in Figure 29.14(b).

These noise events are referred to as fiber waves. We infer that they represent waves that propagate along the wireline in response to sudden slip. The theory for these types of waves has been developed by Leblanc et al. (2022). This type of coherent linear wave is not observed using cemented fiber systems. For wireline fiber, the leading source of fiber decoupling is strain overcoming frictional forces between the wireline cable and the casing (LeBlanc et al., 2022). Figure 29.15 shows crosswell LF-DAS strain data collected in CanDiD-1, and the effect of the sudden slips is obvious. Sudden strain releases in the cable show a stick–slip behavior of the wireline (LeBlanc et al., 2022; Ugueto et al., 2023). The distinguishing feature of these signals is the existence of a section with increasing strain followed by one with decreasing strain. For LF-DAS, the abrupt onset of the slip events can be removed by filtering to provide a cleaner image of the strain changes seen in the wellbore.

To determine the closest point along the fiber for each fiber wave, we picked the apex or the point with the minimum arrival time. This approach enabled a direct comparison (in time and position) with the crosswell strain signal from LF-DAS (Figure 29.16). Sudden strain releases in the cable are visible from the raw LF-DAS data (Figure 29.16(a), (b)), showing a stick–slip behavior of the wireline. The shifts were well removed by a 15 × 15 2D median filter, obtaining clear data to pick frac hits (Figure 29.16(c), (d)). Both microseismic and fiber waves show clear spatiotemporal association with FDIs. A high concentration of fiber waves coincident with the timing and location of several FDIs is well observed. During the treatment stage, clusters of fiber waves appear to correlate closely with localized areas of tensile (extensional) strain

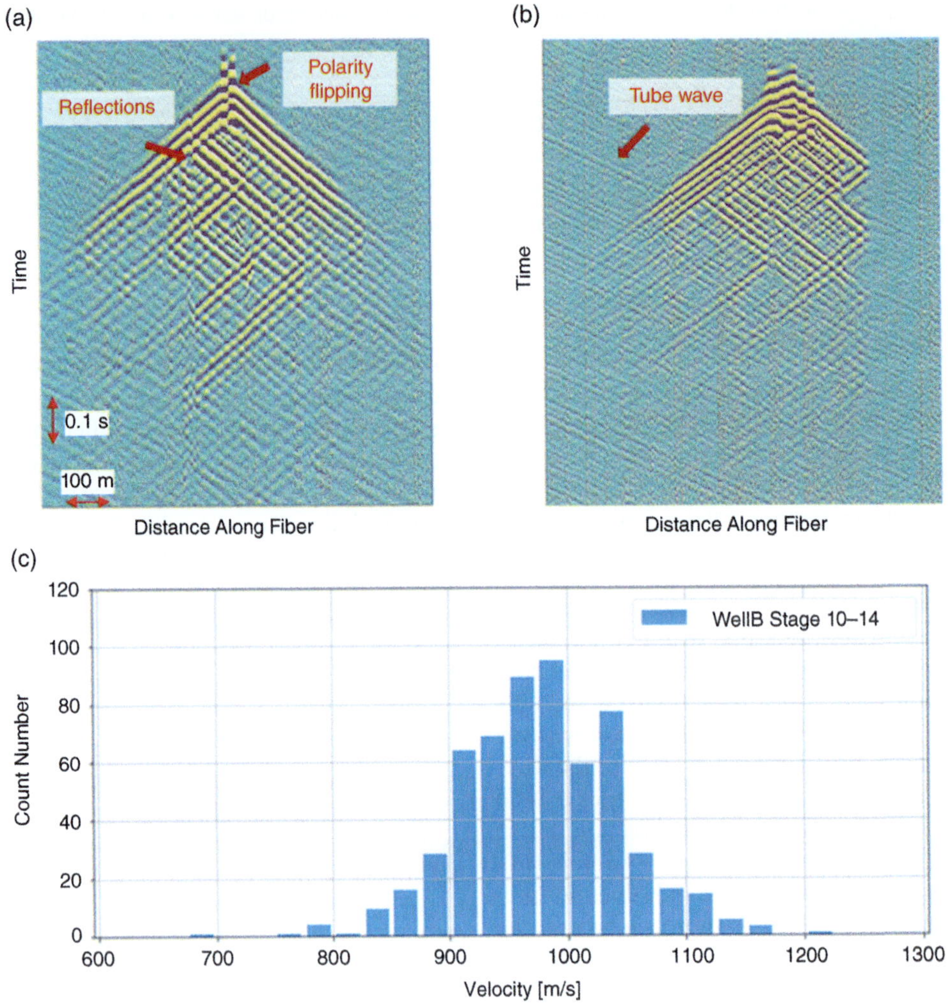

Figure 29.14 (a) A fiber-wave example recorded during CanDiD-1. The wave exhibits linear moveout (apparent velocity) along two main branches that meet at the apex. In this case, a polarity reversal is observed across the apex, as well as reflections. (b) A fiber-wave example with consistent polarity across the apex and strong reflections. (c) Histogram showing measured apparent velocities of fiber-wave signals.

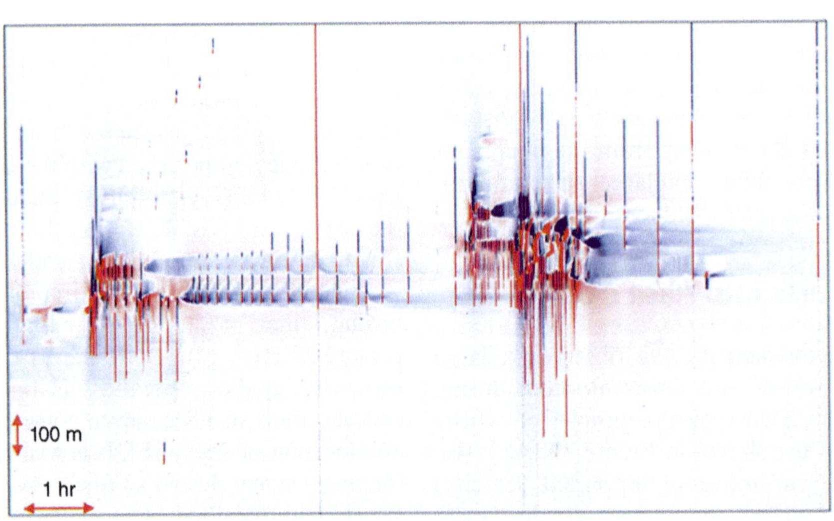

Figure 29.15 Crosswell LF-DAS strain data collected in CanDiD-1. Sudden strain releases in the cable are visible, showing a stick–slip behavior of the wireline.

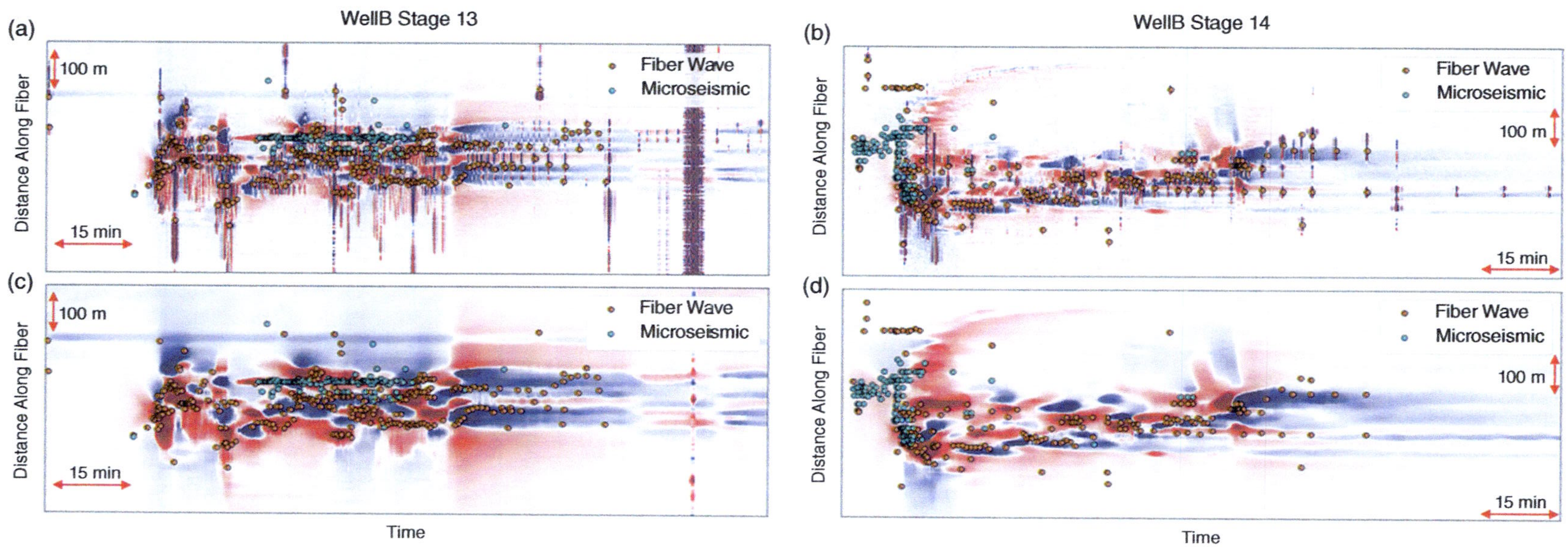

Figure 29.16 Locations of microseismic and fiber waves overlain on both (a, b) raw (top) and (c, d) filtered (bottom) LF-DAS waterfall display for a treatment stage in Well B, stages 13–14. Wang et al. (2022) / with permission of the Society of Exploration Geophysicists.

(Figure 29.16(c), (d)). After the stage is complete, the fiber waves exhibit a pattern of slow spatial migration that is well-aligned with wireline slippage. Therefore, the fiber wave may be useful as an indicator of frac hits.

29.6. DISCUSSION

Analysis of LF-DAS data from both CanDiD projects, such as the construction of fracture connection maps, demonstrates the effectiveness of dip-in fiber for analysis of FDIs signals, capturing fracture initiation, reactivation, and closure visible to a maximum distance of 600 m. In particular, the analysis supports the estimation of the spatial extent of hydraulic fracture systems, both laterally and vertically, showing that some fractures extend across the full lateral and vertical separation between horizontal wells at this pad. In addition, a slight clockwise rotation of fracture azimuths can be observed from the upper part to the lower part of the formation (C. Wang et al., 2022). The observation of the depth-wise progressive rotation of hydraulic fracture azimuths implies some complexity in the local stress field, expressed by vertical variation in SHmax direction. From both experiments, a few anomalous FDIs, detected within the closest treatment well, exhibited a paired extension/compression signature characteristic of shear slip along horizontal planes of weakness in low-frequency DAS (Figure 29.16(b), (d)). These events were also recorded as a minor temperature increase by DTS during CanDiD-2, suggesting interaction with deeper fluids.

Although microseismic signals appear to contain a higher fraction of coherent noise compared to cemented fiber due to poor coupling between wireline and casing, reflections from lithologic interfaces are visible in some events, which may be used to improve focal-depth determination. The travel time of P and S arrivals from vertical and deviated sections (Figure 29.13(a), (b)) could help to reduce the ambiguity of hypocenter location using a single fiber. In addition, a large number of fiber waves caused by wireline slip were observed in CanDiD-1. Fiber-wave characteristics such as velocity and reflections would depend on details of the properties and structure of the wireline, which is currently being investigated. Although fiber waves can be easily treated as noise and neglected during microseismic processing, they can behave as a complementary tool to map fracture geometry and provide insights into the mechanism of fracture opening and closing.

The choice between temporary and permanent deployed fiber for geophysics will depend on the specific requirements and goals of the monitoring project. Temporary fiber can be a more cost-effective and flexible option for short-term monitoring projects, whereas permanent fiber can be more reliable for long-term monitoring projects. In this study, we focus on the performance of wireline fiber systems. Additionally, disposable fiber, another option for temporary fiber deployment, is designed to be cost-effective and has a minimal operational footprint. However, unlike permanent and wireline-deployed fiber, disposable fiber has no protective housing around the fiber and is typically not retrieved after monitoring is complete. Although the wireline-deployed temporary fiber cable used in this study has been found to provide reliable measurement for crosswell strain identification on frac hits and microseismic monitoring, the coupling achieved is not as effective as for cemented fiber. A potential alternative is to merge temporary DAS fiber data with geophone arrays to achieve broad volume coverage for fracture monitoring and completion diagnostics.

29.7. CONCLUSIONS

The CanDiD projects provide case studies for HF monitoring using temporary wireline-deployed DAS systems. Our analysis shows that the wireline DAS system can provide valuable information about fracture geometry, perforation effectiveness, and stress direction by integrating interpretations from high- and low-frequency signals. From the case studies, we derive the following conclusions:

- Fracture connection maps for the stimulation wells suggest hydraulic fractures typically grew at least 300 m horizontally in half-length and 85 m vertically in half-height. LF-DAS data recorded a few long hydraulic fractures growing over 600 m in half-length.
- The wireline-deployed DAS recorded a significant number of fiber waves induced by wireline slippage. The spatiotemporal distribution of fiber waves is consistent with FDIs from LF-DAS data. In principle, fiber waves could therefore provide insights into fracture growth as well as a complementary tool to map fracture geometry.
- Although microseismic signals appear to contain a higher fraction of coherent noise, the signal-to-noise ratios of LF-DAS signals appear to be similar to those of cemented-in DAS fiber. The results suggest that dip-in DAS systems have good potential to enable the mapping of fracture geometry without the need to install permanent optical fibers behind casing.

ACKNOWLEDGMENTS

This research was funded in part through the Canada First Research Excellence Fund (CFREF) by the Global Research Initiative in Sustainable Low Carbon Unconventional Resources (GRI). The authors are grateful to two anonymous companies for providing support for this project, including access to the well pad where the DAS data were collected during HF operations. OptaSense and Silixa Ltd. are thanked for DAS data acquisition for these projects. We are also grateful to ESG Solutions and Altus Interventions for wireline and conveyance services. Sponsors of the Microseismic Industry Consortium are thanked for their ongoing support of this initiative. The authors appreciate Dr. Ge Jin from the Colorado School of Mines for the valuable discussions on the fiber-wave mechanism.

AVAILABILITY STATEMENT

The data used in this study are not available for public release.

REFERENCES

Abadi, M., Agarwal, A., Barham, P., Brevdo, E., Chen, Z., Citro, C., et al. (2015). TensorFlow: large-scale machine learning on heterogeneous distributed systems. Retrieved from https://www.tensorflow.org

Becker, M. W., Ciervo, C., Cole, M., Coleman, T., & Mondanos, M. (2017). Fracture hydromechanical response measured by fiber optic distributed acoustic sensing at milliHertz frequencies. *Geophysical Research Letters*, *44*(14), 7295–7302.

Boone, K., Crickmore, R., Werdeg, Z., Laing, C., & Molenaar, M. (2015). *Monitoring hydraulic fracturing operations using fiber-optic distributed acoustic sensing*. Paper presented at the Society of Petroleum Engineers Unconventional Resources Technology Conference. doi:10.2118/178648-ms

Chollet, F. (2015). Keras. Retrieved from https://github.com/fchollet/keras

Daley, T. M., Freifeld, B. M., Ajo-Franklin, J., Dou, S., Pevzner, R., Shulakova, V., et al. (2013). Field testing of fiber-optic distributed acoustic sensing (DAS) for subsurface seismic monitoring. *Lead Edge Capital Management, LLC.*, *32*, 699–706. doi:10.1190/tle32060699.1

Dehghan, A. N., Goshtasbi, K., Ahangari, K., & Jin, Y. (2015). Experimental investigation of hydraulic fracture propagation in fractured blocks. *Bulletin of Engineering Geology and the Environment*, *74*(3), 887–895. doi:10.1007/s10064-014-0665-x

Eaton, D. W., Ma, Y., Wang, C., & MacDougall, K. (2022). Effectiveness of dip-in DAS observations for low-frequency strain and microseismic analysis: the CanDiD experiment. *SEG Technical Program Expanded Abstracts*, *2022*, 565–569. doi:10.1190/image2022-3745367.1

Garagash, I. A., Osiptsov, A. A., & Boronin, S. A. (2019). Dynamic bridging of proppant particles in a hydraulic fracture. *International Journal of Engineering Science*, *135*, 86–101. doi:10.1016/j.ijengsci.2018.11.004

Hartog, A. H. (2017). *An Introduction to Distributed Optical Fibre Sensors*. CRC Press. doi:10.1201/9781315119014

Hartog, A. H. (2020). Distributed sensors in the oil and gas industry. In I. del Villar & I. R. Matias (Eds.), *Optical Fibre Sensors* (pp. 151–191). Hoboken, NJ: John Wiley & Sons.

He, Q., Suorineni, F. T., Ma, T., & Oh, J. (2017). Effect of discontinuity stress shadows on hydraulic fracture reorientation. *International Journal of Rock Mechanics and Mining Sciences*, *91*, 179–194. doi:10.1016/j.ijrmms.2016.11.021

Henninges, J., Martuganova, E., Stiller, M., Norden, B., & Krawczyk, C. (2021). Wireline distributed acoustic sensing allows 4.2 km deep vertical seismic profiling of the Rotliegend 150 °C geothermal reservoir in the north German Basin. *Solid Earth*, *12*, 521–537. doi:10.5194/se-12-521-2021

Jin, G., & Roy, B. (2017). Hydraulic-fracture geometry characterization using low-frequency DAS signal. *The Leading Edge*, *36*(12), 975–980.

Jin, G., Ugueto, G., Wojtaszek, M., Guzik, A., Jurick, D., & Kishida, K. (2021). Novel near-wellbore fracture diagnosis for unconventional Wells using high-resolution distributed strain sensing during production. *SPE Journal*, *26*(05), 3255–3264. doi:10.2118/205394-PA

Karrenbach, M., Cole, S., Ridge, A., Boone, K., Kahn, D., Rich, J., et al. (2019). Fiber-optic distributed acoustic sensing of microseismicity, strain and temperature during hydraulic fracturing. *Geophysics*, *84*(1), D11–D23.

Karrenbach, M., Ridge, A., Cole, S., Boone, K., Kahn, D., Rich, J., et al. (2017). *DAS microseismic monitoring and integration with strain measurements in hydraulic fracture profiling*. Paper presented at the SPE/AAPG/SEG Unconventional Resources Technology Conference, URTEC-2670716-MS. doi:10.15530/URTEC-2017-2670716

LeBlanc, M., Suh, K., Machovoe, S., Byrd, D., Jaaskelainen, M., Bland, H., et al. (2022). *Theory and practice of a flexible fiber optic cable in a horizontal well used for cross-well and microseismic hydraulic fracture monitoring*. Paper presented at the SPE Hydraulic Fracturing Technology Conference and Exhibition, The Woodlands, Texas, USA. doi:10.2118/209121-MS.

Lellouch, A., Lindsey, N. J., Ellsworth, W. L., & Biondi, B. L. (2020a). Comparison between distributed acoustic sensing and geophones: downhole microseismic monitoring of the FORGE geothermal experiment. *Seismological Research Letters, 1–13*. doi:10.1785/0220200149

Lellouch, A., Mark, A. M., Nemeth, T., & Biondi, B. (2020b). Fracture properties estimation using distributed acoustic sensing recording of guided waves in unconventional reservoirs. *Geophysics*, *85*, M85–M95.

Lindsey, N. J., & Martin, E. R. (2021). Fiber-optic seismology. *Annual Review of Earth and Planetary Sciences*, *49*(1), 309–336. doi:10.1146/annurev-earth-072420-065213

Liu, Y., Wu, K., Jin, G., & Moridis, G. (2020). Rock Deformation and strain-rate characterization during hydraulic fracturing treatments: insights for interpretation of low-frequency distributed acoustic-sensing signals. *SPE Journal*, *25*(5), 2251–2264.

Lumens, P. (2014). *Fiber-optic sensing for application in oil and gas wells* (PhD thesis). Technische Universiteit Eindhoven.

Luo, B., Jin, G., & Staněk, F. (2021). Near-field strain in distributed acoustic sensing-based microseismic observation. *Geophysics*, *86*(5), P49–P60.

Ma, Y., Cao, S., Rector, J. W., & Zhang, Z. (2020). Automated arrival-time picking using a pixel-level network. *Geophysics*, *85*(5), V415–V423. doi:10.1190/geo2019-0792.1

Ma, Y., Eaton, D. W., Igonin, N., & Wang, C. (2023a). Machine learning-assisted processing workflow for multi-fiber DAS microseismic data. *Frontiers in Earth Science, 2023*. doi:10.3389/feart.2023.1096212

Ma, Y., Eaton, D. W., Wang, C., & Aklilu, A. (2023b). Characterizing hydraulic fracture growth using DAS-recorded microseismic reflections. *Geophysics*. doi:10.1190/geo2022-0607.1

Mateeva, A., Lopez, J., Mestayer, J., Wills, P., Cox, B., Kiyashchenko, D., et al. (2013). Distributed acoustic sensing for reservoir monitoring with VSP. *The Leading Edge*, *32*(10), 1278–1283.

Molenaar, M. M., Fidan, E., & Hill, D. J. (2012). *Real-time downhole monitoring of hydraulic fracturing treatments using fiber optic distributed temperature and acoustic sensing*. Paper presented at the SPE/EAGE European Resources Conference, SPE-152981–MS.

Morrill, J. C., & Miskimins, J. L. (2012). *Optimizing hydraulic fracture spacing in unconventional shales*. Paper presented at the SPE Hydraulic Fracturing Technology Conference, The Woodlands, Texas, USA. doi:10.2118/152595-MS

Richter, P., Parker, T., Woerpel, C., Wu, Y., Rufino, R., & Mahmoud, F. (2019). *High-resolution distributed acoustic sensor using engineered fiber for hydraulic fracture monitoring and optimization in unconventional completions*. Paper presented at the SEG International Exposition and Annual Meeting, San Antonio, Texas, USA. doi:10.1190/segam2019-3215860.1

Ronneberger, O., Fischer, P., & Brox, T. (2015). U-net: convolutional networks for biomedical image segmentation. In N. Navab, J. Hornegger, W. Wells, & A. Frangi (Eds.), *Medical Image Computing and Computer-Assisted Intervention – Miccai 2015* (p. 9351). Berlin: Springer. Lecture Notes in Computer Science. doi:10.1007/978-3-319-24574-4_28

Shahri, M., Tucker, A., Rice, C., Lathrop, Z., Ratcliff, D., McClure, M., & Fowler, G. (2021). *High fidelity fibre-optic observations and resultant fracture modeling in support of planarity*. Paper presented at the SPE Hydraulic Fracturing Technology Conference and Exhibition, Virtual. doi:10.2118/204172-MS

Staněk, F., Jin, G., & Simmons, J. (2022). Fracture imaging using DAS-recorded microseismic events. *Frontiers in Earth Science*, *10*, 907749. doi:10.3389/feart.2022.907749

Ugueto, G., Huckabee, P., Wojtaszek, M., Daredia, T., & Reynolds, A. (2019). *New near-wellbore insights from fibre-optics and downhole pressure gauge data*. Paper presented at the SPE Hydraulic Fracturing Technology Conference and Exhibition, The Woodlands, TX.

Ugueto, G., Wojtaszek, M., Mondal, S., Guzik, A., Jurick, D., & Jin, G. (2021). *New fracture diagnostic tool for unconventionals: high-resolution distributed strain sensing via Rayleigh frequency shift during production in hydraulic fracture test 2*. Paper presented at the Unconventional Resources Technology Conference. doi:10.15530/urtec-2021-5408

Ugueto, G., Wu, K., Jin, G., Zhang, Z., Haffener, J., Shahri, M., et al. (2023). *A catalogue of fiber optics strain-rate fracture driven interactions*. Paper presented at the SPE Hydraulic Fracturing Technology Conference and Exhibition, The Woodlands, Texas, USA. doi:10.2118/212370-MS

Wang, C., Ma, Y., Eaton, D., MacDougall, K., & Furlong, C. (2022). *Hydraulic fracture monitoring using distributed acoustic sensing: a case study using dip-in fiber*. Paper presented at the SPE/AAPG/SEG Unconventional Resources Technology Conference. doi:10.15530/urtec-2022-3723692

Wang, J., Tan, Y., Rijken, M., Liu, X., Singh, A., & Li, Y. (2022). *Observations and modeling of fiber-optics strain on hydraulic fracture height growth in HFTS-2*. Paper presented at the SPE/AAPG/SEG Unconventional Resources Technology Conference, Houston, Texas, USA. doi:10.15530/urtec-2021-5295

Zhang, Z., DiSiena, J., Bevc, D., Lim Chen Ning, I. Tan, Y., Swafford, L., et al. (2021). *Hydraulic fracture characterization by integrating multidisciplinary data from the hydraulic fracturing test site 2 (HFTS-2)*. Paper presented at the Unconventional Resources Technology Conference, Houston, Texas, USA.

Zhang, Z., Fang, Z., Stefani, J., DiSiena, J., Bevc, D., Lim Chen Ning, I., et al. (2020). Modeling of fiber-optic strain responses to hydraulic fracturing. *Geophysics*, *85*, A45–A50.

Zhou, J., Chen, M., Jin, Y., & Zhang, G.-q. (2008). Analysis of fracture propagation behavior and fracture geometry using a tri-axial fracturing system in naturally fractured reservoirs. *International Journal of Rock Mechanics and Mining Sciences*, *45*(7), 1143–1152. doi:10.1016/j.ijrmms.2008.01.001

Zhu, W., & Beroza, G. C. (2018). PhaseNet: a deep-neural network-based seismic arrival-time picking method. *Geophysical Journal International*, *216*(1), 261–273. doi:10.1093/gji/ggy423

30

Use of Distributed Acoustic Sensing and Distributed Strain Sensing Technologies for Geomechanical Characterization of Rock Mass Response in Mesoscale Experiments in Underground Laboratories: Lessons Learned

Verónica Rodríguez Tribaldos[1,2], Chet Hopp[2], Antonio P. Rinaldi[3], Katinka B. Tuinstra[3], Federica Lanza[3], Jonathan Ajo-Franklin[4], Timothy Kneafsey[2], Michelle Robertson[2], Yves Guglielmi[2], and Alba Zappone[3,5]

ABSTRACT

Distributed fiber-optic sensing (DFOS) is a promising technology for efficient, high-resolution monitoring and characterization of rock mass response to stimulation during subsurface operations. However, uncertainties in its implementation and the impact of experimental components on the measurements remain. Underground research laboratories (URLs) provide opportunities to test and optimize DFOS approaches through controlled experiments in relevant conditions. Here, we present practical observations gathered from implementing downhole DFOS for geomechanical monitoring during fluid injection experiments conducted in the Sanford Underground Research Facility (SURF) in the United States and the Mont Terri Rock Laboratory (MTRL) in Switzerland, as part of the EGS Collab (at SURF) and the Carbon Sequestration Series D and Fault Slip B (at MTRL) projects. We investigate the performance of low-frequency distributed acoustic sensing (DAS) and distributed strain sensing (DSS) for strain-rate and strain monitoring and evaluate the effect of contrasting cable designs, installation methods, and acquisition parameters. Results suggest that cable construction and installation affect noise levels and measured strain magnitudes. We also discuss experimental and logistical constraints imposed by URLs. Observations are summarized as lessons learned that we hope will guide future investigations and encourage implementing borehole-based DFOS for geomechanical characterization and monitoring in URLs and beyond.

[1]*GFZ German Research Centre for Geosciences, Potsdam, Germany*

[2]*Energy Geosciences Division, Lawrence Berkeley National Laboratory, Berkeley, California, USA*

[3]*Swiss Seismological Service, ETH Zurich, Zurich, Switzerland*

[4]*Department of Earth, Environmental and Planetary Sciences, Rice University, Houston, Texas, USA*

[5]*Department of Mechanical Engineering, ETH Zurich, Zurich, Switzerland*

30.1. INTRODUCTION

The need for actions to mitigate climate change and conduct the energy transition efficiently and safely has sparked interest in the development and utilization of the subsurface for activities such as development of enhanced geothermal systems (EGS) for geothermal energy production, underground storage of CO_2, and disposal of nuclear waste. The success of these operations requires a comprehensive understanding of the conditions

Distributed Acoustic Sensing in Borehole Geophysics, Geophysical Monograph 289, First Edition.
Edited by Yingping Li, Robert Mellors, and Ge Zhan.

DOI: 10.1002/9781394179275.ch30

and behavior of the rock mass and, specifically, of the hydrological, geomechanical, and seismic processes taking place as a result of injection and/or production of fluids (Plenkers et al., 2023). For example, the successful development of EGS relies on increasing the permeability of initially impermeable rocks at depth and enhancing fluid flow therein through fluid injection and consequent fracturing of the rock and/or reactivation of existing faults (Grant, 2016). This process commonly induces low-magnitude seismic events. In some occasions, however, seismic events of magnitudes large enough to be felt on the surface have been induced during EGS operations. These events can pose a threat to nearby infrastructure and communities and reduce the societal acceptance of EGS projects, sometimes leading to their cancellation (Deichmann & Giardini, 2009; Ellsworth et al., 2019; Grigoli et al., 2017, 2018). In the case of CO_2 storage and waste disposal sites, reactivation of nearby faults and associated seismicity, fluid migration, and surface uplift or subsidence are undesired consequences that compromise the integrity of the repository and should therefore be prevented by monitoring these activities and designing operations accordingly (Ringrose et al., 2013; Rutqvist et al., 2016; Zoback & Gorelick, 2012).

Appropriately characterizing and monitoring the geomechanical and seismic response of the rock mass during injection and/or production of fluid or other disturbances, however, is a challenging task. The extents and magnitudes of perturbations are sometimes poorly understood, particularly at significant distances from monitoring wells. Even small changes in pressure and/or temperature of the subsurface have been proven to induce changes in stress and strain, deformation, and seismic events that can occur far beyond the immediate vicinity of operations (Rutqvist, 2012). Microseismicity monitoring is commonly used to assess the risk potential of subsurface operations, and, in the case of EGS and unconventional oil and gas development, it is also the tool of choice for imaging the fracture network being created. However, deformation and fracture opening/fault reactivation can also occur aseismically at a variety of temporal and spatial scales, particularly in clay-rich formations such as caprocks overlying CO_2 storage units and waste disposal sites (Zoback et al., 2012). In this context, remote sensing techniques such as Interferometric Satellite Aperture Radar (InSAR) can be used to detect ground deformation across large spatial areas with sub-centimeter-scale accuracy (Vasco et al., 2018). However, the resolution of InSAR measurements decreases with depth, and temporal sampling is typically 6–12 days at best due to satellite orbital constraints; this might result in missing the early onset of deformation and will likely not resolve sharp features such as fluid leaks at depth. Additional geodetic tools, such as extensometers, tiltmeters, and strainmeters, can be deployed in boreholes and provide accurate, in situ deformation observations. However, these instruments tend to be costly; they need to be deployed in dedicated boreholes to avoid interference with other well activities and commonly provide only a single measurement point (Murdoch et al., 2020).

In the last decade, distributed fiber-optics sensing (DFOS) technologies have emerged as a versatile tool for high-resolution, in situ imaging and monitoring of local physical parameters in a variety of contexts, environments, and applications (Daley et al., 2013; Lellouch & Biondi, 2021; Lindsey & Martin, 2021; Shanafield et al., 2018; Sun et al., 2021; Zhang et al., 2020b). This family of technologies uses laser-based interrogation units (IUs) to convert fiber-optic cables into ultra-dense sensing arrays. Interrogator units continuously inject short laser pulses into the fiber-optic cable, which can be hundreds of kilometers long. The incident light is backscattered at natural impurities within the fiber and recorded back at the IU. Changes in temperature, deformation, or vibrations impacting the fiber-optic cable result in changes in the properties of the scattered field, which can be measured by the IU. Depending on which component of backscattered light is interrogated by the IU—that is, Rayleigh, Brillouin, or Raman backscattering—dynamic strain (distributed acoustic sensing, DAS), absolute strain (distributed strain sensing, DSS), and temperature (distributed temperature sensing, DTS), respectively, can be measured with high sensitivity, at resolution on the order of centimeters to a few meters along distances that range from a few meters to tens of km, depending on the DFOS technology being used (Hartog et al., 2017). In addition to these capabilities, there are clear advantages to using fiber-optic cables as sensing devices: flexibility in deployment configuration (e.g., on surface arrays of multiple configurations, inside boreholes, attached to infrastructure), durability, robustness, resistance to high-pressure and -temperature environments, and ease of permanent deployment.

These capabilities reveal fiber-optics sensing technologies as excellent tools to characterize and monitor the coupled geomechanical and seismic response of the rock mass as a result of subsurface development at unprecedented resolution. In recent years, DAS, and in particular the so-called phase-sensitive DAS, based on phase-sensitive optical domain reflectometry ($\Phi - \text{OTDR}$), has become a popular tool to monitor and characterize induced seismicity in geoenergy applications, such as oil and gas production from unconventional reservoirs (Verdon et al., 2020) and EGS development (Lellouch et al., 2020). In most of these studies, fiber-optic cables are installed in boreholes penetrating the target reservoir and/or surrounding it, providing a continuous, dense recording of seismic events without disturbing other activities in the well (Lellouch & Biondi, 2021). This unprecedented opportunity to deploy the sensing fiber directly inside the rock mass has resulted in monitoring opportunities previously not available. The broadband nature of DAS, with a response that ranges from mHz to kHz (Becker et al., 2017; Lindsey et al., 2020), enables its application not only as a seismic tool but also as a deformation sensor. At low frequencies (<1 Hz), DAS is extremely sensitive to very small variations of strain rate (nε/s). A few studies have demonstrated the potential of low-frequency DAS to detect pressure fronts and the opening and closing of fractures affecting the well where the fiber is installed (Jin et al., 2019; Jin & Roy, 2017; Zhang et al., 2020b). These observations can be combined with DSS monitoring, which provides measurements of absolute strain along the fiber at much higher spatial resolution, although in general with lower temporal resolution (tens of seconds to minutes) and less sensitivity (με) than DAS (Hopp et al., 2022; Sun et al., 2020). Recent field and laboratory studies have demonstrated that DSS measurements based on Rayleigh scattering frequency shift (RFS) can be used to track strain perturbations caused by fluid migration and formation pore-pressure increase with higher stability and accuracy than Brillouin-based DSS measurements (Lei et al., 2019; Xue et al., 2018; Zhang et al., 2020a; Zhang & Xue, 2019).

Despite the success of these studies, some uncertainties remain regarding the full potential of fiber-optic sensing, particularly DAS and DSS, to appropriately characterize seismic and geomechanical processes in these contexts. In particular, questions remain related to data verification and the effects that the different experimental deployment strategies and components have on the final measurements and interpretations. Due to the novelty of DAS and DSS and their application in such investigations, critical aspects such as the choice of cable type and installation procedures that ensure that strain perturbations on the rock mass are accurately transferred to the fiber are still topics of investigation (Sasaki et al., 2021). Additional uncertainties critical for data interpretation are a proper characterization of the system's limitations regarding strain (or strain-rate) sensitivity, as well as the impact of external parameters such as temperature on the measured strain. Although some of these parameters can be tested in laboratory experiments, conditions are usually not representative of the spatial scale and magnitude of strain and strain rate induced by the seismic and geomechanical processes occurring in natural subsurface systems. Full-size field demonstrations in real conditions often only enable acquiring indirect and sparse observations of the processes occurring at depth, which hinders the isolation of the different subsurface mechanisms and installation effects.

Underground rock laboratories (URLs) are an attractive alternative to bridge this gap and provide excellent opportunities to develop, test, and optimize new technologies and sensors to characterize and monitor hydrological, geomechanical, and seismic processes in relevant conditions. URLs offer direct access to the rock mass at depth, and research can be conducted through high-resolution mesoscale experiments under controllable conditions and at relevant temperature and pressure, providing ideal conditions to investigate the behavior of the subsurface during stimulation and utilization. In the last two decades, several URLs have been developed across the world, either by adapting old mines or by excavating new facilities (Blechschmidt et al., 2020; Cho et al., 2008; Kickmaier & McKinley, 1997; Plenkers et al., 2023). Examples of these research infrastructures include the Mont Terri Rock Laboratory, Bedretto Underground Laboratory for Geosciences and Geoenergies and the Grimsel Test Site, all in Switzerland; Äspö Hard Rock Laboratory in Sweden; Reiche Zeche, in Germany; Sanford Underground Research Facility (SURF), in the United States; Underground Research Laboratory, in Canada; Korea Underground Research Tunnel (KURT), in South Korea; and Mizunami Underground Research Laboratory, in Japan.

To date, only a few studies discussing the use and implementation aspects of DFOS techniques in URLs are recorded in the literature. Krietsch et al. (2018) used DSS to monitor a series of high-rate hydraulic fracturing and shearing tests in the Grimsel URL within the framework of the In-situ Stimulation and Circulation (ICS) experiment. They compare strain measurements obtained with Fiber Bragg Gratings (FBGs; Othonos, 1997) and grouted, prestrained fiber-optic cables deployed using two installation procedures: bare cable and a cable packed and fixed with glue. Their analysis revealed that the bare DSS cable was able to recover strain measurements quantitatively comparable to those obtained with FBGs, although with poorer accuracy and temporal resolution, whereas the packed cable did not provide any meaningful measurements. Paitz et al. (2020) report on DAS data recorded during the ICS experiments mentioned earlier, with the objective to calibrate the low-frequency response of their DAS IU. Their results show that the low-frequency DAS response for two DAS channels is comparable to strain measured at two nearby FBG sensors. Piccolo et al. (2023) describe a DFOS experiment conducted in the French National Radioactive Waste Management Agency (Andra) Underground Research Laboratory in France, in which different types of fiber-optic cables were deployed along the walls of small tunnels to assess the potential of both Brillouin-based and Rayleigh-based DSS measurements for monitoring strain changes associated with long-term storage of radioactive waste (Kishida et al., 2014). Their analysis suggested that Rayleigh-based strain measurements are more sensitive than their Brillouin counterparts, whereas Brillouin is better suited for large-magnitude strains.

Here, we report on practical lessons learned on survey design, installation, data acquisition, and data analysis of three borehole-based fiber-optic installations conducted in two URLs within the context of seismic monitoring and geomechanical characterization during fluid injection. In particular, we focus on the advantages, limitations, and complementarities of low-frequency DAS and Brillouin-based DSS for monitoring the deformation of the rock mass. We also comment on the potential and limitations of DAS for microseismic investigations in mesoscale experiments in crystalline rock. First, we summarize general considerations regarding experiment design, sensor installation, and choice of acquisition parameters. This is followed by reporting on two case studies: (a) Experiments 1 and 2 of the EGS Collab project, conducted at Sanford Underground Research Facility (SURF) in South Dakota, USA (Kneafsey et al., 2020, 2023); and (b) The Carbon Sequestration – Series D (CS-D; Zappone et al., 2021) and Fault Slip – B (FS-B) experiments, conducted at the Mont Terri Rock Laboratory (MTRL) in Saint-Ursanne (Switzerland). Our observations will be summarized in a collection of practical lessons learned that we hope are useful for the implementation of future fiber-optics experiments in URLs.

30.2. EXPERIMENT DESIGN AND INSTALLATION CONSIDERATIONS

The deployment of DAS and DSS for monitoring geomechanical processes in URL experiments is constrained by a variety of factors associated not only with the objectives of the study but also with the special conditions imposed by working at an underground facility. Some of the crucial elements that have to be considered when designing a URL DFOS experiment are discussed next:

30.2.1. Main Scientific Questions to be Addressed with the Measurements

The versatility of DFOS, especially DAS, makes for a comprehensive investigation of rock mass response both in terms of static deformation and seismic processes. These objectives, however, will sometimes call for different experiment designs. For example, to track temporal perturbations in strain using low-frequency DAS, long-term, continuous data is critical. The temporal integration of strain-rate data to derive strain will always yield strain from 0 time (i.e., the start of the measurement), and stopping the recording would cause the loss of the reference phase value. For experiments with a focus on low-frequency DAS, recording long-term, continuous data is straightforward because the required sampling rates are low (in the order of only a few Hz) and the generated data volumes will be manageable. For experiments combining strain monitoring with microseismic detection, however, acquisition of continuous data for several weeks or months can be challenging, as seismic investigations usually require higher sampling rates, which can result in GB or TB of data/day depending on the experiment dimensions. Other considerations such as cable type, the importance of cable-rock coupling, cable length, and spatial sampling could also be different for the different applications. It is thus critical that the objectives of the DFOS experiment are well defined prior to experiment design.

30.2.2. Experiment Dimensions and Expected Signals

One of the constraints of URL rock stimulation experiments is the dimensions of the testbed, including the number of boreholes, borehole length and width, and distance between boreholes. These parameters, in combination with site characteristics such as lithology, estimated seismic velocities, and the presence and location of existing structures such as fractures, will determine the required choice of deployment strategies and acquisition parameters such as frequency sampling, spatial sampling, or DAS gauge length. Anticipating, to the degree possible, the characteristics of the processes and signals that we want to record will significantly improve the chances of a successful experiment.

30.2.3. Deployment Logistics

Conducting experiments in URLs can be challenging due to the special conditions and requirements regarding safety concerns; accessibility of both personnel and instrumentation to the site; physical working space to prepare, manage, and deploy long lengths of fiber-optic cable; conditions of temperature and humidity that can affect interrogator units; storage systems and computers; frequency of access to retrieve data; internet connectivity; and remote access for checking instrument state-of-health and for data telemetry. These factors will have important implications for experiment design elements, such as choice of cable type and deployment strategy, decisions on acquisition parameters based on data volumes being generated, choice of storage devices, length of the experiment, or necessary auxiliary equipment (such as tables and weather-tight cabinets to host instrumentation).

This set of constraints will determine the choice of the different components of a DFOS experiment at an underground research laboratory experiment. These components are summarized next.

Cable Type

To use DFOS to measure strain perturbations in a rock formation, it is important that the strain affecting the formation is efficiently transferred to the fiber. However, strain transfer efficiency can strongly depend on cable construction. Other aspects, such as resistance to harsh downhole conditions and cost, and practical considerations, such as ease in handling and deployment, all play an important role when choosing the type of fiber-optic cable to be used in a URL DAS and/or DSS experiment.

One of the main differences between different cable types is the protective structures surrounding the fibers. Fiber-optic cables can be constructed using different fiber packing methods. In so-called loose-tube cables, fibers are loosely threaded in a plastic or metal (stainless steel or nickel alloy) tube (fiber in metal tube, FIMT) that is commonly filled with a viscous optical gel. Because the fiber is not in direct contact with the tube walls, these cables are generally better isolated from mechanical deformation, water ingress, and temperature. However, for multifiber cables with this design, fibers are all bundled together inside the tube, which can make splicing and termination challenging. In so-called tight-buffered cables, fibers are buffered with a two-layer coating; one is a waterproof acrylate tightly bonded to the fiber, and the other one is plastic or metal and surrounded by aramid yarn. This cable construction tends to be robust and flexible but is less resistant to water ingress. In all these designs, an extra protective layer in the form of a metal tube or corrugated steel armor is also common. In addition, the inner cable components are protected by an outer jacket, which is the first line of protection against chemicals, moisture, flames, and mechanical impact. It can also be made of different materials: for example, polyethylene or PVC.

Cable designs with metal elements are especially robust and resistant to harsh conditions, which makes them popular in downhole applications. Loose-tube FIMT designs, for example, are especially common in borehole deployments where water ingress is a concern. However, these cables tend to be heavy and difficult to install due to their lack of flexibility. Tactical cables with no tubing or armoring are more flexible and resistant to bending, lighter, and easier to install, but they are less resistant to temperature extremes and water ingress. Variations exist from these designs, resulting in a wide range of fiber-optic cables available in the market. Specialty strain and acoustic fiber-optic cables for sensing applications are also commercially available.

A limited number of studies have evaluated the effect of fiber-optic cable construction on strain transfer. These studies suggest that, whereas cable design seems to have a limited effect on strain-rate measurements for seismic applications (Dou et al., 2017), when the objective is to measure slow strain response, the cable's protective structure has a significant impact on the magnitude of measured strain, especially in gel-filled cables (Becker et al., 2018; Sasaki et al., 2021).

Installation Conditions

Installation of fiber-optic cables for borehole-based experiments can be conducted in a variety of configurations depending on requirements, including coupling to the rock, the need to retrieve the cable, space available in the borehole, and installation logistics. In URL research experiments, boreholes dedicated to monitoring instrumentation are common, and thus, permanent installations are favored. When possible, fiber-optic cables are deployed behind casing for improved coupling with the rock as well as to accommodate the campaign installation of instruments within casing. However, this deployment method requires installation at an early stage. When the installation of casing is not possible for either logistical or experiment-related reasons, a commonly used alternative is attaching the cable to tubing or conveyance material (e.g., PVC, fiberglass, or metal rod). When permanent deployment is sought, the borehole is grouted to improve coupling. For installations in injection and/or production boreholes, fiber-optic cables can be deployed through straddle packers and attached using clamps or ferrules. These installations are expected to be noisier.

Often, technology development experiments in URLs are characterized by multiparametric investigations in which multiple instruments are deployed in the same borehole to allow for contemporaneous and collocated measurements of different physical parameters that enable calibration and benchmarking. This will introduce an additional constraint in the placement and type of fiber-optic cable that we can deploy for our fiber-optic sensing experiment.

Additionally, the mesoscale dimensions of most URL fiber-optic experiments offer the opportunity to monitor multiple boreholes simultaneously using a single fiber-optic cable. In this configuration, two deployment strategies are possible: installation as a single loop—that is, one single cable looping through all boreholes with no interruptions—or installation using an individual loop for each borehole but connecting all the boreholes by splicing the cable between them. Whereas the single loop option is the most advisable to reduce the number of locations at which light can be lost through splices, it also requires careful cable management practices and strategies to maintain the integrity of the cable. This configuration also offers less flexibility in experiment design and requires careful thought regarding the amount of slack cable to be left between boreholes to account for scenarios in which one of the boreholes needs to be removed from the loop or the cable breaks down in one of the wells. A disadvantage of running separate cables and splicing components together is the potential temperature offset at splice points when conducting DTS measurements; this issue is well-documented and complicates absolute temperature analysis without corrections (Hausner et al., 2011; van de Giesen et al., 2012).

Precise Measurement Timing

One of the most obvious constraints of working underground is ensuring the timing precision of the measurements, given that a GPS signal is seldom available. Particularly for multiparametric investigations, time synchronization between different instruments is crucial. For some applications, such as slow strain monitoring, timing based on Network Time Protocol (NTP), which can usually maintain time to within tens of milliseconds over the internet, is sufficient. For other applications, such as DAS-based high-frequency seismic monitoring, however, better timing precision is needed. An alternative common in URL experiments is the use of portable rubidium atomic clocks, which can be accurate down to the nanosecond if initially synchronized to an accurate absolute time standard at the surface. If such an atomic clock is used, care should be taken to maintain backup power for the clock because absolute synchronization can be lost during power outages.

Data Storage Solutions

One of the critical considerations when choosing acquisition parameters that are constrained by the special characteristics of URL experiments is on-site data storage capacity and access to data retrieval. This is especially important for high-resolution, high-frequency DAS experiments. In long-term, multiwell DAS investigations using fiber lengths on the order of several kilometers, spatial sampling of a few meters, and sampling frequencies of hundreds to thousands of Hz, terabytes/day of data can easily be generated. These experiments will require high storage capacity (which can be limited due to space restrictions) or a site with easy access to retrieve data often, as data telemetry in underground laboratories is usually restricted due to slow or nonexistent internet connectivity.

Choice of Acquisition Parameters

Besides these general considerations, the choice of technology-specific acquisition parameters is also critically conditioned by the characteristics of URL experiments. Based on our experience, the choice of sampling frequency and gauge length in DAS systems can be critically constrained by the dimensions of the testbed. The gauge length—that is, the length of fiber used to calculate the difference in strain rate—will determine the spatial resolution of the DAS measurement. In URL mesoscale experiments, in which borehole length tends to be of a maximum of tens of meters, a large (e.g., 10 m) gauge length will yield a limited number of measurement points along the borehole. Large gauge lengths will also limit the high-frequency end of the seismic events that can be recovered (Dean et al., 2016). In mesoscale experiments, where microseismicity is recorded relatively near the source, high-frequency content (i.e., short wavelength) events are expected. This effect is even more pronounced in crystalline rock, characterized by fast seismic velocities. In this context, short gauge lengths are needed to capture these events. Similarly, high DAS sampling frequencies will be desired. In this scenario, required frequency versus data volume constraints will have to be considered. It is important to note here that whereas for some DAS IUs, the gauge length can be chosen during instrument setup, for some commercial interrogators, this is a fixed parameter that cannot be modified. This feature should also be taken into account when choosing our preferred DAS IU.

Impact of Environmental Conditions on Instrumentation

One final consideration is the effect of environmental parameters on instrumentation durability and performance. Some of the less developed URLs can pose challenging conditions for electronic equipment; dust, humidity, and temperature changes can damage the equipment, and adequate protection strategies should be followed. Additionally, DAS and DSS systems are very sensitive to ambient temperature changes affecting the recording units, which can have an important impact on strain-rate (and strain) measurements. Although the effect of temperature variations around the IU can be neglected for DAS-based seismic measurements, this effect is especially significant in low-frequency DAS measurements. Temperature variations around the DAS IU manifest in the data as consistent strain-rate variations in time that affect the entire fiber length (referred to as "common-mode" noise; Sidenko et al., 2022). Although avoiding these effects entirely is challenging, they can be alleviated by placing the IUs inside ventilated, weather-resistant cabinets that isolate the IU from external conditions as much as possible. Independent monitoring of gallery temperature conditions and analyzing strain-rate changes along exposed portions of the fiber near the IU can help constrain these thermally induced strain-rate variations.

30.3. CASE STUDIES

30.3.1. The EGS Collab Project

The EGS Collab project aimed at improving our understanding of rock response to hydraulic fracturing and stimulation for the development of enhanced or engineered geothermal systems (Kneafsey et al., 2023). This multiyear project, funded by the U.S. Department of Energy, consisted of two experiments conducted at 1.25

and 1.5 km depth within the network of galleries at SURF in Lead, South Dakota, USA (Heise, 2015). These experiments included various injection operations into fractured, crystalline rock mass, which were comprehensively monitored with the objective of using the resulting datasets to improve and validate thermal–hydrological–mechanical–chemical (THMC) modeling approaches. For a comprehensive description of the EGS Collab project activities, we refer the reader to Kneafsey et al. (2020), Kneafsey et al. (2022), Kneafsey et al. (2023), and references therein.

The EGS Collab Experiment 1 took place at a depth of approximately 1.5 km on the 4850 level of SURF over a span of more than 1 year between 2017 and 2019. An injection and production borehole pair were drilled northward from the main drift. Six ~60-m-long dedicated monitoring holes were drilled surrounding the injection/production holes; two northward between injection and production, and two pairs drilled westward, normal to the other boreholes (Figure 30.1(a), (b)). The six monitoring boreholes contained arrays of three-component accelerometers, hydrophones, piezoelectric seismic sources, electrodes used for electrical resistivity tomography (ERT), and a fiber-optic package containing both single mode (SM) and multi-mode (MM) fibers (referred to as a "hybrid cable") for colocated measurements of DAS, DSS, and DTS.

The EGS Collab Experiment 2 took place from February 2022 to September 2022 at a slightly shallower depth of 1.25 km at the 4100 level at SURF. This testbed consisted of nine 101 mm diameter boreholes drilled subhorizontally from the drift walls, with lengths ranging from 55 to 80 m (Figure 30.1(c), (d)). Of these boreholes, two sets of two wells consisted of dedicated monitoring boreholes surrounding the stimulated zone and housed various grouted-in geophysical instruments. The other five boreholes consisted of a central injection well surrounded by four open boreholes that could be used for production and monitoring as needed. As for Experiment 1, the monitoring holes contained arrays of three-component accelerometers, piezoelectric seismic sources, ERT electrodes, and a hybrid fiber-optic cable.

Fiber-Optic Cable Types and Installation Procedure

Fiber-optic cables were deployed along dedicated monitoring boreholes and continuously recorded DAS, DSS, and DTS measurements over the duration of the experiments. In both experiments, the fiber packages were installed as a continuous loop with no splices except for the jumper cables connected to the various interrogator units. This procedure required careful cable management during installation to avoid breaks and kinks but meant that all monitoring holes could be measured simultaneously using techniques that require access to both ends of the fiber (DSS and DTS). This design also improved signal strength by minimizing the number of splices along the fiber, which can constitute points of incident light loss. The total length of the fiber-optic cable was 1,728 m in Experiment 1 and 960 m in Experiment 2. A sketch illustrating the layout of the fiber-optic sensing system for Experiments 1 and 2 is shown in Figure 30.2.

For deployment down the wells, the cable loop was taped to a PVC pipe together with the rest of the borehole instrumentation (Figure 30.3.(b), (c)). Once installed, the boreholes were filled with grout to couple the instruments to the rock. Between boreholes, the cable was free-hanging in coils in the gallery drifts. Either one end (for DAS) or both ends (for DSS/DTS) of each fiber were spliced to jumper cables that were connected to the fiber-optic sensing interrogators. For the single-mode fibers used for DAS, the loose end had to be terminated to reduce the backreflection caused by the fiber end. This was achieved by wrapping and taping the loose end of each fiber to a very small (2 mm) diameter plastic tube. In both experiments, fiber geometry was determined by tapping (for DAS) or heating (using a heat gun, for DSS) the fiber at the entry and exit point of each borehole and identifying the corresponding response in the data. The symmetry of the signal around the turn-around point at the bottom of the borehole was used to refine the channel location inside the borehole.

For both experiments, two water baths were located near the interrogator units for temperature calibration of the DTS measurements. One was maintained at roughly 40 °C; the other was left at ambient temperature. A length of the fiber package (between 15 and 30 m) was coiled and placed into each of these water baths, together with two temperature probes connected to the DTS IU (Figures 30.2. and 30.3.(d)).

Two different fiber-optic cable designs were used for the two experiments, mostly due to constraints imposed by the additional geophysical instrumentation deployed in the monitoring boreholes and as a result of lessons learned during Experiment 1. As mentioned earlier, in both experiments, the cables were chosen to contain both SMFs and MMFs to enable collocated measurements of DAS, DSS (both needing SMF), and DTS (using MMF).

The fiber-optic cable used for the EGS Collab Experiment 1, fabricated by Optical Cable Corporation (OCC), consisted of a polyethylene (PE) jacketed, aramid-packed bundle (tight-buffered) including four SMFs and four MMFs. This is a cost-efficient, standard, telecommunication-grade fiber-optic cable commonly used in fiber experiments. Due to the tight-buffered aramid yarn, these cables are relatively sturdy while being flexible, which makes them easy to handle during installation. The tight buffer also provides additional protection that enables a

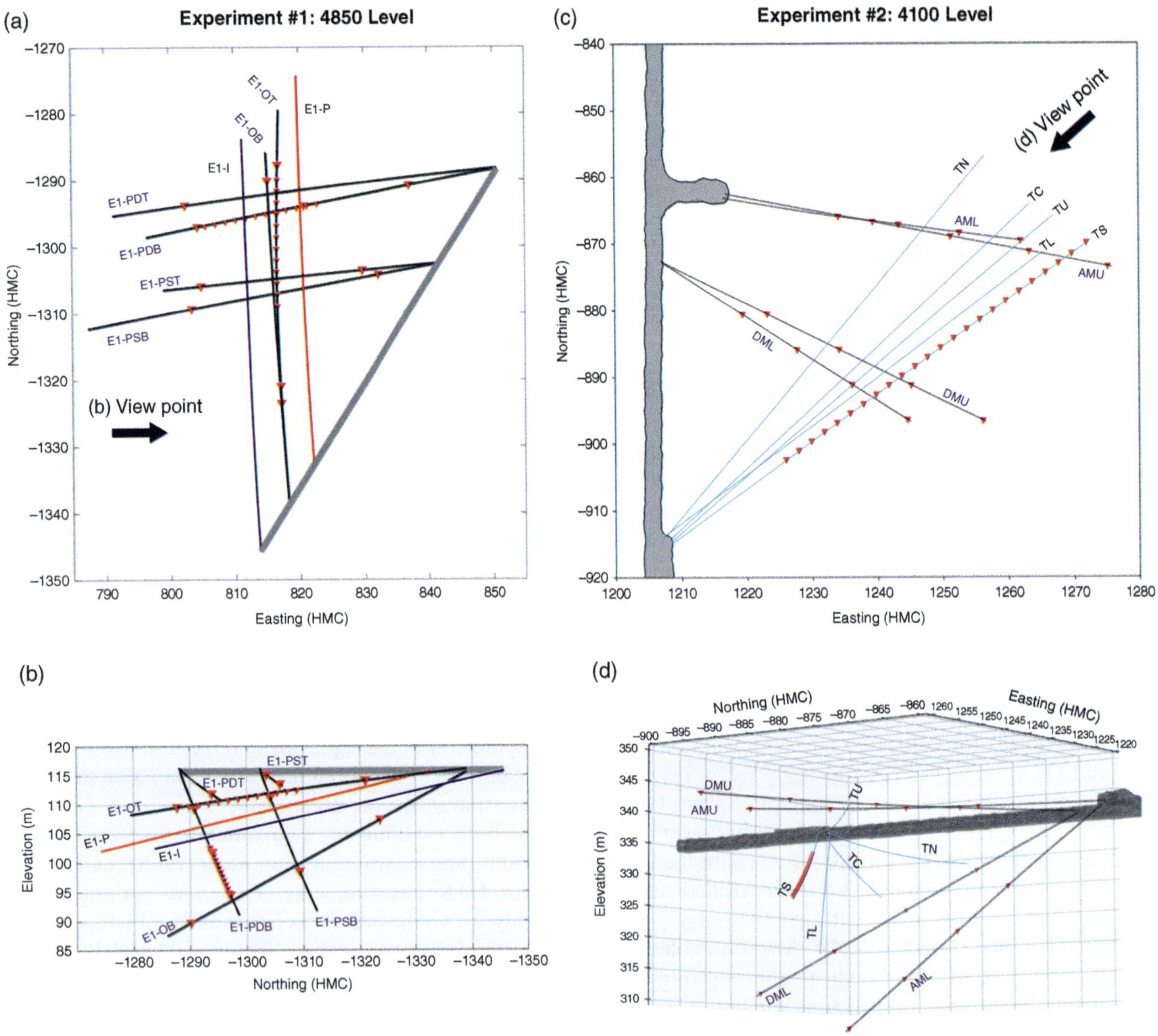

Figure 30.1 Basic layout of the EGS Collab experiments, with (a) and (c) showing map views of Experiments 1 and 2, respectively, and (b) and (d) showing selected 3D perspectives. Black arrows in (a) and (c) show the direction of view for the 3D plots in (b) and (d). Thick, gray line corresponds to the gallery, and thin lines are experiment boreholes. Blue and red boreholes were used for injection and/or production, and black boreholes were dedicated monitoring wells containing grouted instrumentation. Fiber-optic cables were attached to conveyance PVC pipe, lowered into the monitoring boreholes and grouted in as continuous loops. The red triangles indicate the locations of seismic sensors (either hydrophones or accelerometers). HMC stands for Homestake Mine Coordinates, here shown in units of meters.

relatively tight bend radius, useful for the cable turn-around at the bottom of the borehole (Figure 30.3.a).

Despite the good performance of these cables during Experiment 1, a sturdier design was chosen for Experiment 2 based on some of the experiment activities. During the injections conducted in the first testbed, fractures were propagated that intersected the monitoring wells, breaking through the grout and causing water to leak into the borehole. Even though the fiber-optic cable was not damaged, it was decided during the planning of Experiment 2 that a more rugged cable would be used to avoid potential damage. Hence, for Experiment 2, a sensing fiber manufactured by NBG Fiber Optic Corporation was used. The fiber package comprised six individual

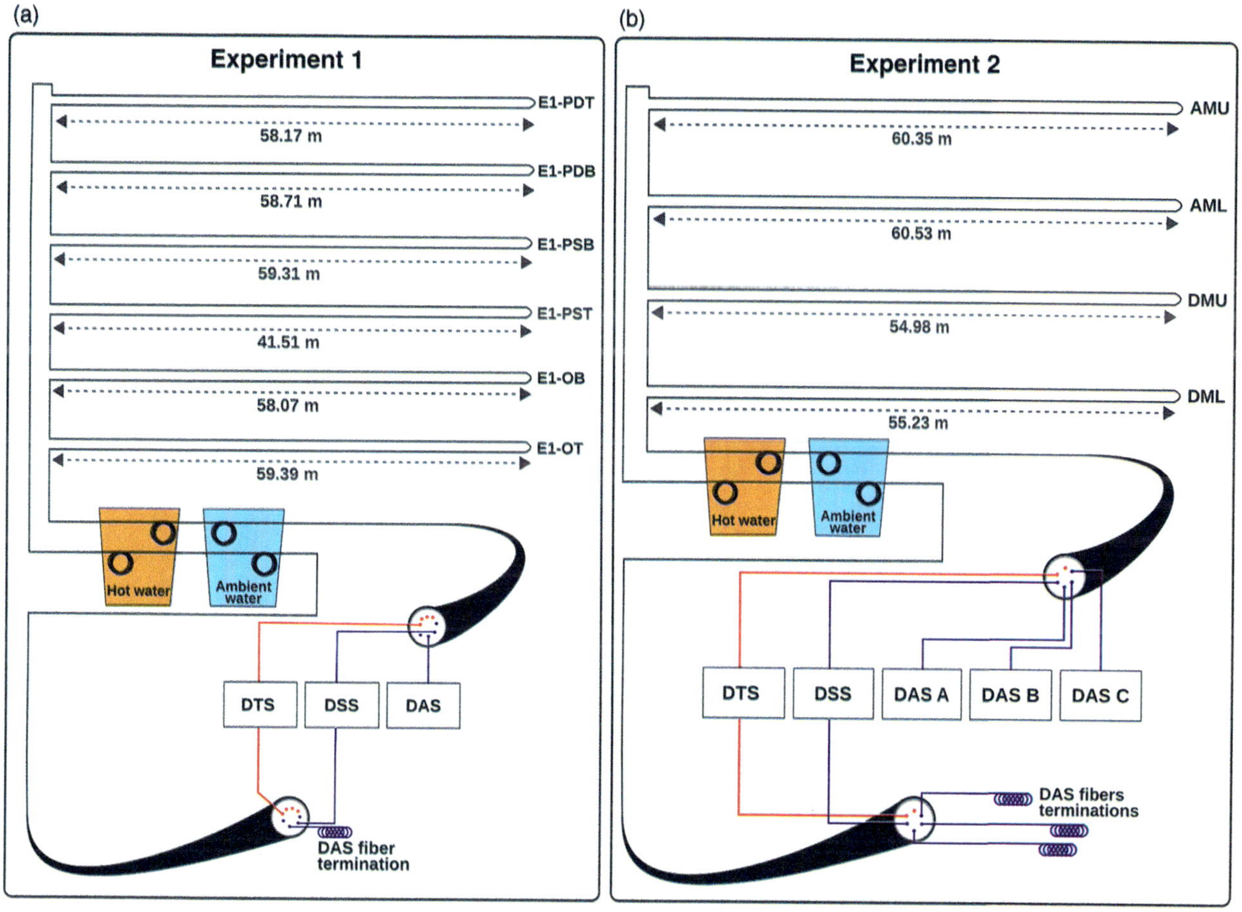

Figure 30.2 Generalized sketch illustrating the layout of the fiber-optic sensing systems deployed in Experiments 1 (a) and 2 (b) of the EGS Collab project. SMF, connected to DAS and DSS interrogator units, are depicted in blue; MMF for DTS measurements are shown in red. Note that the cable was deployed as a single length in a loop configuration, with no splices between boreholes. Small loops at the end of the DAS fibers illustrate the fiber termination, done to avoid a backreflection caused by the loose end of the fiber. The termination was achieved by wrapping and taping the fiber around a very small-diameter plastic tube. Depth of each borehole is indicated.

fibers, two MMFs and four SMFs, suspended in thixotropic gel and encapsulated in a 2.2 mm OD stainless steel casing. The casing was coated in high-density PE, and the entire package had a manufacturer-reported minimum bend radius of 25 mm. On one hand, this cable is sturdier, more resistant to high pressure and temperature conditions, and better protected against water ingress. On the other hand, the rigidity of the cable made it difficult to work with it in the field. Despite being sturdier, the stainless steel casing greatly reduces the bend radius and causes the cable to be less flexible and prone to kinking, which causes the fibers to break. This feature can become a significant problem in the field when having to manage long lengths of cables in space-limited sites, such as is common in the galleries of underground labs, and should be considered during experiment design. As mentioned earlier, an additional potential drawback of this cable design is the presence of the thixotropic gel surrounding the fibers (i.e., loose tube), which can have an impact on the transfer of strain from the rock to the fiber.

DFOS Interrogators, Acquisition Parameters, and General Setup

The DFOS technologies, corresponding IUs, and acquisition parameters are summarized in Table 30.1. Both experiments included DTS measurements to track temperature anomalies caused by fluid flow and perturbations associated with the temperature response of the stimulated rock mass. DSS and low-frequency DAS were deployed for tracking rock deformation associated with fluid injection, hydraulic fracturing, and fracture stimulation, including strain fronts and fracture hits, and for assessing potential far-field strain perturbations away from the injection location. In Experiment 2, two different

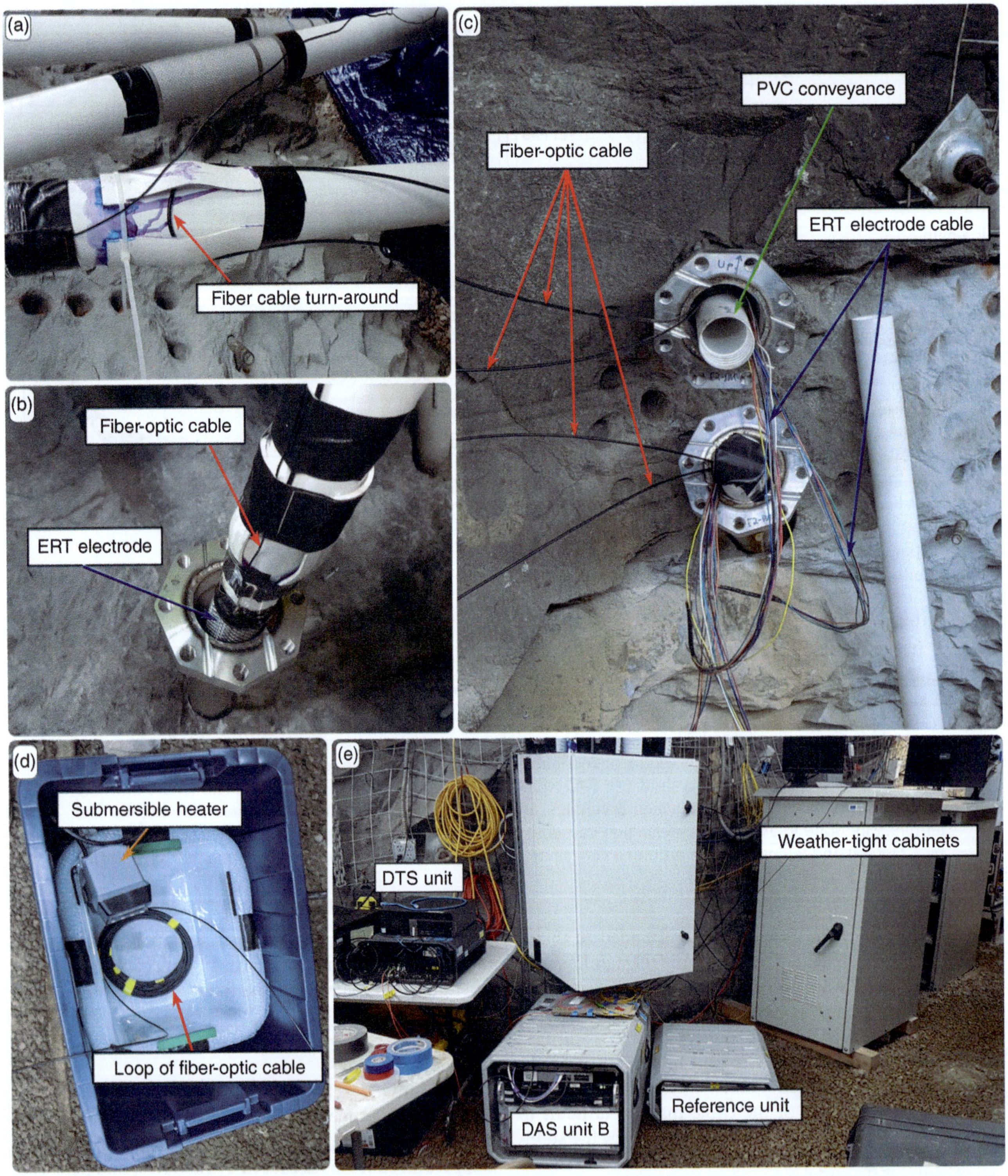

Figure 30.3 System installation in the 4100 level of SURF as part of Experiment 2 of the EGS Collab project. (a) Fiber-optic cable turn-around attached to the bottom-most section of the PVC pipe used to convey monitoring instrumentation down the boreholes. The turn-around was taped to the back of a centralizer to increase the bending radius as much as possible. (b) Conveyance being deployed down the borehole, showing the fiber-optic cables and an ERT electrode taped to the PVC conveyance pipe. (c) Head of two grouted monitoring boreholes (AMU and AML) after installation of the PVC pipe with ERT electrodes and the fiber-optic cable. (d) View of the heated water bath before filling it with water. A loop of fiber was placed inside a small plastic container with a temperature-regulated heater and two temperature probes connected to the DTS IU (not shown here) and subsequently filled with water. (e) View of the alcove where the IUs, data storage servers, and auxiliary equipment were located during the experiment.

Table 30.1 Setup of fiber-optic sensing system and acquisition parameters of each of the DFOS technologies used in Experiments 1 and 2 of the EGS Collab Project.

	Experiment 1			Experiment 2				
General setup								
Cable type	Tight-buffered, PE jacketed, aramid-packed bundle hybrid (4 SMFs and 4 MMFs)			Loose tube with a metal tube filled with thixotropic gel hybrid (4 SMFs and 4 MMFs)				
Installation method	Strapped to PVC casing, grouted in place			Strapped to PVC casing, grouted in place				
Fiber length	1,728 m			960 m				
DFOS acquisition parameters								
	DTS	DSS	DAS	DTS	DSS	DAS IU A	DAS IU B	DAS IU C
Temporal sampling	10 min	~1 min	1 kHz or 10 kHz	10 min	~4 min	500 Hz	200 Hz	36 kHz
Spatial sampling (m)	0.25	0.41	1	0.25	0.51	1	1	2.45
Spatial resolution/gauge length (m)	0.25	1	10	0.25	1	10	7	4.9
Recording period (months)	12	2	2.75	6	4	6	4	3
Total data volume (TB)	0.01	0.0037	23	0.0051	0.031	13.4	2.7	233

DAS IUs with different gauge lengths were deployed for the purpose of low-frequency DAS data analysis, to evaluate the impact of this parameter on the measured strain rate. The units were a Silixa iDAS, with a gauge length of 10 m, also used in Experiment 1, and an OptaSense ODH3, with a gauge length of 7 m (hereafter referred to as Unit A and Unit B, respectively). Unit B was coupled to a strain and temperature reference unit that enabled on-site processing of the data for low frequency using the manufacturer's software.

In both experiments, DAS data were also recorded at high frequency with the aim of capturing microseismicity. However, site characteristics, the size of the experiment, and limitations in acquisition parameters made this challenging. In Experiment 1, Unit A was set to record with alternating sampling frequencies of 1 and 10 kHz at different stages of the experiment, with the higher frequencies being used during significant injection tests when seismicity was expected. However, the complementary arrays of conventional seismic sensors (accelerometers and hydrophones) revealed that the recorded seismicity had most of the energy in the resonance range around 10 kHz (Schoenball et al., 2020), which would not be resolved in the DAS data, which had a maximum Nyquist frequency of 5 kHz. Moreover, an active seismic cross-well survey conducted during site characterization (Schwering et al., 2020) and the analysis of the seismic events revealed a P-wave velocity in the order of 5,900 m/s. These frequencies and seismic velocity ranges indicate a wavelength of these seismic events of less than 1 m, which is significantly smaller than the 10 m gauge length used for DAS acquisition. An attempt at alleviating these issues was made for Experiment 2 by using a third DAS IU, a Terra15 Treble system (hereafter DAS Unit C), a new-generation DAS acquisition system designed specifically for improved response in the low kHz range and short gauge lengths for higher spatial resolution. DAS Unit C recorded seismic data continuously for the duration of the experiment at a frequency of 36 kHz. A gauge length of 4.9 m was chosen, as it provided sufficient spatial resolution without significantly decreasing the signal-to-noise ratio of the data. A total of 233 TB of data were generated by this system alone. At the time of writing, this data has not yet been analyzed.

The DSS interrogator used for strain measurements in both EGS Collab Experiments was an Omnisens DITEST VISION dual temperature and strain unit using the Brillouin Optical Time Domain Analysis (BOTDA) technique. This unit measures Brillouin frequency shift at each point along the fiber, which is linearly related to strain and temperature in the fiber by known strain and temperature coefficients (Horiguchi & Tateda, 1989). The DITEST measures the combined temperature and strain contribution of the Brillouin frequency shift. When independent temperature data along the fiber are available, such as the collocated DTS measurements obtained in these two experiments, these data can be used to correct the DSS measurements for thermally induced Brillouin frequency shifts and isolate the strain signal.

The DTS interrogator used for EGS Collab Experiments 1 and 2 was a Silixa XT-DTS system. Both ends of one of the MMFs were connected to the XT-DTS. This so-called "double-ended" configuration (the fiber is interrogated in both directions) takes longer to make a single measurement than a single-ended setup but results in a more accurate, stable data product.

In Experiment 1, all DFOS interrogators and accessories (monitors, keyboards, etc.) were placed on tables located along the drift and covered with tarps for protection against dust and humidity when not in manual operation. During Experiment 2, a more robust setup was designed. Two of the DAS interrogators and the DSS IU were located inside weather-tight cabinets to minimize the impact of temperature variations. DAS Unit A was also placed on a vibration isolation table to reduce noise. DAS IU B and the accompanying strain and temperature reference unit were suspended inside their own rugged field cases provided by the manufacturer. The DTS unit, which is less sensitive to the environment, was placed on a table. All the instrumentation was placed inside the alcove, where two of the boreholes were drilled (Figure 30.3.(e)).

Different data storage approaches were taken for the two experiments. In Experiment 1, data from all DFOS instruments were stored in 10 TB external hard drives. This strategy was adequate for DTS and DSS, which generated low volumes of data (<1 GB). For DAS datasets, however, manual hard drive swaps had to be conducted every 1–2 weeks. For Experiment 2, three network attached storage (NAS) systems were originally deployed to store the data. Only two swaps were required during the 7-month period of the experiment.

For both experiments, all DFOS and data storage equipment was connected to internet switches that enabled remotely operating the instruments when necessary, evaluating state-of-health and checking the capacity of storage devices for scheduling swaps. Limited bandwidth did not enable streaming full DAS and DSS datasets in real time, but it allowed us to download specific files for quality control and after an event of interest. Because of their smaller size (on the order of 200 KB per file), DTS data were being processed in almost real time by a field computer connected to the DTS unit, and results were posted to a team-shared repository.

Data Analysis and Observations

The datasets acquired in these experiments enable us to analyze the effects that some of the installation and acquisition parameters described earlier have on the characteristics of the data and the final interpretation of the results. In this section, we illustrate these effects by analyzing DAS and DSS data acquired during four sets of hydraulic fracturing and stimulation tests, two for each EGS Collab Experiment. A low (0.4 L/min) and a high (5 L/min) injection flow-rate test is shown for each experiment. In the four tests analyzed here, the fluid injected was ambient-temperature water. For Experiment 1, we will be focusing on strain signals recorded before, during, and after two tests aimed at enlarging a previously created hydraulic fracture. The first dataset (hereafter referred to as Test E1-1), of 3 hr duration, records strain perturbations caused by the hit of a hydraulic fracture into one of the monitoring boreholes (borehole OT in Figure 30.1(a), (b)) during a low flow-rate injection conducted at a depth of 49.98 m (164 ft) in the injection borehole E1-I on 23 May 2018. The second dataset, of 5 hr duration, shows the intersection of a second fracture along the same borehole during a second, higher flow-rate stimulation (Test E1-2, 24 May 2018). For Experiment 2, we analyze 1 day and 5 hr of data recorded along two of the testbed boreholes before, during, and after the first and second stimulation tests, respectively. The first test (hereafter referred to as Test E2-1), conducted on 11–12 April 2022, was performed at a low injection rate, at 58.64–61.05 m (192.4–200.3 ft) in the injection borehole TC. We analyze data recorded along borehole AMU (Figure 30.1(c), (d)), which was intersected by a fracture during the early hours of the experiment. The second test (Test E2-2) was conducted a day later, on 13 April 2022, at higher flow rates at a shallow depth interval of 44.19–46.60 m (145.0–152.9 ft) along the same injection boreholes. Here, we investigate the data recorded along monitoring borehole DML (Figure 30.1(c), (d)), which was intersected by a fracture later in the experiment.

Comparison of DFOS Datasets During Hydraulic Fracturing and Stimulation. Figures 30.4 and 30.5 provide an overview of the DFOS data recorded during the stimulation tests described previously. For all four tests, DTS data are shown for comparison with DAS and DSS strain measurements and as evidence of fracture creation and/or fluid flow into the borehole. Instead of absolute DTS temperatures, we choose to show temperature changes with respect to the start of the measurement period (i.e., with respect to baseline DTS measurements taken before the experiments started) to facilitate data interpretation and highlight the temperature anomalies associated with the fractures. The top panels show injection flow rate and pressure measurements derived from a combination of downhole flowmeters and pressure transducer sensors, and the measurements from the injection pump. Given that our focus in this contribution is on measurements of slow strain rather than seismic signals, the recorded DAS data shown here have been processed to extract the low-frequency component of the signal, which should be mostly composed of the gradual strain perturbations imposed on the rock mass by fluid injection. As mentioned earlier, DAS data were recorded using two different IUs. For the data recorded using IU A (Figure 30.4(c), (d), (h), (i) and Figure 30.5(c), (d), (h), (i)), processing was conducted after the data were recovered from the field. A workflow similar to that described in Jin and Roy (2017) was applied to the raw strain-rate data. This workflow includes (a) decimation of each 30-s-long file from its

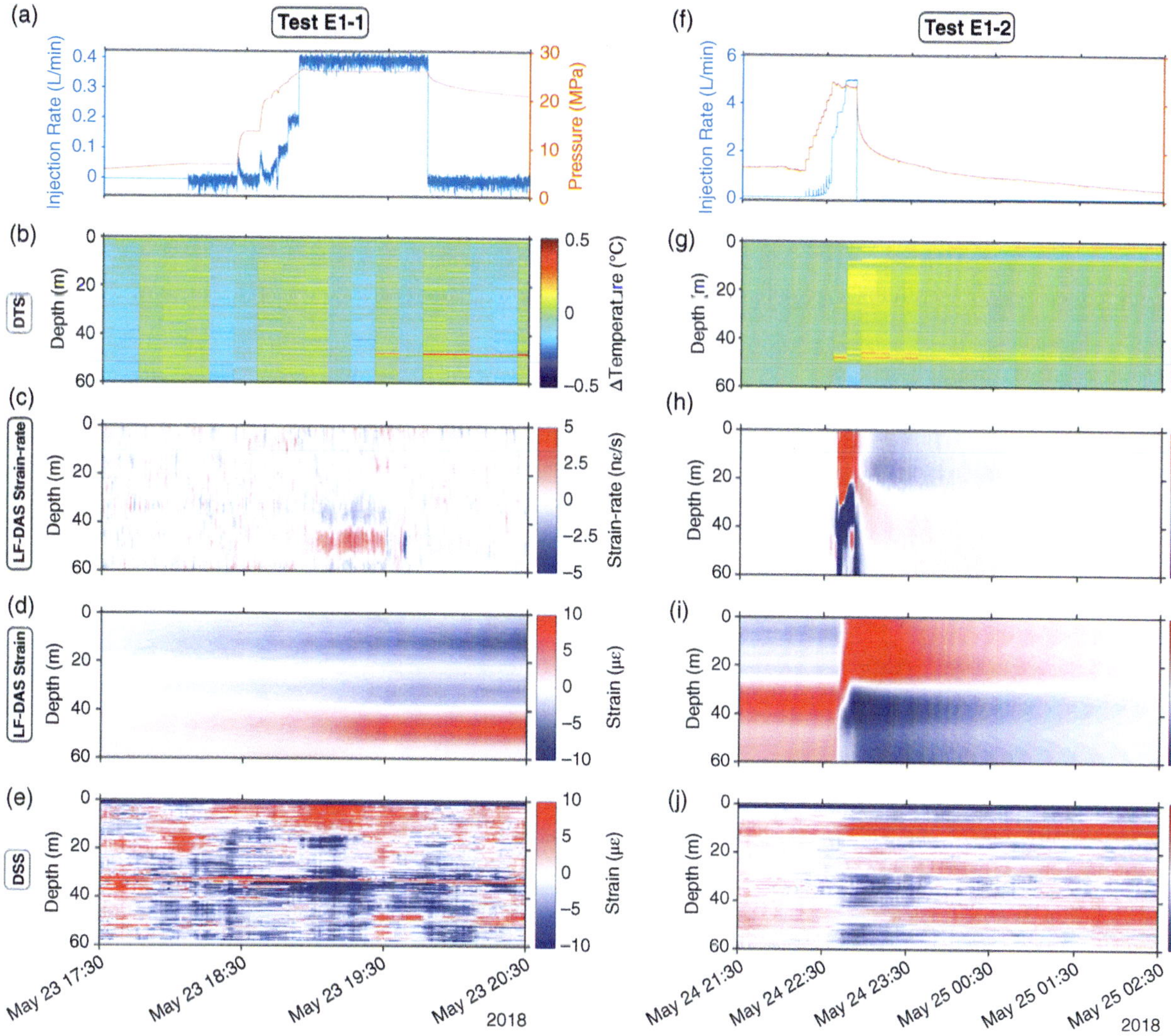

Figure 30.4 Overview of DFOS data recorded along monitoring borehole E1-OT during two tests of EGS Collab Experiment 1. (a, f) Injection rate and pressure. (b, g) Temperature data recorded using DTS. Shown is differential temperature with respect to the start of the measurement period. (c, h) DAS strain-rate data recorded using Unit A after being processed to extract the low-frequency component of the data. (d, i) Low-frequency DAS data shown in panels (c) and (h), respectively, integrated to strain. (e, j) Strain measurements derived from DSS.

original sampling frequency to 1 Hz; (b) concatenation into a file spanning the duration of the test of interest; (c) despiking to remove optical noise by applying a channel-wise median filter; (d) for each borehole, calculating the median of a set of channels outside the area of interest and conduct a channel-wise subtraction for the entire borehole, to remove residual common-mode noise caused by low-frequency vibrations and temperature changes affecting the IU which were not entirely mitigated. This workflow results in the low-frequency strain-rate component of the data (Figure 30.4(c), (h) and Figure 30.5(c), (h)). Two additional steps are conducted to obtain strain: (e) channel-wise detrending to remove the effect of laser drift, and (f) channel-wise temporal integration to obtain strain from the start of the recording (Figure 30.4(d), (i) and Figure 30.5(d), (i)). As introduced in the previous section, the DAS data recorded using IU B in Experiment 2 were processed for low-frequency on-site, providing one day-long record of low-frequency strain downsampled to 1 Hz. Minimal post-processing required channel-wise detrending and median subtraction to alleviate common-mode noise. For the DSS data, post-processing involved correcting the Brillouin frequency shift measurements for temperature effect using the collocated DTS data and temporal averaging in a 20 min moving window to reduce noise.

The data comparisons in Figures 30.4 and 30.5 highlight the potential and complementarity of the different

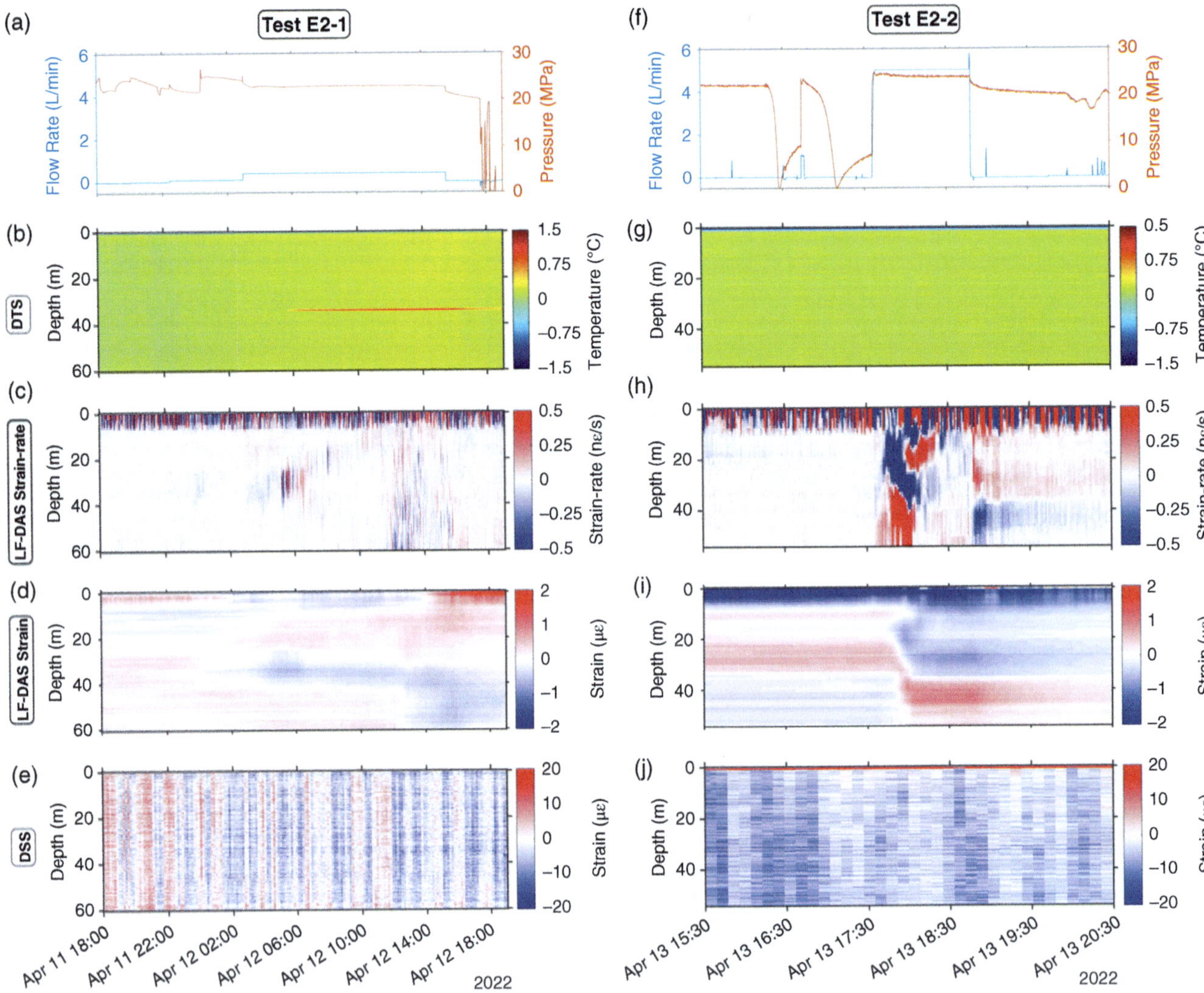

Figure 30.5 Overview of DFOS data recorded along monitoring boreholes AMU and DML during two tests of EGS Collab Experiment 2. (a) Injection rate and pressure during Test E2-1. (b) Temperature data recorded using DTS along borehole AMU. Shown is differential temperature with respect to the start of the measurement period. (c) DAS strain-rate data recorded along borehole AMU using Unit A (10 m gauge length) after being processed to extract the low-frequency component of the data. (d) Low-frequency DAS data shown in panel (c) integrated to strain. (e) Strain measurements derived from DSS data recorded along borehole AMU. (f) Injection rate and pressure during Test E2-2. (g) DTS differential temperature measurements along borehole DML. (h) Low-frequency DAS strain-rate data along borehole DML. (i) Low-frequency DAS data shown in panel (h) integrated to strain. (j) DSS strain recorded along borehole DML.

DFOS techniques for monitoring the response of the rock mass to fluid injection. Although an in-depth interpretation of these signals in the context of EGS operations is beyond the scope of this chapter and will be the subject of future publications, we want to note the most significant events captured by these data and how their signatures differ for the different technologies. In Tests E1-1, E1-2, and E2-1, conspicuous temperature anomalies in the form of sudden increases in temperature of about 0.5 °C to 2 °C are observed in the monitoring boreholes during injection. During Experiment 1, an anomaly appears at a depth of 47 m along borehole E1-OT during Test E1-1, and a second anomaly appears at a shallower depth of 44 m during Test E1-2 (Figure 30.4(b), (g)). In Experiment 2, a small (~1 °C) temperature anomaly develops at a depth of 34 m in borehole AMU during Test E2-1 (Figure 30.5(b)). These anomalies are caused by the intersection of pressurized fractures with the wells, later corroborated with downhole camera footage. The temperature increase is thought to be recording the Joule

Thomson effect, in which pressurized water flowing through the fracture rapidly loses pressure when entering the borehole with a concomitant increase in temperature (Zhang et al., 2018). Because of the high spatial resolution provided by DTS measurements, the depth localization of this anomaly is extremely accurate. In all three tests, these DTS anomalies agree with contemporaneous positive strain rate and strain signals at similar depths, interpreted as extensions due to fracture opening (Figure 30.4(c), (d), (h), (i) and Figure 30.5(c), (d)). Significantly, strain anomalies in Tests E1-1 and E2-1 are only visible in the low-frequency DAS recordings but not in the DSS data, which is comparably noisier (Figures 30.4(e) and 30.5(e)). Especially in Test E2-1, the DSS data are dominated by common-mode noise effects that most likely correspond to thermal strain signals that could not be removed by the temperature correction process (Figure 30.5(e)). This observation illustrates the lower strain sensitivity of DSS measurements (typically in the order of tens of $\mu\varepsilon$) in comparison to low-frequency DAS (typically $< n\varepsilon$). The contamination by common-mode thermal noise is stronger in Experiment 2 than in Experiment 1, despite the fact that the IU was in a weather-tight cabinet during the second experiment. We think that the reason for this difference is larger variations in temperature occurring at the location of Experiment 2 due to the opening and closing of an air door located a few hundred meters away from the experiment location, which was either blocking or allowing external airflow into the gallery. Another interesting observation is that in Test E1-1, the strain recorded by DAS develops about 30 min before the DTS anomaly appears (Figure 30.4(b)–(d)), which suggests that DAS is able to capture rock deformation as the fracture is approaching and before it hits the borehole (i.e., the strain front).

Given the small magnitude of the strain anomaly recorded in Tests E1-1 and E2-1 (in the order of 1 $\mu\varepsilon$), the question arises whether this signal is due to a change in temperature or to rock deformation. Decoupling mechanical and thermal strain measurements in DAS is a challenging task. In experiments with collocated DAS and DTS recordings, DTS measurements may be used to distinguish temperature-induced strain from mechanical strain signals recorded on low-frequency DAS (Miller et al., 2018). However, in the tests analyzed here, both temperature and strain perturbations are expected as a result of fracture opening and fluid flow into the borehole, making it challenging to quantify how much of the strain is thermally induced and how much of it is deformation-related. Theoretically, a temperature-strain relationship should be possible to be established using the loops of fiber inside the water baths described earlier. In practice, however, we found that the unstable temperature of the water baths, due to effects such as thermostat cycling and evaporation of the water causing the thermostat to not be fully submersed, prevented a clear relationship from being determined. Previous work attempting to investigate temperature effects on low-frequency DAS measurements suggests linear temperature-strain relationships ranging from 3 to 10 $\mu\varepsilon/°C$, depending on cable type and installation conditions (Sidenko et al., 2022). Based on these findings, we conclude that it is plausible for these small strain signals to be thermally induced, although no clear conclusion can be drawn.

Significant strain anomalies (30–40 $\mu\varepsilon$) are recorded by both low-frequency DAS and DSS during the high-rate stimulation in Test E1-2 at the depth of the DTS anomalies (Figure 30.4(j)). Here, the differences in spatial resolution between low-frequency DAS and DSS and the effect of the larger DAS gauge length become apparent. The DSS recording shows three well-defined bands of extension centered at ~10, ~27, and ~45 m. The deeper feature agrees with the location of the two hydraulic fractures revealed by DTS. However, the two individual fractures at 44 and 47 m cannot be distinguished, most likely due to the 1 m gauge length. The shallower positive strain bands are interpreted as extension caused by the reopening of natural fractures that intersect the borehole at those depths and were mapped during characterization of the borehole using televiewer logs (Schwering et al., 2020). The shallower zones of deformation seem to correlate with broad temperature anomalies revealed by the differential DTS data Figure 30.4(g). Despite some smearing due to the 1 m gauge length, which is larger than the fracture opening (a few mm), the three localized bands of deformation are better defined in the DSS recording than in the low-frequency DAS recording, in which the three positive strain anomalies are more spatially smeared and their location at depth is more uncertain (Figure 30.4(h)–(j)). Here, the shallower extensional areas are less distinguishable, and the extensional signals expected at the location of the newly created hydraulic fractures overlap with the surrounding compressional anomalies associated with the corresponding stress shadows to the point that a positive anomaly is barely visible in the integrated strain signal at those depths (Figure 30.4(i)). The larger magnitude of the strain anomalies of Test E1-2 with respect to those of Test E1-1 suggests that, whereas a portion of the signal (a few $\mu\varepsilon$) could be thermally induced, most of the strain must be mechanical, thus recording rock mass deformation. This is further supported by the agreement between DAS and DSS signals because DSS has been corrected for temperature effects.

Lastly, the analysis of the Test E2-2 dataset reveals an additional important aspect of the use of DFOS for hydrological and geomechanical characterization of fluid injection experiments. During this test, no temperature anomaly is recorded by the DTS, which suggests that there was no fluid flow into the monitoring borehole and, thus, no fracture intersection (Figure 30.5(g)). However, low-frequency DAS data record two clear positive strain anomalies centered around 15 and 45 m depth a few minutes after injection started (Figure 30.5(h), (i)), which indicates extensional deformation, possibly due to the opening of a fracture. As shown in Kneafsey et al. (2023), the depths of these two anomalies agree very well with DTS anomalies developed during later stimulation tests conducted within the framework of Experiment 2. This suggests that, during the stimulation in Test E2-2, fractures were indeed created (or natural fractures were opened), but they did not conduct any flow at that time. This important observation indicates the potential of low-frequency DAS in monitoring stimulation operations during the creation of an EGS and highlights the importance of multiparametric datasets for the comprehensive characterization of rock mass behavior. These findings could have important implications for EGS development and will be the subject of future publications.

Impact of Gauge Length on Low-Frequency DAS Spatial Resolution. Figure 30.6. shows low-frequency DAS data acquired during Tests E2-1 and E2-2 with DAS Units A and B. As mentioned earlier, Unit A recorded data with a gauge length of 10 m, whereas Unit B used a gauge length of 7 m. Despite a small difference of only 3 m, the impact of the gauge length on the ability to recover short-scale strain variations is evident. Although both units are able to record extensional strain caused by fracture opening, Figure 30.6.(c), (g) reveals that features are better localized and spatially constrained in the 7-m-long gauge length data. In Test E2-1, a well-defined positive strain signal surrounded by compressional zones emerges in the 7 m gauge length recording just before 6 am on 12 April at a depth of 34 m, agreeing with the location of the DTS anomaly (Figure 30.5(b), (c)). In the 10-m-long gauge length dataset, this feature appears as a spatially broad anomaly, and the location of the fracture is less well defined Figure 30.6.(b). A similar comparison can be made for data recorded during Test E2-2. Whereas a well-defined extensional anomaly is observed at 13–16 m depth in the waterfall plot in Figure 30.6.(g) (7 m gauge length), only a broad, faintly positive anomaly is observed in Figure 30.6.(f) (10 m gauge length). These differences become clearer when strain time series are extracted for a single channel located at fracture depths known from DTS data (Figure 30.6.(d), (h)). In general, strain amplitude is lower for the 10 m gauge length data as a result of averaging over a larger area. Despite the better spatial definition of the 7 m gauge length data, it is important to recognize that spatial resolution is still limited, given that 7 m is significantly larger than the width of the targeted fractures (typically only a few millimeters wide). These limitations, coupled with the inevitable uncertainties in the exact channel positioning, introduce some uncertainty in the exact depth location of the observed features, which needs to be carefully considered.

Additional impact of the gauge length parameter in the strain measured using low-frequency DAS can be evaluated when comparing the magnitude of strain anomalies recorded using DAS against strain recorded using DSS. Figure 30.7 shows strain recorded during Tests E1-1 and E1-2 with DAS IU A, with a 10 m gauge length and a channel spacing of 1 m, and with DSS, with a spacing of 0.5 m and a spatial resolution of 1 m, at the location of fracture hits along borehole E1-OT. DSS strain values are relative to the value recorded at the start of DAS recording to enable a direct comparison with cumulative DAS strain, which is an integration of strain rate from time 0. To conduct a "fairer" comparison, we include DSS data after applying a 10 m moving averaging window to resemble the averaging effect of the DAS gauge length. Although DSS is not able to resolve the small strain incurred during the fracture hit in Test E1-1, a similar pattern in the temporal evolution of strain is observed between the two datasets during Test E1-2, and strain values are in the same order of magnitude. A strain perturbation was recorded by DAS on 23 May, just before 11 pm, which seems to correlate with a very small pressure perturbation. For DSS, the signal is almost buried within the noise, but it seems to show a small strain increase. An interpretation of these signals and the apparent difference in polarity would need further investigation, which is beyond the scope of this chapter. We also want to note that some uncertainties exist on the exact geometry mapping between the two DFOS measurements.

Another observation associated with the gauge length spatial averaging effect is that for the low-frequency DAS data, the strain time series of the two individual fractures opening during Test E1-2 have a very similar magnitude (black lines in Figure 30.7). This effect is expected because the fractures are only 3 m apart, a distance much smaller than the span of the gauge length (10 m). The separation is larger for the DSS strain time series displayed in their original spatial resolution (in blue in Figure 30.7), but magnitudes become closer after applying the 10 m averaging filter (red lines).

Although the averaging effect of the gauge length is expected to dampen strain magnitudes for DAS recordings, Figure 30.7(d) shows a larger negative strain recorded by DAS during Test E1-2 just before fracture hits. Both low-frequency DAS and DSS capture compressional strain before extension at the time of fracture

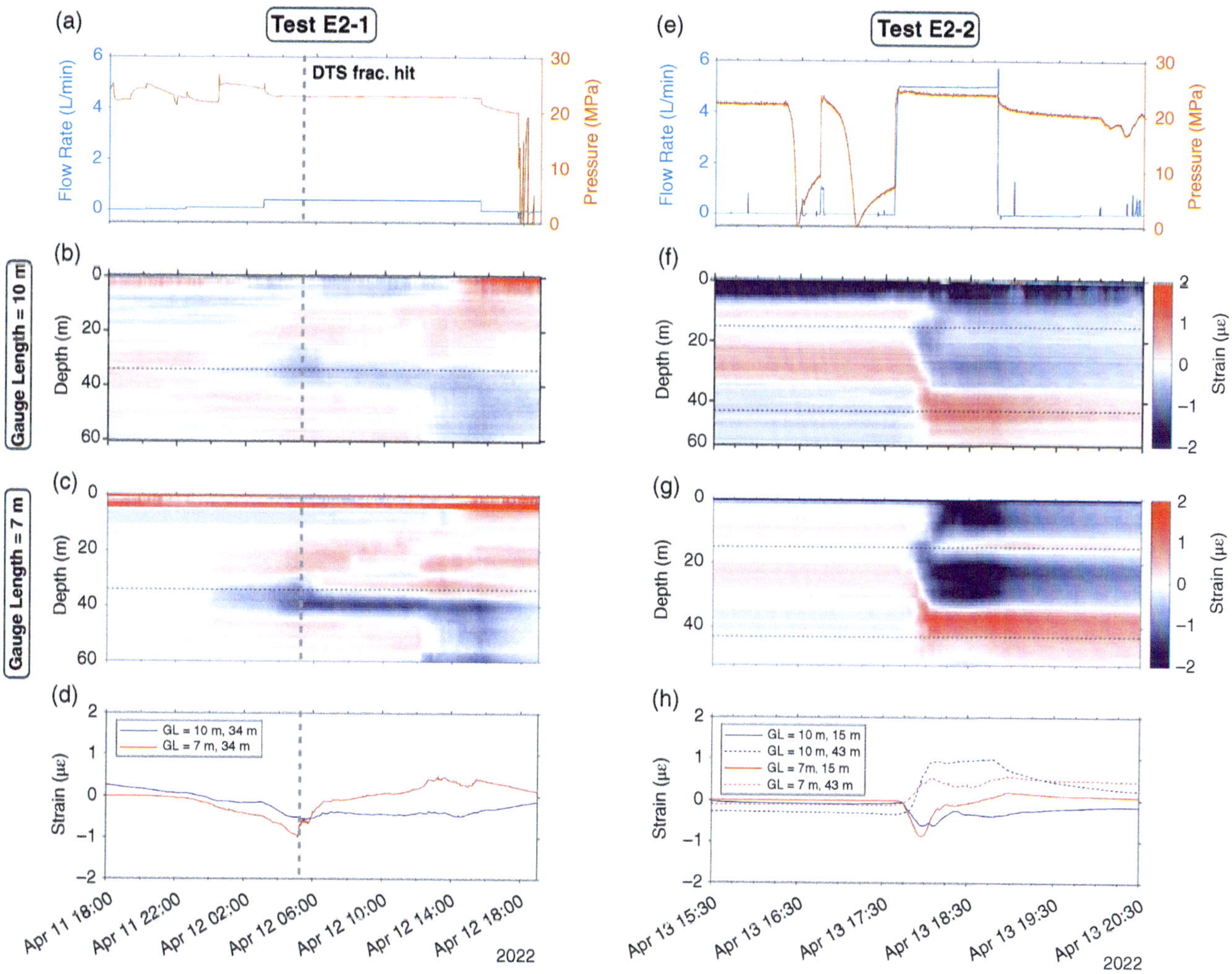

Figure 30.6 Effect of the gauge length on low-frequency DAS recordings during Tests E2-1 and E2-2 of EGS Collab Experiment 2. (a) Injection rate and pressure during Test E2-1. Vertical gray dashed line indicates time of fracture hit as interpreted by DTS data. (b) Low-frequency DAS strain obtained from temporal integration of low-frequency DAS strain-rate measurements recorded with DAS Unit A with a gauge length of 10 m along borehole AMU. Horizontal black dashed line indicates the depth of a fracture hit identified from DTS data (Figure 30.5(b)). (c) Integrated strain recorded with DAS Unit B with a gauge length of 7 m along borehole AMU. (d) Comparison of strain recorded at the depth of the fracture hit in AMU with the two DAS Units. (e) Injection rate and pressure during Test E2-2. (f) As for (b) for integrated strain data recorded during Test E2-2 along borehole DML. Horizontal dashed lines indicate depths of fracture hits identified in DTS data in a later test in May 2023 (Kneafsey et al., 2023, mentioned but not discussed in detail here). (g) As for (c) recorded during Test E2-2 along borehole DML. (h) Comparison of strain recorded with the two DAS IUs at the depths of the two inferred fractures in DML.

opening, which is interpreted as the signature of the approaching pressure front as the fracture propagates toward the borehole. In this case, we hypothesize that the gauge length effect has been overprinted by the ability of DAS to record dynamic signals at much higher resolution. This DSS dataset has an effective temporal sampling of 20 min due to post-processing data averaging necessary to reduce noise. DAS, with a temporal sampling of 1 s, is able to capture the sudden, high-magnitude strain perturbation caused by the approaching fracture, resulting in a large negative strain change in the order of ~80 με. DSS can only record part of the signal, resulting in a much smaller strain perturbation in the order of ~20 με.

These observations demonstrate how the interplay between the different acquisition parameters can play an important role in the strain recorded by both low-frequency DAS and DSS. These impacts need to be considered when designing DFOS experiments for geomechanical applications, as well as during data interpretation, especially when quantification is important.

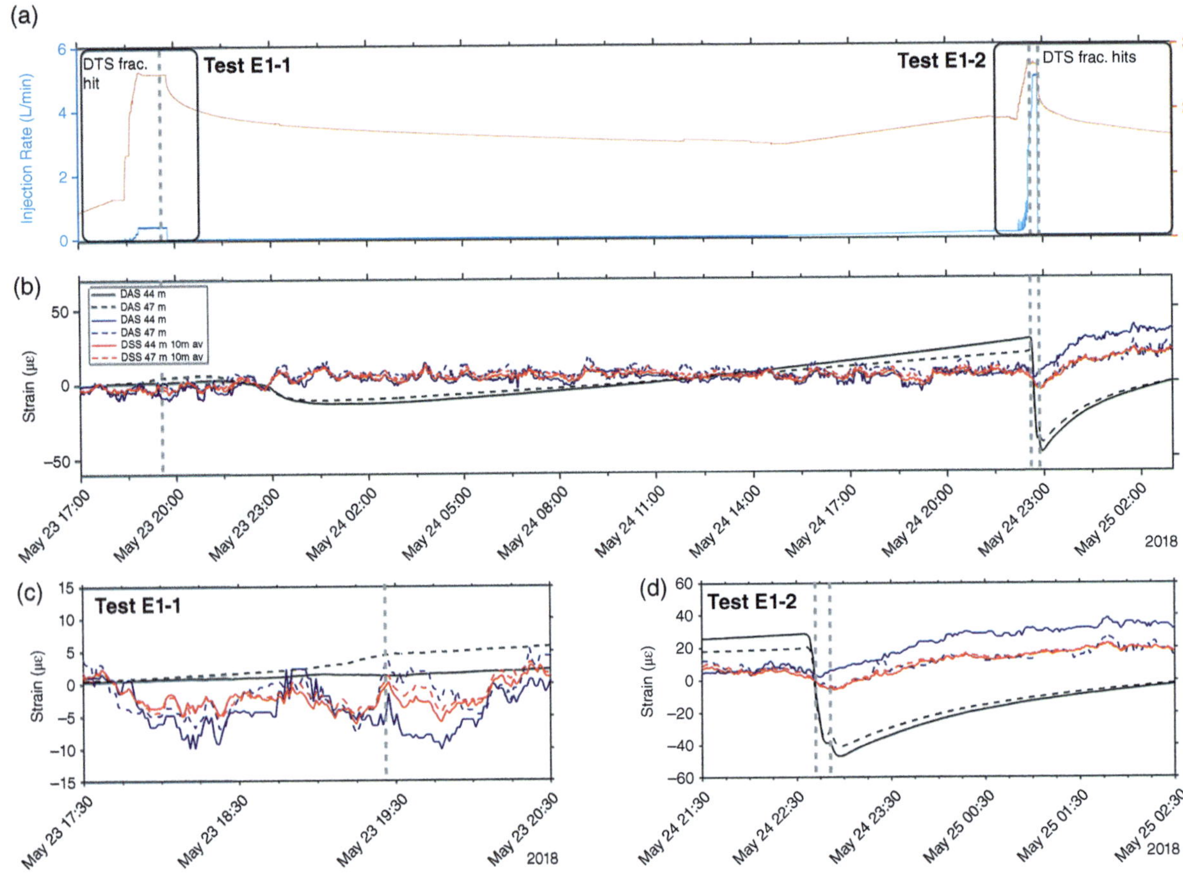

Figure 30.7 Comparison of strain along borehole E1-OT recorded with DSS and derived from temporal integration of low-frequency DAS strain-rate data for Experiment 1 of the EGS Collab project. (a) Injection rate and pressure recorded during the period containing Tests E1-1 and E1-2. Vertical gray dashed lines indicate time of fracture hits as interpreted from DTS data. (b) Strain at 44 and 47 m depths, the location of the two fracture hits identified in DTS data (Figure 30.4(b), (g)). Note that the fracture at 44 m depth only develops on 24 May at 22:40 UTC during Test E1-2. DAS strain (black) is compared to DSS data both at original spatial resolution (blue) and after applying a 10 m spatial averaging moving window to resemble the effect of DAS gauge length (red). (c) Detail of strain measurements during Test E1-1. Note that the DAS-derived strain recorded at 47 m depth increases just before 19:30 UTC, the time at which the DTS anomaly was observed at this depth. Strain recorded at 44 m slightly increases despite the fact that no fracture is observed at this depth during this test. No discernible signal is recorded by DSS. (d) Detail of strain recorded during Test E1-2, when the shallower (44 m) fracture develops and the deeper (47 m) fracture reopens.

In situ testing and calibration are desired to unravel the impact of the different sensing system elements when possible.

Impact of Cable Construction on DAS and DSS Strain Measurements. Decoupling the impact of different installation and acquisition parameters on strain measurements is challenging. That is especially true for deployments in grouted boreholes, where directly accessing the cable is not possible. Nevertheless, the comparison of DAS and DSS data recorded in Experiments 1 and 2, with different cable designs, can provide some indication of the impact of cable construction on strain magnitudes.

Figure 30.8 shows low-frequency DAS data recorded during all four tests of the EGS Collab project analyzed here. The data were acquired using the same IU (Unit A, 10 m gauge length). Despite the fact that these two experiments were conducted in two different testbeds, the rock type, experiment conditions (such as injection rates and pressures), and cable installation were very similar, which enables the comparison of the two datasets. Figure 30.8(a), (b), (e), (f) shows data recorded along the tight-buffered cable in Experiment 1, whereas the data in Figure 30.8(c), (d), (g), (h) were recorded along the FIMT gel-filled cable in Experiment 2. The comparison of strain data recorded during Tests E1-1 and E2-1,

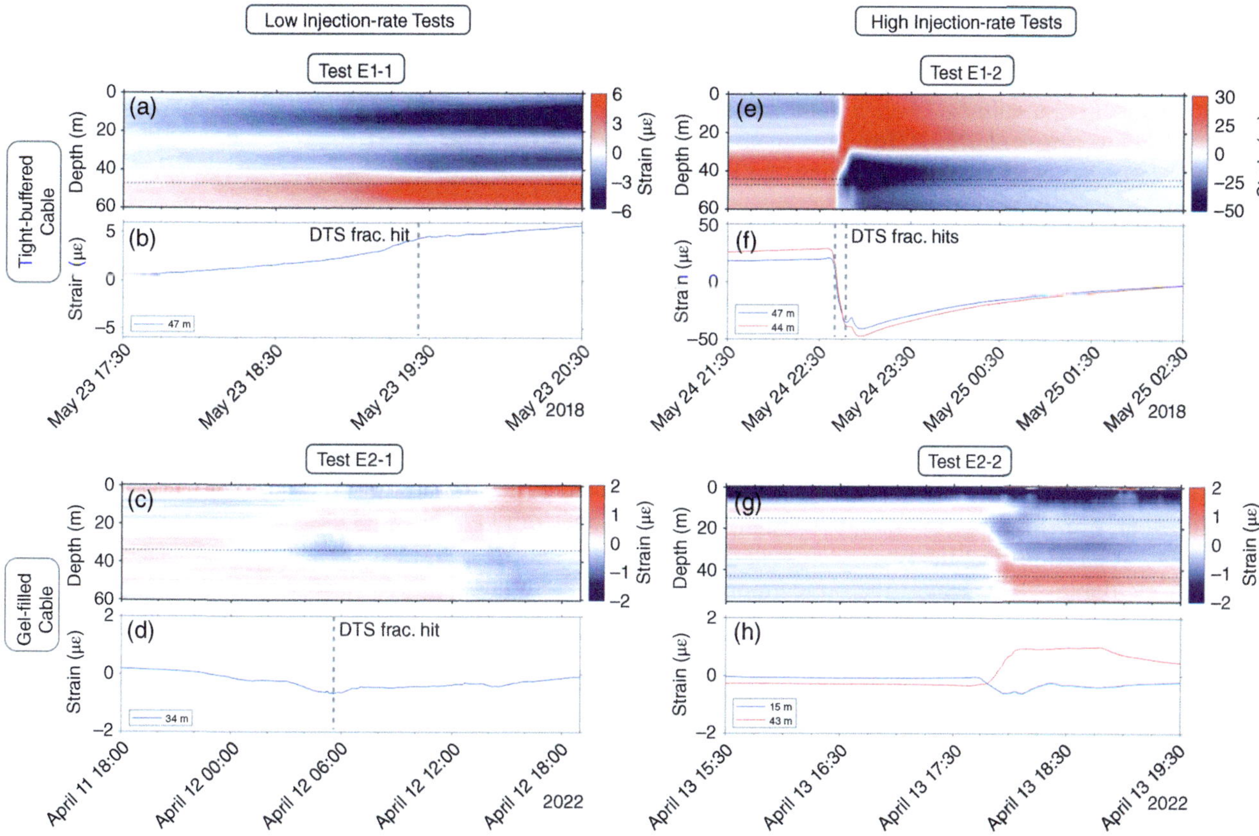

Figure 30.8 Cable construction effect on low-frequency DAS strain measurements in the EGS Collab project for different injection rate conditions. Data shown here are low-frequency DAS strain recorded using DAS Unit A (10 m gauge length). Experiment 1 (Tests E1-1 and E1-2) used a tight-buffered fiber-optic cable, whereas Experiment 2 (Tests E2-1 and E2-2) used a FIMT, gel-filled cable. (a) Low-frequency strain recorded along borehole E1-OT during Test E1-1 (injection rate = 0.4 L/min). Dashed line indicates depth of fracture hit identified from DTS data. (b) Strain evolution at the depth of the fracture hit during Test E1-1. Vertical gray dashed line indicates time of fracture hit as identified from DTS data. (c) Low-frequency DAS strain recorded along borehole AMU during test E2-1 at the same low injection rate. Dashed line shows depth of fracture hit determined by DTS. (d) Strain evolution at the depth of the fracture hit. Vertical gray dashed line indicates time of fracture hit as identified from DTS data. (e) As for (a) for test E1-2, conducted at high flow rate (flow rate = 5 L/min). (f) Strain at the depth of the earlier fracture hit (at 47 m), which was stimulated during Test E1-2, and a new, shallower fracture (at 44 m) created during Test E1-2. Vertical gray dashed line indicates time of fracture hits as identified from DTS data. (g) Low-frequency DAS strain recorded along borehole DML during Test E2-2 at the same high injection rate. Dashed lines indicate the depths of fracture identified in DTS data a month later during another testing phase of Experiment 2, not discussed here (Kneafsey et al., 2023). (h) Temporal evolution of strain at the depth of the two fractures. Note that different strain scales are used for the different tests due to significantly different ranges in strain values between Experiment 1 and Experiment 2, especially during the high injection rate tests (e–h).

conducted at low injection rates of 0.4 L/min, reveals a small difference in the magnitude of the strain recorded during fracture hits, which is slightly lower for Experiment 2 (between −0.2 and 0.6 με; Figure 30.8(c), (d)) than for Experiment 1 (0.5–5.6 με; Figure 30.8(a), (b)). This difference is larger for Tests E1-2 and E2-2, during which water was injected at higher rates of 5 L/min. Whereas values in the order of −46 to +30 με are recorded along borehole E1-OT during Test E1-2 (Figure 30.8(e), (f)), the strain recorded along borehole DML during the high injection-rate Test E2-2 is only slightly larger than that recorded during the low injection-rate Test E2-1, with values of −0.9 to +0.95 με (Figure 30.8(g), (h)). Given that no DTS signal was observed at this time at those depths, one could argue that there was a limited opening of the fractures and thus lower strain than in Test E1-2.

However, we want to note that during the later stages of Experiment 2, in which further fractures are created, and DTS signals emerge at those depths, the recorded strain remains small at magnitudes below 2 $\mu\varepsilon$ (Kneafsey et al., 2023; Schwering et al., 2023). In addition, no visible strain was recorded by the DSS unit during the Experiment 2 tests (Figure 30.5(e), (j)), whereas significant strain was recorded during high flow-rate injections in Test E1-2 (Figures 30.4(j) and Figure 30.7(d)).

Given the very similar conditions of both experiments, these observations suggest that differences in strain magnitude observed between the two experiments (and in particular between Tests E1-2 and E2-2) could be an effect of cable construction. Under conditions of high mechanical stress, the strain response of the Experiment 2 cable is poorer, most likely due to the thixotropic gel in which the fibers are "floating" in this cable design, which is commonly used to protect the fiber against damaging strains during deployment. Under the same amounts of mechanical stress, gel-filled cables are expected to be more resistant to tensile strain than tight-buffered cables, as the fiber is more isolated from external forces. Thus, it is not surprising that the gel-filled FIMT used in Experiment 2 records smaller magnitudes of axial strain inflected on the cable by fracture opening. These observations agree with previous laboratory studies investigating cable construction effect on low-frequency DAS measurements (Becker et al., 2018). Whereas these differences might not be important in seismic applications where strain is larger and more distributed along the cable, cable design seems to have a significant impact on measurements of localized strain rates such as that induced by fractures opening (Becker et al., 2018). Although these observations suggest that gel-filled FIMT might not be adequate for quantitative strain (or strain-rate) sensing, it is important to note that these protective structures are often critical to guarantee the integrity of the installation in harsh environments where fiber breakage is a concern. Hence, an evaluation of the experiment conditions needs to be conducted before choosing a particular fiber-optic cable construction, and strain measurements should always be calibrated before cable deployment when a quantitative analysis is sought. One possible way of calibrating and quantifying mechanical strain before a field experiment is to conduct controlled experiments in a laboratory or test site environment that enables inflicting a known amount of strain on the fiber-optic cable, using step motors or mechanical actuators, recreating the field deployment conditions as faithfully as possible (Becker et al., 2018). Although decoupling the effects of each installation element on the magnitude of strain being recorded by DAS or DSS would still remain challenging, such experiments would enable retrieving an inflicted versus observed strain relationship that could help quantify the data recorded in the field and interpret whether the observed signals are due to mechanical strain affecting the rock or possibly other installation or site-specific factors.

30.3.2. Mont Terri

MTRL (Bossart et al., 2017) in St Ursanne, Canton of Jura, Switzerland, is a research lab dedicated to studying the hydraulic, mechanical, and chemical behavior of a clay formation (Opalinus Clay), which is considered an ideal rock for nuclear waste disposal sites as well as a good analog of common caprock formation in CO_2 storage complexes. The MTRL hosts a variety of experiments, including the Fault Slip – Series B (FS-B) and the Carbon Sequestration – Series D (CS-D) experiments detailed here. The FS-B and CS-D projects consist of complementary experiments studying the effect of fluid migration into a clay-hosted fault zone, the so-called Mont Terri Main Fault (Zappone et al., 2020; Figure 30.9). The CS-D project focuses on long-term, very low flow-rate (<<ml/min) injections of CO_2 brine with the primary goal of studying only leakage flow and to reactivate individual micro-fractures in the Main Fault. The FS-B experiment uses higher flow (5+ L/min) pure water injections to induce fault rupture and investigate the hydro-mechanical behavior of the fault at a scale of tens of meters, including the potential for induced seismicity.

The CS-D and FS-B experiments use a broad array of instruments installed in more than 20 boreholes drilled through or near the Main Fault to measure rock response to the various injection operations. These systems include seismic sources, passive seismic monitoring sensors, in situ displacement sensing probes, electrical resistivity equipment, and various methods of fluid collection and characterization. In addition, both experiments were instrumented with distributed fiber-optics sensing equipment. As in the EGS Collab testbeds, fiber-optic cables were installed as continuous loops, into and out of each borehole, with splices added only for the addition of new boreholes over time. However, between the two experiments, there are a number of differences in the type of cables deployed and their method of installation. We describe these differences and discuss their implications for experiment design and data analysis below. In particular, we focus on analyzing the effects of cable construction and deployment strategies on Brillouin-based DSS data quality.

Cable Types and Installation

The CS-D experiment selected a single-fiber, tight-buffered strain-sensing optic package for their DFOS installation (BRUsens 3.2 mm strain sensing cables).

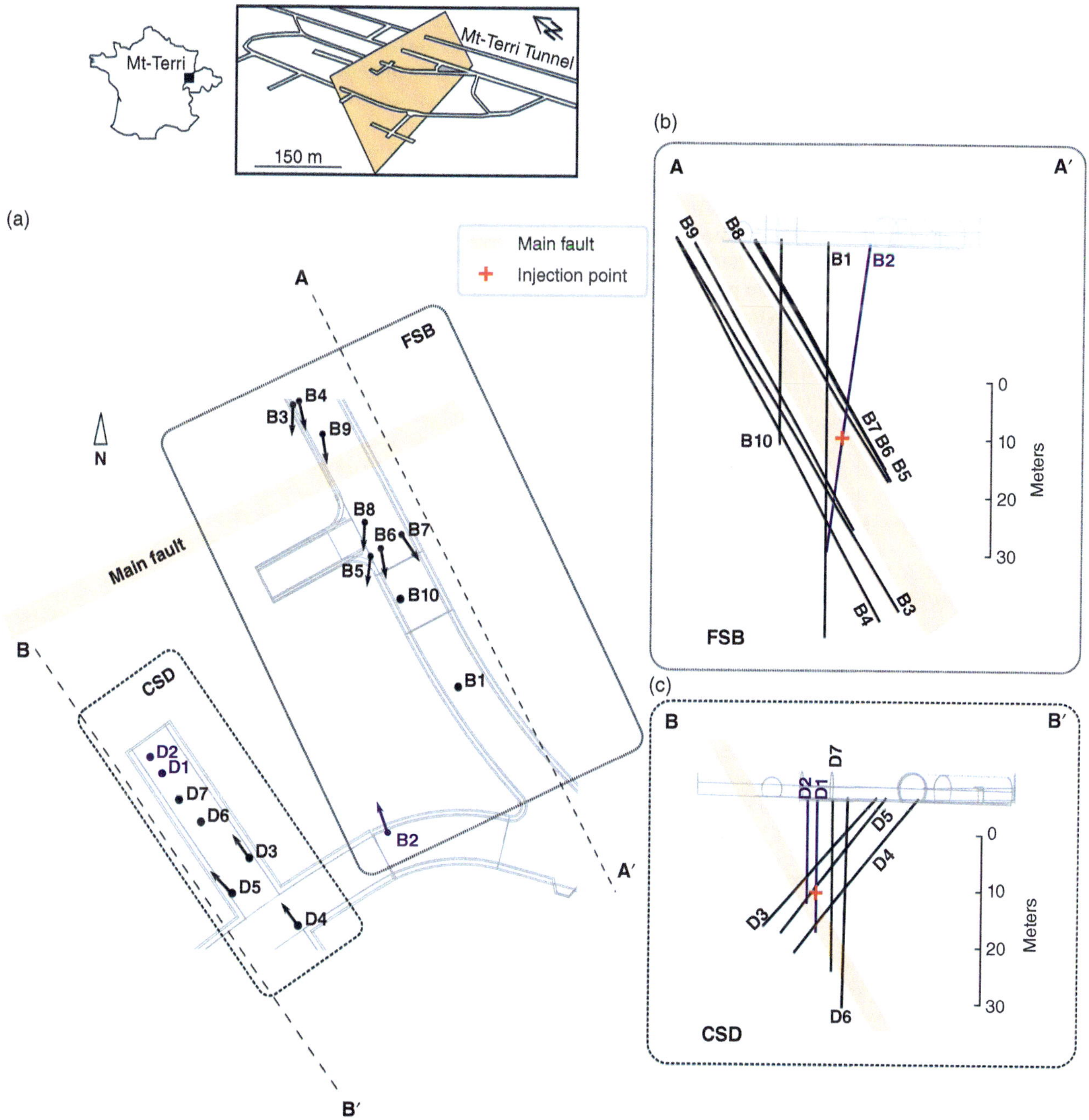

Figure 30.9 Layout of the FS-B and CS-D experiments. (a) Map view of the two experiments, with inset in the upper left showing the distribution of the laboratory's galleries. The approximate orientation of the Main Fault is shown in tan with the general location of the injection point for either experiment indicated by a red cross. Blue and black lines are injection and monitoring boreholes, respectively. (b, c) Cross-sections oriented along the FS-B and CS-D galleries, respectively, with north to the left.

The single fiber is armored with a metal tube, which is coated by different versions of a nylon sheath. This design ensured that any strain transferred to the cable would also be transferred to the fiber itself and result in high-fidelity strain measurements. The downside of this approach is that the system can only be interrogated by a single unit (and therefore technique) at a time.

With the exception of D7, all CS-D boreholes are instrumented with fiber (Figure 30.9). Boreholes D1 and D2 were uncased and contained multilevel straddle packer assemblies that allowed for injection and production of fluid at multiple depths. In these holes, the fiber-optic cable passed through the packers and was anchored to each by a compression ferrule. The cable in these wells was coated with a smooth nylon outer layer. The other boreholes (D3–D6) contained the same fiber but with a textured nylon coating designed to improve coupling to grout. Each of these boreholes was cased with PVC and grouted in place, with the fiber cable on the outside of the casing (Hopp et al., 2022; Figure 30.10(b)). The fiber turn-around was placed inside a protective plastic pod attached to the bottom-most PVC pipe section (Figure 30.10(a)). The grout (a mixture of water, bentonite, and cement) was injected while the fiber was held under tension (Figure 30.10(b)). This technique is often referred to as pretensioning and aims to improve the ability of the system to sense negative strain (effectively contraction). Between boreholes, tactical patch cables were placed loosely into cable trays running along the gallery walls.

At FS-B, all boreholes were instrumented with fiber-optic cable that was grouted behind PVC casing (Figure 30.10(c)). This includes the injection borehole, B2 (Figure 30.9), where the injection probe was placed at slotted intervals inside the liner during the various experiments. The principal differences in the installation from CS-D are (a) that the cables were not pretensioned during grouting and (b) the cables differed in their construction. For FS-B, two cable loops were installed, one for SMFs and one for MMFs. Both cables were multifiber, tight-buffered, aramid-packed packages from Corning. The Corning Ultra SMF28e+ package contained six single-mode fibers, whereas the Corning InfiniCor

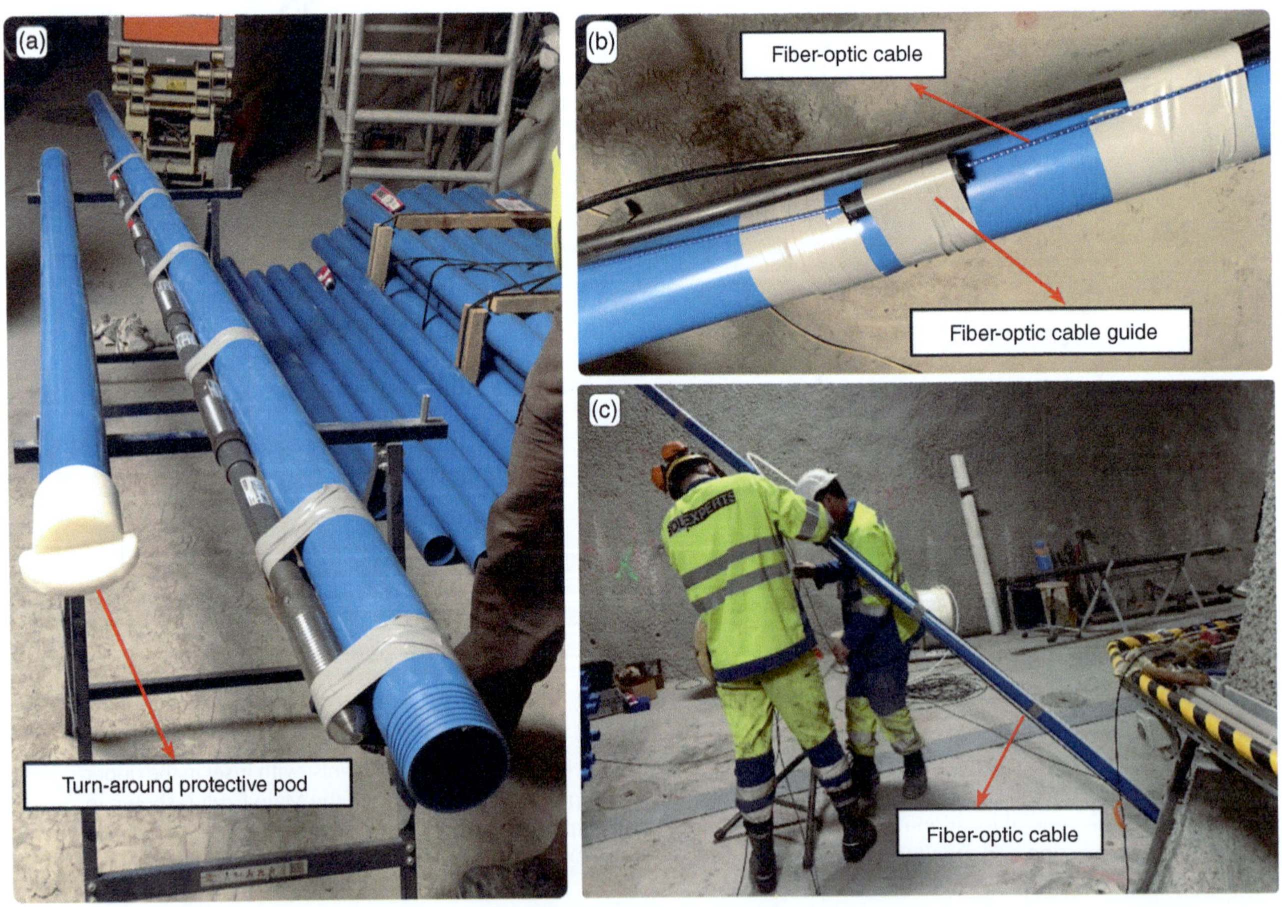

Figure 30.10 Setup and installation of fiber-optic monitoring systems in the CS-D and FS-B Experiments in Mont Terri. (a) CS-D D5 borehole installation. Conveyance PVC pipe taped to fiber-optic cable and additional instrumentation. PVC pipe on the left is the bottom-most section of the conveyance assembly, showing the white plastic cap used to protect the fiber-optic cable turn-around. (b) Detail of fiber-optic guide used in CS-D boreholes to hold the fiber in place and avoid strapping it to the PVC pipe to allow for the fiber to be kept under tension during the deployment. (c) Deployment of PVC conveyance attached to fiber-optic cable in one of the FS-B project monitoring boreholes.

300 MMF package contained six MMFs. Each telecom-standard cable was coated with a PVC jacket.

DSS Interrogator and Acquisition Parameters

The DSS interrogator used for the FS-B and CS-D injection experiments discussed here was the same IU type used in the EGS Collab project, a BOTDA-based Omnisens DITEST VISION dual temperature and strain unit. Figures 30.11 and 30.12 show data recorded during an injection experiment conducted in the FS-B testbed in November 2020, when the configuration parameters in Table 30.2 were used.

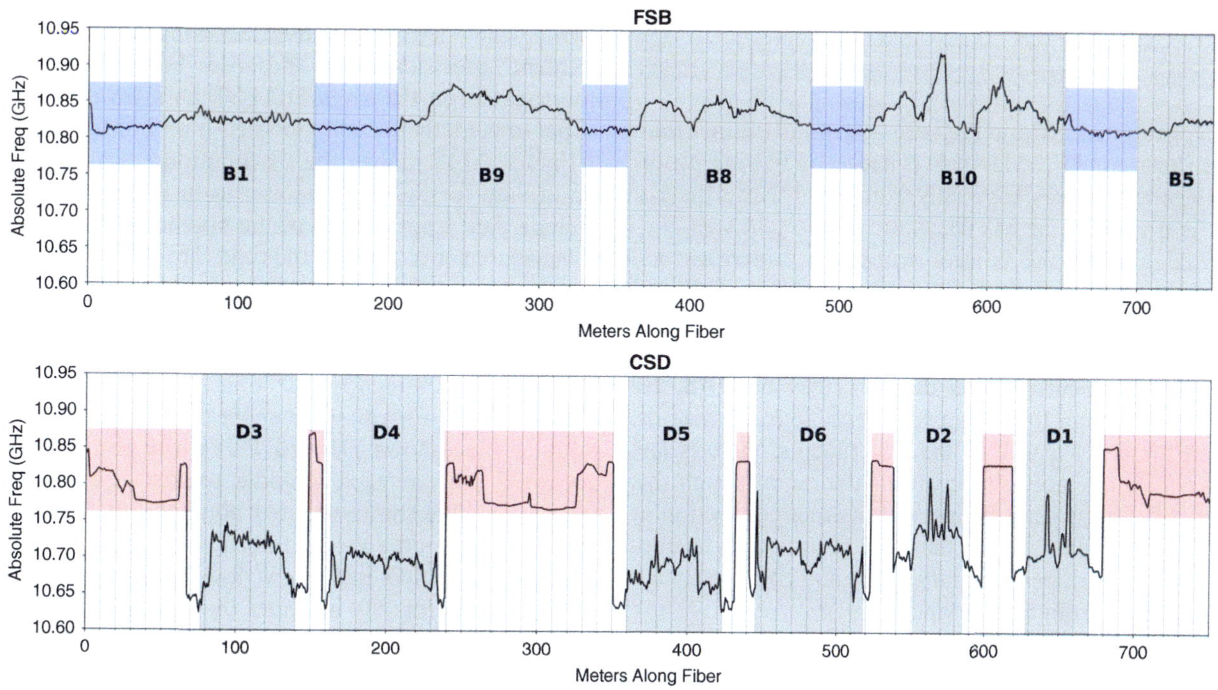

Figure 30.11 Absolute Brillouin frequency shift with distance along the FS-B and CS-D fiber loops. Portions of the fiber installed in boreholes are shaded gray. Portions of the cable lying in cable trays in the gallery are shaded blue (for FS-B) and red (for CS-D).

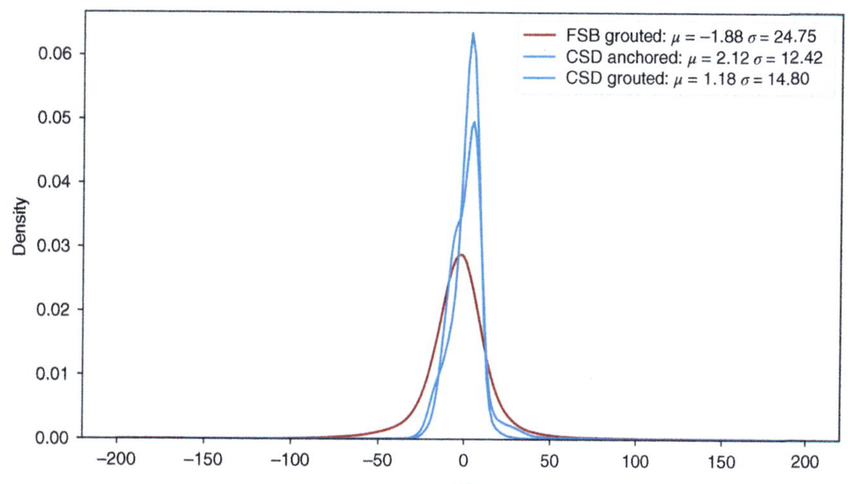

Figure 30.12 Distributions of relative strain data during an injection conducted in the FS-B experiment testbed in November 2020. The two blue curves show relative strain for all CS-D boreholes. The "anchored" boreholes, D1 and D2, are shown as a lighter blue, and all other CS-D boreholes are included in the darker blue distribution. All FS-B boreholes are included in the red distribution. The mean and standard deviation for each dataset are indicated in the legend.

Table 30.2 DITEST configuration parameters during the 2020 FS-B injection experiment.

Spatial resolution (m)	1.5
Spatial sampling (m)	0.51
Temporal sampling (min)	16
Averaging	3000
Frequency start (GHz)	10.6
Frequency stop (GHz)	10.9
Frequency step (MHz)	1.0

The galleries at Mont Terri are larger and cleaner than the drifts at SURF, so less care is needed in weatherproofing the interrogators, especially from water. Nevertheless, dust is a concern at MTRL, so a plastic tent was constructed around the control computers, interrogators, and data storage devices. The experiments at FS-B and CS-D are typically limited in duration to less than 1 week, making data storage simpler than the multimonth EGS Collab experiments. Larger datasets, like the DAS, are still small enough to fit onto semiportable NAS systems. Given the short length of the experiments, there is less need for remote access to instrumentation, although this access does exist via a high-speed internet connection at MTRL. Small datasets can therefore be accessed and copied remotely, whereas the larger (multi-TB) datasets are shipped back from Switzerland.

Data Analysis and Observations

Figure 30.11 shows the absolute Brillouin peak frequency in GHz along both the FS-B and CS-D fibers at the start of the FS-B injection in November 2020. Sections of the cables that are installed inside boreholes are shaded gray, whereas sections in the gallery are colored blue (FS-B) and red (CS-D). The cable type is constant along the entire FS-B fiber loop. However, the cable type is different between the gallery and boreholes at CS-D. Although the absolute frequency shift is higher inside the FS-B boreholes than outside, the exact transition point from borehole to gallery is difficult to determine. At CS-D, the change between tactical cable in the gallery and BRUsens cable in the boreholes is obvious, with the BRUsens having a ~100 MHz negative shift in its absolute Brillouin shift. This is simply a result of the material properties of the fibers themselves, as the Brillouin frequency is related to vibrations of the fiber core lattice. For example, the manufacturing process may include the addition of a "dopant" element that, even in trace concentrations, can alter the Brillouin frequency by hundreds of MHz (Tkach et al., 1986). Variations of this magnitude are possible among manufacturing runs of supposedly identical fibers. It is not indicative of their strain-sensing capability, nor is it particularly useful in mapping the cable geometry, as the sensing fiber could be longer than the actual well. However, it is important to understand the cable properties when configuring the measurements. For example, the CS-D data exhibit a larger range of Brillouin frequency along the fiber loop than the FS-B dataset, and this must be accounted for in the frequency start, stop, and step parameters in Table 30.2, which control the granularity and bounds of the peak fitting algorithm used to find the shift at each channel. A higher frequency start would be appropriate for FS-B, in this case, in which case a smaller frequency step could be used, allowing for higher measurement precision. But because both fibers were interrogated on the same DITEST unit, we chose a set of parameters that accommodated both installations.

Figure 30.12 shows the distributions of the relative strain data for the same dataset as in Figure 30.11. We exclude data from outside of the boreholes and consider all measurements in space and time. The ability of the system to sense real strain is a function of the noise in the data. The standard deviation in the data is one simple estimate of the noise level, with a lower noise value signifying an ability to confidently detect smaller strains. The two blue curves in Figure 30.12 show data from CS-D: the brighter blue shows D1 and D2, where the cable was anchored only at each straddle packer assembly, and the darker blue shows grouted boreholes at CS-D. The red curve shows the grouted portions of the FS-B cable. Both sets of CS-D data show lower noise values than FS-B, which probably indicates that the BRUsens cable is better able to detect strains between 10 and 20 $\mu\varepsilon$ than the tactical cable installed at FS-B. There is little difference in the distribution of the two CS-D curves. There is a minor negative "shoulder" to the anchored cable distribution, but its explanation requires a detailed analysis of the spatiotemporal pattern, which is not shown here. A nonzero mean indicates a cumulative elongation or contraction of the length of the fiber in question. The reason for the negative shift in the FS-B distribution (mean of −1.88), which suggests a small cumulative contraction among the FS-B boreholes, also remains unclear. This negative mean could indicate that the fibers are measuring the bulk contraction of the rock mass triggered by various injection activities. Most of the boreholes do not intersect the area of pressurized fluid, which might be the reason why the fibers record more contraction than extensional strain. In the end, while the BRUsens cable deployed at CS-D certainly shows a lower noise floor, the improvement relative to the tactical cable at FS-B is modest and may be outweighed by the lower cost, easier deployment, and greater number of fibers in the tactical cables.

30.4. DISCUSSION AND LESSONS LEARNED

Observations from four different URL DFOS experiments conducted using various types of DFOS technologies, under different installation conditions, and using different cable types have enabled us to identify the significance of different experiment components and to evaluate the impact these components have on the overall success of such an investigation. Here, we collect a set of lessons learned based on the two case studies introduced earlier. Whereas some of these observations will inevitably be experiment-dependent, we believe that they provide useful insights into the different components of DFOS experiments in URLs, many of which are transferable to field-scale investigations:

30.4.1. Choice of Fiber-Optic Cable

The choice of cable design has both logistical and scientific implications:

- A large variety of fiber-optic cables with different construction designs, fiber packaging types, and outer packaging materials are available in the market. In general, fiber packing (e.g., tight-buffered or gel-filled loose tube) will have an impact on the cable's capability to transfer strain from the rock to the fiber. Cable construction and the presence or lack of strength elements (such as metal tubing or armored structures) will determine the cable's durability, flexibility, weight, and ease of deployment. Fiber-optic cables exist with many possible combinations of these construction elements, and thus, the right combination should be sought based on the application at hand.
- Conventional indoor telecommunication cables commonly consist of tight-buffered designs with a PVC or PE outer jacket surrounding an aramid yarn. These cables offer flexibility, tight bending radius and are lightweight, making them easy to manage and deploy. These designs are economical and perform well in both DAS seismic studies and low-frequency DAS/DSS geomechanical applications in which efficient strain transfer for strain quantification is required. However, these cables are less robust to water ingress than other designs and are less advisable for harsh conditions of pressure and temperature in which water entrance and/or extreme stress are expected.
- Loose-tube cables are better fitted for applications where water ingress and extreme conditions of temperature and pressure are expected, although they also come at a higher price. In these cables, the fibers are isolated by plastic or metal tubes that are commonly filled with a thixotropic gel that absorbs stress. Those constructed with steel or nickel alloy tubing are especially less flexible, and their bent radius is reduced. In our experience, this can cause challenges when working in physically reduced spaces with small teams and in experiments where the cable has to be deployed as a single loop down multiple boreholes. Cable management is complicated by its weight and rigidity, and the cable is more prone to kinking. Moreover, a reduction in the bent radius also complicates the turnaround at the bottom of the borehole. In multifiber cables, splicing individual fibers is more challenging and time-consuming, as the only protection between the fiber itself and the plastic or metal tube is the cladding and the gel. When working with these cables, a fan-out kit is recommended.
- Cable construction can have an impact on the strain rate or strain measured using low-frequency DAS and DSS. Protective structures, such as the thixotropic gel, can effectively dampen strain inflicted on the cable, resulting in poor strain transfer between the rock and the cable. Our observations indicate that, whereas this effect might be less important for low stresses, when high stress is imposed on the rock/cable, the gel absorbs a large part of it, resulting in a much smaller magnitude of strain being recorded (sometimes several orders of magnitude). In situations in which using such cable designs is absolutely necessary, it is important to note that the true strain rate or strain might not be recorded, which can have important implications for geomechanical investigations.
- The use of acoustic and/or strain specialty cables can improve the signal-to-noise ratio of low-frequency DAS and DSS measurements. However, these customized packages tend to be more expensive.
- Independent of the cable packaging, our observations suggest that using multifiber cables that enable collocated and simultaneous acquisition of multiple data streams is very useful to corroborate independent measurements. Hybrid packages with SMFs and MMFs for acquiring DAS/DSS and DTS are critical to identify thermally induced strain in DAS/DSS data. Multiple strains of SMF will enable using absolute strain measurements provided by DSS to corroborate strain estimated using low-frequency DAS measurements and improve spatial resolution. Conversely, our results suggest that low-frequency DAS has comparably higher sensitivity to small magnitude strain than DSS, and thus having both measurements will ensure that small deformation is recorded. Multiple SMFs will also enable acquiring independent DAS measurements using different IUs targeted at different applications (e.g., seismic versus strain) that require different acquisition parameters.

30.4.2. Deployment Methods

- In mesoscale experiments in URLs where experiment dimensions tend to be reduced, and boreholes are closely spaced, deploying the fiber-optic cable as a single, uninterrupted loop when possible is preferred. This strategy eliminates the number of splicing points and provides redundancy inside each borehole, as a downgoing and upgoing leg of cable recording the same event is available. With appropriate geometry mapping, these two legs could be stacked to increase signal-to-noise ratios of the data. This deployment strategy can be challenging in reduced spaces.

- The single-loop deployment strategy can also be challenged by heavy and nonflexible cables, especially those containing strength elements such as metal tubing or armored structures. When using these rigid cables is required, it is advised to have a large field team that can manage long lengths of cable properly and ensure that no extreme bending or kinking occurs. Alternatively, the deployment of individual loops in each borehole can be considered. In that case, high-quality splices between boreholes (fiber-to-fiber splice, without any connectors) will be essential to minimize light losses and data degradation toward the end of the fiber.
- The best coupling is provided by deploying the fiber behind the borehole casing. If that is not possible, attaching the cable to a conveyance rod and grouting it in place inside the borehole is the best next alternative. Temporary deployments with cables going through other instruments, such as packer systems, are possible and can be used, although this deployment is better advised for DAS seismic applications in which larger signals are expected, noise can be reduced during data processing, and strain quantification is not critical.
- Another method to install fiber in boreholes is by means of a splice loop at the bottom of the well instead of a U-turn. This is particularly important when the space for cables is limited due to redundancy with many other instruments. This technique is particularly convenient also for medium/long uncased boreholes that are entirely cemented (e.g., Plenkers et al., 2023).

30.4.3. DFOS Technologies, Acquisition Parameters, and Related Considerations

- Both DSS and low-frequency DAS have the potential for monitoring strain perturbations induced by fluid injection. DSS provides high spatial resolution (<1 m) recordings of absolute strain with a temporal resolution in the order of minutes. Based on our experience, DSS is a good choice to record slow, relatively large (few to tens of $\mu\varepsilon$) near-field strain. Low-frequency DAS provides higher temporal resolution (<1 s) and higher sensitivity to small strain magnitude ($n\varepsilon$), making it useful for monitoring both short-term and long-term strain perturbations both in the near-field and further away from the source (tens of meters based on our experiments). However, the spatial resolution of low-frequency DAS is limited by the gauge length. In URL mesoscale experiments like those introduced here, small-scale structures and somewhat localized signals (such as the opening of small fractures) are expected in the near field, which suggests that small gauge lengths (<5 m) are desired. However, small gauge lengths also result in noisier data, which can pose a challenge when integrating strain rate to strain and to identify small signals. These trade-offs need to be carefully evaluated during experiment design. In the current expansion of DFOS solutions available in the market, additional technologies such as Rayleigh-based DSS can provide strain measurements at spatial resolution down to a few cm or even a few mm for very short (~50 m) fibers (Sasaki et al., 2021; Zhang et al., 2020a).
- Experiment testbed dimensions need to be considered when choosing acquisition parameters. The case studies presented here were characterized by borehole lengths in the order of 30–60 m. In this scenario, DAS measurements with large gauge lengths (e.g., ≥10 m) limit the number of independent measurements along the borehole, which can result in uncertain spatial localization of the signal due to gauge-length-related spatial smearing.
- If the objective is using DAS for induced seismicity detection, an estimation of the expected frequency content is essential. In mesoscale experiments in a crystalline environment, where seismic velocities are fast, and sensors are near the source, high frequencies (kHz range) can be expected. In this scenario, systems that enable high-frequency sampling rates with high signal-to-noise ratios and short gauge lengths (2–4 m) are desirable. For relatively long arrays (≥1 km), these parameters will result in large data volumes (sometimes up to several TB/day). An estimate of data volumes before the experiment will be critical to organize appropriate data storage needs and to avoid data collection interruption during the experiment.
- Temperature isolation of the IU is critical for fiber optics sensing in geomechanical applications. Especially for low-frequency DAS measurements, variations in ambient temperature surrounding the IU introduce common-mode noise affecting the entire fiber, which can be difficult to remove without affecting the targeted signals. This effect can be alleviated by placing the IU inside a weather-tight cabinet and by placing accurate temperature monitoring sensors in the space where the IU is located to track ambient temperature variations. Exposing a length of fiber near the IU can also help track thermally induced strain perturbations.
- Another challenge in DSS and low-frequency DAS experiments is thermally induced strain perturbations affecting the sensing fiber. This is especially important in experiments where fluid interaction is expected. Simultaneous, collocated DTS measurements, if possible along a fiber inside the same cable, will prove useful in detecting temperature variations along the length of

the borehole that could obscure or add to the targeted mechanical strain. However, it is important to note that the strain-temperature relationship is different for DSS and DAS, and it is also dependent on cable construction and deployment methods. Although the impact of the deployment method can be difficult to evaluate, cable effects can be calibrated either in the lab before deployment or at the site by placing a length of cable in a temperature-regulated water bath. Our experience shows us that, when using water baths for long-term experiments, it is critical to monitor the performance of the bath, as effects such as water evaporation and thermostat cycling on the heater can introduce artifacts and make the data unusable. This is especially challenging in URLs where access frequency tends to be limited, and immediate action when a problem is detected might not be possible. Another solution could be to place a section of the fiber inside a thermally insulated box, where the temperature would be kept stable more easily than in a heated water bath.

30.5. CONCLUSIONS

Distributed fiber-optic sensing techniques, particularly DAS and DSS, are emerging as excellent tools for characterizing and monitoring the geomechanical response of the rock mass to stimulation during fluid injection. Due to the relative novelty of this application, however, approaches are still under development. Here, we have collected a set of observations based on four experiments conducted in two different underground research laboratories for investigating EGS development and fault reactivation and leakage due to fluid injection. Our observations demonstrate that mesoscale experiments in underground research laboratories offer an excellent opportunity to learn more about these technologies and evaluate the impact each experiment component has on the observed results at scales and conditions relevant to field-scale demonstrations. These observations also reveal that URL environments pose additional constraints regarding experiment dimensions and logistics that need to be taken into account when designing and implementing DFOS investigations for geomechanical monitoring. A clear definition of the experiment objective is critical for the design of the DFOS experiment, as it will determine the choice of cable, DFOS technology, and acquisition parameters. Cable type and installation procedure can significantly impact strain values being recorded, which is key in DSS and low-frequency DAS experiments in which a quantitative evaluation of strain rate and/or strain is sought. Experiment logistics, such as site access, physical space to work and to place instrumentation, humidity and temperature conditions, remote access to instrumentation, or data storage and data retrieval options, will also play an important role in experiment components, such as the choice of cable design (e.g. sturdier and challenging to deploy versus less robust but easier to work with), the timescale and durability of the installation, the assortment of technologies being used, or the balance between acquisition parameters and data volumes. Although we are aware that the list of lessons learned presented here is not exhaustive and questions still remain, we hope that they help and encourage the advancement and use of DAS and DSS for geomechanical characterization and monitoring of fluid injection experiments in URLs and beyond.

ACKNOWLEDGMENTS

The EGS Collab project and the analysis of the datasets presented here were supported by the U.S. Department of Energy, Office of Energy Efficiency and Renewable Energy (EERE), Office of Technology Development, Geothermal Technologies Office, under Award Number DE-AC02-05CH11231 with LBNL and other awards to other U.S. national laboratories. The authors are deeply grateful to the SURF and its personnel for providing physical access and general logistical and technical support. The authors are also very grateful to the partners of the Mont Terri Project that contributed to the funding of the CS-D and FS-B experiments: the Swiss Federal Office of Topography (Swisstopo), the Swiss Federal Nuclear Safety Inspectorate (ENSI), the Japanese Atomic Energy Agency (JAEA), the Institute of Radioprotection and of Nuclear Safety (IRSN, France), TOTAL SE, CHEVRON, the Federal Institute for Geosciences and Natural Resources (BGR, Hannover), the Swiss Federal Institute of Technology (ETH-Zürich), and the U.S. Department of Energy. The Mont Terri Project is an international research project for the hydrogeological, geochemical, and geotechnical characterizations of a clay formation (Opalinus clay). Funding for Berkeley Lab's analysis of the FS-B data described in this study was provided by the Assistant Secretary for Fossil Energy as part of the Core Carbon Storage and Monitoring Research (CCSMR) and National Risk Assessment Partnership (NRAP) programs of the U.S. Department of Energy under contract FP00007630. Additional analysis support was provided by the U.S. Department of Energy, Office of Fossil Energy and Carbon Management (DOE FECM) under contract DE-FE00032058. The authors thank Thomas Coleman and the Support team at Silixa LLC for helpful technical support with the iDAS DAS IU, Michael Roelens at Terra15 for technical guidance and support on the Treble DAS IU, and Chris Minto, Paul Clarkson, Domininc Graham, Andrés Chavarría and OptaSense Ltd for technical guidance and for providing the ODH3 IU and strain and temperature reference units for the EGS Collab Experiment 2. Finally, the authors thank two anonymous reviewers for their constructive comments, which helped improve this chapter.

AVAILABILITY STATEMENT

The processed distributed fiber-optic sensing data recorded within the framework of the EGS Collab, FS-B, and CS-D projects discussed in this chapter and necessary to reproduce all the figures and support our observations are provided in the following OSF repository: https://osf.io/mxh5n/ (Rodríguez Tribaldos, 2023). The figures were created using MATLAB and Matplotlib.

REFERENCES

Becker, M., Coleman, T., Ciervo, C., Cole, M., & Mondanos, M. (2017). Fluid pressure sensing with fiber-optic distributed acoustic sensing. *The Leading Edge*, *36*(12), 1018–1023.

Becker, M. W., Ciervo, C., & Coleman, T. (2018). *Laboratory testing of low frequency strain measured by distributed acoustic sensing*. Paper presented at the SEG Technical Program, Expanded Abstracts, 4963–4966.

Blechschmidt, I., Martin, A. J., Kober, F., & Vomvoris, S. (2020). *Grimsel test site - a successful international underground research laboratory for many decades*. Paper presented at the WM2020 Conference, Phoenix, Arizona, USA, 20429.

Bossart, P., Bernier, F., Birkholzer, J., Bruggeman, C., Connolly, P., Dewonck, S., et al. (2017). Mont Terri rock laboratory, 20 years of research: introduction, site characteristics and overview of experiments. *Swiss Journal of Geosciences*, *110*(1), 3–22.

Cho, W. J., Kwon, S., & Park, J. H. (2008). KURT, a small-scale underground research laboratory for the research on a high-level waste disposal. *Annals of Nuclear Energy*, *35*(1), 132–140.

Daley, T. M., Freifeld, B. M., Ajo-franklin, J. B., Dou, S., Pevzner, R., Shulakova, V., et al. (2013). Field testing of fiber-optic distributed acoustic sensing (DAS) for subsurface seismic monitoring. *The Leading Edge*, *32*(6), 699–706.

Dean, T., Cuny, T., & Hartog, A. H. (2016). The effect of gauge length on axially incident P-waves measured using fibre optic distributed vibration sensing. *Geophysical Prospecting*, *65*(1), 184–193.

Deichmann, N., & Giardini, D. (2009). Earthquakes Induced by the stimulation of an enhanced geothermal system below Basel (Switzerland). *Seismological Research Letters*, *80*(5), 784–798.

Dou, S., Lindsey, N., Wagner, A. M., Daley, T. M., Freifeld, B., Robertson, M., et al. (2017). Distributed acoustic sensing for seismic monitoring of the near surface: a traffic-noise interferometry case study. *Scientific Reports*, *7*(1), 1–12.

Ellsworth, W. L., Giardini, D., Townend, J., Ge, S., & Shimamoto, T. (2019). Triggering of the Pohang, Korea, earthquake (Mw 5.5) by enhanced geothermal system stimulation. *Seismological Research Letters*, *90*(5), 1844–1858.

Grant, M. A. (2016). Physical performance indicators for HDR/EGS projects. *Geothermics*, *63*, 2–4.

Grigoli, F., Cesca, S., Priolo, E., Rinaldi, A. P., Clinton, J. F., Stabile, T. A., et al. (2017). Current challenges in monitoring, discrimination, and management of induced seismicity related to underground industrial activities: a European perspective. *Reviews of Geophysics*, *55*(2), 310–340.

Grigoli, F., Cesca, S., Rinaldi, A. P., Manconi, A., Clinton, J. F., Westaway, R., et al. (2018). The November 2017 Mw 5.5 Pohang earthquake: a possible case of induced seismicity in South Korea. *Science*, *360*, 1003–1006.

Hartog, A. H., Rao, Y.-J., Ran, Z.-L., Gong, Y., Güemes, A., & Sierra Perez, J. (2017). *An Introduction to Distributed Optical Fibre Sensors*. Boca Raton, Florida: Taylor and Francis Group.

Hausner, M. B., Suárez, F., Glander, K. E., van de Giesen, N., Selker, J. S., & Tyler, S. W. (2011). Calibrating single-ended fiber-optic Raman spectra distributed temperature sensing data. *Sensors*, *11*(11), 10859–10879.

Heise, J. (2015). The Sanford underground research facility at Homestake. *Journal of Physics: Conference Series*, *606*(1), 012015.

Hopp, C., Guglielmi, Y., Rinaldi, A. P., Soom, F., Wenning, Q., Cook, P., et al. (2022). The effect of fault architecture on slip behavior in shale revealed by distributed fiber optic strain sensing. *Journal of Geophysical Research: Solid Earth*, *127*(1), e2021JB022432.

Horiguchi, T., & Tateda, M. (1989). BOTDA-nondestructive measurement of single-mode optical fiber attenuation characteristics using Brillouin interaction: theory. *Journal of Lightwave Technology*, *7*(8), 1170–1176.

Jin, G., Mendoza, K., Roy, B., & Buswell, D. G. (2019). Machine learning-based fracture-hit detection algorithm using LFDAS signal. *Leading Edge*, *38*(7), 520–524.

Jin, G., & Roy, B. (2017). Hydraulic-fracture geometry characterization using low-frequency DAS signal. *Leading Edge*, *36* (12), 975–980.

Kickmaier, W., & McKinley, I. (1997). A review of research carried out in European rock laboratories. *Nuclear Engineering and Design*, *176*(1–2), 75–81.

Kishida, K., Yamauchi, Y., & Guzik, A. (2014). Study of optical fibers strain-temperature sensitivities using hybrid Brillouin-Rayleigh system. *Photonic Sensors*, *4*(1), 1–11.

Kneafsey, T., Blankenship, D., Burghardt, J., Johnson, T., Dobson, P., Schwering, P. C., et al. (2023). *The EGS Collab – discoveries and lessons from an underground experiment series*. Paper presented at the 48th Workshop on Geothermal Reservoir Engineering, Stanford.

Kneafsey, T. J., Blankenship, D., Dobson, P. F., Morris, J. P., White, M. D., Fu, P., et al. (2020). *The EGS Collab Project: learnings from Experiment 1*. Paper presented at the 45th Workshop on Geothermal Reservoir Engineering.

Kneafsey, T. J., Dobson, P. F., Ulrich, C., Hopp, C., Rodriguez Tribaldos, V., Guglielmi, Y., et al. (2022). *The EGS Collab Project – stimulations at two depths*. Paper presented at the 56th U.S. Rock Mechanics/Geomechanics Symposium.

Krietsch, H., Gischig, V., Jalali, M. R., Doetsch, J., Valley, B., & Amann, F. (2018). *A comparison of FBG- and Brillouin-strain sensing in the framework of a decameter-scale hydraulic stimulation experiment*. Paper presented at the 52nd U.S. Rock Mechanics/Geomechanics Symposium.

Lei, X., Xue, Z., & Hashimoto, T. (2019). Fiber optic sensing for geomechanical monitoring: (2)-Distributed strain measurements at a pumping test and geomechanical modeling of deformation of reservoir rocks. *Applied Sciences*, *9*(3), 1–14.

Lellouch, A., & Biondi, B. L. (2021). Seismic applications of downhole DAS. *Sensors*, *21*(9).

Lellouch, A., Lindsey, N. J., Ellsworth, W. L., & Biondi, B. L. (2020). Comparison between distributed acoustic sensing and geophones: downhole microseismic monitoring of the FORGE geothermal experiment. *Seismological Research Letters*, *91*(6), 3256–3268.

Lindsey, N. J., & Martin, E. R. (2021). Fiber-optic seismology. *Annual Review of Earth and Planetary Sciences*, *49*, 309–336.

Lindsey, N. J., Rademacher, H., & Ajo-Franklin, J. B. (2020). On the broadband instrument response of fiber-optic DAS arrays. *Journal of Geophysical Research: Solid Earth*, *125*, 1–16.

Miller, D. E., Coleman, T., Zeng, X., Patterson, J. R., Reinisch, E. C., Cardiff, M. A., et al. (2018). *DAS and DTS at Brady Hot Springs: observations about coupling and coupled interpretations*. Paper presented at the Stanford Geothermal Workshop.

Murdoch, L. C., Germanovich, L. N., DeWolf, S. J., Moysey, S. M., Hanna, A. C., Kim, S., & Duncan, R. G. (2020). Feasibility of using in situ deformation to monitor CO_2 storage. *International Journal of Greenhouse Gas Control*, *93*, 102853.

Othonos, A. (1997). Fiber Bragg Gratings. *Review of Scientific Instruments*, *68*, 4309–4341.

Paitz, P., Edme, P., Gräff, D., Walter, F., Doetsch, J., Chalari, A., et al. (2020). Empirical investigations of the instrument response for distributed acoustic sensing (DAS) across 17 octaves. *Bulletin of the Seismological Society of America*, *111*, 1–10.

Piccolo, A., Lecieux, Y., Lesoille, S., Teixeira, P., Bertrand, J., & Leduc, D. (2023). Performance Assessment of Distributed Strain Sensing Techniques for Convergence Monitoring of Radioactive Waste Repository. *Sensors*, *23*, 398.

Plenkers, K., Reinicke, A., Obermann, A., Gholizadeh Doonechaly, N., Krietsch, H., Fechner, T., et al. (2023). Multidisciplinary monitoring networks for mesoscale underground experiments: advances in the bedretto reservoir project. *Sensors*, *23*(6), 3315.

Ringrose, P. S., Mathieson, A. S., Wright, I. W., Selama, F., Hansen, O., Bissell, R., et al. (2013). The in Salah CO_2 storage project: lessons learned and knowledge transfer. *Energy Procedia*, *37*, 6226–6236.

Rodríguez Tribaldos, V., Hopp, C., Rinaldi, A. P., Tuinstra, K., Lanza, F., Ajo-Franklin, A., et al. (2023). Use of DAS and DSS technologies for geomechanical characterisation of rock mass response in mesoscale experiments in underground laboratories: lessons learned. [Dataset]. OSF. doi:10.17605/OSF.IO/MXH5N.

Rutqvist, J. (2012). The geomechanics of CO_2 storage in deep sedimentary formations. *Geotechnical and Geological Engineering*, *30*, 525–551.

Rutqvist, J., Rinaldi, A. P., Cappa, F., Jeanne, P., Mazzoldi, A., Urpi, L., et al. (2016). Fault activation and induced seismicity in geological carbon storage – lessons learned from recent modeling studies. *Journal of Rock Mechanics and Geotechnical Engineering*, *8*(6), 789–804.

Sasaki, T., Zhang, S., Soga, K., Luo, L., Freifeld, B., Kitayama, Y., et al. (2021). Distributed fiber optic strain sensing of bending deformation of a well mockup in the laboratory. *Journal of Natural Gas Science and Engineering*, *96*(May), 104309.

Schoenball, M., Ajo-Franklin, J. B., Blankenship, D., Chai, C., Chakravarty, A., Dobson, P., et al. (2020). Creation of a mixed-mode fracture network at mesoscale through hydraulic fracturing and shear stimulation. *Journal of Geophysical Research: Solid Earth*, *125*, 12, e2020JB019807. e2020JB019807 2020JB019807.

Schwering, P. C., Doe, T., Roggenthen, W., Neupane, G., Johnston, H., Dobson, P. F., et al. (2020). *Deterministic discrete fracture network (DFN) model for the EGS Collab project on the 4850 level of the Sanford Underground Research Facility (SURF)*. Paper presented at the 54th US Rock Mechanics/Geomechanics Symposium.

Schwering, P. C., Ingraham, M., Vermeul, V. R., Burghardt, J., Johnson, T. C., Strickland, C., et al. (2023). *Shut-in testing on the 4100L - implications on the state of stress, fractures, and wellbores in the second EGS Collab testbed*. Paper presented at the 48th Workshop on Geothermal Reservoir Engineering, Stanford.

Shanafield, M., Banks, E. W., Arkwright, J. W., & Hausner, M. B. (2018). Fiber-optic sensing for environmental applications: where we have come from and what is possible. *Water Resources Research*, *54*(11), 8552–8557.

Sidenko, E., Tertyshnikov, K., Lebedev, M., & Pevzner, R. (2022). Experimental study of temperature change effect on distributed acoustic sensing continuous measurements. *Geophysics*, *87*(3), D111–D122.

Sun, Y., Liu, J., Xue, Z., Li, Q., Fan, C., & Zhang, X. (2021). A critical review of distributed fiber optic sensing for real-time monitoring geologic CO_2 sequestration. *Journal of Natural Gas Science and Engineering*, *88*, 103751.

Sun, Y., Xue, Z., Hashimoto, T., Lei, X., & Zhang, Y. (2020). Distributed fiber optic sensing system for well-based monitoring water injection tests—A geomechanical responses perspective. *Water Resources Research*, *56*(1), 1–19.

Tkach, R., Chraplyvy, A., & Derosier, R. (1986). Spontaneous brillouin scattering for single-mode optical-fibre characterisation. *Electronics Letters*, *19*(22), 1011–1013.

van de Giesen, N., Steele-Dunne, S. C., Jansen, J., Hoes, O., Hausner, M. B., Tyler, S., & Selker, J. (2012). Double-ended calibration of fiber-optic Raman spectra distributed temperature sensing data. *Sensors*, *12*(5), 5471–5485.

Vasco, D. W., Bissell, R. C., Bohloli, B., Daley, T. M., Ferretti, A., Foxall, W., et al. (2018). Monitoring and modeling caprock integrity at the In Salah carbon dioxide storage site, Algeria. In S. Vialle, J. Ajo-Franklin, & J. W. Carey (Eds.), *AGU Geophysical Monograph Series* (Vol. *238*, chapter 12, pp. 243–269).

Verdon, J. P., Horne, S. A., Clarke, A., Stork, A. L., Baird, A. F., & Kendall, J. M. (2020). Microseismic monitoring using a fiber-optic distributed acoustic sensor array. *Geophysics*, *85* (3), KS89–KS99.

Xue, Z., Shi, J. Q., Yamauchi, Y., & Durucan, S. (2018). Fiber optic sensing for geomechanical monitoring: (1)-distributed strain measurements of two sandstones under hydrostatic confining and pore pressure conditions. *Applied Sciences, 8* (11), 2103.

Zappone, A., Rinaldi, A. P., Grab, M., Wenning, Q., Roques, C., Madonna, C., et al. (2020). Fault sealing and caprock integrity for co 2 storage: an in-situ injection experiment. *Solid Earth Discussions*, 1–51.

Zappone, A., Rinaldi, A. P., Grab, M., Wenning, Q. C., Roques, C., Madonna, C., et al. (2021). Fault sealing and caprock integrity for CO_2 storage: an in situ injection experiment. *Solid Earth, 12*(2), 319–343.

Zhang, Y., Doughty, C., Pan, L., & Kneafsey, T. J. (2018). *What could we see at the production well before the thermal breakthrough?* Paper presented at the 43rd Workshop on Geothermal Reservoir Engineering, Stanford.

Zhang, Y., Lei, X., Hashimoto, T., & Xue, Z. (2020a). In situ hydromechanical responses during well drilling recorded by fiber-optic distributed strain sensing. *Journal of Geophysical Research: Solid Earth, 11*(6), 2487–2497.

Zhang, Y., & Xue, Z. (2019). Deformation-based monitoring of water migration in rocks using distributed fiber optic strain sensing: a laboratory study. *Water Resources Research, 55*(11), 8368–8383.

Zhang, Z., Fang, Z., Stefani, J., DiSiena, J., Bevc, D., Lim Chen Ning, I., et al. (2020b). Modeling of fiber optic strain responses to hydraulic fracturing. *Geophysics*, 1–22.

Zoback, M. D., & Gorelick, S. M. (2012). Earthquake triggering and large-scale geologic storage of carbon dioxide. *Proceedings of the National Academy of Sciences, 109*(26), 10164–10168.

Zoback, M. D., Kohli, A., Das, I., & McClure, M. (2012). *The importance of slow slip on faults during hydraulic fracturing stimulation of shale gas reservoirs.* Paper presented at the SPE Americas Unconventional Resources Conference.

31

Distributed Acoustic Sensing in Borehole Geophysics: New Advances and Future Directions

Yingping Li[1,2], Robert Mellors[3], and Ge Zhan[4]

ABSTRACT

This chapter presents conclusions for Chapters 2–30, organized into five sections: (i) borehole distributed acoustic sensing (DAS) instrumentation and modeling, (ii) borehole DAS acquisition and processing, (iii) borehole DAS imaging and inversion, (iv) borehole DAS monitoring, and (v) further reading and future efforts. Although we strive to cover various aspects of borehole DAS geophysical applications, this may not be exhaustive due to the rapid advancements in borehole DAS technology and applications. Therefore, we recommend additional literature for further reading to supplement this monograph. We also discuss future endeavors for DAS technology developments and borehole DAS geophysical applications.

31.1. BOREHOLE DAS INSTRUMENTATION AND MODELING

In this section, we summarize the conclusions of five chapters, Chapters 2–6, on distributed acoustic sensing (DAS) instrumentation and modeling topics.

In Chapter 2, Farhadiroushan et al. (2024) review the latest advancements in integrated distributed strain sensing (DSS) technology for wellbore and reservoir monitoring. The authors utilize an engineered fiber cable with a lower noise floor and an interferometry method to measure the Brillouin frequency shift, achieving integrated strain measurements with a fine resolution of 1 μm. Temperature effects are compensated using Raman distributed temperature sensing (DTS) measurements within the same fiber cable. This development enables the simultaneous measurement of DAS, DSS, and DTS using different modes of optic fibers within a single fiber cable, resulting in a new distributed fiber-optic sensing (DFOS) system with significant scientific merits and cost-effective benefits for wellbore and reservoir monitoring.

Hua et al. (2024), in Chapter 3, introduce a new method called coherence-length-gated microwave photonics interferometry (CMPI) to measure strain for monitoring hydrologic processes in the vadose and saturated zones. CMPI measures distributed strain caused by pressure changes in the vadose zone to characterize hydrologic properties in both the vadose and saturated zones. This method achieves centimeter spatial resolution, up to 10 nε sensitivity, and a kHz strain measurement rate. The authors conclude that constant interrogation by fluctuating barometric pressure provides a new strategy for continuously monitoring the vadose zone accurately and with high resolution.

Haldorsen et al. (2024) recognize that although a near-continuous single-axis DAS array avoids spatial-aliasing issues, it lacks information about the transverse

[1]BlueSkyDAS LLC, Sugar Land, Texas, USA
[2]Department of Earth and Atmospheric Sciences, University of Houston, Houston, Texas, USA
[3]Scripps Institution of Oceanography, University of California San Diego, La Jolla, California, USA
[4]TGS-NOPEC Geophysical Company, Houston, Texas, USA

Distributed Acoustic Sensing in Borehole Geophysics, Geophysical Monograph 289, First Edition.
Edited by Yingping Li, Robert Mellors, and Ge Zhan.

DOI: 10.1002/9781394179275.ch31

components of particle motion. In Chapter 4, the authors introduce newly developed optical three-component (3C) accelerometers to resolve the azimuthal ambiguity of a fiber DAS array. This hybrid sensor array, referred to as a quasi-DAS (qDAS) system, provides adequate data for acoustic monitoring, significantly improves versatility, and reduces the cost for both single and repeat VSP acquisitions. Without downhole electronics, there is no need for downhole power, and the risk of failure for electronics left in the well long-term is eliminated, significantly increasing the lifetime of high-temperature wells, such as geothermal wells. The hybrid optical sensor array is expected to operate reliably at temperatures exceeding 200 °C.

In Chapter 5, Li et al. (2024a) describe the design of a 3D–3C DAS fiber array consisting of a cross and a concentric circular sub-array on the surface, with a portion of the fiber cable installed in a vertical wellbore. The authors calculate profiles of DTT and conduct waveform modeling of DAS amplitudes from selected seismic sources to all fiber sensors in the 3D–3C DAS array using a finite difference method and two velocity models: constant and variable velocity. They demonstrate that the distribution of DTT profiles and variations in DAS amplitudes can effectively determine source azimuths and incidence angles, consistent with those estimated from horizontal and vertical hodograms constructed with DAS amplitudes corrected for fiber directivity. With thousands of fiber sensors in the 3D–3C DAS array, the authors show that it is possible to significantly increase the signal-to-noise ratio (SNR) of the DAS system by employing beamforming and coherent stacking techniques.

The numerical analysis by Zong et al. (2024) in Chapter 6 suggests that noise-free point-wise strain rate data is less sensitive to waves polarized orthogonally to the fiber compared to single-component particle velocity data. However, for a 2D walkaway vertical seismic profiling (VSP) survey with hundreds of shots distributed from near to far offset locations, the final images of the strain rate data and single-component particle velocity data behave similarly due to the greatly improved angular coverage from densely distributed shots. Yet compared to multicomponent geophone records, which provide a vector field of wave propagation, the single-component DAS-VSP strain rate field captures only a subset of the wavefield information, resulting in a less complete image. A better survey design that accounts for the directional sensitivity of DAS measurements can help alleviate this issue. Understanding how DAS responses compare with geophone data and the resultant differences in their final images is crucial for justified imaging evaluation and survey design.

31.2. BOREHOLE DAS ACQUISITION AND PROCESSING

In this section, we present the conclusions of 10 chapters, Chapters 7–16, on borehole DAS acquisition and processing topics.

Bakulin et al. (2024) discuss in Chapter 7 the groundbreaking Smart DAS uphole acquisition system for fiber-based seismic imaging and monitoring. This system, with a network of interconnected shallow upholes and continuous DAS fiber, overcomes near-surface complexities to achieve imaging comparable to surface seismic methods, even in challenging terrains. Its precise near-surface characterization offers detailed subsurface insights, enhancing imaging in difficult overburden areas. Addressing cost and scalability concerns, it facilitates efficient large-scale buried-receiver surveys vital for CO_2 and H_2 storage monitoring. By combining the repeatability and low noise of VSP measurements with extensive coverage of surface seismic surveys, the Smart DAS system revolutionizes onshore reservoir exploration and monitoring.

In Chapter 8, Yu et al. (2024) underscore the value of jointly acquiring 3D DAS-VSP and surface seismic data, showcasing examples from both land and marine joint projects. The dense fiber sensors of 3D DAS-VSP near the reservoir yield direct insights into formation properties like velocities, absorption, and anisotropy, enhancing reflection imaging with a higher number of folds and enabling analyses such as AVO, fracture prediction, and reservoir characterization. Additionally, 3D DAS-VSP data enhances surface seismic processing, aiding in velocity model calibration, deconvolution, demultiple processing, and seismic quality factor (Q) compensation. These attributes of high-resolution images facilitate fluid identification and optimize field development strategies.

Martinez et al. (2024), in Chapter 9, showcase the versatility of DAS-VSP on hybrid cables for specific wellbore seismic applications, offering a potential alternative to conventional VSP surveys using traditional geophone arrays. The cost-effective nature of the DAS-VSP hybrid logging cable method positions it as a viable standard in exploration and appraisal endeavors. This approach allows for seamless routine acquisition of velocity profiles with DAS technology using hybrid cables, minimizing rig downtime and delivering common VSP survey products. The successful utilization of DAS technology in hybrid cables spans various scenarios, including open-cased holes, different well trajectories, shallow and deep wells, multiple tool combinations, and tension profiles. Moreover, this technique is adaptable for both production wells with installed fiber cables and exploration wells without deployed fiber cables yet.

Kasahara et al. (2024a) emphasize in Chapter 10 that the joint acquisition of DAS and DTS enables the generation of velocity and temperature profiles in high-temperature geothermal boreholes, with six DAS-DTS projects conducted in Japan, reaching depths of 3–4 km. The authors demonstrate the application of full-waveform inversion (FWI) and imaging techniques to produce high-quality images of deep structures using deep PP reflections recorded by DAS-VSP surveys. They also observe potential temperature effects on seismic waves. The authors conclude that integrating temperature, velocity, and deep structure images obtained from DTS-DAS joint acquisition and processing with existing geological, geophysical, and drilling data enhances subsurface understanding in geothermal areas, facilitating the development and production of clean geothermal energy.

In Chapter 11, Brisbourne et al. (2024) extensively review the applications of DAS in challenging glacial environments, where fiber cables are either installed in boreholes or deployed on the surface. They conclude that DAS serves as a valuable tool for studying glaciological processes and monitoring changes in the cryosphere in response to rapid climate warming. Particularly in boreholes, DAS offers significant advantages over traditional seismic instrumentation, which would be difficult to deploy in the harsh conditions of glacial settings. Borehole DAS-VSP data allows for the inference of seismic anisotropy in deeper ice layers, observation of basal seismicity and low-amplitude volcanic tremors, and detection of reflectivity variations linked to climatic effects on grain size. The authors also discuss how the relatively simple velocity structure of the cryosphere facilitates the development and validation of DAS data processing techniques that can be applied beyond glaciological research. Furthermore, they stress the importance of ensuring good coupling, suppressing noise, and carefully considering source amplitude and frequency content to fully leverage the advantages of the exceptional channel spacing provided by DAS.

Kean Lim et al. (2024) present in Chapter 12 the processing results from the largest known onshore 3D DAS-VSP survey conducted at a gas hydrate well in Alaska. The study reveals minor faults within the survey area that remain undetectable by surface seismic data, estimates anisotropy parameters, develops a 3D anisotropic velocity model, and generates high-quality images of subsurface structures. These images prove instrumental in pinpointing precise locations for monitoring and producing wells. Despite being single-component sensors with lower sensitivity compared to conventional 3C geophones, DAS offers distinct advantages. Notably, it significantly reduces the time and costs associated with the 3D DAS-VSP survey due to the high spatial resolution of fiber sensors permanently deployed in a well. The authors assert that the success of the onshore DAS 3D VSP survey highlights DAS's potential as a standard tool for VSP applications in the future, especially in reservoir time-lapse monitoring for production operations and CO_2 storage projects. The ongoing advancements in DAS technology and processing techniques are expected to further enhance its capabilities, making it more widely adopted across the industry.

Bouchaala et al. (2024) delve into the investigation of seismic attenuation in carbonate rocks across various seismic scales, ranging from seismic to ultrasonic, with a particular emphasis on DAS data in Chapter 13. The authors affirm the considerable sensitivity of attenuation to the petrophysical properties specific to carbonate rocks. However, they also note distinctions when compared to sandstones, particularly regarding the compressional-to-shear attenuation ratio's sensitivity to fluid saturation. The study uncovers multiple attenuation mechanisms, underscoring the complexity inherent in carbonate rocks due to their intricate pore structures. Nevertheless, achieving an accurate estimate of attenuation requires a trade-off between precision and vertical spatial resolution. Moreover, limitations arise in lateral spatial resolution when assessing attenuation. Although high-density DAS data offers exceptional spatial resolution for attenuation, its current data quality in carbonate media may present challenges in accurately estimating attenuation.

Chapter 14, by Sayed et al. (2024), underscores the pivotal role of DAS-to-velocity transformation in achieving precise borehole seismic imaging. The authors showcase its efficacy in generating consistent and accurate images, highlighting the importance of both model-based and data-driven corrections for aligning with benchmark images. This emphasis on transformational methods resonates throughout the chapter, emphasizing the transformative potential of DAS-to-velocity conversion in bolstering the accuracy of borehole seismic imaging. Through a comparative analysis of various imaging methods using DAS data, the study illuminates differences in frequency and amplitude responses between DAS and geophone sensors. This sheds light on the intricate effects of scaling and correction factors on the data. These insights are paramount, offering a deeper understanding of the challenges and opportunities inherent in estimating Q for both velocity and DAS synthetic data. The chapter advocates for further research in this domain, suggesting a focus on advancing techniques for DAS-to-velocity transformation, gauge length correction, and Q estimation. Additionally, there is a call to explore how local velocities impact DAS imaging in borehole geophysics. Future studies are encouraged to devise innovative approaches aimed at enhancing the accuracy and consistency of borehole seismic imaging using DAS technology.

Li et al. (2024b) highlight in Chapter 15 the crucial need for accurate, precise, and efficient first arrival time picking in both active and passive borehole seismic surveys, particularly when dealing with extensive DAS data due to the high-density fiber sensors spaced as closely as 1 m apart. The authors have developed an automated first-arrival-time picking algorithm using the differential-phase-spectral slope method (DPSSM) and have applied it to process two zero-offset DAS-VSP data sets, each comprising over 1,000 channels recorded at the Utah FORGE geothermal site. The results from the P-wave velocity profiles obtained using the DAS automatic picking method align well with those derived from wireline sonic logging data and reverse VSP data sets, which were generated using perforate shots and induced microearthquakes at various depths. This underscores the effectiveness and efficiency of the DPSSM method in accurately and swiftly picking hundreds to thousands of first arrival times in DAS-VSP data sets. The authors also suggest that the DPSSM automatic picking method holds promise for further development in applications such as microearthquake localization and updating velocity models through travel-time tomography.

Lanza et al. (2024) in Chapter 16 demonstrate that DAS data, with adequate preprocessing, has sufficient quality to detect and locate low-magnitude earthquakes ($M < 0$) occurring several hundred meters to kilometers away. The HADES-R method effectively retrieves the shape of out-of-network clustered seismicity across scales and provides location uncertainty distribution. Azimuthal ambiguity is unavoidable when only a single vertical borehole is available. However, even a slight deviation in the orientation of the wells can significantly enhance the accuracy of the locations. Results from FORGE and Bedretto Lab indicate that DAS signal characteristics and location capabilities perform consistently across different scales. This evaluation supports the use of longer-term experiments to explore the suitability of downhole DAS for passive microseismic monitoring.

31.3. BOREHOLE DAS IMAGING AND INVERSION

This section summarizes the conclusions of seven chapters, Chapters 17–23, on borehole DAS imaging and inversion.

Chen et al. (2024) delve into a VSP imaging technique outlined in Chapter 17, founded on the grid ray tracing method, which amalgamates the strengths of VSP-CDP transformation and depth migration. The authors validate its efficacy through numerical synthetic data, showcasing its ability to produce VSP imaging profiles even in complex structural settings while maintaining cost-effectiveness. They put the method to the test using a walkaway DAS-VSP dataset collected from an oil and gas field in the East China Sea, utilizing air gun sources. The results indicate the method's effectiveness and efficiency in migrating both P-waves and converted S-waves. Noteworthy advantages of this approach include (a) enhanced focus in imaging around reflectors with notably diminished migration artifacts, (b) reduced abnormal imaging amplitudes stemming from uneven coverage, (c) adaptable selection of the imaging aperture to address variable structural complexities, and (d) heightened computational efficiency compared to wave equation migration techniques.

Mizuno et al. (2024), in Chapter 18, delve into a comprehensive workflow aimed at optimizing borehole DAS data acquisition and processing for DAS-VSP imaging projects. The authors focus their attention on two critical aspects of this workflow to extract the maximum technical and economic values from borehole DAS data. First, they emphasize the significance of an algorithmic survey design, which plays a crucial role in quantitatively and automatically designing large borehole DAS surveys. Second, they highlight the importance of DAS amplitude response correction within the imaging algorithm. This correction leads to the production of high-quality images of subsurface structures using much denser fiber sensors, comparable to or even surpassing those generated using sparser 3C geophones. The authors note that the DAS response correction is particularly impactful in deviated wells but less so in vertical wells. Furthermore, the authors acknowledge that challenges persist in DAS migration, primarily due to the limitations of 1C DAS measurements and imaging artifacts arising from spatial aliasing of the wavefield. These challenges are areas for ongoing research and improvement in the field of borehole DAS-VSP imaging projects.

Fuller (2024), in Chapter 19, examines the feasibility of detecting diffractions from fractures within time-lapse borehole DAS data. This is achieved by summing along expected time-offset curves of diffractions using Kirchhoff depth migration images derived from time-lapse borehole DAS-VSPs obtained from a well subjected to hydraulic stimulation. The output from the Kirchhoff-type depth migration program of the time-lapse VSP results in a 2D depth image. Low amplitudes in these images indicate a lack of summed energy from the X-depth space, and high amplitudes signify measurable summed energy. The time-lapse depth images produced in the study reveal significant differences in what is believed to be the diffracted wavefield. This diffraction energy correlates with observed time-lapse reflection variations. Although the imaging results suggest the presence of diffractions as predicted by theoretical modeling, they alone do not conclusively prove the existence of diffractions resulting from thin reservoir fractures. To bolster evidence for or against the existence of such diffractions, repeating time-lapse seismic experiments similar to the one described in the chapter would be beneficial. This repetition should be accompanied by independent data indicating the specific location of fractures, such as formation microresistivity imaging (FMI) logs. Even more convincingly, cores taken from the wellbore adjacent to the stimulation well post-fracture stimulation would provide enhanced evidence.

Ali et al. (2024) detail their processing and imaging methods, particularly focusing on tubing-strapped 3D DAS-VSP data acquired in four wells under flowing conditions in Chapter 20. This chapter explores the multistep, multidomain denoising process, VSP processing, strain-to-velocity conversion, fiber directivity corrections, calibration of the tomographic velocity model, post-image conditioning, and the application of image domain least-squares RTM (reverse time migration) to the DAS-VSP data. The authors demonstrate that their meticulous processing and imaging techniques result in high-quality 3D images of subsurface structures from relatively noisy 3D DAS-VSP data.

In Chapter 21, Martuganova et al. (2024) integrate high-resolution images of subsurface structures from 3D DAS-VSP, acquired using a hybrid wireline DAS system, with existing geological data to characterize reservoir properties in a geothermal area. The resulting 3D geological model, based on these high-resolution site images, provides a comprehensive understanding of the Rotliegend reservoir and facilitates the optimal modification of future drilling plans. The construction of an in situ geothermal research laboratory using the hybrid wireline 3D DAS-VSP technique serves as an exemplary model for enhancing other existing geothermal boreholes.

Kasahara et al. (2024b), in Chapter 22, demonstrate the ability to generate comprehensive, high-resolution images of subsurface structures in a high-temperature geothermal field in northeastern Kyushu, Japan, using a joint DAS and DTS acquisition system. This system involves a 2-km fiber cable installed at the TP-2 geothermal well. The authors measure borehole temperature and Vp velocity profiles down to a depth of 2,000 m. They obtain migration images of subsurface structures by migrating deep reflections of PP and SP-converted waves generated by 12 seismic sources on the surface. Kasahara et al. (2024b) show that their imaging method, using data from the joint DAS-DTS system, can provide commercial services to high-temperature geothermal fields for characterizing subsurface geophysical properties.

In Chapter 23, Cai et al. (2024) adopt an elastic FWI approach to analyze two-azimuth walkaway DAS-VSP datasets and accelerometer data at a carbon capture and storage test site in Newell County, Alberta, Canada, revealing horizontal layering consistent with known geological structures. They have developed a depth registration scheme based on cross-correlation scans of the two datasets to obtain reliable depth estimates for each channel of DAS data. The combination of DAS data and acceleration measurements from accelerometer data has resulted in improved accuracy in subsurface imaging. Validating FWI outcomes using reverse-time migration imaging confirms the reliability and accuracy of their inversion method. The authors emphasize the importance of further research to refine fusion techniques for diverse geological settings, enhance real-time application capabilities, and explore more complex models in elastic FWI.

In addition to the chapters mentioned here, for those interested in borehole DAS imaging and inversion topics, we recommend the current paper by Zhan et al. (2024) as further reading to stay updated on the latest research trends in this field. The authors introduce a tailored FWI imaging workflow to address the specific processing and imaging challenges of large-scale 3D DAS-VSP data. This workflow offers a distinct advantage over conventional DAS-VSP imaging methods, resulting in an optimized image with a significantly reduced turnaround time. Results from field data recorded in an active water injection well demonstrate a geologically consistent DAS-VSP image obtained with minimal manual intervention, showcasing the efficacy of the new workflow.

31.4. BOREHOLE DAS MONITORING

In this section, we include conclusions of seven chapters, Chapters 24–30, on borehole DAS monitoring.

In Chapter 24, Lellouch (2024) reviews current borehole DAS applications in seismological studies. The author notes that higher noise levels in DAS compared to traditional sensors can be mitigated through various array-based approaches. Additionally, Lellouch acknowledges that the directivity of DAS measurements remains a primary limitation in earthquake parameter estimation. Possible mitigation strategies include multiwell recording, combining downhole and surface fibers, using specially designed helically wound cables, and conducting joint analysis with traditional seismometers.

Hasada et al. (2024) discuss making DAS measurements in geothermal wells and characterizing DAS signals by comparing them with surface seismometer measurements in Chapter 25. The authors identify PS and SP conversion phases and conclude that these converted waves are useful for better estimating seismic impedance contrasts in the subsurface of geothermal fields, potentially related to the fluid contents of geothermal reservoirs. The DAS systems, installed in wellbores at different geothermal fields, record numerous natural earthquakes.

This highlights that these DAS systems, primarily intended for geothermal resource exploration, could also serve as borehole DAS seismic networks to monitor natural earthquakes and volcanic tremors, providing possible early warnings to mitigate geohazard risks.

Huang et al. (2024), in Chapter 26, describe the downhole DAS systems installed within two wellbores across an active fault as part of the MiDAS project. These systems provide the opportunity to identify and monitor the active fault zone in situ at depth. The high-resolution DAS systems allow the authors to accurately map a fault zone structure consistent with core samples and directly observe the fault zone's constant strain amplification. Using a DAS amplitude-based analysis method they developed, the authors reveal temporal variations in subsurface media. They believe that such borehole DAS monitoring systems and associated analysis methods could have broader applications in developing geoenergy and groundwater resources, as well as in monitoring earthquakes to mitigate geohazards.

In Chapter 27, Li et al. (2024c) demonstrate the feasibility of using RMS amplitude profiles of borehole DAS ambient noise to identify and locate highly fractured zones, interfaces, and faults at the Utah FORGE. The authors construct DAS noise RMS amplitude profiles, observing noise peaks at certain depths and quiet zones at others. They find that the noise peaks strongly correlate with media characterized by low velocity, low density, high porosity, and a low Poisson's ratio, indicating brittle rather than ductile material. The larger amplitudes of these ambient noise RMS peaks are interpreted as seismic noise energy trapped and accumulated in highly fractured low-velocity zones. Spectral analysis of ambient noise in noisy zones reveals blue noise for frequencies below 450 Hz, transitioning to white noise from 450 to 1000 Hz. In contrast, a time-frequency analysis of ambient noise in quiet zones indicates a clear pattern of daily variation. The authors propose that this method can serve as an accurate, cost-effective, and source-free logging tool for detecting, locating, and continuously monitoring fractures. This has potential applications in the exploration and development of oil, gas, and geothermal energy, as well as in carbon capture, utilization, and storage (CCUS).

Ning and Jin (2024) provide a comprehensive review of LF-DAS and various DSS technologies in Chapter 28. The authors conclude that LF-DAS and DSS measurements offer significant technical and economic value for oil-field applications. The high spatial and temporal resolution of LF-DAS and DSS enables detailed subsurface strain monitoring, facilitating both qualitative and quantitative assessments across hydraulic fracturing, fracture monitoring, geomechanics modeling, and downhole pressure monitoring. Although the chapter focuses on enhancing hydraulic fracturing operations and ensuring the performance of unconventional reservoirs, the methodologies discussed have substantial potential for broader applications. Recent research has demonstrated the adaptability of these techniques in areas such as carbon sequestration for pressure monitoring and geothermal projects for measuring periodic hydraulic responses.

Ma et al. (2024b), in Chapter 29, report the investigation results from two Canadian dip-in DAS (CanDiD) projects, where a temporarily deployed optical fiber cable was used to monitor hydraulic fracturing operations and compared to recordings from cement-in fiber. The authors analyze both microseismic and LF-DAS signals to evaluate the effectiveness of a retrievable optical fiber for fracture monitoring. They discover that the SNR of LF-DAS signals recorded by the dip-in DAS fiber is similar to that of the cement-in DAS fiber. However, microseismic signals captured by the dip-in fiber contain a higher fraction of coherent noise compared to those from the cement-in fiber. The wireline-deployed DAS captured numerous fiber waves induced by wireline slippage, and their spatiotemporal distribution aligned with fracture hits observed in LF-DAS data. The authors conclude that dip-in DAS systems have good potential to map fracture geometry without the need for permanently installed optical fibers behind casings, demonstrating significant technical and economic value.

Observations presented in Chapter 30 by Rodríguez Tribaldos et al. (2024) indicate that different fiber-optic cable constructions and installation methods significantly impact noise levels and the strain rate and strain magnitudes recorded with DAS and DSS. These factors are critical for quantitative studies of borehole DAS and DSS measurements. The choice of these and other experimental components—such as the timescale and durability of the installation, the assortment of technologies used, and the balance between acquisition parameters and data volumes—must be guided by the study's objectives and additional constraints, including testbed dimensions and space, environmental conditions, remote access to instrumentation, and data storage and retrieval options. The authors hope that this study encourages further research into the use of borehole-based DAS and DSS for geomechanical characterization and monitoring of fluid injection experiments in underground research laboratories and beyond.

31.5. FURTHER READING AND FUTURE EFFORTS

For further reading in borehole DAS instrumentation and modeling, we recommend the study by Soulas et al. (2024) on the comparison of VSP responses of hybrid cable 1C DAS, 3C vector fiber sensors (a 3C optical system named BOSS), and 3C geophones. Conducted at the Avalon test site facility in a quarry in England using a seismic vibrator, this study provides valuable insights. The 3C vector fiber sensors are similar to those presented in Chapter 4 by Haldorsen et al. (2024), and the hybrid cable 1C DAS is the same as described by Martinez et al. (2024) in Chapter 9 for temporal DAS deployment. Soulas et al. (2024) discuss their findings in terms of SNR, frequency responses, vector fidelity, and sensitivity and provide commentary on the current state of DAS technology, comparing 3C optical point sensors and 3C geophones.

We also recommend the paper by Cunningham et al. (2023) titled "Three-Dimensional Distributed Acoustic Sensing at the Sanford Underground Research Facility (SURF)" for further reading. The authors investigate the capabilities of a unique 3D-deployed DAS array with spiral-like segments varying in depth and azimuth at SURF, the former Homestake Mine, between 1,250 and 1,488 m in depth. This setup is used to detect local, regional, and teleseismic sources of ground vibrations. The 3D DAS array incorporates multiple types of fiber-optic cables with highly variable coupling to the rock, presenting a significant challenge in recovering 3C DAS data from the 3D DAS array. Calibration with onsite 3C geophones and active seismic sources with known positions is necessary to achieve accurate results.

Joint borehole DAS, DSS, and DTS acquisitions using common engineering fiber cables will become a trend in future DFOS geophysical applications due to their scientific merits and economic advantages. These comprehensive measurements and integrated analyses will enhance our understanding of subsurface structures and reservoirs. The joint acquisition technique will have broad applications in energy development (oil, gas, and geothermal), CO_2 injection monitoring, and geohazard mitigation. For example, a combined DSS, DTS, and DAS tool was used to image and monitor formations and fractures around a borehole in a geologic carbon sequestration and storage project at the San Juan Basin. Kasahara et al. (2024b) enable the generation of velocity and temperature profiles in high-temperature geothermal boreholes with six DAS-DTS projects conducted in Japan. Ma et al. (2024a) employed both borehole DAS and DTS to characterize an active fault associated with large earthquakes in an earthquake-prone area.

3C DAS or qDAS measurements will remain an important topic of continuous attention, with various field trials advancing the maturity of 3C qDAS instruments. Studies on the 3C–3D DAS array (Li et al., 2024a) will address real-world installation and onsite active source calibration issues. The 3C–3D DAS array technique can also be applied in marine geophysics, serving both streamer surveys and joint OBS-DAS array surveys (Li et al., 2023), as well as detecting and monitoring marine mammal activities and ship movements. Additionally, numerical modeling of DAS and geophone responses in complex geological settings (Zong et al., 2024) should be extended to 3C–3D DAS array geometry to test whether 3C DAS data can improve subsurface imaging.

The contributing authors in several chapters of the DAS acquisition and processing part of the book highlight the emerging trend of joint borehole DAS and surface seismic surveys (Bakulin et al., 2024; Yu et al., 2024). This combined approach offers technical and economic advantages for optimizing the processing of both datasets. In previous 3D/4D joint borehole-surface seismic surveys, 3D/4D DAS-VSP was acquired in wells on marine drilling and production platforms, commonly known as "dry tree" wells. For further reading, we recommend the recent paper by van Gestel et al. (2024) titled "Development of a Subsea Distributed Acoustic Sensing Acquisition System." This article introduces a subsea DAS acquisition system with fiber cables permanently installed in subsea wells (wet trees), connected to long-distance fiber cables on the seafloor, and linked back to interrogators on floating platforms. This system enables the efficient and economical joint acquisition of both borehole and surface DAS data, facilitating 4D time-lapse surveys to monitor variations in subsurface reservoirs and optimize production management. The integrated processing of the jointly acquired 3D and 4D borehole and surface DAS data can significantly enhance our ability to accurately image subsurface reservoir structures.

Martinez et al. (2024) highlight the cost-effectiveness of the DAS-VSP hybrid logging cable method with numerous real-world examples. Similar hybrid cable DAS systems were also used in a DAS-VSP survey for geothermal energy exploration (Martuganova et al., 2024) and for monitoring hydraulic fracturing and induced microseismicity (Ma et al., 2024b), underscoring the market demand for such DAS tools in exploratory wells with temporary DAS installations. However, challenges remain in reliably transforming DAS data to velocity, correcting fiber directivity, and applying appropriate Q compensation. Accurately picking first arrival times for DAS-VSP data is crucial for velocity analysis, subsurface imaging, and microearthquake location. The semi-automatic method discussed in Chapter 15 (Li et al., 2024b) could evolve into a fully automatic method with the help of machine learning (ML) and artificial intelligence (AI).

Migration methods are widely applied to 2D and 3D DAS data to generate images of subsurface structures in Chapters 17–21 (Ali et al., 2024; Chen et al., 2024; Fuller, 2024; Martuganova et al., 2024; Mizuno et al., 2024). The major challenges to the migration image are that DAS amplitudes are subject to 1C measurements with fiber directivity effects. More studies are required to overcome those disadvantages and correct the image amplitudes. Fuller (2024) examines the feasibility of detecting diffractions from fractures within time-lapse borehole DAS data and points out that more repeating time-lapse seismic experiments would be beneficial. Both Kasahara et al. (2024b) and Cai et al. (2024) adopt an FWI approach to improve accuracy in subsurface imaging, indicating a trend toward the FWI-migration approach for subsurface structure images.

As pointed out by Lellouch (2024), more borehole DAS applications in seismological studies are expected. Chapters 25–27 (Hasada et al., 2024; Huang et al., 2024; Li et al., 2024c) focus on monitoring earthquakes, active faults, and fractures with various DAS recordings. The authors of Chapters 26 and 27 demonstrate that borehole DAS ambient noise depth profiles can be used to identify and locate an active fault zone (Huang et al., 2024) and highly fractured low-velocity zones (Li et al., 2024c). For further reading, we recommend the paper by Li et al. (2024d) titled "Detecting Fractures and Monitoring Hydraulic Fracturing Processes at the First EGS Collab Testbed Using Borehole DAS Ambient Noise," published in *Geophysics*. In this article, the authors validate the capability of using borehole DAS ambient noise for fracture monitoring by comparing DAS noise RMS amplitude profiles with borehole core photos and core logs. They find that the DAS noise RMS amplitude peaks correspond to locations of fractures or lithological changes indicated in the core photos or core logs. This article provides an excellent example of how borehole DAS passive ambient noise can be used to detect fractures and monitor fracturing processes in EGS reservoirs.

We also recommend another valuable item of reading material: "High-Resolution Analysis of Earthquake Sources and Subsurface Structures Using Downhole Optical Fiber Crossing Active Fault" by Ma et al. (2024a), presented at the 2024 SSA annual meeting in Alaska. The authors use DAS data from numerous earthquakes recorded by borehole DAS systems in two boreholes crossing active faults. These systems, described by Huang et al. (2024) in Chapter 26, have operated continuously and reliably since March 2022. Using seismometer-calibrated DAS data, the authors analyze earthquake nucleation and attenuation features and derive source time functions using the empirical Green function method. Their high spatial resolution and wide frequency-band DAS recordings enable them to map the physical architecture of a weak zone and characterize the slip and stress heterogeneity of the regional fault system. On 2 April 2024, an Mw 7.4 earthquake struck near Hualien, followed by several M > 6 aftershocks and swarms. The systems in two wellbores on either side of the fault captured the entire sequence from about 20 km away. Although there were some saturation issues with the DAS amplitudes during the mainshock due to their proximity, the earlier arrivals offer valuable insights into earthquake initiation. This earthquake sequence highlights the potential of DAS arrays as the next generation of earthquake monitoring and early warning tools to mitigate geohazard risks in earthquake-prone areas. To fully realize this potential, the issue of DAS amplitude saturation should be addressed by expanding the dynamic range of borehole DAS systems.

LF-DAS and various DSS technologies will continue to play crucial roles in hydraulic fracturing, fracture monitoring, geomechanics modeling, and downhole pressure monitoring, as highlighted by the contributing authors of Chapters 28–30 (Ma et al., 2024b; Ning & Jin, 2024; Rodríguez Tribaldos et al., 2024). These applications underscore the need to develop multimeasurement DFOS systems (e.g., Farhadiroushan et al., 2024) to obtain reliable and cost-effective data for addressing geophysical challenges.

31.6. SUMMARY

Section 31.1 covers conclusions from five chapters and two supplementary papers on borehole DAS instrumentation and modeling. It discusses advancements such as integrated DSS, a new CMPI method for monitoring hydrologic processes, hybrid sensor arrays like qDAS systems, and design improvements for 3D-3C DAS fiber arrays. It also touches on the comparative analysis of DAS responses with geophone data and the challenges in recovering 3C DAS data from unique 3D DAS arrays.

Section 31.2 presents insights from 10 chapters on borehole DAS acquisition and processing, covering topics such as Smart DAS uphole acquisition systems, joint acquisition of 3D DAS-VSP and surface seismic data, versatile DAS-VSP applications in various wellbore scenarios, and the integration of DAS and DTS for geothermal exploration. Notable contributions include advancements in seismic imaging, detection of seismicity in glacial environments, processing techniques for large-scale DAS-VSP surveys, and the transformative role of DAS-to-velocity transformation in borehole seismic imaging.

Section 31.3 covers insights from seven chapters on borehole DAS imaging and inversion, focusing on advancements in processing large-scale 3D DAS-VSP data. Topics include innovative imaging techniques using grid ray tracing, algorithmic survey designs, and DAS amplitude response corrections. Additionally, researchers explore the feasibility of detecting fractures and optimizing imaging methods in various geological settings. Notable contributions include enhanced subsurface imaging with reduced artifacts and improved accuracy in subsurface characterization using elastic FWI approaches.

Section 31.4 summarizes the conclusions from seven chapters on borehole DAS monitoring. Key topics covered include the application of DAS in seismology, geothermal wells, fault zone mapping, and its utilization for fracture detection and monitoring hydraulic fracturing using ambient noise. Notable highlights encompass the integration of DAS with traditional sensors to enhance earthquake monitoring and the substantial technical and economic value of LF-DAS in oil-field applications, particularly for detailed subsurface strain monitoring.

Section 31.5 contains several recommended readings, including studies on the 3C DAS-VSP responses of different DAS systems, the capabilities of a 3D-deployed DAS array for detecting ground vibrations, and a subsea DAS acquisition system for efficient joint acquisition of borehole and surface DAS data to enhance subsurface reservoir monitoring. Additionally, it includes a paper on a tailored FWI imaging workflow for large-scale 3D DAS-VSP data. Furthermore, we discuss two recommended papers that focus on borehole DAS ambient noise and core logs for fracture identification and location, as well as high-resolution analysis of earthquake sources for both microearthquakes and large earthquakes using downhole optical fibers at local distances.

Section 31.5 also discusses and anticipates future advancements in borehole DAS technologies and applications. Joint DAS, DTS, and DSS acquisitions using different mode fibers within the same cable, along with simultaneous borehole and surface DAS surveys using the same seismic sources, are expected to become standard practices for 3D and 4D monitoring and imaging of subsurface structures. Hybrid wire-fiber DAS-VSP systems will play a crucial role in exploration wells, which often lack accurate velocity data and high-quality subsurface images. AI and ML methods are expected to facilitate the automatic picking of first arrival times from a large DAS dataset for velocity analysis, tomography, and earthquake location. FWI-RTM migration methods will become popular for processing large-scale 3D and 4D DAS-VSP data, providing high-quality subsurface images with improvements in strain-velocity transformation, Q compensation, and fiber directivity correction. DAS, DTS, and DSS systems will be essential for detecting and monitoring fractures, active faults, and earthquakes, thereby enhancing energy production safety and geohazard mitigation.

This volume offers comprehensive insights into recent advancements in borehole DAS technologies and their diverse applications in geophysics, seismology, and hydrology. Our goal is to present a broad overview of the potential of DAS technologies in various geophysical applications, illustrated with numerous real-world examples throughout this volume. These examples span from borehole DAS applications in energy exploration and production to CCUS initiatives aimed at environmental protection, as well as earthquake monitoring and early warning systems for geohazard mitigation. We recommend additional reading materials to bridge any gaps and complement the information provided in this volume. Our hope is that readers will gain valuable knowledge from this volume, grasp the concepts presented by the contributing authors, and utilize and enhance these methods in their own DAS research projects. Through these efforts, we aim to further advance borehole DAS technologies and their applications in geophysics.

ACKNOWLEDGMENTS

We extend our gratitude to all of the contributing authors whose efforts have brought this monograph to fruition. The authors are particularly thankful to Dr. Taka'aki Taira, an AGU Books Editorial Board Member, for his meticulous review of this manuscript and for providing valuable feedback and suggestions to enhance the chapter. We also express our appreciation to the rest of the AGU Books Editorial Board and the editorial team at Wiley for their ongoing support of this monograph.

REFERENCES

Ali, S., Miranda, F., Emmanuel, S., & Podgornova, O. (2024). Multiwell distributed acoustic sensing 3D vertical seismic profiling under flowing conditions: toward solving the processing and imaging challenges. This volume: Chapter 20.

Bakulin, A., Silvestrov, I., Smith, R., & Golikov, P. (2024). Smart distributed acoustic sensing uphole acquisition system: bridging the gap between surface seismic and borehole geophysics for imaging and monitoring in complex near-surface environments. This volume: Chapter 7.

Bouchaala, F., Rouxel, S., & Zhou, B. (2024). Potential of seismic attenuation for exploring complex media: a focus on carbonate rocks. This volume: Chapter 13.

Brisbourne, A., Hudson, T., & Kendall, M. (2024). Borehole and surface applications of distributed acoustic sensing for characterization of the cryosphere and glacial environments. This volume: Chapter 11.

Cai, X., Kristopher, I., Mattew, E., Scott, K., Hu, Q., & Lawton, D. (2024). Multiazimuth elastic full-waveform inversion of fiber-optic and accelerometer vertical seismic profiling data: a field example from the Carbon Management Canada Newell County Facility, Alberta, Canada. This volume: Chapter 23.

Chen, Y., Zong, J., Xia, S., Yu, G., Fei, W., Hao, Z., & Hu, G. (2024). Distributed acoustic sensing-vertical seismic profiling P-P & P-S wave imaging method: a fast ray-based approach through a multitemplate fast advancement algorithm. This volume: Chapter 17.

Cunningham, E., Lord, N., Fratta, D., Chavarria, A., Thurber, C., & Wang, H. (2023). Three-dimensional distributed acoustic sensing at the Sanford underground research facility. *Geophysics*, *88*(6), WC209–WC220. doi:10.1190/geo2023-0079.1

Farhadiroushan, M., Parker, T. R., Sergey Shatalin, S., Chen Z., Gillies, A., Aghayev, R., et al. (2024). Integrated distributed strain sensing for wellbore and reservoir monitoring. This volume: Chapter 2.

Fuller, B. (2024). Seismic imaging of reservoir fractures using distributed acoustic sensing data. This volume: Chapter 19.

Haldorsen, J., Christensen, C., Metzbower, R., & Berzanskis, A. (2024). A fiber-optic, multicomponent sensor for borehole-seismic formation imaging and for monitoring of natural or induced seismicity. This volume: Chapter 4.

Hasada, Y., Kasahara, J., Mikada, H., & Ohnuma, H. (2024). Investigation of distributed acoustic sensing characteristics by complementary observations of distributed acoustic sensing in geothermal wells and by surface seismometers. This volume: Chapter 25.

Hua, L., Murdoch, L., Zhu, X., Germanovich, L., Wu, Y., Guo, J., et al. (2024). A microwave photonics optical fiber method for measuring distributed strain for hydrologic applications in the vadose and saturated zones. This volume: Chapter 3.

Huang, H. H., Ma, K. F., Wu, E. S., Cheng, Y. Z., Lin, C. J., Ku, C. S., & Su, P. L. (2024). Spatiotemporal monitoring of a frequent-slip fault zone using downhole distributed acoustic sensing at the MiDAS Project. This volume: Chapter 26.

Kasahara, J., Hasada, Y., Furuya, S., Shimazaki, M., Mikada, H., Ohnuma, H., & Fujise, Y. (2024a). Distributed acoustic sensing and DTS geophysical study in the Takigami Geothermal Field in Northeastern Kyushu, Japan. This volume: Chapter 10.

Kasahara, J., Hasada, Y., & Mikada, H. (2024b). Geothermal reservoir characterization using distributed acoustic sensing from vertical seismic profiling in six geothermal fields in Japan. This volume: Chapter 22.

Kean Lim, T., Fujimoto, A., Kobayashi, T., Tcherkashnev, S., & Mondanos, M. (2024). Stratigraphy test well (Hydrate-01) distributed acoustic sensing 3D vertical seismic profiling processing. This volume: Chapter 12.

Lanza, F., Tuinstra, K., Rinaldi, A. P., Rodriguez Tribaldos, V., Hopp, C., Bethmann, F., et al. (2024). A Prospective evaluation of borehole distributed acoustic sensing technologies for detecting and locating out-of-network microseismicity. This volume: Chapter 16.

Lellouch, A. (2024). Seismological applications of downhole distributed acoustic sensing. This volume: Chapter 24.

Li, Y., Li, D., Huang, L., Zheng, Y., (2023). 3D-3C distributed acoustic sensing arrays in marine geophysics. Paper presented at the 3rd EAGE Workshop on Fiber Optics Sensing for Energy Applications, Chengdu, China. doi:10.3997/2214-4609.202376031

Li, Y., Li, D., Huang, L., & Zheng, Y. (2024a). Three-component distributed acoustic sensing arrays with three-dimensional fiber cable deployment. This volume: Chapter 5.

Li, D., Huang, L., Zheng, Y., & Li, Y. (2024b). Automatic picking of distributed acoustic sensing-vertical seismic profiling first-arrival traveltimes using a differential-phase-spectral slope method. This volume: Chapter 15.

Li, D., Huang, L., Zheng, Y., & Li, Y. (2024c). Monitoring granitic and sedimentary rocks at the Utah FORGE site using borehole distributed acoustic sensing ambient noise. This volume: Chapter 27.

Li, D., Huang, L., Zheng, Y., Li, Y., Schoenball, M., Rodriguez-Tribaldos, V., et al. (2024d). Detecting fractures and monitoring hydraulic fracturing processes at the first EGS Collab testbed using borehole DAS ambient noise. *Geophysics*, *89*(2), D131–D138. doi:10.1190/geo2023-0078.1

Ma, K., Liao, J., & Hsiao, L. (2024a). High-Resolution analysis of earthquake sources and subsurface structures using downhole optical fiber crossing active fault. *Seismological Research Letters*, *95*(2B), 1224.

Ma, Y., Wang, C., Maji, V., Salvage, R., MacDougall, K., & Eaton, D. (2024b). Recent advances in the use of temporary optical fiber deployment for downhole hydraulic fracture monitoring. This volume: Chapter 29.

Martinez, A., Useche, M., & Sayed, A. (2024). Distributed acoustic sensing-acquired wellbore seismic data using hybrid wireline cable: field data examples. This volume: Chapter 9.

Martuganova, E., Norden, B., Blöcher, G., & Krawczyk, C.M. (2024). Enhanced geothermal reservoir characterization using wireline distributed acoustic sensing. This volume: Chapter 21.

Mizuno, T., Ali, S., Bettinelli, P., & Le Calvez, J. (2024). Distributed acoustic sensing vertical seismic profiling imaging workflow: transitioning from the geophone era. This volume: Chapter 18.

Ning, Y., & Jin, G. (2024). Enhancing unconventional reservoir development through low-frequency distributed acoustic sensing technology. This volume: Chapter 28.

Rodríguez Tribaldos, V., Hopp, C., Rinaldi, A.P., Tuinstra, K., Lanza, F., Ajo-Franklin, J., et al. (2024). Use of distributed acoustic sensing and distributed strain sensing technologies for geomechanical characterization of rock mass response in mesoscale experiments in underground laboratories: lessons learned. This volume: Chapter 30.

Sayed, A., Khaitan, M., Martinez, A., & Stewart, R.R. (2024). Distributed acoustic sensing to velocity transform for vertical seismic profiling applications. This volume: Chapter 14.

Soulas, S., Tubridy, G., Berry, S., Royds, P., Tubridy, T., Kaack, B., et al. (2024). *A VSP turkey shoot trial on hybrid cable comparing FO DAS, 3-C geophones and FO sensors*. Paper presented at the Fourth EAGE Workshop on Distributed Fibre Optic Sensing, The Hague, Netherlands. doi:10.3997/2214-4609.202430007

Van Gestel, J.-P., Soulas, S., & Garth Naldrett, G. (2024). Development of a subsea distributed acoustic sensing acquisition system. *Geophysics*, *89*(2), P1–P10. doi:10.1190/geo2023-0196.1

Yu, G., Xiong, J., Liu, C., Wu, J., Chen, Y., Zhao, Y., et al. (2024). Joint surface and borehole distributed acoustic sensing-vertical seismic profiling data acquisition and processing. This volume: Chapter 8.

Zhan, G., Gao, F., Malik, R, Preston, M., Macesanu, C., Hu, H., et al. (2024). *Optimizing DAS VSP value through FWI imaging*. Paper presented at the 85th EAGE Annual Conference & Exhibition, Oslo, Norway.

Zong, J., Yang, J., Li, Y. E., Cheng, A., Wu, X., & Willis, M. (2024). Comparing distributed acoustic sensing and geophone vertical seismic profiling imaging – acquisition efficiency of distributed acoustic sensing versus the multi-component advantage of geophone data. This volume: Chapter 6.

INDEX

Note: Locators for tables and figures are marked in *italics*.

Distributed Acoustic Sensing in Borehole Geophysics, Geophysical Monograph 289, First Edition.
Edited by Yingping Li, Robert Mellors, and Ge Zhan.

DOI: 10.1002/9781394179275.index